T5-AFE-751

SAE Dictionary of Aerospace Engineering

William H. Cubberly

Published by:
Society of Automotive Engineers, Inc.
400 Commonwealth Drive
Warrendale, PA 15096-0001

Library of Congress Catalog Card Number 92-64101
ISBN 1-56091-286-3

SAE Order No. M-107

PREFACE

The SAE DICTIONARY OF AEROSPACE ENGINEERING is a technical publication edited for the aerospace engineers who design, test, and manufacture aerospace vehicles, components and parts. Every term and definition has been extensively reviewed by experts, and will be continuously reviewed in the future to incorporate changes as they occur.

This completely new dictionary is, remarkably, the first such dictionary created for engineers who must communicate clearly and accurately in an increasingly more complex technology. This volume has the quality and comprehensiveness that aerospace engineers have come to expect from SAE publications. The engineering terms and definitions are compiled from the following sources:

> 5,188 terms from SAE AEROSPACE STANDARDS and REPORTS... peer reviewed information from these technical publications.
>
> 4,038 terms from the NASA THESAURUS...a publication reviewed and up-dated repeatedly since the 1970's.
>
> 9,541 general engineering and computer terms from ENGINEERING RESOURCES, INC...a leading developer of engineering terms and definitions that are germane to all engineering disciplines.

In sum, here are nearly 20,000 terms and definitions contained in the first dictionary for aerospace experts who actually design and build aircraft and their related components rather than the aerospace "buff."

Obviously, aerospace engineers and those technicians and clerical people who support them, will have a primary interest in the new dictionary. Aerospace engineering students and writers will also welcome this first technical dictionary in aerospace technology.

William H. Cubberly,
Editor

HOW TO USE THIS DICTIONARY

Basic Format

The format for a defined entry provides the term in boldface type and the definition in regular (light) typeface. A term may have more than one definition, in which case the definition is preceded by a number.

SAE Aerospace Standards and Reports definitions always appear first. The standard or report in which the term and definition appeared is referenced and shown in brackets, such as [ARP1201-91]. The last two digits refer to the year in which the standard or report was approved or reaffirmed.

NASA terms and definitions follow the SAE terms, and are so noted with brackets. Lastly, any general or data processing definitions are included.

Abbreviations and acronyms related to aerospace are shown as normally used as well as spelled out, except in those cases where the abbreviation or acronym is so well known as to preclude the need for a spelled-out entry.

Alphabetization

The terms are alphabetized on a letter-by-letter basis. Hyphens, commas, word spacing, etc., in a term are ignored in the sequencing of terms. To aid the user in finding word or abbreviations, the first word on the left-hand page, and the last word on the right-hand page, are shown in bold type at the top of the appropriate page.

a See ampere.

A See Ängstrom.

aberration 1. In astronomy, the apparent angular displacement of the position of a celestial body in the direction of motion of the observer, caused by the combination of the velocity of the observer and the velocity of light. In optics, a specific deviation from perfect imagery, as, for example spherical aberration, coma, astigmatism, curvature of field, and distortion.[NASA] 2. Deviation from ideal behavior by a lens, optical system, or optical component. It exists in all optical systems and designers have to make trade-offs among the different types.

ablated nosetips See PANT program.

ablation The removal of surface material from a body by vaporization, melting, chipping, or other erosive process; specifically the intentional removal of material from a nose cone or spacecraft during high speed movement through a planetary atmosphere to provide thermal protection to the underlying structure.[NASA]

ablative materials Materials, especially coating materials designed to provide thermal protection to a body in a fluid stream through the loss of mass.[NASA]

abnormal electric-system operation The unexpected loss of control of the electric system. The initiating action of the abnormal operation is uncontrolled and the exact moment of its occurrence is not anticipated. However, recovery from this operation is a controlled action. This operation occurs, perhaps, once during a flight as a result of damage, or it may never occur during the life of an aircraft. An example of an abnormal operation is the faulting of electric power to the structure of an aircraft and its subsequent clearing by fault protective devices. [AS1212-71]

abnormal limits Abnormal limits accommodate the trip bands of protective equipment in the primary power generating system.[AS1212-71]

abort 1. To cut short or break off an action, operation, or procedure with an aircraft, space vehicle or the like, especially because of equipment failure, as to abort a takeoff, abort a mission, or abort a launch. 2. An aircraft, space vehicle, or the like that aborts. 3. An act or instance of aborting. 4. During catapulting, the act of suspending a launch and removing the airplane from the catapult.[AIR1489-88] 4. In data processing, the termination of a computer operation before its normal conclusion.

above ground level (AGL) Height of an aircraft, clouds, or the top of an obstruction (for example, building, tower or bridge) above the surface of the earth in the immediate vicinity, usually expressed in feet.[ARP4107-88]

abrade To prepare a surface by roughening the surface by sanding or other means.[AIR4069-90]

abrasion 1. The wearing away by friction, especially hose cover or reinforcement which evidence damage, fraying, etc., due to rubbing or vibrating against another hose or permanent fixture in an installation.[ARP1658A-86] 2. Removal of surface material by sliding or rolling contact with hard particles of the same substance or another substance; the particles may be loose or may be part of another surface in contact with the first. 3. A surface blemish caused by roughening or scratching.

abrasive 1. Particulate matter, usually having sharp edges or points, that can be used to shape and finish workpieces in grinding, honing, lapping, polishing, blasting or tumbling processes; depending on the process, abrasives may be loose, formed into solid shapes, glued to paper or cloth, or suspended in a paste, slurry or air stream. 2. Any substance capable of removing material from a surface by abrasion. 3. A material formed into a solid mass, usually fired or sintered, and used to grind or polish workpieces; common forms are grinding wheels, abrasive discs, honing sticks, cones, and burrs.

abrasion resistance The ability of a material to withstand surface wear. [ARP1931-87]

ABS See Acrylonitrile Butadiene Styrene.

absolute accuracy error The deviation of the analog value at any code from its theoretical value after the full-scale range has been calibrated. Expressed in percent, ppm or fractions of 1 LSB.

absolute address An address which indicates the exact storage location where the referenced operand is to be found or stored in the actual machine code address numbering system. Synonymous with specific address and actual address and related to absolute code.

absolute alarm An alarm caused by the detection of a variable which has exceeded a set of prescribed high- or low-limit conditions.

absolute altimeter See terrain clearance indicator.

absolute altitude Distance from an aircraft or spacecraft to the actual surface of a planet or natural satellite.

absolute angle of attack The angle measured between the chord plane of an airfoil and the position that plane would have if the airfoil were producing zero lift, that is, the sum of the geometric angle of attack and the zero-lift angle of attack; also called an aerodynamic angle of attack. See also angle of attack, zero-lift angle of attack, and critical angle of attack.[ARP4107-88]

absolute ceiling The maximum height above sea level in a standard atmosphere at which a given airplane, under specified operating conditions, can maintain horizontal flight.[ARP4107-88]

absolute code Coding that uses machine instructions with absolute addresses. Synonymous with specific code.

absolute encoder An electronic or electromechanical device which produces a unique digital output (in coded form) for each value of an analog or digital input; in an absolute rotary encoder, for instance, the position following any incremental movement can be determined directly, without reference to the starting position.

absolute feedback In numerical control, assignment of a unique value to each possible position of machine slide or actuating member.

absolute humidity The weight of water vapor in a gas water-vapor mixture per unit volume of space occupied, as, for example, grains or pounds per cubic foot.

absolute instrument An instrument that determines the value of a measured quantity in absolute units by making a simple physical measurement.

absolute measurement A measured value expressed in terms of fundamental standards of distance, mass and time.

absolute pressure 1. Pressure value using absolute vacuum as a reference. [AIR1916-88] 2. Absolute static pressure in psia of the liquid and vapor phase. [AIR1326-91] 3. The pressure measured relative to zero pressure (vacuum). 4. The combined local pressure induced by some source and the atmospheric pressure at the location of the measurement. 5. Gage pressure plus barometric pressure in the same units.

absolute programming In numerical control, using a single point of reference for determining all positions and dimensions.

absolute sealing A level of sealing that requires all seams, slots, holes and fasteners passing through the seal plane to be sealed. (All integral fuel tanks require absolute sealing.)[AIR4069-90]

absolute stability A linear system is absolutely stable if there exists a limiting value of the open-loop gain such that the system is stable for all lower values of that gain, and unstable for all higher values.

absolute value error The magnitude of the error disregarding the algebraic sign or, if a vectorial error, disregarding its direction.

absolute viscosity A measure of the internal shear properties of fluids expressed as the tangential force per unit area at either of two horizontal planes separated by one unit thickness of a given fluid, one of the planes being fixed and the other moving with unit velocity.

absolute zero Temperature of -273.16 °C or -459.69 °F or 0 °K at which molecular motion vanishes and a body has no heat energy.[NASA]

absorbance An optical property expressed as $\log(1/T)$, where T is the transmittance.

absorptance 1. The ratio of the radiant flux absorbed by a body to that incident upon it.[NASA] 2. The fraction of the incident light absorbed.

absorption 1. The process by which radiant energy is absorbed and converted into other forms of energy. In general, the taking up or assimilation of one substance by another. In vacuum technology gas entering the interior of a solid.[NASA] 2. The reduction in intensity of a beam of electromagnetic or particulate radiation as it passes through matter, chiefly due to interactions with atoms or electrons, or with their electric and magnetic fields.

absorption band A region of the electromagnetic spectrum where a given substance exhibits a high absorption coefficient compared to adjacent regions of the spectrum. See also absorption spectra.

absorption coefficient An inherent material property expressed as the fractional loss in radiation intensity per unit mass or per unit thickness determined over an infinitesimal thickness of the given material at a fixed wavelength and band width. See also absorptivity.

absorption cooling Refrigeration in which cooling is effected by the expansion of liquid ammonia into gas and the absorption of the gas by water. The ammonia is reused after the water evaporates.[NASA]

absorption cross sections In radar, cross sections characterized by the amount of power removed from a beam by absorption of radio energy by a target to the power in the beam incident upon the target.[NASA]

absorption curve A graph of the variation of transmitted radiation through a fixed sample while the wavelength material of a given thickness is changed at a uniform rate.

absorption dynamometer A device for measuring mechanical force or power by converting the mechanical energy to heat in a friction mechanism or bank of electrical resistors.

absorption-emission pyrometer An instrument for determining gas temperature by measuring the radiation emitted by a calibrated reference source both before and after the radiation passes through the gas, where it is partly absorbed.

absorption hygrometer An instrument for determining water vapor content of the atmosphere by measuring the amount absorbed by a hygroscopic chemical.

absorption meter An instrument for measuring the quantity of light transmitted through a transparent medium by means of a photocell or other light detecting device.

absorption spectra The arrays of absorption lines and absorption bands which result from the passage of radiant energy from a continuous source through a selectively absorbing medium cooler than the source.[NASA]

absorption spectroscopy The study of the wavelengths of light absorbed by materials and the relative intensities at which different wavelengths are absorbed. This technique can be used to identify materials and measure their optical densities.

absorption tower A vertical tube in which a gas rising through a falling stream of liquid droplets is partially absorbed by the liquid.

absorptive index See absorptivity.

absorptivity The capacity of a material to absorb incident radiant energy, measured as the absorptance of a specimen of material thick enough to be completely opaque, and having an optically smooth surface.[NASA]

accelerated aging A test in which parameters, such as voltage and temperature, are increased above normal operating values to obtain observable or measurable deterioration in a relatively short period of time. The plotted results predict expected service life under normal operating conditions. Also called accelerated life test.[ARP1931-87]

accelerated life test A method of estimating reliability or durability of a product by subjecting it to operating conditions above its maximum ratings.

accelerated stall A stall occurring under acceleration, as in a pullout. Such a stall usually produces more violent motions of the airplane than does a stall occurring in unaccelerated flight. [ARP4107-88]

accelerated test A test in which the applied stress level is chosen to exceed that stated in the reference conditions in order to shorten the time required to observe the stress response of the item, or magnify the response in a given time. To be valid, an accelerated test must not alter the basic modes and/or mechanisms of failure or their relative prevalence.[ARD50010-91]

accelerate-stop distance 1. Runway distance traversed by an aircraft from start to V_1 to stop, assuming failure of the critical engine at V_1.[ARP4107-88] 2. A measure of the field length requirement for aircraft operation. The accelerate-stop distance is the sum of (a) the distance required to accelerate the aircraft from a standing start to the critical engine failure speed and (b) assuming the critical engine to fail at this speed, the distance required to bring the aircraft to a full stop.[AIR1489-88]

accelerating agent 1. A substance which increases a chemical reaction rate. 2. A chemical that hastens the curing of rubber, plastic, cement or adhesives, and may also improve their properties. Also known as accelerator.

accelerating electrode An auxiliary electrode in an electron tube that is maintained at an applied potential to accelerate electrons in a beam.

acceleration 1. The rate of change of velocity with respect to time (as in speed or direction).[AIR1489-88] 2. The act or process of accelerating, or the state of being accelerated. Negative acceleration is called deceleration.[AIR1489-88] 3. The time rate of change of velocity; the second derivative of a distance function with respect to time. 4. The rate of change of velocity. The act or process of accelerating or the state of being accelerated.[NASA]

acceleration cardiovascular effects Reduction in performance capability due to grayout or blackout resulting from high positive G. Analogous negative G effects also apply. See also redout.[ARP4107-88]

acceleration displacement effects Reduction in performance capability due to the physical displacement or restriction in movement of the operator as a result of ± Gz (head to toe), or ± Gy (side to side), or ±Gx (front to back), where G_x, G_y, G_z are the three vector components of the total acceleration to which the operator is being subjected (generally stated in units of the standard value of the gravitational acceleration, at the earth's surface, g=32fps^2 [ARP4107-88]

acceleration effects Reduction in performance capability due to the effects of acceleration on the cardiovascular system or on the vestibular apparatus, or resulting in the restriction of body movement.[ARP4107-88]

acceleration orienting effects Reduction in performance capability due to the effects of acceleration on the vestibular apparatus or visual system. [ARP4107-88]

acceleration time 1. The span of time it takes a mechanical component of a computer to go from rest to running speed. 2. The measurement of time for any object to reach a predetermined speed.

acceleration tolerance The capability to withstand the effects of changing velocity over time on biological processes or spatial orienting mechanisms. [ARP4107-88]

accelerator 1. A chemical additive used to hasten the chemical reaction under specific conditions.[ARP1931-87] 2. An ingredient used in a sealant formulation to accelerate the rate of cure. [AS7200/1-91] 3. The term "accelerator" is used by sealant formulators to denote an ingredient included in the formulation to accelerate the rate of cure. It is also used separately in a mixture to accelerate surface curing only of applied polysulfide sealant.[AIR4069-90]

accelerometer 1. Transducer which measures acceleration or gravitational forces capable of imparting acceleration.[NASA] 2. An instrument for measuring acceleration or an accelerating force such as gravity; if it includes provisions for making a recorded output, it is called an accelerograph.

acceptable message segment (AMS) An AMS condition occurs within a bus monitor when it determines that a message segment contains only valid word(s), and valid data words, and that the words are contiguous where required and that the bus monitor is programmed to accept such a valid message segment.[AS4116-90]

acceptance 1. Acceptance is the act of an authorized representative of the Government by which the Government assumes for itself, or as the agent of another, ownership of existing and identified supplies tendered, or approves specific services rendered, as partial or complete performance of the contract on the part of the contractor.[AS1933-85] 2. Contract acceptance by the airframe manufacturer from the buyer of galleys for installation on production airplanes.[AS1426-80]

acceptance test 1. A test that determines conformance of a product to design specifications as a basis for acceptance.[ARP1931-87] 2. A test conducted under specified conditions by, or on behalf of, the government, using delivered or deliverable items, in order to determine the item's compliance with specified requirements. Includes acceptance of first production units. [ARD50010-91] 3. A test performed on production components to assure the continuance of the quality of that component.[ARP1281A-81] 4. One of a number of tests called out by a sealant specification to assure the conformance of a particular batch of product to the physical and chemical requirements for the material.[AS7200/1-91]

access Pertaining to the ability to place information into, or retrieve information from a storage device.

access control 1. The procedures for providing systematic unambiguous, orderly, reliable and generally automatic use of communication lines, channels, and networks for information transfer. [NASA] 2. Hardware or software features, operating procedures, or management procedures designed to permit authorized access to a computer system.[NASA]

accessibility 1. A design feature which affects the ease of admission to an area for the performance of visual and manipulative maintenance.[ARD50010-91] 2. A measure of the relative ease of admission to the various areas of an item for the purpose of operation or maintenance.[AIR1916-88]

accessible terminal A node in an electronic network that is configured to allow it to be connected to an external circuit.

access method Any of the data-management techniques available to the user for transferring data between main storage and an input/output device.

accessory A part, subassembly, assembly or component designed for use in conjunction with or to supplement another item.[ARD50010-91]

access procedures The procedure by which the devices attached to the network gains access to the medium. The access procedure typically includes provisions to guarantee fairness in sharing the network bandwidth between attached devices. The most common access procedures for LANs are CSMA/CD, Token Bus, Token Ring, and Slotted Ring. See MAC.

access random Pertaining to the process of obtaining data from, or placing data into, storage where the time required for such access is independent of the location of the data most recently obtained or placed in storage.

access time 1.The interval between a request for stored information and the delivery of the information; often used as a reference to the speed of memory. 2. The time interval that is characteristic of a storage unit; a measure of the time required to locate information in a storage position and make it available for processing or to return information from the processing unit to a storage location.

Access Unit Interface The optional interface between a data station using an IEEE 802.3 LAN and a transceiver or modem. The AUI permits transparent connection of a data station to either baseband or broadband media.

accident An occurrence associated with the operation of an aircraft which takes place between the time any person boards the aircraft with the intention of flight and all such persons have disembarked, and in which any person suffers death or serious injury, or in which the aircraft receives substantial damage.[ARP4107-88]

acclimatization The adjustments of a human body or other organism to a new environment; the bodily changes which tend to increase efficiency and reduce energy loss.[NASA]

accommodation coefficient The ratio of the average energy actually transferred between a surface and impinging gas molecules which are scattered by the surface to the average energy which would theoretically be transferred if the impinging molecules reached complete thermal equilibrium with the surface before leaving the surface.[NASA]

accounting The practice and system of recording and summarizing business and financial transactions and reporting as well as verifying and analyzing their results.[NASA]

Accredited Standard Committee A standard committee accredited to ANSI.

accretion disks Rotation disks of matter surrounding an astronomical object, such as a star, galactic nucleus, black hole, etc., which is accumulated gravitationally by the the object.[NASA]

accumulator 1. A closed container for pressure storage of fluid energy. [ARP243B-65] 2. A device or apparatus that accumulates or stores up, as: a contrivance in a hydraulic system that stores fluid under pressure (energy). Also an electric storage battery. [AIR1489-88] 3. In computer technology, device which stores a number and upon receipt of another number add it to the number already stored and store the sum.[NASA] 4. The register and associated equipment in the arithmetic unit of the computer in which arithmetical and logical operations are performed. 5. A unit in the digital computer where numbers are totaled, i.e., accumulated. Often the accumulator stores one operand and upon receipt of any second operand, it forms and stores the result of performing the indicated operations on the first and second operands. Related to adder. 6. A pressure vessel containing water and steam, which is used to store the heat of steam for use at a later period and at some lower pressure. 7. A relatively large-volume chamber or other hydraulic device which receives fluid under low hydraulic power, stores it, and then discharges it at high hydraulic power, after which it is ready to repeat the cycle. 8. A chamber or vessel for storing low-side liquid refrigerant in a refrigeration system. 9. An accumulator is also referred to as a receiver, a reflux receiver or a reflux drum.

accumulator, compensating An accumulator which, in addition to its high pressure volume, incorporates low pressure volumetric capacity which will accommodate a like volume of fluid to that discharged from the high pressure chamber. The sum of the volumes of the high and low pressure chambers remains constant.[ARP243B-65]

accumulator, cylindrical An accumulator in which the fluid is separated from the compressible medium by means of a piston operating in a cylindrical container.[ARP243B-65]

accumulator, flexible separator An accumulator in which the fluid is separated from the compressible medium by means of a flexible bladder or diaphram.[ARP243B-65]

accumulator, hydraulic An accumulator in which the stored operating medium is hydraulic fluid.[ARP243B-65]

accumulator, hydro-pneumatic An accumulator in which the stored operating medium is hydraulic fluid, pressurized by means of compressed gas. [ARP243B-65]

accuracy 1. The closeness with which a measurement approaches the true value established independently.[ARP1256-90] 2. The degree to which an indicated value matches the actual value of a measured variable. 3. Quantitatively, the difference between the measured value and the most probable value for the same quantity, when the latter is determined from all available data, critically adjusted for sources of error. 4. In process instrumentation, degree of conformity of an indicated value to a recognized accepted standard value, or ideal value. 5. The deviation, or error, by which an actual output varies from an expected ideal or absolute output. Each element in a measurement system contributes to errors, which should be separately specified if they significantly contribute to the degradation of total system accuracy. 6. In analog-to-digital converter, accuracy is tied to resolution, a 13-bit A/D, as used in the controller for example, can resolve to one part in 2^{13} or 8192, so best accuracy as a percentage of full scale range is theoretically 1/8192, or about 0.0125 percent.

accuracy, reference See accuracy, rating.

ACEE program A NASA program started in 1975 to reduce fuel consumption for transport aircraft through the study of structural and aerodynamic energy efficiency as well as engine energy efficiency consisting of engine component improvement, new energy efficient engines, and advanced turbopropellers. The acronym stands for aircraft energy efficiency.[NASA]

AC generators Generators for the production of alternating-current power. [NASA]

achromatic Optical elements which are designed to refract light of different wavelengths at the same angle. Typically achromatic lenses are made of two or more components of different refractive index, and are designed for use at visible wavelengths only.

acicular alpha A product of nucleation and growth or an athermal (martensitic) transformation from beta to the lower temperature allotropic alpha phase. It may be needle-like, lenticular, or flattened bar morphology in three dimensions. Its typical aspect ratio is about 10:1.[AS1814-90]

acid cleaning The process of cleaning the interior surfaces of steam generating units by an inhibitor to prevent corrosion, and subsequently draining, washing and neutralizing the acid by a further wash of alkaline water.

acidity Represents the amount of free carbon dioxide mineral acids and salts (especially sulphates of iron and aluminum) which hydrolize to give hydrogen ions in water and is reported as milliequivalents per liter of acid, or p.p.m. acidity as calcium carbonate, or pH the measure of hydrogen ions concentration.

acid pickle Industrial waste consisting of spent liquor from an acidic process for cleaning metal surfaces.

acid rain Low pH rainfall resulting from atmospheric reactions of aerosols containing chlorides and sulfates (or other negative ions).[NASA]

acid-resistant Able to withstand chemical attack by strongly acidic solutions.

acid sludge Oil refinery waste fuel from acid treatment of unrefined petroleum.

acid wash A chemical solution containing phosphoric acid which is used to neutralize residues from alkaline cleaners and to simultaneously produce a phosphate coating that protects a surface of metal from rusting and prepares it for painting.

a-c input module I/O module that converts process switched a-c to logic levels for use in the PC.

ACK See acknowledge.

acknowledge A message sent between peer entities to indicate that data was properly received.

ACLS Automatic Carrier Landing System.

Acme screw thread A type of power-transmission thread made in four series—29° general purpose, 29° stub, 60° stub and 10° modified square; the number of threads per inch is not standardized according to shank diameter.

acoustic Related to sound.

acoustical ohm The unit of measure for acoustic resistance, reactance or impedance; it equals unity when a sound pressure of one microbar produces a volume velocity of one cubic centimetre per second.

acoustic compliance The reciprocal of acoustic stiffness.

acoustic coupler A type of communications device that converts digital signals into audio tones that can be transmitted by telephone.

acoustic delay lines Devices used in a communications link or a computer memory in which the signal is delayed by the propagation of sound waves. [NASA]

acoustic dispersion Separation of a

complex sound wave into its various frequency components, usually due to variation of wave velocity in the medium with sound frequency; usually expressed in terms of the rate of change of velocity with frequency.

acoustic emission The stress and pressure waves generated during dynamic processes in materials and used in assessing structural integrity in machined parts.[NASA]

acoustic excitation The process of inducing vibration in a structure by exposure to sound waves.[NASA]

acoustic generator A transducer for converting electrical, mechanical or some other form of energy into sound waves. See also sound generator.

acoustic holography A technique for detecting flaws or regions of inhomogeneity in a part by subjecting it to ultrasonic energy, producing an interference pattern on the free surface of water in an immersion tank, and reading the interference pattern by laser holography to produce an image of the test object.

acoustic impedance The complex quotient obtained by dividing sound pressure on a surface by the flux through the surface.

acoustic inertance A property related to the kinetic energy of a sound medium which equals $Z_a/2\pi f$, where Z_a is the acoustic reactance and f is sound frequency; the usual units of measure are g/cm4. Also known as acoustic mass.

acoustic interferometer An instrument for measuring either the velocity or frequency of sound pressure in a standing wave established in a liquid or gas medium between a sound source and reflector as the reflector is moved or the frequency is varied.

acoustic levitation Method by which molten materials in space are suspended during processing experiments in the low gravity environment. Also, the use of very intense sound waves to keep a body suspended, thereby eliminating any container contact.[NASA]

acoustic measurement Measurement of properties, quantities, or conditions of acoustical i.e. mechanical waves. [NASA]

acoustic microscopes Instrument which use acoustic radiation at microwave frequencies to allow visualization of microscopic detail exhibited in elastic properties of objects.[NASA]

acoustic radiation See sound waves.

acoustic radiometer An instrument that measures sound intensity by determining unidirectional steady-state pressure when the sound wave is reflected or absorbed at a boundary.

acoustic reactance The imaginary component of acoustic impedance.

acoustic resistance The real component of acoustic impedance.

acoustic retrofitting Modification, especially of aircraft, to effect noise reduction; specifically the introduction of absorber materials and jet noise silencers.[NASA]

acoustics 1. The study of sound, including its production, transmission, and effects. Those qualities of an enclosure that together determine its character with respect to distinct hearing.[NASA] 2. The technology associated with the production, transmission and utilization of sound, and the science associated with sound and its effects. 3. The architectural quality of a room—especially a concert hall, theater or auditorium—that influences the ability of a listener to hear sound clearly at any location.

acoustic signature In sonar applications, the profile that is characteristic of a particular undersea object (or class of objects)—for example, the profile of a school of fish or a sea-bottom formation.

acoustic spectrometer An instrument for analyzing a complex sound wave by determining the volume (intensity) of sound-wave components having different frequencies.

acoustic stiffness A property related to the potential energy of a medium or its boundaries which equals $2\pi fZ_a$, where Z_a is the acoustic reactance and f is sound frequency; the usual units of measure are dyne/cm5.

acoustic streaming Unidirectional flow currents in a fluid that are due to the presence of sound waves.[NASA]

acoustic velocity The speed of propagation of sound waves.[NASA]

acoustic vibrations See sound waves.

acousto-optic An interaction between an acoustic wave and a lightwave passing through the same material. Acousto-optic devices can serve for beam deflection, modulation, signal processing, and Q switching.

acousto-optic glass Glass with a composition designed to maximize the acousto-optic effect.

a-c output module I/O module that converts PC logic levels to output switch action for a-c load control.

ACPL (Spacelab) See atmospheric cloud physics lab.

acronym A word made up of the initial letters of a long or complex technical term, e.g. RAM is the acronym for Random Access Memory.

Acrylonitrile Butadiene Styrene A type of plastic material.

actinicity The ability of radiation to induce chemical change.

actinographs See actinometers.

actinometer(s) 1. The general name for instruments used to measure the intensity of radiant energy, particularly that of the sun.[NASA] 2. An instrument for measuring the actinic quality of radiation—that is, its relative ability to induce chemical change. 3. An instrument for measuring the flux density of solar radiation.

action, air-to-close See fail-open.

action, air-to-open See fail-close.

activated sludge A semiliquid mass removed from the liquid flow of sewage and subjected to aeration and aerobic microbial action. The end product is dark to golden brown, partially decomposed, granular and flocculent, and has an earthy odor when fresh.[NASA]

activation analysis A method of determining composition, especially the concentration of trace elements, by bombarding the composite substance with neutrons and measuring the wavelengths and intensities of characteristic gamma rays emitted from activated nuclides.

activator A chemical additive used to initiate the chemical reaction in a specific chemical mixture.[ARP1931-87]

active 1. An adjective that describes: (a) A system or portion of a system that is in control on line, in contrast to being in standby. (b) The operational status of a servo control device after a failure if the device is activated or remains in control.[AIR1916-88] 2. The general class of devices which control power from a separate supply.[ARP993A-69] See also physical condition.

active alarm point See alarm point.

active built in test A type of built in test which is temporarily disruptive to the prime system operation through the injection of test stimuli into the system.[ARD50010-91]

active control The automatic activation of various control surface functions in aircraft.[NASA]

active controls technology An airplane design concept in which vehicle performance, weight, and economic characteristics are optimized through a reliance on automatic subsystems within the flight control system to augment the airplane's stability, to reduce the designing loads (through load reduction or redistribution and structural mode damping), and to manage the airplane's configuration for aerodynamic efficiency. Active control functions include: pitch stability augmentation, lateral and directional stability augmentation (or both), angle-of-attack limiting, wing-load alleviation, maneuver-load control, gust-load alleviation, flutter-mode control, ride smoothing.[AIR1916-88]

active maintenance time The time during which corrective or preventive maintenance is being done on an item. Active maintenance time is comprised of the following task times: preparation time, fault location time, item obtainment time, fault correction time, adjustment and calibration time, and checkout time.[ARD50010-91]

active medium The material in a laser which produces the amplified stimulated emission. The name of the laser identifies the active medium.

active satellites Satellites which transmit a signal, in contrast to passive satellites.[NASA]

active transducer A transducer whose output waves are produced by power derived from a source other than any of the actuating waves, but whose output power is controlled by the actuating waves.

activity Ratio of escaping tendency of the component in solution to that at a standard state. The ion concentration multiplied by an activity coefficient is equal to the ion activity.

actual address See absolute address.

actual special sortie See special sortie.

actuarial data Refers to the type of information used to define the subcomponents of an engine or system (e.g., serial numbers) and often includes the bookkeeping of data that indicate the life usage on these subcomponents (e.g., operating hours, LCF counts, hot section usage).[ARP1587-81]

actuate To put into action or motion.

actuating error signal See signal, actuating error.

actuating system A system (mechanical, gas, or otherwise) that supplies and transmits energy for the operation of other mechanisms or systems.[AIR1489-88]

actuation signal The setpoint minus the controlled variable at a given instant. Same as error.

actuator(s) 1. A device for converting fluid energy into mechanical energy. [ARP243B-65] 2. The component of the actuation system that does work or dissipates energy to control a load. Actuator output is achieved by conversion of energy from the power system or load into mechanical work, torque, or force. [AIR1916-88] 3. A device that transmits energy and supplies force for the operation of other mechanisms or systems. It may utilize hydraulic, electrical, gaseous, or other sources of energy. [AIR1489-88] 4. Mechanisms to activate process control equipment, e.g., valves. [NASA] 5. A device responsible for actuating a mechanical device such as a control valve. 6. A device that actuates.

actuator, bellows type A fluid powered device in which the fluid acts upon a flexible convoluted component, the bellows.

actuator, diaphragm A fluid powered device in which the fluid acts upon a flexible component, the diaphragm.

actuator, double acting An actuator in which power is supplied in either direction.

actuator, electric A device which converts electrical energy into motion.

actuator, electro-hydraulic type A device which converts electrical energy to hydraulic pressure and into motion.

actuator, electro-mechanical type A device which converts electrical energy into motion.

actuator, hydraulic A fluid device which converts the energy of an incompressible fluid into motion.

actuator, hydro-mechanical A hydraulically actuated mechanism which incorporated mechanical linkage for modification of output.[ARP243B-65]

actuator-locking An actuator with integral provisions for locking the output rod or shaft in a given position or positions.[AIR1489-88]

actuator, piston type A fluid powered device in which the fluid acts upon a movable piston, to provide motion to the actuator stem.

actuator, pneumatic A device which converts the energy of a compressible fluid, usually air, into motion.

actuator, power flight control A flight control actuator using fluid or electrical power to drive the controlled output member, usually a ram or rotary shaft. [ARP1281A-81]

actuator, rotary A device for converting fluid energy into mechanical energy in the form of rotary motion which is limited in angular travel, as distinguished from a motor which can produce continual rotation.[ARP243B-65]

actuator, servo An integral servo valve and hydraulic actuator for use in a control system.[ARP243B-65]

actuator, single acting An actuator in which the power supply acts in only one direction, e.g., a spring diaphragm actuator.

actuator, signal conversion An actuator capable of converting one or more input signals to a single output which is used to drive a power actuator. [ARP1281A-81]

actuator travel time See stroke time.

actuator, vane type A fluid powered device in which the fluid acts upon a pivoted member, the vane, to provide rotary motion.

acute/transient fatigue See fatigue.

Ada A programming language based on PASCAL, originally developed on behalf of the US Department of Defense for use in embedded computer systems. It is named Ada in honor of Augusta Ada Byron, Countess of Lovelace, primarily due to the fact that she was the assistant and patron of Charles Babbage and is considered the world's first programmer.[NASA]

adaptation The adjustment, alteration or modification of an organism to fit it more perfectly for existence in its environment.[NASA]

adapter(s) 1. An intermediate device to provide for connector attachments such as special accessories, special mounting means or special inter-connection means to an electrical termination.[ARP914A-79] 2. Devices or contrivances used or designed primarily to fit or adjust one thing to another. Devices, appliances or the like used to alter something so as to make it suitable for a use for which it was not originally designed.[NASA]

adapting See self-adapting.

adaptive control 1. A control system that maintains optimum system performance by automatically changing system parameters.[AIR1916-88] 2. A control system which adjusts its response to its inputs based on its previous experience. Automatic means are used to change the type or influence (or both) of control parameters in such a way as to improve the performance of the control system. See control, adaptive.

adaptive control system A control system which continuously monitors the dynamic response of the controlled system and automatically adjusts critical system parameters to satisfy the preassigned response criteria, thus producing the same response over a wide range of environmental conditions. [AIR1489-88]

adaptive flight control system A flight control system having the capability to vary its performance parameters in flight and thus adapt to the changing flight conditions so that the vehicle's structural integrity and stability limitations are not exceeded during critical phases of flight.[AIR1916-88]

adaptive gain control A control technique which changes a feedback controller's gain based on measured process variables or controller setpoints.

adaptive optics 1. Real-time optical correction for atmospheric perturbations and other system error sources.[NASA] 2. Optical components which can be made to change the way in which they reflect or refract light. In practice, the term usually means mirrors with surface shapes that can be adjusted.

adaptive tuning 1. In a control system, a way to change control parameters according to current process conditions. 2. The identification of process gains and time that can be used to improve the response of a control loop.

ADC See analog-to-digital converter.

A-D converter (ADC) A hardware device that converts analog data into digital form; also called an encoder.

ADD 1. See OR and false add. 2. See sum.

adder A device which forms, as output, the sum of two or more numbers presented as inputs. Often no data-retention feature is included, i.e., the output signal remains only as long as the input signals are present. Related to accumulator.

adder-subtractor A device whose output is a representation of either the arithmetic sum or difference, or both, of the quantities represented by its operand inputs.

additives Materials or substances added to something else for a specific purpose.[NASA]

address 1. An identification, represented by a name, label or number, of a register or location in storage. Addresses also are part of an instruction word along with commands, tags, and other symbols. 2. The part of an instruction, which specifies an operand for the instruction.

address bus The highway linking subcomponents of the microcomputer system along which address data is transferred.

address field That part of an instruction or word containing an address or operand.

address format The arrangement of the address parts of an instruction.

addressing The means whereby the originator or control station selects the unit to which it is going to send a message.

addressing mode Method for addressing a location used for data storage.

address modification The hardware action of computing an instruction's effective operand address by some sequence of the following operations as prescribed within the instruction: a) Indexing, adding an index to the address; b) Indirect addressing, using the intermediate computed address to obtain another address from memory.

address register A register in which an address is stored.

add time The time required for one addition, not including the time required to get and return the quantities from storage.

adducts Chemical compounds with weak bonds, e.g. occlusive or Van der Waal bonds.[NASA]

ADF Automatic Direction Finder.

adhesion 1. The chemical/mechanical bonding of a material to a surface. [AIR4069-90] 2. A bonding between two surfaces, usually applied to localized welding at high points under substantial contact pressures. 3. Bonding between two surfaces, assisted by an adhesive substance.

adhesion promoter A material applied to a surface for the purpose of chemical enhancing adhesion of a sealant to the surface. Present adhesion promoters for polysulfide integral fuel tank sealant also contains an organic solvent cleaner.[AIR4069-90]

adhesive Any substance capable of bonding two surfaces together.

adhesive bonding 1. Bonding is accomplished by adding an adhesive coating to the surface of wire or cable and curing the adhesive to form a bond. Examples are bonding to potting material at the cable end of a electrical connector and bonding to silicone pressure seals. See potting.[ARP1931-87] 2. A commercial process for fastening parts together in an assembly using only glue, cement, resin or other adhesive.

adhesive sealing The bonding of faying surfaces to each other through the use of a structural adhesive in the form of a tape or a film.[AIR4069-90]

adhesive strength 1. The strength of the bond between the sealant and the substrate. It is desired that the adhesive bond be stronger than the sealant itself. Thus, if an attempt were made to peel the sealant from the surface, the sealant would pull apart before the bond between substrate and sealant would fail.[AS7200/1-91] 2. The strength of an adhesively bonded joint, usually measured in tension (perpendicular to the plane of the bonded joint) or in shear (parallel to the plane of the joint).

ADI Attitude Director Indicator.

adiabatic Referring to a process which takes place without any exchange of heat between the process system and another system or its surroundings.

adiabatic curing Curing concrete or mortar under conditions where heat is neither gained nor lost.

adiabatic demagnetization cooling Use of paramagnetic salts cooled to the boiling point of helium in a strong magnetic field, then thermally isolated and removed from the field to demagnetize the salts and attain temperatures of 10(-3) K.[NASA]

adiabatic engine Any heat engine that produces power without a gain or loss of heat.

adiabatic temperature The theoretical temperature that would be attained by the products of combustion provided the entire chemical energy of the fuel, the sensible heat content of the fuel, and combustion air above the datum temperature were transferred to the products of combustion. This assumes: a) combustion is complete; b) there is no heat loss; c) there is no dissociation of the gaseous compounds formed, and d) inert gases play no part in the reaction.

adjacent channel In FM/FM telemetry, the modulated signal bandwidth immediately below or above the channel of interest.

adjacent conductor An insulated conductor next to any other insulated conductor.[ARP1931-87]

adjacent groove An injection sealing groove machined near to, but not along, the fastener line.[AIR4069-90]

adjustable pitch propeller/variable pitch propeller A propeller the pitch setting of which can be changed in the course of field maintenance, but not when the propeller is rotating. [ARP4107-88]

adjustment The process of altering the value of some circuit element or some component of the mechanism of an instrument, controller or auxiliary device to bring the indication to a desired value, usually a value corresponding to an independently determined value of the measured variable within a specified tolerance.

adjustment error See error.

adjustment hardware Any hardware designed for adjusting the size of a torso restraint system to fit the user, including such hardware that may be integral with a buckle, attachment hardware, or retractor.[AS8043-86]

adjust, span Means provided in an instrument to change the slope of the input-output curve. See span shift.

adsorbents Materials which take up gases by adsorption.[NASA]

adsorption 1. The adhesion of a thin film or liquid or gas to the surface of a solid substance. The solid does not combine chemically with the adsorbed substance.[NASA] 2. The concentration of molecules of one or more specific elements or compounds at a phase boundary, usually at a solid surface bounding a liquid or gaseous medium containing the specific element or compound.

advanced range instrumentation aircraft An EC-135 aircraft configured for reception recording and real-time relay of telemetry data.[NASA]

advanced technology laboratory An all-pallet payload utilizing the Space Shuttle and the European Spacelab and designed to accommodate 8 to 15 experiments per mission.[NASA]

Advanced X Ray Astrophysics Facility See X-Ray Astrophysics Facility.

advancing blade Any rotor blade or wing on a rotary-wing aircraft in horizontal motion, moving into the relative wind.[ARP4107-88]

advection The process of transport of an atmospheric property solely by the mass motion of the atmosphere; also, the rate of change of the value of the advected property at a given point.[NASA]

advection fog Fog resulting from the movement of warm, humid air over a cold surface, especially a cold ocean surface. Also, sometimes, steam fog, which results from the transport of cold air over relatively warm water.[ARP4107-88]

adverse yaw Yaw in the opposite sense to that of the roll of an aircraft, for example, a yaw to the left with the aircraft rolling to the right.[ARP4107-88]

advisory Advice and information provided to assist pilots in the safe conduct of flight and aircraft movement.[ARP4107-88]

advisory indicating system A system which indicates to the pilot, or crew member, a safe or normal configuration, operation of essential equipment, or otherwise attracts attention for routine purposes.[ARP1088-91]

advisory system A system that supplies the crew with information and guidance that they can follow only if they have other reasons that reinforce such information and guidance.[ARP4153-88]

aeolight A type of glow lamp whose intensity of light output varies with an applied signal voltage; its construction employs a cold cathode and an envelope filled with a mixture of gases.

aerator Any device for injecting air into a material or process stream.

aeroassist Changing orbit size by utilizing aerobraking, aerocapture, or aeromaneuvering.[NASA]

aerobatic confidence maneuvers Aerobatics intended to increase pilot skills.[ARP4107-88]

aerobatic demonstration Aerobatics intended to increase or demonstrate pilot skill, or to demonstrate aircraft capabilities.[ARP4107-88]

aerobatics Preplanned, precisely executed flight maneuvers in which the aircraft exceeds either 60 deg of bank or 30 deg of pitch.[ARP4107-88]

aerobiology The study of the distribution of living organisms freely suspended in the atmosphere.[NASA]

aerobraking Changing orbit size by using the upper atmosphere to create drag.[NASA]

aerocapture Making use of the atmosphere of a planet or planetary satellite by capturing the object and reducing the orbit size so that it remains in orbit or lands on the body.[NASA]

aerodynamic angle of attack See absolute angle of attack.

aerodynamic buzz See flutter.

aerodynamic chords See chords

(geometry).

aerodynamic coefficient 1. Any non-dimensional coefficient relating to aerodynamic forces or moments, such as a coefficient of drag, a coefficient of lift, etc.[AIR1489-88]

aerodynamic force The force exerted by a moving gaseous fluid upon a body completely immersed in it.[AIR1489-88]

aerodynamic heating The heating of a body produced by the passage of air or other gases over its surface.[NASA]

aerodynamic lift See lift.

aerodynamics The science that deals with the motion of air and other gaseous fluids, and the forces acting on bodies when the bodies move through such fluids, or when such fluids move against or around the bodies.[NASA]

aerodynamic stability See stability, aerodynamic.

aeroelasticity The study of the response of structurally elastic bodies to aerodynamic loads.[NASA]

aeroelastic research wings Wings that are designed with less than normal stiffness to test devices that suppress flutter.[NASA]

aeroembolism The formation or liberation of gases in the blood vessels of the body, as brought on by a too-rapid change from a high, or relatively high, atmospheric pressure to a lower one. [NASA]

aerograph Any self-recording instrument carried aloft to take meteorological data.

aerology The study of the free atmosphere throughout its vertical extent, as distinguished from studies confined to the layer of the atmosphere adjacent to the earth's surface.[NASA]

aeromagnetic flutter See flutter.

aeromaneuvering Changing orbit size or plane or both by entering the upper atmosphere to create drag or lift or both.[NASA]

aeromaneuvering orbit to orbit shuttle Proposed reusable upper stage for the Space Shuttle superseded by the orbit transfer vehicle.[NASA]

aerometer An instrument for determining the density of air or other gases.

aeronautical beacon A visual NAVAID displaying flashes of white or colored light, or both, to indicate the location of an airport, a heliport, a landmark, a certain point of a Federal airway in mountainous terrain, or an obstruction.[ARP4107-88]

aeronomy The study of the upper regions of the atmosphere where ionization, dissociation, and chemical reactions take place.[NASA]

aerosol A dispersion of fine liquid or solid particles in a gas—for instance, both smoke and fog are aerosols.

aerospace medicine 1. A medical speciality dealing with standards for selection of flight personnel and the determination of their tolerance for safe participation in flight activities. [ARP4107-88] 2. That branch of medicine dealing with the effects of flight through the atmosphere or in space upon the human body and with the prevention or cure of physiological or psychological malfunctions arising from these effects.[NASA]

aerospace safety The engineering assessment and analysis of systems, subsystems, and functions of spacecraft, missiles, advanced aircraft and ground support in order to identify hazards associated with such systems and to design procedures that eliminate those hazards or determine tolerable safety levels.[NASA]

aerospace technology transfer Technology transfer germane to aircraft and space vehicles, their propulsion, guidance, etc.[NASA]

aerospace vehicles Vehicles capable of flight within and outside the sensible atmosphere.[NASA]

aerostats See airships.

aerothermodynamics The study of aerothermodynamic phenomena at sufficiently high gas velocities that thermodynamic properties of gas are important.[NASA]

aerothermoelasticity The study of the response of elastic structures to the combined effects of aerodynamic heating and loading.[NASA]

aerozine A rocket fuel consisting of a mixture of hydrazine and unsymmetrical dimethylhydrazine (UDMH).[NASA]

AFC (control) See automatic frequency control.

affective states Subjective feelings of different types of pleasantness or unpleasantness that a person has about aspects of his/her environment, other people, or himself/herself. Affective states are subdivided into emotions and moods depending on their duration and intensity.[ARP4107-88]

affective states, emotions Affective states which tend to be disruptive of mental physiological, or behavioral processes. Emotions are relatively brief in duration but strong in intensity. A mental state, characterized by strong feeling and accompanied by motor expression, that is related to some object or external situation.[ARP4107-88]

affective states, mood A relatively mild emotional state, enduring or recurrent; an echo of an emotional reaction with or without rememberance of the original stimulus.[ARP4107-88]

afterbodies Companion bodies that trail satellites. Sections or pieces of rockets or spacecraft that enter the atmosphere unprotected behind nose cones or other bodies that are protected for entry. Afterparts of vehicles.[NASA]

afterburners See afterburning.

afterburning Irregular burning of fuel left in the firing chamber of a rocket after cutoff. The function of an afterburner, a device for augmenting the thrust of a jet engine by burning additional fuel in the uncombined oxygen in the gases from the turbine.[NASA]

afterglow 1. Broad, high arches of radiance or glow seen occasionally in the western sky above the highest clouds in deepening twilight, caused by the scattering effect of very fine particles of dust suspended in the upper atmosphere. Also, the transient decay of a plasma after the power has been turned off.[NASA] 2. Luminosity which persists in a gas after an electrical discharge passes through it; the phenomenon is sometimes utilized in flow measurement.

AGC (control) See automatic gain control.

age-control The designation of a specific maximum period of age after cure-date that will assure desired conformance characteristics of an elastomeric material. Age control is based on the premise that elastomers deteriorate upon exposure to ozone, oxygen, heat, sunlight, rain, and other similar environmental factors.[AS1933-85]

aged beta A beta matrix in which alpha, typically fine, has precipitated as a result of aging.[AS1814-90]

age exploration A systematic evaluation of an item based on analysis of collected information from in-service experience, development tests or scientific handbooks. It assesses the item's resistance to a deterioration process with respect to increasing age.[ARD-50010-91]

age hardening Raising the strength and hardness of an alloy by heating a supersaturated solid solution at a relatively low temperature to induce precipitation of a finely dispersed second phase. Also known as aging; precipitation hardening.

age sensitive That characteristic of an elastomer which makes it subject to deterioration by oxygen, ozone, sunlight, eat, rain, and similar factors experienced in the normal environmental exposure subsequent to vulcanization.[AS1933-85]

agglomerate An agglomerate is two or more particles that are in intimate contact and cannot be separated by gentle stirring and from the small shear forces thus generated.[ARP1179A-80]

agglomeration Any process for converting a mass of relatively fine solid material into a mass of larger lumps.

aggregate Natural sand, gravel and crushed stone that is mixed with cement to make mortar or concrete.

aging 1.Alteration of the characteristics of a device due to use. 2. Operating a product before shipping it to stabilize component functions or detect early failures. 3. Any time-dependent change in properties of a material, but especially age hardening at room or slightly elevated temperatures. 4. Curing or stabilizing parts or materials by long-term storage outdoors or under closely controlled storage conditions.

agitator A device for mixing, stirring or shaking liquids or liquid-solid mixtures to keep them in motion.

AGL Above Ground Level.

agricultural aircraft Light aircraft specially equipped for agricultural applications such as crop dusting.[NASA]

agRISTARS project A multiagency program utilizing Landsat remote sensing data to predict crop yields, land use, and detecting pollution.[NASA]

agrophysical units Geographic areas defined for statistical purposes by AgRISTARS personnel whose boundaries are based on natural rather than political lines for the purpose of comparing similar agricultural regions.[NASA]

AGT See automated guideway transit vehicles.

AH-1G helicopter US Army designation for the Bell Model 209 Hueycobra attack helicopter powered by a single Avco Lycoming T53-L-13 turboshaft engine.[NASA]

aiding or opposing load An aiding load is a force or torque on the actuator, provided by load restoration or inertia, or both, that acts in the same direction as the desired direction of load motion. A force or torque that acts in the direction resisting motion is called an opposing load.[AIR1916-88]

air 1. The mixture of gases comprising the earth's atmosphere.[NASA] 2. The mixture of oxygen, nitrogen and other gases, which with varying amounts of water vapor, forms the atmosphere of the earth.

air atomizing oil burner A burner for firing oil in which the oil is atomized by compressed air which is forced into and through one or more streams of oil breaking the oil into a fine spray.

air bearing A device which lubricates motion with flowing air. A linear air bearing, floats a table on air as it travels a straight line.

air bind An air pocket in a pump, conduit or piping system that prevents liquid from flowing past it. Also called a liquid trap.

air binding The inclusion of air in a space hindering the flow of some other gas or liquid.

air blast The flow of air at a high velocity, usually for a short period.

airborne Carried in the atmosphere—either by being transported in an aircraft or by being dispersed in the atmosphere.

airborne integrated reconnaissance system Aerial reconnaissance system incorporating various modes of detection.[NASA]

airborne radar approach The use of airborne radar for aircraft approach control—the radar cursor technique.[NASA]

airborne static electric power inverter Equipment, or a combination of equipment, used in aircraft to convert direct current (DC) electric power to 400

Hz alternating current (AC) electric power.[AS8023-85]

airborne windshear alerting system A device or system which identifies the presence of windshear after the phenomenon is encountered. Alerting devices of this type do not provide guidance information to the pilot. [ARP4109-87]

airborne windshear automatic recovery system A device or system which integrates or couples autoflight systems of the aircraft with an airborne windshear flight guidance system. [ARP4109-87]

airborne windshear detection and avoidance system A device or system which detects a severe windshear phenomenon far enough in advance of the encounter in both the takeoff/climbout profile and the approach/landing profile to allow the pilot to avoid the windshear.[ARP4109-87]

airborne windshear flight guidance system A device or system which provides the pilot with flight guidance to improve recovery probability in an inadvertent windshear encounter. [ARP4109-87]

airborne windshear situational display A display which presents pertinent windshear information such as flight path and stall margins. May be available in conjunction with alerting, guidance and/or detection/avoidance systems. Windshear severity information may be supplied to the pilot on a continuous basis. Does not provide guidance information.[ARP4109-87]

air bottle A container suitable for storing a quantity of air under pressure. May be made of steel, titanium, fiberglass or other suitable material with or without a plastic or aluminum liner. [ARP906A-86]

air breathing boosters Boosters which are possible substitutes for rocket engines and which have inlets for oxygen sources for their engines rather than carrying their own oxygen as in a conventional rocket.[NASA]

air-bubbler liquid-level detector A device for indirectly measuring the level of liquid in a vessel—especially a corrosive liquid, viscous liquid or liquid containing suspended solids; it consists of a standpipe open at the bottom and closed at the top, which is connected to an air supply whose pressure is maintained slightly above maximum head of liquid in the vessel; air bubbles out of the bottom of the pipe, maintaining the internal pressure equal to the head of liquid in the vessel, pressure being measured by a simple gage or transducer. Also known as purge-type liquid-level detector.

air-bubbler specific-gravity meter Any of several devices that measure specific gravity by determining differential pressure between two air-purged bubbler columns; the devices ordinarily use either of two principles for determining specific gravity—comparison of sample density with density of a known liquid, or comparison of pressure between two bubbler columns immersed at different depths in the process liquid.

air carrier A corporate entity or person who undertakes directly, by lease, or by other arrangement, to engage in air transportation.[ARP4107-88]

air compressor A machine that raises the pressure of air above atmospheric pressure and normally delivers it to an accumulator or distribution system.

air condenser 1. A heat exchanger for converting steam to water where the heat-transfer fluid is air. Also known as air- cooled condenser. 2. A device for removing oil or water vapors from a compressed-air line.

air conditioned area See area, air conditioned.

air conditioning Controlling the atmospheric environment in a confined space by measuring and continually adjusting factors such as temperature, humidity, air motion, and concentrations of dust, gases, odors, pollen or microorganisms.

air-cooled engine An engine, such as an internal combustion engine, whose waste heat is removed directly by a flowing stream of air—either a stream blown across the engine's external surfaces, or one blown through internal cooling passages.

air-cooled heat exchanger A device for removing heat from a process fluid by passing it through a bank of finned tubes that are cooled by blowing or drawing a stream of air across the tube exteriors.

aircraft Any machine that can be supported for flight in the air by buoyancy or the effects of the air against its surfaces.[ARP4107-88]

aircraft accident An occurrence associated with the operation of an aircraft which takes place between the time any person boards the aircraft with the intention of flight and the time when all such persons have disembarked, and in which any person suffers death or serious injury, or in which the aircraft receives substantial damage.[ARP4107-88]

aircraft approach category A grouping of aircraft based on a speed of 1.3 times their stall speed in the landing configuration at maximum gross landing weight. For example, aircraft whose stall speed (with landing gear, flaps, etc., in proper position for landing) multiplied by 1.3 is less than 91 knots are classified as Category A aircraft; those for which 1.3 x stall speed is at least 91 knots but less than 121 knots are Category B aircraft; and so on up to Category E aircraft, for which 1.3 x stall speed is 166 knots or more.[ARP-4107-88]

aircraft braking environment The aircraft braking environment is that environment the aircraft experiences during the landing and braking phase of operations including: runway length/width/surface texture/slope/crown/contamination level, wind conditions, temperature/pressure, touchdown velocity and alignment, and touchdown point proficiency.[AS483A-88]

aircraft construction materials A general term designating the materials used in manufacturing an aircraft. [NASA]

Aircraft Energy Efficiency program See ACEE program.

aircraft environmental control A system which provides an environment, controlled within specified operational limits of comfort and safety, for humans, animals, and equipment. These limits may include the following: air pressure, temperature, humidity, velocity, and composition, ventilation rate, wall temperature, audible noise and vibration, etc. The definition of the environmental control system encompasses the air conditioning system but is broader and oes not necessarily rely on air alone as the controlled medium.[AIR746A-83]

aircraft gas turbine engine 1. Any gas turbine engine used for aircraft propulsion or power generation, including those commonly called turbojet, turbofan, turboprop or turboshaft type engines, and afterburning engines in the nonafterburning mode of operation. AIR1533-91][ARP1179A-80][ARP1256-90]

aircraft ground equipment Equipment required on the ground to support the operation and maintenance of an aircraft and its airborne equipment. [AIR1916-88]

aircraft, incipient skid Incipient skid is the point of beginning wheel velocity instability where brake torque begins to exceed resisting tire-runway friction torque and the wheel, therefore, begins an unstable deceleration condition that if continued would result in abrupt wheel lock-up.[AS483A-88]

aircraft, individual wheel control Individual wheel control refers to the feature where each braked wheel is controlled individually, as a function of its wheel speed. For this function each braked wheel requires its own wheel speed transducer, servovalve and control circuit.[AS483A-88]

aircraft integrated data system (AIDS) 1. A term (adopted by ARINC) in general use for airborne recording systems that include acquisition and signal conditioning.[AIR1266-77] 2. The broad term to identify a family of systems that acquires, processes and records data that are used to determine the functional status and condition of various commercial aircraft systems, including engine and engine components.[ARP-1587-81]

aircraft multiplexing Interior communications (signal flow) in an aircraft is conventionally accomplished by point to point wiring between the various electronic and electrical equipment. Multiplexing provides a means of increasing the number of signals transmitted between units while decreasing the number of signal paths. The result is a significant reduction of wire bundles and weight in the aircraft.[AIR-1207-72]

aircraft noise prediction See noise prediction (aircraft).

aircraft on ground (AOG) Cannot fly until replacement parts are received to complete the maintenance action.[AIR-1916-88]

aircraft operational period The time interval between the start of preparation for flight and post flight engine shutdown with consequent deactivation of the aircraft electric system.[AS1212-71]

aircraft power supplies Electrical sources for the normal operation of aircraft.[NASA]

aircraft runup Final engine check prior to takeoff.[NASA]

aircraft, slip ratio Slip ratio is the ratio of reduction in wheel rotational velocity under the influence of braking forces to the equivalent free rolling rotational velocity of the unbraked wheel.[AS483A-88]

aircraft spin A prolonged stall in fixed-wing aircraft characterized by a sustained spiral descent, usually with the nose down.[NASA]

aircraft structural integrity program A time-phased set of required actions performed at the optimum time during the life cycle (design through phase-out) of an aircraft system to ensure the structural integrity (strength, rigidity, damage tolerance, durability and service life capability) of the aircraft.[ARD50010-91]

aircraft subsystems Lesser systems which are components of major aircraft systems. For example, subsystems of the hydraulic system include landing gear, brakes, wing flaps, nosewheel steering, and speed brakes. NOTE: The terms "system" and "subsystem" are often used synonymously.[ARP4107-88]

aircraft systems Major components of the aircraft which operate from a common source of power, provide a common power source to similarly powered components, or perform a major function encompassing lesser functions or components. Examples include hydraulics, electric, flight control, avionics, engine power, fuel, and all-weather systems. [ARP4107-88]

aircraft, wheel speed detectors 1. Inertia Type—A Hub cap or rim driven rotating masses on overriding clutches operating switch(es) or hydraulic/pneumatic valves(s). 2. Generators—AC or DC generators containing bearings and have an output AC or DC voltage signal respectively as a function of wheel speed. 3. Inductive Type—Inductive system contains a sensor and exciter ring. The sensor is a magnetic field device whose reluctance is varied as a function of wheel speed by a rotating exciter ring thus resulting in a frequency proportional to wheel speed. Special attention must be given installation to air gap tolerances, effect of wheel deflection to provide suitable signal amplitude compared to controller input saturation for reliable wheel speed detection.[AS483A-88]

air curtain A stream of high-velocity conditioned air directed downward across an opening such as a door or window to exclude insects and exterior drafts, prevent heat transfer through the opening, and permit the interior space to be air conditioned.

air cushion 1. A mechanical device that uses trapped air to absorb shocks or arrest motion without shock. 2. The partly confined stream of low-pressure, low-velocity air that supports a vehicle known as an air-cushion vehicle, ground-effect machine or hovercraft and allows it to travel equally well over water, ice, marshland, or relatively level ground.

air cushion landing systems Landing systems based on the ground effect principle whereby a stratum of air is utilized as the aircraft ground contacting medium (in place of landing gear). [NASA]

air cylinder A cylindrical body for storing compressed air, for compressing air with a piston, or for driving a piston with compressed air.

air deficiency Insufficient air, in an air-fuel mixture, to supply the oxygen theoretically required for complete oxidation of the fuel.

air dry 1. Air with which no water vapor is mixed. This term is used comparatively, since in nature there is always some water vapor included in air, and such water vapor being a gas, is dry. 2. A papermaking term used to describe "dry" pulp containing about 10% moisture.

air ejector A device for removing air or noncondensible gases from a confined space, such as the shell of a steam condenser, by eduction using a fluid jet.

air entrainment Artificial infusion of a semisolid mass such as concrete or a dense slurry with minute bubbles of air, especially by mechanical agitation.

airfield, expeditionary An extension of SATS that provides a surfaced runway 4000 feet long and 96 feet wide, and parking/maintenance areas for up to four squadrons of aircraft. The field includes catapults and primary recovery systems identical to SATS, as well as three Fresnel Lens Optical Landing Systems (FLOLS), two M-21 emergency recovery systems, and expanded field lighting and communications systems compatible with the expanded capability. See also SATS.[AIR1489-88]

airfield, forward-area Airfields that must support the operation of liaison, observation, and light transport type aircraft, including heavy cargo helicopters, for a period ranging from a few days to three weeks.[AIR1489-88]

airfield, heavy-load an airfield that must support heavy bomber type aircraft. The load-carrying capacity of pavements for this type of airfield is equivalent to a main gear load of 265,000 lb on a four-wheel, dual-twin configuration having tire contact areas of 267 sq in. for each wheel, twin spacing of 37 in. c-c, and inside wheels of twins spaced 62 in. c-c.[AIR1489-88]

airfield, light-load An airfield that must support fighter and medium cargo type aircraft. The load-carrying capacity of pavements for this type of airfield is equivalent to a main gear load of 25,000 lb on a single wheel having a tire contact area of 100 sq in.

airfield, medium load An airfield that must support heavy cargo, tanker, and medium bomber type aircraft. The load-carrying capacity of pavements for this type of airfield is equivalent to a main gear load of 100,000 lb on a two-wheel, twin configuration having tire contact areas of 267 sq in. for each wheel and wheel spacing of 37 in. center to center. [AIR1489-88]

airfield, rear-area Airfields that normally must support the operation of heavy cargo aircraft, medium cargo aircraft, and fighter-bomber aircraft for a period of four to six months. Airfields of this class will be constructed, rehabilitated, extended, and maintained by engineer construction battalions and will usually be located in the Zone of Communications or in the Army rear area. The strength characteristics of the rear-area airfield will normally govern the landing gear flotation design for heavy cargo and fighter-bomber aircraft. The controlling rear-area airfield is characterized as a field having the equivalent of a T11 landing mat surface lying directionally on a 4-CBR subgrade.[AIR1489-88]

airfield, support-area Airfields that normally must support the operation of medium cargo aircraft (and conceivably certain fighter-bomber aircraft designed for close tactical support) for a period of from two weeks to one month. [AIR1489-88]

airfield, theatre of operations Theater-of-Operations airfields (more specifically TO airfield types) are limited-life facilities which represent the maximum construction capability of engineer troop units in the field, considering time limitations imposed by the tactical situation and available construction equipment and surfacing materials. The TO airfield classes are defined as rear-area, support-area, and forward-area airfields, and light VTOL landing areas.[AIR-1489-88]

airfield, zone of interior Zone of Interior airfields are permanent facilities constructed in accordance with the criteria given in Air Force Manual 88-6. Pavements may be either rigid (concrete); flexible (bituminous), or a combination thereof. The ZI airfield classes, as defined for Air Force construction, not only represent a range of flotation capabilities but also directly represent the designs on which most existing military airfields are based. Thus, the relations presented have a direct application to existing airfields in addition to providing a basis for comparison of proposed new aircraft landing gear designs with those of existing aircraft. The ZI airfield classes are defined as heavy-load, medium-load, or light-load airfields.[AIR1489-88]

air filter A device for removing solid particles such as dust or pollen from a stream of air, especially by causing the airstream to pass through layered porous material such as cloth, paper or screening.

air flow parameter A mathematical expression of gas flow, in units of mass flow per unit time, through a pneumatic starter nozzle under choked flow conditions. Also called air flow factor, air flow function, and corrected flow.[ARP-906A-86]

airfoil An aerodynamic surface designed to obtain a reaction from the air through which it moves; for example, aileron, wing, rotor blade, rudder, or similar device.[ARP4107- 88]

airfoil characteristics See airfoils.

airfoils Structures, pieces, or bodies, originally likened to foils or leaves in being wide and thin, designed to obtain a useful reaction on themselves in their motion through the air.[NASA]

airfoil-vane fan A device for creating a stream of moving air by drawing it into a fan casing near the hub and propelling it centrifugally with a rotor whose vanes are curved backward from the direction of rotation.

airframe(s) The fuselage, booms, nacelles, cowlings, fairings, airfoil surfaces (including rotors, but excluding propellers and rotating airfoils or engines), and landing gear of an aircraft, and their associated accessories and controls.[ARP4107-88] 2. The assembled structural and aerodynamic components of an aircraft or rocket vehicle that support the different systems and subsystems integral to the vehicle.[NASA]

airframe manufacturer The manufacturer responsible for providing airframe structure and provisions for galleys in the aircraft of the galley buyer or user. [AS1426-80]

airframe mounted accessory drive (AMAD) An accessory drive package mounted in the airframe remote from the engine but driven by the engine or an APU. Usually shaft or pneumatically driven. Sometimes called airframe mounted accessory gearbox (AMAG). [ARP906A-86]

air-free The descriptive characteristic of a substance from which air has been removed.

air-fuel ratio The ratio of the weight, or volume, of air to fuel.

air furnace Any furnace whose combustion air is supplied by natural draft, or whose internal atmosphere is predominantly heated air.

air gage 1. A device for measuring air pressure. 2. A device for precisely measuring physical dimensions by measuring the pressure or flow of air from a nozzle against a workpiece surface and relating the measurement to distance from the nozzle to the workpiece.

air gap The space between two ferromagnetic elements of a magnetic circuit.

airglow The quasi-steady radiant emission from the upper atmosphere as distinguished from the sporadic emission of the auroras.[NASA]

air-handling unit An assembly of air-conditioning equipment, usually confined within a single enclosure, which treats air prior to distribution and provides the means of propelling the treated air through the distributing ducts.

air-hardening steel A type of tool steel containing sufficient alloying elements to permit it to harden fully on cooling in air from a temperature above its transformation temperature. Also known as self-hardening steel.

air hoist A lifting or hauling tackle whose power is provided by air-driven pistons (for reciprocating motion) or air motors (for rotary motion).

air infiltration The leakage of air into a setting or duct.

air knife A device that uses a thin, flat jet of air to remove excess coating material from sheet stock such as paper.

air lance A device for directing a high-velocity stream of pressurized air into a process vessel, or against a surface such as a boiler wall to remove unwanted deposits.

air law The body of domestic and/or international laws dealing with regulations and liabilities in civil or military aviation.[NASA]

airline transport pilot (ATP) A pilot who has the privileges of a commercial pilot with an instrument rating and is certified to serve as pilot in command for passenger-carrying operations of a turbojet airplane, of an airplane having ten seats or more (excluding the pilot seats), or of a multiengine airplane operated by a commuter air carrier. [ARP4107-88]

air lock(s) 1. A stoppage or diminution of flow in a fuel system, hydraulic system, or the like, caused by pockets of air or vapor. Also chambers capable of being hermetically sealed that provide for passage between two places of different pressure as between an altitude chamber and the outside atmosphere. [NASA] 2. An intermediate chamber between an environmentally controlled confined space and the outside atmosphere that provides for entry of personnel and materials by sealing a door between the chamber and the confined space, opening a door to the outside to admit personnel or materials, closing and sealing that door, changing environmental conditions in the chamber to match those in the confined space, then opening an interior door to permit entry into the confined space; the process is reversed when exiting the confined space. See also air bind.

AIRMET Acronym for AIRman's METeorological Information. In-flight weather advisories issued only to amend the area forecast concerning weather phenomena which are of operational interest to all aircraft and potentially hazardous to aircraft having limited capability because of lack of equipment, instrumentation, or pilot qualifications. AIRMETs concern weather of less severity than that covered by SIGMETs or Convective SIGMETs. AIRMETs cover moderate icing, moderate turbulence, sustained winds of 30 knots or more at the surface, widespread areas of ceilings less than 1000 ft or visibility less than 3 miles, or both, and extensive mountain obscurement. See also SIGMET and Convective SIGMET. [ARP4107-88]

air meter A device for measuring the flow of air or other gas and expressing it as weight or volume per unit time.

air moisture The water vapor suspended in the air.

air monitor A warning device that detects airborne radioactivity or chemical contamination and sounds an alarm when the radiation, gas or vapor level exceeds a preset value.

air motor An engine that produces rotary motion using compressed air or other gas as the working fluid.

air nozzle An air port having direction and appreciable length for directing an air stream.

air permeability A method of measuring the fineness of powdered materials, such as portland cement, by determining the ease with which air passes through a defined mass or volume.

airplane An engine-driven, fixed-wing, heavier-than-air aircraft that is supported in flight by the dynamic reaction of air against its wings.[ARP4107-88]

airplane provisions The airplane structure, equipment, furnishings and related hardware furnished by the airframe manufacturer as part of the airplane. [AS1426-80]

air port An opening through which air passes.

airport advisory area The area within ten miles of an airport that does not have a control tower or where there is a tower which is not in operation and on which a flight service station (FSS) is located. In such cases the FSS provides advisory service to arriving and departing aircraft.[ARP4107-88]

airport marking aids Markings used on runway and taxiway surfaces to identify a specific runway, a runway threshold, a centerline, a holdline, or other designated area. A runway should be marked in accordance with its present usage such as: (a) visual, (b) nonprecision instrument, (c) precision instrument.[ARP4107-88]

airport rotating beacon A visual NAVAID operated at many airports. At civil airports, alternating white and green flashes indicate the location of the airport. At military airports, the beacons flash alternately white and green, but are differentiated from civil beacons by dual peaked (two quick) white flashes between the green flashes. [ARP4107-88]

airport security Organization of trained security personnel, surveillance and screening devices, and procedures used for the protection of airport and airline property, aircraft, passengers, employees, and visitors from injury, air piracy, and other unauthorized acts.[NASA]

airport surveillance radar (ASR) Approach control radar used to detect and display an aircraft's position in the terminal control area.[ARP4107-88]

airport traffic area Unless otherwise specifically designated in FAR Part 93, that airspace within a horizontal radius of five statute miles from the geographical center of any airport at which a control tower is operating, extending from the surface up to, but not including, an altitude of 3000 ft above the elevation of the airport.[ARP4107-88]

air-position indicator An aircraft navigation device which integrates headings and speeds to give a continuous, dead-reckoned indication of position with respect to the surrounding air mass.

air preheater A heat exchanger for transferring some of the waste heat in flue gases from a boiler or furnace to incoming air, which increases the efficiency of combustion.

air-puff blower A soot blower automatically controlled to deliver intermittently jets or puffs of compressed air for removing ash, refuse, or soot from heat absorbing surfaces.

air purge The removal of undesired matter by replacement with air.

air purging Removing airborne contaminants, gases or odors from a confined space by introducing fresh, clean air.

air regulator A device for controlling airflow—for example, a damper to control flow of air through a furnace, or a register to control flow of heated air into a room.

air reheater A device in a forced-air heating system that adds heat to air circulating in the system.

air resistance The opposition offered to the passage of air through any flow path.

air route traffic control center (ARTCC) A facility established to provide ATC service to aircraft operating on IFR flight plans within controlled airspace, principally during the en route phase of flight.[ARP4107-88]

AIRS (reconnaissance sys) See airborne integrated reconnaissance system.

air, saturated Air which contains the maximum amount of the vapor of water or other compound that it can hold at its temperature and pressure.

air separator A device for separating materials of different density, or particles of different sizes, by means of a flowing current of air.

airships Propelled and steerable dirigibles dependent on gases for flotation. [NASA]

air sickness See motion sickness.

air slew missiles Solid propellant rockets utilizing thrust vector control. NASA]

airspace 1. The atmosphere in which aircraft operate, extending upwards from the surface of the earth.[ARP4107-88] 2. The atmosphere above a particular potion of the earth, usually defined by the boundaries of an area on the surface projected perpendicularly upward. NASA]

airspeed Speed of an airborne object with respect to the surrounding air mass; in calm air, airspeed is equal to ground speed; true airspeed is a calibrated airspeed that has been corrected for pressure and temperature effects due to altitude, and for compressibility effects at high airspeeds.

airspeed mach indicator (AMI) An aircraft instrument that displays the vehicle's speed as a ratio of its true airspeed to the speed of sound.[ARP4107-88]

air spring A device commonly used instead of a mechanical spring in heavy vehicles to support the vehicle's body on its running gear; the energy-storage element is an air-filled container with an internal elastomeric bellows or diaphragm.

airstart envelope The region (altitude and airspeed) where airstarts are performed. Its concept is identical to a flight envelope.[ARP906A-86]

air starting Air starting is engine starting while in flight and is obtained by using various starting methods such as spooldown, windmilling or starter-assist.[ARP906A-86]

air supply (AS) 1. The supply of air used in pneumatic instrumentation as a power supply. 2. Plant air supply (PA). 3. Instrument air (IA). 4. The energy supply for pneumatic instrumentation.

air taxi 1. A term used to describe helicopter/VTOL aircraft movement conducted above the surface but normally not above 100 ft above ground level.[ARP4107-88] 2. The carriage in air commerce of persons or property for compensation or hire as a commercial operator in aircraft having a maximum seating capacity of less than 20 passengers or a maximum payload of less than 6000 pounds.[ARP4107-88]

air thermometer A device for measuring temperature in a confined space by detecting variations in pressure or volume of air in a bulb inside the space.

airtight Sealed to prevent passage of air or other gas; impervious to leakage of gases across a boundary.

air traffic clearance/ATC clearance An authorization by air traffic control for an aircraft to proceed within controlled airspace under traffic conditions specified to prevent collisions with known traffic. See also ATC instructions.[ARP4107-88]

air traffic control (ATC) A service operated by appropriate authority, military or civil, to promote the safe, orderly, and expeditious flow of air traffic.[ARP4107-88]

air traffic control radar beacon system (ATCRBS) A radar system in which the object to be detected is fitted with cooperative equipment in the form of a radio receiver/transmitter (transponder). Radar pulses transmitted from the searching transmitter/receiver (interrogator) site trigger a distinctive transmission from the transponder. This reply transmission, rather than a reflected signal, is then received back at the interrogator site for processing and display at an air traffic control facility. In this way, the responding object (usually an aircraft) can be identified as well as detected.[ARP4107-88]

air traffic control specialist/controller A person authorized to provide air traffic control service.[ARP4107-88]

air turbine starter An engine starting device, incorporating a turbine wheel driven by cold or heated compressed air or other gas. The starter also generally includes an inlet scroll, reduction gears, an engaging mechanism or clutch and a switch or speed signal to signal cutoff speed. It may also include an emergency disconnect.[ARP906A-86]

air vent A valve opening in the top of the highest drum of a boiler or pressure vessel for venting air.

air vessel An enclosed chamber, partly filled with pressurized air, that is connected to a piping system to counteract water hammer or promote uniform flow of liquid.

airway/federal airway A control area or portion thereof established in the form of a corridor through airspace with specified width and height, the centerline of which is defined by radio navigational aids. See also colored federal airway.[ARP4107-88]

airway beacon A visual device used to mark airway segments in remote mountain areas. The light flashes Morse Code to identify the beacon site.[ARP4107-88]

airworthiness certificate A certificate issued by the FAA, a designee, or the airworthiness authority of another nation certifying that the aircraft met, at the time of inspection, current airworthiness standards.[ARP4107-88]

airworthy In a condition suitable for safe flight.[ARP4107-88]

airy disk The central bright spot produced by a theoretically perfect circular lens or mirror. The spot is surrounded by a series of dark and light rings, produced by diffraction effects.

Aitken nuclei Microscopic particles in the atmosphere which serve as condensation nuclei for droplet growth during the rapid adiabatic expansion produced by an Aitken dust counter. [NASA]

alarm 1. An abnormal process condition. 2. The sequence state when an abnormal process condition occurs. 3. An instrument, such as a bell, light, printer, or buzzer, which indicates when the value of a variable is out of limits.

alarm severity A selection of levels of priority for the alarming of each input, output, or rate of change.

alarm system An integrated combination of detecting instruments and visible or audible warning devices that actuates when an environmental condition or process variable exceeds some predetermined value.

alarm valve A device that detects water flow and sounds an alarm when an automatic sprinkler system is activated.

albedo The ratio of the amount of electromagnetic radiation reflected by a body to the amount incident upon it, often expressed as a percentage, as, the albedo of the earth is 34%.[NASA]

aldehydes Carbonyl groups to which a hydrogen atom is attached; the first stage of an alcohol; -CHO.[NASA]

alert Indicator (visual, auditory or tactile) which provides information to the crew in a timely manner about a non-normal situation.[ARP4153-88] See also process condition; sequence state.

alert box In data processing, a window that appears on a computer screen to alert the user of an error condition.

alert height The minimum height above ground level at which the pilot is supposed to make the go-around if a system failure is detected.[ARP4107-88]

Alfven waves See magnetohydrodynamic waves.[NASA]

AlGaAs See aluminum gallium arsenides.[NASA]

algae Any plants of a group of unicellular and multicellular primitive organisms that include the Chlorella, Scenedesmus, and other genera.[NASA]

algal bloom See algae.

algebraic adder An electronic or mechanical device that can automatically find the algebraic sum of two quantities.

ALGOL See Algorithmic-Oriented Language.

algorithm 1. A step by step procedure for solving a problem or accomplishing some end.[ARP1587-81] 2. Special mathematical procedures for solving a particular type of problem.[NASA] 3. A prescribed set of well-defined rules or processes for the solution of a problem in a finite number of steps, for example, full statement of an arithmetic procedure for evaluating sin X to a stated precision. 4. Detailed procedures for giving instructions to a computer. 5. Contrast with heuristic and stochastic. 6. A recursive computational procedure.

algorithmic language A language designed for expressing algorithms.

Algorithmic-Oriented Language An international procedure-oriented language.

alias When varying signals are sampled at equally spaced intervals, two frequencies are considered to be aliases of one another if they cannot be distinguished from each other by an analysis of their equally spaced values.

aliasing 1. An error introduced by a sampling measurement system if the sampling rate is not at least double the highest frequency component of the signal being measured. It prevents determination of the frequency of the signal, and can result in a measurement offset.[AIR1872-88] 2. A peculiar problem in data sampling, where data are not sampled enough times per cycle, and the sampled data cannot be reconstructed.

aliasing error An inherent error in time-shared telemetry systems where improper filtering is employed prior to sampling.

alidade 1. An instrument used in the plane-table method of topographic surveying and mapping. 2. Any sighting device for making angular measurements.

alighting gear A general term which includes all equipment or components concerned with alighting on or landing on water, land, shipboard or else, i.e., wheels, shock struts, floats, skis, etc. See also landing gear.[AIR1489-88]

aligning torque or self aligning torque The footprint torque which in a rolling tire resists rolling in a curvilinear path, i.e., tends to align the roll into a straight path.[AIR1489-88]

alignment Performing the adjustments that are necessary to return an item to specified operation.[AIR1916-88]

aliquot An aliquot is a representative sample of the particle suspension in the calibration liquid. Multiple aliquots have identical characteristics.[ARP 1192-87]

alkali metals Metals in group 1A of the periodic system; namely, lithium, sodium, potassium, rubidium, cesium, and francium.[NASA]

alkaline cleaner An alkali-based aqueous solution for removing soil from metal surfaces.

alkalinity 1. The state of being alkaline.[NASA] 2. Represents the amount of carbonates, bicarbonates, hydroxides and silicates or phosphates in the water and is reported as grains per gallon, or p.p.m., as calcium carbonate.

alkali vapor lamps Lamps in which light is produced by an electric discharge between electrodes in an alkali vapor at low or high pressure.[NASA]

Allen screw A screw or bolt that has a hexagonal socket in its head, and is turned by inserting a straight or bent hexagonal rod into the socket.

all fire The minimum stimulus which must be applied to an EED for initiation under specified environmental conditions and reliability requirements.[AIR913-89]

alligatoring 1. Cracking in a film of paint or varnish characterized by broad, deep cracks extending through one or more coats. Also known as crocodiling. 2. Surface roughening of very coarse-grained sheet metal during forming. 3. Longitudinal splitting of flat slabs in a plane parallel to the rolled surface that occurs during hot rolling. Also called fishmouthing.

allobar A form of an element having a distribution of isotopes that is different from the distribution in the naturally occurring form; thus an allobar has a different apparent atomic weight than the naturally occurring form of the element.

allocation The apportionment of numerical requirements to all levels within an equipment which will result in meeting the overall contractual reliability requirement. Synonymous with apportionment.[ARD50010-91]

allowable interface load The maximum capability of a structure (airplane or galley) to react a load without structural failure. Defined as a force in pounds at the structure attachment interface point airplane coordinates. [AS1426-80]

allowable working pressure See design pressure.

allowance Specified difference in limiting sizes—either minimum clearance or maximum interference between mating parts—computed mathematically from the specified dimensions and tolerances of both parts.

alloy 1. A substance having metallic properties and being composed of two or more chemical elements of which at least one is an elemental metal.[ARP 1931-87] 2. Substance having metallic properties and being composed of two or more chemical elements of which at least one is an elemental metal.[NASA] 3. A solid material having metallic properties and composed of two or more chemical elements.

alloy steel An alloy of iron and carbon which also contains one or more additional elements intentionally added to increase hardenability or to enhance other properties.

all-pass network A network designed to introduce phase shift or delay into an electronic signal without appreciably reducing amplitude at any frequency.

Alnico Any of a series of commercial iron-base permanent magnet alloys containing varying amounts of aluminum, nickel and cobalt as the chief alloying elements; the Alnicos are characterized by their ability to produce a strong magnetic field for a relatively small magnet mass, and to retain their magnetism, with relatively insignificant loss in field strength when the magnetizing field is removed.

aloha system A multiple random access communications scheme in which there is a nonfixed allocation of channel capacity, so that the channel is available to any terminal whenever it has a packet ready for transmission.[NASA]

ALNOT Acronym for ALert NOTice. A message sent by a flight service station (FSS) or air route traffic control center (ARTCC) requesting an extensive communication search for overdue, unreported, or missing aircraft.[ARP4107-88]

ALPA Acronym for AirLine Pilots Association.

alpha The allotrope of titanium with a hexagonal, close-packed crystal structure.[AS1814-90]

alpha 2 structure A structure consisting of an ordered alpha phase, such as Ti^3 (Al,Sn) found in highly stabilized alpha. Defined by X-ray diffraction, not optical metallography.[AS1814-90]

alphabet The specific character set used by a computer.

alphabetic word 1. A word consisting solely of letters. 2. A word consisting of characters from the same alphabet.

alpha-beta structure A microstructure which contains both alpha and beta as the principal phases at a specific temperature. It is composed of alpha, transformed beta, and retained beta.[AS1814-90]

alpha case The oxygen, nitrogen, or carbon enriched, alpha-stabilized surface which results from elevated temperature exposure to environments containing these elements. Alpha case is normally hard, brittle, and considered detrimental.[AS1814-90]

alpha counter 1. A system for detecting and counting energetic alpha particles; it consists of an alpha counter tube, amplifier, pulse-height discriminator, scaler and recording or indicating mechanism. 2. An alpha counter tube and necessary auxiliary circuits alone. 3. A term sometimes loosely used to describe just the alpha counter tube or chamber itself.

alpha decay The radioactive transformation of a nuclide by alpha-particle emission.[NASA]

alpha double prime (orthorhombic

martensite) A supersaturated nonequilibrium orthorhombic phase formed by a diffusionless transformation of the beta phase in certain alloys. It occurs when cooling rates are too high to permit transformation by nucleation and growth. It may be strain induced during working operations and may be avoided by appropriate in- process annealing treatments.[AS1814-90]

alpha emitter A radionuclide that disintegrates by emitting an alpha particle from its nucleus.

alphanumeric display/data block Letters and numerals used to show identification, altitude, beacon code, and other information concerning a target on a radar display as used in automated radar terminal systems (ARTS).[ARP 4107-88]

alpha particle 1. A positively charged particle emitted from the nuclei of certain atoms during radioactive disintegration.[NASA] 2. A positively charged, energetic atomic particle consisting of two protons and two neutrons, identical in all measured properties with the nucleus of a helium atom; it may be produced by radioactive decay of certain nuclides or by stripping a helium atom of its electrons.

alpha prime A supersaturated, acicular nonequilibrium hexagonal phase formed by a diffusionless transformation of the beta phase. It occurs when cooling rates are too high to permit transformation by nucleation and growth. It exhibits an aspect ratio of 10:1 or greater. Also known as martensite or martensite alpha.[AS1814-90]

alpha radiation See alpha particles.

alpha-ray spectrometer An instrument used to determine the energy distribution in a beam of alpha particles.

alpha stabilizer An alloying element which dissolves preferentially in the alpha phase and raises the alpha-beta transformation temperature. Aluminum is the most commonly used alpha stabilizer. Interstitial elements such as oxygen and nitrogen are also potent alpha stabilizing elements.[AS1814-90]

alpha-transus The temperature which designates the phase boundary between the alpha and alpha-plus-beta fields. [AS1814-90]

alpine meteorology Wind, precipitation, atmospheric physics, and other climatological phenomena peculiar to the Alps and/or other similar mountainous areas.[NASA]

ALS Approach Light System.

altazimuth A sighting instrument having both horizontal and vertical graduated circles so that both azimuth and declination can be determined from a single reading. Also known as astronomical theodolite; universal instrument.

alteration switch A manual switch on the computer console or a program-simulated switch which can be set on or off to control coded machine instructions.

alternate code complement In a frame synchronization scheme, a frame synchronization pattern is complemented on alternate frames to give better synchronization.

alternate immersion test A type of accelerated corrosion test in which a test specimen is repeatedly immersed in a corrosive medium, then withdrawn and allowed to drain and dry.

alternating current (AC) An electrical current which reverses direction at regular intervals, with the rate expressed as hertz (cycles per second).[ARP1931-87]

alternating-current bridge A bridge circuit that utilizes an a-c signal source and a-c null detector; generally, both in-phase (resistive) and quadrature (reactive) balance conditions must be established to balance the bridge. Some bridges require only one balance (resistive or reactive) and use a phase-sensitive detector.

alternating current generators See AC generators.

alternators See AC generators.

altigraph A recording pressure altimeter.

altimeter 1. Instrument for measuring height above a reference datum.[NASA] 2. An instrument for determining height of an object above a fixed level or reference plane—sea level, for example; the aneroid altimeter and the radio altimeter are the most common types.

altimeter setting The barometric pressure reading used to adjust a pressure altimeter for variations in the local mean sea level reference atmospheric pressure or to adjust to the standard altimeter setting (29.92 inHg).[ARP4107-88]

altitude 1. In astronomy, angular displacement above the horizon. Also height, especially radial distance as measured above a given datum, as average sea level.[NASA] 2. Height above a specified reference plane, such as average sea level, usually given as a distance measurement in feet or metres regardless of the method of measurement.

altitude acclimatization A physiological adaptation to reduced atmospheric and oxygen pressure.[NASA]

altitude restriction An altitude or altitudes stated in the order they are to be flown which are to be maintained until reaching a specific point or time. Such restrictions are issued by ATC to ensure proper altitude separation of traffic or clearances from other known hazards.[ARP4107-88]

altitude sickness In general, any sickness brought on by exposure to reduced oxygen tension and barometric pressure.[NASA]

altitude signals Reflected radio signals returned to an airborne electronic device from the land or sea surface directly underneath the vehicle.

ALU See arithmetic and logical unit.

alum A general name for a class of double sulfates containing aluminum and another cation such as potassium, ammonium or iron.

alumina The oxide of aluminum—Al_2O_3.

aluminides Intermetallic compounds of aluminum and a transition metal.[NASA]

aluminizing 1. Applying a thin film of aluminum to a material such as glass. 2. Forming a protective coating on metal by depositing aluminum on the surface, or reacting surface material with an aluminum compound, and diffusing the aluminum into the surface layer at elevated temperature.

aluminum A soft, white metal that in pure form exhibits excellent electrical conductivity and oxidation resistance; it is the base metal for an extensive series of lightweight structural alloys used in such diverse applications as aircraft frames and skin panels, automotive body panels and trim, lawn furniture, ladders, and domestic cookware.

aluminum arsenides Binary compounds of aluminum with negative, trivalent arsenic.[NASA]

aluminum boron composites Structural materials composed of aluminum alloys reinforced with boron fibers (filaments).[NASA]

aluminum conductor An aluminum wire or group of wires, not insulated from each other, to carry electrical current.[ARP1931-87]

aluminum gallium arsenides Compounds exhibiting characteristics suitable for use in laser devices, light-emitting diodes, solar cells, etc.[NASA]

aluminum graphite composites Structural materials composed of aluminum alloys reinforced with graphite.[NASA]

alveolar air The respiratory air in the alveoli (air sacs) deep within the lungs.

[NASA]

alveoli The terminal air sacs deep within the lungs.[NASA]

amalgamation 1. Forming an alloy of any metal with mercury. 2. A process for separating a metal from its ore by extracting it with mercury in the form of an amalgam; the process was formerly used to recover gold and silver, which are now extracted chiefly with the cyanide process.

Amalthea Innermost satellite of Jupiter.[NASA]

ambient A surrounding or prevailing condition, especially one that is not affected by a body or process contained in it.

ambient air The air that surrounds the equipment. The standard ambient air for performance calculations is air at 80 °F, 60% relative humidity, and a barometric pressure of 29.921 in. Hg., giving a specific humidity of 0.013 lb of water vapor per lb of air.

ambient background 1. Background other than generated background. [AS8034-82] 2. Background light on display face other than generated by the display.[ARP1782-89]

ambient conditions The environment of an enclosure (room, cabinet, etc) surrounding a given device or equipment.

ambient illumination Light that illuminates the face of a display and its surround from sources such as the sky or cockpit and cabin lighting.[ARP1782-89]

ambient orientation A means of maintaining gross orientation in space without "thinking" about it. It is the result of the preconscious level of awareness: keeping track of various sensory inputs, including visual, tactile, kinesthetic, and auditory modes in order to keep the person oriented with respect to the horizon.[ARP4107-88]

ambient pressure The static pressure surrounding a component.[AIR1916-88] See pressure, ambient.

ambient temperature 1. The temperature of the surrounding environment. [ARP1931-87] [ARP914A-79] 2. Temperature of the environment in which the apparatus is working.[AIR1916-88] 3. Ambient temperature is the temperature of the medium, such as air, water, or earth into which the heat of the device is dissipated.[ARP1199A-90] See also temperature, ambient.

ambiguous or erroneous information Information that is capable of being understood in a different sense than intended.[AS8034-82]

American standard pipe thread A series of specified sizes for tapered, straight and dryseal pipe threads established as a standard in the United States. Also known as Briggs pipe thread.

American standard screw thread A series of specified sizes for threaded fasteners, such as bolts, nuts and machine screws, established as a standard in the United States.

American Wire Gauge (AWG) A standard system for identifying the physical size of wire.[ARP1931-87]

AMI Airspeed Mach Indicator.

Amici prism Also known as a roof prism. A right-angle prism in which the hypotenuse has been replaced by a roof, where two flat faces meet at a 90° angle. The prism performs image erection while deflecting the light by 90°; it is the same as rotating the image by 180°—reversing it left to right and at the same time inverting it top to bottom.

ammeter An instrument for determining the magnitude of an electric current.

ammonia A pungent, colorless, gaseous compound of hydrogen and nitrogen—NH3; it is readily soluble in water, where it reacts to form the base, ammonium hydroxide.

AMOOS See aeromaneuvering orbit to orbit shuttle.

Amor asteroid One group of earth-approaching asteroids with orbits between the plants Mars and Jupiter.[NASA]

amorphous film A film of material deposited on a substrate for corrosion protection, insulation, conductive properties or a variety of other purposes. It is non-crystalline and can be deposited by evaporation chemical deposition or by condensation. The method employed would be dictated by its composition and ultimate use.

amp or ampere (A) 1. The standard unit for measuring the strength of an electric current (the amount of current sent by one volt through a resistance of one ohm).[ARP1931-87] 2. Metric unit for electric current produced by one volt acting through a resistance of one ohm. Also the current that will deposit silver at the rate of 0.001118 grams per second, with the current flowing at 1 coulomb per second.

ampere-hour A quantity of electricity equal to the amount of electrical energy passing a given point when a current of one ampere flows for one hour.

ampere-hour meter An integrating meter that measures electric current flowing in a circuit and indicates the integral of current with respect to time.

ampere per metre The SI unit of magnetic field strength; it equals the field strength developed in the interior of an elongated, uniformly wound coil excited with a linear current density in the winding of one ampere per metre of axial distance.

amphiboles A group of dark, rock-forming, ferromagnesian silicate minerals closely related in crystal forma and composition.[NASA]

amplification 1. Increasing the amplitude of a signal by using a signal input to control the amplitude of a second signal supplied from another source. 2. The ratio of the output-signal amplitude from an amplifier circuit to the input-signal amplitude from the control network, both expressed in the same units.

amplification factor The μ factor for plate and control electrodes of an electron tube when the plate current is held constant.

amplifier 1. An active fluidic component which provides a variation in output signal greater than the impressed control signal variation. The polarity of the output signal may be either positive or negative relative to the control signal. The level of the control signal may be greater or less than the output level.[ARP993A-69] 2. Device which enables an input signal to control a source of power whose output is an enlarged reproduction of the essential characteristics of the signal.[NASA] 3. Any device that can increase the magnitude of a physical quantity, such as mechanical force or electric current, without significant distortion of the wave shape of any variation with time associated with the quantity. 4. A component used in electronic equipment to raise the level of an input signal so that the corresponding output signal has sufficient power to drive an output device such as a recorder or loudspeaker.

amplifier (laser) A laser amplifier is a device which amplifies the light produced by an external laser, but lacks the mirrors needed to sustain oscillation and independently produce a laser beam.

amplitude 1. The maximum value of the displacement of a wave or other periodic phenomenon from a reference position. Also, angular distance north or south of the prime vertical; the arc of the horizon, or the angle at the zenith between the prime vertical and a vertical circle, measured north or south from the prime vertical to the vertical circle.[NASA] 2. A measure of the departure of a phenomenon from any given reference.

amplitude distortion A condition in an amplifier or other device when the amplitude of the output signal is not an exact linear function of the input (control) signal.

amplitude-frequency response See frequency response.

amplitude modulation (AM) 1. In general, modulation in which the amplitude of a wave is the characteristic subject to variation.[NASA] 2. The process (or the results of the process) of varying the amplitude of the carrier in synchronism with and in proportion to the variation in the modulating signal.

amplitude noise Random fluctuations in the output of a light source or signal from other generating or detecting means.

amplitude response A measure of the time taken for a defined change of amplitude.

AM rejection The removal of unwanted amplitude modulation of a signal; usually performed by using signal clipping or limiting circuitry.

AMS spec Aerospace Material Specification published by SAE, Aerospace Division.[AIR4069-90]

A/M station (automatic/manual station) In control systems, a device which enables the process operator to manually position one or more valves. A single-loop station enables manual positioning of a single valve; a shared station enables control of multiple valves; and a cascade station provides control of paired loops.

AMTV See automated mixed traffic vehicles.

anaerobic sealant A faying-surface sealant in which cure is inhibited by exposure to air. The sealant can be applied to the faying surface and will not cure until the surfaces are brought together.[AIR4069-90]

analog 1. The representation of numerical quantities by means of physical variables, such as translation, rotation, voltage, or resistance. 2. Contrasted with digital. A waveform is analog if it is continuous and varies over an arbitrary range.

analog back-up An alternate method of process control by conventional analog instrumentation in the event of a failure in the computer system.

analog channel A channel on which the information transmitted can take any value between the limits defined by the channel. Voice channels are analog channels.

analog computer 1. Computer that works on the principle of measuring, as distinguished from counting, in which the input data is analogous to a measurement continuum such as linear lengths, voltages, or resistances which can be manipulated by the computer. [NASA] 2. A computer in which analog representation of data is mainly used. 3. A computer that operates on analog data by performing physical processes on these data. Contrast with digital computer.

analog control Implementation of automatic control loops with analog (pneumatic or electronic) equipment. Contrast with direct digital control.

analog control system Classically, a system that consists of electronic or pneumatic single-loop analog controllers, in which each loop is controlled by a single, manually-adjusted device.

analog data Data represented in a continuous form, as contrasted with digital data represented in a discrete, discontinuous form. Analog data are usually represented by means of physical variables, such as voltage, resistance, rotation, etc.

analog device A mechanism which represents numbers by physical quantities, e.g., by lengths, as in a slide rule, or by voltage or currents as in a differential analyzer or a computer of the analog type.

analog electronic controller Any of several adaptations of analog computers to perform control functions; they may produce an output signal directly related to the difference between a measured value and a predetermined setpoint, or they may produce an output signal modified by rate-of-change or other feedback signals.

analog hardware description language (AHDL) A modeling language capable of representing both the structural and behavioral properties of analog circuits. Structural refers to the connectivity or net-list properties of a circuit; behavioral refers to the mathematical equations for individual components.

analog input 1. A continuously variable input. 2. A termination panel used to connect field wiring from the input device. See input, analog.

analog input module I/O module that converts a process voltage or current signal into a multiple-bit form for use in the PC. The signal is the analog of some process variable.

analog millivoltmeter Galvanometer-type instruments with pointers and scales which may read in millivolts or directly in degrees. Since the deflection of a galvanometer is a function of the current flowing through it, the reading of a particular millivoltmeter-thermocouple system depends upon the total resistance of the circuit including the instrument.[AIR46-90]

analog output A continuously variable output (generally 4-20 ma or 3-15 psi). See also output, analog.

analog output module I/O module that converts a multiple-bit number calculated in the PC to a voltage or current output signal for use in control.

analog signal An analog signal is a continuously variable representation of a physical quantity, property, or condition such as pressure, flow, temperature, etc. The signal may be transmitted as pneumatic, mechanical, or electrical energy. See signal, analog.

analog simulation The calculation of the time or frequency domain response of electrical circuits to input stimulus. It assembles and solves a set of simultaneous equations associated with circuit topology.

analog-to-digital (A/D) 1. A device, or sub-system, that changes real-world analog data (as from transducers) to a form compatible with binary (digital) processing, as done in a microprocessor. 2. The conversion of analog data to digital data. See analog-to-digital converter.

analog-to-digital converter (ADC) 1. Device for converting non-digital information into digits.[NASA] 2. Any unit or device used to convert analog information to approximate corresponding digital information. See converter, analog to digital.

analysis 1. That part of the field of mathematics which arises from calculus and which deals primarily with functions. [NASA] 2. Quantitative determination of the constituent parts.

analysis criticality A procedure by which each potential failure mode is ranked according to the combined influence of severity and probability of occurrence.[ARD50010-91]

analysis program The analysis program is that part of EASY which performs the ECS transient analysis. [AIR1823-86]

analysis, ultimate Chemical analysis of solid, liquid or gaseous fuels. In the case of coal or coke, determination of carbon, hydrogen, sulfur, nitrogen, oxygen, and ash.

analytical balance Any weighing device having a sensitivity of at least 0.1 mg.

analytical curve A graphical representation of some function of relative intensity in spectroscopic analysis plotted against some function of concentration.

analytical gap The separation between the source electrodes in a spectrograph.

analytical line The spectral line of an element used to determine its concentration in spectroscopic analysis.

analytical scale In spectroscopic analysis, the scale that results when an analytical curve is projected onto the intensity axis; it is often used in lieu of an analytical curve to permit direct reading of spectral intensity as element concentration.

analyzer 1. Any of several types of test instruments, ordinarily one that can measure several different variables either simultaneously or sequentially. 2. In an absorption refrigeration system, the component that allows the mixture of water and ammonia vapors leaving the generator to come in contact with the relatively cool ammonia solution entering the generator where the mixture loses some of its vapor content.

anchor bond That portion of the beam lead which adheres to the device dielectric layer and has the required beam lead thickness.[AS1346-64]

AND A logic operator having the property that if P is an expression, Q is an expression, R is an expression..., then the AND of P, Q, R...is true if all expressions are true, false if any expression is false.

Anderson bridge A type of a-c bridge especially suited to measuring the characteristics of extremely low-Q coils.

andesite Volcanic rock composed essentially of andesine and one or more mafic constituents.[NASA]

AND gate A basic electronic circuit used in microprocessor systems. A logical 1 value on output is produced only if all of the inputs have logical 1 values.

anechoic chamber 1. A test room having all surfaces lined with a sound-absorbing material. Also known as dead room. 2. A room lined with a material that absorbs radio waves of a particular frequency or band of frequencies; it is used chiefly for tests at microwave frequencies, such as a radar-beam cross section.

anemobiagraph A recording pressure-tube anemometer, such as a Dines anemometer, in which springs are used to make the output from the float manometer linear with wind speed.

anemoclinometer An instrument for determining the inclination of the wind to a horizontal plane.

anemometer A device for measuring wind speed; if it produces a recorded output, it is known as an anemograph.

anemoscope A device for indicating wind direction.

aneroid Describing a device or system that does not contain or use liquid.

angel In radar meteorology, an echo caused by physical phenomena not discernible to the eye. Such phenomena have been observed when abnormally strong temperature or moisture gradients, or both, were known to exist; they are also sometimes attributed to insects or birds flying in the radar beam. [ARP4107-88]

angle (geometry) The inclination to each other of two intersecting lines, measured by the arc of a circle intercepted between the two lines forming the angle, the center of the circle being the point of intersection.[NASA]

angle beam In ultrasonic testing, a longitudinal wave from an ultrasonic search unit that enters the test surface at an acute angle.

angle modulation A type of modulation involving the variation of carrier-wave angle in accordance with some characteristic of a modulating wave; angle modulation can take the form of either phase modulation or frequency modulation.

angle of attack 1. The acute angle between the wing chord plane and the relative wind; also referred to as geometric angle of attack. See absolute angle of attack and critical angle of attack.[ARP4107-88] 2. The angle between a reference line fixed with respect to an airframe and a line in the direction of movement of the body.[NASA]

angle of attack indicator An instrument which indicates the angle between the wing chord plane and the relative wind.[ARP4107-88]

angle of cord (tires) The angle at which the cords in adjacent layers of fabric are set in the body or carcass of a tire. [AIR1489-88]

angle of elevation The angle between a horizontal plane and the line of sight to an object lying above the plane of the observer.

angle of extinction The phase angle of the stopping instant of anode current flow in a gas tube with respect to the starting instant of the corresponding half cycle of anode voltage.

angle of ignition The phase angle of the starting instant of anode current flow in a gas tube with respect to the starting instant of anode current flow in a gas tube with respect to the starting instant of the corresponding positive half cycle of anode voltage.

angle of incidence 1. The acute angle between a fixed reference, usually the longitudinal axis of the aircraft, and the chord of a wing or other airfoil. [ARP4107-88] 2. The angle between the direction of propagation of a ray of incident radiation and a normal to the surface it strikes; for a reflected wave, the angle of reflection and the angle of incidence are equal.

angle of repose A characteristic of bulk solids equal to the maximum angle with the horizontal at which an object on an inclined plane will retain its position without tending to slide; the tangent of the angle of repose equals the coefficient of static friction.

angle valve A valve design in which one port is colinear with the valve stem or actuator, and the other port is at right angles to the valve stem.

Ångstrom A unit of length defined as 1/6438.4696 of the wavelength of the red line in the Cd spectrum; it equals almost exactly 10^{-10} metre; this unit was once used almost exclusively for expressing wavelengths of light and x-rays, but it has now been largely replaced by the SI unit nanometre, or 10^{-9} metre.

angular acceleration The rate of change of angular velocity.[NASA]

angular accelerometer A device for measuring the rate of change of angular velocity between two objects.

angular frequency A frequency expressed in radians per second; it equals 2 times the frequency in Hz.

angular momentum The product of a body's moment of inertia and its angular velocity.

angular momentum flowmeter A device for determining mass flow rate in which an impeller turning at constant speed imparts angular momentum to a stream of fluid passing through the meter; a restrained turbine located just downstream of the impeller removes the angular momentum, and the reaction torque is taken as the meter output; under proper calibration conditions, the reaction torque is directly proportional to mass flow rate. Also called an axial flowmeter.

angular motion See angular velocity.

angular resolution Specifically the ability of a radar to distinguish between two targets solely by the measurement of angles.[NASA]

angular velocity 1. The change of angle per unit time; specifically the change in angle of the radius vector per unit time.[AIR1489-88] 2. The change of angle per unit time; specifically, in celestial mechanics, the change in angle of the radius vector per unit time. [NASA] 3. Rate of motion along a circular path, measured in terms of angle traversed per unit time.

anhydrous Describing a chemical or other solid substance whose water of crystallization has been removed.

Anik satellites A series of geostationary communication satellites operated by Telesat which is partly owned by the Canadian government and partly owned by private enterprise. The name Anik' is derived from an Eskimo word meaning brother'. It was so designated because of its partial use in the Far North.[NASA]

aniline (phenylamine) A substance produced from coal tar the indigo plant that is used to make inks, dyes and plastics.

ANIP Army-Navy Instrument Program.

anisotropic Exhibiting different properties when characteristics are measured along different directions or axes.

anisotropy Exhibiting different properties or other characteristics—strength or coefficient of thermal expansion, for instance—in different directions with respect to a given reference, such as a specific lattice direction in a crystalline substance.

anneal To heat and then gradually cool in order to relieve mechanical and thermal stress.[ARP1931-87]

annealing 1. Application of heat energy to a material cooling at a suitable rate to relieve stresses, change certain properties, improve machinability, or for realignment of atoms in a distorted lattice as caused, for example, by radiation damage.[NASA] 2. Treating metals, alloys or glass by heating and controlled slow cooling, primarily to soften them and remove residual internal stress but sometimes to simultaneously produce desired changes in other properties or in microstructure.

annotate To add explanatory text to computer programming or any other instructions.

annular ducts Ring-shaped openings for the passage of fluids (gases, etc.) designed for optimum aerodynamic flow properties for the application involved. [NASA]

annular nozzle A nozzle whose inlet opening is ring shaped rather than an open circle.

annular suspension and pointing system In the Shuttle era, high accuracy pointing and stabilization of an experiment payload.[NASA]

annulus 1. Any ring-shaped cavity or opening. 2. A plate that protects or covers a machine.

annunciator An electromagnetic, electronic or pneumatic signaling device that either displays or removes a signal light, metal flag or similar indicator, or sounds an alarm, or both, when occurrence of a specific event is detected; in most cases, the display or alarm is single-acting, and must be reset after being tripped before it can indicate another occurrence of the event.

anode 1. The positive pole or electrode of any electron emitter, such as an electron tube or an electric ell.[NASA] 2. The metal plate or surface that acts as an electron donor in an electrochemical circuit; metal ions go into solution in an electrolyte at the anode during electroplating or electrochemical corrosion. 3. The negative electrode in a storage battery, or the positive electrode in an electrochemical cell. 4. The positive electrode in an x-ray tube or vacuum tube, where electrons leave the interelectrode space.

anode circuit A circuit which includes the anode- cathode path of an electron tube connected in series with other circuit elements.

anode supply voltage The voltage across the terminals of an electric power source connected in series in the anode circuit.

anodic coating An oxide film produced on a metal by treating in an electrolytic cell with the metal as the cell anode.

anodic protection Reducing the corrosion rate of a metal that exhibits active-passive behavior by imposing an external electrical potential on a part.

anodic stripping The removal of metal coatings.[NASA]

anodize To form a protective passive film (conversion coating) on a metal part, such as a film of Al_2O_3 on aluminum, by making the part an anode in an electrolytic cell and passing a controlled electric current through the cell.

anodizing A method of producing film on a metal surface which is particularly well suited for aluminum.

anomalies In general, deviations from the norm.[NASA]

anomalous dispersion Inversion of the derivative of refractive index with respect to wavelength in the vicinity of an absorption band.

anorthosite A group of essentially monomineralic plutonic igneous rocks composed almost entirely of plagioclass feldspar.[NASA]

anoxia 1. Severe hypoxia which may result in permanent damage to the person or one or more of his/her organs.[ARP4107-88] 2. A complete lack of oxygen available for physiological use within the body.[NASA]

ANSI screen control An ANSI standard that specifies a specific set of character sequences which instruct the computer to perform certain actions on the computer screen.

ANSI X3J3 ANSI PL/I Language Standardization Committee.

antecedent events/mishap See mishap.

antenna A device for sending or receiving radio waves, but not including the means of connecting the device to a transmitter or receiver. See also dipole antenna; horn antenna.

antenna array 1. System of antennas coupled together to obtain directional effects, or to increase sensitivity.[NASA] 2. A single mounting containing two or more individual antennas coupled together to give specific directional characteristics.

antennas Conductors or systems of conductors for radiating or receiving radio waves.[NASA]

anthropology The study of the interrelations of biological, cultural, geographical, and historical aspects of man.[NASA]

anthropometrics Measurements of the height, weight, build, and other physical dimensions of a person.[ARP4107-88]

anticlines Geologic formations characterized by folds the core of which contain stratigraphically older rocks; they convex upward.[NASA]

anticlinoria See anticlines.

anticoincidence circuit A circuit with two inputs and one output, which produces an output pulse only if either input terminal receives a pulse within a specified time interval but does not produce a pulse if both input terminals receive a pulse within that interval.

anticorrosive Describing a substance, such as paint or grease, that contains a chemical which counteracts corrosion or produces a corrosion-resistant film by reacting with the underlying surface.

anti-drive end The end of a starter or accessory opposite the end which mates with the engine, or drive device.[ARP 906A-86]

antifouling Measures taken to prevent corrosion or the accumulation of organic or other residues or growths on operating mechanisms, especially in underwater environments.[NASA]

antifriction Describing a device, such

as a bearing or other mechanism, that employs rolling contact with another part rather than sliding contact.

antigravity A hypothetical effect that would arise from cancellation by some energy field of the effect of the gravitational field of the earth or other body.[NASA]

antimagnetic Describing a device which is made of nonmagnetic materials or employs magnetic shielding to avoid being influenced by magnetic fields during operation.

antimisting fuels Fuels which have an additive to reduce misting and thus create safer fuels.[NASA]

antinodes 1. Either of the two points on an orbit where a line in the orbit plane, perpendicular to the line of nodes and passing through the focus intersects the orbit. Also a point line, or surface in a standing wave where some characteristic of the wave field has maximum amplitude.[NASA] 2. The points, lines or surfaces in a medium containing a standing wave where some characteristic of the wave field is at maximum amplitude. Also known as loops.

antioxidant A substance used to prevent or retard the degradation of material through exposure to oxygen or to an oxygen containing environment.[ARP 1931-87]

antiozonant A substance used to prevent or retard the degradation of material through exposure to ozone.[ARP 1931-87]

antiparticles Particles with a charge of opposite signs to the same particles in normal matter.[NASA]

antipodes Anything exactly opposite to something else. Particularly, that point on the earth 180° from a given place.[NASA]

antiquities Man made objects or surviving parts or fragments from the past.[NASA]

antiradiation missiles Missiles that attack radiating targets such as radar transmitters, etc.[NASA]

antireflective coating A coating designed to suppress reflections from an optical surface.

antiresonance A condition existing between an externally excited system and the external sinusoidal excitation when any small increase or decrease in the frequency of the excitation signal causes the peak-to-peak amplitude of a specified response to increase.

antiresonant Describing an electric, acoustic or other dynamic system whose impedance is very high, approaching infinity.

antiresonant frequency A frequency at which antiresonance exists between a system and its external sinusoidal excitation.

anti-rotation, connector Connector design which provides keying or locking provisions to maintain positive orientation for accessory hardware.[ARP 914A-79]

antiskid Describing a material, surface or coating which has been roughened or which contains abrasive particles to increase the coefficient of friction and prevent sliding or slipping. Also known as antislip.

antisurge control Control by which the unstable operating mode of compressors known as "surge" is avoided.

anvil 1. The part of a machine that absorbs the energy of a sharp blow. 2. A heavy block made of wrought iron, cast iron or steel and used to support metal being smith forged. 3. The base of a forging press or drop hammer that supports the die bed and lower die. 4. The stationary contact of a micrometer caliper or similar gaging device.

AOIPS See atmospheric & oceanographic inform sys.

AOPA Aircraft Owners and Pilots Association.

APD See avalanche photodiode.

aperiodic Varying in a manner that is not periodically repeated.

aperiodically damped Reaching a constant value or steady state of change without introducing oscillation.

aperture A hole in a surface through which light is transmitted. Apertures are sometimes called spatial filters, a more descriptive term when placed in the Fourier (focal) plane.

aperture time The time required, in a sample-and-hold circuit, for the switch to open after the "hold" command has been given.

APL (programming language) 'A Programming Language' is a high level interactive computer language primarily designed for mathematical applications. It was developed by Kenneth Iverson in 1962. It is characterized by extensive operators and array handling capability. NASA Goddard was one of the first users and was instrumental in introducing APL to the computer community.[NASA]

apnea See respiration.

Apollo asteroids Earth grazing asteroids in orbits between Mars and Jupiter, and crossing the earth's orbit. This group contains 9 known asteroids. [NASA]

apparent bond width On a bonded device, the maximum width of the beam lead in the bonded area. The apparent bond width is also called the "squash dimension."[AS1346-64]

apparent candlepower The apparent candlepower of an extended source of light measured at a specific distance is the candle power of a point source of light which would produce the same illumination at that distance.[ARP798-91]

apparent density The density of loose or compacted particulate matter determined by dividing actual weight by volume occupied; apparent density is always less than true density of a material comprising the particulate matter because volume occupied includes the space devoted to pores or cavities between particles.

apparent viscosity The resistance to continuous deformation (viscosity) in a non-Newtonian fluid subjected to shear stress.

appearance potential The minimum electron-beam energy required to produce ions of a particular type in the ion source of a mass spectrometer.

application The system or problem to which a computer is applied. Reference is often made to computation, data processing, and control as the three categories of application.

application layer Layer of 7 of the OSI.

application program A program that performs a task specific to a particular end-user's needs; generally, an application program is any program written on a program development operating system that is not part of a basic operating system.

applications software Programs which are unique to a specific process control system installation or other specific installations, rather than general purpose and of broad applicability.

application time The time available for sealant application after mixing or after thawing a premixed and frozen cartridge of sealant. Acceptability limits established for class A brushable sealants are expressed in the time required for the viscosity to increase to a specified level at 77 °F (25 °C) and 50% relative humidity. The acceptability limits for class B extrudable sealants are expressed in terms of the extrusion rate of a sealant from a 6 fl oz (177 mL) cartridge through a nozzle with a 0.125 in (3.18 mm) diameter orifice, using air pressure of 90 psi +5 (621 kPa +34) in a pneumatic sealant gun. The extrusion rate is expressed in grams/minute or (per some specifications) in cubic centimeters (cc)/min. A minimum extrusion rate after the stated application time is given as the acceptable limit.[AIR4069-90]

applied load 1. Weight carried or force sustained by a structural member in service, in most cases the load includes the weight of the member itself. 2. Material carried by the load- receiving member of a weighing scale, not including any load necessary to bring the scale into initial balance.

approach and landing tests (STS) A series of flight maneuvers involving the Space Shuttle.[NASA]

approach clearance Authorization by ATC for a pilot to conduct an instrument approach.[ARP4107-88]

approach light system (ALS) An airport lighting facility which provides visual guidance to landing aircraft by radiating light beams in a directional pattern by which the pilot aligns the aircraft with the extended centerline of the runway on final approach for landing. See also runway edge light system.[ARP4107-88]

approach phase See mishap, phase of flight.

approach sequence The order in which aircraft are positioned while on approach or awaiting approach clearance.[ARP4107-88]

APT See automatically programmed tools.

APU (auxiliary power unit) A propellant powered device used to generate electrical, gas or fluid power independent of the main propulsion unit.[AIR913-89]

aquaplane See hydroplane.

aquatic plants Plants growing in or on water.[NASA]

aquiculture The cultivation (breeding, raising, and harvesting) of fish, mollusks, shellfish, and/or other aquatic life as sources of food.[NASA]

aquifers Bodies of rock that contain sufficient saturated permeable material to conduct ground water and to yield economically significant quantities of ground water to wells and springs. [NASA]

aragonite A white, yellowish, or gray orthorhombic mineral, that contains calcium carbonate.[NASA]

arbitration bar A test bar cast from molten metal at the same time as a lot of castings; it is used to determine mechanical properties in a standard tensile test, which are then evaluated to determine acceptability of the lot of castings.

arbor 1. A piece of material usually cylindrical in shape, generally applied to locate a rotating part, or cutting tool, about a center.[ARP480A-79] 2. In machine grinding, the spindle for mounting and driving the grinding wheel. 3. In machine cutting, such as milling, the shaft for holding and driving a rotating cutter. 4. Generically, the principal spindle or axis of a rotating machine which transmits power and motion to other parts. 5. In metal founding, a bar, rod or other support embedded in a sand core to keep it from collapsing during pouring.

arbor press A mechanical or hydraulic machine for forcing arbors, mandrels, bushings, shafts or pins into or out of drilled or bored holes.

arc 1. The track over the ground of an aircraft flying at a constant distance from a navigational aid by reference to distance measuring equipment (DME). [ARP4107-88] 2. A segment of the circumference of a circle. 3. The graduated scale on an instrument for measuring angles. 4. A discharge of electricity across a gap between electrical conductors.

archaebacteria Organisms belonging to the taxonomic kingdom of the same name which are characterized by distinct t- and r-RNAs, the absence of peptoglyca cell walls and their possible replacement by a proteinaeous coat, ether-linked lipids from phytayl chains, and occurrence in unusually harsh habitats, e.g. methane, halide and thermoacidic environments. These hardy bacteria are significant in the study of the origin of life.[NASA]

architecture 1. In computers, the design of system and logic organization and information flow relationships in a computer rather than the circuit and component features.[NASA] 2. The structure, functional and performance characteristics of a system, specified in an implementation independent way.

archival file In data processing, storage of seldom used data that must be retained for several years.

arcing time Arcing time (as used for fuses) is the time measured from that point when element melt time ends to that point when current is interrupted and permanently becomes zero. If a mechanical indicator (not presently recommended) is utilized which incorporated a secondary element parallel to the fusible element, arcing time will commence from the point at which indicator melt time ends. Arcing time (as used for breakers) is the time measured from that point when contacts first separate to that point when the current is interrupted and permanently becomes zero.[ARP1199A-90]

arc lamp A high intensity lamp in which a direct current electric discharge produces light that is continuous, as opposed to a flashlamp, which produces pulsed light.

arc line A spectral line in spectroscopy.

arc melting Raising the temperature of a metal to its melting point using heat generated by an electric arc; usually refers to melting in a specially designed furnace to refine a metal, produce an alloy or prepare a metallic material for casting.

arc strike See strike.

arc welding A group of welding processes which produce coalescence of metals by heating them with an arc, with or without pressure or the use of filler metal.

area classification The classification of hazardous (classified) locations by Class I, II or III depending upon the presence of flammable gases or vapors, flammable liquids, combustible dust, or ignitable fibers or flyings and by Division 1 or 2 depending upon the existence of these materials to exist in an ignitable concentration under normal or abnormal conditions.

area classification (Class) 1. Class I locations are those in which flammable gases or vapors are or may be present in the air in quantities sufficient to produce explosive or ignitable mixtures. 2. Class II locations are those that are hazardous because of the presence of combustible dust. 3. Class III locations are those that are hazardous because of the presence of easily ignitable fibers or flyings, but in which such fibers or flyings are not likely to be in suspension in the air in quantities sufficient to produce ignitable mixtures.

area classification (Division) 1. Division 1 (hazardous). Where concentrations of flammable gases or vapors exist (a) continuous or periodically during normal operations; (b) frequently during repair or maintenance or because of leakage; or (c) due to equipment breakdown or faulty operation which could cause simultaneous failure of electrical equipment. 2. Division 2 (normally non-hazardous). Locations in which the atmosphere is normally non-hazardous and may become hazardous only through the failure of the ventilating system, opening of pipe lines, or other unusual situations.

area classification (Group) Identified groups of chemicals and compounds whose air mixtures have similar ease of ignition and explosive characteristics, for the purpose of testing, approval, and area classification. Group A: atmospheres containing acetylene. Group B: atmosphere containing butadiene, ethylene oxide, propylene oxide, acrolein, or hydrogen (or gases or vapors equivalent in hazard to hydrogen). Group C: atmospheres such as cyclopropene, ethyl ether, ethylene, or gases or vapors of equivalent hazard. Group D: atmospheres such as acetone, alcohol, ammonia, benzene, benzol, butane, gasoline, hexane, lacquer solvent vapors, naphtha, natural gas, propane, or gases or vapors of equivalent hazard. Group E: atmospheres containing metal dusts. Group G: atmospheres containing combustible dusts having resistivity of $<10^5$ohm -cm^2

area, light VTOL landing This is a special category landing area which will normally require no construction effort other than the clearing of vegetation. [AIR1489-88]

area meter A device for measuring the flow of fluid through a passage of fixed cross-sectional area, usually through use of a weighted piston or float supported by the flowing fluid.

area navigation (RNAV) A method of navigation that permits aircraft operations on any desired course within the coverage of station-referenced navigation signals or within the limits of self-contained system capabilities.[ARP 4107-88]

area navigation high route A route of flight using area navigation within the airspace extending upward from, and including, 18,000 ft MSL to flight level 45.[ARP4107-88]

area navigation low route A route of flight using area navigation within the airspace that extends upward from 1200 ft above the surface of the earth to, but not including, 18,000 ft MSL.[ARP4107-88]

area navigation, random routes/Random RNAV routes Direct routes, based on area navigation capability, between waypoints defined in terms of degree/distance fixes or offset from published or established routes/airways at specified distance and direction. [ARP4107-88]

area navigation, RNAV waypoint/W/P A predetermined geographical position used for route or instrument-approach definition or for progress reporting purposes that is defined relative to a VORTAC station position.[ARP 4107-88]

area, restricted access An area delineated by barrier or floor marking which is clearly posted against unauthorized removal of material.[AS7200/1-91]

argentometer A hydrometer used to find the concentration of a silver salt in water solution.

argument 1. An independent variable, e.g., in looking up a quantity in a table, the number or any of the numbers which identifies the location of the desired value, or in a mathematical function the variable which, when a certain value is substituted for it, determines the value of the function. 2. An operand in an operation on one or more variables. See also parameter. See also independent variables.

Ariel A satellite of Uranus orbiting at a mean distance of 192,000 kilometers. [NASA]

ARIP (impact prediction) See computerized simulation.

arithmetic ability The capability of performing (at least) addition and subtraction in the PC.

arithmetic and logical unit A component of the central processing unit in a computer where data items are compared, arithmetic operations performed and logical operations executed.

arithmetic check See mathematical

check.

arithmetic element The portion of a mechanical calculator or electronic computer that performs arithmetic operations.

arithmetic expression An expression containing any combination of data names, numeric literals, and named constants, joined by one or more arithmetic operators in such a way that the expression as a whole can be reduced to a single numeric flue.

arithmetic operation A computer operation in which the ordinary elementary arithmetic operations are performed on numerical quantities. Contrasted with logical operation.

arithmetic operator Any of the operators, +, and, -, or the infix operators, +, -, *, /, and **.

arithmetic unit The unit of a computing system that contains the circuits that perform arithmetic operations.

arm 1. A rigid member which extends to support or provide contact beyond the perimeter of the basic item.[ARP 480A-79] 2. Allows a hardware interrupt to be recognized and remembered. Contrasted with disarm; see enable.

armature 1. The core and windings of the rotor in an electric motor or generator. 2. The portion of the moving element of an instrument which is acted upon by magnetic flux to produce torque.

arming Provision of pre-setting a skid control function such as touchdown protection by some specific operating sequence of the aircraft (i.e. gear retraction/extension, squat switch activation, etc). Term may also be applied to systems other than skid control.[AIR1489-88]

armor Mechanical protection usually a metallic layer of tape, braid, or wires. [ARP1931-87]

armored cable A wire or cable covered with armor.[ARP1931-87]

armored meter tube Variable area meter tube (rotometer) of all metal construction utilizing magnetic coupling between the float and an external follower.

Army-Navy instrument program (ANIP) A display research and development program initiated in 1952 that advanced the "contact analog" pictorial display concept, including the "highway in the sky" feature.[ARP4107-88]

ARPA computer network The 'Advanced Research Projects Agency' of the Department of Defense nationwide computer network incorporating digital communication between large numbers of dissimilar computers as well as direct access to programs, data storage, etc. shared by all terminals.[NASA]

array 1. An arrangement of elements in one or more dimensions. See also matrix and vector. 2. In a computer program, a numbered, ordered collection of elements, all of which have identical data attributes. 3. A group of detecting elements usually arranged in a straight line (linear array) or in two dimensional matrix (imaging array). 4. A series of data samples, all from the same measurement point. Typically, an array is assembled at the telemetry ground station for frequency analysis.

array dimension The number of subscripts needed to identify an element in the array.

array process A hardware device that processes data arrays; Fast Fourier transforms (FFT) and power-spectral density (PSD) are typical processes.

array processor The capability of a computer to operate at a variety of data locations at the same time.

arrester A device to impede the flow of large dust particles or sparks from a stack, usually screening at the top.

arresting cable A wire rope which is stretched across a deck or runway, and which is engaged by the aircraft arresting hook to decelerate the aircraft. See also arresting gear.[AIR1489-88]

arresting gear Any gear or apparatus designed to arrest something in its motion either all or part of such gear being external to the object being arrested; specifically, any such apparatus used (i.e., in carrier landings) to arrest airplanes in the landing roll.[AIR1489-88]

arresting hook The hook assembly, usually mounted on the aircraft for the purpose of engaging arresting gear mounted on the deck or runway for deceleration (arrestment) of the aircraft. Also see tail hook.[AIR1489-88]

arresting wire See arresting cable.

arrhythmia Absence of rhythm, as, for example, in heart beat.[NASA]

arrival time The time an aircraft touches down on arrival.[ARP4107-88]

arrow keys Keys on a computer keyboard that will move the cursor.

arrow wings Aircraft wings of V-shaped planform, either tapering or of constant chord, suggesting a stylized arrowhead.[NASA]

ARTCC Air Route Traffic Control Center.

articulated On a landing gear, an arrangement where the gear is hinged to permit folding as desired for retraction or operation, i.e., articulated axle.[AIR 1489-88]

articulated arms (waveguides) A beam-direction arrangement in which light passes through a series of jointed pipes containing optics.

articulated structure A structure—either stationary or movable, such as a motor vehicle or train—which is permanently or semipermanently connected so that different sections of the structure can move relative to the others, usually involving pinned or sliding joints.

artificial aging Heat treating a metal at a moderately elevated temperature to hasten age hardening.

artificial gravity A simulated gravity established within a space vehicle by rotation or acceleration.[NASA]

artificial ice Real ice, but formed by artificial means, such as a spray rig or tanker.[AIR1667-89]

artificial intelligence (AI) 1. A characteristic of a knowledge-based concept or process that involves decision making and reasoning based on stored knowledge. A system possessing such a characteristic is sometimes referred to as an "expert system", i.e., it uses knowledge and inference procedures to solve problems, or diagnose conditions by their symptoms.[ARP1181A-85] 2. A subfield of computer science concerned with the concepts and methods of symbolic inference by a computer and the symbolic representation of the knowledge to be used in making inferences.[NASA] 3. The use of computers to simulate the way the human mind operates, as learning or adaptation.

artificial language A language specifically designed for ease of communication in a particular area of endeavor, but one that is not yet natural to that area. This is contrasted with a natural language which has evolved through long usage.

artificial radioactivity Radioactivity induced by bombarding a material with a beam of energetic particles or with electromagnetic radiation.

artificial satellites Man-made satellites.[NASA]

artificial weathering Producing controlled changes in materials, such as surface appearance, under laboratory conditions that simulate outdoor exposure.

ARTS Automated Radar Terminal Systems.

asbestos A fibrous variety of the mineral horneblend, used extensively for its fire-resistant qualities to make insulation and fire barriers; it has been stitched, bonded or woven into blankets, mixed with portland cement and water to make sheet roofing, wall cladding, drainage tiles and corrosion- resistant pipe, and combined with binders such as asphalt or bentonite to make asbestos felt or plaster.

ASC See Accredited Standard Committee.

ASCII 1. A widely-used code (American Standard Code for Information Interchange) in which alphanumerics, punctuation marks, and certain special machine characters are represented by unique, 7-bit, binary numbers; 128 different binary combinations are possible (2^7=128), thus 128 characters may be represented. 2. A protocol.

ASCII file A text file that uses the ASCII character set.

as-fabricated Describing the condition of a structure or material after assembly, and without any conditioning treatment such as a stress-relieving heat treatment; specific terms such as as-welded, as-brazed or as-polished are used to designate the nature of the final step in fabrication.

as-fired fuel Fuel in the condition as fed to the fuel burning equipment.

ash The noncombustible inorganic matter in the fuel.

ash content The incombustible residue remaining after burning a combustible material completely.

ash-free basis The method of reporting fuel analysis whereby ash is deducted and other constituents are recalculated to total 100%.

ash pit A pit or hopper located below a furnace where refuse is accumulated and from which it is removed at intervals.

ASN.1 Abstract Syntax Notation One. An ISO standard (DIS 8824) that specifies a canonical method of data encoding. This standard is an extension of CCITT standard X.409.

aspect ratio 1. ratio of tire section height to section width. General category: low aspect ratio equal to or less than 0.77. [AIR1489-88] 2. The ratio of span to mean chord of a wing or other airfoil.[ARP 4107-88] 3. The ratio of the square of the span of an airfoil to the total airfoil area, or the ratio of its span to its mean chord.[NASA] 4. The ratio of frame width to height for a television picture, which is 4:3 in the United States, Canada and United Kingdom. 5. In any rectangular structure, such as the cross section of a duct or tubular beam, the ratio of the longer dimension to the shorter. 6. A ratio used in calculating resistance to flow in a rectangular elbow and is the ratio of width to depth.

asphalt A brown to black bituminous solid that melts on heating and is impervious to water but soluble in gasoline; it is used extensively in paving and roofing applications, and in paints and varnishes; it occurs naturally in certain oil-bearing rocks, and can be made by pyrolysis of coal tar, lignite tar and certain petroleums.

asphaltenes Components of bitumens that are soluble in carbon disulphide but not in paraffin naphtha, constitute the solid dispersed particles of the bitumens, and consist of high molecular weight hydrocarbons.[NASA]

aspheric For optical elements, surfaces are aspheric if they are not spherical or flat. Lenses with aspheric surfaces are sometimes called aspheres.

aspirating burner A burner in which the fuel in a gaseous or finely divided form is burned in suspension, the air for combustion being supplied by bringing into contact with the fuel, air drawn through one or more openings by the lower static pressure created by the velocity of the fuel stream.

aspiration Using a vacuum to draw up gas or granular material, often by passing a stream of water across the end of an open tube, or through the run of a tee joint, where the open tube or branch pipe extends into a reservoir containing the gas or granular material. See also vacuum.

ASR 1. Acronym for Airport Surveillance Radar. 2. Acronym for Automatic Send/Receive.

ASR approach Surveillance approach.

as-received fuel Fuel in the condition as received at the plant.

assemble To prepare a machine-language program from a symbolic language program by substituting absolute code for symbolic operation codes and absolute or relocatable addresses for symbolic addresses.

assembler A program that translates symbolic source code into machine instructions by replacing symbolic operation codes with binary operation codes, and symbolic addresses with absolute or relocatable addresses.

assembly 1. A unit constructed of many parts or components, and which functions in service as a single device, mechanism or structure. 2. A mid-level computer language.

assembly (fitting) Assembled and torque tightened fittings, nuts, sleeves and tubing.[MA2005-88]

assembly language A computer programming language, similar to computer language, in which the instructions usually have a one-to-one correspondence with computer instructions in machine language, and which utilizes mnemonics for representing instructions.

assembly list A printed list which is the by-product of an assembly procedure. It lists in logical instruction sequence all details of a routine showing the coded and symbolic notation next to the actual notations established by the assembly procedure. This listing is highly useful in the debugging of a routine.

assembly load The total load imposed upon one gear assembly, i.e., one main gear assembly.[AIR1489-88]

assembly program See assembly system.

assembly system An assembly system comprises two elements, a symbolic language and an assembly program that translates source programs written in the symbolic language into machine language.

assembly time Also called "work life" and "open time". A term used in reference to faying-surface sealants. It refers to the amount of time available after a two-part sealant is mixed, before the faying surfaces, to which sealant was applied, must be closed or "squeezed out". If the time is exceeded, the cure will have progressed too far, (i.e., the material will have become too firm) to permit most of the sealant to be squeezed out for the desired surface to surface contact.[AIR4069-90]

assign To designate a part of a system for a specific purpose.

assignment statement A program statement that calculates the value of an expression and assigns it a name.

associated electrical apparatus Electrical apparatus in which the circuits are not all intrinsically safe but which contains circuits that can affect the safety of intrinsically safe circuits connected to it.

association The combining of ions into larger ion clusters in concentrated solutions.

association reactions Gas phase chemical processes in which two molecular species and B react to form a larger molecule AB. In astrophysics these processes are involved in the 'condensation' of small gaseous molecules into larger species.[NASA]

associative processing (computers) Byte-variable computer processing with multifield search, arithmetic, and logic capability.[NASA]

associative storage A storage device in which storage locations are identified by their contents, not by names or positions. Synonymous with content-addressed storage, contrast with parallel search storage.

astatic Without polarity; independent of the earth's magnetic field.

asteroid belts The location of the orbits of most of the minor planets (estimated at a half million asteroids) between Mars and Jupiter; about 2000 asteroids have been assigned numbers and names.[NASA]

asteroid capture The transfer of an asteroid or comet from the influence of a planet into that of another planet or neutral satellite.[NASA]

asteroid missions Space missions for the study of asteroids and related celestial bodies.[NASA]

asteroids Small celestial bodies revolving around the sun, most having orbits between those of Mars and Jupiter. [NASA]

astigmatism A defect in an optical element that causes rays from a single point in the outer portion of a field of view to fall on different points in the focused image.

astrobiology See exobiology.

astrodynamics 1. The practical application of celestial mechanics, astroballistics, propulsion theory, and allied fields to the problem of planning and directing the trajectories of space vehicles.[NASA] 2. Practical application of fundamental science to the problem of planning and controlling the trajectories of space vehicles.

astrolabe 1. Instrument designed to observe the positions and measure the altitudes of celestial bodies.[NASA] 2. An instrument formerly used to find the altitudes of celestial bodies; a predecessor of the sextant.

astronomical coordinates Coordinates defining a point on the surface of the earth, or of the geoid, in which the local direction of gravity is used as a reference.[NASA]

astronomical theodolite See altazimuth.

astronomy The science that treats the location, magnitudes, motions, and constitution of celestial bodies and structures.[NASA]

astrophysics A branch of astronomy that treats the physical properties of celestial bodies, such as luminosity, size, mass, density, temperature, and chemical composition.[NASA]

asymmetric rotor A rotating machine element whose axis of rotation is not the same as its axis of symmetry.

asymmetry potential The difference in potential between the inside and outside pH sensitive glass layers when they are both in contact with 7 pH solutions. It is caused by deterioration of the pH sensitive glass layers or contamination of the internal fill of the measurement electrode.

asymptotic properties Properties of any mathematical relation or corresponding physical system characterized by an approach to a given value as a expression, containing a variable, tends to infinity.[NASA]

asynchronous 1. A mode of operation in which an operation is started by a signal before the operation on which this operation depends is completed. When referring to hardware devices, it is the method in which each character is sent with its own synchronizing information. The hardware operations are scheduled by "ready" and "done" signals rather than by time intervals. This implies that a second operation can begin before the first operation is completed. 2. Not synchronous with the line frequency as applied to rotating a.e. machinery.

asynchronous communication Often called start/stop transmission, a way of transmitting data in which each character is preceded by a start bit and followed by a stop bit.

asynchronous transmission 1. Transmission in which information character, or sometimes each word or small block, is individually synchronized, usually by the use of start and stop elements. The gap between each character (or word) is not of a necessarily fixed length. (Compare with synchronous transmission.) 2. Asynchronous transmission is called start-stop transmission. 3. Data transmission mode in which the timing is self-determined and not controlled by an external clock.

ATA Air Transport Association of America.

ATARS See automatic traffic advisory and resolution.

ATC Air Traffic Control.

ATC advises A phrase used as a prefix to a message of noncontrol information when it is relayed to an aircraft by someone other than an air traffic controller.[ARP4107-88]

ATC assigned airspace (ATCA) Airspace of defined vertical/lateral limits, assigned by ATC, for the purpose of providing air traffic segregation between the specified activities being conducted within the assigned airspace and other IFR air traffic.[ARP4107-88]

ATC clearance Air Traffic Clearance.

ATC instructions Directives issued by air traffic control that require a pilot to take specific actions.[ARP4107-88]

ATCRBS Air Traffic Control Radar Beacon System.

atelectasis Collapsed or airless state of all or part of the lung.[NASA]

athletic See physical condition.

athodyds See ramjet engines.

ATIS Automatic Terminal Information Service.

atmometer A generic name for any instrument that measures evaporation rates; also known as atmidometer; evaporimeter; evaporation gage.

atmospheric air Air under the prevailing atmospheric conditions.

atmospheric and oceanographic inform sys A data system designed primarily for the interactive manipulation of meteorological satellite images. Capabilities include displaying, analyzing, storing, and manipulating digital data in the field of meteorology and earth resources.[NASA]

atmospheric chemistry Study of the production, transport, modification, and removal of atmospheric constituents in the troposphere and stratosphere. [NASA]

atmospheric circulation Global or hemispheric air movements which can be treated by equations of motion in contrast to atmospheric diffusion which is small random movement not amenable to treatment by these equations. [NASA]

atmospheric cloud physics lab (Spacelab) A NASA Spacelab mission involving cloud physics experiments in zero gravity environment. Also known as ACPL.[NASA]

atmospheric communication Sending signals in the form of modulated light through the atmosphere, without the use of fiber optics to contain and direct the beam.

atmospheric conditions See meteorology.

atmospheric correction Removal of the effects of the intervening atmosphere from satellite imagery.[NASA]

atmospheric corrosion Corrosion that occurs naturally due to exposure to climatic conditions; corrosion rates vary depending on specific global location because of variations in average temperature, humidity, rainfall, airborne substances such as sea spray, dust and pollen, and airborne pollutants such as sulfur dioxide, chlorine compounds, fly ash and other combustion products.

atmospheric electricity Electrical phenomena, regarded collectively, which occur in the earth's atmosphere. Also the study of electrical processes occurring within the atmosphere.[NASA]

atmospheric emission See airglow.

atmospheric entry The penetration of any planetary atmosphere by any object from outer space, specifically, the penetration of the earth's atmosphere by a manned or unmanned capsule or spacecraft.[NASA]

atmospheric general circulation experiment Model experiment of the earth's atmospheric circulation as proposed for a Spacelab flight on which a liquid contained between two concentric spheres is subjected to rotation. The thermal driving force will be a stable radial temperature gradient and an unstable latitudinal gradient.[NASA]

atmospheric impurities See air pollution.

atmospheric lasers The theoretical phenomena whereby the upper atmosphere is used as the lasing medium.

atmospheric loading See pollution transport.

atmospheric monochromator A monochromator in which the optical path is through air. This is the standard type used for visible and infrared wavelengths transmitted by air.

atmospheric noise See atmospherics.

atmospheric optics The study of the topical characteristics of the atmosphere and of the optical phenomena produced by the atmosphere's suspensoids and hydrometeors. It embraces the study of refraction, reflection, diffraction, scattering, and polarization of light, but is not commonly regarded as including the study of any other kinds of radiation.[NASA]

atmospheric pressure 1. Absolute pressure of the atmosphere at a given location and time.[AIR1916-88] 2. The pressure at any point in an atmosphere due solely to the weight of the atmospheric gases above the point concerned.[NASA] 3. The barometric reading of pressure exerted by the atmosphere. At sea level 14.7 lb per sq in. or 29.92 in. of mercury.

atmospheric radiation Infrared radiation emitted by or being propagated through the atmosphere.[NASA]

atmospheric refraction Refraction resulting when a ray of radiant energy passes obliquely through an atmosphere.[NASA]

atmospherics The radiofrequency electromagnetic radiations originating, principally, in the irregular surges of charge in thunderstorm lightning discharges. Atmospherics are heard as a quasi-steady background of crackling noise (static) in ordinary amplitude modulated radio receivers.[NASA]

atmospheric shells See atmospheric stratification.

atmospheric sounding Measurement of atmospheric phenomena generally with instruments carried aloft by spacecraft, rockets, etc.[NASA]

atmospheric stratification The presence of strata or layers in the earth's atmosphere.[NASA]

atmospheric tides Defined in analogy to the oceanic tide as an atmospheric motion on a worldwide scale, in which vertical accelerations are neglected (but compressibility is taken into account). [NASA]

ATO (Aborted Takeoff) See abort and rejected takeoff.

ATO (Assisted Takeoff) A procedure in which an assistant is utilized for takeoff operation such as rocket assist (RATO) or jet assist (JATO). A catapult (shipboard and shorebased) is a launching device which can be reused repeatedly and rapidly to provide takeoff assistance.[AIR1489-88]

atomic clocks Timekeeping devices controlled by the frequency of the natural vibrations of certain atoms.[NASA]

atomic mass See atomic weights.

atomic mass unit A unit for expressing

atomic weights and other small masses; it equals, exactly, 1/12 the mass of the carbon-12 nuclide.

atomic number An integer that designates the position of an element in the periodic table of the elements; it equals the number of protons in the nucleus and the number of electrons in the electrically neutral atom.

atomic weight 1. The weight of an atom according to a scale of atomic weight units, awu, valued as one-twelfth the mass of the carbon atom.[NASA] 2. The weight of a single atom of any given chemical element; it is usually taken as the weighted average of the weights of the naturally occurring nuclides, expressed in atomic mass units.

atomization Mechanically producing fine droplets or mist from a bulk liquid or molten substance.

atomizer A device by means of which a liquid is reduced to a very fine spray.

atom probe An instrument, consisting of a field-ion microscope with a probe hole in its screen that opens into a mass spectrometer, used to identify a single atom or molecule on a metal surface.

ATP Airline Transport Pilot.

attachment hardware Any hardware other than retractors designed for terminating the webbing of a torso restraint system.[AS8043-86]

automatic Automatic means self-acting, operating by its own mechanism when actuated by some impersonal influence, as for example, a change in current strength, pressure, temperature, or mechanical configuration.[ARP1199A-90]

attemperation Regulating the temperature of a substance—for instance, passing superheated steam through a heat exchanger or injecting water mist into it to regulate final steam temperature.

attention The active selection of, and emphasis on, one component of a complex experience, and the narrowing of the range of objects to which the organism is responding.[ARP4107-88]

attention, anomalies of Misallocation or untimely interruption of attention to a task.[ARP4107-88]

attention, anomalies of boredom A psychological state resulting from any activity that lacks motivation, or from enforced continuance in an uninteresting situation.[ARP4107-88]

attention, anomalies of channelized attention The focusing of conscious attention upon a limited number of environmental cues which may lead to the exclusion of others of an objectively higher or more immediate priority. Channelized attention is an active anomaly of attention sometimes referred to as fixation.[ARP4107-88]

attention, anomalies of cognitive task saturation That state in which the cognitive task demands exceed the individual's capacity to respond appropriately to all of them. Under such a situation the individual often focuses attention on a subset of the cognitive demands present.[ARP4107-88]

attention, anomalies of complacency A state of adjustment, or a dynamic balance between organism and environment, typified by established habits and responses that are in a quiescent stage. Self-satisfaction accompanied by unawareness of actual dangers or deficiencies.[ARP4107-88]

attention, anomalies of distraction 1. An undesired redirection of the focus of attention by an environmental cue or mental process. (a) External Distraction: Interruption of attention by a nontask related environmental cue. (b) Internal Distraction: Interruption of attention by a nontask related mental process.[ARP4107-88] 2. A stimulus that causes an undesired shift in attention. The distracting stimulus may be either an environmental cue (external distraction) or a mental process (internal distraction).[ARP4107-88]

attention, anomalies of fascination An anomaly of attention in which a person monitors the relevant environmental cues around him/her but fails to respond to them because of a sense of unreality or detachment from events, as if he/she were viewing them from the outside. Fascination is usually associated with a high-stress or crisis situation.[ARP4107-88]

attention, anomalies of habit-pattern interference Reverting to previously learned response modes which are objectively inappropriate to the task at hand. Habit- pattern interference usually occurs at the preconscious level of awareness. See also transfer of training.[ARP4107-88]

attention, anomalies of habituation Adaptation and subsequent inattention to an environmental cue after prolonged or repeated exposure to it.[ARP4107-88]

attention, anomalies of inattention A state of self-reduced conscious attention due to a sense of security, self-confidence, or a perceived absence of threat from the environment. (a) General Inattention: Nonselective inattention typically due to boredom or complacency. (b) Selective Inattention: Insufficient attending to relevant environmental cues due to lack of knowledge or an inappropriate perceptual or attitudinal set.[ARP4107-88]

attention, focus of The part of the span of attention directed toward conscious information processing.[ARP4107-88]

attention, level of The relative proportions of the span, focus, and margin of attention afforded to information processing.[ARP4107-88]

attention, margin of The span of attention minus the focus of attention, or a person's remaining capacity to focus conscious attention.[ARP4107-88]

attenuate To weaken or make thinner—for example, to reduce the intensity of sound or ultrasonic waves by passing them through an absorbing medium.

attenuation 1. Power loss in an electrical system. In cables the loss is expressed in decibels per unit length of cable, at a given frequency.[ARP1931-87] 2. Reducing in intensity.[NASA] 3. The loss of amplitude in a signal as it is transmitted through a conductor. See also gain.

attenuation coefficients A measure of the space rate of attenuation of any transmitted electromagnetic radiation.[NASA]

attenuator 1. An optical device which reduces the intensity of a beam of light passing through it. 2. An electrical component that reduces the amplitude of a signal in a controlled manner.

attitude 1. The orientation of the three major axes of an aircraft (longitudinal, lateral, and vertical) with respect to a fixed reference such as the horizon, the relative wind, or direction of flight. [ARP4017-88] 2. An enduring, learned predisposition to behave in a consistent way toward a given class of objects; a persistent mental or neural, or both, state of readiness to react to a certain object or class of objects, not as they are but as they are conceived to be. [ARP4107-88] 3. The position or orientation of an aircraft, spacecraft, etc., either in motion or at rest, as determined by the relationship between its axes and some reference line or plane or some fixed system of reference axes.[NASA] 4. The position of an object in space determined by the angles between its axes and a selected set of planes.

attitude control The regulation of the attitude of an aircraft, spacecraft, etc. Also a device or system that automatically regulates and corrects attitude, especially of a pilotless vehicle.[NASA]

attitude director indicator (ADI) An integrated flight display in aircraft cockpits which combines the attitude indicator with steering bars (or command bars) that aid the pilot in navigating with VOR or ILS signals.[ARP4107-88]

attitude gyros Gyro-operated flight instruments that indicate the attitude of an aircraft or spacecraft with respect to a reference coordinate system through 360 degrees of rotation about each axis of the craft.[NASA]

attitude indicator An instrument that shows the pitch and roll attitudes of the aircraft with respect to the horizon.[ARP4107-88]

attitudinal set A predisposition, rooted in attitude(s), which may cause the individual to respond in a particular manner to a set of stimuli.[ARP4107-88]

attribute sampling A type of sampling inspection in which an entire production lot is accepted or rejected depending on the number of items in a statistical sample that have at least one characteristic (attribute) that does not meet specifications.

auctioneering device See signal selector.

audible device follower See auxiliary output.

audio Pertaining to audible sound—usually taken as sound frequencies in the range 20 to 20,000 Hz.

audio frequency The band of frequency which is audible to the human ear. Usually 20 to 20,000 Hz.[ARP1931-87]

audio data Useful information at audio signal frequency.[NASA]

audio frequencies Frequencies corresponding to normally audible sound waves.[NASA]

audiometer An instrument used to measure the ability of people to hear sounds; it consists of an oscillator, amplifier and attenuator, and may be adapted to generate pure tones, speech or bone-conducted vibrations.

audio signals Signals with a bandwidth of less than 20 kilohertz.[NASA]

auditory sensation areas In acoustics, the frequency region enclosed by the curves defining the threshold of pain and the threshold of audibility.[NASA]

aufeis (ice) Icing of ground or river water in Arctic areas with continuous permafrost on which the water has continued to flow.[NASA]

auger 1. A woodboring tool consisting of a shank with a T-shaped handle. 2. A feeding device consisting primarily of a set of spiral blades mounted on a central shaft or fastened together to make a spiral rotating assembly; it may rotate in a tube, trough or housing to move powdered, granular or semisolid material axially; in some applications, the auger may be constructed of two counterspiraled augers, which feed material toward the midpoint or outward from the midpoint of the axis depending on the direction of rotation.

AUI See Access Unit Interface.

auroral activity See auroras.

auroral zones Roughly circular bands around either geomagnetic pole above which there is a maximum of auroral activity. The zones lie about 10° to 15° of geometric latitude from the geomagnetic poles.[NASA]

auroras Sporadic radiant emissions from the upper atmosphere over middle and high latitudes.[NASA]

austenite A solid solution of carbon in gamma-iron.[NASA]

austenitic stainless steel 1. Steels having at room temperature a microstructure consisting, at least predominantly, of austenite. Their austenitic microstructure is attained above all by alloying conditions, e.g. manganese and nickel.[NASA] 2. An alloy of iron containing at least 12% Cr plus sufficient Ni (or in some specialty stainless steels, Mn) to stabilize the face-centered cubic crystal structure of iron at room temperature.

austere airfields Those airfields within navigation aids and, in most cases, having short landing areas within paved landing surfaces or other facilities necessary for operation of typical medium/large-size transport aircraft. [ARP4107-88]

autocatalytic degradation The phenomenon whereby the breakdown products of the initial phase of degradation act to accelerate the rate at which subsequent degradation proceeds.[ARP 1931-87]

autoclave An airtight vessel for heating its contents and sometimes agitating them; it usually uses high pressure steam to perform processing, sterilizing or cooking steps using moist or dry heat.

autocollimator A telescopic sight including a light source and a partially reflecting mirror, focused to infinity, for use in measuring small angular motion and checking alignment. See also collimators.

autocorrelation In statistics, the simple linear internal correlation of members of a time series (ordered in time or other domains).[NASA]

AUTOEXEC The name of the file in MS-DOS that has commands to be executed when the computer is booted. Usually named AUTOEXEC.BAT.

autoignition temperature (AIT) Temperature at which the fluid flashes into flame without an external ignition source and continues burning. Actual value is to be determined by one of several approved test methods.[AIR 1916-88]

autokinesis See illusion.

automate 1. To apply the principles of automation. 2. To operate or control by automation. 3. To install automatic procedures, as for manufacturing, servicing, etc.

automated en route ATC An air traffic control technology which allows computers to make decisions about conflict resolution, the generation of clearances, and their automatic transmission, with the operator standing by to take over in an emergency.[NASA]

automated guideway transit vehicles A system of a large number of captive vehicles traveling at relatively close headways on an exclusive guideway controlled by a computer.[NASA]

automated mixed traffic vehicles (AMTV) Low speed surface vehicles automatically operated and controlled in a pedestrian environment by following a buried wire in the roadways sensing obstacles and stopping at predetermined spots for passenger exit and entry.[NASA]

automated pilot advisory system An airport advisory system and an air traffic advisory system designed to improve airport and air traffic advisories at high density uncontrolled airports.[NASA]

automated radar terminal systems (ARTS) 1. A highly automatic radar system that displays, for terminal aircraft controllers, information about aircraft that they are controlling. ARTS gives identification, flight plan data, and other flight-associated information (for example, altitude and speed). (a) ARTS II: A programmable, nontracking, computer-aided display subsystem capable of modular expansion. ARTS II systems provide a level of automated air traffic control capability at terminals having low-to-medium activity. Flight identification and altitude may be associated with the display of secondary radar targets. (b) ARTS III: The Beacon Tracking Level (BTL) of the modular programmable ARTS in use at medium- to high-activity terminals. ARTS III detects, tracks, and predicts secondary radar-derived aircraft targets. These are displayed by computer-generated symbols and alphanumeric characters depicting flight identification, aircraft altitude, ground speed, and flight plan data. (c) ARTS IIIA: The Radar Tracking and Beacon Tracking Level (RT&BTL) of the modular, programmable ARTS. ARTS IIIA detects, tracks, and predicts primary as well as secondary radar-derived aircraft targets. ARTS IIIA is a more sophisticated computer-driven system that upgrades the ARTS III system by providing improved tracking, continuous data recording, and fail-soft operations.[ARP4107-88] 2. Radar tracking system for use in a terminal area. Primary and secondary radar targets are detected and data for the two are correlated for transmission to a central computer.[NASA]

automatic 1. Having the power of self-motion; self-moving; or self-acting: an automatic device. 2. A machine that operates automatically. 3. Pertaining to a process or device that, under specified conditions, functions without intervention by a human operator.

automatically programmed tools A numerical language.

automatic alitude reporting That function of a transponder which responds to Mode C interrogations by transmitting the aircraft's altitude in 100-ft increments.[ARP4107-88]

automatic carrier landing system (ACLS) U.S. Navy final approach equipment, consisting of precision tracking radar coupled to a computer data link, which provides continuous information to the aircraft, monitoring capability to the pilot, and a backup approach system.[ARP4107-88]

automatic control 1. Control of devices and equipment, including aerospace vehicles by automatic means.[NASA] 2. The type of control in which there is no direct action of man on the controlling device.

automatic control engineering The branch of science and technology which deals with the design and use of automatic control devices and systems.

automatic controller Any device which measures the value of a process variable and generates a signal or some controlling action to maintain the value in correspondence with a reference value, or setpoint.

automatic control panel A panel of indicator lights and switches on which are displayed an indication of process conditions, and from which an operator can control the operation of the process.

automatic control system See control system, automatic.

automatic data processing See data processing.

automatic direction finder (ADF) A radio device that is used to locate the direction (bearing) towards a signal on a selected radio frequency; used in navigation, for instrument approaches to airfields, and for locating lost aircraft or persons who might be transmitting a particular radio frequency.[ARP4107-88]

automatic error correction A technique usually requiring the use of special codes or automatic retransmission, which detects and corrects errors occurring in transmission. The degree of correction depends upon coding and equipment configuration.

automatic flight control systems Automatic flight control systems consist of electrical, mechanical, and hydraulic components that generate and transmit automatic control commands which provide pilot assistance or relief through automatic or semiautomatic flight path control or which automatically control airframe response to disturbances. This classification includes automatic pilots, stick or wheel steering, autothrottles, and similar control mechanizations.[AIR1916-88]

automatic frequency control 1. An arrangement whereby the frequency of an oscillator is automatically maintained within specified limits.[NASA] 2. A device or circuit designed to maintain the frequency of an oscillator within a preselected band of frequencies. In an FM radio receiver, the circuitry that senses frequency drift and automatically controls an internal oscillator to compensate for the drift.

automatic gain control 1. A process by which gain is automatically adjusted as a function of input or other specified parameter.[NASA] 2. An auxiliary circuit that adjusts gain of the main circuit in a predetermined manner when the value of a selected input signal varies.

automatic lighter A means for starting ignition of fuel without manual intervention. Usually applied to liquid, gaseous or pulverized fuel.

automatic locking retractor A retractor incorporating adjustment hardware by means of a positive self locking mechanism which is capable, when locked, of withstanding restraint forces.[AS8043-86]

automatic pilot 1. An aircraft subsystem that automatically flies the airplane, maintaining a specified course or heading and altitude. When the automatic pilot is used it substitutes for the pilot as an active component in the flight control loop, and the pilot takes the role of a monitor of the automatic pilot flying the airplane.[ARP4107-88] 2. Equipment which automatically stabilizes the attitude of a vehicle about its pitch, roll, and yaw axes.[NASA] 3. An automatic control system adapted for maintaining an aircraft in stable level flight or for executing selected maneuvers.

automatic reset See reset.

automatic rocket impact predictors See computerized simulation.

automatic self test Self-test to that degree of fault detection and isolation which can be achieved entirely under computer control, without human intervention.[ARD50010-91]

automatic send/receive (ASR) A teletypewriter unit with keyboard, printer, paper tape, reader/transmitter, and paper tape punch. This combination of units may be based on-line or off-line and, in some cases, on-line and off-line simultaneously.

automatic terminal information service (ATIS) The continuous broadcast of recorded noncontrol information in selected terminal areas. Its purpose is to improve controller effectiveness and to relieve radio frequency congestion by automating the repetitive transmission of essential but routine information.[ARP4107-88]

automatic test That performance assessment, fault detection, diagnosis, isolation, and prognosis which is performed with a minimum of reliance on human intervention. This may include BIT.[ARD50010-91]

automatic test equipment (ATE) Equipment that automatically carries out a program of testing for possible malfunction, with the minimum reliance upon human intervention. The ATE should be self checking to ensure that the ATE itself does not lead to additional difficulty.[AIR1916-88]

automatic traffic advisory and resolution (ATARS) Ground based collision avoidance system using the surveillance and data link capabilities of the discrete address beacon system (DABS).[NASA]

automatic utility translator (AUTRAN) A process control language and system offered by Control Data Corporation.

automatic weather stations Weather stations at which the services of observers are not required. They are usually equipped with telemetric apparatus.[NASA]

automatic zero- and full-scale calibration Zero and sensitivity stabilization by servos for comparison of demodulated zero- and full-scale signals with zero- and full-scale references.

automation 1. The implementation of processes by automatic means. 2. The theory, art, or technique of making a process more automatic. 3. The investigation, design, development, and application of methods of rendering processes automatic, self-moving, or self-controlling. 4. The conversion of a procedure, a process, or equipment to automatic operation.

automotive engineering A branch of mechanical engineering that deals with design and construction of landgoing vehicles, especially self-propelled highway vehicles such as automobiles, trucks and buses.

autonomous space clocks Standard Time scale instruments aboard spacecraft with provisions for synchronization with existing satellite-based system (global positioning system, for example).[NASA]

autopilot See automatic pilot.

autoradiography A technique for producing a radiographic image using ionizing radiation produced by radioactive decay of atoms within the test object itself.

autorotation A rotorcraft flight condition in which the lifting rotor is driven entirely by action of the air when the rotorcraft is in motion.[ARP4107-88]

Autosyn A trade name for a type of synchro.

auto throttle(s) A control system that positions and adjusts the throttle(s) of an aircraft to maintain a constant airspeed, which is set (and can be changed) by the pilot.[ARP4107-88]

auto-tracking antenna A receiving antenna which always points to the transmitting site, automatically tracking all movements of the vehicle being telemetered.

autotransformer A type of transformer in which certain portions of the windings are shared by the primary and secondary circuits.

autotrophs Organisms capable of synthesizing organic nutrients directly from simple inorganic substances such as carbon dioxide and inorganic nitrogen.[NASA]

auto-zero logic module A component of a digital controller whose function is primarily to establish an arbitrary zero-reference value for each individual measurement.

AUTRAN See automatic utility translator.

autumn The season of the year between summer an winter. Its beginning is marked by the autumnal equinox and its end by the winter solstice.[NASA]

auxiliary airfoil A secondary airfoil, such as a slat, flap, or tab, that supplements or aids flight in some manner, as by creating an additional force, or by providing a smooth airflow.[ARP 4107-88]

auxiliary contact See auxiliary output.

auxiliary device 1. Generally, any device which is separate from a main device but which is necessary or desirable for effective operation of the system. 2. Specifically, any device used in conjunction with an instrument to extend its range, increase its accuracy, otherwise assist in making a measurement, or perform a function not directly involved in making the measurement.

auxiliary means A device or subsystem, usually placed ahead of the primary detector, which alters the magnitude of the measured quantity to make it more suitable for the primary detector without changing the nature of the measured quantity.

auxiliary output (auxiliary contact) A secondary output.

auxiliary panel 1. A panel which is not in the main control room. The front of an auxiliary panel is normally accessible to an operator but the rear is normally accessible only by maintenance personnel. 2. Located at an auxiliary location.

auxiliary power Those elements of secondary power related to main engine bleed air and shaft power extraction or power generation separate from the main engines. Included are engine bleed air systems, remote engine driven gearboxes, engine starting systems, auxiliary power units, and emergency power systems. See also secondary power. [ARP906A-86]

auxiliary power unit (APU) An internal combustion (usually gas turbine) engine used as the prime mover for equipment which supplies a starter with a desired medium of energy. Also frequently used as a source of bleed air for environmental control systems and as a source of shaft power for electric or hydraulic power generation.[ARP 906A-86]

auxiliary storage A storage device in addition to the main storage of a computer, e.g., magnetic tape, disk, magnetic drum or core. Auxiliary storage usually holds much larger amounts of information than the main storage, and the information is accessible less rapidly. Contrasted with main storage.

auxiliary unit A separate galley unit for food or beverage and liquor service. [AS1426-80]

availability 1. A measure to the degree to which an item is in operable and committable state at the start of a mission when the mission is called for at an unknown (random) point in time. [ARD50010-91] [AIR1916-88] 2. The number of hours in the reporting period less the total downtime for the reporting period divided by the number of hours in the reporting period (expressed in percent).

availability factor The fraction of the time during which the unit is in operable condition.

available draft The draft which may be utilized to cause the flow of air for combustion or the flow of products of combustion.

available energy Energy that theoretically can be converted to mechanical power.

available heat In a thermodynamic working fluid, the amount of heat that could be transformed into mechanical work under ideal conditions by reducing the temperature of the working fluid to the lowest temperature available for heat discard.

available power An attribute of a linear source of electric power defined as $Vrms/4R$, where $Vrms$ is the open circuit rms voltage of the power source and R is the resistive component of the internal impedance of the power source.

available power gain An attribute of a linear transducer defined as the ratio of power available from the output terminals of the transducer to the power available from the input circuit under specified conditions of input termination.

available work The capacity of a fluid or body to do work if applied to an ideal engine.

avalanche Production of a large number of ions by cascade action in which a single charged particle, accelerated by a strong electric field, collides with neutral gas molecules and ionizes them.

avalanche photodiode (APD) A photodiode designed to take advantage of avalanche multiplication of photocurrent. As the reverse-bias voltage approaches the breakdown voltage, hole-electron pairs created by absorbed photons acquire sufficient energy to create additional hole-electron pairs when they collide with substrate atoms, producing a multiplication effect.

average life The mean value for a normal distribution of lives. The term is generally applied to mechanical failures resulting from "wearout".[ARD50010-91]

average outgoing quality limit The average percent of defective units that remain undetected in all lots that pass final inspection; it is a measure of the ability of sampling inspection to limit the probability of shipping defective product; here, a defective unit is considered to be one containing at least one attribute that does not meet specifications.

average-position action A type of control-system action in which the final control element is positioned in either of two fixed positions, the average time at each position being determined from some function of the measured value of the controlled variable.

average value The average root mean square (rms) value of phase quantities is the arithmetical sum of the phase rms values divided by the number of phases.[AS1212-71]

averaging pitot tube An adaptation of the pitot tube in which a multiple-ported pitot tube spans the process tube; total pressure is measured as a composite of the pressures on several ports facing upstream while static pressure is measured using one or more ports facing downstream; the device works best for clean liquids, vapors and gases, but can be used for streams containing suspended solids or viscous contaminants if the purging flow is supplied to the measuring tube.

averaging system The type of fault-tolerant system using two or more active channels wherein the individual channel outputs are summed to provide an average output. All channels are normally operative so performance degradation may occur after a failure. An example of an averaging system is the use of multiple control surfaces on an airplane, each individually actuated.[ARP1181A-85] [AIR1916-88]

aviation meteorology Weather conditions and meteorological studies pertaining to aeronautics.[NASA]

avionics Electrical and electronic equipment used in aviation, principally for navigation and communication.[ARP 4107-88]

avoidance procedure A procedure which attempts avoidance of an area of known or predicted windshear (not yet penetrated).[ARP4109-87]

awards Distinctions that are bestowed upon a person or persons due to their special contributions to a field.[NASA]

awareness, conscious level of A theoretical level of mental awareness at which active information processing or "thinking" takes place. Only one operation at a time can take place at the conscious level.[ARP4107-88]

awareness, level of The theoretical sources of the mental activity which operates in our behavior. The extent to which an organism is conscious of something; the act of "taking account" of an object or state of affairs.[ARP4107-88]

awareness, preconscious level A theoretical level of mental awareness which is the repository of short-term and long- term memory and overlearned response modes and habit patterns. Actions controlled by the preconscious allow us to do more than one thing at a time.[ARP4107-88]

awareness, subconscious level The theoretical repository of information and response modes not available at the conscious level. Reflexes and psychological defense mechanisms operate at the subconscious level.[ARP4107-88]

AWG American Wire Gage.

axial displacement The incremental difference between an initial position and a final position resulting from a force applied along the axis of a component.[ARP914A-79]

AXAF See X Ray Astrophysics Facility.

axes (coordinates) See coordinates.

axial fan Consists of a propeller or disc

type of wheel within a cylinder discharging the air parallel to the axis of the wheel.

axial-flow Describing a machine such as a pump or compressor in which the general direction of fluid flow is parallel to the axis of its rotating shaft.

axial hydraulic thrust In single-stage and multiple- stage pumps, the axial component of the summation of all unbalanced impeller forces.

axial modes Regimes of vibration along a given axis.[NASA]

axial runout For a rotating member, the total amount that a specific surface deviates from a plane perpendicular to the axis of rotation in one complete revolution; it is usually expressed in 0.001 in., or some other suitable unit of measure, taken at a specific radial distance from the axis of rotation.

axisymmetric focused-jet amplifier An amplifier which utilizes control of the attachment of an annular jet to an axisymmetric flow separator, (that is, control of the focus of the jet) to modulate the output.[ARP993A-69]

axle 1. A supporting shaft upon which a wheel turns or which provides support for a rotating member.[AIR1489-88] 2. A rod, shaft or other supporting member that carries wheels and either transmits rotating motion to the wheels or allows the wheels to rotate freely about it.

axle base The dimension between axle centerlines on a multiple axle landing gear.[AIR1489-88]

axle beam The structural member which connects and supports axles on a multiple axle gear arrangement.[AIR1489-88]

axle sleeve A tubular sleeve which fits over the axle and supports the wheel bearings. Function includes: (a) protection of the axle from galling or damage from bearing races, and (b) permits wheel tire assembly buildup and torquing adjustment in the shop and therefore protects against sand, dust, and foreign object damage in installation on the aircraft.[AIR1489-88]

azeotrope A mixture whose evolved vapor composition is the same as the liquid it comes from. This phenomenon occurs at one fixed composition for a given system. At either side of the azeotropic point, the vapors will have different compositions from that of the liquid they evolved from. Such mixtures act as pure substances in distillation and thus are inseparable by standard distillation methods. Azeotropic distillation is necessary to separate such a mixture.

azeotropic distillation A distillation technique in which one of the product streams is an azeotrope. It is sometimes used to separate two components by adding a third, which forms an azeotrope with one of the original two components.

azimuth Horizontal direction or bearing.[NASA]

azimuth angle An angular measurement in a horizontal plane about some arbitrary center point using true North or some other arbitrary direction as a reference direction (0°).

azimuth circle A ring scale graduated from 0° to 360°, and used with a compass, radar plan position indicator, direction finder or other device to indicate compass direction, relative bearing or azimuth angle.

azoles Compounds that contain a five-membered heterocylic ring containing one or more nitrogen atoms.[NASA]

B

B-A-W devices Bulk acoustic wave devices.[NASA]

babbitt Any of the white alloys composed principally of lead or tin which are used extensively to make linings for sliding bearings.

babbitt metal Any of the white alloys composed primarily of tin or lead and of lesser amounts of antimony, copper, and other metals, and used for bearings.[NASA]

babble The composite signal resulting from cross talk among a large number of interfering channels.

backbone The trunk media of a multimedia LAN separated into sections by bridges, routers, or gateways.

backcoating Coating material which is deposited on the side opposite that which is being coated in the evaporated coating process, generally resulting in an objectionable appearing film or haze. [ARP924-91]

back draft A reverse taper on the sidewall of a casting mold or forging die that prevents a casting pattern or forged part from being removed from the cavity.

backfigure antennas Antennas consisting of radiating feeds, reflector elements, and reflecting surfaces such that the antennas function as open resonators, with radiation from the open end of the resonator.[NASA]

background In radiation counting, a low-level signal caused by radiation from sources other than the source of radiation being measured.

background discrimination The ability of a measuring instrument or detection circuit to distinguish an input signal from electronic noise or other background signals.

background luminance The luminance of an area on the face of a display that is NOT illuminated by the graphics presented upon the face of the display. [ARP1782-89]

background noise 1. In recording and reproducing, the total system noise independent of whether or not a signal is present. The signal is not to be included as part of the noise. In receivers, the noise in the absence of signal modulation on the carrier.[NASA] 2. Undesired signals or other stimuli that are always present in a transducer output or electronic circuit, regardless of whether a desired signal or stimulus is also present.

background program A program of the lowest urgency with regard to time and which may be preempted by a program of higher urgency and priority. Contrast to foreground program.

backhand welding Laying down a weld bead with the back of the welder's principal hand (the one holding the torch or welding electrode) facing the direction of welding; in torch welding, this directs the flame backward against the weld bead to provide postheating.

backing pump In a vacuum system using two pumps, the pump discharging directly to the atmosphere which reduces system pressure to an intermediate value, usually 10^{-2} to 10^{-5} psia. Also known as fore pump.

backing ring A ring of steel or other material placed behind the welding groove when joining tubes or pipes by welding, to confine the weld metal.

backings See backups.

backing strip A piece of metal, asbestos or other nonflammable material placed behind a joint prior to welding to enhance weld quality.

backlash 1. In an aircraft control system, a looseness or freeplay in the linkage between crew station controls (input) and the device being controlled (output). In a landing gear system similarly, any looseness, freeplay or slop in the system, i.e., nose wheel steering backlash/ freeplay.[AIR1489-88] 2. In an aircraft control system, a looseness or play in the linkage between the cockpit controls and the control surfaces, or between the cockpit controls and a mechanical feel system.[ARP4107-88] 3. The controlled load motion, due to clearance in actuation elements including the load attach point, and usually expressed in terms of absolute load motion.[AIR1916-88] 4. In a mechanical linkage or gear train, the amount by which the driving shaft must rotate, when reversing direction, in order to merely take up looseness in the linkage or gear train before it begins to transmit motion in the reverse direction. 5. The difference in actual values of a controlled variable when a control dial is brought to the same indicated position from opposite rotational directions.

backlobes Radiation lobes whose axes make angles of approximately 180 degrees with respect to the axes of the major lobes of the antennas; by extension, radiation lobes in the half-space opposed to the direction of peak activity.[NASA]

backmounted A connector installed with its mounting flange positioned behind the mounting surface when looking at the mating face or front side of the connector.[ARP914A-79]

back pressure The pressure level measured in the exhaust collector of the starter or auxiliary power unit.[ARP 906A-91]

backpressure Pressure exerted backward; in a field of fluid flow, a pressure exerted contrary to the pressure producing the main flow, i.e., brake back pressure, net operating pressure is equivalent to gross system pressure less back pressure and other mechanical losses.[AIR1489-88]

backscattering The scattering of light in the direction opposite to the original one in which it was traveling.

backshell A connector accessory or component, which may or may not be supplied with the connector, to provide for strain relief, tighter harness routing in restricted space, shielding from electrical interferences or positive moisture.[ARP914A-79]

backstep sequence A method of laying down a weld bead in which a segment is welded in one direction, then the torch is moved in the opposite direction a distance approximately twice the length of the first segment, and another segment is welded back toward the first; thus the general direction of progress along the joint is opposite to the direction of welding individual segments, with the end point of each segment coinciding with the starting point of the preceding segment.

backtracking A technique used to synchronize mixed- signal simulation systems where an analog simulator is required to back up to a previous time point in order to process a signal originating in the digital simulator.

backup 1. An item kept available to replace an item which fails to perform satisfactorily. An item under development intended to perform the same general function of another item also under development performs.[NASA] 2. An item or system kept available to replace an item or system, i.e., a backup gear extension system.[AIR1489-88] 3. Equipment which is available to complete an operation in the event that the primary equipment fails. 4. A copy of a computer diskette which protects against destruction or loss of the original.

back-up block A back-up block is a piece of metal, usually steel, shaped to support the end of a multi-contact electric connector. Its mass is used to resist the impact of driving taper pins into the taper pin receptacles of the connector.[ARP592-61]

backup copy In data processing, a copy of data or a program that can be used if the original copy is lost or destroyed.

backup seal A seal on the dry side of

the integral fuel tank. It is considered a secondary or redundant seal—never a primary seal.[AIR4069-90]

back-up stiffness The mechanical stiffness of the actuator attach point(s) through which load reaction forces are transmitted.[AIR1916-88]

backup system A mode of control which is engaged upon failure of the primary operational system. Usually used to refer to a system which is as independent as possible from the primary system. Sometimes used as protection against multiple generic failures. [ARP1181A-85]

backward differencing A method of solving a parabolic problem for approximating a time derivative in terms of a previous time step.[NASA]

backward facing steps A step structure which faces an oncoming flow.[NASA]

backward waves In traveling wave tubes, waves whose group velocity is opposite to the direction of electron-stream motion.[NASA]

bactericide Agent that destroys micro-organisms.[NASA]

baffle 1. Plate that regulates the flow of a fluid, e.g., a heat exchanger, boiler flue, or automotive muffler.[NASA] 2. A plate or vane, plain or perforated, used to regulate or direct the flow of fluid. 3. A cabinet or partition used with a loudspeaker produced simultaneously by the front and rear surfaces of the diaphragm.

baffle-nozzle amplifier A device for converting mechanical motion to a pneumatic signal which consists of a supply tube ending in a small nozzle and a movable baffle plate attached to a mechanical arm; the supply tube has a restriction a short distance before the nozzle, so that as the baffle plate moves closer to the nozzle opening the pressure rises in the section of the supply tube between the restriction and the nozzle; arm motion and nozzle clearance are small—on the order of 0.2 mm or less; a baffle-nozzle amplifier serves as the primary detector in almost all pneumatic transmitters and controllers. Often referred to as a flapper-nozzle amplifier because the baffle plate is mounted on a pivoting arm.

baffle plate A tray or partition, solid or perforated, positioned in the flowpath through a process vessel so as to cause the process stream to flow in a certain direction, to reverse its direction of flow, or to slow its velocity.

baffle-type collector A device in gas paths utilizing baffles so arranged as to deflect dust particles out of the gas stream.

bag A deep bulge in the shell or of a furnace or fire-tube boiler.

bag filter A device containing one or more cloth bags for recovering particles from the dust laden gas or air which is blown through it.

bag-type collector A filter wherein the cloth filtering medium is made in the form of cylindrical bags.

bakeout Heating the surfaces of a vacuum system during evacuation to degas them and aid the process of reaching a stable final vacuum level. See also degassing.

balance 1. Generically, a state of equilibrium—static, as when forces on a body exactly counteract each other, or dynamic, as when material flowing into and out of a pipeline or process has reached steady state and there is no discernible rate of change in process variables. 2. An instrument for making precise measurements of mass or weight.

balanced amplifiers See push-pull amplifiers.

balanced field length When the actual field length is equal to the critical field length, then it is designated a balanced field length. See also critical field length.[AIR1489-88]

balanced (to ground) See unbalanced

(to ground).

balance weight A mass positioned on the balance arms of a weighing device so that the arms can be brought to a predetermined position (null position) for all conditions of use.

balancing equipment A combination of the balance machine, proving rotor, tooling, and accessories necessary to accomplish a balancing operation.[ARP 533A-89]

balancing machine A machine that provides a measure of the unbalance in a rotor which can be used for adjusting the mass distribution of that rotor mounted on it so that once per revolution vibratory motion of the journals or force on the bearings can be reduced if necessary.[ARP533A-89]

ballast 1. Relatively dense material placed in the keel of a ship or gondola of a lighter-than-air craft to increase stability or control buoyancy, or both. 2. Crushed stone placed along a railroad bed to help support the ties.

ball bearing A type of antifriction bearing in which the load is borne on a series of hard spherical elements (balls) confined between inner and outer retaining rings (races).

ball burnishing 1. Producing a smooth, dimensionally precise hole by forcing a slightly oversize tungsten-carbide ball through a slightly undersize hole at high speed. 2. A method of producing a lustrous finish on small parts by tumbling them in a wood-lined barrel with burnishing soap, water, and hardened steel balls.

ball bushing A variation of ball bearing that permits axial motion of a shaft instead of rotating motion.

ball check valve A valve that permits flow in one direction only by lifting a spring-loaded ball off its seat when a pressure differential acts in that direction and by forcing the ball more tightly against the seat when a pressure differential acts in the opposite flow direction.

ball-float liquid-level meter A device consisting of a hollow or low-density float attached by means of a linkage to a pointer; in operation, the float rises and falls with the level of liquid in a tank, while the pointer indicates position of the float on a scale outside the tank.

ball, full A closure component that has a complete spherical surface with a flow passage through it.

ball lightning A relatively rare form of lightning, consisting of a reddish, luminous ball, on the order of one foot in diameter, which may move rapidly along solid objects or remain floating in midair. Hissing noises emanate from such balls, and they sometimes explode noisily but may also appear noiselessly. [NASA]

ballistic cameras Ground-based cameras using multiple exposures on the same plate to record the trajectories of rockets.[NASA]

ballistic missiles Missiles designed to operate primarily in accordance with the laws of ballistics.[NASA]

ballistic trajectories Trajectories followed by a body being acted upon only by gravitational forces and the resistance of the medium through which it passes.[NASA]

ballistics The science that deals with the motion, behavior and effects of projectiles, especially bullets, aerial bombs, rockets or the like; the science or art of designing and hurling projectiles so as to achieve a desired performance. [NASA]

balloon 1. A non-power-driven, lighter-than-air aircraft. 2. To rise slightly either just before or after touching down (landing).[ARP4107-88] 3. The circular symbol used to denote and identify the purpose of an instrument or function. It may contain a tag number. 4. Synonym for bubble. See also bubble.

ball, segmented A closure component that is a segment of a spherical surface which may have one edge contoured to yield a desired flow characteristic.

ball sizing See ball burnishing.

ball-type viscometer An apparatus for determining viscosity, especially of high-viscosity oils and other fluids, in which the time required for a ball to fall through liquid confined in a tube is measured.

ball valve A type of shutoff valve consisting of a solid ball with a diametral hole through it which can be rotated within a spherical seat about an axis perpendicular to the axis of the hole; to permit flow, the ball is rotated so that the hole lines up with inlet and outlet ports of the valve, whereas to shut off flow, the ball is rotated so that the hole does not line up with the ports.

Banbury mixer A heavy-duty batch mixer with two counterrotating rotors; it is designed for blending doughy material such as uncured rubber and plastics.

band 1.The gamut or range of frequencies. 2. The frequency spectrum between two defined limits. 3. Frequencies that are within two definite limits and used for a different purpose. 4. A group of channels; see channel. 4. A group of recording tracks on a computer magnetic disk or drum.

band brake A device for stopping or slowing rotational motion by increasing the tension in a flexible band to tighten it around a drum that is attached to the rotating member.

band-elimination filter A wave filter having a single attenuation band whose critical and cutoff frequencies are finite, nonzero values.

bandgap See energy gaps (solid state).

band marking A circular band applied at regular intervals to the insulation of a conductor for the purpose of size designations or circuit identification. [ARP1931-87]

bandpass filter A process or device in which all signals outside a selected band are strongly attenuated, while the signal components lying within the band are passed with a minimum of change.

band spectrum A spectral distribution of light or other complex wave in which the wave components can be separated into a series of discrete bands of wavelengths. See also continuous spectrum.

bandwidth 1. The difference, expressed in hertz, between the two boundaries of a frequency range. 2. A group of consecutive frequencies constituting a band that exists between limits of stated frequency attenuation. A band is normally defined as more than 3.0 decibels greater than the mean attenuation across the band. 3. A group of consecutive frequencies constituting a band that exists between limits of stated frequency delay. 4. The range of frequencies that can be transmitted in an electronic system.

bandwidth (bandpass) The frequency range over which the actuation system has acceptable dynamic response. This spectrum extends from DC up to a specified frequency, which is usually the frequency where the open-loop amplitude ratio has unity gain (0 3 dB), and the closed-loop response is down 3 dB. For a first-order system the closed-loop response is down 3 dB with 45 deg phase lag.[AIR1916-88]

bang-bang control See off-on control.

bank switching A method of equipping a computer with greater memory by giving the same address to added memory chips.

bar 1. A piece of material, long in proportion to its width and thickness. [ARP480A-79] 2. A solid elongated piece of metal, usually having a simple cross section and usually produced by hot rolling or extrusion, which may or may not be followed by cold drawing.

BAR One atmosphere.

Barany chair A kind of chair in which a person is revolved to test his susceptibility to vertigo. It is named after the Swedish physician Robert Barany who lived from 1876 to 1936.[NASA]

barchans See dunes.

bar code A pattern of narrow and wide bars that can be scanned and interpreted into alpha and numeric characters.

bar-code scanner A type of optical scanner developed to read the 12-character Universal Product Code used to identify groceries and other products.

bark A decarburized layer on steel, just beneath oxide scale formed by heating the steel in air.

Barkometer scale A specific gravity scale used primarily in the tanning industry, in which specific gravity of a water solution is determined from the formula: sp gr=1.000 + or - 0.001n where n is degrees Barkometer; on this scale, water has a specific gravity of zero Barkometer.

barn A unit of nuclear cross section where the probability of a specific nuclear interaction, such as neutron capture, is expressed as an apparent area; in this context, one barn equals 10^{-28} m^2

baroclinic instability Hydrodynamic instability arising from the existence of a meridional temperature gradient (and hence a thermal wind) in an atmosphere in quasigeostropic equilibrium and possessing static stability.[NASA]

baroclinity The state of stratification in a fluid in which surfaces of constant pressure (isobaric) intersect surfaces of constant density (isoteric). The number, per unit area, of isobaric-isoteric solenoids intersecting a given surface is a measure of baroclinity.[NASA]

barometer 1. Instrument used to measure atmospheric pressure.[NASA] 2. An absolute pressure gage for determining atmospheric pressure; if it is a recording instrument, it is known as a barograph.

barometric hypsometry Determining elevation above some arbitrary reference plane (usually sea level) through the use of mercury or aneroid barometers.

barometric pressure Atmospheric pressure as determined by a barometer usually expressed in inches of mercury. See also atmospheric pressure.

barometry The study of atmospheric pressure measurement; in particular, determining errors in barometric instrument readings and correcting them.

barostat A device for maintaining constant pressure within a chamber.

barothermograph An instrument for automatically recording both atmospheric temperature and pressure.

barothermohygrograph An instrument for automatically recording atmospheric pressure, temperature and humidity on the same chart.

barotropism The state of a fluid in which surfaces of constant density (or temperature) are coincident with surfaces of constant pressure; it is the state of zero baroclinity.[NASA]

barred galaxies Spiral galaxies whose nuclei are in the shape of bars at the ends of which the spiral arms begin. About one fifth of all spiral galaxies are barred spirals.[NASA]

barrel 1. A cylindrical component of an actuating cylinder, accumulator, etc. in which a piston or sealed separator moves.[ARP243B-65] 2. A unit of volume; for petroleum, it equals 9702 in^3; for fruits, vegetables, other dry commodities and some liquids, a different standard barrel is used.

barrel chamfer The flared entrance or internal bevel at the wire entry end of the termination device intended to facilitate entry of the conductor.[ARP 914A-79]

barrel, conductor The section of the contact, splice or terminal that accommodates the stripped cable conductor. In the case of insulation displacing barrels the cable need not be stripped. [ARP914A-79]

barrel finishing Producing a lustrous surface finish on metal parts by tumbling them in bulk in a barrel partly filled with an abrasive slurry; similar processes are used for cleaning and electroplating using detergent solutions or electrolytes instead of an abrasive slurry.

barrel, insulation The portion of a contact, terminal or splice which accommodates the wire insulation.[ARP 914A-79]

barrel, wire See barrel conductor.

barricade 1. Similar to barrier, for deceleration and recovery of an airplane with a known landing emergency. Generally a strap or webbing arrangement connected to an energy absorbing system. Whereas a barrier often is designed to engage an airplane landing gear, a barricade primarily is designed to engage the wings or other airframe structure, but may also engage the landing gear.[AIR1489-88] See also barrier.

barrier 1. Any material limiting passage through itself of solids, liquids, semisolids, gases, or forms of energy such as ultraviolet light.[NASA] 2. Dielectric material used to insulate electrical circuits from each other or from ground. [ARP914A-79]

barrier injection transit time diodes See Barritt diodes.

barrier, mid-field A type of arresting gear located at the mid-point of the runway to decelerate and stop an aircraft. May be utilized as the aircraft lands in either direction on the runway.[AIR 1489-88]

barrier, overrun A type of arresting gear located near the end of a runway to decelerate and stop an aircraft in case of brake system or other failure. Generally an overrun area is provided off the end of the runway for runout. [AIR1489-88]

barrier, pop-up A type of arresting gear designed such that for normal operations, the aircraft rolls over the barrier. As an emergency requires and a command signal is given, the barrier pops up into a position for engagement as the aircraft passes.[AIR1489-88]

barrier strip A continuous section of dielectric material which insulates electrical circuits from each other or from ground.[ARP914A-79]

Barritt diode Barrier injection transit time diode that operates similarly to IMPATT diode. The operating frequencies are determined by the transit times across the drift.[NASA]

barycenter See center of gravity.

baryon resonance An anomaly found in scattering cross sections indicating the existence of an unstable, excited state baryon.[NASA]

base 1. A structural foundation upon which an item is to be permanently assembled. The base is an integral part of the item for which it is a foundation. See also holder.[ARP480A-79] 2. The foundation or support upon which a machine or instrument rests. 3. The fundamental number of characters available for use in each digital position in a numbering system. 4. A chemical substance that hydrolyzes to yield OH- ions. 5. A reference value. 6. A number that is multiplied by itself as many times as indicated by an exponent. 7. See radix number.

base address 1. A number that appears as an address in a computer instruction but serves as the base, index, initial, or starting point for subsequent addresses to be modified; synonymous with presumptive address and reference address. 2. A number used in symbolic coding in conjunction with a relative address;

an address used as the basis for computing the value of some other relative address.

baseband 1. A single channel signaling technique in which the digital signal is encoded and impressed on the physical medium. 2. The frequencies starting at or near d-c.

base compound The major component of a two-part curing -type sealant containing the prepolymer.[AIR4069-90] [AS7200/1-90]

base flow Fluid flow at the base or extreme aft end of a body.[NASA]

base leg See traffic pattern.

baseline 1. A quantifiable physical condition or level of performance from which changes are measured.[ARP1587-81] 2. Generally, a reference set of data against which operating data or test results are compared to determine such characteristics as operating efficiency or system degradation with time. 3. In navigation, the geodesic line between two stations operating in conjunction with each other.

base load Base load is the term applied to that portion of a station or boiler load that is practically constant for long periods.

base metal 1. The metallic element present in greatest proportion in an alloy. 2. The type of metal to be welded, brazed, cut or soldered. 3. In the welded joint, metal that was not melted during welding. 4. Any metal that will oxidize in air or that will form metallic ions in an aqueous solution. 5. Metal to which a plated, sprayed or conversion coating is applied. Also known as basis metal.

base number Same as radix number.

base pressure In aerodynamics, the pressure exerted on the base, or extreme aft end, of a body, as of a cylindrical or boattailed body or of a blunt-trailing-edge wing, in a fluid flow.[NASA]

BASIC See Beginner's All-purpose Symbolic Instruction Code.

basic element A single component or subsystem that performs one necessary and distinct function in a measurement sequence; to be considered a basic element, the component must perform one and only one of the smallest steps into which the measuring sequence can be conveniently divided.

basic empty weight The empty weight of an airplane with fixed ballast, hydraulic fluid, and other items required by regulatory standards. See also empty weight.[ARP4107-88]

basic frequency In a waveform made up of several sinusoidal components of different frequencies, the single component having the largest amplitude or having some other characteristic that makes it the principal component of the composite wave.

basic input output system (BIOS) That part of a computer operating system that handles input and output.

basic recipe A generic, transportable recipe consisting of header information, equipment requirements, formula and procedure.

basic T A standardized arrangement of four basic flight instruments in a T pattern. Across the top are the airspeed indicator, attitude indicator, and altimeter; the heading indicator is below the attitude indicator to form the T. [ARP4107-88]

basis metal The metal from which a connector, contact, terminal or splice is made.[ARP914A-79]

basis weight For paper and certain other sheet products, the weight per unit area.

basketweave Alpha platelets, with or without interweaved beta platelets, that occur in colonies. Also known as Widmanstatten. Forms during cooling through the beta transus at intermediate cooling rates.[AS1814-90]

batch 1. A mass of base compound or catalyst manufactured at one time. Its identity is maintained by an assigned batch number. [AS7200/1-90] 2. The quantity of material required for or produced by a production operation at a single time. 3. An amount of material that undergoes some unit chemical process or physical mixing operation to make the final product homogeneous or uniform. 4. A group of similar computer transactions joined together for processing as a single unit.

batch distillation A distillation process in which a fixed amount of a mixture is charged, followed by an increase in temperature to boil off the volatile components. This process differs from continuous distillation, in which the feed is charged continuously.

batch mixer A type of mixer in which starting ingredients are fed all at once and the mixture removed all at once at some later time. Contrast with continuous mixer.

batch process A process that manufactures a finite quantity of material by subjecting measured quantities of raw materials to a time-sequential order of processing actions using one or more pieces of equipment.

batch processing 1. Pertaining to the technique of executing a set of programs such that each is completed before the next program of the set is started. 2. Loosely, the execution of programs serially.

bat file A file name ending in .bat which contains a list of commands most often used to initiate a computer program.

bathochrome An agent or chemical group that causes the absorption band of a solution to shift to lower frequencies.

bathometer An instrument for measuring depth in the ocean or other body of water.

bathyclinograph An instrument for measuring vertical ocean currents.

bathyconductograph An instrument for measuring electrical conductivity of sea water as it is towed at various depths behind a moving ship.

bathymeter Instrument that measures the ocean depths and checks the topography of the ocean floor.[NASA]

bathymetry Application of scientific principles to the measurement of ocean depths. See also bathymeter.

bathythermograph An instrument for recording sea temperature versus depth (pressure) as it is towed behind a moving ship. Also known as bathythermosphere.

battery The complete cased assembly of interconnected electrochemical cells, ready for installation in the aircraft. [AS8033-88]

battery power The maximum power discharge current, declared as a minimum of the manufacturer and expressed in amperes, which the fully charged, 23 °C, cell or battery shall be capable of delivering immediately prior to the conclusion of a 15 s maximum power discharge, controlled so as to maintain a constant terminal voltage at 1/2 of the rated value.[ARP906A-91]

battery setting Describes a setting of two or more boilers with common division walls.

baud 1. A unit of signaling speed equal to the number of discrete conditions or signal events per second. (This is applied only to the actual signals on a communication line) 2. If each signal event represents only one bit condition, baud is the same as bits per second. 3. When each signal event represents other than the logical state of only one bit, used for data entry only in the simples of systems. 4. A unit of signaling speed equal to the number of code elements per second. 5. The unit of signal speed equal to twice the number of Morse code dots continuously sent per second; clarified by rate, bit and capacity, and channel.

Baudot code A three-part teletype code consisting of a start pulse (always a space), five data pulses, and a stop pulse (1.42 times the length of the other pulses) for each character transmitted; various combinations of data pulses are used to designate letters of the alphabet, numerals 0 to 9, and certain standard symbols.

baud rate Any of the standard transmission rates for sending or receiving binary coded data; standard rates are generally between 50 and 19,200 bauds.

Baumé scale Either of two specific gravity scales devised by French chemist Antoine Baumé in 1768 and often used to express the specific gravity of acids, syrups and other liquids; for light liquids the scale is determined from the formula: °Bé=(140/sp gr)-130. For heavy liquids it is determined from: °Bé=145-(145/sp gr). 60° F is the standard temperature used.

Bauschinger effect The phenomenon wherein plastic deformation of a metal raises its tensile yield strength but decreases its compressive yield strength.

Bayard-Alpert ionization gage Ionization vacuum gage using a tube with an electrode structure designed to minimize x-ray induced electron emission from the ion collector.[NASA]

bayonet coupling, cylindrical See coupling, bayonet.

BCD See binary coded decimal.

BCOMP See buffer complete.

BDC See buffered data channel.

BDHI Bearing Distance Heading Indicator.

beaching gear Any wheeled device attached to a seaplane or flying boat to permit moving it onto the beach or shore and to move the aircraft about ashore. May also be gear separate from the aircraft normally but attached temporarily for the same purpose.[AIR1489-88]

beacon Light, group of lights, electronic apparatus, or other device that guides, orients, or warns aircraft, spacecraft, etc. in flight.[NASA]

bead 1. Layers of steel (usually) wire imbedded in rubber and wrapped with fabric. They give a base around which the plies are anchored and provide a firm fit on the wheel. To provide increased stiffness in this area, "flippers" are then sometimes wrapped around the enclosed beads. Beads restrict air pressure expansion of the tire and jumping off the bead seat of the wheel.[AIR1489-88] 2. A rolled or folded seam along the edge of metal sheet. 3. A projecting band or rim. 4. A drop of precious metal produced during cupellation in fire assaying. 5. An elongated seam produced by welding in a single pass.

bead bundle The bead wires wrapped together to form the tire bead.[AIR1489-88]

beaded tube end The rounded exposed end of a rolled tube when the tube metal is formed over against the sheet in which the tube is rolled.

bead heel The outer bead edge which fits against the wheel flange.[AIR1489-88]

bead toe The inner bead edge closest to the tire centerline.[AIR1489-88]

bead wire The steel wire from which the tire beads are wound.[AIR1489-88]

beam 1. A member used to provide rigid support. See also adapter, support. [ARP480A-79] 2. An elongated structural member that carries lateral loads or bending moments. 3. A confined or unidirectional ray of light, sound, electromagnetic radiation or vibrational energy, usually of relatively small cross section.

beam center A point located equidistant from the 50% luminance points of the primary beam intensity distribution. [ARP1782-89] [ARP4067]

beam divergence The increase in beam

diameter with increase in distance from a laser's exit aperture. Divergence, expressed in milliradians, is measured at specified points across the beam's diameter.

beam expander An optical system which expands a narrow beam to a larger diameter, ideally without changing the divergence of the beam.

beam injection The introduction of a particle radiation beam into a plasma or ionized gas for the purpose of diagnostics, plasma control, or the study of beam/plasma interactions.[NASA]

beam integrator A device which integrates the energy in a light beam to make it uniform across the beam cross section.

beam lead as-plated surface That surface of a beam lead which is plated last and is bonded to the conductor film metallization during the bonding operation.[AS1346-74]

beam lead bond misalignment On a bonded device, that portion of the unbonded beam lead width not over the conductor film metallization.[AS1346-74]

beam lead length On an unbonded device, the distance from the edge of the dielectric layer to the outer edge of the beam lead.[AS1346-74]

beam lead thickness The thickness of an unbonded (undeformed) beam lead.[AS1346-74]

beam lead width The width of an unbonded (undeformed beam lead.[AS 1346-74]

beam neutralization Neutralization that takes place by means of charge exchange with a neutral gas.[NASA]

beam penetration CRT A tube with multiple phosphors in close association whose color or persistence, or both, can be altered by modulation of the screen voltage. The phosphors may be deposited separately with an electron attenuating layer between them (layer type) or as a single deposit of coated granules using several phosphors and attenuating layers between them (onionskin type).[ARP1782-89]

beam rider guidance System for guiding aircraft, spacecraft, or missiles, along a desired path, by means of a radar beam, light beam, etc. The center of the beam axis forms a line along which the vehicle senses its location and corrects its course relative to the beam axis.[NASA]

beam splitter 1. Partially reflecting mirror which permits some incident light to pass through and reflect the remainder.[NASA] 2. A device which separates a light beam into two beams. Some types affect polarization of the beam.

beam spread The angle of divergence of an acoustic or electromagnetic beam from its central axis as it travels through a material.

bearing 1. The angle, usually expressed in deg (0-360, clockwise) between the direction from an observer (for example, on an aircraft) to an object or point and a reference line, which may be the fore-and-aft axis of an aircraft (relative bearing), true north (true bearing), or magnetic north (compass bearing or magnetic bearing).[ARP4107-88] 2. The angle, in the horizontal plane, between the longitudinal axis of the own aircraft and the relative location of an intruder aircraft measured in the clockwise direction when viewed from above.[ARP4153-88] 3. A part in which a journal, gudgeon, pivot, pin, shaft, or like revolving parts are supported to reduce friction.[ARP480A-79] 4. A machine part that supports another machine part while the latter undergoes rotating, sliding or oscillating motion. 5. That portion of a beam, truss or other structural member which rests on the supports.

bearing can See axle sleeve.

bearing circle A ring-shaped device that fits over a compass or compass repeater to facilitate taking compass bearings.

bearing distance heading indicator A flight instrument that displays bearing to a station as well as the distance to the station (DME).[ARP4107-88]

beat See synchronism.

beat frequencies The frequencies obtained when two simple harmonic quantities of different frequencies 11 and 12 are superimposed. The beat frequency equals 11-12.[NASA]

beat-frequency oscillator An electrical oscillator which generates a frequency which in turn is beat against another frequency to generate a third usually audible frequency. Generally used in communications receivers to provide an audible signal for CW reception or to reinsert a carrier for reception of single side band signals.

beating A resultant pulsating waveform sometimes produced when two or more periodic quantities of different frequencies combine.

beat note The wave of different frequency resulting when two sinusoidal waves whose frequencies differ from each other are supplied to a nonlinear device.

beats Periodic pulsations in amplitude that are created when a wave of one frequency is combined with a wave of a different frequency.

beauty defects Those imperfections of components and elements of an optical system which do not affect the optical characteristics. They are undesirable but may be accepted if they do not cause a significant degradation of image quality or environmental stability. [ARP924-91]

bed 1. The part of a machine having precisely machined ways or bearing surfaces for supporting and aligning other parts such as toolholders or dies. 2. A perforated floor, lining or support structure, often covered with a layer of granular material, in a furnace, chemical processing tank or filtration tank.

Beech 99 aircraft Light, low-wing aircraft manufactured by Beechcraft. [NASA]

Beer's law The law relating the absorption coefficient to the molar density.

Beginner's All-purpose Symbolic Instruction Code A widely used computer language for personal computers.

behavior The way in which an organism, organ, body, or substance acts in an environment or responds to excitation, as the behavior of steel under stress, or the behavior of an animal in a test. [NASA]

behavioral modeling Modeling a device or component directly in terms of its underlying mathematical equations.

behind the panel 1. A term applied to a location that is within an area that contains (a) the instrument panel, (b) its associated rack-mounted hardware, or (c) is enclosed within the panel. 2. Behind the panel devices are not accessible for the operator's normal use. 3. Behind to panel devices are not designated as local or front-of-panel-mounted.

bel A dimensionless unit for expressing the ratio of two power levels; the value in bels equals log (P_2/P_1), where P_1 and P_2 are the two power levels.

belled mouth (bellmouth) See barrel chamfer.

belled tube end See flared tube-end.

Belleville washer See disk spring.

Bell 214A helicopter Sixteen-seat utility helicopter.[NASA]

bellmouth See nozzle.

bellows 1. Mechanical structures with walls like those of an accordion.[NASA] 2. An enclosed chamber with pleated or corrugated walls so that its interior volume may be varied, either to alternately draw in and expel a gas or other fluid, or to expand and contract in response to variations in internal pressure. 3. A pressure transducer that converts pressure into a nearly linear displacement.

bellows expansion joint A type of coupling between two pieces of pipe that uses a flexible metal bellows to prevent leakage while allowing limited linear movement, such as to accommodate thermal expansion and contraction.

bellows gage A pressure-measuring device in which variations in internal pressure within a flexible bellows causes movement of an end plate against spring force; the position of the end plate is directly related to bellows internal pressure.

bellows meter A differential pressure measuring instrument having a measuring element of opposed metal bellows, the motion of which positions the output actuator.

bellows seal 1. A multi-convolution type element used as a protective barrier between the instrument and the process fluid. 2. A seal in the shape of a bellows used to prevent air or gas leakage.

bellows sealed valve A valve utilizing a bellows to replace the conventional packing gland. One end of the bellows is welded to the rising stem; the other is sealed against the valve body.

bellows stem seal A thin wall, convoluted, flexible component which makes a seal between the stem and bonnet or body and allows stem motion while maintaining a hermetic seal.

bell-type manometer A gage for measuring differential pressure which consists essentially of a cup inverted in a container of liquid; pressure from one source is fed to the inside of the cup while pressure from a second source is applied to the exterior of the cup; pressure difference is indicated by the position of the cup in relation to the liquid level.

below minimums Weather conditions below the minimums prescribed by regulation for the particular action involved (for example, landing minimums and takeoff minimums).[ARP4107-88]

bench check 1. A physical inspection or functional test of an item removed for an alleged malfunction to determine if the part or item is serviceable/repairable. It also includes a determination of the extent of maintenance, repair, or possible overhaul required to return it to serviceable status.[ARD50010-91] 2. A laboratory-type test of an assembly, component or subassembly to verify its function or identify a source of malfunction, often done with the unit removed from its housing or system for service or repair. Also known as bench test.

bench mark A natural or artificial object having a specific point marked to identify a reference location, such as a reference elevation.

benchmark program A routine used to determine the performance of a computer or software.

bench (optical) A mounting surface for optical components.

bender A device for use in precision and non-precision bending of material[ARP 480A-79]

bending Applying mechanical force or pressure to form a metal part by plastic deformation around an axis lying parallel to the metal surface; commonly used to produce angular, curved or flanged parts from sheet metal, rod or wire.

bend loss Attenuation caused by high-order modes radiating from the side of a fiber. The two common types of bend losses are: a) those occurring when the fiber is curved around a restrictive radius of curvature and b) microbends caused by small distortions of the fiber imposed by externally induced perturbations, such as poor cabling techniques.

bends See decompression sickness.

bend test 1. Ductility tests in which

specimens are bent through an arc of known radius and angle.[NASA] 2. Ductility test in which a metal specimen is bent through a specified arc around a support of known radius; used primarily to evaluate inherent formability of metal sheet, rod or wire, or to evaluate weld quality produced with specific materials, joint design and welding technique.

bent tube boiler A water tube boiler consisting of two or more drums connected by tubes, practically all of which are bent near the ends to permit attachment to the drum shell on radial lines.

BER See bit error rate.

Bernoulli equation See Bernoulli theorem.

Bernoulli coefficient In any stream, if the area is changed, as by a reducer, there is a change in the velocity and a corresponding change in the static pressure, or "head." This pressure change is measured in units of velocity head. The dimensionless coefficient used for this purpose is the Bernoulli coefficient K_8.

Bernoulli theorem In aeronautics, a law or theorem stating that in a flow of incompressible fluid the sum of the static pressure and the dynamic pressure along a streamline is constant if gravity and frictional effects are disregarded. It is named for Daniel Bernoulli, a Swiss scientist who lived from 1700 to 1782. [NASA]

beryllium A metal lighter than aluminum, non-magnetic, and characterized by good electrical conductivity and high thermal conductivity. Beryllium is used in alloys, especially beryllium copper alloy.[ARP1931-87]

bessel The filter characteristic in which phase- linearity across the pass band, rather than amplitude linearity, is emphasized; known also as constant-delay.

BESS (satellite) A proposed NASA primate biomedical experiment scientific satellite that was never developed. [NASA]

best-straight-line linearity Also called independent linearity; an average of the deviation of all calibration points.

beta The allotrope of titanium with a body-centered cubic crystal structure occurring at temperatures between the solidification of molten titanium and the beta transus.[AS1814-90]

beta emitter A radioactive nuclide that disintegrates by emitting a beta particle.

beta eutectoid stabilizer An alloying element that dissolves preferentially in the beta phase, lowers the alpha-beta to beta transformation temperature, under equilibrium conditions, and results in the beta decomposition to alpha plus a compound. This is a eutectoid reaction and can be very sluggish for some alloys. Commonly used beta eutectoid forming elements are iron, chromium, and manganese.[AS1814-90]

beta factor In plasma physics, the ratio of the plasma kinetic pressure to the magnetic pressure.[NASA]

beta fleck Transformed alpha-lean and/or beta-rich region in the alpha-beta microstructure. This area has a beta transus measurably below that of the matrix. Beta flecks have reduced amounts of primary alpha which may exhibit a morphology different from the primary alpha in the surrounding alpha/beta matrix.[AS1814-90]

beta interactions See weak interactions (field theory).

beta isomorphous stabilizer An alloying element that is soluble in beta titanium in all proportions. It lowers the alpha-beta to beta transformation temperature without a eutectoid reaction and forms a continuous series of solid solutions with beta titanium. Commonly used beta isomorphous forming elements are vanadium and molybdenum.[AS1814-90]

beta particle An electron or positron emitted from the nucleus of a radioactive nuclide.

beta ratio The ratio of the diameter of the constriction to the pipe diameter, B=Dconst/Dpipe.

beta ray A stream of beta particles.

beta-ray spectrometer An instrument used to measure the energy distribution in a stream of beta particles or secondary electrons.

beta test The second stage of testing a new software program.

beta transus The temperature which designates the phase boundary between the alpha plus beta and beta fields. Commercially pure grades transform in a range of 130 to 1760 °F (890 to 960 °C) depending upon oxygen and iron content. In general, aircraft alloys vary in transformation temperature from 1380 to 1900 °F (750 to 1040 °C).[AS1814-90]

betatron Particle accelerator in which magnetic induction is used to accelerate electrons.[NASA]

bevel gear One of a pair of gears whose teeth run parallel to a conical surface so that they can transmit power and motion between two shafts whose axes intersect.

bezel A ring-shaped member surrounding a cover glass, window, cathode-ray tube face or similar area to protect its edges and often to also provide a decorative appearance.

B-H meter An instrument used to determine the intrinsic hysteresis loop of a magnetic material.

bias 1. A constant or systematic error as opposed to a random error. It manifests itself as a persistent positive or negative deviation of the method average from the accepted reference value. [NASA] 2. The departure from a reference value of the average of a set of values; thus, a measurement of the amount of unbalance of a set of measurements or conditions; that is . . . error having an average value that is non-zero. 3. The average d-c voltage or current maintained between a control electrode and the common electrode in a transistor or vacuum tube.

bias load A steady-state load that is unidirectional and constant over full load travel.[AIR1916-88]

bias (tape) The sine wave, typically ten times the amplitude and 3.5 times the top frequency, applied to tape recording heads with a signal in order to eliminate most signal distortion.

BICEPS A General Electric process-oriented language.

bidirectional load cell A column-type strain-gage load cell with female or male fittings at both ends for attaching load hardware; it can be used to measure either tension or compression loading. Also known as universal load cell.

bi-directional printer An electronic printer capable of printing either forward or backward.

bidirectional pulse A wave pulse in which intended deviations from the normally constant values occur in two opposing directions.

Bielby layer An amorphous layer at the surface of mechanically polished metal.

bifurcated contact A spring-type contact with a lengthwise slot to provide two segments which apply contact force in the same direction.[ARP914A-79]

bifurcation (biology) The separation or branching into two parts, areas, aspects or connected segments, of anatomical systems or functions.[NASA]

bilateral tolerance The amount of allowable variation about a given dimension, usually expressed as plus-or-minus a specific fraction or decimal.

bilateral transducer A transducer that can transmit signals simultaneously in both directions between two or more terminations.

billet 1. A semifinished primary mill product ordinarily produced by hot rolling metal ingot to a cylinder or prism of simple cross-sectional shape and limited cross-sectional area. 2. A general term for the starting stock used to make forgings and extrusions.

bimetal A bonded laminate consisting of two strips of dissimilar metals; the bond is usually a stable metallic bond produced by corolling or diffusion bonding; the composite material is used most often as an element for detecting temperature changes by means of differential thermal expansion in the two layers.

bimetallic corrosion A type of accelerated corrosion induced by differences in galvanic potential between dissimilar metals immersed in the same liquid medium (electrolyte) and also in electrical contact with each other.

bimetallic thermometer element A temperature-sensitive strip of metal (or other configuration) made by bonding or mechanically joining two dissimilar strips of metal together in such a manner that small changes in temperature will cause the composite assembly to distort elastically, and produce a predictable deflection; the element is designed to take advantage of the fact that different metals have different coefficients of thermal expansion.

bimetric theories Theories of gravitation.[NASA]

bin activator A vibratory device sometimes installed in the discharge path of a mass-flow bin or storage hopper to promote steady discharge of dry granular material.

binary 1. A computer numbering system that uses two as its base rather than ten. The binary system uses only 0 and 1 in its written form. 2. A device that uses only two states or levels to perform its functions, such as a computer.

binary alloy A metallic material composed of only two chemical elements (neglecting minor impurities), at least one of which is a metal.

binary cell An information-storage element that can assume either of two stable conditions, and no others.

binary code 1. Code composed of a combination of entities each of which can assume one or two possible states. Each entity must be identifiable in time or space.[NASA] 2. A code that uses two distinct characters, usually 0 and 1.

binary coded decimal (BCD) Describing a decimal notation in which the individual decimal digits are represented by a group of binary bits, e.g., in the 8-4-2-1 coded decimal notation each decimal digit is represented by a group of four binary bits. The number twelve is represented as 0001 0010 for 1 and 2, respectively, whereas in binary notation it is represented as 1100. Related to binary.

binary-coded decimal system A system of number representation in which each digit in a decimal number is expressed as a binary number.

binary counter 1. A counter which counts according to the binary number system. 2. A counter whose basic counting elements are capable of assuming one of two stable states.

binary digit 1. In binary notation, either of the characters 0 or 1. 2. Same as bit. 3. See equivalent binary digits.

binary distillation A distillation process that separates only two components.

binary file An electronic term for a file that is not a text file.

binary notation A numbering system using the digits 0 and 1 with a base of 2.

binary number A number composed of the characters 0 and 1, in which each character represents a power of two. The number 2 is 10; the number 12 is 1100; the number 31 is 11111, etc.

binary point The radix point in a binary number system.

binary scaler A signal-modifying device (scaler) with a scaling factor of 2.

binary stars Systems of two stars revolving about a barycenter.[NASA]

binary synchronous A procedure for connecting many terminals that share a single link.

binary synchronous communications A communications procedure using special characters for control of synchronized transmission

binary unit 1. A binary digit. 2. A unit of information content, equal to one binary decision, or the designation of one of two possible and equally likely values or states of anything used to store or convey information. 3. See check bit and parity bit. 4. Same as bit.

binary word A group of binary digits with place values in increasing powers of two.

binder 1. A spirally served tape or thread wrap used for holding in place assembled cable components which are awaiting further manufacturing operations. [ARP1931-87] 2. In metal founding, a material other than water added to foundry sand to make the particles stick together. 3. In powder metallurgy, a substance added to the powder to increase green strength of the compact, or a material (usually of relatively low melting point) added to a powder mixture to bond particles together during sintering that otherwise would not bond into a strong sintered body.

binding post A fixed support, generally screw-type, to which conductors are connected.[ARP914A-79]

Bingham body A non-Newtonian substance that exhibits true plastic behavior—that is, it flows when subjected to a continually increasing shear stress only after a definite yield point has been exceeded.

Bingham viscometer A time-of-discharge device for measuring fluid viscosity in which the fluid is discharged through a capillary tube instead of an orifice or nozzle.

bioastronautics The study of biological, behavioral, and medical problems pertaining to astronautics. This includes systems functioning in the environments expected to be found in space, vehicles designed to travel in space, and the conditions on celestial bodies earth. [NASA]

biochemical oxygen demand The amount of oxygen necessary for the oxidative decomposition of a material by microorganisms. The amount of oxygen consumed in mg/1 of water (or waste water) over a period of 5 days at 20 deg C under laboratory conditions. [NASA]

biochemistry Chemistry dealing with the chemical processes and compounds of living organisms.[NASA]

biocompatibility Compatibility of substances with living tissues and blood components.[NASA]

bioconversion The transformation of algae and/or other biomass materials in successive stages to aliphatic organic acids to aliphatic hydrocarbons to diesel and/or other liquid fuels.[NASA]

biodegradability The characteristic of a substance that can be decomposed by microorganisms.[NASA]

biodynamics The study of the effects of dynamic processes (motion, acceleration, weightlessness, etc.) on living organisms.[NASA]

biofeedback Originally confined to the presenting of a subject with sensory information about his ongoing physiological activities, it now includes the controlling of specific physiological activities through trained mental effort.[NASA]

bioinstrumentation Instruments that can be attached to humans or animals to record biological parameters, such as pulse rate, breathing rate or body temperature.

biological analysis See bioassay.

biological corrosion Deterioration of

metal surfaces due to the presence of plant or animal life; deterioration may be caused by chemicals excreted by the life form, or by concentration cells such as those under a barnacle, or by other interactions.

biological models See bionics.

biological models (mathematics) Mathematical models for living systems.[NASA]

biomagnetism Magnetic fields surrounding parts or the whole of a living biological system; also, the effects of magnetism on parts or the whole of a biological entity.[NASA]

biomass The dry weight of living matter in a given area expressed in terms of mass or weight per unit of volume or area.[NASA]

biomechanics See biodynamics.

biomedical engineering The application of engineering principles to the solution of medical problems, including the design and fabrication of prostheses, diagnostic instrumentation and surgical tools.

biomedical experiment scientific satellite See BESS (satellite).

bionics The study of systems, particularly electronic systems, which function after the manner characteristic of, or resembling living systems.[NASA]

bioreactors Biological processors to remove or produce certain chemicals or a particular chemical.[NASA]

bioregenerative life support systems See closed ecological systems.

biosatellites Artificial satellites which are specifically designed to contain and support man, animals, or other living material in a reasonably normal manner for an adequate period of time and which, particularly for man and animals, possesses the proper means for safe return to the earth.[NASA]

biosimulation See bionics.

biosphere That transition zone between earth and atmosphere within which most forms of terrestrial life are commonly found; the outer portion of the geosphere and inner or lower portion of the atmosphere.[NASA]

Biot number A standard heat transfer dimensionless number.[NASA]

biotechnology The application of engineering and technological principles to the life sciences.[NASA]

biotelemetry The remote sensing and evaluation of life functions, as, e.g., in spacecraft and artificial satellites.[NASA]

biotite A widely distributed and important rock-forming mineral of the mica group.[NASA]

bi-phase A method of bit encoding for serial data transmission or recording whereby there is a signal transition every bit period.

bipolarity Capability of assuming negative or positive values.[NASA]

bipolar technology Technology that uses two different polarity electrical signals to represent logic states of 1 and 0.

bipolar transistor A transistor created by placing a layer of P- or N-type semiconductors between two regions of an opposite type of semiconductor.

bipropellants See liquid rocket propellants.

biquinary code A method of coding decimal digits in which each numeral is coded in two parts—the first being either 0 or 5, and the second any value from 0 to 4; the digit equals the sum of the two parts.

birdcage A defect in stranded wire where the strands have separated from the normal lay.[ARP1931-87]

birefringent element A device that has a refractive index which is different for lightwaves of different orthogonal polarizations. Because of this difference, light of the two orthogonal polarizations travels at different speeds and is refracted slightly differently.

Birmingham wire gage A system of standard sizes used in the United States for brass wire, and for strip, bands, hoops and wire made of ferrous and nonferrous metals; the decimal equivalent of standard Bwg sizes is generally larger than for the same gage number in both the American wire gage and U.S. steel wire gage systems.

biscuit 1. A piece of pottery that has been fired but not glazed. 2. An upset blank for drop forging. 3. A small cake of primary metal, generally one produced by bomb reduction or a similar process.

bistable The capability of assuming either of two stable states, hence of storing one bit of information.

bistable amplifiers See flip-flops.

bistable and tristable control A control system in which the power to control the load is fully "on" in either polarity (bistable), or fully "on" in one polarity; "off", or fully "on" in the other polarity (tristable). These systems are sometimes called "on-off" systems or "bang-bang" systems. When the time duration of the application of power is modulated by the input, the system is called pulse width modulated (PWM). [AIR1916-88]

bistatic radar See multistatic radar.

bistatic reflectivity The characteristic of a reflector which reflects energy along a line, or lines, different from, or in addition to, that of the incident ray.[NASA]

BISYNC See binary synchronous communications.

bit 1. A manual or power driven boring tool.[ARP480A-79] 2. A cutting tool for drilling or boring. 3. The blade of a cutting tool such as a plane or ax. 4. A removable tooth of a saw, milling cutter or carbide-tipped cutting tool. 5. The heated tip of a soldering iron. 6 An abbreviation of binary digit. 7. A single character in a binary number. 8. A single pulse in a group of pulses. 9. A unit of information capacity of a storage device. The capacity in bits is the logarithm to the base two of the number of possible states of the device. Related to storage capacity. 10. The smallest unit of information that can be recognized by a computer.

bit density A measure of the number of bits recorded per unit of length or area.

BITE Built-in test equipment.[ARP 4102/13-90]

bit error rate 1. The number of erroneous bits or characters received from some fixed number of bits transmitted. [NASA] 2. The ratio of bits received in error to bits sent.

bit error rate tester A system which measures the fraction of bits transmitted incorrectly by a digital communication system.

bit map A table that describes the state of each member of a related set; bit map is most often used to describe the allocation of storage space; each bit in the table indicates whether a particular block in the storage medium is occupied or free.

bit pattern A combination of n binary digits to represent 2 to the n possible choices, e.g., a 3-bit pattern represents 8 possible combinations.

bit rate 1. The speed at which bits are transmitted, usually expressed in bits per second. (Compare with baud). 2. The rate at which binary digits, or pulses representing them, pass a given point on a communications line or channel. Clarified by baud.

bits per second In a serial transmission, the instantaneous bit speed within one character, as transmitted by a machine or a channel. See baud.

bit stream A binary signal without regard to grouping by character.

bit string A string of binary digits in which each bit position is considered as an independent unit.

bit synchronizer A hardware device that establishes a series of clock pulses in synchronism with an incoming bit stream and identifies each bit.

bituminous Describing a substance that contains organic matter, mostly in the form of tarry hydrocarbons (described as bitumen).

black body 1. A physical object that absorbs incident radiation, regardless of spectral character or directional preference of the incident radiation; a perfect black body is most closely approximated by a hollow sphere with a small hole in its wall—the plane of the hole being the black body; a perfect black body is used as an ideal reference concept in the study of radiant energy. 2. Denotes a perfectly absorbing object, none of the incident energy is reflected. It radiates (perfectly) at a rate expressed by the Stefan-Boltzmann Law; the spectral distribution of radiation is expressed by Planck's radiation formula. When in thermal equilibrium, a black body absorbs and radiates at the same rate.

black body radiation The electromagnetic radiation emitted by an ideal black body; it is the theoretical maximum amount of radiant energy of all wavelengths which can be emitted by a body at a given temperature.[NASA]

blackbody temperature The true temperature of a blackbody source. When used to calibrate a radiation pyrometer, the radiation pyrometer will measure brightness temperature of sources other than the blackbody. To obtain true temperature of non-blackbodies using a radiation pyrometer, multiply the brightness temperature by the emissivity of the observed source.

black box A generic term used to describe an unspecified device which performs a special function or in which known inputs produce known outputs in a fixed relationship.

black-bulb thermometer A thermometer whose sensitive element is covered with lampblack to make it approximate a black body.

black data Data that do not require safeguards.[AIR4271-89]

Black Hawk assault helicopter See H-60 Helicopter.

black liquor 1. The solution remaining after cooking pulpwood in the soda or sulfite papermaking process. 2. A black, iron-acetate solution containing 5 to 5.5% Fe, and sometimes tannin or copperas, used in dyes and printing inks.

blackout A temporary loss of vision and sometimes even consciousness resulting from stagnant hypoxia, commonly induced by positive G forces of severe intensity or duration, or both. See also grayout.[ARP4107-88]

blade Arms of propeller and rotating wings. Specifically, restrictive, those parts of propellers or of rotating wings from the shank outward, i.e., those parts having efficient airfoil shapes and that cleave the air. Vanes such as rotating vanes or stationary vanes in rotary air compressors, or vanes of turbine wheels.[NASA]

blade (knife blade) A fuse or limiter terminal having a substantially rectangular cross-section.[ARP1199A-90]

blade angle The acute angle between the chord of a section of a propeller, or of a rotary wing system, and a plane perpendicular to the axis of rotation. [ARP4107-88]

blade slap noise Impulsive noise (short high pressure sound waves) of rotating blades primarily helicopter blades. [NASA]

blade-type consistency sensor A pneumatic device for determining changes in consistency of a flowing non-Newtonian substance such as a slurry; it senses the force required for a shaped blade to shear through the flowing stock, and transmits a pneumatic output signal proportional to changes in consistency; its normal operating range is 1.75 to 6.0% suspended solids, with a sensitivity of -0.02% in many applications.

blank In computer programming, the character used to represent a space.

blank alarm point See alarm point.

blankets (fission reactors) Damper materials for fusion reactors.[NASA]

blanking 1. Inserting a solid disc at a pipe joint or union to close off flow during maintenance, repair or testing. 2. Using a punch and die to cut a shaped piece from sheet metal or plastic for use in a subsequent forming operation. 3. Using a punch and die to make a semifinished powder-metal compact.

blast deflectors Devices used to divert the exhaust of a rocket fired from a vertical position.[NASA]

blast furnace gas Lean combustible by-product gas resulting from burning coke with a deficiency of air in a blast furnace.

blasting 1. Detonating an explosive. 2. Using abrasive grit, sand or shot carried in a strong stream of air or other medium to remove soil or scale from a surface.

blazars Strongly optical polarized active galactic nuclei objects exhibiting BL lacertae-like and quasar-like characteristics.[NASA]

bleed and burn A phrase to describe a process whereby air is extracted from a compressor or air accumulator, and to which fuel is added and is burned to increase the air temperature.[ARP906A-91]

bleeding 1. Allowing a fluid to drain or escape to the atmosphere through a small valve or cock; used to provide controlled slow reduction of slight overpressure, to withdraw a sample for analysis, to drain condensation from compressed air lines, or to reduce the airspace above the liquid level in a pressurized tank. 2. Withdrawing steam from an intermediate stage of a turbine to heat a process fluid or boiler feedwater. 3. Natural separation of liquid from a semisolid mixture—such as oil from a lubricating grease or water from freshly poured concrete.

bleeding cycle A type of steam cycle where steam is withdrawn from the turbine at one or more intermediate stages and used to heat feedwater before it enters the boiler.

blend 1. To mix ingredients so that they are indistinguishable from each other in the mixture. 2. To produce a smooth transition between two intersecting surfaces, such as at the edges of a radiused fillet between a shaft and an integral flange or collar.

blind hole A hole in a piece of material that does not completely penetrate to the back surface.

blind nipple A short piece of pipe or tubing with one end closed and sealed.

blind pressure transmitter A pressure transmitter not having an integral readout device.

blind speed The rate of departure or closing of a target relative to the radar antenna at which cancellation of the primary radar target by moving target indicator (MTI) circuits in the radar equipment causes a reduction or complete loss of signal from that target. [ARP4107-88]

blip 1. On radar screens, a streak of light caused by an object, vehicle, or some electronic disturbance passing through the path of the radar beam.[ARP4107-88] 2. Any erratic signal on a computer screen.

blister 1. A small area on the surface of metal or plastic where a thin layer of the material has been separated from underlying material and is raised due to gas trapped between the layers, yet remains attached around the edges of the raised area. 2. An enclosed macroscopic cavity in a glaze or other fired ceramic coating. 3. A raised area where a paint, electroplate or other coating has become detached from the substrate due to accumulation of gas or moisture at the coating-substrate interface.

BL lacertae objects One of a class of astronomical objects exhibiting: (a) rapid variations in intensity at radio, infrared, and optical wavelengths, (b) energy distributions largely at infrared wavelengths, (c) absence of discrete features in low dispersion spectra, and (d) strong and rapidly varying polarization at visual and radio wavelengths. [NASA]

block 1. A piece of material such as wood, stone, or metal, usually with one or more plane or approximately plane faces, used to strengthen or sustain.[ARP480A-79] 2. A set of things, such as words, characters, or digits, handled as a unit. 3. A collection of contiguous records recorded as a unit, blocks are separated by interblock gaps, and each block may contain one or more records. 4. In data communication, a group of contiguous characters formed for transmission purposes. The groups are separated by interblock characters. 5. A group of physically adjacent words or bytes of a specified size particular to a device. The smallest system-addressable segment on a mass-storage device in reference to I/O. See also cylinder block; block-and-tackle.

blockage seal A type of isolation seal used with a faying-surface seal. It is a prepak or injection seal that must join the primary seal.[AIR4069-90]

block-and-tackle A hoisting gear consisting of a rope or cable and one or more independently rotating frictionless pulleys. Also known as block and fall.

block, data A set of associated characters or words handled as a unit.

block diagram 1. A graphical representation of the hardware in a computer system. The primary purpose of a block diagram is to indicate the paths along which information or control flows between the various parts of a computer system. It should not be confused with the term flow chart. 2. A coarser and less symbolic representation than a flow chart. 3. A graphical representation of a computer program. 4. Used to provide a simple pictorial representation of a control system. Block diagrams have two basic symbols, the circle and the function block. The arrows entering and leaving the circle represent the flow of information and the head of each arrow has an algebraic sign associated with it, either plus or minus. 5. Block diagrams show the graphical representation of the hardware in a system. The primary purpose of a block diagram is to indicate the paths along which information or control flows between various parts of the system.

blocked impedance Of an electromechanical transducer; the electrical impedance at the input terminals when the mechanical system is "blocked", or prevented from moving.

blockend interrupt (BIN) A signal in TELEVENT that indicates that a buffer is completely filled with data.

blocker-type forging A shape forging designed for easy forging and extraction from the die through the use of generous radii, large draft angles, smooth contours and generous machining allowances; used as a preliminary stage in multiple-die forging or when machining to final shape is less costly than forging to final shape.

block flow The distance freshly mixed, uncured sealant will sag while hanging from a vertical surface for approximately one-half hour at standard conditions.[AS7200/1-90]

blocking 1. Producing a semifinished forging of approximate shape suitable for further forging or machining to final size and shape. 2. Reducing the oxygen content of the bath in an open-hearth furnace. 3. Undesired adhesion between plastics surfaces during storage or use. 4. Of computer records, see grouping.

block sequence A welding sequence in which separated lengths of a continuous multiple-pass weld are built up to full cross section before gaps between the segments are filled in. Compare with cascade sequence.

block switching A two-level multiplexing technique used in data transmission, whereby one level selects the input channel to be transmitted and the second level selects the group of first-level input channels to be addressed; the chief advantage of block switching is reduction of leakage currents from "off" channels which interfere with data signals being transmitted. Also known as submultiplexing.

blocky alpha Alpha phase which is considerably larger and more polygonal in appearance than the primary alpha present. It is induced by unidirectional metal working and has an aspect ratio of 3:1 or higher. It may result from extended exposure high in the alpha-beta phase field following rapid cooling through the beta transus during forging or heat treating operations. It may be removed by beta recrystallization or by all-beta working followed by further alpha-beta work. May accompany grain boundary alpha. Microhardness not significantly different from surrounding normal alpha-beta matrix.[AS1814-90]

bloedite A mineral consisting of hydrous sodium magnesium sulfate that is colorless. Also known as astrakanite or astrochanite.[NASA]

blood-brain barrier A mechanism which maintains the constancy of the neurons in the central nervous system by preventing certain substances from leaving the bloodstream and entering the neural tissue.[NASA]

bloom 1. A semifinished metal bar of large cross section (usually a square or rectangle exceeding 36 sq in.) hot rolled or sometimes forged from ingot. 2. Visible fluorescence on the surface of lubricating oil or an electroplating bath. 3. A bluish fluorescent cast to a painted surface caused by a thin film of smoke, dust or oil. 4. A loose, flower-like corrosion product formed when certain nonferrous metals are exposed in a moist environment. 5. To apply an antireflection coating to glass. 6. To hammer or roll metal to brighten its surface.

blowback The difference between the pressure at which a safety valve opens and at which it closes, usually about 3% of the pressure at which valve opens.

blowby Leakage of fluid through the clearance between a piston and its cylinder during operation.

blowdown 1. In a safety valve, the difference between opening and closing pressures. 2. In a steam boiler, the practice of periodically opening valves attached to the bottom of steam drums and water drums, during boiler operation, to drain off accumulations of sediment.

blow down valve A valve generally used to continuously regulate concentration of solids in the boiler, not a drain valve.

blower A fan used to force air under pressure.

blowhole A pocket of air or gas trapped during solidification of a cast metal.

blow-off valve A specially designed, manually operated, valve connected to the boiler for the purpose of reducing the concentration of solids in the boiler or for draining purposes.

blowout Term used to indicate the combuster becomes unlit.[ARP906A-91]

blowout disk See rupture disk device.

blowtorch action Impingement of a localized jet of hot gas on a surface.

blue brittleness In some steels, loss of ductility associated with tempering or service temperatures in the blue heat range, 400 to 600 °F.

blue stars Stars of spectral type , B, A, or F according to the Draper catalog. [NASA]

blue vitriol A solution of copper sulfate sometimes applied to metal surfaces to make scribed layout lines more visible.

bluff bodies Bodies having a broad, flattened front, as in some reentry vehicles.[NASA]

bluing Also spelled blueing. 1. Forming a bluish oxide film on steel by exposing it to steam, air or other agents at a suitable temperature, thus giving scale-free surfaces an attractive appearance and improved corrosion resistance. 2. Heating formed springs after fabrication to improve their properties and reduce residual stress. 3. A thin blue oxide formed on polished metal surfaces when exposed briefly to air at high temperatures.

blunt leading edges The obtuse cross sections of certain front edges of airfoils or wings.[NASA]

blunt trailing edges The rounded or obtuse angled trailing edges of wings and/or control surfaces designed to enhance aerodynamic characteristics. [NASA]

BNI See Bureau ďOrientation de la Normalisatin en Informatique.

board In computers, a flat sheet in which integrated circuits are mounted. See panel.

boattails The rear portions of elongated bodies, as in rockets, having decreasing cross-sectional area toward the rear.[NASA]

BOD See biochemical oxygen demand.

Bode plot 1. A Bode plot presents system frequency response data plotted in rectangular coordinate form. The magnitude ratio (output-to-input) and phase angle (time relationship of output-to-input) are plotted against frequency of a change in input.[AIR1823-86] 2. A graph of transfer function versus frequency wherein the gain (often in decibels) and phase (in degrees) are plotted against the frequency on log scale. Also called bode diagram.

bodies of revolution Symmetrical bodies having the form described by rotating a plane curve about an axis in its plane.[NASA]

body, connector The main portion of a connector consisting of the housing and insert assembly to which contacts and accessories are attached.[ARP914A-79]

body, encapsulated A body with all surfaces covered by a continuous surface layer of a different material, usually an elastomeric or polymeric material.

body, split A valve body design in which trim is secured between two segments of a valve body.

body temperature (non-biological) See temperature.

body temperature regulation See thermoregulation.

body, wafer A thin annular section body whose end surfaces are located and clamped between the piping flanges by bolts extending from flange to flange.

body, wafer, lugged A thin annular section body whose end surfaces mount between the pipeline flanges, or may be attached to the end of a pipeline without any additional flange or retaining parts, using either through-bolting and/or tapped holes.

body, weir type A body having a raised contour contacted by a diaphragm to shut off fluid flow.

Boeing 757 aircraft Boeing's twin turbofan short/medium range transport aircraft that made its first flight on February 19, 1982.[NASA]

Boeing 767 aircraft Boeing's wide-bodied medium range commercial transport aircraft that made its first flight on September 26, 1981.[NASA]

bogie Also spelled bogey; bogy. 1. A low,

strongly built cart. In landing gear usage, two or more wheels or sets of wheels on axles connected by means of an axle beam (or bogie beam) and pivoted at the lower end of a shock strut assembly or structure to permit pitching motion as it rides over the runway or other supporting surface.[AIR1489-88] 2. A type of aircraft landing gear consisting of two sets of wheels in tandem with a central strut. 3. A supporting and aligning idler wheel or roller on the inside of an endless track. 4. A swivel-mounted axle or truck that supports a railroad car, the leading end of a locomotive, or the end of a vehicle such as a gun carriage. 5. The drive-wheel assembly and supporting frame for the two rear axles of a three-axle motor truck, mounted so that the wheels are kept in contact with the road surface, especially around curves and over rough roads.

bogie beam See axle beam.

bogie gear A type of landing gear configuration utilizing a shock strut or structural support assembly and at the lower end, a pivoted bogie assembly. [AIR1489-88]

bogs See marshlands.

Bohr magneton A constant equivalent to the magnetic moment of an electron. [NASA]

boiler horsepower The evaporation of 34 1/2 lbs of water per hour from a temperature of 212 °F into dry saturated steam at the same temperature. Equivalent to 33,475 Btu.

boiler water A term construed to mean a representative sample of the circulating boiler water, after the generated steam has been separated and before the incoming feed water or added chemical becomes mixed with it so that its composition is affected.

boiling The conversion of a liquid into vapor with the formation of bubbles.

boiling out The boiling of a highly alkaline water in boiler pressure parts for the removal of oils, greases, etc.

boilup Vapors that are generated in the column reboiler.

bolides Brilliant meteors—especially ones which explode—detonating fireballs.[NASA]

bolograms See bolometers.

bolometer 1. Instrument which measures the intensity of radiant energy by employing thermally sensitive electrical resistors; a type of actinometer. [NASA] 2. A sensitive infrared detector whose operation is based on a change in temperature induced by absorbing infrared radiation. It is made of two thin, blackened gratings of platinum, one illuminated and the other kept in the dark. The absorption of heat changes the electrical resistance which is detected by comparing the resistances of the two gratings in an electrical circuit.

bolster A steel block or plate used to support dies and attach them to a press bed; in drop forging, a bolster is also used to attach dies to the ram.

bolt 1. A pin or rod to fasten or hold something in place, having a wrench head designed to be held or turned at one end and a screw thread on the other.[ARP480A-79] 2. A threaded fastener consisting of a rod, usually made of metal, having threads at one end and an integral round, square or hexagonal head at the other end; short bolts usually have threads running the entire length below the head, and longer bolts often have an unthreaded shank between the head and threaded end.

bolted joint 1. Joint fastened with bolts. Usually designed for heavy loads. [NASA] 2. An assembly of two or more parts held together by a bolt and nut, with or without washers, or by a bolt that threads into a tapped hole in one of the parts.

bolter An attempted arrested landing wherein (a) the arresting hook fails to engage the arresting wire due to hook skip, (b) the aircraft touches down beyond the wire area, (c) the arresting hook fails to retain the wire, (d) the arresting hook or wire fails after engagement. The aircraft then becomes airborne and resumes normal flight.[AIR 1489-88]

bolting 1. A collective term for threaded fasteners, especially bolts, nuts, screws and studs. 2. Assembling parts together using threaded fasteners.

bomb calorimeter An apparatus for measuring the quantity of heat released by a chemical reaction; it consists of a strong-walled metal container (bomb) immersed in about 2.5 litres of water in an insulated container; a sample is sealed in the bomb, the bomb immersed, the sample ignited (or a reaction started) by remote control, and the heat released measured by observing the rise in temperature of the water bath.

bombs (ordnance) Explosive devices designed to be detonated under specified conditions.[NASA]

bond 1. An electrical connection between conductive parts which provides the required electrical conductivity.[ARP1870-86] 2. A wire rope that attaches a load to a crane hook. 3. Adhesion between cement or mortar and masonry. 4. In an adhesive bonded or diffusion bonded joint, the junction between faying surfaces. 5. In welding, brazing or soldering, the junction between assembled parts; where filler metal is used, it is the junction between fused metal and heat-affected base metal. 6. In grinding wheels and other rigid abrasives, the material that holds abrasive grains together. 7. Material added to molding sand to hold the grains together. 8. The junction between base metal and cladding in a clad metal product.

bonded Conductive parts that are considered to be bonded when they are mechanically interconnected to maintain a common electrical potential. [ARP1870-86]

bonded liner In a butterfly valve body, a liner vulcanized or cemented to the body bore.

bonded strain gage A device for measuring strain which consists of a fine-wire resistance element, usually in zigzag form, embedded in nonconductive backing material such as impregnated paper or plastic, which is cemented to the test surface or sensing element.

bonded transducer A pressure sensor that uses a bonded strain gage to generate the output signal.

bonding Specifically, a system of connections between all metal parts of an aircraft or other structure forming a continuous electrical unit and preventing jumping or arching of static electricity. Gluing or cementing together for structural strength.[NASA]

bonding connector A device used to connect exposed metal to ground. It normally carries no current but is used as a current path to eliminate shock or spark hazards and insures the operation of circuit protective devices in cases of insulation breakdown.[ARP914A-79]

bond length On a bonded device, the distance from the outer edge of the bonded beam lead to the apparent inner edge of the bonded area.[AS1346-74]

bone dry A papermaking term used to describe pulp fibers or paper from which all water has been removed. Also known as oven dry; moisture free.

Bonne projection A type of conical map projection in which meridians are plotted as curves and the parallels are spaced along them at true distances. [NASA]

bonnet, seal-welded A bonnet welded to a body to provide a zero leakage joint.

Boolean Pertaining to logic quantities.

Boolean add See OR.

Boolean algebra 1. The study of the manipulation of symbols representing operations according to the rules of logic. Boolean algebra corresponds to an algebra using only the number 0 and 1, therefore can be used in programming digital computers which operate on the binary principle.[NASA] 2. A process of reasoning, or a deductive system of theorems using a symbolic logic, and dealing with classes, propositions, or on-off circuit elements. It employs symbols to represent operators such as and, or, not, except, if, then, etc., to permit mathematical calculation. Named after George Boole, famous English mathematician.

Boolean expression A quantity expressed as the result of Boolean operations such as and, or, and not upon Boolean variables.

Boolean functions A system of mathematical logic often executed in circuits to provide digital computations such as "OR", "AND", "NOR", "NOT", etc.

Boolean operator A logic operator each of whose operands and whose result have one of two values.

Boolean variable See logical variable.

boost See acceleration (physics).

booster 1. An intermediate explosive charge to augment the initiating component of an explosive train and cause detonation of deflagration of the main explosive charge.[AIR913-89] 2. Propulsion system employed in the launching phase of a vehicle flight.[AIR913-89]

booster fan A device for increasing the pressure or flow of a gas.

booster relay A volume or pressure amplifying pneumatic relay that is used to reduce the time lag in pneumatic circuits by reproducing pneumatic signals with high volume and/or high pressure outputs.

boostglide vehicles Vehicles designed to glide in the atmosphere following a rocket-powered phase. Portions of the flights may be ballistic, out of the atmosphere.[NASA]

boot 1. A connector accessory, usually made from a flexible or semi-rigid insulating material, designed to house wire/cable terminations as a protective device, provide harness direction, moister seal when bonded or used as a potting form.[ARP914A-79] 2. A computer routine in which a few instructions are loaded which then cause the rest of the system to be loaded.

bootstrap A technique for loading the first few instructions of a routine into storage, then using these instructions to bring in the rest of the routine. This usually involves either the entering of a few instructions manually or the use of a special key on the console.

bootstrap loader A routine whose first instruction is sufficient to load the remainder of itself into memory from an input device; normally used to start a complete system of programs.

bore 1. The inner cavity in a pipe or tube. 2. The diameter of the cylinder of a piston-cylinder device such as a reciprocating compressor, engine or pump, or a hydraulic or pneumatic power cylinder. 3. To penetrate or pierce a workpiece with a rotating cutting tool. 4. To increase the size of an existing hole, generally with a single-point cutting tool, while either the work or the cutting tool rotates about the central axis of the hole. 5. The inner surface of a gun tube. 6. The central hole in a laser or other type of tube (a capillary, waveguide, or hole in a micro-channel plate).

boredom See attention, anomalies of boredom).

boreholes Holes made by drilling into the ground to study stratification, to search for or to obtain natural resources, or to release underground pressures.[NASA]

bore Reynolds number Calculated Reynolds number including R_d using V_{bore}, P_{bore}, μ_{bore}, d_{bore}; also $R_d=R_D/\beta$.

borescope A viewing device used to visually inspect items such as cannon bores, internal combustion engine cylinder walls, propellant perforations of rocket motors or similar items, for defects of manufacture, corrosion, scars, erosion, etc., without the necessity of disassembling the item(s). It may or may not have extension tubes to cover the various ranges of inspection. The item is basically a straight tube telescope using a mirror or prism.[ARP480A-79]

boresight error Linear displacement between two parallel lines of sight.[NASA]

boron fibers Fibers produced by vapor deposition methods; used in various composite materials to impart a balance of strength and stiffness.[NASA]

boresighting Initially aligning a gun, directional antenna or other device by optical means or by observing a return signal from a fixed target at a known location; the term is derived from an early military practice of looking down the bore of an artillery piece to obtain an initial line of sight to a target.

boron counter tube A type of radiation counter tube used to detect slow neutrons; the tube has electrodes coated with a boron compound, and it also may be filled with BF_3; a slow neutron is easily absorbed by a B^{10} nucleus, with subsequent emission of an alpha particle.

borosilicate glass 1. Low expansion heat resistant glass.[NASA] 2. A type of heat-resisting glass that contains at least 5% boric acid.

Borsic (tradename) Trademark of United Aircraft Products, Inc. for its boron aluminum composite materials.[NASA]

bort Industrial diamonds or diamond fragments.

boss 1. A raised portion of metal or small area and limited thickness on flat or curved metal surfaces. 2. A short projecting section of a casting, forging or molded plastics part, often cylindrical in shape, used to add strength or to provide for alignment or fastening of assembled parts.

bottle A hollow vessel usually having a neck smaller than the body and a narrow mouth for a stopper or other type closure. See also tank.[ARP480A-79]

bottom contraction The vertical distance from the crest to the floor of the weir box or channel bed.

bottom dead center The position of a piston and its connecting rod when at the extreme downstroke position.

bottoming drill A flat-end twist drill that converts the conical bottom of a blind hole into a cylinder.

bottoming tap A tap designed for cutting full threads all the way to the bottom of a blind hole; usually used to finish the bottom of a hole tapped with a regular, tapered-end tap.

bottoms The higher boiling product streams usually taken from the bottom of a distillation column—sometimes from the reboiler and sometimes from a separate surge vessel.

Bouger law A relationship describing the rate of decrease of flux density of a plane-parallel beam of monochromatic radiation as it penetrates a medium which both scatters and absorbs at that wavelength.[NASA]

boundary element method Technique for solving two- and three-dimensional boundary value problems in thermodynamics, mechanics, etc.[NASA]

boundary integral method Technique related to the boundary element method, and used for laminar and turbulent flow problems.[NASA]

boundary layer 1. The layer of air immediately adjacent to a moving surface, such as an airfoil. The movement of air in the boundary layer relative to the surface is relatively small.[ARP 4107-88] 2. In a flowing fluid, a low-velocity region along a tube wall or other boundary surface.

boundary layer amplifier An amplifier which utilizes the control of the separation point of a power stream from a curved or plane surface to modulate the output.[ARP993A-69]

boundary layer plasmas Plasmas resulting from the frictional heat of hypersonic spacecraft entering the earth's atmosphere.[NASA]

boundary lubrication A condition in sliding contact when contact pressures are high enough and sliding velocities low enough that hydrodynamic lubrication is completely absent; mating surfaces slide across each other on a multimolecular layer of lubricant, often with some solid-to-solid surface contact; for liquid lubricants, a bearing-characteristic (Sommerfield) number of 0.01 is considered to be the upper limit of boundary lubrication.

boundary structure The fuel tight primary structure of an integral fuel tank including skin panels, bulkheads and spars—all elements that form the tank boundaries.[AIR4069-90]

boundary value problems Physical problems completely specified by a differential equation in an unknown, valid in a certain region of space, and certain information (boundary condition) about the unknown, given on the boundaries of that region. The information required to determine the solution depends completely and uniquely on the particular problem.[NASA]

bound water In a moist solid to be dried, that portion of the water content which is chemically combined with the solid matter.

Bourdon tube A flattened tube, twisted or curved, and closed at one end, which is used as the pressure-sensing element in a mechanical pressure gage or recorder; a process stream pressure is routed to the open end of the tube, and the tube flexes or untwists in relation to the internal pressure, with the change in shape of the tube being used to operate a mechanical pointer or pen positioner. Also known as Bourdon element; Bourdon pressure gage.

Boussinesq approximation The assumption (frequently used in the theory of convection) that the fluid is incompressible except insofar as the thermal expansion produces a buoyancy. [NASA]

bow shock waves See bow waves, shock waves.

bow waves Shock waves in front of a body, such as an airfoil, or apparently attached to the forward tip of the body. [NASA]

box 1. A container having a cover (usually attached) to be used for storage or shipment of a specific item. See also case.[ARP480A-79] 2. A flow-chart symbol.

boxcar averager A signal processing instrument which averages selected portions of repetitive signals to improve signal quality. Sometimes called a gated integrator because it passes or gates portions of the signal, then integrates them.

box girder A hollow girder or beam, usually having a spare or rectangular cross section. Also known as box beam.

box header boiler A horizontal boiler of the longitudinal or cross drum type consisting of a front and rear inclined rectangular header connected by tubes.

box wrench A closed-end wrench designed to fit a single size and shape nut; different wrench ends are needed for different nut sizes and different nut shapes. Also known as box end wrench.

B power supply An electrical power supply connected in the plate circuit of a vacuum tube electronic device.

BPS See bits per second. Also see baud.

brace See drag brace, jury brace, side brace.

bracket An item of rigid construction

which is attached to or projects from a main body, for the purpose of sustaining a secondary item in a predetermined suspended position, relative to the main body.[ARP480A-79]

Bragg curve A curve showing the average specific ionization of an ionizing particle of a particular kind as a function of its kinetic energy, velocity, or residual range.[NASA]

Bragg law A principle describing the apparent reflection of x-rays (and De-Broglie waves associated with certain particulate beams) from atomic planes in crystals; maximum reflected intensity occurs along the family of directions defined by θ=arcsin λ n l/2d, where θ is the Bragg angle (angle of reflection and of incidence), n is an integer, λ is the wavelength of monochromatic radiation reflected from the crystal, and d is the interplanar spacing of the reflecting parallel planes in the crystal.

braid 1. A woven or braided sheath made from conductive or nonconductive material.[ARP914A-79] 2. An assembly of fibrous or metallic filaments woven to form a protective and/or conductive covering over one or more wires, or as a flexible metallic conductive cable such as a grounding strap.[ARP1931-87]

braid angle The angle of the braided filaments or fibers in relation to the longitudinal axis of the wire or cable.[ARP1931-87]

braid carrier The yarn or strand, or group of yarns or strands, laid parallel in the braid by a single bobbin of the braider.[ARP1931-87]

braid end An individual yarn or strand in a braid carrier.[ARP1931-87]

braille A system of writing that uses characters made up of raised dots. It was named after Louis Braille.[NASA]

brake A machine element for applying frictional force to slow or stop relative motion.

brake, annular piston A type of brake utilizing for operation an annular or ring shaped piston/cylinder arrangement and powered by hydraulic or air pressure to apply force for squeezing a set of alternate rotating/stationary discs together and produce torque and a decelerating force.[AIR1489-88]

brake back plate The structural end plate on a multiple disc brake opposite the actuators. Its purpose is to restrain the discs and resist the force exerted by the actuators.[AIR1489-88]

brake carrier See brake housing.

brake chatter A self induced brake vibration of less than 100 cycles per second, excited by the friction characteristics of the rubbing surfaces.[AIR1489-88]

brake control system The total system for control of a brake or brakes. May be a mechanical system, hydraulic system, electrical/electronic system, etc., but generally includes a combination of types i.e., mechanical pilot input to brake pedals, cable system to a metering valve and to the brakes actuators via an electronically/mechanically controlled antiskid system.[AIR1489-88]

brake disc A single disc of a brake. May be a rotating disc (rotor) in a single or multiple disc brake or stationary disc (stator) in a multiple disc brake. Functions in a heat absorbing capacity and to produce a frictional retarding force. May be a solid disc or an assembly of segments.[AIR1489-88]

brake, disc A type of brake utilizing one or more rotating discs with a clamping force applied to produce a frictional retarding force.[AIR1489-88]

brake displacement The volumetric displacement required in a brake actuation system to actuate from full off to full on.[AIR1489-88]

brake drag Any residual resistance to motion produced by a brake assembly when in the full off condition.[AIR1489-88]

brake drag force Any resistance to motion produced by a brake assembly. May apply to an on or off condition but generally considered as the decelerating force produced by the brake when powered to on condition.[AIR1489-88]

braked roll The portion of an air vehicle landing roll during which the brakes are applied for deceleration of the vehicle.[AIR1489-88]

brake drum Cylindrical, rotating part fastened to the wheel. Acts as a friction surface, heat sink, or transmits torque.[AIR1489-88]

brake, drum A type of brake utilizing a brake drum and brake shoe or expander tube arrangement.[AIR1489-88] See also drum.

brake energy (kinetic energy) The energy due to the motion of the vehicle mass which must be converted by the brake assembly to some other form of energy and either dissipated or absorbed.[AIR1489-88]

brake, expander tube A drum type brake using a rubber like tube to expand and press the brake lining segments against the friction surface of the drum.[AIR1489-88]

brake horsepower 1. The horsepower (time rate of doing work) delivered to the brake.[AIR1489-88] 2. The power of an engine or other motor as calculated from the force exerted on a friction brake or absorption dynamometer applied to the flywheel or the shaft. See also shaft horsepower.[ARP4107-88]

brake housing Also brake carrier, power plate, actuator housing. The brake part, fixed to the axle and used to contain hydraulic or pneumatic fluid passages and the actuating pistons and cylinders.[AIR1489-88]

brake lining 1. Specially compounded materials applied to rotor and/or stator to create a predictable coefficient of friction and act as a wear surface.[AIR 1489-88] 2. A material having high coefficient of friction that is used as the principal friction element in a mechanical brake; it usually is made of fabric or molded asbestos, and usually can be readily replaced to extend the brake's service life and restore braking efficiency.

brake linkage Links or levers, or structural members used to connect an axle mounted, free rotating brake assembly to the fixed structure of the gear. Also a linkage system for the purpose of actuating a brake assembly.[AIR1489-88]

brake master cylinder In a fluid operated brake system, the cylinder assembly which receives an input command force and in turn transmits a fluid (hydraulic/pneumatic) force to one or more actuating cylinders, usually located at or on the individual brake assemblies.[AIR1489-88]

brake, multiple disc A brake assembly utilizing a multiplicity of alternate rotating and stationary discs (rotors and stators).[AIR1489-88]

brake, multiple piston A brake configuration utilizing two or more actuating cylinders and pistons, interconnected and circumferentially spaced around the brake housing.[AIR1489-88]

brake overheat indicator A device or system for the purpose of detecting and indicating to the pilot that an overheat condition exists in the brakes, i.e., caution and/or warning.[AIR1489-88]

brake pressure metering valve A fluid metering valve in the brake control system. Upon receiving an input command (i.e., mechanical/electrical) the valve meters fluid pressure and flow (in proportion to the input command) to the brakes.[AIR1489-88]

brake pressure plate Also primary disc, retractor plate. A stationary disc with a prescribed cross-section used to distribute the applied piston actuating loads or forces to the brake discs.[AIR 1489-88]

brake rotor Rotating disc. A part normally keyed to the wheel and rotating with it relative to the fixed axle.[AIR 1489-88]

brake shoe A structural part of the brake assembly which is actuated by the system force input and which presses the brake lining against the rotating brake drum.[AIR1489-88] See also shoe.

brake, single disc A brake configuration utilizing a single rotating disc. [AIR1489-88]

brake, spot type A brake configuration utilizing a single cylinder and piston arrangement with a "spot" type lining carrier to rub against the rotating disc. [AIR1489-88]

brake squeal A self-induced brake vibration mode with a frequency greater than 100 cycles per second.[AIR1489-88]

brake stator Stationary disc. A part normally keyed to a stationary part of the brake assembly.[AIR1489-88]

brake torque The torque developed by the friction elements of the brake assembly. This torque is transmitted through the wheel and tire where it is reacted by a drag force at the tire/runway interface. Other resultant forces are reacted by the air vehicle structure. See also brake torque compensating linkage.[AIR1489-88]

brake torque compensating linkage Also brake torque reaction linkage. A mechanical linkage for the purpose of transmitting brake torque reaction loads to the fixed structure of a landing gear structure. Especially in a bogie type landing gear configuration where brake drag loads reacted at the tire/runway interface would cause excessive pitching of the bogie assembly if uncompensated with the linkage to fixed gear structure. The linkage system may also contain hydromechanical and/or automatic electronic control elements.[AIR 1489-88]

brake torque tube A tubular shaped part of the brake assembly which fits around the axle and to which the stators (stationary discs) are keyed. The torque tube transmits the stator torque to the brake housing and/or other fixed brake structure and to the axle, compensating linkage, or gear structure.[AIR1489-88]

braking action A report of conditions on the airport surface area providing a pilot with a degree/quality of braking that might be expected. Braking action is reported as good, fair, poor, or nil. [ARP4107-88]

braking, antiskid The process of braking an air vehicle and utilizing an antiskid system as an integral part of the brake control system.[AIR1489-88]

braking, automatic Control of air vehicle braking with an automatic system and the pilot out of the control loop. An input of the desired rate of deceleration is made prior to landing. The system generally includes an antiskid system and manual selection of a deceleration rate. Eliminates pilot technique during the braked roll.[AIR1489-88]

Brale A 120° conical diamond indenter used in Rockwell hardness testing of relatively hard metals.

branch The selection of one of two or more possible paths in the control of flow based on some criterion. The instructions which mechanize this concept are sometimes called branch instructions; however the terms transfer of control and jump are more widely used. Related to conditional transfer.

branch circuit That portion of a wiring system extending beyond the final overcurrent device protecting the circuit. [ARP1199A-90]

branch instruction An instruction that performs a branch.

branchpoint A point in a routine where one of two or more choices is selected under control of the routine. See conditional transfer.

brass Any of the many alloys based on the binary system copper-zinc; most brasses contain no more than 40 wt% zinc.

Brayton cycle A thermodynamic cycle consisting of two constant-pressure processes interspersed with two constant-antropy cycles. Named after George B. Brayton, American engineer.[NASA]

braze welding A joining process similar to brazing but in which the filler metal is not distributed in the joint by capillary action.

Brazilian space program The space program of Brazil which is under the jurisdiction of the Instituto de Pesquisas Espaciais (INPE).[NASA]

brazing A joining process whereby a filler material which melts at a temperature in excess of 800 °F but less than the melting point of the base metal, combines with the base metal.[ARP 1931-87]

breadboard model 1. Assembly of preliminary circuits or parts used to prove the feasibility of a device, circuit, system, or principle without regard to the final configuration or packaging of the parts.[NASA] 2. A prototype or uncased assembly of an instrument or electronic device whose parts are laid out on a flat surface and connected together to demonstrate or check its operation.

break 1. An interruption in computer processing. 2. To interrupt the sending end and take control of the circuit of the receiving end.

breakdown 1. Initial hot working of ingot-cast or slab-cast metal to reduce its size prior to final working to finished size. 2. A preliminary press-forging operation.

breakdown voltage The electrical potential necessary to cause the passage of a specified electric current through an insulator or insulating material. [ARP1931-87]

break-in An initial run procedure required to produce desired surface conditions on component parts (e.g., gears, piston and cylinder) operating relative to each other.[ARP906A-91]

breakout The point other than at the end, along the length of a harness or other multiconductor configuration, at which a wire or group of wires leaves the configuration.[ARP1931-87]

breakout pressure The pressure required to overcome static friction in a component.[AIR1916-88]

break point A location at which program operation is suspended so that partial results can be examined; a preset point in a program where control passes to a debugging routine.

breakpoint instruction 1. An instruction which will cause a computer to stop or to transfer control, in some standard fashion, to a supervisory routine which can monitor the progress of the interrupted program. 2. An instruction which, if some specified switch is set, will cause the computer to stop or take other special action.

break rate The percent of time an aircraft will return from an assigned mission with one or more previously working system/subsystems on the Mission Essential Subsystem List (MESL) inoperable (code 3 including ground and air aborts).[ARD50010-91]

breaks Creases or ridges, usually appearing in aged sheet or strip, where the yield point has been locally exceeded; depending on the origin of the break it may be termed a coil break, cross break, edge break or sticker break.

breech The container provided on cartridge starters to receive and fire the cartridge.[ARP906A-91]

breech cap The removable portion of the breech which contains the igniting mechanism.[ARP906A-91]

breech chamber The fixed portion of

the breech which directs the cartridge gas to the starter.[ARP906A-91]

breeching A duct for the transport of the products of combustion between parts of a steam generating unit or to the stack.

Bremsstrahlung 1. Electromagnetic radiation produced by the rapid change in the velocity of an electron or another fast, charged particle as it approaches an atomic nucleus and is deflected by it. In German it means braking radiation.[NASA] 2. X-rays having a broad spectrum of wavelengths, which are formed due to deceleration of a beam of energetic electrons as they penetrate a target. Also known as white radiation.

Brewster-angle window A window inserted into an optical path at Brewster's angle—the angle at which unpolarized light must be incident upon a nonmetallic surface for the reflected radiation to acquire maximum plane polarization. At Brewsters angle, the reflected plane polarized beam and the refracted beam through the window are at 90°.

bricks Solid masonry units of clay or shale, usually formed into a rectangular prism while plastic and burned or fired in a kiln. Bricks are ceramic products.[NASA]

bridge 1. A device that interfaces and transfers data between two or more similar communication systems.[AIR 4271-89] 2. A network device that interconnects two local area networks that use the same LLC but may use different MACs. A bridge requires only OSI Level 1 and 2 protocols. See gateway and router. 3. The strain-to-voltage converter in many measurement systems (actually, a Wheatstone bridge).

bridge amplifier A type of amplifier circuit used extensively in instrumentation to provide gains up to 1000 at bandwidths up to 50 kHz; it is generally configured as a direct-coupled amplifier constructed of four subamplifiers and suitable fixed resistances.

bridge circuit An electronic network in which an input voltage is applied across two parallel elements and the output voltage—to an indicating device or load—is taken across two intermediate points on the parallel elements.

bridged-T network A T-network having a fourth branch connected in parallel with the two series branches of the T, the fourth branch termination at one input and one output terminal.

bridgewall A wall in a furnace over which the products of combustion pass.

bridging 1. Premature solidification of metal across a mold section before adjacent metal solidifies. 2. Welding or mechanical jamming of the charge in a downfeed furnace. 3. Forming an arched cavity in a powder metal compact. 4. Forming an unintended solder connection between two or more conductors, either a secure connection or merely an undesired electrical path without mechanical strength. Also known as crossed joint; solder short.

bridgewire A resistance wire which thermally or explosively initiates the charge in an EED.[AIR913-89]

bridle, catapult A wire rope or synthetic fiber assembly designed to connect an aircraft to a catapult for the purpose of accelerating the aircraft for takeoff. The bridle connects to two catapult hooks on the aircraft and passes around the catapult shuttle spreader. The bridle falls off the aircraft (sheds) as it overtakes the shuttle and is generally restrained by a bridle arrester system on the deck. See also pendant, catapult. [AIR1489-88]

Briggs pipe thread See American standard pipe thread.

bright dipping Producing a bright surface on metal, such as by immersion in an acid bath.

brightness 1. The psychometric attribute

of visual perception by which an area appears to emit more or less light (luminance is the physical measurement of this parameter). The term refers to the visual sensation which results from viewing surfaces or spaces. This sensation is determined in part by the measurable luminance and in part by the conditions of observation (such as the color being observed or the state of adaptation of the eye).[ARP1782-89] 2. The attribute of visual perception in accordance with which an area appears to emit more or less light.[NASA] 3. A term used in nonquantitative statements with reference to sensations and perceptions of light, in quantified statements with reference to the description of brightness by photometric units.

brightness distribution The statistical distribution based on brightness, or the distribution of brightness over the surface of an object.[NASA]

brightness ratio The ratio of the brightnesses of any two surfaces. When the two surfaces are adjacent, the brightness ratio is called the brightness contrast.[ARP798-91]

brightness temperature In astrophysics, the temperature of a black body radiating the same amount of energy per unit area at the wavelengths under consideration as the observed body. The temperature of a nonblack body determined by measurement with an optical pyrometer.[NASA]

bright plating Electroplating to yield a highly reflective coated surface.

bright switch A solid-state switch consisting of two bipolar transistors connected in an inverted configuration to achieve a low offset voltage; used in only limited applications today.

brine Water saturated or strongly impregnated with common salt.[NASA]

Brinell test A standard bulk hardness test in which a 10-mm-diameter ball is pressed into the surface of a test piece and a hardness number determined by dividing applied load in kg by area of the circular impression in sq mm.

briquetting Producing relatively small lumps or block of compressed granular material, often incorporating a binder to help hold the particles together.

brisance The shattering effect of a high explosive.[AIR913-89]

British thermal unit The mean British thermal unit is 1/180 of the heat required to raise the temperature of 1 lb of water from 32 °F to 212 °F at a constant atmospheric pressure. It is about equal to the quantity of heat required to raise 1 lb of water 1 °F. A Btu is essentially 252 calories.

brittle fracture Separation of solid material with little or no evidence of macroscopic plastic deformation, usually by rapid crack propagation involving less energy than for ductile fracture of a similar structure.

brittleness The tendency of a material to fracture without apparent plastic deformation.

Brix scale A specific gravity scale used almost exclusively in sugar refining; the degrees Brix represent the weight percent pure sucrose in water solution at 17.5 °C.

broaching Cutting a finished hole or contour in solid material by axially pulling or pushing a bar-shaped, toothed, tapered cutting tool across a workpiece surface or through a pilot hole.

broadband A medium based on CATV technology where multiple signals are frequency division multiplexed. Due to the use of CATV technology, a broadband cable is unidirectional (within any given block of frequencies). As a result, two types of broadband systems are in common use, single cable and dual cable. In a single cable system, stations transmit and receive on the same cable but at different frequencies. The station transmits on one frequency, travels down the network to the head end, gets translated into a different frequency, and sent back down the network where it is received by all stations. In a dual cable system, the stations transmit and receive at the same frequency but on different cables. The end of the transmit cable is connected to the beginning of the receive cable, forming a double loop through the plant.

broadband pyrometer See wideband radiation thermometer.

broadband transmission (fiber optic) Transmission of signals with a large bandwidth, such as video transmission or higher.

broadcast 1. The transmission of a message from one station to all other stations.[AIR4271-89] 2. The simultaneous dissemination of information to one or more stations, one-way, with no acknowledgment of receipt. 3. A message addressed to all stations connected to a LAN.

broken clamps Loose, broken, or damaged firesleeve clamps which evidence a failure to properly secure firesleeving to a hose assembly.[ARP1658A-86]

broken symmetry Phenomena where a loss of symmetry is present such as in piezoelectricity.[NASA]

broken wires On hose with wire-braided jacket or cover, a strand of wire, or multiple strands of wire forming a plait, which is severed or broken in two due to abrasion, flexing or stress fatigue. [ARP1658A-86]

bronze 1. A copper-rich alloy of copper and tin, with or without small amounts of additional alloying elements. 2. By extension, certain copper-base alloys containing less tin than other elements, such as manganese bronze and leaded tin bronze, and certain other copper-base alloys that do not contain tin, such as aluminum bronze, beryllium bronze and silicon bronze. 3. Trade names for certain copper-zinc alloys (brasses), such as architectural bronze (Cu-40Zn-3Pb) and commercial bronze (Cu-10Zn).

bruise resistance The capability of a tire to resist bruise damage i.e., from rocks and foreign objects.[AIR1489-88]

Brunt-Vaisala frequency The frequency at which an air parcel will oscillate when subjected to an infinitesimal perturbation in a stably stratified atmosphere.[NASA]

brush A device composed of bristles, wire, etc., set in a suitable back or handle and used for applying liquids, cleaning, scrubbing, removing burrs, etc.[ARP 480A-79]

brushcoat The thin layer of a curing-type sealant usually applied over fasteners, seams, and various parts and small openings prior to the application of the primary class B fillet sealant. [AIR4069-90]

brushes (electrical contacts) Conductive metal or carbon blocks used to make sliding electrical contact with a moving part as in an electric motor.[NASA]

brush plating An electroplating process in which the surface to be plated is not immersed, but rather rubbed with an electrode containing an absorbent pad or brush which holds (or is fed) a concentrated electrolyte solution or gel.

Btu See British thermal unit.

bubble 1. A small volume of steam enclosed within a surface film of water from which it was generated. 2. Any small volume of gas or vapor surrounded by liquid; surface-tension effects tend to make all bubbles spherical unless they are acted upon by outside forces.

bubblegas Any gas selected to bubble from the end of a tube immersed in liquid for level measurement from the hydrostatic back pressure created in the tube.

bubble memory See magnetic bubble memory.

bubble point The temperature at which a liquid mixture begins to boil and evolve vapors.

bubbler-type specific-gravity meter See air-bubbler specific-gravity meter.

bubble sort In data processing, a method of arranging a group of numbers in some order.

bubble tight A nonstandard term used to refer to control valve seat leakage. Refer to ANSI/FC1 70-2 for specification of seat leakage classifications.

bubble tube A length of pipe or tubing placed in a vessel at a specified depth to transport a gas injected into the liquid to measure level from hydrostatic back pressure in the tube.

bubble-type viscometer A device similar to a ball-type viscometer, except viscosity is determined from timed rise of a standard-size bubble through the sample liquid instead of timed fall of a ball.

buckle 1. Localized waviness in a metal bar or sheet, usually transverse to the direction of rolling. 2. An indentation in a casting due to expansion of molding sand into the cavity.

Buckley gage A device that measures very low gas pressure by sensing the amount of ionization produced by a prescribed electric current.

buckling 1. An unstable state of equilibrium of a thin-walled body stemming from compressive stresses in walls. The lateral deflection of a thin-walled body resulting from such instability.[NASA] 2. Producing a lateral bulge, bend, bow, kink or wavy condition in a beam, bar, column, plate or sheet by applying compressive loading.

buckstay A structural member placed against a furnace or boiler wall to restrain the motion of the wall.

buffer 1. An internal portion of a data processing system serving as intermediate storage between two storage or data-handling systems with different access times or formats; usually to connect an input or output device with the main or internal high-speed storage. Clarified by storage buffer. 2. An isolating component designed to eliminate the reaction of a driven circuit on the circuits driving it, e.g., a buffer amplifier.

buffer, catapult In nose gear catapulting, a device or system for decelerating the aircraft to zero velocity at hookup condition, by the engagement of the holdback bar with the slider of the catapult deck hardware.[AIR1489-88]

buffer complete (BCOMP) In TELEVENT, the signal that indicates when the computer buffer is complete.

buffered computer A computing system with a storage device which permits input and output data to be stored temporarily in order to match the slow speeds of input and output devices. Thus, simultaneous input-output and computer operations are possible. A data transmission trap is essential for effective use of buffering since it obviates frequent testing for the availability of a data channel.

buffered data channel (BDC) A device that provides high-speed parallel data interfaces into and out of the computer memory.

buffered I/O channel A computer I/O channel that controls the movement of data between an external device and memory, under the control of self-contained registers (i.e., independent of the operating program). See BDC.

buffer storage In computer operations, storage used to compensate for a difference in rate of flow or time of occurrence when transferring information from one device to another.[NASA]

buffeting The beating of an aerodynamic structure or surfaces by unsteady flow, gusts, etc.; the irregular shaking or oscillation of a vehicle component owing to turbulent air or separated flow.[NASA]

buffing Producing a very smooth and

bright surface by rubbing it with a soft wheel, belt or cloth impregnated with fine abrasive such as jeweler's rouge.

bug An error, defect or malfunction in a computer program.

buildup 1. Excessive electrodeposition on areas of high current density, such as at corners and edges. 2. Small amounts of work metal that adhere to the cutting edge of a tool and reduce its cutting efficiency. 3. Deposition of metal by electrodeposition or spraying to restore required dimensions of worn or undersize machine parts.

built in test (BIT) An operational status checkout or test system which is integrated into a control system or function. Usually used to verify operational status of as many aspects of a control function as possible within the limits of integrated test capability. May be initiated automatically or on command.[ARP1181A-85] 2. An integral capability of the mission equipment which provides an on-board, automated test capability to detect, diagnose, or isolate system failures. The fault detection and, possibly, isolation capability is used for periodic or continuous monitoring of a system's operational health, and for observation and, possibly, diagnosis as a prelude to maintenance action.[ARD50010-91] 3. An automated internal test which interrupts normal operation and requires participation of either aircrew or maintenance personnel.[ARP1782-89] 4. A testing routine that is initiated to assist in diagnosing a suspected system fault.[ARD50020-91]

built in test equipment (BITE) 1. Equipment built into a unit to provide a self test capability for the unit.[ARP 1587-81] 2. Any device permanently mounted in the prime equipment and used for the express purpose of testing the prime equipment, either independently or in association with external test equipment.[ARD50010-91]

bulge A local distortion of swelling outward caused by internal pressure on a tube wall or boiler shell caused by overheating. Also applied to similar distortion of a cylindrical furnace due to external pressure when overheated provided the distortion is of a degree that can be driven back.

bulk acoustic wave devices Acousto-optic devices utilizing bulk sound waves at megahertz frequencies in thin film transducers.[NASA]

bulk density Mass per unit volume of a bulk material, averaged over a relatively large number of samples.

bulk memory See secondary storage.

bulk modulus 1. The reciprocal of the coefficient of compressibility.[NASA] 2. An elastic modulus determined by dividing hydrostatic stress by the associated volumetric strain (usually computed as the fractional change in volume).

bulk storage A hardware device in a computer system that supplements computer memory; typically, a magnetic tape or disk.

bulk storage memory Any non-programmed large memory. For example, discs, drums, or magnetic tape units.

bull block A machine with a power-driven rotating drum for pulling wire through a drawing die.

bull gear A bull wheel with gear teeth around its periphery.

bullion 1. A semirefined alloy containing enough precious metal to make its recovery economically feasible. 2. Refined gold or silver, ready for coining.

bull wheel 1. The main wheel or gear of a machine, usually the largest and strongest. 2. A cylinder with a rope wound around it for lifting or hauling.

bump A raised or flattened portion of a boiler drum head or shell formed by fabrication, generally used for nozzle or pipe attachments.

bumpless transfer Change from

manual mode to automatic mode of control, or vice versa, without change in control signal to the process.

bumpy toruses The shapes (doughnuts) of certain plasmas.[NASA]

Buna-N A nitrile synthetic rubber known for resistance to oils and solvents.

bunched lay In a bunched lay conductor or cable, the strands or wires are twisted together in the same direction without regard to geometrical arrangement, with the same lay length.[AS1198-82]

bunched stranding Any number of conductor strands of the same diameter twisted together in the same direction without regard to geometric arrangement of the individual strands. [ARP1931-87]

bundle 1. A group of wires fastened or held together by auxiliary means such as straps, ties, clamps, lacing tape/twine or flexible wrappings (jackets) or sheaths. Also called "cable".[ARP914A-79] 2. A number of individually insulated conductors (wires), and/or cables, groups, harnesses, routed together. [ARP1931-87]

bundle (fiber optic) A group of fibers packaged together which collectively transmit light, in a coherent bundle, the end fibers are in a fixed relationship to each other and can transmit an image.

bungee A spring, elastic cord, or other tension (or compression) device used in a system to balance an opposing force or in a landing gear system to assist in retracting or extending the gear or to absorb shock, etc., i.e., downlock bungee, uplock bungee, overcenter bungee, etc.[AIR1489-88]

bunker C oil Residual fuel oil of high viscosity commonly used in marine and stationary steam power plants. (No. 6 fuel oil).

buoyancy The tendency of a fluid to lift any object submerged in the body of the fluid; the amount of force applied to the body equals the product of fluid density and volume of fluid displaced.

buoyancy displacers The technique of measuring liquid level by measuring the buoyant force on a partially immersed volumetric displacing device.

buoyancy-type liquid-level detector Any of several designs of level gage that depend for their operation on the buoyant force acting on a float or similar device located inside the tank or vessel.

burden 1. The amount of power consumed in the measuring circuit of an instrument, usually given as the volt-amperes consumed under normal operating conditions. 2. The property of a circuit connected to the secondary winding of an instrument transformer which determines active and reactive power at the transformer output terminals.

Bureau d'Orientation de la Normalisatin Informatique The French national standards body for computer related standards.

burner 1. Any device for producing a flame using liquid or gaseous fuel. 2. A device in the firebox of a fossil-fuel-fired boiler that mixes and directs the flow of fuel and air to give rapid and complete combustion. 3. A worker who cuts metal using an oxyfuel-gas torch.

burner windbox A plenum chamber around a burner in which an air pressure is maintained to insure proper distribution and discharge of secondary air.

burner windbox pressure The air pressure maintained in the windbox or plenum chamber measured above atmospheric pressure.

burn-in 1. The operation of an item to stabilize its characteristics. Basically, a reliability conditioning procedure which is a method of aging an item by operating it under specified environmental and test conditions in accordance with an established procedure in order to eliminate early failures and age or stabilize the item prior to final test and shipment.[ARD50010-91] 2. The operation of an item under stress to stabilize its characteristics.[AIR1916-88]

burning See combustion.

burning process See combustion.

burning rate The rate at which a solid propellant burns normal to its surface at a specified chamber pressure and propellant temperature.[AIR913-89] 2. The velocity at which a solid propellant in a rocket is consumed.[NASA]

burnish 1. To polish or make shiny. 2. Specifically, to produce a smooth, lustrous surface finish on metal parts by tumbling them with hardened metal balls or rubbing them with a hard metal pad.

burnout The termination of combustion in a rocket engine because of exhaustion of the propellant.[NASA]

burr 1. A thin, turned over edge or fin produced by a grinding wheel, cutting tool or punch. 2. A rotary tool having teeth similar to those on a hand file.

bursting In data processing, the act of separating continuous forms into single sheets.

burst pressure The pressure required to rupture or burst. Generally a design includes allowance for maximum expected pressure in service times a reasonable safety factor.[AIR1489-88]

burst pressure, actual Pressure at which a component bursts, exhibits massive leakage due to permanent or non-permanent structural failure, or due to structural deflection.[AIR1916-88]

burst pressure, minimum (ultimate pressure) Pressure during burst pressure testing up to which no externally visible bursting and no significant external leakage occurs. Deformation and permanent set are permitted. Function may be impaired.[AIR1916-88]

bus 1. A generic term describing a channel along which signals travel from one or several sources (stations) to one or more destinations (stations).[AIR4271-89] 2. A group of wires or conductors, considered as a single entity, which interconnects parts of a system. 3. In a computer, signal paths such as the address bus, the data bus, etc. 4. A circuit over which data or power is transmitted; often one which acts as a common connection among a number of locations. Synonymous with trunk. 5. A communications path between two switching points. 6. A common connector circuit, usually multiwire, for transfer of power, data, timing, etc., between the several modules or units on the bus.

bus cycle The transfer of one word or byte between two devices.

bushing 1. A replaceable part, cylindrical in shape, hollow, and designed primarily to be inserted in a hole to reduce the effective inside diameter of the hole, and to protect the body structure about the hole from damage resulting from stress, strain, and vibration.[ARP480A-79] 2. A removable piece of soft metal or impregnated sintered-metal sleeve used as a bearing or guide. 3. A ring-shaped device made of ceramic or other nonconductive material used to support an electrical conductor while preventing it from becoming grounded to the support structure.

business flying Flying by an individual in his/her own, company-owned, rented, or leased aircraft in connection with his/her profession or occupation, or in furtherance of company business.[ARP4107-88]

busing The joining of two or more circuits to provide a common electrical connection.[ARP914A-79]

bus interface unit (BIU) The BIU implements the physical and data link layers of the OSI Model.[AIR4271-89]

bus request The DEC PDP-11 priority

system for determining which external device will obtain control of the UNIBUS to interrupt the CPU for service; there are seven bus requests and one bus grant.

butt braze Joining of two conductors end-to-end, with no overlap and with axis in line, using the process of brazing.[ARP1931-87]

butterfly valve A valve consisting of a disc inside a valve body which operates by rotating about an axis in the plane of the disc to shut off or regulate flow in a piping system; a similar device used in heating or ventilating ductwork is called a butterfly damper.

buttering Coating the faces of a weld joint prior to welding to preclude cross contamination of a weld metal and base metal.

butterworth The filter characteristic in which constant amplitude across the pass band is the objective; known also as constant amplitude (CA).

butting dies Crimping dies so designed that the opposing die faces touch at the closed condition of the crimping cycle. Also called "bottoming dies".[ARP914A-79]

butt joint A joint between two members lying approximately in the same plane; in welded joints, the edges may be machined or otherwise prepared to create any of several types of grooves prior to welding.

buttock line A vertical fore and aft plane used for identifying inboard/outboard locations within the airplane.[AS1426-80]

button samples The small amount of sealant extruded from each mixed sealant cartridge onto an indexed card before the cartridge is frozen. It is used as an indicator of the condition of the sealant in each tube, i.e., uniformity of mix, presence of air, cure rate, hardness on cure.[AIR4069-90]

butt splice 1. A device for joining conductors end-to-end with their axis in line and not overlapping. See splice.[ARP 914A-79] 2. A splice wherein two conductors are joined end-to-end with their longitudinal axis in line and not overlapping. A device for accomplishing such a splice.[ARP1931-87]

buttstrap A narrow strip of boiler plate overlapping the joint of two butted plates, used for connecting by riveting.

butt weld A weld that joins the edges or ends of two pieces of metal having similar cross sections, without overlap or offset along the joint line.

butt wrap A spirally wrapped tape over a cable core where the trailing edge of one wrap just meets the leading edge of the preceding wrap with neither overlap nor spacing.[ARP1931-87]

butylene oxides See tetrahydrofuran.

buyer The operator or airline purchasing galleys or an aircraft containing galleys.[AS1426-80]

buyer furnished equipment The equipment (galleys and inserts) purchased by the buyer and furnished to the airframe manufacturer for installation in production airplanes.[AS1426-80]

by-pass A passage for a fluid, permitting a portion or all of the fluid to flow around its normal pass flow channel.

bypass capacitor A capacitor connected in parallel with a circuit element to provide an alternative a-c current path of relatively low impedance.

bypass ratio Ratio of the secondary to the primary inlet airflows for a turbofan engine.[NASA]

by-product Incidental or secondary output of a chemical production or manufacturing process that is obtained in addition to the principal product with little or no additional investment or allocation of resources.

byte 1. Generally accepted as an eight-bit segment of a computer word. 2. Eight contiguous bits starting on an addressable byte boundary; bits are numbered from the right, 0 through 7, with 0 the low-order bit. When interpreted arithmetically, a byte is a two's complement integer with significance increasing from bits 0 through 6; bit 7 is the sign bit. The value of the signed integer is in the range of -128 to 127 decimal. When interpreted as an unsigned integer, significance increases from bits 0 through 7 and the value of the unsigned integer is in the range 0 to 255 decimal. A byte can be used to store one ASCII character. 3. A collection of eight bits capable of representing an alphanumeric or special character.

C_1 The numerical value of the minimum expected capacity in units of ampere hours, at a discharge rate of C_1 amperes (the one-hour rate), and a battery temperature of 23 °C. The C_1 is the rating which indicates the relative amount of available emergency electrical energy capability of a particular cell/battery design.[AS8033-88]

C-8A augmentor wing aircraft NASA's research, short haul, jet aircraft[NASA]

CA See constant amplitude.

cabin pressurization The process of producing pressures in an aircraft that are higher than ambient pressures outside the aircraft.[ARP4107-88]

cable 1. Two or more wires contained in a common jacket or covering; or two or more wires twisted or molded together without a common covering; or one or more wires contained in a shield or in a shield and jacket.[AR1931-87] Unless otherwise specified, a cable or wire is one on which all manufacturing operations have been completed. See also wire.[AS1198-82] 2. A large, strong rope made of fiber or wire. 3. A rope or chain used to restrain a vessel at its mooring. 4. A composite electrical conductor consisting of one or more solid or stranded wires usually capable of carrying relatively large currents, covered with insulation and the entire assembly encased in a protective overwrap.

cable assembly A completed cable and its associated hardware.[AR1931-87]

cable clamp A connector accessory or portion of a component which is designed to grip the wire or cable to provide strain relief and absorb mechanical stress which would otherwise be transmitted to the termination.[ARP914A-79]

cable core The portion of an insulated cable which lies under the protective covering or coverings.[AR1931-87]

cable core binder A wrapping of tapes or cords around the several conductors of a multiple-conductor cable used to hold them together. Cable core binder is usually supplemented by an outer covering of braid, jacket or sheath. [AR1931-87]

cable, fiber-optic That portion of an interconnection plant, carrying signals between two subsystems, that consists of the signal carrying medium plus one or more mechanical protective members such as a sheath plus strength member and scuff jacket. A fiber or multiple fibers in a structure fabricated to meet optical, mechanical and environmental specifications.[ARD50020-91]

cable filler Material used in multiple conductor cables to occupy the interstices formed by the assembly of the insulated conductors, thus forming a cable core of the desired shape.[AR1931-87]

cable shielding clamp A connector accessory device consisting of a sealing member and cable support designed to terminate the shield of the electrical cable at the connector.[ARP914A-79]

cabling the act of twisting together two or more insulated conductors to form a cable.[AR1931-87]

cache memory A small, high-speed memory placed between the slower main memory and the processor. A cache increases effective memory transfer rates and processor speed. It contains copies of data recently used by the processor and fetches several bytes of data from memory in anticipation that the processor will access the next sequential series of bytes.

CAD See computer aided design.

cadmium mercury tellurides See mercury-cadmium tellurides.

cadmium plating An electroplated coating of cadmium on a steel surface which resists atmospheric corrosion. Applications include nuts, bolts, screws, and many hardware items in addition to enclosures.

cage A circular frame for maintaining uniform separation between balls or rollers in a rolling-element bearing. Also known as separator.

caking Producing a solid mass from a slurry or mass of loose particles by any of several methods involving filtration, evaporation, heating, pressure, or a combination of these.

calcine 1. To heat a material such as coke, limestone or clay without fusing it, for the purpose of decomposing compounds such as carbonates and driving off volatiles such as moisture, trapped gases and water of hydration. 2. To heat a material under oxidizing conditions. 3. The product of a calcining or roasting process.

calculating action A type of control system action in which one or more feedback signals are combined with one or more actuating signals to provide an output signal which is some function of the combination.

calculation A group of numbers and mathematical symbols that is executed according to a series of instructions.

calculation, V/L Involves determining the amount of air that will come out of solution between two partial pressures (P initial - P_{TVP}) and (P final - P_{TVP}) while flowing in a closed system.[AIR 1326-91]

calculus of variations The theory of maxima and minima of definite integrals whose integrand is a function of the dependent variables, the independent variables, and their derivatives.

calderas Large, basin-shaped volcanic depressions, more or less circular in form, the diameter of which is many times greater than that of the included vent or vents.[NASA]

calefaction 1. A warming process. 2. The resulting warmed condition.

calendars Orderly arrangements of days, weeks, months, etc. to suit a particular need such as civil life.[NASA]

calender 1. To pass a material such as rubber or paper between rollers or plates to make it into sheets or to make it smooth and glossy. 2. A machine for performing such an operation.

calibrated airspeed (CAS) The indicated airspeed of an aircraft corrected for position and instrument error. Calibrated airspeed is equal to true airspeed in standard atmosphere at sea level. [ARP4107-88]

calibrating tank A liquid vessel of known capacity which is used to check the volumetric accuracy of positive-displacement meters. Also known as meter-proving tank.

calibration 1. The comparison of a measurement system or device of unverified accuracy to a measurement system or device of known and greater accuracy, to detect and correct any variation from required performance specifications of the measurement system or device.[ARD50010-91] 2. The comparison of a standard with a higher ranking reference standard.[AS7200/1-90] 3. Determination of the experimental relationship between the quantity being measured and the output of the device which measures it; where the quantity measured is obtained through a recognized standard of measurement.

calibration curve A plot of indicated value versus true value used to adjust instrument readings for inherent error; a calibration curve is usually determined for each calibrated instrument in a standard procedure and its validity confirmed or a new calibration curve determined by periodically repeating the procedure.

calibration cycle The frequency that a device is due for calibration. This cycle could be dependent on calendar, cycles, or hours.

calibration gas A mixture of gases of specified and known composition used as the basis for interpreting analyzer response in terms of the concentration of the gas to which the analyzer is responding.[ARP1256-90]

calibration record A record of the measured relationship of the transducer output to the applied measurand over the transducer range.

caliper A gaging device with at least one adjustable jaw used to measure linear dimensions such as lengths, diameters and thicknesses.

caliper, brake A "C" shaped structural part of a brake assembly which fits around the edge and adjacent to both sides of a rotating disc. The caliper usually contains an actuator(s) on both sides to provide opposing forces against the rotating disc to produce the decelerating force.[AIR1489-88]

call 1. To transfer control to a specified closed subroutine. 2. In communications, the action performed by the calling party, or the operations necessary to making a call, or the effective use made of a connection between two stations.

calling sequence A specified arrangement of instructions and data necessary to set up and call a given subroutine.

Callisto A satellite of Jupiter orbiting at a mean distance of 1,884,000 kilometers. Also called Jupiter IV.[NASA]

call-up Initial voice contact between a facility and an aircraft, using the identification of the unit being called and the unit initiating the call.[ARP4107-88]

calorie The mean calorie is 1/100 of the heat required to raise the temperature of 1 gram of water from 0 °C to 100 °C at a constant atmospheric pressure. It is about equal to the quantity of heat required to raise one gram of water 1 °C. A more recent definition is: a calorie is 3600/860 joules, a joule being the amount of heat produced by a watt in one second.

calorific value The number of heat units liberated per unit of quantity of a fuel burned in a calorimeter under prescribed conditions.

calorimeter 1. An instrument designed to measure heat evolved or absorbed. [NASA] 2. A device for determining the amount of heat liberated during a chemical reaction, change of state or dissolution process. 3. Apparatus for determining the calorific value of a fuel. 4. An instrument or detector which measures the amount of heat in a light beam—used to measure incident radiation if the percentage of absorbed radiation is known.

calorimetric detection A detector which operates by measuring the amount of heat absorbed—incident radiation must be absorbed as heat to be detected.

calorize To produce a protective coating of aluminum and aluminum-iron alloys on iron or steel (or, less commonly, on brass, copper or nickel); the calorized coating is protective at temperatures up to about 1800 °F.

CALS Computer-aided Acquisition and Logistic Support.

cam 1. An irregularly shaped device, that revolves around an axis, or slides within limits, which actuates other parts by contact to change the direction, speed and/or timing of a part in motion. [ARP480A-79] 2. A machine element that produces complex, repeating translational motion in a member known as a follower that slides or rolls along a shaped surface or in a groove that is an integral part of the cam; a cam is usually a rotating plate, eccentrically mounted on an axis perpendicular to the plate surface, with the follower resting against the contoured periphery of the plate; alternatively, it may be a rotating cylinder or reciprocating plate with a groove cut into its surface for the follower to rest in, or it may be some other shape.

CAM See computer aided manufacturing.

camber 1. The angle which the plane of a wheel or tire makes with a plane normal to the ground. Usually considered positive when the angle vector moves up and outboard relative to the centerline of the vehicle.[AIR1489-88] 2. The maximum distance between the mean line and the chord of an airfoil. The convexity or rise of the curve of an airfoil from its chord, and the ratio of the maximum departure of that curve from the chord to the length of the chord. [ARP4107-88] 3. Deviation from a straight line, most often used to describe a convex, edgewise sweep or curve. 4. The angle of deviation from the vertical for the steerable wheels of an automobile or truck.

camera tube An electron-beam tube in which an optical image is converted to an electron-current or charge-density image, which is scanned in a predetermined pattern to provide an electrical output signal whose magnitude corresponds to the intensity of the scanned image.

cam follower The output link of a cam mechanism.

Campbell bridge A type of a-c bridge used to measure mutual inductance of coil or other inductor in terms of a mutual inductance standard.

camshaft The rotating member that drives a cam.

cam-type timer Any of several designs of timing device using a single contoured cam to continually adjust a process parameter, such as a setpoint, or employing several cams mounted on a single timer shaft to provide interlocked sequence control of a complex operation without using relays.

can A metal vessel or container, usually cylindrical, and usually having an open top or removable cover.

canard An aerodynamic vehicle in which trim surfaces used for longitudinal (or pitch) control are forward of the main lifting surface. Also, more commonly, such a forward-placed control surface itself.[ARP4107-88]

canard configurations Pertaining to an aerodynamic vehicle in which horizontal surfaces used for trim and control are forward of the main lifting surface; the horizontal trim and control surfaces in such an arrangement. [NASA]

candela Metric unit for luminous intensity. The unit used to express the intensity of light visible to the human eye. It corresponds to the emission from 1/60th of a square centimeter of a black body operating at the solidification temperature of platinum, and emitting one lumen per steradian.

candela/m^2 Luminous intensity per unit area of a radiating source with an extended area rather than a point source: the unit area is normal to the direction of observation. The cd/m^2 is the metric unit for luminance (photometric brightness) and is equal to 0.2919 fL (1 fL = 3.426 cd/m^2).[ARP1782-89]

candelas (cd) The cd is the metric unit of luminous intensity. Luminous intensity is used to measure light coming from a point source in a given direction, such as the light coming from a landing light or anticollision light on an aircraft. It is roughly equal to the obsolete units of candles and candlepower.[ARP1782-89]

candle The unit of luminous intensity. It is defined as 1/60 of the intensity of one square centimeter of a black body radiator at the temperature of solidification of platinum (2,046 K).[ARP798-91]

candlepower 1. Luminous intensity expressed in candles.[ARP798-91] 2. An obsolete unit of measure for luminous intensity.

canned 1. Describing a pump or motor enclosed within a watertight casing; in the case of a motor, it is usually enclosed within the same casing as the driven element (such as a pump) and designed so that its bearings are lubricated by the pumped liquid. 2. Describing a composite billet or slab consisting of a reactive metal core encased in metal that is relatively inert so that the reactive metal may be hot worked in air by rolling, forging or extrusion without excessive oxidation.

cannibalize 1. Removal of serviceable parts from one aircraft for installation on another.[ARD50010-91] 2. To disassemble or remove parts from one assembly and use the parts to repair other, like assemblies.

cannot duplicate A fault indicated by BIT or other monitoring circuitry which cannot be confirmed at the next level of maintenance.[ARD50010-91]

canopies The topmost layers of leaves and branches of forest trees or other plants.[NASA]

cantilever A beam or other structural member fixed at one end and hanging free at the other end.

cap A protecting and/or closing part, basically circular, designed with an integral means of securing itself and must partially enclose some protruding, external portion of the item to which it is attached. See also guard, cover, protector.[ARP481A-79]

capacitance 1. That property of a system of conductors and dielectrics which permits the storage of electricity when potential differences exist between the conductors. Its value is expressed as the ratio of the electrostatic charge on a conductor to the potential difference between the conductors required to maintain that charge. Capacitance is measured in farads.[AR1931-87] 2. The ability of a condensor to store a charge before the terminals reach a potential difference of one volt. The greater the capacitance the greater the charge that can be stored.

capacitance meter An instrument for determining electrical capacitance of a circuit or circuit element. See also microfaradmeter.

capacitance-voltage characteristics The characteristics of a metal semiconductor contact or a semiconductor junction that manifests a measured capacitance as a function of a dc bias voltage with small, superimposed ac voltage applied to that junction or contact. [NASA]

capacitive coupling Electrical interaction between two conductors caused by the capacitance between them. [AR1931-87]

capacitive instrument A measuring device whose output signal is developed by varying the capacitive reactance of a sensitive element.

capacitor 1. A passive fluidic element which, because of fluid compressibility, produces a pressure across the device which lags net flow into it by essentially 90°.[ARP993A-69] 2. A device used for storing an electrical charge.

capacitor discharge system An ignition system in which the spark energy is primarily the result of a capacitor discharge.[AIR784-88]

capacity The dischargeable ampere hours available from a fully charged cell/battery at any specified discharge rate/temperature condition.[AS8033-88]

capacity factor The ratio of the average load carried to the maximum design capacity.

capacity lag In any process, the amount of time it takes to supply energy or material to a storage element at one point in the process from a storage point elsewhere in the process. Also known as transfer lag.

capillary 1. Having a very small internal diameter. 2. A tube with a very small diameter.

capillary action 1. The phenomenon of liquid rising in a small interstice due to surface tension.[AR1931-87] 2. Spontaneous elevation or depression of a liquid level in a fine hair-like tube when it is dipped into a body of the liquid. 3. Capillary action is induced by differences in surface energy between the liquid and the tube material.

capillary drying Progressive removal of moisture from a porous solid by evaporation at an exposed surface followed by movement of liquid from the interior to the surface by capillary action until the surface and core reach the same stable moisture concentration.

capillary tube A tube sufficiently fine that capillary action is significant.

cap screw A threaded fastener similar to a bolt, but generally used without a nut by threading it into a tapped hole in one part of an assembly.

capstan A vertical-axis drum used for pulling or hauling; it may be power driven or it may be turned manually by means of a bar extending radially from a hole in the drum.

capsules See space capsules.

captive device A multi-part fastener, usually screw-type, whose components are retained without separation when loosened from its base assembly. [ARP914A-79]

captive device-fastener A fastener, usually screw-type, whose components are retained without separation when loosened from its base assembly. [ARP914A-79]

captive tests Holddown tests of a propulsive subsystem, rocket engine or motor as distinguished from a flight test.[NASA]

capture cross sections See absorption cross sections.

capture effect An effect in frequency-modulation (FM) reception where the stronger signal of two stations on the same frequency completely suppresses the weaker signal.[NASA]

carbenes An organic radical containing divalent carbon.[NASA]

carbides Compounds of carbon with one or more metallic elements.[NASA]

carbide tool A cutting tool whose working edges and faces are made of tungsten, titanium or tantalum carbide particles, compacted and sintered into a hard, heat-resistant and wear-resistant solid by powder metallurgy; the heat-resistant properties of the material are derived in part from a matrix alloy, usually cobalt, which cements the carbide particles together.

carbon An element; the principal combustible constituent of all fuels.

carbon cycle The path of carbon in living beings in which carbon dioxide is fixed by photosynthesis to form organic nutrients and ultimately restored to the inorganic state by respiration and protoplasmic decay.[NASA]

carbon equivalent An empirical relationship that is used to estimate the ability to produce gray cast iron, or one that is used to rate weldability of alloy steels; for cast iron, the formula is $CE=TC+1/3(Si+P)$, where CE is the carbon equivalent, TC is the total carbon content, Si is the silicon content and P is the phosphorus content, all in wt%; for weldability, the formula is $CE=C+Mn/6+(Cr+Mo+V)/5+(Ni+Cr)/15$ where each symbol stands for the concentration of the indicated element in wt%.

carbonitriding A surface-hardening process in which a suitable ferrous material is heated at a temperature above the lower transformation temperature in an atmosphere that will cause simultaneous absorption of carbon and nitrogen at the surface and, by diffusion, create a concentration gradient; final properties are achieved by controlled cooling from temperature, and sometimes by subsequent tempering.

carbonization The process of converting coal to carbon by removing other ingredients.

carbon loss The loss representing the unliberated thermal energy occasioned by failure to oxidize some of the carbon in the fuel.

carbon-pile pressure transducer A resistive-type pressure transducer that depends for its operation on the change in resistance that occurs when irregular carbon granules or smooth carbon disks are pressed together; because of its low resistance, it can often provide sufficient output current to actuate electrical instruments without amplification.

carbon potential A measure of the ability of an environment to alter or maintain the surface-carbon content of ferrous alloys; the specific effect that occurs depends on temperature, time and steel composition as well as on carbon potential.

carbon steel An alloy of carbon and iron containing not more than 2% carbon, and which does not contain alloying elements other than a small amount of manganese.

carbon suboxides Colorless lacrimatory gases having unpleasant odors and boiling points of approximately -7 °C. [NASA]

carbonaceous materials Substance composed of or containing carbon or carbon compounds.[NASA]

carburetor A component of a spark-ignition internal combustion engine that mixes fuel with air, in proper proportions, and delivers a controlled quantity of the mixture to the cylinders.

carburizing A surface-hardening process in which a suitable ferrous material is heated at a temperature above the transformation range in the presence of carbon-rich environment, which may be produced from solid carbon, vaporized liquid hydrocarbons or gaseous hydrocarbons; following production of a carbon concentration gradient in the alloy, it is either quenched from the carburizing temperature and tempered, or reheated, quenched and tempered, to achieve desired properties in both the carbon-rich outer case and the carbon-lean inner core.

carcass The structural body of a tire. Generally would include all tire components except the tread.[AIR1489-88]

card A circuit board within a computer or other electronic instrument or system.

card hopper See hopper, card.

cardinal altitudes or flight levels "Odd" or "Even" thousand-foot altitudes or flight levels, for example, 1000 ft, 12,000 ft, or FL 350, as distinct from 1500 ft, 15,276 ft, or FL 355. See also flight level.[ARP4107-88]

cardinal heading A heading toward one of the cardinal points of the compass (that is, north, south, east, or west). [ARP4107-88]

cardiovascular system The system of an animal pertaining the the heart and blood vessels.[NASA]

card reader (Hollerith cards) A hardware device for reading computer-standard punched cards for computer entry.

card stacker See stacker, card.

Caribbean region The region that consists of all or parts of the islands of the Caribbean Sea, the Bahamas, the British dependent territories, the Virgin Islands, and the mainland areas of the three Guianas and Belize.[NASA]

Carnot cycle An idealized reversible thermodynamic cycle. The carnot cycle consists of four stages: (a) an isothermal expansion of the gas at temperature T1; (b) an adiabatic expansion to temperature T2, (c) an isothermal compression at temperature T2, (d) an adiabatic compression to the original state of the gas to complete the cycle.[NASA]

carriage 1. A mechanism that moves along a predetermined path in a machine to carry and position another component. 2. A mechanism designed to hold paper in the active portion of a printing or typing machine, and to advance the paper as necessary; sometimes, the mechanism also provides for automatically feeding new sheets of paper on demand.

carriage bolt A threaded fastener with a plain (unslotted) head and a square shoulder below the head which keeps the bolt from turning as the nut is tightened; this type of bolt is designed primarily for bolting wood members, but can be used with metal members if the one next to the bolt head has a square bolt hole to accommodate the bolt's shoulder.

carriage return The operation that causes printing to be returned to the left margin with or without line advance. Sometimes used to signify completion of manual data entry.

carriage return character (CR) A format effector that causes the location of the printing or display position to be moved to the left margin with or without line advance.

carriage stop A device attached to the outer way of a lathe bed which permits accurate and repeatable positioning of the tool carriage for cutting grooves, turning multiple diameters and lengths, and cutting off pieces of specific lengths.

carrier 1. The element or combination of several elements laid parallel in the braid by a single bobbin of a braider. [AR1931-87] 2. A carrier is the yarn or strand, or a combination of several yarns or strands laid parallel in the braid by a single bobbin of the braiding machine.[AS1198-82] 3. A continuous frequency signal capable of being modulated to carry information.

carrier band A single channel signaling technique in which the digital signal is modulated on a carrier and transmitted.

carrier density The charge carrier concentration of holes and/or electrons in a semiconductor which determines its electronic characteristics and function.[NASA]

carrier frequency The basic frequency or pulse repetition rate of a transmitted signal, bearing no intrinsic intelligence until it is modulated by another signal that does bear intelligence.

carrier modulation See modulation.

carrier sense multiple access with collision detect (CSMA/CD) An access procedure where a device with data to transmit first listens to the medium. When the medium is not busy, the device starts transmitting. While the device is transmitting, it listens for collisions (simultaneous transmission by another station). If a collision occurs, the node stops transmitting, waits, and tries again.

carrier-to-noise ratio 1. RF signal power input to the receiver divided by the noise power input.[NASA] 2. Carrier amplitude divided by noise amplitude or carrier power divided by noise power.

carrier transport The mobility of conduction electrons or holes in semiconductors.[NASA]

carrier waves Waves generated at a point in the transmitting system and modulated by the signal.[NASA]

carryover The chemical solids and liquid entrained with the steam from a boiler.

Cartesian coordinates A coordinate system in which the locations of points in space are expressed in reference to three planes, called coordinate planes, no two of which are parallel.[NASA]

cartridge 1. A gas generation device, packaged to include ignition, propellant and other required items. Monopropellant: A cartridge using a single propellant. Bi- Propellant: A cartridge using a fuel and an oxidizer, generally in liquid form. Solid Propellant: A cartridge which incorporates a fuel and oxidizer in a solid form.[AIR906A-91] 2. A small unit used for storing computer programs or data values. The amount of information stored tends to be small and access times large compared to discs. However, they are widely used in hobbyist applications.

cartridge actuated devices See actuators.

cartridge disk A relatively low-capacity data-or program-storage medium; generally removable.

cartridge tape Small magnetic tape for digital program storage; stores discrete records.

CAS Calibrated Airspeed; Collision Avoidance System.

CASA/SME Computer and Automated Systems Association of the Society of Manufacturing Engineers.

cascade circuit A circuit in which more than one protector is connected in series between the power source and the load. See also coordination.[ARP1199A-90]

cascade control 1. A control system composed of two loops where the setpoint of one loop (the inner loop) is the output of the controller of the other loop (the outer loop). 2. A control technique that incorporates a master and a slave loop. The master loop controls the primary control parameters and establishes the slave-loop set point. The purpose of the slave loop is to reduce the effect of disturbances on the primary control parameter and to improve the dynamic performance of the loop. See control, cascade.

cascade control action Control action where the output of one controller is the setpoint for another controller.

cascaded Describing a series of machines, devices, or machine elements, so arranged that the output from one feeds directly into the next.

cascade sequence A welding sequence in which a continuous multiple-pass weld is built up by depositing weld beads in overlapping layers; usually, weld beads are laid in a backstep sequence, starting with the root bead that extends only part way along the joint length, then starting successive beads a short distance farther along the joint from the start of the previous bead. Compare with block sequence.

case 1. A container designed to hold a specific item(s) in a fixed position by virtue of conforming dimensions and/or attachments. The item(s) which it contains is complete in itself for removal and use outside the container. However, the container may be constructed so as to permit the use of the item(s) without removal. It does not include the item(s). Excludes shipping containers designed to be discarded after shipment of equipment. See also box.[ARP481A-79] 2. An enclosure designed to hold one or more components in a fixed position, usually by nestling into a conforming recess or resting on fixed supports; in some instances, components are attached directly to the enclosure; the entire unit may be kept in storage or taken to a jobsite, and the contents removed as needed; sometimes as with certain portable instruments, the contents can be used by merely opening the cover of the enclosure and making appropriate connections to the device inside. 3. A hardened outer layer on a ferrous alloy produced by suitable heat treatment, which sometimes involves altering the chemical composition of the outer layer before hardening.

CASE See Common Applications Service Elements.

case-bonded grain A solid propellant grain which is cast in place in the surrounding rocket case.[AIR913-89]

cased glass Glass composed of two or more layers of different glasses, usually a clear, transparent layer to which is added a layer of white or colored glass. The glass is sometimes referred to as flashed, multilayer, or polycased glass. [ARP798-91]

case drain flow Cooling flow and internal leakage of a hydraulic pump or motor to the housing and out of the case drain port, not provided in all models. [AIR1916-88]

case hardening Producing a hardened outer layer on a ferrous alloy by any of several surface-hardening processes, including carburizing, carbonitriding, nitriding, flame hardening and induction hardening. Also known as surface hardening.

casing A covering of sheets of metal or other material such as fire resistant composition board used to enclose all or a portion of a steam generating unit.

cassette 1. A light-tight container for holding photographic or radiographic film, or a photographic plate, and positioning it within a camera or other device for exposure. 2. A small, compact container holding magnetic tape along with supply and takeup reels so that it can be inserted and removed as a unit for quick loading and unloading of a tape recorder or playback machine; different sizes and styles of tape and cassette are used for audio, video and computer applications, depending on the hardware being used.

cassette tape Magnetic tape for digital data storage.

cast 1. To produce a solid shape from liquid or semisolid bulk material by allowing it to harden in a mold. 2. A tinge of a specific color; a slight overtint of a color different from the main color—for instance, white with a bluish cast.

castellated nut A hexagonal nut with a slotted cylindrical projection above one of the hexagonal sides; it is used in conjunction with a cotter pin or safety wire that passes through a lateral hole in the bolt or stud which is aligned with two of the slots in the nut; the cotter pin or safety wire keeps the nut from turning so the joint stays tight.

caster Also castor. See also trail. 1. To swivel about an eccentric vertical axis. Also the dimension which defines this characteristic, i.e., distance from the point where the swivel axis intersects the ground and the centroid of the tire contact area. Generally considered positive when the tire centroid point falls aft of the swivel axis intersect point. [AIR1489-88] 2. The fore-and-aft angle of deviation from the vertical of the kingpin (or its equivalent) in an automobile or truck steering gear. 3. A wheel, usually small in diameter, which is mounted so it is free to swivel about a vertical axis; it is commonly used to support hand trucks, machinery or furniture.

castering gear A gear configuration which has a designed-in capability to caster such as a nose gear or tail gear. See also cross wind gear.[AIR1489-88]

castering wheel(s) Landing wheel(s) mounted in a frame that can rotate about a vertical axis. Castering wheels are of two types: (a) free-castering wheels, and (b) those whose vertical alignment is controlled from the cockpit.[ARP4107-88]

casting 1. The process of making a solid shape by pouring molten metal into a cavity, or mold, and allowing it to cool and solidify. 2. A near-net-shape object produced by this process; a rough casting, cylindrical, square or rectangular in cross section and intended for subsequent hot working or remelting, is called an ingot.

casting alloy An alloy having suitable fluidity when molten and having suitable solidification characteristics to make it capable of producing shape castings; most casting alloys are not suitable for rolling or forging and can only be shaped by casting.

castings A designation for low-quality drill diamonds.

casting shrinkage 1. Total reduction in volume due to the three stages of shrinkage—during cooling from casting temperature to the liquidus, during solidification, and during cooling from the solidus to room temperature. 2. Reduction in volume at each stage in the solidification of a casting.

casting slip A slurry of clay and additives suitable for casting into molds to make unfired ceramic products.

casting wheel A large turntable with molds positioned around its periphery so that each can be moved, in turn, into position for receiving molten metal.

cast iron Any iron-carbon alloy containing at least 1.8% carbon and suitable for casting to shape.

cast tape (film) A material which is formed directly into a tape (or film) by means of flowing or casting a solution or dispersion of the film forming material onto a suitable carrier, then removing the solvents.[AR1931-87]

CAT Clear Air Turbulence.

CAT II Category II. See instrument landing system/ILS category.

catalog In data processing, the contents of a computer disk or tape.

catalyst The component of a two-part curing-type sealant that causes the prepolymer to polymerize.[AIR4069-90] [AS7200/1-90]

catapult 1. A device or machine used to accelerate an airplane to takeoff sped. The capability for rapid and repeated re-use distinguishes it from rocket or JATO assisted takeoff devices. Also, a device (usually explosive) for ejecting a person from an aircraft.[AIR1489-88] 2. A power-actuated machine or device for hurling forth something as an airplane or missile, at a high initial speed; also a device usually explosive, for ejecting a person from an aircraft.[NASA]

catapult bridle See bridle.

catapult hook A hook or hooks built into an aircraft for the purpose of engaging the catapult, pendent, or bridle for the purpose of catapulting the aircraft.[AIR1489-88]

catapult shuttle The deck-mounted of field-mounted arrangement or device for the purpose of towing the aircraft for catapulting. Usually includes an inverted hook (spreader) for engagement with the catapult launching bridle or pendant.[AIR1489-88]

cataphoresis Movement of suspended solid particles in a liquid medium due to the influence of electromotive force.

catastrophic failure 1. A sudden failure that occurs without prior warning, as opposed to a failure that occurs gradually by degradation. 2. Failure of a mechanism or component that renders an entire machine or system inoperable.

category See instrument landing system/ILS categories.

category "A" utilization equipment Those utilization equipments whose installation in aircraft will be controlled so that line drops will be limited to 2 volts ac line drop or 1 volt dc, or both. The line drop is the voltage difference between the point of voltage regulation and the power input terminals of the equipment. Use of this category should be held to a minimum, and its use will be subject to approval by the procuring activity.[AS1212-71]

category "B" utilization equipment Those utilization equipments destined for aircraft for which the line drops will be less than 4 volts ac or 2 volts dc, or both. When a detail equipment specification does not designate a category, the equipment will be considered a category "B" equipment. This category will include the majority of aircraft electric equipments and is the preferred category.[AS1212-71]

category "C" utilization equipment Those equipments which are intermittently operated. During operation, voltage limits include allowance for 8 volts ac line drop or 3 volts dc line drop, or both.[AS1212-71]

catenary The shape produced by holding a rope or cable at its ends and allowing the center section to sag under its own weight.

cathetometer An optical instrument for measuring small differences in height -— for instance, the difference in height between two columns of mercury.

cathode 1. In electron tubes, electrodes through which a primary stream of electrons enters the interelectrode space. [NASA] 2. The metal plate or surface that acts as an electron acceptor in an electrochemical circuit; metal ions in an electrolytic solution plate on the cathode during electroplating, and hydrogen may be formed at the cathode during electroplating or electrochemical corrosion. 3. The positive electrode in a storage battery, or the negative electrode in an electrolytic cell. 4. The negative electrode in an x-ray tube or vacuum tube, where electrons enter the interelectrode space.

cathode burn The burning sensation of the eyes resulting from prolonged staring at cathode ray tubes (electronic displays).[ARP4107-88]

cathode corrosion 1. Corrosion of the cathode in an electrochemical circuit, usually involving the production of alkaline corrosion products. 2. Corrosion of the cathodic member of a galvanic couple.

cathode follower A type of electronic circuit in which the output load is connected in the cathode circuit of an electron tube or equivalent transistor and the input signal is impressed across a terminal pair where one is connected directly to the control grid and the other to the remote end of the output load.

cathode ray In an electron tube or similar device, a stream of electrons emitted by the cathode.

cathode-ray oscillograph An instrument that produces a record of a waveform by photographing its graph produced on a cathode-ray tube, or by otherwise recording such an image.

cathode-ray oscilloscope An instrument that indicates the shape of a waveform by producing its graph on the screen of a cathode-ray tube.

cathode ray tube (CRT) 1. An electron beam tube in which the beam, or beams, can be focused to a desired cross section on a surface and varied in position and intensity to produce a visible or otherwise detectable pattern. Unless otherwise stated, the term cathode-ray tube is reserved for devices in which the screen is cathodoluminescent and in which the output information is presented in the form of a pattern of light. [ARP1782-89] 2. Vacuum tubes consisting essentially of an electron gun producing a concentrated electron beam (or cathode ray) which impinges on a phosphorescent coating on the back of a viewing face (or screen). The excitation of the phosphor produces light, the intensity of which is controlled by the flow of electrons. Deflection of the beam is achieved either electromagnetically by currents in coils around the tube, or electrostatically by voltages on internal deflection plates.[NASA] 3. An electronic vacuum tube containing a screen on which information may be stored for visible display by means of a multigrid modulated beam of electrons from the thermionic emitter, storage is effected by means of charged or uncharged spots. 4. A storage tube. 5. An oscilloscope tube. 6. A picture tube. 7. A computer terminal using a cathode ray tube as a display device.

cathodic coating 1. Material forming a continuous film on a base metal by mechanical coating or by electroplating.[NASA] 2. A mechanical plate or electrodeposit on a base metal, with the coating being cathodic to the underlying base metal.

cathodic protection Preventing electrochemical corrosion of a metal object by making it the cathode of a cell using either a galvanic or impressed current.

cathodoluminescence 1. Luminescence produced when high velocity electrons bombard a metal in a vacuum, thus vaporizing small amounts of the metal which, in an excited state, emit radiation characteristic of the metal.[NASA] 2. Luminescence induced by exposure of a suitable material to cathode rays.

CATT devices Controlled avalanche transit time triodes which use avalanche multiplication in the collector depletion region of a silicon, bipolar, transistor-like structure to increase the gain and thereby achieve a higher frequency operation of silicon bipolar transistors.[NASA]

CATV Community Antenna Television. See broadband.

caulk 1. A heavy paste such as a mixture of a synthetic or rubber compound and a curing agent, or a natural product such as oakum, used to seal cracks or seams and make them airtight, steam-tight or watertight. Also known as caulking compound; calk. 2. To seal a crack or seam with caulk.

caustic dip A strongly alkaline solution for immersing metal parts to etch them, to neutralize an acid residue, or to remove organic material such as grease or paint.

caustic embrittlement Intergranular cracking of carbon steel or Fe-Cr-Ni alloy exposed to an aqueous caustic solution at a temperature of at least 150 °F while stressed in tension; a form of stress-corrosion cracking. Also known as caustic cracking.

caustic lines The locations of wave front interactions induced by the maneuvers of supersonic aircraft in changing direction and/or attitude.[NASA]

caustics The envelope of rays diffracted by surface defects in materials.[NASA]

caustic soda The most important of the commercial caustic materials—it consists of sodium hydroxide that contains 76 to 78% sodium oxide.

caution alert Abnormal operational or aircraft system condition that requires immediate crew awareness and subsequent corrective or compensatory crew action.[ARP4153-88]

Cavendish balance A torsional instrument for determining the gravitational constant by measuring the displacement of two spheres of known small mass, mounted on opposite ends of a thin rod suspended on a fine wire, when two spheres of known large mass are brought near the small spheres.

cavitation Formation of cavities, either gaseous or vapor within a liquid stream, which occurs where the pressure is locally reduced to vapor pressure of the liquid. It may include gas coming out of solution in the liquid as pressure is reduced (soft cavitation).[AIR1916-88] See also cavitation flow.

cavitation erosion Progressive removal of surface material due to localized hydrodynamic impact forces associated with the formation and subsequent collapse of bubbles in a liquid in contact with the damaged surface. Also known as cavitation damage; liquid-erosion failure.

cavitation flow The formation of bubbles in a liquid, occurring whenever the static pressure at any point in the fluid flow becomes less than the fluid vapor pressure.[NASA]

cavitons Density cavities created by localized oscillating electric fields. [NASA]

cavity resonator A space normally enclosed by an electrically conducting surface, which is used to store electromagnetic energy and whose resonant frequency is determined by the shape of the enclosure.

cavity-type wavemeter An instrument used to determine frequency in a waveguide system; typically, the position of a piston inside a cylindrical cavity is tuned to resonance, which is determined by a drop in transmitted power; the meter is then detuned for normal operation.

CBR (California Bearing Ratio) A measure of the bearing capacity of soil. The CBR of a soil is expressed in terms of a percentage of the bearing capacity of a standard crushed limestone surface.[AIR1489-88]

CBW See constant bandwidth.

CCD See charge coupled devices.

CCD star tracker Navigation instrument designed for the NASA space transportation system.[NASA]

CCITT International Consultative Committee on Telegraphy and Telephony.

CCR See control complexity ratio.

CCV Control-Configured Vehicle (aircraft).

CD Constant delay. See bessel.

CDI Course Deviation Indicator.

CDMA See code division multiple access.

CD-ROM A compact disk used for computer data storage. The letters stand for Compact Disk Read-Only Memory.

ceiling 1. The height above the earth's surface of the lowest layer of clouds or other obscuring phenomena that are reported as "broken," "overcast," or "obscuration," and not classified as "thin" or "partial". 2. The upper operating limit(s) of an aircraft. See absolute ceiling and service ceiling.[ARP 4107-88]

ceilometer A device or apparatus for measuring the height of a cloud ceiling or determining the vertical visibility to an obscuration.[ARP4107-88]

celestial bodies Any aggregations of matter in space constituting a unit for astronomical study, as the sun, moon, a planet, comet, star, or nebula. Also called heavenly bodies.[NASA]

celestial geodesy The determination of the form of the earth, of the earth's gravitational field and of relative positions of satellite trajectories.[NASA]

celestial mechanics The study of the theory of motions of celestial bodies under the influence of gravitational fields.[NASA]

celestial navigation The process of directing a craft from one point to another by reference to celestial bodies of known constants.[NASA]

celestial observation See astronomy.

celestial sphere An imaginary sphere of infinite radius concentric with the earth, on which all celestial bodies except the earth are assumed to be projected.[NASA]

cell 1. The basic battery building block. An electrochemical storage device consisting of positive and negative plates, separator, gas barrier and electrolyte contained in a cell case (jar) with suitable electrical terminals for connection to additional cells.[AS8033-88] 2. One of a series of chambers in which a chemical or electrochemical reaction takes place—for example, the chambers of a storage battery or electrolytic refining bath. 3. One of the cavities in a honeycomb structure. 4. The storage of one unit of information, usually one character or one word. 5. A location specified by whole or part of the address and possessed of the faculty of store. Specific terms such as column, field, location, and block, are preferable when appropriate. See storage cell.

Celsius A scale for temperature measurement based on the definition of 0 °C and 100 °C as the freezing point and boiling point, respectively, of pure water at standard pressure.

cement 1. A dry, powdery mixture of silica, alumina, magnesia, lime and iron oxide which hardens into a solid mass when mixed with water; it is one of the ingredients in concrete and mortar. 2. An adhesive for bonding surfaces where intimate contact cannot be established and the adhesive must fill a gap over all or part of the faying surfaces.

cementation 1. High temperature impregnation of a metal surface with another material. 2. Conversion of wrought iron into steel by packing it in charcoal and heating it at about 1800 °F for 7 to 10 days.

cemented carbide A powder-metallurgy product consisting of granular tungsten, titanium or tantalum carbides in a temperature-resistant matrix, usually cobalt; used for high- performance cutting tools, punches and dies; the proportion of matrix material is small compared to the amount of carbide.

cementite An intermetallic compound containing iron and carbon.[NASA]

cent The interval between two sound frequencies, where the ratio of the two frequencies is the twelve-hundredth root of 2. Also equal to one-hundredth of a semi-tone.

center gage A gage used to check angles, such as the angle of a cutting-tool point or screw thread.

centerline groove An injection-sealing groove machined along the fastener line.[AIR4069-90]

center of gravity The center of mass of a system of masses, as the barycenter of the earth-moon system.[NASA]

center of mass A point of a material body or system of bodies which moves as though the system's total mass existed at that point and all external forces were applied at the point.[NASA]

centigrade A nonpreferred term formerly used to designate the scale now referred to as the Celsius scale.

centimeter waves Electromagnetic radiation in the 3,000 to 30,000 MHz range.[NASA]

centimetre-gram-second (CGS) A standard metric system of units used largely for scientific work prior to adoption of the international SI system currently preferred for both scientific and engineering work.

central 80% of useful display area Area within the locus of points 80% of the distance from display center to the edge of the useful display area.[ARP 1782-89] [ARP4067-89] [AS8034-82]

central flow control function (CFCF) The air traffic control command center function that is responsible for coordination and approval of all major intercenter flow control restrictions on a system basis in order to obtain maximum utilization of the airspace. See fuel advisory departure, quota flow control. [ARP4107-88]

centralized To bring under one control.

centralized control In data communications systems, a method of media access control in which a designated station exercises positive control over all message traffic.[AIR4271-89]

centralized maintenance shops One maintenance shop that has responsibility to maintain all equipment in the facility. Usually several crafts work out of this one centralized maintenance shop.

central processing unit (CPU) 1. The unit of a computing system that includes the circuits controlling the interpretation of instructions and their execution. [NASA] 2. The brain of the computing machine, usually defined by the arithmetic and logic units (ALU) plus a control section, often called a processor. 3. The part of a computing system that contains the arithmetic and logical units, instruction control unit, timing generators, and memory and I/O interfaces. See also unit, central processing.

central station A power plant or steam heating plant generating power or steam for sale.

centrifugal fan Consists of a fan rotor or wheel within a housing discharging the air at right angle to the axis of the wheel.

centrifugal force 1. The apparent force in a rotating system, deflecting masses radically outward from the axis of rotation, with magnitude per unit mass ω^2R where ω is the angular speed of rotation; and R is the radius of curvature of the path. This magnitude may also be written V^2/R, in terms of the linear speed V. This force (per unit mass) is equal and opposite to the centripetal acceleration.[AIR1489-88] 2. The apparent force in a rotating system, deflecting masses radially from the axis or rotation.[NASA] 3. A force acting in a direction along and outward on the radius of turn for a mass in motion.

centrifugal growth The growth in diameter of a rotating tire due to centrifugal force.[AIR1489-88]

centrifugal tachometer An instrument that measures the instantaneous angular speed of a rotating member such as a shaft by measuring the centrifugal force on a mass that rotates with it.

centrifuge 1. Specifically in aerospace, large motor driven apparatus with long arms at the end of which human and animal subjects or equipment can be revolved and rotated at various speeds to simulate (very closely) the (prolonged) acceleration in high performance aircraft, rockets, and spacecraft. Sometimes called astronautic centrifuges. [NASA] 2. A rotating device that separates suspended fine or colloidal particles from a liquid, or separates two liquids of different specific gravities, by means of centrifugal force.

cepstra The Fourier transformation of the logarithm of a power spectrum. [NASA]

cepstral analysis The application of cepstral methods to wave or signal phenomena in seismology, speech analysis, echoes, underwater acoustics, etc. [NASA]

ceramal protective coatings. See cements.

ceramic 1. A heat-resistant natural or synthetic inorganic product made by firing a nonmetallic mineral. 2. A shape made by baking or firing a ceramic material, such as brick, tile or labware.

ceramic coating A protective coating made by thermal spraying a material such as aluminum or zirconium oxide, or by cementation of a material such as aluminum disilicide, on a metal substrate.

ceramic fibers Fibers composed of ceramic materials. They are usually used for reinforcement.[NASA]

ceramic matrix composites Composite materials consisting of a reinforced ceramic matrix.[NASA]

ceramics Inorganic compounds or mixtures requiring heat treatment to fuse them into homogeneous masses usually possessing high temperature strength but low ductility. Types and uses range from china for dishes to refractory liners for nozzles.[NASA]

ceramic tool A cutting tool made from fused, sintered or cemented metallic oxides.

ceramic transducer See electrostriction transducer.

Cerenkov effect See Cerenkov radiation.

Cerenkov radiation 1. The radiation from a charged particle whose velocity is greater than the phase velocity that an electromagnetic wave would have if it were propagating in the medium. The particle will continue to lose energy by radiation until the velocity is less than this phase velocity.[NASA] 2. Visible light produced when charged particles pass through a transparent medium at a speed exceeding the speed of light in the medium.

cermet A body consisting of ceramic particles bonded with a metal; used in aircraft, rockets, and spacecraft for high strength, high temperature applications. The name is derived from a combination of CERamic and METal.[NASA] 2. A powder-metallurgy product consisting of ceramic particles bonded together with a metal matrix.

certification 1. The act of certifying. 2. The state of being certified. 3. The attainment of certification is a means for individuals to indicate to the general public, co-workers, employers, and others that an impartial, nationally-recognized organization has determined that they are qualified to perform specific technical tasks by virtue of their technical knowledge and experiences.

certification pressure The authorized pressure for formal acceptance tests. [AIR1916-88]

certified takeoff weight The maximum takeoff weight stated on the type certificate data sheet of an aircraft (or, for aircraft manufactured prior to 1957, on the type specification sheet).[ARP4107-88]

certify 1. To confirm formally as true, accurate, or genuine. 2. To guarantee as meeting a standard. 3. To issue a license or certificate.

Cessna 402B aircraft A lighter, twin-engine, short-haul cargo/passenger aircraft manufactured by the Cessna Aircraft Company.[NASA]

CFCF Central Flow Control Function.

CFD See charge flow devices.

CFR engine Cooperative Fuel Research engine—a standard test engine for determining the octane number of motor fuels.

CGM Computer Graphics Metafile.

CGS See centimetre-gram-second.

chafer strip The strips of rubber coated fabric which reinforce the bead area and protect the carcass plies against wheel chafing and against damage when mounting or demounting the tire.[AIR 1489-88]

chafing Repeated relative motion between wires, cables, groups, harnesses or bundles, or between these wiring system components and structure or equipment, which results in deleterious wear.[AR1931-87]

chafing fatigue See fretting.

chain 1. A nonrigid series of metal links or rings that are interlinked with each other, or are pinned or otherwise held together, to make an elongated flexible member suitable for pulling, hauling, lifting, supporting or restraining objects, or for transmitting power. 2. A mesh of rods or plates used in place of a belt to convey objects or transmit power. 3. An organization in which records or other items of data are strung together by means of pointers.

chain-balanced density meter A submerged-float meter, using an iron-core float that moves up and down within a pickup coil; a slack chain attached to the bottom of the float applies more weight as the float rises and establishes a definite equilibrium position for any given fluid density within the range of the instrument.

chain block A lifting tackle, often suspended from an overhead track, which uses a chain instead of rope to lift heavy weights and is hand driven by pulling on an endless chain; some models are power driven. Also known as chain fall; chain hoist.

chain drive A device for transmitting power and motion without slipping which consists of an endless chain that meshes with driving and driven sprockets; chain drives are used on bicycles and motorcycles to provide the motive power, on conveyors to drive the belts and in hoisting mechanisms to provide the lifting power.

chain-float liquid-level gage A device for indicating liquid level in a tank which consists of a float connected to a counterweight by a chain running over a sprocket; as the float rises and falls with liquid level in the tank, the chain rotates the sprocket which in turn positions a pointer to indicate liquid level.

chain-link-fence illusion See illusion, visual.

chalking 1. The formation of a powdery surface condition due to disintegration of firesleeving or hose cover material by weathering or other destructive environments.[ARP1658A-86] 2. A defect of coated metals caused by formation of a layer of powdery material at the metal-coating interface.

chamfer 1. A beveled edge that relieves an otherwise sharp corner. 2. A relieved angular cutting edge at a tooth corner on a milling cutter or similar tool.

Chandler motion See polar wandering.

change detection A process of examining imagery to detect changes on a planetary surface or astronomical body.[NASA]

channel 1. One signal or control path of a redundant set. A channel is an entity within itself and contains elements individual to that channel. To achieve true fault-tolerance, each channel must be designed so that a failure in one channel does not cause failure in another. A model may be used as a reference channel in a detection-correction system.[AIR1916-88] [ARP1181-85] 2. A passage formed by a structural discontinuity such as the opening under a joggle or other void.[AIR4069-90] 3. A groove machined in a faying surface to accept a uniform bead or section of sealant.[AIR4069-90] 4. A path along which information, particularly a series of digits or characters, may flow. 5. One or more parallel tracks treated as a unit. 6. In a circulating storage, a channel is one recirculating path containing a fixed number of words stored serially by word. 7. A path for electrical communication. 8. A band of frequencies used for communication.

channel buffering A technique used to minimize the possibility of a failure in one channel from inducing a failure in another channel.[ARP1181A-85] [AIR1916-88]

channeling leak A leak that develops at a source located some distance from the leak exit point.[AIR4069-90]

channelized attention See attention, anomalies of channelized attention.

channel noise In communications, bursts of interruptive pulses caused mainly by contact closures in electromagnetic equipment or by transient voltages in electric cables during transmission of signals or data. Impulsive noise is the frequent cause of transmission errors.[NASA]

channel priority The order of authority of the various channels in a redundant system where the channels are not equivalent. Examples of channel prioritizing are: primary/secondary, active/standby, normal/alternate.[ARP1181A-85] [AIR1916-88]

channel sampling rate The number of times a given data input is sampled during a specified time interval.

channel seal See injection seal, groove seal.

channel selector In an FM discriminator, the plug-in module which causes the device to select one of the channels and demodulate the subcarrier to recover data.

channel summing The combining of multiple channels to provide a control function. Examples of channel summing techniques are: flow summing, flux summing, force summing, position summing, torque summing, velocity summing.[ARP1181A-85] [AIR1916-88]

Chapman-Jouget flame See detonation.

CHAR See character.

character codes The binary code patterns used to create characters in a computer.

characteristic 1. Specifically, distinguishing quality, property, feature, or capability of an entity.[NASA] 2. The integral part of a common logarithm, i.e., in the logarithm 2.5, the characteristic is 2, the mantissa is 0.5. 3. Sometimes, that portion of a floating point number indicating the exponent. 4. A distinctive property of an individual, document, item, etc.

characteristic curve 1. A curve expressing a relation between two variable properties of a luminous source, as candlepower and volts, candlepower and rate of fuel consumption, etc.[ARP798-91] 2. Of a photographic or radiographic film, the graph of relative transmittance of the emulsion versus exposure, or a graph of functions of these two quantities. Also known as characteristic emulsion curve.

characteristic impedance (Zo) Characteristic impedance of a uniform line is the ratio of an applied potential difference to the resultant current at the point where the potential difference is applied, when the line is of infinite length. The term is applied only to a uniform line. Coaxial cable is such a uniform line.[AR1931-87]

charge 1. A given quantity of explosive.[AIR913-89] 2. A defined quantity of an explosive. 3. The starting stock loaded into a batch process. 4. Material loaded into a furnace for melting or heat treating. 5. A measure of the accumulation or depletion of electrons at any given point. 6. The amount of substance loaded into a closed system, such as refrigerant into a refrigeration system. 7. The quantity of excess protons (positive charge) or excess electrons (negative charge) in a physical body, usually expressed in coulombs.

chargeable Within the responsibility of a given organizational entity (applied to terms such as failures, maintenance time, etc.).[ARD50010-91]

charge coupled devices Semiconductor devices arrayed so that the electric charge at the output of one provides the input stimulus to the next.[NASA]

charge efficiency The efficiency of electric cell recharging.[NASA]

charge exchange The collisional transfer of an electron from a neutral atom or molecule to an ion.[NASA]

charge flow devices Metal oxide semiconductor (MOS) devices used for fire detectors and humidity sensors.[NASA]

charging The return of electrical energy to a battery and its storage in electrochemical form.[AS8033-88]

charm A quantum number which has been proposed to account for an apparent lack of symmetry in the behavior of hadrons relative to that of leptons, to explain why certain reactions of elementary particles do not occur, and to account for the longevity of the J particle.[NASA]

Charon Natural satellite of the planet Pluto, discovered and named by Dr. James W. Christy.[NASA]

chart A sheet or plate giving printed information in tabular and/or graphic form. It is not intended for, neither does it have provisions for, recording additional information thereon. Excludes maps.[ARP481A-79]

chart recorder A device for automatically plotting a dependent variable against an independent variable; the dependent variable is proportional to the input signal from a transducer; the independent variable may be proportional to a transducer signal, also, but is most often time or a time-dependent variable that can be produced by controlling the rate of advance of rolled chart paper.

chase 1. A vertical passage in a building that contains the pipes, wires and ducts which provide heat, ventilation, electricity, running water, drains and other building services. 2. The main body of a mold that contains one or more mold cavities. 3. To make a series of cuts, each following the path of a preceding cut, such as is done to produce a thread in lathe turning using a single-point tool. 4. To straighten and clean damaged or debris-filled threads on a screw or pipe end.

chassis 1. A frame or box-like sheet-metal support for mounting the components of an electronic device. 2. A frame for a wheeled vehicle that provides most of the stiffness and strength of the vehicle body, and supports the body, engine, and passenger or load compartment on the running gear.

chatter Servoactuator chatter is a low amplitude (usually) high frequency oscillation of the output.[ARP1281A-81]

CHEAPERNET An IEEE 802.3 standard for a low cost, 10 Mbit LAN, compatible with TOP.

check A process of partial or complete testing of the correctness of machine operations. The existence of certain prescribed conditions within the computer, or the correctness of the results produced by a program. A check of any of these conditions may be made automatically by the equipment or may be programmed. Related to marginal check.

check, bench A functional check of an item in the shop to determine whether the item may be returned to service, or whether it requires adjustment, repair, overhaul, or replacement.[AIR1916-88]

check bit A binary check digit; often a parity bit. Related to parity check.

check, C A heavy maintenance check. See also letter check.[AIR1916-88]

check digit In data transmission, one or more redundant digits appended to a machine word, and used in relation to the other digits in the word to detect errors in data transmission.

checker work An arrangement of alternately spaced brick in a furnace with openings through which air or gas flows.

checking 1. Short, shallow cracks in the surface of an elastomeric hose cover material resulting from damaging action of environmental conditions. [ARP1658A-86] 2. A network of fine cracks in a coating or at the surface of a metal part; they may appear during processing but are more often associated with service, especially when it involves thermal cycling.

checkout 1. Man/machine task to determine that the equipment is operating satisfactorily and ready for return to service.[ARD50010-91] 2. A sequence of actions taken to test or examine a thing as to its readiness for incorporation into a new phase of use, or for the performance of its intended function. The sequence of steps taken to familiarize a person with the operation of an airplane or other piece of equipment. [NASA] 3. Determination of the working condition of a system. 4. A test or preliminary operation intended to determine whether a component or system is ready for service or ready for a new phase of operation.

checkpoint A point in time in a machine run at which processing is temporarily halted, to make a record of the condition of all the variables of the machine run, such as the status of input and output devices and a copy of working storage. Check points are used in conjunction with a restart routine to minimize reprocessing time occasioned by functional failures. A checkpoint also may be a particular point in a program at which processing is halted for checking.

check problem A problem used to test the operation of a computer or to test a computer program; if the result given by the computer does not match the known result, it indicates an error in programming or operation.

checksum 1. A routine for checking the accuracy of data transmission by dividing the data into small segments, such as a disk sector, and computing a sum for each segment. 2. Entry at the end of a block of data corresponding to the binary sum of all information in the block. Used in error-checking procedures.

check valve A valve that allows flow in a single direction only.[ARP986-90]

chemical affinity 1. The relative ease with which two elements or compounds react with each other to form one or more specific compounds. 2. The ability of two chemical elements to react to form a stable valence compound.

chemical analysis Determination of the principal chemical constituents.

chemical clouds Artificial clouds of chemical compounds released in the ionosphere for observation of dispersion and other characteristics.[NASA]

chemical conversion coating A decorative or protective surface coating produced by inducing a chemical reaction between surface layers of a part and a specific chemical environment, such as in chromate treatment or phosphating.

chemical defense All actions and counteractions designed for the protection of personnel and material against offensive chemical agents.[NASA]

chemical energy Energy produced or absorbed in the process of a chemical reaction. In any such a reaction, energy losses or gains usually involve only the outermost electrons of the atoms or ions of the system undergoing change; here a chemical bond of some type is established or broken without disrupting the original atomic or ionic identities of the constituents.[NASA]

chemical engineering A branch of engineering that deals with the design, operation and maintenance of plants and equipment for chemically converting raw materials into bulk chemicals, fuels and other similar products through the use of chemical reaction, often accompanied by a change in state or in physical form.

chemical evolution The theory of the creation or production of living matter from nonliving matter.[NASA]

chemical feed pipe A pipe inside a boiler drum through which chemicals for treating the boiler water are introduced.

chemical fuels Fuels that depend upon an oxidizer for combustion or to development of thrust, such as liquid or solid rocket fuel or internal combustion engine fuel; distinguished from nuclear fuel.[NASA]

chemical release modules Shuttle launched, free-flying spacecraft containing canisters for injecting chemicals into the upper atmosphere and the measurement of the reactions.[NASA]

chemiluminescence Any luminescence produced by chemical action.[NASA]

chemisorption The binding of a liquid or gas on the surface or in the interior of a solid by chemical bonds or forces. [NASA]

chemosphere The vaguely defined region of the upper atmosphere in which photochemical reactions take place. It is generally considered to include the stratosphere (or the top thereof) and the mesosphere, and sometimes the lower part of the thermosphere.[NASA]

cherry picker Any of several types of small traveling cranes, especially one consisting of an open passenger compartment at the free end of a jointed boom.

chimney A brick, metal or concrete stack.

chimney core The inner cylindrical section of a double wall chimney, which is separated from the outer section by an air space.

chimney lining The material which forms the inner surface of the chimney.

chip Single large scale integrated circuit.

chip breaker An attachment or a relieving channel behind the cutting edge of a lathe tool to cause removed stock to break up into pieces rather than to come off as long, unbroken curls.

chipping 1. Using a manual or pneumatic chisel to remove seams, surface defects or excess metal from semifinished mill products. 2. Using a hand or pneumatic hammer with chisel-shaped or pointed faces to remove rust, scale or other deposits from metal surfaces.

chips Areas from which glass has broken away from the surface, edge or bevel of an optical element.[ARP924-91]

chips (electronics) Integrated microcircuits mounted on substrates and performing significant numbers of functions.[NASA]

chips (memory devices) Integrated microcircuit devices used collectively to perform the functions of data storage: accepting, retaining, and emitting bits of data.[NASA]

Chiron Minor planet 2060, a solar system asteroid discovered by Charles T. Kowal of Hale Observatories.[NASA]

chirp An all encompassing term for the various techniques of pulse expansion-pulse compression applied to pulse radar; a technique to expand narrow pulses to wide pulses for transmission, and compress wide received pulses to the original narrow pulse width and wave shape, to gain improvement in signal-to-noise ratio without degradation to range resolution and range discrimination.[NASA]

chitin A polysaccharide which is the principal constituent of the shells of crabs and lobsters and of the shards of beetles. It is also found in certain fungi.[NASA]

chlorate candles Sodium chlorate candles can store oxygen in a compact volume approaching that attainable with liquid oxygen. The candles have desirable weight, cost and storage life characteristics. The liberated oxygen contains a small amount of fine sodium chloride which can be readily removed by filters.[AIR1246-77]

Chlorella A genus of unicellular green algae to be adapted to converting carbon dioxide into oxygen in a closed ecological system.[NASA]

chlorocarbons All compounds containing chlorine and carbon with or without other elements.[NASA]

choke 1. A flow restriction in which the length of the reduced area passage is of sufficient magnitude that fluid viscosity becomes of major importance in determining pressure drop.[ARP243B-65] 2. A valve which increases suction to draw in an excess proportion of fuel and facilitate starting a cold internal combustion engine.

choke coil An inductor that allows direct current to pass but presents relatively large impedance to alternating current.

choked flow 1. Flow of a compressible fluid (gas), limited by the speed of sound in the fluid at the throat of a control section.[AIR1916-88] 2. The condition that exists when, with the upstream conditions remaining constant, the flow through a valve cannot be further increased by lowering the downstream pressure.

chokes See decompression sickness.

Cholesky factorization A numerical algorithm used to solve linear systems of equations.[NASA]

chondrites Meteoritic stones characterized by small rounded grains or spherules.[NASA]

chop (fuel) Shutoff of fuel when engine is running.[AIR906A-91]

chop (throttle) Rapid throttle movement from any power setting to idle or cutoff.[AIR906A-91]

chopper Any device for periodically interrupting a continuous current or flux.

chord/chord line In aeronautics, a straight line parallel to the plane of symmetry connecting the leading and trailing edges of an airfoil. The ordinates and angles of the airfoil are measured from the chord line.[ARP4107-88]

chords Straight lines intersecting circles

or other curves, or straight lines connecting the ends of arcs. In aeronautics, straight lines intersection or touching airfoil profiles at two points, specifically, those parts of lines between two points of intersections.[NASA]

Christiansen filter A device for admitting monochromatic radiation to a lens system; it consists of coarse powder of a transparent solid confined between parallel windows, with the spaces between particles being filled with a liquid whose refractive index is the same as that of the powder for a certain wavelength; only that wavelength is transmitted by the filter without deviation.

chromadizing Improving paint adhesion on aluminum and its alloys by treating the surface with chromic acid.

chromate treatment Applying a solution of hexavalent chromic acid to produce a protective conversion coating of trivalent and hexavalent chromium compounds.

chromatic aberration The focusing of light rays of different wavelengths at different distances from the lens. This is not a significant effect with a single wavelength laser source, but can be when working at different or multiple wavelengths.

chromaticity 1. The expression of color in terms if CIE codents.[ARP798-91] 2. The color quality of a color stimulus definable by its chromaticity coordinates.[ARP1782-89]

chromaticity coordinate The ratio of each of a set of three tristimulus values to their sum. They are given in ordered paris, for example, (x,y), u[1] [7]1 etc.[ARP 1782-89]

chromaticity diagram 1. A diagram in which distance along suitable axes represent chromaticity coordinates.[ARP 1782-89] 2. A graph of one of the chromaticity coordinates against another.

chromaticity difference (CD) Distance between two color points.[ARP4067-89]

chromatography 1. The separation of chemical substances by making use of differences in the rates at which the substances travel through or along a stationary medium.[NASA] 2. An instrumental procedure for separating components from a mixture of chemical substances which depends on selective retardation and physical absorption of substances by a porous bed of sorptive media as the substances are transported through the bed by a moving fluid; the sorptive bed (stationary phase) may be a solid or a liquid dispersed on a porous, inert solid; the moving fluid (moving phase) may be a liquid solution of the substances or a mixture of a carrier gas and the vaporized sample; a wide variety of detection techniques are used, some of which can be automated or microprocessor driven.

chrome See Munsell chroma.

chromium plating Electrodeposition of either a bright, reflective coating or a hard, less-reflective coating of chromium on a metal surface. Also known as chrome plating; chromium coating.

chromium steels Steels containing chromium as the main alloying element.[NASA]

chromizing Producing an alloyed layer on the surface of a metal by deposition and subsequent diffusion of metallic chromium.

chromophore The group of atoms within a molecule that contributes most heavily to its light-absorption qualities.

chromosphere A thin layer of relatively transparent gases above the photosphere of the sun.[NASA]

chronograph An instrument used to record the time at which an event occurs or the time interval between two events.

chronotron A device for measuring elapsed time between two events in which the time is determined by measuring the position of the superimposed loci of a pair of pulses initiated by the events. See also time lag.

chuffing Intermittent or irregular burning in a solid propellant rocket motor with corresponding low frequency pressure oscillations.[AIR913-89]

chugging An irregular combustion of liquid fuels in a rocket engine with corresponding low frequency pressure oscillations.[AIR913-89]

Chukchi Sea Part of the Arctic Ocean north of the Bering Strait between Asia and North America.[NASA]

chute, deceleration See chute, drag.

chute, drag A parachute attached to an air vehicle and deployed usually after touchdown for the purpose of decelerating the air vehicle by means of aerodynamic drag.[AIR1489-88]

chute, drogue A parachute, deployed for the purpose of providing the necessary force for extracting and deploying a main drag chute or chutes.[AIR1489-88]

chute, pilot A parachute deployed to initiate a sequence of events in a drag chute system; i.e., the pilot chute is mechanically (or otherwise) ejected into the airstream. The pilot chute provides enough drag to pull the drogue chute which in turn provides adequate power to extract or deploy the main chute(s). [AIR1489-88]

Ci See curie.

CID See computer interface device.

cinder A particle of gas borne partially burned fuel larger than 100 microns in diameter.

cinder trap A dust collector having staggered elements in the gas passage which concentrates larger dust particles. A portion of the gas passes through the elements with the concentrated dust into a settling chamber, where change in direction and velocity drops out coarser particles.

Cipolletti weir An open-channel flow-measurement device similar to a rectangular weir but having sloping sides, which results in a simplified discharge equation.

circadian desynchronization That state in which the body's "normal" 24-h rhythmic biological cycle (circadian rhythm) is disturbed; typically caused by movement across several time zones and generally having an adverse effect upon pilot performance. Colloquially referred to as "Jet Lag".[ARP4107-88]

circadian rhythm 1. The tendency for some biological processes to occur at approximately the same time in each 24-hour period.[ARP4107-88] 2. Regular change in phsyiological function occurring in approximately 24 hour cycles. [NASA]

circle of confusion A circular image in the focal plane of an optical system which is the image formed by that system of a distant point object.

circle-to-land maneuver/circling maneuver A maneuver initiated by the pilot to align the aircraft with a runway for landing when a straight-in landing from an instrument approach is not possible or is not desirable. This maneuver is made only after ATC authorization has been obtained and the pilot has established the required visual reference to the airport.[ARP4107-88]

circling minimums See landing minimums.

circuit 1. A complete path over which electrons can flow from the negative terminals of a voltage source through parts and wires to the positive terminals of the same voltage source.[AR 1931-87] 2. Network providing one or more closed paths.[NASA] 3. Any group of related electronic paths and components which electronic signals will pass to performa a specific function.

circuit analyzer A multipurpose assembly of several instruments or instrument circuits in one housing which are to be used in measuring two or more operating characteristics of an electronic circuit.

circuit breaker 1. A device designed to open and close a circuit by non-automatic means, as well as to open the circuit automatically on a predetermined overload of current without injury to itself when properly applied within its rating.[ARP1199A-90] 2. A resettable circuit-protective device. Circuit breakers can be divided by function into three classes: (a) control circuit breaker: A breaker whose function is to protect the wiring used to operate control devices, such as relays. (b) power circuit breaker: A breaker, as distinguished from a control circuit breaker, whose function is to protect the wiring carrying the power to using equipment. (c) remote circuit breaker: A breaker which is not accessible to the crew during flight.[AS486B-78]

circuit diagram A line drawing of an electronic/electrical system which identifies components and diagrams how they are connected.

circuit-noise meter An instrument which uses frequency-weighting networks and other components to measure electronic noise in a circuit, giving approximately equal readings for noises that produce equal levels of interference.

circular-chart recorder A type of recording instrument where the input signal from a temperature, pressure, flow or other transducer moves a pivoted pen over a circular piece of chart paper that rotates about its center at a fixed rate with time.

circularity In data processing, a warning message that the commands for two separate but interdependent cells in a program cannot proceed until a value for one of the cells is determined.

circularly polarized light Light in which the polarization vector rotates periodically, but does not change magnitude, describing a circle. It can also be stated as the superposition of two plane-polarized (or linearly polarized) lightwaves, of equal magnitude, one 90° in phase behind the other.

circular mil 1. A unit of measure used to express the cross-sectional area of a conductor. It is the unit of area equal to $\pi/4$ times a square mil.[AR1931-87] 2. A circular mil is a unit of area equal to $\pi/4$ times 78.54% of a square mil. the cross-section area of a circle in circular mils is, therefore, equal to the square of its diameter in mils.[AS1198-82] 3. A wire-gage measurement equal to the cross-sectional area of a wire one mil (0.001 in.) in diameter; actual area is 7.8540×10^{-7} in.

circular mil area (CMA) 1. The cross-sectional area of the current carrying portion of a conductor expressed in circular mils.[AR1931-87] 2. The circular mil area, abbreviated "CMA", of a strand or conductor is the cross-section area of the actual current carrying portion, expressed in circular mils. In a stranded conductor, the area of the voids or interstices between the strands are not included in the conductors circular mil area.[AS1198-82]

circular polarized wave An electromagnetic wave for which the electric field vector, magnetic field vector, or both, describe a circle.

circularvection See illusion, vection.

circular waveguides Small hollow tubes that are designed to transmit a specific wavelength along the length of the tube.[NASA]

circulating memory In an electronic memory device, a means of delaying information combined with a means for regenerating the information and reinserting it into the delaying means.

circulation 1. The flow or motion of a fluid in or through a given area or volume. A precise measure of the average flow of a fluid along a given closed curve.[NASA] 2. The movement of water and steam within a steam generating unit.

circulation control airfoils Airfoils in which a high lift capability is produced by supercirculation where control of the stagnation points by the jet sheet produces high lift coefficients.[NASA]

circulation control rotors Rotors that provide STOL capability on high performance aircraft by means of tangential blowing over a rounded trailing edge and mass flow characteristic of turbine engine bleed.[NASA]

circulation distribution The line integral of the velocity component around a curve along the closed contour.[NASA]

circulation ratio The ratio of the water entering a circuit to the steam generated within that circuit in a unit of time.

circulator A pipe or tube to pass steam or water between upper boiler drums usually located where the best absorption is low. Also used to apply to tubes connecting headers of horizontal water tube boilers with drums.

circumferential crimp A type of crimp where the crimping dies completely surround a barrel resulting in a symmetrical reshaping of the barrel. Some circumferential crimps are oval, hexonical, circular, etc.[ARP914A-79]

circumsolar radiation Radiation from small angle scattering of direct sunlight from atmospheric aerosols with dimensions on the order of or greater than the wavelength of light.[NASA]

circumsolar telescopes Optical instruments for measuring the circumsolar radiation for application to solar energy systems. Mirrors and lenses are utilized for incident sunlight concentration.[NASA]

cislunar space Of or pertaining to phenomena, projects, or activity in the space between the earth and the moon, or between the earth and the moon's orbit.[NASA]

CL-600 challenger aircraft Canadair turbofan aircraft with supercritical wings.[NASA]

cladding 1. A method of applying a layer of metal over another metal whereby the junction of the two metals is continuously bonded.[AR1931-87] 2. A coating placed on the surface of a material and usually bonded to the material.[NASA] 3. Covering one piece of metal with a relatively thick layer of another metal and bonding them together; the bond may be produced by corolling or coextrusion at high temperature and pressure, or by explosive bonding. 4. The low refractive index material which surrounds the core of a fiber and protects against surface contaminant scattering.

cladding strippers Chemicals or devices which remove the cladding from an optical fiber to expose the light-carrying core. The term might sometimes be misapplied to chemicals or devices which remove the protective coating applied over cladding to protect the fiber from the environmental stress.

clamp A device which, by rigid compression, holds a piece or part in position, or retains units in close proximity or parts in alignment, its compression quality depending upon an integral screw mechanism or screws, bolts, and like mechanical fasteners.[ARP481A-79]

clamping circuit 1. Circuit which maintains either extremity of a waveform at a prescribed potential. Networks for adjusting the absolute voltage level of waveforms.[NASA] 2. A circuit which maintains either the maximum or minimum amplitude level of a waveform at a specific potential.

clamping plate A plate for attaching a mold to a plastics-molding or die-casting machine.

clamping pressure In die casting, injection molding and transfer molding, the force (or pressure) used to keep the mold closed while it is being filled.

clasp A nonthreaded fastener, usually

hook-like, with a releasable catch.

class 1. With respect to the certification, ratings, privileges, and limitations of airmen, a classification of aircraft within a category having similar operating characteristics, such as single engine, land, water, and free balloon. 2. With respect to the certification of aircraft, a broad grouping of aircraft having similar characteristics of propulsion, flight, or landing, such as airplane, rotorcraft, balloon, landplane, and seaplane.[ARP 4107-88]

class A amplifier An amplifier in which the grid bias and alternating grid voltages are such that plate current always flows in a specified tube.

class AB amplifier An amplifier in which the grid bias and alternating grid voltages are such that plate current in a specified tube flows considerably more than one-half but less than the entire electrical cycle.

class B amplifier An amplifier in which the grid bias is approximately equal to the cutoff value, therefore making the plate current in a specified tube approximately zero when the grid voltage is zero.

class C amplifier An amplifier in which the grid bias is considerably more negative than the zero plate current value.

Class I location A location in which flammable gases or vapors are or may be present in the air in quantities sufficient to produce ignitible mixtures.

Class II location A location that is hazardous because of the presence of combustible dust.

Class III location A location in which easily ignitible fibers or materials producing combustible flyings are handled, manufactured or used.

classification 1. Implies a decision rule where data or information may be identified and grouped (e.g., waveforms or signatures) to be an indication of a particular status, discrepancy or failure mode.[ARP1587-81] 2. Sorting particles or objects by specific criteria, such as size or function. 3. Separating a mixture into its constituents, such as by particle size or density. 4. Segregating units of product into various adjoining categories, often by measuring characteristics of the individual units, thus forming a spectrum of quality. Also termed grading.

classification of a location The assignment of a rating such as Division 1, Division 2, or non-hazardous.

classification of nosewheel steering systems Nosewheel steering (NWS) systems may be classified by their importance and operational usage as follows: Primary (or Class A) Steering Systems—A primary system is one which is essential to safe ground operation of the aircraft. Redundancy (fail operative) is an implied requisite. Secondary (or Class B) Steering Systems—A secondary system is one that is normally in full time use during ground operation, but is not essential for safe ground operation of the aircraft. Fail-safety (fail passive) is an implied requisite. Tertiary (or Class C) Steering Systems—A tertiary NWS system is one that is used primarily for taxi-parking-catapult spotting, and is not normally required or used for take-off or landing operations. A totally passive disengaged mode is an implied requisite.[ARP1595-82]

class of hydraulic system A pressure standard for a military aircraft hydraulic system based on the nominal pump output pressure or other supply pressure, defined in ISO 6771.[AIR1916-88]

clay atmometer A simple device for determining evaporation rate to the atmosphere, which consists of a porous porcelain dish connected to a calibrated reservoir filled with distilled water.

clean fuels Energy sources from which

pollutants and other impurities have been removed by refining, purification, and other means, to produce fuels less conducive to pollution.[NASA]

cleanout door A door placed so that accumulated refuse may be removed from a boiler setting.

cleanup 1. Removing small amounts of stock by an imprecise machining operation, primarily to improve surface smoothness, flatness or appearance. 2. The time required for an electronic leak-testing instrument to reduce its output signal to 37% of the initial signal transmitted when tracer gas is first detected. 3. The gradual disappearance of internal gases during operation of a discharge tube.

clear To erase the contents of a storage device by replacing the contents with blanks, or zeros. Clarified by erase.

clear-air turbulence (CAT) Turbulence encountered in air where no clouds are present. See wind shear, jet stream. [ARP4107-88]

clearance 1. Authorization by a traffic control facility for an aircraft to proceed within controlled airspace, taking into account the location of other known aircraft.[ARP4107-88] 2. The lineal distance between two adjacent parts that do not touch. 3. Unobstructed space for insertion of tools or removal of parts during maintenance or repair.

clearance, brake running The clearance provided in a brake assembly in the "full off" condition between the actuators and the pressure plate, i.e., the actuator travel required from "full off" to initial application of force to the friction surfaces.[AIR1489-88]

clearance fit A type of mechanical fit in which the tolerance envelopes for mating parts always results in clearance when the parts are assembled.

clearance, lateral The clearance provided at the side of a wheel or tire to ensure against interference during rotation and/or variations in a predetermined location, i.e., lateral runout, looseness in the system, etc.[AIR1489-88]

clearance limit The fix, point, or location to which an aircraft is cleared when issued an air traffic clearance.[ARP 4107-88]

clearance, radial The clearance provided radially to ensure against interference during rotation and/or variations in a predetermined location. Tire radial clearance should include allowances for new tire growth due to pressure, service relaxation, and centrifugal growth.[AIR1489-88]

cleared for approach Phrase denoting ATC authorization for an aircraft to execute any standard or special instrument approach procedure for that airport.[ARP4107-88]

cleared for the option Phrase denoting ATC authorization for an aircraft to make a touch-and-go, low approach, missed approach, stop and go, or full stop landing at the discretion of the pilot.[ARP4107-88]

cleared to land Phrase denoting ATC authorization for an aircraft to land. [ARP4107-88]

clear icing/clear ice Generally, the formation of a layer or mass of ice which is relatively transparent because of its homogeneous structure and small number and size of air spaces; synonymous with glaze, particularly with respect to aircraft icing.[ARP4107-88]

clearway An area beyond the takeoff runway under the control of airport authorities within which terrain or fixed obstacles may not extend above certain limits.[ARP4107-88]

cleaver A device used to cut or break optical fibers in a precise way so the ends can be connected with low loss.

clevis A U-shaped metal fitting with holes at the open ends of the legs for insertion of a pin or bolt to make a closed link for attaching or suspending a load.

climbout phase See mishap, phase of flight.

clinical thermometer A thermometer for accurately determining the temperature of the human body; most often, it is a mercury in glass maximum thermometer.

clinker A hard compact congealed mass of fused furnace refuse, usually slag.

clinometer A divided-circle instrument for determining the angle between mutually inclined surfaces.

clip A mechanical device (usually quickly removable) which clasps, holds, fastens; its holding or gripping quality depending entirely upon the spring action of the material.[ARP481A-79]

clipboard In data processing, an area of information can be stored in order to use it later in a different application.

clipping circuit 1. A circuit that prevents the peak amplitude of a signal from exceeding some specific level. 2. A circuit that eliminates the tail of a signal pulse after some specific time. 3. A circuit element in a pulse amplifier that reduces the pulse amplitude at frequencies less than some specific value.

CLK See clock.

clock (CLK) 1. A master timing device used to provide the basic sequencing pulses for the operation of a synchronous computer. 2. A register which automatically records the progress of real time, or perhaps some approximation to it, records the number of operations performed, and whose contents are available to a computer program. 3. A timing pulse that coincides with or is phase-related to the occurrence of an event, such as bit rate or frame rate.

clock frequency The master frequency of periodic pulses which schedules the operation of the computer.

clock mode A system circuit that is synchronized with a clock pulse, that changes states only when the pulse occurs, and will change state no more than once for each clock pulse.

clock pulse A synchronization signal provided by a clock.

clock rate The time rate at which pulses are emitted from the clock. The clock rate determines the rate at which logical arithmetic gating is performed with a synchronous computer.

clock skew A phase shift between the clock inputs of devices in a single clock system; the result of variations in gate delays and stray capacitance in a circuit.

clone In data processing, an exact duplication of another computer device or software.

closed bomb A fixed volume chamber used for testing the pressure-time and chemical reaction characteristics of combustible materials.[AIR913-89]

closed center In a hydraulic system, when no service is actuated the system is closed to flow, as distinguished from an open center system. A closed center system may utilize a continuous pressure supply, as with a variable displacement pump, or an intermittent pressure supply by means of unloader valve, motor driven pump or pump by-passing provisions.[ARP243B-65]

closed circuit 1. Any device or operation where all or part of the output is returned to the inlet for further processing. 2. A type of television system that does not involve broadcast transmission, but rather involves transmission by cable, telephone lines or similar method.

closed die A forming or forging operation in which metal flow takes place only within the die cavity.

closed ecological systems Systems that provide for the maintenance of life in an isolated living chamber through complete reutilization of the material available, in particular, by means of a cycle wherein exhaled carbon dioxide, urine, and other waste matter are converted chemically or by photosynthesis into oxygen, water, and food.[NASA]

closed end splice A splice, open at one end only, designed to terminate two or more conductors. See also splice.[ARP 914A-79]

closed entry A socket contact or insert cavity design feature which prevents the entry of over-size pin contacts, test probes or other insertable components. [ARP914A-79]

closed faults See geological faults.

closed-fireroom system A forced draft system in which combustion air is supplied by elevating the air pressure in the fireroom.

closed loop 1. A combination of control units in which the process variable is measured and compared with the desired value (or set point). If the measured value differs from the desired value, a corrective signal is sent to the final control element to bring the controlled variable to the proper value. 2. An hydraulic or pneumatic system where flow is recirculated following the power cycle; the system contains a limited amount of fluid, which is continually reused. 3. Pertaining to a system with feedback type of control, such that the output is used to modify the input. 4. An operation by which the computer applies control action directly to the process without manual intervention. 5. A signal path which includes a forward path, a feed-back signal, and a summary point, and forms a closed circuit. See loop, closed.

closed-loop control See closed loop.

closed-loop control system A control system in which the command is compared with a measurement of system output and the resulting error signal is used to drive the load towards the desired output.[AIR1916-88]

closed-loop frequency response The frequency response between command input and control system output with the feedback signal summed with command. Actuation system response for a closed-loop system is usually specified as closed-loop frequency response. [AIR1916-88]

closed loop gain See gain, closed loop.

closed loop numerical control A type of numerical-control system in which position feedback, and often velocity feedback as well, is used to control the dynamic behavior and successive positions of machine slides or equivalent machine members.

closed-loop system 1. A system in which the output is used to control the input. See feedback control loop.[AIR 1489-88] 2. A control system that includes feedback, a reference mechanism, a capability to detect error, and a means of correcting error so that the output of the system can be modified in progress. For example, the system of a pilot manually flying an aircraft and adjusting its altitude involves the controller's (pilot's) reception of feedback of his/her performance (altimeter, vertical velocity indicator, and visual perception of height), comparison with desired or commanded altitude, detection of any difference (error), and output to the flight control system to adjust the aircraft's altitude so that the result will be zero error. In this situation the pilot is said to be "in-the-loop". See also in-the-loop. [ARP4107-88]

closed pass A metal rolling arrangement in which a collar or flange on one roll fits into a groove on the opposing roll thus permitting production of a flash-free shape.

closed position A position that is zero percent closed.

closed runway A runway that is unusable for aircraft operations.[ARP4107-88]

closed traffic successive operations involving takeoffs and landings or low approaches where the aircraft does not depart from the traffic pattern.[ARP 4107-88]

close-grained Consisting of fine, closely spaced particles or crystals.

close-tolerance forging Hot forging in which draft angles, forging tolerances and cleanup allowances are considerably smaller than those used for commercial-grade forgings.

closing See polarization.

closing plate A plate used to cover or close openings in non-pressure parts.

closing pressure In a safety relief valve, the static inlet pressure at the point where the disc has zero lift off the seat.

closure component, characterized Closure component with contoured surface, such as the "vee plug," to provide various flow characteristics.

closure component, cylindrical A cylindrical closure component with a flow passage through it (or a partial cylinder).

closure component, eccentric Closure component face is not concentric with the shaft centerline and moves into seat when closing.

closure component, eccentric spherical disk Disk is spherical segment, not concentric with the disk shaft.

closure component, linear A closure component that moves in a line perpendicular to the seating plane.

closure component, rotary A closure component which is rotated into or away from a seat to modulate flow.

closure component, tapered Closure component is tapered and may be lifted from seating surface before rotating to close or open.

cloud base The lower surface of a cloud. [ARP4107-88]

cloud chamber 1. Device for observing the paths of ionizing particles, based on the principle that supersaturated vapor condenses more readily on ions than on neutral molecules.[NASA] 2. An enclosure filled with supersaturated vapor that can indicate the paths of energetic particles when vapor condenses along the trail of ionized molecules created as the particle passes through the enclosure.

cloud deck The upper surface of a cloud.[ARP4107-88]

cloud physics A subdivision of physical meteorology concerned with physical properties of clouds in the atmosphere and the processes occurring therein.[NASA]

cloud seeding Any technique carried out with the intent of adding to a natural cloud in a planetary atmosphere certain substances that will alter the natural development of that cloud.[NASA]

clusec A unit of power used to express the pumping power of a vacuum pump; it equals about 1.333×10^{-6} watt, or the power associated with a leak rate of 10 ml/sec at a pressure of 1 millitorr.

cluster analysis The analysis of data with the object of finding natural groupings within the data either by hand or with the aid of a computer.[NASA]

clutch A machine element which allows a shaft in an equipment drive to be connected and disconnected from the power train, especially while the shaft is running. See also engaging mechanism.

clutter Atmospheric noise, extraneous signals, etc. which tend to obscure the reception of a desired signal in a radio receiver, radarscope, etc.[NASA]

C-M diagram See color-magnitude diagram.

CMOS 1. The combination of a PMOS (p-type channel metal oxide semiconductor) with an NMOS (in-type channel metal oxide semiconductor).[NASA] See also Complementary Metal Oxide Semiconductor.

CN emission Radio waves emitted from incandescent gaseous cyanide (CN) in space under low pressures at wavelengths characteristic of the elements comprising the gas.[NASA]

cnoidal waves Finite amplitude progressive waves in shallow water having a wave profile represented by the Jacobian elliptic function CN'.[NASA]

CO_2 welding See gas metal-arc welding.

coal chemicals A group of chemicals used to make antiseptics, dyes, drugs and solvents that are obtained initially as by-products of the conversion of coal to metallurgical coke.

coal derived gases The gases which are derived from various coal gasification processes.[NASA]

coal derived liquids Fluid hydrocarbons derived from the liquefaction of coal.[NASA]

coalescence A term used to describe the bonding of materials into one continuous body, with or without melting along the bond line, as in welding or diffusion bonding. See also coalescing.

coal gas Gas formed by the destructive distillation of coal.

Coanda effect A phenomenon of fluid attachment to one wall in the presence of two walls.

coarse aggregate Crushed stone or gravel, used in making concrete, which will not pass through a sieve with 1/4-in. (6 mm) holes.

coarse grained 1. Having a coarse texture. 2. Having a grain size, in metals, larger than about ASTM No. 5.

coarse vacuum An absolute pressure between about 1 and 760 torr.

coastal dunes See dunes.

coastal fix The navigation aid or intersection where an aircraft transitions between the domestic route structure and the oceanic route structure.[ARP 4107-88]

coastal marshlands See marshlands.

coasting flight The flight of a rocket between burnout of thrust cutoff of one stage and ignition of another, or between burnout and summit altitude or maximum horizontal range.[NASA]

coating A continuous film of some material on a surface.

coating (fiber optic) A layer of plastic or other material applied over the cladding of an optical fiber to prevent environmental degradation and to simplify handling.

coating holes (voids) Areas void of coating arising from dust, dirt, lint, or improperly cleaned surfaces beneath the film.[ARP924-91]

coating (optics) A thin layer or layers applied to the surface of an optical component to enhance or suppress reflection of light, and/or to filter out certain wavelengths.

coaxial 1. A construction of two (usually cylindrical) entities sharing a common axis.[ARP1931-87] 2. Having coincident axes; for example as in a cable where a central insulated conductor is surrounded by one or more metallic sheaths that act as ground leads or secondary conductors.

coaxial cable 1. A cable in which one conductor completely surrounds the other, the two being coaxial, and separated by a continuous (usually solid) insulating material.[ARP1931-87] 2. Waveguide consisting of two concentric conductors insulated from each other.[NASA] 3. Cable with a center conductor surrounded by a dielectric sheath and an external conductor. Has controlled impedance characteristics that make it valuable for data transmission.

coaxial nozzles Class of nozzle configurations in jet aircraft for reducing noise.[NASA]

coaxial propellers (shafts) The concentric arrangement in which each component of the propeller rotates in an opposite direction wherein the speed of each component can vary independently because no fixed gear ratio is incorporated between the shafts.[ARP355-84]

coaxial transmission See coaxial

cables, transmission.

COBE See Cosmic Background Explorer satellite.

COBOL See Common Business-Oriented Language.

cobra dane (radar) Radar installation for monitoring Soviet missiles.[NASA]

cock A valve or other mechanism that starts, stops or regulates the flow of liquid, especially into or out of a tank or other large-volume container.

cockpit voice recorder An approved device for recording electronically detected voice communications within the aircraft cockpit and between the aircraft and others. The recorder must operate continuously from the use of the checklist before the flight to completion of the final check at the end of the flight. The device may have erasure features, but the most recent 30 min of recording must be retained.[ARP4107-88]

COD (Carrier Onboard Delivery) Operation involving aircraft delivery of personnel or material on board an aircraft carrier.[AIR1489-88]

COD (cracks) See crack opening displacement.[NASA]

CODAB See configuration data block.

code 1. A system of symbols for meaningful communication. Related to instruction. 2. A system of symbols for representing data or instructions in a computer or a tabulating machine. 3. To translate the program for the solution of a problem on a given computer into a sequence of machine language, assembly language or pseudo instructions and addresses acceptable to that computer. Related to encode. 4. A machine language program.

code division multiple access Multiple access system in which users are segregated by means of pseudorandom signal coding and bandwidth spreading so that the complete time and frequency axes are occupied and only the power is shared.[NASA]

code division multiplexing The separation of two or more simultaneous radio transmissions over a common path by signal coding and bandwidth spreading.[NASA]

codes In PCM telemetry, the manner in which ones and zeros in each binary number are denoted.

codes/transponder codes The number assigned for a particular electronic multiple-pulse reply signal transmitted by the transponder in an aircraft which makes it easier for the airtraffic controller to identify that particular aircraft on his/her radar display screen.[ARP 4107-88]

CODIL See Control Diagram Language.

coding The ordered list, in computer code or pseudo code, of the successive computer instructions representing successive computer operations for solving a specific problem.

coding sheet A fill-in form on which computer programming instructions are written.

coefferdam An earthwork or piling structure that prevents water from filling an excavation or keeps it from surrounding and undermining a pier or foundation. 2. A raised projection surrounding a hatch or trapdoor to keep water out of the opening.

coefficient of discharge The ratio of actual flow to theoretical flow. It includes the effects of jet contraction and turbulence.

coefficient of expansion The fractional change in dimension of a material for a unit change in temperature.[ARP1931-87]

coefficient of friction A number indicating the amount of some change under certain specified conditions, often expressed as a ratio. Specifically, coefficient of friction is the ratio of the tangential force at a common boundary of two bodies (i.e., tire and runway) in contact that resists the motion or tendency to motion of one relative to the other to the normal force acting on the two bodies at that boundary.[AIR1489-88]

coefficient of friction, effective The net effective value for coefficient of friction within a specified operating range (i.e., of velocity, temperature, etc.)[AIR 1489-88]

coefficient of friction, maximum instantaneous A peak value for coefficient of friction which usually could not be sustained over a period of time. [AIR1489-88]

coefficient of variation (CV) The difference in size (diameter) between the largest and smallest referee particles in a given size sample expressed as a percent deviation from the nominal size is the coefficient of variation and can be expressed as the ratio of the standard deviation to the mean.[ARP1192-87]

coefficient, side friction Coefficient of friction in a lateral direction for a tire. May be affected by tread pattern.[AIR 1489-88]

coercimeter An instrument for measuring the magnetic intensity of a magnet or electromagnet.

coesite A polymorph of silicon dioxide. [NASA]

coextrusion 1. A process for bonding two metal or plastics materials by forcing them simultaneously through the same extrusion die. 2. The bimetallic or bonded plastics shape produced by such a process.

Coffin-Manson law A relationship which enables one to estimate the fatigue life from the cyclic plastic strain range. The specific life for a given metal or alloy is determined by its tensile ductility.[NASA]

cog A tooth on the edge of a wheel.

cogeneration The generation of electricity or shaft power by an energy conversion system and the concurrent use of the rejected thermal energy from the conversion system as an auxiliary energy source.[NASA]

cognitive disorientation A situation in which a person has lost proper perspective within his/her environment and which results in confusion as to the sequence or priority of tasks to perform. This is referred to colloquially as "getting behind the power curve" or "losing situational awareness".[ARP 4107-88]

cognitive engineering The application of knowledge from cognitive psychology (psychology of information processing) to the engineering design of systems. [ARP4107-88]

cognitive flexibility An individual's ability to shift from one mental task to another or to effectively time-share between several tasks while maintaining situational awareness.[ARP4107-88]

cognitive psychology The study of acquisition, storage, and retrieval of information; the processing of that information; and the consequent decision-making processes.[ARP4107-88]

cognitive task saturation See attention, anomalies of cognitive task saturation.

cogwheel A wheel with radial teeth on its rim.

coherence A property of electromagnetic waves that are all the same wavelengths and precisely in phase with each other.

coherence length The distance over which light from a laser retains its coherence after it emerges from the laser.

coherent fiber bundle A bundle of optical fibers with input and output ends in the same spatial relationship to each other, allowing them to transmit an image.

coherent radar A type of radar that employs circuitry which permits comparison of the phase of successive received target signals.[NASA]

coherent scattering Scattering of electromagnetic or particulate rays in which definite phase relationships exist between the incident and scattered waves; coherent waves scattered from two or more scattering centers are capable of interfering with each other.

cohesion/cohesive strength 1. The strength of internal forces holding a sealant together.[AS7200/1-90] 2. Refers to the internal forces holding a sealant together. For example, when peel tests are conducted on samples of sealant bonded to a surface, a strip of canvas cloth or a strip of wire screen is embedded in the sealant. The specimen is clamped in a testing machine. The panel is secured and the strip of canvas or wire screen is pulled at 180°. Assuming the adhesive bond of the sealant to the panel doesn't fail, the force necessary to tear the sealant, making it fail within itself, is the cohesive strength.[AIR4069-90]

coil breaks Creases or ridges in metal sheet or strip that appear as parallel lines across the direction of rolling, generally extending the full width of the material.

coil spring A flexible, elastic member in a helical or spiral shape which stores mechanical energy or provides a pulling or restraining force directly related to the amount of elastic deflection.

coincidence Existence of two phenomena or occurrence of two events simultaneously in time or space, or both.

coincidence circuits Circuits that produce a usable output only when each of two or more input circuits receive pulses simultaneously or within an assignable time interval.[NASA]

coincidence error Coincidence error is a statistical function of particle concentration and the sensing zone volume. A coincidence error of 10% is the maximum that is considered acceptable in using APC's. Maximum particle concentration introduced into the particle counter shall be limited to values resulting in coincidence error below 10%. Coincidence is defined as the probability of more than one particle being present in the optical sensing zone at any one time. When coincidence occurs, the APC cannot distinguish individual particles and will record one or more particles larger than those present in the sensing zone.[ARP1192-87]

coining Squeezing a metal blank between closed dies to form well-defined imprints on both front and back surfaces, or to compress a sintered powder-metal part to final shape; the process is usually done cold, and involves relatively small amounts of plastic deformation.

coke The solid residue remaining after most of the volatile constituents have been driven out by heating a carbonaceous material such as coal, pitch or petroleum residues; it consists chiefly of coherent, cellular carbon with some minerals and a small amount of undistilled volatiles.

coke oven gas Gas produced by destructive distillation of bituminous coal in closed chambers. Heating value 500-550 Btu/cu ft.

cold bend A test used to determine the affect of low temperatures on the insulation system of wire and cable when the wire or cable is flexed. Failure is characterized by the appearance of cracks or other defects in the insulation system.[ARP1931-87]

cold cathode fluorescent lamps Cold cathode fluorescent lamps are similar in design and construction to hot cathode lamps except that no filaments are provided in the electrode ends of the lamps. To start conduction of "strike the arc" in a cold cathode lamp, a minimum voltage is required of 2 to 4 times greater than that required for starting a hot cathode tube. The electrode in each end is shaped with a cavity and is coated with a special barium oxide compound to optimize conduction and minimize impedance.[AIR512B-91]

cold cathodes Cathodes whose operation does not depend on its temperature being above the ambient temperature. [NASA]

cold cathode tubes Electron tubes containing cold cathodes.[NASA]

cold drawing Pulling rod, tubing or wire through one or more dies that reduce its cross section, without applying heat either before or during reduction.

cold extrusion Striking a cold metal slug in a punch-and-die operation so that metal is forced back around the die. Also known as cold forging; cold pressing; extrusion pressing; impact extrusion.

cold-finished Referring to a primary-mill metal product, such as strip, bar, tubing or wire, whose final shaping operation was performed cold; the material has more precise dimensions, and usually higher tensile and yield strength, than a comparable shape whose final shaping operation was performed hot.

cold flow Permanent deformation of wire insulation due to mechanical forces, without the aid of heat softening. [ARP1931-87]

cold flow tests Tests of liquid rockets without firing them to check or verify the efficiency of a propulsion subsystem providing for the conditioning and flow of propellants (including tank pressurization, propellant loading, and propellant feeding).[NASA]

cold forging See cold extrusion.

cold forming 1. Any operation to shape metal which is performed cold. 2. Shaping sheet metal, rod or wire by bending, drawing, stretching or other stamping operations without the application of heat. See also cold working.

cold galvanizing Painting a metal with a suspension of zinc particles in a solvent, so that a thin zinc coating remains after the organic solvent evaporates.

cold heading Cold working a metal by application of axial compressive forces that upset metal and increase the cross sectional area over at least a portion of the length of the starting stock. Also known as upsetting.

cold joint In soldering, making a soldered connection without adequate heating, so that the solder does not flow to fill the spaces, but merely makes a mechanical bond; a cold joint typically exhibits poor to nonexistent electrical conduction across the joint, is not leak tight, and may break loose under vibration or other mechanical forces.

cold junction See reference junction.

cold neutrons Neutrons of less velocity than thermal neutrons; at 152 °C their energy is below 0.01 eV.[NASA]

cold plate A mounting plate for electronic components which has tubing or internal passages through which liquid is circulated to remove heat generated by the electronic components during operation. Also known as liquid-cooled dissipator.

cold pressing See cold extrusion.

cold rolling Rolling metal at about room temperature; the process reduces thickness, increases tensile and yield strengths, improves fatigue resistance, and produces a smooth, lustrous or semilustrous finish.

cold shut A cold shut is a portion of a part that is partially separated from the main body of metal by oxide, or by the failure of two streams of metal to unite.[AS3071A-77]

cold solder joint A solder connection exhibiting poor wetting and grayish, porous appearance due to insufficient heat, inadequate cleaning prior to soldering, or excessive impurities in the solder solution.[ARP1931-87]

cold start temperature The temperature at which the hydraulic system will start to operate, but need not necessarily meet full performance.[AIR1916-88]

cold trap A length of tubing between a vacuum system and a diffusion pump or instrument which is cooled by liquid nitrogen to help remove condensable vapors.

cold treatment Subzero treatment of a metal part—usually at -65 °F, -100 °F or liquid-nitrogen temperature—to induce metallurgical changes that either stabilize dimensions, complete a phase transformation or condition the metal and prepare it for further processing.

cold working 1. Deforming of metal plastically at a temperature lower than its recrystallization temperature. Examples: fillet rolling, thread rolling, shot peening, cold upsetting (below the metal's recrystallation temperature) hot upsetting, hot-cold upsetting, wire drawing, extruding.[ARP700-65] 2. Deforming metal plasticity at a temperature lower than the recrystallization temperature.[NASA] 3. Any plastic deformation of a metal carried out below its recrystallization temperature; the process always induces strain hardening to a degree directly related to the percent reduction in cross section.

collar A rigid, ring-shaped machine element that is forced onto or clamped around a shaft or similar member to restrict axial motion, provide a locating surface or cover an opening.

collating sequence In data processing, the order of the ASCII numeric codes for the characters.

collator 1. A mechanical device at the output of a printing machine or copier which sorts multiple-page documents and arranges them into sets. 2. In data processing, a device for combining sets of data cards or other information-bearing elements into a desired sequence. 3. In data processing using electronic files, a program or routine used to merge two or more files into a single, ordered output file.

collector 1. Any of a class of instruments for determining electrical potential at a point in the atmosphere, and ultimately the atmospheric electric field; all collectors consist of a device for bringing a conductor rapidly to the potential of the surrounding air and an electrometer for measuring its potential with respect to the earth. 2. A device used for removing gas borne solids from flue gas. 3. One of the functional regions in a transistor. See also accumulators.

collimate To make parallel.

collimation Producing a beam of light or other electromagnetic radiation whose rays are essentially parallel.

collimator 1. Optical device which renders rays of light parallel.[NASA] 2. An optical system which focuses a beam of light so all the rays form a parallel beam.

collision A close approach of two or more bodies (including energetic particles) that results in an interchange of energy, momentum or charge. See also elastic collision; inelastic collision.

collision parameters In orbit computation, the distances between centers of attraction of central force fields and the extension of velocity vectors of moving objects at great distances from the centers. In gas dynamics and atomic physics, any of several parameters such as cross section, collision rate, mean free path, etc. which provide a measure of the probability of collision.[NASA]

collision rates Ratios defined by the average number of collisions per second suffered by a molecule of other particle moving through a gas.[NASA]

colloid 1. A dispersion of particles of one phase in a second phase, where the particles are so small that surface phenomena play a dominant role in their chemical behavior; typical colloids include mists or aerosols (liquid dispersed phase in gaseous dispersion medium), smoke (solid in gas), foam (gas in liquid), emulsions (liquid in liquid), suspensions (solid in liquid), solid foam such as pumice (gas in solid), and solid solution such as colloidal gold in glass (solid in solid). 2. A finely divided organic substance which tends to inhibit the formation of dense scale and results in the deposition of sludge, or causes it to remain in suspension, so that it may be blown from the boiler.

Colmonoy A series of high nickel alloys (manufactured by Wall-Colmonoy Corp.) used for hard facing of surfaces subject to erosion.

colonies Regions within prior beta grains with alpha platelets having nearly identical orientations. In commercially pure titanium, colonies often have serrated boundaries. Colonies arise as transformation products during cooling from the beta field at cooling rates slow enough to allow platelet nucleation and growth.[AS1814-90]

color Color is that characteristic of mental percepts variously stimulated by and corresponding to the quantitative aspects of visible light energy as sensed through the shaped response of the eyes as sensors of vision. It is that aspect of visual perception by which an observer may distinguish differences between two structure-free fields of view of the same size and shape, such as may be caused by differences in the spectral composition of the radiant energy concerned in the observation. Color may be stated in terms of luminance, dominant, wavelength, and purity. Neutral color qualities such as black, white and grey that possess a zero saturation (or chroma) are called achromatic colors. Colors having a finite saturation or chroma are chromatic or colored.[ARP1782-89]

color (particle physics) See quantum chromodynamics.

colorants Substances which are used to produce the colors of objects are called colorants (dyes, pigments, inks, paints, and decorative coatings).[ARP798-65]

color code 1. A color identification mark appearing on the surface of a wire or cable component (sometimes in the form of a stripe) to identify a given component within a complex.[ARP1931-87] 2. Any system of colors used to identify a specific type or class of objects from other, similar objects—for example, to differentiate steel bars of different grades in a warehouse. 3. A system of colors used to identify different piping systems from each other in a factory or other building—for example, red for fire protection, yellow for hazardous chemical, blue for potable water and green for compressed air.

color coding A system for the identification of components, materials, tools and related devices by means of color. [ARP914A-79] 2. Any system of colors used for purposes of identification. [NASA]

color-color diagram A two-axis coordinate graph showing the distribution of stars or other objects with reference to difference color indices.[NASA]

colored federal airways L/MF airways depicted in brown on aeronautical charts and identified by color name and number (for example, Amber One). Green and Red airways are plotted east and west. Amber and blue airways are plotted north and south. (The term colored airways is no longer used in the U.S.A.) [ARP4107-88]

color enhancement See color coding.

color filter A filter containing a colored dye, which absorbs some of the incident light and transmits the remainder.

colorimetry Any analytical process that uses absorption of selected bands of visible light, or sometimes ultraviolet radiation, to determine a chemical property such as the end point of a reaction or the concentration of a substance whose color is indicative of product purity or uniformity.

color infrared photography A representation of temperature differences using false colors.[NASA]

color-magnitude diagram The plot of the absolute or apparent magnitude against the color index for a group of stars. Also known as C-M diagram. [NASA]

color of objects The capacity of the object to modify the color of the light incident upon it.[ARP798-65]

color shift A change in the dominant wavelength or the color purity of light which has been reflected from or has passed through the glass.[ARP924-91]

color stimulus Radiant power of given magnitude and spectral composition entering the eye and producing a sensation of color.[ARP1782-89]

color temperature A source of light is the temperature at which a black body must be operated to give a color matching that of the source in question. [ARP798-65]

Columbus space station The European Space Agency's manned orbital platform.[NASA]

column A vertical structural member of substantial length designed to bear axial compressive loads.

column loading A factor that takes into consideration the quantity of liquid descending in the column and the quantity of vapor ascending in the column. If either the liquid or the vapor flow rate becomes too high, column flooding will occur.

coma A lens aberration in which light rays from an off-axis source which pass through the center of a lens arrive at the image plane at different distances from the axis than do rays from the same source which pass through the edges of the lens.

combination automatic controller A type of control system arrangement in which more than one closed control loop are coupled through primary feedback or through any of the controller elements.

combination die A forging, forming or casting die with more than one cavity.

combination pliers A pliers whose jaws are designed for holding objects in one portion, and for cutting and bending wire in another portion.

combination scale An instrument scale consisting of two or more concentric or colinear scales, each graduated in equivalent values with two or more units of measure.

combination square A measuring and rough layout tool consisting of a special head and a short steel rule that, when used together, can check angles of both 90° and 45°.

combination wrench A fixed-size wrench having an open-end wrench at one end and box wrench at the other, usually both intended to fit the same size bolthead.

combined cycle power generation Power generation which combines an open-cycle gas turbine and a closed-cycle steam turbine.[NASA]

combined flight control/utility system A combined flight control/utility hydraulic system (combined system) is a system that supplies a portion of the power required to operate the flight control system and also supplies power to the utility system.[ARP578-69]

combustibility See flammability.

combustible The heat producing constituents of a fuel.

combustible loss The loss representing the unliberated thermal energy occasioned by failure to oxidize completely some of the combustible matter in the fuel.

combustion 1. An exothermic chemical reaction which liberates heat and usually produces high temperature gases and light.[AIR913-89] 2. The rapid chemical combination of oxygen with the combustible elements of a fuel resulting in the production of heat.

combustion chamber 1. Container in which the actual burning of fuel takes place.[NASA] 2. Any chamber or enclosure designed to confine and control the generation of heat and power from burning fuels.

combustion chemistry The study of the exothermic oxidation reactions occurring immediately before and during combustion.[NASA]

combustion control Control of factors (temperature, preheating, draft, excess or deficient air, etc.) which affects combustion efficiency.[NASA]

combustion efficiency 1. The percentage ratio of the energy actually released by the combustion process to the energy which would be realized if all the carbon in the fuel were oxidized to CO_2 and the hydrogen to water vapor.[AIR 1533-91] 2. The efficiency with which fuel is burned, expressed as the ratio of the actual energy released by the combustion to the potential chemical energy of the fuel.[NASA]

combustion engine An energy conversion machine that operates by converting to motion heat from the burning of a fuel.

combustion (flame) safeguard A system for sensing the presence or absence of flame and indicating, alarming or initiating control action.

combustion rate The quantity of fuel fired per unit of time, as pounds of coal per hour, or cubic feet of gas per minute.

combustion safety control—programming type A combustion safety control that provides for various operations at definite periods of time in predetermined sequences.

combustors See combustion chambers.

come-along 1. A lever-operated chain or wire-rope hoist for lifting or pulling at any angle, which has a reversible ratchet in the handle to permit using short strokes for tensioning or relaxing the fall. Also known as puller. 2. A device for gripping and applying tension to a length of cable, wire rope or chain by means of jaws that close when the user pulls on a ring.

cometary atmospheres The region of the coma of a comet as well as the gaseous part surrounding the coma that often is a hydrogen atmosphere that contains particulate matter.[NASA]

comets Luminous members of the solar system composed of a head, or coma, and often with a spectacular gaseous tail extending a great distance from the head.[NASA]

COM file A computer file name ending in .COM which most often contains a machine code program. It is short for "command" file.

comfort curve A line on the graph of dry-bulb temperature versus wet-bulb temperature or relative humidity that represents optimum comfort for the average person who is not engaged in physical activity.

comfort zone The respective ranges of indoor temperature, relative humidity and ventilation rate (air-movement rate) that most persons consider acceptable for their normal degree of physical activity and mode of dress.

command 1. A signal or input whereby functions are performed as the result of a transmitted signal.[AIR1489-88] 2. A signal that causes a computer to start, stop or to continue a specific operation.

command bars Needles on the attitude director indicator that assist the pilot in intercepting and maintaining a glide slope or course, or both. The needles present "command" information in that the pilot must fly to the bars to bring them to a neutral position (zero displacement), and keeping the bars in the neutral position the system will cause the pilot to intercept and maintain the course (and the glide slope, if on an ILS approach). Also, known as the steering bars.[ARP4107-88]

command/control The orderly distribution of authority and responsibility designed to systematically accomplish a mission; and the continuous-feed-back-loop communications network connecting all levels of command so that decisions can be made, efforts coordinated, and discipline maintained.[ARP4107-88]

command control A system whereby functions are performed as a result of a transmitted signal.[AIR1489-88]

command guidance The guidance of a spacecraft or rocket by means of electronic signals sent to receiving devices in the vehicle.[NASA]

command input An input that represents the desired output of the control system.[AIR1916-88]

command language 1. Vocabulary to interactively execute activities such as computer retrieval or input.[NASA] 2. A source language consisting primarily of procedural operations, each capable of invoking a function to be executed.

command post A place at which the commander of a unit receives orders from his/her superiors and from which command is exercised over a unit. [ARP4107-88]

command resolution The maximum change in the value of a command signal which can be made without inducing a change in the controlled variable.

command response (data communication systems) A method of media access control where the remote controllers receive and transmit data only when commanded by the bus controller.[AIR4271-89]

command systems Se command guidance.

comment An expression which explains or identifies a particular step in a routine, but which has no effect on the operation of the computer in performing the instructions for the routine.

commercial aircraft Any civilian aircraft being used in the transportation of persons or property for compensation or hire.[ARP4107-88]

commercial spacecraft Commercial satellites and other spacecraft operated by the private sector.[NASA]

Commission Internationale de l'Eclairage (CIE) The principal international body in the field of illumination. Its purpose is to establish agreeable standards for industry such as the Standard Colorimetric Observer (or standard observer). The standard observer is the basis for most trichromatic systems and has not changed since 1931 when it was adopted. The standard observer was expanded in 1964 to fields of view larger than 4 degrees. (The currently available data sets are for 2 deg (1981) and 10 deg (1964).)[ARP1782-89]

common A reference within a system having the same electrical potential throughout. Usually connected to ground at one point. Often different commons are used throughout a system such as power common, signal common, etc. depending on the accuracy to which the reference is held.

commonality The factors which are common in equipment or systems. [NASA]

Common Applications Service Elements (CASE) One of the application protocols specified by MAP.

common area A section in memory that is set aside for common use by many separate programs or modules.

Common Business Oriented Language (COBOL) A specific language by which business data-processing procedures may be precisely described in a standard form. The language is intended not only as a means for directly presenting any business program to any suitable computer, for which a compiler exists, but also as a means of communicating such procedures among individuals.

common field A field that can be accessed by two or more independent routines.

common machine language In data processing, coded information which is in a form common to a related group of data-processing machines.

common mode In analog data, an interfering voltage from both sides of a differential input pair (in common) to ground.

common mode interference A form of interference which appears between the terminals of any measuring circuit and ground. See common mode voltage. See also interference, common mode.

common-mode rejection ratio Abbreviated CMRR. A measure of the ability of a detector to damp out the effect of a common-mode-generated interference voltage; usually expressed in decibels.

common mode voltage (CMV) In-phase, equal-amplitude signals that are applied to both inputs of a differential amplifier, usually referred to as a guard shield or chassis ground. See voltage, common mode.

common trip A common-trip multipole circuit breaker is one in which an overload on any pole will cause all poles to open simultaneously.[ARP1199A-90]

communication Transmission of intelligence between points of origin and reception without alteration of sequence or structure of the information content.

communication link 1. The physical means of connecting one location to another for the purpose of transmitting and receiving information. 2. The physical realization of a specified means by which stations communicate with each other. The specification normally cover the interfaces and some aspects of functional capability. 3. A link may provide multiple channels for communications.

communication networks Organization of facilities for the rapid reception of, transmission of, and/or relaying of electrical impulses for reproduction as printed messages, pictures, or other data.[NASA]

communication satellites Satellites designed to reflect or relay electromagnetic signals used for communication.[NASA]

communications media The communications system that is used by the data link system to effect up and down link communications, e.g., VHF, UHF, HF, Mode S, SATCOMM, MLS.[ARP4102/13-90]

commutation 1. Sequential sampling, on a repetitive timesharing basis, of multiple data sources for transmitting or recording, or both, on a single channel.[NASA] 2. Cyclic sequential sampling on a time-division basis of multiple data sources.

commutation duty cycle A channel dwell period, expressed as a percentage of a channel interval.

commutation frame period The time required for sequential sampling of all input signals; this would correspond to one revolution of a simple multicontact rotary switch.

commutation rate The number of commutator inputs sampled per specified time interval.

commutator 1. Device used to accomplish time division multiplexing by repetitive sequential switching.[NASA] 2. A segmented ring, usually constructed of hard-drawn copper segments separated by an insulator such as mica, which is used to energize only the correct windings of a d-c generator or motor at any given instant.

compact 1. A powder-metallurgy part made by pressing metal powder, with or without a binder or other additives; prior to sintering it is known as a green compact, and after sintering as a sintered compact or simply a compact. 2. To consolidate earth or paving materials by weight, vibration, impact or kneading so that the consolidated material can sustain more load than prior to consolidation.

companding A process in which compression is followed by expansion, as in noise reduction systems.[NASA]

comparative tracking index (CTI) The numerical value of the maximum voltage in volts at which the material withstands 50 drops without tracking.

comparator 1. In computer operations, device or circuit for comparing information from two sources.[NASA] 2. A device for inspecting a part to determine any deviation from a specific dimension by electrical, optical, pneumatic or mechanical means.

compass 1. Instrument for indicating a horizontal reference direction, specifically a magnetic compass.[NASA] 2. A drafting or layout tool for drawing circles, arcs or fillets which consists of a bow or radius bar connecting a pin center and a pan or drafting pencil. 3. An instrument for indicating relative direction on the earth's surface, usually measured with respect to a reference direction such as magnetic or true North.

compass bearing See bearing.

compass locator A low-power, low- or medium-frequency (L/MF) radio beacon installed at the site of the outer or middle marker of an instrument landing system (ILS). It can be used for navigation at distances of approximately 15 miles or as otherwise authorized in the approach procedure.[ARP4107-88]

compass locator, middle (LLM) A compass locator installed at the site of the middle marker of an instrument landing system.[ARP4107-88]

compass locator, outer (LOM) A compass locator installed at the site of the outer marker of an instrument landing system.[ARP4107-88]

compass rose A circle, graduated in deg, printed on some charts or marked on the ground at an airport. It is used as a reference to either true or magnetic direction.[ARP4107-88]

compatibility 1. A characteristic ascribed to a major subsystem that indicates it functions well in the overall system. Also applied to the overall system with reference to how well its various subsystems work together, as in 'the vehicle has good compatibility'. Also applied to materials which can be used in conjunction with other materials and not react with each other under normal operating conditions.[NASA] 2. The ability for two devices to communicate with each other in a manner that both understand.

compatibility interface A point at which hardware, logic, and signal levels are defined to allow the interconnection of independently designed and manufactured components.

compatible The state in which different kinds of computers or equipment can use the same programs or data.

compensated pendulum A pendulum made of two materials having different coefficients of linear expansion, and so constructed that the distance between the center of oscillation and the point of suspension remains the same over the normal range of ambient temperatures.

compensation signals In telemetry, a set of reference signals recorded on tape along with the data, and used during playback to automatically compensate for any nonuniformity in tape speed.

compensator A device which compensates, makes up for, or offsets. For example a hydraulic pressure compensator (nose wheel steering system, etc.) which would maintain a positive pressure on the unit in case of loss of system pressure. Device may also be of mechanical or electrical design.[AIR1489-88]

compensatory leads An arrangement

of connecting elements between an instrument and a transducer or other observation device such that variations in the properties of any of the connecting elements—such as temperature effects that induce changes in resistance—are compensated so that they do not affect instrument accuracy.

compile 1. A computer function that translates symbolic language into machine language. 2. To prepare a machine-language program from a computer program written in another programming language by making use of the overall logic, structure of the program, or generating more than one machine instruction for each symbolic statement, or both, as well as performing the function of an assembler.

compiler 1. A program that translates a high-level source language (such as FORTRAN IV or BASIC) into a machine language suitable for a particular machine. 2. A computer program more powerful than an assembler. In addition to its translating function, which is generally the same process as that used in an assembler, it is able to replace certain items of input with a series of instructions, usually called subroutines. Thus, an assembler translates item for item, and produces as output the same number of instructions or constants which were put into it; a compiler will do more than this. The program which results from compiling is a translated and expanded version of the original. Synonymous with compiling routine and related to assembler.

compile time In general, the time during which a source program is translated into an object program.

compiling routine Same as compiler.

complacency See attention, anomalies of complacency.

complement 1. An angle equal to 90° minus a given angle. The true complement of any quantity in positional notation, i.e. the quantity which, when added to the first quantity, gives the least quantity containing one more place. The base-minus-one complement of any quantity in positional notation; i.e., the quantity which when added to the first quantity containing the same number of places.[NASA] 2. A quantity expressed to the base n, which is derived from a given quantity by a particular rule. Frequently used to represent the negative of the given quantity. 3. A complement on n, obtained by subtracting each digit of the given quantity from n-1, adding unity to the least significant digit, and performing all resultant carries, e.g., the twos complement of binary 11010 is 00110. The tens complement of decimal 456 is 544. 4. A complement on n-1, obtained by subtracting each digit of the given quantity from n-1, e.g., the ones complement of binary 11010 is 00101. The nines complement of decimal 456 is 543.

Complementary Metal Oxide Semiconductor (CMOS) 1. One type of computer semiconductor memory. The main feature of CMOS memory is its low power consumption. 2. A type of semiconductor device not specifically memory.

complementary operator The logic operator which is the NOT of a given logic operator.

complementary wavelength The monochromatic wavelength of light which matches a standard reference light when combined with the sample color in suitable proportions as applied to colorimetry.

complete combustion The complete oxidation of all the combustible constituents of a fuel.

complete contraction A combination of both end and bottom contractions in a weir.

completion network In a strain gage signal conditioner, the one to three resistors which must be added to make a four-arm bridge (the transducer being the active arm or arms).

complex compounds Chemical compounds in which part of the molecular boding is of the coordinate type.[NASA]

complex frequency A complex number used to characterize exponential or damped sinusoidal waves in the same way as an ordinary frequency is used to characterize a simple harmonic wave.

complex lens A lens system consisting of more than one optical element.

complex tone A sound wave produced by combining simple sinusoidal component waves of different frequencies.

compliance (elasticity) See modulus of elasticity.

component An article which is a self-contained element of a complete operating unit and performs a function necessary to the operation of that unit. [NASA]

composite A material or structure made up of physically distinct components that are mechanically, adhesively or metallurgically bonded together; examples include filled plastics, laminates, filament-wound structures, cermets, and adhesive-bonded honeycomb-sandwich structures. See also composite material.

composite flight plan A flight plan which specifies VFR operation for one portion of flight and IFR for another portion.[ARP4107-88]

composite joint A connection between two parts that involves both mechanical joining and welding or brazing, and where both contribute to total joint strength.

composite materials Structural materials of metals, ceramics, or plastics with built-in strengthening agents which may be in the form of filaments, foils, powders, or flakes of a different compatible material.[NASA]

composite propellants Solid rocket propellants consisting of a fuel and an oxidizer neither of which would burn without the presence of the other. [NASA]

composite route system An organized oceanic route structure, incorporating reduced lateral spacing between routes, in which composite separation [of aircraft] is authorized.[ARP4107-88]

composite separation A method of managing route and altitude assignments separating aircraft so that a combination of half the lateral and half the vertical minimums specified for the area concerned is applied.[ARP4107-88]

composite subcarrier Two or more subcarriers that are combined in a frequency-division multiplexing (FDM) scheme.

composite wave filter A selective transducer made up of two or more filters—the filters being any combination of high- pass, low-pass, band-pass or band-elimination types.

compound A homogeneous substance, composed of two or more essentially different chemicals which are present in definite proportions, having properties which are different from those of its constituent elements.[ARP1931-87]

compound angle The surface contour formed by two intersecting mitered angles.

compound die Any die so constructed that it performs more than one operation on a given part with a single stroke of the punch.

compound document A digital representation of a paper document consisting of multiple object types such as text, graphics and photo images.[AS4159-88]

compound engine A multicylinder engine in which the working fluid—steam, air or hot gas—expands successively as it passes from one cylinder to another through the engine.

compound lever A device consisting of two or more levers, where force or motion is transferred from the arm of one lever to the next lever in the train.

compound screw A screw having threads of different pitches or opposite helixes on opposite ends of the shank.

compound semiconductor A semiconductor such as gallium arsenide which is made up of two or more materials, in contrast to simple single element materials such as silicon and germanium.

compressibility 1. The property of a substance, as air, by virtue of which its density increases with increase in pressure.[NASA] 2. Volumetric strain per unit change in hydrostatic pressure.

compressibility factor (Z) A factor used to compensate for deviation from the laws of perfect gases. If the gas laws are used to compute the specific weight of a gas, the computed value must be adjusted by the compressibility factor Z to obtain the true specific weight.

compressible Capable of being compressed. Gas and vapor are compressible fluids.

compressible flow 1. In aerodynamics, flow at speeds sufficiently high that density changes in the fluid cannot be neglected.[NASA] 2. Fluid flow under conditions which cause significant changes in density.

compressional wave A wave in an elastic medium which causes an element of the medium to undergo changes in volume without rotating.

compression failure Buckling, collapse or fracture of a structural member that is loaded in compression.

compression member A beam, column or other structural component that is loaded in such a way as to be under predominantly compressive stress.

compression mold A type of plastics mold that is opened to introduce starting material, closed to shape the part, and reopened to remove the part and restart the cycle; pressure to shape the part is supplied by closing the mold.

compression ratio 1. In internal combustion engines, the ratio between the volume displaced by the piston plus the clearance space, to the volume of the clearance space.[NASA] 2. In powder metallurgy, the ratio of the volume of loose powder used to make a part to the volume of the pressed compact.

compression ring A separate ring, within the backshell assembly that is chamfered to provide an environmental seal by compressing the rear grommet.[ARP914A-79]

compression set The amount of compression an elastomer retains. Expressed as a percentage of original dimension. [ARP1931-87]

compression spring An elastic member, usually made by bending metal wire into a helical coil, that resists a force tending to compress it.

compression test A destructive test for determining fracture strength, yield strength, ductility and elastic modulus by progressively loading a short-column specimen in compression.

compression waves In acoustics, waves in an elastic medium which cause an element of the medium to change its volume without undergoing rotation. Mathematically, a compression wave is one whose velocity wave has zero curl.[NASA]

compressor 1. Machine for compressing air or other fluids.[NASA] 2. A machine—usually a reciprocating-piston, centrifugal, or axial-flow design—which is used to increase pressure in a gas or vapor. Also known as compression machine. 3. A hardware or software process for removing redundant or otherwise uninteresting words from a stream, thereby "compressing" the data quantity.

compressor blades Blades which are either rotor blades or stator blades in axial-flow compressors; sometimes used restrictively (and ambiguously) for compressor rotor blades.[NASA]

compressor bleed valve A valve used

to control air flow from a driven compressor to provide: (a) compressor stall margin during starting and/or acceleration, (b) compressed air for starting and/or other purposes.[ARP906A-91]

Compton effect The decrease in frequency and increase in wavelength of x-rays or gamma-rays when scattered by free electrons.[NASA]

Compton scattering A form of interaction between x-rays and loosely-bound electrons in which a collision between them results in deflection of the radiation from its previous path, accompanied by random phase shift and slight increase in wavelength.

compulsators Compensated pulsed alternators i.e., single phased alternators designed for pulsed power duty with air gap armature windings and air gap compensating windings.[NASA]

compulsory reporting points Reporting points, designated on aeronautical charts by solid triangles or filed in a flight plan as fixes selected to define direct routes, which must be reported to ATC as each is reached.[ARP4107-88]

computational chemistry A complementary method for determining properties of gases, solids, and their interactions from first principle calculations. It extends testing capabilities to realms that are too dangerous or too costly to obtain experimentally.[NASA]

computational fluid dynamics The application of large computer systems for the numerical solutions of complex fluid dynamics equations.[NASA]

computational process An instance of execution of a segment by a processor using a data area.

computational stability The degree to which a computational process remains valid when subjected to effects such as errors, mistakes, or malfunctions.

computer 1. A data processor that can perform substantial computation, including numerous arithmetic or logic operations, without intervention by a human operator during the run. 2. A device capable of solving problems by accepting data, performing described operations on the data, and supplying the results of these operations. Various types of computers are calculators, digital computers, and analog computers. 3. See analog computer, digital computer, general purpose computer, hybrid computer, and stored program computer.

computer aided design The use of the computer in design work.[NASA]

computer aided engineering See computer aided design.[NASA]

computer aided manufacturing Interactive computing in support of manufacturing.[NASA]

computer aided mapping Creating data bases of topographic and manmade features for the production of traditional maps and digital maps. Resultant digital maps have great flexibility and can be easily updated. The user can select the appropriate scale, view selected features, and view any desired area.[NASA]

Computer and Automated Systems Association of the Society of Manufacturing Engineers (CASA/SME) A professional engineering association dedicated to the advancement of engineering technology. CASA/SME supports the administrative functions of the MAP/TOP users group.

computer code A machine code for a specific computer.

computer compatible tapes Machine readable tapes.[NASA]

computer control Computer control is a device in which control and/or display actions are generated for use by other system devices. When used with other control devices on the communication link the computer normally performs or functions in a hierarchical relationship to the other control devices.

computer-dependent language A relative term for a programming language whose translation can be achieved only by a specific model (or models) of computer.

computer graphics 1. The representation of a picture in terms of computer graphic primitives such as lines, markers, polygons, text and cell arrays. [AS4159-88] 2. The technique of combining computer calculations with various display devices, printers, plotters, etc. to render information in graphical or pictorial format.[NASA] 3. Any display in pictorial form on a computer monitor that can be printed.

computer-independent language A language in which computer programs can be created without regard for the actual computers which will be used to process them. Related to transportability.

computer information security Protective measures to prevent destruction, larceny, and/or unauthorized use of information in computerized files. [NASA]

computer instruction A machine instruction for a specific computer.

Computer Integrated Manufacturing (CIM) A central computer gathers all types of data, provides information stored in the data base for decisions, and controls production input and output.

computer interface Serves as the interface device between the host computer and other devices on the data highway. It converts data from the protocol of the computer and that of the highway, and vice versa.

computer interface device (CID) Hardware that allows a general-purpose computer to share data with the rest of the distributed control system.

computer-limited Pertaining to a situation in which the time required for computation exceeds the time available.

computer network 1. The interconnection of two or more computers for the mutual or individual processing of data to and from a multitude of terminals or stations by utilizing appropriate switching techniques, transmission systems, or miniprocessors.[NASA] 2. A complex consisting of two or more interconnected computing units.

computer networking Interconnection of two or more geographically separated computers so that information can be exchanged between them, usually under the direction of individual, autonomous control programs. See also distributed processing.

computer operator A person who performs standard system operations such as adjusting system operation parameters at the system console, loading a tape transport, placing cards in a card reader, and removing listings from the line printer.

computer part-programming In numerical control, the preparation of a part program to obtain a machine program using the computer and appropriate processor and post processor.

computer program A series of instructions or statements in a form acceptable to a computer prepared in order to achieve a certain result.

computer program integrity The completeness of a program to execute its intended function.[NASA]

computer security See computer information security.

computer simulation A logical-mathematical representation of a simulation concept, system, or operation programmed for solution on an analog or digital computer. See also computerized simulation.

computer systems performance The efficiency and reliability that characterize the real operation of the system. [NASA]

computer systems simulation Forecasting of computer requirements by the use of predictive modeling and estimating computer workloads.[NASA]

computer vision Capability of computers to analyze and act on visual input.[NASA]

computer word A sequence of bits or characters treated as a unit and capable of being stored in one computer location. Synonymous with machine word.

computing instrument See instrument, computing.

computerized design See computer aided design.

computerized simulation Computer-calculated representation of a process, device, or concept in mathematical form. Used for ARIP (impact prediction), automatic rocket impact predictors, computer simulation, and IP (impact prediction).[NASA]

COMSEC Communication Security.

ComStar C The third in a series of Comsat domestic communications satellites launched in a transfer orbit by NASA for COMSAT.[NASA]

ComStar satellites Series of domestic Comsat communication satellites. [NASA]

conc See concentrated.

concatenate To combine several files into one file, or several strings of characters into one string, by appending one file or string after another.

concatenated codes Two or more codes which are encoded and decoded in series.[NASA]

concave A term describing a surface whose central region is depressed with respect to a flat plane approximately passing through its periphery.

concentrate 1. To separate metal-bearing minerals from the gangue in an ore. 2. The enriched product resulting from an ore-separation process. 3. An enriched substance that must be diluted, usually with water, before it is used.

concentration 1. The volume fraction of the component of interest in the gas mixture—expressed as volume percentage or as parts per million.[ARP1256-90][AIR1533-91] 2. The weight of solids contained in a unit weight of boiler or feed water. 3. The number of times that the dissolved solids have increased from the original amount in the feedwater to that in the boiler water due to evaporation in generating steam.

concentricity 1. The measurement which is the location of the center of the conductor with respect to the geometric center of the circular insulation. [ARP1931-87] 2. The quality of two or more geometric shapes having the same center—usually, the term is applied to plane shapes or cross sections of solid shapes that are approximately circular.

concentric lay A concentric lay conductor or cable is composed of a central core surrounded by one or more layers of helically wound strands or wires. Unless otherwise specified, it is optional for the successive layers to be alternately reversed in direction of lay (true concentric lay) or to be in the same direction (unidirectional lay). See also unidirectional lay.[AS1198-82]

concentric lay conductor A conductor with one or more layers of helically wound strands in a fixed round geometric arrangement. It is optional for the direction of lay for successive layers to be alternately reversed or in the same direction. If the direction of the lay for successive layers is the same, the lay length shall increase with each successive layer. The standard direction of the lay of the outer layer is left hand. [ARP1931-87]

concentric orifice plate A fluid-meter orifice plate having a circular opening whose center coincides with the axis of the center of the pipe it is installed in.

conceptual phase The identification and exploration of alternative solutions or solution concepts to satisfy a validated need.[ARD50013-91]

concentric spheres Structures in which the space between the spheres is utilized for experiments involving fluid flow, etc.[NASA]

concrete A mixture of aggregate, water and a binder, usually Portland cement, that cures as it dries and becomes rock hard.

concrete structures Buildings, dams, stadiums, etc. constructed entirely of a mixture of aggregates, water, and portland cement.[NASA]

concurrent processing Two (or more) computer operations that appear to be processed simultaneously when in fact the CPU is rapidly switching between them.

cond See conductivity.

condensate 1. The liquid product of a condensing cycle. Also known as condensate liquid. 2. A light hydrocarbon mixture formed by expanding and cooling gas in a gas-recycling plant to produce a liquid output.

condensate pot A section of pipe (4 in. diameter) installed horizontally at the orifice flange union to provide a large-area surge surface for movement of the impulse line fluid with instrument element position change to reduce measurement error from hydrostatic head difference in the impulse lines.

condensate trap 1. A device to separate saturated water from steam in a pipe or piece of process equipment. 2. A device used to trap and retain condensate in a measurement impulse line to prevent hot vapors from reaching the instrument.

condensation test This test is used to evaluate the rust preventive properties of greases covering steel objects under alternate low temperature and moderate temperature, high humidity conditions.[S-5C,40-55]

condensation The physical process by which a vapor becomes a liquid or solid; the opposite of evaporation. Specifically, in meteorology, the transformation from vapor to liquid.[NASA]

condensation nuclei Liquid or solid particles upon which condensation of water begins in the atmosphere.[NASA]

condensation-type hygrometer Any of several designs of dew-point instruments that operate by detecting the equilibrium temperature at which dew or frost forms on a thermoelectrically, mechanically or chemically cooled surface; surface condensation may be detected by optical, electrical or nuclear techniques.

condenser The heat exchanger, located at the top of the column, that condenses overhead vapors. For distillation, the common condenser cooling media are water, air, and refrigerants such as propane. The condenser may be partial or total. In a partial condenser only part of the vapors are condensed, with the remainder usually withdrawn as a vapor product.

condenser boiler A boiler in which steam is generated by the condensation of a vapor.

conditional branch See conditional transfer.

conditional jump See conditional transfer.

conditional stability 1. A linear system is conditionally stable if it is stable for a certain interval of values of the open-loop gain, and unstable for certain lower and higher values. 2. The property of a controlled process by which it can function in either a stable or unstable mode, depending on conditions imposed.

conditional transfer An instruction which, if a specified condition or set of conditions is satisfied, is interpreted as an unconditional transfer. If the conditions are not satisfied, the instruction causes the computer to proceed in its normal sequence of control. A conditional transfer also includes the testing of the condition. Synonymous with conditional lump and conditional branch and related to branch.

condition monitoring (CM) 1. Condition monitoring (CM) is a data gathering, not a preventative maintenance, process. It allows failures to occur, and relies on analysis of operating experience information to indicate the need for appropriate action. If the removal rate is deemed to be excessive, the component may temporarily be assigned a hard time removal limit until corrective action can restore its removal performance to an acceptable level. 2. A CM item has neither hard-time removal nor scheduled test to determine its continued serviceability. All removals are for unscheduled cause. All units removed must, upon arrival in the shop, be given a check in accordance with the procedures in the appropriate shop manual. If the unit "passes", it is returned to service. Failure to pass the check requires the unit to be repaired to a state that it will pass and be serviceable. [AIR1916-88]

condition monitoring system A system designed to monitor the condition of a machine or process.

conductance A measure of the ability of any material to conduct an electric charge. Conductance is a ratio of the current flow to the potential difference causing the current flow.[ARP1931-87] See also resistance.

conducted susceptibility The tendency of a piece of equipment to have its performance degraded in response to interference on its connecting wires. [ARP1931-87]

conducting See conduction.

conducting media See conductors.

conducting polymer A plastics material having electrical conductivity approaching that of metals.

conduction 1. The transfer of energy within and through a conductor by means of internal particle of molecular activity and without any net external motion.[NASA] 2. Flow of heat through or across a conductor.

conduction band(s) A range of states in the energy spectrum of a solid in which electrons can move freely.[NASA] 2. A partially filled or empty energy band in which electrons are free to move easily.

conduction pump A device for pumping a conductive liquid, such as a liquid metal, by passing an electric current across the stream of liquid and applying a magnetic field at right angles to the electrical current.

conductive elastomer An elastomeric material that conducts electricity; usually made by mixing powdered metal into a silicone before it is cured.

conductivity 1. The capability of a material to carry an electric charge. Conductivity of metals is usually expressed as a percentage of copper conductivity—copper being one hundred percent (100%).[ARP1931-87] 2. The ability to transmit, as electricity, heat, sound, etc. A unit measure of electrical conduction; the facility with which a substance conducts electricity, as represented by the current density per unit electrical-potential gradient in the direction of flow.[NASA] 3. The amount of heat (Btu) transmitted in one hour through one square foot of a homogeneous material 1 in. thick for a difference in temperature of 1 °F between the two surfaces of the material. 4. The electrical conductance, at a specified temperature, between the opposite faces of a unit cube; usually expressed as ohm^{-1} cm^{-1}

conductivity bridge A simple four-arm a-c bridge circuit in which a conductivity cell is the unknown circuit element; electrically, the cell is equivalent to a resistance and a capacitance in series; higher a-c frequencies lead to lower cell-polarization errors, but introduce greater errors due to capacitance impedance; the latter can be reduced by using a phase-sensitive detector.

conductivity-type moisture sensor An instrument for measuring moisture content of fibrous organic materials such as wood, paper, textiles and grain at moisture contents up to saturation.

conductometer An instrument that measures thermal conductivity, especially one that does so by comparing the rates at which different rods conduct heat.

conductor 1. A conductor is a stand, or a group of strands not usually insulated from each other, suitable for carrying a single electric current. While a cable shield may actually fulfill this definition and be used as a conductor, it is termed "shield" and not "conductor". [AS1198-82] 2. An electric current-carrying material; the conductive element in an electrical wire.[ARP914A-79] 3. Substance or entity which transmits electricity, heat, or sound.[NASA] 4. Any material through which electrical current can flow.

conductor, card edge A rectangular connector into which the edge of a printed wiring board is inserted so as to make electrical contact with conductive traces located on the circuit board. [ARP914A-79]

conductor size For a solid conductor (one strand), the size is the AWG size number; that is "Size 20", not "Size 20 AWG" or "20 AWG". For a stranded conductor, the size is that number of which the conductor's circular mil area most closely corresponds to the even numbered AWG circular mil area. For this purpose, even numbered AWG's are: 40, 38, 36, 34, 32, 30, 28, 26, 24, 22, 20, 18, 16, 14, 12, 10, 8, 6, 4, 2, 1, 0, 00, 000, and 0000. Odd numbered sizes, such as "Size 23" are not used for stranded conductors. A 760 CMA stranded conductor (of 19 Size 34 strands) is designated "Size 22", not "Size 21".[AS1198-82]

conductor stop A device or design feature on a terminal, splice, contact or tool which correctly positions the conductor in the conductor barrel.[ARP 914A-79]

conduit 1. A rigid or semi-rigid tube within which the signal carrying medium may be placed. It also serves the functions of strength member and scuff jacket.[ARD50020-91] 2. Any channel, duct, pipe or tube for transmitting fluid along a defined flow path. 3. A thin-wall pipe used to enclose wiring.

cone bearing A tapered sleeve bearing in the shape of a truncated cone that runs in a correspondingly tapered bearing block.

cone of ambiguity An inverted cone extending upward from the site of a VOR/TACAN facility in which navigational signals tend to be unreliable. [ARP4107-88]

cone-plate viscometer An instrument for routinely determining the absolute viscosity of fluids in small sample volumes by sensing the resistance to rotation of a moving cone caused by the presence of the test fluid in a space between the cone and a stationary flat plate.

cones Geometric configurations having a circular bottom and sides tapering off to an apex (as in nose cone).[NASA]

confidence An attitudinal set in which a person is predisposed to think that he/she can perform a task. (a) Overconfidence: An attitudinal set in which a person assumes that he/she can perform a task even though he/she has not successfully performed it in the past, has not successfully performed it recently, or has performed it in the past but under different circumstances. (b) Underconfidence: An attitudinal set in which a person assumes that he/she cannot perform a task, even though he/she has performed it successfully and sufficiently often in the recent past.[ARP 4107-88]

confidence level 1. The probability that a given statement is correct, usually associated with statistical predictions. [AIR1916-88] 2. The probability that the interval quoted will include the true value of the quantity being measured. 3. In acceptance sampling, the probability that accepted lots will be better than a specific value known as the rejectable quality level (RQL); a confidence level of 90% indicates that 90 out of every 100 lots accepted will have a quality better than the RQL. 4. In statistical work, the degree of assurance that a particular probability applies to a specific circumstance.

confidence limits In statistics, the upper and lower extremes of the confidence interval.[NASA]

confidence test A test primarily performed to increase the confidence that the unit under test is operating acceptably.[AIR1916-88]

CONFIG.SYS A basic computer file that outlines how that particular device is designed to operate.

configuration 1. The functional and/or physical characteristics of hardware/software as set forth in technical documentation and achieved in a product. [ARD50013-91] 2. The arrangement of the parts or elements of something. 3. A low-level, fill-in-the-blank form of programming a process control device. 4. A particular selection of hardware devices or software routines and/or programs that function together. 5. A term applied to a device or system whose functional characteristics can be selected or rearranged through programming or other methods. 6. The hardware, firmware and/or software combinations make up a system.

configuration data block (CODAB) In TELEVENT, the data section that identifies the "personality" of a hardware-software combination.

configuration interaction In physical chemistry, the interaction between two different possible arrangements of the electrons in an atom or molecule.[NASA]

configure The installation procedure that sets up software to operate on a particular computer and printer.

confined detonating fuse (CDF) A detonating cord with a flexible outer sheath which retains the products of detonation.[AIR913-89]

confined flow Flow of a continuous stream of fluid within a process vessel or conduit.

conflict alert A feature of certain automated air traffic control systems designed to alert radar controllers to existing or pending conflict situations recognized by the computer program parameters.[ARP4107-88]

conflict situation A state of affairs that exists when an aircraft comes within certain vertical, lateral, or longitudinal distances from each other. These distances may vary depending on the density of the airspace and the type of aircraft.[ARP4107-88]

confluence See convergence.

conformance A device conforms to the manufacturers specifications. See accuracy and error band.

confusion Loss of situational awareness that is recognized by the individual concerned. A state characterized by bewilderment, emotional disturbance, lack of clear thinking, and (sometimes) perceptual disorientation.[ARP4107-88]

conical flare See cones.

conical orifice An orifice having a 45° bevel on the inlet edge to yield more constant and predictable discharge coefficient at low flow velocity (Reynolds number less than 10,000).

conical scan antenna An automatic-tracking antenna system in which the beam is driven in a circular path such that it forms a cone. The antenna is steered automatically so that the telemetry source is kept at the center of the cone.

conical scanning Scanning in which the direction of maximum radiation generates a cone whose vertex angle is of the order of the beam width. Such scanning may be either rotating of nutating, according as the direction of polarization rotates or remains unchanged.[NASA]

coniscope See koniscope.

conjugate bridge An arrangement of electrical or electronic components in which the supply circuit and detector circuit are interchanged as compared with the normal arrangement for that type of bridge.

conjugated circuits Branches of an electrical network configured so that a change in the electromotive force in either branch does not result in a current change in the other.[NASA]

conjugate gradient method An interactive method for solving a system of linear equations of dimension N which terminates in at most N steps if no rounding errors are encountered. Each iterate will bring one closer to the solution.[NASA]

conjugate impedances An impedance pair having the magnitudes of resistance and reactive components of one equal to the corresponding values of the other, but whose reactive components are of opposite signs.

connect Establish linkage between an interrupt and a designated interrupt servicing program. See disconnect.

connecting rod Any straight link that transmits power or motion from one part of a mechanism to another, especially one that links a rotating member to a reciprocating member—for example, the link that attaches a piston to the crankshaft in a reciprocating internal-combustion engine.

connector 1. Not recommended, except to describe provision for an electrical connection. See also fitting.[ARP243B-65] 2. Any detachable device for providing electrical continuity between two conductors. 3. In fiber optics, a device which joins the ends of two optical fibers together temporarily.

connector, electrical A conductor terminating device which allows for the separation of one or more electrical circuits to facilitate servicing.[ARP914A-79]

connector, hermaphroditic A connector which has features enabling it to be mated with an identical connector.[ARP914A-79]

connector, plug An electrical connector, intended to be attached to the free end of a conductor, wire, cable or bundle, which couples or mates to a receptacle connector.[ARP914A-79]

connector, receptacle An electrical connector, intended to be mounted or installed into a fixed structure such as a panel, electrical case or chassis, which couples or mates to a plug connector. [ARP914A-79]

connector, right angle A rectangular connector which is generally mounted onto a printed wiring board whose contacts are inserted into a matching pattern or plated through holes in the circuit board and soldered in place.[ARP 914A-79]

connector set, electrical Two or more separate plug and receptacle connectors designed to be mated together. The set may include mixed connectors mated together, such as one plug connector and one dummy receptacle connector or one receptacle connector and one dummy plug connector.[ARP914A-79]

connector, umbilical An electrical connector, used to connect a cable to a vehicle such as an aircraft or rocket, which is unmated prior to or during initial movement or launching of the vehicle.[ARP914A-79]

conscious level See awareness, level of conscious.

consecutive access A method of data access that is characterized by the sequential nature of the I/O device involved; for example, a card reader is an example of a consecutive access device; each card must be read one after another, and no distinction is made between logical sets of data in or among the cards in the input hopper.

consecutive starts Start cycles which follow each other at close intervals. The engine may or may not have come to rest before the second or successive starts.[ARP906A-91]

consistency A qualitative means of classifying substances, especially semisolids, according to their resistance to dynamic changes in shape.

console 1. Array of controls and indicators for the monitoring and control for a particular sequence of actions, as in the checkout of a rocket, a countdown action, or a launch procedure.[NASA] 2. A main control desk for an integrated assemblage of electronic equipment. Also known as control desk. 3. A grouping of control devices, instrument indicators, recorders and alarms, housed in a freestanding cabinet or enclosure, to create an operator's work station. 4. The cabinet or enclosure for a floor-model radio or television receiver, or similar electronic device.

constant 1. A value that remains the same throughout the distinct operation; opposite of variable. 2. A data item which takes as its value its name (hence, its value is fixed during program execution).

constant-amplitude (filter) A reference to the characteristic of a Butterworth filter. See Butterworth.

constantan An alloy of 55% copper and 45% nickel used with copper in thermocouples in the temperature range 169 °C to 386 °C. Temperature coefficient of resistivity, 0.0002/°C. Normally, the copper is the positive wire, and the constantan is the negative wire.[ARP1931-87]

constant-bandwidth (CBW) The spacing of FM subcarriers equally with relation to each other; see proportional bandwidth.

constant-current potentiometer A type of null-balance instrument for determining an unknown d-c voltage, usually less than 10 V, under conditions that maintain constant current in the detector circuit; resolution up to one part in 103 can be achieved with a single potentiometer slidewire, and up to one part in 107 with a multidecade device.

constant-current transformer A type of transformer which automatically adjusts its output of its secondary circuit to maintain a constant current under varying load impedances when its primary windings are connected to a constant-voltage power supply.

constant-delay (filter) See Bessel.

constant-head meter A flow measurement device that maintains a constant pressure differential by varying the cross section of a flowpath through the meter, such as in a piston meter or rotameter.

constant horsepower line The locus of points on a curve where the product of torque and speed is constant.[ARP 906A-91]

constant-load balance A single-pan weighting device, having a constant load, in which the sample weight is determined by hanging precision weights from a counterpoised beam.

constant-resistance potentiometer A type of null-balance instrument for determining an unknown d-c voltage, usually less than 10 V, using a constant scaling resistor in parallel with the potentiometer circuit.

constant-speed propeller A propeller

designed to maintain engine speed at a constant RPM, automatically increasing or decreasing pitch as engine speed tends to increase or decrease.[ARP4107-88]

constant volume balloon See superpressure balloon.

constant-volume gas thermometer A device for detecting and indicating temperature based on Charles' Law—the pressure of a confined gas varies directly with absolute temperature; in practical instruments, a bulb immersed in the thermal medium is connected to a Bourdon tube by means of a capillary; changes in temperature are indicated directly by movement of the Bourdon tube due to changes in bulb pressure.

constellation Originally conspicuous configuration of stars; now regions of the celestial sphere marked by arbitrary boundary lines.[NASA]

constrained mechanism A mechanical device in which all members move only along predetermined paths.

constraint 1. A speed or altitude or time restriction usually related to some downstream waypoint.[ARP1570-86] 2. The limit of normal operating range. 3. Anything which keeps a member under longitudinal tension from contracting laterally, which sets up a condition of biaxial tension in the member; the term is used most often in connection with welded joints that cannot shrink laterally as the weld solidifies and cools.

consumable electrode An arc-welding electrode that melts during welding to provide the filler metal.

consumable insert A piece of metal placed in the root of a weld prior to welding, and which melts during welding to supply part of the filler metal.

consumables (spacecraft) All supplies for spacecraft and spacecrews that will be consumed during a mission.[NASA]

consumer's risk (BETA) The probability of accepting equipment with a true mean time between failures (MTBF) equal to the lower test MBF (THETA 1). The probability of accepting equipment with a true MTBF less than the lower test MTBF (THETA 1 will be less than (BETA).[ARD50010-91]

contact 1. To establish communications with a facility and, if appropriate, to specify the frequency that should be used. For example: "Contact departure control." or "Contact tower on one nineteen point niner". 2. A flight condition wherein the pilot ascertains the attitudes of his/her aircraft and navigates by maintaining visual reference (contact) with the surface.[ARP4107-88] 3. In hardware, a set of conductors that can be brought into contact by electromechanical action and thereby produce switching. In software, a symbolic set of points whose open or closed condition depends on the logic status assigned to them by internal or external conditions.

contact analog A flight display that takes the essential visual cues from the contact view that the pilot uses in landing an aircraft and in other ground-referenced maneuvers and incorporates them in a vertical situational display that may present a highly stylized contact view in which the dynamic responses of the pictured elements are analogous to those of their visual-world counterparts in contact flight. A true contact analog display remains pictorial in that all elements obey the same laws of motion perspective as their visual-world counterparts.[ARP4107-88]

contact approach An approach wherein an aircraft on an IFR flight plan, operating clear of clouds with at least 1 mile flight visibility, having received an air traffic control authorization, may deviate from the prescribed instrument approach procedure and proceed to the airport of destination by visual reference to the surface.[ARP4107-88]

contact area 1. The area of a loaded tire which is actually in contact with the supporting surface. Generally elliptical in shape.[AIR1489-88] 2. The designated surface required to contact a mating surface within the limits specified. The contact area shall be distributed uniformly over the surface.[AS 291D-64] 3. The area in contact between two conductive elements through which current flow can take place.[ARP914A-79]

contact arrangement The position and layout pattern of contacts in a connector.[ARP914A-79]

contact back wipe Usually refers to an actuated contact surface where a contact travels on the surface of its mating contact during the actuation cycle then moves back to a clean wiped surface at the completion of the actuation or engagement cycle.[ARP914A-79]

contact, electrical The conductive element in a connector or other device which mates with a corresponding element to provide an electrical path or circuit.[ARP914A-79]

contact, female See contact, socket.

contact float The allowable, free axial, lateral or angular movement of a contact in a connector.[ARP914A-79]

contact, hermaphroditic An electrical contact which has features that enable it to be mated with an identical contact.[ARP914A-79]

contact input See input, contact.

contact inspection In ultrasonic testing, a method of scanning a test piece which involves placing a search unit directly on a test piece surface covered with a thin film of couplant.

contact loads Dynamic loading by contact between two bodies.[NASA]

contact, male See contact, pin.

contactor A mechanical or electromechanical device for repeatedly making and breaking electrical continuity between two branches of a power circuit, thereby establishing or interrupting current flow.

contact output See output, contact.

contact, pin A contact intended to make electrical contact on its outer surface upon engagement with a socket contact.[ARP914A-79]

contact rectifier A device for converting a-c electrical power to d-c power that is constructed of two different solids in contact with each other; rectification is accomplished because the selected combination of solids yields greater electrical conductivity in one direction across the interface between them than in the other direction.

contact resistance The electrical resistance through a pair of engaged contacts or terminals. Resistance may be measured in ohms or millivolt drop at a specified test current. This resistance measurement does not include the resistance of the terminating conductor joint or the conductor.[ARP914A-79]

contact retainer A device either on the contact or in the connector insert whose purpose is to retain the contact in the insert.[ARP914A-79]

contact retention The provision or means in an electrical connector by which the contacts are retained; the ability of the connector to retain contacts. See also force, contact retention.[ARP914A-79]

contacts The electrically conducting parts in a contactor that repeatedly come in contact or separate to make or break electrical continuity.

contact sense module A device which monitors and converts program-specified groups of field-switch contacts into digital codes for input to the computer.

contact size Either a single number designator based on the AWG size number most closely corresponding in circular mil area (CMA) to the CMA of the pin contact of a given contact set, or a double number designator, similarly based, whereby the first number corresponds to the CMA of the pin contact, and the second number corresponds to the max wire size accommodated by the contacts conductor barrel.[ARP914A-79]

contact, socket A contact having an engagement end that will accept entry of a pin contact with a point of electrical contact on the inside diameter of contacting surface.[ARP914A-79]

contact symbology Representation of logic schemes in contact or ladder diagram form.

contact thermography A method of measuring surface temperature in which the surface of an object is covered with a thin layer of luminescent material and then viewed under ultraviolet light in a darkened room; the brightness viewed indicates surface temperature.

contact tube In gas metal arc welding and flux cored arc welding, a metal part with a hole in it that provides electrical contact between the welding machine and the wire electrode that is fed continuously through the hole.

contact-type membrane switch A disk-shaped momentary- contact switch of multilayer construction; the active element consists of two conductive buttons separated by an insulating washer; finger pressure on one face of the disk brings the buttons into contact, completing the electrical circuit; when the pressure is released, the contacts separate, breaking the electrical circuit.

contact-wear allowance The thickness that may be lost due to wear from either of a pair of mating electrical contacts before they cease to adequately perform their intended function.

contact wipe The distance a contact travels on the surface of its mating contact during engagement or separation.[ARP914A-79]

containment The ability of the starter of APU to retain within its envelope all energy laden fragments of the turbine(s) or other rotating components if these parts are caused to fail.[ARP906A-91]

contaminant Any solid or liquid substance that can interfere with the function of the sealant, coating, adhesion promoter, etc.[AIR4069-90]

contaminate That which contaminates to make impure or corrupt by contact or mixing.

contamination 1. Foreign matter contained in, or potentially capable of being contained in, a system fluid (liquid or gas). This foreign matter is capable of moving with the system fluid and of being deposited upon or forming upon system filters. It may be in a solid, gelatinous, liquid, or gaseous form.[AIR787-65] 2. Presence of an unwanted substance—usually, a substance that causes an undesired effect or interferes with a desired effect.

content-addressed storage See associative storage.

contention 1. A condition on a multidrop communication channel when two or more locations try to transmit at the same time. 2. Unregulated bidding for a line or other device by multiple users.

context The composition, structure, or manner in which something is put together. Also refers to the situation or environment of an event.[NASA]

contiguous file A file consisting of physically adjacent blocks on a mass-storage device.

continental control area See controlled airspace.

continental margins See continental shelves.

continental shelves The ocean floor that is between the shoreline and the abyssal ocean floor, including various provinces; the continental shelf; continental borderland; continental slope; and the continental rise.[NASA]

contingency One or more unplanned

occurrences which make the performance of a task more difficult, but are not inherently hazardous (for example, unforecast bad weather, ATC delays, etc.).[ARP4107-88]

continuity tester A device for testing that a fiber optic communication system forms a continuous optical path between two points.

continuous blowdown The uninterrupted removal of concentrated boiler water from a boiler to control total solids concentration in the remaining water.

continuous dilution A technique of supplying a protective gas flow continuously to an enclosure, housing electrical circuitry, containing an internal potential source of flammable gas or vapor for the purpose of diluting any flammable gas or vapor which could be present to a level well below the lower-explosion-limit.

continuous duty A requirement of service that demands operation at a substantially constant load current for an indefinitely long time.[ARP1199A-90]

continuous-duty rating The maximum power or other operating characteristic that a specific device can sustain indefinitely without significant degradation of its functions.

continuous flow electrophoresis See electrophoresis.

continuous furnace A type of reheating furnace where the charge is loaded at one end, moves through the furnace to accomplish the intended treatment, and is discharged at the other end.

continuous mixer A type of mixer in which starting ingredients are fed continuously and the final mixture is withdrawn continuously, without stopping or interrupting the mixing process; generally, unmixed ingredients are fed at one end of the machine and blended progressively as they move towards the other end, where the mixture is discharged. Contrast with batch mixer.

continuous operation A process that operates on the basis of continuous flow, as opposed to batch, intermittent or sequenced operations.

continuous-path numerical control A type of numerical- control system involving not only specification of successive end positions of machine slides or equivalent machine members, but also automatic generation of the linear, circular or parabolic path to be followed in moving from one end position to the next. Also known as contouring numerical control.

continuous rating A defined power input or set of operating variables that represent the maximum values for operating a device continuously for an indefinite time without reducing its normal service life.

continuous sampling The presentation of a flowing sample to the analytical analyzer so as to obtain continuous measurement of concentrations of the components of interest.[ARP1256-90]

continuous spectra Spectra in which wavelengths, wave numbers, and frequencies are represented by the continuum of real numbers or a portion thereof, rather than by a discrete sequence of numbers. For electromagnetic radiation, spectra that exhibit no detailed structure and represent a gradual variation of intensity with wavelength from one end to the other, as the spectra of incandescent solids. For particles, spectra that exhibit a continuous variation of the momentum or energy. [NASA]

continuous spectrum A distribution of wavelengths in a beam of electromagnetic radiation in which the intensity varies continuously with wavelength, exhibiting no characteristic structure such as a series of bands where the intensity does not abruptly change at discrete wavelengths. See also band spectrum.

continuous weld A welded joint where the fusion zone is continuous along the entire length of the joint.

continuum Things that are continuous which have no discrete parts as the continuum of real numbers as opposed to the sequence of discrete integers, as the background continuum of a spectrogram due to thermal radiation. [NASA]

contour control system A system of control in which two or more controlled motions move in relation to each other so that a desired angular path or contour is generated.

contouring numerical control See continuous-path numerical control.

contour sensors The sensing of image coincidences by means of optical processing techniques.[NASA]

contraction The narrowing of the stream of liquid passing through a notch of a weir.

contract maintenance 1. Maintenance not normally done by plant personnel. 2. A maintenance service organization which contracts to do specific maintenance.

contractor 1. Any individual, company, partnership, association, or corporation holding a contract or letter of intent to supply items for which this standard is specified.[AS478G-79] 2. A contractor is an individual, partnership, company, corporation, or association having a contract with a procuring activity for the design, development design and manufacture, manufacture, maintenance, modification or supply of items under the terms of the contract.[AS1933-85]

contrahelical A method of applying two or more layers of spirally twisted, served, or wrapped materials, where each successive layer is wrapped in the opposite direction of the preceding layer. [ARP1931-87]

contrarotating propellers Two propellers mounted on concentric shafts having a common drive and rotating in opposite directions.[NASA]

contrast 1. Contrast (C) = $(L_t\text{-}L_b)/L_b$. This quantity ranges from 0.0 (no contrast) to infinity, and is commonly used in instrument and control panel specifications. It is used in lieu of the contrast ratio definition only because it starts at zero, which is more logical to some people.[ARP1782-89] 2. Contrast as Modulation (C_m) = $(L_t\text{-}L_b)/L_t\text{+}L_b)$. This quantity ranges from 0.0 (no contrast) to 1.0, and is often found in human factors research such as in discussions of contrast sensitivity of the eye. It has been called contrast, contrast ratio, modulation, luminance modulation, or, when multiplied by 100, percent contrast. This quantity is consistent with "modulation" as defined in communications theory.[ARP1782-89] 3. Luminance contrast = $(L_t\text{-}L_b)/L_t$. This quantity ranges from 0.0 (no contrast) to 1.0, and is called Luminance Contrast in MIL-STD-1472C. It is luminance difference divided by (normalized to) max luminance, rather than mean luminance or min luminance, and is rarely used. NOTE: L_t, the total luminance of the symbol or image, including any background or reflected light as measured in the specified lighting conditions. L_b, the luminance of the background, or dimmer area, including any reflected light and any stray display emissions measured in the specified lighting conditions.[ARP1782-89] 4. In general, the degree of differentiation between different tones in an image.[NASA] 5. In a photographic or radiographic image, the ability to record small differences in light or x-ray intensity as discernible differences in photographic density.

contrast factor The slope of the central portion of a graph of photographic density versus exposure for a given photographic or radiographic emulsion.

contrast ratio (CR) The contrast ratio of a symbol in the presence of incident light is the ratio of symbol luminance (Bs) to background luminance (Bb) of the symbol.[ARP4067-89]

control 1. (Mechanical) A device used to govern a machine or mechanism in operation.[ARP480A-79] 2. (Electrical) A component which governs the operation of another component or grouping of components.[ARP480A-79] 3. Frequently, one or more of the components in any mechanism responsible for interpreting and carrying out manually initiated directions. 4. In some applications, a mathematical check. 5. Instructions which determine conditional jumps often are referred to as control instructions, and the time sequence of execution of instructions is called the flow of control. 6. Any manual or automatic device for the regulation of a machine to keep it at normal operation. If automatic, the device is motivated by variations in temperature, pressure, water level, time, light, or other influences.

control accuracy The degree to which a controlled process variable corresponds to the desired value or setpoint.

control action The act of changing some variable (position, power, etc.) to effect a correction of the controlled temperature.[ARP89C-70]

control agent The energy or material comprising the process element which is controlled by manipulating one or more of its attributes—the attribute(s) commonly termed the controlled variable(s).

control algorithm A mathematical representation of the control action to be performed.

control amplifier The device which mixes the signals from the input and feedback devices, detects an error if there is one, and causes a correction to be made.[ARP89C-70]

control apparatus An assembly containing one or more control devices which acts to manipulate a controlled variable.

control area See controlled airspace.

control augmentation system (CAS) 1. A vehicle flight control system wherein the control system responds to the error between the commanded vehicle motion and the actual vehicle motion.[ARP1181A-85] 2. A function of the flight control system, including command shaping, sensors, actuators, etc., that perform in such a manner so as to augment the static and dynamic stability, and maneuver response of the aircraft. When considered as an entity it is essentially a closed-loop tracking control system, responding to pilot commands.[AIR1916-88]

control authority The amount of control surface or force effector deflection that can be produced by AFCS signals relative to the total available control deflection. This phrase is often preceded by the word electrical or abbreviations such as CAS or SAS, so as to be more explicit.[ARP1181A-85] [AIR1916-88]

control bandpass See bandwidth.

control block A storage area through which a particular type of information required for control of the operating system is communicated among its parts.

control board A panel that contains control devices, instrument indicators and sometimes recorders which display the status of a system or subsystem, and from which switches, dials and controllers can be manipulated to alter system operating variables. Also known as control panel; panel board.

control bus The data highway used for carrying control signals.

control-by-light (CBL) (fly-by-light) system A control system wherein control information is transmitted by light through a fiberoptic cable. A true CBL system does not have CBW or mechanical backup, nor CBW or mechanical override.[AIR1916-88] [ARP1181A-85]

control-by-wire (fly-by-wire/CAS-by-wire) system A control system wherein in control information is transmitted completely by electrical means. [AIR1916-88] [ARP1181A-85]

control calculations Installation-dependent calculations that determine output signals from the computer to operate the process plant. These may or may not use generalized equation forms such as PID forms.

control card A card which contains input data or parameters for a specific application of a general routine.

control character A character whose purpose is to control an action rather than to pass data to a program; ASCII control characters have an octal code between 0 and 37; normally typed by holding down the CTRL key on a terminal keyboard while striking a character key.

control chart A plot of some measured quantity, such as a dimension, versus sample number, time, or quantity of goods produced, which plot can be used to determine a quality trend or to make adjustments in process controls as necessary to keep the measured quantity within prescribed limits.

control circuit 1. A circuit in a control apparatus which carries the electrical signal used to determine the magnitude or duration of control action; it does not carry the main power used to energize instrumentation, controllers, motors or other control devices. 2. A circuit in a digital computer which performs any of the following functions—directs the sequencing of program commands, interprets program commands, or controls operation of the arithmetic element and other computer circuits in accordance with the interpretation.

control column A lever or post having a wheel, half- wheel, or other device which is manipulated in controlling the attitude of the aircraft.[ARP4107-88]

control complexity ratio (CCR) A measure of the complexity of a particular control system's logic configuration.

control computer A process computer which directly controls all or part of the elements in the process. See process computer.

control-configured vehicle (CCV) A fixed-wing aircraft designed to allow modulation of positive aerodynamic or reactive forces, or both, along and about all three axes, thereby providing limited independent maneuverability in all six deg of freedom. A control-configured vehicle can be moved vertically without changing pitch and can be moved laterally without changing bank or heading.[ARP4107-88]

control counter A physical or logical device in a computer that records the storage locations of one or more instruction words which are to be used in sequence, unless a transfer or special instruction is encountered.

control device Any device—such as a heater, valve, electron tube, contactor, pump, or actuator—used to directly effect a change in some process attribute.

Control Diagram Language (CODIL) A process-oriented language and system offered by Leeds and Northrup Company.

control electrode In an electron tube or similar device, an electrode whose potential can be varied to induce variations in the current flowing between two other electrodes.

control element A component of a control system that reacts to manipulate a process attribute when stimulated by an actuating signal.

control feel The "feel," or reaction, that a pilot perceives through the cockpit controls, either from the aerodynamic forces acting on the control surfaces, or from artificial input simulating these aerodynamic forces.[ARP4107-88]

control-force reversal A reversal or disappearance of the conventional forces acting on the aircraft control system, for example, when, owing to abnormal conditions, a forward movement of the stick or yoke causes the nose of the plane to rise instead of the normal response of dropping.[ARP4107-88]

control function See control operation.

control grid An element of an electron tube ordinarily positioned between the anode and cathode to act as a control electrode.

control initiation The signal introduced into a measurement sequence to regulate any subsequent control action as a function of the measured quantity.

control instruction A computer instruction that directs the sequence of operations.

control key A computer control key that when pressed with another key gives that key a different meaning.

controllability The capability of an aircraft rocket, or other vehicle to respond to control, especially in direction or attitude.[NASA]

controllable-pitch propeller/variable pitch propeller A propeller whose blade angle may be changed from the cockpit while the propeller is rotating.[ARP4107-88]

controlled airspace Airspace designed as continental control area, control area, control zone, terminal control area, transition area, or positive control area within which some or all aircraft may be subject to air traffic control.[ARP 4107-88]

controlled airspace, continental control area The airspace of the 48 contiguous states, the District of Columbia and Alaska (excluding the Alaska peninsula west of Longitude 160W) at and above 14,500 ft MSL, but not including the airspace less than 1500 ft above the surface of the earth.[ARP4107-88]

controlled airspace, control area airspace designated as colored federal airways, VOR Federal Airways, control areas associated with jet routes outside the continental control area, additional control areas, control area extensions, and area low routes.[ARP4107-88]

controlled airspace, control zone Controlled airspace which extends upward from the surface and terminates at the base of the continental control area.[ARP4107-88]

controlled airspace, terminal control area (TCA) Controlled airspace extending upward from the surface or higher to specified altitudes, within which all aircraft are subject to operating rules and pilot and equipment requirements.[ARP4107-88]

controlled airspace, transition area Controlled airspace extending upward from 700 ft or more above the surface of the earth when designated in conjunction with an airport for which an approved instrument approach procedure has been prescribed, or from 1200 ft or more above the surface of the earth when designated in conjunction with airway route structures or segments.[ARP4107-88]

controlled avalanche transit time devices See CATT devices.

controlled cooling Cooling a part from elevated temperature in a specific medium to produce desired properties or microstructure, or to avoid cracking, distortion or high residual stress; the usual cooling mediums, in descending order of severity, are brine, water, soluble oil, fused salt, oil, fan-blown air and still air.

controlled device The final element which is under the direct control of the control amplifier, such as a valve, damper, programmer, or a resistance heater.[ARP89C-70]

controlled medium The process fluid or other substance containing the controlled variable.

controlled variable 1. The variable which the control system attempts to keep at the setpoint value. The setpoint may be constant or variable. 2. The part of a process which you want to control (flow, level, temperature, pressure, etc.). 3. A process variable which is to be controlled at some desired value by means of manipulating another process variable.

controller 1. The component of the actuation system that controls the power modulator as a function of the command or error signal. Many of the electronic controllers have dynamic compensation as well as significant gain, so they must be recognized as separate components in the actuator loop.[AIR1916-88] See air traffic control specialist. 2. A device which permits the human pilot to initiate turn, pitch, and bank signals in the automatic pilot.[ARP419-57] 3. A device for interfacing a peripheral unit or subsystem in a computer; for example, a tape controller or a disk controller. 4. Device which contains all the circuitry needed for receiving data from external devices, both analog and digital, processes the data according to preselected algorithms, then provides the results to external devices.

control limit An automatic safety control responsive to changes in liquid level, pressure or temperature or position for limiting the operation of the controlled equipment.

control limits In statistical quality control, the upper and lower values of a measured quantity that establish the range of acceptability; if any individual measurement falls outside this range, the part involved is rejected and if the sample average for the same measurement falls outside the range, the entire lot is rejected.

controlling extensions A controller which derives its input from the motion of the float can be installed within the extension housing.

controlling means The components of an automatic controller that are directly involved in producing an output control signal or other controlling action.

controlling system See system, controlling.

control lock A securing device that prevents movement of control surfaces. [ARP4107-88]

control logic The sequence of steps or events necessary to perform a particular function. Each step or event is defined to be either a single arithmetic or a single Boolean expression.

control loop 1. A combination of two or more instruments or control functions arranged so that signals pass from one to another for the purpose of measurement and/or control of a process variable. See closed loop and open loop.

control motor A reversible motor which is designed to be operated from two independent voltage sources of the same frequency in such a way that variations in the voltage source will determine the direction, speed, and torque of the motor. A high torque to inertia ratio and a straight line speed torque curve are inherent characteristics of this type of motor.[ARP667-82]

control operation An action performed by a single device, such as the starting or stopping of a particular process. Conventionally, carriage return, fault change, rewind, end of transmission, etc., are control operations, whereas the actual reading and transmission of data are not.

control output module A device which stores commands from the computer and translates them into signals which can be used for control purposes. It can generate digital outputs to control on-off devices or to pulse setpoint stations, or it can generate analog output (voltage or current) to operate valves and other process control devices.

control panel 1. A part of a computer console that contains manual controls. 2. See plugboard. 3. See console and automatic control panel.

control point 1. The actual value of the controlled temperature at which the system is controlling.[ARP89C-70] 2. The setpoint or other reference value that an automatic controller acts to maintain as the measured value of a process variable under a given set of conditions.

control precision The degree to which a given value of a controlled variable can be reproduced for several independent control initiations using the same control point and the same system operating conditions.

control pressure (pilot pressure) The pressure required to control or influence any motion or change in motion.[AIR 1916-88]

control program 1. A group of programs that provides such functions as the handling of input/output operations, error detection and recovery, program loading and communication between the program and the operator. IPL, supervisor, and job control make up the control program in the disk and tape operating systems. 2. Specific programs which control an industrial process.

control programming Writing a user program for a computer which will control a process in the sense of reacting to random disturbances in time to prevent impairment of yield, or dangerous conditions.

control resolution The smallest increment of change that can be induced in the controlled process variable as a result of control-system action.

control rockets Vernier engines, retrorockets, or other such rockets, used to change the attitude of, guide, or make small changes in the speed of a rocket, spacecraft, or the like.[NASA]

control rod A long piece of neutron-absorbing material that fulfills one or both of the functions of controlling the number of neutrons available to trigger nuclear fission or of absorbing sufficient neutrons to stop fission in case of an emergency.

control room area See area, control room.

control, safety Control (including relays, switches, and other auxiliary equipment used in conjunction therewith to form a safety control system) which are intended to prevent unsafe operation of the controlled equipment.

control, safety combustion See combustion (flame) safeguard.

control sector An airspace area of defined horizontal and vertical dimensions for which a controller, or group of controllers, has air traffic control responsibility; normally, such an area is within an air route traffic control center or an approach control facility.[ARP4107-88]

control spring A spring designed to produce a torque equal and opposite to the torque produced by an instrument's moving element for any position of the moving element within the limits of its operating range.

control system 1. A system in which deliberate guidance or manipulation is used to achieve a prescribed value of a variable. A control system has at least one input and one output.[AIR1916-88] 2. A system consisting of a controller, power source(s), control junction, control effector, and feedback sensor(s) that interact to attain a goal(s). See closed-loop system.[ARP4107-88]

control unit 1. Those parts of computers that cause the arithmetic unit, storage, and transfer of a computer to operate in proper sequence.[NASA] 2. The portion of a computer which directs the sequence of operations, interprets the coded instructions, and initiates the proper commands to the computer circuits preparatory to execution. 3. A device designed to regulate the fuel, air, water, or electrical supply to the controlled equipment. It may be automatic, semi-automatic, or manual.

control valve 1. A valve used to control the fluid flow of a hydraulic type servo. It operates in response to signals from a controller, selector, or sensor. Also called a boost control valve, or power boost control valve.[ARP419-57] 2. A final controlling element, through which a fluid or gas passes, which adjusts the size of flow passage as directed by a signal from a controller to modify the rate of flow.

control variable 1. The variable which the control system attempts to keep at the setpoint value. 2. The part of a process which you want to control (flow, level, temperature, pressure, etc.) 3. A process variable which is to be controlled at some desired value by means of manipulating another process variable.

control zone See controlled airspace.

CONUS Acronym for the CONtinental United States.

convection 1. In general, mass motion within a fluid resulting in transport and mixing of the properties of that fluid. Specifically, in meteorology, atmospheric motions that are predominately vertical.[NASA] 2. The transmission of heat by the circulation of a liquid or a gas such as air. Convection may be natural or forced.

convection cooling Removing heat from a body by means of heat transfer using a moving fluid as the transfer medium, usually involving only the motion caused by differences in heat content between fluid near the hot surface and fluid at some distance from the surface.

convection-type superheater See superheater.

convective SIGMET/convective significant meteorological information A weather advisory concerning convective weather significant to the safety of all aircraft. Convective SIGMETs are issued for tornadoes, lines of thunderstorms, embedded thunderstorms of any intensity level, areas of thunderstorms greater than or equal to radar weather echo intensity level 4 with an area coverage of 4/10 (40 percent) or more, and hail 3/4 inches or more in diameter. See also SIGMET.[ARP4107-88]

convergence 1. Approach to a limit, e.g., by an infinite sequence.[NASA] 2. The condition in which all the electron beams of a multibeam (color) cathode ray tube intersect at a specific point. See also misconvergence.

convergence, dynamic The incorporation of a means (electrostatic or magnetic) for automatically converging the multiple beam spots of a color CRT over the entire screen as a function of scanning.[ARP1782-89]

conversational mode Communication between a terminal and a computer in which each entry from the terminal elicits a response from the computer and vice versa.

conversion 1. The process of changing from one type of control or operational state to another; e.g. from an active to a standby control or from a primary to a secondary system.[ARP1181A-85]

conversion coating A protective surface layer on a metal that is created by chemical reaction between the metal and a chemical solution.

conversion time The time required for a complete measurement by an analog-to-digital converter.

conversion (to engineering units) Scaling signals from their raw input form to the form used internally, usually into floating point engineering units.

conversion transducer Any transducer whose output-signal frequency is different from its input-signal frequency.

convertaplanes See V/STOL aircraft.

converter 1. Rotary device for changing alternating current to direct current. Transducer whose output is a different frequency from its input.[NASA] 2. A type of refining furnace where impurities are oxidized and removed by blowing air or oxygen through the molten metal. 3. A/D analog to digital; D/A digital to analog; I/P current to pneumatic pressure converter; P/I pneumatic pressure to current converter; P/V pneumatic pressure to voltage converter; V/P voltage to pneumatic pressure converter. 4. A converter is also referred to as a transducer; however, transducer is a completely general term, and its use specifically for signal conversion is not recommended.

convex A term describing a surface whose central region is raised with respect to a flat plane approximately passing through its periphery.

convex programming In operations research, a particular case of nonlinear programming in which the function to be maximized or minimized and the constrains are appropriately convex or concave functions of the controllable variables. Contrast with dynamic programming, integer programming, linear programming, mathematical programming, and quadratic programming.

conveyor A continuously moving materials-handling device for transferring large numbers of individual items or quantities of bulk solids from one location to another over a relatively short distance along a fixed path.

cook-off The detonation or deflagration of an explosive device caused by externally applied heat.[AIR913-89]

coolant 1. Liquids of gases used to cool something, as a rocket combustion chamber.[NASA] 2. Any fluid used primarily to remove heat from an object and carry it away. 3. In a machining operation, any cutting fluid whose chief function is to keep the tool and workpiece cool.

Coolidge-type x-ray tube A high-vacuum tube in which electrons emitted from a high-voltage cathode impinge on a water-cooled metal target inclined with respect to the tube axis; x-rays emitted from the focal spot on the target are directed through a side window in the metal tube enclosure, where a material relatively transparent to x-rays—beryllium foil, mica, aluminum or special low-absorption glass—allows them to escape.

cooling effect The ability of a fluid stream to carry away heat. For a given fluid, this is a function of temperature and mass flow.[ARP89C-70]

coordinate Set of measures defining points in space.[NASA]

coordinate system See coordinate.

coordination Coordination defines the ability of the protector with the lowest rating (in a cascade arrangement) to open before protectors with higher ratings when a fault occurs downstream from the lowest rated protector. See also cascade circuit.[ARP1199A-90]

copilot A licensed pilot serving in any piloting capacity other than as pilot-in-command, but excluding a pilot who is onboard the aircraft for the sole purpose of receiving flight instruction. [ARP4107-88]

copilot syndrome An attitude resulting in ineffective crew coordination based on the comforting premise that one or more the other crewmembers have the situation under control and are looking out for one's best interest. Implicit in the term "other crewmembers" are non-flight members such as personnel in the ARTCC, the command post, or a RAPCON facility.[ARP4107-88]

copolymer A compound resulting from the polymerization of two or more different monomers.[ARP1931-87]

copper A basic element, atomic number 29, which is widely used for electrical conductors.[ARP1931-87]

copper alloy An alloy in which copper is the predominant element. Generally, the addition of sulfur, lead, or tellurium improves machinability. Cadmium improves tensile strength and wearing qualities. Chromium gives very good mechanical properties at temperatures well above 200 °C. Zirconium provides harness, ductility, strength, and relatively high electrical conductivity at temperatures where copper, and common high conductivity copper alloys tend to weaken. Nickel improves corrosion resistance, while silicon offers much higher mechanical properties. Beryllium, when present in an approximate 2% content in copper alloys, permits maximum strength, while about 0.5% content offers high conductivity.[ARP1931-87]

copper, ETP Electrolytic tough pitch copper (ETPC) has a minimum copper content of 99.9%. Annealed conductivity averages 101% with a 100% minimum.[ARP1931-87]

copper, OFHC Oxygen-free high conductivity copper (OFHC) has a 99.95% minimum copper content with an average annealed conductivity of 101%. It is suitable for apparatus that is welded or exposed to reducing gasses at high temperatures. This copper has no residual deoxidant.[ARP1931-87]

copper, silver bearing Silver bearing copper with a 99.9% copper content provides nearly the same electrical conductivity as ETP copper, but offers a higher softening point, greater resistance to creep, and higher strength at elevated temperatures. It also offers higher resistance to wear and oxidation, and improved machinability. [ARP1931-87]

coprocessor A device added to a CPU that performs special functions more efficiently than the CPU alone.

copy In data processing, reproducing data from one storage device to another.

copy protection 1. The inability to copy a disk, particularly program disks, by the addition of codes on the disk. 2. The inability to copy data on a disk by the addition of an adhesive cover on a side slot of the disk.

corbinotron A device consisting of a corbino disc, made of high-mobility semiconductor material, and a coil that produces a magnetic field perpendicular to the plane of the disc.

cord body The main body of the tire which includes layer(s) of cord fabric (plies).[AIR1489-88]

cord fabric The fabric from which the plies, breakers, and chafer strips are made. May be rayon, nylon, fibreglass, etc.[AIR1489-88]

cordite See double base propellant.

core 1. A component or assembly of components over which other components, such as a shield, jacket, sheath, or armor are applied in order to form a cable. See cable core.[ARP1931-87] 2. A strongly ferromagnetic material used to concentrate and direct lines of flux produced by an electromagnetic coil. 3. The inner layer in a composite material or structure. 4. The central portion of a case-hardened part, which supports the hard outer case and gives the part its toughness and shock resistance. 5. An insert placed in a casting mold to form a cavity, recess or hole in the finished part. 6. A rod or closed tube inserted in a tube to reduce the flow area. See magnetic core. 7. Magnetic memory elements; typically the main memory in a computer system.

cored electrode A tubular welding electrode containing flux or some other material in the central cavity.

cored solder Wire solder having a flux-filled central cavity.

core dump See storage dump.

core (fiber optic) The inner portion of an optical fiber which carries light along the length of the fiber. Light is confined to the core by a difference in refractive index between core and cladding, with the latter having a lower index.

core iron A grade of soft steel suitable for making cores used in electromagnetic devices such as chokes, relays and transformers.

core memory The most common form of main memory storage used by a central processing unit, in which binary data are represented by switching the polarity of magnetic cores.

core resident A term pertaining to programs or data permanently stored in core memory for fast access.

core storage See magnetic core.

core wire Copper wire having a steel core, often used to make antennas.

coring A metallurgical condition where individual grains or dendrites vary in composition from center to grain boundary due to nonequilibrium cooling during solidification in an alloy that solidifies over a range of temperatures.

Coriolis effect 1. The physiological effect felt by a person moving radially in a rotating system, as a rotating space station resulting in nausea, vertigo, dizziness, etc.[NASA] 2. An accelerating force acting on any body moving freely above the earth's surface due to the fact that the earth is rotating with respect to a given axis through its center; it is the Coriolis effect that causes, for instance, a level bubble carried in an airplane to be deflected perpendicular to the direction of flight, and a river in the Northern Hemisphere to scour its right bank more than its left bank whereas a river in the Southern Hemisphere scours its left bank more than its right. Coriolis effect is the basis for mass-flow meters. See also illusion, vestibular.

Coriolis force Results from Coriolis acceleration acting on a mass moving with a velocity radially outward in a rotating plane.

Coriolis-type mass flowmeter An instrument for measuring mass flowrate by determining the torque from radial acceleration of the fluid.

Corliss valve A type of valve used to admit steam to, or exhaust it from, a reciprocating engine cylinder.

corner-cube prism A prism in which three flat surfaces meet at right angles, as they would if they were the corner of the cube. Incident light through a planar face is reflected back to the source.

cornering force Lateral or cornering force of a tire (perpendicular to the direction of motion).[AIR1489-88]

corner taps The differential pressure signal location in an orifice flange union defined by the corner formed between the orifice plate and the internal diameter of the flange.

corona An electrically detectable (usually luminous) field-intensified ionization that occurs in an insulating system due to a potential gradient which exceeds a certain critical level.[ARP1931-87]

corona discharge See electric corona.

coronal holes Solar areas where extreme UV and x-ray coronal emission is abnormally low or absent. These are coronal regions apparently associated with diverging magnetic fields.[NASA]

coronal loops Loop like structures revealed in soft x-ray images of the solar limb and believed to evolve from the introduction of energy and density perturbations at the top of an arched, cylindrical magnetic flux tube initially in equilibrium in the coronal plasma. [NASA]

corona onset point The critical value of electrical potential where corona is first detected. Also known as ignition voltage.[ARP1931-87]

corona voltmeter A type of voltmeter that uses the inception of corona to determine the crest value of voltage in an a-c electric current.

corotating See wheels, corotating.

corporate aircraft Aircraft flown by professional pilots and operated by a corporation or business in furtherance of the firm's business.[ARP4107-88]

Corporation for Open Systems (COS) An organization formed in 1985 to coordinate member company efforts in the selection of standards and protocols, conformance testing, and the establishment of certification.

corpuscular radiation Nonelectromagnetic radiation consisting of energetic charged or neutral particles.[NASA]

corrected speed The corrected speed of a starter or APU is utilized to present performance corrected for temperature. It is usually expressed as $N/\sqrt{T}$ or $N/\sqrt{0}$ where T or 0 are defined at the inlet of the starter or APU.[ARD50013-91]

correction A quantity, equal in absolute magnitude to the error, added to a calculated or observed value to obtain a true value.[NASA]

correction time See time, settling.

corrective action 1. A documented design, process, procedure, or materials change implemented and validated to correct the cause of failure or design deficiency.[ARD50010-91] 2. Controller output which results in a change in controlled temperature in the direction of the control point.[ARP89C-70] 3. A documented design, process, procedure, or materials change implemented and validated to correct the cause of failure or design deficiency.[ARD50013-91] 4. The change produced in a controlled variable in response to a control signal.

corrective maintenance Maintenance specifically intended to eliminate an existing fault. Synonymous with emergency maintenance. Contrast with preventive maintenance.

corrective network An electronic network incorporated into a circuit to improve its transmission or impedance properties, or both.

correlation 1. In statistics, a relationship between two occurrences which is expressed as a number between minus one (-1) and plus one (+1).[NASA] 2. Measurement of the degree of similarity of two images as a function of detail and relative position of the images. It is obtained by multiplying the Fourier transforms of the two images, then taking the the Fourier transform of the product.

correlation detection A method of detection in which a signal is compared, point-to-point, with an internally generated reference.[NASA]

correlation function See correlation.

correlator 1. Device that detects weak signals in noise by performing an electronic operation.[NASA] 2. A logic device which compares a series of bits in a data stream with a known bit sequence and puts out a signal when correlation is achieved. One use of the correlator is as a PCM frame synchronizer.

corresponding states A principle that states that two substances should have similar properties at corresponding conditions with reference to some basic properties, e.g., critical pressure and critical temperature.

Corrodekote test An accelerated corrosion test for electrodeposits in which a specimen is coated with a slurry of clay in a salt solution, and then is exposed for a specified time in a high-humidity environment.

corrosion 1. A physical discoloration of hose cover or reinforcement indicative of material degradation due to extreme heat, chemical attack, weathering, oxidation, etc.[ARP1658A-86] 2. The chemical deterioration of a portion of a part by oxidation. Can be caused by oxygen, moisture, or attack by other oxidizing chemicals.[AIR4069-90] 3. The deterioration of a metal by chemical or electrochemical reaction with its environment.[NASA] 4. The wasting away of metals due to chemical action in a boiler; usually caused by the presence of O_2, CO_2 or an acid.

corrosion coating A material applied to integral tanks to coat the surface and supply protection to the metal, preventing corrosion. The term is used widely, but actually means, a corrosion preventive coating.[AIR4069-90]

corrosion fatigue A synergistic interaction of the failure mechanisms corrosion and fatigue such that cracking occurs much more rapidly than would be predicted by simply adding their separate effects; failure by corrosion fatigue requires the simultaneous presence of a cyclic stress and a corrosive environment.

corrosion protection Preventing corrosion or reducing the rate of corrosive attack by any of several means including coating a metal surface with a paint, electroplate, rust-preventive oil, anodized coating or conversion coating; adding a corrosion-inhibiting chemical to the environment; using a sacrificial anode; or using an impressed electric current.

corrosive Any substance or environment that causes corrosion.

corrosive flux A soldering flux that removes oxides from the base metal when the joint is heated to apply solder; the flux is usually composed of inorganic salts and acids which are corrosive and must be removed before placing the soldered components in service to ensure maximum service life.

corrosiveness The degree to which a substance causes corrosion.

corrugated fastener A thin, corrugated strip of steel used to fasten two pieces of wood together by hammering it into the wood approximately at right angles to the joint line.

corrugating Forming sheet metal into a series of alternating parallel ridges and grooves; forming may be done by rolling the metal between matched grooved rolls or by forming it in a press brake equipped with a special-shaped punch and die.

corrupt In data processing, the inclusion of errors in programs or data.

COS See Corporation for Open Systems.

cosmetic defect A variation from the conventional appearance of an item, such as a slight deviation from its usual color, which is not detrimental to the items performance.[ARP1931-87]

Cosmic Background Explorer satellite A NASA satellite designed to measure background radiation in order to confirm or deny the big bang theory.[NASA]

cosmic dust Finely divided solid matter with particle sizes smaller than a micrometeorite, thus with diameters much smaller than a millimeter, moving in interplanetary space.[NASA]

cosmic gamma ray bursts See gamma ray bursts.

cosmic noise Interference caused by cosmic radio waves.[NASA]

cosmic radiation See cosmic rays.

cosmic rays 1. The aggregate of extremely high energy subatomic particles which travel the solar system and bombard the earth from all directions. Cosmic ray primaries seem to be mostly protons, hydrogen nuclei, but also contain heavier nuclei. On colliding with atmospheric particles they produce many different kinds of lower energy secondary cosmic radiation.[NASA] 2. Penetrating ionizing radiation whose ultimate origin is outside the earth's atmosphere; some of the constituents of cosmic rays can penetrate many feet of material such as rock.

cosmochemistry The branch of chemistry that deals with the chemical composition and changes in the universe. [NASA]

COSPAS The USSR satellite of the COSPAS-SarSat project which is a satellite-aided project for the search and rescue of distressed vehicles, administered by USSR, US, French, and Canadian agencies.[NASA]

cost index (CI), flight index The operator's cost factor for that specific flight that takes into account such variables as refueling fuel cost, time of day, gate scheduling, etc.[ARP1570-86]

cotter A tapered part similar to a wedge or key that can be driven into a tapered hole to hold an assembly together.

cottered joint A joint in which power is transferred across the joint via shear force transverse to the longitudinal axis of a bar (usually tapered along one side to ensure a tight fit) known as a cotter, which holds the joint together.

cotter pin A split pin, usually formed by folding a length of half-round wire back on itself; the pin is inserted into a hole and then is bent to keep a castle nut from turning on a bolt, to hold a cotter securely in place, to hold hinge plates together, or to pin various other machine parts together. Also known as cotter key.

Cottrell precipitator A device for removing dust or mist from a gas by passing the gas through a vertical, electrically grounded pipe where the particulates become ionized by corona discharge from an axial wire maintained at a high negative voltage; the ionized particles migrate to the pipe's inner wall where they collect for later removal by mechanical means.

coulomb Metric unit for quantity of electricity.

Coulomb collisions The collisions of sets of two particles both of which are charged.[NASA]

coulomb damping The dissipation of energy that occurs when a particle in a vibrating system is resisted by a force whose magnitude is a constant independent of displacement and velocity, and whose direction is opposite to the direction of the velocity of the particle. Also called dry friction damping.[AIR1489-88]

coulomb friction load A constant friction load opposing motion.[AIR1916-88]

coulombmeter An instrument for measuring the quantity of electricity (in coulombs) by integrating a stored charge in a circuit that has a high impedance.

coulometer An electrolytic cell constructed and operated to measure a quantity of electricity in terms of the electrochemical action it produces.

coulometric titration A method of wet chemical analysis in which the amount of an unknown substance taking part in a chemical reaction is determined by measuring the number of coulombs required to reach the end point in electrolysis.

count In computer programming, the total number of times a given instruction is performed.

countdown A step-by-step process that culminates in a climatic event, each step being performed in accordance with a schedule marked by a count in inverse numerical order; specifically, this process is used in leading up to the launch of a large or complicated rocket vehicle, or in leading up to a captive test, a readiness firing, a mock firing or other firing test.[NASA]

counter 1. A device or register in a digital processor for determining and displaying the total number of occurrences of a specific event. 2. In the opposite direction. 3. Device or PC program element that can total binary events and perform ON/OFF actions based on the value of the total. 4. A device, register, or location in storage for storing numbers or number representations in a manner which permits these numbers to be increased or decreased by the value of another number, or to be changed or reset to zero or to an arbitrary value.

counterbore 1. An end cutting tool having two or more cutting edges with flutes or grooves adjacent thereto for the passage of cuttings and an integral or inserted pilot, or centrally located hole for insertion of a pilot, for guiding the tool. The flutes or grooves are usually helical to give a positive rake at the cutting edge. It is used to form a flat-bottomed enlargement of the mouth of a previously formed cylindrical hole. The tool may have a straight shank, tapered shank, or special shank requiring a special holder, for holding in a power- or hand-operated machine. [ARP480A-79] 2. A drilled or bored flat-bottomed hole, often concentric with another, smaller hole.

countercurrent flow Flow of two fluids in opposite directions within the same device, such as a tube-in-shell heat exchanger. Contrast with counterflow.

counterflow Flow of a single fluid in opposite directions in adjacent portions of the same device, such as a U-bend tube. Contrast with countercurrent flow.

counter rotating propellers (shafts) The airplane arrangement in which shafts and propellers rotate in opposite directions but in which the shafts are not concentric.[ARP355-84]

counter rotation Movement of sets of bodies or fluids around a common axis where movement in own rotational direction is opposed by movement in the opposite direction.[NASA]

countershaft A secondary shaft, driven by the main shaft of a machine, and used to supply power to one or more machine parts.

countersink 1. An angular cutting tool, usually made with angular relief, having two or more flutes with specific size angle cutting edges. It is used for chamfering and countersinking holes. The tool may have a straight shank, tapered shank, bit stock shank or special shank requiring a special holder, for holding in a power- or hand-operated machine. It may be designed to be driven or turned by means of a wrench or carpenter's brace.[ARP480A-79] 2. A chamfer around the edge of a circular hole, which removes burrs, provides a seat for a flat-head screw or other fastener, or provides a tapered surface for a machine center to rest in.

counterweight 1. A mass which counterbalances the weight of the lifting device or load platform of an elevator or hoist so that the engine must only work against the payload, friction, and any remaining unbalanced machine loads. 2. Any mass incorporated into a mechanism to compensate for an out-of-balance condition and maintain static equilibrium. Also known as counterbalance; counterpoise.

counting rate The average number of ionizing events that occur per unit time, as determined by a counting tube or similar device.

counting-rate meter An instrument whose indicated output is related to the average rate of occurrence of ionizing events.

counting scale Any of several designs of weighing device where the total weight of a large number of identical parts is compared with the weight of one part or the weight of a small, easily counted number of parts, and the number of parts in the unknown quantity determined by automatic indication, readout or calculation.

counts 1. An alternate form of representing raw data corresponding to the numerical representation of a signal received from or applied to external hardware. 2. The accumulated total of a series of discrete inputs to a counter.

3. The discrete inputs to an accumulating counter. See digitized signal.

couplant A substance used to transmit sound waves from an ultrasonic search unit to the surface of a test piece, thus reducing losses and improving test accuracy; usual couplants include water, oil, grease, paste or other liquid or semisolid substances.

coupled See wheels.

coupled control-element action A type of control system action in which two or more actuating signals or control element actions are used in concert to operate one control device.

coupled mode Mode of vibration that is not independent but which influences one mode to the other.[NASA]

coupled reference input See cascade action.

coupler 1. In data processing, a device that joins similar items. 2. In fiber optics, a device which joins together three or more fiber ends—splitting the signal from one fiber so it can be transmitted to two or more other fibers. Directional, star, and tee couplers are the most common.

couple unbalance That condition of unbalance for which the central principal axis intersects the shaft axis at the center of gravity. NOTE 1: The quantitative measure of couple unbalance can be given by the vector sum of the movements of the two dynamic unbalance vectors about a certain reference point in the plane containing the center of gravity and the shaft axis. NOTE 2: If static unbalance in a rotor is corrected in any plane other than that containing the reference point, the couple unbalance will be changed. [ARP588A-89]

coupling 1. A device or contrivance for joining adjacent ends or parts of anything.[AIR1489-88] 2. A device permitting transfer of energy from one electrical circuit to another, or from one mechanical device to another.[AIR1489-88] 3. A device that serves to couple or connect the ends of adjacent parts. [ARP480A-79] 4. Any device that connects the ends of adjacent parts; the connection may be rigid, allowing little or no relative movement, or it may be flexible, accommodating misalignment and other sources of relative movement. 5. A mechanical fastening between two shafts that provides for the transmission of power and motion. Also known as shaft coupling. 6. Interdependence in a computer system.

coupling, bayonet, cylindrical A coupling mechanism utilizing spiral ramps in one cylindrical connector half to engage projections in the mating half so as to provide jacking and locking together of the mating halves through limited rotation of the coupling ring. [ARP914A-79]

coupling, breech A coupling mechanism which distributes the coupling load over large solid metal engaging and locking lands for positive coupling alignment and complete connector mating with a limited rotation of the coupling ring. [ARP914A-79]

coupling, quick disconnect A design feature which permits relatively rapid joining and separation of mating parts. [ARP914A-79]

coupling ring That portion of a cylindrical plug connector housing which, by rotation, aids in the mating, captivation or unmating of the plug to the receptacle connector. See also coupling, bayonet; coupling, quick disconnect; coupling, threaded.[ARP914A-79]

coupling, threaded A coupling mechanism utilizing matching screw threads for mating and unmating of cylindrical connectors or other devices.[ARP 914A-79]

coupling, threaded self lockingA coupling mechanism utilizing matching screw threads for mating and unmating of cylindrical connectors or devices incorporating automatically actuated locking mechanism to prevent the coupling ring from disengaging under vibration conditions.[ARP914A-79]

coupling torque The force required to rotate a coupling ring or jackscrew when engaging a mating pair of connectors.[ARP914A-79]

coupling triple start, self locking A coupling mechanism using a triple start thread for quick connector mating with one full turn of the coupling ring which is also designed with an anti-coupling device.[ARP914A-79]

course 1. The intended direction or path of flight in the horizontal plane measured in deg from north. The actual course or path of an aircraft should be distinguished from its heading. They may coincide, but usually do not, the difference being a function of heading, sideslip, and drift. See also heading. 2. The ILS localizer signal pattern usually specified as front course or back course. (a) Front Course (ILS): The approach course of the localizer, which is used with other functional parts (for example, glide slope, marker beacons, etc.). Its signal is transmitted from the far end of the runway (opposite from the runway approach threshold) and is adjusted so that the distance between full-scale deflections (left to right) of the course deviation indicator needle should equate to approximately 700 ft of linear width at the runway threshold. (b) Back Course (ILS): The course line along the extended centerline of a runway, in the opposite direction to the front course (ILS).[ARP4107-88]

course deviation indicator (CDI) A display device for presenting the magnitude (deg, ft, miles) of displacement that a vehicle is from a selected course. [ARP4107-88]

course selector A manually operated navigation instrument used in conjunction with a VHF omnidirectional radio range to ascertain station direction. [ARP4107-88]

cover A protective device, rigid or flexible, to enclose an opening against foreign material where mutilation is not involved. See cap, guard, mask, protector.[ARP480A-79]

coverage 1. The calculated percentage which defines the completeness with which a metal braid covers the underlying surface. The higher percentage of coverage, the greater the protection against external interference.[ARP1931-87] 2. Also load repetition factor. The number of passes of loaded tires of an air vehicle in adjacent tire paths sufficient to just cover a given width of pavement one time. Determining factors are: width of traffic lane, number of wheels, width of tire contact area, and the traffic distribution.[AIR1489-88]

cover blisters Raised areas or spots on an elastomeric hose cover surface usually forming a void or air-filled space in the material.[ARP1658A-86]

cover, dust A covering device or material used during stowage and transit to protect connectors, harnesses or electronic assemblies against dust and other foreign matter. It may be of a design which attaches to a connector or may completely envelop a connector, harness or electronic assembly. See also cover, protective.[ARP914A-79]

covering power The ability of an electroplating solution to give a satisfactory plate at low current densities, such as occur in recesses, but not necessarily to build up a uniform coating. Contrast with throwing power.

cover plate 1. Any flat metal or glass plate used to cover an opening. 2. Specifically, a piece of glass used to protect the tinted glass in a welder's helmet or goggles from being damaged by weld spatter.

cover, protective An accessory used to

cover the mating portion of a connector for mechanical, environmental and/ or electrical protection.[ARP914A-79]

cover the six A request to protect a person's or vehicle's vulnerable area(s); those that cannot be monitored by the operator of the system. A phrase generally used by pilots of combat aircraft requesting their wingman to guard the area to their rear (6 o'clock position). [ARP4107-88]

covolume The space occupied by a gas when compressed to its limit.[AIR913-89]

cowling A metal cover, usually one that provides a streamlined enclosure for an engine.

CPA Closest point of approach of two aircraft in conflict.[ARP4153-88]

CP/M An operating system for microcomputers.

C power supply An electrical power supply connected between the cathode and grid of a vacuum tube to provide a grid-bias voltage.

CPU See central processing unit; see also unit, central processing.

CPU-bound A state of program execution in which all operations are dependent on the activity of the central processor, for example, when a large number of calculations are being performed; compare to I/O-bound.

crack 1. A crack is a clean (crystalline) fracture passing through or across the grain boundaries and may possibly follow inclusions of foreign elements. Cracks are normally caused by overstressing the metal during forging or other forming operations, or during heat treatments.[MA1568-87] Where parts are subjected to significant reheating, cracks usually are discolored by scale.[MA2005-88] 2. A fissure in a part where it has been broken but not completely severed into two pieces. 3. The fissure or chink between adjacent components of a mechanical assembly. 4. To incompletely sever a solid material, usually by overstressing it. 5. To open a valve, hatch, door or other similar device a very slight amount.

crack closure Phenomenon which occurs when the cyclic plasticity of a material gives rise to the development of residual plastic deformations in the vicinity of a crack tip, causing the fatigue crack to close at positive load. [NASA]

cracked residue The fuel residue obtained by cracking crude oils.

cracked sleeve A firesleeve which evidences surface cracks or fissures caused by strain and environmental conditions.[ARP1658A-86]

crack geometry The shape and size of partial fractures or flaws in materials. [NASA]

cracking (chemical engineering) 1. A process used to reduce the molecular weight of hydrocarbons by breaking molecular bonds by thermal, catalytic, or hydrocracking methods.[NASA] 2. The thermal decomposition of complex hydrocarbons into simpler compounds or elements.

cracking pressure The pressure at which a valve becomes unseated and begins to pass fluid.[AIR1916-88]

cracking process A method of manufacturing gasoline and other hydrocarbon products by heating crude petroleum distillation fractions or residues in the presence of a catalyst so that they are broken down into lighter hydrocarbon products, some of which can be distilled off.

crack opening displacement The displacement at the mouth of a crack in a material.[NASA]

crack tip The boundary between cracked and uncracked material.[NASA]

cradle A device to cradle a part by engaging the contour of the part. See also saddle.[ARP480A-79]

crane A hoisting machine with a power-driven horizontal or inclined boom and lifting tackle.

crane hoist A mobile hoisting machine used principally for lifting loads by means of cables; it consists of a mobile undercarriage and support structure, a power unit and winch enclosed in a cab or house (often one that swivels on the undercarriage), a movable boom and various lifting, boom positioning and support cables.

crane scale A type of lifting device integral with or attached to a crane hook and having an internal load cell that automatically weighs a load as it is lifted; where a strain-gage load cell is used, weight can be indicated or recorded remotely.

crank 1. To rotate the main shaft of an internal combustion or gas turbine engine by means of a starter.[ARP906A-91] 2. A mechanical link that can revolve about a center of rotation.

Crank-Nicholson method A method for solving parabolic partial differential equations, whose main feature is an implicit method which avoids the need for using very small time steps.[NASA]

crankpin A cylindrical projection on a crank for attaching a connecting rod.

crankshaft 1. A straight shaft to which one or more cranks are attached. 2. A cast, forged or machined shaft with integral cranks, such as is used in a reciprocating automobile engine.

crank throw 1. The web or arm of a crank. 2. The radial displacement of the crankpin from the crankshaft axis.

crank web The portion of a crank that connects a crankpin to the crankshaft or to another adjacent crankpin. Also known as crank throw.

crash A computer hardware or software malfunction that causes the system to be reset or restarted.

crashworthiness The ability of a vehicle to withstand a crash.[NASA]

crater 1. A spot on the face of a cutting tool where it has been worn by contact with chips. 2. A depression at the finishing end of a weld bead.

crazing 1. Minute cracks on or near the surface of materials.[ARP1931-87] 2. A surface effect on elastomeric hose cover material characterized by multitudinous minute cracks.[ARP1658A-86] 3. A network of fine, shallow cracks at the surface of a coating, solid metal or plastics material. 4. Development of such a network.

Cray computer Supercomputer built by Cray Research Inc. that requires the supporting services of another front end general purpose computer for operation. It incorporates very fast scalar and vector hardware, is used primarily for the simulation of physical phenomena, and is programmed in FORTRAN. [NASA]

CRC See cyclic redundancy check.

create To open, write data to, and close a file for the first time.

creep 1. The time and temperature dependent deformation that occurs in a material as a result of the application of a constant load. Although this can occur at any temperature in some materials, it is of concern in metals at higher temperature, generally above one-half of the melting temperature (absolute). The calculation of life usage due to creep requires accurate time and temperature information since a change of 15 °C, at high temperatures, can affect the life of a part by a factor of two.[AIR1872-88] 2. Time-dependent plastic strain occurring in a metal or other material under stress, usually at elevated temperature.

creep (leakage) distance The shortest distance on the surface of an insulator separating two electrically conductive surfaces.[ARP914A-79]

creep rate The creep rate is always measured at a specified temperature and stress and is the rate at which creep deformation occurs. It is classified in four stages and can further be expressed as the slope of the creep/time curve. [ARP700-65]

creep resistance See creep strength.

creep strength The constant nominal stress that will cause a specified quantity of creep in a given time at constant temperature.[NASA]

crest 1. The top of a screw thread. 2. The bottom edge of a weir notch, sometimes referred to as the sill.

crestatrons See traveling wave tubes. [NASA]

crest voltmeter An instrument whose indicated value is the average positive peak amplitude of a sinusoidal a-c electric voltage.

crest width The distance along the crest between the sides.

crevice corrosion A type of concentration-cell corrosion associated with the stagnant conditions in crevices, fissures, pockets and recesses away from the flow of a principal fluid stream, where concentration or depletion of dissolved salts, ions or gases such as oxygen leads to deep pitting.

crew coordination The systematic division of subtasks between or among crew or flight members so as to accomplish a larger task more efficiently. Crew coordination is the most basic level of command/control.[ARP4107-88]

crewmember A person assigned to perform duty in an aircraft during flight time. Flight crewmember refers to the pilot, copilot, navigator, or (where applicable) flight engineer.[ARP4107-88]

crew procedure (inflight) Operations performed by crews aboard aircraft or spacecraft during flight. Includes flight operations as well as spaceborne experiment procedures.[NASA]

crew procedure (preflight) Operations performed by crews aboard aircraft or spacecraft and by ground support crews before flight or launching.[NASA]

crew size The number of people in a crew.[NASA]

crew systems Those portions of aircraft systems/subsystems that are affected by the aircrew, such as the flight control system, flight display system, radar systems, and environmental control systems.[ARP4107-88]

crimp The physical compression or reshaping of a conductor barrel or ferrule around a conductor in order to obtain a mechanical and an electrical connection.[ARP914A-79]

crimp anvil (nest) That portion of a crimping die which supports a barrel or ferrule during crimping.[ARP914A-79]

crimper A device to flute, corrugate; or compress or otherwise deform a part to change its initial shape.[ARP480A-79]

crimp indentor That portion of the crimping die which indents or reshapes the barrel or ferrule.[ARP914A-79]

crimping 1. Forming small corrugations in order to set down and lock a seam, create an arc in a metal strip or reduce the radius of an existing arc or circle. 2. Causing something to become wavy, crinkled or warped. 3. Pinching or pressing together to seal or unite, especially the longitudinal seam of a tube or cylinder.

crimping dies That portion of a crimping tool that compresses and reshapes the conductor barrel or ferrule to form the crimp.[ARP914A-79]

crimping tool 1. A crimping tool is a mechanical device having specially shaped dies for reforming the wire barrel of a taper pin around a wire.[ARP592-90] 2. The device used to perform a crimp.[ARP914A-79]

crimp pot adapter A sleeve that fits around the stripped conductor and allows for a small wire to fit into a large gage crimp pot.[ARP914A-79]

crimp tensile strength The axial force required to separate the wire from the crimped conductor barrel. The wire may pull out of or break in the crimped area of the conductor barrel.[ARP914A-79]

crimp termination A termination accomplished by the controlled reforming of the wire barrel portion of a terminating device, through physical compression exerted by appropriate tooling. [ARP1931-87]

criterion referenced instruction (CRI) 1. A training system methodology in which an individual's achievement in the instructional program is measured in terms of a predetermined set of absolute criteria rather than relative to the performance of other individuals. 2. The concept of using a criterion test to measure the effectiveness of the instruction rather than to measure student proficiency.[ARP4107-88]

critical altitude The maximum altitude in standard atmospheric conditions at which it is possible to maintain a specified engine power or a specified manifold pressure at a specified rotational speed (engine RPM).[ARP4107-88]

critical angle of attack The minimum angle of attack of a given airfoil or airfoil section at which extensive flow separation occurs, with consequent loss of lift and increase of drag; generally results in stalling of the airfoil.[ARP4107-88]

critical cooling rate The minimum cooling rate that will suppress undesired transformations during a hardening heat treatment.

critical damping The minimum damping that will allow a displaced system to return to its initial position without oscillation.[AIR1489-88] See also damping

critical defect A critical defect is a defect that judgment and experience indicate is likely to result in hazardous or unsafe conditions for individuals using, maintaining, or depending upon the product; or a defect that judgment and experience indicate is likely to prevent performance of the tactical function of a major end item such as an aircraft or missile. Synonymous with failure.[ARD50010-91]

critical dimension 1. Generally, any physical measurement whose value or accuracy is considered vital to the function of the involved component or assembly. 2. In a waveguide, the cross-sectional dimension which determines the waveguide's critical frequency.

critical engine The engine whose failure would most adversely affect the performance or handling qualities of an aircraft.[ARP4107-88]

critical engine failure speed A minimum speed at which it is considered safe to attempt to complete the takeoff with one engine inoperative.[AIR1489-88]

critical field length Critical field length is the length of field (runway) required to accelerate a multiengined aircraft to the critical engine failure speed, experience failure of the critical engine, and either (1) continue takeoff and have just enough runway left to execute a safe takeoff, or (2) elect to reject the takeoff and be able to stop the aircraft exactly at the end of the runway.[AIR1489-88]

critical flow This is a somewhat ambiguous term that signifies a point at which the characteristics of flow suffer a finite change. In the case of a liquid, critical flow could mean the point at which the flow regime changes from laminar to transitional. It more often is used to mean choked flow. In the case of a gas, critical flow may mean the point at which the velocity at the vena contracta attains the velocity of sound, or it may mean the point at which the flow is fully choked.

critical frequency 1. The limiting frequency below which magnetoionic wave components are reflected and above which they penetrate through an ionized medium (plasma) at vertical incidence.

[NASA] 2. The frequency below which a traveling wave of a given mode cannot be maintained in a given waveguide.

critical item 1. Any item: (a) the failure of which would critically affect system safety, cause the system to become unavailable or unable to achieve mission objectives, or cause extensive/expensive maintenance and repair. NOTE: High value are reliability critical for design to life cycle cost. (b) the failure of which would prevent the acquisition of data to evaluate system safety, availability, mission success, or need for maintenance/repair. (c) which has stringent performance requirement(s) in its intended application relative to state-of-the-art techniques for the item. (d) the sole failure of which causes system failure. (e) which is stressed in excess of specified derating criteria. (f) which has a known operating life, shelf life, or environmental exposure such as vibration, thermal, propellant; or a limitation which warrants controlled surveillance under specified conditions. (g) which is known to require special handling, transportation, storage, or test precautions. (h) which is difficult to procure or manufacture relative to state-of-the-art techniques. (i) which has exhibited an unsatisfactory operating history. (j) which does not have sufficient history of its own, or similarity to other items having demonstrated high reliability, to provide confidence in its reliability. (k) which has past history, nature, function, or processing with a deficiency warranting total traceability. (l) which is used in large quantities (typically, at least 10 per cent of the configured items electronic parts count).[ARD50010-91] 2. An item that has a limitation to warrant controlled surveillance under specified conditions.[AIR1916-88]

criticality A relative measure of the consequences of a failure mode and its frequency of occurrences.[ARD50010-91] [AIR1916-88]

critical Mach number See critical velocity.

critical mass The amount of concentrated fissionable material that can just support a self-sustaining fission reaction.[NASA]

critical part A life-limited part, whose failure would likely impair the ability of the aircraft to continue safe flight either through damage to the structure or controls or injury to the crew.[AIR 1872-88]

critical point The thermodynamic state in which liquid and gas phases of a substance coexist in equilibrium at the highest possible temperature. At higher temperature than the critical no liquid phase can exist.[NASA]

critical pressure 1. In rocketry, the pressure in the nozzle throat for which the isentropic weight flow rate is maximum. The pressure of a gas at the critical point, which is the highest pressure under which a liquid can exist in equilibrium with its vapor.[NASA] 2. The equilibrium pressure of a fluid that is at its critical temperature.

critical-pressure ratio The ratio of downstream pressure to upstream pressure which corresponds to the onset of turbulent flow in a moving stream of fluid.

critical Reynolds number See critical velocity, Reynolds number.

critical speed 1. A speed of a rotating system that corresponds to a resonance frequency of the system.[AIR1489-88] 2. The speed of angular rotation at which a shaft becomes dynamically unstable due to lateral resonant vibration. See also critical velocity.

critical strain The amount of prior plastic strain that is just sufficient to trigger recrystallization when a deformed metal is heated.

critical temperature 1. The temperature above which a substance cannot exist in the liquid state regardless of the pressure. As applied to reactor overheat or afterheat, the temperature at which the least resistant component of the reactor core begins to melt down. As applied to materials, the temperature at which a change in phase takes place causing an appreciable change in the properties of the material.[NASA] 2. The temperature of a fluid above which the fluid cannot be liquefied by pressure alone.

critical velocity 1. In rocketry, the speed of sound at the conditions prevailing at the nozzle throat. Used for critical Mach number, critical Reynolds number, and critical speed.[NASA] 2. For a given fluid, the average linear velocity marking the upper limit of streamline flow and the lower limit of turbulent flow at a given temperature and pressure in a given confined flowpath.

critical viewing sector Those geometric sectors of an instrument which include the normal viewing angles of the various crew members for the particular instrument.[ARP1161-91]

crop calendars Schedules for the maturation and harvesting of seasonal crops.[NASA]

crop dusting The application of fungicides or insecticides in powder form to a crop, usually from a low flying aircraft.[NASA]

crop inventories Numerical estimates of vegetable, fruit, and other commercial farm products based on the analysis of photography or imagery from aircraft or satellites made during periodic passes during the growth cycle. [NASA]

Crop Inventories by Remote Sensing See AgRISTARS project.

cross-assembler An assembler program run on a larger host computer and used for producing machine code to be executed on another usually smaller, computer.

cross-axis acceleration See transverse acceleration.

crossbar micrometer An instrument for determining differences in right ascension and declination of celestial objects; it consists of two bars mounted perpendicular to each other in the focal plane of a telescope and inclined at 45° to the east-west path of the stars.

cross-bleed Ducted air from the compressor section of an operating engine to another engine such that this compressed air can be used for staring or other purposes.[ARP906A-91]

cross bleed starting The staring of one engine utilizing cross bleed air from another operating engine.[ARP906A-91]

cross-compiler A computer program run on a larger host computer and used for translating a high level language program into the machine code to be executed on another computer.

cross controls To position aircraft controls in an uncoordinated fashion (for example, to deflect the right aileron downward while holding right rudder). [ARP4107-88]

cross drum boiler A section header or box boiler in which the axis of the horizontal drum is at right angles to the center lines of the tubes in the main bank.

cross faults See geological faults.

cross feed The feeding or transfer of fuel or oil from engine to engine or from tank to tank on a multiengine aircraft. [ARP4107-88]

cross flow A flow going across another flow, as a spanwise flow over a wing. [NASA]

crosshair An inscribed line or a thin hair, wire or thread used in the optical path of a telescope, microscope or other optical device to obtain accurate sightings or measurements; sometimes, a pair of hairs at right angles are used, which is the original source of the term.

crosshead 1. A sliding block that moves back and forth between guides and that contains a wrist pin for converting reciprocating motion to rotary motion. 2. A device designed to extrude material at an angle which is used most extensively at the discharge end of an extruder in a wire-coating operation.

cross-linking The establishing of chemical links between the molecular chains in polymers through the process of chemical reaction, electron bombardment, or vulcanization.[ARP1931-87]

cross-modulation Carrier and signal harmonics of one or more channels appearing in other channels of a system; in the case of a large number of cross-modulation products, the resultant cross-talk noise approaches the characteristics of fluctuation noise (AM).

crossover frequency 1. The frequency at which a dividing network delivers equal power to upper-band and lower-band channels. 2. The frequency at which the asymptotes to the constant-amplitude and constant-velocity portions of the frequency-response curve of an acoustic recording system intersect. Also known as transition frequency; turnover frequency.

crossover network A selective network that divides the audio-frequency output of an amplifier into two or more bands of frequencies to supply two or more loudspeakers. Also known as dividing network; loudspeaker dividing network.

cross-pointer indicator An aircraft instrument having two crossing needles that indicate the position of the aircraft with respect to an instrument landing system localizer and glide slope.[ARP 4107-88]

cross polarization The component of the electric field vector normal to the desired polarization component.[NASA]

cross section 1. Measure of the effectiveness of a particular process expressed either as area (geometric cross section) which would produce the observed results, or as ratio.[NASA] 2. For a given confined flowpath or a given elongated structural member, the dimensions, shape or area determined by its intersection with a plane perpendicular to its longitudinal axis. 3. In characterizing interactions between moving atomic particles, the probability per unit flux and per unit time that a given interaction will occur.

cross-sectional area of a conductor The sum of the cross-sectional areas of its component wires, that of each wire being measured perpendicular to its axis.[ARP1931-87]

cross sensitivity, cross-axis sensitivity See transverse sensitivity.

crosstalk 1. Electrical disturbances in a communication channel as a result of coupling with other communication channels.[NASA] 2. The unwanted signals in a channel that originate from one or more channels in the same communication system. 3. Signals electrically coupled from another circuit, usually undesirably, but sometimes for useful purposes.

crosswind 1. When used concerning the traffic pattern, the word means "crosswind leg". See also traffic pattern. 2. When used concerning wind conditions, the word means a wind not parallel to the runway or the path of an aircraft. See also crosswind component.[ARP 4107-88]

crosswind component The wind component measured in knots at 90 deg to the longitudinal axis of the runway. [ARP4107-88]

cross wind gear A landing gear configuration which permits alignment of the wheels to compensate for crab angle of the air vehicle (due to crosswinds) at touchdown and for rollout. A fixed angle may be set in to the gear or freedom to caster and align with the direction of aircraft travel may be utilized.

[AIR1489-88]

cross-wire weld A resistance weld made by passing a controlled electric current through the junction of a pair of crossed wires or bars; used extensively to make mesh or screening.

crown 1. The part of a drill bit that is inset with diamonds. 2. The vertex of a structural arch or arched surface. 3. The domed top of a furnace or kiln. 4. The central portion of sheet material that is slightly thicker than at the edges. 5. Any raised central portion of a nominally flat surface.

crown glass An optical glass of alkali-lime-silica composition with index refraction usually 1.5 to 1.6.

crown sheet In a firebox boiler, the plate forming the top of the furnace.

CRT See Cathode Ray Tube.

CRT display 1. Cathode ray tube (video screen). 2. Hardware display on a cathode-ray-tube alphanumerically and/or graphically.

CRT line width The width of a luminous intensity distribution at 50% of its maximum amplitude.[ARP1782-89]

crucible A pot or vessel made of a high-melting-point material, such as a ceramic or refractory metal, used for melting metals and other materials.

crude oil Unrefined petroleum.

cruise phase See mishap, phase of flight.

crush 1. A casting defect caused by displacement of sand as the mold is closed. 2. Buckling or breaking of a section of a casting mold caused by incorrect register as the mold is closed.

crustal dynamics See geodynamics.

cryochemistry The study of chemical phenomena in very low temperature environment.[NASA]

cryogenic Any process carried out at very low temperature, usually considered to be -60 °F (-50 °C) or lower.

cryogenic cooling Use of cryogenic fluids to reach temperatures near absolute zero.[NASA]

cryogenic fluid A liquid which boils below -123 °Kelvin (-238 °F, -150 °C) at one atmosphere absolute pressure.

cryogenic rocket propellants Rocket fuels, oxidizers, or propulsion fluids which are liquid only at very low temperatures.[NASA]

cryogenics The study of the methods of producing very low temperatures. The study of the behavior of materials and processes at cryogenic temperatures. [NASA]

cryogenic wind tunnel Wind tunnel employing a cryogenic environment and utilizing independent control over Mach number, Reynolds number, aeroelastic effects, and model-tunnel interactions. [NASA]

cryometer A thermometer for measuring very low temperatures.

cryoscope A device for determining the freezing point of a liquid.

cryosorption See sorption.

cryostat An apparatus for establishing the very low-temperature environment needed for carrying out a cryogenic operation.

cryotron Device based upon the principle that superconductivity established at temperatures near absolute zero is destroyed by the application of a magnetic field.[NASA]

cryptography The science of preparing messages in a form which cannot be read by those not privy to the secrets of the form.[NASA]

crystal dislocations Types of lattice imperfections whose existence in metals is postulated in order to account for the phenomenon of crystal growth and of slip, particularly for the low value of shear stress required to initiate slip. [NASA]

crystal lattices Three-dimensional, recurring patterns in which the atoms of crystals are arranged.[NASA]

crystalline fracture A type of fracture

surface appearance characterized by numerous brightly reflecting facets resulting from cleavage fracture of a polycrystalline material.

crystal oscillator A device for generating an a-c signal whose frequency is determined by the properties of a piezoelectric crystal.

crystal spectrometer An instrument that uses diffraction from a crystal to determine the component wavelengths in a beam of x-rays or gamma rays.

CSMA/CD See carrier sense multiple access with collision detect.

CSU (cumulative sum) algorithm See compressor.

CTI See comparative tracking index.

cubicle 1. Any small room or enclosure. 2. An enclosure, usually free standing, that houses high-voltage electrical equipment.

cultural resources Archaeological and historical sites.[NASA]

cumulative/chronic fatigue See fatigue.

cumulative dose The total amount of penetrating radiation absorbed by the whole body, or by a specific region of the body, during repeated exposures.

cup fracture A mixed mode fracture in ductile metals, usually observed in round tensile specimens, in which part of the fracture occurs under plane-strain conditions and the remainder under plane-stress conditions, such that in a round tensile bar one of the mating fracture surfaces looks like a miniature cup and the other like a truncated cone. Also known as cup-and-cone fracture.

cupping 1. The first step in deep drawing. 2. The fracture of severely worked rod or wire where one of the fracture surfaces is roughly conical and the other cup-shaped.

cure To change the physical properties of a material by chemical reaction, by the action of heat and catalysts, alone or in combination with or without pressure.[ARP1931-87]

cured angle The angle at which the cords are set in the tire after cure. In practice it may be measured from centerline of axle or from the plane of symmetry.[AIR1489-88]

cure-date The date the compounded, uncured elastomer is vulcanized to produce an elastomeric product.[AS1933-85]

cured end count Cords per inch in the tire ply fabric after cure.[AIR1489-88]

cure rate A measure of the rate of polymerization based upon the increasing hardness of the sealant with time. The time required to reach the specified hardness varies widely with the type of sealant and with the temperature and relative humidity. Testing is performed at standard conditions.[AS7200/1-90]

cure time 1. A measure of the time required for the polymerization to advance to a given hardness at standard conditions.[AS7200/10] 2. The time required for a sealant to polymerize and develop its full physical/mechanical properties. In practice, however, it is the time required to reach a hardness called for by a specification.[AIR4069-90]

Curie Abbreviated Ci. The standard unit of measure for radioactivity of a substance; it is defined as the quantity of a radioactive nuclide that is disintegrating at the rate of 3.7×10^{10} disintegrations per second.

Curie temperature the temperature in a ferromagnetic material above which the material becomes substantially nonmagnetic.[NASA]

curing 1. Allowing a substance such as a polymeric adhesive or poured concrete to rest under controlled conditions, which may include clamping, heating or providing residual moisture, until it undergoes a slow chemical reaction to reach final bond strength or hardness. 2. In thermoplastics molding, stopping all movement for an interval prior to releasing mold pressure so that the molded part has sufficient time to stabilize.

curing type sealants Sealants that polymerize either by mixing two parts together or which polymerize by exposure to moisture from the air—to form a cured, nonreversible polymeric elastomer.[AIR4069-90]

curl (vectors) A vector operation upon a vector field which represents the rotation of the filed, related to the circulation of the field at each point.[NASA]

current 1. Rate of transfer of electricity expressed in amperes.[ARP1931-87] 2. The rate of flow of an electrical charge in an electric circuit analogous to the rate of flow of water in a pipe.

current amplification For a given amplifier, the ratio of current delivered to the output circuit to the corresponding current supplied to the input circuit.

current carrying capacity The maximum current which a wire or cable with a given circular mil area is capable of carrying without exceeding its temperature limit.[ARP1931-87]

current limitation The ability of a protective device to reduce the short-circuit peak current to a value less than that which would be available if no protective device were in the circuit.[ARP 1199A-90]

current loop (20mA) A serial transmission standard widely used for v.d.u.'s and teletypes. 0 and 1 are represented by the absence or presence of a current (20mA).

current meter Any of a wide variety of devices for measuring a-c or d-c electric current—including moving-coil, moving iron, electronic and electrodynamic instruments. See velocity-type flowmeter.

current rating 1. The maximum continuous electrical flow of current recommended for a given wire in a given configuration, expressed in amperes.[ARP1931-87] 2. The maximum current which a device is designed to conduct for a specified time at a specified operating temperature.[ARP914A-79]

current-responsive element (fusible element) A current-responsive element is that part of the fuse or limiter which carries current and melts when the current exceeds a predetermined value.[ARP1199A-90]

currents (oceanography) See water currents.

current-to-pressure transducer (I/P) A device which receives an analog electrical signal and converts it to a corresponding air pressure.

current transformer An instrument transformer designed to have its primary winding connected in series with a circuit carrying the current being measured or controlled.

current word address (CWA) The memory address of a word that is currently being operated on.

curvature of field A defect in an optical lens or system which causes the focused image of a plane field to lie along a curved surface rather than a flat plane.

curve-fit The process of determining the coefficients in a curve by mathematically fitting a given set of data to that curve class; for example, linear curve-fit, or nth-order polynomial curve-fit.

curvilinear coordinates See spherical coordinates.

custody transfer The act of transferring ownership of a fluid for money or the equivalent.

custom LSI A large scale integrated circuit designed for a specific purpose and which hence has a dedicated function.

cutoff 1. An act or instance of shuffling something off. Specifically, in rocketry, an act or instance of shuffling off the propellant flow in a rocket or stopping the combustion of the propellant.[NASA] 2. The parting line on a compression-molded plastics part. Also known as flash groove; pinch-off. 3. The point in the stroke of an engine where admission of the working fluid to the cylinder is shut off. 4. The time required to shut off the flow of working fluid into a cylinder.

cutoff tool A lathe tool with a narrow cutting edge used to sever a finished piece from remaining bar stock. Also known as parting tool.

cutoff valve A quick-acting valve used to stop the flow of working fluid into an engine cylinder.

cutoff wheel A thin abrasive wheel used to cut stock or to make slots in a part.

cutout, connector A hole or group of holes cut in a panel, case or chassis for the purpose of mounting a connector. [ARP914A-79]

cut-out pressure The pressure at which the sequence of reduced flow of a component or system begins.[AIR1916-88]

cutout speed See speed, starter cutoff.

cutout switch A speed sensing device, located on either the starter or the engine side of the engaging mechanism, used to terminate starter operation at a predetermined cutout speed.[ARP906A-91]

cutter 1. A rotary cutting tool of cylindrical form with either single or double cutting ends and with either a straight or tapered shank. The teeth may be straight or helical and be either right hand or left hand. It is used for milling slots, keyways and pockets where the ordinary arbor type of milling cutter cannot be used.[ARP480A-79] 2. A cutting tool, especially a rotary, toothed cutting wheel.

cutter bar A supporting member for the cutting tool in a lathe or other machine tool.

cut-through resistance The ability of a material to withstand mechanical pressure, usually a sharp edge, without penetration of the impinging item through the material.[ARP1931-87]

cutting angle The angle between the face of a cutting tool and the uncut stock surface.

cutting edge 1. In a diamond or ceramic tool, the point or edge of the insert material that actually cuts the work. 2. Generally, the sharpened edge of any cutting tool that contacts the work during machining.

cutting fluid In a metal-cutting operation, any liquid that is introduced into the area where the tool contacts the work, especially a liquid used to provide lubrication at the cutting edge, to carry away the heat generated during machining, and to flush out chips or other machining debris. Some cutting fluids have chemical compounds which react with the tool and material being cut to enhance cutting action.

cutting speed The relative velocity between cutting tool and workpiece along the main direction of cutting. Also known as peripheral speed.

cutting tool A sharp-edged single-point or toothed tool that comes in contact with the workpiece and removes stock in a machining operation. Also known as cutter.

cutting torch A device for producing a controlled flame which has an additional supply line for introducing a jet of oxygen into the flame; it cuts metal and other materials by first heating a small area, then rapidly oxidizing and melting the material along a thin line when the jet of oxygen is turned on. Usually a special plasma torch is needed for stainless steel because of its oxidation resistance.

CVR Cockpit Voice Recorder.

CVT (current value table) See multiplex processor.

CWA See current word address.

cyanide emission See CN emission.

cyaniding A surface-hardening process similar to carbonitriding that produces a carbon- and nitrogen-rich surface layer on steel by immersing parts in a bath of molten cyanide salts; can also be done in the gas phase.

cybernetics 1. The study of methods of control and communication which are common to living organisms and machines.[NASA] 2. The branch of learning which brings together theories and studies on communication and control in living organisms and machines.

cycle 1. The complete sequence including reversal of the flow of an alternating current.[ARP1931-87] 2. The complete sequence of values of a periodic quantity that occur during a period.[NASA] 3. An interval of space or time in which one set of events or phenomena is completed. 4. Any set of operations that is repeated regularly in the same sequence. The operations may be subject to variations in each repetition. 5. In any repetitive variable process, variation of a given variable through one complete range of values. 6. To run a machine through a complete set of operating steps. 7. The fundamental time interval for operations inside the computer. 8. A condition in a sequential circuit; from an initial, unstable state the circuit passes through more unstable states before reaching a stable state.

cycle, aircraft operating A completed take-off and landing sequence. NOTE: Touch and go landings are counted as aircraft operating cycles.[ARD50010-91]

cycle, engine operating A completed engine thermal cycle including the application of takeoff power.[ARD50010-91]

cycle, gear extension In a retractable landing gear system, the transition cycle from the up and locked (stowed) position to the down and locked (extended) position.[AIR1489-88]

cycle, gear retraction In a retractable landing gear system, the transition cycle from the gear down and locked (extended) position to the up and locked (stowed) position.[AIR1489-88]

cycle-index The number of times a cycle has been executed or the difference, or the negative of the difference, between the number that has been executed and the number of repetitions desired.

cycle redundancy check (CRC) An error detection scheme, usually hardware implemented, in which a check character is generated by taking the remainder after dividing all the serialized bits in a block of data by a predetermined binary number. This remainder is then appended to the transmitted date and recalculated and compared at the receiving point to verify data accuracy.

cycle stealing Data transferred over the data bus during a direct memory access while little disruption occurs to the normal operation of the microprocessor.

cycle time 1. The time required by a computer to read from or write into the system memory. If system memory is core, the read cycle time includes a write-after-read (restore) subcycle. 2. Cycle time is often used as a measure of computer performance, since this is a measure of the time required to fetch an instruction.

cyclic 1. Helicopter control mechanism for periodically varying the blade angle of each rotor, producing a tilt in the tip-path plane and effecting motion in a desired direction.[ARP4107-88] 2. A condition of either steady-state or transient oscillation of a signal about the nominal value.

cyclic adenosine monophosphate See cyclic AMP.

cyclic AMP A nucleotide which is implicated as an intracellular messenger in a wide variety of cellular processes. Prototypically it acts as a molecular transducer of nonsteroid signals from outside the cell to relevant cellular enzymes by a series of reactions.[NASA]

cyclic code A form of gray code, used

for expressing numbers in which, when coded values are arranged in the numeric order of real values, each digit of the coded value assumes its entire range of values alternately in ascending and descending order.

cyclic compounds In organic chemistry, compounds containing a ring of atoms. [NASA]

cyclic redundancy check (CRC) An error-checking technique in which a checking number is generated by taking the remainder after dividing all the bits in a block (in serial form) by a predetermined binary number. Can easily be achieved by shift operations.

cyclic redundancy check character (CRC) A character used in a modified cyclic code for error detection and correction.

cyclic shift A shift in which the data moved out of one end of the storing register are re-entered into the other end, as in a closed loop.

cycling Periodic repeated variation in a controlled variable or process action. See also cycles.

cyclograph A device for electromagnetically sorting or testing metal parts by means of the pattern produced on a cathode-ray tube when a sample part is placed in an electromagnetic sensing coil; the CRT pattern is different in shape for different values of carbon content, case depth, core hardness or other metallurgical properties.

cyclones (equipment) See centrifuges.

cyclotron A device that utilizes an alternating electric field between electrodes positioned in a constant magnetic field to accelerate ions or charged subatomic particles to high energies.

cyclotron frequency Frequency at which a charged particle orbits in a uniform magnetic field. It depends on the charge to mass ratio of the particle times the magnetic field. While the frequency is independent of the particle energy, Lamor orbit increases with energy.[NASA]

cyclotron radiation The electromagnetic radiation emitted by charged particles as they orbit in a magnetic field. The radiation arises from the centripetal acceleration of the particle as it moves in a circular orbit.[NASA]

cyclotron resonance Energy transfer to charged particles in a magnetic field from an alternating-current electric field whose frequency is equal to the cyclotron frequency.[NASA]

cyclotron resonance device Microwave amplifier based n the interaction between electromagnetic waves and transverse electron streams moving along helical trajectories.[NASA]

cylinder 1. A domed, closed storage tank for hot water. Also known as storage calorifier. 2. A strong, thick-walled container for storing and transporting compressed gases. 3. A round, straight-walled cavity, closed at one or both ends, that a piston rides in to convert the potential energy in pressurized gas to linear mechanical motion and power, or to utilize mechanical power to compress a gas.

cylinder, actuating Sometimes called jack, ram, or strut. A linear motion device in which the thrust or force is proportional to the effective cross-sectional area and the pressure differential. [ARP243B-65]

cylinder, balanced An actuating cylinder in which the effective thrust producing area is equal in both directions. [ARP243B-65]

cylinder block A massive piece of metal, usually made by casting, that contains the piston chambers of a multicylinder engine or compressor. Also known as block; engine block.

cylinder bore The inside diameter of a piston chamber.

cylinder, disk All like-numbered tracks on a disk pack; a portion of the disk which can be recorded or reproduced without moving the heads.

cylinder, double acting A cylinder with provisions for applying fluid pressure at each end, and thus capable of exerting a force in either direction. [ARP243B-65]

cylinder, fixed end A cylinder which is held in a rigid position.[ARP243B-65]

cylinder, flight control An actuating cylinder designed for in flight actuation of aerodynamic surfaces.[ARP243B-65]

cylinder head The cap, which usually has a specially shaped recess, used to close the end of a piston chamber in a reciprocating engine, pump or compressor; usually, it provides valve openings, spark-plug taps and other penetrations necessary for machine operation.

cylinder, landing gear A tubular structural member which houses the piston, orifice, oil, gas, and other elements of an air-oil shock absorber or the mechanical elements of a mechanical system. Carries lugs and/or provisions for various attachments on the exterior.[AIR 1489-88]

cylinder liner A separate cylindrical sleeve that is inserted into a piston chamber to provide a cylinder wall with properties different from those of the cylinder block. Normally used to furnish a better wearing material for piston rings than the block, i.e. a cast iron liner in an aluminum block.

cylinder, rotating end A cylinder mounted to permit limited rotary movement about a fixed point.[ARP243B-65]

cylinder, single acting A cylinder in which fluid pressure is introduced in one end so that fluid force is exerted in one direction only. Gravity, spring forces, or other means are used to accomplish the return stroke.[ARP243B-65]

cylinder, swivel end A cylinder with one or both ends provided with a joint which not only allows oscillation of the cylinder but which also incorporates stationary fluid connections.[ARP243B-65]

cylinder, transfer A device for transmitting fluid pressure from one circuit to another without intermixture of fluid between the circuits.[ARP243B-65]

cylindrical afterbodies See afterbodies.

cylindrical cam A mechanism consisting of a cylinder which rotates on its longitudinal axis and causes linear motion parallel to that axis in a cam follower which rolls in a groove cut in the cylindrical surface.

cylindrical lens A lens which is cylindrical in cross section, so it is curved in one direction but not in the perpendicular direction, used to expand a laser beam into a plane of light.

cylindrical plasmas Magnetic self-attraction of parallel electric currents causing constriction of a conducting plasma through which a large current is flowing.[NASA]

cylindrical waves Waves in which the wave fronts are coaxial cylinders. [NASA]

D

D/A See digital-to-analog.

DAEMO (data analysis) See data processing; data reduction.

daisy chain 1. A serial interconnection of devices. Signals are passed from one device to another, generally in the order of high priority to low priority. 2. A method of propagating signals along a bus, often used in applications in which devices are connected in series.

daisywheel printer Using a rotating wheel to type, a printer providing slow but good quality print output.

Dalton's law The empirical generalization that for many so called perfect gases, a mixture of these gases will have a pressure equal to the sum of the partial pressures that each of the gases would have as a sole component with the same volume and temperature, provided there is no chemical interaction. [NASA]

DAMA See demand assignment multiple access.

damage, accidental Physical deterioration of an item caused by contact or impact with an object or influence which may/may not be a part of the aircraft, or by improper manufacturing or maintenance practices.[ARD50010-91]

damage assessment Estimate of injury and loss to components, subsystems, or entire systems, as well as the cost of repairs or replacement to restore serviceability.[NASA]

damage, environmental Physical deterioration of an item's strength or resistance to failure as a result of chemical interaction with its climate or environment.[ARD50010-91]

damage factor A relative number assigned to indicate a defined amount or unit of engine component or piece part life usage; e.g., LOF counts, hot section factors.[ARP1587-81]

damage tolerance The ability of the airframe and engine structure to resist failure due to the presence of flaws, cracks, or other damage for a specific period of unrepaired usage.[ARD50010-91]

damp To suppress oscillations or disturbances.[AIR1489-88]

damped frequency See frequency, damped.

damped wave A wave in which the source amplitude diminishes with each succeeding cycle.

dampener A device for progressively reducing the amplitude of spring oscillations after abrupt application or removal of a load.

damper 1. A device or system for the purpose of suppressing oscillations or disturbances. May be hydraulic or mechanical.[AIR1489-88] 2. A device for introducing a variable resistance for regulating the volumetric flow of gas or air: a) butterfly type damper—A single blade damper pivoted about its center; b) curtain type damper—A damper, composed of flexible material, moving in a vertical plane as it is rolled; c) flat type damper—A damper consisting of one or more blades each pivoted about one edge.; d) louvre type damper—A damper consisting of several blades each pivoted about its center and linked together for simultaneous operation; e) slide type damper—A damper consisting of a single blade which moves substantially normal to the flow.

damper, drag A damper for the purpose of damping drag forces. Sometimes built into a landing gear drag strut. [AIR1489-88]

damper, hop A damper designed into a landing gear system for the purpose of suppressing pitching oscillations of the bogie assembly in the bogie type gear. [AIR1489-88]

damper loss The reduction in the static pressure of a gas flowing across a damper.

damper, shimmy A damper designed into a landing gear system for the purpose of suppressing shimmy or oscillation of the wheel system. Especially on gear configuration with castering provisions such as nose or tail wheel.[AIR 1489-88]

damping 1. The suppression of oscillations or disturbances; the dissipation of energy with time. Used for damping factor, damping in pitch, damping in roll, damping in yaw, elastic stability, and jet damping.[NASA] 2. Reducing or eliminating vibrations, especially reducing noise or reverberations by using sound-absorbing materials.

damping behavior The characteristic of a tire which tends to damp out periodically applied forces.[AIR1489-88]

damping factor In any damped oscillation, the ratio of the amplitude of any given half-cycle to the amplitude of the succeeding half-cycle. See also damping.

damping, friction See coulomb damping.

damping, hydraulic A damping system in which a hydraulic fluid is utilized. External applied forces cause the oil to be moved through an orifice or restriction. Resisting force is proportional to the square of the velocity.[AIR1489-88]

damping in yaw See damping.

damping magnet A permanent magnet used in conjunction with a moving conductor to produce an opposing torque when there is relative motion between the magnet and the conductor; a secondary function is to dissipate kinetic energy resulting from eddy currents that may be induced in the moving conductor.

damping pitch See damping; pitch (inclination).

damping ratio 1. The ratio of actual damping to critical damping. It may be expressed as the ratio of output under static conditions to twice the output at the lowest frequency where a 90° phase shift is observed.[AIR1489-88] 2. The ratio of the deviations of the indicator following an abrupt change in the measurand in two consecutive swings from the position of equilibrium, the greater deviation being divided by the lesser. The deviations are expressed in angular measure.

damping, structural Also solid damping. Damping due to internal friction within the material itself. It is independent of frequency and proportional to the maximum stress of the vibration cycle.[AIR1489-88]

damping tachometer generator A generator which is designed to be energized from a single phase voltage source and to deliver a generated output voltage essentially proportional to speed and energizing voltage. In practice residual voltages at zero speed exist. The frequency of this output voltage is the same as the fundamental energizing voltage frequency.[ARP667-82]

damping, viscous This is damping encountered by bodies moving a moderate speed through a fluid. The resisting force is proportional to the velocity. [AIR1489-88]

dark adaptation The process by which the iris and retina of the eye adjust to allow maximum vision in dim illumination, following exposure of the eye to a relatively brighter illumination. [NASA]

dark ambient Any ambient light level that contributes less than 1% of the display luminance value measured by a luminance measurement device focused on the display surface. This ambient contribution can result from reflections off the display surface or any other means.[ARP4067-89]

dark current The current that flows in photosensitive detectors when there is no incident radiant flux (total darkness).

d'Arsonval galvanometer A galvanometer made by suspending a light coil of wire on thin gold or copper ribbons in the field of a permanent magnet; when current is carried to the coil via the suspending ribbons, the coil rotates, and the amount of rotation is indicated by reflecting a beam of light from a small mirror carried on the coil onto a fixed linear scale. Also known as light-beam galvanometer.

d'Arsonval movement The mechanism of a permanent-magnet moving-coil instrument such as a d'Arsonval galvanometer.

DAS See data acquisition system.

dashpot 1. A snubbing device within a fluid operated unit which operates at the extremity of the stroke by displacing operating fluid through a restricted passage, as sometimes used in an actuating cylinder.[ARP243B-65] 2. A type of shock absorber which utilizes resistance of oil being forced through an orifice or restriction to oppose an impact load. Differs from air-oil shock absorber in that no air charge system is utilized.[AIR1489-88] 3. A fluid-filled cylinder containing a loose-fitting piston that is used to damp vibratory motion or to change the effect of a sharp change in load from an instantaneous change in position to a more gradual change.

dashpot relay A timing device relying upon the restrictive action of an orifice upon a fluid to provide the delay. When the relay coil is energized, the armature piston moves against a reservoir of fluid, forcing it through a restriction. This slows down the action. Timing is achieved by variations in orifice size.

Dassault Mystere 50 aircraft See Mystere 50 aircraft.

DAST program A NASA program which uses the Firebee 2 target drone aircraft as a test bed for getting flight data on research wings. The drone is launched from the wing of a B52 and recovered by parachute. The program's purpose is the study of flight loads and load control.[NASA]

data 1. Information of any type. 2. A common term used to indicate the basic elements that can be processed or produced by a computer.

data acquisition The function of obtaining data from sources external to a microprocess or a computer system, converting it to binary form, and processing it.

data acquisition system A system used for acquiring data from sensors via amplifiers and multiplexers and any necessary analog to digital converters.

data adaptive evaluator/monitor See data processing; data reduction.

data analysis See data processing; data reduction.

data averaging An optional mode of operation for an automatic data logger which allows readings from two or more data acquisition channels to be averaged in each scan or, alternatively, readings from each of several channels to be averaged over a preselected number of successive scans.

data bank A comprehensive collection of data, for example, several automated files, a library, or a set of loaded disks. Synonymous with database.

database 1. Any body of information. 2. A specific set of information available to a computer. 3. A collection of interrelated data stored together with controlled redundancy to serve one or more applications; the data are stored so that they are independent of programs that use the data; a common and controlled approach is used in adding new data and in modifying and retrieving existing data within a data base. A system is said to contain a collection of databases if they are disjointed in structure.

database management A system that provides meaningful information from the data included in a database.

data base management systems Software products that control data structures containing interrelated data stored so as to optimize accessibility and control, minimize redundancy, and offer multiple views of the data to various applications programs.[NASA]

data block In TELEVENT, a short section of memory in which data relating to events or operating programs are stored. See also alphanumeric display.

data bus The highway connecting the various microcomputer components carrying the data signals.

data capture (logging) The systematic collection of data to use in a particular data processing routine, such as monitoring and recording temperature changes over a period of time.

data channel A bidirectional data path between I/O devices and the main memory of a digital computer. Data channels permit one or more I/O operations to proceed concurrently with computation thereby enhancing computer performance.

data code A structured set of characters used to represent the data items of a data element, for example, the data codes 1,2,...,7 may be used to represent the data items Sunday, Monday,...,Saturday.

data collection The act of bringing data from one or more points to a central point.

data communication The transmission of data from one point to another.

data compression The elimination of redundant data without loss of information; a few standard telemetry data compression algorithms are ZFN, ZVP, ZVA, FFN, FFP, FFA, FVP, and FVA.

data converter Any of numerous devices for transforming analog signals to digital signals, or vice versa.

data directory A listing of data stored in a database.

data display module A device which stores computer output and translates this output into signals which are distributed to a program-determined group of lights, annunciators, numerical indicators, and cathode ray tubes in operator consoles and remote stations.

data distributor A manually or automatically controlled unit that is used to distribute specific data channels to quick-look devices.

data element A scalar, array, or structure.

data error A deviation from correctness in data, usually an error, which occurred prior to processing the data.

data file In a computer, a portion of memory allocated to a specific set of organized data, including codes that identify the file name and sometimes the file type. Also referred to as data set.

data gathering See data collection.

data graph An exhibit of relationship between sets of numeric data, either as points with coordinates or lines, used to extract data values by interpolating or extrapolating or to simply illustrate a trend or formula.[AS4159-88]

data handling See data processing.

data highway A communication link between separate stations tied with a multidrop cable and/or optical connections. It eliminates a need for separate, independently wired data links. Each station on a highway can function independently.

data input/output unit (DI/OU) A device that interfaces to the process for the sole purpose of acquiring or sending data.

data integration Taking data from multiple sources and merging the data into a single data file.[NASA]

data link 1. Digital communication of data which may be initiated automatically or manually from the aircraft or ground.[ARP4102/13-90] 2. Communication channel or circuit used to transmit data from a sensor to a computer, a readout device or a storage device.[NASA] 3. Equipment which permits the transmission of information in data format. 4. Facility for transmission of information. Also used to refer to layer 2 of the Open Systems Interconnection definitions. 5. A fiber optic signal transmission system which carries information in digital or analog form. The term usually refers to short distance communications, spanning distances less than a kilometer.

data logger 1. A system or subsystem with a primary function of acquisition and storage of data in a form that is suitable for later reduction and analysis, such as computer- language tape. 2. A computer system designed to obtain data from process sensors and to provide a log of the data. Many data loggers can carry out some filtering and linearising of the data.

data logging Recording of data about events that occur in time sequence. See also data collection.

data management A general term that collectively describes those functions of the control program that provide access to data sets, enforce data storage conventions, and regulate the use of input/output devices.

data processing 1. Application of procedures, mechanical, electrical, computational, or other whereby data are changed from one form to another. [NASA] 2. The execution of a systematic sequence of operations performed upon data. Synonymous with information processing.

data processing equipment Machines for handling information in a sequence of reasonable apportions.[NASA]

data processing system A network of machine components capable of accepting information, processing it according to plan, and producing the desired results.

data processor A device capable of performing data processing, for example, a desk calculator, punched-card machine, or computer. See also data processing equipment.

data protection Any method to preserve computer data from destruction or misuse. Backing up computer files is one example.

data reduction 1. Transformation of observed values into useful, ordered, or simplified information.[NASA] 2. The process of transforming masses of raw test or experimentally obtained data, usually gathered by automatic recording equipment, into useful, condensed, or simplified intelligence.

data set (DS) 1. A collection of data in one of several prescribed arrangements to which the system has access. 2. A device which performs the modulation/demodulation and control functions necessary to provide compatibility between data processing equipment and communications facilities. See also subset.

data signaling rate In communications, the data transmission capacity of a set of parallel channels. The data signaling rate is expressed in bits per second.

data simulation The use of statistical or physical models to produce synthetic data for testing purposes.[NASA]

data sink In communications, a device capable of accepting data signals from a transmission device. It also may check these signals and originate error control signals. Contrast with data source.

data smoothing The mathematical process of fitting a smooth curve to dispersed data points.[NASA]

data source In communications, a device capable of originating data signals for a transmission device. It also may accept error control signals. Contrast with data sink.

data stream The movement of a group of measurements in one multiplexer.

data structures 1. The organization of computer memory used to represent information in a computer program or data base.[NASA] 2. The storage of related data in computer memory by use of arrays, records or data lists.

data terminal equipment Either a data source or a data sink, or both.

data transfer (computers) The technique used by the hardware manufacturer to transmit data from computer to storage device or from storage device to computer, usually under specialized program control.[NASA]

data transmission The sending of data from one part of a system to another part.

data type Any one of several different types of data, such as integer, real, double precision, complex, logical, and Hollerith. Each has a different mathematical significance and may have different internal representation.

datum 1. A point, direction or level used as a convenient reference for measuring angles, distances, heights, speeds or similar attributes. 2. Any value that serves as a reference for measuring other values of the same quantity.

datum plane A permanently established reference level, usually average sea level, used for determining the value of a specific altitude, depth sounding, ground elevation or water-surface elevation. Also known as chart datum; datum level; reference level; reference plane.

daughter A nuclide formed as a result of nuclear fission or radioactive decay.

dawsonite A mineral consisting of aluminum sodium carbonate.[NASA]

day A unit of time whose exact value depends on the system of time measurement being used—apparent solar time, mean solar time, apparent sidereal time, universal time, ephemeris time, or atomic time; except for atomic time, the basis of the definition is the period during which the earth makes one rotation on its axis; for general purposes, one day equals 24 h or 86,400 s.

dB See decibel.

DBS (satellites) See direct broadcast satellites.

d controller See controller, derivative (D).

DDC See Direct Digital Control.

DDCMP See Digital Data Communications Message Protocol.

DDN Defense Data Network[AS4159-88]

deacclimatization See acclimatization.

dead band 1. Free motion at the wheel(s) under control (output) over which the controlling system (input) has no authority. See also backlash. May also apply to other mechanical systems. May also be defined as movement of the input control which produces no response in the output.[AIR1489-88] 2. The complete range of values of the controlled temperature in which no corrective action will be taken by the controller.[ARP89C-70]

dead center 1. Either of two positions of a crank where the turning force between the crank and its connecting rod are zero; it occurs when the centerline of the crank and the centerline of the connecting rod lie in the same plane. 2. A nonrotating center for holding a rotating workpiece.

dead-end shutoff A nonstandard term used to refer to control valve leakage. Refer to ANSI/FCI 70-2 for specifications of leakage classifications.

dead-end tube A tube with a closed end—for example, a tube in a porcupine boiler.

dead-front switchboard A switching panel constructed so that all of the live terminations are made on the rear of the panel.

dead length, actuator Nominal length between attach points of an actuator when in the fully retracted condition less the available stroke.[AIR1489-88]

deadman's brake A safety device that automatically stops a vehicle when the driver does not have his foot on the pedal; it is also used on other operator-controlled mechanisms such as cranes and lift trucks.

deadman's handle A hand grip or handle that an operator must squeeze or press on continuously to keep a machine running.

dead man timer (DMT) Circuit monitors operation of the processor cards and signals if a failure occurs.

dead reckoning (DR) 1. Navigational method for determining the location of an aircraft based upon time, heading, and estimated true airspeed (with allowances made for winds and compass errors) flown since last observed position.[ARP4107-88] 2. In navigation, determination of position by advancing a previous known position for courses and distances.[NASA]

dead room See anechoic chamber.

dead-stick Without power, as in dead-stick landing, or into land dead-stick.[ARP4107-88]

dead time 1. The interval of time between initiation of an input change or stimulus and the start of the resulting response. 2. Any definite delay deliberately placed between two related actions in order to avoid overlap that might cause confusion or to permit a particular different event, such as a control decision, switching event or similar action to take place. See time, dead.

dead-time correction A correction applied to an instrument reading to account for events or stimuli actually occurring during the instrument's dead time.

deadweight gage A device used to generate accurate pressures for the purpose of calibrating pressure gages; freely balanced weights (dead weights) are loaded on a calibrated piston to give a static hydraulic pressure output.

dead zone Also called dead band. A range of values around the set point. When the controlled variable is within this range, no control action takes place. See zone, dead.

deaeration Removing a gas—air, oxygen or carbon dioxide, for example—from a liquid or semisolid substance, such as boiler feedwater or food.

debug 1. To locate and correct any errors in a computer program. 2. To detect and correct malfunctions in the computer itself. Related to diagnostic routine. 3. To submit a newly designed process, mechanism or computer program to simulated or actual operating conditions for the purpose of detecting and eliminating flaws or inefficiencies.

debuggers System programs that enable computer programs to be debugged.

debugging 1. A process to detect or remedy inadequacies, preferably prior to operational use.[ARD50010-91] 2. A process to detect and remedy inadequacies. With respect to software, it is the development process to locate, identify, and correct programming mistakes, including omissions in the software.[AIR 916-88] 3. The process of detecting, diagnosing and then correcting program faults. See also checkout.

debugging aid routine A routine to aid programmers in the debugging of their routines. Some typical aid routines are storage printout, tape print-out, and drum print-out routines.

debugging on-line See on-line debugging.

deburr To remove burrs, fins, sharp edges and the like from corners and edges of parts or from around holes, by any of several methods, often involving the use of abrasives.

Debye length A theoretical length which describes the maximum separation at which a given electron will be influenced by the electric field of a given positive ion.[NASA]

Debye temperature. See specific heat.

decade A group or assembly of ten units,

e.g., a counter which counts to ten in one column or a resistor box which inserts resistance quantities in multiples of powers of 10.

decade scaler A scaling device that produces one output pulse for each ten input pulses.

decalescence Darkening of a metal surface upon undergoing a phase transformation on heating; the phenomenon is caused by isothermal absorption of the latent heat of transformation.

decanting Boiling or pouring off liquid near the top of a vessel that contains two immiscible liquids or a liquid-solid mixture which has separated by sedimentation, without disturbing the heavier liquid or settled solid.

decarburization The loss of carbon from the surface of a ferrous alloy as a result of heating in a medium that reacts with the carbon at the alloy surface. [ARP700-65]

decarburizing Removing carbon from the surface layer of a steel or other ferrous alloy by heating it in an atmosphere that reacts selectively with carbon; atmospheres that are relatively rich in water vapor or carbon dioxide are typical deoxidizing atmospheres.

decay 1. Decrease of a radioactive substance because of nuclear emission of alpha or beta particles, positrons, or gamma rays.[NASA] 2. The spontaneous transformation of a nuclide into one or more other nuclides either by emitting one or more subatomic particles or gamma rays from its nucleus or by nuclear fission; radioactive decay of a specific nuclide is characterized by its half life—the time it takes for one-half of the original mass to spontaneously transform.

decay time The time in which a voltage or current pulse will decrease to one tenth of its maximum value. Decay time is proportional to the time constant of the circuit.

Decca navigation A long range, ambiguous, two dimensional navigation system using continuous wave transmission to provide hyperbolic lines of position through the radio frequency phase comparison techniques from four transmitters.[NASA]

decelerating electrode An intermediate electrode in an electron tube which is maintained at a potential that induces decelerating forces on a beam of electrons.

deceleration 1. The act or process of moving, or of causing to move, with decreasing speed. Sometimes called negative acceleration.[AIR1489-88] [NASA] See also impact deceleration.

decelerometer An instrument for measuring the rate at which speed decreases.

decentralized 1. To distribute the functions among several authorities. 2. Decentralized maintenance distributes maintenance functions among areas of responsibilities or areas of the physical plant.

decibel (dB) 1. The unit used to express differences of power level. The decibel is ten times the common logarithm of the power ratio. It is used to express power loss in cables. A 3dB loss approximates a 50% decrease. A 2dB loss approximates a 27% decrease.[ARP1931-87] 2. A unit of measure used to express amplitude ratio of control input and output. Decibels = 20 $\log_{10}$ (amplitude out/amplitude in).[AIR1916-88] 3. A unit for measuring relative strength of a signal parameter, such as power, voltage, etc. The number of decibels is twenty (ten for power ratio), times the logarithm (base 10) of the ratio of the measured quantity to the reference level. The reference level must always be indicated, such as 1 milliamp for current ratio. See also power level.

decibel meter An instrument calibrated in logarithmic steps and used for measuring power levels, in decibel units, of audio or communication circuits.

decimal 1. Pertaining to a characteristic or property involving a selection, choice, or condition in which there are ten possibilities. 2. Pertaining to the numeration system with a radix of ten. 3. See binary code decimal.

decimal balance A type of balance having one arm ten times as long as the other, so that heavy objects can be balanced with light weights.

decimal coded digit A digit or character defined by a set of decimal digits, such as a pair of decimal digits specifying a letter or special character in a system of notation.

decimal digit In decimal notation, one of the characters 0 through 9.

decimal notation A fixed radix notation, where the radix is ten; for example, in decimal notation, the numeral 576.2 represents the number 5x10 squared plus 7x10 to the first power, plus 6x10 to the zero power, plus 2x10 to the minus 1 power.

decimal number A number, usually of more than one figure, representing a sum, in which the quantity represented by each figure is based on the radix of ten. The figures used are 0, 1, 2, 3, 4, 5, 6, 7, 8, and 9.

decimal numbering system A system of reckoning by 10 or the powers of 10 using the digits 0-9 to express numerical quantities.

decimal numeral Aa decimal representation of a number.

decimal point The radix point in decimal representation.

decimal-to-binary conversion The process of converting a number written to the base ten, or decimal, into the equivalent number written to the base two, or binary.

decision The selection of a response designed to achieve a desired goal after having made a judgment as to the significance and priority of available information.[ARP4107-88]

decision delay Failure to select a response in a timely manner due to an anomaly of attention or motivation. [ARP4107-88]

decision elements. See logical elements.

decision height (DH) The height at which a decision must be made during an ILS or PAR instrument approach either to continue the approach or to execute a missed approach.[ARP4107-88]

decision instruction An instruction that effects the selection of a branch in a program, for example, a conditional jump instruction.

decision, poor Selection of an inappropriate response (assuming adequate information and time to decide) due to an anomaly of attention or motivation. [ARP4107-88]

decision table A table of all contingencies that are to be considered in the description of a problem, together with the actions to be taken. Decision tables are sometimes used in place of flow charts for problem description and documentation.

DECK See digital to analog converter, an electronic device that converts a digital signal, often from a computer, into a proportional analog voltage or current.

deck A collection of cards, commonly a complete set of cards which have been punched for a definite service or purpose.

deck run, catapult The distance from the end of the catapult power stroke to the end of the deck.[AIR1489-88]

deck scale A low-profile weighing device used for moderate to heavy loads—up to 20,000 lb; because the load platform is 2 to 10 in. above floor level, loads must be lifted onto the scale or ramps must be provided to enable wheeled vehicles to move onto the platform and off again; the frame of a deck scale rests directly on the existing floor, rather than in a pit, and most models can be moved to different locations as needed.

declaration As used in many programming languages, a statement that is not to be executed, but usually is used for descriptive purposes.

declination Angular distance north or south of the celestial equator; the arc of an hour circle between the celestial equator and a point on the celestial sphere, measured northward or southward from the celestial equator through 90 degrees, and labeled N or S to indicate the direction of measurement. [NASA]

declinometer An instrument similar to a surveyor's compass used for determining the variation of magnetic directions from true directions; the horizontal circle is constructed so that the line of sight can be aligned with the magnetic needle or with any other desired setting.

decode 1. To apply a code so as to reverse some previous encoding. 2. To determine the meaning of individual characters or groups of characters in a message. 3. To determine the meaning of instructions from the status of bits which describes the instruction, command, or operation to be performed.

decoder 1. Device for translating electrical signals into predetermined functions. In computer operations, networks or devices in which one or two or more possible outputs results from a prescribed combination of inputs.[NASA] 2. A device which determines the meaning of a set of signals and initiates a computer operation based thereon. 3. A matrix of switching elements which selects one or more output channels according to the combination of input signals present. Contrasted with encoder and clarified by matrix. 4. A device used to change computer data from one coded format to another.

decollate The separation of multi-part computer forms.

decommissioning Disposal or deactivation of equipment or sites whose usefulness has diminished to a point where it is no longer required for its original purpose.[NASA]

decommutation A reversal of the commutation process; separation of information in a commutated data stream into as many independent information channels as were originally commutated.

decommutator 1. Equipment for separation, demodulation, or demultiplexing commutated signals.[NASA] 2. Equipment for the separation or demultiplexing of commutated signals.

decompression Any method for relieving pressure.

decompression sickness Effects produced by the evolution of gas (usually nitrogen) from tissues and fluids in the body due to changes in barometric pressure. (a) Bends: Manifestations of mild to severe pain, usually in the larger joints, due to nitrogen evolving from the blood. (b) Chokes: Deep and sharp pain centrally located under the sternum due to nitrogen evolving from the blood and locating in the smaller blood vessels of the lungs and producing a dry, nonproductive cough. (c) Neurological Manifestations: The effects of nitrogen evolving from the blood and locating in the brain or spinal cord. Symptoms may include blurred vision, blind spots, flickering lights, headaches, or unilateral numbness or tingling. (d) Skin Manifestations: The effects of nitrogen evolving from the blood and locating in subcutaneous tissue. Symptoms may include itching, hot or cold sensations, tingling, or the appearance of a mottled rash.[ARP4107-88] 2. A disorder experienced by deep sea divers and aviators caused by reduced atmospheric pressure and evolved gas bubbles in the body, marked by pain in the extremities, pain in the chest (chokes), occasionally leading to severe central nervous symptoms and neurocirculatory collapse. See also bends (physiology).[NASA]

decontamination Removing or neutralizing an unwanted chemical, biological or radiological substance.

decoupling The technique of reducing process interaction through coordination of control loops.

decoupling control A technique in which interacting control loops are automatically compensated when any one control loop takes a control action.

decrement 1. The quantity by which a variable is decreased. 2. A specific part of an instruction word in some binary computers, thus a set of digits.

decremeter An instrument for measuring the damping of a train of waves by determining its logarithmic decrement.

decryption Translating computer data from an unreadable format to a readable format.

dedicated In data processing, a device that performs only one function.

deep drawing A press operation for forming cup-shaped or deeply recessed parts from sheet metal by forcing the metal to undergo plastic deformation between dies without substantial thinning.

deep stall A stabilized high angle of attack assumed by an aircraft after it reaches the stall angle.[ARP4107-88]

deep well injection (wastes) Storage of liquid wastes, particularly chlorohydrocarbons, by injection into subsurface geologic strata for long term isolation from the environment.[NASA]

default 1. The value of an argument, operand, or field assumed by a program if a specific assignment is not supplied by the user. 2. The alternative assumed when an identifier has not been declared to have one of two or more alternative attributes.

default directory In MS-DOS, the directory in which the computer looks for files if no directory is specified.

default drive In MS-DOS, the disk drive the computer will use to search for files if no disk drive is specified.

defect 1. Any nonconformance of the unit or product with specified requirements. Defects will normally be grouped into one or more of the following classes but may be grouped into other classes or subclasses with these classes: (a) Critical Defect—A defect that constitutes a hazardous or unsafe condition, or as determined by experience and judgment could conceivably become so, relative to its deleterious effect on the prime intended function, or mission capability of the aircraft or its operating personnel. (b) Major Defect—A defect, other than critical, that could result in failure or materially reduce the usability of the unit or part for its intended purpose. (c) Minor Defect—A defect that does not materially reduce the usability of the unit or part for its intended purpose, or is a departure from standard but which has no significant bearing on the effective use or operation of the unit or part.[ARD50010-91] 2. A departure of any quality characteristic from its specified or intended value that is severe enough to constitute cause for rejecting the object or service.

Defense Meteorological Satellite Program See DMSP satellites.

definition 1. The resolution and sharpness of an image, or the extent to which an image is brought into sharp relief. 2. The degree with which a communication system reproduces sound images or messages.

deflagration 1. The chemical decomposition (burning) of a material in which the reaction front advances into the unreacted material at less than sonic velocity.[AIR913-89] 2. A sudden or rapid burning, as opposed to a detonation or explosion.[NASA]

deflashing Removing fins or protrusions from the parting line of a die casting or molded plastics part.

deflecting electrode An intermediate electrode in an electron tube whose surrounding electric field induces constant or variable deflecting forces on an electron beam.

deflecting force In a direct-acting recording instrument, the force produced at the marking device, for any position of the scale, by its positioning mechanism acting in response to the electrical quantity being measured.

deflecting yoke An assembly of one or more coils that induce a magnetic field to deflect an electron beam in a manner related to the oscillating frequency and magnitude of the current flowing through the coils.

deflection 1. The radial compression or deflection of a tire under load. This may be expressed by a finite value or as a ratio (percentage of the actual deflection measurement to the total available [(undeflected outside diameter less the wheel rim flange diameter) divided by two.][AIR1489-88] 2. Movement of a pointer away from its zero or null position. 3. Elastic movement of a structural member under load. 4. Shape change or change in diameter of a tubular member without fracturing the material.

deflection factor The reciprocal of the instrument sensitivity.

deflection polarity In an oscilloscope, the relationship between direction of electron-beam displacement and polarity of applied signal voltage.

deflectometer An instrument for determining minute elastic movements that occur when a structure is loaded.

deflector 1. Plate, baffle, or the like that divert something in its movement or flow.[NASA] 2. A device for changing direction of a stream of air or of a mixture of pulverized fuel and air.

defocus To cause a beam of electrons, light, x-rays or other type of radiation to depart from accurate focus at a specific point in space, ordinarily the surface of a workpiece or test object.

deformed beam lead thickness On a bonded device, the mean thickness of the beam lead in the bonded area. [AS1346-74]

defrost To remove ice from a surface, usually by melting or sublimation.

deg or ° See degree.

degas To remove dissolved, entrained or adsorbed gas from a solid or liquid.

degasification Removal of gases from samples of steam taken for purity tests. Removal of CO2 from water as in the ion exchange method of softening.

degasifier 1. An element or compound added to molten metal to remove dissolved gases. 2. A process or type of vessel that removes dissolved gases from molten metal.

degassing The deliberate removal of gas from a material, usually by application of heat under high vacuum. See also bakeout.[NASA]

degenerate matter A state of matter found in white dwarf stars and other ultrahigh-density objects in which the electrons follow Fermi-Dirac statistics, i.e. the matter reaches a density high enough so that the pressure increases more and more rapidly to the point where it becomes independent of the temperature and is a function of the density only, thereby departing from the classical laws of physics.[NASA]

degenerate waveguide modes A set of waveguide modes having the same propagation constant for all frequencies of interest.

degeneration 1. A gradual impairment in ability to perform.[ARD50010-91] 2. Negative feedback.

degenerative feedback See negative feedback.

degradation 1. The condition or status indicating impaired or deteriorating condition, function or physical state.[ARP1587-81] 2. Gradual deterioration in performance.[NASA]

degrees of freedom A mode of motion, either angular or linear, with respect to a coordinate system, independent of any other mode. A body in motion has six possible degrees of freedom, three linear and three angular.[NASA]

degradation failure Gradual shift of an attribute or operating characteristic to a point where the device no longer can fulfill its intended purpose.

degreasing An industrial process for removing grease, oil or other fatty substances from the surfaces of metal parts, usually by exposing the parts to condensing vapors of a polyhalogenated hydrocarbon solvent.

degree rise The amount of increase in temperature caused by the flow of electrical current through a wire.[ARP1931-87]

dehumidification Reducing the moisture content of air, which increases its cooling power.

deicing Using heat, chemicals or mechanical rupture to remove ice deposits, especially those that form on motor vehicles and aircraft at low temperatures or high altitudes.

Deimos A satellite of Mars orbiting at a mean distance of 23,500 kilometers.[NASA]

deionization time The time it takes for the grid in a gas tube to regain control of tube output after the anode current has been interrupted.

delamination 1. The separation of layers in a laminate through failure of the adhesive bond.[ARP1931-87] 2. Separation of a material into layers, especially a material such as a bonded laminate.

delay A pyrotechnic device which introduces a controlled time delay between initiation and functioning of an explosive device.[AIR913-89]

delay distortion A form of distortion in a transmitted radio wave that occurs when the rate of change of phase shift with frequency is not constant over the transmission-frequency range.

delayed combustion A continuation of combustion beyond the furnace. See also secondary combustion.

delayed perception See perception.

delayed response See response.

delay-interval timer A timing device which is electrically reset to delay energization or deenergization of a circuit for an interval of time up to 10 min following a specific event such as restoration of power after a power failure or turning a manual switch off.

delay line 1. A cable constructed so as to provide very low velocity of propagation with a specific electrical delay for transmitted signals.[ARP1931-87] 2. (computer storage) In electronic computers, devices for producing a time delay of a signal.[NASA] 3. A transmission medium which delays a signal passing through it by a known amount of time; typically used in timing events.

delay-line memory A type of circulating memory having a delay circuit as the chief element in the path of circulation.

delay-line register An acoustic or electric delay line, one or more words long, combined with appropriate input, output and circulation circuits.

delay modulation A method of data encoding for serial data transmission or recording; a logic ONE (or ZERO) is represented by a signal transition at midbit time and a logic ZERO (or ONE) followed by a logic ZERO (or ONE) is represented by a transition at the end of the first ZERO (or ONE) bit.

delay-on-make timer A timing device that holds its main contacts open for a preset period of time after it receives an initiating signal, then closes the contacts and allows current to flow in the main circuit; when the timer receives a stopping signal, the contacts open and after a short interval the timer automatically resets so it can repeat the cycle.

delimiter A character that separates, terminates, or organizes elements of a character string, statement, or program.

delta network A set of three circuit branches connected in series, end-to-end, to form a mesh having three nodes.

delta wing(s) 1. A symmetrical triangular wing having a low aspect ratio, tapered leading edge, and straight trailing edge.[ARP4107-88] 2. Triangularly shaped wings of aircraft.[NASA]

demand assignment multiple access A technique of assigning communication resources on an as needed basis' such as in satellite communications. [NASA]

demand meter Any of several types of instruments used to determine the amount of electricity used over a fixed period of time.

demodulation The process of retrieving intelligence (data) from a modulated carrier wave. The reverse of modulation.

demodulator 1. Electronic device which operates on an input of a modulated carrier to recover the modulating wave as an output.[NASA] 2. A device which recovers information from a carrier or subcarrier. A telemetry receiver has a demodulator; an FM discriminator is a demodulator.

demography Statistical study of human populations especially with reference to size, density, distribution, and vital data.[NASA]

demonstrated That which has been proven by the use of concrete evidence gathered under specified conditions. [ARD50013-91] [ARD50010-91]

demonstrated MTBF interval (THETA D) The probable range of true MTBF under test conditions; that is, an interval estimate of MTBF at a stated confidence level.[ARD50010-91]

demonstration The joint contractor and procuring activity effort to determine whether specific maintainability contractual requirements have been achieved.[ARD50010-91]

demultiplexer 1. The device which enables the telemetry operator to observe individual measurements from within a multiplexer. The opposite of a "multiplexer." 2. A device which separates two or more signals which have been multiplexed together for transmission through a single optical fiber. 3. A reverse multiplexer which allows the transfer of data from one microprocessor port to a number of output devices such as actuators.

demultiplexing Separation of two or more signals that were previously combined by a compatible multiplexer and transmitted over a single channel. [NASA]

Dendrochronology The use of annual growth rings in plant tissue to determine the age of the plant or tree. Used for tree ring dating.**densimeter** 1. Instrument for measuring the density of specific gravity of liquids, gases, or solids.[NASA] 2. An instrument for determining the density of a substance in absolute units, or for determining its specific gravity—that is, its relative density with respect to that of pure water. Also known as density gage; density indicator; gravitometer.

densitometer 1. Instrument for measuring the density of specific gravity of liquids, gases or solids.[NASA] 2. Instruments for the measurement of optical density (photographic transmission, photographic reflection, visual transmission, etc.) of a material, generally of a photographic image.[NASA] 3. An instrument for determining optical density of photographic or radiographic film by measuring the intensity of transmitted or reflected light.

density 1. Weight per unit volume of a substance.[ARP1931-87] 2. A physical property of materials measured as mass per unit volume. 3. The weight of a substance for a specified volume at a definite temperature, for example, grams per cubic centimeter at 20 °C. 4. Closeness of texture or consistency. 5. Degree of opacity, often referred to as optical density.

density bottle See specific gravity bottle.

density correction Any correction made to an instrument reading to compensate for the deviation of density from a fixed reference value; it may be applied because the fluid being measured is not at standard temperature and pressure, because ambient temperature affects density of the fluid in a fluid-filled instrument, or because of other similar effects.

density (rate/area) See flux density.

density transmitter An instrument used to determine liquid density by measuring the buoyant force on an air-filled float immersed in a flowing liquid stream.

Department of Defense Flight Information Publications (DoD FLIP) Publications used for flight planning, en route, and terminal operations. FLIPs are produced by the Defense Mapping Agency for worldwide use. En route charts and instrument approach procedure charts are incorporated in DoD FLIP for use in the National Airspace System (NAS).[ARP4107-88]

departure control A function of an approach control facility providing air traffic control service for departing IFR (and, under certain conditions, VFR) aircraft.[ARP4107-88]

dependability 1. A measure of the degree to which an item is operable and capable of performing its required function at any (random) time during a specified mission profile, given item availability at the start of the mission. (Item state during a mission includes the combined effects of the mission-related system R&M parameters but exclude non-mission time; see availability).[ARD50010-91]

dependent variables Variables considered as a function of other variables, the latter being called independent. [NASA]

depolarizers Optical components which scramble the polarization of light passing through them, effectively turning a polarized beam into an unpolarized beam.

deposit 1. Any substance intentionally laid down on a surface by chemical, electrical, electrochemical, mechanical, vacuum or vapor transfer methods. 2. Solid or semisolid material accumulated by corrosion or sedimentation on the interior of a tube or pipe.

deposited metal In a weldment, filler metal added to the joint during welding.

deposit gage Any instrument used for assessing atmospheric quality by measuring the amount of particulate matter that settles out on a specific area during a defined period of time.

deposition rate 1. The amount of filler metal deposited per unit time by a specific welding procedure, usually expressed in pounds per hour. 2. The rate at which a coating material is deposited on a surface, usually expressed as weight per unit area per unit time, or as thickness per unit time.

deposition sequence The order in which increments of a weld deposit are laid down.

depth gage An instrument or micrometer device capable of measuring distance below a reference surface to the nearest 0.001 in.; it is most often used to measure the depth of a blind hole, slot or recess below the normal part surface surrounding it, or to measure the height of a shoulder or projection above the adjacent part surface.

depth of engagement The radial contact distance between mating threads.

depth of fusion The distance from the original surface that the molten zone extends into the base metal during welding.

depth of thread The radial distance from crest to root of a screw thread.

depth perception See space perception.

derandomizer The circuit which removes the effect of data randomizing, thereby recovering data which had been randomized for tape storage.

derating 1. Using an item in such a way that applied stresses are below rated values. It is an intentional reduction of the stress/strength ratio in the application of an item, usually for the purpose of achieving a "reliability margin" in design which should reduce the occurrence of stress related failures. [ARD50010-91] 2. Using an item in such a way that applied stresses are below rated values.[AIR1916-88] 3. The lowering of the rating of an item in one stress field to allow an increase in another stress field.[AIR1916-88]

derivative 1. Mathematically it is the reciprocal of rate. 2. This control action will cause the output signal to change according to the rate at which input signal variations occur during a certain time interval.

derivative action A type of control-system action in which a predetermined relation exists between the position of the final control element and the derivative of the controlled variable with respect to time.

derivative control Change in the output that is proportional to the rate of change of the input. Also called rate control. See control action, derivative (D).

derivative control action (rate action) Control action in which the output is proportional to the rate of change of the input. See control action; see also control action, derivative.

derivative controller See controller, derivative.

derivative control mode A controller mode in which controller output is directly proportional to the rate of change of controlled variable error.

derivative time The time interval by which rate action advances the effect of proportional action on the final control element.

denier A term which describes the weight of a yarn, which in turn determines its physical size.[ARP1931-87]

depth filter A filter medium that retains contaminant, primarily within tortuous passages, at different levels within the filtration media. Other materials are frequently used in combination with wire to improve filtration.[AIR888-89]

depth of crimp The distance the crimp die indentor indents the conductor barrel or ferrule.[ARP914A-79]

derating factor A factor used to determine the acceptable reduced current carrying capacity of a wire when that wire is used in an environment or application other than that for which its original current carrying capacity was determined.[ARP1931-87]

descaling Removing adherent deposits from a metal surface, such as thick oxide from hot rolled or forged steel, or inorganic compounds from the interior of boiler tubes; it may be done by chemical attack, mechanical action, electrolytic dissolution or other means, alone or in combination.

descent phase See mishap, phase of flight.

describing function For a nonlinear element in sinusoidal steady state, the frequency response obtained by taking only the fundamental component of the output signal. The describing function depends on the frequency and on the amplitude of the input signal, or only on the amplitude of the input signal.

desertification The formation of a desert or the gradual expansion of a desertline into previously usable land, due to man-made or natural causes. [NASA]

design development test Tests conducted to establish or verify design concepts for items which have not been proved by previous use.[AS1426-80]

design eye position (DEP) A point fixed in relation to the aircraft structure (neutral seat reference point) at which the midpoint of the pilot's eyes should be located at the normal position. The DEP is the principal dimensional reference point for the location of flight deck panels, controls, displays, and external vision.[ARP4067-89]

design load 1. A specified load that a structural member or part should withstand without failing. It is determined by multiplying some particular load by an appropriate factor, usually the limit load multiplied by a factor of safety. [ARP4107-88] 2. The load for which a steam generating unit is designed, considered the maximum load to be carried.

design mission scenario(s) Those portions of the total mission scenario(s) selected for use in designing a particular system. Segments of the total mission scenario have been eliminated because they are contained within other segments, determined to be noncritical, determined to be redundant or for other similar reasons. The design mission scenario may be described in the same variety of ways as the total mission scenario (that is, summary-of-mission narrative, mission narrative, ribbon-in-the-sky, altitude/timeline curves, or design scenario timeline). Typically, it is described in all of these ways during the process.[ARP4107-88]

design pressure The maximum allowable working pressure permitted under the rules of the ASME Construction Code. See pressure, design.

design steam temperature The temperature of steam for which a boiler is designed.

design stress The maximum permissible load per unit area a given structure can withstand in service, including all allowances for such things as unexpected or impact loads, corrosion, dimensional variations during fabrication and possible underestimation of service loading.

design thickness The sum of thickness required to support service loads. This method of specifying material thickness is used particularly when designing boilers, chemical process equipment, and metal structures that will be exposed to atmospheric environments, soils or seawater.

design to cost A process whereby cost factors are determined and calculated for the life cycle of a product as an integral part of its design.[NASA]

desired course (CRS) The intended horizontal direction of travel, expressed as an angular distance from a reference direction (usually true or magnetic north).[ARP1570-86]

desired track The imaginary line on the earth's surface connecting successive points over which flight is desired. This line describes the great circle course between successive waypoints and is further defined by the intersection of a plane and the earth's surface when the plane passes through (2) successive waypoints and the center of the earth. See also ground track.[ARP1570-86]

desired track angle (DTK) (DSRTK) The clockwise angle from true north to an imaginary line or path on the earth's surface connecting successive points over which flight is desired. This line describes the great circle course between successive waypoints and is further defined by the intersection of a plane and the earth's surface when the plane passes through (2) successive waypoints and the center of the earth. Consequently, DTK is most meaningful when the airplane present position is close to the great circle path defined by the FROM-TO points.[ARP1570-86]

desired value See value, desired.

desk top publishing The computer merging of text and graphics to produce manuals and leaflets.

desorption 1. The process of removing sorbed gas.[NASA] 2. Removing adsorbed material.

destructive testing 1. Prolonged endurance testing of equipment or a specimen until it fails in order to determine service-life or design weakness.[ARD 50013-91] 2. Testing in which the preparation of the test specimen or the test itself may adversely affect the life expectancy of the unit under-test (UUT) or render the sample unfit for its intended use.[ARD50013-91] 3. Any method of determining a material property, functional attribute or operational characteristic which renders the test object unsuitable for further use or severely impairs its intended service life.

desynchronization (biology) The loss of synchronization between two or more rhythms so that they show independent periods.[NASA]

detachment A particular state of isolation in which man is separated or detached from his accustomed behavioral environment by inordinate physical and psychological distances. This condition may compromise his performance.[NASA]

detectability The quality of a measured variable in a specific environment that is determined by relative freedom from interfering energy or other characteristics of the same general nature as the measured variable.

detection-correction system 1. The type of fault-tolerant system wherein a failure or out-of-operating tolerance condition is detected and corrective action is taken automatically. This may involve switching to a standby system; or, if two or more systems are normally operating, correction may involve switching-out the failed channel. Inherent in this type of system is the existence of a finite time for detection and correction. With detection-correction systems, it is possible to use a model of an active system as a reference in order to extend the failure correction capability of the total system.[ARP1181A-85]

detector 1. Sensor, also an instrument employing a sensor to detect the presence of something in the surrounding environment i.e., wheel speed sensor; proximity sensor, etc.[AIR1489-88] 2. Sensor or instrument employing a sensor.[NASA] 3. A device which detects light, generating an electrical signal which can be measured or otherwise processed. See transducer.

detector-amplifier A device in which an optical detector is packaged together with electronic amplification circuitry.

detent A catch or lever that initiates or prevents movement in a mechanism, especially an escapement.

detergent A natural material or synthetic substance having the soaplike quality of being able to emulsify oil and remove soil from a surface.

deterioration Decline in the quality of a device, mechanism or structure over time due to environmental effects, corrosion, wear or gradual changes in material properties; if allowed to continue unchecked, deterioration often leads to degradation failure.

determination See measurement.

detonating cord Flexible tube containing a core of high explosive.[AIR913-89]

detonation 1. The extremely rapid chemical decomposition (explosion) of a material in which the reaction front advances into the unreacted material at greater than sonic velocity.[AIR913-89] 2. A rapid chemical reaction which propagates at a supersonic velocity. Used for Chapman-Jouget flame. [NASA]

detonation waves Shock waves that accompany detonation and have a shock front followed by a region of decreasing pressure in which the reaction occurs.[NASA]

detonator An explosive train component capable of initiating high order detonation in a subsequent high explosive component.[AIR913-89]

deuterium A heavy isotope of hydrogen having one proton and one neutron in the nucleus. Used for hydrogen 2.[NASA]

deuterium fluoride lasers See DF lasers.

deuterium fluorides Fluorides of deuterium, a heavy isotope of hydrogen. [NASA]

deuterium oxides See heavy water.

deuterons The nuclei of deuterium atoms.[NASA]

deutron detector A type of specialized radiation detector used in some nuclear reactors to detect the concentration of deuterium nuclei present.

developed boiler horsepower The boiler horsepower generated by a steam generating unit.

development system A system used to develop both the hardware and software for a microcomputer system. The development system may contain an editor, assembler and/or high level language, compiler, debugging and in-circuit emulation facilities.

development test A test performed by the developing agency to verify the operation or performance of a system or component design, or to produce data which will permit improving the design of the item under test.[ARD50013-91]

deviation 1. The difference between control or set point, and a value of process variable. 2. In quality control, any departure of a quality characteristic from its specified value. 3. A statistical quantity that gives a measure of the random error which can be expected in numerous independent measurements of the same value under the same conditions of measurement.

deviation alarm 1. Alarm that is set whenever the deviation exceeds the preset limits. 2. An alarm caused by a variable departing from its desired value by a specified amount.

deviation controller A type of automatic control device which acts in response to any difference between the value of a process variable and the instrument setpoint, independent of their actual values.

deviation ratio The ratio given by $M=f/f_{max}$, where f is the maximum frequency difference between the modulated carrier and the unmodulated carrier, and f_{max} is the maximum modulation frequency.

device 1. A component or assembly designed to perform a specific function by harnessing mechanical, electrical, magnetic, thermal or chemical energy. 2. Any piece of machinery or computer hardware that can perform a specific task. 3. A component in a control system, such as; primary element, transmitter, controller, recorder, or final control element.

device control character One of a class of control characters intended for the control of peripheral devices associated with a data processing or telecommunication system, usually for switching devices "on" or "off".

device controller A hardware unit that electronically supervises one or more of the same types of devices; acts as the link between the CPU and I/O devices.

device driver A program/routine that controls the physical hardware activities on a peripheral device; a device driver is generally the device-dependent software interface between a device and the common, device-independent I/O code in an operating system.

device flags One-bit registers which record the current status of a device.

device handler A program/routine that drives or services an I/O device; a device handler is similar to a device driver but provides more control and interfacing functions than a device driver.

device independence The ability to request input/output operations without regard to the characteristics of the input/output devices.

device lift On a bonded device, the vertical distance from the conductor film metallization to the dielectric surface of the semiconductor device. The device lift is also called the "bugging distance."[AS1346-74]

dewars Insulated thermos-like containers for cryogenic liquids, which can be designed to house detectors or lasers requiring cooling.

dewatering 1. Removal of water by draining, pumping, or other means. [NASA] 2. Removing water from solid or semisolid material—for instance, by centrifuging, filtering, settling or evaporation. 3. Removing water from a riverbed, pond, caisson or other enclosure by pumping or evaporation.

dew cell An instrument consisting of two bare electrical wires wound spirally around an electrical insulator and covered by wicking wetted with an aqueous solution containing an excess of LiCl; dew point of the surrounding atmosphere is determined by passing an electric current between the two wires, which raises the temperature of the LiCl solution until its vapor pressure is the same as that of the ambient atmosphere.

dewetting 1. Generally, loss of surface attraction between a solid and a liquid. 2. Specifically, flow of solder away from a soldered joint upon reheating.

dew point 1. The dew point is the temperature at which the air would become saturated (with respect to water) if cooled at constant pressure and without the addition or removal of water vapor. It is expressed in degrees Fahrenheit.[AIR1335-75] 2. Temperature at which water vapor beings to condense. [NASA]

dew-point recorder An instrument that determines dew- point temperature by alternately heating and cooling a metal plate and using a photocell to automatically detect and record the temperature at which condensed moisture appears and disappears on the target. Also known as mechanized dew-point meter.

dew-point temperature That temperature at which condensation of moisture from the vapor phase begins.

DF See deuterium fluorides.

DF lasers Gas lasers in which the active material is deuterium fluoride. [NASA]

DFT See diagnostic function test.

DG Directional Gyro.

DH Decision Height.

diagnostic 1. Pertaining to the detection and isolation of a malfunction or mistake. 2. Program or other system feature designed to help identify malfunctions in the system. An aid to debugging.

diagnostic alarm Alarm that is set whenever the diagnostic program reports a malfunction.

diagnostic function test (DFT) A program to test overall system reliability.

diagnostic message An error message in a programming routine to help the programmer identify the error.

diagnostic programs 1. A troubleshooting aid for locating hardware malfunctions in a system. 2. A program to aid in locating coding errors in newly developed programs. 3. Computer programs that isolate equipment malfunctions or programming errors.

diagnostic routine 1. A sequence of tests or fault tree logic designed to use data inputs and predetermined standards or operational limits to establish condition status and locate a malfunction or discrepancy.[ARP1587-81] 2. A routine used to locate a malfunction in a computer, or to aid in locating mistakes in a computer program. Thus, in general, any routine specifically designed to aid in debugging or trouble shooting. Synonymous with malfunction routine and related to debug.

diagnostic(s) 1. An analysis result pertaining to the detection and isolation of a malfunction or discrepancy.[ARP 1587-81] 2. Data analysis techniques which lead to the detection and identification of a deterioration, malfunction, or failure.[AIR1873-88] 3. Information concerning known failure modes and their characteristics that can be used in troubleshooting and failure analysis to help pinpoint the cause of a failure and aid in defining suitable corrective measures.

diagnostic sensitivity A measure of the threshold level at which a change of condition or functional status yields symptomatic indications with a given diagnostic routine or technique. The threshold level is an accumulation of all error contributions which input into the diagnostic routine or technique. [ARP1587-81]

diagnostics, or diagnostic software The program by which a computer or other programmable device or system can literally "check itself," to diagnose any defects that may be present.

diagnostic test The running of a machine program or routine for the purpose of discovering a failure or a potential failure of a machine element, and to determine its location or its potential location.

diagonal stay A brace used in fire-tube boilers between a flat head or tube sheet and the shell.

dial 1. Generally, any circular scale. 2. The graduated scale adjacent to a control knob that is used to indicate the value or relative position of the control setting.

dial indicator 1. Any meter or gage with a graduated circular face and a pivoted pointer to indicate the reading. 2. A type of measuring gage used to determine fine linear measurements, such as radial or lateral runout of a rotating member, by resting a feeler against a surface and noting the change in position of a pivoted pointer relative to the calibrated gage face as the part is rotated; the gage also can be adapted to other setups where precise relative position is to be determined.

diamagnetic material A substance whose specific permeability is less than 1.00 and is therefore weakly repelled by a magnetic field.

diamond-pyramid hardness A material hardness determined by indenting a specimen with a diamond-pyramid indenter having a 136° angle between opposite faces then calculating a hardness number by dividing the indenting load by the pyramidal area of the impression. Also known as Vickers hardness.

diamond-turned mirror A mirror in which the surface has been formed by machining away material with a diamond tool.

diamond wheel A grinding wheel for cutting very hard materials which uses synthetic diamond dust as the bonded abrasive material.

diaphragm 1. A thin, flexible disc that is supported around the edges and whose center is allowed to move in a direction perpendicular to the plane of the disc; it is used for a wide variety of purposes, such as detecting or reproducing sound waves, keeping two fluids separate while transmitting pressure or motion between them, or producing a mechanical or electrical signal proportional to the deflection produced by differential pressure across the diaphragm. 2. A partition of metal or other material placed in a header, duct or pipe to separate portions thereof.

diaphragm (anatomy) Musculomembranous partition separating the abdominal and thoracic cavities.[NASA]

diaphragm motor A diaphragm mechanism used to position a pneumatically operated control element in response to the action of a pneumatic controller or pneumatic positioning relay.

diaphragm seal A thin flexible sheet of material clamped between two body halves to form a physical barrier between the instrument and process fluid.

dichroic filter A filter which selectively transmits some wavelengths of light and reflects others. Typically such filters are based on multilayer interference coatings.

dichromate treatment A technique for producing a corrosion-resistant conversion coating on magnesium parts by boiling them in a sodium dichromate solution.

dictionary A list of code names used in a computer routine or system and their intended meaning in that routine or system.

didymium A mixture of rare earth elements that is freed from cerium. It was once regarded as an element but contains chiefly neodymium and praseodymium and is usually associated with lanthanum. It is used in coloring glass for optical filters.[NASA]

die 1. The movable block of characteristic shape by which solid material is formed or shaped in a forming operation.[ARP480A-79] 2. A base containing appropriate contours used to shape, form or establish a piece from a parent metal sheet.[ARP480A-79] 3. A tool, usually containing at least one cavity, that imparts shape to solid, molten or powdered metal, or to elastomers or plastics, primarily because of the shape of the tool itself; a die is used together with a punch or a matching die in such operations as stamping, forging, forming, blanking, die casting, plastics molding and coining; in certain operations—die casting, powder metallurgy and plastics forming, for instance—dies are sometimes referred to as molds.

die block A heavy block, usually of tool steel, into which the desired impressions are sunk, formed or machined, and which is bolted to the bed of a press.

die body The stationary part of a powder pressing or extrusion die.

die casting 1. A casting process in which molten metal is forced under pressure into the cavity of a metal mold. 2. A part made by this process.

die chaser One of the cutting parts of a threading die.

die clearance The amount of lateral clearance between mated die parts when the dies are closed; commonly expressed as clearance per side.

die cushion A press accessory located beneath or within a bolster or die block, and actuated by air, oil, rubber or springs, to provide additional motion or pressure during stamping.

die forging 1. The process of forming shaped metal parts by pressure or impact between two dies. 2. A part formed in this way.

die holder A plate or block mounted between the die block and press bed.

die insert A removable part of a die or punch.

dielectric 1. A material having electrical insulting properties.[ARP914A-79] 2. An insulator or nonconductor. (a) An insulating medium which intervenes between two conductors. (b) A material having the property that energy required to establish an electric field may be stored, and later recovered in whole or in part, as electrical energy.[ARP 1931-87] 3. An insulating material, or a material that can sustain an electric field with very little dissipation of power.

dielectric breakdown The voltage required to cause an electrical failure or breakthrough of the insulation. See breakdown voltage.[ARP1931-87]

dielectric coating An optical coating made up of one or more layers of dielectric (nonconductive) materials. The layer structure determines what fractions of incident light at various wave lengths are transmitted and reflected.

dielectric constant 1. Permittivity, capacitivity, specific inductivity capacity. (a) The dielectric constant K^1, is a measure of the stored energy impressed in the molecular structure of the dielectric material by an external electric field. (b) The ratio of the permittivity of the dielectric material to the permittivity of a vacuum. (c) The ratio of the capacitance of a capacitor filled with a given dielectric to that of the same capacitor having a vacuum for a dielectric.[ARP 1931-87] 2. A material characteristic expressed as the capacitance between two plates when the intervening space is filled with a given insulating material divided by the capacitance of the same plate arrangement when the space is filled with air or is evacuated.

dielectric loss The time rate at which electric energy is transformed into heat in a dielectric when it is subjected to a changing electric field.[ARP1931-87]

dielectric materials See dielectrics.

dielectrics Substances that contain few or no free charges and which can support electrostatic stresses.[NASA]

dielectric strength The voltage which an insulating material can withstand before breakdown occurs. It is usually expressed as a voltage gradient such as "volts per mil".[ARP1931-87] See also breakdown voltage rating and insulation resistance.

dielectric test Test which consists of the application of a voltage higher than the rated voltage for a specified time for the purpose of determining the adequacy against breakdown of the insulation under normal conditions.[ARP 1931-87]

dielectronic satellite lines See resonance lines.

die scalping Improving the surface quality of bar stock, rod tubing or wire by drawing it through a sharp-edged die to remove a thin surface layer containing minor defects.

die set A tool or tool holder consisting of a die base and punch plate for attaching matched upper and lower dies, and that can be inserted into a press and removed from it as a single unit.

diesinking Making a shaped recess in a working face of a die, usually by mechanical, electrochemical or spark discharge machining.

die slide A device that slides into and out of the bed of a power press, carrying the lower die and providing for easy access and improved safety in feeding stock or removing stamped parts.

die welding Forge welding using shaped dies.

difference The output equals the algebraic difference between the two inputs.

difference limen The increment in a stimulus which is barely noticed in a specified fraction of independent observations where the same increment is imposed.

differential 1. When applied to two position control action, it is the difference between the value of the controlled temperature at which the controller operates to one position and that value of controlled temperature at which it operates to the other position. When applied to a control with a deadband, it is the difference between the value of the controlled temperature at which the controller action in a given direction is started and the value at which it is stopped. The differential is not necessarily the same on both sides of null. [ARP89C-70] 2. Any arrangement of epicyclic gears that allows two driven shafts to revolve at different speeds, with the speed of the main driving shaft being the algebraic mean of the speeds of the driven shafts. Also known as differential gear.

differential amplifier A device which compares two input signals and amplifies the difference between them.

differential analyzer 1. Analog computer designed and used primarily for solving differential equations.[NASA] 2. A computer (usually analog) designed and used primarily for solving many types of differential equations.

differential delay The difference between the maximum and the minimum frequency delays occurring across a band.

differential gap The smallest increment of change in a controlled variable required to cause the final control element in a two-position control system to move from one position to its alternative position.

differential gap control See control, differential gap.

differential input The difference between the instantaneous values of two voltages both being biased by a common mode voltage.

differential input (to a signal conditioner) 1. An input in which both sides are isolated from the chassis and power supply ground. The signal is applied as a differential voltage across the two sides. 2. Allows an analog-to-digital converter to measure the difference between two input signals.

differential instrument Any instrument that has an output signal or indication proportional to the algebraic difference between two input signals.

differential mode interference See interference, normal mode.

differential motion A mechanism in which the net motion of a single driven element is the difference between motions that would be imparted by each of two driving elements acting alone.

differential (of a control) The difference between cut in and cut out points.

differential pressure 1. The difference in value between two functionally related pressures occurring simultaneously at different points, such as at opposite sides of an actuator piston. [AIR1916-88] 2. The difference in pressure between two points of measurement. 3. The static pressure difference generated by the primary device when there is no difference in elevation between the upstream and downstream pressure taps.

differential-pressure gage Any of several instruments designed to measure the difference in pressure between two enclosed spaces, independent of their absolute pressures.

differential-pressure transmitter Any of several transducers designed to measure the pressure difference between two points in a process and transmit a signal proportional to this difference, without regard to the absolute pressure at either point.

differential-pressure-type liquid-level meter Any of several devices designed to measure the head of liquid in a tank above some minimum level and produce an indication proportional to this value; alternatively, the head below some maximum level can be measured and similarly displayed.

differential pulse code modulation An efficient signal encoding method of reducing the transmission rate of digital signals. The basic principle of DPCM is to quantize code and transmit the difference between the actual sample and prediction value.[NASA]

differential quantum efficiency Used

in describing quantum efficiency in devices having nonlinear output/input characteristics, the slope of the characteristic curve is the differential quantum efficiency.

differential screw A type of compound screw which produces a motion equal to the difference in motion between the two components of the compound screw.

differential thermal analysis See thermal analysis.

differential windlass A windlass that has a barrel with two sections of different diameter; the pulling rope passes around one section, then through a pulley and around the other section; the pulley is attached to the load.

differentiator 1. In computer operations, a device whose output is proportional to the derivative of an input signal. In electronics, a transducer whose output wave form is the time derivative of its input waveform.[AIR1489-88] 2. A device whose output function is proportional to the derivative, i.e., the rate of change, of its input function with respect to one or more variables (usually with respect to time).

diffracted beam In x-ray crystallography, a beam of radiation composed of a large number of scattered rays mutually reinforcing one another.

diffracted wave The wave component existing in the primary propagation medium after an interaction between the wave and a discontinuity or a second medium; the diffracted wave coexists in the primary medium with incident waves and with waves reflected from suitable plane boundaries.

diffraction 1. The process by which the direction of radiation is changed so that it spreads into the geometric shadow region of an opaque or refractive object that lies in a radiation field.[NASA] 2. Deviation of light from the paths and foci prescribed by rectilinear propagation; phenomenon responsible for bright and dark bands found within a geometrical shadow. 3. A phenomenon associated with the scattering of waves when they encounter obstacles whose size is about the same order of magnitude as the wavelength; in effect, each scattering point produces a secondary wave superimposed on the unscattered portion of the incident wave, the intensity of the scattered wave varying with direction from the scattering point; diffraction effects form the basis for x-ray crystallography, and they also tend to produce aberrations that must be dealt with in the design and construction of high-quality acoustical and optical systems.

diffraction grating An array of fine, parallel, equally spaced reflecting or transmitting lines which diffract light into a direction characteristic of the spacing of the lines and the wavelength of the diffracted light.

diffraction-limited beam A beam with a far-field spot size dependent only on the theoretical diffraction limit, which is the function of output wavelength divided by output aperture diameter.

diffraction propagation Wave propagation around objects, or over the horizon, by diffraction.[NASA]

diffraction radiation Electromagnetic radiation excited by an electron flux passing near a diffractive, periodic structure, such as a wiggler magnet in a free electron laser.[NASA]

diffraction x-ray machine An apparatus consisting of an x-ray tube, power supply, controls and auxiliary equipment used in the study of crystals, semiconductors and polymeric materials.

diffused-semiconductor strain gage A component used in manufacturing transducers, principally diaphragm-type pressure transducers, that consists of a slice of silicon about 2.5 to 22 mm in diameter into which an impurity element such as boron has been diffused;

modern photolithographic-masking techniques make it possible to simultaneously produce hundreds of full four-arm Wheatstone bridge patterns, complete with leadwire soldering pads, on a single slice of silicon about 50 to 75 mm (2 to 3 in.) in diameter.

diffuser 1. Specially designed duct, chamber, or section, sometimes equipped with guide vanes, that decreases the velocity of a fluid, as air, and increases its pressure, as in jet engines, wind tunnels, etc.[NASA] 2. A duct, chamber or enclosure in which low-pressure, high-velocity flow of a fluid, usually air, is converted to high-pressure, low-velocity flow. 3. As applied to oil or gas burners, a metal plate with openings so placed as to protect the fuel spray from high velocity air while admitting sufficient air to promote the ignition and combustion of fuel. Sometimes termed impeller.

diffuse radiation Radiant energy propagating in many different directions through a given small volume of space; to be contrasted with parallel radiation.[NASA]

diffuse reflection That in which the light is reflected in all directions.[ARP 798-91]

diffuse reflection factor The ratio of the diffusely reflected light to the incident light.[ARP798-91]

diffuse-specular Surfaces which are essentially diffuse but contain an outer layer of glazed material which reflects specularity. Porcelain-enamel is a common example.[ARP798-91]

diffuse transmission That in which the transmitted light is emitted in all directions from the transmitting body.[ARP798-91]

diffuse transmission factor The ratio of the diffusely transmitted light to the diffuse incident light.[ARP798-91]

diffusing surfaces and media Surfaces which break up the incident light and distribute it more or less in accordance with lambert's cosine law of emission, as for example, rough plaster and white glass.[ARP798-91]

diffusion 1. In an atmosphere, or in any gaseous system, the exchange of fluid parcels between regions, in apparently random motions of a scale too small to be treated by the equations of motion. In materials, the movement of atoms of one material into the crystal lattice of an adjoining material, e.g., penetration of the atoms in a ceramic coating into the lattice of the protected metal. In ion engines, the migration of neutral atoms through a porous structure incident to ionization at the emitting surface. Used for diffusion effect and perfusion.[NASA] 2. Conversion of gas-flow velocity into static pressure, as in the diffuser casing of a centrifugal fan. 3. The movement of ions from a point of high concentration to low concentration. 4. Migration of atoms, molecules or ions spontaneously, under the driving force of compositional differences, and using only the energy of thermal excitation to cause atom movements.

diffusion coefficient The absolute value of the ratio of the molecular flux per unit area to the concentration gradient of a gas diffusing through a gas or a porous medium where the molecular flux is evaluated across a surface perpendicular to the direction of the concentration gradient.[NASA]

diffusion effect See diffusion.

diffusion pump A vacuum pump in which a stream of heavy particles such as oil or mercury vapors carries gas molecules out of the vacuum chamber.

diffusivity A measure of the rate of diffusion of a substance, expressed as a diffusivity coefficient K.[NASA]

digital 1. A method of measurement using precise quantities to represent variables. 2. Binary. 3. A reference to the representation of data by discrete pulses, as in the presence or absence of a signal level to indicate the 1's and 0's of binary data. 4. A type of readout in which the data is displayed as discrete, fully-informed alphanumeric characters.

digital back-up An alternate method of digital process control initiated by use of special purpose digital logic in the event of a failure in the computer system.

digital circuits See digital electronics.

digital computer 1. Computers which operate with information, numerical or otherwise, represented in a digital form.[NASA] 2. A computing device that uses numerical digits to represent discretely all variables. 3. A computer in which discrete representation of data is mainly used. 4. A computer that operates on discrete data by performing arithmetic and logic processes on these data. Contrast with analog computer.

digital controller A control device consisting of a microprocessor plus associated A/D input converters and D/A output converters; it receives one or more analog inputs related to current process variables, uses the digitized information to compute an output signal using a predetermined control algorithm, and converts the result to an analog signal which operates the final control element; the device also may be adapted to furnish additional outputs such as alarms, totalizer signals and displays.

digital data Data represented in discrete discontinuous form, as contrasted with analog data represented in continuous form. Digital data is usually represented by means of coded characters, for example, numbers, signs, symbols, etc.

Digital Data Communications Message Protocol (DDCMP) A character-oriented communications protocol standard.

digital delay generator An electronic instrument which can be programmed digitally to delay a signal by a specific interval—time delay generator.

digital differential analyzer 1. An incremental computer in which the principal type of computing unit is a digital integrator whose operation is similar to the operation of an integrating mechanism. 2. A differential analyzer that uses digital representation for the analog quantities.

digital electronics The use of circuits in which there are usually only two states possible at any point. The two states can represent any of a variety of binary digits (bits) of information. [NASA]

Digital Equipment Corporation (DEC) Manufacturer of the PDP-11 series computer systems and peripheral devices.

digital filter 1. Computational means of attenuating undesired frequencies in sets of time-dependent data.[NASA] 2. An algorithm which reduces undesirable frequencies in the signal.

digital indicator A device that displays the value of a measured variable in digitized form; in most instances, the measurement range is not displayed simultaneously, which is considered an inherent disadvantage.

digital input A number value input. See input, digital.

digital logic A signal level is represented as a number value with a most significant and least significant bit. Binary digital logic uses numbers consisting of strings of 1's and 0's.

digital manometer A manometer equipped with a sonar device which measures column height and produces a digitized display.

digital millivoltmeter Many modern aircraft use high impedance digital millivoltmeters. The high impedance eliminates the need for controlling input resistance.[AIR46-90]

digital motor See stepping motor.

digital multiplexer A data selection device that permits sharing a common information path between multiple groups of digital devices, such as from a computer CPU to any of several groups of digital output devices.

digital readout An electrically powered device which interprets a continuously variable signal and displays its amplitude, or another signal attribute, as a series of numerals or other characters that correspond to the measured value and can be read directly; the accuracy of measurement is limited by the decimal position of the rightmost character in the display rather than by characteristics of the measurement circuit alone.

digital resolution The value of the least significant digit in a digitally coded representation.

digital signal A discrete or discontinuous signal, one whose various states are discrete intervals apart. See signal, digital.

digital speed transducer See digital tachometer.

digital subset See data set.

digital tachometer Any of several instruments designed to determine rotational speed and display the indication in digital form.

digital television Television in which picture redundancy is reduced or eliminated by transmitting only the data needed to define motion in the picture, as represented by changes in the areas of continuous white or black.[NASA]

digital-to-analog converter (D/A or DAC) 1. A device, or sub-system that converts binary (digital) data into continuous analog data, as, for example, to drive actuators of various types, motor-sped controllers, etc. 2. An electronic device that converts a binary-coded word to an analog voltage proportional to the binary value of that word. See converter, digital to analog.

digital valve A single valve casing containing multiple solenoid valves whose flow capacities vary in binary sequence (1, 2, 4, 8, 16, ...); to regulate flow, the control device sends operating signals to various combinations of the solenoids; applications are limited to very clean fluids at moderate temperatures and pressures, but within these limitations precise flow control and rapid response are possible—an eight-element valve, for example, yields flow resolution of 0.39% (1 part in 256).

digitize To convert an analog measurement of a physical variable into a numerical value, thereby expressing the quantity in digital form. See analog-to-digital converter.

digitized signal Representation of information by a set of discrete values, in accordance with a prescribed law. Every discrete value represents a definite range of the original undigitized signal. See analog-to-digital converter.

digitizer A device which converts an analog measurement into digital form. See also analog to digital converter.

digs Breaks of the polished surface of a round, oval, square, etc. shape including pits, holes and surface-broken bubbles.[ARP924-91]

dihedral The spanwise inclination of wing or other surface relative to horizontal.[ARP4107-88]

dihydroxyphenylalanine See dopa.

dikes (geology) See rock intrusions.

dilatant substance A material which flows under low shear stress but whose rate of flow decreases with increasing shear stress.

dilatometer An apparatus for accurately measuring thermal expansion of materials. See also extensometer.

diluent A substance which reduces the viscosity and pour point of an engine's lubrication oil.[S-4, 6-84]

dilution 1. Adding solvent to a solution to lower its concentration. 2. Melting low-alloy base metal or previously deposited weld metal into high-alloy filler metal to produce a weld deposit of intermediate composition.

dimensional stability The ability of a material to retain its size and shape over an extended period of time under a defined set of environmental conditions, especially temperature.

dimetcote An inorganic zinc coating composed of two materials, (1) a reactive liquid and (2) a finely divided powder which are mixed together. The mixture reacts in place with a steel surface to form an insoluble coating.

diminished radix complement A number obtained by subtracting each digit of the given number from one less than the radix; typical examples are the nine's-complement in decimal notation and one's-complement in binary notation.

DIN Abbreviation for the standards institution of the Federal Republic of Germany.

diode A two-electrode electronic component containing merely an anode and a cathode.

diode laser A laser in which stimulated emission is produced at a p-n junction in a semiconductor material. Only certain materials are suited for diode-laser operation, among them gallium arsenide, indium phosphide, and certain lead salts.

diode laser array A device in which the output of several diode lasers is brought together in one beam. The lasers may be integrated on the same substrate, or discrete devices may be coupled optically and electronically.

Dione One of the natural satellites of Saturn orbiting at a mean distance of 378,000 kilometers.[NASA]

diopter A measurement of refractive power of a lens equal to the reciprocal of the focal length in meters. A lens with 20-centimeter focal length has power of five diopters, while one with a 2-meter focal length has a power of 0.5 diopter.

DI/OU See data input/output unit, a device that interfaces to the process for the sole purpose of acquiring or sending data.

DIP See Dual In-line Package.

dip brazing Producing a brazed joint by immersing the assembly in a bath of hot molten chemicals or hot metal; a chemical bath may provide the brazing flux; molten metal, the brazing alloy.

dip coating Covering the surface of a part by immersing it in a bath containing the coating material.

dip needle A device for indicating the angle, in a vertical plane, between a magnetic field and the horizontal plane.

dipole System composed of two, separated, equal electric or magnetic charges of opposite sign.[NASA]

dipole antenna 1. A straight radiator, usually fed in the center, and producing a maximum of radiation in the plane normal to its axis. The length specified is the overall length. SN (single dipole antennas).[NASA] 2. A center-fed antenna which is approximately half as long as the wavelength of the radio waves it is primarily intended to transmit or receive.

dip soldering A process similar to dip brazing, but using a lower-melting filler metal.

dip tube See bubble tube.

dir In MS-DOS, the command that will cause file directories to be displayed.

direct access The retrieval or storage of data by a reference to its location on a volume, rather than relative to the previously retrieved or stored data.

direct access device See random access device.

direct access storage device (DASD) A data storage unit on which data can be accessed directly at random without having to progress through a serial file such as tape; a disk unit is a direct-access storage device.

direct acting controller A controller in which the value of the output signal increases as the value of the input (measured variable or controlled variable) increases. See controller, direct acting.

direct acting recorder A recorder in which the pen or other writing device is directly connected to, or directly operated by, the primary sensor.

direct action A controller in which the value of the output signal increases as the value of the input (measured variable or controlled variable) increases.

direct address An address which indicates the location where the referenced operand is to be found or stored with no reference to an index register. Synonymous with first-level address.

direct addressing An addressing mode in which the instruction operand specifies the location of the data to be used.

direct broadcast satellite Domestic satellite used for direct TV transmission to home receivers.[NASA]

direct code A code which specifies the use of actual computer command and address configurations.

direct-connected An arrangement whereby a meter or other driving mechanism is connected to a driven mechanism without intervening gears, pulleys or other speed-changing devices.

direct coupling The association of two circuits which is accomplished by capacitance, resistance or self-inductance common to both circuits.

direct current (DC) An electrical current which travels uniformly in one direction.[ARP1931-87]

direct-current amplifier An amplifier designed to amplify signals of infinitesimally small frequency.

direct current resistance The resistance offered by any circuit or circuit component to the flow of direct current. [ARP1931-87]

direct digital control (DDC) 1. A computer control technique that sets the final control-elements position directly by the computer output. 2. A control system in which the computer carries out the functions normally performed by conventional controllers, for example, three term control. 3. A term used to imply that a digital controller is connected directly to a final control element or actuator in a manufacturing process, e.g., a valve in a process stream, an electric drive motor mechanically operating on a process. Used to distinguish from analog control. 4. A method of control in which all control outputs are generated by the computer directly, with no other intelligence between the central computer and the process being controlled. See control, direct digital.

direct drive Any powered mechanism where the driven portion is on the same shaft as the driving portion, or is coupled directly to the driving portion.

direct entry In data processing, the input of data directly to computer memory and disk, in contrast to earlier methods of keying to punched cards which were then read into a computer.

direct extensions A device that provides flow rate indication by means of viewing the position of the extension of the metering float within a glass extension tube.

direct glare The sensation produced by brightnesses within the visual field that are sufficiently greater than the luminance to which the eyes are adapted to cause annoyance, discomfort, or loss in visual performance and visibility.[AIR 1151-70]

directional antennas Antennas that radiate or receiver radio signals more efficiently in some directions than in others.[NASA]

directional control valve A valve whose chief function is to control the direction of flow within a fluid system.

directional coupler 1. A device for separately sampling either the forward or backward oscillations in a transmission line. 2. A fiber optic coupler is directional if it preferentially transmits light in one direction.

directional gyroscope A navigational instrument for indicating direction; it contains a free gyroscope which holds its position in azimuth, thus allowing the instrument scale to indicate deviation from the reference direction.

directional property Any mechanical or physical property of a material whose value varies with orientation of the test axis within the test specimen.

directional solidification (crystals) Controlled solidification (crystal growth) of molten metal in a casting so as to provide feed metal to the solidifying front of the casting.[NASA]

directional stability The properties of an aircraft, rocket, tire, etc. enabling it to restore itself from a yawing or sideslipping condition. Also called weathercock stability.[AIR1489-88]

direction finder (radio) See radio direction finder.

direction finding A procedure or process for locating or localizing the origin of radar, acoustical, or optical emissions.[NASA]

direction of lay The direction of lay is the lateral direction, either right-hand or left-hand, in which a strand or wire passes over the top as it recedes from an observer looking along the axis of the conductor or cable.[AS1198-82]

direction of polarization The direction of the electric field vector of an electromagnetic wave.

direction of propagation The direction of average energy flow with respect to time at any point in a homogeneous, isotropic medium.

directive An operator command that is recognized by computer software.

directivity The ability of an antenna to radiate or receive more energy in some directions.[NASA]

directly controlled system See system, directly controlled.

directly controlled variable See variable, directly controlled.

direct maintenance man hours The total time in direct man-hours required to restore or maintain an item in serviceable condition.[ARD50010-91]

direct memory access (DMA) 1. A method of fast data transfer between the peripherals and the computer memory. The transfer does not involve the CPU. 2. Pertains to hardware that enables data to be entered into computer memory without involving the CPU; this is the method used by most telemetry/computer systems.

direct multiplex control See control, direct multiplex.

direct numerical control (DNC) A distributed numerical control system in which the supervisory computer controls several CNC or NC machines.

directory 1. Alphabetical, geographical, or classified listing by field of persons, organizations, programs and/or objects such as instruments, devices, and products. Use of this term excludes directories in computers.[NASA] 2. A file with the layout for each field of the record which it describes. 3. The layout of a record within a file. 4. A table that contains the names of, and pointers to, files on a mass-storage device.

directory device A mass-storage retrieval device, such as disk or DECtape, that contains a directory of the files stored on the device.

directory service The network management function that provides all addressing information required to access an application process. See PSAP address.

direct power generation Any method of producing electric power directly from thermal or chemical energy without first converting it to mechanical energy; examples include thermopiles, primary batteries, and fuel cells.

direct process Any method for producing a commercial metal directly from metal ore, without an intervening step such as roasting or smelting that produces semirefined metal or another intermediate product.

direct-reading gage Any instrument that indicates a measured value directly rather than by inference—for instance, indicating liquid level by means of a sight glass partly filled with liquid from the tank or by means of a pointer directly connected to a float in the tank.

direct record In instrumentation tape, the mode in which tape magnetization is directly related to data voltage level.

direct storage access (DSA) See access, direct memory.

direct wave A wave that is propagated through space without relying on the properties of any gas or other substance occupying the space.

direct-writing recorder A pen-and-ink recorder in which the position of the pen on the chart is controlled directly by a mechanical link to the coil of a galvanometer, or indirectly by a motor controlled by the galvanometer.

dirigibles See airships.

DIS See Draft International Standard.

disability glare Glare which reduces visual performance and visibility and often is accompanied by discomfort. [AIR1151-70]

disable 1. To remove or inhibit a computer hardware or software feature. 2. Disallow the processing of an established interrupt until interrupts are enabled. Contrasted with enable. See disarm.

disarm Cause an interrupt to be completely ignored. Contrasted with arm. See disable.

disassemble To reduce an assembly to its component parts by loosening or removing threaded fasteners, pins, clips, snap rings or other mechanical devices—in most instances, for some purpose such as cleaning, inspection, maintenance or repair followed by reassembly.

disaster Large-scale drought, glacier movement, flood, fire, storm, etc.[NASA]

discernible Discernible is intended to separate the time constant associated with jitter from that associated with drift or dither. The drift and dither time constants are long enough that symbol movement would not be discernible. [AS8034-82]

discharge head The pressure at which a pump discharges freely to the atmosphere, usually measured as feet of water above the intake level.

discharging The removal of electrical energy from a battery.[AS8033-88]

discomfort glare Glare which produces discomfort. It does not necessarily interfere with visual performance or visibility.[AIR1151-70]

disconnect 1. A conductive device designed to be separated from its mated part.[ARP914A-79] 2. To disengage the apparatus used in a connection and to restore it to its ready condition when not in use. 3. Disengaging the linkage between an interrupt and a designated interrupt servicing program. See connect.

disconnect, AMAD A device used to decouple the AMAD from the engine. [ARP906A-91]

disconnect switch An electrical switch for interrupting power supplied to a machine; it is usually separate from the machine controls (often mounted nearby on the wall) and serves mainly to deenergize the equipment for safety during setup or maintenance.

discontinuity 1. A discontinuity is an interruption in the normal physical structure or surface configuration of a part material and is considered to be a defect when its nature, degree, frequency and location is detrimental to the quality, appearance, or performance of the part.[AS3071A-77] 2. A break in sequence or continuity of anything. [NASA] 3. Any feature within a bulk solid that acts as a free surface; it may be a crack, lap, seam, pore or other physical defect, or it may be a sharp boundary between the normal structure and an inclusion or other second phase; a discontinuity may or may not impair the usefulness of a part.

Discos (satellite attitude control) A satellite orbit Disturbance COmpensation System' designed to maintain an object (proof object) in correct orbit by detecting forces and compensating for them by using thrusters.[NASA]

discrepancy 1. Any difference of inconsistency between a requirement for a characteristic or a material or an item, as specified in a contract, drawing specification standard, test procedures or other document and the actual characteristic of the material or item.[ARD 50010-91][AIR1916-88] 2. Deviation from an expected condition.[ARP1587-81]

discrete 1. Pertaining to distinct elements or to representation by means of distinct elements, such as characters. 2. In data processing, data organized in specific parts. 3. An individual bit from a selected word. 4. Discrete manufacturing refers to the manufacture of distinct products or parts.

discrete address beacon system Radar beacon system with discretely addressable transponders and a ground-air-ground data link for automated air traffic control (FAA).[NASA]

discrete component circuit A circuit implemented by uses of individual transistors, resistors, diodes, capacitors, etc. Contrasted with integrated circuit.

discrete frequency A particular radio frequency for use in direct pilot-controller ATC communications which is selected to reduce radio frequency congestion by controlling the number of aircraft operating on a particular frequency at one time.[ARP4107-88]

discrete instrument Pertaining to distinct elements or to representation by means of distinct elements.

discrete programming See integer programming.

discriminant analysis (statistics) A linear combination of a set of N variables that will classify (into two different classes) the events or items for which the measurements of the N variables are available, with the smallest proportion of misclassifications.[NASA]

discriminant functions See discriminant analysis (statistics).

discrimination ratio One of the standard test plan parameters, it is the ratio of the upper test MTBF (THETA 0) to the lower MTBF (THETA 1); that is, d = (THETA 0)/(THETA 1).[ARD50010-91]

discriminator 1. In general, a circuit in which output depends upon the difference between an input signal and a reference signal.[NASA] 2. A hardware device used to demodulate a frequency-modulated carrier or subcarrier to produce analog data.

disdrometer An apparatus capable of measuring and recording the size distribution of raindrops in the atmosphere.

disengage To intentionally pull apart two normally meshing or interlocking parts, such as gears or splines, especially for the purpose of interrupting the transmission of mechanical power.

disengagement The uncoupling of the starter from the engine by means of the starter clutch, or engaging device. Normally occurs at starter cutoff speed or starter maximum speed.[ARP906A-91]

disengaging surface The surface of the boiler water from which steam is released.

dish See parabolic reflector.

dish antenna An antenna in which a parabola-shaped "dish" serves as the reflector to increase antenna gain.

dishing A metalforming operation that forms a shallow concave surface.

disinfectant A chemical agent that destroys microorganisms, bacteria, and viruses or renders them inactive.

disintegrated Separated or decomposed into fragments; loss of original form. [ARP4107-88]

disk A high-speed rotating magnetic platter for storing computer data.

disk area The area of the circle described by the blade tips of a rotating propeller or rotor.[ARP4107-88]

disk brake A mechanical brake in which the friction elements, normally called pads, press against opposite sides of a spinning disk attached to the rotating element to slow or stop its motion.

disk cam A flat cam with a contoured edge that rotates about an axis perpendicular to the plane of the cam, communicating radial linear motion to a follower that rides on the edge of the cam.

disk clutch A device for engaging or disengaging a connection between two shafts where the chief clutch element is a pair of disks, one coupled to each shaft, and which transmit power when engaged by means of disk-face linings made of friction materials.

disk coupling A flexible coupling in which power and motion is transmitted by means of a disk made of elastomeric or other flexible material.

disk directory Table for storing the location of files held on the disk.

disk drive 1. The mechanism which moves the disk in a disk storage unit, usually including the spindle, drive motor, read-record heads, and head actuating mechanism. The term is sometimes used to include the logic control unit and other electronic circuits included in the drive unit. 2. A device that reads and writes computer data on disks.

diskette A round, flat, flexible platter coated with magnetic material and used for storage of software or data.

disk formatting See format.

disk galaxies Galaxies consisting of a central bulge of a spheroidal aggregation of stars and a surrounding disk of stars fanning outward in a thin layer. [NASA]

disk map The organization of information stored on disks.

disk meter A flow-measurement device that contains a nutating disk mounted in such a way that each time the disk nutates, a known volume of fluid passes through the meter.

disk operating system (DOS) 1. A program with which the computer performs such mundane but useful tasks as storing, locating, and retrieving files on disk, reading the keyboard, and issuing display and print information. [NASA] 2. A collection of system programs for operating the microcomputer system.

disk pack A large disk with very high storage capacity.

disk spring A mechanical spring consisting of a dished circular plate and washer supported in such a way that one opposing force is distributed uniformly around the periphery and the second acts at the center. Washer-type disk springs are sometimes known as Belleville washers.

dispatching priority A number assigned to tasks, and used to determine precedence for the use of the central processing unit in a multitask situation.

dispersing prism A prism designed to spread out the wavelengths of light to form a spectrum.

dispersion 1. Finely divided particles in suspension in another substance.[ARP 1931-87] 2. Any process that breaks up an inhomogeneous, lumpy mixture and converts it to a smooth paste or suspension where particles of the solid component are more uniform and small in size. 3. Breaking up globs of oil and mixing them into water to make an emulsion. 4. Intentionally breaking up concentrations of objects of substances and scattering them over a wide area. 5. The process by which an electromagnetic signal is distorted because the various frequency components of that signal have different propagation characteristics. 6. The relationship between refractive index and frequency (or wavelength). 7. In wave mechanics, linear dispersion is the rate of change of distance along a spectrum with frequency, whereas reciprocal linear dispersion is the rate of change of frequency with distance along a spectrum.

dispersion limited operation Denotes operation when the dispersion of the pulse, rather than its amplitude, limits the distance between repeaters. In this regime of operation, waveguide and material dispersion preclude an intelligent decision on the presence or absence of a pulse.

displaced threshold A threshold that is located somewhere other than at the designated beginning of the runway. [ARP4107-88]

displacement 1. A vector quantity that specifies the change of position of a body or particle usually measured from the mean position or position of rest. [NASA] 2. The volume swept out by a piston as it moves inside a cylinder from one extreme of its stroke to the other extreme. 3. For a reciprocating engine, pump or compressor, the volume swept out by one piston as it moves from top dead center to bottom dead center, multiplied by the number of cylinders. 4. Forcing a fluid or granular substance to move out of a cavity or tube by forcing more of the substance in, or by means of a piston or inflatable bladder that moves or expands into the space.

displacement antiresonance A condition of antiresonance where the external sinusoidal excitation is a force and the specified response is displacement at the point where the force is applied.

displacement meter A meter that measures the amount of a material flowing through a system by recording the number of times a vessel or cavity of known volume is filled and emptied.

displacement resonance A condition of resonance where the external sinusoidal excitation is a force and the specified response is displacement at the point where the force is applied.

displacement-type density meter A device that measures liquid density by means of a float and balance beam used in conjunction with a pneumatic sensing system; the float is confined within a small chamber through which the test liquid continually flows, so that density variations with time can be determined.

displacer-type liquid-level detector A device for determining liquid level by means of force measurements on a cylindrical element partly submerged in the liquid in a vessel; as the level in the vessel rises and falls, the displacement (buoyant) force on the cylinder varies and is measured by the lever system, torque tube or other force measurement device.

displacer-type meter An apparatus for detecting liquid level or determining gas density by measuring the effect of the fluid on the buoyancy of a displacer unit immersed in it.

display 1. A component which converts a liquid signal into an equivalent visual output.[ARP993A-69] 2. The cockpit or crew station presentation of information regarding condition of a component or system such as landing gear position (up and locked, down and locked, in transit), break overheat warning, etc.[AIR1489-88] 3. A visual presentation of data. 4. In data processing, the visible representation of data on a screen.

display center The center of the useful

display area.[ARP1782-89] [AS8034-82]

display surface The outermost surface of the display unit through which the display image is projected. The display surface includes the CRT imaging surface, envelope glass, and all contrast and anti-reflective filter surfaces.[ARP 4067-89]

display tube A cathode ray tube used to display information.

display unit A device which provides a temporary visual representation of data. See cathode ray tube.

dissector tube A camera tube which produces an output signal by moving the electron-optical image formed by photoelectric emission on a continuous photocathode surface past an aperture.

dissipation factor A measure of the AC power loss. The ratio of the loss index to its relative permittivity. The ratio of the energy dissipated to the energy stored in the dielectric per cycle. The tangent of the loss angle. Dissipation factor is proportional to the power loss per cycle (f) per potential gradient (E^2 per unit volume, as follows: Dissipation Factor = power loss/(E^2 x f x volume x constant).[ARP1931-87]

dissociation 1. The separation of a complex molecule into constituents by collision with a second body or by absorption of a photon. The product of dissociation of a molecule is two ions, one positively charged and one negatively charged.[NASA] 2. The process by which a chemical compound breaks down into simpler constituents, as the CO_2 and H_2O at high temperature.

dissolved gases 1.Gases in solution.[NASA] 2. Gases which are "in solution" in water.

dissolved solids Those solids in water which are in solution.

dissymmetrical transducer A transducer in which interchanging at least one pair of specified terminals will change the output signal delivered when the input signal remains the same.

distance measuring equipment (DME) Equipment (airborne and ground) used to measure, in nautical miles, the slant-range distance of an aircraft from the position at altitude to the DME navigational aid on the ground. See also TACAN, VORTAC.[ARP4107-88] 2. A radio aid to navigation which provides distance information by measuring total round trip time of transmission from an integrator to a transponder and return.[NASA]

distance perception See space perception.

distillate 1. The distilled product from a fractionating column. 2. The overhead product from a distillation column. When a partial condenser is used, there may be both a liquid and a vapor distillate stream. 3. In the oil and gas industry the term distillate refers to a specific product withdrawn from the column, usually near the bottom.

distillate fuel Any of the fuel hydrocarbons obtained during the distillation of petroleum which have boiling points higher than that of gasoline.

distillation 1. A unit operation used to separate a mixture into its individual chemical components. 2. Vaporization of a substance with subsequent recovery of the vapor by condensation. 3. Often used in less precise sense to refer to vaporization of volatile constituents of a fuel without subsequence condensation.

distilled water Water produced by vaporization and condensation with a resulting higher purity.

distortion 1. Inhomogeneities in the glass or irregularities in the surface of the element causing displacement of images.[ARP924-91] 2. An undesired change in waveform. In a system used for transmission or reproduction of sound, a failure by the system to transit or reproduce a received waveform with exactness. An undesired change in the dimensions or shape of a structure as, distortion of a fuel tank due to abnormal stresses or extreme temperature gradients.[NASA] 3. A lens defect that causes the images of straight lines to appear geometrically other than straight lines. See harmonic content.

distortion meter An instrument that visually indicates the harmonic content of an audio-frequency signal.

distraction See attention, anomalies of distraction.

distributed In a control system, refers to control achieved by intelligence that is distributed about the process to be controlled, rather than by a centrally located single unit.

distributed control A method of media access control in which responsibility for control is distributed among all stations.[AIR4271-89]

distributed control system 1. Comprised of operator consoles, a communication system, and remote or local processor units performing control, logic, calculations and measurement functions. 2. Two meanings of distributed shall apply: a) Processors and consoles distributed physically in different areas of the plant or building, b) Data processing distributed such as several processors running in parallel, (concurrent) each with a different function. 3. A system of dividing plant or process control into several areas of responsibility, each managed by its own controller (processor), with the whole interconnected to form a single entity usually by communication buses of various kinds.

distributed database Relational computer data that can be stored in more than one networked computer, but accessed entirely by one computer.

distributed digital control systems (DDCS) See distributed control system.

distributed feedback lasers Lasers containing a periodic medium which provides the necessary feedback for laser action.[NASA]

distributed processing 1. Processing with multiple small computers that are capable of operating independently but can communicate over a network with each other/and or a central computer. [NASA] 2. Interconnection of two or more computers so that they can work together on the same problem, not necessarily under the direction of a single control program. See also computer networking.

distributed system An arrangement whereby the computer processing power is distributed instead of being centralized.

distribution functions The density functions or number of particles per unit volume of phase space. The distribution functions are a function of the three space coordinates and the three velocity coordinates.[NASA]

distributor 1. A block type unit with an inlet and several outlets to provide force to multiple points, generally hydraulic or air. See manifold.[ARP480A-79] 2. Any device for apportioning current or flow among various output paths. 3. In an automotive engine, a device for sending an ignition spark to the individual cylinder in a fixed order at a rate determined by engine speed.

disturbance resolution The minimum change caused by a disturbance in a measured variable which will induce a net change of the ultimately controlled variable.

disturbance variable A measured variable that is uncontrolled and that affects the operations of the process.

dither The deliberate slow periodic motion of symbology on the face of a CRT intended to prevent degradation of phosphors.[ARP1782-89]

diurnal rhythm See circadian rhythm.

dive brake/speed brake Movable aerodynamic devices on aircraft that reduce airspeed during descent and landing. [ARP4107-88]

divergence 1. The expansion of spreading out of a vector field; also a precise measure thereof. A static instability of a lifting surface or of a body on a vehicle wherein the aerodynamic loads tending to deform surface or body are greater than the elastic restoring forces.[NASA] 2. The spreading out of a laser beam with distance, measured as an angle.

divergence loss The portion of energy in a radiated beam which is lost due to nonparallel transmission, or spreading.

diversion valve A type of fluidic control device that uses the Coanda effect to either switch flow from one outlet port to another or proportion flow between two divergent outlet ports.

diversity combiner The device that accepts two radio signals from a single source that have been received with polarization, frequency, or space diversity, and combines them to yield an output that is better than either original signal.

diversity reception The use of two or more radio receivers, each being connected to different antennas, to improve the signal level. The antennas have diversity in space, phasing, and polarity.

divider A layout tool resembling a draftsman's compass which is used in toolmaking or sheetmetal work to draw circles or arcs, or to scribe hole spacings or other linear dimensions.

dividing network See crossover network.

DMA See direct memory access.

DME Distance Measuring Equipment. The abbreviation DME is also used in audio communications to mean simply distance.

DME separation Spacing of aircraft in terms of distances (nautical miles) determined by distance measuring equipment (DME).[ARP4107-88]

DMSP satellites Satellites of the defense meteorological satellite program, a program sponsored by the United States Air Force System Command's Space Division which provides timely global imagery and specialized meteorological data for supporting a variety of Department of Defense operations. Used for Defense Meteorological Satellite Program.[NASA]

docking The act of coupling two or more orbiting objects. The operation of mechanically connecting together, or in some manner bringing together orbital payloads.[AIR1489-88] See also spacecraft docking.

document 1. A medium and the data recorded on it for human use, for example, a report sheet, a book. 2. By extension, any record that has permanence and that can be read by man or machine.

documentation 1. The assembling, coding, and disseminating of recorded knowledge.[NASA] 2. The creating, collecting, organizing, storing, citing, and disseminating of documents, or the information recorded in documents. 3. A collection of documents or information on a given subject. 4. Often used in specific reference to computer program explanation.

DoD Department of Defense.

Dod FLIP Department of Defense Flight Information Publications.

Dodge-Romig tables A set of standard tables with known statistical characteristics that are used in lot-tolerance and AOQL acceptance sampling.

dog 1. A machine tool accessory used as a clamp for gripping a piece of work and conveying motion to it.[ARP480A-79] 2. Any of several simple devices for fastening, gripping or holding.

doghouses (electronics) Small enclosures placed at the base of transmitting antenna towers to house antenna tuning equipment.[NASA]

dolly A low truck with one or more wheels, rollers, or casters, having either an open or solid platform for moving heavy objects. It does not have superstructure, handles, tongues, stakes, or the like. It may have provisions to permit it to be pulled. NOTE: Use only if stand, trailer or truck does not apply. [ARP480A-79]

dolly, catapult A low mobile platform that rolls on casters. A wheeled apparatus used to support an aircraft for launching from a catapult. A low wheeled platform used to support the nose wheel of an airplane for launching from a SATS catapult, and to which the aircraft is connected for launching. The airplane lifts off from the dolly at the end of the power stroke, and the dolly remains on the ground and is braked to a halt.[AIR1489-88]

dolomite (mineral) A common rock-forming rhombohedral material consisting of calcium, magnesium, and carbonates. It is used for refractory products.[NASA]

DO loop A FORTRAN statement which directs the computer to perform that sequence to which it is keyed.

dome nuts Dome nuts are plate nuts with a mechanical seal at the base and a cap over the top which together provide a fuel tight seal. Commonly used to permit screw or bolt attachments to inaccessible areas and/or access doors. [AIR4069-90]

dominant wavelength 1. The wavelength of homogeneous light, which when combined with white light in suitable proportions, matches a color.[ARP 798-91] 2. The wavelength of monochromatic light which matches a given color when combined in suitable proportions with a standard reference light.

door, strut fairing A door in the landing gear installation for the purpose of fairing the gear (strut) well to the contours of the air vehicle after the strut is retracted into the air vehicle.[AIR1489-88]

door, wheel fairing A door in the landing gear installation for the purpose of fairing the wheel well to the contours of the air vehicle after the wheels are retracted inside the vehicle. [AIR1489-88]

dopa An intermediate organic compound produced by oxidation of tyrosine by tyramine; also, an intermediate product in the synthesis of both epinephrine and melanin.[NASA]

dope A cellulose ester lacquer used as an adhesive or coating.

doped germanium A type of detector in which impurities are added to germanium to make the material respond to infrared radiation at wavelengths much longer than those detectable by pure germanium.

doping 1. Adding a small amount of a substance to a material or mixture to achieve a special effect. 2. Coating a mold or mandrel to prevent a molded part from sticking to it. See also additive.

Doppler effect The change in frequency with which energy reaches a receiver when the receiver and the energy source are in motion relative to each other. [NASA]

Doppler-effect flowmeter A device that uses ultrasonic techniques to determine flow rate; a continuous ultrasonic beam is projected across fluid flowing through the pipe, and the difference between incident-beam and transmitted-beam frequencies is a measure of fluid flow rate.

Doppler-Fizeau effect The Doppler effect applied to a source of light. When the distance between the observer and the source of light is diminishing, the lines of the spectrum are displaced towards the violet, and, when the distance is increasing, they are displaced toward the red, the displacement being proportional to the relative velocity of approach of recession.[NASA]

Doppler navigation Dead reckoning performed automatically by a device which gives a continuous indication of position by integrating the speed derived from measurement of the Doppler effect of echoes from directed beams of radiant energy transmitted from the craft.[NASA]

Doppler radar Radar which utilizes the Doppler effect to determine the radial components of relative radar target velocities or to select targets having particular radial velocities.[NASA]

Doppler shift 1. A phenomenon that causes electromagnetic or compression waves emanating from an object to have a longer wavelength if the object moves away from an observer than would be the case if the object were stationary with respect to the observer, and to have a shorter wavelength if the object moves toward the observer; it is the physical phenomenon that forms the basis for analyzing certain sonar data and certain astronomical observations. 2. A change in the wavelength of light caused by the motion of an object emitting (or reflecting) the light. Motion toward the observer causes a shift toward shorter wavelengths, while motion away causes a shift toward longer wavelengths.

DOS See disk operating system.

dose The amount of radiation received at a specific location per unit area or unit volume, or the amount received by the whole body.

dose meter Any of several instruments for directly indicating radiation dose.

dose rate Radiation dose per unit time.

dose-rate meter Any of several instruments for directly indicating radiation dose rate.

dosimeter Instrument for measuring the ultraviolet in solar and sky radiation. Device worn by persons working around radioactive material, which indicate the dose of radiation to which they have been exposed.[NASA]

DoT Department of Transportation.

dot matrix Characters formed by a matrix of dots.

dot-matrix printer A printer that produces letters, numbers and symbols from a two-dimensional group of dot patterns.

double acting Acting in two directions—for example, as in a reciprocating compressor where each piston has a working chamber at both ends of the cylinder, in a pawl that drives in both directions, or in a forging hammer that is raised and driven down by air or steam pressure.

double-action forming A metalforming process in which one stroke of the press performs two die operations.

double amplitude The peak-to-peak value.

double base powder (propellant) A powder or propellant containing nitrocellulose and another principle explosive ingredient, usually nitroglycerin. [AIR913-89]

double base propellants Solid rocket propellants using two unstable compounds, such as nitrocellulose and nitroglycerin. The unstable compounds used in a double based propellant do not require a separate oxidizer.[NASA]

double-buffered I/O An input or output operation that uses two buffers to transfer data; while one buffer is being used by the program, the other buffer is being read from or written to by an I/O device.

double compression A weave in which the shute wires are compressed tightly by the comb of the loom so that they are deformed in the machine direction. The shute wires are in contact with each other.[AIR888-89]

double-density A type of computer diskette that has twice the storage capacity of a single-density diskette.

double groove weld A weldment in

which the joint is beveled or grooved from both sides to prepare the joint for welding.

double pole A type of device such as a switch, relay or circuit breaker that is capable of either closing or opening two electrical paths.

double precision 1. Pertaining to the use of two computer words to represent a number. 2. In floating-point arithmetic, the use of additional bytes or words representing the number, in order to double the number of bits in the mantissa.

double sampling A type of sampling inspection in which the lot can be accepted or rejected based on results from a single sample, or the decision can be deferred until the results from a second sample are known.

double sided A computer diskette that stores data on both sides.

double standard A state of affairs in which rules are not applied uniformly to everyone, and violations committed by certain persons or groups are condoned or disregarded. Also, the perception that such a situation exists.[ARP 4107-88]

double stars Stars which appear as single points of light to the eye but which can be resolved into two points by a telescope. A double star is not necessarily a binary, a two star system revolving about a common center, but may be an optical double, two unconnected stars in the same line of sight.[NASA]

doublet lens A lens with two components of different refractive index—generally designed to be achromatic.

double-welded joint A weldment in which the joint is welded from both sides.

double window fibers Optical fibers which are designed for transmission at two wavelength regions, 0.8 to 0.9 micrometer and around 1.3 micrometers.

doughnut shape wheel See toroidal wheel.

DOVAP See Doppler effect.

dowel 1. A headless, cylindrical pin used to locate parts in an assembly or to hold them together. 2. A round wood stick or metal rod used to make dowel pins.

dowel screw A dowel that is threaded at both ends.

down 1. Any machinery or equipment that is not operating. 2. In data processing, computer hardware that is not running.

downhand welding See flat-position welding.

downing event The event which causes an item to become unavailable to initiate its mission (the transition from up-time to down-time).[ARD50010-91]

downlink Message or data transmitted by the aircraft to the ground network. [ARP4102/13-90]

downlinking The transmission of signals (data, information, etc.) from satellites to ground terminals.[NASA]

download Data or program transfer, usually from a larger computer to a PC.

downlock The downlock is a mechanism or device for locking the landing gear in the down or extended position preparatory to landing. It locks the structure in the proper position for taking loads imposed by ground operations. Some types used are: (a) "on center" or "past center" side or drag brace held in position hydraulically and/or mechanically. (b) secondary "jury brace" held on or past center hydraulically and/or mechanically. (c) lock pin engagement. (d) hook and roller engagement, etc. [AIR1489-88]

downrange The airspace extending downstream on a given rocket test range.[NASA]

downtime 1. That element of time during which the item is not in condition to perform its intended function.[ARD 50010-91] 2. A period during which equipment is not operating correctly because of machine failure.[NASA] 3.

The time when a piece of equipment is not available due to various causes, such as maintenance, set up, power failure or equipment malfunction.

downtime, logistics That portion of downtime during which repair is delayed solely because of the necessity for waiting for a replacement part or other subdivision of the system.[ARD50010-91]

downtime, mean The average elapsed time between loss of mission capable status and restoration of the system to mission capable status.[ARD50010-91]

downwash A flow of air deflected or forced downward, as by the passage of a wing or by the action of a rotor or a rotor blade. Also referred to as rotor wash. See also wake turbulence.[ARP 4107-88]

dowtherm A constant boiling mixture of phenyl oxide and diphenyl oxide used in high-temperature heat transfer systems (boiling point 494 °F, 257 °C).

DP See draft proposal.

DP cell A pressure transducer that responds to the difference in pressure between two sources. Most often used to measure flow by the pressure difference across a restriction in the flow line.

DPCM (modulation See differential pulse code modulation.

DR Dead Reckoning.

draft Also spelled draught. 1. The side taper on molds and dies that makes it easier to remove finished parts from the cavity. 2. The depth to which a boat or other vessel is submerged in a body of water; the value varies with vessel weight and water density. 3. Drawing a product in a die. 4. The small, positive pressure that propels exhaust gas out of a furnace and up the stack. 5. The difference between atmospheric pressure and some lower pressure existing in the furnace or gas passages of a steam generating unit. 6. A preliminary document.

draft differential The difference in static pressure between two points in a system.

draft gage 1. A type of manometer used to measure small gas heads, such as the draft pressure in a furnace. 2. A hydrostatic indicator used to determine a ship's depth of submergence.

Draft International Standard (DIS) The second stage of the ISO standard process.

draft loss A decrease in the static pressure in a boiler or furnace due to flow resistance.

draft proposal (DP) The first stage of the ISO standard process.

drag 1. A retarding force acting upon a body in motion through a fluid reacting opposite and parallel to the direction of motion of the body.[ARP4107-88] 2. A retarding force acting upon the direction of motion of the body. It is a component of the total fluid forces acting on the body.[NASA] 3. The bottom part of the flask for a casting mold. See also cope. 4. Resistance of a vehicle body to motion through the air due to total force acting parallel to and opposite to the direction of motion. 5. Generally, any resistance to the motion of a solid shape through a body of fluid. 6. In data processing, the movement of an object on a screen by using a mouse.

drag balance See lift drag ratio.

drag-body flowmeter A device that measures the net force on a submerged solid body in a direction parallel to the direction of flow, and converts this value to an indication of flow or flow rate.

drag brace A structural brace of a landing gear system whose primary function is to react drag loads imposed on the system. This member may be a folding member and often serves a secondary function of locking the gear system in the extended and/or retracted position.[AIR1489-88]

drag brake A device for the purpose of providing a retarding force upon a body in motion through a fluid, parallel to the direction of motion of the body. See also ground spoilers, speed brake.[AIR 1489-88]

drag chute Any of various types of parachutes attached to high performance aircraft that can be deployed, usually during landings, to decrease speed and also, under certain flight conditions to control and stabilize the aircraft.[AIR 1489-88]

drag coefficients The ratios of drag to the products of dynamic pressures and reference areas.[NASA]

drag effect See drag.

drag efficiency The measure of efficiency of a skid control system expressed as a ratio (percentage) of the actual drag performance produced over a given period of time (or distance) relative to the maximum (100%) drag performance obtainable for the given set of conditions (i.e., coefficient tire to runway, wet, dry, temp., etc.)[AIR1489-88]

drag force anemometer Instrument for measuring both the static and dynamic velocity head and flow in high frequency, unsteady flow.[NASA]

drag stay See drag brace.

drain 1. A pipe that carries away waste solutions or effluent. 2. To empty a tank or vessel by means of gravity flow into a waste system or auxiliary holding vessel. 3. A valved connection at the lowest point for the removal of all water from the pressure parts.

draw 1. To pull a load. 2. To form cup-shaped parts from sheet metal. 3. To reduce the size of wire or bar stock by pulling it through a die. 4. To remove a pattern from a sand-mold cavity. 5. A fissure or pocket in a casting caused by inadequate feeding of molten metal during solidification. 6. A shop term referring to temper.

draw bead 1. A bead or offset used for controlling metal flow during sheet-metal forming. 2. A contoured rib or projection on a draw ring or holddown to control metal flow in deep drawing.

drawbench The stand that holds a die and draw head used for reducing the size of wire, rod, bar stock or tubing.

drawdown The curvature of the liquid surface upstream of the weir plate.

drawdown ratio The ratio of die opening to product thickness in a deep drawing operation.

drawhead 1. The die holder on a drawbench. 2. A group of rollers through which strip, tubing or solid stock is pulled to form angle stock.

drawing A process used in the manufacture of wire. It consists of pulling the metal through a die, or series of dies, for the purpose of reducing the diameter.[ARP1931-87]

drawing back 1. A shop term for tempering. 2. Reheating hardened steel below the critical temperature to reduce its hardness.

drawing compound A lubricating substance such as soap or oil applied to prevent draw marks, scoring or other defects caused by metal-to-metal contact during a stamping, wiredrawing or similar metalforming operation.

drawing tower Equipment for making optical fibers, in which optical fibers are drawn from heated glass preforms.

draw mark Any surface flaw or blemish that occurs during drawing, including scoring, galling, pickup or die lines.

draw radius The curvature at the edge of the cavity in a deep-drawing die.

draw ring A ring-shaped die part which the punch pulls the draw blank over during a drawing operation.

dredged material Sand, mud, silt, gravel, etc. recovered from the bottom of harbors, canals, etc. during dredging operations.[NASA]

dredging Mechanical or hydraulic excavation of underwater material. Used in maintaining and building of channels and ports as well as underwater mining of sand, gravel, and minerals. [NASA]

dress 1. To shape a tool such as a grinding wheel. 2. To restore a tool to its original contour and sharpness.

drift 1. The lateral displacement of an aircraft from its intended course or heading induced by the crosswind component of the ambient air movement relative to the earth.[ARP4107-88] 2. A bar tube or shaft type tool used for specific purpose of assembling or separating two or more parts by axially applying impact or pressure by means of an external force.[ARP480A-79] 3. Drift is usually expressed as the change in output over a specified time with fixed input and operation conditions. Drift is used in connection with analog transducers, analyzers, etc.

drift angle The angle between the airplane centerline and ground track or the angular difference between true heading and ground track angle. Drift angle is right when ground track angle is greater than true heading and left when ground track angle is less than true heading.[ARP1570-86]

driftpin A round, tapered metal rod that is driven into matching holes in mating parts to stretch them and bring them into alignment, such as for riveting or bolting.

drift plug A tapered rod that can be driven into a pipe to straighten it or flare its end.

drift rate The amount of drift, in any of its several senses, per unit time. Drift rate has many specific meanings in different fields. The type of drift rate should always be specified.[NASA]

drill 1. An end cutting tool having one or more cutting edges, and having straight or helical flutes or grooves adjacent thereto for the admission of coolant and the ejection of cuttings or chips, used for drilling holes in various materials.[ARP480A-79] 2. A cylindrical tool with one or more cutting edges on one end, which is used to make or enlarge holes in solid material by rotating it about its longitudinal axis and applying axial force.

drill drift A flat, tapered piece of steel used to remove taper shank drills and other tools from their tool holders.

drill gage A flat, thin steel plate with numerous holes of accurate sizes that can be used to check the size of drills.

drill jig A tool constructed to guide a drill during repeated drilling of the same size holes, either at many locations in a given piece or at the same location in many identical pieces, especially where exceptional straightness or accuracy of location is desired.

drill press A vertical drilling machine so constructed as to hold the work stationary and apply vertical force to press a rotating drill into the work.

drill sleeve A hollow, tapered cylinder used as an adapter to fit the shank of a taper-shank drill or other tool into the spindle of a drill press or similar machine tool.

drive In data processing, a device that manipulates a diskette, disk, or magnetic tape so the computer can read or write data to them. See disk drive and tape drive.

drive coupling See coupling.

drive fit A type of interference fit requiring light to moderate force to assemble.

driven gear The member(s) of a gear train that receive power and motion from another gear.

driver 1. A hand tool for assembling or removing components. See also extractor, puller, pusher.[ARP480A-79] 2. A software element that converts operator instructions into suitable language to drive a hardware device (unit or stream drivers, for example). 3. A small program or routine that handles the control of an external peripheral device or executes other programs.

drive shaft A shaft which transmits power and motion from a motor or engine to the other elements of a machine.

drive stiffness (actuator stiffness) The stiffness of the actuator between the mounting and output motion attach points. It is the resultant of the mechanical stiffness of the load carrying elements and of the actuation system stiffness.[AIR1916-88]

driving pinion The gear in a gear train that receives power and motion by means of a shaft connected to the source of power and transmits the power and motion via its teeth to the next gear in the train.

driving-point impedance The complex ratio of applied sinusoidal voltage, force or pressure at the driving point of a transducer to resulting current, velocity or volume velocity, respectively, at the same point, all inputs and outputs being terminated in some specified manner.

driving-point reactance The imaginary component of driving-point impedance.

driving-point resistance The real component of driving- point impedance.

drizzle Fairly uniform precipitation, composed exclusively of fine drops of water (diameter less than 1/50 of an inch) very close to one another. Drizzle drops are too small to cause appreciable ripples on the surface of still water. [AIR1335-75]

drone A remotely controlled, self-powered aircraft or missile.

drone aircraft Remotely controlled aircraft.[NASA]

drone helicopter See drone aircraft.

drones for aerodynamic and struct test See DAST program.

droop The difference between the control point and the set point due to some inherent control characteristic. Also called offset or deviation.[ARP89C-70] See also offset.

drooped airfoils A baseline airfoil with an abrupt change in cross-section at about midspan from the fuselage. The outboard portion of the wing has a cross-section with a nearly flat bottom and a drooped (downward) leading edge in relation to the inboard baseline wing.[NASA]

drop leg The section of measurement piping below the process tap location to the instrument.

dropout Discrete variation in signal levels during the reproduction of recorded data which result in data reduction error.[NASA]

dropsondes Radiosondes equipped with a parachute, dropped from an aircraft to transmit measurement of atmospheric conditions as it descends. [NASA]

drop test A type of test to determine the dynamic capabilities of a landing gear shock absorber system to accept the loads and kinetic energies specified and imposed upon it. The shock absorber system may be mounted in a test rig and dropped at specified loads and velocities, or a complete air vehicle may be free dropped. The test is for the purpose of verifying the energy absorption capability and for optimization of the energy control system (hydraulic, pneumatic, mechanical, etc.).[AIR1489-88]

drop tester May be either of two types. It may consist of a hammer assembly that drops upon a test item, or it may be a device by means of which the test item can be dropped in a prescribed manner onto a suitable surface.[ARP 476A-76]

drop tower Large device for low gravity processing of molten material which consists of either a capsule which is dropped, or a drop tube where containerless low gravity studies are conducted or both.[NASA]

drop tube See drop tower.

drosometer An instrument for measuring the amount of dew that condenses on a given surface.

drugs Any chemical compound(s) taken for prevention of disease, treatment of disease, weight management, mood alteration, birth control, sleep management or other purposes. The effects may be direct or residual, but either may reduce performance capability.[ARP 4107-88]

drum 1. Any machine element consisting essentially of a thin-walled, hollow cylinder. 2. A thin-walled, cylindrical container, especially a flat-ended shipping container holding liquids or bulk solids and having a capacity of 12 to 110 gallons (50 to 400 litres). 3. The cylindrical member around which a hoisting rope is wound. 4. A high capacity computer storage device.

drum baffle A plate or series of plates or screens placed within a drum to divert or change the direction of the flow of water or water and steam.

drum brake A mechanical brake in which the friction elements, normally called shoes, press against the inside surface of a cylindrical member (the drum) attached to the rotating element to slow or stop its motion.

drum course A cylindrical section of a drum.

drum head A plate closing the end of a boiler drum or shell.

drum internals All apparatus within a drum.

drum operating pressure The pressure of the steam maintained in the steam drum or steam-and-water drum of a boiler in operation.

dry air Air with which no water vapor is mixed. This term is used comparatively, since in nature there is always some water vapor included in air, and such water vapor, being a gas is dry.

dry ash Refuse in the solid state, usually in granular or dust form.

dry back The baffle provided in a firetube boiler joining the furnace to the second pass to direct the products of combustion, that is so constructed to be separate from the pressure vessel and constructed of heat resistant material, (generally refractory and insulating material).

dry-back boiler The baffle provided in a firetube boiler joining the furnace to the second-pass to direct the products of combustion, that is so constructed to be separate from the pressure vessel and constructed of heat resistance material. (Generally refractory and insulating material).

dry basis A method of expressing moisture content where the amount of moisture present is calculated as a percentage of the weight of bone-dry material; used extensively in the textile industry.

dry blast cleaning Using a dry abrasive medium such as grit, sand or shot to clean metal surfaces by driving it against the surface with a blast of air or by centrifugal force.

dry-bulb temperature The temperature of the air indicated by thermometer not affected by the water vapor content of the air.

dry corrosion Atmospheric corrosion taking place at temperatures above the dew point.

dry assay Determining the amount of a metal or compound in an alloy, ore or metallurgical residue by means that do not involve the use of liquid to separate or analyze for constituents.

dry gas Gas containing no water vapor.

dry-gas loss The loss representing the difference between the heat content of the dry exhaust gases and their heat content at the temperature of ambient air.

drying oven A closed chamber for driving moisture from surfaces or bulk materials by heating them at relatively low temperatures.

dry pipe A perforated pipe in the steam

space above the water level in a boiler which helps keep entrained liquid from entering steam outlet lines.

dry steam Steam containing no moisture. Commercially dry steam containing not more than one half of one percent moisture.

dry steam drum A pressure chamber, usually serving as the steam offtake drum, located above and in communication with the steam space of a boiler steam-and-water drum.

dry test meter A type of meter used extensively to determine gas flow rates for billing purposes and to calibrate other flow-measuring instruments; it has two chambers separated by a flexible diaphragm which is connected to a dial by means of a gear train; in operation, the chambers are filled alternately, with a flow control valve switching from one chamber to the other as the first becomes completely filled, while flow rate is indicated indirectly from movement of the diaphragm.

DSL (delta slope) algorithm See compressor.

DTA (analysis) See thermal analysis.

dual-axis tracking antenna A tracking antenna which is steered automatically in both azimuth and elevation.

dual-beam analyzer A type of radiation-absorption analyzer that compares the intensity of a transmitted beam with the intensity of a reference beam of the same wavelength.

dual-fail operative (DFO) A condition or requirement wherein an active control device or system can sustain any two failures within the system and remain operative. It is implicit with DFO that the system be able to accept identical but non- simultaneous failures in two of its channels and continue to operate with no nominal loss of performance. Unless specifically stated, it is understood that no nominal loss of performance occurs after one or two failures.[ARP1181A-85]

Dual In-line Package (DIP) A common way of packaging semiconductor components.

duality principle Principle that for any theorem in electric circuit analysis there is a dual theorem in which quantities are replaced with dual quantities. Examples are current and voltage or impedance and admittance.[NASA]

duality theorem Theorem which states that if either of the two dual linear programming problems has a solution, then so does the other.[NASA]

dual load path A type of mechanical paralleling wherein two separate load carrying paths exist from the control system input to the system output. Each load path is capable of carrying sufficient load such that failure of any one member will not jeopardize system performance.[ARP1181A-85]

dual ramp ADC Technique for converting analog data into digital format. The unknown voltage is input to a ramp generator and integrated for a specified time. At the end of this time a counter is started and a reference voltage applied to cause a controlled ramp down. The counter is stopped when the voltage becomes zero. The count gives the digital number output.

dual ratio Dual purpose gear box which permits engine starting at one speed and pumping or a generating capability from the starter driven at another speed. The two gear systems are isolated from one another by overrunning clutches.[ARP906A-91]

dual rotation propellers (shafts) The concentric arrangement in which each component of the propeller rotates in an opposite direction at the same speed because of a fixed gear ratio between shafts.[ARP355-84]

dual-slope converter An integrating analog-to-digital converter in which the unknown signal is converted to a proportional time interval.

dual system Special configurations that use two computers to receive identical input and execute the same routines, with the results of such parallel processing subject to comparison. Exceptional high-reliability requirements usually are involved.

dual-tandem valve A tandem valve having two separate control sections. [ARP1181A-85] [AIR1916-88]

dual wing configuration A configuration of two wings of nearly the same planform and area, one behind the other.[NASA]

duct 1. A single duct designed to be attached to the rear of a turbine engine, providing an uninterrupted discharge for the exhaust gases. Not to be used when exhaust gases are to be controlled for proper velocity. See also nozzle. [ARP480A-79] 2. Specifically a tube or passage that confines and conducts fluids, as passage for the flow of air to compressor of a gas turbine engine, or pipe leading air to a supercharger. [NASA] 3. An enclosed fluid-flow passage, which may be any size up to several feet in cross section, usually constructed of galvanized sheet metal and not intended to sustain internal pressures of more than a few psi; the term is most often applied to passages for ventilating air, and to intakes and exhausts for engines, boilers and furnaces.

ducted fan Fan enclosed in a duct. [NASA]

ducted fan engine Aircraft engine incorporating a fan or propeller enclosed in a duct; especially, jet engines in which an enclosed fan or propeller is used to ingest ambient air to augment the gases of combustion in the jetstream.[NASA]

duct geometry The shape and dimension of ports or other openings designed for passage of fluid (gas, liquid, or mixtures) in or external to engines.[NASA]

ductile iron The term preferred in the United States for cast iron containing spheroidal nodules of graphite in the as-cast condition. Also known as nodular cast iron; nodular iron; spherulitic-graphite cast iron.

ductility The property of a metal that indicates its relative ability to deform without fracturing; it is usually measured as percent elongation or reduction of area in a uniaxial tensile test.

dud An explosive device that has failed to initiate as intended.[AIR913-89]

dullness See luster.

Dumet wire Wire made of Fe-42Ni, covered with a layer of copper, which is used to replace expensive platinum as the seal-in wire in incandescent lamps and vacuum tubes; the copper coating prevents gassing at the seal.

dummy 1. A device constructed physically to resemble another device, but without the operating characteristics. 2. A cathode, usually corrugated to give varying current densities, which is plated at low current densities to preferentially remove impurities from an electroplating solution. 3. A substitute cathode used during adjustment of the operating conditions in electroplating. 4. An artificial address, instruction, or record of computer information inserted solely to fulfill prescribed conditions, such as to achieve a fixed word length or block length, but without itself affecting machine operations except to permit the machine to perform desired operations.

dummy argument A variable such as the one which appears in the argument list of a function definition but which is replaced by the actual argument when the function is used.

dummy block A thick plate the same shape as the extrusion billet which is placed between billet and ram to prevent the latter from overheating.

dummy connector, plug A connector device designed to mate with a receptacle connector so as to perform protective, environmental and/or electrical shorting functions.[ARP914A-79]

dummy connector, receptacle A connector device designed to mate with a plug connector so as to perform protective, environmental, and cable and harness routing/fitting and stowage functions.[ARP914A-79]

dummy instruction An artificial instruction or address inserted in a list to serve a purpose other than for execution as an instruction.

dummy load See impedance.

dump 1. A printout of computer memory or a file in hexadecimal and character form. 2. The transfer of data without regard for its significance. Same as storage dump.

dump combustor Combustor having a means of reducing flow velocity and forming recirculation zones through the sudden enlargement area between the inlet duct and the combustion chamber.[NASA]

dump valve A large valve in the bottom of a tank or container that can quickly empty the tank in an emergency.

dune Low mound, ridge, bank or hill of loose, windblown granular material, usually sand, capable of movement. [NASA]

duodecimal number A number, consisting of successive characters, representing a sum, in which the individual quantity represented by each character is based on a radix of twelve. The characters used are 0, 1, 2, 3, 4, 5, 6, 7, 8, 9, T (for ten) and E (for eleven). Related to number systems.

duplex 1. An adjective meaning twofold, as a duplex valve, duplex actuator, etc. [ARP1181A-85] [AIR1916-88] 2. Pertaining to a twin, pair, or a two-in-one situation, e.g., a channel provided simultaneous transmission in both directions or a second set of equipment to be used in event of the failure of the primary device. 3. Referring to any item or process consisting of two parts working in connection with each other.

duplex cable A cable which contains two optical fibers in a single cable structure. Light is not coupled between the two fibers; typically one is used to transmit signals in one direction and the other to transmit in the opposite direction.

duplex connector A connector which simultaneously makes two connections, joining one pair of optical fibers with another.

duplex control A control in which two independent control elements share a common input signal for the operation of separate final control elements both of which influence the value of the controlled condition.

duplexed system A system with two distinct and separate sets of facilities, each of which is capable of assuming the system function while the other assumes a standby status. Usually, both sets are identical in nature.

duplexer Device which permits a single antenna system to be used for both transmitting and receiving. Duplexers should not be confused with diplexers, devices permitting an antenna system to be used simultaneously or separately by two transmitters.[NASA]

duplex mode The communication link which allows simultaneous transmission and receipt of data.

duplex operation 1. The operation of associated transmitting and receiving apparatus in which the processes of transmission and reception are concurrent.[NASA] 2. Operating an associated transmitter and receiver which are designed for concurrent transmission and reception.

duplex process Any integrated process in which a manufacturing operation is carried out by two procedures in series—for example, refining steel by the Bessemer process followed by producing ingots or continuously cast slabs by the basic oxygen or electric furnace process.

duplex pump A reciprocating or diaphragm pump having two parallel flow paths through the same housing, with a common inlet and a common outlet.

durability A measure of useful life (a special case of reliability).[ARD50010-91]

duration See time.

durometer A mechanical device used to measure the hardness of a sealant. A Shore A or REX A reading is produced. Some coatings are measured on a Shore D scale, which encompasses a higher hardness range.[AS7200/1-90]

dust Particles of gas borne solid matter larger than one micron in diameter.

dust counter A photoelectric instrument that measures the number and size of dust particles in a known volume of air. Also known as Kern counter.

dust loading The amount of dust in a gas, usually expressed in grains per cu ft or lb per thousand lb of gas.

dutch oven A furnace that extends forward of the wall of a boiler setting. It usually is of all refractory construction, although in some cases it is water cooled.

dutch roll Oscillating motion of an aircraft combining rolling and yawing (roll-induced yaw) so named for the resemblance to the characteristic rhythm of an ice skater.[ARP4107-88]

Dutch weave A weave in which the warp wires are straight and heavier than the shute wires.[AIR888-89]

duty The specification of service conditions which defines the type, duration and constancy of applied load or driving power.

duty cycle 1. The operating cycle required of the ignition system. It is expressed as a function of time ON and time OFF or continuous, as applicable, and is generally associated with the ignition exciter specification.[AIR784-88] 2. The period when a component is performing its designed function. The duty cycle may consist of multiple cycles.[ARP906A-91] 3. A description of the load throughout the total mission time with sufficient detail to determine load velocity requirements, frequency of occurrence, and dynamic load characteristics. A complete duty cycle description defines the actuation energy required for the total mission.[AIR1916-88] 4. For a device that operates repeatedly, but not continuously, the time intervals involved in starting, running and stopping plus any idling or warm-up time. 5. For a device that operates intermittently, the ratio of working time to total time, usually expressed as a percent. Also known as duty factor. 6. The percent of total operating time that current flows in an electric resistance welding machine.

duty cyclometer A meter for directly indicating duty cycle.

dwarf galaxy Galaxy with low luminosity.

dwarf novae Short period binary systems in which a red quasi-main sequence star fills its Roche lobe and transfers matter, via an accretion disk, onto a white dwarf.[NASA]

dwell 1. A contour on a cam that causes the follower to remain at maximum lift for an extended portion of the cycle. 2. In a hydraulic or pneumatic operating cycle, a pause during which pressure is neither increased nor decreased.

dwell period The time spent by a commutator at a given channel position.

dwell time In any variable cycle, the portion of the cycle when all controlled variables are held constant—for example, to allow a parameter such as temperature or pressure to stabilize, or to allow a chemical reaction to go to completion.

dye penetrant A low-viscosity liquid containing a dye used in nondestructive examination to detect surface discontinuities such as cracks and laps in both magnetic and nonmagnetic materials.

dynamically relocatable coding Coding for a computer which has special hardware to perform the derelativization. With an appropriately designed computer system, coding can be loaded into various sections of core, appropriate addresses changed, and the program executed.

dynamic calibration A calibration procedure in which the quantity of liquid is measured while liquid is flowing into or out of the measuring vessel.

dynamic compensation A technique used in control to compensate for dynamic response differences to different input streams to a process. A combination of lead and lag algorithms will handle most situations.

dynamic gain See gain, dynamic.

dynamic impedance (stiffness) The impedance associated with the output deflections of an active closed-loop actuation system caused by externally applied dynamic forces (usually sinusoidal) over a specified frequency range. Dynamic impedance includes the load mass or inertia, load friction, system stiffness, and any other load-related compliance effects. Since it is a complex quantity, the term dynamic impedance is preferred to the term dynamic stiffness.[AIR1916-88]

dynamic load 1. A load imposed by dynamic action, as distinguished from a static load. Specifically, with respect to aircraft, rockets, or spacecraft, a load due to an acceleration, as imposed by gusts, by maneuvering, by landing, by firing rockets, by braking, etc.[AIR1489-88][NASA] 3. That portion of a service load which varies with time, and cannot be characterized as a series of different, unvarying (static) loads applied and removed successively.

dynamic luminance (Shades of Gray) Dynamic luminance is the difference between adjacent luminance values that is readily distinguishable by the human eye. This relationship between perceived brightnesses is a power law relationship to luminance and is limited to non highly saturated colors. A minimum ratio between two adjacent luminance values of 1.414 (square root of 2) is normally used.[ARP1782-89]

dynamic memory Same as dynamic storage.

dynamic model 1. Model of aircraft or other objects having their linear dimensions and its weight and moments of inertia reproduced in scale in proportion to the original.[NASA] 2. A model in which the variables are functions of time. Contrast with static model and steady-state model.

dynamic optimization A type of control, frequently multivariable and adaptive in nature, which optimizes some criterion function in bringing the system to the setpoints of the controlled variables. The sum of the weighted, time-absolute errors is an example of a typical criterion function to be minimized. Contrast with steady-state optimization.

dynamic pressure 1. That portion of the local pressure in a fluid which is recovered when the fluid is brought to rest.[AIR1916-88] 2. The pressure of a fluid resulting from its motion, equal to one half the fluid density times the fluid velocity squared. In incompressible flow, dynamic pressure is the difference between total pressure and static pressure.[NASA] 3. The increase in pressure above the static pressure that results from complete transformation of the kinetic energy of the fluid into potential energy.

dynamic programming In operations research, a procedure for optimization of a multi-stage problem wherein a number of decisions are available at each stage of the process. Contrast with convex programming, integer programming, linear programming, mathematical programming, nonlinear programming, and quadratic programming.

dynamic RAM Random access memory that needs to be refreshed at regular time intervals. It involves the extra complexity of refresh circuits but higher densities can be achieved.

dynamic range 1. The dynamic range of an APC is the ratio of smallest to the largest particle that can be processed by the instrument; for example, a size range of 2.0-120 m would yield a ratio or dynamic range of 60:1.[ARP1192-87] 2. The range of signals which is accepted by a device without manual adjustment.

dynamic response The behavior of an output in response to a changing input.

dynamics Study of the motion of a system of material particles under the influence of forces, especially those which originate outside the system under consideration.[NASA]

dynamic sensitivity In leak testing, the minimum leak rate that a particular device is capable of detecting.

Dynamics Explorer satellites Two satellites that have been designed to occupy different orbits and supply comparative data for studying the boundary region between earth and space. Of the 24 goals of the program, one half require both satellite's data, one fourth one satellite's data and one fourth the other satellite's data. The satellites were launched together in August of 1981. [NASA]

dynamic stability 1. The characteristics of a body, such as an aircraft, that causes it, when disturbed from an original state of steady flight or motion, to damp the oscillations set up by restoring moments and gradually return to its original state; specifically, the aerodynamic characteristics.[AIR1489-88] 2. The property which permits the response of a positively damped physical system to asymptotically approach a constant value when the level of excitation is constant. Compare with static stability. See stability, dynamic.

dynamic stop A loop stop consisting of a single jump instruction.

dynamic storage The storage of data on a device or in a manner that permits the data to move or vary with time, and thus the data is not always available instantly for recovery, e.g., acoustic delay line, magnetic drum, or circulating or recirculating of information in a medium. Synonymous with dynamic memory.

dynamic storage allocation A storage allocation technique in which the location of programs and data is determined by criteria applied at the moment of need.

dynamic subroutine A subroutine which involves parameters, such as decimal point position or item size, from which a relatively coded subroutine is derived. The computer itself is expected to adjust or generate the subroutine according to the parametric values chosen. Contrasted with static subroutine.

dynamic test A test of a device or mechanism conducted under variable loading or stimulation.

dynamic torque The torque which a brake will produce for deceleration dynamically as opposed to static torque. Dynamic torque is effected by velocity, temperature, lining coefficient, and other factors.[AIR1489-88]

dynamic unbalance 1. That condition in which the central principal axis is not coincident with the shaft axis. NOTE: The quantitative measure of dynamic unbalance can be given by two complementary unbalance vectors in two specified planes (perpendicular to the shaft axis) which completely represent the total unbalance of the rotor. [ARP588A-89] 2. A condition in rotating equipment where the axle of rotation does not exactly coincide with one of the principal axes of inertia for the mechanism; it produces additional forces and vibrations which, if severe, can lead to failure or malfunction.

dynamic variable Process variables that can change from moment to moment due to unspecified or unknown sources.

dynamometer 1. Instrument for measuring power or force; specifically, instrument for measuring the power, torque, or thrust of aircraft engines or rockets.[NASA] 2. An electrical instrument in which current, voltage or power is measured by determining the force between a fixed coil and a moving coil. 3. A special type of rotating machine used to measure the output or driving torque of rotating equipment.

dynamometer, brake test A device for measuring (testing) forces and power or capability of brakes. Generally the system includes a large rotating steel wheel (mass) which is brought up to speed by external power and represents an inertia equivalent to which the brake is to be tested. The test brake/wheel/ tire assembly is loaded against the dynamometer wheel and the power of the brake is utilized to stop the rotating mass.[AIR1489-88]

dynamometer, tire test See also dynamometer, brake test. A device for measuring (testing) capabilities of tires. Differs from a brake test dynamometer in that an inertia equivalent is generally not required. The dynamometer wheel is powered to rotate and the tire is loaded radially against it. Tire test parameters of interest are usually load, speed, time, and distance.[AIR1489-88]

dyspnea Difficult or labored breathing. [NASA]

Dzus Fastener A trade name for a quick-release type fastener, designed to permit rapid removal of inspection plates or aircraft cowling.[ARP4107-88]

e The base of natural logarithms.

E See modulus of elasticity.

ear blocks See trapped gas effects.

earing Forming a scalloped edge around a deep-drawn sheet-metal part due to directional properties in the blank material.

EAROM See electrically alterable read-only memory.

earphone(s) Electroacoustic transducers operating from an electrical system to an acoustical system and intended to be closely coupled acoustically to the ear. Synonymous with headsets.[NASA]

earth axis Any one of a set of mutually perpendicular reference axes established with the upright axis (the Z axis) pointing to the center of the earth, used in describing the position or performance of an aircraft or other body in flight. The earth axes may remain fixed or may move with the aircraft or other object.[NASA]

earth currents See telluric currents.

earth figure See geodesy.

earth hydrosphere That part of the earth that consists of the oceans, seas, lakes, and rivers.[NASA]

earth mantle The zone of the earth below the crust and above the core (to a depth of 3480 km), which is divided into the upper mantle and the lower mantle, with a transition zone between. [NASA]

earth observations (from space) The acquisition of earth surface data from aircraft or spacecraft.[NASA]

earth observing system (EOS) NASA's orbital multisensor observatory system for the long term acquisition of earth sciences data to be operated in conjunction with an integrated ground-based science information system. This international system will become operational in 1995 when the first of four polar platforms will be launched. The first and third will be launched under U.S. auspices. The second under ESA auspices and the last under Japanese auspices.[NASA]

earthquake resistance Structural strength of natural geological formations reacting to seismic forces.[NASA]

earthquake resistant structures Buildings and other structures designed for maximum safety and protection from the effects of earthquakes.[NASA]

earth reference or inertial (I) Referenced with respect to the earth, as opposed to referenced with respect to air mass.[ARP1570-86]

earth shape See geodesy.

earth terminal measurement system NBS system for measuring electromagnetic parameters of communication satellites and ground stations relative to antenna gain, ratio of carrier power to operating noise temperature, and satellite effective isotropic power.[NASA]

earth terminals Portable or stationary ground based equipment used to transmit and receive signals and other data via satellites in communications networks.[NASA]

EAS Equivalent Airspeed.

easily cleanable Surfaces which are accessible without tools for hand cleaning, and of such material and finish and so fabricated that residue may be completely removed by normal cleaning methods.[AS1426-80]

easily interpretable Values or information, or both, displayed that can be perceived and understood with a high degree of accuracy by users without additional measuring devices/scales. [ARP4107-88]

easily removable Capable of being removed from its normally used or installed position with the use of tools. [AS1426-80]

EASY Environmental Analysis SYstem computer program.[AIR1823-86]

EBCDIC See Extended Binary Coded Decimal Interchange Code.

ebullition The act of boiling or bubbling.

eccentric Describing any rotating mechanism whose center of rotation does not coincide with the geometric center of the rotating member.

eccentricity A measure of the center of a conductor's location with respect to the circular cross section of the insulation. Expressed as a percentage of center displacement of one circle within another.[ARP1931-87]

eccentric orifice An orifice whose center does not coincide with with the centerline of the pipe or tube; usually, the eccentricity is toward the bottom of a pipe carrying flowing gas and toward the top of a pipe carrying liquid, which tends to promote the passage of entrained water or air rather than allowing entrained water or gas to build up in front of the orifice.

echelon faults See geological faults.

echo(es) 1. Waves that have been reflected or otherwise returned with sufficient magnitude and delay to be detected as a wave distinct from that directly transmitted. In radar, a pulse of reflected radiofrequency energy, the appearance on a radar indicator of the energy returned from a target.[NASA] 2. In MS-DOS, the echo command prints the text of the command on the screen as they are executed.

echo check A check of accuracy of transmission in which the information which was transmitted to an output device is returned to the information source and compared with the original information to insure accuracy of output.

echoencephalography A diagnostic technique in which pulses of ultrasonic waves are beamed through the head from both sides, and echoes from the midstructures of the brain are recorded as graphic tracings.[NASA]

echo ranging A form of active sonar, in which the sonar equipment generates pulses of sound, then determines the distance to underwater objects by precisely measuring the time it takes for a pulse to reach an object, be reflected and return to a known location, usually one adjacent to the transmitter; by using a narrow, focused sound beam, direction to the object also can be found.

echosonogram A graphic display, such as an echocardiogram, determined with ultrasonic pulse-echo techniques.

echo sounding Determining the depth of water below a vessel or platform by sonic or ultrasonic pulse-echo techniques.

eclipses The reductions in visibility or disappearances of nonluminous bodies by passing into the shadows cast by another nonluminous body. The apparent cutting off, wholly or partially, of the light from a luminous body by a dark body coming between it and the observer.[NASA]

ecliptic The apparent annual path of the sun among the stars; the intersection of the plane of the earth's orbit and with the celestial sphere. The eliptic is a great circle of the celestial sphere inclined at an angle of about 23 degrees 27 minutes to the celestial equator.[NASA]

ECM See electrochemical machining.

ecological systems See ecology.

ecology The study of the environmental relations of organisms.[NASA]

econometrics The application of mathematics and statistical techniques to the testing and quantifying of economic theories and the solution of economic problems.[NASA]

economic impact The impact on the economy from whatever cause.[NASA]

economic lot size The number of items to be manufactured in each setup, or the number to be purchased on each order, that will keep the costs of manufacturing, purchasing, setup, inspection and warehousing to a minimum over a certain time, usually a year or longer.

economizers Heat exchangers used to recover excess thermal energy from process streams. Economizers are used for feed preheat and as column reboilers. In some systems the reboiler for one column is the condenser for another.

eddy A whirlpool of fluid. See also vortices.

eddy current An electric current set up in the near-surface region of a metal part by induction resulting from the electromagnetic field of an external coil carrying an alternating current; eddy currents are used to generate heat or electromagnetic fields for use in such applications as induction heating, electromagnetic sorting and testing of materials, vibration damping in spacecraft, and various types of instrumentation.

eddy-current tachometer A device for measuring rotational speed which has been used extensively in automotive speedometers; it consists of a permanent magnet revolving in close proximity to an aluminum disk which is pivoted to turn against a spring; as a magnet revolves, it induces eddy currents in the disk, setting up torque that acts against the spring; the amount of disk deflection is indicated by a moving pointer directly coupled to the disk.

eddy viscosity The turbulent transfer of momentum by eddies giving rise to an internal fluid friction, in a manner analogous to the action of molecular viscosity in laminar flow, but taking place on a much larger scale.[NASA]

edge filter An interference filter which abruptly shifts from transmitting to reflecting over a narrow range of wavelengths.

edit To organize data for subsequent processing on a computer.

editor 1. A routine which performs editing operations. 2. A system program for amending the source code programs in high level language or assembly languages.

EDM See electrical discharge machining.

EDP See electronic data processing.

eductor 1. A device that withdraws a fluid by aspiration and mixes it with another fluid. 2. Using water, steam or air to induce the flow of other fluids from a vessel. See injector.

EEPROM See electrically erasable and programmable read-only memory.

effective address 1. A modified address. 2. The address actually considered to be used in a particular execution of a computer instruction.

effective bandwidth An operating characteristic of a specific transmission system equal to the bandwidth of an ideal system whose uniform pass-band transmission equals maximum transmission of the real system and whose transmitted power is the same as the real system for equal input signals having a uniform distribution of energy at all frequencies.

effective cutoff frequency A transducer characteristic expressed as the frequency where the insertion loss between two terminating impedances exceeds the loss at some reference frequency in the transmission band by a specified value.

effectiveness, built in test A measure of the system's automated capability to (a) correctly ascertain the operating condition of a subsystem/function and (b) isolate defective item(s) at a designated ambiguity level without the use of equipment which is external to the system. The BIT function may be contained within the item being evaluated, may be a separate piece of system equipment, or a combination of both elements.[ARD50010-91]

effective value The root-mean-square value of a cyclically varying quantity; it is determined by finding the average of the squares of the values throughout one cycle and taking the square root of the average.

efficiency 1. The ratio of output to the input. The efficiency of a steam generating unit is the ratio of the heat absorbed by water and steam to the heat in the fuel fired. 2. In manufacturing, the average output of a process or production line expressed as a percent of its expected output under ideal conditions. 3. The ratio of useful energy supplied by a dynamic system to the energy supplied to it over a given period of time.

efficiency of a light source The ratio of the total luminous flux to the total power input. In an electrical lamp it is expressed in lumens per watt.[ARP798-91]

efficiency, power (turbine) Ratio of actual power produced to theoretical power. For an air turbine starter, it can be expressed as the ratio of shaft power out to air horsepower in.[ARP906A-91]

efficiency, retraction A measure of efficiency for retraction of a landing gear system represented by a ratio (percentage) i.e., area under the load/stroke curve for the actuator compared to the maximum attainable load/stroke curve area for the actuator.[AIR1489-88]

efficiency, shock absorber A measure of efficiency for a shock absorber represented by a ratio (percentage) i.e., area under the load stroke curve for the shock absorber compared to a theoretical straight line (horizontal) load/stroke curve representing maximum attainable output.[AIR1489-88]

efficiency, volumetric Percentage ratio of actual flow to that exact flow established by design parameters of record (e.g., speed, displacement) for a specific item.[ARP906A-91]

effluent Liquid waste discharged from an industrial processing facility or waste treatment plant.

effluvium Waste by-products of food or chemical processing.

E format In FORTRAN, an exponential type of data conversion denoted by Ew.d where w is the number of characters to be converted as a floating point number with d spaces reserved for the digits to the right of the decimal point, e.g., E11.4 yields 000.5432E03 as input, 543.2 internally and 0.5432E+03 as output.

E glass A low alkali lime borosilicate glass made into glass fiber filaments used in composite materials.[NASA]

EGT Literally, exhaust gas temperature. As used, EGT refers to the temperature measurement defined by the engine manufacturer as a limiting parameter (for example, MGT, TIT, TOT, etc.). [AIR1873-88]

E.I.A. Electronics Industry Association who provide standards for such things as interchangeability between manufacturers.

EIA interface Serial word transfer in ASCII characters with RS-232C logic levels, as between the computer and a manual operator.

eigenvalues Eigenvalues are the roots (or solution parameters) of a system's characteristic equation. The roots are complex numbers, with a real and an imaginary part.[AIR1823-86]

eigenvalue sensitivity Eigenvalue sensitivity is a quantitative measure of the change in system eigenvalues for a given change in the magnitude of a design parameter.[AIR1823-86]

Einstein Observatory See HEAO 2.

EISCAT radar system (Europe) The European incoherent Scatter Radar system.[NASA]

ejecta Matter ejected during impact cratering processes, usually meteoritic. [NASA]

ejection 1. Physical removal of an object from a specific site—such as removal of a cast or molded product from a die cavity by hand, compressed air or mechanical means. 2. Emergency expulsion of a passenger compartment from an aircraft or spacecraft. 3. Withdrawal of fluid from a chamber by the action of a jet pump or eductor.

ejector 1. A device used in conjunction with the mechanism of a machine, fixture, or the like; which automatically sorts or throws out completed, accepted, or rejected items from the working station.[ARP480A-87] 2. Device consisting of a nozzle, mixing tube, and diffuser utilizing the kinetic energy of a fluid from a low pressure region by direct mixing and ejecting both streams. [NASA] 3. A device which utilizes the kinetic energy in a jet of water or other fluid to remove a fluid or fluent material from tanks or hoppers.

ejector condenser A direct-contact condenser in which vacuum is maintained by a jet of high-velocity injection water, which simultaneously condenses the steam and discharges water, condensate and noncondensible gases to the atmosphere or to the next stage ejector.

ejector half The movable portion of a diecasting or plastics-forming mold.

ejector pin A pin or rod that is driven into a hole in the rear of a mold cavity to remove the finished piece. Also known as knockout pin.

ejector plate The plate in a die-casting or plastics molding machine that backs up the ejector pins and holds the ejector assembly together.

ejector rod A rod or rodlike member that automatically operates the ejector assembly when a mold is opened.

Ekman layer The layer of transition between the surface boundary layer of the atmosphere, where the shearing stress is constant, and the free atmosphere, which is treated as an ideal fluid in approximate geostraphic equilibrium.[NASA]

elastic chamber The portion of a pressure-measuring system that is filled with the medium whose pressure is being measured, and that expands and collapses elastically with changes in pressure; examples include Bourdon tube, bellows, flat or corrugated diaphragm, spring-loaded piston, or a combination of two or more single elements, which may be the same or different types.

elastic collision A collision between two or more bodies in which the internal energy of the participating bodies remains constant, and in which the kinetic energy of translation for the combination of bodies is conserved.

elastic deformation The changes in dimensions of items, caused by stress, provided a return to original dimensions occurs when stress is removed.[ARP700-65]

elastic modulus See modules of elasticity.

elastic scattering A collision between two particles, or between a particle and a photon, in which total kinetic energy and momentum are conserved.

elastic stability See damping.

elastomer 1. A material which possesses elastic properties similar to those of natural rubber in the vulcanized state. At room temperature an elastomer can be stretched repeatedly to at least twice its original length and will, upon release of stress, return to its approximate original length.[AS1933-85] 2. Any elastic, rubber-like substance, such as natural or synthetic rubber.[ARP1931-87]

Elber equation In fatigue crack propagation studies, the effective stress range ratio $U = 0.5 + 0.4R$, where R is the stress ratio.[NASA]

elbow 1. A fitting that connects two pipes at an angle, usually 90° but may be any other angle less than 100°. 2. Any sharp bend in a pipe.

elbow meter A pipe elbow that is used as a flow measurement device by placing a pressure tap at both the inner and outer radius and measuring

electrical actuation systems (EAS) All electrical, mechanical, optical, or fluidic components necessary to convert a command signal and electrical power from the vehicle into a controlled linear or rotary force while using an electromagnetic prime mover.[ARP4255-91]

electrical apparatus 1. All items applied as a whole or in part for the utilization of electrical energy. These include, among others, items for generation, transmission, distribution storage measurement, control and consumption of electrical energy and items for telecommunications. 2. Intrinsically safe electrical apparatus and intrinsically safe parts of associated electrical apparatus shall be placed in one of two categories, 'ia' or 'ib'.

electrical apparatus category 'ia' An electrical apparatus that is incapable of causing ignition in normal operation, with a single fault and with any combination of two faults applied, with the following safety factors: 1.5—in normal operation and with one fault; 1.0—with two faults.

electrical apparatus category 'ib' An electrical apparatus that is incapable of causing ignition in normal operation and with a single fault applied, with the following safety factors: 1.5—in normal operation and with one fault; 1.0—with one fault, if the apparatus contains no unprotected switch contacts in parts likely to be exposed to a potentially explosive atmosphere and the fault is self-revealing.

electrical circuits—low-voltage circuit A circuit involving a potential of not more than 30 volts and supplied by a primary battery or by a standard Class 2 transformer or other suitable transforming device, or by a suitable combination of transformer and fixed impedance having output characteristics in compliance with what is required for a Class 2 transformer. A circuit derived from a source of supply classified as a high-voltage circuit, by connecting resistance in series with the supply circuit as a means of limiting the voltage and current, is not considered to be a low-voltage circuit.

electrical conductivity A material characteristic indicative of the relative ease with which electrons flow through the material—usual units are %IACS, which relates the conductivity to that of annealed pure copper; it is the reciprocal of electrical resistivity.

electrical discharge machining A machining method in which stock is removed by melting and vaporization under the action of rapid, repetitive spark discharges between a shaped electrode and the workpiece through a dielectric fluid flowing in the intervening space, often referred to by its abbreviation, EDM; process variations include electrical discharge grinding and electrical discharge drilling. Also known as electroerosive machining; electron discharge machining; electrospark machining.

electrical engineering A branch of engineering that deals with practical applications of electricity, especially the generation transmission and utilization of electric power by means of current flow in conductors.

electrical logic Logic for mode switching or failure detection and correction performed with electronic or electrical components.[ARP181A-85][AIR1916-88]

electrically actuated valves Hose valves that rely only on electrical power to position the valve.[ARP986-90]

electrically alterable read-only memory (EAROM) A type of computer memory that is normally unchangeable, its contents can be changed only under special conditions.

electrically erasable and programmable read-only memory (EEPROM) A later version of EAROM that is simpler to use.

electrically operated extensions

Usually a highly sensitive induction type device for signaling high or low flows or deviations from any set flow. The device consists of a sensing coil positioned around the extension tube of the rotameter. Movement of the metering float into the field of the coil causes a low level signal change which is usually amplified to a level suitable for performing annunciator or control functions.

electrical pumping Deposition of energy into a laser medium by passing an electrical current or discharge, or a beam of electrons through the material.

electrical resistivity A material characteristic indicative of its relative resistance to the flow of electrons—usual units are ohm-m (SI) or ohms per circular-mil foot (U.S. customary); it is the reciprocal of electrical conductivity.

electrical steel Low carbon steel that contains 0.5 to 5% Si or other material; produced specifically to have enhanced electromagnetic properties suitable for making the cores of transformers, alternators, motors and other iron-core electric machines. Contrast with electric steel.

electric boiler A boiler in which electric heating means serve as the source of heat.

electric chart drive A clocklike mechanism driven by an electric motor which advances a circular or strip chart at a preset rate.

electric circuits See circuits.

electric contact Either of two opposing, electrically conductive buttons or other shapes which allow current to flow in a circuit when they touch each other; they are usually attached to a spring-loaded mechanism that is mechanically or electromagnetically operated to control whether or not the contacts touch.

electric controller An assembly of devices and circuits which turns electric current to an electrically driven system off and on in response to a stimulus; in most instances, the assembly also monitors and regulates one or more characteristics of the electric supply—voltage or amperage, for example.

electric corona A luminous, and often audible, electric discharge that is intermediate in nature between a spark discharge (with, usually, its single discharge channel) and a non point discharge (with it diffuse, quiescent, nonluminous character).[NASA]

electric discharge The flowing of electricity through a gas, resulting in the emission of radiation that is characteristic of the gas and the intensity of the current.[NASA]

electric/electronic extensions A system that converts float position to a proportional electric signal (either a-c or d-c), or to a proportional shift or unbalance in impedance which is balanced by a corresponding shift in impedance in the receiving instrument.

electric field A condition within a medium or evacuated space which imposes forces on stationary or moving electrified bodies in direct relation to their electric charges.

electric field strength The magnitude of an electric field vector.

electric furnace Furnace whose heat is derived from electrical energy, generally achieved through resistance heating. Materials research and space processing are research uses.[NASA]

electric heating Any method for converting electric energy into heat, but especially those methods involving resistance to the passage of electric current.

electric hybrid vehicle Surface vehicle which utilizes propulsion systems of both electric motors and conventional internal combustion engines.[NASA]

electric hygrometer An instrument that uses an electrically powered sensing means to determine ambient atmospheric humidity.

electric instrument An indicating device for measuring electrical attributes of a system or circuit. Contrast with electric meter.

electricity meter A device for indicating the time integral of an electrical quantity.

electric meter A recording or totalizing instrument that measures the amount of electric power generated or used as a function of time. Contrast with electric instrument.

electric potential In electrostatics, the work done in moving unit positive charge from infinity to the point whose potential is being specified.[NASA]

electric propulsion A general term encompassing all the various types of propulsion in which the propellant consists of charged electrical particles which are accelerated by electrical or magnetic fields, or both; for example, electrostatic propulsion, electromagnetic propulsion, and electrothermal propulsion.[NASA]

electric-resistance-type liquid-level detector A device for detecting the presence of liquid at a given point; it consists of an electric probe which is insulated from the side of a vessel and positioned so that the end of the probe is at the desired liquid level; in operation, a small electric voltage is impressed between the probe and the vessel; if liquid exists at the probe level, current flows in the circuit, but does not flow if the liquid is below probe level; simple electrical systems are used when the solution has a resistivity less than 20,000 ohm-cm, but electronic systems can extend the range to 20-million ohm-cm; two probes, widely separated, can be used to control level between high and low limits or to provide high-level and low-level alarms.

electric steel Any steel melted in an electric furnace, which allows close control of composition. Also known as electric-furnace steel. Contrast with electrical steel.

electric stroboscope A device that uses an electric oscillator or similar element to produce precisely timed pulses of light; oscillator frequency can be controlled over a wide range so that the device can be used to determine the frequency of a mechanical oscillation—rpm of a rotating shaft or frequency of a mechanical vibration, for instance—by determining light-pulse frequency at which the object appears motionless.

electric tachometer An electrically powered instrument for determining rotational speed, usually in rpm.

electric telemeter An apparatus for remotely detecting and measuring a quantity—including the detector intermediate means, transmitter, receiver and indicating device—in which the transmitted signal is conducted electrically to the remote indicating or recording station.

electric thermometer An instrument that uses electrical means to measure and indicate temperature.

electric transducer A type of transducer in which all input, output and intermediate signals are electric waves.

electroacoustic transducer 1. Transducer for receiving waves from an electric system and delivering waves to an acoustic system, or vice versa. Microphones and earphones are electroacoustic transducers.[NASA] 2. A type of transducer in which the input signal is an electric wave and the output a sound wave, or vice versa.

electrochemical cleaning Removing soil by the chemical action induced by passing an electric current through an electrolyte. Also known as electrolytic cleaning.

electrochemical coating A coating formed on the surface of a part due to chemical action induced by passing an electric current through an electrolyte.

electrochemical corrosion Corrosion of a metal due to chemical action induced by electric current flowing in an electrolyte. Also known as electrolytic corrosion.

electrochemical machining A machining method in which stock is removed by electrolytic dissolution under the action of a flow of electric current between a tool cathode and the workpiece through an electrolyte flowing in the intervening space. Abbreviated ECM. Also known as electrochemical milling; electrolytic machining.

electrochemical recording A type of recording system in which a signal-controlled electric current passes through a sensitized recording medium, usually in sheet form, inducing a chemical reaction to occur in the medium.

electrochemical transducer A device that uses a chemical change to measure an input parameter, and that produces an output electrical signal proportional to the input parameter.

electrochromism A phenomenon whereby a select number of solid materials will change color when an electric field is applied.[NASA]

electrode 1. Terminal at which electricity passes from one medium into another. The positive electrode is called the anode; the negative electrode is called the cathode. In a semiconductor device, an element that performs one or more of the functions of emitting or collecting electrons or holes, or of controlling their movements by an electric field. In electron tubes, a conducting element that performs one or more of the functions of emitting, collecting or controlling, by an electromagnetic field, the movements of electrons or ions.[NASA] 2. An electrically conductive member that emits or collects electrons or ions, or that controls movement of electrons or ions in the interelectrode space by means of an electric field.

electrode characteristic The relation between electrode voltage and electrode amperage for a given electrode in a system, with the voltages of all other electrodes in the system held constant; the electrode characteristic is usually shown as a graph.

electrode force The force that tends to compress the electrodes against the workpiece in electric-resistance spot, seam or projection welding. Also known as welding force.

electrodeposition Any electrolytic process that results in deposition of a metal from a solution of its ions; it includes processes such as electroplating and electroforming. Also known as electrolytic deposition.

electrode voltage The electric potential difference between a given electrode and the system cathode or a specific point on the cathode—the latter is especially applicable when the cathode is a long wire or filament.

electrodynamic instrument An electrical instrument having both a fixed and a moving coil, both of which carry all or part of the current to be measured; if the coils are connected in series, interaction of the fields induced by the coils produces a torque proportional to the square of the current (as in an a-c voltmeter), and if connected in parallel, proportional to the product of the two coil currents (as in an a-c ammeter); in both cases, the indication is an effective (rms) value.

electrodynamics The science dealing with the forces and energy transformations or electric currents and the magnetic fields associated with them.[NASA]

electrodynamometers See dynamometers.

electroemissive machining See electrical discharge machining.

electroepitaxy Crystal growth process achieved by passing an electric current through the substrate solution.[NASA]

electro-explosive device (EED) Any cartridge, squib, ignitor, detonator, etc., which is initiated electrically.[AIR913-89]

electroforming Shaping a component by electrodeposition of a thick metal plate on a conductive pattern; the part may be used as formed, or it may be sprayed on the back with molten metal or other material to increase its strength.

electrogalvanizing Coating a metal with electrodeposited zinc.

electrograph 1. A tracing produced on prepared sensitized paper or other material by passing an electric current or electric spark through the paper. 2. A plot or graph produced by an electrically controlled stylus or pen.

electrohydraulic actuator A self-contained EAS that utilizes an electric motor driving a hydraulic pump to operate a hydraulic actuation system. [ARP4255-91]

electrojets Laterally limited relatively intense electric currents located in the ionosphere.[NASA]

electroless deposition Controlled autocatalytic reduction method of depositing coatings.[NASA]

electroless plating Deposition of a metal from a solution of its ions by chemical reduction induced when the basis metal is immersed in the solution, without the use of impressed electric current.

electroluminescence Emission of light caused by an application of electric fields to solids or gases. In gas electroluminescence, light is emitted when the kinetic energy of electron or ions accelerated in an electric field is transferred to the atoms or molecules of the gas in which the discharge takes place.[NASA]

electroluminescent lamps See electroluminescence.

electrolysis The production of chemical changes by passage of an electric current through an electrolyte.[ARP1931-87]

electrolyte The ionically conductive alkaline solution used in the nickel cadmium cell.[AS8033-88]

electrolytic cleaning See electrochemical cleaning.

electrolytic corrosion Corrosion caused by electrochemical reactions. [ARP1931-87] See also electrochemical corrosion.

electrolytic deposition See electrodeposition.

electrolytic etching Engraving a pattern on a metal surface by electrolytic dissolution.

electrolytic grinding A combined grinding and electrochemical machining operation in which an electrically conductive grinding wheel is made the cathode and the workpiece the cathode, and an electric current is impressed between them in the presence of a chemical electrolyte.

electrolytic hygrometer An apparatus for determining water-vapor content of a gas by directing it at known flow rate through a teflon or glass tube coated on the inside with a thin film of P_2O_5 (phosphorus pentoxide), which absorbs water from the flowing gas; the water is dissociated by a d-c voltage impressed on a winding embedded in the hygroscopic film dissociates the water, and the resulting current represents the number of molecules dissociated; a calculation based on flow rate, current and temperature yields water concentration in ppm.

electrolytic machining See electrochemical machining.

electrolytic pickling Removal of scale and surface deposits by electrolytic action in a chemically active solution.

electrolytic polishing See electropolishing.

electrolytic powder Metal powder that is produced directly or indirectly by electrodeposition.

electromagnet Any magnet assembly whose magnetic field strength is determined by the magnitude of an electric current passing through some portion of the assembly.

electromagnetic See electromagnetism.

electromagnetic acceleration The use of perpendicular components of electric and magnetic fields to accelerate a current carrier.[NASA]

electromagnetic control See remote control

electromagnetic environment experiment Shuttleborne radio frequency experiment.[NASA]

electromagnetic instrument Any instrument in which the indicating means or recording means is positioned by mechanical motion controlled by the strength of an induced electromagnetic field.

electromagnetic interference Any electrical or electronic disturbance phenomenon, signal or emission (man made or natural) which causes undesirable responses, unacceptable responses, malfunctions, degradation of performance, or premature and undesired location, detection or discovery by enemy forces, except deliberately generated interference.[ARP1281A-81] See also interference, electromagnetic.

electromagnetic pulse (EMP) A type of disturbance that leads to noise in radio-frequency electric or electronic circuits.

electromagnetic radiation 1. Energy propagated through space or through material media in the form of an advancing disturbance in electric and magnetic fields existing in space or in media. The term radiation, alone, is used commonly for this type of energy, although it actually has a broader meaning.[NASA] 2. An all-inclusive term for any wave having both an electric and a magnetic component; the spectrum of electromagnetic waves includes—in order of increasing photon energy, increasing frequency and decreasing wavelength—radio waves, infrared, visible light, ultraviolet, x-rays, gamma rays and cosmic rays.

electromagnetic spectra Spectra of known electromagnetic radiations, extending from the shortest cosmic rays, through gamma rays, x-rays, ultraviolet radiation, visible radiation, and including microwave and all other wavelengths of radio energy.[NASA]

electromagnetic wave A wave in which both the electric and magnetic fields vary periodically, usually at the same frequency. See also electromagnetic radiation.

electromagnetism Magnetism produced by an electric current. The science dealing with the physical relations between electricity and magnetism.[NASA]

electromechanical actuator An EAS that utilizes an electric motor mechanically coupled to the load.[ARP4255-91]

electromechanical transducer A type of transducer in which the input signal is an electric wave and the output mechanical oscillation, or vice versa.

electromechanics The technology associated with mechanical devices and systems that are electromagnetically or electrostatically actuated or controlled.

electrometallurgy The technology associated with recovery and processing of metals using electrolytic and electrical methods.

electrometer 1. Instrument for measuring differences in electric potential. [NASA] 2. An instrument for measuring electric charge, usually by means of the forces exerted on one or more charged electrodes in an electric field.

electrometer tube A high-vacuum tube that can measure extremely small d-c voltages or amperages because of its exceptionally low control-electrode conductance.

electromotive force Force capable of maintaining a potential difference, and thus a current, within a circuit. These forces can be established by chemical action or by mechanical work.[NASA]

electromyogram See electromyography.

electromyograph See electromyography.

electromyography The study of the response of a muscle to an electric stimulation.[NASA]

electron An elementary subatomic particle having a rest mass of $9.107 x 10^{-28}$ g and a negative charge of $4.802 x 10^{-10}$ statcoulomb. Also known as negatron. A subatomic particle of identical weight and positive charge is termed a positron.

electron acceleration The acceleration of electrons by action of solar cosmic rays.[NASA]

electron avalanche The process in which a relatively small number of free electrons in a gas that is subjected to a strong electric field accelerate, ionize gas atoms by collision, and thus form new free electrons to undergo the same process in cumulative fashion.[NASA]

electron beam 1. Specifically, focused stream of electrons used for neutralization of the positively charged ion beam in an ion engine. Also used to melt or weld materials with externally high melting points.[NASA] 2. A narrow, focused ray of electrons which streams from a cathode or emitter and which can be used to cut, machine, melt, heat treat or weld metals.

electron-beam instrument Any instrument whose operation depends on using variable electric or magnetic fields, or both, to deflect a beam of electrons; the electron beam may be of constant intensity or it may vary in intensity according to a control signal.

electron-beam tube Any of several types of electron tube whose performance depends on formation and control of one or more electron beams.

electron cyclotron heating A type of radio frequency plasma heating in which high-power microwave energy is introduced into the plasma region. [NASA]

electron device Any device whose operation depends on conduction by the flow of electrons through a vacuum, gas-filled space or semiconductor material.

electron emission Ejection of free electrons from the surface of an electrode into the adjacent space.

electron flux See flux (rate).

electron gun An electron-tube subassembly that generates a beam of electrons, and may additionally accelerate control, focus or deflect the beam.

electron-hole drops Exciton condensations exhibiting the properties of electrically conducting plasmas which form in germanium and silicon crystals at sufficiently low cryogenic temperatures.[NASA]

electronic aircraft Designation for tactical electronic warfare aircraft. [NASA]

electronic amplifier See amplifier.

electronic attitude director indicator (EADI) See primary flight display.

electronic data processing (EDP) Data processing performed largely by electronic equipment.

electronic engineering A branch of engineering that deals chiefly with the design, fabrication and operation of electron-tube or transistorized equipment, which is used to generate, transmit, analyze and control radio-frequency electromagnetic waves or similar electrical signals.

electronic equipment Equipment in which electricity is conducted principally by electrons moving through a vacuum, gas or semiconductor.[NASA]

electronic heating Producing heat by the use of radio-frequency current generated and controlled by an electron-tube oscillator or similar power source.

Also known as high-frequency heating; radio-frequency heating.

electronic horizontal situation indicator (EHSI) See nav-display.

electronic level See energy level.

electronic mail The use of a large centralized computer to store messages for users of the electronic mail network.

electronic photometer See photoelectric photometer.

electronics That branch of physics that treats of the emission, transmission, behavior, and effects of electrons. [NASA]

electronic switch A circuit element causing a start and stop action or a switching action electronically, usually at high speeds.

electronic transition A transition in which an electron in an atom or molecule moves from one energy level to another.

electronic warfare Military action involving the use of electromagnetic energy to determine, exploit, reduce, or prevent hostile use of the electromagnetic spectrum, and action which retains friendly use of the electromagnetic spectrum.[NASA]

electron ionization See ionization.

electron metallography Using an electron microscope to study the structure of metals and alloys.

electronmicroprobe analysis A technique for determining concentration and distribution of chemical elements over a microscopic area of a specimen by bombarding the specimen with high-energy electrons in an evacuated chamber and performing x-ray fluorescent analysis of secondary x-radiation emitted by the specimen.

electron microscope Any of several designs of apparatus that use diffracted electron beams to make enlarged images of tiny objects.

electron microscopy The interpretive application of an electron microscope for the magnification of materials that cannot be properly seen with an optical microscope.[NASA]

electron multiplier See photomultiplier tube.

electron multiplier tube A type of electron tube that uses cascaded secondary emission to amplify small amperages.

electron runaway (plasma physics) High acceleration of electrons in a collisional plasma caused by a suddenly applied electric field (which greatly reduces the collision cross section of the electrons).[NASA]

electron spectroscopy The study and interpretation of atomic, molecular, and solid state structure based on x-ray induced electron emission from substances.[NASA]

electron tube Any device whose operation depends on conduction by the flow of electrons through a vacuum or gas-filled space within a gastight envelope.

electron volt Abbreviated eV (preferred) or EV. A unit of energy equal to the work done in accelerating one electron through an electric potential difference of one volt.

electro-optic effect A change in the refractive index of a material under the influence of an electric field. Kerr and Pockels effects are respectively quadratic and linear in electric field strength.

electropainting Electrodeposition of a thin layer of paint on metal parts which are made anodic. Also known as electrophoretic painting.

electrophonic effect A hearing sensation that results when a-c electric current of suitable amplitude and frequency passes through an animal's tissues.

electroplate The application of a metallic coating on a surface by means of electrolytic action.[ARP1931-87]

electroplating Electrodeposition of a thin layer of metal on a surface of a part that is in contact with a solution, or electrolyte, containing ions of the deposited metal; in most electroplating processes, the part to be plated is the cathode, and the concentration of metal ions in the solution is maintained by placing a sacrificial anode of the deposited metal in the electrolyte.

electropneumatic controller An electrically powered controller in which some or all of its basic functions are performed by pneumatic devices.

electropolishing Smoothing and polishing a metal surface by closely controlled electrochemical action similar to electrochemical machining or electrolytic pickling.

electroscope An instrument for detecting an electric charge by observing the effects of mechanical force exerted between two or more electrically charged bodies.

electroseismic effect See seismic waves.

electrospark machining See electrical discharge machining.

electrostatic bonding Use of the particle-attracting property of electrostatic charges to bond particles of one charge to those of the opposite charge.[NASA]

electrostatic instrument Any instrument whose operation depends on forces of electrostatic attraction or repulsion between charged bodies.

electrostatic lens A set of electrodes arranged so that their composite electric field acts to focus a beam of electrons or other charged particles.

electrostatic memory A memory device which retains information by means of electrostatic charge, usually involving a special type of cathode-ray tube and its associated circuits.

electrostatic microphone An electroacoustic transducer for converting sound into an electrical signal by means of variation in electrostatic capacitance of the active transducer element.

electrostatic painting A spray painting process in which the paint particles are charged by spraying them through a grid of wires which is held at a d-c potential of about 100 kV; the parts being painted are connected to the opposite terminal of the high-voltage circuit so that they attract the charged paint particles; the process yields more uniform coverage than conventional spray painting, especially at corners, edges, recesses and oblique surfaces.

electrostatic plasma See plasma (physics).

electrostatic precipitator A device for removing dust and other finely divided matter from a flowing gas stream by electrostatically charging them and then passing the gas stream over charged collector plates which attract and hold the particles.

electrostatic voltmeter An instrument for measuring electrical potential by means of electrostatic forces between elements in the instrument.

electrostriction The phenomenon wherein some dielectric materials experience an elastic strain when subjected to an electric field, this strain being independent of polarity of the field. [NASA]

electrostriction transducer A device which consists of a crystalline material that produces elastic strain when subjected to an electric field, or that produces an electric field when strained elastically. Also known as piezoelectric transducer.

electrothermal process Any process that produces heat by means of an electric current—using an electric arc, induction or resistance method—especially when temperatures higher than those obtained by burning a fuel are required.

electrothermal recording A form of electrochemical recording in which chemical change is induced by thermal effects associated with the passage of electrical current.

electrothermic instrument Any instrument whose operation depends on the heating effect associated with the passage of electric current.

element 1. In data processing, one of the items in an array. 2. A substance which cannot be decomposed by chemical means into simpler substances.

element melt time The time elapsed from the moment a fusing current begins to flow to the moment the current sharply drops in value and arcing commences.[ARP1199A-90]

elements The general class of devices in their simplest form used to make up fluidic components and circuits; for example, fluidic restrictors and capacitors, a proportioned amplifier or an OR-NOR logic gate. These are the least "common denominators" of the fluidics technology.[ARP993A-69]

elements of industrial process measurement and control systems Functional units or integrated combinations thereof which ensure the transducing, transmitting or processing of measured values, control quantities or variable, and reference variables. A valve actuator in combination with a current to pressure transducer, valve positioner, or a booster relay is considered an element which receives the standard pneumatic transmission signal or standard electric current transmission signal.

elevated range See range, suppressed-zero.

elevated span See range, suppressed-zero.

elevated-zero range See range, elevated-zero.

elevation Vertical distance above a reference level, or datum, such as sea level. See range, suppressed-zero.

elevation error A type of error in temperature-measuring or pressure-measuring systems that incorporate capillary tubes partly filled with liquid; the error is introduced when the liquid-filled portion of the system is at a different level than the instrument case, the amount of error varying with distance of elevation or depression.

elevator A movable horizontal airfoil, usually attached to the horizontal stabilizer, that is used to control pitch. [ARP4107-88]

elevator illusion See illusion, vestibular.

elinvar An iron-nickel-chromium alloy that also contains varying amounts of manganese and tungsten and that has low thermal expansion and almost invariable modulus of elasticity; its chief uses are for chronometer balances, watch balance springs, instrument springs and other gage parts.

ellipses Plane curve constituting the locus of all points the sum of whose distances from two fixed points called focuses or foci is constant; an elongated circle.[NASA]

ellipsoid Surface whose plane sections (cross sections) are all ellipses or circles, or the solid enclosed by such a surface.[NASA]

ellipsometer An optical instrument which measures the constants of elliptically polarized light. It is most often used in thin-film measurements.

elliptically polarized light Light in which the polarization vector rotates periodically, changing in magnitude with a period of 360° so it describes an ellipse. The result of two plane polarized beams of light (each approximately a sine wave) perpendicular to each other and having a constant phase difference; the resultant planes polarized wave in the direction of the common beams well describe an ellipse. A special case called circular polarization occurs when the amplitudes of the two planes' polarized waves are equal and the phase difference is an odd multiple of $\pi/2$.

elliptically polarized wave Any electromagnetic wave whose electric or magnetic field vector, or both, at a given point describes an ellipse.

elliptical plasma Confined non-circular plasma.[NASA]

elliptical polarization The polarization of a wave radiated by an electric vector rotating in a plane and simultaneously varying in amplitude so as to describe an ellipse.[NASA]

ellipticity The amount by which a spheroid differs from a circle, calculated by dividing the difference in the length of the axes by the length of the major axis.[NASA]

elongated alpha The hexagonal crystal phase appearing as stringer-like arrays, considerably larger in appearance than the primary alpha. Commonly exhibits an aspect ratio of 3:1 or higher. [AS1814-90]

elongation 1. The permanent extension in length of a material which has been mechanically stressed by tension or exposure to heat.[ARP1931-87] 2. Elongation is the permanent extension in the gage length of a test specimen, measured after rupture. Ultimate elongation is the extension measured at the moment of rupture. Unless otherwise specified, elongation is expressed as a percentage of the original gage length. For example, if a 1-inch gage is marked on an unstretched specimen and the specimen is stretched until the gage marks are 7 inches apart, the elongation is 6 inches or 600 percent.[AS1198-82]

ELT Emergency Locator Transmitter.

elutriation Separation of fine, light particles from coarser, heavier particles by passing a slow stream of fluid upward through a mixture so that the finer particles are carried along with it.

embedded computer systems Computer systems physically incorporated into larger systems whose primary function is not data processing.[NASA]

embolism Large amount of air in the blood stream which, when reaching the heart, causes it to fail; small amounts are resorbed and cause no symptoms. [NASA]

embossing Raising in relief on a surface.[NASA]

embrittlement cracking A form of metal failure that occurs in steam boilers at riveted joints and at tube ends, the cracking being predominantly intercrystalline.

emergency A sudden unplanned occurrence which jeopardizes the safe completion of a task and requires specific and timely action to avoid damage or injury.[ARP4107-88]

emergency electric-system operation That condition of the electric-system during flight when the primary electric-system becomes unable to supply sufficient or proper electric power, thus requiring the use of a limited independent source(s) of emergency power for essential utilization equipments. [AS1212-71]

emergency locator transmitter (ELT) A radio transmitter attached to the aircraft structure which operates from its own power source on 121.5 MHz and 243.0 MHz. It aids in locating downed aircraft by radiating a downward-sweeping audio tone between two and four times per second. It is designed to function independently after an accident.[ARP4107-88]

emergency locking retractor (inertia reel) A retractor incorporating adjustment hardware by means of a locking mechanism that is activated by aircraft acceleration, webbing movement relative to the aircraft, or other automatic action during an emergency and is capable, when locked, of withstanding restraint forces.[AS8043-86]

emergency maintenance An urgent need for repair or upkeep that was unpredicted or not previously planned work. See corrective maintenance.

emergency power unit (EPU) A device which can provide short term emergency electric, pneumatic or hydraulic power for engine starting, flight controls, etc. in the even of primary source failure. [ARP906A-91]

emergency safe altitude See minimum safe altitude.

emery An abrasive material composed of pulverized, impure corundum—used in various forms including cloth or paper with an adhesive-bonded layer of emery grains, and compacted emery-binder mixtures shaped into cakes, sticks, stones, grinding wheels and other implements.

EMI Electromagnetic interference. See interference, electromagnetic.[ARP1161-91]

emission characteristic The relation between rate of electron emission and some controlling factor—temperature, voltage or current of a filament or heater, for instance—for a specific element of a system, all other factors being held constant.

emission index The mass of emissions of a given constituent per unit mass of fuel, multiplied by 1000.[AIR1533-91]

emission spectra The spectra of wavelengths and relative intensities of electromagnetic radiation emitted by a given radiator. Each radiating substance has a unique, characteristic emission spectrum, just as every medium of transmission has its individual absorption spectrum.[NASA]

emissivity 1. A property of a material, measured as the emittance of a specimen of the material that is thick enough to be completely opaque and has an optically smooth surface.[NASA] 2. A material characteristic determined as the ratio of radiant-energy emission rate due solely to temperature for an opaque, polished surface of a material divided by the emission rate for an equal area of a blackbody at the same temperature. 3. The rate at which electrons are emitted from a solid or liquid surface when additional energy is imparted to the system by radiant energy such as heat or light or by energetic particles such as a beam of electrons.

emissograph See actinometer.

emittance An alternative term for emissivity.

emotion See affective states.

empennage See tail assembly.

empty field myopia See illusion, visual.

empty weight The weight of an aircraft with fixed ballast, unusable fuel, and full operating fluids including oil, hydraulic fluid, and other fluids required for normal operation of the airplane's systems. See also basic empty weight.-[ARP4107-88]

EMR The earlier name of the Data Systems Division within Fairchild-Weston, a Schlumberger corporation, supplier of telemetry/computer systems.

E-MSAW En Route Minimum Safe Altitude Warning.

emulate To imitate one system with another such that the imitating system accepts the same data, executes the same programs, and achieves the same results as the imitated system.

emulator 1. A device or program that emulates, usually done by microprogramming the imitating system. Contrasted to simulate. 2. A computer that behaves very much like another computer by means of suitable hardware and software.

emulsifier A substance that can be mixed with two immiscible liquids to form an emulsion. Also known as disperser; dispersing agent.

emulsion characteristic curve A graph of relative transmittance of a developed photographic or radiographic emulsion versus exposure; alternatively, a graph of a function of transmittance versus a function of exposure.

emulsion corrosion test A test used

to evaluate the rust preventive property of grease-water emulsions when coated steel objects are exposed to the severely corrosive atmosphere of the salt spray test.[S-5C,40-55]

enable 1. To restore a computer system to ordinary operating conditions. 2. To "arm" a software or hardware element to receive and respond to a stimulus. 3. Allow the processing of an established interrupt. 4. Remove a blocking device, i.e. switch, to permit operation. Contrasted with disable. See arm.

enamel 1. Thin ceramic coating, usually of high glass content, applied to a substrate, generally a metal.[NASA] 2. A type of oil paint that contains a finely ground resin and that dries to a harder, smoother, glossier finish than other types of paint. 3. Any relatively glossy coating, but especially a vitreous coating on metal or ceramic obtained by covering it with a slurry of glass frit and firing the object in a kiln to fuse the coating. Also known as glaze; porcelain enamel.

encapsulated microcircuit Microelectronic circuit enclosed in plastic.[NASA]

Enceladus A satellite of Saturn orbiting at a mean distance of 238,000 kilometers.[NASA]

encipher See encode.

Encke comet A very faint comet with a periodicity of 3.3 years which is the shortest of any known comet.[NASA]

encode 1. To apply a code, frequently one consisting of binary numbers, to represent individual characters or groups of characters in a message. Synonymous with encipher. 2. To represent computer data in digital form. 3. To substitute letters, numbers, or characters for other numbers, letters, or characters, usually to intentionally hide the meaning of the message except to certain individuals who know the enciphering scheme.

encoder 1. A device capable of translating from one method of expression to another method of expression, e.g., translating a message, "add the contents of A to the contents of B," into a series of binary digits. Contrasted with decoder. 2. A device that transforms a linear or rotary displacement into a proportional digital code. 3. A hardware device that converts analog data into digital representations.

encrustation The buildup of slag, corrosion products, biological organisms such as barnacles, or other solids on a structure or exposed surface.

encryption Converting data into codes that cannot be read without a key or password.

end 1. An end is an individual yarn or strand in a braid or shield.[AS1198-82] 2. In computer programming, a word indicating the completion of a program structure.

endangered species Living organisms (except plants) whose populations have diminished to such low levels that survival may require extraordinary conservation procedures. Changes in size and quality of the ecology are considered the cause of the possible extinction of some species.[NASA]

end around carry A carry from the most significant digit place to the least significant digit place.

end burning A term used to describe a solid propellant grain which is inhibited so that it burns from one end only so that burning progresses in the direction of the longitudinal axis.[AIR913-89]

end connections, flanged End connections incorporating flanges that mate with corresponding flanges on the piping.

end connections, split clamp End connections of various proprietary designs using split clamps to apply gasket or mating surface loading.

end connections, threaded End connections incorporating threads, either male or female.

end connections, welded End connections that have been prepared for welding to the line pipe or other fittings. May be butt weld (BW), or socket weld (SW).

end device The last device in a chain of devices that performs a measurement function, the last device being the one performing final conversion of a measured value into an indication, record or control-system input signal.

end device, end instrument See transducer.

end effect The consequence(s) a failure mode has on the operation, function, or status of the highest indenture level. [ARD50010-91]

end of descent (EOD) The metering fix, clearance, or other scheduled end point for the main descent from enroute cruise. If a level flight deceleration segment is programmed at the end flight level, the EOD or BOD (bottom of descent) will be the downstream end of that segment. The EOD or BOD includes a geographical position, altitude and speed. It may also include a required or estimated time of arrival.[ARP1570-86]

end-of-file (EOF) A magnetic marker on tape that signifies where a data file ends.

end of tape (EOT) A unique reflective marker near the end of a reel of magnetic tape to warn the computer that the end is approaching.

endothermic reaction A reaction which occurs with the absorption of heat.

end play Axial movement in a shaft-bearing assembly due to clearances within the assembly.

end point In titration, an experimentally determined point close to the equivalence point, which is used as the signal to terminate titration; it is used instead of equivalence point in most calculations, and corrections for the error between end point and equivalence point usually are not applied.

endpoint control The exact balancing of process inputs required to satisfy its stoichiometric demands.

endpoint linearity The linearity of the object taken between the end points of calibration.

ends See braid ends.

end scale value The value of an actuating electrical quantity that corresponds to the high end of the indicating or recording scale on a given instrument.

end-to-end data system Comprehensive data system which demonstrates the processing of sensor data to the user thus reducing data fragmentation. [NASA]

endurance limit The maximum stress below which a material can presumably withstand an infinite number of stress cycles; if the stress is not completely reversed, the minimum stress also should be given. See also fatigue strength.

energetic particles Charged particles having energies equaling or exceeding a hundred Mev.[NASA]

energize To apply rated voltage to a circuit or device in order to activate it. [ARP1931-87]

energy 1. Any quantity with dimension which can be represented as mass times length squared divided by time squared. [NASA] 2. The capacity of a body for doing work or its equivalent—it may be classified as potential or kinetic, depending on whether it is associated with bodies at rest or bodies in motion; or it may be classified as chemical, electrical, electromagnetic, electrochemical, mechanical, radiant, thermal or vibrational, or any other type, depending on its source or nature.

energy balance The balance relating the energy in and energy out of a column. In control applications the energy balance manipulative variables are reflux and boilup.

energy beam An intense ray of electromagnetic radiation, such as a laser beam, or of nuclear particles, such as electrons, that can be used to test materials or to process them by cutting, drilling, forming, welding or heat treating.

energy, brake The portion of total energy of an air vehicle which in deceleration process goes to the brake. (Other portions go to aerodynamic drag, rolling resistance, etc.)[AIR1489-88]

energy budget Quantitative description of the total energy exchange into and out of a given physical or ecological system; may include radiation heat, kinetic, and biological process.[NASA]

energy density Light energy per unit area, expressed in joules per square meter—equivalent to the radiometric term "irradiance". See also flux density.

energy, drag chute The portion of total energy of an air vehicle which in the deceleration process goes into the drag chute in the form of aerodynamic drag.[AIR1489-88]

energy efficiency transport program See ACEE program.

energy exchanger A generic term for any of several devices whose primary function is to transfer energy from one medium to another—examples include heat exchangers, boilers, and electrical transformers.

energy gap (solid state) A range of forbidden energies in the band theory of solids.[NASA]

energy level Any one of different values of energy which a particle, atom, or molecule may adopt under conditions here the possible values are restricted by quantizing conditions.[NASA]

energy limited A starter that can provide only a fixed amount of energy on a given start. Examples include cartridge starter, fuel-air starter, hydraulic starter operating from an accumulator, and battery operated electric starter.[ARP906A-91]

energy losses Tire loss of energy due to rolling resistance of the tire which includes consideration of forces required to compress the tire as it rolls, as well as the work required to flex the tire carcass.[AIR1489-88]

energy, shock absorber The portion of the total air vehicle energy which in the landing process goes into the shock absorber(s). Generally, it is the part due to the vertical sink speed of the vehicle at touchdown although some of this energy is damped out in the air vehicle structure, and tires.[AIR1489-88]

energy, tire The portion of the total air vehicle energy which in the landing process goes into the tires. The tires also absorb energy during the braking process.[AIR1489-88]

engagement The connecting of a starter shaft to the engine. Sometimes recalled re-engagement if the engine is not at rest.[ARP906A-91]

engagement, crash The connecting of the starter shaft to the engine while the engine is decelerating and the starter is running at a no load speed which is greater than that of the engine. The differential speed that exists at engagement can cause a destructive crash engagement. Sometimes called crash re-engagement.[ARP906A-91]

engagement, running The connecting of the starter shaft to the engine while the engine is running (usually decelerating or steady windmilling) at some speed below starter cutoff speed. Except for the effects of backlash in the drive train, the engagement takes place when the starter speed is the same as that of the engine. Sometimes called running re-engagement.[ARP906A-91]

engaging mechanism A device for connecting the starter shaft to the engine. [ARP906A-91]

engaging mechanism, jaw type An engaging mechanism that uses matching elements with jaw teeth formed on their faces. One of the pair is typically attached to the engine-side accessory shaft. The other is attached to the starter and is caused to move axially into engagement with the engine-side jaw. Separation is automatic and the mechanism permits the starter elements to come to rest between start cycles.[ARP 906A-91]

engaging mechanism, overrunning clutch A type of clutch mechanism that employs sprags, rollers, or pawls, etc., with appropriate driving and driven members. This clutch drives in one direction only, overrunning in the other direction.[ARP906A-91]

engine 1. Machine or apparatus that converts energy, especially heat energy, into work.[NASA] 2. A machine whose chief purpose to to convert various forms of energy, such as heat or chemical energy, into mechanical power, and to perform work by imparting mechanical force and motion to other mechanisms.

engine accessory A part or assembly usually driven by the engine but not required to operate the engine.[ARP 1587-81]

engine airframe integration Physics of the interface between the engine and the airframe.[NASA]

engine bleed systems Bleed gas is normally defined as "high pressure" air that is extracted from the compressor section of turbine engine. The term "high pressure" embraces engine compressor bleed systems used for engine starting, auxiliary power and wing anti-icing. Pressures may range to 250 psig and 1000 F in present day aircraft.[AIR 744A-72]

engine block See cylinder block.

engine duty cycle A composite cycle (or cycles) derived from the mission profiles and mission mix. The engine duty cycle is usually expressed in terms of power lever position versus time.[ARD 50010-91]

engine component A part or assembly that is fundamental to the operation of the engine.[ARP1587-81]

engine control Any control for regulating the power and speed of an engine, such as the throttle, mixture control, manifold pressure regulator, fuel pressure control, or supercharger control. [NASA]

engineering drawing Two-dimensional geometry, usually representing 3D parts, assemblies, etc., and associated text annotations. They are historically the medium of engineering design or product definition.[AS4159-88]

engineering plastics Plastics materials that are suitable for making into structural members and machine elements.

engineering psychology The study of human behavior in using tools and machines and of machine design in relation to the human's behavioral capacities, abilities, and motivations.[ARP 4107-88]

engineering time The total machine downtime necessary for routine testing, good or bad, for machine servicing due to breakdowns, or for preventive servicing measures. This includes all test time, good or bad, following breakdown and subsequent repair of preventive servicing. Synonymous with servicing time.

engineering units Terms of data measurement, as degrees Celsius, pounds, grams, and so on.

engine failure Engine stoppage attributable to the engine structure. See also power failure.[ARP4107-88]

engine failures, chargeable All engine failures observed less the excluded failures.[ARD50010-91]

engine failures, observed The inability of the engine to perform within engine specification limits or service manual limits.[ARD50010-91]

engine health monitoring The general

discipline or technique for indication of status of the mechanical or functional condition of an engine or engine components; sometimes referred to as engine condition monitoring.[ARP1587-81]

engine lathe A manually operated lathe whose headstock is driven by a gear train, by a stepped pulley mechanism, or by a combination of gears and pulleys.

engine monitoring system (EMS) An EMS is a complete system approach to define engine, engine component and sub-system health status through the use of sensor inputs, data collection, data processing, data analysis and the human decision process. This system approach can consist of an integrated set of hardware and software and several separate engine monitoring system elements and can be manual, computer aided or automated.[AIR1873-88]

engine pod A streamlined structure or nacelle on an airplane, usually mounted beneath the wing or attached to the wing tip, housing one or more jet engines.[ARP4107-88]

engine structural integrity program A time-phased set of required actions performed at the optimum time during the life cycle (design through phase-out) of an aircraft engine, to ensure the structural integrity (strength, rigidity, damage tolerance, durability, and service life capability) of the engine.[ARD50010-91]

Engler viscosity A standard time-based viscosity scale used primarily in Europe.

Enhanced Performance Architecture (EPA) EPA is an extension to MAP that provides for low delay communication between nodes on a single segment. See MAP/EPA and MINI-MAP.

enhancement, serial data A method whereby a continuous string of logical ONEs or ZEROs is modified to introduce bit transitions to enable bit synchronization for recording purposes; also preserves bandwidth. For example, in an incoming serial data stream, a number of words are all logic ZEROES, and therefore a DC level that the bit synchronizer cannot synchronize on; the data are enhanced by making the LSB of the words a logic ONE.

en route descent Descent from the en route cruising altitude which takes place along the route of flight.[ARP4107-88]

en route flight advisory service/flight watch A service specifically designed to provide, upon pilot request, timely weather information pertinent to his/her type of flight, intended route of flight, and altitude. See also flight watch.[ARP4107-88]

en route minimum safe altitude warning (E-MSAW) A function of the NAS State A en route computer that aids the controller by alerting him/her when a tracked aircraft is below (or is predicted by the computer to go below) a predetermined minimum IFR altitude (MIA). [ARP4107-88]

enter key The key on a computer terminal that is pressed to enter data into a computer.

enthalpy A mathematically defined thermodynamic function of state. [NASA]

entity An active element within an OSI layer (e.g. Token Bus MAC is an entity in the Layer 2).

entrainment The conveying of particles of water or solids from the boiler water by the steam.

entropy A measure of the extent to which the energy of a system is unavailable.[NASA]

entropy (statistics) A factor or quantity that is a function of a mechanical system and is equal to the logarithm of the probability of the particular arrangement in that state.[NASA]

entry Any item of computer data to be stored and processed.

entry conditions The initial data and control conditions to be satisfied for successful execution of a given routine.

entry data The initial data required for successful execution of a given routine. See entry conditions.

entry guidance (STS) THe precise steering commands for trajectory from initial penetration of the earth's atmosphere until the terminal area guidance is activated at an earth-relative speed (about 2500 fps).[NASA]

entry name The alphanumeric name given to an entry point. See entry point.

entry point In a routine, any place to which control can be passed.

envelope 1. Generally, the boundaries of an enclosed system or mechanism. 2. Specifically, the glass or metal housing of an electron tube, or the glass enclosure of an incandescent lamp. Also known as bulb.

environment The aggregate of all external and internal conditions (such as temperature, humidity, radiation, magnetic and electric fields, shock, vibration, etc.) either natural or manmade, or self-induced, that influences the form, performance, reliability, or survival of an item.[ARD50013-91]

environmental area See area, environmental.

environmental chamber See test chamber.

environmental chemistry Collective term comprising the complex chemical relationships involving the atmosphere, climatology, air and water pollution, fuels, pesticides, energy, biochemistry, geochemistry, etc.[NASA]

environmental damage/deterioration Physical deterioration of an item's strength or resistance to failure as a result of interaction with climate or environment.[ARD50010-91]

environmental engineering A branch of engineering that deals with the technology related to control of the surroundings which humans live in, especially the control or mitigation of contamination or degradation of natural resources such as air quality and water purity.

environmental influence See operating influence.

environmental stress screening A series of tests conducted under environmental stresses to disclose weak parts and workmanship defects for correction. [ARD50010-91]

environmentally sealed The provision or characteristic of a device which enables it to be protected against the entry of moisture, fluids and foreign, particulate contaminants which could otherwise affect the performance of the device.[ARP914A-79]

environmental temperature See ambient temperature.

environmental test Any laboratory test conducted under conditions that simulate the expected operating environment in order to determine the effect of the environment on component operation or service life.

EOF marker In data processing, a code written after the last record of a file to indicate the end of that file.

eosinophils A type of white blood cell or leukocyte which stains a red color with resin stain; normally about 2 or 3 percent of white cells in the blood but tending to decrease during stressful situations and thus usable as an index for stress.[NASA]

EPA See Enhanced Performance Architecture.

ephemeride Periodical publication tabulating the predicted positions of celestial bodies at regular intervals, such as daily, and containing other data of interest to astronomers. A publication giving similar information useful to a navigator is called an almanac.[NASA]

ephemeris time The uniform measure of time defined by the laws of dynamics and determined in principle from the orbital motions of the planets, specifically the orbital motion of the earth as represented by Newcomb's Tables of the Sun.[NASA]

EP lubricant See extreme-pressure lubricant—an oil or grease containing additives that enhance the ability of the lubricant to adhere to a surface and reduce friction under high bearing loads.

epoxy adhesive An adhesive made of epoxy resin.

epoxy matrix composite High strength composition consisting of epoxy resin and a reinforcing matrix of filaments or fibers of glass metal, or other materials.[NASA]

Eppley pyrheliometer A thermoelectric device for measuring direct and diffuse solar radiation; radiation is directed onto two concentric silver rings, the outer covered with MgO and the inner covered with lampblack, and a thermopile is used to determine the difference in temperature between the two rings.

equalization (channel balancing) The use of feedback to achieve close coincidence between the outputs of two or more elements or channels in a fault-tolerant control system. Equalization may be necessary to reduce the transient that could occur while shutting off a failed channel, or it may be necessary to minimize the adverse effects of normal tolerances.[ARP181A-85]

equalizer 1. A device that connects parts of a boiler to equalize pressures. 2. The electronic circuit in a tape reproducer whose gain across the spectrum of interest compensates for the unequal gain characteristic of the record/reproduce heads, thereby providing "equalized" gain across the band.

equal percent characteristic See characteristic, equal percentage.

equations of motion A set of equations which give information regarding the motion of a body or of a point in space as a function of time when initial position and initial velocity are known. [NASA]

equations of state Equations relating temperature, pressure, and volume of a system in thermodynamic equilibrium. [NASA]

equator The primary great circle of a sphere or spheroid, such as the earth, perpendicular to the polar axis; or a line resembling or approximating such a circle.[NASA]

equatorial atmosphere The composition and characteristics of the earth's atmosphere at and/or near the equator. [NASA]

equatorial regions Areas on or near the earth's equator, regions between the Tropic of Cancer and the Tropic of Capricorn (23 degrees 27 minutes North or south of the Equator).[NASA]

equiaxed structure A polygonal or spheroidal microstructural feature having approximately equal dimensions in all directions. In alpha-beta titanium alloys, such a term commonly refers to a microstructure in which most of the alpha phase appears spheroidal, primarily in the transverse direction. [AS1814-90]

equilay stranding Stranding composed of more than one layer of helically laid strands, with a reversed direction of lay, and the same length of lay for each successive layer.[ARP1931-87]

equilibrium flow Gas flow in which energy is constant along streamlines and the composition of the gas at any point is not time dependent.[NASA]

equilibrium state Any set of conditions that results in perfect stability—mechanical forces that completely balance each other and do not produce acceleration, or a reversible chemical reaction in which there is no net increase or decrease in the concentration of reactants or reaction products, for instance.

equinox One of two points of intersection of the ecliptic and the celestial equator occupied by the sun when its declination is zero degrees.[NASA]

equipment A generic term for any apparatus, assembly, mechanism or machine, or for a group of units constructed similarly, or for a group of units performing similar functions.

equipment, automatic test Equipment that is designed to automatically conduct analysis of functional or static parameters and to evaluate the degree of unit under test (UUT) performance degradation; and may be used to perform fault isolation of UUT malfunction. The decision making, control, or evaluative functions are conducted with minimum reliance on human intervention and usually done under computer control.[ARD50010-91]

equipment compatibility The characteristic of computers by which one computer may accept and process data prepared by another computer without conversion or code modification.

equipment failure A fault in the equipment, excluding all external factors, which prevents continued performance.

equipment temperature Temperature of the unit at a specified position and measured at a specified point, normally at the surface.[AIR1916-88]

equipotential For all practical purposes, an identical state of electrical potential for conducting item(s).[ARP1870-86]

equivalence point Point on the titration curve where the acid ion concentration equals the base ion concentration.

equivalent airspeed The calibrated airspeed of an aircraft corrected for adiabatic compressible flow for the particular altitude. Equivalent airspeed is equal to calibrated airspeed in standard atmosphere at sea level.[ARP4107-88]

equivalent binary digits The number of binary digits required to express a number in another base to the same precision, e.g., approximately 3 1/3 binary digits are required to express in binary form each digit of a decimal number. For the case of binary coded decimal notation, the number of binary digits required is usually four times the number of decimal digits.

equivalent evaporation Evaporation expressed in pounds of water evaporated from a temperature of 212 °F to dry saturated steam at 212 °F.

equivalent network A network that can perform the functions of another network under certain conditions; the two networks may be of different forms—one mechanical and one electrical, for instance.

equivalent optical diameter The size perceived by the APC is the equivalent optical diameter. It is based on calibration data for the particular instrument. Normally, the diameter reported is that of a sphere with an equal projected area, and specific optical properties.[ARP 1192-87]

equivalent single wheel load A theoretical calculated load which, if applied to a single tire, with a contact area equal to that of one tire of the assembly, would produce the same effect on the airfield as does a multiple wheel assembly. [AIR1489-88]

equivalent static wheel load A static wheel or tire load rating based on a dynamic load reduced by a specified factor. May be higher than the actual required static load rating for a specific application, in which case the tire/wheel static rating is dictated by this consideration rather than the actual static rating requirement.[AIR1489-88]

erasable storage 1. A storage device whose data can be altered during the course of a computation, e.g., magnetic tape, drum and cores. 2. An area of storage used for temporary storage.

erase To obliterate information from a storage medium, for example, to delete or to overwrite.

ERBE See earth radiation budget experiment.

erg The unit of energy in the CGS system; it is the amount of energy consumed (work) when a force of one dyne is applied through a distance of one centimeter.

ergometer Instrument for measuring muscular work.[NASA]

ergonomics 1. Human factors engineering which deals with machine design and workspace environment to make them compatible with human capacities and limitations.[ARP4107-88] 2. The science of designing machines and work environments to suit the needs of people. See also human factors engineering.

Erichsen test A cupping test for determining the suitability of metal sheet for use in a deep drawing operation; it is expressed as the depth in millimeters of a cup-shaped impression in a sheet of metal, supported on a ring, which is deformed at the center by a spherical tool until it breaks or tears.

erosion 1. In a solid rocket, burning of the propellant at a rate greater than the rate normally associated with the existing motor pressure and propellant temperature, usually due to high gas velocity parallel to the burning surface.[AIR913-89] 2. The wearing away of refractory or of metal parts by gas borne dust particles. 3. Progressive destruction of a structural member by the abrasive action of a moving fluid, often one that contains solid particles in suspension; if the fluid is a gas, erosion may be caused by liquid droplets carried in the moving gas stream.

erosion-corrosion Progressive destruction of a structural member by the combined effects of corrosion and erosion acting simultaneously.

erosive burning Combustion of solid propellants accompanied with nonsteady, high velocity flows of product gases across burning propellant surfaces.[NASA]

err See error.

error 1. An unintended and objectively inappropriate physical or mental operation. A deviation from a desired, specified, or standard type of performance. Human error is often the result of omission or untimeliness of the proper action or the commission of an inappropriate action.[ARP4107-88] 2. A mistake in specification, design, production, maintenance, or operation that causes an undesired performance of a function.[AIR1916-88] 3. The occurrence of a difference between the information transmitted and the information received.[AIR4271-89] 4. The difference between the set point and the control point.[ARP89C-70] 5. In mathematics, the difference between the true value and a calculated or observed value.[NASA] 6. The general term referring to any deviation of a computed or a measured quantity from the theoretically correct or true value. 7. The part of the error due to a particular identifiable cause, e.g., a truncation error, or a rounding error. In a restricted sense, that deviation due to unavoidable random disturbances, or to the use of finite approximations to what is defined by an infinite series. 8. In a single automatic control loop, the setpoint minus the controlled variable measurement.

error, adjustment Operating a control too slowly or too rapidly, moving a control/switch to the wrong position, or following the wrong sequence in operating several controls/switches.[ARP4107-88]

error band See accuracy.

error burst In data transmission, a sequence of signals containing one or more errors but counted as only one unit in accordance with some specific criterion or measure. An example of a criterion is that if three consecutive correct bits follow an erroneous bit, an error burst is terminated.

error checking Data quality assurance

usually attempted by calculating some property of the data block before transmission. The resulting property or check character is also sent to the receiver, where it may be inspected and compared with a recalculated value based on the received data.

error correcting code A code in which each acceptable expression conforms to specific rules of construction that also define one or more equivalent nonacceptable expressions, so that if certain errors occur in an acceptable expression, the result will be one of its equivalents, and thus the error can be corrected.

error detecting code A code in which each expression conforms to specific rules of construction, so that if certain errors occur in an expression, the resulting expression will not conform to the rules of construction, and thus the presence of the errors is detected. Synonymous with self-checking code.

error detection routine A routine used to detect whether or not an error has occurred, usually without special provision to find or indicate its location.

error, forgetting Failing to check, set, or operate a control/switch at the proper time.[ARP4107-88]

error, hysteresis See hysteresis.

error, hysteretic See hysteresis.

error indication Ideal value.

error message An audible or visual indication of a software or hardware malfunction, or a non-acceptable data entry attempt.

error range 1. The range of all possible values of the error of a particular quantity. 2. The difference between the highest and the lowest of these values.

error ratio The ratio of the number of data units in error to the total number of data units.

error, reversal Moving a control/switch in a direction opposite to that necessary to produce the desired result.[ARP4107-88]

error signal 1. The output of the summing point where algebraic summation of two or more control loop variables is performed.[AIR1916-88] 2. Voltages the magnitude of which are proportional to the difference between an actual and a desired position.[NASA] 3. The output of a comparing element. See signal, error.

error, substitution Confusing one control/switch with another or failing to identify a control/switch when it is needed.[ARP4107-88]

error, unintentional activation Accidentally operating a control/switch. [ARP4107-88]

ERTS-C See Landsat 3.

ESA spacecraft Spacecraft of the European Space Agency.[NASA]

escape Of a particle of large body; to achieve an escape velocity and a flightpath outward from a primary body so as neither to fall back to the body nor to orbit it.[NASA]

escape key A key on a computer keyboard that returns the operator to the prior step in a command sequence.

escape maneuver A computed maneuver to prevent a potential collision. It can be any single or combination of maneuvers which resolve a conflict. [ARP4153-88]

escapement A ratchet device that permits motion only in one direction, such as the device that controls motion in the works of a mechanical watch or clock.

escape rocket Small rocket engine attached to the leading end of an escape tower, which may be used to provide additional thrust to the capsule to obtain separation of the capsule from the booster vehicle in an emergency.[NASA]

escape velocity The radial speed which a particle of larger body must attain in order to escape from the gravitation of a planet, satellite, or star.[NASA]

escutcheon A decorative shield, flange,

or border around a panel-mounted part such as a dial or control knob. Also known as escutcheon plate.

essential flight information A message that contains information relevant to the safety of the flight.[ARP4102/13-90]

Etalon A type of Fabry-Perot interferometer in which the distance between two highly reflecting mirrors is fixed. It is used to separate light in different wavelengths when the wavelengths are closely spaced.

etalons Two adjustable parallel mirrors mounted so that either one may serve as one of the mirrors in a mechelson interferometer; used to measure distance in terms of wavelengths of spectral lines.[NASA]

etch cleaning Removing soil by electrolytic or chemical action that also removes some of the underlying metal.

etch cracks Shallow cracks in the surface of hardened steel due to hydrogen embrittlement that sometimes occurs when the metal comes in contact with an acidic environment.

etched wire insulation A process applied to a fluoroplastic insulated wire in which the wire is passed through a sodium bath to create a rough surface to allow a material to bond to the surface of the fluoroplastic.[ARP1931-87]

etching 1. Controlled corrosion of a metal surface to reveal its metallurgical structure. 2. Controlled corrosion of a metal part to create a design; the design may consist of alternating raised and depressed areas, or it may consist of alternating polished and roughened areas, depending on the conditions and corrodent used.

ethics The standards of conduct and moral judgment of a group, religion, profession, etc.[NASA]

ethnic factors The complex patterns of behavior which distinguish an ethnic group.[NASA]

etiology The doctrine of causes, particularly the causes and reasons for diseases.[NASA]

Europa A satellite of Jupiter orbiting at a mean distance of 761,000 kilometers. Also called Jupiter II.[NASA]

European Incoherent Scatter Radar See EISCAT radar system (Europe).

European Large Telecomm Satellite See L-Sat.

European Workshop on Industrial Computer Systems (EWICS) The European industrial computer control standards group.

eutectic 1. A process by which a liquid solution undergoes isothermal decomposition to form two homogeneous solids—one richer in solute than the original liquid, and one leaner. 2. The composition of the liquid which undergoes eutectic decomposition and possesses the lowest coherent melting point of any composition in the range where the liquid remains single- phase. 3. The solid resulting from eutectic decomposition, which consists of an intimate mixture of two phases.

eutectic composites Composite materials with a metal matrix of a mixture of solids including eutectoids.[NASA]

eutectoid A decomposition process having the same general characteristics as a eutectic, but taking place entirely within the solid state.

EUVE See extreme ultraviolet Explorer satellite.

evacuating (transportation) The organized withdrawal or removal of people from a place or area as a protective measure.[NASA]

evaluation kit A small microcomputer system used for learning the instruction set of a given microcomputer. It usually includes light emitting diodes, a keyboard, a monitor/debugger in ROM, a small amount of RAM and some input-output ports.

evaluation scenario Portions of the

design mission scenario that are selected for use in testing/evaluating the system. Segments are chosen to represent most "worst case" uses/operations and to condense the total mission scenario so that less time is required during the testing phase. The evaluation scenario may be described in the same variety of ways as the total mission scenario; however, it is typically only described in the forms of an evaluation scenario timeline and an experimenter's script. (For example, long inflight cruise segments may be eliminated, similar type maneuvers may be flown only once rather than repeatedly, and mission segments may be condensed or combined to test only the portions entailing high workloads for the aircrew.) [ARP4107-88]

evaporated make-up Distilled water used to supplement returned condensate for boiler feed water.

evaporation 1. The physical process by which a liquid or solid is transformed into the gaseous state; the opposite of condensation.[NASA] 2. The change of state from a liquid to a vapor.

evaporation gage See atmometer.

evaporation rate 1. The mass of material evaporated per unit time from unit surface of a liquid or solid. The number of molecules of a given substance evaporated per second per square centimeter from the free surface of the condensed phase.[NASA] 2. The number of pounds of water evaporated in a unit of time.

evaporative cooling 1. Lowering the temperature of a mass of liquid by evaporating part of it, using the latent heat of vaporization to dissipate a significant amount of heat. 2. Cooling ambient air by evaporating water into it. 3. See vaporization cooling.

evaporator Any of several devices where liquid undergoes a change of state from liquid to gas under relatively low temperature and low pressure.

evaporimeter See atmometer.

event 1. The occurrence of some programmed action within a process which can affect another process. 2. In TELEVENT, an occurrence recognized by telemetry hardware such as frame synchronization, buffer complete, start, halt, and the like.

event oriented Pertaining to a physical occurrence.

event recorder An instrument that detects and records the occurrence of specific events, often by recording on-off information against time to show when an event starts and stops, and how often it recurs.

EWICS See European Workshop on Industrial Computer Systems.

exactness See precision.

exception reporting An information system which reports on situations only when actual results differ from planned results. When results occur within a normal range they are not reported.

excess air Air supplied for combustion in excess of that theoretically required for complete oxidation.

excessive motivation See motivation, anomalies of excessive motivation.

excimer laser A laser in which the active medium is an "excimer" molecule—a diatomic molecule which can exist only in its excited state. The internal physics are conducive to high powers in short pulses, with wavelengths in the ultraviolet.

excimers Molecules characterized by repulsive or very weakly bound ground electronic states.[NASA]

excitation 1. Addition of energy to a nuclear, atomic or molecular system transferring it to another energy state. [NASA] 2. Voltage supplied by a signal conditioner to certain types of physical measurement transducers (bridges, for example).

excitation voltage A precision voltage applied to transducers; when pressure, strain, or the like are sensed by the transducer, a small portion of this voltage appears on the signal lines; the value of this signal voltage is proportional to the stimulus applied.

excited state See excitation.

exciter/regulator The devices necessary to provide voltage regulation and to provide or control the excitation of the alternator.[AS8011A-85]

exclusions See exclusive OR.

exclusive OR A logical operator which has the property that if P and Q are two statements, then the statement P*, where the * is the exclusive OR operator, is true if either P or Q, but not both are true, and false if P and Q are both false or both true.

executable statement Constituent of a program specifying the action of the program. Contrasted with a nonexecutable statement which describes the use of the program, the characteristics of the operands, editing information, statement functions, or data arrangement.

execute (EXEC) A computer term usually associated with safeguarding by requiring a second insert ("Execute").[ARP 1570-86] 2. To interpret machine instructions or higher-level statements and perform the indicated operations on the operands specified. 3. In computer terminology, to run a program.

execution The act of performing programmed actions. Execution time is the interval required for a specified action to be performed.

execution of an instruction The set of elementary steps carried out by the computer to produce the result specified by the operation code of the instruction.

execution time 1. Time required to execute a program. 2. The period during which a program is being executed. 3. The time at which execution of a program is initiated. 4. The period of time required for a particular machine instruction. See also instruction time.

executive The controlling program or set of routines in an operating system; the executive coordinates all activities in the system including I/O supervision, resource allocation, program execution, and operator communication. See also monitor.

executive commands In TELEVENT, the several commands that establish modes of operation, such as SETUP, END, CONNECT, and the like.

executive mode A central processor mode characterized by the lack of memory protection and relocation by the normal execution of all defined instruction codes.

executive program A program which controls the execution of all other programs in the computer based on established hardware and software priorities and real time or demand requirements.

executive software That portion of the operational software which controls online, response-critical events, responding to urgent situations as specified by the application program. This software is also known as the real-time executive.

executive system 1. A system that supplies the crew with guidance that they are required to follow unless they have good reason to believe it should be ignored.[ARP4153-88] 2. An integrated collection of service routines for supervising the sequencing of programs by a computer.

executive systems (computers) See operating system (computers)[NASA]

EXE file In MS-DOS, the designation .EXE follows most application program file names.

exfoliation corrosion A type of corrosion that proceeds parallel to the surface of a material, causing thin outer layers to be undermined and lifted by corrosion products.

exhaust 1. Discharge of working fluid from an engine cylinder or from turbine vanes after it has expanded to perform work on the piston or rotor. 2. The fluid discharged. 3. A duct for conducting waste gases, fumes or odors from an enclosed space, especially the discharge duct from a steam turbine, gas turbine, internal combustion engine or similar prime mover; gas movement may be assisted by fans.

exhaust clouds Clouds formed from the exhaust aerosols of launch vehicle engines and boosters at liftoff.[NASA]

exhaust emission The movement of gaseous or other particles and radiation from the nozzle of a rocket or other reaction engine.[NASA]

exhaust-gas analyzer An instrument that measures the concentrations of various combustion products in waste gases to determine the effectiveness of combustion.

exhaust steam Steam discharged from a prime mover.

exhaust stroke The portion of the cycle in an engine, pump or compressor that expels working fluid from the cylinder.

exhaust valve A valve in the headspace of a cylinder that opens during the exhaust stroke to allow working fluid to pass out of the cylinder.

exhaust velocity The velocity of gases or particles (exhaust stream) that exhaust through the nozzle or a reaction engine, relative to the nozzle.[NASA]

exit The time or place at which the control sequence ends or transfers out of a particular computer program or subroutine.

exobiology That field of biology which deals with the effects of extraterrestrial environments on living organisms and with the search for extraterrestrial life. [NASA]

exosphere The outermost, or topmost, portion of the atmosphere. It's lower boundary is the critical level of escape, variously estimated at 500 to 1000 kilometers above the earth's surface. [NASA]

exothermic reaction A reaction which occurs with the evolution of heat.

exotic fuels High-energy fuels, especially the hydroborons, which have higher calorific values than the corresponding hydrocarbons, and which at one time were proposed for use in high-performance aircraft and missiles.

expandable tire A tire which in the free, uninflated state assumes a shape which is smaller in diameter and cross section than when in the inflated state. For the purpose of conserving stowage volume in the air vehicle. Requires deflation after takeoff and prior to retraction, and reinflation after gear extension and prior to landing.[AIR1489-88]

expanded EMS An EMS where the instrumentation is improved and supplemented such that the engine components and modules can be individually monitored. Data processing and analysis are more extensive and may require integration with the control of individual engine components and modules. [AIR1873-88]

expanded joint The pressure tight joint formed by enlarging a tube end in a tube seat.

expanded memory The ability to add usable memory to a computer.

expanded metal A form of coarse screening made by lancing sheet metal in alternating rows of short slits, each offset from the adjacent rows, then stretching the sheet in a direction transverse to the rows of slits so that each slit expands to give a roughly diamond-shaped opening.

expanded plastic A light, spongy plastics material made by introducing air or gas into solidifying plastic to make it foamy. Also known as foamed plastic; plastic foam.

expander 1. A device which has main function of temporarily increasing an expansible type part to a desired dimension. See also spreader.[ARP480A-87] 2. The tool used to expand tubes.

expanding Increasing the diameter of a ring-shaped or cylindrical part, usually by placing it over a circular segmented die and forcing the segments to move radially in a controlled manner to stretch the part circumferentially.

expansion 1. Increasing the volume of a working fluid, with a corresponding decrease in pressure, and usually with an accompanying decrease in temperature, as in an engine, turbine or other prime mover. 2. Generally, any increase in volume or dimension which causes a body to occupy more physical space.

expansion factor (Y) Correction for the change in density between two pressure-measurement stations in a constricted flow.

expansion joint The joint to permit movement due to expansion without undue stress.

expansion ratio In rocketry, the ratio of nozzle exit area to the nozzle throat area.[AIR913-89]

expectancy A mental set in which environmental conditions are assumed prior to their occurrence. This may lead to a perceptual response, or attitudinal set. An acquired disposition whereby a response to a certain sign, object, or cue stimulus is expected to bring about a certain other situation.[ARP4107-88]

expedite A command used by ATC when prompt compliance is required to avoid the development of a dangerous situation.[ARP4107-88]

experience Applicable work in design, construction, preoperational and startup testing activities, operation, maintenance, onsite activities, or technical services. Observation of others performing work in the above areas is not experience. This experience can be obtained during startup or operations in a nuclear power plant, in fossil power plant, in other industries, or in the military.

expert system 1. Computer program that manipulates symbolic information to produce the same results as human experts would. They deal with uncertain data and make decisions on that data. Input and design relies on human experts.[NASA] 2. A computer program that uses stored data to reach conclusions, unlike a data base which presents data unchanged. See also artificial intelligence.

exploding bridge wire (EBW) A bridge wire designed to be exploded by a high energy electrical discharge.[AIR913-89]

exploding conductor circuit See circuit.

explosion 1. A rapid chemical reaction with the generation of a high temperature and usually a large quantity of gas.[AIR913-89] 2. The sudden production of large quantities of gases, usually hot, from much smaller amounts of gases, liquids, or solids.[NASA] 3. Combustion which proceeds so rapidly that a high pressure is generated suddenly.

explosion door A door in a furnace or boiler setting designed to be opened by a pre-determined gas pressure.

explosion proof apparatus Apparatus enclosed in a case that is capable of withstanding an explosion of a specified gas or vapor which may occur within it and of preventing the ignition of a specified gas or vapor surrounding the enclosure by sparks, flashes, or explosion of the gas or vapor within, and which operates at such an external temperature that a surrounding flammable atmosphere will not be ignited thereby.

explosion suppression Any method used to confine or suppress an explosion.[NASA]

explosion welding A solid-state process for creating a metallurgical bond by driving one piece of metal rapidly against another with the force of a controlled explosive detonation.

explosive cladding Producing a bimetallic material by explosion welding a thin layer of one metal on a substrate; it is used most advantageously to yield a material with one surface having a unique property, such as resistance to corrosion by certain strong chemicals, while the bulk of the material possesses good fabrication and structural properties.

explosive decompression Rapid reduction of air pressure inside an aircraft, coming to a new static condition of balance with the external pressure. A change in cabin pressure faster than the lungs can decompress. Generally, any decompression which occurs in less than 0.5 seconds.[ARP4107-88]

explosive forming Shaping parts in dies through the use of explosives to generate the forming pressure; most often, a sheet metal part is placed over an open die and covered with a sheet of explosive, which is then detonated to drive the metal into the die.

explosive train A series of explosive elements.[AIR913-89]

exponent In floating-point representation, one of a pair of numerals representing a number that indicates the power to which the base is raised. Synonymous with characteristic.

exponential notation A way to express very large or small numbers in data processing.

exponentiation A mathematical operation that denotes increases in the base number by a previously selected factor.

EXPOS (Spacelab payload) X-ray spectropolarimetry payload for Spacelab.[NASA]

exposure 1. For a photographic or radiographic emulsion, the product of incident radiation intensity and interval of time it is allowed to impinge on the emulsion. 2. A term loosely used to indicate time of exposure in photography.

exposure time 1. The period (in hours or cycles) during which a system, subsystem, unit, or part is exposed to failure, measured from when it was last verified as functioning to when its proper performance is or may be required.[AIR1916-88] 2. The elapsed time during which radiant energy is allowed to impinge on photographic or radiographic emulsion.

expression 1. A combination of operands and operators that can be evaluated to a distinct result by a computing system. 2. Any symbol representing a variable or a group of symbols representing a group of variables possibly combined by symbols representing operators in accordance with a set of definitions and rules. 3. In computer programming, a set of symbols that can have a specific value.

extars See x-ray stars.

Extended Binary Coded Decimal Interchange Code (EBCDIC) An 8-bit code that represents an extension of a 6- bit "BCD" code, which has been widely used in computers of the first and second generations. EBCDIC can represent up to 256 distinct characters and is the principal code used in many of the current computers.

extended duration space flight See long duration space flight.

extended instruction set (EIS) The software in the DEC system that provides hardware with fixed-point arithmetic and direct implementation of multiply, divide, and multiple shifting.

extended position The configuration of a retractable landing gear when it is down and locked and ready for landing and supporting the air vehicle. Also used in describing the state of a shock absorber when it is in the unloaded or uncompressed position.[AIR1489-88]

extensible flap A flap that can be extended and rotated downward, effectively increasing both the area and the camber of the wing.[ARP4107-88]

extension 1. An attachment for extending the length of a boring bit, socket wrench or handle, tow bar, and like items. The body of the item may be either a coil spring or a solid bar, as applicable. The items permit the connection of ends that are designed for direct mating to each other.[ARP480A-87] 2. A multiple character set that follows a computer filename that further clarifies the filename.

extension furnace See dutch oven.

extension spring A tightly coiled helical spring designed to resist a tensile force.

extensometer 1. An apparatus for studying seismic displacements by measuring the change in distance between two reference points that are separated by 20 to 30 metres or more. 2. An instrument for measuring minute elastic and plastic strains in small objects under stress, especially the strains prior to fracture in standard tensile-test specimens.

external distraction See attention, anomalies of external distraction.

external leakage Leakage from the interior of a device to the exterior, other than out of the fluid ports.[AIR1916-88]

external loads Aerodynamic loads imposed on the actuator output. (They vary generally as a function of surface displacement, rate of displacement, and operating conditions such as Mach number and altitude.)[ARP1281A-81]

externally fired boiler A boiler in which the furnace is essentially surrounded by refractory or water-cooled tubes.

externally quenched counter tube A radiation counter tube equipped with an external circuit that inhibits reignition of the counting cycle by internal ionizing events.

external memory See external storage.

external-mix oil burner A burner having an atomizer in which the liquid fuel is struck, after it has left an orifice, by a jet of high velocity steam or air.

external multiplexors Scanivalves, switching temperature indicators, and other devices that permit input of several signals on one computer input channel.

external party line (XPL) A logic level from telemetry equipment that causes the buffered data channel to switch input ports (to merge time, for example).

external start (XST) The hardware-generated pulse that causes the system to start receiving data (concurrent with word 1 of a frame, and often frame 1 of a subframe).

external storage 1. The storage of data on a device which is not an integral part of a computer but in a form prescribed for use by the computer. 2. A facility or device, not an integral part of a computer, on which data usable by a computer is stored, such as off-line magnetic tape units, or punch card devices. Synonymous with external memory and contrasted with internal storage.

external treatment Treatment of boiler feed water prior to its introduction into the boiler.

extract instruction An instruction that requests the formation of a new expression from selected parts of given expressions.

extraction tool 1. A hand operated mechanical device consisting of a hand grip and a tip designed to hold a taper pin. It is used to remove taper pins from taper pin receptacles and may be identical to an insertion tool.[ARP592-90] 2. A device used for removing removable contacts from their retaining cavity or mechanism.[ARP914A-79]

extractive distillation A distillation technique (employing the addition of a solvent) used when the boiling points of the components being separated are very close 'within 3 °C (5 °F)' or the components are constant boiling mixtures. In extractive distillation, which is a combination of fractionation and solvent extraction, the solvent is generally added to the top of the column and recovered from the bottom product by means of subsequent distillation. The chemical added is a solvent only to the less-volatile components. See azeotrope.

extractor A device which has main function of removing threaded or sleeved part from blind holes and requires hand gripping. For other removing categories, see driver, puller.[ARP480A-87]

extragalactic light See extraterrestrial radiation; light (visible radiation).

extra hard temper A level of hardness and strength in nonferrous alloys and some ferrous alloys corresponding approximately to a cold worked state one-third of the way from full hard to extra spring temper.

extra spring temper A level of hardness and strength for nonferrous alloys and some ferrous alloys corresponding to a cold worked state above full hard beyond which hardness and strength cannot be measurably increased by further cold work.

extraterrestrial intelligence Intelligent life existing elsewhere than on earth.[NASA]

extraterrestrial life Life forms evolved and existing outside the terrestrial biosphere.[NASA]

extra terrestrial radiation In general, solar radiation received just outside the earth's atmosphere.[NASA]

extremely improbable For airworthiness purposes, the likelihood of a failure is less than once in a billion flight hs (10^{-9}).[ARP4107-88]

extreme ultraviolet Explorer satellite An explorer satellite carrying scientific instruments for scanning the sky in the 100-900 Angstrom region of the spectrum to study the very hot celestial bodies (white dwarfs, for example).[NASA]

extreme ultraviolet radiation Ultraviolet emission in the 100-1000 Angstrom range.[NASA]

extremum values In statistics, the upper or lower bound of the random variable which is not expected to be exceeded by a specified percentage of the population within a given confidence interval.[NASA]

extrusion 1.Method of forcing plastic or elastomer material through an orifice to apply insulation or jacketing to a conductor or cable.[ARP1931-87] 2. A process for forming elongated metal or plastic shapes of simple to moderately complex cross section by forcing ductile, semisoft solid material through a die orifice. 3. A length of product made by this process.

extrusion billet A slug of metal, usually heated into the forging temperature range, which is forced through a die by a ram in an extrusion process.

extrusion pressing See cold extrusion.

eye A device which has main function of lifting with a hoist.[ARP480A-87]

eyebar A metal bar having a hole through an enlarged section at each end of the bar.

eyebolt A bolt with a loop formed at one end in place of a head.

eyelet 1. A reinforced conductive device or hole into which conductors are passed/routed or terminated.[ARP914A-79] 2. A small ring or barrel-shaped piece of metal used to reinforce a hole, especially in fabric.

eyeleting Forming a lip around the rim of a hole in sheet metal.

F

F See farad and Fahrenheit.

FAA Federal Aviation Administration.

FAA aircraft Any aircraft used in the service of the FAA. This includes aircraft owned, leased, rented, held under military bailment, or otherwise in possession of the FAA for the purpose of flight, ground test, or school use. It does not include private, club, or rented aircraft used for travel purposes.[ARP4107-88]

fabrication 1. A general term for parts manufacture, especially structural or mechanical parts. 2. Assembly of components into a completed structure.

fabry-perot A pair of highly reflecting mirrors, whose separation can be adjusted to select light of particular wavelengths. When used as a laser resonator, this type of cavity can narrow the range of wavelengths emitted by the laser.

Fabry-Perot lasers See lasers.

face 1. The portion of the glass envelope through which the luminous pattern is viewed.[ARP1782-89] 2. An exposed structural surface. 3. In a weldment, the exposed surface of the fusion zone.

face curtain A sheet of heavy fabric designed to be pulled in front of the face for protection against wind blast during ejection from an aircraft.[ARP 4107-88]

faceplate 1. A circular plate attached to the spindle of a lathe with the plane of the plate perpendicular to the spindle axis; it is used to attach and align certain types of workpieces. 2. A protective cover for holes in an equipment enclosure. 3. A glass or plastic window in personal protective gear such as welding helmets, respirator masks or diving masks. 4. A two-dimensional array of separate optical fibers, fused together, serving to strongly direct light forward.

facet The plane surface of a crystal or fracture surface.

facing 1. Machining a flat planar surface in lathe turning by positioning a single-point tool against the workpiece at the axis of rotation and moving the tool radially outward so that it cuts a spiral path in a plane perpendicular to the axis of rotation. 2. Fine molding sand applied to the surface of the mold cavity.

facsimile A system for utilizing telephone transmission apparatus to send written or pictorial information to a remote location; it consists of a transmitter, which scans the hard-copy record and converts its image into an electrical signal wave, and a receiver, which converts the electrical wave into its final pictorial form and registers it on a record sheet.

factorization Process of instance of factoring.[NASA]

faculae Large patches of bright material forming a veined network in the vicinity of sunspots. The appear to be more permanent than sunspots and are probably due to elevated clouds of luminous gas.[NASA]

FAD Fuel Advisory Departure.

fading A drop in signal intensity, or a slow undulation, caused by changes in the properties of the transmission medium.

FAF Final Approach Fix.

Fahrenheit A temperature scale where the freezing point of pure water occurs at 32 °F and the span between freezing point and boiling point of pure water at standard pressure is defined to be 180 scale divisions (180 degrees).

fail closed A condition wherein the valve closure component moves to a closed position when the actuating energy source fails.

fail-functional A more limited case of fail-operative wherein performance is degraded following a failure.[ARP1181 A-85] [AIR1916-88]

fail-hardover The type of failure wherein the output of the failed element is at an extreme condition (e.g. position, force, etc.). In cases having a polarity of output, a hardover failure may be of either polarity.[ARP1181A-85] [AIR 1916-88]

fail-neutral A failure mode where the control device or system fails to a passive null or locked-at-null condition. [ARP1181A-85] [AIR1916-88]

fail-open The type of failure wherein the failed element disconnects the normal control path within a device. Such a failure either prevents the signal from passing or seriously alters the signal that passes through a system.[ARP 1181A-85] [AIR1916-88]

fail operational A design feature that enables a system to continue to operate despite the malfunction or failure of one or more components. See redundant design.[ARP4107-88]

fail-operative A quality wherein a control device or system can continue operation after a failure or failures. A more explicit description is given by SFO or DFO. In a true fail-operative situation, a failure will cause no nominal loss of performance.[ARP1181A-85]

fail-passive A quality wherein the failed device or system ceases to create any active output. In the purest sense a device that fails passively would simply remove its presence from the control system. However, a device is still considered fail-passive if it remains a part of the system but acts only as an additional load. Sometimes referred to as fail-soft.[ARP1181A-85] [AIR1916-88]

fail-safe (FS) 1. A quality wherein the control device or system ceases to function but the conditions or consequences resulting from the failure are not hazardous and do not preclude continued safe flight. The condition following failure may be completely passive, or it may involve driving to a predetermined nonactive condition.[ARP1181A-85] 2. A design or systems used to minimize risk in case of a malfunction. For example, a steering control system designed to fail in a free caster mode rather than a wheel hard over condition.[AIR1489-88] 3. A design feature of a system that permits malfunction or failure of the component(s) most at risk without resulting in a dangerous or catastrophic outcome. See also redundant design. [ARP4107-88]

fail-safe device A component, system or control device so designed that it places the controlled parameter in a safe condition in case of a power interruption, controller malfunction or failure of a load-carrying member.

fail-safe systems Systems used to minimize risk in case of malfunction.[NASA]

fail soft With reference to electronic equipment, a design feature that enables the equipment to compensate automatically when a partial failure occurs. [ARP4107-88]

failure 1. A noun describing the state of having failed. In dealing with fault-tolerant flight control systems a failure occurs when a device within the system fails to function within prescribed limits without regard to the cause of the failure. Thus a failure may be: (a) any loss of function of any element within the control system, (b) loss of supply power to the system, (c) erroneous hardover conditions or loss of control intelligence at the signal input, (d) any out-of-tolerance condition that exceeds normal operating limits.[ARP-1181A-85] 2. The inability of an item to perform within previously specified limits.[ARD50010-91] 3. The inability of a system, subsystem, unit or part to perform within previously specified limits. Note that some failures may have no effect on the capability of the airplane and therefore are not failure conditions.[AIR1916-88] 4. A functional status or physical condition characterized by the inability of an engine, engine component or sub-assembly to fulfill its design purpose; the most severe degree of malfunction.[ARP1587-81]

failure analysis 1. The logical, systematic examination of an item or its diagram(s) to identify and analyze the probability causes, and consequences of potential and real failures.[ARD 50010-91] 2. Subsequent to a failure, the logical systematic examination of an item, its construction, application, and documentation to identify the failure mode and determine the failure mechanisms and its basic course.[ARD50013-91]

failure, catastrophic A failure that can cause item loss.[ARD50013-91]

failure cause The circumstance that induces or activates a failure mechanism; e.g., defective soldering, design weakness, assembly techniques, software error, etc.[ARD50010-91] [ARD 50013-91]

failure condition A consequential airplane state that has an impact on the functional capability of the airplane or the ability of the crew to cope with adverse operating conditions, or that would prevent continued safe flight and landing. A failure condition can result from the occurrence of a specific single event or a combination of related faults, failures, errors, operating conditions, or environment. Postulated failure conditions are assessed for their impact on safety and assigned an appropriate probability classification. A defined failure condition provides the criteria for classifying system functions as nonessential, essential, or critical.[AIR1916-88]

failure condition, catastrophic Failure conditions that would prevent continued safe flight and landing. A failure condition is considered catastrophic if more than a relatively few occupants incur serious injuries or fatalities.[AIR 1916-88]

failure condition, major (failure condition, hazardous) Failure conditions that would reduce the capability of the airplane or the ability of the crew to cope with adverse operating conditions to the extent that there would be (a) a significant reduction in the safety margin, (b) a significant increase in crew workload, or in conditions that impair crew efficiency, or (c) in more severe cases, adverse effects on occupants.[AIR 1916-88]

failure condition, minor Failure conditions that would not significantly reduce airplane safety, and which involve crew actions that are well within their capabilities. Minor failure conditions may include (a) a slight reduction in safety margins, (b) a slight increase in workload, such as routine flight plan changes, (c) some possible discomfort to occupants.[AIR1916-88]

failure, critical A failure, or combination of failures, that prevents an item from performing a specified mission. [ARD50010-91] [ARD50013-91]

failure, dependent Failure of an item caused either directly or indirectly by the failure of another item. A dependent failure is one which occurs as a byproduct of an independent failure. Dependent failures are not necessarily present when simultaneous failures occur.[ARD 50010-91]

failure detection The process or technique of identification of engine, engine component or subsystem failure.[ARP 1587-81]

failure effect 1. The consequence(s) a failure mode has on the operation, function, or status of an item. Failure effects are classified as local effect, next higher level, and end effect.[ARD50010-91] 2. The consequences of item failure on system operation.[AIR1916-88]

failure, dormant (hidden) A defect that is not visible, which cannot be detected and which does not cause failure under normal conditions, but can cause failure within the design envelope. In airline parlance this is usually called "hidden failure".[AIR1916-88]

failure, expected Expected failures are based on failure rates and exposures. The number is obtained by multiplying the official failure rate for each age interval by the exposures in the interval and then adding the results.[AIR1916-88]

failure isolation A man/machine task to isolate the cause of a malfunction or failure.[ARD50010-91]

failure mechanism The physical, chemical, electrical, thermal, or other process which results in failure.[ARD50013-91]

failure, mission Failure to complete the intended mission as a consequence of equipment failure.[AIR1916-88]

failure mode A manner in which a device can or did fail. Simple devices may have only one failure mode; whereas, more complex devices can have several failure modes.[ARP1181A-85]

failure mode and effects analysis (FMEA) A procedure by which each potential failure mode in a system is analyzed to determine the results or effects thereof on the system and to classify each potential failure mode according to its severity.[ARD50010-91] [AIR1916-88] [ARD50013-91]

failure mode and effects criticality analysis—maintainability information A procedure by which each potential failure is analyzed to determine how the failure is detected and the actions to be taken to repair the failure. [ARD50010-91]

failure, nonchargeable 1. A non-relevant failure. 2. A relevant failure caused by a condition previously specified as not within the responsibility of a given organizational entity. (All relevant failures are chargeable to one organizational entity or another.)[ARD50010-91]

failure, nonrelevant Failure to be excluded in interpreting test results or in calculating the value of a reliability characteristic.[ARD50010-91]

failure path The chain of events or set of circumstances that result in an engine, engine component or sub-system failure because of interrelationships between components and sub-systems.-[ARP1587-81]

failure, primary A failure which occurs without being related to the failure of associated items, distinguished from dependent failure.[ARD50010-91] [ARD50013-91]

failure probability The probability of failure in a specified period of time. [ARD50013-91]

failure, random Any failure whose occurrence is unpredictable in an absolute sense but which is predictable only in a probabilistic or statistical sense. [ARD50010-91] [ARD50013-91]

failure rate 1. The total number of failures within an item population, divided by the total number of life units expended by that population during a particular measurement interval under stated conditions.[ARD50013-91] Symbol used (Lamda). A reliability measure related to MTBF.[ARD50010-91] 2. The probable number of times that a system or component fails during a given time or cyclic period of operation.[AIR 1916-88]

failure rate, initial An initial estimate of the expected failure frequency of an item, which may later be adjusted, as actual experience is gained.[AIR1916-88]

failure rate, smooth Rates determined when sufficient failures and exposures per age intervals are available; enough to eliminate peaks and valleys from the data.[AIR1916-88]

failure, relevant Failure to be included in interpreting test results or in calculating the value of a reliability characteristic.[ARD50010-91]

failure, secondary 1. A non-specific condition of a system that has manifested a number of failures, but still provides most of its functional capability.[ARD50010-91] 2. Failure of an item caused either directly or indirectly by the failure of another item. A dependent failure is one which occurs as a byproduct of an independent failure. Dependent failures are not necessarily present when simultaneous failures occur.[ARD50010-91] [ARD50013-91]

failure, wearout A failure that occurs as a result of deterioration process or mechanical wear and whose probability of occurrence increases with time. [ARD50010-91]

failures, excluded 1. Failures resulting from transportation, storage, inspection, maintenance, repair, installation, overhaul or replacement improperly performed by Using Service personnel contrary to currently applicable instruction or reasonable standards of aircraft quality workmanship. This exclusion does not apply to improper actions by contractor personnel.[ARD50010-91] 2. Failures resulting from operation of an engine beyond engine specification defined environment conditions and time cycle limitation, or with fuels or lubricants not conforming to the applicable specification.[ARD50010-91] 3. Failures which are the result of fuel system contaminations, where the contamination levels are outside the limits specified in the engine specification, unless evidence exists that the contamination was engine generated.[ARD50010-91] 4. Failures for which a corrective engine design change or an operational procedure change has been verified and engineering approved by the appropriate Using Service will not be counted after date of approval, unless the failures are identical and the corrective change was in place, or used, on the failed engine.[ARD50010-91] 5. Failures where the primary failure cause was not directly attributable to the design or quality of the engine: such as failures attributed to foreign object damage (FOD) which is in excess of engine specification requirements.[ARD50010-91] 6. Failure of equipment not furnished by the engine contractor, which do not occur as a result of failure of the engine to provide a proper function or interface within the limits stated in the engine specification.[ARD50010-91] 7. Any reported malfunctions which cannot be verified by subsequent investigation and do not occur in subsequent operation.[ARD50010-91] 8. Discrepancies due to manufacturing defects discovered during green, final and penalty runs, provided these defects are corrected prior to acceptance of the engine by the Using Service.[ARD50010-91] 9. Filter changes, chip detector inspection and cleaning, borescope inspections, engine wash procedures, or removal of engine or components due to life and/or cycle limitations. Engine failures detected by any of these procedures shall be included, unless excluded for other reasons.[ARD50010-91] 10. Multiple part removals performed to correct a single failure mode or secondary damage from a single failure are counted as one failure against the engine.[ARD50010-91]

faint object camera One of the five components of the first scientific payload of the Hubble Space Telescope. The faint object camera will be used to observe extremely faint astronomical objects with wavelengths between 120 and 700 nm.[NASA]

fairing or feathering 1. A shape that produces a smooth transition from one angular direction to another; or the act of producing this smooth contour. Its purpose in tooling sealants is to ensure good contact with the surfaces and to minimize air entrapment.[AIR4069-90] 2. A stationary member or structure, whose primary function is to produce a smooth contour. It serves to cover projecting parts that would offer resistance to air flow.[ARP480A-87]

fall The chain, rope or wire rope used in lifting tackle.

fall block In lifting tackle, a pulley block attached to the load and rising or descending with it.

fall time The time required for the output voltage of a digital circuit to change from a logical high level (1) to a logical low level (0).

false add To form a partial sum, that is, to add without carries.

false alarm In general, the unwanted detection of input noise. In radar, an indication of a detected target even though one does not exist, due to noise or interference levels exceeding the set threshold of detection.[NASA]

false alert 1. An alert, caused by a false track or a system malfunction, that is given when no threat exists in the TCAS operational envelope.[ARP4153-88] 2. An alert which occurs when the design windshear threshold conditions do not exist.[ARP4109-87]

false axle See axle sleeve.

false brinelling Fretting between the rolling elements and races of ball or roller bearings.

false set Rapid hardening of freshly mixed cement, mortar or concrete with a minimum evolution of heat; plasticity can be restored by mixing without adding more water.

fan A rotating mechanism, usually consisting of a paddle wheel or screw, with or without a casing, used to induce movement (currents) in air or other gas, such as in a circulation, ventilation or exhaust system where large volumes must be delivered.

fan blade One or more revolving vanes attached to a rotary hub and operated by a motor.[NASA]

fang bolt A bolt having a triangular head with sharp projections at the corners which is used primarily to attach metal parts to wood.

fan inlet area The inside area of the fan outlet.

fan jet A jet engine having a ducted fan in its forward end that draws in extra air, the compression and expulsion of which provides additional thrust.[ARP 4107-88]

fan performance A measure of fan operation in terms of volume, total pressures, static pressures, speed, power input, mechanical and static efficiency, at a stated air density.

fan performance curves The graphical presentation of total pressure, static pressure, power input, mechanical and static efficiency as ordinates and the range of volumes as abscissa, all at constant speed and air density.

fan requirements—recommended or specified The fan requirements recommended by the manufacturer of the steam generating equipment which will include necessary tolerances to overcome unfavorable operating conditions.

FAR Federal Aviation Regulations.

farad (F) 1. Unit of electrical capacitance. The capacitance of a capacitor which, when charged with one coulomb, gives a difference of potential of one volt.[ARP1931-87] 2. Metric unit of electrical capacitance.

Faraday rotation A rotation of the plane of polarization of light caused by the application of a magnetic field to the material transmitting the light.

Faraday rotator A device which relies on the Faraday effect to rotate the plane of polarization of a beam of light passing through it. Faraday rotator glass is a type of glass with composition designed to display the Faraday effect.

far field Distant from the source of light. This qualification is often used in measuring beam quality, to indicate that the measurement is made far enough away from the laser that local aberrations in the vicinity of the laser have been averaged out.

far-infrared laser Generically, this term could be taken to mean any laser emitting in the far infrared, a vaguely defined region of wavelengths from around 10 micrometers to 1 millimeter. This family of lasers require optical pumping by an external laser—usually carbon dioxide.

fascination See attention, anomalies of fascination.

fast break In magnetic particle testing of ferromagnetic materials, interrupting the current in the magnetizing coil to induce eddy currents and strong magnetization as the magnetizing field collapses.

fastener 1. Any of several types of devices used to hold parts firmly together in an assembly; some fasteners hold parts firmly in position, but allow free or limited relative rotation. 2. A device for holding a door, gate or similar structural member closed.

Fast-Fourier Transform A type of frequency analysis on data that can be done by computer using special software, or by an array processor or by a special-purpose, hardware device.

fatal injury Any injury which results in death within 30 days of occurrence. [ARP4107-88]

fatigue 1. The progressive decrease in performance ability due to prolonged mental or physical activity, extreme mental or physical activity, sleep deprivation, disrupted diurnal cycles, or life-event stress.[ARP4107-88] 2. The progressive failure of materials (especially metals) as a result of repeated cyclic loading that causes cracks to form and grow, eventually resulting in complete failure if loading is continued. [AIR1872-88] 3. The tendency of a metal to break under repeated cyclic stressing under load, the highest of which must be below the metal's ultimate tensile strength.[ARP700-65] 4. Progressive fracture of a material by formation and growth of minute cracks under repeated or fluctuating stresses whose maximum value is less than the material's tensile strength, and is often wholly within the elastic-stress range.

fatigue, acute/transient The type of fatigue associated with physical or mental activity between two regular sleep periods. Acute/transient fatigue is eliminated after a regular sleep period. [ARP4107-88]

fatigue (biology) State of the human organism after exposure to any time of physical or psychological stress (e.g. pilot fatigue).[NASA]

fatigue, cumulative/chronic fatigue The type of fatigue resulting from an inadequate recovery from successive periods of acute/transient fatigue. One regular sleep period will not eliminate cumulative fatigue; however several sleep periods and reduced interim activity will eliminate it.[ARP4107-88]

fatigue life 1. The number of applied stress reversals, at a particular stress level, at which a material fails.[AIR 1872-88] 2. The number of stress cycles that can be sustained under given load conditions.[ARP700-65] 3. The number of stress cycles that a material can sustain prior to fracture for a given set of fatigue conditions.

fatigue limit (fatigue endurance limit) The level of applied stress at which an infinite number of reversals can be endured without the material failing. This is also defined as the number of cycles that can be accumulated by a part before being retired from service.[AIR1872-88]

fatigue (materials) A weakening or deterioration of metal or other material occurring under load, especially under repeated cyclic, or continued loading. [NASA]

fatigue notch factor The ratio of the fatigue strength of an unnotched specimen to the fatigue strength of a notched specimen of the same material and condition; the notch used is of a specified size and contour, and the strengths are compared at the same number of stress cycles.

fatigue notch sensitivity An estimate of the effect of a notch or hole on the fatigue properties of a material; it is expressed as $q=(K_f\text{-}1)/K_t\text{-}1)$, where q is the fatigue notch sensitivity, K_f is the fatigue notch factor, and K_t is the stress concentration factor for a specimen of the material containing a notch of a specific size and shape.

fatigue, physical The effects of prolonged physical activity or the effects of brief but extreme physical activity, or both, either of which taxes a person's capacity.[ARP4107-88]

fatigue strength The maximum stress that ordinarily leads to fatigue fracture in a specified number of stress cycles; if the stress is not completely reversed during each stress cycle, the minimum stress also should be given. See also endurance limit.

fatigue, subjective the type of fatigue associated with the wearing effects of such psychosocial problems as unresolved conflicts, prolonged frustration, or constant worrying. Subjective fatigue is not eliminated by any number of sleep periods without first resolving the conflict or removing the frustrations.[ARP 4107-88]

fault 1. A physical condition that causes a device, component, or element to fail to perform in a required manner; for example, a short circuit or a broken wire.[ARD50010-91] 2. An undesired anomaly in the functional operation of a system, subsystem, unit, or part.[AIR 1916-88] 3. An occurrence that causes operational characteristics that are in conflict with those specified.[AIR4271-89] 4. A defect of any component upon which the intrinsic safety of a circuit depends. 6. The failure of any part of a computer system.

fault current See short-circuit current.

fault detection The process, technique or capability of identification of a discrepancy.[ARP1587-81]

fault isolation 1. The process of determining the location of a fault to the extent necessary to effect repair.[ARD 50010-91] [AIR1916-88] 2. The process and action, following equipment failure, that are involved in the identification of the unit, assembly, or piece part that has failed.[AIR1266-77]

fault isolation analysis (FIA) A systematic evaluation of failure causes, indications and probabilities to determine the level of fault isolability provided for in the system design.[AIR1266-77]

fault localization The process of determining the approximate location of a fault.[AIR1916-88]

fault tolerance 1. The ability of a system to perform a designated set of functions after having suffered one of a specified set of faults.[AIR4271-89] 2. The capability of systems to function despite one or more critical failures, by use of redundant circuits or functions and/or reconfigurable elements.[NASA] 3. That property of a system which permits it to carry out its assigned function even in the presence of one or more faults in the hardware or software components.

fault tree 1. An expression for a logic path used to establish engine, engine component or subsystem functional status and condition.[ARP1587-81] 2. Acyclic directed graphs used in the analysis or prediction of faults and defects.[NASA]

fault tree analysis A method for block diagramming constituent elements of a critical problem area in a logic network for analysis of failure modes and failure effects on the system. In advanced models, the probabilities of failure events are also determined.[ARD 50010-91] [ARD50013-91]

FAX Sending digitized copies of documents by telephone.

faying surface Either of two surfaces in contact with each other in a welded, fastened or bonded joint, or in one about to be welded, fastened or bonded.

faying-surface seal A preassembly seal installed between two mating (overlapping) surfaces. Faying-surface sealants are used to prevent corrosion, and, in conjunction with fillet seals, to prevent a leak path from extending through a faying surface to another area. When modified by a groove, a faying-surface seal has been used as a primary seal. [AIR4069-90]

FCC See frame code complement.

F-center laser A solid state laser in which optical pumping by light from a visible-wavelength laser produces tunable near-infrared emission from defects—called "color centers" or "F centers"—in certain crystals.

FDDI See fiber distributed data interface.

FDM See Frequency-Division Multiplex.

FDMA See frequency division multiple access.

feasibility study Any evaluation of the worth of a proposed project based on specific criteria.

feather See fairing.

feather angle The lade angle setting at which the nonrotating propeller produces the least drag.[ARP4107-88]

feed 1. The act of supplying material to a process or to a specific processing unit. 2. The material supplied. Also known as feedstock. 3. Forward motion tending to advance a tool or cutter into the stock in a machining operation.

feedback 1. The return of a portion of the output of a device to the input; positive feedback adds to the input, negative feedback subtracts from the input. [AIR1489-88] 2. Information, as to progress, results, etc., returned to an originating source.[AIR1489-88] 3. Process signal used in control as a measure of response to control action. 4. The part of a closed-loop system which automatically brings back information about the condition under control. 5. Part of a closed loop system which provides information about a given condition for comparison with the desired condition.

feedback control An error driven control system in which the control signal to the actuators is proportional to the difference between a command signal and a feedback signal from the process variable being controlled. See control, feedback.

feedback control loop A closed transmission path (loop), which includes an active transducer and which consists of a forward path, a feedback path, and one or more mixing points arranged to maintain a prescribed relationship between the loop input signal and the loop output signal.[AIR1489-88]

feedback control system A control system, comprising one or more feedback control loops, which combines functions of the controlled signals with functions of the commands to tend to maintain prescribed relationships between the commands and the controlled signals.[AIR1489-88]

feedback control signal The output signal which is returned to the input in order to achieve a desired effect, such as fast response.

feedback element The component of the actuation system that measures the system output and transforms it in to a form compatible with the command signal.[AIR1916-88] See also elements, feedback.

feedback loop The components and processes involved in correcting or controlling a system by using part of the output as input. See loop, closed (feedback loop).

feedback-loop control elements Elements in a closed-loop control system that provide the feedback signal of the output, or a function of the output that can be compared with the reference input.[AIR1916-88]

feedback oscillator An amplifier circuit in which an oscillating output signal is coupled in phase with the input signal, the oscillation being maintained at a frequency determined by frequency-selective parameters of the amplifier and its feedback circuits.

feedback ratio In a control system, the ratio of the feedback signal to a corresponding reference input.

feedback signal 1. The signal used to cancel further action when the command is carried out.[AIR1916-88] 2. A signal derived from some attribute of the controlled variable, or from a control-system output, which is combined with one or more input or reference signals to produce a composite actuating signal. See signal, feedback.

feeder 1. A feeder is a circuit conductor originating at the power source bus from which the branch circuit loads are served.[ARP1199A-90] 2. A conveyor adapted to control the rate of delivery of bulk materials, packages or objects to a specific point or operation. 3. A device for the controlled delivery of materials to a processing unit. 4. In metalcasting, a runner or riser so placed that it can deliver molten metal to the contracting mass of metal as it cools and solidifies, thus preventing voids, porosity or shrinkage cavities.

feed-forward 1. An industry standard process control program, in which mathematically predicted errors are corrected before they occur; usually mainly for process loops with long lags or response times. 2. Open loop control.

feedforward control A method of control that compensates for a disturbance before its effect is felt in the output. It is based on a model that relates the output to the input where the disturbance occurs. In distillation the disturbances are usually feed rate and feed compositions. Steady-state feedforward models are usually combined with dynamic compensation functions to set the manipulative variables and combined with feedback adjustment (trim) to correct for control model-accuracy constraints. See control, feedforward.

feedforward control action Control action in which information concerning one or more external conditions that can disturb the controlled variable is converted into corrective action to minimize deviations of the controlled variable. Feedforward control is usually combined with other types of control to anticipate and minimize deviations of the controlled variable.

feedhead A reservoir of molten metal that extends above a casting to supply additional molten metal and compensate for solidification shrinkage. Also known as riser; sinkhead.

feed pipe A pipe through which water is conducted into a boiler.

feed rate The relative velocity between tool holder and workpiece along the main direction of cutting in a machining operation.

feedscrew An externally threaded rod used to control the advance of a tool or tool slide on a lathe, diamond drilling rig, percussion drill or other equipment.

feedstock Material delivered to a process or processing unit, especially raw material delivered to a chemical process or reaction vessel.

feedthrough error A signal caused by coupling from reference input to output when the digital-to-analog converter logic inputs are all low. Expressed in mV or dB relative to V REF.

feed-thru A connector, terminal block or terminal device having conductive elements accessible from opposite sides of an insulator or a partition for termination or connection with mating devices.[ARP914A-79]

feed trough A trough or pan from which feed water overflows in the drum.

feedwater Process water supplied to a vessel such as a boiler or still, as opposed to circulating water or cooling water.

feed-water treatment The treatment of boiler feed water by the addition of chemicals to prevent the formation of scale or eliminate other objectionable characteristics.

feet (anatomy) The lower, pedal, extremities of the legs.[NASA]

feldspars A group of abundant rock-forming minerals of the family of anhydrous silicates.[NASA]

felsite A light colored, fine grained igneous rock composed chiefly of quartz or feldspar.[NASA]

female fitting An element of a connection in pipe, tubing, electrical conductors or mechanical assemblies that surrounds or receives the mating (male) element; for example, the internally threaded end of a pipe fitting is termed female.

fermat principle 1. The principle which states that the path along which electromagnetic radiation travels between any two points will be that path for which the elapsed time for the travel is in minimum.[NASA] 2. Also called the principle of least time; a ray of light traveling from one point to another, including reflections and refractions which may be suffered, follows that path which requires the least time. Stated another way, the optical path is an extreme path, in the terminology of the calculus of variations.

Fermi-Dirac statistics The statistics of an assembly of identical half-integer spin particles; such particles have wave functions antisymmetrical with respect to particle interchange and satisfy the Pauli exclusion principle.[NASA]

ferric percentage Actual ferric iron in slag, expressed as percentage of the total iron calculated as ferric iron.

ferrite Solid solution of carbon in alpha-iron.[NASA]

ferroalloy An alloy, usually a binary alloy, of iron and another chemical element which contains enough of the second element for the alloy to be suitable for introducing it into molten steel to produce alloy steel, or in the case of ferrosilicon or ferroaluminum to produce controlled deoxidation.

ferrodynamic instrument An electrodynamic instrument in which the presence of ferromagnetic material (such as an iron core for an electromagnetic coil) enhances the forces ordinarily developed in the instrument.

ferrography 1. A technique for the isolation and analysis of wear particles in a lubricant.[NASA] 2. Wear analysis conducted by withdrawing lubricating oil from an oil reservoir and using a ferrograph analyzer to determine the size distribution of wear particles picked up as the oil circulates between moving mechanical parts; the technique also may be used to assess deterioration of human joints or joint-replacement prostheses by analyzing for the presence of bone, cartilage and prosthetic-material fragments in human synovial fluid.

ferromagnetic material Any material that exhibits the phenomena of magnetic hysteresis and saturation, and whose permeability depends on the magnetizing force, all of which are exhibited by the chemical element iron.

ferrometer An instrument for measuring magnetic permeability and hysteresis in iron, steel and other ferromagnetic materials.

ferrous alloy Any alloy containing at least 50% of the element iron by weight.

ferrule 1. A short tube or sheath of conductive material used to make connections to shielded or coaxial cable; a specially formed metal ring used in connector accessories to reduce the transmission of torque to the connector grommet.[ARP914A-79] 2. The ferrule is a fuse or limiter terminal of a cylindrical shape which enclosed the end of a fuse or limiter.[ARP1199A-90] 3. A metal ring or cap that is fitted onto the end of a tool handle, post or other similar member to strengthen and protect it. 4. A bushing inserted in the end of a boiler flue to spread and tighten it. 5. A tapered bushing used in compression-type tubing fittings to provide the wedging action that creates a mechanical seal. 6. An element of a fiber optic connector, typically used to house or align fibers.

fetch The process of obtaining the data from an address memory location.

F format In FORTRAN Fw.d indicates that w characters are to be converted to a floating point mixed number with d spaces reserved for digits to the right of the decimal point.

FFT See Fast-Fourier Transform.

fiber 1. The characteristic of wrought metal that indicates directionality, and can be revealed by etching or fractography. 2. The pattern of preferred orientation in a polycrystalline metal after directional plastic deformation such as by rolling or wiredrawing. 3. A filament or filamentary fragment of natural or synthetic materials used to make thread, rope, matting or fabric. 4. In stress analysis, a theoretical element representing a filamentary section of solid material aligned with the direction of stress; usually used to characterize nonuniform stress distributions, as in a beam subjected to a bending load.

fiber composite Structural material consisting of combinations of metals or alloys or plastics reinforced with one or more types of fibers.[NASA]

fiber metal A material composed of metal fibers which have been pressed or sintered together, and which may also have been impregnated with resin, molten metal or other material that subsequently hardened.

fiber optic gyroscope A device in which changes in the wavelength of light going in different directions through a long length of optical fiber wound many times around a ring is used to measure rotation speed.

fiber optics 1. The technique of transmitting light through long thin, flexible fibers of glass, plastic, or other transparent materials.[NASA] 2. A medium that uses light conducted through glass or plastic fibers for data transmission, optical measurements, or optical observations.

fiber optic system 1. A relatively new method of data transmission. Light transmitting fibers are used for connecting sensors to the computer. Fiber optic systems have very good immunity from noise. 2. Any system employing fiber optics to provide analytical observation of measurements employing the transmission properties of glass, plastic, polycrystalline, or crystals materials.

fiber release The release of carbon or graphite when graphite reinforced composites are burned, especially in aircraft crashes or fires.[NASA]

fiber sensor A sensing device in which the active element is an optical fiber or an element attached directly to an optical fiber. The quantity being measured changes the optical properties of the fiber in a way that can be detected and measured.

fibrous composite A material consisting of natural, synthetic or metallic fibers embedded in a matrix, usually a matrix of molded plastics material or hardenable resin.

fibrous fracture A type of fracture surface appearance characterized by a smooth, dull gray surface.

fibrous structure 1. In fractography, a ropy fracture-surface appearance, which is generally synonymous with silky or ductile fracture. 2. In forgings, a characteristic macrostructure indicative of metal flow during the forging process, which is revealed as a ropy appearance on a fracture surface or as a laminar appearance on a macroetched section; a ropy appearance on the fracture surface of a forging does not carry the same implication as a ropy fracture of other wrought metals, and should not be considered the same as a silky or ductile fracture. 3. In wrought iron, a microscopic structure consisting of elongated slag fibers embedded in a matrix of ferrite.

fidelity The degree to which a system, subsystem or component accurately reproduces the essential characteristics of an input signal in its output signal. See also accuracy.

field The part of a computer record containing a specific portion of information.

field aligned current Ionospheric and magnetospheric current aligned along the electric field of a planet.[NASA]

field bus A standard under development in ISA SP50 for a bus to interconnect process control sensors, actuators, and control devices.

field coil A stationary or rotating electromagnetic coil.

field contact follower See auxiliary output.

field curvature Formation of an image that lies on a curved surface rather than a flat plane. For single and double element lenses, curvature is always inward, but for other types the curvature can be in either direction.

field emission Induced electron emission from an unheated metal surface resulting from application of a strong electric field.

field excitation Controlling the speed of a series-wound electric motor or a diesel-electric locomotive engine by changing the relationship between armature current and field strength, either through the use of shunts to reduce field current or through the use of field taps.

field-free emission current The electron current flowing from a cathode when the electric gradient at the cathode surface is zero.

field installable Nominally, a fiber optic splice or cable is field installable if it can be mounted by technicians working in the field, without a lab full of equipment at hand.

field of view The area or solid angle that can be viewed through or scanned by an optical instrument.[NASA]

field-replaceable unit Computer hardware modules that are easily replaced.

field strength For any physical field, the flux density, intensity, or gradient of the field at the point in question. [NASA]

field weld A weld made at a construction or installation site as opposed to a weld made in a fabrication shop.

FIFO See first in, first out.

filament A very fine single strand of metal wire, extruded plastic or other material. See also solar prominence.

filament winding Fabricating a composite structure by winding a continuous fiber reinforcement on a rotating core under tension; the reinforcement usually consists of glass, boron or silicon carbide thread, either previously impregnated with resin or impregnated during winding.

filament-wound structure A composite structure made by fabricating one or more structural elements by filament winding, then curing them and assembling them, or assembling them first and curing the entire structure.

filar micrometer An attachment for a microscope or telescope consisting of two parallel fine wires or knife edges in an eyepiece, one of them in a fixed position and the other capable of being moved in a direction perpendicular to its length by means of a very accurate micrometer screw; the device is used to make accurate measurements of linear distances in the optical field of view; actual distances are determined by dividing the micrometer reading by the magnification of a microscope, although in some cases the micrometer scale is calibrated for direct reading at a specific magnification.

file An organized structure consisting of an arbitrary number of records, for storing information on a bulk storage device, e.g., disc, drum, core, or tape.

filed A term normally used in conjunction with flight plans; it means that a flight plan has been submitted to air traffic controller (ATC).[ARP4107-88]

file gap An interval of space or time associated with a file to indicate or signal the end of the file; related to gap.

file handling The manipulation of data files by various methods. It generally involves read, write, and compare.

file maintenance program The file maintenance program is that part of the EASY program which is used to load component models or to modify the standard component models.[AIR 1823-86]

file name extension An addition to a computer file name that indicates the file type, such as .BAT, .COM, etc.

file specification A name that uniquely identifies a file maintained by an operating system. A file specification generally consists of at least three components; a device name identifying the volume on which the file is stored, a file name, and a file name extension. In addition, depending on the system, a file specification can include a user file directory name, or UIC, and a version number.

file structured device A device on which data are organized into files; the device usually contains a directory of the files stored on the device.

file transfer access and management (FTAM) One of the application protocols specified by MAP and TOP.

filled composite A plastics material made of short-strand fibers or a granular solid mixed into thermoplastic or thermosetting resin prior to molding.

filled-system thermometer Any of several devices consisting of a temperature-sensitive element (bulb), an element sensitive to changes in pressure or volume (Bourdon tube, bellows or diaphragm), capillary tubing and an indicating or recording device; the bulb, capillary tube and pressure- or volume-sensitive element are partly or completely filled with a fluid that changes its volume or pressure in a predictable manner with changes in temperature.

filler 1. A material used in cable to fill large interstices.[ARP1931-87] 2. A substance, often inert, added to a compound to improve properties and/or reduce cost.[ARP1931-87] 3. An inert material added to paper, resin, elastomers and other materials to modify their properties or improve quality in end products. 4. A material used to fill holes, cracks, pores and other surface defects before applying a decorative coating such as paint. 5. A metal or alloy deposited in a joint during welding, brazing or soldering; usually referred to as filler metal.

fillet 1. A concave transition surface between two surfaces that meet at an angle. 2. A molding or corner piece placed at the junction of two perpendicular surfaces to lessen the likelihood of cracking.

fillet seal A primary seal (post assembly) applied at the juncture of two adjoining parts or surfaces and along the edges of faying surfaces as a continuous bead of sealing material. It can be applied over, along the edges of, and between installed parts.[AIR4069-90]

fillet weld A roughly triangular weld that joins two members along the intersection of two surfaces that are approximately perpendicular to each other.

fill-in-the-blank programming language A nonprocedural programming language in which programs are developed by filling out data sheets for an existing program. Examples: BICEPS, PROSPRO, CODIL.

fill time The time required to completely fill the combuster fuel manifold after fuel flow is initiated.[ARP906A-91]

film 1. Thin sheeting. The finished form of a material which is (a) processed by casting a fluid material on a large surface (usually a rotating drum) and exposing it to a curing process, or (b) melt extruded directly into the sheets. See also cast tape.[ARP1931-87] 2. A flat, continuous sheet of thermoplastic resin or similar material that is extremely thin in relation to its width and length. 3. A very thin coating, deposit or reaction product that completely covers the surface of a solid.

film cooling The cooling of a body or surface, such as the inner surface of a rocket combustion chamber, by maintaining a thin fluid layer over the affected area.[NASA]

film edge—bond spacing For a bonded device, the distance from the end of the substrate conductor film metallization to the closest edge of the bonded area. [AS1346]

filmogen The material or binder in paint that imparts continuity to the coating.

film strength 1. Generally, the resistance of a film to disruption. 2. In lubricants, a measure of the ability to maintain an unbroken film over surfaces under varying conditions of load and speed.

filter 1. A device serving to remove solid particles from a flowing fluid by passing it through a porous element.[ARP 243B-65] 2. In electronic, acoustic and optical equipment, a device that allows signals of certain frequencies to pass, while rejecting signals having frequencies in another range. 3. A device used in a frequency transmission circuit to exclude unwanted frequencies and to keep the channels separate. 4. A device to suppress interference which would appear as noise. 5. A machine word that specifies which parts of another machine word are to be operated upon, thus the criterion for an external command. Synonymous with mask. 6. A porous material or structural element designed to allow fluids to pass through it while collecting and retaining solids of a certain particle size or larger.

filter aid An inert powdery or granular material such as diatomaceous earth, fly ash or sand which is added to a liquid that is about to be filtered in order to form a porous bed on the filter surface, thereby increasing the rate and effectiveness of the filtering process.

filter, bandpass A circuit which is tuned to pass all frequencies between certain points in the spectrum.

filter cake The solid or semisolid material retained on the surface of a filter after a liquid containing suspended solids has passed through.

filter capacitor A capacitor used as an element of an electronic filter circuit.

filter, clean pressure drop The differential pressure across a new filter element at specified flow rates and temperatures with a specified fluid. As a specification requirement, clean pressure drop provides little measure of element performance compared to the dirt holding capacity at a specified terminal pressure drop.[AIR887A-85]

filter, collapse pressure The minimum differential pressure across the element for which the element must not collapse, cause permanent degradation of filtration efficiency or fail in any other manner.[AIR887A-85]

filter, dirt capacity A measure of the "life" of the element, i.e. the amount of contaminant added in a laboratory test before the differential pressure due to clogging becomes so high that the element must be replaced or cleaned.[AIR 887A-85]

filter, flow fatigue The integrity of a filter element to withstand pulsating flow by subjecting the element to a flow cycle that normally ranges from zero to a specified percentage of rated flow and back to zero.[AIR887A-85]

filter inductor An inductor used as an element of an electronic filter circuit.

filter, line type A filter designed for installation in a fluid line and having a case capable of withstanding the operating pressure of the line, usually system pressure.[ARP243B-65]

filter, low pass A circuit which is tuned to pass all frequencies lower than a specified cutoff point.

filter, material compatibility The integrity of the filter element after artificially aging it in the fluid in which it is to be used at an elevated temperature. [AIR887A-85]

filter, media migration The amount of material that a filter releases from its own media and materials of construction. The release is caused by failure of the media to maintain its structural integrity under intended service conditions.[AIR887A-85]

filter medium The portion of a filter or filtration system that actually performs the function of separating out the solid material; it may consist of metal or nonmetal screening, closely woven fabric, paper, matted fibers, a granular bed, a porous ceramic cup or plate, or other porous component.

filter micron rating Unit of measure associated with filtration quality where one micron equals 1.0×10^{-6} m (3.937 ⁹ 10^{-5} in).[ARP906A-91]

filter, optical Optical coating, glass or plastic panel placed on or near the display face to provide selective spectral transmission of light or to reduce the amount of light transmitted. Reasons for its use are improvement of contrast ratio, color selection, anti-reflection or absorption, or both. For example, a neutral density filter improves contrast ratio by doubly attenuating the ambient light (compared to the single attenuation of the display luminance).[ARP 1782-89]

filter, reservoir type A filter to be housed within the shell of a reservoir for purification of the return flow to the reservoir, usually applied to the filter element alone.[ARP243B-65]

filter, reverse flow In some systems, the filter element may be subject to reverse flow, i.e., fluid flow through the filter housing is in a direction the reverse of normal.[AIR887A-85]

filter, vent type A filter for installation in an atmospheric vent line.[ARP243B-65]

fin 1. Fixed or adjustable airfoil or vane attached longitudinally to an aircraft, rocket, or a similar body to provide a stabilizing effect. Also, a flat plate of structure, as a cooling fin.[NASA] 2. A thin, flat or curved projecting plate, typically used to stabilize a structure surrounded by flowing fluid or to provide an extended surface to improve convective or radiative heat transfer. 3. A defect consisting of a very thin projection of excess material at a corner, edge or hole in a cast, forged molded or upset part, which must be removed before the part can be used.

final approach fix (FAF) The designated fix from or over which the final approach (IFR) to an airport is executed. It identifies the beginning of the final approach segment of the instrument approach. See segments of an instrument approach procedure.[ARP4107-88]

final approach-IFR See traffic pattern.

final approach segment See segments of an instrument approach procedure. [ARP4107-88]

final approach-VFR See traffic pattern.

final control element 1. An instrument that takes action to adjust the manipulated variable in a process. This action moves the value of the controlled variable back towards the set point. 2. The last system element that responds quantitatively to a control signal and performs the actual control action. Examples include valves, solenoids and servometers. 3. The device that exerts a direct influence on the process. 4. Unit of a control loop (such as a valve) which manipulates the control agent.

final controller The controller providing information and final approach guidance during precision approach radar (PAR) and surveillance approach radar (SAR) approaches utilizing radar equipment.[ARP4107-88]

final controlling element The element in a control system that directly changes the value of the manipulated variable. See element, final controlling.

fine grinding 1. Mechanical reduction of a powdery material to a final size of at least -100 mesh, usually in a ball mill or similar grinding apparatus. 2. In metallography or abrasive finishing, producing a surface finish of fine scratches by use of an abrasive having a particle size of 320 grit or smaller.

fineness Purity of gold or silver expressed in parts per thousand; for instance, gold having a fineness of 999.8 has only 0.02%, or 200 parts per million, of impurities by weight.

fineness ratio The ratio of the length of a body to its maximum diameter, or, sometime to some equivalent dimension—said especially of a body such as an airship hull or rocket.[NASA]

fines 1. In a granular substance having mixed particle sizes, those particles smaller than the average particle size. 2. Fine granular material which passes through a standard screen on which the coarser particles in the mixture are retained. 3. In a powdered metal, the portion consisting of particles smaller than a specified particle size.

finish 1. A chemical or other substance applied to the surface of virtually any solid material to protect it, alter its appearance, or modify its physical properties. 2. The degree of reflectivity of a lustrous material, especially metal; it is usually described by one of the following imprecise terms, listed in order of increasing luster and freedom from scratches—machined, ground brushed, matte, dull lustrous, bright, polished and mirror. 3. Generally, the surface quality, condition or appearance of a metal or plastic part.

finish grinding The final step in a grinding operation, which imparts the desired surface appearance, contour and dimensions.

finishing temperature In a rolling or forging operation, the metal temperature during the last reduction and sizing step, or the temperature at which hot working is completed.

finite impulse response filter See FIR filter.

finite volume method A moving mesh method for analyzing transonic flow over airfoils.[NASA]

fin tube A tube in a boiler having water on the outside and carrying the products of combustion on the inside.

fiord Arm of the sea having steep sides, deep bottom, and shallow sills separating it from the sea.

FIPS Federal Information Processing Standard.[AS4159-88]

fire assay Determining the metal content of an ore or other substance through the use of techniques involving high temperatures.

fire box The equivalent of a furnace. A term usually used for the furnaces of locomotive and similar types of boilers.

fire crack A crack starting on the heated side of a tube, shell, or header resulting from excessive temperature stresses.

fire detection, aircraft cargo compartment Type I: Carbon Monoxide, an instrument which will actuate an alarm signal when the concentration of carbon monoxide in air exceeds a specified value. TYPE II: Smoke Detector, Electronic, an instrument operating on the principle of smoke particles modifying the relationship between a light beam and electronic light sensor which will actuate an alarm signal when the concentration of smoke in air exceeds a specified value. Type III Smoke Detector, Visual, an instrument which by visual means will show in a positive manner the presence of smoke when the concentration of smoke in air exceeds a specified value.[AS446-84]

fired pressure vessel A vessel containing a fluid under pressure exposed to heat from the combustion of fuel.

fired torque The torque produced by an engine during the start cycle after engine light off has occurred.[ARP906A-91]

fire point The lowest temperature at which, under specified conditions, fuel oil gives off enough vapor to burn continuously when ignited.

fire polish A specular reflective finish that is produced by flame or moulding or an equivalent surface that may be produced by mechanical means.[ARP 924-91]

fireproof Resistant to combustion or to damage by fire under all but the most severe conditions.

fire resistance See flammability.

fire-resistant Resistant to combustion and to heat of standard intensity for a specified time without catching fire or failing structurally.

fire retardant 1. Treated by coating or impregnation so that a combustible material—wood, paper or textile, for instance—catches fire less readily and burns more slowly than untreated material. 2. The substance used to coat or impregnate a combustible material to reduce its tendency to burn.

fire tube A tube in a boiler having water on the outside and carrying the products of combustion on the inside.

fire tube boiler A boiler with straight tubes, which are surrounded by water and steam and through which the products of combustion pass.

firewall Also fire wall. 1. A fireproof or fire-resistant wall or bulkhead separating an engine from the aircraft structure; designed to prevent the spread of any fire originating at the engine. 2. To move the throttle forward toward the firewall; this movement produces an increase in power from the engine. [ARP4107-88]

FIR filter Physically unrealizable nonrecursive digital filter.[NASA]

firing rate control A pressure, temperature or flow controller which controls the firing rate of a burner according to the deviation from pressure or temperature set point. The system may be arranged to operate the burner on-off, high-low or in proportion to load demand.

firmware 1. Hardwired software which often encompasses microcodes.[NASA] 2. Programs or instructions that are permanently stored in hardware memory devices (usually read-only memories) that control hardware at a primitive level.

first alert See first out.

first-in, first-out (FIFO) An ordered queue. A discipline wherein the first transaction to enter a queue is also the first to leave it. Contrast with last-in, first-out.

first-level address See direct address.

first-order system A system definable

by a first-order differential equation.

first out reset See reset.

first word address (FWA) A program/routine.

Fisher loop test One of several Wheatstone bridge test arrangements commonly used to determine the distance to a fault (grounded or crossed wires) in a communications cable.

fisheye 1. An area on a fracture surface having a characteristic white crystalline appearance, usually caused by internal hydrogen cracking. 2. A small globular mass in a blended material such as plastic or glass which is not completely homogeneous with the surrounding material; it is particularly noticeable in transparent or translucent materials.

fishing tool An elongated or telescoping tool with a magnet, hook or grapple at one end which is used to retrieve objects from inaccessible places.

fish plate Either of the two plates bolted or riveted to the webs of abutting rails or beams on opposite sides to secure a mechanical joint.

fishtail Excess metal at the trailing end of an extrusion or a rolled billet or bar, which is generally cropped and either discarded or recycled into a melting operation.

fissile material See fissionable material.

fission The splitting of an atomic nucleus into two more-or-less equal fragments.[NASA]

fissionable material Material containing nuclides capable of undergoing fission only by fast neutrons with energy greater than 1MeV, e.g., thorium-232 and uranium-238.[NASA]

fissure A small, cracklike surface discontinuity, often one whose sides are slightly opened or displaced with respect to each other.

fissures (geology) Extensive cracks in rocks.[NASA]

fit The closeness of mating parts in an assembly, as determined by their respective dimensions and tolerances; fits may be classified as running (sliding), locational, transition or force (shrink) fits, depending on the size and direction (positive for running or negative for force fits) of the dimensional allowance; fits may also be termed clearance or interference depending on whether there is always a gap between mating parts or always interference, as long as the parts are within specified tolerances.

fitting 1. Not recommended, except in relation to fluid passage attaching devices of the nature used external to operating components. See fitting, tube. [ARP243B-65] 2. An auxiliary part of standard size and configuration that can be used to facilitate assembly; in constructing a system of pipe or tubing, for example, connections are more easily made if standard elbows, tees, unions and couplings are used to connect straight lengths of pipe, rather than bending the pipe or making special preparations before welding lengths of pipe together.

fitting, tube A self-contained detachable device including a fluid passage for attaching or connecting fluid carrying lines.[ARP243B-65]

fix A geographical position determined by visual reference to the surface, by reference to one or more radio NAVAIDs, by celestial plotting, or by another navigational device.[ARP4107-88]

fixed carbon The carbonaceous residue less the ash remaining in the test container after the volatile matter has been driven off in making the proximate analysis of a solid fuel.

fixed length record A record where the number of characters is constant.

fixed pitch A typeface in which all letters are the same width.

fixed point 1. (mathematics) Positional notation in which corresponding places in different quantities are occupied by coefficients of the same power of the base. Notation in which the base point is assumed to remain fixed with respect to one end of the numeric expressions. [NASA] 2. A reproducible standard value, usually derived from a physical property of a pure substance, which can be used to standardize a measurement or check an instrument calibration. 3. Pertaining to a numeration system in which the position of the radix point is fixed with respect to one end of the numerals, according to some convention. See fixed-point arithmetic.

fixed-point arithmetic 1. A method of calculation in which operations take place in an invariant manner, and in which the computer does not consider the location of the radix point. This is illustrated by desk calculators or slide rules with which the operator must keep track of the decimal point, and similarly with many automatic computers, in which the location of the radix point is the programmer's responsibility. Contrasted with floating-point arithmetic. 2. A type of arithmetic in which the operands and results of all arithmetic operations must be properly scaled so as to have a magnitude between certain fixed values.

fixed-point data In data processing, the representation of information by means of the set of positive and negative integers. It is faster than floating point data and requires fewer circuits to implement.

fixed point notation In data processing, numbers that are expressed by a set of digits with the decimal point in the correct position.

fixed-point part In a floating-point representation, the numeral of a pair of numerals representing a number, that is the fixed-point factor by which the power is multiplied. Synonymous with mantissa.

fixed-program computer A computer in which the sequence of instructions are permanently stored or wired in, and perform automatically and are not subject to change either by the computer or the programmer except by rewiring or changing the storage input. Related to wired program computer.

fixed restrictor A fixed physical restriction to fluid flow.

fixed storage A storage device that stores data not alterable by computer instructions, for example, magnetic core storage with a lockout feature, or photographic disk.

fixed wing A wing, which may be permanently fixed, foldable or adjustable, that is fixed to the airplane fuselage and outspread in flight; that is, a nonrotating wing.[ARP4107-88]

fixed word length Having the property that a machine word always contains the same number of characters or bits.

fixture 1. Use only when other categories do not suffice. A device for rigidly holding or positioning a part for a specific operation. EX: fixture, checking; fixture, reaming.[ARP480A-87] 2. An auxiliary component or operator aid attached to a structure or machine enclosure—a light or tool shelf, for instance. 3. A special holder that positions the work in a machining operation but does not guide the tool.

FL Flight Level.

flag 1. A bit of information attached to a character or word to indicate the boundary of a field. 2. An indicator frequently used to tell some later part of a program that some condition occurred earlier. 3. An indicator used to identify the members of several sets which are intermixed. 4. A storage bit whose location is usually reserved to indicate the occurrence or nonoccurrence of some condition, e.g., a Halt/Run flag would be 1 when the processor was halted and 0 when in the RUN condition.

flag/flag alarm A warning device incorporated in certain airborne navigation and flight instruments indicating that: (a) instruments are inoperative or otherwise not operating satisfactorily, or (b) the signal strength or quality of the received signal falls below acceptable values.[ARP4107-88]

flag register This is an 8-bit register in which each bit acts as a flag.

flag terminal A terminal having a tongue protruding from the side of its barrel.[ARP914A-79]

flake 1. Dry, unplasticized cellulosic plastics base material. 2. Plastics material in chip form used as feed in a molding operation. 3. An internal hydrogen crack such as may be formed in steel during cooling from high temperature. Also known as fisheye; shattercrack; snowflake. 4. Metal powder in the form of fish-scale particles. Also known as flaked powder.

flame A luminous body of burning gas or vapor.

flame cutting Using an oxyfuel-gas flame and an auxiliary oxygen jet to sever thick metal sections or blanks.

flame deflector In a vertical launch, any of variously designed obstructions that intercept hot gases of rocket engines so as to deflect them away from the ground or from a structure. In captive tests, elbows in the exhaust conduits or flame buckets that deflect the flame into the open.[NASA]

flame detector A device which indicates if fuel, such as liquid, gaseous, or pulverized, is burning, or if ignition has been lost. The indication may be transmitted to a signal or to a control system.

flame hardening A form of surface hardening that uses the inherent hardenability of a steel or other hardenable alloy to produce a hardened surface layer by spot-heating the metal with a fuel-gas flame to a shallow depth and then rapidly cooling the heated metal.

flame ionization detector A hydrogen-air diffusion flame detector that produces a signal nominally proportional to the mass-flow rate of hydrocarbons entering the flame per unit of time—generally assumed responsive to the number of carbon atoms entering the flame.[ARP1256-90]

flameout Unintended loss of combustion in turbine engines resulting in the loss of engine power.[ARP4107-88]

flame photometer An instrument for determining compositions of solutions by spectral analysis of the light emitted when the solution is sprayed into a flame.

flame plate A baffle of metal or other material for directing gases of combustion.

flame propagation rate Speed of travel of ignition through a combustible mixture.

flame quenching See quenching (cooling).

flame resistance A characteristic of material which is demonstrated by the tendency of the material, when burning to self-extinguish once the ignition source is removed.[ARP1931-87]

flame spraying 1. Applying a plastic coating on a surface by projecting finely powdered plastic material mixed with suitable fluxes through a cone of flame toward the target surface. 2. Thermal spraying by feeding an alloy or ceramic coating material into an oxyfuel-gas flame; compressed gas may or may not be used to atomize the molten material and propel it onto the target surface.

flame-spray strain gage A fine-wire strain gage element attached to a substrate by flame spraying a ceramic encapsulation over the element, which attaches it without damaging either the gage or the substrate; this attachment technique produces a bond suitable for operating over the temperature range -270 to 820 °C (-450 to 1500 °F).

flame treating Making inert thermoplastics parts receptive to inks, lacquers, paints or adhesives by bathing them in open flames to promote surface oxidation.

flammability Susceptibility to combustion.

flammable A characteristic of a material which is demonstrated by the tendency of the material to ignite and burn when an ignition source is brought sufficiently close.[ARP1931-87]

flammable liquid A liquid, usually a liquid hydrocarbon, that gives off combustible vapors.

flange, connector A projection extending from or around the periphery of a connector for the purpose of attaching the connector to a rigid surface.[ARP 914A-79]

flanged spade tongue terminal A slotted tongue terminal having the ends of the tongue formed up or down to the tongue plane, so as to form a degree of protection against the terminal slipping out from under its captive hardware. [ARP914A-79]

flange taps See orifice flange taps.

flank 1. On a cutting tool, the end surface adjacent to the cutting edge. 2. On a screw thread, the side of the thread.

flap A hinged, pivoted, or sliding airfoil or plate, or a combination of the former, regarded as a single surface, normally located at the trailing edge of the wing, and designed to add camber well aft on the chord or increase wing area, or both. See also Fowler flap, leading edge flap. [ARP4107-88]

flaperons Airplane control surfaces that serve the function of both aileron and flap.[NASA]

flap valve A valve with a hinged flap or disc that swings in only one direction.

flare The increase in pitch angle of an aircraft just before touchdown. This change in pitch attitude on final approach to touchdown attitude allows the airspeed to dissipate and the aircraft to settle down to the runway in the proper attitude.[ARP4107-88]

flareback A burst of flame from a furnace in a direction opposed to the normal flow, usually caused by the ignition of an accumulation of combustible gases.

flared tube-end The projecting end of a rolled tube which is expanded or rolled to a conical shape.

flare stars Members of a class of dwarf stars that show sudden intensive outbursts of energy.[NASA]

flaring Increasing the diameter at the end of a pipe or tube to form a conical section.

flash 1. In plastics molding, elastomer molding or metal die casting, a portion of the molded material that overflows the cavity at the mold parting line. 2. A fin of material attached to a molded, cast or die forged part along the parting line between die halves, or attached to a resistance flash welded, upset welded or friction welded part along the weld line.

flashback Backward burning of a flame into the lip of a burner or torch.[NASA]

flash converter A converter in which all bit choices are made at the same time.

flashing 1. A desired and usually controlled variation in the luminance of a symbol or group of symbols.[AS8034-82] 2. (vaporizing) The evaporation of a heated liquid as a consequence of rapid pressure reduction.[NASA] 3. Steam produced by discharging water at saturation temperature into a region of lower pressure.

flashlamp A gas-filled lamp which is excited by an electrical pulse passing through it to emit a short bright flash of light. A broad range of wavelengths are produced, with their precise nature depending on the gas or gases used.

flash line A raised line on the surface of a molded or die cast part that corresponds to the parting line between mold faces.

flash plating Electrodeposition of a very thin film of metal, usually just barely enough to completely cover the surface.

flash point 1. Temperature at which a fluid gives off sufficient vapor to cause it to ignite when a small flame is applied, under controlled conditions.[AIR 1916-88]

flash welding A resistance welding process commonly applied to wide, thin members, irregularly shaped parts and tube-to-tube joints, in which the faying surfaces are brought into close proximity, electric current is passed between them to partly melt the surfaces by combined arcing and resistance heating, and the surfaces are then upset forged together to complete the bond.

flask A wood or metal frame for holding a sand mold in foundry work; it is open ended, and usually consists of two halves—the cope (upper half) and the drag (lower half), although three or more flask sections are occasionally used.

flat cable A cable with geometric configuration which is flat or essentially flat rather than round. In a multiconductor flat cable the conductors are laid side by side in the same plane.[ARP 1931-87]

flat conductor A conductor with a rectangular cross section.[ARP1931-87]

flat conductor cable A cable constructed using flat conductors.[ARP 1931-87]

flat die forging Shaping metal by compressing or hammering it between simple flat or regularly contoured dies.

flat pattern Shape of a part or parts put in 3 space in its undefined condition.[NASA]

flat-position welding Welding from above the work, with the face of the weld in the horizontal plane. Also known as downhand welding.

flat spin A spin in which the longitudinal axis of an aircraft inclines downward at an angle less than 45 deg. [ARP4107-88]

flatspotting The wearing of a flat spot on a tire by means of skidding due to overbraking or faulty skidding due to overbraking or faulty skid control.[AIR 1489-88]

flattening Straightening metal sheet by passing it through a set of staggered and opposing rollers that bend the sheet slightly to flatten it without reducing its thickness.

flattening test A test that evaluates the ductility, formability and weld quality of metal tubing by flattening it between parallel plates to a specified height.

flatting agent A chemical additive that promotes a nonglossy, matte finish in paints and varnishes.

flavor (particle physics) The specific identifiers of quarks which distinguish various combinations of electric charge and mass.[NASA]

flaw A discontinuity or other physical attribute in a material that exceeds acceptable limits; the term flaw is nonspecific, and more specific terms such as defect, discontinuity or imperfection are often preferred.

fleet satellite communication system Global communication system utilizing satellites.[NASA]

FLEETSATCOM See fleet satellite communication system.

flettner tab See servo tab.

flexibility A 4-D term representing the FMCS total range of control capability over arrival time at a designated downstream waypoint or waypoints. It may be expressed as an TRA Max and TRA Min in respective GMT hrs:min:sec. [ARP1570-86]

flexible manufacturing systems (FMS)- A manufacturing system under computer control with automatic material handling. The system is primarily designed for batch manufacturing.

flexible pavement See pavement, flexible.

flexible spacecraft Space vehicles (usually space structures or rotating satellites) whose surfaces and/or appendages may be subject to elastic flexural deformations (vibrations).[NASA]

flexivity Temperature rate of flexure for a bimetal strip of given dimensions and material composition.

flex life The number of flexes that can be achieved when flexing an item under given conditions (temperature, radius of bend, load, arc, etc.) before the failure point is reached.[ARP1931-87]

flicker 1. Undesired temporal variation of luminance in a portion or the total display at a rate detectable by the eye.[ARP1782-89] 2. Undesired rapid temporal variation in part or total display luminance.[AS8034-82]

flicker vertigo See illusion, visual.

flight The movement of an object through the atmosphere or through space, sustained by aerodynamic, aerostatic, or reaction forces, or by orbital speed; especially, the movement of a man operated or man controlled device, such as a rocket, a space probe, a space vehicle, or an aircraft.[NASA]

flight characteristic Characteristic exhibited by an aircraft, rocket, or the like in flight, such as a tendency to stall or to yaw, or an ability to remain stable at certain speeds.[NASA]

flight check 1. In-flight investigation and evaluation of a navigational aid (NAVAID) to determine whether it meets established tolerances. 2. An in-flight evaluation of a flight crewmember's ability to perform assigned duties. 3. A call sign prefix used by Federal Aviation Administration (FAA) aircraft engaged in flight inspection/certification of navigational aids and flight procedures.[ARP4107-88]

flight control component A component used in a system controlling in flight actuation of aerodynamic surfaces. (Actuator, cylinder, valve, etc.)[ARP 243B-65]

flight control system (FCS) A system which includes all aircraft subsystems and components used by the pilot or other sources to control one or more of the following: aircraft flight path, attitude, airspeed, aerodynamic configuration, ride and structural modes.[ARP 1181A-85]

flight crewmember A pilot, copilot, flight engineer, or flight navigator assigned to an aircraft during flight time.[ARP4107-88]

flight discipline Adherence to established procedures throughout the course of a sortie. This includes not pursuing irrational or impulsive courses of action, actions that are inconsistent with established procedure, or actions not prebriefed.[ARP4107-88]

flight envelope The bounds within which a certain flight system can operate, especially a graphic representation of these bounds showing interrelationships of operational parameters.[NASA]

flight inspection See flight check.

flight level A level of constant atmospheric pressure related to a reference datum of 29.92 in of mercury. It is stated in digits representing hundreds of feet. For example, flight level 250 represents a barometric altimeter indication of 25,000 ft; flight level 255 indicates 25,500 feet.[ARP4107-88]

flight monitor package A special software system that enables several operators to monitor data from an aircraft or other source in real time.

flight operation Collective term for ground support operations by flight crew or supported personnel preparatory to space flight, or tasks performed by crew during flight.[NASA]

flight path 1. A line, course, or track along which an aircraft is flying or intended to be flown.[ARP4107-88] 2. Path made or followed in the air or in space by an aircraft or rocket; the continuous series of positions occupied by a flying body; more strictly, the path of the center of gravity of the flying body, referred to the earth or other fixed reference. [NASA]

flight-path angle The angle between the flight path of the aircraft and the horizontal.[ARP4107-88]

flight performance See flight characteristic.

flight plan Specified information relating to the intended flight of an aircraft that is filed orally or in writing with a flight service station (FSS) or an air traffic control (ATC) facility.[ARP4107-88]

flight plan—progress display The pilot's primary navigation display. The display may include, but is not limited to, the following: aircraft relative position and orientation; altitude intercept range; course, desired track or localizer; course or localizer deviation; drift angle and/or track angle error; ETA's and RTA's; heading; predictive track; track; time control speed reference; waypoints; weather.[ARP1570-86]

flight profile A graphic vertical-plane portrayal of an aircraft's flight path. [ARP4107-88]

flight recorder An instrument or device that records information about the performance of an aircraft in flight or about conditions encountered during flight. [ARP4107-88]

flight service station (FSS) Air traffic facilities which (a) provide pilot briefing, en route communications, and VFR search and rescue services; (b) assist lost aircraft and aircraft in emergency situations; (c) relay ATC clearances; (d) originate Notices to Airmen (NOTAM); (e) broadcast aviation weather and NAS information; (f) receive and process IFR flight plans; and (g) monitor NAVAIDs. Selected flight service stations provide en route flight advisory service (flight watch), take weather observations, issue airport advisories, and advise Customs and Immigration of transborder flights.[ARP4107-88]

flight simulator Training device or apparatus that simulates certain conditions of flight or of flight operations. [NASA]

flight test Test by means of actual or attempted flight to see how an aircraft, spacecraft, space-air vehicle, or missile flies. Tests of a component part of a flying vehicle, or of an object carried in such a vehicle, to determine its suitability or reliability in terms of its intended function by making it endure actual flight.[NASA]

flight test vehicle Test vehicle for the conduct of flight tests either to test its own capabilities or to carry equipment requiring flight tests.[NASA]

flight visibility The average forward horizontal distance from the cockpit of an aircraft in flight at which prominent unlighted objects may be seen and identified by day, and prominent lighted objects may be seen and identified by night.[ARP4107-88]

flight watch A term for use in air-ground contacts on frequency 122.0 MHz to identify the flight service station providing en route flight advisory service. See also en route flight advisory service.[ARP4107-88]

flinching In quality control inspection, failure of an inspector to call a borderline defect a defect.

flint glass An optical glass which contains lead or other elements which raise its refractive index between 1.6 and 1.9; higher than other types of optical glass.

FLIP See DoD FLIP.

flip-flop 1. A digital component or circuit with two stable states and sufficient hysteresis so that it has "memory". Its state is changed with a control pulse; a continuous control signal is not necessary for it to remain in a given state. [ARP993A-69] 2. Device having two stable states and two input terminals (or types of input signals) each of which corresponds with one of the two states. The circuits remain in either state until caused to change to the other state by application of the corresponding signal. Similar bistable devices with an input which allows it to act as a single-stage binary counter.[NASA] 3. A bistable device, i.e., a device capable of assuming two stable states. 4. A bistable device which may assume a given stable state depending upon the history of pulses of one or more input points and having one or more output points. The device is capable of storing a bit of information. 5. A control device for opening or closing gates, i.e., a toggle.

FLIR detector Forward-looking infrared detector for sensing all emissions of heat or light.[NASA]

float 1. A device or component part of an air vehicle which permits it to remain suspended upon the surface of a body of water without sinking: also to maneuver upon, takeoff, and land upon a body of water.[AIR1489-88] 2. Any component having positive buoyancy—for example, a hollow watertight body that rests on the surface of a liquid, partly or completely supported by buoyant forces. 3. See plummet.

float chamber A vessel in which a float regulates the liquid level.

float control A type of control apparatus in which the control signal is regulated by a float riding up and down with liquid level. Contrast with floating control.

float gage Any of several types of devices that use pulleys, levers or other mechanisms to transmit the position of a float to a scale that indicates liquid level in a tank or vessel.

floating A condition of a line in a logic circuit that is not grounded or tied to any established potential.

floating action A type of control-system action in which a fixed relationship exists between a measured deviation and the rate of motion of the final control element.

floating control 1. A control which has no unique position of the final controlled device. In any given stable condition the control point and the set point are the same or within the deadband. When an error exists, continuous control action will be exercised until the error is eliminated at the control element(s) (i.e., cabin sensing element). The rate of corrective action may or may not be a linear function of error and is often modified by velocity feedback, pulse modulation or anticipation.[ARP 89C-70] 2. A control device in which the output control signal is proportional to the difference between an indicator signal and the controller's setpoint; this difference is often referred to as an error signal; in operation, floating control reduces the tendency to overshoot the setpoint because it reduces power input to the system as the controlled variable approaches the setpoint value. Contrast with float control.

floating control action See control action, floating.

floating controller See controller, floating.

floating control mode A controller mode in which an error in the controlled variable causes the output of the controller to change at a constant rate. The error must exceed preset limits before controller change starts.

floating plug A short nosed mandrel attached to a rod which is inserted into pipe or tubing during reduction by drawing. Also known as plug die.

floating point 1. An arithmetic notation in which the decimal point can be manipulated; values are sign, magnitude, and exponent ($+.833\text{x}10^2$). 2. A form of number representation in which quantities are represented by a bounded number (mantissa) and a scale factor (characteristic or exponent) consisting of a power of the number base, e.g., 127.6=0.1276*10**3 where the bounds are 0 and 1 for the mantissa and the base is ten.

floating point arithmetic A method of calculation which automatically accounts for the location of the radix point. This usually is accomplished by handling the number as a signed mantissa times the radix raised to an integral exponent, e.g., the decimal number +88.3 might be written as +.883*10**2. The binary number -.0011 as -.11*2**-2. Contrasted with fixed-point arithmetic.

floating point base In floating-point representation, the fixed positive integer that is the understood base of the power. Synonymous with floating-point radix.

floating point notation In data processing, numbers that are expressed as a fraction coupled with an interger exponent of the base.

floating-point routine A set of subroutines which causes a computer to execute floating-point arithmetic. These routines may be used to simulate floating-point operations on a computer with no built in floating-point hardware.

floating rate The rate of motion of the final control element in a proportional-speed floating-control system which corresponds to a given deviation.

floating speed The rate of motion of the final control element in a single-speed or multispeed floating-control system.

floatless level control Any device for measuring or controlling liquid level in a tank or vessel without the use of a float—methods include manometers, electrical probes, capacitance devices, radiation instruments, and sonic or ultrasonic instruments.

float switch An on-off switch activated by the position of a float.

float valve An on-off-type valve whose action is triggered by the rise or fall of a float.

FLOLS Fresnel Lens Optical Landing System.

flooding The consequence of excessive column liquid loading where, in effect, the liquid on trays becomes too deep for the vapor to pass through or where the vapor flow rate is too high, creating an excessive differential pressure or a decrease in the differential temperature across the column.

flood lighting Lighting similar to normal home lighting where a light source floods a general area or is directed by reflectors to light a particular area. [ARP1161-91]

flood lit systems Integrally lit displays in which the light source or a light distributing material directs light to the front of the display where it is reflected to the observer. Examples of flood lit systems are wedge lighting, ring lighting, and parallel plate lighting.[ARP 1161-91]

floppy disk Any flexible platter with a magnetic coating that can accept computer data. Also called a diskette.

flospinning Forming cylindrical, conical or curvilinear parts from light plate by power spinning the metal over a rotating mandrel.

flotation A process for separating particulate matter in which differences in surface chemical properties are used to make one group of particles float on water while other particles do not; it is used primarily to separate minerals from gangue but is also used in some chemical and biological processes. In mining engineering, also known as froth flotation.

flotation, ground The process of supporting an air vehicle on the surface of the ground without sinking into the ground. Also used to describe capability of pavement and other surfaces on the surface of the ground to support a vehicle or object.[AIR1489-88]

flow 1. Rate of fluid movement, usually expressed in GPM.[ARP243B-65] 2. Quantity of a fluid (volume or mass) crossing the transverse plane of a flow path per unit of time. Gas volume flow may be expressed at standard reference conditions of sea level atmospheric pressure and ambient temperature.[AIR 1916-88] 3. A stream or movement of air or other fluid, or the rate of fluid movement, in the open or in a duct, pipe, or passage; specifically an airflow. [NASA] 4. The order of events in the computer solution to a problem. 5. The movement of material in any direction. See also block flow.

flowability A general term describing the ability of a slurry, plasticized material or semisolid to behave like a fluid.

flow amplifier A component designed specifically for amplifying flow signals. [ARP993A-69]

flow brazing A brazing process in which the joint is heated by pouring hot molten nonferrous filler metal over the assembled parts until brazing temperature is attained.

flow, calibrated In a unit which controls or limits rate or quantity of flow, that rate or quantity of flow for which the unit is calibrated or adjusted. When calibration is related to specific conditions, the condition of calibration must be stated. Example, a restrictor may be calibrated to pass 1 GMP at 1500 psi pressure differential.[ARP243B-65]

flow chart 1. Graphical representation of sequences of operations using symbols to represent the operations. Flow charts are more detailed representations than diagrams.[NASA] 2. A system-analysis tool that provides a graphical presentation of a procedure. Includes block diagrams, routine sequence diagrams, general flow symbols, and so forth. 3. A chart to represent, for a problem, the flow of data, procedures, growth, equipment, methods, documents, machine instructions, etc. 4. A graphical representation of a sequence of operations by using symbols to represent the operations such as COMPUTE, SUBSTITUTE, COMPARE, JUMP, COPY, READ, WRITE, etc.

flow coat To apply a coating by pouring liquid over an object and allowing the excess to drain off.

flow compensation Using secondary signals to correct flow values for changes in density or viscosity.

flow control 1. Measures used by ATC to adjust the flow of traffic into a given airspace, along a given route, or bound for a given airport to ensure the most effective utilization of the airspace. [ARP4107-88] 2. Any method for controlling the flow of a material through piping, ductwork or channels.

flow diagram See flow chart.

flow divider A flow proportioner which operates only with dividing flow.[ARP 243B-65]

flow equalizer A flow proportioner in which the portions are equal.[ARP243B-65]

flow line(s) 1. The texture revealed by etching a metal surface or section showing direction of metal flow.[ARP700-65] 2. The connecting line or arrow between symbols on a flow chart. 3. A mark on a molded plastic part where two flow fronts met during molding. Also known as weld mark. 4. In mechanical metallurgy, a path followed by minute volumes of metal during forming.

flow marks Wavy surface marks on a molded thermoplastic part resulting from improper flow of resin during molding.

flowmeter 1. An instrument used to measure linear, non-linear or volumetric flow rate or discharge rate of a fluid flowing in a pipe. Also known as fluid meter.

flowmeter secondary device The device that responds to the signal from the primary device and converts it to a display or to an output signal that can be translated relative to flow rate or quantity.

flow mixer A device for mixing two solids, liquids, or gases together in which the mixing action occurs as the materials pass through the device. Also known as line mixer.

flow nozzle A type of differential pressure producing element having a contoured entrance. Characterized by its ability to be mounted between flanges and have a lower permanent pressure loss than an orifice plate.

flow proportioner A device which automatically maintains a relatively constant ratio between the portions of dividing or combining flow passing through it regardless of differences in pressure between the portions. A flow proportioner may operate only with combining flow, dividing flow or with both. [ARP243B-65]

flow rate The quantity of fluid which moves through a pipe or channel within a given period of time.

flow, rated The nominal maximum operating flow, applicable where control of that flow rate is not basic. Example, a one way restrictor valve with a calibrated flow of 1 GMP at 1500 psi pressure differential may have a rated flow of 3.5 GMP in the reverse or free flow direction.[ARP243B-65]

flow-rate range Range of flow rates bounded by the minimum and maximum flow rates.

flow soldering See wave soldering.

flow surge (surge) Temporary rise and fall of flow.[AIR1916-88]

flow transmitter A device that senses the flow of liquids in a pipe and converts the sensor output into electric signals proportional to flow rate which can be transmitted to a remote indicator or controller.

FLTSATCOM See fleet satellite communication system.

flue A conduit or duct for conveying combustion products from a furnace chamber or firebox to the point of discharge to the atmosphere.

flue dust The particles of gas-borne solid matter carried in the products of combustion.

flue gas Gaseous combustion products from a furnace.[NASA]

flue-gas analyzer An instrument that monitors the composition of flue gas as it passes out of a boiler or heating unit; the readout is used to guide adjustment of combustion controls to achieve maximum combustion efficiency or heat output.

flueric An adjective which can be applied to fluidic devices and systems performing sensing, logic, amplification, and control functions if no moving mechanical elements whatsoever are used. [ARP993A-69]

fluid A gas or liquid, both of which have the property of undergoing continuous deformation when subjected to any finite shear stress as long as the shear stress is maintained.

fluid coupling A device for transmitting rotational motion and power between shafts by means of the acceleration or deceleration of oil or another suitable liquid. Also known as hydraulic coupling.

fluid filled shell Shell of revolution containing a gas or liquid.[NASA]

fluidics The general field of fluid devices or systems performing sensing, logic, amplification and control functions employing primarily no-moving-part (flueric) devices.[ARP993A-69]

fluidity The degree to which a substance flows freely.

fluidized bed A dynamic mixture of a gas and/or vapor and minute solid particles of such a size that the mixture resembles a fluid in motion.

fluid management The isolation and separation of liquids from gas in a storage vessel which operates in a reduced or zero gravity environment using liquid acquisition devices such as those used in the Space Shuttle RCS tankage.[NASA]

fluid meter See flowmeter.

fluid, operating The medium or fluid to be used in a unit or system.[ARP243B-65]

fluid-solid interaction The interaction of a rigid or elastic structure with an incompressible or compressible fluid. Airblast loading and response, acoustic interaction, aeroelasticity, and hydroelasticity comprise its major divisions.[NASA]

fluid stiffness The pressurized fluid chamber on each side of the piston in an actuator contains a trapped volume of fluid when the control valve is at null. The trapped fluid exerts stiffness, opposing change in fluid volume, when an external load is applied to the piston.

fluid temperature Temperature of the fluid measured at a specified point in the system.[AIR1916-88]

fluid transpiration See transpiration.

flume An adaptation of the venturi concept of flow constriction applied to open-channel flow measurement.

fluorescence 1. Emission of light or other radiant energy as a result of and only during absorption of radiation of a different wavelength from some other source.[NASA] 2. Emission of electromagnetic radiation from a surface upon absorption of energy from other electromagnetic or particulate radiation, the emission being sustained only so long as the stimulating radiation impinges on the material. 3. The electromagnetic radiation produced by the above process. 4. Characteristic x-rays produced due to absorption of higher-energy x-rays.

fluorescence spectroscopy The study of materials by light which they emit when irradiated by other light. Many materials emit visible light after they have been illuminated by ultraviolet light. The intensity and wavelengths of the emitted light can be used to identify the material and its concentration.

fluorescent emission See fluorescence.

fluorescent lamps In general, fluorescent lamps are used for area lighting, including cabin, galley and lavatory and decorative lighting. The fluorescent lamp is an electric discharge light source, in which light is predominantly produced by fluorescent powders activated by ultraviolet energy generated by a mercury arc. There are two basic types of fluorescent lamps, hot cathode and cold cathode.[AIR512B-91]

fluorocarbon 1. A polymer or gas containing only carbon and fluorine.[ARP 1931-87] 2. A compound containing fluorine and carbon (including other elements).[NASA]

fluorometer An instrument for measuring the fluorescent radiation emitted by a material when excited by monochromatic incident radiation—usually filtered radiation from a mercury-arc lamp or from a tungsten or molybdenum x-ray tube. Also spelled fluorimeter.

fluoroplastics A family of plastics resins based on fluorine substitution of hydrogen atoms in certain hydrocarbon molecules. See also fluoropolymers.

fluoropolymer(s) 1. A polymer which contains fluorine as one of its elements.[ARP1931-87] 2. A family of polymers based on fluorine replacement of hydrogen atoms in hydrocarbon molecules. Compounds are characterized by chemical inertness, thermal stability, and low coefficient of friction.[NASA]

fluoroscopy X-ray examination similar to radiography, but in which the image is produced on a fluorescent screen instead of on radiographic film.

fluorosilicone sealant A sealant based upon a fluorosilicone polymer. There are two types used in integral fuel tanks: (a) A one part, moisture cured polymerizing elastomer used in brushcoat and fillet (extrusion type) applications inside the fuel tank. (b) A one part noncuring channel or groove sealant. It can be prepacked into the grooves in the faying surface in assembly or injection through external ports.[AIR4069-90]

flushing Removing debris, deposits, wear particles, or used lubricating oil from a piping system, chamber or mechanism by circulating a liquid such as a solvent oil or water, then draining the system to carry off unwanted substances.

flushing connection A connection on the instrument, manifold or piping to permit periodic back-flow of an external fluid for clearing purposes.

flushing pressure Pressure required to flush a system at defined conditions (for instance at defined flow).[AIR1916-88]

flush left In a printout, the alignment of type so that the left start-point of each line is the same.

flush right In a printout, the alignment of type so that the end-point of each line is the same. Most books are typeset with lines flushed both left and right.

flute 1. In a drill, reamer or tap, a channel or groove in the body of the tool which exposes the cutting edge and provides a passage for cutting fluid and chips. 2. In a milling cutter or hob, the chip space between the back of one cutting tooth and the face of the following tooth.

fluted-rotor flowmeter A type of flow-measurement device in which fluid is trapped between two fluted rotors which are dynamically balanced but hydraulically unbalanced so that they turn at a rate proportional to the volume rate of fluid flow.

flutter 1. An aeroelastic self excited vibration in which the external source of energy is the airstream and which depends on the elastic, inertial and dissipative forces of the system in addition to the aerodynamic forces.[NASA] 2. In tape recorders, the higher-frequency variations in record and/or reproduce speed which cause time base errors in the record/reproduce process. 3. Irregular alternating motion of a control surface, often due to turbulence in a fluid flowing past it. 4. Repeated speed variation in computer processing.

flux 1. The rate of flow of some quantity, often used in reference to the flow of some form of energy. In nuclear physics generally, the number of radioactive particles per unit volume times their mean velocity.[NASA] 2. In metal refining, a substance added to the melt to remove undesirable substances such as sand, dirt or ash, and sometimes to absorb undesirable elements or compounds such as sulfur in steelmaking or iron oxide in copper refining. 3. In welding, brazing and soldering, a substance preplaced in the joint or fed into the molten zone to prevent formation of oxides or other undesirable compounds, or to dissolve them and make it easy to remove them. 4. In magnetic or electromagnetic applications, the integral of magnetic field strength over the cross sectional area of the field.

flux-cored arc welding Abbreviated FCAW. A form of electric arc welding in which the electrode is a continuous tubular wire of filler metal whose central cavity contains welding flux; welding may be performed with or without a shielding gas such as CO_2 or argon.

flux density The flux (rate of flow) of any quantity, usually a form of energy, through a unit area of specified surface. (Note that this is not a volumetric density like radiant density).[NASA]

flux gate A detector that produces an electric signal whose magnitude and phase are proportional to the magnitude and direction of an external magnetic field aligned with the detector's axis.

flux guide A shaped piece of metal used in magnetic or electromagnetic applications to direct magnetic flux along preferred paths or to prevent it from spreading beyond specific boundaries.

flux mapping See flux density.

fluxmeter An instrument for measuring the intensity of magnetic flux.

flux pinning In superconductors, the interaction between the magnetic and the metallurgical microstructures. It controls the critical current density in a given superconducting material. [NASA]

flux pump Cryogenic dc generator. [NASA]

flux (rate) The total emanation of energy, material or particles from a single source per unit time. Used for electron flux, neutron flux, and particle flux.[NASA]

flux (rate per unit area) See flux density.

flux vector splitting The splitting of the nonlinear flux vectors of the conservation law form of the inviscid gas-dynamic equations into subvectors by similarity transformations so that each subvector has associated with it a specified eigenvalue spectrum.[NASA]

fly A fan with two or more blades that is used in timepieces or light machinery to control rotational speed by means of air resistance.

fly ash Fine particulate, essentially noncombustible refuse, carried in a gas stream from a furnace.[NASA]

fly-by-light See control-by-light.

flyby mission Interplanetary mission in which the vehicle passes close to the target planet but does not impact it or go into orbit around it.[NASA]

fly by tube control A fluidic light control for aircraft in which a hydraulic control signal link connects the pilot's controls to the control surface actuators.[NASA]

fly-by-wire (FBW) See control-by-wire.

fly cutting Machining using a rotating single-point tool or a milling cutter having only one tooth.

fly heading (degrees) A controller command that informs the pilot of the heading that he/she should fly.[ARP 4107-88]

flying See flight.

flying qualities See flight characteristics.

flying tab See servo tab.

flywheel A balanced rotating element attached to a shaft which utilizes inertial forces to maintain uniform rotational speed and damp out small variations in power generated by the driving elements.

FM discriminator A device that converts frequency variations to proportional variations in voltage or current.

FM/FM Frequency modulation of a carrier by subcarriers that are frequency modulated by information.

FM (frequency modulation) The process (or the result of the process) in which the frequency deviates from the unmodulated carrier in proportion to the instantaneous value of the modulating signal.

FM/PM (modulation) Phase modulation of a carrier by subcarrier(s) which is (are) frequency modulated by information.[NASA]

FMS See flexible manufacturing system.

FM (tape record/reproduce) The tape record/reproduce process whereby data modulate an FM oscillator for recording, and are demodulated by an FM discriminator.

F number The ratio of the principle length of a lens to its diameter.

foaming 1. Any of various methods of introducing air or gas into a liquid or solid material to produce a foam. 2. The continuous formation of bubbles which have sufficiently high surface tension to remain as bubbles beyond the disengaging surface.

foam-in-place A method widely used to apply foamed insulation to homes and industrial equipment, in which two or more reactive substances are deposited onto a surface to be covered where the foaming reaction takes place.

focal length The distance from the focal point of a lens or lens system to a reference plane at the lens location, measured along the focal axis of the lens system.

focal plane array See focal plane device.

focal plane device Radiation sensitive device positioned at the focal area of electromagnetic detectors.[NASA]

focal point 1. The location on the opposite side of a lens or lens system where rays of light from a distant object meet at a point. Also known as focus. 2. The point in space where a beam of electromagnetic energy (such as light, x-rays or laser energy) or of particles (such as electrons) has its greatest concentration of energy; it corresponds to the point where a converging beam of energy undergoes a transition to become a diverging beam.

focal spot The area of the target in an x-ray tube where the stream of electrons from the cathode strikes the target.

focus 1. To adjust the position of a lens with respect to an imaging surface so that sharp features of the object appear sharp in the image. 2. A focal point.

focusing coil An assembly containing one or more electromagnetic coils which is used to focus an electron beam.

focusing electrode An electrode configured so that its electric field acts to control the cross-sectional area of an electron beam.

focusing magnet An assembly containing one or more permanent magnets or electromagnets which is used to focus an electron beam.

focus of attention See attention, focus of.

FOD Acronym for Foreign Object Damage; damage to a jet engine caused by the ingestion of debris (foreign objects); also, the debris which damages a jet engine when it is ingested.[ARP4107-88]

fog 1. A suspension of very small water droplets in the air.[AIR1335-75] 2. A defect in developed radiographic, photographic or spectrographic emulsions consisting of uniform blackening due to unintentional exposure to low-intensity light or penetrating radiation.

fogged metal A metal surface whose luster has been greatly reduced by the creation of a film of oxide or other reaction products.

fog quenching Rapidly cooling an item by subjecting it to a fine mist, usually of water.

fog time The period between a fuel manifold full condition and lightoff during a ground or airstart attempt. It is typified by a light fuel mist exiting the exhaust nozzle.[ARP906A-91]

foil Very thin metal sheet, usually less than 0.006 in. (0.15 mm) thick.

foil strain gage A type of metallic strain gage usually made in the form of a back-and-forth grid by photoetching a precise pattern on foil made of a special alloy having high resistivity and low temperature coefficient of resistivity.

fold A fold is a doubling over of metal which may occur during the forging operation. Folds may occur at or near the intersection of diameter changes, and are especially prevalent with noncircular necks, shoulders and heads. [MA2005-88]

folding error An error in sampling an electronic signal arising from failure to sample at a high enough rate (sampling rate should be at least double the maximum signal frequency), so that the sampling device perceives high-frequency components of the signal as low-frequency components. Also known as aliasing error.

follow-up control unit A device which provides an output signal proportional to the displacement, or rate of movement, of that which is being driven by the servo. Also called a feedback or repeatback control unit.[ARP419-57]

follow-up control unit A device which provides an output signal proportional to the displacement, or rate of movement, of that which is being driven by the servo. Also called a feedback or repeatback control unit.[ARP419-57]

follow-up signal A signal produced by a follow-up control unit. Also feedback or repeatback control unit.[ARP419-57]

font Typesetting characters of a particular style.

food chain The scheme of feeding relationships by tropic levels which unites member species of a biological community.[NASA]

food processing The transformation of foodstuffs into forms for easy packaging, greater palatability, longer storage, etc.[NASA]

foot A fundamental unit of length in the British and U.S. Customary systems of measurement equal to 12 inches.

footcandle A measure of the intensity or level of illumination. One footcandle (fc) is the intensity of illumination at a point on a surface one foot from a uniform point source of one standard candle.[AS264D-91] [ARP1161-91]

footlambert (fL) A unit of brightness equal to the uniform brightness of a perfectly diffusion surface emitting or reflecting light at the rate of one lumen per square foot. The average brightness of any reflecting surface in footlamberts is the product of the illumination in footcandles by the reflection factor of the surface.[ARP798-91]

foot-pound A force of one pound applied to a lever one foot long.

footprint 1. The area of airplane floor space reserved for locating the galleys. Also the lower surface of galley envelope areas shown in plan view.[AS1426-80] 2. Ground pattern or contour of an acoustical or microwave nature that are predictable and measurable.[NASA]

footprint area See tire footprint.

Forbush decease The observed decrease in cosmic ray activity in the earth's atmosphere about a day after a solar flare.[NASA]

Forbush effect See Forbush decrease.

force The cause of the acceleration of material bodies measured by the rate of change of momentum produced on a free body.[NASA]

force-balance transmitter A transmitter design technique utilizing feedback of the output signal to balance the primary input signal from the measuring element. The balanced output signal is proportional to the measured variable.

force, contact engaging The force required to fully engage a pair of mating contacts.[ARP914A-79]

force contact retention The maximum allowable force which, if applied axially in either direction on a contact, does not displace the contact permanently from its normal position in the connector or jeopardize or damage the contact or retention provision.[ARP914A-79]

force, contact separation The force required to separate a pair of fully mated contacts.[ARP914A-79]

forced circulation Using a pump or fan to move fluid through a conduit or process vessel—for instance, air or gases through a furnace or combustion chamber (often referred to as forced draft), ambient or conditioned air through ductwork (often referred to as forced ventilation), or a mixture of water and steam through tubes in a boiler.

forced draft fan A fan supplying air under pressure to the fuel burning equipment.

forced oscillation Oscillation of a system attribute where the period of oscillation is determined by an external periodic force. See also forced vibration.

forced vibration An oscillation of a system in which the response is imposed by the excitation. If the excitation is periodic and continuing, the oscillation is steady state.[NASA]

forced vibratory motion equation See forced vibration.

force factor 1. The complex ratio of the force required to block the mechanical system of an electromechanical transducer to the corresponding current in the electrical system. 2. The complex ratio of open-circuit voltage in the electrical system of an electromechanical transducer to corresponding velocity in the mechanical system.

force fit A class of interference fit involving relatively large amounts of negative allowance, which requires large amounts of force to assemble and results in relatively large induced stresses in the assembled parts.

force vector recorder Instrument for recording force displacements in a variety of disciplines.[NASA]

forcing Applying control signals greater than those warranted by a given deviation in the controlled variable in order to induce a more rapid rate of adjustment in the controlled variable.

force, insert retention The maximum allowable force which, if applied to the mating face of a connector insert, does not displace the insert permanently from its normal position in the connector housing or jeopardize or damage the insert or connector housing retention provision.[ARP914A-79]

Ford cup viscometer A time-to-discharge apparatus used primarily for determining the viscosity of paints and varnishes.

foreground The area in memory designated for use by high-priority programs; the program, set of programs, or functions that gain the use of machine facilities immediately upon request.

foreground/background A control system that uses two computers, one performing the control functions and the other used for data logging, off-line evaluation of performance, financial operations, and so on. Either computer is able to perform the control functions.

foreground/background processing A computer system organized so that primary tasks dominate computer processing time when required, and secondary tasks fill the remaining time.

foreground program A time-dependent program initiated via request, whose urgency preempts operation of a background program. Contrast with background program.

forehand welding Welding in which the palm of the welder's torch or electrode hand faces the direction of weld travel; it has special significance in oxyfuel-gas welding, where the welding flame is directed ahead of the weld puddle and provides preheating. Contrast with backhand welding.

foreign object damage Damage to any portion of the aircraft caused by impact or ingestion of birds, stones, hail or other debris.[ARD50010-91]

fore pump A vacuum pump operated in series with another vacuum pump to produce vacuum at the discharge of the latter, where the second pump is not capable of discharging gases at atmospheric pressure.

fore vacuum A space on the exhaust side of a vapor jet or pump where the ambient static pressure is below atmospheric.

forgetting error See error, forgetting.

forging 1. Using compressive force to plastically deform and shape metal; it is usually done hot, in dies or between rolls. 2. A shaped part made by impact, compression or rolling; if by rolling, the part is usually referred to as a roll forging.

forging range In hot forging, the optimum temperature range for shaping the metal.

forging stock A piece of semifinished metal used to make a forging. Also known as forging billet.

fork The structural member of a shock strut assembly which supports the axle and extends alongside the wheel and tire and connects to the basic shock strut.[AIR1489-88]

fork, full A configuration of landing gear fork which supports the axle on both ends, extends adjacent to the wheel on both sides, and joins around the tire to attach to the basic shock strut.[AIR 1489-88]

fork, half A configuration of landing gear fork which supports the axle on one end only, lies adjacent to the wheel on one side only, and attaches to the shock strut assembly.[AIR1489-88]

format 1. A symbol or group of symbols arranged in a specific manner to portray/display information.[ARP4107-88] 2. To prepare a diskette for acceptance of computer data. 3. A specific arrangement of computer data. 4. The arrangement of a tape record, buffer, or the like for compatibility with processing or storage standards. 5. The arrangement of programming elements comprising any field, record, file, or volume. 6. The basic parameters of a telemetry/data-acquisition sample plan; for example: number of words-per-frame, frames-per-subframe in a sample plan.

formation flight More than one aircraft which, by prior arrangement between the pilots, operate as a single aircraft with regard to navigation and position reporting. Separation between aircraft within the formation is the responsibility of the flight leader and the pilots of the other aircraft in the flight. This includes transition periods when aircraft within the formation are maneuvering to attain separation from each other to effect individual control, and during join-up and breakaway.[ARP4107-88]

formatted ASCII A mode in which data are transferred; a file containing formatted ASCII data is generally transferred as strings of seven-bit ASCII characters (bit eight is zero) terminated by a line-feed, formfeed, or vertical tab. Special characters, such as null, RUBOUT, and tab may be interpreted specially.

formatted binary A mode in which data are transferred; formatted binary is used to transfer check-summed binary data (eight-bit characters) in blocks. Formatting characters are start-of-block indicators, byte count and check-sum values.

formatter A hardware or software process of arranging data on tape or disk, or in a buffer.

formed taper pin receptacles Formed receptacles are taper pin receptacles having the tapered hole formed by rolling a flat strip of metal into the appropriate shape. This type of receptacle has an open longitudinal seam.[ARP592-90]

formed taper pins Formed taper pins are pins formed by rolling a flat strip of metal into the appropriate shape. [ARP592-90]

form grinding Producing a contoured surface on a part by grinding it with an abrasive wheel whose face has been shaped to the reverse of the desired contour.

forming Applying pressure to shape a material by plastic deformation without intentionally altering its thickness.

forming die A die for producing a contoured shape in sheet metal or other material.

form perception See space perception.

form tool A single-edge, nonrotating cutting tool that produces its inverse or reverse form on a workpiece.

formula A set of parameters that distinguish the products defined by procedures. It may include types and quantities of ingredients, along with information such as the magnitude of process variables. It may effect procedures.

formula translating system (FORTRAN) A procedure-oriented language for solution of arithmetic and logical programs.

FORTRAN compiler A processor program for FORTRAN.

forward controlling elements See elements, forward controlling.

forward-loop control elements Elements situated between the error signal and the output of the control system.[AIR1916-88]

forward scattering The scattering of radiant energy into the hemisphere of space bounded by a plane normal to the direction of the incident radiation and lying on the side toward which the incident radiation was advancing: the opposite of backward scatter.[NASA]

forward slip A slip in which the airplane's direction of motion continues the same as before the slip was begun. The primary purpose of the forward slip is to dissipate altitude without increasing the airplane's airspeed, particularly in airplanes not equipped with flaps. [ARP4107-88]

forward thrust See thrust.

fossil fuel Coal or petroleum hydrocarbon fuel as distinguished from nuclear fuel.

fossil meteorite crater See fossils.

fossils Remains, traces, or imprints of an organism preserved in the earth's crust some time in the geologic past. [NASA]

fouling 1. Growth of adherent plant or animal life on submerged structures, which often leads to biological corrosion or degradation of performance, such as reduction of heat transfer or increase of fluid friction. 2. The accumulation of refuse in gas passages or on heat absorbing surfaces which results in undesirable restrictions to the flow of gas or heat.

foundry A commercial enterprise, plant or portion of a factory where metal or glass is melted and cast.

four-ball tester An apparatus for determining lubrication efficiency by driving one ball against three stationary balls clamped together in a cup filled with the test lubricant; effectiveness of lubrication is expressed relatively in terms of wear-scar diameters on the stationary balls.

Fourier analysis The representation of physical or mathematical data by the use of the Fourier series of Fourier integral.[NASA]

Fourier optics 1. Optical components used in making Fourier transforms and other types of optical processing operations. 2. A prism or grating monochromator which essentially performs a Fourier transform on the light incident upon the entrance slit.

fovea The central part of the retina, which contains a high concentration of the color sensitive receptors known as cones.[NASA]

foveal visual cues Visual stimuli occurring within an approximately 60-degree cone from a person's normal sight line. Visual cues in this region are typically detected phototropically (with cones). Foveal vision is mostly used for discerning fine detail, estimating depth and distance, and differentiating colors.[ARP4107-88]

Fowler flap Named after Harlan D. Fowler, American aeronautical engineer—a type of extendible trailing-edge flap that effectively increases both the camber and wing area.[ARP4107-88]

fractals Highly irregular geometrical figures such as snowflakes or the boundary of a cloud whose capacity dimension is not an integer. The capacity dimension characterizes the measuring of the number of different size superimposed squares needed to cover the geometric shape. By the use of differing size boxes, one is able to determine the capacity dimension.[NASA]

fraction 1. In classification of powdered or granular solids, the proportion of the sample (by weight) that lies between two stated particle sizes. 2. In chemical distillation, the proportion of a solution of two liquids consisting of a specific chemical substance.

fractional distillation A thermal process whereby a mixture of liquids that boil at different temperatures is heated at a series of increasing temperatures, and the distillates boiled off at each temperature are collected separately.

fractionating column An apparatus for fractional distillation in which rising vapor and falling liquid are brought into intimate contact.

fraction defective In quality control, the average number of units of product containing one or more defects for each 100 units of product in a given lot.

fractography The study of fracture surfaces, especially for the purpose of determining the causes of failure and relating these causes to macrostructural and microstructural characteristics of parts and materials.

fracture mechanics A solid mechanics analytical method dealing with the state of stress in a cracked, scratched, or otherwise flawed material, used to describe the condition under which it may grow, or to predict the growth rate of the crack, scratch or flaw or both.[AIR 1872-88]

fracture test A method for determining composition, grain size, case depth or material soundness by breaking a test specimen and examining the fracture surface for certain characteristic features.

fragmentation In data processing, the effect of often used files that are growing in size to become non-contiguous when stored on a soft or hard disk.

frame 1. The image in a computer display terminal. 2. In time-division multiplexing, one complete commutator revolution that includes a single synchronizing signal or code.

frame code complement (FCC) The subframe synchronization method whereby the frame synchronization code is complemented to signal the beginning of each subframe.

frame rate (FRATE) The rate, or the pulses that clock that rate, of rotation of a data multiplexer "wheel."

frame synchronization pattern A unique code, coded pulse, or interval to mark the start of a commutation frame period.

frame synchronizer (FSY) Telemetry hardware that recognizes the unique signal that indicates the beginning of a frame of data; a typical frame synchronizer "searches" for the code, "checks" the recurrence of the code in the same position for several frame periods, and then "locks" on the code.

framework The load carrying members of an assembled structure.

framing error An error resulting from transmitting or receiving data at the wrong speed. The character of data will appear to have an incorrect number of bits.

Franhofer lines Dark lines in the absorption spectrum of solar radiation due to absorption by gases in the outer portions of the sun and in the earth's atmosphere.[NASA]

fraying The unraveling of a material, usually a woven fibrous braid.[ARP 1931-87]

free ash Ash which is not included in the fixed ash.

free atmosphere That portion of the earth's atmosphere, above the planetary boundary layer, in which the effect of the earth's surface friction on the air is negligible, and in which the air is usually treated (dynamically) as an ideal fluid. The base of the free atmosphere is usually taken as the geostrophic wind level.[NASA]

free-electron laser Multifrequency laser utilizing optical radiation amplification by a beam of free electrons passing through a vacuum in a transverse periodic magnetic field, as opposed to conventional lasers in which the oscillating electrons are bound to atoms and molecules and have a specific wavelength.[NASA]

free electrons Electrons which are not bound to an atom.[NASA]

free fall 1. Descriptive term for a retractable landing gear configuration capable of moving from the stowed (up and locked) position to the extended (down and locked) position by means of gravitational forces.[AIR1489-88] 2. The fall or drop of a body, such as a rocket, not guided, not under thrust, and not retarded by a parachute or a braking device.[ARP4107-88] The free and unhampered motion of a body along a Keplerian trajectory, in the force of gravity is counterbalanced by the force of inertia.[NASA]

free field Ideally, a wave field or potential-energy field in a homogeneous, unbounded medium, but practically, a field where boundary effects are negligible over the useful portion of the medium.

freefit A type of clearance fit having a relatively large allowance; it is used when accuracy of assembly is not essential, or when large temperature variations may occur, or both. Also known as free-running fit.

free flight Unconstrained or unassisted flight, as in the flight of a rocket after consumption of its propellant or after motor shutoff, in the flight of an unguided projectile, and in the flight in certain kinds of wind tunnels of unmounted models.[NASA]

free flow A condition in which the liquid surface downstream of the weir plate is far enough below the crest so that air has free access beneath the nappe.

free gyroscope A gyro wheel mounted in two or more gimbal rings so that its spin axis can maintain a fixed position in space.

free impedance A transducer characteristic equal to the input impedance when the load impedance is zero.

free jet Fluid jet without solid boundaries, such as a jet discharging into the open.[NASA]

free-machining A material description that indicates some alteration of chemical composition to substantially improve machinability—such as by the addition of sulfur, phosphorus or lead to steel, or lead to nonferrous metals. Also termed free-cutting.

free oscillation Periodic variation of some system variable when externally applied forces consist of either those that do no work or those that are derived from an invariant potential. See also free vibration.

free-piston engine Engine in which the pistons are not connected to the crank.[NASA]

free rolling A free rolling wheel/tire is one that is neither powered (driven) nor braked. The peripheral velocity in the contact zone then is the same as ground velocity.[AIR1489-88]

free radicals Atoms or groups of atoms broken away from stable compounds by application of external energy, and although containing unpaired electrons, remaining free for transitory or longer periods.[NASA]

free vibration Oscillation of a system in the absence of external forces.[NASA]

free water The amount of water released when a wet solid is dried to its equilibrium moisture content.

freeze To hold the contents of a register

(time, for example) until they have been transferred to another device.

freeze-up 1. Abnormal operation of a refrigeration unit because ice has formed on the heat absorbing elements. 2. Stoppage of rotational motion due to radial expansion or adhesive welding between a bearing and its journal. Also called seizure.

freezing drizzle Drizzle, the drops of which freeze on impact with the ground or with other objects at or near the earth's surface.[AIR1335-75]

freezing point The temperature at which equilibrium is attained between liquid and solid phases of a pure substance; the term also is applied to compounds and alloys that undergo isothermal liquid-solid phase transformation.

freezing rain Rain, the drops of which freeze on impact with the ground or with other objects at or near the earth's surface.[AIR1335-75]

french coupling A coupling with a right and left hand thread.

frequencies The harmonics of a periodic variable.

frequency 1. The number of times an alternating current repeats its cycle in one second.[ARP1931-87] 2. The number of cycles a periodic variable passes through per unit time. 3. Rate of signal oscillation in Hertz.

frequency assignment The specific frequency or frequencies authorized by competent authority; expressed for each channel by: (a) the authorized carrier frequency, the frequency tolerance, and the authorized emission bandwidth, (b) the authorized emission bandwidth in reference to a specific assigned frequency (when a carrier does not exist), or (c) the authorized frequency band (when a carrier does not exist).[NASA]

frequency band The continuum between two specified limiting frequencies. See also frequency.

frequency departure The amount that a carrier frequency or center frequency varies from its assigned value.

frequency deviation The peak difference between the instantaneous frequency of a modulated wave and the frequency of the unmodulated carrier wave.

frequency discriminator Electronic circuit which delivers output voltages proportional to the deviations of signals from predetermined frequency values. [NASA]

frequency distortion A form of distortion in which the relative magnitudes of the components of a complex wave are changed during transmission.

frequency divider An electronic circuit or device whose output-signal frequency is a proper fraction of its input-signal frequency.

frequency-divisional multiplexing The combination of two or more signals at different frequencies so they can be transmitted as one signal. This can be done electronically, or it can be done optically by using two or more light sources of different wavelengths.

frequency division multiple access Multiple access communication system in which the user has a specific frequency allocation and uses all of the time axis while sharing the available power.[NASA]

frequency-division multiplex (FDM) A system for the transmission of information about two or more quantities (measurands) over a common channel, by dividing the available frequency bands; amplitude, frequency, or phase modulation of the subcarriers may be employed.

frequency drift The slow and random variation of the controlled frequency level within the steady-state limits occurring, for example, as a result of environmental effects and wear on the electric power-drive system.[AS1212-71]

frequency drift rate The rate of change

of frequency owing to frequency drift when plotted against time.[AS1212-71]

frequency hopping Random changing of frequencies in transmission to mislead or prevent interception by unauthorized equipment.[NASA]

frequency meter An instrument for determining the frequency of a cyclic signal, such as an alternating current or radio wave.

frequency modulation 1. The cyclic or random variation, or both, of instantaneous frequency about a mean frequency during steady-state electric-system operation. The frequency modulation is normally within narrow frequency limits and occurs as a result of speed variations in a generator rotor owing to the dynamic operation of the rotor coupling and drive speed regulation. It is frequently nonsinusoidal.[AS1212-71] 2. Angle modulation of a sine wave carrier in which the instantaneous frequency of the modulated wave differs from the carrier frequency by an amount proportional to the instantaneous value of the modulating wave.[NASA] 3. A type of electronic circuit which produces an output signal whose frequency has been modified by one or more input signals. See also modulated wave. 4. In telemetry, modulation of the frequency of an oscillator to indicate data magnitude.

frequency modulation repetition rate The reciprocal of the period of the modulation waveform.[AS1212-71]

frequency monitor An instrument that determines the amount that a frequency deviates from its assigned value.

frequency multiplication The generation of harmonics of the frequency of a lightwave by nonlinear interactions of the lightwave with certain materials. Frequency doubling is equivalent to dividing the wavelength in half. High power beams are needed for the nonlinear interaction to occur.

frequency multiplier An electronic circuit or device whose output-signal frequency is an exact multiple of its input-signal frequency.

frequency of occurrence A referee determination by viewing 50 fields, 4 x 5 in (102 x 127 mm), projected at 100X. The number of fields containing the feature of interest is divided by the total number of fields viewed to represent the lot, thus arriving at a percentage. [AS1814-90]

frequency response 1. Frequency response characteristics of a servoactuator are defined as the steady state relationship of the output amplitude to the input amplitude and the output to input phase difference, when the input is subjected to constant amplitude sinusoidal signals of various frequencies. [ARP1281A-81] 2. The portion of the frequency spectrum which can be sensed by a device within specified limits of amplitude error. Response of a system as a function of the frequency of excitation.[NASA] 3. A measure of the effectiveness with which a circuit or device transmits signals of different frequencies, usually expressed as a graph of magnitude or phase of an output signal as a function of frequency. Also known as amplitude-frequency response; sine-wave response. 4. For a linear system in sinusoidal steady-state, the ratio of the Fourier-transform of the output signal to the Fourier-transform of the corresponding input signal. 5. The response of a component, instrument, or control system to input signals at varying frequencies.

frequency response (Bode plot) The complex ratio of the actuation system output to the command input, while the input is cycled sinusoidally and the frequency is varied. Frequency response is usually presented as a combined plot of normalized amplitude ratio in dB, and of the input to output phase angle, in degrees, versus the logarithm of frequency (called a Bode plot).[AIR1916-88]

frequency response method A method of tuning a process control loop for optimum operation by proper selection of controller settings. This method is based on a study of the frequency response of the open process control loop.

frequency reuse A digital satellite communication technique which features the reuse of frequency bands in a downlink transmission to provide high power utilization and flexible accommodations of dynamic source destination traffic patterns.[NASA]

frequency shift keying (FSK) 1. The form of frequency modulation in which the modulating wave shifts the output frequency between predetermined values, and the output wave is coherent with no phase discontinuity.[NASA] 2. Modulation accomplished by switching from one discrete frequency to another discrete frequency.

frequency stability 1. A measurement of how well the output frequency (or equivalently emitted wavelength) of a laser stays constant. In some types the emitted wavelength tends to drift because of factors such as changing temperature of the laser itself. 2. Statement of deviation with time, temperature or supply voltages of an electronic oscillator when compared to a standard.

frequency swing A characteristic of a frequency-modulation system equal to the difference between the maximum and minimum design values of instantaneous frequency in the modulated wave.

frequency telemetering A system for transmitting measurements where the information values are represented by frequencies within a specific band, the specific frequency being determined by the percent of full scale equivalent to the current value of the measured variable.

fresh water Water in rivers, lakes, springs, etc. containing no significant amounts of dissolved salts.[NASA]

Fresnel lens(es) 1. Thin lenses constructed with stepped setbacks so as to have the optical properties of much thicker lenses. 2. A lens in which the surface is composed of a number of concentric lens sections with the same focal length desired for the larger lens. Typically for high quality optical applications, the smaller lenses are concentric circles. This technique is used to compress a short focal length optical component into a thickness much less than a plane-convex lens of the same material and focal length.

Fresnel reflector Device characterized by a set of mirrors with varying orientation arranged so as to have the optical properties of a smooth reflector e.g., parabolic reflector.[NASA]

Fresnel region The region between the antenna and the Fraunhofer region. [NASA]

fretting 1. Wear caused by oscillatory slip of small amplitude between two surfaces. The amplitude below which wear is to be considered fretting will be that which is large enough to maintain a lubricant film of finite thickness between the surfaces at all times. Class 1: Fretting that occurs in parts of a mechanism that were never intended to have relative motion. Class 2: Fretting that occurs in parts of a mechanism intended to undergo relative motion some or all of the time, but not through the action of antifriction bearings. Class 3: Fretting that occurs in anti-friction bearings.[S-2,47-91] 2. A form of wear that occurs between closely fitting surfaces subjected to cyclic relative motion of very small amplitude; it is usually accompanied by corrosion, especially of the very fine wear debris. Also known as chafing fatigue; fretting corrosion; friction oxidation; molecular attrition; wear oxidation; and in rolling-element bearings, false brinelling.

fretting corrosion Corrosion resulting from fretting.[S-2,47-91]

fretting, molecular attraction theory The theory holds that fretting is a form of molecular attrition (or transfer) caused when alternating close approach and subsequent separation of the molecules of the two mating surfaces occurs. On close approach, the fields of force of the molecules overlap, and on separation the molecules are torn out of both surfaces as a result of breaking the cohesional bonds.[S-2,47-91]

fretting, weld-breaking theory This theory is based on the modern concepts of friction, and holds that "When two solids come in contact, the force bringing them together is concentrated on a small number of high points in both surfaces. These high points flow plastically under the load, and cold-weld, equilibrium being finally reached when the cross-sectional area of the welded junctions is sufficiently large to support the load. Friction and wear result from breaking of the welds and from plowing of the asperities of one material through the other, the weld-breaking process being the dominant one for most practical purposes".[S-2,47-91]

friable Capable of being easily crumbled, pulverized or otherwise reduced to powder.

friction See friction error.

frictional error See error, frictional.

friction feed printer A printer using the pressure of a platen to advance the paper.

friction oxidation See fretting.

friction saw A toothless circular saw that cuts materials largely by fusion due to frictional heat along the line of contact with the piece being cut.

friction tape A type of cotton tape impregnated with a sticky, moisture-resistant compound, which is used to cover and insulate exposed electrical connections or terminations; it has been largely replaced by electrical tape made of polyvinyl chloride resin backed with a sticky adhesive.

friction-tube viscometer A device for measuring viscosity by determining the pressure drop across a friction tube as the fluid is pumped through it.

friction unit force A measure of the shearing force on the brake friction material. Equals brake torque divided by the product of brake radius times lining area.[AIR1489-88]

frigorimeter A thermometer for measuring low temperatures.

fringe multiplication The duplicating effect of a family of curves superimposed on another family of curves so that the curves intersect at angles less than 45 degrees. A new family of curves appears which pass through intersections of the original curves.[NASA]

frit 1. A powdered ceramic prepared by fusing a physical mixture of oxides into a uniform melt, which is then quenched and milled into a fine, homogeneous powder.[NASA] 2. Fusible ceramic mixture used to make ceramic glazes and porcelain enamels.

frit seal A hermetical seal for enclosing integrated circuits and other electronic components, which is made by fusing a mixture of metallic powder and glass binder.

from-to tester A type of electronic test equipment for checking continuity between two points in a circuit.

front-end processor 1. A device that receives computer data from other input devices, organizes such data as specified, and then transmits this data to another computer for processing. 2. The computer equipment used to receive plant signals, including analog-to-digital converters and the associated controls.

front mounted A connector mounted with its mounting flange positioned in front of the mounting surface when looking at the mating face or front side of the connector.[ARP914A-79]

frost plug A device for determining liquid level when the contents of a tank are at a temperature below 0 °C; a side tube resembling a sight glass but having a series of closed tubes (plugs) at different levels instead of the glass; the tubes below liquid level are cooled so that moisture from the atmosphere forms frost on them, while the tubes above liquid level remain frost free.

frothing Production of a layer of relatively stable bubbles at an air-liquid interface; it can be accomplished by any of several methods, including aeration, agitation or chemical reaction; in many instances it is an undesired side effect of an operation, but sometimes it is an essential element of the operation, as in froth flotation to separate a mineral from its ore.

Froude number The nondimensional ratio of the inertial force to the force of gravity for a given fluid flow; the reciprocal of the Reech number.[NASA]

frozen soil See permafrost.

FSS Flight Service Station.

FTAM 1. File Transfer, Access and Management.[AS4159-88] See also file transfer access and management.

fuel Any material that will burn or otherwise react to release heat energy—common fuels include coal, charcoal, wood and petroleum products (fossil fuels), which burn, and uranium, which undergoes nuclear fission.

fuel advisory departure (FAD) Procedures (for example, postponement of takeoff) to minimize engine running time for aircraft destined for an airport that is experiencing prolonged arrival delays.[ARP4107-88]

fuel-air mixture Mixture of fuel and air.

fuel-air ratio 1. The mass rate of fuel flow to the engine divided by the mass rate of dry airflow through the engine.[ARP1256-90][AIR1533-91] 2. The ratio of the weight, or volume, or fuel to air.

fuel cell power plant Power generating device that directly produces electrical energy from chemical energy and consists of fuel processors, stacked fuel cells, and dc to ac converters. The main types, distinguished by electrolytes which are heated to different temperatures, are base, phosphoric acid, molten carbonate, and solid oxide.[NASA]

fuel cell Device which converts chemical energy directly into electrical energy. The overall functioning of a fuel cell is similar to that of a conventional engine. When supplied with fuel and oxidant (usually hydrogen and oxygen), the fuel cell produces electric power and, as by-products, pure water and waste heat. Instead of the chemical to heat to electricity conversion processes required by conventional engines, fuel cells convert the hydrogen and oxygen reactants directly into electricity. Because of this, fuel cell efficiency is relatively high (in the order of 60 to 70%), which results in low reactant consumption. Fuel cell power systems can provide energy densities greater than 1,000 watt-hours per pound.[AIR744A-72] 2. Device which converts chemical energy directly into electrical energy but differing from a storage battery in that the reacting chemicals are supplied continuously as needed to meet output requirements.[NASA]

fuel consumption The using of fuel by an engine or power plant; the rate of this consumption, measured, e.g. in gallons or pounds per minute.[NASA]

fuel gas A combustible gaseous substance that is used as a fuel.

fuel oil Any oily hydrocarbon liquid having a flash point of at least 100 °F (38 °C) which can be burned to generate heat.

fuel production Producing of conventional and/or alternative fuels by various technologies.[NASA]

fuel temperature (t) Final equilibrium test temperature in °F of the liquid and vapor phase.[AIR1326-91]

fulchronograph An instrument for recording lightning strikes electromagnetically.

fulgurator An atomizer used in flame analysis to spray the salt solution to be analyzed into the flame.

full adder A computer logic device that accepts two addends and a carry input, and produces a sum and a carry output.

full annealing An imprecise term that implies heating to a suitable temperature followed by controlled cooling to produce a condition of minimum strength and hardness.

full cycling control Controls placed on the crimping cycle of crimping tools forcing the tool to be closed to its fullest extent requiring completion of the crimping cycle before the tool can be opened. See also ratchet control.[ARP-914A-79]

full duplex 1. Communications that appear to have information transfer in both directions (transmit and receive) at the same time. 2. The electronic transmission of data simultaneously in two directions. See duplex, full.

Fuller's earth A highly absorbent, clay-like material formerly used to remove grease from woolen cloth, but now used principally as a filter medium.

full hard temper A level of hardness and strength for nonferrous alloys and some ferrous alloys corresponding to a cold worked state beyond which the material can no longer be formed by bending.

full-height drive A 5 1/4-inch disk drive that is 3 1/4-inches or 1 1/2-inches wide when installed.

full mission capable Systems and equipment shall be considered FMC when they are safe and have all mission-essential subsystems installed and operating as designated by the Military Service.[ARD50010-91]

full-scale development phase The period when the system and the principal items necessary for its support are designed, fabricated, tested, and evaluated.[ARD50013-91]

full-scale value 1. The largest value of a measured quantity that can be indicated on an instrument scale. 2. For an instrument whose zero is between the ends of the scale, the sum of the absolute values of the measured quantity corresponding to the two ends in the scale.

full-wave rectifier An electronic circuit that converts an a-c input signal to a d-c output signal, with current flowing in the output circuit during both halves of each cycle in the input signal.

function 1. A specific purpose of an entity or its characteristic action. 2. In communications, a machine action such as a carriage return or line feed. 3. A closed subroutine which returns a value to the calling routine upon conclusion. 4. The operation called for in a computer software instruction.

functional design The specification of the working relations between the parts of a system in terms of their characteristic actions.

functional design specifications Those levels of design in which all subtasks are specified and their relationships defined so that the total collection of subsystems will perform the intended task of the entire system.[NASA]

functional diagram A diagram that represents the functional relationships among the parts of a system.

functional program A routine or group of routines which, when considered as a whole, completes some task with a minimum of interaction of other functional programs other than to obtain data and signal completion of its task. For example, a group of routines which take data from an analog scanner and store it on a bulk storage device might be considered to be a functional program.

functional requirements A specification of required functional behavior, operation, performance, or purpose.

functional specification A document that tells exactly what the system should do, what will be supplied to the system, and what is expected to come out of it.

functional test See test.

functioning time In an EED, the lapsed time between application of initiating energy and some later function such as bridge wire beak, case opening or start of pressure rise, peak pressure, etc.[AIR913-89]

function keys Special keys on a computer keyboard that instruct the computer to perform a specific operation.

function scan Function scan is a feature of the EASY program that provides a graphic output of algebraic functional relationships or of tabulated data input by a user.[AIR1823-86]

function selector A device which permits the human pilot to select automatic pilot functions or modes of operation. [ARP419-57]

function subprogram The function subprogram is an independently written program and is treated as such by the compiler. It may consist of any number of statements which are executed when it is called. See also subroutine and subprogram.

function switch A circuit having a fixed number of inputs and outputs designed such that the output information is a function of the input information, each expressed in a certain code, signal configuration, or pattern.

function table 1. The two or more sets of information so arranged that an entry in one set selects one or more entries in the remaining sets. 2. A dictionary. 3. A device constructed of hardware, or a subroutine, which can either decode multiple inputs into a single output or encode a single output into multiple outputs. 4. A tabulation of the values of a function for a set of values of the variable.

fundamental frequency The frequency of a sinusoidal function having the same period as a complex periodic quantity.

fundamental mode 1. The mode of a waveguide having the lowest critical frequency. 2. A type of sequential circuit in which there is only one input change at a time and no further change occurs until all states are stabilized.

fundamental natural frequency The lowest frequency in a set of natural frequencies.

furans Organic heterocyclic compounds containing diunsaturated rings of four carbon atoms and one oxygen atom; also known as furfuran or tetrol.[NASA]

furnace An apparatus for liberating heat and using it to produce a physical or chemical change in a solid or liquid mass; most often, the heat is produced by burning a fossil fuel, passing electric current through a heavy-duty resistance element, generating and sustaining an electric arc, or electromagnetically inducing large eddy currents in the charge.

furnace draft The draft in a furnace, measured at a point immediately in front of the highest point at which the combustion cases leave the furnace.

furnace volume The cubical contents of the furnace or combustion chamber.

fuse 1. A device which automatically shuts off flow in a line in event of downstream system rupture.[ARP243B-65] 2. A fuse is a replaceable circuit protecting device depending on the melting of a conductor for circuit interruption. [AS486B-78] 3. An igniting device used to communicate fire or other initiation stimuli consisting of a flexible tube containing a core of explosive, usually black power.[AIR913-89] 4. Any of several devices for detonating an explosive—

for example, by elapsed time, command, impact, proximity or thermal effects.

fuse clips The contacts of the fuseholder which support the fuse or limiter and connect their terminals with the circuit.[ARP1199A-90]

fused fiber optics A number of separate fibers which are melted together to form a rigid fused bundle to transmit light. Fused fiber optics may be used for transmitting images or simply illumination; they are not necessarily coherent bundles of fibers.

fused silica The term usually applied to synthetic fused silica, formed by the chemical combination of silicon and oxygen to produce a high purity silica. Optical glass is made by the melting of high purity sands, while fused quartz is made by crushing and melting natural quartz. See silica glass.

fused slag Slag which has coalesced into a homogeneous solid mass by fusing.

fuse, hydraulic A hydraulic device, designed to shut off flow of fluid if the flow exceeds a specified value. Used in brake systems to stop flow and depletion of the A/V fluid system if a line rupture or similar failure occurs.[AIR-1489-88]

fuse or limiter holder A mounting device with contacts and terminals for the purpose of accepting fuses or limiters for easy connection within a circuit.[ARP1199A-90]

fuse pull-out A removable fuseholder which can be removed to replace fuses or to open an electrical circuit.

fuse, quantity measuring A fuse which closes when more than a predetermined quantity of fluid has passed through it.[ARP243B-65]

fuse, return flow A fuse which closes both pressure and return lines when the ratio between the flows deviates beyond a predetermined value.[ARP243B-65]

fuse tube A tube of insulting material which surrounds the current-responsive element.[ARP1199A-90]

fusibility Property of slag to fuse and coalesce into a homogeneous mass.

fusible alloy An alloy with a very low melting point, in some instances approaching 150 °F (65 °C), usually based on Bi, Cd, Sn or Pb; the fusible alloys have varied uses, the most widely known being solders and fusible links for automatic sprinklers, fire alarms and other safety devices.

fusible plug 1. A thermal sensitive pressure release device used to prevent tubeless tire or wheel assembly failure when an overheat condition occurs which would cause degradation of wheel and/or tire structural properties.[AIR1489-88] 2. A hollowed threaded plug having the hollowed portion filled with a low melting point material, usually located at the lowest permissible water level.

fusiform shape See cone.

fusion The combining of atoms and consequent release of energy.[NASA]

fusion welding Any welding process that involves melting of a portion of the base metal.

fusion zone In a weldment, the area of base metal melted, as determined on a cross section through the weld.

future alarm point See alarm point.

fuze A device designed to initiate a munition.[AIR913-89]

fuzzy sets Mathematical models coupled with a provision for the effect of human factors and construction process and experience.[NASA]

fuzzy systems Systems that involve fuzzy sets.[NASA]

FWA See first word address.

FW-SIFR Fixed-Wing Special IRF.

g See gram.

G The basic unit of acceleration. 1G = 32.2ft/s^2 (9.81 cm/s^2).[ARP4107-88] See also specific gravity.

GAAMA General Aviation Aircraft Manufacturers Association.

G-adaptation illusion See illusion, kinesthetic.

gadolinium alloy Mixture of gadolinium, a rare earth metal, with other metals.[NASA]

gage Also spelled gauge. 1. Device to ascertain a specific or comparative dimension.[ARP480A-87] 2. The thickness of metal sheet, or the diameter of rod or wire. 3. A device for determining dimensions such as thickness or length. 4. A visual inspection aid that helps an inspector to reliably determine whether size or contour of a formed, stamped or machined part meets tolerances.

gage block A rectangular chromium steel block having two flat parallel surfaces, with flatness and parallelism guaranteed within a few millionths of an inch; they are usually manufactured and sold in sets for use as standards in linear measurement. Also known as Johanssen block; Jo block; precision block; size block.

gage cock A valve attached to a water column or drum for checking water level.

gage glass A glass or plastic tube for measuring liquid level in a tank or pressure vessel, usually by direct sight; it is usually connected directly to the vessel through suitable fittings and shutoff valves.

gage length In materials testing, the original length of an elongated specimen over which measurements of strain, thermal expansion or other properties are taken.

gage point A specific location used to position a part in a jig, fixture or qualifying gage.

gage pressure 1. Pressure measured relative to ambient pressure. 2. The difference between the local absolute pressure of the system and the atmospheric pressure at the place of the measurement. 3. Static pressure as indicated on a gage.

gain 1. The relative degree of amplification in an electronic circuit. 2. The ratio of the change in output to the change in input which caused the change. 3. In a controller, the reciprocal of proportional band—for example, if the proportional band is set at 25%, the controller gain is .25. Proportional band can be expressed as a dimensionless number (gain) or as a percent.

gain, antenna By common definition, the difference in signal strengths between a given antenna and an isotropic antenna.

gain, crossover frequency See frequency, gain crossover.

gain crossover, phase-margin frequency The point on the Bode plot of the transfer function at which the magnitude crosses over unit: $Log_m G(j\omega) = 0$ dB where ω is the oscillation frequency, in radians per second, and j is the square root of minus one. The frequency at gain crossover is called the phase-margin frequency.[AIR1916-88]

gain margin 1. A measure of system stability defined as the reciprocal of the magnitude of the open loop gain at 180 deg of phase lag crossover frequency usually expressed in dB.[AIR1916-88] 2. The gain margin is the factor by which the loop gain must be multiplied for a system to reach the stability boundary.[AIR1823-86]

gain, open loop See gain, loop.

gain, zero frequency See gain, static (zero-frequency gain).

gal See gallon.

Gal A unit of acceleration equal to 1 cm/s^2. The milligal is frequently used because it is about 0.001 times the earth's gravity.

galactic cosmic rays Energetic particles

that come from outside the solar systems. They generally come from within our galaxy.[NASA]

galactic mass The total amount of matter contained in a galaxy.[NASA]

galactic radio waves Radio waves emanating from our galaxy.[NASA]

galaxies Vast assemblages of stars or nebulae, composing island universes separated from other such assemblages by great distances.[NASA]

Galilean satellites The four largest and brightest satellites of Jupiter (Io, Europa, Ganymede, and Callisto).[NASA]

Galileo probe The NASA Jupiter atmospheric entry probe to be deployed from the Galileo spacecraft. The probe will make in situ measurements while descending from a parachute.[NASA]

Galileo spacecraft A NASA orbiter spacecraft which will carry the Galileo probe and following deployment at Jupiter, will become an orbiting platform for remote sensing of Jupiter and its satellites.[NASA]

galvanic cell See electrolytic cell.

galley(s) All galley hardware items furnished for installation on the airplane, exclusive of airplane provisions. This includes galley complex structure, decorative panels, ceilings, furnishings, lighting, galley units, modules, food, beverages, utensils, ovens, coffee makers, floor tie down fittings, refrigeration equipment, electrical, water and drain system connections from the galley equipment to the airplane interface connectors.[AS1426-80]

galley complex A structural integral installation of galleys at a designated galley envelope area within the airplane. A complex includes items such as one or more galley units, work space, partitions, curtains, walls, ceilings, lights and structural attachment devices.[AS 1426-80]

galley envelope The three-dimensional outside surface limits of the volume(s) within the airplane reserved for locating the galleys.[AS1426-80]

galley unit A major subsection of a galley complex, or an independent galley structure which houses various items such as ovens, food trays, coffee makers, refrigeration, liquor, and service items.[AS1426-80]

galling Localized adhesive welding with subsequent spalling and roughening of rubbing metal surfaces as a result of excessive friction and metal-to-metal contact at high spots.

gallon A unit of capacity (volume) usually referring to liquid measure in the British or U.S. Customary system of units. The capacity defined by the British (Imperial) gallon equals 1.20095 U.S. gallons; one U.S. gallon equals four quarts or 3.785×10^{-3}+3

galvanic corrosion Electrochemical corrosion associated with current in a galvanic cell, which is set up when two dissimilar metals (or the same metal in two different metallurgical conditions) are in electrical contact and are immersed in an electrolytic solution.

galvanizing Coating a metal with zinc, using any of several processes, the most common being hot dipping and electroplating.

galvanometer An instrument for measuring small electric currents using electromagnetic or electrodynamic forces to create mechanical motion, such as changing the position of a suspended moving coil.

galvanometer recorder A sensitive moving-coil instrument having a small mirror mounted on the coil; a small signal voltage applied to the coil causes a light beam reflected from the mirror to move along the length of a slit, producing a trace on a light-sensitive recording medium that moves transverse to the slit at constant speed.

game theory 1. Application of mathematics to a game, business situation, or other problem to maximize gain or minimize loss.[NASA] 2. A mathematical process of selecting an optimum strategy in the face of an opponent who has a strategy of his own.

gamma A measure of the contrast properties of a photographic or radiographic emulsion which equals the slope of the straight-line portion of its HD curve.

gamma counter An instrument for detecting gamma radiation— either by measuring integrated intensity over a period of time or by detecting each photon separately.

gamma radiation See gamma rays.

gamma ray 1. Quantum of electromagnetic radiation emitted by nuclei, each such photon being emitted as the result of a quantum transition between two energy levels of the nucleus. Gamma rays have energies usually between 10 thousand electron volts and 10 million electron volts with correspondingly short wavelengths and high frequencies.[NASA] 2. A term sometimes used to describe any high-energy electromagnetic radiation, such as x-rays exceeding about 1 MeV or photons of annihilation radiation.

gamma-ray astronomy Astronomy based on the detection of gamma-ray emission and interactions from supernova remnants, neutron stars, flare stars, galactic core and disc, black holes, etc.[NASA]

gamma-ray bursts Short (about 0.1-4 sec.) intense low-energy (about 0.1-1.2 MeV) bursts recorded by the Vela satellite system in 1967. Their isotropic distribution suggests an extragalactic origin, but a galactic disk origin cannot be ruled out.[NASA]

gamma-ray lasers Stimulated emission devices producing coherent gamma radiation.[NASA]

gamma-ray observatory A late 1980's NASA mission to explore the gamma-ray window to the universe from 0.06 MeV to 30 GeV.[NASA]

gamma-ray spectra The energy distribution of gamma rays emitted by nuclei.[NASA]

gamma-ray spectrometer 1. Instrument for deriving the physical constants of materials by using induced gamma radiation as the emission source.[NASA] 2. An instrument for measuring the energy distribution in a beam of gamma rays.

gamma structure An ordered structure of titanium-aluminum compound with a stoichiometric ratio TiAl and face-centered tetragonal crystal structure. [AS1814-90]

gamma-ray telescope Special telescope for the observation (and recording) of astronomical phenomena in the gamma-ray spectrum.[NASA]

gang disconnect A connector that permits the simultaneous connection or disconnection of two or more electrical circuits.[ARP914A-79]

Gantt chart A style of bar chart used in production planning and control to display both work planned and work done in relation to time.

gantries See gantry cranes.

gantry cranes Large cranes mounted on platforms that usually run back and forth on parallel tracks astride the work area.[NASA]

Ganymede A satellite of Jupiter orbiting at a mean distance of 1,071,000 kilometers. Also called Jupiter III.[NASA]

gap 1. An interval of space or time that is used as an automatic sentinel to indicate the end of a word, record, or file of data on a tape; for example, a word gap at the end of a word, a record or item gap at the end of a group of words, or a file gap at the end of a group of records or items. 2. The absence of information for a specified length of time, or space on a recording medium, as contrasted with marks and sentinels that indicate the presence of specific information to achieve a similar purpose. 3. The space between the reading or recording head and the recording medium, such as tape, drum, or disk; related to gap, head. 4. In a weldment, the space between members, prior to welding, at the point of closest approach for opposing faces.

gap scanning In ultrasonic examination, projecting the sound beam through a short column of fluid produced by pumping couplant through a nozzle in the ultrasonic search unit.

garbage In data processing, meaningless or incorrect data.

garnets Groups of minerals that are silicates of cubic crystalline form.[NASA]

garter spring A closed ring made by welding the ends of a closely wound helical spring together.

gas amplification A counter-tube or ionization-chamber characteristic equal to the charge collected divided by the charge produced in the active volume by a given ionizing event.

gas analysis The determination of the constituents of a gaseous mixture.

gas atomization Atomization of fluids by high velocity gas jets.[NASA]

gas bearing A journal or thrust bearing that uses a film of gas to lubricate the running surfaces. Also known as gas-lubricated bearing.

gas barrier The ionically conductive film, between the cell plates, which forces all plate generated gases to the top of the electrolyte and prevents them from reaching plates of opposite polarity. [AS8033-88]

gas burner A burner for use with gaseous fuel.

gas carburizing A surface hardening process in which steel or an alloy of suitable alternative composition is exposed at elevated temperature to a gaseous atmosphere with a high carbon potential; hardening of the resulting carbon-rich surface layers is done by quenching the part from the carburizing temperature or by reheating and quenching.

gas counter A type of counter tube in which a gaseous sample whose radiation is to be measured is introduced directly into the counter tube itself.

gas current A current of positive ions flowing to a negatively biased electrode, the positive ions being produced when electrons flowing between two other electrodes collide with residual gas molecules.

gasdynamic pumping The production of a population inversion by a gasdynamic process, in which a hot, dense gas is expanded into a near vacuum, causing the gas to cool rapidly. If the gas cools faster than energy can be redistributed, a population inversion is generated.

gaseous cavitation See cavitation flow.

gaseous emission Substance emitted in the form of gas downstream of the combustion chamber and limited to carbon monoxide, carbon dioxide, nitric oxide, nitrogen dioxide and hydrocarbons. [AIR1533-91] [ARP1256-90]

gas etching Removing material from a semiconductor material by reacting it with a gas to form a volatile compound.

gas generator 1. A device in which a propellant is burned to produce a sustained flow of pressurized gas.[AIR913-89] 2. A device used to generate gases in the laboratory; a chemical plant for producing gas from coal, for example, water gas.[NASA]

gas generator engine See engine; gas generator.

gas giant planets The giant planets, Jupiter, Saturn, Uranus, and Neptune, of our solar system.[NASA]

gas house tar By-product from the distillation of coal for illuminating gas.

gasification The process of converting solid or liquid fuel into a gaseous fuel such as the gasification of coal.

gasket 1. The flexible sealing element in a stationary or static fluid seal.[ARP 243B-65] 2. A sealing member, usually made by stamping from a sheet of cork, rubber, metal or impregnated synthetic material and clamped between two essentially flat surfaces to prevent pressurized fluid from leaking through the crevice; typical applications include flanged joints in piping, head seals in a reciprocating engine or compressor, casing seals in a pump, or virtually anywhere a pressure-tight joint is needed between stationary members. Also known as static seal.

gas lift The technique of raising a liquid in a vertical flow line by injecting a gas below a portion of the liquid column causing upward flow.

gas metal-arc welding (GMAW) A form of electric arc welding in which the electrode is a continuous filler metal wire and in which the welding arc is shielded by supplying a gas such as argon, helium or CO_2 through a nozzle in the torch or welding head; the term GMAW includes the methods known as MIG welding.

gas meter An instrument for measuring and recording the volume or mass of a gaseous fluid that flows past a given point in a piping system.

gasohol Synthetic fuel consisting of a mixture of gasoline and grain alcohol (ethanol).[NASA]

gasometer A piece of apparatus typically used in analytical chemistry to hold and measure the quantity of gas evolved in a reaction; similar equipment is used in some industrial applications.

gas path analysis Mathematical process of determining overall engine performance, individual module performances and sensor performances from any specific set of engine related measurements.[NASA]

gas plasma display A data display screen used on some laptop computers. Characters are easier to read than those on liquid crystal display screens, but the unit is more expensive.

gas pliers A pinchers-type tool for grasping round objects such as pipes, tubes and rods.

gas pocket A cavity within a solid or liquid body that is filled with gas.

gas seal A type of shaft seal that prevents gas from leaking axially along a shaft where it penetrates a machine casing.

gas-shielded arc welding An all-inclusive term for any arc welding process that utilizes a gas stream to prevent direct contact between the ambient atmosphere and the welding arc and weld puddle.

gassing 1. Absorption of gas by a material. 2. Formation of gas pockets in a material. 3. Evolution of gas during a process—for example, evolution of hydrogen at the cathode during electroplating, gas evolution from a metal during melting or solidification, or desorption of gas from internal surfaces during evacuation of a vacuum system; the last is sometimes referred to as outgassing.

gas specific gravity balance A weighing device consisting of a tall gas column with a floating bottom; a pointer mechanically linked to the floating bottom indicates density or specific gravity directly, depending on scale calibration.

gas thermometer A temperature transducer that converts temperature to pressure of gas in a closed system. The relation between temperature and pressure is based on the gas laws at constant volume.

gastrointestinal gas expansion See trapped gas effects.

gas tube An electron tube whose operating characteristics are substantially affected by the presence of gas or vapor within the tube envelope.

gas tungsten-arc welding (GTAW) A

form of metal-arc welding in which the electrode is a nonconsumable pointed tungsten rod; shielding is provided by a stream of inert gas, usually helium or argon; filler metal wire may or may not be fed into the weld puddle, and pressure may or may not be applied to the joint; the term GTAW includes the method known as Heliarc or TIG welding.

gas turbine Turbine rotated by expanding gases, as in a turbojet engine or in a turbosupercharger.[NASA]

gas-solid interaction Effect of the impingement of gases (particles) on solid surfaces in various environments.[NASA]

gas welding See preferred term oxyfuel-gas welding.

gate 1. The area from which cargo or persons, or both, are put on or taken off an aircraft. At some airports, fueling also takes place in the gate area; at others, there is a separate area for fueling. 2. A particular point in the air route structure from which aircraft proceed to enter the terminal control area for approach to an airfield. A limiting point of the en route phase of a flight.[ARP 4107-88] 3. A movable barrier. 4. A device such as a valve or door which controls the rate at which materials are admitted to a conduit, pipe or conveyor. 5. A device for positioning film in a movie camera, printer or projector. 6. The passage in a casting mold that connects the sprue to the mold cavity. Also known as in-gate. 7. An electronic component that allows only signals of predetermined amplitudes, frequencies or phases to pass.

gate-hold procedures Procedures at selected airports to hold aircraft at the gate or other ground location whenever departure delays exceed or are anticipated to exceed 5 minutes.[ARP4107-88]

gate valve A type of valve whose flow-control element is a disc or plate that undergoes translational motion in a plane transverse to the flow passage through the valve body.

gateway 1. Device that interfaces and transfers data between two or more dissimilar communication systems.[AIR 4271-89] 2. A network device that interconnects two networks that may have different protocols.

gauge A term used to denote the physical diameter of a wire or conductor.[ARP1931-87] See also gage.

gauge pressure Absolute pressure minus atmospheric pressure.[AIR1916-88]

gauge theory A field theory in which symmetrics of the theory are implemented locally in space and time. This leads to theories where forces are generally carried by vector bosons. Some gauge theories are electrodynamics, quantum chromodynamics, and Yang Mills theory.[NASA]

gauss The CGS unit of magnetic flux density or magnetic induction; the SI unit, the tesla, is preferred.

Gaussian beam A laser beam in which the intensity has its peak at the center of the beam, then drops off gradually toward the edges. The intensity profile measured across the center of the beam is a classical Gaussian curve.

Gaussian Distribution A Gaussian distribution has an equal deviation on each side of a median value.[ARP1192-87]

Gaussian elimination A technique for solving linear equations by progressive differencing.[NASA]

Gaussian noise See random noise.

gaussmeter A magnetometer for measuring only the intensity, not the direction, of a magnetic field; its scale is graduated in gauss or kilogauss. See also magnetometer.

gauze 1. A sheer, loosely woven textile fabric; one of its widest uses is for surgical dressings, but it also has some industrial uses such as for filter media.

2. Plastic or wire cloth of fine to medium mesh size.

GCA Ground-Controlled Approach-Landing.

G-differential illusion See illusion, kinesthetic.

GDOP See geometric dilution of precision.

gear 1. A toothed machine element for transmitting power and motion between rotating shafts whose axes are relatively close to each other or are intersecting. 2. A collective term for equipment that performs a specific functions—lifting gear, for example. 3. A collective term for the portion of a machine that transmits motion from one mechanism to another. 4. A specific combination of gears in a transmission or adjustable gear train that determines mechanical advantage, speed and direction of rotation.

gear configuration, bicycle Configuration in which there are two gear assemblies, arranged in tandem design, one behind the other. Generally includes outrigger gears also for ground stability, i.e., B-47 aircraft.[AIR1489-88]

gear configuration, dual twin Configuration in which the axle arrangement is such that two wheel/tire assemblies are located on either side of the shock strut or supporting member, i.e., C5A nose gear.[AIR1489-88]

gear configuration, dual twin tandem Configuration in which there are two axles, one behind the other, connected by an axle beam, and each of which is a dual twin wheel/tire arrangement, i.e., B-58 main gears.[AIR1489-88]

gear configuration, quadracycle Configuration in which four main landing gear assemblies are located, one in each aircraft quadrant. An arrangement of main gear two, side by side, with one set behind or in tandem with the first set. May also include outrigger gears for ground stability, i.e., B-52 aircraft.[AIR1489-88]

gear configuration, single wheel Configuration utilizing only one wheel/tire assembly on the gear assembly.[AIR-1489-88]

gear configuration, tandem (also single tandem) Configuration on which the two wheel/tire assemblies are mounted on separate axles one behind the other or "in tandem", i.e., Avro Arrow.[AIR1489-88]

gear configuration, tricycle General descriptive term for an arrangement with three gear assemblies, one in front and two main gears located aft, in a tricycle arrangement. Each gear may have its own configuration depending upon the number of wheel/tire assemblies and the individual arrangement, i.e., B-707, DC8, DC10.[AIR1489-88]

gear configuration, triple wheel Configuration in which there is one axle on which are located three wheel/tire assemblies, i.e., RS-71 Aircraft main gear.[AIR1489-88]

gear configuration, tri-tandem Configuration in which three wheel/tire assemblies are arranged one behind the other or "in tandem" on a single gear assembly.[AIR1489-88]

gear configuration, tri-twin tandem Configuration in which there are three axles, one behind the other, connected by means of an axle beam, and each of which carries two wheel/tires assemblies, i.e., Tupolev TU-144 main gear. [AIR1489-88]

gear configuration, twin tandem Configuration in which there are two axles, one behind the other, connected by an axle beam, and each axle having two wheel/tire assemblies, i.e., DC8, DC10, B-707, L1011 main gears.[AIR 1489-88]

gear configuration, twin tricycle Configuration of an individual landing gear assembly on which there are three axles, each mounting two wheel/tire assemblies. The axle arrangement is one in front, and two aft in a tricycle arrangement, i.e., C5A main gears.[AIR 1489-88]

gear configuration, twin wheel Configuration in which there is one axle on which is mounted two wheel/tire assemblies.[AIR1489-88]

gear down To arrange a gear train so that the driven shaft rotates at a slower speed than the driving shaft.

gear drive A mechanism for transmitting power (torque) and motion from one shaft to another by means of direct contact between toothed wheels.

gear level To arrange a gear train so that the driving and driven shafts rotate at the same speed.

gear, lever suspension A gear arrangement in which the axle is mounted at the aft end of a lever (structural member) which is pivoted at the forward end to permit freedom for the wheels to ride up and over bumps and/or obstructions. The lever motion is controlled by a shock absorber connected between the lever and the strut or other fixed structure. [AIR1489-88]

gear, main The main gear forms the principle support of the aircraft on land or water. It may include any combination of wheels, tires, floats, skis, shock absorbing mechanisms, brakes, retracting mechanisms, and structural provisions necessary for attachment and operation. The combination may be in varied configurations.[AIR1489-88]

gear meter A positive-displacement fluid meter in which two meshing gear wheels provide the metering action.

gearmotor 1. A device consisting of an electric motor and a direct-coupled gear train; the arrangement allows the motor to run at optimum speed—usually 1800 or 3600 rpm—while delivering rotational motion at a substantially lower speed.

gear, neutral A gear assembly which may be utilized on either right hand or left hand positions on the aircraft.[AIR 1489-88]

gear, nose The nose gear is an auxiliary landing gear unit. In addition to supporting some weight of the aircraft, it provides balance, controls the aircraft attitude and gives ground stability. In many instances, it includes provisions for steering the aircraft. It is located forward on the aircraft or at the nose. [AIR1489-88]

gear, outrigger The outrigger is an auxiliary gear which supports some weight of the aircraft under some conditions. It is located at the outer extremities (i.e., wing tip, etc.) of the airplane and is designed to alleviate overturning and to provide stability during ground operations, i.e., B-52 outrigger or wing tip gear.[AIR1489-88]

gear pump A pump in which fluid is fed to one side of a set of meshing gears, which entrain the fluid and discharge it on the other side. 2. A gear pump supplied with pressurized fluid which converts fluid flow to rotary motion.

gear ratio The ratio of the revolutions per unit time of one shaft to the revolutions in the same unit time of another shaft connected to it by a gear device. Starter gear ratio usually expressed as ratio of starter turbine speed to starter output speed.[ARP906A-91]

gear, scooter A gear arrangement in which there are usually two axles located one behind the other and connected by an axle beam. Main support for the bogie assembly is a pin ended shock absorber attached between the axles and connected to the gear structure at the upper end. The forward end of the axle beam is pin jointed to a post which is free to slide within a structural cylinder assembly. The post assembly carries drag, side and torque loads. Vertical load is carried basically by the shock absorber. In the side view the forward located post with one axle under the post and the other aft, gives the appearance of a "scooter."[AIR1489-88]

gear, tail The tail gear is an auxiliary gear which supports a small portion of the aircraft weight, and usually consists of a shock strut, wheel/tire assembly, (or skid) and controlling devices. It is sometimes steerable. The tail gear is located in the aft or tail portion of the aircraft.[AIR1489-88]

gear train A combination of two or more gears, arranged to transmit power and motion between two rotating shafts or between a rotating shaft and a member that moves linearly.

gear up To arrange a gear train so that the driven shaft rotates at a higher speed than the driving shaft.

gear walk Cyclic fore and aft motion of the landing gear strut assembly about a normally static vertical strut centerline. Caused by drag loads applied at the tire/ground interface and the natural spring rate of the gear structure. Sometimes aggravated by anti-skid braking action (cycling) and resonant frequency of vibration of the strut.[AIR 1489-88]

gear wheel A wheel with integral gear teeth that mesh with another gear, a rack, or a worm.

gegenschein A round or elongated spot of light in the sky at a point 180 degrees from the sun. Also called counterglow.[NASA]

Geiger counter Instrument for detecting and measuring radioactivity. In full, Geiger-Mueller counter.[NASA]

Geiger-Mueller counter A radiation-measuring instrument whose active element is a gas-filled chamber usually consisting of a hollow cathode with a fine-wire anode along its axis; in operation, the voltage between anode and cathode is high enough that the discharge caused by a primary ionizing event spreads over the entire anode until stopped when the space charge reduces the electric-field magnitude. Also known as Geiger counter.

Geiger-Mueller tube See Geiger counter.

Geiger threshold The lowest voltage applied to a counter tube which results in output pulses of essentially equal amplitude, regardless of the magnitude of the ionizing event.

gel coat A resin gelled on the internal surface of a plastics mold prior to filling it with a molding material; the finished part is a two-layer laminate, with the gel coat providing improved surface quality.

general adaptation syndrome The heightened physiological state, automatically assumed by the body when faced with a crisis, to prepare for "flight or fight." This heightened physiological state may detract from rational processes and cause a person to overreact, overcontrol, or overlook significant cues.[ARP4107-88]

general aviation That portion of civil aviation which encompasses all facets of aviation with the exception of air carriers holding a certificate of public convenience and necessity, and large commercial aircraft operators.[ARP 4107-88]

general illustration Two-dimensional line art with labels, titles, etc., which is not inherently useful for extracting computational data values or geometric information.[AS4159-88]

general inattention See attention, anomalies of inattention.

general processor In numerical control, a computer program which carries out computations on the part program and prepares the cutter location data (CL data) for a particular part without reference to the machine on which it might be made.

general-purpose computer A computer designed to solve a large variety of problems, e.g., a stored program computer which may be adapted to any of a very large class of applications.

general-purpose simulation system A generic class of discrete, transaction-oriented simulation languages based on a block (diagramming) approach to problem statement. Abbreviated GPSS.

generated background Internally generated imagery upon which symbology may be superimposed.[AS8034-82][ARP 1782-89]

generating electric field meter An instrument for measuring electric field strength in which a flat conductor is alternately exposed to the field and shielded from it; potential gradient of the field is determined by measuring the rectified current through the conductor.

generating magnetometer An instrument for measuring magnetic field strength by means of the electromotive force generated in a rotating coil immersed in the field being measured.

genetic engineering The intentional production of new genes and alteration of genomes by the substitution or addition of new genetic material.[NASA]

geoastrophysics See astrophysics; geophysics.

geodesy The science which deals mathematically with the size and shape of the earth, and the earth's external gravity field, and with surveys of such precision that overall size and shape of the earth must be taken into consideration.[NASA]

geodetic accuracy The degree to which point positions or boundaries indicated on maps or imagery correspond with true geodetic positions.[NASA]

geodetic coordinates Quantities which define the position of a point on the spheroid of reference with respect to the planes of the geodetic equator and of a reference meridian.[NASA]

geodetic survey Survey which takes into account the size and shape of the earth. [NASA]

Geodynamic Experimental Ocean Satellite See GEOS-D satellite.

geodynamics Study of the dynamic forces or processes within the earth. [NASA]

geofabrics See geotechnical fabrics.

geofractures See geological faults.

geographic disorientation The type of spatial disorientation in which a person is correctly oriented with reference to the horizon, but not oriented in relation to known ground references or navigational fixes. Simply stated, the person is lost.[ARP4107-88]

geographic information system Computer assisted system that acquires, stores, manipulates, and displays geographic data. Some systems are not automated.[NASA]

geoids The figure of the earth as defined by the geopotential surface which most nearly coincides with mean sea level over the entire surface of the earth. [NASA]

geological fault A surface or zone of rock fracture along which there has been displacement from a few centimeters to a few kilometers in scale.[NASA]

geomagnetically trapped particle See radiation belt.

geomagnetic crotchet See sudden ionospheric disturbance.

geomagnetic equator See magnetic equator.

geomagnetic field See geomagnetism.

geomagnetic latitude Angular distances from the geomagnetic equator, measured northward of southward through 90 degrees and labeled N or S to indicate the direction of measurement.[NASA]

geomagnetic storm See magnetic storm.

geomagnetism The magnetic phenomena, collectively considered, exhibited by the earth and its atmosphere and by extension the magnetic phenomena in interplanetary space. The study of the magnetic field of the earth.[NASA]

geometric accuracy The internal geometric fidelity of an imaging system. [NASA]

geometrical acoustics The study of the behavior of sound under the assumption that sound transversing a homogeneous medium travels along straight line or rays.[NASA]

geometrical hydromagnetics See magnetohydrodynamics.

geometrical optics The geometry of paths of light rays and their imagery through optical systems.[NASA]

geometrical theory of diffraction A ray theory of diffraction process.[NASA]

geometric angle of attack See angle of attack.

geometric dilution of precision A navigation and positioning system performance index expressing the dilution of range measurement precision due to the geometric relationship between user and satellites. It is formulated as the square root of the sum of the variances of position estimates in the three orthogonal directions and can be employed to determine the optimal locations for network satellites and in the selection of optimal satellite signals sources.[NASA]

geometric distortion Geometric distortion is an aberration which causes a displayed pattern to be contorted from the desired pattern. Geometrical distortion includes the linearity errors in a display. Linearity errors are defined as on-axis horizontal and vertical deflection non-uniformities, whereas geometrical distortion is a general term applying to deflection non-uniformities over the entire phosphor screen.[ARP 1782-89]

geometric-perspective illusion See illusion, visual.

geometric pitch The tangent of the acute angle between the extended chord plane of a propeller airfoil and its plane of rotation. See also standard pitch. [ARP4107-88]

geometric rectification (imagery) The correction of image distortions due to sensor view angle, platform attitude, or target surface features.[NASA]

geometrodynamics See relativity.

geomorphology A science that deals with the land and submarine relief features of the earth's surface and genetic interpretation of them through using the principles of physiography in its descriptive aspects and of dynamic and structural geology in its explanatory phases.[NASA]

geophysical fluid General term for the liquids and gases on or in the earth (from water in all forms, to petroleum and hydrocarbons in liquid and gaseous form, and molten rock material within the earth).[NASA]

geophysical fluid flow cell Apparatus used in model experiments for deep solar convection and Jovian atmospheric circulation for Spacelab 1 and Spacelab 3.[NASA]

geophysics The physics of the earth and its environment, i.e. , earth, air, and (by extension) space. Classically, geophysics is concerned with the nature of and physical occurrences at and below the surface of the earth including, therefore, geology, oceanography, geodesy, seismology, and hydrology. The trend is to extend the scope of geophysics to include meteorology, geomagnetism, astrophysics, and other sciences concerned with the physical nature of the universe.[NASA]

geopotential The potential energy of a unit mass relative to sea level, numerically equal to the work that would be done in lifting the unit mass from sea level to the height at which the mass is located; commonly expressed in terms of dynamic height or geopotential height.[NASA]

geopotential height The height of a given point in the atmosphere in units proportional to the potential energy of unit mass (geopotential) at this height, relative to sea level.[NASA]

geopotential research mission A NASA Gravity field mapping mission utilizing the low-low satellite tracking concept to measure the Doppler shift between two coorbiting polar satellites. [NASA]

geopressure Pressure that exceeds the normal hydrostatic pressure of about 0.465 psi per foot of depth.[NASA]

Geosari project Launch of GEOS on second development flight of Ariane launcher into a geostationary elliptical orbit in 1979. The name is derived from a combination of GEOS and ARiane. [NASA]

GEOS-D satellite Another in a series of the European Space Agency's geostationary scientific satellites launched by NASA for long-term cosmic radiation studies.[NASA]

Geostational Operati Environ Satellite B See Goes 2.

Geostationary Operational Environ Sats See Goes satellites.

geostationary platform See synchronous platform.

geostationary satellite See synchronous satellite.

geostrophic wind The horizontal wind velocity for which the coriolis acceleration exactly balances the horizontal pressure force.[NASA]

geotechnical engineering The science and practice of that part of civil engineering involving the inter-relationship between a geologic environment and the works of man.[NASA]

geotechnical fabric Generic term for a variety of artificial fiber products used in engineering construction of civil works such as embankments. Also called geofabrics, filter cloth, geotextiles, and civil engineering fabrics.[NASA]

geotechnology Application of science and engineering to problems involved in utilizing natural resources.

geotemperature Internal temperature of the planet earth.[NASA]

geotextile See geotechnical fabric.

geothermal energy extraction The removal for storage and/or utilization of heat from natural sources within the earth (hot springs, geysers, hot rocks, etc.)[NASA]

geothermal energy utilization Any application of energy derived from sources within the earth.[NASA]

geothermal technology The gamut of operations involved in the exploration, exploitation, and conversion of energy derived from geothermal sources.[NASA]

geothermometry See geotemperature.

German infrared laboratory A proposed infrared telescope for Spacelab that was discontinued in 1985. It superseded the LIRTS (telescope).[NASA]

germicides See bactericides.

germinator See phytotron.

gesso A mixture of chalk and either gelatine or casein glue, which is painted on panels to provide a suitable surface for tempera work or for polymer-based paints.

get away specials (STS) Low-cost, man-independent Space Shuttle experimental payloads.[NASA]

getter(s) 1. Materials which are included in a vacuum system or device for removing gas by sorption.[NASA] 2. A material exposed to the interior of a vacuum system in order to reduce the concentration of residual gas by absorption or adsorption.

getter-ion pump A type of vacuum pump that produces and maintains high vacuum by continuously or intermittently depositing chemically active metal layers on the wall of the pump, where they trap and hold inert gas atoms which have been ionized by an electric discharge and drawn to the activated pump wall. Also known as sputter-ion pump.

gewel hinge A hinge consisting of a

hook inserted in a loop.

geyser Hot spring that intermittently erupts jets of hot water and steam. [NASA]

G force See acceleration (physics).

ghost point A term used in boiler water testing with soap solution. A lather appears to form but will disappear upon the addition of more soap solution. This point represents total calcium hardness and the final lather total hardness.

GHz Gigahertz.

giant-hand illusion See illusion, vestibular.

gib A removable plate that holds other parts or that acts as a bearing or wear surface.

gilbert The CGS unit of magnetomotive force; the SI unit, the ampere (or ampere-turn) is preferred.

gimbal Device with two mutually perpendicular and intersecting axes of rotation, thus giving free angular movement in two directions, on which engines or other objects may be mounted in gyros, supports which provide the spin axes with degrees of freedom.[NASA]

gimbal lock A position in a gyro having two degrees of freedom where the spin axis becomes aligned with an axis of freedom, thus depriving it of a degree of freedom and therefore depriving it of its useful properties.

gimbal mount An optical mount which allows position of a component to be adjusted by rotating it independently around two orthogonal axes.

gimlet A small tool for boring holes in wood, leather and similar materials; it consists of a threaded point, spiral-fluted shank and cross handle; a tool without the handle and adapted for use in a drill is known as a gimlet bit.

gin A hoisting machine consisting of a windlass, pulleys and ropes in a tripod frame.

Giotto mission The European Space Agency's mission to fly through the head of Halley's Comet in order to make in site measurements of the composition and physical state as well as the structures of the head.[NASA]

gland 1. The cavity or space provided for the accommodation and operation of an elastic packing or gasket for sealing of a fluid vessel or compartment. [ARP243B-65] 2. A device for preventing a pressurized fluid from leaking out of a casing at a machine joint, such as at a shaft penetration. Also known as gland seal. 3. A movable part that compresses the packing in a stuffing box. See packing follower; see also lantern ring.

glare The sensation produced by luminances from extraneous sources within the visual field that are sufficiently greater than the emitted luminance, to which the eyes are adapted, or cause loss in visual performance.[ARP1782-89]

glare source Any light emitting source of reflective surface of similar brightness; normally associated with integral flood lit systems.[ARP1161-91]

glass A hard, brittle, amorphous, inorganic material, often transparent or translucent, made by fusing silicates (and sometimes borates and phosphates) with certain basic oxides and cooling rapidly to prevent crystallization.

glass fiber A glass thread less than 0.001 in. (0.025 mm) thick; it is used in loose, matted or woven form to make thermal, acoustical or electrical insulation; in matted, woven or filament-wound form to make fiber-reinforced composites; or in loose, chopped form to make glass-filled plastics parts.

glassine A thin, dense, transparent, super-calendered paper made from highly refined sulfite pulp; it is used industrially as insulation between layers of iron-core transformer windings.

glass laser High power laser used in laser fusion technology research.[NASA]

glassmaker's soap A substance such as MnO_2, which is added to glass to eliminate the green color imparted by the presence of iron salts.

glass paper 1. An abrasive material made by bonding a layer of pulverized glass to a paper backing. 2. Paper made of glass fibers.

glass sand The raw material for glass-making; it normally consists of high-quartz sand containing small amounts of the oxides of Al, Ca, Fe and Mg.

glassware Laboratory containers, vessels, graduated cylinders, tubing and the like which are made from glass.

glass wool A relatively loose mass of glass fibers used chiefly in insulating, packing and filtering applications.

glassy alloy A metallic material having an amorphous or glassy structure. Also known as metallic glass.

glassy carbon Form of carbon with unique properties and characteristics. Formed by carbonizing phenolic resins made by reacting phenols with cellulosics, aldehydes, and ketones.[NASA]

Glauert coefficient See aerodynamic forces; Mach number.

glaze 1. A coating of ice, generally clean and smooth, formed by freezing of supercooled water on a surface.[ARP4107-88] 2. A glossy, highly reflective, glass-like, inorganic fused coating. See enamel.

glazed surface An undesirable, hard, glossy surface developed where rubbing action (friction) breaks down the parts of the surface or the lubricant.[ARP 4107-88]

glazing 1. Cutting and fitting glass panes into frames. 2. Smoothing the exposed solder of a wiped pipe joint with a hot iron.

glazing compound A caulking compound, such as putty, used to seal the edges of a pane of glass where it fits into its frame.

glide angle See glide path.

glide path 1. The path used by an aircraft in approach procedures as defined by an instrument landing facility.[ARP 419-57] 2. Flight path of aeronautical vehicles in a glide, seen from the side. The path used by aircraft or spacecraft in approach procedure and which are generated by instrument landing facilities.[NASA]

glide path, (on/above/below) Terms used by ATC to inform an aircraft making a precision approach radar (PAR approach) of its vertical position (elevation) relative to the desired descent profile to the runway.[ARP4107-88]

glide slope (GS) That which provides vertical guidance for aircraft during approach and landing. The glide slope consists of the following: (a) electronic components emitting signals which provide vertical guidance by reference to airborne instruments during instrument approaches such as ILS, or (b) visual ground aids, such as VASI, which provide vertical guidance for VFR approach or for the visual portion of an instrument approach and landing.[ARP4107-88] 2. An included path in space which is defined by radio signals emanating from an instrument landing facility. [ARP419-57] See also glide path.

glide slope intercept altitude The minimum altitude of the intermediate approach segment prescribed for a precision approach which assures required obstacle clearance. It is depicted on instrument approach procedure charts. See segments of an instrument approach procedure and instrument landing system.[ARP4107-88]

Glimm method Numerical technique for solving gas dynamics problems involving hyperbolic systems of conservation laws.[NASA]

glitch Undesirable electronic pulses that cause processing errors.

glitter Decorative flaked powder having a particle size large enough so that the individual flakes produce a visible reflection or sparkle; used in certain decorative paints and in some compounded plastics stock.

global 1. Any name that is declared global has as its scope the entire system in which it resides. 2. A computer instruction that causes the computer to locate all occurrences of specific data. 3. A value defined in one program module and used in others; globals are often referred to as entry points in the module in which they are defined, and externals in the other modules that use them.

global array A set of data listings that can be referenced by other parts of the software.

global common An un-named data area that is accessible by all programs in the system. Sometimes referred to as blank common.

global positioning system A satellite navigation system which will display many (up to 24) satellites in three sets of orbits by means of a precise time standard and three-dimensional information on position and velocity.[NASA]

global variable Any variable available to all programs in the system. Contrast with reserved variable.

globe An enclosing device of clear or diffusing material to protect the lamp, to diffuse or redirect its light, or to modify its color.[ARP798-91]

globe valve A type of flow-regulating valve consisting of a movable disc and a stationary-ring seat in a generally spherical body. In the general design, the fluid enters below the valve seat and leaves from the cavity above the seat.

globular alpha A spheroidal form of equiaxed alpha.[AS1814-90]

glossimeter An instrument for measuring the 'glossiness' of a surface—that is, the ratio of light reflected in a specific direction to light reflected in all directions—usually by means of a photoelectric device. Also known as glossmeter.

glow See luminescence.

glow discharge 1. Electrical discharge that produces luminosity.[NASA] 2. A discharge of electrical energy through a gas, in which the space potential near the cathode is substantially higher than the ionization potential of the gas.

glue 1. Generally, a term often used improperly to describe an adhesive. 2. Specifically, a crude, impure form of commercial gelatine that softens to a gel consistency when wetted with water and dries to form a strong adhesive layer.

glued A mixed signal simulation system that combines existing analog and digital simulation software into a hybrid analog or digital simulation system.

gluons The carriers of the strong force which holds atomic nuclei together (holding together groups of quarks making up stable particles, which in turn are bound together in the atomic nuclei).[NASA]

GMAW See gas metal-arc welding.

GMT Greenwich Mean Time. The mean solar time of the meridian of Greenwich, England used as the prime basis of standard time throughout the world. Expressed in hs GMT or hs Z (Zulu phonetically).[ARP4107-88]

gnomonic projection A projection on a plane tangent to the surface of a sphere having the point of project at the center of the sphere.[NASA]

gnotobiotics The study of germ free animals.[NASA]

GNP See gross national product.

go around 1. Instructions for a pilot to abandon his/her approach to landing. A pilot on an instrument approach should execute the published missed approach procedure or follow instructions provided by ATC. 2. The discontinuance of the final approach to landing for any reason. See also missed approach.[ARP4107-88]

Goes 1 The first in a series of geostational operational environmental satellites launched in October 1975. It ceased operation in June 1977.[NASA]

Goes 2 The second in a series of geostationary operational environmental satellites launched in June 1977.[NASA]

Goes 3 The third in a series of geostationary operational environmental satellites launched in June 1978.[NASA]

Goes 4 The fourth in a series of geostationary operational environmental satellites launched in September 1980.-[NASA]

Goes 5 The fifth in a series of geostationary operational environmental satellites launched in May 1981.[NASA]

Goes 6 The sixth in a series of geostationary operational environmental satellites launched in April 1983.[NASA]

Goes-G Satellite which was to have been the seventh in a series of geostationary operational environmental satellites. The May 1986 launch failed.[NASA]

Goes satellites Geostationary operational environmental satellites.[NASA]

Golay cell An infrared detector in which the incident radiation is absorbed in a gas cell, thereby heating the gas. The temperature induced expansion of the gas deflects a diaphragm, and a measurement of this deflection indicates the amount of incident radiation.

goniometer 1. Instrument for measuring angles.[NASA] 2. Specifically, an instrument used in crystallography to determine angles between crystal planes, using x-ray diffraction or other means. 3. An instrument used to measure refractive index and other optical properties of transparent optical materials or optically scattering in materials at UV, VIZ or IR wavelengths.

go/no-go gage A composite gaging device that enables an inspector to quickly judge whether specific dimensions or contours are within specified tolerances; in many instances, the device is so constructed that the part being inspected will fit one part of the gage easily and will not fit another part if it is within tolerance, and will pass both parts or pass neither if it is not within tolerance.

go/no-go test A test in which one or more parameters are determined, but which can only result in acceptance or rejection of the test object, depending on the value(s) measured.

goodness of fit The degree to which the observed frequencies of occurrence of events in an experiment correspond to the probabilities in a model of the experiment.[NASA]

GOSIP Government Open Systems Interconnection Procurement.[AS4159-88]

gouging Forming a groove in an object by electrically, mechanically, thermomechanically or manually removing material; the process is typically used to remove shallow defects prior to repair welding.

governor A device for automatically regulating the speed or power of a prime mover—especially a device that relies on centrifugal force in whirling weights opposed by springs or gravity to actuate the controlling element.

GPSS See general-purpose simulation system.

grabens See geological faults.

grab sampling A method of sampling bulk materials for analysis, which consists of taking one or more small portions (usually only imprecisely measured) at random from a pile, tank, hopper, railcar, truck or other point of accumulation.

graceful degradation A system attribute wherein when a piece of equipment fails, the system falls back to a degraded mode of operation rather than failing catastrophically and giving no response to its users.

grade 1. To move earth, making a land surface of uniform slope. 2. A classification of materials, alloys, ores, units of product or characteristic according to some attribute or level of quality. 3.

To sort and classify according to attributes or quality levels. 4. Oil classification according to quality.

graded index fiber (GRIN) An optical fiber in which the refractive index changes gradually between the core and cladding, in a way designed to refract light so it stays in the fiber core. Such fibers have lower dispersion and broader bandwidth than step-index fibers.

graded refractive index lens A lens in which the refractive index of the glass is not uniform. Typically the index will differ with distance from the center of the lens.

gradient The rate of change of some variable with respect to another, especially a regular uniform or stepwise rate of change.

gradient index optics Optical systems with components whose refractive indexes vary continuously within the material used for the optical elements. [NASA]

graduation Any of the major or minor index marks on an instrument scale.

grain 1. A single piece of solid propellant regardless of size or shape.[ARP 913-89] 2. The appearance or texture of wood, or a woodlike appearance or texture of another material. 3. In paper or matted fibers, the predominant direction most fibers lie in, which corresponds to directionality imparted during manufacture. 4. In metals and other crystalline substances, an individual crystallite in a polycrystalline mass. 4. In crumbled or pulverized solids, a single particle too large to be called powder.

grain boundary The plane of mismatch between adjacent crystallites in a polycrystalline mass, as revealed on a polished and etched cross section through the material.

grain boundary alpha Primary or transformed alpha outlining prior beta grain boundaries. It may be continuous unless broken up by subsequent work. Also may accompany blocky alpha. Occurs by slow cooling from the beta field into alpha-beta field.[AS1814-90]

grain flow Fibrous appearance on a polished and etched section through a forging, which is caused by orientation of impurities and inhomogeneities along the direction of working during the forging process.

grain growth An increase in the average grain size in a metal, usually as a result of exposure to high temperature.

graininess Visible coarseness in a photographic or radiographic emulsion, which is caused by countless small grains of silver clumping together into relatively large masses visible to the naked eye or with slight magnification.

graining Working a translucent stain while still wet to simulate the grain in wood or marble; tools such as special brushes, combs and rags are used by hand to create the desired irregular patterns.

grain size 1. For metals, the size of crystallites in a polycrystalline solid, which may be expressed as a diameter, number of grains per unit area, or standard grain size number determined by comparison with a chart such as those published by ASTM; in most instances the grain size is given as an average, unless there are substantial proportions which can be given as two distinct sizes; if two or more phases are present, grain size of the matrix is given. 2. For abrasives, see preferred term grit size.

grains per cu ft The term for expressing dust loading in weight per unit of gas volume (7000 grains equals one pound).

grains (water) A unit of measure commonly used in water analysis for the measurement of impurities in water (17.1 grains=1 part per million—ppm).

gram The CGS unit of mass; it equals 0.001 kilogram, which has been adopted as the SI unit of mass.

grand unified theory A theory describing the unification of gravity with the other elementary forces in physics, i.e., the weak force, the strong force and the electromagnetic force.[NASA]

grant Asset bestowed or transferred, such as money or land, for a particular purpose.[NASA]

granular fracture A rough, irregular fracture surface, which can be either transcrystalline or intercrystalline, and which often indicates that fracture took place in a relatively brittle mode, even though the material involved is inherently ductile.

granular structure Nonuniform appearance of molded or compressed material due to the presence of particles of varying composition.

graphic Pertaining to representational or pictorial material, usually legible to humans and applied to the printed or written form of data such as curves, alphabetic characters, and radar scope displays.

graphical display unit An electronics device that can display both text and pictorial representations.

graphic character See graphic.

graphic panel A master control panel which, pictorially and usually colorfully, traces the relationship of control equipment and the process operation. It permits an operator, at a glance, to check on the operation of a far-flung control system by noting dials, valves, scales, and lights.

graphics A computer technique using lines and symbols to display information rather than letters and numbers.

graphite-epoxy composites Structural materials composed of epoxy resins reinforced with graphite.[NASA]

graphite flake A form of graphite present in gray cast iron which appears in the microstructure as an elongated, curved inclusion.

graphite-polyimide composite Composite material utilizing graphite reinforcing fibers in a resin matrix.[NASA]

graphite rosette A form of graphite present in gray cast iron which appears in the microstructure as graphite flakes extending radially outward from a center of crystallization.

graphitic carbon Free carbon present in the microstructure of steel or cast iron; it is an essential feature of most cast irons, but is almost always undesirable in steel.

graphitic corrosion Corrosion of gray cast iron in which the iron matrix is slowly leached away, leaving a porous structure behind that is largely graphite but that may also be held together by corrosion products; this form of corrosion occurs in relatively mild aqueous solutions and on buried pipe and fittings.

graphitic steel Alloy steel in which some of the carbon is present in the form of graphite.

graphitization Formation of graphite in iron or steel; it is termed primary graphitization if it forms during solidification, and secondary graphitization if it forms during subsequent heat treatment or extended service at high temperature.

graphitizing Annealing a ferrous alloy in such a way that at least some of the carbon present is converted to graphite.

graphoepitaxy The use of artificial surface relief structures to induce crystallographic orientation in thin films. [NASA]

graph theory The mathematical study of the structure of graphs and networks. [NASA]

Grashof number A nondimensional parameter used in the theory of heat transfer. The Grashof number is associated with the Reynolds number and the Prandtl number in the study of convection.[NASA]

graveyard spin A sequence of repeated

spins occurring because, in the proper recovery from a spin, the pilot's motion-sensing system tends to create an illusion of spinning in the opposite direction. Responding to this somatogyral illusion, the pilot returns the plane to its original spin.[ARP4107-88]

graveyard spiral A progressively steepening spiral resulting from the somatological illusion during a coordinated constant rate turn that has ceased to stimulate the motion sensing system of the pilot. When, in this situation, the pilot observes a loss of altitude, the tendency is to pull back on the controls, thus tightening the spiral and increasing the loss of altitude.[ARP4107-88]

gravimeter A device for measuring the relative force of gravity by detecting small differences in weight of a constant mass at different points on the earth's surface. Also known as gravity meter.

gravireceptors Highly specialized nerve endings and receptor organs located in skeletal muscles, tendons, joints, and in the inner ear which furnish information to the brain with respect to body position, equilibrium and the direction of gravitational forces.[NASA]

gravitation The acceleration produced by the mutual attraction of two masses, and of magnitude inversely proportional to the square of the distance between the two centers of mass.[NASA]

gravitational constant 1. The coefficient of proportionality in Newton's law of gravitation.[NASA] 2. A dimensionless conversion factor in English units which arises from Newton's second law (F=ma) when mass is expressed in pounds-mass (lb_m).

gravitational wave antennas Devices for receiving propagating gravitational fields produced by some change in the distribution of matter.[NASA]

gravitometer See densimeter.

gravitons The hypothetical elementary units of gravitation which are equivalent in the electronics in electromagnetic theory.[NASA]

gravity Weight index of fuels; liquid, petroleum products expressed either as specific, Baumé or A.P.I. (American Petroleum Institute) gravity; weight index of gaseous fuels as specific gravity related to air under specified conditions; or weight index of solid fuels as specific gravity related to water under specified conditions. See also gravitation.

gravity meter 1. A device that uses a U-tube manometer to determine specific gravities of solutions by direct reading. 2. An electrical device for measuring variations in gravitational forces through different geological formations. 3. A gravimeter.

gravity probe B An experiment designed to measure general relativistic induced torques on a gyroscope in orbit about the earth.[NASA]

gravity wave Wave in an interface between fluids of different density in which the restoring force is gravity.[NASA]

Gravsat satellites See geopotential research mission.

gray Metric unit for absorbed dose.

gray body An object having the same spectral emissivity at every wavelength, or one whose spectral emissivity equals its total emissivity.

gray code A generic name for a family of binary codes which have the property that a change from one number to the next sequential number can be accomplished by changing only one bit in a code for the original number. This type of code is commonly used in rotary shaft encoders to avoid ambiguous readings when moving from one position to the next. See also cyclic code and shaft encoder.

gray iron Cast iron containing free graphite in flake form; so named because a freshly broken bar of the alloy appears gray.

grayout A temporary condition in which vision is hazy, restricted, or otherwise impaired, owing to insufficient oxygen supply to the brain commonly induced by positive G forces of moderate intensity and duration. See also blackout. [ARP4107-88]

gray scale Images that are not colored or multispectral.[NASA]

grazing incidence Incidence at a small glancing angle.[NASA]

grazing incidence solar telescope See GRIST (telescope).

grease 1. Rendered, inedible animal fat. 2. A semisolid to solid lubricant consisting of a thickening agent, such as metallic soap, dispersed in a fluid lubricant, such as petroleum oil.

grease seal ring See lantern ring.

great circle(s) A circle on the surface of a sphere, especially the earth, whose plane passes through the center of the sphere.[ARP4107-88]

great circle course The direction of the great circle through the point of departure and the destination expressed as an angular distance from a reference direction (usually north) to the great circle. The angle varies from point to point along the great circle. A great circle is an imaginary line of intersection on the earth's surface of any plane through the earth's center.[ARP1570-86]

greaves Armor to protect the legs. In landing gear, usage, armor to protect the struts or equipment mounted thereon from arresting cables, barrier straps, etc.[AIR1489-88]

green Unfired, uncured or unsintered.

greenhouse effect The heating effect exerted by the atmosphere upon the earth by virtue of the fact that the atmosphere (mainly, its water vapor) absorbs and reemits infrared radiation. [NASA]

green strength The mechanical strength of a ceramic or powder metallurgy part after molding or compacting but before firing or sintering; it represents the quality needed to maintain sharpness of contour and physical integrity during handling and mechanical operations to prepare it for firing or sintering.

greenware Unfired ceramic ware.

grid 1. A network of lines, typically forming squares, used in layout work or in creating charts and graphs. 2. A crisscross network of conductors used for shielding or controlling a beam of electrons.

grid circuit An electronic circuit which includes the grid-cathode path of an electron tube in series with other circuit elements.

grid control A method of controlling anode current in an electron tube by varying the potential of the grid electrode with respect to the cathode.

grid emission Emission of electrons or ions from the grid electrode of an electron tube.

grid nephoscope A device for determining the direction of cloud motion by sighting through a grid work of bars and adjusting the angular position of the grid until some feature of the cloud in the field of view appears to move along the major axis of the grid.

grid spaced contacts Contacts in a multiple contact connector spaced in a geometric pattern.[ARP914A-79]

GRIN See graded index fiber.

grinding 1. Removing material from the surface of a workpiece using an abrasive wheel or belt. 2. Reducing the particle size of a powder or granular solid.

grinding aid Something added to the charge in a rod or ball mill to accelerate the grinding process.

grinding burn Localized overheating of a workpiece surface due to excessive grinding pressures, or inadequate supply of coolant or both.

grinding check Grinding checks are fine thermal cracks that develop from overheating of the area being ground.

Such cracks are generally at right angles to the direction of grinding but may appear as a complete network.[AS3071A-77]

grinding cracks Shallow cracks in the surface of a ground workpiece; they appear most often in relatively hard materials due to excessive grinding friction or high material sensitivity.

grinding fluid A cutting fluid used in grinding operations, primarily to cool the workpiece but also to lubricate the contacting surfaces and carry away grinding debris.

grinding medium Any material—including balls, rods, and quartz or chert pebbles—used in a grinding mill.

grindstone A stone disc mounted on a revolving axle and used for grinding or tool sharpening.

GRIST (telescope) An ESA Spacelab payload designed for grazing incidence solar phenomena that is still in the study phase.[NASA]

grit A particulate abrasive consisting of angular grains.

grit blasting Abrasively cleaning metal surfaces by blowing steel grit, sand or other hard particulate against them to remove soil, rust and scale. Also known as sandblasting.

grommet 1. An elastomeric or plastic sealing device which supports and protects terminations and wires/cables from adverse mechanical and environmental conditions.[ARP914A-79] 2. A metal washer or eyelet, often used to reinforce a hole in cloth or leather. 3. A rubber or soft plastic eyelet inserted in a hole through sheet metal, such as an electronic equipment chassis or enclosure, to prevent a wire from chafing against the side of the hole, damaging its insulation or shorting out to the chassis. 4. A circular piece of fibrous packing material used under a bolt head or nut to seal the bolthole.

grommet nut A blind nut with a round head that is sometimes used with a screw to attach a hinge to a door.

groove 1. A long narrow channel or furrow in a solid surface. 2. In a weldment, a straight-sided, angled, or curved gap between joint members prior to welding which helps confine the weld puddle and ensure full joint penetration to produce a sound weld.

grooved drum A windlass drum whose face has been grooved, usually in a helical fashion, to support and guide the rope or cable wound on it.

grooved tube-seat A tube seat having one or more shallow groves into which the tube may be forced by the expander.

groove seal A postassembly seal formed by injecting a noncuring sealant material into a grove machined in one faying surface of the mating or overlapping structure.[AIR4069-90]

gross national product The total value of the goods and services produced in a nation during a specific period and also comprising the total expenditures by consumers and government plus gross private investment.[NASA]

gross porosity In weld metal or castings, large or numerous gas holes, pores or voids that are indicative of substandard quality or poor technique.

ground 1. The primary aircraft structure is the referenced ground for the negative of the dc and neutral of the ac in the power generation and power utilization systems.[AS1212-71] 2. A conducting connection, whether intentional or accidental, by which an electric current or equipment is connected to the earth, or to a conducting structure that serves a function similar to that of an earth ground (that is, a structure such as a frame of an air, space or land vehicle that is not conductively connected to earth).[ARP1870-86] [ARP1199A-90] 3. A (neutral) reference level for electrical potential, equivalent to the level of electrical potential of the earth's crust. 4.

A secure connection to earth which is used to reference an entire system. Usually the connection is in the form of a rod driven or buried in the soil or a series of rods connected into a grid buried in the soil.

ground clouds See exhaust clouds.

ground clutter A pattern produced on the radar scope by reflections from local ground features. Ground clutter may make it difficult to detect other radar returns in the affected area.[ARP4107-88]

grounding conductor A conductor which provides a current return path from an electrical device to ground. [ARP914A-79]

ground-controlled approach/landing A radar approach system operated from the ground by air traffic control personnel transmitting instructions to the pilot by radio.[ARP4107-88]

ground effect 1. The apparent increase in aerodynamic lift experienced by an aircraft when flying near the ground and observed up to a distance above the ground approximately equal to the wingspan of the aircraft.[ARP4107-88] 2. Increase in the lift of an aircraft operating close to the ground caused by reaction between high-velocity downwash from its wing or rotor and the ground.[NASA]

ground effect (communications) The effect of ground conditions on radio communications.[NASA]

ground flotation Characteristic definition used in describing the capability of an aircraft to operate satisfactorily on or from a specific airfield with respect to bearing and support parameters of that airfield.[AIR1489-88]

ground lead See work lead.

ground loads Any load imposed on the landing gear assembly by contact with the ground during any ground operation, i.e., drag, side, vertical, torsion, spin up, springback, towing, turning, etc.[AIR1489-88]

ground loop 1. The generation of undesirable current flow within a ground conductor, owing to the circulation currents which originate from a second source of voltage—frequently as a result of connecting two separate grounds to a single circuit.[ARP1931-87] 2. An uncontrolled violent turn of an airplane while taxiing, or during the landing or takeoff run.[ARP4107-88]

ground plane A surface, all points of which are assumed to be at the same potential, usually the zero reference potential for the system. (Note: A true, equipotential ground plane does not exist in practice. The deviations from the ideal increase with the frequency of the signals appearing on the ground plane conductor and can become a very important consideration in system design.)[ARP1870-86]

ground power A power source that remains on the ground and is not airborne.[ARP906A-91]

ground safety lock A device designed to prevent inadvertent retraction of the gear when the aircraft is on the ground and supported by the gear. Generally installed manually after landing and removed prior to flight. Also used on landing gear doors and arresting hooks to provide safety for ground maintenance personnel.[AIR1489-88]

ground service Airplane servicing operations during which rotatable galley equipment, such as containers, carts, modules and similar inserts are unloaded from galleys on the aircraft, transported to ground kitchens, washed and cleaned, stored or recycled with food and supplies for installation in galleys ready for flight.[AS1426-80]

ground speed The speed of an aircraft relative to the surface of the earth, typically expressed in knots or statute miles per hour.[ARP4107-88]

ground spoilers A device or system designed to "spoil" or reduce the aerodynamic lift on the aircraft quickly after touchdown, thus applying weight to the gears and enabling braking action for deceleration of the aircraft. See also speed brake.[AIR1489-88]

ground support equipment That equipment on the ground, including all implements, tools, and devices (mobile or fixed), required to inspect, test, adjust, calibrate, appraise, gage, measure, repair, overhaul, assemble, transport, safeguard, record, store, or otherwise function in support of a rocket, space vehicle, or the like, either in the research and development phase or in the operational phase, or in support of the guidance system used with the missile, vehicle, or the like.[NASA] See also aircraft ground equipment.

ground track (track, actual track) The imaginary line or path on the earth's surface connecting successive points over which the aircraft has flown. See also desired track.[ARP1570-86]

ground track angle (track angle, actual track angle) The clockwise angle from true north to an imaginary line on the earth's surface connecting successive points over which the aircraft has flown (ground track). The normal usage implies present track angle—i.e., using the most recent successive points over a relatively short period of time.[ARP1570-86]

ground truth Data obtained on the ground concerning the significance of anomalies observed in remote sensing to help interpretation.[NASA]

group velocity The velocity of a wave disturbance as a whole, i.e. of an entire group of component simple harmonic waves.[NA]

ground visibility The prevailing horizontal visibility near the earth's surface as reported in the United States by the National Weather Service or an accredited observer.[ARP4107-88]

group A number of wires and/or cables secured together and routed to a single item or set-up of equipment.[ARP1931-87]

grouping Combining two or more computer records into one block of information to conserve storage space or disk or tape. Also known as blocking.

group velocity The velocity corresponding to the rate of change of average position of a wave packet as it travels through a medium.

grouting Placing or injecting a fluid mixture of cement and water (or of cement, sand and water) into a grout hole, crevice, seam or joint for the purpose of forming a seepage barrier, consolidating surrounding earth or rock, repairing concrete structures, or sealing the joint where an equipment base rests on a concrete floor.

growth chambers See phytotrons.

grub screw A headless screw that is slotted at one end to receive a screw driver.

GS Glide Slope.

GTAW See gas tungsten-arc welding.

guanosines Guanine riboside; a nucleoside composed of guanine and ribose.[NASA]

guard 1. A protective or safety device to protect against injury, soiling, loss or the like. See also, cap, cover, mask, protector.[ARP480A-89] 2. A shield or cowling that surrounds moving parts to prevent workers from being injured or to prevent incidental equipment damage from foreign objects.

guard bit A bit contained in each word or groups of words of memory which indicates to computer hardware or software whether the content of that memory location may be altered by a program. See protected location.

guard ring An auxiliary ring-shaped electrode in a counter tube or ionization chamber whose chief functions are to control potential gradients, reduce insulation leakage or define the active region of the tube.

guard vacuum An enclosed evacuated space between a primary vacuum system and the atmosphere whose primary purpose is to reduce seal leakage into the primary system.

guayule A desert shrub native to southwestern United States and north Mexico that produces polymeric isoprene essentially identical to that made by Hevea rubber trees in southeast Asia. [NASA]

guidance (motion) The process of directing the movements of an aeronautical vehicle or space vehicle, with particular reference to the selection of a flight path.[NASA]

guide 1. A device with prime function of attaching to engine parts for the purposes of locating two or more parts for assembly or disassembly.[ARP480A-89] 2. A pulley, idler roll or channel member that keeps a rope, cable or belt traveling in a predetermined path. 3. A runway in which a conveyor travels. 4. A stationary machine element—a beam, bushing, rod or pin, for instance—whose primary function is to maintain one or more moving elements confined to a specific path of travel.

guide bearing A plain bushing used to prevent lateral movement of a machine element while allowing free axial translation, with or without (usually without) simultaneous rotation. Also known as guide bushing.

guide bushing See bushing.

guided bend test A bend test in which the specimen is bent to a predetermined shape in a jig or around a grooved mandrel.

guided missile An unmanned airborne vehicle whose flight path or trajectory can be altered by some mechanism within or attached to the vehicle in response to either a preprogrammed control sequence or a control sequence transmitted to the vehicle while in flight.

guided wave A wave whose energy is confined by one or more extended boundary surfaces, and whose direction of propagation is effectively parallel to the boundary.

guide pin A pin or rod, extending beyond the mating force or body of a connector, designed to position and guide connectors during mating so as to ensure proper engagement of the contacts. [ARP914A-79]

guide socket A socket or hole in a connector designed to accept a guide pin of a mating connector and thereby position and guide the connectors during mating so as to ensure proper engagement of the contacts.[ARP914A-79]

guide vanes Control surfaces that may be moved into or against a rocket's jetstream, used to change the direction of the jet flow for thrust vector control. [NASA]

gun launcher Ordnance device for firing missiles and rockets with initial attitude control.[NASA]

gusset The transition between the terminal tongue and conductor barrel. [ARP914A-79]

gust A sudden brief increase in the wind; according to United States Weather Bureau practice, gusts are reported when the variation in wind speed between peaks and lulls is at least 10 knots.[ARP 4107-88]

GUT See grand unified theory.

gutter 1. A drainage trough or trench, usually surrounding a raised surface. 2. A groove around the cavity of a forging or casting die to receive excess flash.

guy A wire, rope or rod used to secure a pole, derrick, truss or temporary structure in an upright position, or to hold it securely against the wind.

guyed-steel stack A steel stack of insufficient strength to be self-supporting which is laterally stayed by guys.

gyratory screen A sieving machine

having a series of nested screens whose mesh sizes are progressively smaller from top to bottom of the stack; the mechanism shakes the stacked screens in a nearly circular fashion, which causes fines to sift through each screen until an entire sample or batch has been classified.

gyro See gyroscope.

gyrodamper Single-gimbal control moment gyro actively controlled to extract the structural vibratory energy through the local rotational deformations of a structure; used in large space structures.[NASA]

gyrofrequency The natural period of revolution of a free electron in the earth's magnetic field.[NASA]

gyro horizons Artificial horizons or attitude gyroscopes.[NASA]

gyromagnetic ratio The magnetic moment of a system divided by its angular momentum.

gyroscope 1. Device which utilizes the angular momentum of a spinning mass (rotor) to sense angular motion of its base about one or two axes orthogonal to the spin axis.[NASA] 2. An instrument that maintains a stable, angular reference direction by virtue of the application of Newton's second law of motion to a mechanism whose chief component is a rapidly spinning heavy mass.

gyroscopic couple The turning moment generated by a gyroscope to oppose any change in the position of its axis of rotation.

gyroscopic drift See gyroscope.

gyroscopic horizon A gyroscopic instrument that simulates the position of the natural horizon and indicates the attitude of an aircraft with respect to this horizon.

gyrostat See gyroscope.

gyrotron See cyclotron resonance device.

gyro wheel The heavy rotating element of a gyroscope, which consists of a wheel whose rather large mass is distributed uniformly around its rim; in precision gyroscopes, the gyro wheel is specially constructed to have nearly perfect balance.

H See henry.

H-60 helicopter The Black Hawk (Sikorsky) assault helicopter.[NASA]

HAA Height Above Airport.

habitats The areas or types of environment in which plants or animals normally occur or live.[NASA]

habit pattern interference See attention, anomalies of habit pattern interference.

habit pattern substitution See attention, anomalies of habit pattern interference.

habituation See attention, anomalies of habituation.

hair-line cracks Fine, random cracks in a coating such as paint or any rigid surface.

HAL Height Above Landing.

half-adder A logic circuit that accepts two binary input signals and produces corresponding sum and carry outputs; two half-adders and an OR gate can be combined to realize a full- adder. See also full adder.

half-adjust To round a number so that the least significant digit(s) determines whether or not a "one" is to be added to the digit next higher in significance that the digit(s) used as criterion for the determination. After the adjustment is made, if required, the digit(s) used as criterion will be dropped. e.g. 432.784 using the terminal 4 as criterion yields 432.78 as the half-adjusted value. The number 432.785 half-adjusts to 432.79, since the terminal digit is "one half, or more."

half-and-half solder A lead-tin alloy (50Pb-50Sn) used primarily to join copper tubing and fittings.

half cycle In alternating circuits, the time to complete one-half of a full cycle at the operating frequency.

half duplex Communications in both directions (transmit and receive), but in only one direction at a given instant in time. See duplex, half; see also full duplex.

half-height drive A 5 1/4-inch disk drive that is 1 5/8-inches wide when installed.

half-life The average time required for one half the atoms in a sample of radioactive element to decay.[NASA]

half-thickness The thickness of an absorbing medium that will depreciate the intensity of radiation beam by one-half.

half-wave plate A polarization retarder which causes light of one linear polarization to be retarded by a half wavelength 180° relative to the phase of the orthogonal polarization.

half-wave rectifier 1. An electronic circuit that converts an a-c input signal to a d-c output signal, with current flowing in the output circuit during only one half of each cycle in of the input signal. 2. A rectifier that feeds current during the half cycle when the alternating current voltage is in the polarity at which the rectifier has low resistance. Whereas during the other half cycle the rectifier passes no current.

Hall coefficient See Hall effect.

Hall current See Hall effect.

Hall effect 1. The electrical polarization of a horizontal conducting sheet of limited extent, when that sheet moves laterally through a magnetic field having a component vertical to the sheet. The Hall effect is important in determining the behavior of the electrical currents generated by winds in the lower atmosphere.[NASA] 2. An electromotive force developed as a result of interaction when a steady-state current flows in a steady-state magnetic field; the direction of the emf is at right angles to both the direction of the current and the magnetic field vector, and the magnitude of the emf is proportional to the product of current intensity, magnetic force, and sine of the angle between current direction and magnetic field vector.

Halley's comet A ember of the solar system with an orbit and a period of about 76 years.[NASA]

halocarbon Compound consisting of halogen atoms and carbon atoms. [NASA]

HALOE See halogen occultation experiment.

halogen occultation experiment Shuttle experiment to provide global stratospheric vertical concentration profiles of key chemical species involved in the catalytic destruction of ozone due to chlorine compounds.[NASA]

HAL/S (language) Programming language developed for the flight software of the NASA Space Shuttle program. [NASA]

hammer 1. A hand tool used for striking a workpiece to shape it or to drive it into another object. 2. A machine element consisting of an arm and a striking head, such as for ringing a bell, or consisting of a guided striking head, often carrying one-half of a die set, for shaping metals by forging.

hammerhead stall An abrupt maneuver in which the airplane zooms into a vertical climb, stalls and yaws simultaneously, and goes into a dive from which recovery is made opposite to the direction of entry.[ARP4107-88]

Hamming code An error-correcting code, with or without parity, that allows a data device to detect and correct single-bit errors in coded digital data.

Hamming distance A characteristic of any given data code that indicates the ability to detect single-bit errors; it equals the number of bits in any given character that must be changed to produce another legitimate character.

H and D curve The measurement of photographic emulsion shown as a curve in which density is expressed as a function of the logarithm of exposure.

handhole An opening in a pressure part for access, usually not exceeding 6 in. in longest dimension.

handhole cover A handhole closure.

hand lance A manually manipulated length of pipe carrying air, steam, or water for blowing ash and slag accumulations from heat absorbing surfaces.

handle An item designed to be held in the hand to provide proper leverage. EX: handle, wire brush; handle, socket wrench.[ARP480A-87]

handling qualities See controllability.

handoff An action taken to transfer the radar identification of an aircraft from one controller to another if the aircraft will enter the receiving controller's airspace and radio communications with the aircraft will be transferred to that controller.[ARP4107-88]

handshake The recognition between two computers that they are able to communicate.

hang fire An undesired delay in the functioning of an explosive device after initiating energy is applied.[AIR913-89]

hang glider Ultralight, unpowered aircraft in which the pilot controls the flight attitude and glide path by shifting his position on a suspended seat (swing seat).[NASA]

hard card A type of computer hard disk on a card rather than a spinning disk.

hard clad silica fibers Silica optical fibers which are coated with hard plastic material, not with the soft materials typically used in plastic clad silica.

hard copy Readable data from the computer to a printer which is produced on paper.

hard disk A computer storage medium with a large storage capacity as compared to floppy disks. 80 megabytes of storage is now common.

hard-disk management Since hard disk life is limited, there are four basic things that will enhance disk life and use: (A) Use subdirectories rather than have all work files in one directory; (B)

Delete files that are no longer needed; (C) Run CHKDSK periodically to check for lost clusters; (D) Run a defragmentation program every three months.

hard drawn copper wire Copper wire that has been drawn to size and not annealed.[ARP1931-87]

hard-drawn wire Cold-drawn metal wire of relatively high tensile strength and low ductility.

hardening Producing increased hardness in a metal by quenching from high temperature, such as hardening steel, or by precipitation-hardening (aging) a dilute alloy, such as hardening certain aluminum or other nonferrous alloys.

hardening (system) Technique for decreasing the susceptibility or vulnerability of weapon systems and components.[NASA]

hard landing An impact landing of a spacecraft on the surface of a planet or natural satellite destroying all equipment except possibly a very rugged package.[NASA]

hard lead Any of a series of lead-antimony alloys of low ductility; typically, hard lead contains 1 to 12% Sb.

hardness 1. Resistance of metal to plastic deformation usually by indentation. However the term may also refer to stiffness or temper, or to resistance to scratching, abrasion, or cutting.[NASA] 2. A measure of the amount of calcium and magnesium salts in a boiler water. Usually expressed as grains per gallon or p.p.m. as $CaCO_3$. See also SHORE A and REX hardness.

hardover Response or movement of an aircraft control system at the maximum rate toward its maximum deflection. [ARP4107-88]

hardware 1. Any part of the torso restraint system, other than webbing. [AS8043-86] 2. Physical equipment as contrasted to ideas or design that may exist only on paper.[NASA] 3. In data processing, hardware refers to the physical equipment associated with the computer. 4. The electrical, mechanical and electromechanical equipment and parts associated with a computing system, as opposed to its firmware and software.

hardware priority interrupt See priority interrupt and software priority interrupt.

hard water Water which contains calcium or magnesium in amounts which require an excessive amount of soap to form a lather.

harmonic Having a frequency that is a multiple of the basic cyclical quantity to which it is related.

harmonic analysis A statistical method for determining the amplitude and period of certain harmonic or wave components in a set of data with the aid of Fourier series.[NASA]

harmonic analyzer 1. An instrument for measuring the magnitude and phase of harmonic segments of a cyclical function from a graph. 2. An electronic instrument which measures the amplitude and frequency of an a.c. signal; including those of its harmonics.

harmonic conversion transducer A transducer in which the output frequency is a multiple of the input frequency.

harmonic distortion Distortion caused by the presence of harmonics of a desired signal.

harmonic function Any solution of the Laplace equations.[NASA]

harmonic generation 1. The multiplication of the frequency of a lightwave by nonlinear interactions of the lightwave with certain materials. Generating the second harmonic is equivalent to dividing the wavelength in half. 2. Electronic means of multiplying frequency, usually accomplished with the assistance of nonlinear devices.

harmonic motion The projection on a diameter of the circle of such motion.

[NASA]

harmonics Eigenfrequency oscillations excited in a vibrating system.[NASA]

harness 1. A multi-conductor cable with leads spaced along its length.[ARP480A-87] 2. A group of wires or cables routed together with or without attached components and secured in a manner to provide a preshaped electrical wire or cable assembly.[ARP914A-79] 3. An assembly of wires and/or cables arranged so it may be installed and removed as a unit.[ARP1931-87]

harness—high density A harness designed to save weight and space.[ARP 1931-87]

Hartley information unit In information theory, a unit of logarithmic measurement of the decision content of a set of 10 mutually exclusive events, expressed as the logarithm to the base 10; for example, the decision content of an eight-character set equals log 8, or 0.903 Hartley.

Hartree-Fock-Slater method A refined approximation method for the calculation from wave function of electron total energies, kinetic energies, etc. for chemical elements.[NASA]

hashing The generation of a meaningless number from a group of records that can be used as a location address.

Hastelloy B An International Nickel Co. alloy having a nominal composition of nickel (Ni) 66.7%; iron (Fe) 5%; molybdenum (Mo) 28%; vanadium (V) 0.3%.

Hastelloy C An International Nickel Co. alloy having a nominal composition of nickel (Ni) 59%; iron (Fe) 5%; molybdenum (Mo) 16%; tungsten (W) 4%; chromium (Cr) 16%.

HAT Height Above Touchdown.

Hawkeye 1 satellite See Explorer 52 satellite.

HAX (metallurgy) See heat affected zone.

Hay bridge A general-purpose a-c bridge circuit in which two opposing sides of the bridge are fixed resistances; the unknown leg is a combination of resistance and inductance, and the remaining side consists of a variable resistor and a variable capacitor.

hazardous area An area in which explosive gas/air mixtures are, or may be expected to be, present in quantities such as to require special precautions for the construction and use of electrical apparatus.

hazardous atmosphere 1. A combustible mixture of gases and/or vapors. 2. An explosive mixture of dust in air.

hazardous material Any substance that requires special handling to avoid endangering human life, health or well being. Such substances include poisons, corrosives, and flammable, explosive or radioactive chemicals.

hazardous material disposal (in space) The disposal in space of hazardous material. When radioactive materials are involved the expected lifetime of orbit exceeds the lifetime of the radioactivity.[NASA]

haze A suspension of extremely small, dry particles invisible to the naked eye and sufficiently numerous to give the air an opalescent (milky or pearly) appearance.[AIR1335-75]

hazemeter See transmissometer.

HCL argon laser Gas laser in which the active material is gaseous hydrogen chloride and argon.[NASA]

HCL laser Gas laser in which the active material is gaseous hydrogen chloride.[NASA]

HDDR See high density digital recording.

HDLC See high-level data link control.

head The portion of a computer disk drive that reads, writes, or erases any magnetic storage medium.

head crash In data processing, the malfunction of the read and write head in a disk drive.

header 1. A conduit or chamber that

receives fluid flow from a series of smaller conduits connected to it, or that distributes fluid flow among a series of smaller conduits. 2. In data processing, data placed at the beginning of a file for identification. 3. The portion of a batch recipe that contains information about the purpose, source and version of the recipe such as recipe and product identification, originator, issue date and so on.

head gap 1. The space between the reading or recording head and the recording medium, such as tape, drum, or disk. 2. The space or gap intentionally inserted into the magnetic circuit of the head in order to force or direct the recording flux into the recording medium.

heading 1. The upsetting of wire, rod or bar stock in dies to form parts having some of the cross-sectional area larger than the original.[ARP700-65] 2. The direction—usually expressed in deg relative to true or magnetic north—in which the longitudinal axis of an aircraft point. (For example, "Fly heading 270 magnetic"). See also course.[ARP4107-88]

head loss Pressure loss in terms of a length parameter such as inches of water or millimeters of mercury.

head pressure Expression of a pressure in terms of the height of fluid. $P=y\rho g$, where ρ is fluid density and y is the fluid column height.

headset See earphones.

head up display See HUD.

health physics The technology associated with the measurement and control of radiation dose in humans.

HEAO A See HEAO 1.

HEAO B See HEAO 2.

HEAO C See HEAO 3.

HEAO 1 The first of three NASA high energy astronomy observatories launched during 1977 for the study of cosmic rays and earth's magnetic field to study the x ray and gamma ray sky.[NASA]

HEAO 2 The second of three NASA high energy astronomy observatories. It was launched during 1978 for the study of specific x ray objects, quasars, x ray pulsars, and candidate black holes.[NASA]

HEAO 3 The third of three NASA high energy astronomy observatories. It was launched during 1979 for the study of cosmic rays and elemental and isotropic composition as a corollary to a search of narrow gamma ray lines.[NASA]

heat 1. Energy transferred by a thermal process.[NASA] 2. Energy that flows between bodies because of a difference in temperature; same as thermal energy.

heat absorbing filter A glass filter which transmits most visible light, but strongly absorbs infrared light.

heat affected zone That portion of the base metal the structure or properties of which have been altered by the heat of welding or gas-cutting operation.[NASA]

heat aging A test used to determine the ability of a wire or material to withstand a specific temperature for a specific length of time. The test temperature is usually higher than the related temperature of the test specimen. Sometimes referred to as "life cycle".[ARP1931-87]

heat available The thermal energy above a fixed datum that is capable of being absorbed for useful work. In boiler practice, the heat available in a furnace is usually taken to be the higher heating value of the fuel corrected by subtracting radiation losses, unburned combustible, latent heat of the water in the fuel formed by the burning of hydrogen, and adding the sensible heat in the air for combustion, all above ambient temperatures.

heat balance 1. The equilibrium which exists on the average between the radiation received by a planet and its atmosphere from the sun and that emitted by the planet and the atmosphere. The equilibrium which is known to exist when all sources of heat gain and loss for a given region of body are accounted for. In general this balance cludes advective or evaporative terms as well as a radiation term.[NASA] 2. An accounting of the distribution of the heat input and output.

heat capacity See specific heat.

heat content The amount of heat per unit mass that can be released when a substance undergoes a drop in temperature, a change in state or a chemical reaction. See also enthalpy.

heat damage Charring, chalking, or hardening and cracking of hose cover caused by excessive or prolonged exposure to heat. Other evidence may include fluid leakage from socket where fitting or coupling is attached to hose. [AIR1658A-86]

heat equation See thermodynamics.

heater A device which provides controlled heat to a specific area.[ARP480A-87]

heat exchanger 1. Device for transferring heat from one fluid to another without intermixing the fluids, as a regenerator and, an apparatus for cooling or heating the air in a wind tunnel. [NASA] 2. A vessel in which heat is transferred from one medium to another.

heat flow See heat transmission.

heating surface Heating surface shall be expressed in square feet and shall include those surfaces which are exposed to products of combustion on one side and water on the other. This surface shall be measured on the side receiving the heat, which is as provided in the ASME Power Test Code.

heat of fusion The increase in enthalpy accompanying the conversion of one mole, or a unit mass, of a solid to a liquid at its melting point at constant pressure and temperature.[NASA]

heat pack See heat sink.

heat rate The ratio of heat input to work output of a thermal power plant. It is a measure of power plant efficiency.

heat release The total quantity of thermal energy above a fixed datum introduced into a furnace by the fuel, considered to be the product of the hourly fuel rate and its high heat value, expressed in Btu per hour per cubic foot of furnace volume.

heat resistance See thermal resistance.

heat resistant alloys Alloys developed for very high temperature service here relatively high stresses (tensile, thermal, vibratory, and shock) are encountered and where oxidation resistance is frequently required.[NASA]

heat shielding The use of devices that protect something from heat. Specifically, the protective structure necessary to protect a reentry body from aerodynamic heating.[NASA]

heat shock A test used to determine the stability of a material by sudden exposure to a high temperature for a short period of time.[ARP1931-87]

heatsink Any device, usually a static device, used primarily to absorb heat and thereby protect another component from damage due to excessive heat.

heat sink The portion of the brake which provides the primary storage medium for the heat generated in deceleration of the aircraft. (Aircraft kinetic energy converted to heat energy.) Generally includes the rotors, stators, and sometimes portions of the back plate pressure plate, etc.[AIR1489-88]

heat sink loading A measure of the energy absorbed per unit weight by the brake heat sink during a single stop. Equals the total kinetic energy divided by the heat sink weight (foot pounds per pound or joules per kilogram.)[AIR 1489-88]

heat transfer The transfer or exchange of heat by radiation, conduction, or convection with a substance and between the substance and its surroundings.[NASA]

heat transfer, coefficient of 1. The rate of heat transfer per unit area per unit temperature difference, a quantity having the dimensions of reciprocal length.[NASA] 2. Heat flow per unit time across a unit area of a specified surface under the driving force of a unit temperature difference between two specified points along the direction of heat flow. Also known as over-all coefficient of heat transfer.

heat tracing The technique of adding heat to a process or instrument measurement line by placing a steam line or electric heating element adjacent to the line.

heat transmission Heat transmitted from one substance to another.[NASA]

heat treatment Heating and cooling a solid metal or alloy in such a way as to obtain desired conditions or properties.[NASA]

heavier-than-heavy key The remaining components in the bottoms stream other than the heavy key.

Heaviside bridge A type of a-c bridge for making mutual inductance measurements when the inductance of the primary winding is already known.

heavy duty cable Generally a type of fiber optic or electrical cable designed to withstand unfriendly conditions, such as those encountered outdoors. Some varieties are armored to withstand hostile conditions.

heavy ends The fraction of a petroleum mixture having the highest boiling point.

heavy fraction The final products retrieved by distilling crude oil.

heavy key The component in multicomponent distillation that is removable in the bottoms stream and that has the highest vapor pressure of the components at the bottoms. If more reboiler heat is added, the heavy key component is the first component to be put in the overhead product.

heavy lift airships Airships designed to lift heavy materials.[NASA]

heavy oil A viscous fraction of petroleum or coal-tar oil having a high boiling point.

heavy water 1. Water in which the hydrogen of the water molecule consists entirely of the heavy hydrogen isotope of mass 2 (deuterium).[NASA] 2. A liquid compound D_2O whose chemical properties are similar to H_2O (light water), and occurs in a ratio of 1 part in 6000 in fresh water.

hectare A metric unit of land measure equal to 10,000 m^2, or approximately 2.5 acres.

hectare-meter A metric unit of volume, commonly used in irrigation work, which equals 10,000 m^3; it represents the amount of water needed to cover an area of one hectare to a depth of one metre.

heel block A block or plate attached to a die that keeps the punch from deflecting too much.

height Vertical distance; the distance above some reference point or plane, as, height above sea level. The vertical dimension of anything; the distance which something extends above its foot or root, as blade height.[NASA]

height above airport (HAA) The height of the minimum descent altitude above the published airport elevation. See also minimum descent altitude.[ARP4107-88]

height above landing (HAL) The height above a designated helicopter landing area used for helicopter instrument approach procedures.[ARP4107-88]

height above touchdown (HAT) The height of the decision height or minimum descent altitude above the highest runway elevation in the touchdown zone (the first 3000 ft of the runway). See also decision height, minimum descent altitude.[ARP4107-88]

height gage A mechanical device, usually having a vernier or micrometer scale, used for measuring precise distances above a reference plane.

heliarc welding See gas tungsten-arc welding.

helical antennas Antennas used where circular polarization is required. The driven element consists of a helix supported above a ground plane.[NASA]

helical stripe A continuous, colored spiral stripe applied over the outer perimeter of an insulated conductor for circuit identification purposes.[ARP 1931-87]

helicopter A rotorcraft that depends principally on its engine-driven rotors for movement as well as lift.[ARP4107-88]

helicopter impulsive noise. See blade slap noise.

heliograph See spectroheliograph.

heliography See spectroheliograph.

heliometry See pyroheliometer.

heliosphere The region around the sun whose plasma processes are dominated by solar wind.[NASA]

heliostat instrument consisting of mirrors moved by clockwork for reflecting the sun's rays in a fixed direction. [NASA]

helipad A particular location on an airfield, building, or other area specifically designated as the place of helicopter landings and takeoffs.[ARP4107-88]

heliport An area of land or water, or a structure, used or intended to be used for the landing and takeoff of helicopters; it includes the helipad and other associated structures and facilities, such as fences, terminal buildings, and fire suppression equipment.[ARP4107-88]

helix A spiral winding.[ARP1931-87]

helix tube See traveling wave tube.

Hellige turbidimeter A variable-depth instrument for visually determining the cloudiness of a liquid caused by the presence of finely divided suspended matter.

Helmholtz resonator An enclosure having a small opening consisting of a straight tube of such dimension that the enclosure resonates at a single frequency determined by the geometry of the resonator.[NASA]

help In data processing, an on-screen information resource that a user can activate to answer questions.

hematite A common iron mineral; ferric oxide.[NASA]

hemoperfusion Type of poison treatment in which the patient's blood is passed over a bed of absorbent material (activated carbon, resin, etc.) to remove the toxin from the bloodstream. [NASA]

henry (H) 1. Unit of inductance. The inductance of a circuit in which an electromotive force (emf) of one volt is produced when the current in the circuit changes uniformly at the rate of one ampere per second.[ARP1931-87] 2. Metric unit for inductance.

Henry's law A principle of physical chemistry that relates equilibrium partial pressure of aa substance in the atmosphere above a liquid solution to the concentration of the same substance in the liquid; the ratio of concentration to equilibrium partial pressure equals the Henry's-law constant, which is a temperature-sensitive characteristic; Henry's law generally applies only at low liquid concentrations of a volatile component.

hepatitis An inflammation of the liver, commonly of viral origin, but also associated with other diseases.[NASA]

herbicide Chemical agent used for the eradication of undesirable plants or for the inhibition of their growth.[NASA]

Herbig-Haro objects Celestial objects having many of the characteristics of a T Tauri star (e.g., their spectra show a weak continuum with strong emission lines), believed to be stars in the very early stages of development. All known Herbig-Haro objects have been found within the boundaries of dark clouds.

Those strong infrared sources are characterized by mass loss.[NASA]

hertz (Hz) Unit of frequency equal to one cycle per second.[ARP1931-87]

Hessian matrices Given a real value function of N variables, an N by N symmetric of all second order partial derivatives.[NASA]

heterodyne A combination of a.c. signals of two different frequencies, coupled so as to produce beats whose frequency is the sum or difference of the frequencies of the original signals.

heterodyne conversion transducer A transducer in which output frequency is the sum and difference of the input frequency and the local oscillator frequency.

heterodyning Mixing two radio signals of different frequencies to produce a third signal which is of lower frequency, i.e., to produce beating.[NASA]

heterogeneous radiation A beam of radiation containing rays of several different wavelengths or particles of different energies or different types.

heterojunction A junction between semiconductors that differ in doping levels and also in their atomic compositions. Example: a junction between layers of GaAs and GaAIAs—a double heterojunction laser contains two such junctions, a single heterojunction laser contains only one.

heterojunction devices Electron devices utilizing junctions between different semiconducting materials. The characteristics and performance of the devices are dependent on the relative lineup of the energy bands at the junctions.[NASA]

heterojunctions Boundaries between two different semiconductor materials, usually with a negligible discontinuity in the crystal structure.[NASA]

heterosphere The upper portion of a two part division of the atmosphere according to the general homogeneity of atmospheric composition; the layer above the homosphere. The heterosphere is characterized by variation in composition and mean molecular weight of constituent gases. This region starts at 80 to 100 kilometers above the earth, and therefore closely coincides with the ionosphere and the thermosphere. [NASA]

heuristic Pertaining to a method or problem solving in which solutions are discovered by evaluation of the progress made toward the final solution, such as a controlled trial and error method. An exploratory method of tackling a problem, or sequencing of investigation, experimentation, and trial solution in closed loops, gradually closing in on the solution. A heuristic approach usually implies or encourages further investigation, and makes use of intuitive decisions and inductive logic in the absence of direct proof known to the user. Thus, heuristic methods lead to solutions of problems or inventions through continuous analysis of results obtained thus far, permitting a determination of the next step. A stochastic method assumes a solution on the basis of intuitive conjecture or speculation and testing the solution against known evidence, observations, or measurements. The stochastic approach tends to omit intervening or intermediate steps toward a solution. Contrast with stochastic and algorithmic.

heuristic program A program that monitors its performance with the objective of improved performance.

hex A number of representation system of base 16. The hex number system is very useful in cases where computer words are composed of multiples of four bits (that is, 4-bit words, 8-bit words, 16-bit words, and so on).

hexadecimal Number system using base 16 and the digit symbols from 0 to 9 and A to F. See hex.

hexadecimal notation A numbering system using 0, 9, A, B, C, D, E, and F with 16 as a base.

hexagonal-head bolt A standard threaded fastener with an integral hexagon-shaped head.

hexagonal nut A hexagon-shaped fastener with internal threads used with a mating, externally threaded bolt, stud or machine screw.

hex code A low-level code in which the machine code is represented by numbers using a base of 16.

Heydweiller bridge A type of a-c bridge circuit suitable for determining the mutual inductance between two interacting windings, both having unknown inductances.

hidden function 1. An item whose function is normally active and whose failure is not evident to the operating crew during performance of normal duties.[ARD50010-91] 2. An item whose function is normally dormant and whose readiness to perform, prior to it being needed, is not evident to the operating crew during performance of normal duties.[ARD50010-91]

hierarchical distributed control A hierarchy of computer systems in which one computer acting as supervisor controls several lower level computers.

hierarchy Specified rank or order of items, thus, a series of items classified by rank or order.

high-alloy steel An iron-carbon alloy containing at least 5% by weight of additional elements.

high altitude flight See flight.

high aluminum defect (HAD) An aluminum-rich alpha stabilized region containing an abnormally large amount of aluminum which may extend across a large number of beta grains. It contains an inordinate fraction of primary alpha but has a microhardness only slightly higher than the adjacent matrix. These are also known as Type II defects.[AS1814-90]

high brass A commercial wrought brass containing 65% copper and 35% zinc.

high-carbon steel A plain carbon steel with a carbon content of at least 0.6%.

high cycle fatigue Failure caused by cyclic loading to levels less than the material elastic limit, which results in complete failure in more than 50,000 loading cycles. High cycle fatigue life usage is difficult to calculate as it tends to be very localized on engine components where measurements are not made during service.[AIR1872-88]

high density digital recording (HDDR) The technique which combines the good features of NRZ and biphase codes, to achieve a packing density of up to 33,000 bits per inch per track in instrumentation tape recording.

high density inclusion (HDI) A region with a concentration of elements, usually tungsten or columbium, having a higher density than the matrix. Regions are readily detectable by x-ray and will appear brighter than the matrix.[AS 1814-90]

high electron mobility transistors A recently developed field effect transistor based on the technique of modulation doping of GaAs/Al(x)Ga(1-x) as heterojunctions. This technique achieves high mobility in part by introducing carriers into high purity GaAs from donor ions in an adjacent A1GaAs layer, the electrons and ions being separated by the built in heterojunction potential.[NASA]

High Energy Astronomy Observatory See HEAO.

high frequency (HF) The radio frequency band between 3 and 30 MHz.[ARP4107-88]

high-frequency bias A sinusoidal signal that is mixed with the data signal during the magnetic tape direct recording process for the purpose of increasing the linearity and dynamic range of the recording medium. Bias frequency is usually three to four times the highest data frequency to be recorded.

high-frequency heating See electronic heating.

high gas pressure switch A switch to stop the burner if the gas pressure is too high.

high-heat value See calorific value.

high intensity laser See high power laser

high-intensity runway light system (HIRL) See runway edge light system.

high interstitial defect (HID) Interstitially stabilized alpha phase region of substantially higher hardness than surrounding material. It arises from very high local nitrogen, oxygen, or carbon concentrations which increase the beta transus and produce the high hardness, often brittle, alpha phase. Their boundaries are always diffused. They are commonly called Type I defects of low-density inclusions (LDI). Defects often associated with voids and cracks.[AS1814-90]

high-level computing device (HLCD) A microprocessor-based device used to perform computer-like functions.

high-level data link control (HDLC) A type of data link protocol.

high-level human interference (HLHI) A device that allows a human to interact with the total distributed control system over the shared communications facility.

high-level language(s) 1. Computer languages whose instructions or statements each correspond to several machine language instructions.[NASA] 2. A programming language whose statements are translated into more than one machine-language instruction. Examples of high-level languages are BASIC, FORTRAN, COBOL, and TELEVENT.

high-level operator interface (HLOI) A type of HLHI designed for use by a process operator.

high-lift device Any device, such as a flap, slat, or boundary-layer-control device, used to increase the maximum lifting capacity (or maximum lift coefficient) of a wing.[ARP4107-88]

high limiting control See control, high limiting.

high-low bias test Same as marginal check.

high order Pertaining to the weight or significance assigned to the digits of a number, e.g., in the number 123456, the highest order digit is 1, the lowest order digit is 6. One may refer to the three higher-order bits of a binary word, as another example.

highpass filter 1. Wave filter having a single transmission band extending from some critical or cutoff frequency, not zero, up to infinite frequency.[NASA] 2. A filter which passes high frequencies above the cut-off frequency, with little attenuation.

high pitch The setting of a propeller blade at a high angle relative to the plane of rotation other than the feather angle. The specific angle considered as high pitch may vary from one model of propeller to another. High pitch settings are used to conserve fuel by reducing engine rpm when power requirements are low to moderate.[ARP4107-88]

high power laser Stimulated emission device having high energy flux density outputs.[NASA]

high-pressure boiler A boiler furnishing steam at pressure in excess of 15 pounds per square inch or hot water at temperatures in excess of 250 °F or at pressures in excess of 160 pounds per square inch.

high pressure stored gas system Consists of four elements: the fluid medium or gas; a container; an actuator to open the container upon command; and a pressure reducer which reduces the container gas pressure down to the required system discharge level. The more simple systems may operate without the use of a pressure reducer, thereby providing unregulated output.

high resolution graphics A finely defined graphical display on a computer monitor screen.

high Reynolds number A Reynolds number above the critical Reynolds number of a sphere.[NASA]

high speed flight See flight.

high-speed taxiway/exit/turnoff A long-radius taxiway provided with lighting or marking to define the path of aircraft, traveling at speeds up to 60 knots, from the runway center to the point on the center of a nearby taxiway that is clear of the active runway. The high-speed taxiway is designed to expedite aircraft turning off the runway after landing, thus reducing runway occupancy time.[ARP4107-88]

high-strength alloy A metallic material having a strength considerably above that of most other alloys of the same type or classification.

high strength alloy conductor A conductor which shows a maximum of 20% increase in resistance and a minimum of a 70% increase in breaking strength over the equivalent construction in pure copper while exhibiting a minimum elongation of 5% in 10 inches. As required, the alloy should be capable of sustaining continuous exposure to temperatures as high as 300 °C without suffering an appreciable permanent change in properties.[ARP1931-87]

high-temperature alloy A metallic material suitable for use at 500 °C (930 °F) or above. This classification includes iron-base, nickel-base and cobalt-base superalloys, and the refractory metals and their alloys, which retain enough strength at elevated temperature to be structurally useful and generally resist undergoing metallurgical changes that weaken or embrittle the material. See also heat resistant alloy.

high-temperature hot-water boiler A water heating boiler operating at pressure exceeding 160 psig or temperatures exceeding 250 °F.

high temperature superconductor New superconducting material consisting of mixed metal oxide ceramics that maintain their superconductivity at higher temperature ranges (above 24 K) than the more traditional superconductors.[NASA]

high tension system Ignition system capable of delivering voltages in excess of 5 kV to the firing tip of the spark igniter.[AIR784-88]

highway in the sky A pictorial representation of the path that the pilot is to fly, generally presented on a forward-looking vertical situation display.[ARP 4107-88]

hinge A mechanical device that connects two members across a joint yet allows one member to pivot about an axis that runs along the joint.

hinge moments See torque.

HIP (process) See hot isostatic pressing.

Hipparcos satellite A planned ESA astronometric satellite to determine trigonometric parallaxes, proper motions, and positions of 100,000 stars, mainly for stars brighter than magnitude 10.[NASA]

HIRL High-Intensity Runway Light System. See also runway edge light system.

hiss Random noise in the audiofrequency range, having subjective characteristics analogous to prolonged sibiland sounds.[NASA]

histochemical analysis In biochemistry, the analysis of chemical components in tissues.[NASA]

histotoxic hypoxia See hypoxia, histotoxic.

hit In data processing, the isolation of a matching record.

hitch pin See cotter pin.

hit rate The number of successful matches in a computer search.

HLCD See high-level computing device.

HLHI See high-level human interference.

HLOI See high-level operator interface.

hohlraums In radiation thermodynamics, cavities whose walls are in radiative equilibrium with the radiant energy with the cavity.[NASA]

hohmann trajectory See transfer orbits.

hoist A mechanical device consisting of a supporting frame and integral mechanism specifically designed to raise or lower a load by tensile force.[ARP480A-87]

holdback A device or assembly designed for restraining the aircraft during engine run up.[AIR1489-88]

holdback and release assembly A device or assembly designed to restrain the aircraft against engine thrust, ship motion and tensioning forces prior to catapulting. Once the catapult is fired, the holdback and release assembly is designed to release at a specific load, usually by means of a frangible link (tension bar), thus permitting the aircraft to be accelerated by the catapult.[AIR1489-88]

holdback bar A rigid holdback and release assembly used to restrain the aircraft prior to launch specifically from a nose gear launch catapult system.[AIR1489-88]

holdback, engine run up A device or assembly designed to restrain the aircraft during engine trim procedures and full power checkouts.[AIR1489-88]

holdback fitting, aircraft The attachment point or device on the aircraft to which the holdback is coupled for engine runup or for catapulting.[AIR1489-88]

holdback, repeatable A device or assembly designed for catapult holdback, and release which may be used repeatedly without dependence on or replacing a frangible link.[AIR1489-88]

hold down A device or assembly; part of the arresting hook installation of an aircraft, designed to hold down the hook after deployment and contact with the deck (runway) to prevent hook bounce and insure engagement with the arresting gear cable. See also dashpot and snubber.[AIR1489-88]

holder A device specifically designed to accommodate and position another item to facilitate quick replacement of the item held. See also base, stand.[ARP 480A-87]

hold/holding procedure A predetermined maneuver which keeps aircraft within a specified airspace while awaiting further clearance from air traffic control.[ARP4107-88]

holding beam An electron beam for reactivating the charge on the surface of an electronic device.

holding fix A specified fix, identifiable to a pilot by NAVAIDs or visual reference to the ground, that is used as a reference point in establishing and maintaining the position of an aircraft while holding.[ARP4107-88]

hold time In any process cycle, an interval during which no changes are imposed on the system. Hold time is usually used to allow a chemical or metallurgical reaction to reach completion, or to allow a physical or chemical condition to stabilize before proceeding to the next step.

hole burning A laser process that depletes, spatially or spectrally, the electron/hole pair density in a region of space or frequency of high coherent light, being spatial hole burning and spectral hole burning respectively.[NASA]

hole geometry (mechanics) The sizes, locations, and shapes of perforations created in materials.[NASA]

Hollerith card A punched card used in digital computing; it is named for Herman Hollerith, who developed a computing method using punched cards to compile the 1890 U.S. census.

Hollerith code A widely used system of encoding alphanumeric information onto cards, hence Hollerith cards are the same as punch cards. Such cards were first used in 1890 for the U.S. Census and were named after Herman Hollerith, their originator. See code, Hollerith.

holographic diffraction grating Diffraction grating in which the pattern of light diffracting lines was recorded holographically rather than mechanically ruled into the surface.

holographic optical elements Holograms which have been made to diffract light in the same pattern as other optical components. It is possible to produce (usually by computer synthesis) a hologram which mimics the function of the lens. In some applications, such holographic optical elements are less costly than conventional optics.

holographic subtraction A holographic technique by which two dissimilar optical fields can be subtracted to yield only their difference.[NASA]

home In personal computers, a key which places the cursor at the upper left-hand position on the screen or the upper left-hand position of the entire file.

homing The following of a path of energy waves to or toward their source or point of reflection.[NASA]

homogeneous glass Glass of essentially uniform composition throughout its structure.[ARP798-91]

homogeneous radiation A beam of radiation containing rays whose wavelengths all fall within a narrow band of wavelengths, or containing particles of a single type having about the same energy.

homojunction 1. Solar cell where both sides of the cell are made of the same material.[NASA] 2. A junction between semiconductors that differ in doping levels but not in atomic composition. Example: A junction between n-type and p-type GaAs.

homologous pair In optical spectroscopy, two lines so chosen that the ratio of their radiant powers has minimal change with variations in the input conditions.

homopolar generator Rotating electric machine for converting mechanical power into pure direct current by utilizing poles having the same polarity of the armature.[NASA]

homosphere The lower portion of a two part division of the atmosphere according to the general homogeneity of atmospheric composition; opposed to the heterosphere. The region in which there is no gross change in atmospheric composition, that is, all the atmosphere from the earth's surface to about 90 kilometers. The homosphere is about equivalent to the neutrosphere, and includes the torposphere, stratosphere, and mesosphere; it also includes the ozonosphere and at least part of the chemosphere. [NASA]

hone To remove a small amount of material using fine-grit abrasive stones and thereby obtain an exceptionally smooth surface finish or very close dimensional tolerances.

honeycomb core lightweight strengthening material of structures resembling honeycomb meshes.[NASA]

hook, downlock A hook designed to engage a roller or restraining device and lock the gear assembly in the down position, ready for touchdown and ready to resist applied ground loads.[AIR1489-88]

Hookean behavior A condition in liquid expansion when the fractional change in volume is proportional to the hydrostatic stress, if under such stress it evidences ideal elastic behavior.

hook gage An instrument consisting of a pointed metal hook mounted on a micrometer slide that is used to measure the level of a liquid in an evaporation pan. The level with respect to a reference height is determined when the point of the hook just breaks the liquid surface.

hook point The point of a hook as designed for proper engagement of operation in a mechanism. In an arresting hook installation, that portion of the arresting hook assembly which engages the arresting gear cable. Usually detachable from the arresting hook shank for wear replacement purposes.[AIR1489-88]

hook tongue terminal A terminal with a hook-shaped tongue.[ARP914A-79]

hookup The process or operation of preparation of the aircraft for launch from a catapult. Includes engagement of the holdback and attachment of the catapult bridle or launch bar. The condition in which the aircraft is spotted on the catapult ready for tensioning. [AIR1489-88]

hook, uplock A hook designed to engage a roller or restraining device and lock the gear and/or door assembly in the up or stowed position in a safe condition for all modes of flight maneuvering of the aircraft.[AIR1489-88]

hopper scale A weighing device consisting of a bulk container or hopper suspended on load cells or a lever system and used to batch-weigh bulk solids, often in connection with automated batch processing or with continuous receiving or shipping operations.

horizon That great circle of the celestial sphere midway between the zenith and nadir, or a line resembling of approximating such a circle.[NASA]

horizon misplacement See illusion, visual.

horizontal boiler A water tube boiler in which the main bank of tubes are straight and on a slope of 5 to 15 degrees from the horizontal.

horizontal branch stars Horizontal strips of stars on the Hertzsprung-Russell diagram of globular clusters to the left of the red giant branch.[NASA]

horizontal orientation The attitude of an object in reference to the plane which is perpendicular to the direction of gravity.[NASA]

horizontal return-tubular boiler A fire-tube boiler consisting of a cylindrical shell, with tubes inside the shell attached to both end closures. The products of combustion pass under the bottom half of the shell and return through the tubes.

horn A device for directing and intensifying sound waves that consists of a tube whose cross section increases from one end to the other.

horn antenna 1. Antenna shaped like a horn.[NASA] 2. The flared end of a radar waveguide, whose dimensions are chosen to give efficient radiation of electromagnetic energy into the surrounding environment.

horn (antenna) A moderate-gain wide-beamwidth antenna, generally limited to use in manually steerable applications.

horn relay contact See auxiliary output.

hose assembly A length of hose with a coupling attached to one or each end. [AS1933-85]

hose assembly date The date the hose end fittings are applied to the hose to form a hose assembly.[AS1933-85]

host (computer) The primary computer in a multielement system; the system that issues commands, has access to the most important data, and is the most versatile processing element in a system. Compare with target computer or object computer.

hot atoms Atoms with high internal or kinetic energy as a result of a nuclear process such as beta decay or neutron capture.[NASA]

hot cathode Cathode that functions primarily by the process of thermionic emission.[NASA]

hot cathode fluorescent lamps Hot cathode fluorescent lamps employ coiled tungsten filaments as electrodes. These are coated with one or more of the alkaline earth oxides. This electron-emissive coating provides an abundance of free electrons when hot. By suitable circuit arrangements, these cathodes can be heated to a satisfactory electron emitting temperature before the arc is struck (preheat, trigger or rapid start) or they may be required to act momentarily as cold cathodes until heated by the electron stream after starting (instant start or slimline).[AIR512B-91]

hot corrosion The corrosion at high temperatures as a result of the reduction of protective oxide coatings and scales and the subsequent accelerated oxidation.[NASA]

hot dip A method of coating whereby the item to be coated is immersed in a molten bath of the coating material. [ARP1931-87]

hot dip galvanizing A process for rust-proofing iron and steel products by the application of a coating of metallic zinc.

hot dipping A process for coating parts by briefly immersing them in a molten metal bath, then withdrawing them and allowing the metal to solidify and cool.

hot isostatic pressing A thermomechanical process for forming metal-powder compacts or ceramic shapes by use of isostatically applied gas pressure in order to achieve high density in the treated material.[NASA]

HOTOL launch vehicle A British unmanned horizontal takeoff and landing single-stage-to-orbit launch vehicle. Later launches will be manned.[NASA]

hot start An engine start during which allowable engine turbine gas temperature limits are exceeded.[ARP906A-91]

hot-wire instrument A measuring device that depends on the heating reaction of a wire carrying a current for its operation.

housing The outer component of an assembly designed to enclose, support, and/or protect the internal mechanism. [ARP480A-87]

housing, electrical connector The portion of a connector into which the insert is assembled. Also called shell. [ARP914A-79]

housekeeping Administrative or overhead operations or functions that are necessary in order to maintain control of a situation; for example, for a computer program, housekeeping involves the setting up of constants and variables to be used in the program; synonymous with "red tape."

housing A protective enclosure or case.

hover Maintenance of a relatively stable position over a point on the surface at a particular height above that surface; a capability of helicopters, hovercraft, hummingbirds, and many insects. [ARP4107-88]

hover check Term used to indicate that a helicopter/VTOL aircraft requires a stabilized hover to conduct a performance/power check prior to hover taxi, air taxi, or takeoff. The altitude of the hover will vary based on the purpose of the check.[ARP4107-88]

hover taxi Term describing a helicopter/VTOL aircraft movement conducted above the surface and in ground effect at airspeeds less than approximately 20 knots. The actual height may vary, and some helicopters may require hover taxi above 25 ft AGL to reduce ground-effect turbulence or provide clearance for cargo slingloads.[ARP4107-88]

HPLC See high-performance liquid chromatography.

HSDB High speed data bus.[AIR4271-89]

HSI Horizontal Situation Indicator.

HSRB High speed ring bus.[AIR4271-89]

HTPB propellant Solid rocket propellant containing hydroxyl terminated polybutadiene as bonding material.

[NASA]

Hubble constant The rate at which the velocity of recession of the galaxies increases with distance.[NASA]

HUD Acronym for Head Up Display. A method of presenting images to the pilot of an aircraft while he/she is looking forward through the windscreen. These images are generated from a device out of the pilot's field of view and reflected from a transparent surface in front of the pilot. Thus, the pilot looks through this transparent surface and sees the information superimposed on his/her vision of the real world.[ARP 4107-88]

hue The characteristic of color that determines whether the color is basically yellow, blue, red, etc.

Huggenberger tensometer A magnifying extensometer that employs a compound lever system to intensify changes taking place in a 10 to 20 mm gage length about 1200 times.

hum 1. Electrical disturbance at the power supply frequency or harmonics thereof.[NASA] 2. An undesirable by-product in an alternating current power supply.

human-computer interface. See man-computer interface.[NASA]

human engineering See human factor engineering.

human factors The study of the physical, physiological, psychological, psychosocial, and pathological variables which affect humans' performance. [ARP4107-88]

human-factors engineering 1. Application of the knowledge of human factors to the design of devices, systems, and environments to optimize the safety, efficiency, and the general well-being of the persons who interact with them.-[ARP4107-88] 2. Application of information on physical and psychological characteristics of man to the design of devices and systems for human use. [NASA]

humidification Artificially increasing the moisture content of a gas.

humidistat An instrument for measuring and controlling relative humidity.

humidity A measure of the water vapor content of the air.[AIR1335-75]

humidity, absolute The moisture content of air on a mass or volumetric basis.

humidity element The part of a hygrometer that senses the amount of water vapor in the atmosphere.

humidity, relative The moisture content of air relative to the maximum that the air can contain at the same pressure and temperature.

humidity test A corrosion test for comparing relative resistance of specimens to a high humidity environment at constant temperature.

hung start The operation of starting a gas turbine engine wherein the combined torque output of the engine and starter are insufficient to provide positive acceleration during an attempt to start. [ARP906A-91]

Huygens principle A very general principle applying to all forms of wave motion which states that every point on the instantaneous position of an advancing phase front (wave front) may be regarded as a source of secondary spherical wavelets. The position of the phase front a moment later is then determined as the envelope of all the secondary wavelets (ad infinitum).[NASA]

hybrid cable A multiconductor cable containing two or more types of conductors. Such a cable could include optical fiber and electrical conductors. [ARP1931-87]

hybrid computer 1. A computer for data processing using both analog representation and discrete representation of data. 2. A computing system using an analog computer and a digital computer working together.

hybrid T Series T and shunt T junctions

located at the same point in a waveguide and designed to restrict energy flow to specified channels.

hybrids (biology) See genetic engineering.

hybrid structure An assembly constructed of interconnected rigid and flexible structural shapes; designed to sustain dynamic, static, and other loads.[NASA]

hydraulic Referring to any device, operation or effect that uses pressure or flow of oil, water or any other liquid of low viscosity.

hydraulically actuated valves Those valves that utilize an incompressible fluid such as hydraulic oil or fuel for operation. This includes valves that are solenoid controlled, but hydraulically actuated.[ARP986-90]

hydraulic actuator See actuator.

hydraulic booster The use of hydraulic power actuation to reduce the pilot effort needed for control of a vehicle wherein the actuator output force is in direct proportion to the manual input force.[AIR1916-88]

hydraulic circuit A fluid-flow circuit that operates somewhat like an electric circuit.

hydraulic engineering A branch of civil engineering that deals with the design and construction of such structures as dams and other flood-control devices, sewers and sewage-disposal plants, water-driven electric power stations, and water treatment and distribution systems.

hydraulic fluid A light oil or other low-viscosity liquid used in a hydraulic circuit.

hydraulic gage A gage designed for service at extremely high pressure.

hydraulic power transfer unit A device which transfers power by mechanical interconnection but does not transfer fluid from one system to another. The power may be either unilateral or bilateral.[ARP578-69]

hydraulic system A closed system using the same fluid throughout the entire circuit and composed of components and elements to generate, transmit, and control hydraulic power as well as convert fluid energy to mechanical power.[ARP578-69]

hydride phase The phase formed in titanium when the hydrogen content exceeds the solubility limit. Hydrogen and, therefore, hydrides tend to accumulate at areas of high stresses.[AS1814-90]

hydroaeromechanics See aerodynamics.

hydroburst The procedure used for burst testing of tires in which water is used as the pressurizing medium rather than air. Avoids the explosive decompression effect of air.[AIR1489-88]

hydrocarbon A chemical compound of hydrogen and carbon.

hydrocracker A chemical reactor in which large hydrocarbon molecules are fractured in the presence of hydrogen.

hydrocracking Technique for the catalytic conversion of coal into liquid fuels.[NASA]

hydrodynamic coefficients The factors producing motions in floating objects in liquids.[NASA]

hydrodynamic ram effect The physical effect (force) transmitted to the walls of a liquid filled container by the action of a projectile penetrating the container and transferring its energy to the liquid as kinetic energy. The fluid, in turn, transfers this kinetic energy to the walls of the container, causing excessive structural damage.[NASA]

hydroelectricity Electric power produced by water power using water wheels, turbogenerators, or other conversion equipment.[NASA]

hydroelectric plant An electric power generating station where the power is produced by generators driven by hydraulic turbines.

hydrogen chloride laser See HCL laser.

hydrogen damage Any of several forms of metal failure caused by dissolved hydrogen, including blistering, internal void formation, and hydrogen-induced delayed cracking.

hydrogen deuterium oxide See heavy water.

hydrogen embrittlement A decrease in fracture strength of metals due to the incorporation of hydrogen in the metal lattice.[NASA]

hydrogen engine Internal combustion engine utilizing gaseous hydrogen as the fuel.[NASA]

hydrogen masers A stimulated emission device in which hydrogen gas provides an output signal with a high degree of stability and spectral purity.[NASA]

hydrogen metabolism The physical and chemical processes by which an organism transforms the complex hydrogen components of foodstuffs into simple hydrogen compounds by disassimilation and catabolism in the production of energy.[NASA]

hydrogen oxygen engine Engine using liquid hydrogen as fuel and liquid oxygen as oxidizer.[NASA]

hydrogen production Production of hydrogen for fuel purposes by photosynthetic, chemical, electrical, thermal, electrochemical, or other means.[NASA]

hydrokineter A device for recirculating or causing flow of water by the use of a jet of steam or water at higher pressure than the water caused to flow.

hydrology model Mathematical or physical representation by which the circulation distribution, and properties of the waters of the earth can be studied.[NASA]

hydromagnetics See magnetohydrodynamics.

hydromagnetic wave See magnetohydrodynamic wave.

hydromagnetism See magnetohydrodynamics.

hydromatic propeller A constant-speed hydraulically operated propeller.[ARP 4107-88]

hydromechanical logic Logic for mode switching or failure detection and correction, performed only with mechanical elements using hydraulic information in the form of hydraulic pressures or flows.[ARP1181A-85]

hydrometer 1. Instrument used for measuring the specific gravity of a liquid.[NASA] 2. An instrument for directly indicating density or specific gravity of a liquid.

hydrophone 1. Microphone suitable for use in water or other liquid.[NASA] 2. A transducer that reacts to water-borne sound waves.

hydroplane Tire hydroplaning is the name given to a pneumatic tire operating condition in which the water on a wet runway or surface is not displaced from the nominal tire-ground contact area by a rolling tire or by a moving but non-rotating (full skidding) tire at a rate fast enough to allow the tire to make contact with the ground surface over its complete nominal footprint area, as would be a case of operation on a dry ground surface. When hydroplaning occurs, the tire rides on a wedge or film of water over a part or all of its footprint area, depending upon conditions.[AIR1489-88]

hydroplaning, dynamic See hydroplane. The process of hydroplaning associated with movement of a body over water. To skim over water at high speed.[AIR1489-88]

hydroplaning speed Tire hydroplaning develops progressively from low speed through an intermediate range where the tire progressively loses contact with the runway and, finally, at sufficient forward speed, total hydroplaning develops.[AIR1489-88]

hydroplaning, total A hydroplaning

situation in which the total tire footprint (contact area) is supported by a film of water.[AIR1489-88]

hydroplaning, viscous Another term for total hydroplaning i.e., the tire contact footprint is supported by a film of viscous fluid (water).[AIR1489-88]

hydropneumatic Referring to a device operated by both liquid and gas power.

hydroponics Growing of plants in a nutrient with the mechanical support of an inert medium such as sand.[NASA]

hydropyrolysis A coal-to-liquid process in which bituminous coal, lignite, tars, sand and related materials are rapidly heated to 1000-1100 degrees K in pressurized hydrogen gasification reactors to generate pure methane.[NASA]

hydroscopic Refers to any material that easily absorbs and retains moisture.

hydro ski A type of landing gear designed to enable an aircraft to take off and land on water. (i.e., water ski)[AIR 1489-88]

hydrosphere (earth) See earth hydrosphere.

hydrothermal stress analysis The evaluation of the combined effects of temperature-humidity cycling.[NASA]

hydrostatic head The pressure created by a height of liquid above a given point.

hydrostatic-head gage A pressure gage that is unique from others in the graduation of scale; usually in feet.

hydrostatic test Determining the burst resistance or leak tightness of a fluid component or system by imposing internal pressure.

hydrox engine See hydrogen oxygen engine.

hygral properties The affinity of something for moisture.[NASA]

hygrometer An instrument for directly indicating humidity.

hygrometry Any process for determining the amount of moisture present in air or another gas.

hydrothermal system Energy system utilizing hot water from geysers, hot springs, solar heating, and other sources.[NASA]

hygrothermograph An instrument that records both temperature and humidity on the same chart.

hypemic hypoxia See hypoxia, hypemic.

hyperbolas Open curves with two branches, all points of which have a constant difference in distance from two fixed points called focuses.[NASA]

hyperbolic navigation Radio navigation in which a hyperbolic line of position is established by signals received from two stations at a constant time difference.[NASA]

hypercube multiprocessor Distributed-memory, message- passing multiprocessor designed to reduce the number of interconnections compared to the number of processors. Other simple geometries such as rings, meshes, or trees of processors can be embedded in hypercubes.[NASA]

hypergolic 1. A multi-phase propellant system which will spontaneously ignite upon mixing of the constituents.[AIR 913-89] 2. A term related to spontaneous ignition upon contact.

Hyperion One of the natural satellite of Saturn orbiting at a mean distance of 1,481,000 kilometers.[NASA]

hyperkinesia Excessive exercise, that is often accompanied by uncontrollable muscular movement.[NASA]

hypermodel A behavioral model of the analog-digital interface in a mixed-mode simulator.

hyperon In the classification of subatomic particles according to mass, the heaviest of such particles. Some large and highly unstable components of cosmic rays are hyperons.[NASA]

hyperoxia A condition in which the total oxygen content of the body is increased above that normally existing at sea level.[NASA]

hypersonic flow In aerodynamics, flow of a fluid over a body at speeds much greater than the sped of sound and in which the shock waves start at a finite distance from the surface of the body.[NASA]

hypersonic glider Unpowered vehicle, specifically reentry vehicle, designed to flow at hypersonic speeds.[NASA]

hypersonics That branch of aerodynamics that deals with hypersonic flow.[NASA]

hypervelocity Extremely high velocity. Applied by physicists to speeds approaching the speed of light, but generally implies speeds of in the order of satellite speed and greater.[NASA]

hyperventilation An abnormal increase in the volume of air breathed in and out of the lungs, which can occur subconsciously when a stressful situation is encountered in flight. As pilot "blows off" excessive carbon dioxide from his/her body, symptoms of lightheadedness, suffocation, drowsiness, tingling of the extremities, and coolness may occur. Continued hyperventilation may result in incoordination, disorientation, painful muscle spasms, and, finally, unconsciousness.[ARP4107-88]

hypocapnia Deficiency of carbon dioxide in the blood and body tissues, which may result in dizziness, confusion, and muscular cramps.[NASA]

hypoventilation A respiratory minute volume, or pulmonary ventilation that is less than normal. Also called underbreathing.[NASA]

hypoxemia The condition of reduction of the normal oxygen tension in the blood.[NASA]

hypoxia 1. A state of oxygen deficiency in the body sufficient to impair functions of the brain and other organs. See also anoxia.[ARP4107-88] 2. Oxygen want of deficiency; any state wherein a physiologically inadequate amount of oxygen is available to, or utilized by, tissue without respect to cause or degree.[NASA]

hypoxia, histotoxic Inability of tissues to accept oxygen.[ARP4107-88]

hypoxia, hypemic Inability of the blood to carry sufficient oxygen.[ARP4107-88]

hypoxia, hypoxic Insufficient inspired oxygen.[ARP4107-88]

hypoxia, stagnant Insufficient flow of blood to the brain.[ARP4107-88]

hypoxic hypoxia See hypoxia, hypoxic.

hypsometer An instrument that determines elevation above a reference plane (such as sea level) by measuring the boiling point of a liquid and from that measurement finding atmospheric pressure.

hysteresimeter A device for measuring a lagging effect related to physical change, such as the relationship between magnetizing force and magnetic induction.

hysteresis 1. The difference in actuation system input that yields the same output command level during a complete cycle of input command when cycled throughout the full range of travel. The cycling rate must be significantly below the control bandpass so that velocity error signals are not included in this parameter.[AIR1916-88] 2. Hysteresis is the difference of the output of the same input for increasing and decreasing inputs.[ARP1281A-81] 3. Any of several effects resembling a kind of internal friction, accompanied by the generation of heat within the substance affected. The delay of an indicator in registering a change in a parameter being measured.[NASA] 4. A phenomenon demonstrated by materials which make their behavior a function of the history of the environment to which they have been subjected. 5. The tendency of an instrument to give a different output for a given input, depending on whether the input resulted from an increase or decrease from the previous value. 6. The lagging in the response of a unit of a system behind an increase or a decrease in the strength of signal.

Hz See hertz.

I

I See current; also moment (definition 2.)

IACS Abbreviation for "International Annealed Copper Standard". A method of rating the conductivity of copper. [ARP1931-87]

IAE See integral absolute error.

IAF Initial Approach Fix.

Iapetus A satellite of Saturn orbiting at a mean distance of 3,562,000 kilometers. [NASA]

IAS Indicated Airspeed.

IC See integrated circuit.

ICAO International Civil Aviation Organization.

ice crystal Frozen supercooled water droplets. May exist in the form of snow crystals or ice nodules such as sleet. [AIR1667-89]

ice, freezing rain Precipitation in the form of large above-freezing water droplets which become supercooled and freeze upon contact with a below-freezing surface within a below-freezing air mass.[AIR1667-89]

ice, glaze or clear Transparent ice formed during flight in clouds by the slower freezing of supercooled water droplets. This is most likely to occur at ambient temperatures near freezing when the droplets may flow along the surface or remain liquid before freezing occurs. The ice formed during freezing rain is also an example of glaze ice. This type of ice may occur at conditions above the Ludlam Limit, reducing the apparent LWC.[AIR1667-89]

ice, glime A mixture of glaze and rime, generally with rough surfaces and runback ice.[AIR1667-89]

ice, hoarfrost Ice crystals deposited directly from water vapor onto surfaces that are below freezing.[AIR1667-89]

ice pellets Precipitation of transparent or translucent pellets of ice which are spherical or irregular, rarely conical, having a diameter of 1/5 of an inch or less. Ice pellets are subdivided into two main types: (a) frozen raindrops; or snowflakes which have largely melted and then refrozen; the freezing process usually taking place near the ground, (b) pellets of snow encased in a thin layer of ice, which has formed from the freezing either of droplets intercepted by the pellets or of water resulting from the partial melting of the pellets.[AIR 1335-75]

ice prisms A fall from the air of unbranched ice crystals, in the form of needles, columns or plates, often so tiny that they seem to be suspended in the air.[AIR1335-75]

ice, rime Opaque ice formed during flight in clouds by the rapid freezing of small supercooled water droplets producing a streamlined spear shape. This type of ice occurs below the Ludlam Limit.[AIR1667-89]

ice-warning indicator An instrument that detects the presence of ice on the aircraft or of icing conditions.[ARP4107-88]

icing The formation of any type of ice which adheres to a structure, especially on airfoils or other parts of an airframe. [ARP4107-88]

icing intensity The relationship of the icing intensity terms of trace, light, moderate, and severe to the corresponding cloud liquid water content (LWC). [AIR1667-89]

icing, light The rate of accumulation may create a problem if flight is prolonged in this environment (over 1 h). Occasional use of deicing/anti-icing equipment removes/prevents accumulation. It does not present a problem if the deicing/anti-icing equipment is used.[AIR1667-89]

icing, moderate The rate of accumulation is such that even short encounters become potentially hazardous and use of deicing/anti-icing equipment or diversion is necessary.[AIR1667-89]

icing, natural Icing that occurs during flight in a cloud formed by nature.[AIR

1667-89]

icing, severe The rate of accumulation is such that deicing/anti-icing equipment fails to reduce or control the hazard. Immediate diversion is necessary. [AIR1667-89]

icing, trace Ice becomes perceptible. Rate of accumulation slightly greater than rate of sublimation. It is not hazardous even though deicing/anti-icing equipment is not utilized, unless encountered for an extended period of time (over 1 h).[AIR1667-89]

ICL computers Family of British digital computers produced by International Corporation, Ltd.[NASA]

icon In data processing, a picture that represents a particular command that is used with a mouse.

I controller See controller, integral (reset) (I).

ICP See integrated circuit piezoelectric.

ideal elastic behavior A material characteristic, under given conditions, when the strain is a unique straight-line function of stress and is independent of previous stress history.

ideal gas 1. A gas which conforms to Boyle's law and has zero heat of the expansion (or also obeys Charles' law).[NASA] 2. A hypothetical gas characterized by its obeying precisely the equation for a perfect gas, PV=nRT.

idealized system See system, idealized.

ideal transducer A hypothetical passive transducer which produces the maximum possible output for a given input.

ideal value See value, ideal.

ident A request for a pilot to activate the aircraft transponder identification feature to help the controller identify an aircraft.[ARP4107-88]

ident feature The special feature in the air traffic control radar beacon system (ATCRBS) equipment used to immediately distinguish one displayed beacon target from other beacon targets.[ARP 4107-88]

identification Any marking applied to an item or its package for the purpose of engineering, manufacturing or inspection control.[AS478G-79]

identification plate See data plate.

identifier A symbol used in data processing whose purpose is to identify, indicate or name a body of data.

idle characters Control characters interchanged by a synchronized transmitter and receiver to maintain synchronization during non-data periods.

idle time 1. That part of available time during which computer hardware is not being used. Contrast with operating time. 2. That part of uptime in which no job can run because all jobs are halted or waiting for some external action such as I/O data transfer.

idling pressure Pressure required to maintain a system or component at the idling speed, or flow.[AIR1916-88]

ID synchronization A count contained in one word of a telemetry frame to indicate which subframe is being sampled at any given time.

ID synchronizer A method of PCM telemetry subframe recognition in which a specific word in the format activates a counter that identifies the number of the subframe word being received.

ID (Time Code) A three-numeral identification that can be inserted into time code manually in the place of the "day of the year" information.

IEC See International Electrotechnical Commission.

IEE 488 A parallel transmission standard for connecting instruments to a computer. An industry standard byte serial, bit parallel system handling 8-bit words.

IF Intermediate Fix. See also intermediate frequency.

IF amplifier An intermediate-frequency stage in a typical superhetrodyne radio receiver.

if and only if (IFF) A conditional statement implying that an action is to be taken or that a result is true if, and only if, stated prerequisite conditions are satisfied.

IFIP See International Federation for Information Processing.

I format In FORTRAN Iw indicates that w characters are to be converted as a decimal integer, e.g., 17 yields -24680 as input, +24680 internally, and -24680 as output.

IFR Instrument Flight Rules.

IFR aircraft/IFR flight An aircraft conducting flight in accordance with instrument flight rules.[ARP4107-88]

IFR conditions Weather conditions below the minimum for flight under visual flight rules.[ARP4107-88]

IFR departure procedure See IFR takeoff minimums and departure procedures.

IFR takeoff minimums and departure procedures FAR, Part 91, prescribes standard takeoff rules for certain civil aviation users. At some airports, obstructions or other factors require the establishment of nonstandard takeoff minimums, departure procedures, or both, to assist pilots in avoiding obstacles during a climb to the minimum en route altitude. Pilots should be familiar with the departure procedures and must assure that their aircraft can meet or exceed any climb gradients specified by the procedure.[ARP4107-88]

if-then See inclusion.

igniter 1. An explosive device specifically designed to initiate burning of a fuel mixture or a propellant.[AIR913-89] 2. Device used to begin combustion, such as a spark plug in a combustion chamber of a jet engine, or a squib used to ignite the fuel in a rocket.[NASA] 3. A device for initiating an explosion or combustion in a fuel-air mixture.

ignitiator The primary stimulus component in all explosive and pyrotechnic devices.[AIR913-89]

ignition The initiation of combustion.

ignition capable equipment and wiring Equipment and wiring which in its normal operating condition releases sufficient electrical or thermal energy to cause ignition of a specific hazardous atmosphere, under normal operating conditions.

ignition exciter An assembly of component parts which provides a means of changing low voltage alternating current or low voltage direct current to a condition suitable to provide (with or without additional devices) a spark discharge for ignition purposes.[ARP667-82]

ignition lag The time interval between spark discharge and fuel ignition in an internal combustion engine. Also known as ignition delay.

ignition lead, high tension A definite length of electrical cable having at least one end terminated in a single, common fitting. Its construction, materials used, etc. must be such as to conduct the discharge energy from a high tension ignition exciter (in excess of 5 kV) to a high tension spark igniter.[ARP667-82]

ignition lead, low tension A definite length of electrical cable having at least one end terminated in a single, common fitting. Its construction, materials used, etc. must be such as to conduct the discharge energy from a low tension ignition exciter (less than 5 kV) to a low tension spark igniter.[ARP667-82]

ignition period See trial for ignition.

ignition system 1. The system associated with rocket engines which provides for igniting the propellant.[AIR913-89] 2. That portion of the electrical subsystems of an internal combustion engine that produces a spark to ignite the fuel.

ignition temperature Lowest temperature of a fuel at which combustion becomes self-sustaining.

ignitor A flame or high energy spark which is utilized to ignite the fuel at the main burner.

ignitor intermittent An electric-ignited pilot which is automatically lighted each time there is a call for heat. It burns during the entire period that the main burner is firing.

ignitor interrupted An electric-ignited pilot which is automatically lighted each time there is a call for heat. The pilot fuel is cut off automatically at the end of the trial-for-ignition period of the main burner.

IGY (geophysical year) See International Geophysical Year.

ILI (In Limits) algorithm See compressor.

illuminance 1. At a point on a surface is the quotient of the luminous flux that is incident on an infinitesimal element of the surface containing the point under consideration, by the area of that surface element. As applied to electronic displays, illuminance is the metric of measurement of light from a source, such as the sun, that impinges upon a surface such as the face of an electronic display. The units of illuminance are the footcandle (fcd) and lux or lumen per square meter (lm/m^2)[ARP1782-89] 2. The total luminous flux received on a unit area of a given real or imaginary surface, expressed in such units as the footcandle, lux, or phot. Illuminance is analogous to irradiance, but is to be distinguished from the latter in that illuminance refers only to light and contains the luminous efficiency weighting factor necessitated by the nonlinear wavelength response of the human eye. [NASA] 3. Luminous flux per unit area over a uniformly illuminated surface.

illuminant C A source of illumination having an energy distribution similar to that adopted by the International Commission on Illumination as "average daylight." Normally produced by a tungsten lamp and an appropriate filter. Also similar to "North skylight" illumination.[ARP1161-91]

illuminants Light oil or coal compounds that readily burn with a luminous flame such as ethylene, propylene and benzene.

illuminated dial A transparent, semitransparent or nontransparent circular scale that is artificially illuminated.

illumination The density of luminous flux on a surface; it is equal to the flux divided by the area when the latter is uniformly illuminated.[ARP798-91]

illusion An erroneous perception of sensory input due to limitations of sensory receptors or the manner in which sensory information is presented, or both. The incorrect perception of an object(s). Often, the laws of physics explain the erroneous perception.[ARP4107-88]

illusions, kinesthetic An erroneous perception of somatosensory stimuli to the ligaments, muscles, or joints of the body. (a) G-adaptation illusion: An erroneous perception that motion has ceased after continued exposure to a sustained velocity. For example, movement in an elevator is only perceived at the beginning and end of the ascent or descent. (b) G-differential illusion: An erroneous perception of aircraft attitude based on "seat of the pants" sensations. For example, without other sensory inputs, a 30-deg-bank level turn feels the same as a 60-deg-bank turn.[ARP4107-88]

illusions, vection Visual illusions of motion, erroneously detected peripherally, in which a person perceives that he/she is moving when in fact an external object is moving. (a) Circularvection: An erroneous sensation of rotation due to movement detected in the visual field, especially peripherally. (b) Linearvection: An erroneous perception of linear movement due to motion detected in the visual field, especially peripherally.[ARP4107-88]

illusions, vestibular Erroneous perceptions of orienting stimuli to the semicircular ducts or otolith organs of the vestibular apparatus. (a) Coriolis Illusion: An erroneous sensation of rotation due to the movement of the head into a plane of angular or linear acceleration which induces fluid movement in the semicircular ducts. (b) Elevator Illusion: An erroneous sensation of pitch-up after level off from a steep descent, or pitch-down after level off from a steep climb, or when in turbulence. (c) Giant-Hand Illusion: The erroneous sensation that controls will not respond to inputs, even with seemingly great effort, when the source of resistance is in fact the operator himself/herself attempting to respond to conflicting sensory cues. (d) Leans: An illusion of angular displacement (bank) due to an undetected, subthreshold angular acceleration followed by a detected, transthreshold angular acceleration. (e) Somatogravic Illusion: An erroneous sensation of tilt in the vertical plane due to linear acceleration. This illusion is most common during rapid acceleration or deceleration. (f) Somatogyral Illusion: An erroneous perception that rotation has ceased because the semicircular canal fluid has stabilized after angular acceleration. The graveyard spin and graveyard spiral are results of the somatogyral illusion. [ARP4107-88]

illusions, visual Erroneous perceptions of stimuli to the visual system. (a) Autokinesis: An erroneous perception of movement of a light when stared at for a length of time in a dark visual field. (b) Chain-Link-Fence Illusion: The blending into the foreground of nearby objects when focusing on a distant object. (c) Empty Field Myopia: The tendency for the eyes to focus at a distance of about one meter when viewing a visually non-stimulating field. (d) Flicker Vertigo: The disruptive psychological effects of cyclic visual stimulation of about 10 to 15 cycles per second. (e) False Horizon Illusion: An illusion created by sloping cloud formations, an obscure horizon, a dark scene with ground lights and stars, or certain geometric patterns of ground light, which results in the pilot placing the aircraft in a dangerous attitude because of the perception of not being aligned properly with the actual horizon. (f) Geometric-Perspective Illusion: An erroneous perception of being nearer to, or farther away from, an object than one actually is, due to equating retinal image size to distance or angular displacement of familiar objects. For example, an 8000-ft runway viewed from 1000 ft up may appear the same size as a 10,000-ft runway viewed from 1500 ft up; another example is the tendency to flare high on a wider than usual runway.[ARP4107-88]

ILS (landing systems) See instrument landing systems.

IM Inner Marker.

image analysis Technique for understanding or quantification of digital data as presented in a two-dimensional format.[NASA]

image converter camera A camera which converts images from one wavelength region to another, typically from the infrared to the visible.

image digitizer A device which measures light intensity at each point in an image and generates a corresponding digital signal which indicates that intensity. It converts an analog image to a digital data set.

image impedances Of a transducer, the impedances that will simultaneously produce equal impedances in both directions at each of its inputs and outputs.

image intensifier A viewing system which functions as a light amplifier, taking a faint image and amplifying it so that it can be viewed more easily.

image inverter A fused fiber optic bundle which is permanently twisted during manufacture to turn the image it transmits upside down. The same can be done with conventional optics, but a fiber optic image inverter can do it in a distance of less than an inch.

image orthicon A camera tube whose output is generated using a low-velocity electron beam to scan the reverse side of a storage target containing an image produced by focusing the electron image from a photoemitting surface on it.

image processing Conversion of optical images into digital data form for storage and reconstruction by computer techniques.[NASA]

image reconstruction The reproduction of the original scene from data stored or transmitted after scanning by an electron beam. In reprography, the recreation of graphic images from digital data stored in a computer.[NASA]

image resolution In optics, a measure of the ability of an optical instrument to produce separable images of different points on an object.[NASA]

image rotation Mechanized or digital rotation of an image.[NASA]

IMC Instrument Meteorological Conditions. See also Institute of Measurement and Control.

immediate-access storage A device, usually consisting of an array of storage elements, in which stored information can be read in one microsecond or less.

immediate address Incorporating an operand, instead of merely the address of an operand, in the address portion of a digital computer instruction.

immediate mode In data processing, the ability to interrupt a program sequence to perform another function.

immersion length Of a thermometer, the distance along the thermometer body from the boundary of the medium whose temperature is being determined to the free end of the well, bulb or element, if unprotected.

immersion test This test is used to determine the relative abilities of greases to prevent corrosion on metal surfaces when the grease-coated metal object is immersed in water for an extended period of time.[S-5C,40-55]

immunity An inherent or induced electrochemical condition that enables a metal to resist attack by a corrosive solution.

immunoassay An assay that utilizes antigen-antibody reactions for the determination of biochemical substances. [NASA]

I_{mp} The numerical value of discharge current capability of a batter at maximum power delivery, partway through a simulated turbine engine start. The I_{mp} rating indicates relative engine-starting power capability of the battery, and is the minimum value of I_{mp} expected from a fully charged battery at 23 °C. For engine start systems design purposes, the use of a straight line approximation between a terminal voltage of 1.2N volts at zero load current and 0.6N volts at I_{mp} Amps, tends to be slightly conservative in the first few seconds of discharge toward engine light-off.[AS8033-88]

impact acceleration The acceleration generated by very sudden starts or stops of a vehicle. The term is usually applied in the context of physiological acceleration.[NASA]

impact angle The angle at which an aircraft or object strikes the terrain relative to the slope of the impact site terrain.[ARP4107-88]

impacted airport An airport affected by adverse weather or runway conditions, equipment failure, personnel shortages, or other phenomena which impair its ability to accommodate its normal flow of aircraft traffic.[ARP 4107-88]

impact deceleration See deceleration;

impact acceleration.

impact extrusion See cold extrusion.

impact fusion The conversion of the kinetic energy of a fast moving, initially stationary, macroparticle projectile into the internal energy of fusile material using a particle accelerator. Impact fusion is generally an inertial confinement fusion concept.[NASA]

impact ice Ice which forms when snow, sleet, or supercooled water droplets impinge upon aircraft surfaces, which are at or below freezing temperature.[ARP 4107-88]

impact melts Molten material resulting from hypervelocity impact.[NASA]

impact modulator An amplifier which utilizes the control of the intensity of two directly opposed, impacting power jets thereby controlling the position of the impact plane to modulate the output.[ARP993A-69]

impact pressure The pressure a moving stream of fluid produces against a surface which brings part of the moving stream abruptly to rest; it is approximately equal to the stagnation pressure for subsonic flow in the fluid medium.

impact strength 1. A test for determining the resistance of an insulating material or system to damage caused by impacting with a given weight dropped from a given distance in a controlled environment.[ARP1931-87] 2. The amount of energy required to fracture a material. The type of specimen and the testing conditions affect the values and therefore should be specified. [NASA] 3. A material property that indicates its ability to resist breaking under extremely rapid loading, usually expressed as energy absorbed during fracture.

impact temperature The temperature of a gas, after impact with a solid body which converts some of the kinetic energy of the gas to heat and thus raises the gas temperature above ambient.

impact tube A small diameter tube, immersed in a fluid, and oriented so that the fluid stream impinges normally on its open end.

impedance (Z) 1. The total opposition that a circuit offers to the flow of alternating current or to any other varying current at a particular frequency. It is a combination of resistance and reactance, measured in ohms.[ARP1931-87] 2. The complex ratio of a forcelike parameter to a related velocity-like parameter—for instance, force to velocity, pressure volume velocity, electric voltage to current, temperature to heat flow, or electric field strength to magnetic field strength.

impedance bridge A four-arm bridge circuit in which one or more of the arms have reactive components instead of purely resistive components; an impedance bridge must be excited by an a-c signal to yield complete analysis of the unknown bridge element.

impeller 1. Device that imparts motion to a fluid; specifically in centrifugal compressors, rotary disks which, faced on one or both sides with radial vanes, accelerate the incoming fluid outward into diffusers.[NASA] 2. As applied to pulverized coal burners, a round metal device located at the discharge of the coal nozzle in circular type burners, to deflect the fuel and primary air into the secondary air stream. As applied to oil burners, same as diffuser. 3. The driven portion of a centrifugal pump or blower.

impingement 1. The striking of moving matter, such as the flow of steam, water, gas or solids, against similar or other matter. 2. A method of removing entrained liquid droplets from a gas stream by allowing the stream to collide with a baffle plate.

impingement attack A form of accelerated corrosion in which a moving corrosive liquid erodes a protective surface layer, thus exposing the underlying metal to renewed attack.

impingement starting A starting method wherein nozzle(s) are built into the engine such that the starting fluid works directly on one of the engine turbine or compressor rotors.[ARP906A-91]

implication See inclusion.

implosion The rapid inward collapsing of the walls of vacuum systems or devices as the result of failure of the walls to sustain the ambient pressure.[NASA]

impregnated bit A diamond cutting tool made of fragmented bort or screened whole diamonds in a sintered powder-metal matrix.

improbable For airworthiness purposes, the likelihood of a failure that is equal to or less than once in 1,000,000 flight hours (10^{-9}).[ARP4107-88]

improvement maintenance Efforts to reduce or eliminate the need for maintenance. Reliability engineering efforts should emphasize elimination of failures that require maintenance. Includes modification, retrofit, redesign, or change-order.

impulse The integral of the thrust with respect to time over the entire burning period.[AIR913-89]

impulse excitation A method of producing oscillations in which the duration of stimulus is relatively short in relation to the duration of oscillation.

impulse line The conduit that transfers the pressure signal from the process to the measuring instrument.

impulse pressure A rapidly occurring pressure rise, peaking at a prescribed multiple of the nominal or operating pressure. After the impulse peak the pressure trace follows a prescribed curve, with a hold at nominal and zero pressure during one impulse pressure cycle.[MA2005-88]

impulses The products of the forces and the times during which the forces are applied.[NASA]

impulse (specific) The impulse per unit weight of propellant consumed measured under standard gravity conditions.[AIR913-89]

impulse strength The voltage breakdown of insulation under voltage surges on the order of microseconds in duration.[ARP1931-87]

impulse (total) The integral of the thrust with respect to time over the entire burning period.[AIR913-89]

impulse-type telemetering Employing intermittent electrical impulses to transmit instrument readings to remote locations.

IMS See International Magnetospheric Study.

inaccessible area An area is considered inaccessible if entry may be dangerous without special controls to enter.

inaccuracy See error.

inactive See physical condition.

inattention See attention, anomalies of inattention.

incandescence 1. Emission of light due to high temperature of the emitting material. Any other emission of light is called luminescence.[NASA] 2. Spontaneous radiation of light energy from a hot object.

incandescent filament lamp In general, incandescent filament lamps used in aircraft cabin lighting comprise a range from 0.3 watt, 0.03 candela to 36 watt, 50 candelas. They are used in every facet of cabin lighting, including indications, signs, area, reading and decorative. Their efficacy of light production ranges from approximately 1.2 lumens per watt to 15 lumens per watt, depending on size, wattage and design life.[AIR512B-91]

incandescent lamp A light source consisting of a glass bulb containing a filament electrically maintained at incandescence.[ARP798-91]

inches water gage ("w.g.) Usual term

for expressing a measurement of relatively low pressures or differentials by means of a U-tube. One inch w.g. equals 5.2 lb per sq ft or 0.036 lb per sq in.

incidence Partial coincidence, as a circle and a tangent line. The impingement of a ray on a surface.[NASA]

incident An occurrence other than an aircraft accident, associated with the operation of an aircraft, which adversely affects or could affect the safety of operations.[ARP4107-88]

incident wave A wave in a given medium that impinges on a discontinuity or a medium of different propagation characteristics.

incipient failure A functional status or condition which is existing at the beginning of a failure of engine, engine component of subsystem.[ARP1587-81]

incipient skid The point of wheel instability where brake torque exceeds resisting tire-runway friction torque an the wheel, therefore, begins a deceleration condition, which if continued, would result in abrupt wheel lock-up. [AIR1489-88]

in-circuit emulcation (ICE) A development aid for testing the software in computer hardware. It involves an umbilical link between a development system and the target hardware being plugged into the microprocessor socket.

inclination The angle between the plane of an orbit and the reference plane. The equator is the reference plane for geocentric orbits and the elliptic is the reference plane for heliocentric orbits. Also the magnetic dip.[NASA]

inclination error See error, inclination.

inclined-tube manometer A glass-tube manometer having one leg inclined from the vertical to give more precise readings.

inclinometer 1. An instrument for determining the angle of the earth's magnetic field vector from the horizontal. 2. A device for finding the direction of the earth's magnetic field with respect to the horizon. 3. An instrument on a ship which indicates the angular deviation of the ships attitude to the true vertical.

inclusion 1. Non-metallic particles inherent in the material when it was made. These particles may be isolated or distributed in the form of longitudinal stringers.[MA1568-87] 2. Inclusions are nonmetallic impurities such as slag, oxide, and sulphides which were present in the original ingot.[AS3071A-77] 3. Foreign matter trapped within the glass substrate or the surface by a coating.[ARP924-91] 4. A logic operator having the property that if P is a statement and Q is a statement, then P inclusion Q is false if P is true and Q is false, true if P is false, and true if both statements are true. P inclusion Q is often represented by P>Q. Synonymous with if-then and implication.

inclusive OR See OR.

incoherent fiber optics A bundle of fibers in which the fibers are randomly arranged at each end. The pattern may be truly random to achieve uniform illumination, or the manufacturer may simply not bother to align individual fibers. In either case, the fiber bundle cannot transmit an image along its length.

incoherent scatter radar Radar used in the study of the ionosphere, thermosphere, etc.[NASA]

incomplete combustion The partial oxidation of the combustible constituents of a fuel.

incompressible Liquids are referred to as being incompressible since their change in volume due to pressure is negligible.

incompressible flow Fluid flow under conditions of constant density.

Inconel A series of International Nickel Co. high-nickel, chromium and iron alloys characterized by inertness to certain corrosive fluids.

increased safety A type of protection by which measures are applied so as to prevent with a higher degree of security the possibility of excessive temperatures and of the occurrence of sparks in the interior and on the external parts of electrical apparatus which does not produce them in normal service and which is intended for use in hazardous locations defined by the IEC as Zone 1.

increaser A pipe fitting identical to a reducer except specifically referred to for enlargements in the direction of flow.

increment The specific amount in which a variable is changed.

incremental See incremental representation.

incremental backup A computer routine that copies only those files that have not yet been backed up.

incremental compiler Computer software that compiles programs as they are entered into a computer rather than compiling a program upon completion.

incremental cost The cost of the next increment of output from a process.

incremental encoder An electronic or electromechanical device which produces a coded digital output based on the amount of movement from an arbitrary starting position; the output for any given position with respect to a fixed point of reference is not unique.

incremental feedback In numerical control, assignment of a value for any given position of machine slide or actuating member based on its last previous stationary position.

incremental plotter A discrete X-Y plotter.

incremental representation A method of representing a variable in which changes in the value of the variables are represented, rather than the values themselves.

indentor See crimp indentor.

independent conformity See conformity, independent.

independent linearity See linearity, independent.

independent variable 1. Any of the variables of a problem, chosen according to convenience, which may arbitrarily be specified, and which then determines the other or dependent variables of the problem. The independent variable is often called the coordinate, particularly in problems involving motion in space. Dependent and independent variables can be interchanged, e.g., height and pressure.[NASA] 2. A process or control-system parameter that can change only due to external stimulus. 3. A parameter whose variations, intentional or unintentional, induce changes in other parameters according to predetermined relationships.

index 1. An ordered reference list of the contents of a computer file or document, together with keys or reference notations for identification or location of those contents. 2. To prepare a list as in 1. 3. A symbol or number used to identify a particular quantity in an array of similar quantities, for example, the terms of an array represented by X(1), X(2),..., 100 respectively. 4. Pertaining to an index register. 5. To move a machine part to a predetermined position, or by a predetermined amount, on a quantized scale.

index address modification (indexing) See address modification.

indexed address 1. An address in a computer instruction that indicates a location where the address of the reference operand is to be found. In some computers, the machine address indicated can itself be indirect. Such multiple levels of addressing are terminated either by prior control or by a termination symbol. Synonymous with second-level address. 2. An address that is to be modified or has been modified by an index register or similar device. Synonymous with variable address.

indexed addressing A method of addressing computer data whereby the address is obtained by adding the instruction operand to the address in the index register.

indexed sequential files Collection of related computer records stored on discs. The records are arranged in the same sequence as the key number and an index or table is used to define the actual location of these records on the disc.

index graduations The heaviest or longest division marks on a graduated scale, opposite the scale numerals.

indexing A technique of address modification often implemented by means of index registers.

index matching fluid A liquid with refractive index that matches that of the core or cladding of an optical fiber. It is used in coupling light into or out of optical fibers, and can help in suppressing reflections at glass surfaces.

index register A register which contains a quantity which may be used to modify addresses.

index-word A computer storage position or register, the contents of which may be used to modify automatically the effective address of any given instruction.

indicated airspeed The speed of an aircraft as shown on the aircraft's airspeed indicator. This is the speed used in pilot/controller communications under the general term "airspeed".[ARP 4107-88]

indicating gage Any measuring device whose output can be read visually but is not automatically transcribed on a chart or other permanent record.

indicating instrument See instrument, indicating.

indicating scale On a recording instrument, a scale that allows a recorded quantity to be simultaneously observed.

indication In nondestructive testing, any visible sign or instrument reading that must be interpreted to determine whether or not a flaw exists.

indicator 1. A mechanism for amplifying and measuring the displacement of a movable contact point, to be measuring a determination or variation from a standard determination. It consists essentially of a case with means for mounting the indicator, a spindle carrying the contact point, an amplifying mechanism, a point, and a graduated dial. May include accessories, and/or attachments.[ARP480A-87] 2. An instrument which graphically shows a value of a variable. 3. The pointer on a dial or scale that provides a visual readout of a measurement. 4. An instrument for diagramming pressure-volume changes during the working cycle of a positive-displacement compressor, engine or pump.

indicator card A chart for recording an indicator diagram.

indicator diagram A graphic representation of work done by or on the working fluid in a positive-displacement device such as a reciprocating engine.

indicator, gear position An instrument or indicator for the purpose of providing information to the pilot or crew regarding position of the landing gear. May be located at the pilot or crew station or remotely (i.e., pop up indicator projecting above upper surface of wing above the gear). Green lights are used for safe configurations and red lights for unsafe or in transition conditions and "up" and "down" to indicate up and locked or down and locked.[AIR 1489-88]

indicator melt time The elapsed time from the moment element melt time ends and arcing commences.[ARP1199A]

indicator tube An electron-beam tube in which useful information is conveyed by variations in beam cross section at a luminescent target.

indirect-acting recording instrument An instrument in which the output level of the primary detector is raised through intermediate mechanical, electric, electronic or photoelectric means to actuate the writing or marking device.

indirect address An address that specifies a computer storage location that contains either a direct address or another indirect address. Synonymous with multi-level address.

indirect addressing A method of addressing computer data in which the operand of the instruction is a location address which contains the address of the data.

indirect commands In data processing, commands to the system from previously recorded inputs, rather than from the operator terminal; the operator can call a sequence of indirect commands by file name.

indirect file In data processing, a file that contains commands that are processed sequentially, yet could have been entered interactively at a terminal.

indirectly controlled system See system, indirectly controlled.

indirectly controlled variable See variable, indirectly controlled.

indirectly heated cathode A cathode in a thermionic tube that is heated by an independent heating element.

indium-tin-oxide semiconductors See ITO (semiconductors).

individual wheel control One type of control used in antiskid system where each wheel utilizes its own valve and control circuit.[AIR1489-88]

indoor air pollution Pollution found in enclosed spaces often compounded by insufficient air mixing which intensifies the concentration of pollutants caused by outdoor and/or indoor sources. [NASA]

induced draft Airflow through a device such as a firebox or drying unit which is produced by placing a fan or suction jets in the exit duct.

induced draft fan A fan exhausting hot gases from the heat absorbing equipment.

induced drag In subsonic flow over a finite airfoil or other body, that part of the drag induced by lift. Induced drag is inversely proportional to the airspeed. The direct result of the aerodynamic force resulting from the downward velocity imparted to the air as the airfoil moves through the air.[ARP4107-88]

inductance (L) 1. The property of a circuit or circuit element that opposes a change in current flow. Inductance causes current changes to lag behind voltage changes. Inductance is measured in henrys.[ARP1931-87] 2. In an electrical circuit, the property that tends to oppose changes in current magnitude or direction. 3. In electromagnetic devices, generating electromotive force in a conductor by means of relative motion between the conductor and a magnetic field such that the conductor cuts magnetic lines of force.

inductance-type pressure transducer Any of several designs of pressure sensor where motion of the primary sensor element, such as a bourdon tube or diaphragm, is detected and measured by a variable-inductance element and measuring circuit.

induction heating Raising the temperature of an electrically conductive material by electromagnetically inducing eddy currents in the material.

induction instrument A type of meter whose indicated output is determined by the reaction between magnetic flux in fixed windings and flux in a moving coil where the two fluxes are induced by electric currents from different sources.

induction motor meter A type of meter resembling an induction motor, in which the rotor moves in direct relation to the reaction force between a magnetic field and currents induced in the rotor.

inductive bridge position transducer A device for measuring linear position by means of induction between a fixed member slightly longer than the limits of motion and a movable member approximately half as long; position is determined by selecting appropriate taps from the longer member that are connected in a successive decade with external inductors to form a bridge circuit, and relating the configuration that balances the bridge with actual position of the movable member; the chief advantage is the relatively high output voltage developed for a relatively small change in position.

inductive coupling 1. Crosstalk resulting from the action of the electromagnetic field of one conductor on the other. [ARP1931-87] 2. Using common or mutual inductance to cause signals in one circuit to vary in accordance with signals in another.

inductive plate position transducer A device for measuring rotary position by means of induction between a stationary and rotary plate, each having an etched winding projected onto a nonconductive surface, or for measuring linear motion by means of induction between a stator plate and a sliding member, each also having etched windings; advantages include eliminating wear and backlash as well as providing good resolution, often within 0.001 in. or less.

inductive system An ignition system in which the spark energy is primarily the result of a rapid variation in magnetic flux in an induction coil.[ARP667-82]

inductor A wire coil that will store energy in the form of a magnetic field.

industrial computer A computer used on-line in various areas of manufacturing including process industries (chemical, petroleum, etc.), numerical control, production lines, etc. See process computer and numerical control.

industrial computer language A computer language for industrial computers. A language used for programming computer control applications and system development, e.g., assembly language, FORTRAN, RTL, PROSPRO, BICEPS, and AUTRAN.

industrial controls A collective term for control instrumentation used in industry.

industrial engineering A branch of engineering that deals with the design and operation of integrated systems of personnel, equipment, materials and facilities.

Industrial Technology Institute (ITI) A non-profit organization founded by the University of Michigan dedicated to computer integrated manufacturing. ITI offers MAP conformance testing and certification.

industry standards A standard which specifies state-of-art, complete industry standards which will be useful to industry for the design and procurement of equipment.[AIR818C-91]

inelastic collision 1. Collision between two particles in which changes occur both in the internal energy of one or both of the particles and in the sums, before and after collision, of their kinetic energies.[NASA] 2. A collision between two or more bodies in which there is a net change in internal energy of at least one of the participating bodies and a net change in the sum of their kinetic energies.

inelastic stress A force acting on a solid and producing a deformation such that the original shape and the size of the solid are not restored after the force is removed.[NASA]

inert Description of condition of a rocket system or component thereof which contains no explosive, pyrotechnic or other reactive material.[AIR913-89]

inert atmosphere A gaseous medium that because of its lack of chemical reaction is used to enclose tests or equipment.[NASA]

inert gas See rare gas.

inert gaseous constituents Incombustible gases such as nitrogen which may be present in a fuel.

inertia 1. Resistance to acceleration or deceleration.[AIR1489-88] 2. Inherent resistance of a body to changes in its state of motion.

inertia bonding The joining of materials with friction and pressure.[NASA]

inertia equivalent In brake testing; the equivalent amount of the total air vehicle energy which goes to a single brake assembly. This equivalent energy is translated into dynamometer energy which is absorbed by the brake in the test process.[AIR1489-88]

inertial confinement fusion The process of using intense beams of heavy ions to convey the energy needed to compress and heat small pellets containing deuterium-tritium fuels to achieve ignition of the pellets.[NASA]

inertial force See inertia.

inertial fusion (reactor) Reactors in which pellet fusion is initiated by high energy sources including lasers.[NASA]

inertial guidance Guidance by means of the measurement and integration of acceleration from within the craft. [NASA]

inertial navigation Dead reckoning performed automatically by a device which gives a continuous indicational position by integration of accelerations since leaving a starting point.[NASA]

inertia load A load opposing any change in the state of motion and proportional to load acceleration and inertia.[AIR 1916-88]

inertia, moment of Tire, wheel, etc. (Symbol "I") of a body about an axis, ΣMR^2, where M is the mass of a particle of the body and R is its distance from the axis.[AIR1489-88]

inertia-type timer Any of several types of relay devices that incorporate extra weights or flywheels to achieve brief time delay in normal relay action by providing additional inertia to be overcome; delays are usually on the order of 80 to 120 milliseconds.

infant mortality The initial phase of the lifetime of a population of a particular component when failures occur as a result of manufacturing errors, etc. Infant mortalities are screened out by burn-in.[ARD50010-91]

infiltration Casing molten metal to be drawn into void spaces in a powder-metal compact, foamed-metal shape, or fiber-metal layup.

infinite frequency stiffness (frequency-independent dynamic stiffness) The stiffness associated with the output deflections of an active actuation system caused by externally applied loads where the frequency of the load disturbance is significantly above the bandpass of the actuation system. Since the system cannot actively react against high frequency load disturbances, this stiffness is identical to stiffness to ground at frequencies well beyond the system bandwidth.[AIR1916-88]

infinite loop In data processing, a routine that can be ended only by terminating the program.

infinity A point, line, or region, beyond measurable limits.[NASA]

inflation pressure The pressure (gage) to which the tire is to be inflated for a specific load and deflection, usually the rated pressure for a given service application.[AIR1489-88]

in-flight engine status In-flight indications (real time or near real time) of potential failures and warnings of a cautionary or advisory nature; e.g., high vibration. (Warnings to the cockpit should be those only to which the flight crew can react. Event detection and exceedance documentation should be provided).[ARP1587-81]

in-flight shutdown Cessation of engine operation during flight for any reason other than training procedure.[ARP 4107-88]

influence The change in an instrument's indicated value caused solely by a difference in value of a specified variable or condition from its reference value or condition when all other variables are held constant.

information Any facts or data which can be used, transferred, or communicated.[NASA]

information adaptive system The spaceborne portion of the NASA End-to-End Data System.[NASA]

information processing 1. The mental process of receiving incoming information from the environment, assessing its meaning, and deciding on an appropriate response. A general term for the presumed operations whereby the raw sense-data are refashioned into items of knowledge and utilized for decision making that may lead to action(s). Among these operations are perceptual organization, comparison with items stored in memory, and the making of decisions as to the response to be made. The mental processes from sensory input to evoked response.[ARP4107-88] 2. The organization and manipulation of data usually by a computer. See data processing.

information processing (biology) An approach to the study of perception, memory, language and/or thought that considers organisms to be complex systems that receive, transform, store and transmit information.[NASA]

information theory The mathematical theory concerned with information rate, channels, channel width, noise, and other factors that affect information transmission; initially developed for electrical communications, it is now applied to business systems and other phenomena that deal with information units and flow of information in networks.

infrared Any electromagnetic wave whose wavelength is 0.78 to 300 μm.

infrared absorption The taking up of energy from infrared radiation by a medium through which the radiation is passing.[NASA]

infrared absorption moisture detector An instrument for determining moisture content of a material such as sheet paper; moisture content can be read directly by determining the ratio of two beam intensities, one at a wavelength within the resonant-absorption band for water and the other at a wavelength just outside the band.

infrared imaging device Any device that receives infrared rays from an object and displays a visible image of the object.

infrared photometry Photometry in the infrared region.[NASA]

infrared radar Radar covering a range from the limit of the visible spectrum to the shortest microwaves.[NASA]

infrared radiation Electromagnetic radiation lying in the wavelength interval from 75 microns to an indefinite upper boundary sometimes arbitrarily set at 1000 microns (0.01 centimeter).[NASA]

infrared signatures The infrared spectral characteristics of an object or uniform land surface which uniquely defines it.[NASA]

infrared source (astronomy) Celestial bodies or astronomical regions emitting a large amount of radiation in the infrared portion of the electromagnetic spectrum.[NASA]

Infrared Space Observatory (ISO) An astronomical satellite observatory funded by ESA operating at wavelengths from 3 to 2000 microns. The observatory is comprised of a 60 cm Cassegrain telescope, a CCD infrared camera, two Michelson interferometers, and a photo-polarimeter.[NASA]

infrared spectroscopy A technique for determining the molecular species present in a material, and measuring their concentrations, by detecting the characteristic wavelengths at which the material absorbs infrared energy and measuring the relative drop in intensity associated with each absorption band.

infrared suppression The shielding and/or protection of aircraft engines and exhausts from heat-seeking missiles and/or detecting devices.[NASA]

infrared telescope Special optical instrument for astronomical observations in the range from one micron to one millimeter.[NASA]

infrared windows A frequency region in the infrared where there is good transmission of electromagnetic radiation through the atmosphere.[NASA]

infrasonic frequencies A sound-wave frequency lower than the audio-frequency range.

inhalation See respiration.

inherent Achievable under ideal conditions, generally derived by analysis, and potentially present in the design.[ARD 50010-91]

inherent damping Using mechanical hysteresis of materials such as cork or rubber to reduce vibrational amplitude.

inherent error The error in quantities that serve as initial conditions at the beginning of a step in a step-by-step set of operations. Thus, the error carried over from the previous operation from whatever source or cause.

inherent failure causes Failures due to integral operating factors of engine usage such as creep and low cycle, high cycle and thermal fatigue.[AIR1872-88]

inherent regulation See self-regulation.

inherent reliability A measure of reliability that includes only the effect of an item design and its application and assumes an ideal operation and support environment.[ARD50010-91]

inherent stability See stability, inherent.

inhibitor(s) 1. A material applied to the surface(s) of propellant grains to prevent burning on the coated surface(s).[AIR913-89] 2. Things that inhibit; specifically substances bonded, taped, or dip dried onto a solid propellant to restrict the burning surface and to give direction to the burning process.[NASA] 3. A substance which selectively retards a chemical action. An example in boiler work is the use of an inhibitor, when using acid to remove scale, to prevent the acid from attacking the boiler metals.

in-house maintenance 1. Maintenance performed by plant maintenance personnel. 2. Not contract maintenance.

initial approach fix (IAF) The fixes depicted in instrument approach procedure charts that identify the beginning of the initial approach segment(s).[ARP4107-88]

initial approach segment See segments of an instrument approach procedure.

initial isolation level of ambiguity The number of possible equipment/system subunits defined by the built-in-test, external tests equipment, or manual test procedure, which might contain the failed component. It is possible that a combination of built-in-test, external special purpose test equipment, and manual procedures may be necessary to effect isolation. For example, if an equipment test subsystem (built-in, external, manual) isolates a fault to one or two subunits, the level of ambiguity is equal to two; if it isolates it to one of three subunits the level of ambiguity is equal to three.[ARD50010-91]

initialize In data processing, to send a rest command to clear all previous or extraneous information, as when starting a new operating sequence.

initial set The start of a hardening reaction following water addition to a powdery material such as plaster or portland cement.

initial value problem See boundary value problem.

initiation The first action in the first element of an explosive train.[AIR913-89]

injection laser diode A semiconductor device in which lasing takes place within the P-N junction. Light is emitted from the diode edge.

injection molding A forming process in which a heat softened or plasticized material is forced from a cylinder into a relatively cool cavity which gives the product a desired shape. A similar process is used for forming solid propellants from quick cure ingredients.[NASA]

injection seal A seal accomplished by injecting sealant into holes, joggles, channels, grooves, and other voids caused by build-up of structure in the fuel tank boundaries. This seal is used to provide continuity where fillet seals are interrupted by the structure and also to fill cavities completely.[AIR4069-90]

injector 1. Device that propels fuel or propellant into a combustion chamber under pressure other than atmospheric. [NASA] 2. Any nozzle or nozzle-like device through which a fluid is forced into a chamber or passage.

ink A liquid or semisolid material consisting of a pigment or dye and a carrier, and used to produce a design or mark on a material such as paper or cloth, after which the carrier evaporates leaving behind a colored residue of pigment or dye.

ink-jet printer A printer that forms characters by shooting tiny dots of ink onto paper. See also dot matrix printer.

ink-vapor recording A type of electromechanical recording in which the trace is produced by depositing vaporized particles of ink directly on the chart paper.

inlet A passage or opening where fluid enters a conduit or chamber.

inlet airframe configuration Optimum location of engine inlet for various purposes.[NASA]

inlet box An enclosure at or near the entrance to a chamber or duct system for attaching a fan to the system.

inlet filter A fluid contamination filter located upstream of the skid control valve (or other hydraulic unit).[AIR 1489-88]

inlet pressure In connection with performance data on pumps, when not otherwise specified, the total static pressure measured in a standard testing chamber by a vacuum gage located near the inlet port.[NASA]

inlet pressure (supply pressure) Pressure at the inlet of a component.[AIR 1916-88]

inlet temperature 1. Fluid temperature at the plane of the inlet port.[AIR1916-88] 2. A location for measuring the temperature of fluids, particles, etc. entering a heat system, an engine, or other machine.[NASA]

inlet valve A valve for admitting the working fluid to the cylinder of a positive-displacement device such as a reciprocating pump or engine.

in line 1. Centered on an axis. 2. Having several features, components or units aligned with each other. 3. In a motor-driven device, having the motor shaft parallel to the device's driven shaft and approximately centered on each other.

inner marker (IM) inner marker beacon A marker beacon used with an ILS (CAT II) precision approach which is located between the middle marker and the approach end of the ILS runway, transmitting a radiation pattern keyed at six dots per second, and indicating to the pilot, both aurally and visually, that he/she is at the designated decision height (DH) (normally, 100 ft above touchdown zone elevation on the ILS CAT II approach).[ARP4107-88]

inner tube A rubber tube located within a pneumatic tire for the purpose of sealing in the air.[AIR1489-88]

in-plant system 1. A system whose parts, including remote terminals, are all situated in one building or localized area. 2. The term is also used for communication systems spanning several buildings and sometimes covering a large distance, but in which no common carrier facilities are used.

input 1. Signals taken in by an input interface as indicators of the condition of the process being controlled. 2. Data keyed into a computer or computer peripherals. See excitation or measurand.

input area An area of computer storage reserved for input. Synonymous with input block.

input block See input area.

input channel A channel for impressing a state on a device or logic element. See channel, input.

input counter See counter, input.

input device In data processing, the device or collective set of devices used for conveying data into another device.

input interface Any device that connects computer hardware or other equipment for the input of data.

input-output (I/O) 1. The interface to a unit that provides data or signals used or generated by that unit.[AS8034-82] 2. A general term for the equipment used to communicate with a computer and the data involved in the communications. Synonymous with I/O.

input-output control system (IOCS) A set of flexible routines that supervises the input and output operations of a computer at the detailed machine-language level.

input/output curve The graphic representation of actuation system output versus command input. This is usually a continuous plot throughout a complete cycle between plus and minus rated commands. The cycling rate must be significantly below control bandpass so that velocity error signal is not included in this parameter.[AIR1916-88]

input-output limited Pertaining to a computer system or condition in which the time for input and output operation exceeds other operations.

input-output (I/O) software That portion of the operational software which organizes efficient flow of data and messages to and from external equipment.

input-output (I/O) statement A statement that controls the transmission of information between the computer and the input/output units.

input ports In computer hardware, terminals for connection in external devices which input data to the computer.

input resistance See resistance, input.

input signal A signal applied to a device, element, or system. See signal, input.

input state The state occurring on a specified computer input channel.

input strobe (INSTRB) A signal that enters set-up data into registers.

input work queue A list of summary information of job-control statements maintained by the job scheduler, from which it selects the jobs and job steps to be processed.

inquiry A technique whereby the interrogation of the contents of a computer's storage may be initiated at a keyboard.

insensitive time See dead time.

insensitivity See sensitivity.

insert 1. Any design feature of a cast or molded component that is made separately and placed in the mold cavity prior to the casting or molding step. 2. A removable part of a die, mold or cutting tool.

insert, electrical connector The insulating element of a connector which supports and positions the contracts.[ARP914A-79]

insertion gain The ratio of the power delivered to the portion of a transmission system following a transducer to the power delivered to the same portion without the transducer in place.

insertion loss The loss in load power resulting from the insertion of a cable. It is expressed in decibels as the ratio of power received at the load before insertion to the power received at the load after insertion.[ARP1931-87]

insertion point Usually indicated by a computer cursor, the place where characters will appear when an operator starts typing.

insertion tool 1. An insertion tool is a mechanical device with a tip to hold a taper pin and force it into a taper pin receptacle.[ARP592-90] 2. A tool used to insert contacts into their retaining device.[ARP914A-79]

inside caliper A caliper having outward-turned feet on each leg for measuring inside dimensions.

inside diameter The maximum dimension across a cylindrical or spherical cavity. Ideally, this is a line passing through the exact center of the cavity and perpendicular to the cavity's inner surface.

inside gage 1. A fixed-dimension device for checking inside diameters. 2. The inside diameter of a bit, measured between opposing cutting points.

inside micrometer A micrometer caliper designed for measuring inside diameters and similar inside dimensions between opposing surfaces.

insolation In general, solar radiation received at the earth's surface. The rate at which direct solar radiation is incident upon a unit horizontal surface at any point on or above the surface of earth. Contracted from INcoming SOLar radiATION.[NASA]

inspection 1. An examination of an item against a specified standard.[ARD 50010-91] 2. A deliberate critical examination to determine whether or not an item meets established standards. Inspection may involve measuring dimensions, observing visible characteristics, or determining inherent properties of an object, but usually does not involve determining operating characteristics. The last is more properly termed testing.

inspection door A small door in the outer enclosure so that certain parts of the interior of the apparatus may be observed.

inspection, final Quality control inspector certification performed after the last in-plant fabrication or assembly operation and prior to preparation for shipment.[AS7200/1-90]

inspection hole A hole located in the conductor barrel that permits inspection to determine that the conductor is properly located before crimping and that the conductor is properly located after crimping thus ensuring a proper termination.[ARP914A-79]

inspection, nondestructive (NDT) A family of methods for investigating the quality, integrity, properties, and dimensions of materials and components, without damaging or impairing their serviceability through the use of dye penetrant, magnetic eddy current, ultrasonic, radiographic, infrared, etc. devices. Nondestructive testing (NDT) is widely used in aircraft structural inspection.[AIR1916-88]

inspection, periodic The periodic inspection is a thorough and close check of the overall aircraft. Each periodic inspection comprises all intermediate pre-flight, thru-flight and post-flight inspections. It is repeated at regular intervals of calender time or hours of operation.[AIR1916-88]

inspection, quality control A critical comparison of characteristics to standards required by the customer.[AS 7200/1-90]

inspection, sample The monitoring or withdrawal, or both, of random selected devices from assembly line or service to permit determination of their condition at predetermined progressive intervals.[AIR1916-88]

inspection tests Those tests performed upon a production article to determine acceptability prior to shipment. Also referred to as production acceptance tests.[ARP906A-91]

instability See also stability.

installation Putting equipment or software in place prior to commencing operation.

installation drawings Drawings which define the exterior size of the unit and include data on installation interfaces. [ARP906A-91]

installed Hose assemblies joined to other parts to form assemblies or components offered for acceptance against procurement contracts.[AS1933]

instantaneous frequency In an angle-modulated wave, the derivative of the angle with respect to time.

instantaneous trip (opening) "Instantaneous" indicates that delay is not purposely introduced into the action of the device.[ARP1199A-90]

instantaneous sampling Taking a series of readings of the instantaneous values of one or more wave parameters.

instantons Field configurations of Yang-Mills theory which are localized in space and time. These configurations are solutions of the Yang-Mills field equations in Euclidean space time which allow the transitions (tunneling) from one vacuum state to another.[NASA]

Institute of Measurement and Control A British professional organization.

instruction In data processing, a statement that specifies an operation and the values or locations of its operands. In that context, the term instruction is preferable to the terms command or order, which are sometimes used synonymously. Command should be reserved for electronic signals, and order should be reserved for sequence, interpolation and related usage.

instruction address An instruction's computer memory address. An asterisk is frequently used to designate this address.

instruction area 1. A part of computer storage allocated to receive and store the group of instructions to be executed. 2. The storage locations used to store the program.

instruction buffer An eight-bit byte buffer in the computer processor that is used to contain bytes of the instruction currently being decoded and to prefetch instructions in the instruction system.

instruction code See operation code.

instruction counter A counter that indicates the location of the next computer instruction to be interpreted.

instruction format The bits or characters of a computer instruction allocated to specific functions.

instruction register In data processing, a storage register which contains the address of the instruction.

instruction repertory 1. The set of instructions which a computing or data processing system is capable of performing. 2. The set of instructions which an automatic coding system assembles.

instruction set In computer software, the particular set of instructions that are implemented on a microcomputer.

instruction time The portion of an instruction cycle during which the computer control unit is analyzing the instruction and setting up to perform the indicated operation.

instrument 1. The word "instrument" shall be considered to mean the specific device when the device is totally self contained including mechanization and display (i.e. rate-of-climb instrument). [AIR818C-91] 2. A device for measuring the value of an observable attribute; the device may merely indicate the observed value, or it may also record or control the value. 3. Measuring, recording, controlling, and similar apparatus requiring the use of small to moderate amounts of electrical energy in normal operation.

instrumental analysis Any analytical procedure that uses an instrument to measure a value, detect the presence or absence of an attribute, or signal a change or end point in a process.

instrumentation amplifiers High precision amplifiers with high noise rejection capabilities.

instrumentation tape Analog magnetic tape, ungapped, for continuous data (as PCM or FM telemetry).

instrument correction A quantity added to, subtracted from, or multiplied into an instrument reading to compensate for inherent inaccuracy or degradation of instrument function.

instrument flight rules (IFR) Rules governing the procedures for conducting flight by reference to instruments. Also a term used by pilots and controllers to indicate a type of flight plan. [ARP4107-88]

instrument landing system (ILS) 1. A precision instrument approach system which normally consists of the following electronic components and visual aids defined under separate alphabetized headings: (a) localizer, (b) glide slope, (c) outer marker, (d) middle marker, (e) approach lights. See also airport lighting.[ARP4107-88] 2. A system which provides, in the aircraft, a display of the lateral, longitudinal, and vertical references necessary for a landing.[NASA]

instrument landing system (ILS) categories Instrument landing systems (ILS) procedures are classified as follows: (a) ILS Category I: An ILS approach procedure which provides for approach to a height above touchdown of not less than 200 ft and with runway visual range of not less than 1800 feet. (b) ILS Category II: An ILS approach procedure which provides for approach to a height above touchdown of not less than 100 ft and with runway visual range of not less than 1200 feet. (c) ILS Category III: This category subsumes the following: 1. IIIA: An ILS approach procedure which provides for approach without a decision height minimum and with runway visual range of not less than 700 feet. 2. IIIB: An ILS approach procedure which provides for approach without a decision height minimum and with runway visual range of not less than 150 feet. 3. IIIC: An ILS approach procedure which provides for approach without a decision height minimum and without runway visual range minimum.[ARP4107-88]

instrument loop diagram A loop diagram contains the information needed to understand the operation of the loop and also show all connections to facilitate instrument startup and maintenance of the instruments. The loop diagram must show the components and accessories of the instrument loop, highlighting special safety and other requirements.

instrument meteorological conditions (IMC) Meteorological (weather) conditions expressed in terms of visibility, distance from cloud, and ceilings that are less than the minima specified for visual meteorological conditions.[ARP 4107-88]

instrument, millivoltmeter type Operated and actuated by varying E.M.F. output of a thermocouple; the varying E.M.F. input to the instrument being obtained by temperature changes of the temperature sensing thermocouple.[AS413B-91]

instrument oil A special grade of lubricating oil for instruments and other delicate mechanisms. It is formulated to resist oxidation and gumming, to be compatible with electric insulation, and to inhibit metals from tarnishing.

instrument, ratiometer type Actuated by changes in electrical resistance of a temperature sensing electrical resistance element; the resistance changes being obtained by temperature changes of the temperature sensing resistance element.[AS413B-91]

instrument reading time The time lag between an actual change in an attribute and stable indication of that change on a continuous-reading instrument.

Instrument Society of America (ISA) A U.S. society of instrument and controls professionals.

instrument specification A detailed and exact statement of particulars, especially a statement prescribing performance, dimensions, construction, tolerances, bill of materials, features, and operating conditions.

instrument system An "instrument system" is a device which consists of more than one assembly or device (i.e. electronic display system).[AIR818C-91] See also instrumentation.

instrument torque The turning moment on an instrument's moving element produced directly or indirectly by the quantity being measured.

instrument transformer A precision transformer capable of reproducing a signal in a secondary circuit that is suitable for use in measuring, control or protective devices.

insulated terminal A terminal having its conductor barrel and insulation support, if any, covered with a dielectric material.[ARP914A-79]

insulation 1. Material having a high resistance to the flow of electric current, which is used to prevent leakage of current from a conductor.[ARP1931-87] 2. Insulation is nonconducting material used to separate a conductor or shield from other conductors and shields and from ground.[AS1198-82] 3. A material of low thermal conductivity used to reduce heat loss. 4. A material of specific electrical properties used to cover wire and electrical cable.

insulation crimp The physical reshaping of an insulation sleeve to close or compress around the wire insulation. [ARP914A-79]

insulation grip That portion of an insulation barrel which when closed or compressed around the conductor insulation makes contact with the insulation on the cable. See barrel, insulation. [ARP914A-79]

insulation piercing terminal A terminal having a barrel with a design that displays the wire insulation and makes contact with the conductor.[ARP914A-79]

insulation resistance 1. The ratio of applied voltage to the total current between two electrodes in contact with a specific insulation, usually expressed in megohms for 1000 feet.[ARP1931-87] 2. The insulation resistance of a wire or cable is the electrical resistance offered by its insulation to an impressed direct-current potential tending to produce a leakage of current through the insulation.[AS1198-82]

insulation support he portion of an insulation barrel which extends around but is not necessarily compressed or closed to the point of making contact with the cable insulation.[ARP914A-79]

insulator 1. A prefabricated item specifically designed to prevent any undesirable flow of energy between a conductor and/or other objects.[ARP 480A-87] 2. A material through which electrical current cannot flow.

insurance (contracts) Coverage by contract whereby one party undertakes to indemnify or guarantee another against loss by a specified contingency or peril. [NASA]

intake 1.An opening where a fluid enters a chamber or conduit; an inlet. 2. The amount of fluid entering through the opening.

integer A whole number signified by a binary "word."

integer programming 1. In operations research, a class of procedures for locating the maximum or minimum of a function subject to constraints, where some or all variables must have integer values. Contrast with convex programming, dynamic programming, linear programming, mathematical programming. 2. Loosely discrete programming.

integers Whole numbers; numbers that are not a fraction.[NASA]

integral 1. This control action will cause the output signal to change according to the summation of the input signal values sampled at regular intervals up to the present time. 2. Mathematically it is the reciprocal of reset.

integral absolute error (IAE) A measure of controller error defined by the integral of the absolute value of a time-dependent error function; used in tuning automatic controllers to respond properly to process transients. See also integral time absolute error.

integral action A type of controller function where the output (control) signal or action is a time integral of the input (sensor) signal.

integral action time constant See time constant, integral action.

integral blower A blower built as an integral part of a device to supply air thereto.

integral-blower burner A burner of which the blower is an integral part.

integral control 1. A system that uses integration in the control loop elements to provide an output in response to an error signal. This is referred to as a Type N system, where N is the order of the integration.[AIR1916-88] 2. Form of control action that returns the value of the controlled variable to the set point when sustained offset occurs without this action. Also called reset control.

integral control action (reset) Control action in which the output is proportional to the time integral of the error input, i.e., the rate of change of output is proportional to the error input. See control action, integral.

integral controller See controller, integral (reset).

integral control mode A controller mode in which the controller output increases at a rate proportional to the controlled variable error. Thus, the controller output is the integral of the error overtime with a gain factor called the integral gain.

integral flange A flange on a length of pipe, a nozzle or a pressure vessel which is cast or forged with the item itself, or is permanently attached to it by welding.

integral fuel tank A load-carrying structure of an aircraft absolutely sealed to provide for fuel containment. It exists as a cavity in a wing or in the fuselage or both.[AIR4069-90]

integrally lighted instrument An instrument where the lighting system is designed such that the light sources are contained within the instrument enclosure, thus not relying on any outside source of illumination; the instrument may employ reflected or transilluminated light.[ARP924-91]

integral orifice A differential pressure measuring technique for small flow rates in which the fluid flows through a miniature orifice plate integral with a special flow fitting.

integral rocket ramjets A combination of a solid propellant rocket and a ramjet which uses the empty booster case as a ramjet combustor.[NASA]

integrals Of or pertaining to an integer.

[NASA]

integral time absolute error (ITAE) A measure of controller error defined by the integral of the product of time and the absolute value of a time-dependent error function; whereas the absolute value prevents opposite excursions in the process variable from canceling each other, the multiplication by time places a more severe penalty on sustained transients. See also integral absolute error.

integrated actuator package (IAP) An actuator package wherein a power converter is contained within the package. Usually a servoactuator with an electric-motor-driven pump and fluid reservoir in a single LRU.[AIR1916-88]

integrated circuit (IC) A complete electronic circuit containing active and passive elements fabricated and assembled as a single unit, usually as a single piece of semiconducting material, resulting in an assembly that cannot be disassembled without destroying it.

integrated circuit piezoelectric (ICP) A type of pressure-sensitive sensor that combines a piezoelectric element with isolation amplifier and signal conditioning microelectronics inside the sensor housing so that the output signal can be transmitted over ordinary two-wire cable instead of special low-noise cable.

integrated energy systems Community systems for energy generation and distribution.[NASA]

integrated optics Thin film devices containing tiny lenses, prisms, and switches to transmit very thin laser beams, which serve the same purposes as the manipulation of electrons in thin film devices of integrated electronics.[NASA]

integrated software A computer program that combines several functions for ease of use.

Integrated Systems Digital Network (ISDN) A suite of protocols being defined by CCITT to provide voice and data services over wide area networks (WANs).

integrating accelerometer A device that measures acceleration of an object, and converts the measurement to an output signal proportional to speed or distance traveled.

integrating ADC A type of analog-to-digital converter where the analog input is integrated over a specific time with the advantages of high resolution, noise rejection and linearity.

integrating control A control system in which rate of change of valve position is proportional to temperature error, or valve position is the integral of temperature error.[ARP89C-70]

integrating extensions An integrator which derives its input from the motion of the float can be installed within the extension housing.

integrating frequency meter A master frequency meter for an electric power system that measures the actual number of cycles of alternating voltage for comparison with the theoretical number of cycles for the same time at the prescribed frequency.

integrating meter 1. A totalizing meter, such as for electric energy consumed. 2. An instrument whose output is proportional to the single (or higher order) integral of the quantity measured.

integrating network A transducer circuit whose output waveform is a time integral of its input waveform.

integrating sphere A sphere used in optical measurements which is intended to integrate the input light over the output aperture to provide uniform illumination.

integrating temperature compensated tachometer generator A generator used in computing applications requiring integration of a variable with respect to time. This type of tachometer generator is characterized by very small deviations of the output voltage as a function of temperature and a minimum warm-up time. Temperature control and compensation networks are usually integral parts of integrating generators. [ARP667-82]

integrator(s) 1. A device whose output is proportional to the integral of an input signal. 2. In digital computers, a device for accomplishing a numeric approximation of the mathematical process of integration.[AIR1489-88] 3. A device which continually totalizes or adds up the value of a quantity for a given time. 4. A device whose output is proportional to the integral of the input variable with respect to time.

integrity In data processing, a word to describe data that has not been corrupted.

intelligence In data processing, the processing capability of a computer.

intelligent terminal A computer terminal with some local processing capability.

intensifier A device which converts an input pressure to a higher output pressure which is proportional to its input pressure.[ARP243B-65]

intensifying screen A sheet of material placed in contact with radiographic film that undergoes secondary fluorescence when struck with x-rays or gamma rays, thereby increasing image density for a given exposure.

intensity 1. In general, the degree or amount, usually expressed by the elemental time rate or spatial distribution of some condition or physical quantity, such as electric field, sound, magnetism, etc. With respect to electromagnetic radiation, a measure of the radiant flux per unit solid angle emanating from some source. Frequently, it is desirable to specify this as radiant intensity in order to distinguish it clearly from luminous intensity.[NASA] 2. The amount of light incident per unit area. For human viewing of visible light, the usual term is illuminance; for electromagnetic radiation in general, the term is radiant flux.

intensity level The amplitude of a sound wave, commonly measured in decibels.

interaction A phenomena, characteristic of a multivariable process, in which the effect of a manipulative variable change in one control loop not only affects its own controlled variable, but also the controlled variable in another loop. In distillation the primary consideration is interaction between the overhead composition control loop and the bottoms composition control loop.

interaction analysis A technique used in determining the pairing of manipulative and controlled variables in a control loop.

interactive In data processing, a technique of user/system communication in which the operating system immediately acknowledges and sets upon a request entered by the user at a terminal; compare with batch.

interactive computing See conversational mode.

interactive control The sending of multiple commands that are selected on the basis of data received from an experiment in real time.[NASA]

interactive graphics See computer graphics.

interblock gap Blank space on a computer storage medium between two adjacent blocks of data.

intercalation Production of layer type semiconducting as well as other conducting materials. Also called synthetic metals.[NASA]

intercept method A method for estimating the quantity of particles or number of grains within a unit area of a microscopic image by counting the number intercepted by a series of straight lines through the image. This is one of the standard methods of determining grain size of a polycrystalline metal.

interchange Removing the item that is to be replaced, and installing the replacement item.[ARD50010-91]

interchangeability 1. That quality which allows an assembly or part to substitute or be substituted for another of the same part number designation and meet all physical, functional, and structural requirements and be installed by the application of the attaching means only. This specifically excludes trimming, cutting, filing, drilling, reaming, shimming and forming during installation. No tools other than those normally available to service mechanics are require for installation of the item. No operations or alterations except designed adjustments are required on supporting and surrounding structure in order to install the item.[AS1426-80] 2. The ability to interchange, without restriction, like equipments or portions thereof in manufacture, maintenance, or operation.[ARD50010-91]

interchangeable item One, which (a) possesses such functional and physical characteristics as to be equivalent in performance, reliability, and maintainability, to another item of similar or identical purposes; and (b) is capable of being exchanged for the other item 1) without selection for fit or performance, and 2) without alteration of the items themselves or of adjoining items, except for adjustment.[ARD50010-91]

interchangeable, one way Introduces a new item and places restrictions on the use of the old item. The new item may be used in place of either the old or the new, but the old item can only be used in an application where it has previously been installed. The new item is considered to supersede the old. For example, the interchangeability characteristics of items A and B where B may freely be used in applications where A is specified, but A must not be used in applications where B is specified. [ARD50010-91]

intercom A voice-communication system among different stations in an aircraft. [ARP4107-88]

interconnect wire A type of wire used in general purpose application in aerospace vehicles for interconnecting individual electrical or electronic devices. [ARP1931-87]

intercooler A heat exchanger in the path of fluid flow between stages of a compressor to cool the fluid and allow it to be further compressed at lower power demand.

interelectrode capacitance 1. The capacitance between electrodes of a vacuum tube. 2. A capacitance determined by measuring the short-circuit transfer admittance between two electrodes.

interface 1. A shared boundary.[ARD 50020-91] 2. That part of a sensor system that includes the receiver plus circuits to reformat the signal (electrically) for the end user plus an emitter to generate a transducer driving signal.[ARD 50020-91] 3. Those parts of a data bus system where optical/electrical conversion takes place.[ARD50020-91] 4. The physical, functional, or procedural relationships established as a basis for division of responsibilities between two or more independent design, manufacturing or test activities as related to the hardware for which they are jointly responsible.[AS1426-80] 5. The two surfaces of mating connectors which face each other when mated.[ARP914A-79] 6. A common boundary between two parts of a system, whether material or non material. Specifically, in a rocket vehicle or other mechanical assembly, a common boundary between two components. Specifically, in fluid dynamics, a surface separating two fluids across which there is a discontinuity of some fluid property such as density or velocity or of some derivative of these properties in a direction normal to the interface. The equations of motion do not apply at the interface but are replaced by the boundary conditions. [NASA] 7. A specific electronic circuit that is a boundary between other circuits or devices.

interface-functional The relationship between functions at the hardware physical interface, such as electrical voltage and current, fluid pressure and flow, temperature, acceleration, acoustic and other environments.[AS1426-80]

interface-physical The hardware physical relationship at the juncture between items designed, manufactured, or tested by different activities jointly responsible for the hardware. It includes dimensional geometric relationship, tolerances, materials, finishes at the juncture.[AS1426-80]

interface-procedural The matters related to successful conduct of the program for hardware which has physical or functional interfaces, such as design reviews, tests, inspections, approvals, government certification, data preparation, schedules, and related procedural activities.[AS1426-80]

interfaces The systems, external to the system being analyzed, which provide a common boundary or service and are necessary for the system to perform its mission in an undegraded mode, e.g., systems that supply power, cooling, heating, air services, or input signals. [ARD50013-91]

interfacial seal See seal.

interference 1. Instrument response due to presence of components other than the gas (or vapor) that is to be measured.[AIR1533-91] 2. Any undesirable electromagnetic emission or any electrical or electromagnetic disturbance, phenomenon, signal or emission, man-made or natural which causes an undesirable response, malfunction or degradation of the performance of electrical or electronic equipment.[ARP1931-87] 3. The waveform resulting from superimposing one wave train on another. 4. In signal transmission, spurious or extraneous signals which prevent accurate reception of desired signals.

interference, common mode A form of interference which appears between measuring circuit terminals and ground.

interference (electrical) Any spurious voltage or current rising from external sources and appearing in the circuits of a device. See noise.

interference, electrostatic field See interference, electromagnetic.

interference filter An optical filter which selectively transmits specific wavelengths of light because of interference resulting from dielectric coatings on the surface of the material. Multilayer interference coatings may include metallic layers.

interference fit 1. The condition where the diameter of the fastener is larger than the hole that it is to fit in.[NASA] 2. Any combination of pin or shaft diameter and mating hole diameter where the tolerance envelope of the hole overlaps or is smaller than the tolerance envelope of the pin.

interference, longitudinal See interference, common mode.

interference, magnetic field See interference, electromagnetic.

interference monochromatization See diffraction.

interference pattern The pattern of some characteristic of a stationary wave produced by superimposing one wave train on another—it may be the distribution in space of energy density, energy flux, particle velocity, pressure or some other characteristic.

interference seal A seal produced between a fastener and its hole when a fastener of a given diameter is driven into a hole of smaller diameter (the hole diameter is approximately 0.003 in (0.08 mm) smaller to be effective). An interference seal is also produced when a fastener shank is expanded by the installation process.[AIR4069-90]

interference, transverse See interference, normal mode.

interferometer 1. Apparatus used to produce and measure interference from two or more coherent wave trains from the same source. Interferometers are used to measure wavelengths, to measure angular width of sources, to determine the angular position of sources (as in satellite tracking), and for many other purposes.[NASA] 2. An instrument so designed that the variance of wavelengths and light path lengths within the mechanism allows very accurate measurement of distances.

interferometric pressure transducer A type of pressure sensor developed to read pressure differentials on the order of 200 Pa (.030 psi) with a resolution of 1 Pa (0.00015 psi) by detecting very small deflections of a fragile diaphragm through optical interferometry.

interferon A protein (lymphokine) released by cells in response to virus infection. When taken up by other cells, interferon inhibits the replication of viruses within them.[NASA]

intergranular beta Beta phase situated between alpha grains. It may be at grain corners as in the case of equiaxed alpha type of microstructures in alloys having low beta stabilizer content.[AS 1814-90]

interior ballistics That branch of ballistics that deals with the propulsion of projectiles, i.e., the motion and behavior of projectiles in a gun barrel, the temperatures and pressures developed inside a gun barrel or rocket.[NASA]

interleaving 1. The act of accessing two or more bytes or streams of data from distinct computer memory banks simultaneously. 2. The alternating of two or more operations or functions through the overlapped use of a computer facility.

interlock 1. Instrument which will not allow one part of a process to function unless another part is functioning. 2. A device such as a switch that prevents a piece of equipment from operating when a hazard exists. 3. To join two parts together in such a way that they remain rigidly attached to each other solely by physical interference. 4. A device to prove the physical state of a required condition, and to furnish that proof to the primary safety control circuit.

interlock, motor start A connection made through contacts on the motor controller which is wired in series with the safety circuit so that the motor must be energized before the system is allowed to proceed.

intermediate addressing Method of addressing data stored in a computer memory. The instruction operand is the data to be used with the instruction.

intermediate approach segment See segments of an instrument approach procedure.

intermediate band A mode of recording and playback in which the frequency response at a given tape speed is "intermediate."

intermediate fix (IF) The fix that identifies the beginning of the intermediate approach segment of an instrument approach procedure. The fix is not normally identified on the instrument approach chart as an intermediate fix (IF). See also segments of an instrument approach procedure.[ARP4107-88]

intermediate frequency 1. The beat frequencies used in heterodyne receivers, usually the difference between the received radiofrequency signal and a locally generated signal.[NASA] 2. In a superhetrodyne receiver, the stage where a down-converted carrier is passed through a bandpass filter and amplifier.

intermediate level maintenance That maintenance performed in direct support of using organizations. Tasks normally consist of calibration, repair, or replacement of damaged or unserviceable parts, components, or assemblies; the emergency manufacture of non-available parts; and the provision of technical assistance to using organizations. (Level 2 maintenance). Synonymous with intermediate maintenance. [ARD50010-91]

intermediate means In an instrumentation or control system, all system elements between the primary detector and the end device which transmit or modify the output of the former to make it compatible with input requirements of the latter.

intermediate mode Method of operating a computer, with interpretive languages such as BASIC, whereby an individual instruction or a small number of instructions, not forming part of a program, are executed.

intermediate phase 1. A distinguishable homogeneous phase whose composition range does not extend to any of the pure components of the system, such as TiH and TiO.[AS1814-90] 2. A distinct compound or solid solution in an alloy system whose composition limits do not extend to any of the pure constituents.

intermediate position A specified position that is greater than zero and less than 100 percent open.

intermediate zone See zone, intermediate.

intermetallic compound A phase in an alloy system which usually occurs at a definite atomic ratio and exhibits a narrow solubility range. Nearly all such phases are brittle.[AS1814-90]

intermittency effect In photography or radiography, a departure from the reciprocity law when the emulsion is exposed in a series of discrete increments, compared to the response when it is exposed continuously to the same total energy level.

intermittent blowdown The blowing down of boiler water at intervals.

intermittent duty 1. A requirement of service that demands operation for alternate intervals of load and no load. [ARP1199A] 2. An operating cycle that consists of alternating periods of use and idle time—for example, on and off, load and no-load, load and rest, or load, no-load and rest; in most instances, successive periods of use or idle time vary widely in length, although some intermittent-duty cycles follow well-defined patterns.

intermittent firing A method of firing by which fuel and air are introduced into and burned in a furnace for a short period, after which the flow is stopped, this succession occurring in a sequence of frequent cycles.

intermittent precipitation Precipitation must have stopped and recommenced at least once during the hour (60 minutes) preceding the actual time of observation.[AIR1335-75]

intermodulation 1. The modulation of the components of a complex wave by each other in a nonlinear system. [NASA] 2. The modulation of the components of a complex wave by each other, producing new waves whose frequencies are equal to the sums and differences of integral multiples of the component frequencies of the original complex wave.

intermodulation distortion (IMD) Defined as 20 log (rms sum of the sum and difference distortion products)/(rms amplitude of the fundamental).

internal In PC ladder programs, a coil or contact whose reference is a logical element in the program and not directly concerned with I/O. May also refer to the storage location used for the logical status of such an element.

internal combustion engine A mechanical prime mover that uses exhaust gases resulting from the burning of fuel within the engine as the thermodynamic working fluid.

internal energy 1. A mathematically defined thermodynamic function of state, interpretable through statistical mechanics as a measure of the molecular activity of the system.[NASA] 2. Ability of a working fluid to do its work based on the arrangement and motion of its molecules.

internal furnace A furnace within a boiler consisting of a straight or corrugated flue, surrounded with water.

internal gear Any ring-type or annular gear whose teeth are on the inner surface of the rim.

internalized unit values A system of values, motives, and prioritized goals held by a unit and adopted by a member of that unit. Such a person is referred to colloquially as a "team player".[ARP 4107-88]

internal leakage Leakage between internal cavities of a device.[AIR1916-88]

internally fired boiler A fire tube boiler having an internal furnace such as a Scotch, Locomotive Fire-Box, Vertical Tubular, or other type having a water-cooled plate-type furnace.

internal-mix oil burner A burner having a mixing chamber in which high velocity steam or air impinges on jets of incoming liquid fuel which is then discharged in a completely atomized form.

internal oxidation A form of degradation of a material involving absorption of oxygen at the surface and diffusion of oxygen to the interior, where it forms subsurface scale or oxide inclusions.

internal pressure 1. The pressure inside a system or component.[AIR1916-88] 2. The pressure inside a portion of matter due to the attraction between molecules. [NASA]

internal standard In chemical analysis, especially instrumental analysis, a material present in or added to a sample in known amounts to serve as a reference in determining composition.

internal storage Addressable storage directly controlled by the central processing unit of a digital computer.

internal stress See residual stress.

internal treatment The treatment of boiler water by introducing chemicals directly into the boiler.

internal wave In fluid mechanics, wave motions of stably stratified fluids in which the maximal vertical motions occur below the surface of the fluids. [NASA]

International Civil Aviation Organization (ICAO) A specialized agency of the United Nations whose objective is to develop the principles and techniques of international air navigation and to foster planning and development of international civil air transport. [ARP4107-88]

International Computers Limited See ICL computers.

International Electrotechnical Commission An international standards development and certification group in the area of electronics and electrical engineering.

International Federation for Information Processing An international group of technical societies.

International Geophysical Year By international agreement, a period during which greatly increased observation of world-wide geophysical phenomena is undertaken through the cooperative effort of participating nations. July 1957 to December 1958 was the first such year; however, precedent was set by the International Polar Years of 1882 and 1932.[NASA]

International Magnetospheric Study

Joint US, ESA, Japanese, and Canadian effort (1976-1979) for observation and measurement of magnetospheric and ionospheric phenomena and involving spacecraft, aircraft, balloons, and rockets, as well as ground based equipment.[NASA]

International Solar Polar Mission See Ulysses mission.

International Standard (IS) The third (and highest) stage of the ISO standard process. Prospective ISO standards are balloted three times. The first stage is a Draft Proposal (DP). After a Draft Proposal has been in use a period of time (typically six months to a year) the standard, frequently with corrections and changes, is re-balloted as a Draft International Standard (DIS). After the Draft International Standard has been in use for a period of time (typically one to two years) it is reballoted as an International Standard (IS).

International System of Units The metric system of units based on the meter, kilogram, second, ampere, Kelvin degree, and candela. Other SI units are hertz, radian, newton, joule, watt, coulomb, volt, ohm, farad, weber and tesla.[NASA]

interpass temperature The lowest temperature reached by weld metal before the next pass is deposited in a multiple-pass weld.

interplanetary propulsion See rocket engines.

interpreter A system program which allows the execution of computer programs using a step by step translation of individual instructions instead of translating the complete program before execution.

interprocessor communication Communication between two or more processors in a computer system.[NASA]

Inter-Range Instrumentation Group (IRIG) The telemetry working group of IRIG is responsible for specifying the industrywide standards and practices of telemetry.

inter-record gap (IRG) On magnetic tape, the blank gap between records; the tape can stop and start within this gap.

interrogator The ground-based surveillance radar beacon transmitter-receiver which normally scans in synchronism with a primary radar system. This interrogator transmits signals which cause properly functioning transponders located in aircraft to reply with a unique signal. The reply signals from the interrogated aircraft are mixed with the primary radar returns and displayed on the controller's radar scope. Also applied to the airborne element of the TACAN/DME system.[ARP4107-88]

interrupt (INT) In data processing, a signal that, when activated, causes a transfer of control to a specific location in memory, thereby breaking the normal flow of control of the routine being executed. An interrupt is normally caused by an external event such as a "done" condition in a peripheral. It is distinguished from a trap, which is caused by the execution of a processor instruction.

interrupting capacity The rated interrupting capacity is the maximum short-circuit current at rated voltage which a protective device is required to interrupt under the operating duty specified and with a normal frequency recovery voltage not less than rated voltage. See also rupture capacity.[ARP1199A-90]

interrupt service routine In data processing, a unique address that points to two consecutive memory locations containing the start address of the interrupt service routine and priority at which the interrupt is to be serviced.

interrupt vector In data processing, an address generated by an interrupt. It points to the start of the interrupt service routine.

interrupt vector register In data processing, a register for storing the interrupt vector.

intersection(s) In Boolean algebra, the operation in which concepts are described by stating that they have all the characteristics of the classes involved. Intersection is expressed as AND.[NASA]

intersection departure/intersection takeoff A takeoff or proposed takeoff on a runway commencing from the intersection of a taxiway with a runway or the intersection of two runways.[ARP 4107-88]

interstellar chemistry Molecular formation/dissociation in interstellar space due to radiation, collision, and other forces.[NASA]

interstice The space or void left between or around the cabled or stranded components.[ARP1931-87]

interstitial element An element with relatively small atomic diameter that can assume position in the interstices of the titanium crystal lattice. Common examples are oxygen, nitrogen, hydrogen, and carbon.[AS1814-90]

interval The number of word times that occur between successive repetitive samples of the same channel; synonymous with supercommutation and strapping interval.

interval timer A device that provides an interrupt signal upon completion of a predetermined or programmed interval of time. See timer.

in-the-loop An expression which indicates that a component is a necessary part of the closed control loop and that its removal, inaction, failure, or malfunction would interrupt the control-action-feedback loop of the system. For example, when an aircraft is flying on autopilot, the autopilot subsystem is in-the-loop, and the pilot is no longer in-the-loop. The pilot is merely monitoring the closed-loop flight control system. If the autopilot fails, the pilot must get back in-the-loop (that is, resort to manual reversion). See also closed-loop system.[ARP4107-88]

intraorbit transfer vehicle Small scooter type tugs that would move men and materials within an orbit.[NASA]

intrinsic joint loss A loss intrinsic to the fiber that is caused by fiber parameter mismatches when joining two nonidentical fibers.

intrinsic safety 1. A method to provide safe operation of electric process control instrumentation where hazardous atmospheres exist. The method keeps the available electrical energy so low that ignition of the hazardous atmosphere cannot occur. 2. A protection technique based upon the restriction of electrical energy within apparatus and of interconnecting wiring, exposed to a potentially explosive atmosphere, to a level below that which can cause ignition by either sparking or heating effects. Because of the method by which intrinsic safety is achieved, it is necessary to ensure that not only the electrical apparatus exposed to the potentially explosive atmosphere but also other electrical apparatus with which it is interconnected is suitably constructed.

intrinsic safety barrier A device inserted in wire between process control instrumentation and the point where the wire passes into the hazardous area. It limits the voltage and current on the wire to safe levels.

intruder Any aircraft in the airspace of another aircraft that is tracked by the collision avoidance system (including all threat aircraft).[ARP4153-88]

intuitively obvious Term applied to that which can be described or operated correctly without training or explanation.[ARP4107-88]

invalidity See errors.

inverse response The dynamic characteristic of a process by which its output responds to an input change by moving initially in one direction but finally in the other.

inverse scattering Method of analyzing some classic wave scattering. [NASA]

inverse-time See time-inverse.

inversion temperature In a thermocouple, the temperature of the "hot" junction when the thermoelectric emf of the circuit is equal to zero.

inverted spin A spin throughout which the airplane is upside down.[ARP4107-88]

inverter A NOT element. The output signal is the reverse of the input signal.

involute splines, wear limit The amount of tooth thickness decrease for an external involute spline, or the amount of space width increase for an internal involute spline which may be considered allowable without expectation of failure.[ARP1200-89]

I/O See input/output.

Io A satellite of Jupiter orbiting at a mean distance of 421,800 kilometers. Also called Jupiter I.[NASA]

I/O-bound A state of program execution in which all operations are dependent on the activity of an I/O device; for example, when a program is waiting for input from a terminal. See also CPU-bound.

I/O hardware Computer hardware used to carry signals into and out of the processing hardware.

I/O isolation Usually refers to the electrical separation of field circuits from computer internal circuits. Accomplished by optoelectronic devices. Occasionally refers to the ability to have input or output field wiring on isolated circuits, i.e., with one return for each.

I/O limited See input-output limited.

I/O module Basic set of I/O interfaces sharing a common computer unit housing. Can be a set of discrete I/O or a smart control I/O.

ion A charged atom or radical which may be positive or negative.

ion chamber See ionization chamber.

ion density (concentration) In atmospheric electricity, the number of ions per unit volume of a given sample of air; more particularly, the number of ions of a given type (positive small ion, negative small ion, positive large ion, or negative large ion) per unit volume of air.[NASA]

ion engine Reaction engine in which ions, accelerated in an electrostatic field, are used as propellants.[NASA]

ion exchange A chemical process for removing unwanted dissolved ions from water by inducing an ion-exchange reaction (either cation or anion) as the water passes through a bed of special resin containing the substitute ion.

ion-exchange resin A synthetic organic compound (resin) that can remove unwanted ions from a dilute solution by combining with them or by exchanging them for ions that produce desirable or neutral effects.

ion gage See ionization gage.

ionic mobility In gaseous electric conduction, the average velocity with which a given ion drifts through a specified gas under the influence of an electric field of unit strength. Mobilities are commonly expressed in units of centimeters per second per volt per centimeter.[NASA]

ionic propellant See ion engine.

ionic strength Effective strength of all ions in a solution that is equal to the sum of one half of the product of the individual ion concentration and their ion valence or charge squared for dilute solutions.

ion implantation A process for enhancing surface properties of a solid by bombarding it with a beam of high-energy ions, which are absorbed into the material's surface layer.

ionization 1. Generally the disassociation of an atom or molecule into positive or negative ions or electrons. Restructively, the state of an insulator whereby it facilitates the passage of current due to the presence of charged particles, usually induced artificially. [ARP1931-87] 2. The process of splitting a neutral molecule into positive and negative ions, or of detaching one or more electrons from a neutral atom.

ionization chamber 1. Apparatus used to study the production of small ions in the atmosphere by cosmic ray and radioactive bombardment of air molecules.[NASA] 2. An enclosure filled with gas that is ionized when radiation enters the chamber; it contains two or more electrodes that sustain an electric field and collect the charge resulting from ionization.

ionization constant A measure of the degree of dissociation of a polar compound in dilute solution at equilibrium; it equals the product of the concentrations of the dissociated compound (ions) divided by the concentration of the undissociated compound.

ionization counter See ionization chamber; radiation counter.

ionization gage 1. Vacuum gage with a means of ionizing the gas molecules and a means of correlating the number and type of ions produced with the pressure of the gas. Various types of ionization gages are distinguished according to the method of producing the ionization.[NASA] 2. A pressure transducer based on conduction of electric current through ionized gas of the system whose pressure is to be measured. Useful only for very low pressures (for example, below 10^{-3} atm).

ionization potential The energy required to ionize an atom or molecule. The energy is usually given in terms of electron volts.[NASA]

ionization time In a gas tube, the interval between the time when conduction conditions are established and the time when conduction actually begins at some stated value of tube potential.

ionization vacuum gage An instrument for measuring very low pressures (high vacuums) by means of a current of positive ions produced in the gas by electrons emitted from a hot cathode and accelerated across a portion of the evacuated space toward another electrode.

ionization voltage (corona level) The minimum value of falling r.m.s. voltage which sustains electrical discharge within the vacuous or gas-filled spaces in the cable construction or insulation. [ARP1931-87]

ionized plasma See plasmas (physics).

ionizer Filament, grid, or porous body in ion engines or other devices which strip electrons from the outer shells of neutral atoms to form positively charged ions.[NASA]

ionizing event Any interaction between an atom or molecule and an energy beam, particle, atom or molecule that causes one or more ions to be generated.

ionizing radiation 1. Any electromagnetic or particulate radiation capable of producing ions, directly or indirectly in its passage through matter. [NASA] 2. Any electromagnetic or particulate radiation that can produce ions, either directly or indirectly, when it interacts with matter.

ion laser A laser in which the active medium is an ionized gas, typically one of the rare gases, argon or krypton, or a mixture of the two.

ionopause The upper boundary of the ionosphere of certain planets (excluding the earth) and comets where electrons decline sharply.[NASA]

ionosphere That portion of the earth's atmosphere where ionization takes place due to ultra-violet radiation of the sun or from bombardment by hydrogen bursts from sunspots. The various layers, identified as B, C, D, E and F, have characteristics that reflect and refract radio waves according to their frequency, time of day, sunspot cycle and earth weather.

ionospheric storms Disturbances of the ionosphere, resulting in anomalous variations in its characteristics and effects on radio communication.[NASA]

ionospheric tilts Ionospheric conditions where the variability of the number of the electrons as a function of altitude is present. Ionospheric tilts are sometimes created by traveling ionospheric disturbances (TID's) and ionospheric tilts deflect radio waves in unexpected directions adversely affecting radio reception.[NASA]

ion pair The combination of a positive ion and a negative ion having the same magnitude of charge, and formed from a neutral molecule due to absorption of the energy in radiation.

ions Charged atoms or molecularly bound groups of atoms; sometimes also free electrons or other charged subatomic particles. In atmospheric electricity, any of several types of electrically charged submicroscopic particles normally found in the atmosphere. Atmospheric ions are of two principal types, small ions and large ions, although a class of intermediate ions has occasionally been reported. In chemistry, atoms or specific groupings of atoms which have gained or lost one or more electrons, as the chloride ion or ammonium ion. Such ions exist in aqueous solutions and in certain crystal structures.[NASA]

ion spectrometer See mass spectrometer.

ion storage Ions within an electromagnetic trap and cooled to sub-Kelvin temperatures with lasers. Potential uses are for frequency standards.[NASA]

ion stripping A procedure following the focusing of ion beams in the target chamber of a reactor to be used for particle beam pellet fusion.[NASA]

I/O page That portion of computer memory in which specific storage locations are associated directly with I/O devices.

I/O rack Chassis for mounting computer I/O modules. May be local or remote from CPU/memory unit.

IP (impact prediction) See computerized simulation.

I/P converter A device that linearly converts electric current into gas pressure (for example, 4-20 mA into 3-15 psi).

IRIG See Inter-Range Instrumentation Group.

irradiance The power per unit area incident upon a surface. Also called radiant flux density.

iron 58 A radioactive isotope of iron. [NASA]

irradiance The detection rate per unit area of radiation.[NASA]

irradiation 1. The exposure of an insulating or jacketing material to high energy emissions for the purpose of favorably altering the molecular structure by providing cross-linking of molecular chains.[ARP1931-87] 2. Exposing an object or person to penetrating ionizing radiation such as x-rays or gamma rays. 3. Exposing an object or person to ultraviolet, visible, or infrared energy.

irregular galaxies Galaxies with amorphous structure and with relatively low mass (10 to the 8th to 10 to the 10th solar masses). Fewer than 10% of all galaxies are classified as irregular.[NASA]

IS See International Standard.

ISA See Instrument Society of America.

ISDN See Integrated Systems Digital Network.

isentrope A line of equal or constant pressure, with respect to either space or time.[NASA]

isentropic Proceeding at constant entropy.

isentropic exponent A ratio defined by the specific heat at constant pressure divided by the specific heat at constant volume.

isobaric Proceeding at constant pressure.

isobaric flame temperature The temperature of a propellant flame under constant pressure conditions.[AIR913-89]

isobars (pressure) Lines of equal or constant pressure, specifically such lines on a weather map.[NASA]

isochoric flame temperature The temperature of a propellant flame under constant volume conditions.[AIR913-89]

isochronous governor A device that maintains rotational speed of an engine constant, regardless of load.

isolation A technique used in fault tolerant systems that removes the effects of a failure or prevents a failure from propagating or affecting the continued operation of the system.[ARP1181A-85] See also burst pressure, proof pressure, or reference pressure.

isolation level The functional level to which a failure can be isolated using accessory test equipment at designated test points.[ARD50010-91]

isolation seal A secondary seal used to isolate potential fuel leakage paths. To prevent channeling of fuel along a leak path between structural members. It can be a repair seal installed to reestablish seal continuity in areas of direct contact with the fuel.[AIR4069-90]

isomer nuclide having the same mass number A and atomic number Z, but existing for measurable times in different quantum states with different energies and radioactive properties. Molecules having the same atomic composition and molecular weight, but differing in geometrical configuration. [NASA]

isomerization Process of converting hydrocarbon or other organic compound to an isomer.[NASA]

ISO OSI International Organization for Standardization's Reference Model of Open System Interconnect.[AIR4271-89]

isoparametric finite elements The basis for the calculation of physical properties of structural shapes including stress analyses.[NASA]

isopleths Se nomographs.

isopotential point Point on the millivolt versus pH plot at which a change in temperature has no effect. It is at 7 pH and zero millivolts unless shifted by the standardization and meter zero adjustments or an electrode asymmetry potential.

isostasy A supposed equality existing in vertical sections of th earth, whereby the weight of any column from the surface of the earth to a constant depth is approximately the same as that of any other column of equal area, the equilibrium being maintained by plastic flow from one part of the earth to another.[NASA]

isotensoid structure Filamentary structure in which the filaments are uniformly stressed throughout from the design loading conditions.[NASA]

isothermal Proceeding at constant temperature.

isothermal process Thermodynamic change of state of a system that takes place at constant temperature.[NASA]

isotherms Lines connecting points of equal temperature.[NASA]

isotope(s) Any of two or more nuclides that have the same number of protons in their nuclei but different numbers of neutrons; such atoms are of the same element, and thus cannot be separated from each other by chemical means, but because they have different masses can be separated by physical means.

isotope effect The effect of nuclear properties other than the number of protons on the nonnuclear physical and chemical behavior of the nuclides.

isotopic enrichment Process by which the relative abundance of the isotopes of a given element are altered in a batch to produce a form of the element enriched in a particular isotope.[NASA]

isotropic turbulence Turbulence in which the products and squares of the velocity components and their derivatives are independent of direction, or more precisely, invariant with respect to rotation and reflection of the coordinate axes in a coordinate system moving with the mean motion of the fluid. [NASA]

ITAE See integral time absolute error.

item A nonspecific term used to denote any product, including systems, materials, parts, subassemblies, sets, accessories, etc.[ARD50013-91][ARD50010-91]

item, completed Items of equipment (including basic or end items, components, assemblies) that have been overhauled, modified, renovated, and completed in accordance with terms of contracts, project orders, or other work directives and authorizations, and are ready for their intended use after receiving final mechanical acceptance inspection.[AIR1916-88]

item, consumable An item that is used only once.[ARD50010-91]

item, critical See critical item.

item, expendable Items for which no authorized repair procedure exists, and for which cost of repair would normally exceed that of replacement.[ARD50010-91]

item levels Item levels from the simplest to the more complex are as follows: part, subassembly, assembly, unit, group, set, subsystem, system.[AIR1916-88]

item, life limited 1. An item that has a limited and predictable useful life and could be considered for replacement on a preplanned basis for reliability, safety or economic reasons.[ARD50010-91] 2. An item that must be removed from service and discarded before a specified time is achieved. It is referred to as a XXX time life part or component.[AIR-1916-88]

item, line maintenance Any item which can be readily changed on an aircraft during the maintenance operations.[ARD50010-91]

item, maintenance significant (MSI) Items identified by the manufacturer whose failure: (a) could affect safety (ground or flight), and/or (b) is undetectable during operations, and/or (c) could have significant operational economic impact, and/or (d) could have significant nonoperational economic impact.[ARD50010-91]

item, mandatory replacement An item that, if disturbed or removed during the course of maintenance or overhaul, must be replaced to comply with specifications and procedures.[ARD50010-91]

item, repairable An item comprising or including replaceable parts, commonly economical to repair, and subject to being rehabilitated to a fully serviceable condition over a period less than the life of the flight equipment to which it is related.[ARD50010-91]

item, serviceable An item that can be returned to service with appropriate airworthiness documentation.[AIR1916-88]

iterate To repeatedly execute a loop or series of steps. For example, a loop in a routine.

iterative Describing a procedure or process which repeatedly executes a series of operations until some condition is satisfied. An iterative procedure can be implemented by a loop in a routine.

ITI See Industrial Technology Institute.

ITO (semiconductor) Semiconductor device consisting of a layer of tin sandwiched between an indium layer and an oxide layer.[NASA]

IVD See Integrated Voice Data LAN.

Izsak ellipsoid See ellipsoids; geodesy.

J

J See joule.

jack 1. A device usually hydraulic (but may be pneumatic or mechanical), for the purpose of lifting or raising heavy weights such as aircraft. Various types and configurations are used such as axle jacks, fuselage jacks, etc. See actuator. [AIR1489-88] 2. A connecting device to which a wire or wires of a circuit may be attached and which is arranged for the insertion of a plug.

jacket 1. The outermost separable layer of insulating material on a wire or cable. [ARP1931-87] 2. The outermost layer of insulating material on a wire or cable. [ARP914A-79] 3. Covering or casing of some kind. Specifically, a shell around the combustion chamber of a liquid fuel rocket, through which the propellant is circulated in regenerative cooling. Coatings of one material over another to prevent damage such as oxidation or micrometeroid penetration.[NASA] 4. A plastic layer applied over the coating of an optical fiber, or sometimes over the bare fiber. Used for color coding in optical cables, to make handling easier, or for protection of the fiber against mechanical stress and strain. 5. Stiff plastic protective material that encases a floppy disk with slots for a disk drive to access the data. 6. The layer of plastic, fiber or metal surrounding insulated electrical wires to form a cable. This outer cover may be for mechanical or environmental protection of the wires contained therein.

jack pad A pad or surface built on a landing gear for the purpose of accepting jacking loads or a separate attachable jack fitting.[AIR1489-88]

jack point A raised point usually spherical on a landing gear for the purpose of accepting the lifting forces imposed by a jack.[AIR1489-88]

jackscrew 1. A captive screw and nut assembly, attached to connectors, for use in mating and unmating connectors; may also facilitate connector mounting and polarization.[ARP914A-79] 2. A device which is type of puller which separates two parts of utilizing threaded action by one of its details. Used to life or exert pressure.[ARP480A-87] 3. A portable device for lifting a heavy load a short distance by means of a screw mechanism. 4. The screw of such a device.

jackshaft A countershaft, especially an auxiliary shaft used between two other shafts.

Jahn-Teller effect The effect whereby, except for linear molecules, degenerate orbital states in molecules are unstable.[NASA]

jamming 1. Electronic or mechanical interference which may disrupt the display of an aircraft on radar or the transmission/reception of radio communications/navigation.[ARP4107-88] 2. Intentional transmission or reradiation of radio signals in such a way as to interfere with reception of desired signals by the intended receiver.[NASA] 3. Intentional transmission of radio-frequency waves for the purpose of interfering with transmissions from another station.

Janus One of the natural satellites of Saturn.[NASA]

Japanese Map Users Group See World Federation.

JATO Acronym for Jet-Assisted Takeoff. A method of enabling aircraft to take off in a shorter distance than otherwise possible by the use of boosters (rockets) temporarily attached, usually to the underside of the wings.[ARP4107-88]

jaw 1. A mechanical device which engages a mating component for the purpose of holding or exerting force to transmit motion.[ARP480A-87] 2. A mechanical device resembling or suggesting the jaw of an animal in form or action; either of two or more opposing parts movable so as to open or close, for grasping or crushing anything between them; as jaws of a vice, or lathe. [ARP480A-87]

JCL See job control language.

jerk A sudden, abrupt motion.

jet Rapid flow of fluid from a nozzle or orifice.

jet airstream See jet streams (meteorology)

Jet Assisted Take-Off (JATO) A rocket motor used to assist the take-off of an aircraft. Also known as Rocket Assisted Take-Off (RATO).[AIR913-89]

jetavators See guide vanes.

jet blast The equivalent of a wind gust produced by the jet engines of an aircraft.[ARP1328-74]

jet damping See damping.

jet engine(s) 1. An aircraft engine that produces thrust by the rearward expulsion of combustion products from the burning of fuel with air.[ARP4107-88] 2. Broadly, engines that eject jets or streams of gas or fluids, obtaining all or most of their thrust by reaction to the ejection. Specifically, aircraft engines that derive all or most of their thrust by reaction to their ejection of combustion products (or heated air) in a jet and that obtains oxygen from the atmosphere for the combustion of their fuel (or outside air for heating, as in the case of the nuclear jet engine), distinguished in this sense from a rocket engine. Jet engines of this kind may have compressors, commonly turbine driven to take in and compress air (turbojets), or they may be compressorless, taking in and compressing air by other means (pulsejets, ramjets).[NASA]

jet fuel starter A small gas turbine engine mounted on the engine gearbox or AMAD to provide cranking power for starting. Also referred to as gas turbine starter.[ARP906A-91]

jet interaction amplifier An amplifier which utilizes control jets to deflect a power jet and modulate the output. Usually employed as an analog amplifier. [ARP993A-69]

jet lag Desynchronization of biological rhythms because of transmeridian flight. [NASA]

jet membrane process Method for separating or enriching isotopes of the same element by using a condensible vapor as the carrier fluid. A process gas containing the isotopes enters a chamber into which a heavy condensible gas (the jet) flows. The lighter of the two isotopes is enriched relative to the heavier species and is collected by a probe downstream for further enrichment or analysis.[NASA]

jet pump A type of pump that uses a jet of fluid to induce flow in another fluid.

jet reference fluid (JRF) A combination of aliphatic and aromatic liquids, also containing mercaptans, used as a harsh environment to test the resistance of integral fuel tank sealants to jet fuels.[AS7200/1-90]

jet route A route designed to serve aircraft operations from 18,000 ft MSL on up to and including flight level 450. The routes are referred to as "J" routes with numbering to identify the designated route (for example, J 105).[ARP4107-88]

jet stream A narrow, shallow, meandering river of strong winds which usually extends around the temperate zone of the earth. A jet stream is considered to exist whenever winds of 50 knots or stronger, embedded in the high tropospheric or lower stratospheric general wind flow, are concentrated in a band at least 300 miles long. It is generally found in segments of 1000 to 3000 miles in length, 100 to 400 miles in width, and 3000 to 7000 ft in depth with winds generally between 100 to 150 knots. [ARP4107-88]

jet thrust The thrust of a fluid, especially as distinguished from the thrust of a propeller.[NASA]

jet vanes Vanes either fixed or movable,

used in a jetstream, especially in the jetstream of a rocket, for purposes of stability or control under conditions where external aerodynamic controls are ineffective. Also called blast vane. [NASA]

jet wash See wake turbulence.

jewels Recessed bearings of glass, sapphire or diamond that support the ends of a pivot pin in an instrument or a fine mechanical watch or clock.

JFET Junction field effect transistors in which semiconductor channels of low conductivity join the source and drain and in which these channels are reduced and cut off by the junction depletion regions, which reduce the conductivity and cause a voltage to be applied between the gate electrodes.[NASA]

jig A device which attaches and positions on the part for purposes of locating and controlling another tool for machining. [ARP480A-87]

J integral A contour energy integral formulated by Rice and used for evaluating fracture toughness of elastoplastic materials.[NASA]

jitter 1. Undesired, rapid movement of symbology upon the display face that is discernible to a human eye located at the design eye position.[ARP1782-89] 2. A computer signal instability. See also vibration.

job A group of computer data and control statements that do a unit of work, such as a program and all its related subroutines, data, and control statements; also a batch control file.

job control language (JCL) A language for identifying a job and requesting action from a computer operating system.

job satisfaction A person's subjective evaluation of the extent to which he/she is performing and progressing satisfactorily in the occupation of his/her choice and which meets his/her professional needs.[ARP4107-88]

Jodrell Bank Observatory A large radio telescope, located near Manchester, England.[NASA]

joggle A displacement machined or formed in a structural member to accommodate the base of an adjacent member. Although joggles are sealed by prepacking during preassembly whenever possible, in some cases they must be sealed by injection during post-assembly operations.[AIR4069-90]

Johansson curved crystal spectrometer A type of spectrometer having a reflecting crystal whose face is concave so that x-rays that diverge slightly after passing through the primary slit are refocused at the detector slit.

joint A separable or inseparable juncture between two or more materials.

joule Metric unit for energy, quantity of heat, or work.

Joule-Thomson effect A change of temperature in a gas undergoing Joule-Thomson expansion.[NASA]

JOVIAL See Jules' Own Version of International Algorithmic Language.

joy stick A device by which an individual can communicate with an electronic information system through a cathode ray tube.

judder A mode of vibration—the oscillation of the landing gear leg due to externally applied forces, and the inertia forces due to the combined inertia of the wheel, brake, axle, and the leg.[AIR 1489-88]

judgment Assessment of the significance and priority of data from the environment in terms of how they relate to the task at hand. The result of this process forms the basis upon which decisions are made. See also decision.[ARP4107-88]

judgment delay Failure, due to an anomaly of attention or motivation, to assess the significance and priority of information from the environment in a timely manner, assuming adequate quality and quantity of information.[ARP 4107-88]

judgment, poor Failure, due to an anomaly of attention or an anomaly of motivation, to realistically assess the significance and priority of information from the environment, assuming adequate quality and quantity of information.[ARP4107-88]

Jules' Own Version of International Algorithmic Language (JOVIAL) A compiler language based on the International Algorithmic Language, ALGOL.

jump An instruction which causes a new address to be entered into a computer program counter; the program continues execution from the new program counter address.

jumper(s) 1. Short lengths of conductors used to complete electrical circuits, usually temporary, between terminals, or bypassing an existing circuit.[NASA] 2. A temporary wire used to bypass a portion of an electrical circuit or to attach an instrument or other device during testing or troubleshooting.

jumper tube A short tube connection for by-passing, routing, or directing the flow of fluid as desired.

junction The portion of a transistor where opposite types of semiconductor material meet.

junction box A protective enclosure around connections between electric wires or cables.

junction field effect transistors See JEFT.

junctions In semiconductor devices, regions of transition between semiconducting regions of different electrical properties.[NASA]

Jupiter rings Ring structures around the planet Jupiter discovered on March 4, 1979 by Voyager 1.[NASA]

jury rig A makeshift or temporary assembly.

jury strut (or brace) A secondary toggle kinematic arrangement to provide a lock for the gear assembly in a specific configuration. A secondary brace which locks the primary brace in position, both of which are "on center" or near "on center" condition. Also, a piece of ground support equipment used to provide extra bracing for movable or folded assemblies against abnormally high loads; i.e., to brace folded wings on carrier-based aircraft against heavy weather conditions.[AIR1489-88]

justification 1. The act of adjusting, arranging, or shifting digits to the left or right, to fit a prescribed pattern. 2. The use of space between words in printed material that causes each line to be of equal length.

justify To align computer data about a specified reference.

K Symbol used to indicate a kilobyte which is a measure of a computer memory. One Kilobyte of memory can store 1024 characters. See also kelvin; see also ratio of specific heats.

Kalman filter 1. The Kalman filter is an optimal estimator or filtering technique for estimating the state of a linear system.[AIR1823-86] 2. A technique for calculating the optimum estimates of process variables in the presence of noise; the technique, which generates recursion formulas suitable for computer solutions, also can be used to design an optimal controller.

kaolinite A hydrous silicate of aluminum. It constitutes the principle mineral in kaolin.[NASA]

Karl Fischer technique A titration method for accurately determining moisture content of solid, liquid or gas samples using Karl Fischer reagent—a solution of iodine, sulfur dioxide and pyridine in methanol or methyl Cellusolve; the titration is highly suitable for automation, and has high sensitivity (5 ppm) and good accuracy (+ or - 1%) over a wide range of moisture content (10 ppm to 100%).

Karman vortex street A double trail of vortices formed alternately on both sides of a cylinder of similar body moving at right angles to its axis through a fluid, the vortices in one row rotating in a direction opposite to that of the other row.[NASA]

Karnaugh map A tubular arrangement that facilitates the combination and elimination of logical functions by listing similar logical expressions, thereby taking advantage of the human brain's ability to recognize visual patterns to perform the minimization.

K_b See Bernoulli coefficient.

KByte $1024(2^{10})$bytes.

Kelvin Metric unit for thermodynamic temperature. An absolute temperature scale where the zero point is defined as absolute zero (the point where all spontaneous molecular motion ceases) and the scale divisions are equal to the scale divisions in the Celsius system; in the Kelvin system, the scale divisions are not referred to as degrees as they are in other temperature-measurement systems; 0 °C equals approximately 273.16 K.

Kelvin bridge A type of d-c bridge circuit similar to a Wheatstone bridge, but incorporating two extra resistances in parallel with two of the known resistances to minimize the inaccuracies introduced because of finite lead and contact resistances in the circuit.

Kelvin-Varney voltage divider(KVVD) A resistive-type voltage divider used in some d-c bridge circuits to provide greater sensitivity at low values of the unknown resistance.

Kennison nozzle A specially shaped nozzle designed for measuring flow through partially filled pipes; because of its self-scouring, nonclogging design, it is especially useful for measuring flow of raw sewage, raw and digested sludge, final effluent, trade wastes, and other liquids containing suspended solids or debris; it also functions well at low flow rates or when flow rates vary widely.

Kepler laws The three empirical laws governing the motions of the planets in their orbits, discovered by Johannes Kepler (1571-1630). These are: a) the orbits of the planets are ellipses, with the sun at a common focus; b) as a planet moves in its orbit, the line joining the planet and the sun sweeps over equal areas in equal intervals of time (also called law of equal areas); c) the squares of the periods of revolution of any two planets are proportional to the cubes of their mean distances from the sun.[NASA]

Kern counter See dust counter.

Kerr cell A device in which the Kerr effect is used to modulate light passing through the material. The modulation depends on rotation of beam polarization caused by the application of an electric field to the material. The degree of rotation determines how much of the beam can pass through a polarizing filter.

Kevlar A Dupont synthetic textile material, lightweight and nonflammable, and with high impact resistance.[NASA]

key 1. A projection on a connector which engages a keyway in a mating connector so as to guide the connectors halves during mating; may be positioned and/or used in multiples to provide a polarization feature.[ARP914A-79] 2. A machine part inserted into a groove (or keyway) to lock two parts together, such as a shaft and a gear or pulley. 3. One of a set of control levers used to operate a machine such as a typewriter or computer processing unit. 4. A device that moves or pivots to secure or tighten components in an assembly. 5. A component, usually having notches or grooves in its working face, that is inserted into a lock to engage or disengage the locking mechanism. 6. In data processing, characters that identify a record.

keyboard An orderly arrangement of keys for operating a machine such as a typewriter.

keyboard entry 1. An element of information inserted manually, usually via a set of switches or marked punch levers, called keys, into an automatic data processing system. 2. A medium for achieving access to or entrance into an automatic data processing system.

keyboard lockout An interlock feature which prevents sending from the keyboard while the tape transmitter of another station is sending on the same circuit.

keyboard monitor A computer program that provides and supervises communication between the user at the system console and an operating system.

keyboard perforator A machine that operates somewhat like a typewriter and produces punched paper tape for automatically operating computers or communications equipment.

key disk A disk required to start certain programs.

keypunch 1. A special device to record information on cards or tape by punching holes in the cards or tape to represent letters, digits, and special characters. 2. To operate a device for punching holes in cards or tape.

key-to-disk device Input equipment designed to accept keyboard entry directly on magnetic disks.

keyway A slot or groove into which a key engages.[ARP914A-79]

keyword One of the significant and informative words in a title or document which describe the content of that document.

kg See kilogram.

kiln An oven or similar heated chamber for drying, curing or firing materials or parts.

kilo A decimal prefix denoting 1,000.

kilogram Metric unit of mass.

kilometric waves Electromagnetic waves with wavelengths between 1,000 and 10,000 meters.[NASA]

kilovolt ampere (kVA) 1000 volt x amperes.[ARP1931-87]

kilovolts (kV) 1000 volts.[ARP1931-87]

kilowatt (kW) A unit of power equal to one thousand watts.[ARP1931-87]

kimberlite See biotite.

kinematics The branch of mechanics dealing with the description of the motion of bodies or fluids without reference to the forces producing the motion.[NASA]

kinematic viscosity Absolute viscosity of a fluid divided by its density.

kinesthetic illusion See illusions, kinesthetic.

kinetic energy 1. The energy which a body possesses as a consequence of its motion.[NASA] 2. The energy of a working fluid caused by its motion. 3. Energy related to the fluid of dynamic pressure, $1/2 \rho V^2$.

kinetic theory The derivation of the bulk properties of fluids from the properties of their constituent molecules, their motions and interactions.[NASA]

kinetic vacuum system A vacuum system capable of attaining and sustaining limiting pressures of 5×10^{-5} to 5×10^{-7} torr despite a relatively high outgassing load or the presence of small leaks.

kink A tight loop in wire or wire rope that results in permanent damage due to deformation.

kinking A temporary or permanent distortion of a hose, induced by winding or doubling upon itself and consequently exceeding the recommended minimum bend radius established for that hose.[ARP1658A-86]

Kirchhoff-Huygens principle See diffraction.

Kirchhoff law of radiation The radiation law which states that at a given temperature the ratio of the emissivity to the absorbtivity for a given wavelength is the same for all bodies and is equal to the emissivity of an ideal black body at that temperature and wavelength.[NASA]

Kirchhoff's Law The sum of the voltage across a device in a circuit series is equal to the total voltage applied to the circuit.

kit 1. A container or containers in which a one-part or a two-part sealant is packaged. Two component kits consist of the base compound in one container and the catalyst in the other.[AS7200/1-90] 2. A group of related but non-homogeneous items used for service or modification purposes.[ARP480A-87]

klystrons Electron tubes for converting direct current energy into radio frequency energy by alternately speeding up and slowing down the electrons.[NASA]

"K" Modulus 1. The load in psi on a loaded area of the subgrade divided by the deflection in inches of the subgrade under that load. 2. The total load in pounds divided by the total volume displaced in cubic inches.[AIR1489-88]

kneeling gear A landing gear assembly or installation designed to "kneel" the aircraft for ground handling purposes, such as parking, loading, maintenance, catapulting, etc. The gear may retract to an intermediate position or fold in a manner different from normal retraction. A mechanism for raising or lowering the aircraft while parked on the ground.[AIR1489-88]

knockout pin See ejector pin.

knot A velocity of one nautical mile per hour, 1.150779 statute miles per hour; not a measure of distance. Although the term "knot" is defined in some dictionaries (based on lay usage) as "1 nautical mile," this usage (as in "knots per h") is not acceptable in professional aviation or nautical contexts.[ARP4107-88]

knowledge Familiarity with theory and concepts, and detailed understanding of job-related topics.

knowledge engineering See expert systems.

knowledge representation The use of symbolic data structures to represent knowledge so that a computer can manipulate them.[NASA]

Knudsen cells See Knudsen gages.

Knudsen flow Gas flow in a long tube at pressures such that the mean free path of a gas molecule is significantly greater than the tube radius.

Knudsen gages Gages which measure pressure in terms of the net rate of transfer of momentum by molecules between two surfaces maintained at different temperatures and separated by a distance smaller than the mean free path of the gas molecules.[NASA]

kohm See kilo-ohm.

kondo effect Change in superconductivity characteristics resulting from magnetic impurities in the compounds involved.[NASA]

konimeter A device for determining dust concentration by drawing in a measured volume of air, directing the air jet against a coated glass surface thereby depositing dust particles for subsequent counting under a microscope.

koniscope An indicating instrument for detecting dust in the air.

Korteweg-Devries equation The mathematical representation describing the propagation of long waves of small but finite amplitude.[NASA]

kraft process (woodpulp) Woodpulping process in which sodium sulfate is used in the caustic soda pulp-digestion liquor. Also known as sulfate pulping or kraft pulping.[NASA]

Kramers-Kronig formula The relationship between the attenuation coefficient and the dispersion (frequency dependent phase velocity) for viscoelastic waves.[NASA]

kreep A yellow-brown glassy lunar mineral enriched in potassium, rare earth elements, and phosphate.[NASA]

kriging A method of providing unbiased estimates of variables in regions where the available data exhibit spatial autocorrelation, and these estimates are obtained in such a way that they have minimum variance.[NASA]

krypton fluoride lasers Rare gas halide ultraviolet stimulated emission devices in which krypton fluoride is the active lasing medium.[NASA]

kurtosis In statistics, the extent to which a frequency distribution is peaked or concentrated about the mean; it is sometimes defined as the ratio of the fourth moment of the distribution to the square of the second moment.[NASA]

Kurtosis number Figure of merit, K, used for monitoring impulsive type vibrations of ball bearings.

KVVD See Kelvin-Varney voltage divider.

Kynar Tradename of Polyvinylidene fluoride, manufactured by Pennwalt Corp.

L

L See litre; also see length.

La Chemical symbol for lanthanum.

label In data processing, a set of symbols used to identify or describe an item, record, message, or file. Occasionally it may be the same as the address in storage.

labeled common Named data areas that are accessible to all computer programs declaring the named area.

labeled molecule A molecule of a specific chemical substance in which one or more of its component atoms is an abnormal nuclide—that is a nuclide that is radioactive when the molecules normally are composed of stable isotopes, or vice versa.

laboratory ambient conditions 18-24 °C (65-75 °F) and 45-55 percent relative humidity.[AS8043-86]

labyrinth seals Minimum leakage seals that offer resistance to fluid flow while providing radial or axial clearance. [NASA]

lacquer A liquid material applied to fibrous braid to prevent fraying, wicking or moisture absorption in the braid. A saturant.[ARP1931-87]

lactometer A hydrometer designed for measuring the specific gravity of milk.

ladder diagram Symbolic representation of a control scheme. The power lines form the two sides of a ladder-like structure, with the program elements arranged to form the rungs. The basic program elements are contacts and coils as in electromechanical logic systems.

lag 1. The delay between change of conditions and the indication of the change on an instrument.[AIR1489-88] 2. Delay in human reaction.[AIR1489-88] 3. The amount one cyclic motion is behind another, expressed in degrees. The opposite is lead.[AIR1489-88] 4. A relative measure of the time delay between two events, states, or mechanisms. 5. In control theory, a transfer function term in the form, 1/(Ts+1). 6. The dynamic characteristic of a process giving exponential approach to equilibrium. See also time lag.

lagging 1. In an a-c circuit, a condition where peak current occurs at a later time in each cycle than does peak voltage. 2. A thermal insulation, usually made of rock wood and magnesia plaster, that is used to prevent heat transfer through the walls of process equipment, pressure vessels or piping systems.

Lagrange coordinates Systems of coordinates by which fluid parcels are identified for all times by assigning them coordinates which do not vary with time. Examples of such coordinates are: a) the values of any properties of the fluid conserved in the motion; or b) more generally, the positions in space of the parcels at some arbitrarily selected moment. Subsequent positions in space of the parcels are then the dependent variables, functions of time and of the Lagrange coordinates.[NASA]

lambert A unit of luminance; it equals the uniform luminance of a perfectly diffusing surface emitting or reflecting light at one lumen per square centimeter.

Lambert law See Bouguer law.

Lambert's cosine law The radiance of certain surfaces, known as Lambertian reflectors, Lambertian radiators, or Lambertian sources, is independent of the angle from which the surface is viewed.

lamella (metallurgy) Crystalline materials whose grains are in the form of thin sheets.[NASA]

laminar boundary layer 1. In fluid flow, layer next to the fixed boundary. The fluid velocity is zero at the boundary layer but the molecular viscous stress is large because the velocity gradient normal to the wall is large.[NASA] 2. A layer of a moving turbulent stream adjacent to the wall of a pipe or other conduit, where the motion approximates streamline flow.

laminar boundary layer separation

See laminar boundary layer.

laminar flames See laminar flow.

laminar flow 1. Flow of a fluid characterized by the gliding of fluid layers (laminae) past one another in orderly fashion.[AIR1916-88] 2. In fluid flow, a smooth flow in which no cross flow of fluid particles occurs between adjacent stream lines; hence, a flow conceived as made up of layers—commonly distinguished from turbulent flow.[NASA]

laminar flow control See laminar boundary layer.

laminar jets See laminar flow.

laminated materials See laminates.

laminated tape A tape consisting of two or more layers of different materials bonded together.[ARP1931-87]

laminations Laminations, found only in plate steel, are thin flat discontinuities seen only at the edge or end of the plate.[AS3071A-77] See also laminates.

lamp 1. The generic term for an artificial source of light.[ARP798-91] 2. Any device for producing light, usually one that converts electric energy to light.

lamp follower See auxiliary output.

lamp test See test.

LAN (computer networks) See local area networks.

Landau damping The damping of a space charge wave by electrons which move at the phase velocity of the wave and gain energy transferred from the wave.[NASA]

landfills Disposal sites for solid wastes which are buried in layers of earth.[NASA]

landing gear The apparatus comprising those components of an aircraft or spacecraft that support and provide mobility for the craft on land, water, or other surfaces. The landing gear consists of wheels, floats, skids, bogies, and treads, or other devices, together with all associated struts, bracing, shock absorbers, etc. Landing gear includes all supporting components, such as the tail wheel or tail skid, outrigger wheels or pontoons, etc., but the term is often conceived to apply only to the principal components, i.e., to the main wheels, floats, etc., and the nose gear if any. For military aircraft, catapulting provisions and the arresting hook installation are often considered an ancillary component of the landing gear.[AIR 1489-88]

landing-gear-extended speed The maximum speed at which an aircraft can be safely flown with the landing gear extended.[ARP4107-88]

landing-gear operating speed The maximum speed at which the landing gear can be safely extended or retracted.[ARP4107-88]

landing hot Landing an aircraft at a speed substantially greater than its stalling speed.[ARP4107-88]

landing mat A ground cover laid down for the purpose of supporting aircraft operations, i.e., parking, taxi, takeoff, landing. Several types are used such as pierced metal strips or aluminum planks.[AIR1489-88]

landing minimums/IFR landing minimums The minimum ceiling and visibility prescribed for landing civil aircraft while using an instrument approach. These minimums apply along with other limitations set forth in FAR Part 91 with respect to the minimum descent altitude (MDA) or decision height (DH) prescribed in the instrument approach procedures as follows: (a) Straight-in Landing Minimums. The MDA and visibility, or the DH and visibility, required for straight-in landing on a specified runway, or (b) Circling Minimums: The MDA and visibility required for the circle-to-land maneuver. Descent below the established MDA or DH is not authorized during an approach unless the aircraft is in a position from which a normal approach to the runway of intended landing can be made, and adequate visual reference to required visual cues is maintained.[ARP4107-88]

landing phase See mishap, phase of flight.

landing roll/rollout The distance from point of touchdown to the point where the aircraft can be brought to a stop or exit the runway.[ARP4107-88]

land mobile satellite service A proposed radio relay satellite system for serving thinly populated or large geographical areas.[NASA]

language In data processing, a set of representations, conventions and rules used to convey information. See algorithmic language, artificial language, machine language, natural language, object language, problem-oriented language, procedure-oriented language, programming language, source language, and target language.

language extendibility The ability to change a programming language through source statements written in that language.

language translator A general term for any assembler, compiler, or other routine that accepts statements in one language and produces equivalent statements in another language.

lap 1. A lap is similar to a seam and may result from improper rolling practices or, in the case of forging, the metal being folded over but failing to weld into a single piece. Rolling laps generally run lengthwise with the bar but can extend into the bar at an angle. [AS3071A-77] 2. A controlled surface, used to remove insignificant amounts of metal.[ARP480A-87]

lap joint 1. Two conductors joined by placing them side-by-side so that they overlap.[ARP914A] 2. A connection between two parts made by overlapping members at the junction and welding, riveting or bolting them together.

Laplace transform 1. The Laplace transform is a mathematical relationship that is used to transform a function from the time domain to a transformed domain.[AIR1823-86] 2. In control theory, a mathematical method for solution of differential equations.

lapping Smoothing or polishing a surface by rubbing it with a tool made of cloth, leather, plastic, wood or metal in the presence of a fine abrasive.

lapping in A process of mating contact surfaces by grinding and/or polishing.

lapse rate The decrease of an atmospheric variable with height, the variable being temperature unless otherwise specified. The term applies ambiguously to the environmental lapse rate and the process lapse rate and the meaning must often be ascertained from the context. [NASA]

laptop A small, portable computer usually with a flip-up screen.

lap weld A lap joint made by welding.

large aircraft Aircraft having a maximum certified takeoff weight of more than 12,500 pounds.[ARP4107-88]

large core fiber An optical fiber with a comparatively large core, usually a step index type. There is no standard definition of "large" but for the purposes here, diameters of 400 micrometers or more are designated as "large".

Large Infrared Telescope on Spacelab See LIRTS (telescope).

large scale integration (LSI) 1. A computer chip containing a large number of digital circuits in a small area. 2. Integrated circuits with more than 100 logic gates.

larmor radius For a charged particle moving transversely in a uniform magnetic field, the radius of curvature of the projection of its path on a plane perpendicular to the field.[NASA]

laser Device for producing light by emission of energy stored in a molecular or atomic system when stimulated by an input signal (From Light Amplification by Stimulated Emission of Radiation).

[NASA]

laser anemometer Measuring instrument in which the wind being measured passes through two perpendicular light beams and the resulting change in velocity of one or both beams is measured. [NASA]

laser annealing Rapid heating of metals and/or alloys with the use of lasers. [NASA]

laser cutting The cutting of material by means of lasers.[NASA]

laser diode array A device in which the output of several diode lasers is brought together in one beam. The lasers may be integrated on the same substrate, or discrete devices coupled optically and electronically.

laser doppler flowmeter An apparatus for determining flow velocity and velocity profile by measuring the Doppler shift in laser radiation scattered from particles in the moving fluid stream; contaminants such as smoke may have to be introduced into a gas stream to provide scattering centers; the technique can be used to measure velocities of 0.01 to 5000 in./s (0.25 mm/s to 125 m/s).

laser glass An optical glass doped with a small concentration of a laser material. When the impurity atoms are excited by light, they are stimulated to emit laser light.

laser guidance Guidance system for rockets or projectiles, utilizing a laser beam for a precise trajectory to a designated target.[NASA]

laser gyroscope Ring-laser angular rotation sensor for stabilizing and controlling large space structures, for space vehicle guidance, etc.[NASA]

laser induced fluorescence Emission of electromagnetic radiation that is caused by the flow of laser radiation into the emitting body and which ceases abruptly with the excitation.[NASA]

laser interferometer A type of optical interferometer using a laser as the source of monochromatic light; accuracies of better than 20 microinch (1.25 μm) are achieved when measuring lengths up to 200 in. (5.08 m).

laser interferometry The design and use of interferometers in which a laser is the light source. The monochromaticity and brilliance of the laser light enables the differentiation between interfering beams of hundreds of meters, in contrast to a maximum of 20 centimeters for the classical interferometers.[NASA]

laser line filter A filter which transmits light in a narrow range of wavelengths centered on the wavelength of a laser. Light at other wavelengths is reflected. Such filters are used to remove light from non-laser sources, which could interfere with operation of a laser system.

laser microscopy The application of a laser microscope having a ceramic tube in which a metal vapor is formed at 1600 degrees C. Copper (or other metal atoms) are excited and amplify light so that when used with a projection microscope, the object to be magnified is illuminated. The power of the emitted beam on the screen remains constant. [NASA]

laser plasma interactions The results of the actions of laser beams on electrically ducting fluids, such as plasmas or ionized gases.[NASA]

laser printer A print-quality printer that uses a laser beam to electrostatically transfer an image to paper.

laser propulsion The use of high power lasers for aircraft, rocket, or spacecraft propulsion by indirect conversion of laser heated propellants or working fluids to produce thrust; direct thrust generation with laser light pressure on the vehicle; direct conversion of laser energy into electricity for propulsion. [NASA]

laser pumping The application of a laser

beam of appropriate frequency to a laser medium so that absorption of the radiation increases the population of atoms or molecules in higher energy states. [NASA]

laser simulator A light source which simulates the output of a laser. In practice, the light source is a 1.06 micrometer LED which simulates the output of a neodymium laser at much lower power levels.

laser spectrometer Spectrometer that uses a laser.[NASA]

laser spectroscopy The use of lasers for spectroscopic analysis; particularly in Raman spectroscopy.[NASA]

laser stability Characteristic of a laser beam free from oscillations.[NASA]

laser target designator Laser equipment aboard spacecraft for identifying satellites, missiles, and objects in space. [NASA]

laser target interactions Interactions where lasers are used to produce heating, fusion, or damage in targets.[NASA]

laser targets Objects subjected to laser radiation, especially for laser fusion applications.[NASA]

laser weapons Military applications of high power lasers (mainly gasdynamic and chemical mixing lasers).[NASA]

laser welding Microspot welding with a laser beam.[NASA]

lasing Generation of visible or IR light waves having very nearly a single frequency by pumping or exciting electrons into high energy states in a stimulated emission device (laser).[NASA]

last-in, first-out (LIFO) In an ordered pushdown stack, a discipline wherein the last transaction to enter a stack is also the first to leave it. Contrast with first-in, first-out.

latches Devices that fasten one thing to another, as a rocket to a launcher, but are subject to ready release so that things may be separated.[NASA]

latching digital output A contact closure output that holds its condition (set or reset) until changed by later execution of a computer program. See momentary digital output.

latching relay Real device or program element that retains a changed state without power. In a computer program element, the power removal is only in terms of the logic power expressed in the diagram. Real power removal will affect the PC outputs according to some scheme provided by the manufacturer.

latch switch A control to prevent fuel valve opening if the burner is not secured in the firing position.

latch-up A pnpn self-sustaining low impedance state which is a type of electronic malfunction.[NASA]

latency 1. A measure of time delay.[AIR 4271-89] 2. In data processing, the time between the completion of the interpretation of an address and the start of the actual transfer from the addressed location. Latency includes the delay associated with access to storage devices such as drums and delay lines.

latent defect An inherent or induced weakness, not detectable by ordinary means, which will either be precipitated to early failure under environmental stress screening conditions or eventually fail in the intended use environment.[ARD50010-91]

latent heat 1. The unit quantity of heat required for isothermal change in a state of a unit mass of matter. Latent heat is termed heat of fusion, heat of sublimation, heat of vaporization, depending on the change of state involved.[NASA] 2. Heat that does not cause a temperature change.

latent heat of fusion See heat of fusion.

lateral axis An imaginary lateral line at right angles to the longitudinal axis, passing through the center of gravity of the airplane, and lying within a plane normal to the plane of symmetry. Angular movement about this axis is called pitching.[ARP4107-88]

lateral navigation (LNAV) Those functions which provide navigation guidance in the horizontal plane and which provide command signals to the roll channel of the auto flight system and displays. Currently used interchangeably with R-NAV (random navigation), a term created by ARINC.[ARP1570-86]

lateral separation The lateral spacing of aircraft at the same altitudes which is achieved by requiring operation on different routes or in different geographical locations.[ARP4107-88]

lateral spring constant Also lateral spring rate. The number of unit load required to produce a unit deflection in a lateral direction (normal to plane of the wheel).[AIR1489-88]

lateral stability See stability, lateral.

Latin square method In mathematics, the use of an n x n square array of n different symbols, each symbol appearing once in each row and once in each column.[NASA]

latitude 1. Angular distance from a primary great circle or plane.[NASA] 2. Of a photographic emulsion, the ratio of the exposure limits between which the film density curve (Hurter and Driffield curve, or H & D curve) is essentially linear.

lattice network An electronic network composed of four branches connected end-to-end to form a mesh, and in which two nonadjacent junctions are the input terminals and the two remaining nonadjacent junctions are the output terminals.

lattice parameter In crystallography, the length of any side of the unit cell in a given space lattice; if the sides are unequal, all unequal lengths must be specified.

launch bar In a nose gear launch system, the link which provides means of steering the nose gear during tracking and couples the aircraft to the catapult during the power run; part of the nose gear assembly.[AIR1489-88]

launch clouds See exhaust clouds.

launch complexes See launching bases.

launcher 1. Specifically, a structure or device often incorporating a tube, a group of tubes, or a set of tracks, from which self-propelled missiles or aircraft are sent forth and by means of which the missiles or aircraft are aimed or imparted inertial guidance—distinguished in this specific sense from a catapult.[AIR1489-88] 2. Broadly, a structure, machine, or device including the catapult, by means of which airplanes, rockets, or other vehicles are directed, hurled or set forth.[AIR1489-88]

launching bases Areas such as Cape Kennedy or Vandenburg Air Force Base that has several launch sites.[NASA]

launching device See launcher.

launching pad The load-bearing base or platform from which a rocket vehicle is launched.[NASA]

launching run During catapulting, the extent of aircraft travel from "release" to the end of "deck run". It is the sum of the power run or power stroke and the deck run.[AIR1489-88]

launching site Defined area from which a rocket vehicle is launched, either operationally or for test purposes; specifically, at Cape Kennedy or Vandenburg, any of the several areas equipped to launch a rocket.[NASA]

launch time See launch windows.

launch windows The postulated openings in the continuum of time or of space, through which a spacecraft or missile must be launched in order to achieve a desired encounter, rendezvous, impact or the like.[NASA]

Lauritsen electroscope An electroscope in which the sensitive element is a metallized quartz fiber.

lava A general term for a molten extrusive; also for the rock that is solidified from it.[NASA]

law of exception In the investigation of aircraft accidents, the principle according to which, if all other possible causes have been ruled out, it is concluded that the operative cause was pilot behavior.[ARP4107-88]

lay The axial length of a turn of the helix made by a helical element of a conductor or cable.[ARP1931-87]

layer 1. In reference to sky cover, clouds, or other obscuring phenomena whose bases are approximately at the same level. The layer may be continuous or composed of detached elements. The term layer does not imply that a clear space exists between the layers or that the clouds or obscuring phenomena composing them are of the same type.[ARP 4107-88] 2. A subdivision of the OSI architecture.

lay-up Production of reinforced plastics by positioning the reinforced material (such as glass) in the mold prior to impregnation with resin.[NASA]

lb See pound.

L-band In telemetry, the radio spectrum which is available for manned vehicles; 1435-1540 MHz.

LCD See liquid crystal display.

LCU See Local Control Unit.

LCVASI Low Cost Visual Approach Slope Indicator.

L/D (reflux-to-distillate ratio) A quantity used in analyzing column operations. See reflux ratio.

LDA Localizer-type Directional Aid.

lead 1. The amount one cyclic motion is ahead of another, expressed in degrees. The opposite is lag.[AIR1489-88] 2. A definite length of one-conductor electrical wire, wire braid, or other conductive material, except cable or cord, one or both ends of which are processed or terminated. The lead may be of any size or shape, insulated or uninsulated.[ARP-480A-87] 3. A column of high explosive used as one component of an explosive train.[AIR913-89] 4. In control theory, a transfer function term in the form, (Ts+1). 5. The distance a screw mechanism will advance along its axis in a single rotation.

lead acid batteries The common automobile batteries in which the electrodes are grids of metallic lead containing lead oxides that change in composition during charging and discharging. The electrolyte generally is dilute sulfuric acid.[NASA]

lead angle 1. In welding, the angle between the axis of the electrode and the axis of the weld. 2. The angle between the tangent to a helix and a plane perpendicular to the axis of the helix.

lead equivalent The radiation-absorption rating of a specific material expressed in terms of the thickness of lead that reduces radiation dose an equal amount under given conditions.

leader 1. A blank section of tape at the beginning of a reel of magnetic tape or at the beginning of a paper tape. 2. A system program which enables other programs to be loaded into the computer.

leading 1. In an a-c circuit, a condition where peak current occurs earlier in each cycle than does peak voltage. 2. In printing, the insertion of additional space between lines.

leading edge The first transition of a pulse going in either a positive (high) or a negative (low) direction.

leading edge flaps Control surfaces at the leading edges of airfoils. Hinged panels deflected downward to induce and control separation of the air flow. [NASA]

leading edge thrust The increase in lift produced by highly swept, low-aspect ratio wings which develop a strong separation vortex; however, an even larger increase in drag is produced.[NASA]

lead-lag compensator A dynamic compensator combining lead action (the inverse of lag) with lag.

lead time In industrial engineering, the

amount of time required to design and develop a piece of equipment before it is ready for use.

lead wire Any wire connecting two points in an electrical circuit, but especially a wire connecting an electric device to a source of power or connecting an indicating or controlling instrument to a sensor.

lead zirconate titanates Dense ceramics with high piezoelectric coefficients and a high relative permitivity.[NASA]

leakage 1. Flow through a passage which is in a nominally closed position or at a location which normally should permit no flow, usually of relatively small magnitude.[ARP243B-65] [AIR1916-88] 2. A type of streamline flow most often observed in viscous fluids near solid boundaries, which is characterized by the tendency for fluid to remain in thin, parallel layers to maintain uniform velocity. 3. A nonturbulent flow regime in which the stream filaments glide along the pipe axially with essentially no transverse mixing. Also known as viscous or streamline flow. 4. Flow under conditions in which forces due to viscosity are more significant than forces due to inertia. 5. Undesirable loss or entry across the boundary of a system. The term is usually applied to slow passage of a fluid through a crack or fissure, but may also be used to describe passage of small quantities of particles, radiation, electricity or magnetic lines of force beyond desired boundaries.

leakage rate The amount of leakage across a defined boundary per unit time.

leak detector An instrument such as a helium mass spectrometer used for detecting small cracks or fissures in a vessel wall.

leaker 1. A crack of hole in the tube which allows fluids to escape.[ARP 1658A-86] 2. A hose assembly which allows fluids to escape at the fittings or couplings.[ARP1658A-86]

leak exit The point outside the tank where the leak appears.[AIR4069-90]

leak path The path a fuel leak follows from the leak source to its external exit. [AIR4069-90]

leak pressure See pressure, leak.

leak source The point inside a fuel tank where the leak starts.[AIR4069-90]

lean mixture An insufficient amount of fuel in the fuel-to-air mixture (weight-ratio basis) required by an internal combustion engine; a relatively large amount of air compared to fuel, generally greater than 16:1. See also rich mixture.[ARP 4107-88]

leans See illusion, vestibular.

learning Long-term adjustments to a person's behavior as a result of reinforcement and practice. These adjustments may be either physical or mental. A relatively permanent change in behavior or knowledge which comes as a result of experience and is not the direct result of a body state such as fatigue or illness.[ARP4107-88]

learning ability The innate capacity to acquire new skills or knowledge and apply them practically.[ARP4107-88]

learning rate The relative speed at which new information is acquired and permanent adjustments made in one's behavior.[ARP4107-88]

learning reinforcement The process of strengthening prior experiences through their recall and review (rehearsal) in order to retain them in long-term memory. The strengthening of a response when that response leads to a satisfying state of affairs.[ARP4107-88]

learning transfer The ability of a person to apply the experience acquired in learning situations to "real world" situations. The change, positive or negative, in the ability of a person to perform a given act as a direct consequence of prior learning of a related kind.[ARP 4107-88]

least significant bit (LSB) The smallest

bit in a string of bits, usually at the extreme right.

least significant digit (LSD) The right-most digit of a number.

least squares method Any statistical procedure that involves minimizing the sum of squared differences.[NASA]

LED A semiconductor diode, the junction of which emits light when passing a current in the forward (junction on) direction. See also light emitting diode.

Ledoux bell meter A type of manometer whose reading is directly proportional to flow rate sensed by a head producing measuring device such as a pitot tube.

left justified A field of numbers (decimal, binary, etc.) which exists in a memory cell, location or register, possessing no zeroes to its left.

leg 1. A single landing gear structure or assembly.[AIR1489-88] 2. One of the members of a branched object or system. 3. The distance between the root of a fillet weld and the toe. 4. Any structural member that supports an object above the horizontal.

length A fundamental measurement of the distance between two points, measured along a straight or curved path.

length of lay The length of lay of any helically wound strand of wire or cable is the axial length of one complete turn of the helix, usually expressed in inches.[AS1198-82]

leptons In the classification of subatomic particles according to mass, the lightest of all particles; examples of leptons are the electron and positron.[NASA]

letdown The descent of an aircraft from cruising altitude in preparation for an approach or landing.[ARP4107-88]

letdown procedure Procedure used for descending from cruising altitude to the airport.[ARP4107-88]

letter check In airline industry, the alphabetic designations given to scheduled-maintenance packages. A maintenance package is a group of maintenance tasks scheduled for accomplishment at the same time. See check, C.[AIR1916-88]

let-through current The current that actually passes through the protective device after initiation of a fault.[ARP 1199A-90]

level 1. Any bubble-tube device used to establish a horizontal line or plane. 2. To make the earth's surface even and roughly horizontal.

level, confidence The probability that a given statement is correct.[ARD 50010-91]

level indicator 1. An indicating instrument for determining the position of a liquid surface within a vessel. 2. An instrument containing a meter, neon lamp or cathode-ray tube which shows audio voltage level in an operating sound-recording system.

leveling saddle A pipe clamp anchoring a swivel joint 2 in. threaded socket allowing a 2 in. pipestand to be properly positioned.

level (logic) A signal that remains at the "O" or "I" level for long amounts of time.

level-off To make the flight path of an airplane horizontal after a climb, glide, or dive.[ARP4107-88]

level of repair analysis A technique which establishes (a) whether an item should be repaired or discarded, (b) at what maintenance level, that is, organizational, intermediate, or depot.[ARD 50010-91]

levered suspension A type of landing gear arrangement in which an axle is supported on a trailing arm from a structural post or support. The trailing arm rotates about the pivot as a lever to provide the vertical wheel motion for shock absorption.[AIR1489-88]

levitation melting A metallurgical process in which a piece of metal placed above a coil carrying a high frequency current can be supported against gravity by the Lorentz force caused by the induced surface currents in the metal. At the same time, the heat produced by Joule dissipation melts the metal. [NASA]

lexical analysis In data processing, a stage in the compilation of a program in which statements, such as IF, AND, END, etc. are replaced by codes.

LF Low Frequency.

L/F (reflux-to-feed ratio) A quantity used in analyzing column operations.

LFR Low Frequency Range.

Liapunov's second method A method analogous to the rate-of-change-of energy method for mechanical systems whereby stability or instability or a process control system can be determined. Also referred to as the indirect method.

liberation See heat release.

library 1. A collection of information and standard programs available to a computer, usually on auxiliary storage. 2. A file that contains one or more relocatable binary modules which are routines that can be incorporated into other programs.

libration A real or apparent oscillatory motion, particularly the apparent oscillation of the moon. Because of libration more than half of the moon's surface is revealed to an observer on the earth even though the same side of the moon is always toward the earth, because the moon's periods of rotation and revolution are the same. Other motions regarded as librations are long period orbital motions and periodic perturbations in orbital elements.[NASA]

Lichtenberg figure camera A device for indicating the polarity and approximate crest value of a voltage surge; it consists of a photographic film or plate backed by an extended plane electrode, with its emulsion contacting a small electrode that is connected to the circuit in which a surge occurs.

LIF (fluorescence) See laser induced fluorescence.

life (biology) See life sciences.

life cycle The total life span of an aeronautical system beginning with the concept formulation phase and extending through the operational phase up to retirement from the inventory. Synonymous with life cycle profile.[ARD50010-91] See "heat aging".

life cycle costs The sum of the acquisition costs and maintenance costs for the life of a system.[NASA]

life cycle maintenance (support) cost The total cost of an item maintenance during its useful life including organizational, intermediate, depot, and contractor maintenance, spares and repair parts provisioning, test equipment, maintenance personnel salaries and subsistence, training, etc.[ARD50010-91]

life sciences The field of scientific disciplines encompassing biology, physiology, psychology, medicine, sociology, and other related areas.[NASA]

life, service The life of an item at which it is no longer physically or economically feasible to repair or overhaul the items to acceptable standards.[ARD 50010-91]

life, storage The length of time an item can be stored under specified conditions and still meet specified requirements. [ARD50010-91]

life test Life test is the process of placing the "unit of product" under a specified set of test conditions and measuring the time it takes until failure.[ARD 50010-91]

life units A measure of use duration applicable to the item (such as operating hours, cycles, distance, rounds fired, attempts to operate).[ARD50010-91]

life, useful The length of time a population of items is expected to operate with a constant failure rate. This excludes any infant mortality and wear-out periods.[ARD50010-91]

life test A destructive test in which a device is operated under conditions that simulate a lifetime of use.

life usage A method of tracking either the useful life remaining or life consumed on a life-limited part in a gas turbine engine.[AIR1873-88]

LIFO See last-in, first-out.

lift 1. The force component acting perpendicular to the line of flight (and in general usage, parallel to the plane of symmetry of the aircraft), and produced primarily by the pressure forces acting on the surfaces of the aircraft.[ARP4107-88] 2. That component of the total aerodynamic force acting on a body perpendicular to the undisturbed airflow relative to the body. To lift off, to take off in vertical ascent.[NASA] See also travel.

lift coefficients See aerodynamic coefficients; lift.

lift dampers See ground spoilers.

lift distribution See lift.

lift drag ratio The ratio of lift to drag obtained by dividing the lift by the drag of the lift coefficient by the drag coefficient.[NASA]

lift forces See lift.

liftoff The initial separation of an aircraft or other vehicle from the ground, especially the vertical takeoff of a rocket or VTOL aircraft.[ARP4107-88]

lift ratio The ratio of the tire outside diameter to the rim diameter of the wheel (bead seat diameters).[AIR1489-88]

ligament The minimum cross section of solid metal in a header, shell or tube sheet between two adjacent holes.

light 1. Visible radiation (about 0.4 to 0.7 in wavelength) considered in terms of its luminous efficiency, i.e., evaluated in proportion to its ability to stimulate the sense of sight.[NASA] 2. Electromagnetic radiation having a wavelength in the range over which it can be detected with the unaided human eye.

light-beam galvanometer A type of sensitive galvanometer whose null-balance point is indicated by the position of a beam of light reflected from a mirror carried in the moving coil of the instrument. Also known as d'Arsonval galvanometer.

light-beam instrument A measurement device that indicates measured values by means of the position of a beam of light on a scale.

light-coupled switch A switch in which the switching signal is transmitted to the activating device by means of a light beam.

light crude Crude oil rich in low-viscosity hydrocarbons of low molecular weight.

light curtain An arrangement whereby a wide, thin beam of invisible modulated light is used to detect passage of objects through a plane up to about 8 by 78 in. (200 mm by 2 m).

light duration See pulse duration.

light duty cable Generally a type of fiber optic cable designed to withstand conditions encountered in a building—not outdoor conditions.

lighted dial A dial or indicating scale and pointer which has a small lamp within the assembly so that the scale and pointer are illuminated for viewing in darkness. Compare with luminous dial.

light emitting diode (LED) A semiconductor diode which emits visible or infrared light. Light from an LED is incoherent spontaneous emission, as distinct from the coherent stimulated emission produced by diode lasers and other types of lasers. The indicator lights on most I/O modules are LED's.

light ends The fraction of a petroleum mixture having the lowest boiling point.

lighter-than-air aircraft Aircraft that can rise and remain suspended by using contained gas such that the total weight of the aircraft including the gas is less than the weight of the total volume of air it displaces at the given altitude.[ARP4107-88]

lighter-than-light key The remaining components in the overhead stream other than the light key.

lighting off torch A torch used for igniting fuel from a burner. The torch may consist of asbestos wrapped around an iron rod and saturated with oil or may be a small oil or gas burner.

light intensity See luminous intensity.

light ions Ions of helium, boron, and other elements used in implantation experiments.[NASA]

light key The component in multicomponent distillation that is removed in the overhead stream and that has the lowest vapor pressure of the components in the overhead. If the reboiler head is decreased or the reflux flow increased, the light key component is the first component to fall into the bottoms product.

light leaks Light emitted by a surface intended to be opaque; normally associated with integral transilluminated systems.[ARP1161-91]

light meter A small, hand-held instrument for measuring intensity of illumination.

light modulator An apparatus that produces a sound track by means of a source of light, an appropriate optical system, and a device for inducing controlled variations in light-beam characteristics.

light-off The initiation of combustion in the combustor of a gas turbine.[ARP 906A-91]

light oil Any oil whose boiling point is in the temperature range 110 to 210 °C, especially a coal tar fraction obtained by distillation.

light pen A device by which an individual can communicate with an information system through a cathode ray tube.

light pressure See illuminance.

light transport aircraft A classification of multiengine airplanes having a maximum passenger capacity of 30 seats and a gross weight of about 35,000 pounds.[NASA]

light valve(s) 1. Optical shutters which, when activated by light, become either transparent or opaque.[NASA] 2. A device whose ability to transmit light can be made to vary by applying an external electrical quantity such as a current, voltage, electric field, electron beam or magnetic field.

light water Water in which both hydrogen atoms in each molecule are of the isotope protium.[NASA]

light water reactors Nuclear reactors using ordinary (rather than heavy) water as moderator.[NASA]

lignite Coal of relatively recent origin, an intermediate between peat and bituminous coal.[NASA]

likelihood ratio The probability of a random drawing of a specified sample from a population, assuring a given hypothesis abut the parameters of the population, divided by the probability of a random drawing of the same sample, assuring that the parameters of the population are such that this probability is maximized.[NASA]

limb brightening The increase in the intensity of radio or x-ray brightness of the sun or other stars from its center to its limb.[NASA]

limb darkening A condition, sometimes observed on celestial bodies, in which the brightness of the object decreases as the edges or limbs of the object are approached. The sun and Jupiter exhibit limb darkening.[NASA]

limen Threshold; a psychophysical concept denoting the lowest detectable intensity of any sensory stimulus.[NASA]

limit-check The comparison of data from a specific source with pre-established allowable limits for that source.

limit checking Internal program checks for high, low, rate-of-change, and deviation from a reference. These checks are to detect signals indicating undesirable or unsafe plant operation.

limit control A sensing device that shuts down an operation or terminates a process step when a prescribed limiting condition is reached.

limit cycle A sustained oscillation of finite amplitude.

limit cycling A repetitive on-off cycling of brake pressure by the locked wheel circuit.[AIR1489-88]

limited EMS An EMS where the measurement parameters are usually limited to those that are provided as part of the standard aircraft instrumentation. [AIR1873-88]

limiter(s) 1. A fuse designed specifically with a high temperature melting point to provide protection for electric power distribution systems against fault short-circuit current. A limiter is relatively insensitive to ambient temperature.[ARP 1199A-90] 2. (fusion reactors) Material aperture in fusion power reactors which collect particles from the outer surfaces of the plasmas to control their transport to regions of low density.[NASA] 3. A device which applies limits to a signal.

limit exceedances Parameter excursions beyond pre-established values. [ARP1587-81]

limiting The action which causes a transducer output to become constant even though its input continues to rise above a certain value.

limit load The highest load factors which can be expected in normal operation under various operational situations. FAR requires that aircraft structures be capable of supporting 1-1/2 times the limit load factor without failure. [ARP4107-88]

limit of detection In any instrument or measurement system, the smallest value of the measured quantity that produces discernible movement of the indicator.

limit of error In an instrument or control device, the maximum error over the entire scale or range of use under specific conditions.

limit of measurement In any instrument or measurement system, the smallest value of the measured quantity that can be accurately indicated or recorded.

limit priority A priority specification associated with every task in a multitask operation, representing the highest dispatching priority that the task may assign to itself or to any of its subtasks.

limits The prescribed maximum and minimum values of a dimension or other attribute.

limits, confidence The values, upper and lower, between which a true value can be expected to fall, with a pre-established level of confidence.[ARD50010-91]

limit switch An electromechanical device that is operated by some moving part of a power-driven machine to alter an electrical circuit associated with the machine. See also position switch.

limnology The physical, chemical, meteorological, and especially the biological and ecological conditions in inland waters.[NASA]

lin See linear.

Lindemann electrometer An electrometer in which a metallized quartz fiber mounted on a quartz torsion fiber perpendicular to its axis is positioned within a system of electrodes to produce a visual indication of electric potential.

line 1. A tube, pipe or hose which acts as a conductor of fluid.[ARP243B-65] 2. In word processing, a string of characters that terminates with a vertical tab, form-feed, line-feed, or carriage return. 3. In communications, a line provides a data transmission link. 4. In process plants, a collection of one or more assciated units and equipment modules, arranged in serial and/or parallel paths, used to make a complete batch of material or finished product. Also "production line" or "train".

lineal scale length The distance from one end of an instrument scale to the other, measured along the arc if the scale is curved or circular.

linear The type of relationship that exists between two variables when the ratio of the value of one variable to the corresponding value of the other is constant over the entire range of possible values.

linear accelerator Device for accelerating charged particles employing alternate electrodes and gaps arranged in a straight line, so proportioned that when their potentials are varied in the proper amplitudes and frequency, particles passing through them receive successive increments of energy.[NASA]

linear actuator A device for converting power into linear motion.

linear (or nonlinear) analysis Nonlinear systems may be analyzed by using a linear approximation of the nonlinear equations about a specific point, or by using a number of digital computer techniques available for solving nonlinear differential equations.[AIR1823-86]

linear arrays Antenna arrays whose elements are equally spaced along a straight line.[NASA]

linear control system A control system in which the transfer function between the controlled condition and the command signal is independent of the amplitude of the command signal.

linear evolution equations Denotes a large class of differential or integral differential equations which are used to describe the evolution in time of some physical systems from an initial state. The equation is said to be linear if the unknown functions and their derivatives appear linearly.[NASA]

linearity 1. A measure of the linear relationship between input/output over the entire operating range.[ARP1281A-81] 2. The degree to which the normal output curve conforms to a straight line under specified load conditions, usually expressed as percentage of rated command.[AIR1916-88]

linearity, differential Any two adjacent digital codes should result in measured output values that are exactly 1 LSB apart. Any deviation of the measured "step" from the ideal difference is called differential nonlinearity expressed in multiples of 1 LSB.

linearization The process of converting a nonlinear (nonstraight-line) response into a linear response.

linear meter An instrument whose indicated output is proportional to the quantity measured.

linear optimization See linear programming.

linear polarization 1. Polarization of an electromagnetic wave in which the electric vector of a fixed point in space remains pointing in a fixed direction although varying in magnitude.[NASA] 2. Light in which the electric field vector points in only a single direction.

linear position sensing detector An optical detector which can measure the position of a light spot along its length.

linear potentiometer A variable resistance device whose effective resistance is a linear function of the position of a control arm or other adjustment; most often, the device is constructed so that a single length of straight or coiled wire whose resistance varies uniformly along its length is in contact with a shoe or similar sliding member; effective resistance is varied by connecting the circuit to one end of the wire and to the shoe, and then varying the position of the shoe; use of a wire-wound resistor, thin film or printed circuit element allows greater voltage drop per unit length along the potentiometer, and therefore stronger and more useful output signals.

linear programming (LP) A method of solution for problems in which a linear function of a number of variables is subject to a number of constraints in the form of linear inequalities.

linear quadratic Gaussian control A type of optimal-state feedback control whose design considers noise. It is primarily used to control aircraft and spacecraft systems.[NASA]

linear quadratic regulator A type of optimal-state feedback controller that does not consider noise. It is primarily used to control aircraft and spacecraft. [NASA]

linear regulator A linear regulator is a special case for minimization of the performance index to obtain optimal performance.[AIR1823-86] See also linear quadratic regulator.

linear system See system, linear.

linear transducer A type of transducer for which a plot of input signal level versus output signal level is a straight line.

linear variable differential transformer (LVDT) A type of position sensor consisting of a central primary coil and two secondary coils wound on the same core; a moving-iron element linked to a mechanical member induces changes in self induction that are directly proportional to movement of the member.

linear variable reluctance transducer (LVRT) A type of position sensor consisting of a center-tapped coil and an opposing moving coil attached to a linear probe; the winding is continuous over the length of the core, instead of being segmented as in an LVDT; the chief disadvantage of an LVRT is that the overall length must be at least double the stroke, whereas the chief advantage is its excellent linearity over an effective stroke up to 24 in. (610 mm).

linearvection See illusion, vection.

linear velocity A vector quantity whose magnitude is expressed in units of length per unit time and whose direction is invariant; if the direction varies in circular fashion with time, the quantity is known as angular or rotational velocity, and if it varies along a fluctuating or noncircular path the quantity is known as curvilinear velocity.

line, bleed A line, selectively open to overboard, which serves only for removing foreign substances from a system or unit, as for removal of entrapped air from a hydraulic circuit.[ARP243B-65]

line-class valve A valve qualified by its design characteristics to be used as the first valve off the process line.

line combustor A fuel burning device in the air line of a pneumatic starting system utilized to heat the air directed to the starter.[ARP906A-91]

line, drain A line returning leakage fluid independently to reservoir or return circuit. Also, a line selectively open to overboard for removing fluid from the system.[ARP243B-65]

line mixer See flow mixer.

line of sight 1. The straight line between the observer and a target or the observed point. 2. The straight line from a transmitting radar antenna in the direction of the beam, especially toward a target. [ARP4107-88] 3. An aim or observation taken with mechanical or optical aid to establish a direct path to an objective, target, etc.[NASA]

line of sight communication Electromagnetic wave propagation, usually microwaves, in a straight line between the transmitter and receiver. The useful transmission distance is generally limited to the horizon as sighted from the elevation of the transmitter.[NASA]

line-of-sight-transmission Transmissions that follow a straight line rather than the curvature of the earth. UHF and VHF transmissions tend to be more line of sight than LF or MF transmissions.[ARP4107-88]

line oriented flight training See LOFT.

line, pilot A line which acts as a conductor of control actuating fluid.[ARP 243B-65]

line pressure A line which conducts fluid from the pressure source to a control unit or units.[ARP243B-65] See reference pressure.

line printer A computer printer that operates on a line-by-line (rather than character-by-character) basis for high-speed systems.

liner A material applied to the inside of a solid rocket case which adheres to both the case and the propellant.[AIR 913-89]

line replaceable unit (LRU) 1. A item which is replaced at the organizational maintenance level.[ARD50010-91] 2. A component or assembly of components that by design, is packaged to be removed and replaced as a single item. [AIR1266-77] 3. Propulsion component or assembly that may be replaced at the lowest level of maintenance, sometimes called a weapon replacement assembly (WRA).[ARP1587-81]

line, return A line which conducts working fluid back to the reservoir.[ARP 243B-65]

line spectra The spontaneous emission of electromagnetic radiation from the bound electrons as they jump from high to low energy levels in an atom.[NASA]

line spectrum The spectrum of a complex wave consisting of several components having discrete frequencies.

line, supply A line which conducts a fluid supply, as from a reservoir to a pump.[ARP243B-65]

line, vent A line which is continuously open to atmosphere.[ARP243B-65]

line width 1. Width at 50 percent of peak luminance of the line luminance distribution.[AS8034-82] [ARP4067-89] 2. On a CRT, the width of a luminance distribution at 50% of the maximum amplitude of the luminance distribution. [ARP1782-89]

lining The material used on the furnace side of a furnace wall. It is usually of high grade refractory tile or brick or plastic refractory material.

lining, brake A specially compounded material applied to rotor and/or stator to create a predictable coefficient of friction and act as a wear surface in the brake assembly.[AIR1489-88]

lining loading A measurement of the total amount of energy absorbed through the interface of each square unit of lining and its mating surface over a short period of time or a single stop.[AIR 1489-88]

lining power A measure of the average amount of energy absorbed by a square unit of lining area and its mating surface during each second.[AIR1489-88]

link 1. An intermediate rod or piece of material that transmits force or motion.[ARP480A-87] See also communication link.

linkage 1. A technique for providing interconnections between routines. 2. A mechanism consisting of bars, slides, pivots and rotating members, which transfers motion from one part of a machine to another.

linkage editor A computer program that produces a load module by transforming object modules into a format that is acceptable to fetch, combining separately produced object modules and previously processed load modules into a single load module, resolving symbolic cross references among them, replacing, deleting, and adding control sections automatically on request, and providing overlay facilities for modules requesting them.

linked list In data processing, a method of organizing data so that it is retrievable in an order that is not always the same order as the data is stored.

linker A computer program that binds together independently assembled programs. The program is developed in modules which are then linked together to form the whole.

link library A generally accessible partitioned computer data set which, unless otherwise specified, is used in fetching load modules referred to in execute (EXEC) statements and in attach, link, load, and transfer control (XCTL) macro instructions.

link test Test conducted to establish the integrity of the data link.[ARP4102/13-90]

liquid barometer A simple device for measuring atmospheric pressure, which can be constructed by filling a glass tube having one closed end with a liquid such as mercury, then temporarily plugging the open end, inverting the tube into a container partly filled with the liquid, and unplugging the open end; if the liquid is mercury the tube must be at least 30 in. (76.2 mm) long; liquids of different densities require tubes of different lengths.

liquid cooled dissipator See cold plate.

liquid crystal display A type of digital display device.

liquid crystal light valve A device used in optical processing to convert an incoherent light image into a coherent light image.

liquid drops See drops (liquids).

liquid-filled thermometer Any of several designs of temperature-measurement devices that depend for their operation on predictable change in volume with temperature of a liquid medium confined in a closed system.

liquid knockout See impingement.

liquid level control A device for sensing and regulating the position of a liquid surface within a vessel.

liquid-level manometer A differential pressure gage in which the reading is obtained by viewing the change in level of one or both of the free surfaces of a liquid column spanning both gage legs.

liquid-metal embrittlement A decrease in strength or ductility of a solid metal caused by contact with a liquid metal.

liquid phase epitaxy A liquid phase transformation during crystal growth.[NASA]

liquid plus solid zones See mushy zones.

liquid propellant auxiliary power system In its simplest form it consists of propellant, propellant tankage, propellant expulsion system and flow control devices, and the combustion/decomposition chamber.[AIR744A-72]

liquid propellant rocket engines Rocket engines using a propellant or propellants in liquid form.[NASA]

liquid rocket propellants Specifically, rocket propellants in liquid form. Examples of liquid propellants include fuels such as alcohol, gasoline, aniline, liquid ammonia, and liquid hydrogen; oxidants such as liquid oxygen, hydrogen peroxide (also applicable as a monopropellant), and nitric acid; additives such as water; and monopropellants such as nitromethane.[NASA]

liquids Substances in a state in which the individual particles move freely with relation to each other and take the shape of the container, but do not expand to fill the container.[NASA]

liquid sloshing The back and forth movement of a liquid fuel in its tank. creating problems of stability and control in the vehicle.[NASA]

liquid spring A type of shock absorber which is completely filled with fluid and which utilizes the compressibility of the fluid for the shock absorption function. It provides a relatively high spring rate compared to an air-oil type shock absorber.[AIR1489-88]

liquid wastes The liquid counterpart of solid wastes from industrial, chemical, metabolic, and/or mineral sources.[NASA]

LIRL Low Intensity Runway Lights. See runway edge light system.

LIRTS (telescope) A proposed large infrared telescope for Spacelab superseded by the German infrared laboratory.[NASA]

lissajous figures Figures where the path of a particle moving in a plane when the components of its position along two perpendicular axes each undergo simple harmonic motions and the ratio of their frequencies is a rational number.[NASA]

list An ordered set of items contained within an electronic memory in such a way that only two items are readily program addressable. These items are the earliest appended (beginning item) and the most recently appended (ending item). Items stored into the list are "appended" following the ending item. Items read from the list are "removed." Same as push-up list.

listing The hard copy generated by a line printer; may also refer to a visual CRT display generated in lieu of hard copy.

list processing A method of processing data in the form of lists. Usually, chained lists are used so that the logical order of items can be changed without altering their physical location.

literal An element of a programming language that permits the explicit representation of character strings in expressions and command and function elements; in most languages, a literal is enclosed in either single or double quotes to denote that the enclosed string is to be taken "literally" and not evaluated.

lithium iodates Salts of iodic acid containing the 10 to the third power radical.[NASA]

lithium sulfur batteries Primary cells for producing electrical energy using lithium metal for one electrode and sulfur for the other.[NASA]

lithology Description of the physical character of a rock as determined by eye or with a low-power magnifier and based on color, structure, mineralogic components, and grain size.[NASA]

litmus A blue, water-soluble powder derived from lichens and used as an acid-base indicator; it is blue at pH 8.3 and above, and is red at pH 4.5 and below.

litre Also spelled liter. Abbreviated L. The SI unit of volume; it equals 0.001 m^3 or 1.057 quarts.

live center A lathe center held in the headstock that rotates with the headstock and part being turned.

live front An assembly arrangement which has all moving or energized parts exposed on the front of the panel, framework or cabinet.

live part A part which is considered capable of rendering an electric shock.

live room An enclosed space characterized by unusually small capacity for absorbing sound.

live steam Steam which has not performed any of the work for which it was generated.

live zone See zone, live.

lixiscopes Portable light weight battery-operated low intensity x-ray imaging systems with medical, industrial, and scientific applications.[NASA]

LLC See logical link control.

LLEI See low-level engineering interface.

LLHI See low-level human interface.

LLOI See low-level operator interface.

LLWSAS Low-Level, Wind-Shear Alert System.

lm See lumen.

L/MF Low to Middle Frequency.

L/MF airway(s) Airways whose NAVAIDs utilize low or medium frequencies.[ARP4107-88]

LMM Compass Locator at the Middle Marker. See compass locator.

L network An electronic network composed of two branches in series, with the junction and the free end of one branch being connected to one pair of terminals and the free ends of both branches being connected to another pair of terminals.

load 1. The quantity that can be carried at one time (for example, cargo load or passenger load); the total weight of a vehicle (for example, as in the calculation of wing loading—the load supported by the wings). 2. The stress or forces placed upon a structure either by external weights or air pressures.[ARP4107-88] 3. To store a computer program or data into memory. 4. To mount a magnetic tape on a device so that the read point is at the beginning of the tape. 5. To place a removable disk in a disk drive and start the drive. 6. The amount of force applied to a structural member in service. 7. The quantity of parts placed in a furnace, oven or other piece of process equipment. 8. The quantity or mass of bulk material placed in a hopper, railcar or truck. 9. The power demand on an electrical distribution system. 10. The amount of power needed to start or maintain motion in a power-driven machine. 11. The process load is a term to denote the nominal values of all variables in a process that affect the controlled variable. 12. In an electric power circuit, the resistive and reactive components which comprise the device being powered by the circuit. 13. In a physical structure the externally applied force, or the sum of external forces and the weight of the structure borne by a single member or by the entire structure. See also load impedance.

load-and-go In data processing, an automatic coding procedure that not only compiles the program, creating machine language, but also proceeds to execute the created program; load-and-go procedures are usually part of a monitor.

load, assembly The total load imposed on a landing gear assembly or leg. May be further broken down to individual wheel loads.[AIR1489-88]

load cell A transducer for the measurement of force or weight. Action is based on strain gages mounted within the cell on a force beam.

load circuit A circuit or a branch of a network which carries the main portion of current flow.

load classification number A rating number which indicates the capability of an aircraft with specific tire/wheel equipment to operate from a given type of airbase. Intercontinental express = 100, light service = less than 14.[AIR 1489-88]

load deflection curve For a tire, a plot of applied load (usually the abscissa) versus the deflection (usually the ordinate). Also, usually provided for a given (or multiple) inflation pressure.[AIR 1489-88]

load, dynamic A load imposed by dynamic action as distinguished from a static load. Specifically with respect to aircraft, rockets, or spacecraft, a load due to an acceleration of the craft, as imposed by gusts, by maneuvering, by landing, taxiing, take off, turning, etc. [AIR1489-88]

loaded galley weight The weight of all galleys including maximum allowable weight of food, carts, beverages, inserts, galley service items.[AS1426-80]

load, equivalent single wheel The calculated load which, if applied to a single tire, would produce the same effect on the airfield as does the multiple wheel installation under consideration.[AIR 1489-88]

load factor 1. A number which yields the inertial load (g load) when multiplied by the weight of the object.[AIR 1489-88] 2. The ratio of two loads (basic load is the denominator). The load factor for a landing gear assembly is obtained by dividing the sum of the external forces acting upon it by the static load. The force of gravity does not appear in the sum of external forces because on each particle of mass, the gravity force is cancelled by the inertial force of free-fall acceleration.[AIR1489-88] 3. The ratio of a specified load to the total weight of the aircraft. The specified load is expressed in terms of any of the following: aerodynamic forces, inertia forces, ground or water reactions.[ARP 4107-88] 4. The ratio of the average load in a given period to the maximum load carried during that period.

load, holdback The load applied to the aircraft holdback fitting. The load results from the buffing force during deceleration of the aircraft from taxi-in velocity, or from the simultaneous application of catapult tensioning force, aircraft full takeoff power, and ship motion before firing the catapult.[AIR 1489-88]

loading Buildup of material along the cutting edge of a bit or other tool. Similarly, buildup of grinding debris on the working face of a grinding wheel or abrasive disc.

loading density The weight of explosive per unit volume.[AIR913-89]

load module A program prepared in a format that is ready for loading and executing.

load natural frequency The undamped resonant frequency of the load mass coupled with the load related stiffness. [AIR1916-88]

load point (mag tape) The point, near the beginning of the tape, at which the computer can start to record data.

load range The range of load over which a tire is capable of operating. For a tire, variation of loads would usually be accompanied by a proportional increase or decrease of inflation pressure within design limitations.[AIR1489-88]

load, rated The specific load at which a tire (or wheel, etc.) is rated or qualified. For a tire, the rated load would be at a specific inflation pressure and tire deflection.[AIR1489-88]

load, release The maximum load applied to the aircraft holdback fitting by application of the catapult firing force in addition to the existing holdback load. The release load of predetermined magnitude ruptures a release element or disengages a repeatable release fitting, thereby allowing the aircraft to be released.[AIR1489-88]

load repetition factor Sufficient passes of load carrying tires in adjacent tire paths to just cover a given width of pavement one time. The term is identical with "coverage" as used in pavement design.[AIR1489-88]

load sensor See weigh carriage. Also called load receiving element.

load, speed, time curves Curves plotted against time as the abscissa which show the relationships of load and speed vs. time.[AIR1489-88]

load, ultimate The load that should cause destructive failure according to stress analysis; the load that causes failure during a test of strength.[ARP 4107-88]

LOC Localizer Only Approach.

Local Area Network (LAN) 1. Network, generally microcomputer based, that enables users in the same location to use the same programs and equipment such as printers.[NASA] 2. A communications mechanism by which computers and peripherals in a limited geographical area can be connected. They provide a physical channel of moderate to high data rate (1 to 20 Mbit) which has a consistently low error rate (typically 10^{-9}).

Local Control Unit (LCU) A control device that performs closed-loop control and interfaces directly with the process.

local group (astronomy) The cluster of galaxies to which our galaxy belongs. It is a poor, irregular cluster with some 20 certain members including the Milky Way Galaxy, the Andromeda Galaxy, the Triangulum, four irregular galaxies, and about 13 intermediate or dwarf ellipticals. [NASA]

localizer 1. A radio facility which provides signals for use in lateral guidance of aircraft with respect to a runway centerline.[ARP419] 2. The component of an ILS which provides course guidance to the runway.[ARP4107-88]

localizer only approach (LOC) A nonprecision instrument approach procedure utilizing the localizer signal for course guidance to the runway; the glide slope is not available.[ARP4107-88]

localizer-type directional aid (LDA) A NAVAID used for nonprecision instrument approaches with utility and accuracy comparable to a localizer but which is not part of a complete ILS and is not aligned with the runway.[ARP 4107-88]

local oscillator An oscillator whose output is combined with another frequency to generate a sum or difference frequency either of which may be easier to amplify and use, as in a superhetrodyne receiver.

local processing unit Field station with input/output circuitry and the main processor. These devices measure analog and discrete inputs, convert these inputs to engineering units, perform analog and logical calculations (including control calculations) on these inputs and provide both analog and discrete (digital) outputs.

local reference A copper bar mounted on the cabinets of a subsystem which become the signal reference point for the entire subsystem. All power commons and signal commons of a subsystem are tied to the local reference. Each local reference is tied to the master reference, by a separate wire.

local traffic Aircraft operating in the local traffic pattern or within sight of the tower; or, aircraft known to be departing for or arriving from flight in local practice areas; or, aircraft executing simulated instrument approaches at the airport.[ARP4107-88]

location An address in computer storage or memory where a unit of data or an instruction can be stored.

location counter 1. In data processing, the control-section register which contains the address of the instruction currently being executed. 2. A register in which the address of the current instruction is recorded. Synonymous with instruction counter and program address counter.

locator A device for positioning a contact, terminal or splice in the crimping dies.[ARP914A]

locator beacon, personnel A portable, lightweight, manually operated beacon, designed to be carried on the person, in the cockpit of an aircraft, or attached to a parachute, which operates from its own power source on 121.5 MHz or 243 MHz, or both, (preferably on both emergency frequencies) transmitting a distinctive downward swept audio tone for homing purposes, which may or may not have voice capability; and which is capable of operation by unskilled persons.[ARP4107-88]

lock 1. In a forging, having the flash line in more than one plane. 2. A device for securing a door, drawer or hatch that features a movable bolt operated by a key. 3. To prevent a movable part from moving; to seize.

locked-in liner In a butterfly valve body, a liner retained in the body bore by a key ring or other means.

locked wheel protection The temporary removal of brake pressure by a secondary control circuit when the controlled wheel(s) reaches a low predetermined wheel velocity when the aircraft is at a significantly higher velocity.[AIR1489-88]

lock-in amplifier An amplifier which selects signals at one prespecified frequency and amplifies them, while discriminating against signals at other frequencies.

locking Pertaining to code extension characters that change the interpretation of an unspecified number of following characters. Contrast with nonlocking.

locking spring See contact retainer.

lockout Any condition which prevents any or all senders or receivers from communicating.

lock-step A method to synchronize a mixed-signal simulation system where each simulator progresses one time step and passes all interacting signals to the other simulator.

locomotive boiler A horizontal fire-tube boiler with an internal furnace the rear of which is a tube sheet directly attached to a shell containing tubes through which the products of combustion leave the furnace.

locus of control An attitudinal set in which a person believes either that he/she is in control of his/her destiny (internal locus of control), or that outside influences control his/her destiny (external locus of control).[ARP4107-88]

LOFT Acronym for Line Oriented Flight Training. A technique of conducting aircrew flight training in simulators by having the crew fly full mission scenarios together in the simulator as they would in an operational aircraft. This type of flight training emphasizes the whole mission concept and is aimed at developing effective communications among crewmembers, effective human resource utilization, and good crew coordination.[ARP4107-88]

log 1. A record of everything pertinent to a machine run including: identification of the machine run, record of alteration switch settings, identification of input and output tapes, copy of manual key-ins, identification of all stops, and a record of action taken on all stops. 2. To record occurrences in a chronological sequence.

logarithmic amplifier An amplifier whose output is a logarithmic function of its input.

logarithmic decrement In an exponentially damped oscillation, the natural logarithm of the ratio of one peak value to the next successive peak value in the same direction.

logarithms The power to which a fixed number, called the base, usually 10 or e (2.7182818) must be raised to produce the value to which the logarithm corresponds.[NASA]

logger A device which automatically records physical processes and events, usually chronologically.

logging (industry) The business of felling trees, cutting them up into logs and transporting the logs to sawmills or to a place of sale.[NASA]

logic 1. A means of solving complex problems through the repeated use of simple functions which define basic concepts. Basic logic functions are "AND", "OR", "NOT", etc. 2. The science dealing with the criteria or formal principles of reasoning and thought. 3. The systematic scheme which defines the interactions of signals in the design of an automatic data processing system. 4. The basic principles and application of truth tables and interconnection between logical elements required for arithmetic computation in an automatic data processing system. Related to symbolic logic.

logical addressing A mode of addressing in which message content or type is labeled.[AIR4271-89]

logical block An arbitrarily-defined, fixed number of contiguous bytes used as the standard I/O transfer unit throughout a computer operating system. For example, the commonly used logical block in PDP-11 systems is 512 bytes long. An I/O device is treated as if its block length is 512 bytes, although a device's actual (physical) block length may be different. Logical blocks on a device are numbered from block 0 consecutively up to the last block on the volume.

logical connectives The computer operators or words, such as and, or, or else, if then, neither, nor, and except, which make new expressions from given expressions and which have the property that the truth or falsity of the new expressions can be calculated from the truth or falsity of the given expressions and the logical meaning of the operator.

logical decision 1. The choice or ability to choose between alternatives. Basically, this amounts to an ability to answer yes or no with respect to certain fundamental questions involving equality and relative magnitude. 2. The utilization of a logic instruction.

logical device name An alphanumeric name assigned by a user to represent a physical device; the name can be used synonymously with the physical device name in all references to the device. Logical device names are used in device-independent systems to enable a program to refer to a logical device name that can be assigned to a physical device at run time.

logical difference All elements belonging to Class A but not to Class B, when two classes of elements, Class A and Class B, are given.

logical element(s) In computers or data processing systems, the smallest building blocks which can be represented by operators in an appropriate system of symbolic logic. Typical logical elements are the AND gate and flip-flop, which can be represented as operators in a suitable symbolic logic.[NASA]

logical expression A logical expression consists of logical constants, variables, array elements, function references, and combinations of those operands, separated by logical operators and parentheses.

logical link control (LLC) The upper sublayer of the Data Link Layer (Layer 2) used by all types of IEEE 802 Local Area Networks. LLC provides a common set of services and interfaces to higher layer protocols. Three types of services are specified: a) Type 1 - Connectionless: A set of services that permit peer entities to transmit data to each other without the establishment of connections. Type 1 service is used by both MAP and TOP; b) Type 2 - Connection Oriented: A set of services that permit peer entities to establish, use and terminate connections with each other in order to transmit data; c) Type 3 - Acknowledged Connectionless: A set of services that permit a peer entity to send messages requiring immediate response to another peer entity. This class of service can also be used for polled (master-slave) operation.

logical operation 1. An operation in which logical (yes or no) quantities form the elements being operated on, e.g., AND, OR. 2. The operations of logical shifting, masking, and other nonarithmetic operations of a computer. Contrasted with arithmetic operation.

logical operator See logical connectives.

logical product Same as AND.

logical record A logical unit of data within a file whose length is defined by the user and whose contents have significance to the user; a group of related fields treated as a unit.

logical sum A result, similar to an arithmetic sum, obtained in the process of ordinary addition, except that the rules are such that a result of one is obtained when either one or both input variables is a one, and an output of zero is obtained when the input variables are both zero. The logical sum is the name given the result produced by the inclusive OR operator.

logical unit number A number associated with a physical device unit during a task's I/O operations; each task in the system can establish its own correspondence between logical unit numbers and physical device units.

logical variable A variable that may

have only the value true or false. Also called Boolean variable.

logic analyzer A device used to analyze the logical operation of the microcomputer. It is a test device used for debugging systems.

logic design The specification of the working relations between the parts of a computer system in terms of symbolic logic and without primary regard for hardware implementation.

logic device The general category of digital fluidic components which perform logic functions; for example AND, NOT, OR, NOR, and NAND. They can gate or inhibit signal transmission with the application, removal, or other combinations of control signals.[ARP993A-69]

logic diagram 1. In data processing, a diagram that represents a logic design and sometimes the hardware implementation. 2. Graphic method of representing a logic operation or set of operations.

logic gate A device that takes binary bits as input and produces an output bit to some specification.

logic instruction A computer instruction that executes an operation that is defined in symbolic logic, such as AND, OR, NOR.

logic levels Electrical convention for representing logic states. For TTL systems, the logic levels are nominally 5V for logic 1, and 0V for logic 0.

logic network In data processing, an arrangement of logic gates designed to achieve specific outputs.

log-in See log-on.

logistics The science of planning and carrying out the movement and maintenance of forces. For its most comprehensive sense, those aspects of military operations that deal with: (a) design and development, acquisition, storage, movement, distribution, maintenance, evaluation, and hospital inspection of personnel; (b) acquisition of construction, maintenance, operation, and disposition of facilities; and (c) acquisition or furnishing of services.[ARD50010-91]

logistics support The materials and services required to enable the operating forces to operate, maintain, and repair the end item within the maintenance concept defined for that end item. Logistics support encompasses the identification, selection, procurement, scheduling, stocking, and distribution of spares, repair parts, facilities, ground support equipment, trainers, technical publications, contractor engineering and technical services, and personnel training as necessary to provide the operating forces with the capability needed to keep the end item in a functioning status.[ARD50010-91]

logistic time Logistic time is all replacement procurement time, except that time when the maintenance man is engaged in the procurement activity.[ARD50010-91]

log-on In data processing, to enter into a system or network.

log-out In data processing, to exit from a system or network.

LOM Outer Compass Locator. See compass locator.

long duration space flight Space flight involving interplanetary and/or interstellar travel.[NASA]

long flame burner A burner in which the fuel emerges in such a condition, or one in which the air for combustion is admitted in such a manner, that the two do not readily mix, resulting in a comparatively long flame.

longitude Angular distance, along a primary great circle, from the adopted reference point; the angle between a reference plane through the polar axis and a second plane through that axis.[NASA]

longitudinal axis An imaginary straight line lying in the plane of symmetry extending from the nose to the tail of an aircraft; and passing through the center of gravity. Movement about this axis is called roll.[ARP4107-88]

longitudinal drum boiler A sectional header or box header boiler in which the axis on the horizontal drum or drums is parallel to the tubes in a vertical plane.

longitudinal interference See common mode interference.

longitudinal redundancy check (LRC) A system of error control based on the formulation of a block check following preset rules. The check formation rule is applied in the same manner to each character.

longitudinal separation The longitudinal spacing of aircraft at the same altitude by a minimum distance expressed in units of time or miles. The amount of separation required is a function of the availability of radar coverage and the distance the aircraft is from the radar antenna site.[ARP4107-88]

longitudinal stability See stability, longitudinal.

longitudinal wave(s) 1. Waves in which the direction of displacement at each point of the medium is normal to the wave front.[NASA] 2. A wave in which the medium is displaced in a direction perpendicular to the wave front at all points along the wave.

long period variables See Mira variables.

long range navigation See loran.

long term trending Tracking of engine, engine components or subsystem degradation on a periodic basis, often by flight. This type of tracking indicates a deviation of monitoring data from an established trend.[ARP1587-81]

long waves (meteorology) See planetary waves.

look angles (electronics) The solid angle in which an instrument operates effectively, generally used to describe radars, optical instruments, and space radiation detectors.[NASA]

look angles (tracking) The elevation and azimuth at which a particular satellite is predicted to be found at a specified time.[NASA]

look-up table Same as table, and not to be confused with the verb form, table look up.

loop 1. The signal path in a closed-loop control system beginning with the error signal and ending with the resultant feedback signal that functionally interconnects the forward-loop and feedback-loop elements.[AIR1916-88] 2. A sequence of instructions that is executed repeatedly until a terminal condition prevails. 3. Synonymous with control loop. See closed loop and open loop. 4. The doubled part of a cord, wire, rope or cable; a bight or noose. 5. A complete hydraulic, electric, magnetic or pneumatic circuit. 6. A length of magnetic tape or motion picture film that has been spliced together, end-to-end, so it can be played repeatedly without interruption. 7. In data processing, a closed sequence of instructions that are repeated. 8. All the parts of a control system: process, or, sensor, any transmitters, controller, and final control element. 9. In a computing program: A sequence of instructions that is written only once but executes many times (iterates) until some predefined condition is met.

loop (computing) Instructions that actually perform the primary function of a loop, as distinguished from loop initialization, modification, and testing.

loop diagram A schematic representation of a complete hydraulic, electric, magnetic or pneumatic circuit.

loop, feedback See loop, closed (feedback loop).

loop gain The product of the gains of all the elements in a loop. See also gain, loop.

loop identification 1. Consists of a first-letter and a number of an instrument loop. Each instrument within a loop has assigned to it the same loop number and, in the case of parallel numbering, the same first letter. 2. Each instrument loop has a unique loop identification.

loop (initialization) Instructions that immediately precede the loop proper, that set addresses, counters, or data to initial values.

loop load The algebraic sum of the applied loads at the anchorages of a torso restraint system segment. A balanced loop load is achieved when the reaction loads at each lap belt anchorage are equal.[AS8043-86]

loop-modification Instructions of a loop that alter instruction addresses, counters, or data.

loop-testing Instructions of a loop that determine whether the loop is complete.

loose cover A separation of the cover from the carcass or reinforcements.[ARP1658A-86]

LORAN 1. Acronym for LOng RAnge Navigation. An electronic navigational system by which hyperbolic lines of position are determined by measuring the difference in the time of reception of synchronized pulse signals from two fixed transmitters. The vehicle's position is determined by the intersection of these two hyperbolic lines of position generated by the two transmitters. Loran A operates in the 1750 to 1950 kHz frequency band; Loran C and D operate in the 100 to 110 kHz frequency band.[ARP4107-88] 2. A two dimensional pulse synchronized radio navigation system to determine hyperbolic lines of position through pulse time differencing from a master compared to two slave stations.[NASA]

Lorentz force The force affecting a charged particle due to the motion of the particle in a magnetic field.[NASA]

loss 1. Energy dissipated without accomplishing useful work.[ARP1931-87] 2. Dissipation of power, which reduces the efficiency of a machine or system. 3. Dissipation of material or energy due to leakage.

loss factor A factor of an insulating material which is equal to the product of its dissipation factor and dielectric constant.[ARP1931-87]

lossless materials Dielectric materials that do not dissipate energy or that do not dampen oscillations.[NASA]

lossy line A cable having large attenuation per unit of length.[ARP1931-87]

lossy media A material that dissipates energy of electromagnetic or acoustic energy passing through it.[NASA]

lost cluster A group of one or more disk sectors that are not available for storage use.

lot (2-part material) A quantity of material packaged from a batch of base compound and a batch of catalyst, given a specific identity, and offered for acceptance. This definition also pertains to one-part material except that it is produced from a single batch.[AS 7200/1-90]

loudness 1. The intensive attribute of an auditory sensation, in terms of which sounds may be ordered on a scale extending from soft to loud. Loudness is measured in sones. Loudness depends primarily upon the sound pressure of the stimulus, but it also depends upon the frequency and waveform of the stimulus.[NASA] 2. The relative auditory intensity of a sound wave.

loudness level A measurement of sound intensity numerically equal to the sound pressure, in decibels, relative to 0.0002 microbar, of a simple tone whose frequency is 1000 Hz and which is judged by the listeners to be equivalent in loudness; the units of measure determined in this way are called phons.

loudspeaker An electroacoustic transducer usually constructed to effectively radiate sound of varying frequencies into the air.

low-alloy steel An iron-carbon alloy which contains up to about 1% C, and less than 5% by weight of additional elements.

low altitude airway structure/federal airways The network of airways serving aircraft operation up to but not including 18,000 ft MSL.[ARP4107-88]

low approach An approach over an airport or runway following an instrument (IFR) approach, or a visual flight rules (VFR) approach including the go-around maneuver, where the pilot intentionally does not make contact with the runway. Low approaches are often performed during training flights where a landing is not required at the completion of each approach or in those circumstances in which the pilot feels that it would be unsafe to continue the approach for a landing.[ARP4107-88]

low brass A binary copper-zinc alloy containing about 20% zinc.

low-carbon steel 1. An iron-carbon alloy containing about 0.05 to 0.25% C, and up to about 0.7% Mn. 2. Iron alloys containing carbon in low percentages that display temper and malleability characteristics not found in ordinary carbon steels.[NASA]

low cost visual approach slope indicator (LCVASI) A visual approach slope indicator system consisting of painted plywood panels (normally black and white or fluorescent orange) whose alignment as seen by the pilot indicates the aircraft's relative position with regards to a fixed glide path to a touchdown point on the runway.[ARP4107-88]

low cycle fatigue (LCF) 1. Failure caused by cyclic loading to levels resulting in non-elastic deformation that results in complete failure of the component in less than 50,000 loading cycles. Most rotating gas turbine components are prone to low cycle fatigue failure as a result of centrifugal and thermally induced loads, and their life usage can be measured with reasonable accuracy. [AIR784-88] 2. Component material life usage incurred by cyclic stress excursions.[ARP1587-81]

low draft switch A control to prevent burner operation if the draft is too low. Used primarily with mechanical draft.

lower atmosphere Generally, and quite loosely, that part of the atmosphere in which most weather phenomena occur (i.e., the troposphere and lower stratosphere); hence, used in contrast to the common meaning for the upper atmosphere.[NASA]

lower body negative pressure Application and/or measurement of reduced pressure in the portion of the body below the illiac crests. Used as a simulator or orthostatic stress or as an indicator of cardiovascular deconditioning in weightless environment.[NASA]

lower limit The signal corresponding to the minimum value of the transmitted input.

lower range-limit See range-limit, lower.

lower range-value See range-value, lower.

lower test MTBF (THETA 1) That value which is unacceptable. The standard test plans will be reject, with high probability, equipment with a true MTBF that approaches (THETA 1).[ARD50010-91]

low-fire start The firing of a burner with controls in a low-fire position to provide safe operating condition during light-off.

low frequency (LF) The frequency band between 30 and 300 kHz.[ARP4107-88]

low frequency range (LFR) A directional NAVAID, in the frequency band between 30 and 300 kHz, no longer used in the United States.[ARP4107-88]

low frequency stiffness (frequency dependent dynamic stiffness) The stiffness associated with the output deflection of an active actuation system caused by externally applied loads where the frequency of the load disturbance is within the bandpass (bandwidth) of the actuation system.[AIR1916-88]

low gas pressure switch A control to stop the burner if gas pressure is too low.

low gravity See reduced gravity.

low head boiler A bent tube boiler having three drums with relatively short tubes in a vertical plane.

low-heat value The high heating value minus the latent heat of vaporization of the water formed by burning the hydrogen in the fuel.

low-hydrogen electrode A covered welding electrode that provides an atmosphere around a welding arc which is low in hydrogen.

low intensity readability The visual clarity of a lighted display when compared with other lighted displays, both energized at a selected low level of excitation.[ARP1161-91]

low intensity runway light system (LIRL) See runway edge light system.

Low Intensity X-Ray Imaging Scopes See lixiscopes.

low-level human interface (LLHI) A device that allows a human to interact with a local control unit.

low-level language In data processing, a program instruction that usually has a single machine instruction.

low-level operator interface (LLOI) A type of LLHI designed for use by a process operator.

low-level, wind-shear alert system A system of five or six anemometers around the periphery of an airport the readouts of which are automatically compared with that of the center-field anemometer. A wind vector difference of 15 knots or more between the center-field anemometer and any peripheral anemometer is indicative of potential wind shear, and the tower will advise pilots of the potential for wind shear.[ARP4107-88]

low limiting control See control, low limiting.

low loss Term applied to a dielectric material or cable that has a small amount of power loss over long lengths making it suitable for transmission of radio frequency energy.[ARP1931-87]

low mass See mass.

low noise cable A cable configuration specially constructed to eliminate spurious electrical disturbances caused by capacitance changes or self generated noise.[ARP1931-87]

low oil temperature switch (cold oil switch) A control to prevent burner operation if the temperature of the oil is too low.

low order Pertaining to the weight or significance assigned to the digits of a number, e.g. in the number, 123456, the low order digit is six. One may refer to the three low order bits of a binary word, as another example. See order.

lowpass filter(s) 1. Wave filters having a single transmission band extending from zero frequency up to some critical or bounding frequency, not infinite. [NASA] 2. A filter which passes frequencies below its cut-off frequency with little attenuation.

low-pass output filter (LPOF) In a subcarrier discriminator, the filter which rejects subcarrier components and all extraneous noise while passing the frequencies which are known to contain data.

low-pressure hot-water and low-pressure steam boiler A boiler furnishing hot water at pressures not exceeding 160 pounds per square inch or at temperatures not more than 250 °F or steam at pressures not more than 15 pounds per square inch.

low resolution graphics In data processing, the ability of a dot-matrix printer to reproduce simple forms or pictures.

low Reynolds number A Reynolds number below the critical Reynolds number of a sphere.[NASA]

low-temperature hygrometry The

measurement of water vapor at low temperatures; requires special techniques because of the small amounts of moisture typically present and because of unusual instrument operating characteristics at such temperatures.

low tension An item incorporating an electrode(s) across which an electric spark is discharged to ignite a combustible mixture in a continuous burning cycle engine and categorized by "shunted surface gap." This type requires less than 5 kV potential to create a spark between the electrodes. General practice dictates that a "new" spark igniter shall spark when 1000 V is applied.[AIR784-88]

low tension system Ignition systems capable of delivering voltages up to 5 kV inclusive to the firing tip of the spark igniter.[AIR784-88]

low to middle frequency The frequency band of nondirectional radio beacons between 200 and 1750 kHz.[ARP4107-88]

low pitch A propeller setting in which the chord of the blade is at a relatively acute angle to the plane of rotation, resulting in a high propeller speed. The specific angle considered to be low pitch will vary from one model propeller to another. Low pitch settings are used when very high or maximum power is required at low airspeeds, as in takeoff or go-around.[ARP4107-88]

low vacuum The condition in a gas filled space at pressures less than 760 torr corresponding approximately to the vapor pressure of water at 25 deg. C and to 1 inch of mercury.[NASA]

low water cutoff A device to stop the burner on unsafe water conditions in the boiler.

lox-hydrogen engines See hydrogen oxygen engines.

LP See linear programming.

L_{po} Sound pressure level at a point four feet downstream of a valve and three feet from the surface of the pipe.

LPOF See low-pass output filter.

LQG control See linear quadratic Gaussian control.

LQR See linear quadratic regulator.

L-Sat A communications satellite designed by European Space Agency member states to meet future communications satellite market needs such as European broadcast services, global telecommunications trunk services, and mobile services.[NASA]

LSB See least significant bit.

LSI See large-scale integration.

lubricant ring See lantern ring.

lubricator A device for automatically applying lubricant.

ludlam limit The point at which some supercooled water droplets no longer freeze within their catchment area and the forward growth of ice is diminished.[AIR1667-89]

lug Any projection, like an ear, used for supporting or grasping. See also terminal.

lugged body See body, wafer, lugged.

lumen(s) 1. The unit of luminous flux equal to the flux in a unit solid angle (one steradian) from a uniform point source of one candela.[ARP1782-89] 2. Metric unit for measuring the flux or power of light visible to the human eye; the photometric equivalent of the watt.

lumens per square meter (lm/m^2) The lm/m^2, or lux (light flux) is the metric unit for illuminance (commonly called illumination). Illuminance is used to measure light falling on a surface such as a display surface. A lux is approximately equal to 0.0929 foot candles: (1 fc approx - 10.8 lux).[ARP1782-89]

luminaire A complete lighting unit consisting of a light source together with its direct appurtenances, such as the globe, reflector, housing, and such support as is integral with the housing. [ARP798-91]

luminance 1. A measure of what the human eye perceives as brightness of a display. Luminance is defined as the luminous intensity per unit area that is emitted by a surface in a given direction. Luminance is measured in units of footlamberts (l/pi candelas per square foot) or Nits (candela per square meter). [ARP1782-89] 2. In photometry, a measure of the intrinsic luminous intensity emitted by a source in a given direction; the illuminance produced by light from the source upon a unit surface area oriented normal to the line of sight at any distance from the source, divided by the solid angle subtended by the source at the receiving surface.[NASA] 3. The luminous intensity of any surface in a given direction per unit of projected area in a plane perpendicular to that direction.

luminance intensity Luminous intensity is the luminous flux emitted into a given solid angle by a source. It is measured in units of candelas.[ARP1782-89]

luminance uniformity The luminance uniformity of a display is a measure of the variation of luminance across the display surface. It is defined in terms of a ratio or percent of luminance measured on the display.[ARP1782-89]

luminescence Light emission by a process in which kinetic heat energy is not essential for the mechanism of excitation.[NASA]

luminescent intensity See luminous intensity.

luminosity Emissive power with respect to visible radiation.

luminosity coefficients The constant multipliers for the respective tristimulus values of any color such that the sum of the three products is the luminance of the color.

luminous Emitting radiation in the form of visible light.

luminous dial A dial or indicating scale and pointer whose scale divisions, numerals and pointer are made of or coated with a light-emitting substance such as luminous paint so that they can be seen in the dark. Compare with lighted dial.

luminous efficiency Luminous flux divided by radiant flux.

luminous flux 1. Luminous flux is luminous power (that is, radiant power that has been adjusted for human eye response) that is emitted by a source. The process of correcting radiant power for the human eye response has been defined by the CIE. The unit of luminous flux is the lumen.[ARP1782-89] 2. The amount of light passing a given point per unit time.

luminous flux density See luminous intensity.

luminous intensity Luminous energy per unit time per unit solid angle; the intensity (flux per unit solid angle) of visible radiation weighted to take into account the variable response of the human eye as a function of the wavelength of light; usually expressed in candles.[NASA]

lumped-constant wavemeter A device for determining frequency using a tunable resonant LC circuit coupled to a crystal detector; the circuit generally utilizes plug-in coils of various inductances and a continuously variable capacitor with a dial calibrated in frequency.

lumped parameter systems Systems in which the parameters may be considered to represent, for purposes of analysis, a single inductance, capacitance, resistance, etc. throughout the frequency range of interest.[NASA]

LUNA lunar probes See lunik lunar probes.

lunar craters A depression, usually circular, on the surface of the moon, usually with a raised rim called a ringwall. [NASA]

lunar eclipses The phenomenon observed when the moon enters the shadow of the earth.[NASA]

lunar probes Probes for exploring and reporting on conditions on or about the moon.[NASA]

lunar scattering See diffuse radiation.

lunation See month.

lung, expiratory reserve volume The maximal volume of gas that can be expired from the end-expiratory level.[AIR 825B-91]

lung, functional residual capacity The volume of gas remaining in the lungs at the resting expiratory level. The resting end-expiratory position is used here as a base line because it varies less than the end-inspiratory position.[AIR825B-91]

lung, inspiratory capacity The maximal volume of gas that can be inspired from the resting expiratory level.[AIR 825B-91]

lung, inspiratory reserve volume The maximal amount of gas that can be inspired from the end-inspiratory position.[AIR825B-91]

lung, tidal volume The volume of gas inspired or expired during each respiratory cycle.[AIR825B-91]

lung, total capacity The amount of gas contained in the lung at the end of a maximal inspiration.[AIR825B-91]

lung, vital capacity The maximal volume of gas that can be expelled from the lungs by forceful effort following a maximal inspiration.[AIR825B-91]

lunik lunar probes Russian term for a space probe launched to the moon's vicinity or to impact on the moon.[NASA]

lux Metric unit of illuminance.

LVDT See linear variable differential transformer.

LVRT See linear variable reluctance transducer.

Lx See lux.

Lyman alpha radiation The radiation emitted by hydrogen at 1216 angstrom, first observed in the solar spectrum by rocket borne spectrographs. Lyman alpha is very important in the heating of the upper atmosphere thus affecting other atmospheric phenomena.[NASA]

lysimeters Instruments for measuring the water percolating through soils and determining the materials dissolved by the water.[NASA]

m See metre.

M Mach number.

MAA Maximum Authorized IFR Altitude.

MAC See media access control.

mach See machine.

Mach angle The angle between the path of a body moving with supersonic velocity and a corresponding Mach line; the speed of sound divided by the body's velocity equals the sine of the Mach angle.

Mach cones The cone shaped shock waves theoretically emanating from an infinitesimally small particle moving at supersonic speed through a fluid medium. It is the locus of the Mach lines. The cone shaped shock waves generated by a sharp pointed body, as at the nose of a high speed aircraft.[NASA]

machine Any device capable of performing useful work, especially a device for producing controlled motion or for regulating the effect of a given force.

machine address An absolute, direct unindexed address expressed as such, or resulting after indexing and other processing has been completed.

machine code The lowest level of computer language in the form of the digital code that can be directly executed by the computer.

machine code instruct A code that defines a particular computer operation that can be used without further translation.

machine-dependent program A program that operates on only one type of computer.

machine element Any standard mechanical part used in constructing a machine, such as a bearing, fastener, cam, gear, lever, link, pin or spring.

machine error A deviation from correctness in computer data resulting from an equipment failure.

machine-independent Pertaining to procedures or programs created without regard for the actual devices which will be used to express them.

machine-independent program A program that operates on a variety of different computers.

machine instruction An instruction that a machine can recognize and execute.

machine language 1. A language that is used directly by a machine. 2. Binary words (on the PDP-11 family, sixteen-bit) that are required to make the computer perform. 3. In software, the language which a computer understands; ones and zeros.

machine-language code Same as computer code and contrasted with symbolic code.

machine-language programming The term basically means programming using machine language. See machine-language code.

machine operator The person who manipulates the computer controls, places information media into the input devices, removes the output and performs other related functions.

machine-oriented language 1. A language designed for interpretation and use by a machine without translation. 2. A system for expressing information which is intelligible to a specific machine, e.g., a computer or class of computers. Such a language may include instructions which define and direct machine operations, and information to be recorded by or acted upon by these machine operations. 3. The set of instructions expressed in the number system basic to a computer, together with symbolic operation codes with absolute addresses, relative addresses, or symbolic addresses. Synonymous with machine language. Clarified by language. Related to object language and contrasted with problem-oriented language.

machine program 1. A program that is to be loaded in a computer and executed by it. 2. In numerical control, an ordered set of instructions in automatic control language and format and based on the part program, recorded on appropriate input media and sufficiently complete to effect the direct operation of an automatic control.

machine readable Data that will be accepted by a computer through an input device.

machine-readable medium A medium that can convey data to a given sensing device.

machine recognition See artificial intelligence.

machinery One or a group of machines; an apparatus or system constructed of machines.

machine word A unit of information of a standard number of bits or characters which a machine regularly handles in each transfer, e.g. a machine may regularly handle numbers or instructions in units of 36 binary digits.

machining center A versatile CNC machine tool with multi-axis control and usually automatic tool loading. The machining centers are designed to carry out a range of operations.

machining tear A pattern of short, jagged individual cracks, generally at right angles to the direction of machining, and are the result of improperly set cutting tools, or dull cutting tools.[AIR1667-89] [AS3071A-77]

mach meter An instrument that indicates the ratio of aircraft speed to the speed of sound at a particular altitude and temperature.[ARP4107-88]

Mach number (M) 1. The ratio of true airspeed to the speed of sound. A number expressing the ratio of the speed of a body or of a point on a body with respect to the surrounding air or other fluid, or the speed of a flow, to the speed of sound in the medium; the speed represented by this number. Named after Ernst Mach (1838-1916), Austrian scientist.[ARP4107-88]

Mach reflection The reflection of a shock wave from a rigid wall in which the shock strength of the reflected wave and the angle of reflection both have the smaller of the two values theoretically possible.[NASA]

macro Directions for expanding abbreviated text; a boilerplate that generates a known set of instructions, data, or symbols. A macro is used to eliminate the need to write a set of instructions that are used repeatedly; for example, an assembly-language macro instruction enables the programmer to request the assembler to generate a predefined set of machine instructions.

macro-assembler An assembler which allows the use of macros and converts them to machine code.

macro instruction The more powerful instructions which allow a programmer to refer to several instructions as though they were a single instruction. When a programmer uses the name of a macro instruction, all of the instructions are inserted at that point in his coding by the macroprocessor.

macro modeling The representation of a component or device in terms of a net-list description of an equivalent circuit. Standard components, such as resistors or capacitors, are typically employed.

macromolecules See molecules.

macroprocessor 1. A program which translates a single symbolic statement into one or more assembly language statements. 2. A phase of an assembler that has the capability of translating selected mnemonic or symbolic instructions into multiple machine-language instructions.

macro-program A program containing macros.

macroprogramming Programming with macro instructions.

macroscopic stress Load per unit area distributed over an entire structure or over a visible region of the structure.

macrostructure The features of a polycrystalline metal revealed by etching and visible at magnifications of 10 diameters or less.

Magellan Mission (ESA) See Magellan ultraviolet astronomy satellite.

Magellan project (NASA) A Venus exploratory mission to acquire radar imagery and topographic profiles of the planet surface and determine the characteristics of the Venusian gravity field.[NASA]

Magellan spacecraft (NASA) A Venus probe incorporating Voyager and Galileo hardware designs equipped with a synthetic aperture radar system to acquire surface imagery, altimetric profiles, and surface radiothermal emissivities. Earth-based Doppler radio tracking of the spacecraft will be used to derive gravimetric data.[NASA]

Magellan ultraviolet astronomy satellite This ESA mission will provide high resolution spectra of celestial sources down to sixteenth magnitude over the extreme ultraviolet wavelength range (between 50 and 150 nm).[NASA]

magic tees Compound waveguides or coaxial tees with four arms which exhibit directional characteristics, when properly matched, so that a signal entering one arm will be split between two of the other arms but not the third. A signal entering another arm is likewise split with half the energy entering one of the arms common to the other input but not its second arm and the other half of the energy entering the arm not used by the other input. Magic tees are used in radar as transmitter receiver duplexers.[NASA]

magma Naturally occurring mobile rock materials, generated within the earth and capable of intrusion and extrusion, from which igneous rocks are thought to have been derived by solidification and related processes.[NASA]

magnetically actuated extensions A device attached to the meter body which contains an electrical switch and which is magnetically actuated by the metering float extension to signal a high or low flow. The switch is adjustable with respect to the float position over a range equal to the travel of the metering float. Standard switch ratings are usually 0.3 amperes for 110 volt, 60 cycle a-c supply (five amperes or more if relays are used).

magnetic amplifier An electronic amplification or control device that functions through the use of saturable reactors, either alone or in combination with other circuit elements.

magnetic bearing(s) 1. Any application of the principle in which something capable of rotation and translation is held by the use of electromagnetic force without touching it. Applications range from small instruments to very large forces.[NASA] 2. The angle between the line of sight to an object and the direction from the observer to magnetic North, measured in a plane parallel to the earth's surface. See also bearing.

magnetic biasing Simultaneous conditioning of a magnetic recording medium by superimposing a second magnetic field on the magnetic signal being recorded.

magnetic blowout switch A special type of switch designed to switch high d-c loads; a small permanent magnet contained in the switch housing deflects the arc to quench it when the contacts open.

magnetic bubble memory A high-density information storage device composed of a magnetic film only a few micrometres thick deposited on a garnet substrate; information is stored in small magnetized regions (bubbles) whose magnetic polarity is opposite to that of the surrounding region.

magnetic card A card with a magnetic surface on which data can be stored by selective magnetization of portions of the flat surface.

magnetic compass Any of several devices for indicating the direction of the horizontal component of a magnetic field, but especially for indicating magnetic North in the earth's magnetic field.

magnetic compression The force exerted by a magnetic field on an electrically conducting fluid or on a plasma. [NASA]

magnetic contactor A device for opening and closing one or more sets of electrical contacts which is actuated by either energizing or deenergizing an electromagnet within the device.

magnetic cooling Keeping a substance cooled to about 0.2 K by using a working substance (paramagnetic salt) in a cycle of processes between a high-temperature reservoir (liquid helium) at 1.2 K and a low temperature reservoir containing the substance to be cooled. [NASA]

magnetic core 1. A configuration of magnetic material that is, or is intended to be, placed in a spatial relationship to current-carrying conductors and whose magnetic properties are essential to its use. It may be used to concentrate an induced magnetic field, as in transformer, induction coil, or armature, to retain a magnetic polarization for the purpose of storing data, or for its nonlinear properties as in a logic element. It may be made of such material as iron, iron oxide, or ferrite and in such shapes as wires, tapes, toroids, or thin film. 2. A storage device in which binary data is represented by the direction of magnetization in each unit of an array of magnetic material, usually in the shape of toroidal rings, but also forms such as wraps on bobbins. Synonymous with core.

magnetic damping Progressive reduction of oscillation amplitude by means of current induced in electrical conductors due to changes in magnetic flux.

magnetic disk A flat, circular plate with a magnetic surface on which data can be stored by selective magnetization of portions of the flat surface.

magnetic drum Memory device used in computers; rotating cylinders on which information may be stored as magnetically polarized areas, usually along several parallel tracks around the periphery.[NASA]

magnetic equator That line on the surface of the earth connecting all points at which the magnetic dip is zero.[NASA]

magnetic extensions A device that provides flow rate indication by means of a magnetic coupling between the extension of the metering float and an external indicator follower surrounding the extension tube.

magnetic field intensity See magnetic flux.

magnetic field interference A form of interference induced in the circuits of a device due to the presence of a magnetic field. It may appear as common mode or normal mode interference in the measuring circuits. See also interference, electromagnetic.

magnetic field reconnection A change in topology of the magnetic field configuration resulting from a localized breakdown of the requirement for 'connection' of fluid elements at one time on a common magnetic field line. Alternatively, it occurs when an electric field exists with a component parallel to a locally two-dimensional X-type magnetic neutral line which is equivalent to a breakdown in connection.[NASA]

magnetic fields Regions of space wherein magnetic dipoles would experience a magnetic force of torque; often represented as the geometric array of the imaginary magnetic lines of force that exist in relation to magnetic poles. [NASA]

magnetic float gage Any of several designs of liquid- level indicator that use a magnetic float to position a pointer or change the orientation of bicolor wafers.

magnetic float switch A device for operating a mercury switch by repositioning a magnetic piston with respect to a small permanent magnet attached to the pivoting mercury switch capsule; in the usual configuration, a float attached to the piston positions it near the small magnet when liquid level is high, and drops the piston out of proximity when the level is low, allowing a light spring to retract the magnet and pivot the mercury capsule.

magnetic flux The magnetic force exerted on an imaginary unit magnetic pole placed at any specified point of space. It is a vector quantity. It's direction is taken as the direction toward which a north magnetic pole would tend to move under the influence of the field. If the force is measured in dynes and the unit pole is a cgs unit pole, the field intensity is given in oersteds.[NASA]

magnetic focusing Causing an electron beam to become diverging or converging to position an image or beam on an object—usually CRT screen—by interacting with a magnetic or electromagnetic field.

magnetic hardness comparator A device for determining hardness of a steel part by comparing its response to electromagnetic induction with the response of a similar part of known hardness.

magnetic head A transducer for converting electrical signals into magnetic signals suitable for storing on magnetic recording media, for converting stored magnetic signals into electrical output signals, or for erasing stored magnetic signals.

magnetic ink An ink which contains magnetic particles. Characters printed in magnetic ink can be read both by humans and by machines designed to read the magnetic pattern.

magnetic lens Electric coils electromagnets or permanent magnets assembled into a configuration that can accomplish magnetic focusing.

magnetic memory See magnetic storage.

magnetic mirrors Magnetic fields so arranged that they will theoretically confine a hot plasma.[NASA]

magnetic moments The quantities obtained by multiplying the distances between two magnetic poles by the average strength of the poles. Measures of the magnetic flux set up by the gyration of an electric field in a magnetic field. Moments are negative, indicating they are diagrammatic, and equal to the energy of rotation divided by the magnetic field. In atomic and nuclear physics, moments, measured in Bohr magnetrons, are associated with the intrinsic spin of the particle and with the orbital motion of the particle in a system. [NASA]

magnetic north North, as determined by the earth's magnetic lines of force; the reference direction for measurement of magnetic directions. See true north. [ARP4107-88]

magnetic poles Either of the two places on the surface of the earth where the magnetic dip is 90 deg., that in the northern Hemisphere (at, approximately, latitude 73 deg. 8 N, longitude 1001 deg. W in 1955) being designated north magnetic pole, and that in the Southern Hemisphere (at, approximately, latitude, 68 deg. S, longitude 144 deg. E in 1955) being designated south magnetic pole. Either of those two points of a magnet where the magnetic force is the greatest. In magnetic theory, a fictitious entity analogous to a unit charge of electrostatic theory. In nature only dipoles, magnetic poles exist.[NASA]

magnetic printing Permanently transferring a recorded signal from one magnetic recording medium to another magnetic recording medium (or to another portion of the same medium) by bringing the two sections into close proximity.

magnetic proximity sensor Any of several devices that are activated when a magnetized or ferromagnetic object passes within a defined distance of the active element; there are four types—variable-reluctance sensors, hermetically sealed dry-reed switches, Hall-effect switches, and Weigand-effect sensors.

magnetic recorder A device for producing a stored record of a variable electrical signal as a variable magnetic field in a ferromagnetic recording medium.

magnetic separator A machine that uses strong magnetic fields to remove pieces of magnetic material from a mixture of magnetic material and nonmagnetic or less strongly magnetic material.

magnetic shield A metal shield which insulates the contents from external magnetic fields. Such shields are often used with photomultiplier tubes.

magnetic storage 1. In computer terminology, any device which makes use of the magnetic properties of materials for the storage of information.[NASA] 2. A device or devices which utilize the magnetic properties of materials to store information.

maintenance conditions Conditions under which maintenance is performed.

maintenance, corrective All actions performed as a result of failure, to restore an item to a specified condition. Corrective maintenance can include any or all of the following steps: localization, isolation, disassembly, interchange, reassembly, alignment and checkout. [ARD50010-91]

maintenance, deferred Maintenance not having any bearing on flight safety, which is deferred to a convenient time and/or location for accomplishment. [ARD50010-91]

maintenance, depot Maintenance which is the responsibility of and performed by designated maintenance activities, to augment stocks of serviceable material, and to support organizational maintenance and intermediate maintenance activities by the use of more extensive shop facilities, equipment, and personnel of higher technical skill than are available at the lower levels of maintenance. Its phases normally consist of repair, modification, alteration, modernization, overhaul, reclamation, or rebuilding of assemblies, subassemblies, units, and equipment; the emergency manufacture of nonavailable parts; and providing technical assistance to using activities and intermediate maintenance organizations. Depot maintenance is normally accomplished in fixed shops, shipyards and shore based facilities. [ARD50010-91]

maintenance, direct That effort expended by maintenance personnel in the actual performance of maintenance on aircraft, aeronautical equipment, or Support Equipment (SE) in accordance with the applicable technical manual. It applies equally to both contractor and Government Furnished Equipment (GFE).[ARD50010-91]

maintenance, direct (man hours per maintenance action) A measure of maintainability that is the sum of direct maintenance man hours, divided by the total number of maintenance actions during a stated period of time. [AIR1916-88]

maintenance event One or more maintenance actions required to effect corrective and preventative maintenance due to any type of failure or malfunction, false alarm or scheduled maintenance plan.[ARD50010-91]

maintenance, hard time (HT) See maintenance, scheduled.

maintenance indirect costs Maintenance labor and material costs, not considered to be direct maintenance costs, that contribute to the overall maintenance program costs through overhead operations, administration, record keeping, supervision, tooling, test equipment, facilities, etc.[AIR1916-88]

maintenance, in shop Work that requires the use of shop facilities and cannot be normally performed outside the shop. (Bench test and component disassembly and repair are examples of in-shop maintenance work.)[ARD 50010-91]

maintenance level Division of maintenance, based on difficult and requisite technical skill, in which jobs are allocated to organizations in accordance with the availability of personnel, tools, supplies, and the time within the organization. Maintenance levels include organizational, intermediate, and depot. Organizational maintenance embraces the maintenance performed by a using organization on its own equipment. This includes inspection, cleaning, servicing, preservation, lubrication, adjustment, minor repair not requiring detailed disassembly, and replacement not requiring high technical skill. Intermediate maintenance is performed by designated maintenance activities in direct support of using organizations. This category will normally be limited to maintenance consisting of replacement of unserviceable parts, subassemblies, or assemblies. Depot maintenance refers to the maintenance required for major overhaul or complete rebuilding of parts, subassemblies, assemblies, and other end items. Such maintenance is intended to augment stocks of serviceable equipment or to support lower levels of maintenance by use of more extensive shop equipment and personnel of higher technical skill than available in organizational or field maintenance activities.[ARD 50010-91]

maintenance, line Routine check, inspection and malfunction rectification performed enroute and base stations during transit, turnaround or night stop. Synonymous with maintenance, line station.[ARD50010-91]

Maintenance Management System A part of the Management Information System (MIS) that is useful for maintaining the companies equipment. It accesses equipment information, spare parts availability and location, maintenance work order systems, preventive maintenance systems, maintenance personnel qualifications, equipment maintenance history, and any other information that will help the maintenance engineer, supervisor, technician or mechanic be more proficient in his job.

maintenance, on condition 1. A primary maintenance process having repetitive inspections or tests to determine the condition of units, systems, or portions of structure with regard to continued serviceability (corrective action is taken when required by item condition.)[ARD50010-91] 2. A maintenance concept whereby an engine has no fixed time limitation on repair or replacement of the engine or any of its components. Repair or replacement of the engine or any of its components shall be determined by the condition of the unit. The unit shall be subjected to periodic diagnostic checks and inspections to insure its continued ability to perform its function within specified limits.[ARD 50010-91] 3. A maintenance process, where a component's suitability for continued service is determined by periodic inspection or test, or both, in situ on the airplane. Such a plan gives full recognition to the fact that failure is random in nature and cannot be completely obviated by any known maintenance scheme.[AIR1916-88]

maintenance plan A document containing technical data, tailored to a specific weapon system maintenance concept, which identifies maintenance and support resource requirements to maintain aeronautical systems, equipment, and Support Equipment (SE) in an operationally ready state. The maintenance plan provides the interface between maintenance engineering and supply for provisioning purposes and communicates necessary (but incomplete) inputs to enable other logistic element managers to develop their hardware support requirements. The maintenance plan is designed as a tool for the shore community for Integrated Logistic Support (ILS) planning and is prepared in accordance with NAVAIRINST 4790.4A (NOTAL).[ARD50010-91]

maintenance, preventive Tests, measurement, replacements, adjustments, repairs and similar activities carried out with the intention of preventing faults or malfunctions from occurring during subsequent operation. Preventive maintenance is designed to keep hardware and software in proper operating condition and may be performed on a scheduled basis.[ARD50010-91]

maintenance ratio A measure of the total maintenance manpower burden required to maintain an item. It is expressed as the cumulative number of manhours of maintenance expended in direct labor during a given period of the life units divided by the cumulative number of end item life units during the same period.[ARD50010-91]

maintenance, scheduled That maintenance performed at defined intervals to retain an item in a serviceable condition by systematic inspection, detection, replacement of wearout items, adjustment, calibration, cleaning, etc. Synonymous with maintenance, preventive routine.[ARD50010-91]

maintenance significant Maintenance items of equipment or components which are judged to be relatively the most important for safety, reliability, or economic impact.[AS1426-80]

maintenance steering group (MSG) An ATA (Air Transportation Association) sponsored study group which publishes recommended methodologies and analytical procedures for developing a maintenance plan for aircraft, engines and systems.[ARP1587-81]

maintenance task The maintenance effort necessary for retaining an item in, or changing/restoring it to a specified condition.[ARD50010-91]

maintenance time Time used for equipment maintenance. It includes preventive maintenance time and corrective maintenance time.

maintenance time, elapsed For the purposes of Maintenance Data Reporting (MDR), EMT is defined as the actual clock time, in hours and tenths, that maintenance was being performed on a job. EMT does not include the clock hours and tenths for cure time, charging time, or leak test when they are being conducted without maintenance personnel actually monitoring the work. Although the EMT is directly related to job man-hours, it is not to be confused with total manhours required to complete a job. For example, if five persons complete a job in 2.0 hours of continuous work, the EMT=2.0 hours and the man-hours=10.0.[ARD50010-91]

maintenance, unscheduled 1. That maintenance performed to restore an item to a satisfactory condition by providing correction of a known or suspected malfunction and/or defect.[ARD50010-91] 2. Corrective maintenance performed, as required by item condition.[AIR 1916-88]

major alteration A change not listed in the aircraft, aircraft engine, or propeller specifications that might appreciably affect weight, balance, structural strength, performance, powerplant operation, flight characteristics, or other characteristics affecting airworthiness.

[ARP4107-88]

major diameter The largest diameter of a screw thread; it is measured at the crest of an external thread and at the root of an internal thread.

major frame With reference to telemetry formats, the time period where all data of a multiplex are sampled at least once; includes one or more minor frames. Major frame length is determined as (N)(Z) words, where N = the number of words per minor (prime) frame and Z = the number of words in the longest submultiple frame.

major graduations Intermediate graduation marks on a scale which are heavier or longer than other graduation marks but which are not index graduations.

majority A logic operator having the property that if P is a statement, Q is a statement, and R is a statement, then the majority of P, Q, R is true if more than half the statements are true, false if half or less are true.

majority voting system A fault-tolerant system wherein the outputs of three or more signals are summed to provide a single signal representative of the majority of the individual signals, often providing detection logic for identifying a failed channel. See also voter.[ARP 1181A-85]

major repair A repair that, if improperly done, might adversely affect weight, balance, structural strength, performance, powerplant operation flight characteristics, or other qualities affecting airworthiness.[ARP4107-88]

major time In telemetry computer systems, two sixteen- bit words: minutes/seconds, and hours/days.

make-up The water added to boiler feed to compensate for that lost through exhaust, blowdown, leakage, etc.

male fitting An element of a connection in pipe, tubing, electrical conductors or mechanical assemblies that fits into the mating (female) element; for example, the externally threaded end of a pipe fitting is termed male.

malfunction 1. A general term used to denote the occurrence of failure of a product to give satisfactory performance. It need not constitute a failure if readjustment or operator's controls can restore an acceptable operating condition.[ARD50010-91] 2. Abnormal condition or status of an engine, component or sub-system.[ARP1587-81] 3. Improper functioning of components, causing improper operation of a system.[NASA] 4. The effect of a fault.

malfunction routine Same as diagnostic routine.

malleable iron A somewhat ductile form of cast iron made by heat treating white cast iron to convert the carbon-containing phase from iron carbide to nodular graphite.

MALS Medium intensity Approach Light System. See approach light system.

MALSR Medium intensity Approach Light System with Runway alignment indicator lights. See approach light system.

man See manual.

Management Information System (MIS) A computerized system using a large database containing information on: a)customers, b) equipment, c) supplies, d) spare parts, e) personnel, f) process, g) sales forecast, h) history, i) costs, j) profits, etc. Selected information is available to those persons making decisions.

manatees Large plant eating aquatic mammals living in shallow tropical waters near the coasts of North and South America.[NASA]

man-computer interface The interface between man and the computer and its interrelationships including ergonomic factors.[NASA]

mandrel A cylindrical-shaped tool with a slight taper on the overall length, with or without a flexible sleeve, centered on each end with a flat milled on the ends for a holder. Used for holding work for machining operations. Also used as a round anvil to form metal.[ARP480A-87]

maneuverability That property of any vehicle which determines the rate at which its attitude and direction of movement can be changed.[ARP4107-88]

manhead The head of a boiler drum or other pressure vessel having a manhole.

manhole The opening in a pressure vessel of sufficient size to permit a man to enter.

manhours The total number of accumulated direct labor hours (in hours and tenths) expended in performing a maintenance action. Direct maintenance man-hours are man-hours expended by assigned personnel to complete the work described on the source document. This includes the functions of preparation, inspection, disassembly, adjustment, fault correction, replacement or reassembly of parts, and calibration/tests required in restoring the item to a serviceable status. It also includes such associated tasks as checking out and returning tools, looking up part numbers in the Illustrated Parts Breakdown (IPB), transmitting required information to material control, and completing documentation of the Visual Information Display System/Maintenance Action Form (VIDS/MAF) or Support Action Form (SAF).[ARD50010-91]

manhours, maintenance The manhours required to complete the maintenance task.[ARD50010-91]

manhours per flying hours A performance figure calculated by dividing the direct manhours expended to maintain a particular aircraft fleet during a given period, by the flying hours (airborne) during that period.[ARD50010-91]

manifold 1. A preformed item with two or more inlet ports and passages, and a common outlet. See also distributor. [ARP480A-87] 2. A pipe or header for collecting a fluid from, or the distributing of a fluid to a number of pipes or tubes. 3. A branch pipe which distributes intake or exhaust fluids to a series of valve ports, as in a multicylinder engine such as an automobile engine.

manifold equalizing line The conduit within a manifold which connects the high and low differential pressure impulse lines.

manifold (instrumentation) Any configuration of valves which can be manipulated to create zero differential pressure at the measuring instrument.

manifold pressure 1. Absolute pressure as measured at the appropriate point in the induction system and usually expressed in inches of mercury.[ARP4107-88] 2. The fluid pressure in the intake manifold of an internal combustion engine.

manifold variable A quantity or condition which is varied so as to change the value of the controlled variable.

manipulated variable 1. In a process that is desired to regulate some condition, a quantity or a condition that is altered by the control in order to initiate a change in the value of the regulated condition. 2. The part of the process which is adjusted to close the gap between the set point and the controlled variable. See also variable, manipulated.

manipulative variable In a control loop, the variable that is used by the controller to regulate the controlled variable.

manipulators Mechanical devices for the remote handling of hazardous materials; they are usually hand operated, often from behind a shield, and may or may not be power assisted.

man machine system Systems in which the functions of the man and the machine are interrelated and necessary for the operation of the system.[NASA]

manometer 1. Instrument for measuring pressure of gases and vapors above and below atmospheric pressure.[NASA] 2. A gage for measuring pressure or a pressure difference between two fluid chambers. A U-tube manometer consists of two legs, each containing a liquid of known specific gravity.

manometric equivalent The length of a vertical column of a given liquid at standard room temperature which indicates a pressure differential equal to that indicated by a 1-mm-long column of mercury at 0 °C.

mantissa See floating point.

mantle (earth structure) See earth mantle.[NASA]

manual backup An alternate method of process control by means of manual adjustment of final control elements in the event of a failure in the computer system.

manual control The operation of a process by means of manual adjustment of final control elements.

manual controller A control device whose output signals, power or motions are all varied by hand.

manual data entry module A device which monitors a number of manual input devices from one or more operator consoles and/or remote data entry devices and transmits information from them to the computer.

manual flight control system (MFCS) A system which transmits manual pilot commands directly or generates and conveys commands which augment manual pilot control commands and thereby accomplishes flight control functions. The system may include electrical, mechanical and hydraulic components which provide means for transmission of manual pilot commands to the control function. This classification includes the longitudinal, lateral-directional, life, drag and variable geometry control systems. Associated scheduling, limiting and control devices are included.[ARP1181A-85]

manual input 1. The entry of data by hand into a device at the time of processing. 2. The data entered as in definition 1.

manually actuated valves Those valves that are operated by hand motion applying a force through a mechanical lever or cable system.[ARP986-90]

manual operation Processing of data in a system by direct manual techniques.

manual override The capability of a flight control system to enable the pilot to override the AFCS through a cable and/or linkage system and exert control in excess of the AFCS authority or in opposition to the AFCS command.[AIR1916-88]

manual rest See reset.

manual reversion The action of reverting to manual control because of failure to the automatic or semiautomatic system.[ARP4107-88]

manual station 1. Synonymous with manual loading station. 2. A single loop hard manual control to operate the final control devices in case of control system failure. 3. Provides for bypassing normal controller operation to manually vary an analog output signal in a controller. Used primarily in an emergency, or possibly during a maintenance shutdown of the controller.

manufactured gas Fuel gas manufactured from coal, oil, etc., as differentiated from natural gas.

manufacturer software A complex program package that develops the user's application and organizes computer procedures to obtain efficient response to the application program. Often this software is referred to as an operating system.

Manufacturing Automation Protocol (MAP) A specification for a suite of communication standards for use in manufacturing automation developed under the auspices of General Motors Corporation. The development of this specification is being taken over by the MAP/TOP Users Group under the auspices of CASA/SME.

Manufacturing Messaging Format Standard (MMFS) One of the application protocols specified by MAP.

map 1. To establish a correspondence or relationship between the members of one set and the members of another set and perform a transformation from one set to another, for example, to form a set of truth tables from a set of Boolean expressions. Information should not be lost or added when transforming the map from one to another. 2. See memory map.

MAP Missed Approach Point. See also Manufacturing Automation Protocol.

MAP/EPA Part of the EPA architecture, a MAP/EPA node contains both the MAP protocols and the protocols required for communication to MINI-MAP. It can communicate with both MINI-MAP nodes on the same segment and full MAP nodes anywhere in the network.

mapped system A system that uses the computer hardware memory management unit to relocate virtual memory addresses.

Mapsat A proposed stereoscopic system for mapping the earth from space to replace Landsat D as defined by the US Geological Survey.[NASA]

MAP/TOP Users Group United States and Canada's MAP/TOP Users Group. See CASA/SME.

Marangoni convection Convective flow induced by surface tension gradients. This important in both ground and space processing where a free surface is present.[NASA]

Marecs maritime satellites The European Space Agency's system of two satellites provides maritime communications links between ships and coast earth stations. Originally known as Marots, the system operates with one satellite over the Atlantic Ocean and one over the Pacific Ocean.[NASA]

marginal check A preventive-maintenance procedure in which certain operating conditions are varied about their normal values in order to detect and locate incipient defective units, e.g., supply voltage or frequency may be varied. Synonymous with marginal test and high-low bias test, and related to check.

marginal test Same as marginal check.

margin of attention See attention, margin of.

margin of safety The ratio between maximum service load (allowable design load) for a structure and the load that would cause the structure to deform, collapse or break.

marine chemistry The study of the chemical processes in oceanic environments.[NASA]

marine engineering A branch of engineering that deals with the design, construction and operation of shipboard propulsion systems and associated auxiliary machinery.

Marisat 1 satellite The first commercial maritime communication satellite.[NASA]

Marisat satellites A class of maritime commercial communication service satellites designed to provide telephone, telegraph, radio, distress messages and facsimile services to merchant ships, etc.[NASA]

mark A sign or symbol used to signify or indicate an event in time or space, e.g., end of word or message mark, a file mark, a drum mark, or an end of tape mark.

marker beacon An electronic navigation facility transmitting a 75 MHz vertical fan or one-shaped radiation pattern. Marker beacons are identified by their modulation frequency and keying code, and when received by compatible airborne equipment, indicate to the pilot, both aurally and visually, that he/she is passing over the facility.[ARP4107-88]

marking pointer An adjustable stationary pointer, usually of a color different from that of the indicating pointer, that can be positioned opposite any location on the scale of interest to the user.

Markov chain A probabilistic model of events, in which the probability of an event is dependent only on the event that precedes it.

mark-sense To mark a position on a punch card with an electrically conductive pencil, for later conversion to machine punching.

mark-sense device An electronic machine that will read mark-sensed forms.

mark sensing A technique for detecting special pencil marks entered in special places on a punch card and automatically translating the marks into punched holes. See also sensing, mark.

marshes See marshlands.

marshlands Transitional land-waste areas, covered at least part of the time by estuarine or coastal waters and characterized by aquatic and grasslike vegetation. Used for bogs, coastal marshlands, marshes, and swamps.[NASA]

martensite See alpha prime.

Martensitic transformation A phase transformation occurring in some metals and resulting in formation of martensite.[NASA]

Marx generators A high voltage electrical pulse generator in which capacitors are charged in parallel, then discharged in series to generate a voltage much higher than the charging voltage.

MASER See Microwave Amplification by the Stimulated Emission of Radiation.

mask 1. A protective covering used to limit the area to be affected by a surface treatment. See also cover, guard, protector.[ARP480A-87] 2. A protective face covering which usually provides for filtration of breathing air or for attachment to an external supply of breathing air. 3. A frame or similar device to prevent certain areas of a workpiece surface from being coated, as with paint. 4. A frame that conceals the edges of a cathode-ray tube, such as a television screen. 5. A machine word or register that specifies which parts of another machine word or register are to be operated on.

masking 1. The process of extracting a nonword group or a field of characters from a word or a string of words. 2. The process of setting internal program controls to prevent transfers which otherwise would occur upon setting of internal machine latches.

mask programmed memory Computer memory dedicated to the storage of a particular set of data. A mask containing the particular pattern of bits is used in the manufacture of the memory.

mass 1. A quantity characteristic of a body, which relates the attraction of this body toward another body. Since the mass of a body is not fixed in magnitude, all masses are referred to the standard kilogram, which is a lump of platinum.[NASA] 2. Amount of matter an object contains.

MASSBUSS The thirty-two-bit direct-memory-access bus on the PDP-11/70 and VAX-11 computers.

mass drivers (payload delivery) Proposed method for payload delivery into earth orbit from the moon by electromagnetic acceleration; also for deliveries to Lagrange equilibrium points. [NASA]

mass flow The amount of fluid, measured in mass units, that passes a given location or reference plane per unit time.

mass-flow bin A bin with steep, smooth sides which allow its contents to flow, without stagnant regions, whenever some of the contents are withdrawn.

mass flowmeter An instrument for

measuring the rate of flow in a pipe, duct or channel in terms of mass per unit time.

mass flow rate The mass of fluid moving through a pipe or channel within a given period of time.

mass number The sum of the number of protons and the number of neutrons in the nucleus of a specific nuclide.

mass ratio The ratio of the mass of the propellant charge of a rocket to the total mass of the rocket when charged with the propellant.[NASA]

mass spectrograph A mass spectroscope which records intensity distributions on a photographic plate.

mass spectrometer A mass spectroscope which uses an electronic instrument to indicate intensity distribution in the separated ion beam.

mass spectroscope An instrument for determining the masses of atoms or molecules, or the mass distribution of an ion mixture, by deflecting them with a combination of electric and magnetic fields which act on the particles according to their relative masses.

mass spectrum In a mixture of ions, the statistical distribution by mass or by mass-to-charge ratio.

mass storage Pertains to a computer device that can store large amounts of data so that they are readily accessible to the central processing unit; for example, disks, DEC tape, or magnetic tape.

mass to light ratio The ratio of the mass of celestial body to its luminosity. [NASA]

mass velocity Mass flow per unit cross-sectional area.

master 1. A device or mechanism that controls the operation of different mechanisms or establishes a standard.[ARP 480A-87] 2. A device which controls other devices in a system. 3. A precise pattern for making replicate workpieces, as in certain types of casting processes.

master change Documentation prepared by the airframe manufacturer which defines and coordinates a negotiable change to the contract specifications after contract signing.[AS1426-80]

master clock A device which functions as the primary source of timing signals.

master file directory The system-maintained file on a volume that contains the names and addresses of all the files stored on the volume.

master gage A device with fixed locations for positioning parts or holes in three dimensions.

master recipe A basic recipe which has been made site specific.

master reference A signal point which is the signal reference point for an entire system. Usually a ground rod or grid. All local references are tied back to the master reference point.

master-slave A mode of operation where one data station (the master) controls the network access of one or more data stations (the slaves).

master/slave manipulator A remote manipulator which mechanically, hydromechanically or electromechanically reproduces hand or arm motions of an operator.

mate The joining, engaging, connecting or coupling of two connectors or devices designed to be utilized together.[ARP 914A-79]

material balance 1. The procedure of accounting for the mass of material going into a process versus the mass leaving the process. 2. The balance relating the material in and material out of a distillation column. Material-balance manipulative variables are overhead flow, bottoms flow, sidestreams flow, and feed flow.

material dispersion Light pulse broadening due to differential delay of various wavelengths of light in a waveguide material. This group delay is aggravated by broad bandwidth light sources.

materials handling Transporting or conveying materials, parts or assemblies, including all aspects of loading, unloading, moving, storing and shipping them, both within a facility and between facilities.

materials recovery The treatment of a material to reclaim one or more of its components.[NASA]

materials science The study of materials used in research, construction and manufacturing; includes the fields of metallurgy, ceramics, plastics, rubber and composites.

mathematical check A check which uses mathematical identities or other properties, occasionally with some degree of discrepancy being acceptable, e.g., checking multiplication by verifying that AxB=BxA. Synonymous with arithmetic check.

mathematical logic Same as symbolic logic.

mathematical model The general characterization of a process, object, or concept, in terms of mathematics, which enables the relatively simple manipulation of variables to be accomplished in order to determine how the process, object, or concept would behave in different situations.

mathematical programming In operations research, a procedure for locating the maximum or minimum of a function subject to constraints. Contrast with convex programming, dynamic programming, integer programming, linear programming, nonlinear programming, and quadratic programming.

mat, landing See landing mat.

matrix 1. The constituent which forms the continuous or dominant phase of a two or more phase microstructure.[AS 1814-90] 2. In mathematics, an n dimensional rectangular array of quantities. Matrices are manipulated in accordance with the rules of matrix algebra. 3. In computers, a logic network in the form of an array of input leads and output leads with logic elements connected at some of their intersections. 4. The principal microstructural constituent of an alloy. 5. The binding agent in a composite or agglomerated mass.

matrix management An organized approach to administration of a program by defining and structuring all elements to form a single system with components united by interaction.[NASA]

matrix material The ingredients used as binding agents to produce composite materials.[NASA]

matrix printer A type of computer device that forms letters and symbols by printing a pattern of dots.

mat-surface glass Glass whose surface has been altered by etching, sand-blasting, grinding, etc., to increase the diffusion. Either one or both surfaces may be so treated.[ARP798-65]

matte 1. A smooth but relatively nonreflective surface finish. 2. An intermediate product in the refining of sulfide ores by smelting.

max See maximum.

maximum allowable working pressure The highest gage pressure that can safely be applied to an internally pressurized system under normal operating conditions. It is usually well below the design bursting pressure and the hydrostatic test pressure for the system, and is the pressure at which relief valves are set to lift.

maximum authorized altitude (MAA) The highest altitude on a federal airway, jet route, area navigation low or high route, or other direct route for which a minimum en route altitude is designated in FAR Part 95, at which adequate reception of navigation aid signals is assured.[ARP4107-88]

maximum continuous load The maximum load which can be maintained for a specified period.

maximum entropy method Procedure

used in estimating high resolution power spectra from short data lengths.[NASA]

maximum, error See error, maximum.

maximum flow The maximum flow rate is given in the instrument specifications. The accuracy and validity of calibration depends on maintaining a flow rate within specifications.[ARP1192-87]

maximum fluid temperature Highest fluid temperature in the system at which the fluid is intended to be operated. [AIR1916-88]

maximum instantaneous demand The sudden load demand on a boiler beyond which an unbalanced condition may be established in the boiler's internal flow pattern and/or surface release conditions.

maximum limit of ultimate trip The minimum current which will cause a circuit breaker to open under a given set of ambient conditions. Also known as the "rated trip current".[ARP1199A-90]

maximum permissible pressure The highest pressure that is permitted for safety reasons.[AIR1916-88]

maximum pointer A movable pointer that is repositioned as the indicating pointer of an instrument moves upscale, but remains stationary at the highest point reached when the indicating pointer moves downscale.

maximum pressure The highest transient pressure that can occur temporarily.[AIR1916-88]

maximum thermometer A thermometer that indicates maximum temperature reached during a given interval of time; a clinical thermometer used to determine a patient's body temperature is one type of maximum thermometer.

maximum time to repair The maximum time required to complete a specified percentage of all maintenance actions. [ARD50010-91]

maximum usable frequency For a given distance from a transmitter, the highest frequency at which sky waves can be received.[NASA]

maximum working pressure See pressure, maximum working (MWP).

Maxwell The CGS unit of magnetic flux.

Maxwell bridge A type of a-c bridge circuit in which the impedance of an unknown inductor is measured in terms of an adjustable resistor and adjustable inductor; since the latter may be difficult to obtain, an alternative bridge arrangement uses an adjustable resistor and capacitor in parallel with the unknown inductor.

Maxwellian distribution The velocity distribution of the moving molecules of a gas in thermal equilibrium, as determined by applying the kinetic theory of gases.

maypole antennas A class of antennas which use the deployable reflector concept for large space systems applications.[NASA]

MBit Million bits per second.

MBM See magnetic bubble memory.

MBM junctions Diode devices using metal-barrier-metal layers.[NASA]

MByte 1,048,576(2^{20})bytes.

MCA Minimum Crossing Altitude.

McLeod vacuum gage A common type of mercury filled pressure gage whose design is a special case of a liquid manometer used as a pressure amplifier; the design enables use of a manometer-type instrument for measuring vacuum on the order of 10^{-6} torr instead of the 10^{-2} torr usually achieved with precision manometers.

MDA Minimum Descent Altitude.

MEA Minimum En route IFR Altitude.

mean accuracy See accuracy, mean.

mean chord The average length of the chord line from the wing tip to the wing root.[ARP4107-88]

mean corrective maintenance time The mean time required to complete a maintenance action, i.e., total maintenance down time divided by total maintenance actions, over a given period of time. Mean time to repair (often denoted as MTTR) is the sum of all maintenance downtime during a given period divided by the number of maintenance actions during the same period of time.[ARD50010-91]

mean effective pressure The average net pressure difference across a piston in a positive displacement machine such as a compressor, engine or pump. It is commonly used to evaluate performance of such a machine.

mean free path 1. Of any particle, the average distance that a particle travels between successive collisions with the other particles of an ensemble. Specifically, the average distance traveled by the molecules of a perfect gas between consecutive collisions with one another. For any process the reciprocal of the cross section per unit volume for that process.[NASA] 2. In a gas, liquid or colloid, the average distance traveled by an individual atom, molecule or particle between successive collisions with other particles.

mean horizontal candlepower The average candlepower in the horizontal plane passing through the luminous center of the lamp. It is assumed that the lamp or other light source is mounted in the usual manner, as in the case of a filament lamp, with its axis of symmetry vertical.[ARP798-65]

mean life The arithmetic mean of the times to failure of a group of nominally identical items.[ARD50010-91]

mean line A line lying in a plane parallel to the plane of symmetry that is equidistant from the upper and the lower surfaces from leading edge to the trailing edge of an airfoil.[ARP4107-88]

mean maintenance time The measure of item maintainability taking into account maintenance policy. The sum of preventive and corrective maintenance times, divided by the sum of scheduled and unscheduled maintenance events, during a stated period of time.[ARD 50010-91]

mean sea level (MSL) Sea level between mean high tide and low water.[ARP 4107-88]

mean square values In statistics, values representing the average of the sum of the squares of the deviations from the mean value.[NASA]

mean time between critical failure MTBCF is the average time between failure of mission essential system functions.[ARD50010-91]

mean time between demands A measure of the system reliability parameter related to demand for logistic support: The total number of system life units divided by the total number of item demands on the supply system during a stated period of time. e.g. Shop Replaceable Unit (SRU), Weapon Replaceable Unit (WRU), Line Replacement Unit (LRU), and Shop Replaceable Assembly (SRA).[ARD50010-91]

mean time between failure (MTBF) 1. A basic measure of reliability for repairable items: The mean number of life units during which all parts of the item perform within their specified limits, during a particular measurement interval under stated conditions.[ARD 50010-91] 2. Total system operating time divided by the number of system failures that have occurred during that period; the average time one could expect a given system to operate before experiencing a system failure.[ARP4107-88] 3. The limit of the ratio of operating time of equipment to the number of observed failures as the number of failures approaches infinity. The total operating time divided by the quantity (n+1), where n is the number of failures during the time considered.

mean time between maintenance A measure of the reliability taking into account maintenance policy. The total number of life units expended by a given time, divided by the total number of maintenance events (scheduled and unscheduled) due to that item.[ARD50010-91]

mean time between maintenance actions A measure of the system reliability parameter related to demand for maintenance manpower: The total number of system life units, divided by the total number of maintenance actions (preventive and corrective) during a stated period of time.[ARD50010-91]

mean time between removals A measure of the system reliability parameter related to demand for logistic support: The total number of system life units divided by the total number of items removed from that system during a stated period of time. This term is defined to exclude removals performed to facilitate other maintenance and removals for product improvement.[ARD 50010-91]

mean time between unscheduled removals A performance figure calculated by dividing the total unit flying hours (airborne) accrued in a period by the number of unscheduled unit removals that occurred during the same period.[ARD50010-91]

mean-time-to-failure (MTTF) The average or mean-time between initial operation and the first occurrence of a failure or malfunction, as the number of measurements of such time on many pieces of identical equipment approaches infinity.

mean time to repair The total corrective maintenance time divided by the total number of corrective maintenance actions during a given period of time. [ARD50010-91]

mean time to restore system A measure of the system maintainability parameter related to availability and readiness: The total corrective maintenance time, associated with downing events, divided by the total number of downing events, during a stated period of time. (Excludes time for off-system maintenance and repair of detached components.)[ARD50010-91]

mean time to unscheduled removal A performance figure calculated by dividing the summation of times to unscheduled removal for a sample of removed items by the number of removed items in the sample. NOTE: This is different from mean time between unscheduled removal (MTBUR) since no allowance is given to items that have not been removed.[ARD50010-91]

measurand A physical quantity, force, property or condition which is to be measured.[ARP1587-81]

measured accuracy See accuracy, measured.

measured ceiling In U.S. aviation weather observations, the ceiling classification that is applied when the ceiling value has been determined by means of: (a) a ceiling light or ceilometer, or (b) the known heights of unobscured portions of objects or other natural landmarks within 1.5 nautical miles of any runway of the airport. It applies only to clouds and obscuring phenomena aloft, and is identified by the ceiling designator "M".[ARP4107-88]

measured signal See signal, measured.

measured value See value, measured.

measured variable 1. The physical quantity, property, or condition which is to be measured. Common measured variables are temperature, pressure, rate of flow, thickness, speed, etc. 2. The part of the process that is monitored to determine the actual condition of the controlled variable. See also variable, measured.

measured variable modifier The second letter when first-letter is used in combination with modifying letters D (differential), F (ratio), M (momentary), K (time rate of change), Q integrate or totalize), (A could be used for absolute), or any combination of these is intended to represent a new and separate measured variable, and the combination is treated as a first-letter entity.

measurement A data point which is or can be converted into a suitable signal for telemetry transmission.

measurement component A general term indicating the components or subassemblies in a specific device that together determine the value of a quantity and produce the indicated or recorded output.

measurement device A self-contained assembly comprised of all the necessary components to perform one or more measuring operations.

measurement energy The energy, usually obtained from the measurand or the primary detector required to operate a measurement device or system.

measurement equipment A general term used to describe components, devices, assemblies or systems capable of performing measuring operations.

measurement mechanism A mechanical device that performs one or more operations in a measuring sequence.

measurement range The portion of the total response range of an instrument over which specific standards of accuracy are met.

measurement system Any set of interconnected components, including one or more measurement devices, that performs a complete measuring function, from initial detection to final indication, recording or control-signal output.

measuring See measurement.

measuring instrument See instrument, measuring.

measuring junction The electrical connection between the two legs of a thermocouple which is attached to the body, or immersed in the medium, whose temperature is to be measured.

measuring means The components of an automatic controller which determine the value of a controlled variable and communicate that value to the controlling means.

measuring modulator A component in a measuring system which modulates a direct-current or low-frequency alternating-current input signal to produce an alternating-current output signal whose amplitude is related to the measured value, usually as a preliminary step to producing an amplified output signal.

measuring range The extreme values of the measured variable within which measurements can be made within the specified accuracy. The difference between these extreme values is called "span."

mechanical Referring to tools or machinery.

mechanical atomizing oil burner A burner which uses the pressure of the oil for atomization.

mechanical chart drive A spring-driven clock mechanism which feeds continuous chart paper past a recorder head at a predetermined speed.

mechanical classification Any of several methods for separating mixtures of particles or aggregates according to size or density, usually involving the action of a stream of water.

mechanical compliance Displacement of a mechanical element per unit force; it is the mechanical equivalent of capacitance in an electrical circuit.

mechanical damping Attenuating a vibrational amplitude by absorption of mechanical energy.

mechanical draft The negative pressure created by mechanical means.

mechanical efficiency The ratio of power output to power input.

mechanical engineering A branch of engineering that deals with the generation and use of thermal and mechanical energy, and with the design, manufacture and use of tools and machinery.

mechanical hygrometer A hygrometer that uses an organic material, such as a bundle of human hair, to sense changes in humidity. In operation, the organic material expands and contracts with changes in moisture content in the air, and the change in length alters the position of a pointer through a spring-loaded mechanical linkage.

mechanical impedance The complex quotient of alternating force applied to a system divided by the resulting alternating linear velocity in the direction of the force at its point of application.

mechanical linkage A set of rigid bars, or links, that are joined together at pivot points and used to transmit motion. Frequently they are used in a mechanism along with a crank and slide to convert rotary motion to linear motion.

mechanical properties The properties of a material that can be determined through application of a force and measuring the material's response.

mechanical reactance The imaginary component of mechanical impedance.

mechanical register A mechanical or electromechanical recording or indicating counter.

mechanical resistance The real component of mechanical impedance.

mechanical reversion The capability of reverting from fly-by-wire control to a state wherein the pilot's control is mechanically coupled to the actuator control valves.[AIR1916-88]

mechanical transmission An assembly of mechanical components suitable for transmitting mechanical power and motion.

mechanical scale A weighing device in which objects are balanced through a system of levers against a counterweight or counterpoise.

mechanical seal A seal produced by an elastomeric O-ring or other molded or extruded shape when it is deformed by pressure mechanically as in access doors.[AIR4069-90]

mechanism 1. Generally, an arrangement of two or more mechanical parts in which motion of one part compels the motion of the others. 2. Specifically the arrangement of parts in an indicating instrument which control motion of the pointer or other indicating means, excluding those parts which form the enclosure, scale, or support structure, or which adapt the instrument to the quantity being measured. 3. In a recording instrument, the arrangement of parts which control motion of the marking device, the marking device itself, the device for driving the chart, and the parts which carry the chart.

mechanized dew-point meter See dew-point recorder.

mechanoreceptor Nerve ending that reacts to mechanical stimuli, as touch, tension, and acceleration.[NASA]

media 1. The physical interconnection between devices attached to the LAN. Typical LAN media are twisted pair, baseband coax, broadband coax, and fiber optics. 2. The plural of medium. 3. A name for the various materials used to hold or store electronic data, such as printer paper, disks, magnetic tape or punched cards.

media access control (MAC) The lower sublayer of the Data Link Layer (Layer 2) unique to each type of IEEE 802 Local Area Networks. MAC provides a mechanism by which users access (share) the network.

median corrective maintenance time The downtime within which 50% of all corrective maintenance actions can be completed under the specified maintenance conditions. The median value, Mct, is often referred to as the geometric mean (MTTRG) or equipment repair time (ERT) in some maintainability documents.[ARD50010-91]

median preventive maintenance time

The equipment downtime required to perform 50% of all scheduled preventive maintenance actions on the equipment under the specified conditions.[ARD 50010-91]

medical certificate Acceptable evidence of physical fitness on a form prescribed by the Federal Aviation Administrator. [ARP4107-88]

medium In data processing, the material on which data is recorded and stored.

medium-carbon steel An alloy of iron and carbon containing about 0.25 to 0.6% C, and up to about 0.7% Mn.

medium intensity approach light system (MALS) See runway edge light system.[ARP4107-88]

medium scale integration (MSI) 1. A medium level of chip density, lower than for LSI circuits but more than small scale integration. 2. An integrated circuit with 10 to 100 logic gates.

mega Prefix denoting 1,000,000.

megabit One million bits.

megabyte (MByte) A unit of computer memory size. One million bytes. See MByte.

megahertz One million hertz or cycles per second.

megohmmeter A device which measures the electrical resistance of a sealant (and other materials as well).[AS7200/1-90]

Meissner effect See superconductivity.

melt extrusion An extrusion process in which the insulation material is heated above its melting point and forced through a die.[ARP1931-87]

melt index Extrusion rate of a thermoplastic material through an orifice of specified diameter and length under specified condition of time, temperature and pressure. Also known as melt flow number.[ARP1931-87]

melting point The temperature at which a solid substance becomes liquid; for pure substances and some mixtures it is a single unique temperature; for impure substances, solutions and most mixtures it is a temperature range.

melts (crystal growth) Molten substances from which crystals are formed during the cooling or solidifying process.[NASA]

melt spinning A material process by which polymers such as nylon and polyesters and glass are melted to permit extrusion into fibers through spinnerets. [NASA]

melt time See element melt time.

membrane 1. A thin tissue which covers organs, lines cavities, and forms canal walls in the body of an animal. 2. A thin sheet of metal, rubber or treated fabric used to line cavities or ducts, or to act as a semirigid separator between two fluid chambers.

membrane analogy See membrane structures.

membrane structures Shell structures, often pressurized, that do not take wall bending or compression loads.[NASA]

memory 1. The mental activity of recalling past experience. Experience includes any information a person receives through any means, any cognitive functions he/she performed on that information, and any response he/she made as a result of the information. The general function of reviving or reliving past experience, with more or less definite realization that the present experience is a revival.[ARP4107-88] 2. The component of a computer control system, guidance system, instrumented satellite, or the like, designed to provide ready access or data or instructions previously recorded so as to make them bear upon an immediate problem, such as the guidance of a physical object, or the analysis and reduction of data.[NASA] 3. Any form of computer data storage, including main memory and mass storage, in which data can be read and written; in its strictest sense, memory refers to main memory.

memory access time See access time.

memory address The address in computer memory of the location containing an instruction or operand.

memory addressability A measure of capability and ease of programming used in evaluating computers. The maximum number of locations specifiable by a nonindexed instruction using the instruction's minimum execution time.

memory bus The computer bus (or buses) which interconnects the processor, memory, and peripherals on a high-speed data processing highway.

memory capacity Same as storage capacity.

memory chip An electronic device that accepts data for computer use or storage.

memory cycle time The minimum time between two successive data accesses from a memory.

memory dump A listing of the contents of a storage device, or selected parts of it.

memory image A replication of the contents of a portion of memory.

memory latency time See latency.

memory, long-term The recall or recognition of experience days, months, or years after its occurrence. A system that retains past experiences for long periods, has a very large capacity, and stores items in relatively processed forms. [ARP4107-88]

memory management A function of a PDP-11 computer that enables it to operate with a larger memory than 32k words.

memory map Graphic representation of the general functional assignments of various areas in memory. Areas are defined by ranges of addresses.

memory mapping A map showing the usable and unusable (or protected) areas of memory.

memory protect A technique of protecting the contents of sections of memory from alteration by inhibiting the execution of any memory modification instruction upon detection of the presence of a guard bit associated with the accessed memory location. Memory modification instructions accessing protected memory are usually executed as a no-operation and a memory protect violation program interrupt is generated.

memory protection A scheme for preventing read and/or write access to certain areas of memory.

memory resident A program that remains in RAM memory even when other programs are operating, and be called up by interrupting the currently running program.

memory, short-term The recall or recognition of experience within a few minutes or hours of occurrence. A hypothesized memory system that keeps material for intervals of a minute or so, is very dependent upon rehearsal, has a small storage capacity (sometimes said to be 7 + or - 2 pieces or chunks of information), and holds material in relatively less processed form than long-term memory.[ARP4107-88]

meniscus The concave or convex surface, caused by surface tension, at the top of a liquid column, as in a manometer tube.

meniscus lens A lens with one concave surface and one convex surface.

mental workload The total cognitive demands upon a person in a particular time unit. See cognitive psychology. [ARP4107-88]

menu In data processing, a list from which an operator can select the tasks to be done.

mercury cadmium tellurides Compounds of tellurium exhibiting photovoltaic characteristics and used for photodiodes and photodetectors in the 3 to 12 micrometer wavelengths at cryogenic temperatures.[NASA]

mercury ion engines Machines providing thrust by expelling accelerated or high velocity mercury ions and often using energy provided by nuclear reactors.[NASA]

mercury meter A differential pressure measuring device utilizing mercury as the seal between the high and low chambers.

mercury switch A type of switch consisting of two wires sealed into the end of a glass capsule containing a bead of mercury; if the capsule is tipped one way, the mercury covers the exposed ends of the wires and completes the circuit; if tipped the other way, the mercury exposes the wires and breaks the circuit.

mercury vapor lamp A type of ion discharge lamp widely used in ultraviolet analyzers because it emits several strong monochromatic lines with characteristic wavelengths such as 254, 313, 360, 405 etc.; lamp emission can be made almost completely monochromatic by using special filters.

mercury-vapor tube A gas tube in which the active gas is mercury vapor.

mercury-wetted relay A device using mercury as the relay contact closure substance.

merge In data processing, to combine two or more groups of records into a single file.

merge sort In data processing, an operation of combining data and then sorting it in some prescribed manner.

meridian plane Any plane which contains the optical axis.

mesh 1. A measure of screen size equal to the number of openings per inch along the principal direction of the weave. 2. The size classification of particles that pass through a sieve of the stated screen size. 3. Engagement of a gear with its mating pinion or rack. 4. A closed path through ductwork in a ventilation survey.

mesons In the classification of subatomic particles by mass, the second lightest of such particles. Their mass is intermediate between that of the lepton and the nucleon.[NASA]

mesopause The base of the inversion at the top of the mesosphere, usually found at 80 to 85 kilometers.[NASA]

mesoscale phenomena Meteorological phenomena extending approximately one to a hundred kilometers (mesoscale cloud pattern, for example).[NASA]

mesosphere The atmospheric shell, in which temperature generally decreases with heights, extending from the stratopause at about 50 to 55 kilometers to the mesopause at about 80 to 85 kilometers.[NASA]

message An arbitrary amount of information whose beginning and end are defined or implied.

message exchange A device placed between a communication line and a computer to take care of certain communication functions and thereby free the computer for other work.

message latency category A quality of service in which a message transfer is guaranteed to be initiated within a certain period of time.[AIR4271-89]

message processing In communication operations, the acceptance, preparation for transmission, receipt and/or delivery of a series of words or symbols intended for conveying information.[NASA]

message routing The function performed at a central message processor of selecting the route, or alternate route if required, by which a message will proceed to the next point in reaching its destination.

message switching The technique of receiving a message, storing it until the proper outgoing circuit is available, and then retransmitting it.

metabolites Products of biological synthesis and/or metabolism.[NASA]

metal A chemical element that is crystalline in the solid state, exhibits relatively high thermal and electrical conductivity, and has a generally lustrous or reflective surface appearance.

metal-barrier-metal junctions See MBM junctions.

metal corrosion See corrosion.

metal foams Foamed materials formed under low gravity conditions in space from sputtered metal deposits. This experimental space processing was completed in the second NASA SPAR flight. [NASA]

metal-insulator-metal diodes See MIM diodes.

metallic Exhibiting characteristics of a metal.

metallic coating A thin layer of metal applied to an optical surface to enhance reflectivity.

metallic glass(es) Amorphous alloys (glassy metals) produced by extremely rapid quenching of molten transition-metal alloys (e.g., iron, nickel, and/or cobal). These metallic glasses exhibit unique mechanical magnetic, and electrical properties, superconductive behavior, and anticorrosion resistance, depending on the alloys, their formation and quenching techniques.[NASA] See also glassy alloy.

metallicity The abundance index of a metal or metals for a celestial body. [NASA]

metallic superoxides The more common superoxides are those of potassium and sodium with KO_2. In addition to serving as oxygen sources, the superoxides remove carbon dioxide, water and odors. They weigh less, cost less and occupy a smaller volume than oxygen or lithium hydroxide (for carbon dioxide removal) systems. They provide a degree of sterilization and have an excellent shelf life.[AIR1246-77]

metallography The study of the structure of metals—the most common techniques are optical microscopy, electron microscopy and x-ray diffraction analysis.

metal-nitride-oxide semiconductor 1. Class of semiconductors utilizing silicon nitride and silicon oxide dielectrics. [NASA] 2. One type of computer semiconductor memory used in EAROMs.

metal-semiconductor-metal semiconductors See MSM (semiconductors).

metal units Concentration units defined as the number of gm-moles per 1000 gm of solvent.

metal vapor lasers Stimulated emission devices the active materials of which are vaporized metals.[NASA]

metamorphism (geology) The mineralogical and structural adjustment of solid rocks to physical and chemical conditions which have been imposed at depths below the surface zones of weathering and cementation, which differ from the conditions under which the rocks in question originated.[NASA]

metastable beta A nonequilibrium phase composition that can be partially or completely transformed to martensite, alpha, or eutectoid decomposition products with thermal or strain energy activation during subsequent processing or service exposure.[AS1814-90]

meteor bursts See meteoroid showers.

meteorite compression tests See meteorites.

meteorites Meteoroids which have reached the surface of the earth without being completely vaporized.[NASA]

meteoritic ionization See meteor trails.

meteorograph A recording instrument for measuring meteorological data—temperature, barometric pressure and humidity, for example.

meteoroids Solid objects moving in interplanetary space, of a size considerably smaller than asteroids and considerably large than atoms or molecules. [NASA]

meteoroid showers Groups of meteoroids with approximately parallel trajectories.[NASA]

meteorological instrumentation

Equipment for measuring weather data.

meteorological rockets See sounding rockets.

meteorology The study dealing with the phenomena of the atmosphere. This includes not only the physics, chemistry, and dynamics of the atmosphere, but is extended to include many of the direct effects of the atmosphere upon the earth's surface, the oceans, and life in general. A distinction can be drawn between meteorology and climatology, the latter being primarily concerned with average not actual weather conditions.[NASA]

meteors See meteoroids.

meteor trails Anything, such as light or ionization, left along the trajectory of the meteor after the head of the meteor has passed.[NASA]

meter 1. A device for measuring and indicating the value of an observed quantity. 2. An international metric standard for measuring length, equivalent to approximately 39.37 in. in the U.S. customary system of units. Spelled metre in the International Standard of Units (SI).

metered brake pressure The fluid pressure (usually hydraulic) which is metered to the brakes by pilot demand and control to provide the desired decelerating force for the aircraft.[AIR1489-88]

meter factor 1. A constant used to multiply the actual reading on a scale or chart to produce the measured value in actual units. 2. A correction factor applied to a meter's indicated value to compensate for variations in ambient conditions such as a temperature correction applied to a pressure indication.

metering 1. A method of time-regulating the traffic flow into a terminal area so as not to exceed a predetermined terminal acceptance rate (the maximum inbound traffic flow to the field as determined by the FAA).[ARP4107-88] 2. Regulating the flow of a fluid so that only a measured amount is permitted to flow past a given point in the system. 3. Measuring any variable (flow rate, electrical power, etc.).

metering fix (MF) 1. The airport arrival flow control points used by Air Traffic Control to meter spacing for landing. Usually the fix is associated with a required flight level, speed and ETA.[ARP 1570-86] 2. A fix along an established route over which aircraft will be metered prior to entering terminal airspace.-[ARP4107-88]

metering pin A pin within the shock absorber which operates through the fluid metering orifice as stroking occurs. Variation of the diameter of the pin provides a preprogrammed metering of the fluid through the orifice to obtain maximum efficiency for a given landing condition and optimum balance of efficiency for all conditions.[AIR1489-88]

metering tube An arrangement within the shock absorber for metering fluid for control of energy absorption. The metering tube has drilled passages through the wall to program the metering of fluid as stroking occurs.[AIR 1489-88]

meter prover A device for checking the accuracy of a gas meter.

meter proving tank See calibrating tank.

meter run A flowmeter installed and calibrated in a section of pipe having adequate upstream and downstream length to satisfy standards of flowmeter installation. See also orifice run.

meter sensitivity The accuracy with which a meter can measure a value; it is usually expressed as percent of the meter's full scale reading.

methanation The conversion of various organic compounds to produce methane.[NASA]

method of moments A method of estimating the parameters of a distribution by relating the parameters to moments.[NASA]

metre Metric unit of length (SI). See meter.

metrication The conversion on an industry and/or nationwide basis of English units of measurement into the International System of Units, including engineering and manufacturing standards, tools and instruments, and all affected areas in the government and private sectors.[NASA]

metric conversion See metrication.

metric photography The recording of events by means of photography (either singly or sequentially), together with appropriate metric coordinates to form the basis for accurate measurements. [NASA]

metric system See International System of Units.

metrology The science of dimensional measurement; sometimes includes the science of weighing.[NASA]

MeV Mega-electron-volts; a unit of energy equivalent to the kinetic energy of a single electron accelerated through an electric potential of 1-million volts.

MF Middle or Medium Frequency.

Mf The temperature at which the martensite reaction is complete.[AS1814-90]

mho 1. Unit of conductance. Reciprocal of an ohm. One ampere of current passing through a material under a potential difference of one volt provides one mho of conductance.[ARP1931-87] 2. A customary unit of conductance and admittance generally defined as the reciprocal of one ohm, or the conductance of an element whose resistance is one ohm; the equivalent SI unit Siemen is preferred.

MIA Minimum IFR Altitudes.

mica An inorganic material which separates into layers and has high insulation resistance, dielectric strength and heat resistance. It is used as a insulation wrap in wires and cables to a limited degree where resistance requirements are severe and for high temperature work demanding good heat resistance. [ARP1931-87]

micro 1. (μ) Prefix denoting one-millionth. [ARP1931-87] 2. A common term meaning very small.

microamanometer See manometer.

microbalance A small analytical balance for weighing masses of 0.1 g or less to the nearest μg.

microballoons Very small glass spheres (50 to 100 micrometers in diameter) used as targets in the laser fusion programs. [NASA]

microbar A unit of pressure equal to one dyne per square centimeter.

microburst A localized but very severe weather phenomenon resulting in abrupt changes in wind direction and velocity. [ARP4107-88]

microcalorimeter See calorimeter.

microchannel plate(s) 1. An array of microchannels formed into plates and contained in a photomultiplier tube. [NASA] 2. A glass device with many tiny, parallel holes passing through it. It is used, with suitable biasing, as an electron amplifier, primarily for use in imaging detectors.

micro code 1. A system of coding making use of suboperations not ordinarily accessible in programming, e.g., coding that makes use of parts of multiplication or division operations. 2. A list of small program steps; combinations of these steps, performed automatically in a prescribed sequence from a macrooperation like multiply, divide, and square root. See multiprocessor.

microcomputer(s) 1. Complete digital computers utilizing a microprocessor consisting of one or more integrated circuit chips as the central arithmetic and logic unit, and added chips to provide timing, program memory, random access memory interfaces for input and output signals and other functions. Some microcomputers consist of a single integrated-circuit chip.[NASA] 2. A computer based on the use of a microprocessor integrated circuit. The entire computer often fits on a small printed circuit board and works with a data word of 4, 8, or 16 bits. 3. A complete computer in which the CPU is a microprocessor.

microcurie A unit of radioactivity equal to one millionth (10^{-6}) curie.

microdensitometer A device for measuring the density of photographic films or plates on a microscopic scale; the small scale version of a densitometer.

microfarad (μf)One-millionth of a farad, a unit of capacitance.[ARP1931-87]

microfaradmeter A capacitance meter calibrated in microfarads.

micro-floppy disks 3 1/2-inch disks that have greater storage capacity than a 5 1/4-inch floppy disk.

microgravity See reduced gravity.

microinstruction Controls the operations of the various primitive resources of a computer: main and local store registers (both general and special purpose), arithmetic and logic units (ALUs), data paths, and so on. Microinstructions are stored as words in a control store that is traditionally (but not necessarily) separate from the main storage.

micromanipulator A positioning device for making small adjustments to the position of an optical component or other device.

micromechanics The study of the constraints, the grain size, and their interrelationship in materials.[NASA]

micrometeorites Very small meteorites or meteoritic particles with a diameter in general less than a millimeter.[NASA]

micrometer 1. Instrument for making precise linear measurements in which the displacements measured correspond to the travel of a screw of accurately known pitch.[NASA] 2. A metric measure with a value of 10^{-6} meters or 0.000001 meter, previously referred to as "micron" 3. Any device incorporating a screw thread for precisely measuring distances or angles, such as is sometimes attached to a telescope or microscope. 4. A type of calipers that incorporates a precision screw thread and is capable of measuring distance between two opposing surfaces to the nearest 0.001 or 0.0001 in.

micron One millionth of a meter, or 0.000039 in. The diameter of dust particles is often expressed in microns.

micron cubic meter of gas That quantity of a gas that will occupy a one cubic meter volume at a pressure of one micron (one millionth of a meter of mercury).[AIR818C-91]

microphone An electroacoustic transducer which transmits an electrical output signal that is directly related to the loudness and frequency distribution of sound waves that strike the active element.

microphonism 1. In an electron tube, modulation of one or more electrode currents as a direct result of mechanical vibrations of a tube element. 2. An undesirable electrical output signal in response to mechanical or acoustic vibration of an electronic or electrical devise.

microphotometer See photometer.

microprocessor 1. A usually monolithic, large-scale-integrated (LS) central processing unit (CPU) on a single chip of semi-conductor material; memory, input/output circuits, power supply, etc. are needed to turn a microprocessor into a microcomputer. 2. A large-scale integrated circuit that has all the functions of a computer, except memory and input/output systems. The IC thus includes the instruction set, ALU, registers and control functions. 3. Sometimes abbreviated as MPU, uP, etc.

microprogramming A method of operating the control unit of a computer, wherein each instruction initiates or calls for the execution of a sequence of more elementary instructions. The microprogram is generally a permanently stored section of nonvolatile storage. The instruction repertory of the microprogrammed system can thus be changed by replacing the microprogrammed section of storage without otherwise affecting the construction of the computer.

microradiography Production of a magnified radiographic image.

microradiometer A device for detecting radiant power which consists of a thermopile supported on and directly connected to the moving coil of a galvanometer.

microscopic stress Load per unit area over a very short distance, on the order of the diameter of a metal grain, or smaller. A term usually reserved for characterizing residual stress patterns.

microsecond (μs)One-millionth of a second (10^{-6}) second).[ARP1931-87]

microwave 1. Sources do not agree on the length of microwaves and their frequency, but it is generally considered to be in the frequency band between approximately 500 MHz and 300 GHz. [ARP4107-88] 2. Of, or pertaining to, radiation in the microwave region. [NASA] 3. Electromagnetic radiation having a wavelength of 1 to 300 mm.

Microwave Amplification by the Stimulated Emission of Radiation (MASER) The microwave equivalent and predecessor of the laser. It produces coherent microwaves.

microwave landing system (MLS) An instrument landing system operating in the microwave spectrum which provides lateral and vertical guidance to aircraft having compatible avionics equipment. This system provides precise continuous three-dimensional position information anywhere within the approach zone, potentially allowing unrestricted choice of approach paths. [ARP4107-88]

microwave radiation See microwaves.

microwave scanning beam landing system Primary position sensor of Space Shuttle Orbiter's navigation system during the autoland phase of the flight.[NASA]

microwave spectrum The portion of the electromagnetic spectrum of frequencies lying between infrared waves and radio waves.

microyield strength Stress at which a microstructure (single crystal, for example) exhibits a specified deviation in its stress-strain relationship.[NASA]

midaltitude The average of many measurements of altitudes as with satellite instruments for the compiling of planetary maps.[NASA]

middle atmosphere The portion of the earth atmosphere extending from the troposphere to 100 kilometers.[NASA]

middle compass locator (LMM) See compass locator.

middle frequency (MF) The frequency band between 300 kHz and 3 MHz.[ARP 4107-88]

middle marker (MM) A marker beacon that defines a point along the glide slope of an ILS normally located at or near the point of decision height (ILS Category I). It is keyed to transmit alternate dots and dashes, with the alternate dots and dashes keyed at the rate of 95 dot/dash combinations per minute, on a 1300 Hz tone, which is received aurally and visually by compatible airborne equipment. See marker beacon.[ARP4107-88]

mid RVR See runway visual range.

mid-value logic system A fault-tolerant system having an odd number of active channels (usually three) where the system output is determined by the middle of the (three) input signals.[AIR 1916-88]

Mie scattering Any scattering produced by spherical particles without special regard to comparative size of radiation wavelength and particle diameter.

[NASA]

MIE theory See Mie scattering.

MIG aircraft Any of a series of Soviet fighter aircraft, fighter-bombers, interceptors, and air supremacy aircraft, designed by Mikoyan.[NASA]

migration The movement of ions from an area of the same charge to an area of opposite charge.

MIG welding Metal inert-gas welding; see gas metal-arc welding.

mil 1. One one-thousandth of an inch.[ARP1931-87] 2. A unit of angular measurement commonly used in the military for setting artillery elevations.

mild detonating fuse (MDF) A flexible metal tube, usually lead, containing a much smaller core of high explosive than the normal detonating cord.[AIR 913-89]

mile A British and U.S. unit of length commonly used to specify distances between widely separated points on the earth's surface; a statute mile, used for distances over land, is defined as 5280 ft; a nautical mile, used for distances along the surface of the oceans, is defined as one minute of arc measured along the equator, which equals 6080.27 ft or 1.1516 statute miles.

military climb corridor A restricted area established in the vicinity of certain military bases used by military aircraft to climb out from an airfield to their desired operating altitude.[ARP 4107-88]

Milky Way Galaxy The galaxy to which the sun belongs.[NASA]

millilambert A measure of the brightness of a surface which emits or reflects light. A perfectly reflecting surface illuminated by one footcandle has a brightness of one footlambert (fl).[AS264D-91]

millimeter Also spelled millimetre. 1. A unit of length equal to 0.001 meter. 2. A millimeter of mercury, abbreviated mm Hg, is a unit of pressure equivalent to the pressure exerted by a column of pure liquid mercury one mm high at 0 °C under a standard gravity of 980.665 cm/s^2; it is roughly equivalent to 1/760th of standard atmospheric pressure.

milliradian (mr) An angular measurement equal to 0.0573 deg defined as 0.001 of an arc whose length equals the circle radius.[ARP4067]

millisecond One thousandth of a second (10^{-3}) second.[ARP1931-87]

MIL-STD-1533 The military standard that defines serial data communications protocol on modern military vehicles, especially aircraft.

Mimas A satellite of Saturn orbiting at a mean distance of 186,000 kilometers.[NASA]

MIMD (computers) A type of parallel processor that is essentially two or more individual computers with facilities for interaction and work sharing.[NASA]

MIM diodes Junction diodes each consisting of an insulating layer sandwiched between two metallic surface layers and exhibiting a negative differential resistance in its V-1 characteristics conceivably because of stimulated inelastic tunneling of electrons.[NASA]

min See minute.

mini In data processing, a term used to describe a smaller computer of 12- to 32-bit word length and memory sizes of 16K-8M bytes.

miniature boiler Fired pressure vessels which do not exceed the following limits: 16 in. inside diameter of shell; 42 in. over-all length to outside of heads at center; 20 sq ft water heating surface; or 100 psi maximum allowable working pressure.

miniaturization The design and production of a scaled-down version of a device or mechanism that is capable of performing all of the same functions as the larger-sized original.

minicomputer A medium size computer for more dedicated applications than mainframe computers. It generally has a larger instruction set, wider range of languages and better support than microcomputers.

minimal surfaces Surfaces for which the first variation of the area integral vanish.[NASA]

MINI-MAP A subset of the MAP protocols extended to provide higher performance for application whose communications are limited to a single LAN. A MINI-MAP node contains only the lower two layers (physical and link) of the MAP protocols. It can only communicate directly with MAP/EPA or MINI-MAP nodes on the same segment.

mini-micro In data processing, a very small microcomputer containing a CPU, memory, and I/O interfaces for data exchange and timing circuits to control the flow of data.

minimum bend radius The smallest radius around which a piece of sheet metal, wire, bar stock or tubing can be bent without fracture, or in the case of tubing, without collapse.

minimum crossing altitude (MCA) At certain fixes, the lowest altitude at which an aircraft must cross when proceeding in the direction of a higher minimum en route IFR altitude (MEA). For example, when the minimum en route IFR altitude changes from 6000 to 8000 ft at a particular intersection when flying from west to east, the minimum crossing altitude for that intersection eastbound would be set at 8000 feet.[ARP 4107-88]

minimum descent altitude (MDA) The lowest altitude, expressed in feet above mean sea level, to which descent is authorized on final approach or during circle-to-land maneuvering in execution of a standard instrument approach procedure where no electronic glide slope is provided.[ARP4107-88]

minimum en route IFR altitude (MEA) The lowest published altitude between radio fixes that assures acceptable navigational signal coverage and meets obstacle clearance requirements between those fixes.[ARP4107-88]

minimum entropy method. Application of entropy in statistical mechanics. [NASA]

minimum equipment list (MEL) An approved list of equipment that must be operative for full mission capability. Any systems that are inoperative result in partial mission capable status. See minimum equipment and dispatch procedures (MEDP).[AIR1916-88]

minimum fusing current The smallest value of current a circuit breaker must hold without tripping under a given set of ambient conditions. Also known as the "rated hold current".[ARP1199A-90]

minimum holding altitude (MHA) The lowest altitude prescribed for a holding pattern that assures navigational signal coverage and communications, and that meets obstruction clearance requirements.[ARP4107-88]

minimum IFR altitudes (MIA) minimum altitudes for IFR operations as prescribed in FAR Part 91.[ARP4107-88]

minimum limit of ultimate trip The maximum current a circuit breaker must hold without tripping under a given set of ambient conditions. Also known as the "rated hold current".[ARP1199A-90]

minimum obstruction clearance altitude (MOCA) The lowest published altitude in effect between radio fixes on VOAR airways, off-airway routes, or route segments, that meets obstacle clearance requirements for the entire route segment and that assures acceptable navigational signal coverage only within 25 statute miles of a VOR station.[ARP4107-88]

minimum operating pressure The lowest pressure at which a system or component must function.[AIR1916-88]

minimum pressure The lowest transient pressure than can occur temporarily. [AIR1916-88]

minimum reception altitude The lowest altitude at which one can receive signals to determine specific VOR/VORTAC/TACAN fixes.[ARP4107-88]

minimum reflux The quantity of reflux required to perform a specified separation in a column that has an unlimited number of trays. At minimum reflux, no products are withdrawn.

minimum safe altitude (MSA) 1. The minimum safe altitude specified in FAR Part 92. for various operations. 2. Any altitude depicted on approach charts which provides at least 1000 ft obstacle clearance for emergency use within a specified distance from the navigation facility upon which an approach procedure is predicated. MSAs are identified as minimum sector altitudes or emergency safe altitudes and are established as follows: (a) Minimum Sector Altitudes: Altitudes depicted on approach charts which provide at least 1000 ft of obstacle clearance within a 25-mile radius of the navigation facility upon which the approach procedure is predicated. Sectors depicted on approach charts are at least 90 deg in scope radially and extend outward from the facility for 25 miles. These altitudes are for emergency use only and do not necessarily assure acceptable navigational signal coverage. (b) Emergency Safe Altitudes: Altitudes depicted on approach charts which provide at least 1000 ft of obstacle clearance in nonmountainous areas and 2000 ft of obstacle clearance in designated mountainous areas within a 1000-mile radius of the navigation facility upon which the approach procedure is predicated. Emergency Safe Altitudes are normally used only in military procedures.[ARP 4107-88]

minimum safe altitude warning (MSAW) A function of the ARTS III computer that alerts the controller when an aircraft equipped with an operating Mode C transponder is being tracked by the radar facility and is below or is predicted by the computer to go below a predetermined minimum safe altitude. [ARP4107-88]

minimum safe performance standard A standard which specifies the minimum instrument design requirements as established by the operational and environmental conditions encountered during normal flight (take off, climb, cruise, descent and landing). It also specifies the minimum instrument performance necessary for safe operation of the aircraft during normal flight.[AIR818C-91]

minimums/minima Weather condition requirements established for a particular operation or type of operation.[ARP 4107-88]

minimum thermometer A thermometer that indicates the lowest temperature reached during a given interval of time.

minimum vectoring altitude (MVA) The lowest altitude, expressed in feet above mean sea level, at which aircraft will be vectored by a radar controller. This altitude assures communications and radar coverage, and meets obstruction clearance criteria.[ARP4107-88]

mining engineering A branch of engineering that deals with the discovery, extraction and initial processing of minerals—usually metal ores or coal—found in the earth's crust.

minitrack optical tracking system See minitrack system.

minitrack system A satellite tracking system consisting of a field of separate antennas and associated receiving equipment interconnected so as to form interferometers which track a transmitting beacon in the payload itself.[NASA]

minor frame The period between frame synchronization words that include one complete cycle of a commutator having the highest rate; normally does not exceed 8192 bit internals; synonymous with prime frame.

minor graduations The shortest or lightest division marks on a graduated scale, which indicate subdivisions lying between successive major graduations or between an index graduation and an adjacent major graduation.

Minor Planet 1221 See Armor asteroid.

Minor Planet 2060 See Chiron.

minor time In data processing systems, one sixteen-bit word: binary milliseconds.

minute 1. A measure of angle equal to 1/60th of one degree. 2. A measure of time equal to 60 s.

mips (million instructions per second) The measure of computer machine code instructions per second.

Mirage aircraft Collective term for a class of French attack aircraft.[NASA]

Miranda A satellite of Uranus orbiting at a mean distance of 124,000 kilometers.[NASA]

Mira variables Long-period (80 to over 600 days) variable stars of red giant or red supergiant type, exemplified by the star Mira Ceti.[NASA]

MIRL Medium Intensity Runway Lights. See runway edge light system.

mirror fusion An open-ended configuration which traps low beta plasmas. It is realized by associating two identical magnetic mirrors having the same axis.[NASA]

mirror scale An instrument scale and a mirror, so arranged that the indicating pointer and its reflection are aligned when the observer's eye is in the correct position to read the instrument without parallax error.

misalignment load A force component perpendicular to the actuator output motion, which is the resultant of a load not in line with the actuation output. [AIR1916-88]

misconvergence The degree to which the midpoint of the line widths of two or three primary colors are misregistered at the phosphor surface of a CRT.[ARP 1782-89]

misfire Failure of an explosive device to fire after initiating energy is applied. [AIR913-89]

mishap An unplanned, unintended event that results in damage to equipment or injury to personnel.[ARP4107-88]

mishap, antecedent events Those events or conditions which occurred prior to flight but which relate to the conditions making the mishap more likely (for example, fatigue or "get-home-itis").[ARP4107-88]

mishap, maneuver A sub-element of the mishap phase of flight described by the sequence of tasks required to perform the maneuver (for example, turnout of traffic, formation crossover, or egress from a weapons delivery pass). Maneuver mishaps may occur in any of the eight phases of flight described as follows: 1. Approach Phase—From the final approach fix to the missed approach point for an instrument approach; from reaching traffic- pattern altitude until crossing the runway threshold for a visual approach. A go-around is considered part of the approach phase if it occurs prior to the missed approach point for an instrument approach or prior to crossing the runway threshold for a visual approach. 2. Climbout Phase—From crossing the field boundary to attaining cruise altitude. 3. Cruise Phase—From reaching cruise altitude to arriving at the area of range activity; or from leaving the area of the range activity to beginning descent into the base of intended landing. 4. Descent Phase—From the initial approach fix to the final approach fix for an instrument descent; from beginning descent from cruise altitude to arriving at the final approach fix for an en route descent to an instrument approach; from beginning descent from cruise altitude until reaching traffic pattern altitude for an en route descent to a visual approach. Holding is considered part of the descent phase of flight. 5. Landing Phase—From the missed approach point until touchdown for an instrument approach; from crossing the runway threshold until touchdown from a visual approach. A go-around is considered part of the landing phase if it occurs after the missed approach point for an instrument approach or after crossing the runway threshold for a visual approach. After touchdown, a touch-and-go is considered a takeoff. 6. Range Phase—From the time the aircraft enters the area designated for practicing/conducting mission activities until completion of those activities and departure from the designated area. This Range Phase may be a low-level route, military operating area, gunnery range, warning area, or a refueling track. 7. Takeoff Phase—From runway holdline to the point when the aircraft is airborne and has passed the field boundaries. 8. Taxi Phase—From engine start to runway holdline, and from clearing the active runway to having parked the aircraft.[ARP4107-88]

mishap, point of That point in the mishap sequence of events at which no preventive or evasive action by the operator would have avoided the mishap.[ARP4107-88]

mishap, predisposing events Those events or conditions more general in nature or more longstanding than mishap antecedent events but which are predisposing to mishap occurrence (for example, risk-taking tendencies, lax supervision, etc.)[ARP4107-88]

mishap, sequence of events Those events or conditions related to the mishap which begin with demonstration of intent for flight as defined in AFR 127-4, and end when damage or injury has occurred and ceased.[ARP4107-88]

mishap, task A sub-element of the mishap maneuver which describes each specific action required of the operator to accomplish that maneuver (for example, switchology, target tracking, or aircraft positioning).[ARP4107-88]

mismatch 1. (electrical) Condition in which the impedance of a source does not match or equal the impedance of the connected load or transmission line.[NASA] 2. Lateral offset between two halves of a casting mold or forging die, which produces distortion in shape across the parting line.

misperception See perception.

misplaced motivation See motivation, misplaced.

missed alert A system alert which is not given even though an aircraft is in the TCAS operational envelope and the threat of collision or potential collision exists.[ARP4153-88]

missed approach 1. A term used by the pilot to inform the ATC that he is executing the missed approach. See go-around. 2. A maneuver conducted by a pilot when an instrument approach cannot be completed to a landing. The route of flight and altitude to be flown in the execution of a missed approach are shown on the instrument approach procedure charts. A pilot executing a missed approach prior to the missed approach point (MAP) must continue along the final approach to the MAP. The pilot may climb immediately to the altitude specified in the missed approach procedure but may not alter course until the missed approach point has been crossed. At locations where ATC radar service is provided, the pilot should conform to radar vectors issued by the controller in lieu of the published missed approach procedure. See go-around and segments of an instrument approach procedure.[ARP4107-88]

missed approach point (MAP) A point prescribed in each instrument approach procedure at which a missed approach shall be executed if the required visual reference with the outside world does not exist.[ARP4107-88]

missed approach segment See segments of an instrument approach procedure.

missile(s) Any object that is designed to be thrown, dropped, projected or propelled for the purpose of making it strike a target.[AIR913-89]

missing mass (astrophysics) A problem related to a cluster of galaxies. The dynamical mass, is substantially larger than the mass estimated by the mass-to-luminosity ratio of the visible parts of the galaxies, the visible mass.[NASA]

mission 1. That period beginning with the start of the engine prior to flight and ending at engine shutdown at the completion of the flight.[ARD50010-91] 2. A flight operation of an aircraft in the performance of an assigned task. Missions may consist of such tasks as search and rescue of persons from a capsized boat, television coverage of local traffic problem areas, delivery of munitions on/at a target, and the transportation of people or cargo, or both, from one airport to another.[ARP4107-88]

mission capable MC status data shall consist of the sum of full mission capable (FMC) and par mission capable (PMC) for purposes of reporting to Office of the Secretary of Defense (OSD).[ARD50010-91]

mission capable rate The percent of possessed time that a system is capable of performing at least one of its assigned missions.[ARD50010-91]

mission effects Failure effects which preclude the completion of the aircraft mission. These failures cause delays, cancellations, ground or flight interruptions, high drag coefficients, flight envelope restrictions, etc.[ARD50010-91]

mission essential equipment Mission essential equipment is that interdependent equipment in a missile weapon system which is assigned a mission on the battlefield in support of forces in contact with the enemy and without which the assigned mission cannot be successfully completed.[ARD50010-91]

mission essential functions Those subsystem functions required to enable an aircraft to perform its designated mission(s).[ARD50010-91]

mission essential subsystem Anything authorized and assigned to approved combat and combat support forces which would be immediately employed to wage war and provide support for combat actions.[ARD50010-91]

mission mix The relative frequency that each mission profile is encountered during a specified time period.[ARD 50010-91]

mission narrative A report of the planned use of the system (for example, an aircraft) told in chronological order. Description is presented in terms of the real sequence and describes roles, activities, relations, and events. Example: "The aircraft will depart Boston as leader in a 3-ship formation, carrying a 23-man special forces team, 2 jeeps, and a 1-ton truck. The flight will proceed at an altitude of 27,000 ft to Newfoundland and rendezvous with a KC-135 tanker." (The narrative continues with route, destination, en route weather, en route mission tasks, threat, recovery procedures, etc.) See also system.[ARP 4107-88]

mission profile A time-phased description of the events and environments an item experiences from initiation to completion of a specified mission, to include the criteria of mission success or critical failures.[ARD50010-91]

mission reliability The ability of an item to perform its required functions for the duration of a specified mission profile.[ARD50013-91]

mission time The time during such a system or equipment is actually operating (in an "up" status). Operating time is usually divisible among several operating periods or conditions. These include "standby time", filament "on-time", pre-flight "checkout" time, flight. [ARD50010-91]

mission time between critical failures A measure of mission reliability: The total amount of mission time, divided by the total number of critical failures during a stated series of missions.[ARD 50010-91]

mission time to restore functions A measure of mission maintainability: The total corrective failure maintenance time, divided by the total number of critical failures, during the course of a specified mission profile.[ARD50010-91]

miter valve A valve in which the disc is at an angle of approximately 45° to the axis of the valve body.

MIUS See modular integrated utility system.

mixed level A simulation system combining both low-level transistor and gate circuit descriptions with high-level behavioral circuit representations.

mixed mode See mixed signal.

mixed oxides Mixture of oxides, particularly of radioactive metals.[NASA]

mixed radix Pertaining to a numeration system that uses more than one radix, such as the biquinary system.

mixed signal A simulation system combining both analog and digital circuit representations.

mixer In sound recording or reproduction equipment, a device capable of combining two or more input signals into a single linearly-proportioned output signal, usually with the additional capability of adjusting the levels of any of the inputs.

mixed icing conditions Mixture of supercooled water droplets and ice crystals existing within the same cloud environment.[AIR1667-89]

mixing depth See mixing height.

mixing valve A valve having more than one inlet but only one outlet port; it is used to blend two or more fluids to give a mixture of predetermined composition.

MLA See multispectral linear arrays.

MLS Microwave Landing System.

MMA (maximum-minimum) algorithm See compressor.

MMFS See Manufacturing Messaging Format Standard.

MMS (Manufacturing Message Specification) ISO/IEC 9506 A set of international standards developed to facilitate the interconnection of information processing systems. The first part is to define the service provided by the MMS. The second part specifies the protocol that supports the MMS. See also multimission modular spacecraft.

mnemonic Pertaining to assisting, or intending to assist, human memory; thus a mnemonic term, usually an abbreviation, that is easy to remember, e.g., MPY for multiply and ACC for accumulator.

mnemonic operation code An operation code in which the names of operations are abbreviated and expressed mnemonically to facilitate remembering the operations they represent. A mnemonic code normally needs to be converted to an actual operation code by an assembler before execution by the computer. Examples of mnemonic codes are ADD for addition, CLR for clear storage, and SQR for square root.

mnemonics An assembly language instruction, defined by a symbol, that has some resemblance to the operations carried out. Mnemonics are easier to remember and use than the equivalent Hex code or machine code.

MNOS See metal-nitride-oxide semiconductor (one type of computer semiconductor memory used in EAROMs).

mobile communication systems Any configuration of mobile or transportable voice and data communication equipment which allows for communication between combinations of mobile/fixed points with or without the aid of satellites.[NASA]

mobile telemetering Any arrangement for transmitting instrument readings from a movable data-acquisition station to a remote stationary or movable indicating or recording station without the use of interconnecting wire.

mobility 1. The average drift velocity of a charged particle induced by a unit electrical potential gradient. 2. In gases, liquids, solids or colloids, the relative ease with which atoms, molecules or particles can move from one location to another without external stimulus.

MOCA Minimum Obstruction Clearance Altitude.

mockup 1. The reassembly of an aircraft following its breakup in an accident. This procedure may supply a clue as to accident cause. 2. A structural scale model of an aircraft or other artifact used as part of the design/fabrication process or as part of a simulator system. 3. A model of a piece of equipment or a system, frequently full size, used for experiments, performance testing or training.

mode 1. The letter or number assigned to a specific pulse spacing of radio signals transmitted to aircraft or received by ground interrogator from airborne transponder components of the air traffic control radar beacon system (ATCRBS).[ARP4107-88] 2. A single component in a computer network. 3. Real or complex (number system). 4. A stable condition of oscillation in a laser. A laser can operate in one mode (singlemode) or in many modes (multimode).

mode C The automatic altitude reporting capability which converts the aircraft's altitude in 100-ft increments to coded digital information that is transmitted, together with the aircraft's transponder code, to the interrogating facility. The altitude of Mode C-equipped (and operating) aircraft can be automatically presented on the ATC controller's radar display.[ARP4107-88]

mode changer A device for changing the characteristics of a guided wave from one mode of propagation to another.

mode coupling See coupled modes.

mode filter In a waveguide circuit, an arrangement of waveguide elements which pass waves that are being propagated in certain mode(s) and exclude waves being propagated in other modes.

model A device used in a failure detection-correction system to simulate the performance of a component or a channel used for control. Typical models are electrically implemented.[ARP1181A-85] See also mathematical model.

model basin A large tank of water for design experiments and performance studies of ship hulls using scale models. Also known as model tank, towing tank.

model dispersion That component of pulse spreading caused by differential optical path lengths in a multimode fiber.

model generation program The model generation program (MGP) of EASY assembles the ECS by organizing standard components or new components, and components interconnections, based on user inputs.[AIR1823-86]

model reference adaptive control This deals with three parameters: an ideal adaptive control system whose response is agreed to be optimum; computer simulation in which both the model system and the actual system are subjected to the same stimulus; and parameters of the actual system, which are adjusted to minimize the difference in the outputs of the model and the actual system.[NASA]

model specification A specification covering the essential detail and technical requirements of a specific unit design, including a description of the procedures by which it will be demonstrated that the requirements have been met. [ARP906A-91]

modem 1. A device that provides both combining (modulation) and separation (demodulation) of data. Typically used to connect a node to a broadband network. See also modulator-demodulator; see also transceiver. 2. An electronic device for serial transmission of digital data in the audio frequency spectrum over a voice-grade telephone line. 3. A device that converts signals in one form to another form compatible with another kind of equipment. In particular, a circuit board that changes digital data being transmitted within a particular device into a form suitable to be transmitted over a data highway and vice versa. (From MOdulator + DEModulator).

mode of vibration See vibration mode.

moderation Reducing the kinetic energy of neutrons, usually by means of successive collisions with hydrogen, carbon or other light atoms.

moderators Materials that have a high cross section for slowing down fast neutrons with a minimum of absorption, e.g., heavy water, beryllium, used in reactor cores.[NASA]

modes The PID algorithms operate in several modes. These are determined by the operator and/or by states of other instructions inside the controller. 2. AUTO, CASCADE, MANUAL, etc.

MODFETS Heterojunction field effect transistor device structures in which only the larger (Al, GaAs) bandgap is doped with donors, while the GaAS layer is left undoped. This results in high electron mobilities due to spatially separated electrons and donors.[NASA]

modular engine Those engines consisting of several independent assemblies called modules, which by design can be removed/replaced without major disassembly of the engine or other modules, for example, compressor, combustion, turbine, after burner, gearbox, torquemeter, or combination thereof.[ARD 50010-91]

modular integrated utility system A joint NASA-HUD concept incorporating various utilities—electric power plant, water supply, heating and air conditioning, sewage treatment, and waste disposal—into a single system having increased efficiency and economy.[NASA]

modularity The degree to which a system of programs is developed in relatively independent components, some of which may be eliminated if a reduced version of the program is acceptable.

modularization Designing a series of components, subassemblies or devices for interchangeability of physical location, so that different assemblies can be easily constructed on a standard frame or mounted in standard enclosures.

modular programming Programming in which tasks are programmed in distinct sections or sub-sections resulting in the ability to modify one section without reference to other sections.

modulated brake pressure Metered brake pressure modulated by a skid control valve and control system.[AIR 1489-88]

modulated wave A radio-frequency wave in which amplitude, phase or frequency is varied in accordance with the waveform of a modulating signal.

modulation 1. The variation in the value of some parameter characterizing a periodic oscillation. Specifically, variation of some characteristic of a radio wave, called the carrier wave, in accordance with instantaneous values of another wave, called the modulating wave. [NASA] 2. The process of impressing information on a carrier for transmission (AM, amplitude modulation; PM, phase modulation; FM, frequency modulation). 3. Regulation of the fuel-air mixture to a burner in response to fluctuations of load on a boiler.

modulation doped fets See MODFETS.

modulation doping The process of doping only the larger bandgap of a heterojunction device with donors, while the other layer is left undoped. Since the electrons and donors are spatially separated, ionized impurity scattering is avoided and extremely high electron mobilities are obtained.[NASA]

modulation factor The ratio of peak variation actually used in a given type of modulation to the maximum design variation possible.

modulation index In frequency modulation with a sinusoidal waveform, the modulation index is the ratio of the peak (not peak-to-peak) frequency deviation to the frequency of the modulating wave.

modulation meter An instrument for measuring modulation factor of a wave train, usually expressed in percent.

modulation noise The noise in an electronic or acoustic circuit caused by the presence of a signal, but not including the waveform of the signal itself.

modulator Device to effect the process of modulation.[NASA]

module 1. A combination of assemblies, subassemblies and parts, contained in one package, or so arranged as to be installed in one maintenance action. [ARD50010-91] 2. A subsection of a galley unit or complex that is easily removable and transferable to ground equipment for cycling through ground kitchens for cleaning and reloading with food or service items.[AS1426-80] 3. A computer program unit that is discrete and identifiable with respect to compiling, combining with other units, and loading, for example the input to, or output from, an assembler, compiler, linkage editor, or executive routine.

modulo A mathematical operation that yields the remainder function of division; thus, 39 modulo 6 equals 3.

modulo N check 1. A check that makes use of a check number that is equal to the remainder of the desired number when divided by n, e.g., in a modulo 4 check, the check number will be 0, 1, 2, or 3 and the remainder of the desired number when divided by 4 must equal the reported check number, otherwise an equipment malfunction has occurred. 2. A method of verification by congruences, e.g., casting out nines.

modulus of elasticity 1. The ratio of stress to strain in material that is elastically deformed.[ARP1931-87] 2. In any solid, the slope of the stress-strain curve within the elastic region; for most materials, the value is nearly constant up to some limiting value of stress known as the elastic limit; modulus of elasticity can be measured in tension, compression, torsion or shear; the tension modulus is often referred to as Young's modulus.

moire A moire pattern results from interference or light blocking due to beat frequencies between two or more superimposed spatially periodic structures. On a color CRT display, moire patterns take the form of light and dark areas on raster; on stroke-written lines, moire takes the form of periodic light and dark modulation of the line.[ARP1782-89]

Moire fringes The bands which appear in the Moire effect.[NASA]

Moire interferometry The use of intersecting families of curves as instruments for making precise measurement, the study of indices of refractions, etc. by utilizing the interference patterns. [NASA]

moisture Water in the liquid or vapor phase.

moisture absorption Generally, the amount of moisture, in percentage, that an insulation will absorb under specified conditions.[ARP1931-87]

moisture barrier A material or coating that retards the passage of moisture through a wall made of more permeable materials.

moisture-free See bone dry.

moisture in steam Particles of water carried in steam usually expressed as the percentage by weight.

moisture loss The loss representing the difference in the heat content of the moisture in the exit gases and that at the temperature of the ambient air.

moisture resistance The ability of a material to resist absorbing moisture. [ARP1931-87]

mol See mole.

molar units Concentration units defined as the number of gm-moles of the component per liter of solution.

mold, potting An accessory used as a form for containing the potting compound around the terminations of a connector. See also boot.[ARP914A-79]

mold skid depth The depth of the grooves in a tire (mold), or conversely, the height of the tread on the tire. Variation in this parameter increases or decreases wear life of the tire. Tread depth may be limited by speed requirement and centrifugal force/stresses. [AIR1489-88]

mole Metric unit for amount of a substance.

molecular attrition See fretting.

molecular beam A unidirectional beam of neutral-charge molecules passing through a vacuum.

molecular beam epitaxy Ultrahigh vacuum technique for growing very thin epitaxial layers of semiconductor crystals.[NASA]

molecular clouds Thickest and densest interstellar clouds consisting mainly of molecular hydrogen but also a high concentration of dust grains.[NASA]

molecular dissociation See dissociation.

molecular flow 1. The flow of gas through a duct under conditions such that the mean free path is greater than the largest dimension of a transverse section of the duct.[NASA] 2. Gas flow in a tube at a pressure low enough that the mean free path of the molecules is greater than the inside diameter of the tube.

molecular shields Furlable devices used in space vacuum research to permit deployment and retrieval of instruments and the performance of experiments without contamination.[NASA]

molecular weight The weight of a given molecule expressed in atomic weight units.[NASA]

molecule(s) 1. Aggregates of two or more atoms of a substance that exists as a unit.[NASA] 2. The smallest division of a unique chemical substance which maintains its unique chemical identity.

mole fraction The volume concentration of a gas per unit volume of the gas mixture of which it is a part.[AIR1533-91]

moles Number of molecular weights which is the weight of the component divided by its molecular weight.

Moliere formula See secondary cosmic rays.

Moll thermopile A type of thermopile used in some radiation-measuring instruments. It consists of multiple manganan-constantan thermocouples connected in series. Alternate junctions are embedded in a shielded nonconductive plate of large heat capacity; the remaining junctions are blackened and exposed directly to the radiation. The voltage across the thermopile is directly proportional to the intensity of radiation.

molten salts High temperature inorganic salt or mixtures of salts used for thermal energy storage, heat exchangers, high power electric batteries, heat treatment of alloys, etc.[NASA]

moment 1. Of force, the effectiveness of a force in producing rotation about an axis; it equals the product of the radius perpendicular to the axis of rotation that passes through the point of force application and the tangential component of force perpendicular to the plane defined by the radius and axis of rotation. 2. Of inertia, the resistance of a body at rest or in motion to changes in its angular velocity.

momentary 1. An alarm that returns to normal before being acknowledged. 2. Returns to normal state when pressure or signal is removed.

momentary alarm See alarm.

momentary digital output A contact closure, operated by a computer, that holds its condition (set or reset) for only a short time. See latching digital output.

momentary switch A spring-loaded switch whose contacts complete a circuit only while an actuating force is applied; for a typical momentary pushbutton, electric current flows through the switch only while the operator has a finger on the button.

moment of inertia A measure of the resistance offered by a body to angular acceleration. The mass moment of inertia is usually expressed in slug ft^2 or when multiplied by the gravity constant (g), lb ft^2. Its point of application must be expressed, as it varies as the square of a gear ratio.[ARP906A-91] See also inertia.

momentum 1. Quantity of motion. 2. The product of a body's mass and its linear velocity.

momentum energy See kinetic energy.

Monel A series of International Nickel Co. high-nickel, high-copper alloys used for their corrosion resistant properties to certain conditions.

monitor 1. A device used for sensing the operation of a component or channel such that failures may be detected. Inline and cross-channel are two forms of monitoring. In-line monitoring compares output performance to the command input or a model. Cross-channel monitoring compares equivalent performance features of two or more channels. [ARP1181A-85] 2. To measure a quantity continuously or at regular intervals so that corrections to a process or condition may be made without delay if the quantity varies outside of prescribed limits. 3. Software or hardware that observes, supervises, controls, or verifies the operations of a system. 4. In data processing, a high-resolution viewing screen.

monitor command An instruction issued directly to a monitor from a user.

monitor console The system control terminal.

monitoring The act or technique of establishing functional status or conditions.[ARP1587-81]

monitoring, inflight Monitoring by on-board equipment during the period from engine start to engine shut down. [ARP1587-81]

monitoring, on site Utilization of data on site. Ground test of engines using on-board or portable ground test equipment.[ARP1587-81]

monitoring, remote Utilization of data at a remote (off site) location.[ARP1587-81]

monitor light See pilot light.

monitor routine See executive program.

monitor software That portion of the operational software which controls on-line and off-line events, develops new on-line applications and assists in their debugging. This software is also known as a batch monitor.

monitor system Same as operating system.

monochromatic A single wavelength or frequency. In reality, light cannot be purely monochromatic, and actually extends over a range of wavelengths. The breadth of the range determines how monochromatic the light is.

monochromatic radiation Any electromagnetic radiation having an essentially single wavelength, or in which the photons all have essentially the same energy.

monochromator An optical device

which uses a prism or diffraction grating to spread out the spectrum, then pass a narrow portion of that spectrum through a slit; it generates monochromatic light from a non-monochromatic source.

monomer Any molecule that can be chemically bound as a unit of a polymer.[ARP1931-87]

monopropellant A liquid propellant which contains an oxidizing agent and combustible matter (fuel) in a single phase.[AIR913-89]

monostable Pertaining to a device that has one stable state.

monotectic alloys Metallic composite materials having a dispersed phase of solidification products distributed within a matrix. The dispersed components can be selected to provide characteristics such as superconductivity or lubricity.[NASA]

monotonic A digital-to-analog converter is monotonic if the output either increases or remains constant as the digital input increases.

mono wheel An aircraft landing gear configuration in which there is a single wheel.[AIR1489-88]

Monte Carlo method A trial-and-error method of repeated calculations to discover the best solution of a problem. Often used when a great number of variables are present, with interrelationships so extremely complex as to forestall straightforward analytical handling.

month The period of the revolution of the moon around the earth. The month is designated as sidereal, tropical, anomalistic, dracontic, or synodical, according to whether the revolution is relative to the stars, the vernal equinox, the perigee, the ascending node, or the sun. The calendar month is a rough approximation of the synodical month.[NASA]

mood See affective states.

mooring Tie-down or securing of the aircraft by cables, ropes, straps or chains for protection against wind, propeller or rotor wash, jet blast, ship motion, etc. Also sometimes called picketing. [AIR1489-88]

most significant bit (MSB) The bit in a digital sequence which defines the largest value. It is usually at the extreme left.

most significant digit The leftmost non-zero digit.

mother board A circuit board that includes the primary components of a microcomputer.

motion The act, process or instance of change of position. Also called movement, especially when used in connection with problems involving the motion of one craft relative to another.[NASA]

motion balance instrument An instrument design technique utilizing the motion of the measuring element against a spring to reach a balance of forces representing the magnitude of the measured variable.

motion converter The component of the actuation system that converts the motion of the actuator into the motion of the load. For example, a ball screw converts rotary motion into linear and a crank arm converts linear motion into rotary.[AIR1916-88]

motion equations See equations of motion.

motion sickness The syndrome of pallor, sweating, nausea, and vomiting which is induced by unusual acceleration. [NASA]

motion simulation Replication of exact motion or replication of part of a motion to provide the sensation of the motion. [NASA]

motivation A person's prioritized value system which influences his/her behavior. The nonstimulus variables controlling behavior; the general name for the fact that an organism's acts are partly determined in direction and strength by its own nature or internal state, or both. A specific hypothesized personal determiner of the direction and strength of action or of a line of action.[ARP4107-88]

motivation, anomalies of Characteristics of a person's value system which may result in unsafe acts. **motivation, excessive** Attributing a higher value to successfully performing the mission than is actually warranted by the importance of the mission.[ARP4107-88]

motivation, misplaced A situation in which the factors that influence a person's selection of a course of action are either remotely related or not related at all to the object requirements of the mission.[ARP4107-88]

motivation, under- Attributing a lower value to successfully performing the mission than is actually warranted by the importance of the mission.[ARP 4107-88]

motor (rocket) A generic term for a solid propellant rocket consisting of the assembled propellant, case, ignition system, nozzle, and appurtenances. NOTE: The term rocket engine is usually used for liquid fueled rockets.[AIR913-89]

motor (verb) To rotate an engine with the starter, but without providing fuel or ignition to the engine.[ARP906A-91]

motor, hydraulic A device for converting liquid fluid energy into mechanical energy in the form of continual rotary motion of a mechanical member.[ARP 243B-65]

motor meter A type of integrating meter; it consists of a rotor, one or more stators, a retarding device which makes the speed of the rotor directly proportional to the integral of the quantity measured (usually power or electric current), and a counter or set of dials that indicates the number of rotor revolutions.

motor operator The electric or hydraulic power mechanism that receives a control signal and repositions a valve or other final control element.

motors Machines supplied with external energy which is converted into force and/or motion.[NASA]

motor vehicles Automotive vehicles that do not run on rails, generally having rubber tires.[NASA]

MOTS (tracking system) See minitrack system.

mount A device to be attached directly to a part to suspend or support the part at a given position.[ARP480A-87]

mounting pad The pad upon which the starter or accessory is attached. Consists of a fixed pad with studs or suitable attaching means concentrically located with respect to a shaft which has a spline or similar power transmitting means.[ARP906A-91]

mounting strain error See error, mounting strain.

mouse In data processing, an input device capable of moving a cursor on a computer screen.

movement See motion.

movement area The runways, taxiways, and other areas of an airport which are utilized for taxiing, takeoff, and landing of aircraft (exclusive of loading ramp and parking areas). At those airports with a tower, specific approval for entry onto the movement area must be obtained from ATC.[ARP4107-88]

moving-coil instrument An instrument whose output is related to the reaction between the magnetic field set up by current flow in one or more movable coils and the magnetic field of a fixed-position permanent magnet. Also known as permanent-magnet moving-coil instrument.

moving-dial indicator A type of indicator where a flat, circular scale (dial) is attached to the moving element and the instrument scale is continually repositioned with respect to a fixed pointer to indicate changing values of a measured variable.

moving-drum indicator A type of indicator where a circular member (drum) with a scale along its periphery revolves in relation to a fixed pointer to indicate changing values of a measured variable.

moving element Of an instrument, the parts which move as a direct result of a variation in the quantity being measured.

moving-iron instrument An instrument which depends on the reaction between one or more pieces of magnetically soft material, at least one of which moves, and the magnetic field set up by electric current flowing in one or more fixed coils, to produce its output indication or signal.

moving-magnet instrument An instrument which depends on the reaction between a movable permanent magnet that aligns itself with the resultant field produced by another permanent magnet interacting with one or more current-carrying coils, or by two or more coils interacting with each other, to produce its output indication or signal.

moving-scale indicator Any of several designs of instrument indicator where the scale moves in relation to a fixed pointer to indicate changing values of a measured variable.

moving target indicator (MTI) An electronic device that permits radar scope presentation only from targets that are in motion. This is done by eliminating signals from nonmoving targets (such as buildings and high terrain), signals from targets at less than a predetermined threshold velocity, and random noise. Use of the moving target indicator is a partial remedy for ground clutter.[ARP4107-88]

MPL A high-level language suitable for the development of microprocessor application software.

MRAC (systems) See model reference adaptive control.

Ms The maximum temperature at which a martensite reaction begins upon cooling from the beta phase.[AS1814-90]

MSAT A joint Canada United States mobile satellite system which is being developed with a voice and data communication link between mobile units and the switched telephone network or between mobile units and other mobile units via a satellite. Each country will have a satellite capable of mutual backup.[NASA]

MSAW Minimum Safe Altitude Warning.

MSB See most significant bit.

MSBLS See microwave scanning beam landing system.

MS-DOS The most widely used operating system for microcomputers.

MSI See medium scale integration.

MSL Mean Sea Level. Used in conjunction with an altitude, MSL means "above mean sea level"; for example, "16,000 ft MSL" means "16,000 ft above mean sea level."[ARP4107-88]

MSM (semiconductors) Semiconductor devices consisting of a semiconductor layer sandwiched between two layers of metal.[NASA]

M synchronization A type of linking between a camera shutter and a flash unit that gives a 15 millisecond delay so that the metal foil flash lamp reaches peak brightness before the shutter actually trips.

MTBF Mean Time Between Failure.

MTI radar See moving target indictors.

MTTF See mean-time-to-failure.

multi-access A multiprogramming system which permits a number of users to simultaneously make on-line program changes.

multi-address Same as multiple address.

multi-anode microchannel arrays A family of photoelectric, photon counting array detectors being developed for use in instruments on both ground based and spaceborne telescopes.[NASA]

multibeam antennas Antennas that have the ability to form more than one beam from a single radiating aperture.

[NASA]

multi-bus Intel's proprietary link between single board systems used for industrial systems.

multicast 1. The transmission of a message from one station to selective stations.[AIR4271-89] 2. A message addressed to a group of stations connected to a LAN.

multichannel plates See microchannel plates.

multichannel spectral analyzer A measurement system which sorts signals into a number of different channels, then counts and analyzes the signals channel by channel.

multiconductor A cable containing more than one component wire.[ARP 1931-87]

multicoupler A device for coupling several receivers or transmitters to one antenna; it also enables the proper impedance match between same.

multicraft 1. Maintenance personnel who are proficient in more than one craft such as Instrument Technician, Electrician and Instrument Mechanic. 2. Responsible to maintain a variety of equipment used in control systems.

MULTICS The time-sharing system developed at Project MAC (Man and Computer Project at M.I.T.). See task, definition 3.

multi-element control system A control system utilizing input signals derived from two or more process variables for the purpose of jointly affecting the action of the control system. Examples are input signals from pressure and temperature or from speed and flow, etc. See also multi-variable control; see also control system, multi-element (multivariable).

multifiber cable A fiber optic cable containing many fibers which transmit signals independently and are housed in separate substructures within the cable.

multifuel burner A burner by means of which more than one fuel can be burned, either separately or simultaneously, such as pulverized fuel, oil or gas.

multifunction multiloop controller A type of microprocessor based controller that combines the process control functions of a dedicated loop controller with many of the logic functions of a programmable logic controller to provide the control strategy of an entire unit operation.

multilayer A type of printed circuit board which has several layers of circuit etch or pattern, one over the other and interconnected by electroplated holes. Since these holes can also receive component leads, a given component lead can connect to several circuit points, reducing the required dimensions of a printed circuit board.

multilayer coating Optical coatings in which several layers of different thicknesses of different materials are applied to an optical surface. Interference affects light passing through the layers. Reflection and transmission are influenced differently at different angles of incidence and different wavelengths.

multilayer structures See laminates.

multilevel address See indirect address.

multimeter See volt-ohm-milliammeter.

multimission modular spacecraft Future spacecraft to be operated in conjunction with the Space Shuttle orbiter vehicle and serviced by its module exchange mechanism.[NASA]

multimode fiber An optical fiber capable of carrying more than one mode of light in its core.

multipath transmission The process, or condition, in which radiation travels between source and receiver via more than one path. Since there can be only one direct path, some process of reflection, refraction or scattering must be involved.[NASA]

multiphoton absorption Ionization

and dissociation of a molecule under the action of powerful laser radiation. Laser-flux dependent light intensities are emitted by different excited states of the molecule to indicate the various absorption processes.[NASA]

multiple access The allocation of communication system resources (output) among multiple users by means of power, bandwidth, and power assignment singly or in combination.[NASA]

multiple action A control-system action that is a composite of the actions of two or more individual controllers.

multiple address A type of instruction which specifies the addresses of two or more items which may be the addresses of locations of inputs or outputs of the calculating unit or the addresses of locations of instructions for the control unit. The term multi-address is also used in characterizing computers, e.g., two-, three-, or four-address machines. Synonymous with multi-address.

multiple failures The simultaneous occurrence of two or more independent failures. When two or more failed parts are found during troubleshooting and failures cannot be shown to be dependent, multiple failures are presumed to have occurred.[ARD50010-91]

multiple grain An assembly of solid propellant grains inside an explosive device or motor.[AIR913-89]

multiple input See reflash.

multiple-output system A system which manipulates a plurality of variables to achieve control of a single variable.

multiple-purpose meter See volt-ohm-milliammeter.

multiple sampling A type of statistical quality control in which several samples, each consisting of a specified number of items, are withdrawn from a lot and inspected. The lot is accepted, rejected, or resampled, depending on the number of unacceptable items found.

multiple wheel control Control of two or more braked wheels utilizing one valve and control of circuit.[AIR1489-88]

multiplexer 1. Optical multiplexers combine signals at different wavelengths. Electronic multiplexers combine signals electronically before they are converted into optical form. 2. A device for combining two or more signals, as for multiplex, or for creating the composite color video signal from its components in color television. 3. A device which samples input and/or output channels and interleaves signals in frequency or time. 4. A device that allows selection of one of many input channels of analog data under computer control. The device is often an integral part of a DAS. 5. A device which mixes several measurements for transmission and/or tape recording; time-division (PAM or PCM) or frequency-division (FM). 6. A device or circuit that samples many data lines in a time ordered sequence, one at a time, and puts all sampled data onto a single bus. (A demultiplexer does the reverse job).

multiplexer channel 1. An input/output channel which serves several input/output units. 2. A single path that is capable of transferring data from multiple sources or to multiple destinations through the use of time multiplexing.

multiplexing 1. The simultaneous transmission of two or more signals within a single channel. The three basic methods of multiplexing involve the separation of signals by time division, frequency division, and phase division. [NASA] 2. Utilizing a single device for several similar purposes or using several devices for the same purpose, e.g., a duplexed communications channel carrying two messages simultaneously. 3. Technique of selecting from many inputs to provide a specified output. Allows a single A/D to serve several voltage sources by selecting them one at a time.

multiplex transmission See multiplexing.

multiplier phototubes See photomultiplier tubes.

multipliers Devices which have two or more inputs and whose output is a representation of the product of the quantities represented by the input signals. [NASA]

multiplier tube A phototube in which secondary emission from auxiliary electrodes produces an internally amplified output signal. Also known as multiplier phototube; photomultiplier tube.

multipole circuit breaker A multipole circuit breaker has two or more poles controlled by a single-actuating member.[ARP1199A-90]

multiport burner A burner having a number of nozzles from which fuel and air are discharged.

multiposition action A type of controller action in which the final control element is positioned in one of three or more preset configurations, each corresponding to a definite range of values for the controlled variable.

multi-position controller See controller, multi-position.

multiprocessing 1. Pertaining to the simultaneous execution of two or more programs or sequences of instructions by a computer or computer network. 2. Loosely, parallel processing.

multiprocessor A machine with multiple arithmetic and logic units for simultaneous use.

multi-processors A number of independent central processing units each having access to a common memory. One unit is usually an information interchange controller while others carry out distinct defined parts of a task.

multiprogramming A computer processing method in which more than one task is in an executable state at any one time.

multipropellants See rocket propellants.

multiradar tracking See radar networks.

multirange Having two or more specific ranges of values over which an instrument or control device can be used; changing from one range to another usually involves simply repositioning a switch and does not require removing or replacing any internal parts.

multiskilled Maintenance personnel who are skilled in more than one craft.

multispectral linear arrays Large number of interconnected solid state detectors in a pushbroom mode wherein the forward motion of the vehicle (spacecraft) sweeps the assembly of detectors which are oriented perpendicular to the ground track.[NASA]

multispectral resource sampler An experimental remote sensing instrument for satellites to measure both intensity and polarization at several wavelengths.[NASA]

multi-speed floating controller See controller, multiple-speed floating.

multistage Occurring in a sequence of separate steps; in a multistage pump or compressor, for instance, pressure is raised by passing the working fluid through a series of impellers or pistons, each of which raise the pressure above the outlet pressure from the previous stage.

multistage rocket vehicles Vehicles having two or more rocket units, each unit firing after the one in back of it has exhausted its propellant. Normally, each unit, or stage, is jettisoned after completing its firing.[NASA]

multistatic radar System in which successive lobes of the antenna are sequentially engaged to provide a tracking capability without physical movement of the antenna.[NASA]

multitasking The facility that allows the programmer to make use of the multiprogramming capability of a computer system.

multitemporal analysis See temporal resolution.

multivariable control A control system involving several measured and controlled variables where the interdependences are considered in the calculation of the output variables.

multi-variable control system See control system, multi-element (multivariable).

multivibrators Two-stage regenerative circuits with two possible states and an abrupt transition characteristic.[NASA]

multi wheel Any landing gear assembly which utilizes two or more wheel assemblies. More appropriate for description of an aircraft or assembly with many wheels.[AIR1489-88]

"Mu" Meter An instrument, device, or apparatus for measuring the coefficient of friction of a runway surface at various conditions of damp, wet, flooded, icy, etc.[AIR1489-88]

Munsell chroma The dimension of the Munsell system of color corresponding most closely to saturation.

muon spin rotation Particle spin depolarization caused by sensitivity of muon spin to the presence of defects in certain metals.[NASA]

muscovite An important mineral of the mica group.[NASA]

mushy zones Regions of liquid plus solid phases in alloys that solidify over a range of temperatures.[NASA]

mutagens Agents that raise the frequency of mutations above the spontaneous rate.[NASA]

mutual capacitance Capacitance between two conductors when all other conductors, including ground, are connected together and then regarded as an ignored ground.[ARP1931-87]

mush To settle or to gain little or no altitude while flying in a semistalled condition or at a relatively high angle of attack.[ARP4107-88]

MUX See multiplexer.

MX missile United States strategic intercontinental ballistic missile.[NASA]

Mystere 50 aircraft A tri-engine business jet aircraft (Dassault).[NASA]

N

N See newton.

nacelle A partially separate streamlined enclosure attached to an aircraft.[ARP 4107-88]

naked singularities Singularities in spacetime that will be visible and communicable to the outside world, i.e., singularities that are not shielded by an event horizon from infinity.[NASA]

NAND 1. A logical operator having the property that if P is a statement, Q is a statement, R is a statement, ..., then the NAND of P, Q, R, ... is true if at least one statement is false; false if all statements are true. Synonymous with NOT-AND. 2. Logical negation of AND. Supplies a logic 0 when all inputs are at logic 1.

nano A prefix which means one billionth.

nanosecond (ns) One billionth of a second (10^{-9} second).[ARP1931-87]

nap of the earth (flight) An operational tactic whereby helicopters are flown as close to the surface of the earth as possible, taking advantage of natural and cultural terrain cover to escape detection or minimize exposure. Aircraft becomes hidden in the surface irregularities (nap) of the earth.[ARP4107-88]

nap-of-the-earth navigation Low altitude flight of helicopters during night or day utilizing electronic means for detection and recognition of landmarks and targets.[NASA]

nappe A sheet of liquid passing through the notch and falling over the weir crest.

narrow-angle diffusion That in which light is scattered in all directions from the diffusing medium but in which the intensity is notably greater over a narrow angle in the general direction which the light would take by regular reflection or transmission.[ARP798-91]

narrowband A description of frequency measurement whose frequency band of energy is smaller relative to the rest of the band.[NASA]

narrowband radiation thermometer A type of temperature-measuring instrument that responds accurately only over a given, relatively narrow band of wavelengths, often a band chosen to meet a special requirement of the intended application.

National Airspace System (NAS) The common network of U.S. airspace; air navigation facilities, equipment and services, airports and landing areas; aeronautical charts, information and services; rules, regulations and procedures; technical information; and manpower and material. Included are system components shared jointly with the military.[ARP4107-88]

national airspace system stage A/NAS stage A The en route ATC system's radar, computers and computer programs, controller plan view displays (PVDs/radar scopes), input/output devices, and related communications equipment which are integrated to form the basis of the automated IFR air traffic control system.[ARP4107-88]

National Electrical Manufacturers Association (NEMA) A group of U.S. manufacturers in the electrical industry who support committee activity for the purpose of standardization and preparation of voluntary standards and specifications of wire and cable.[ARP1931-87]

National Electric Code (NEC) A set of regulations governing construction and installation of electrical wiring and apparatus, established by the National Fire Protection Association. It is widely used by state and local authorities within the United States.

National Oceanic Satellite System Joint NASA (Goddard)-DOD venture.[NASA]

National Operational Environmental Sat Sys See NOESS.

natural circulation The circulation of water in a boiler caused by differences in density.

natural contour of tire The uninflated tire cross sectional contour.[AIR1489-88]

natural draft Convective flow of a gas—as in a boiler, stack or cooling tower—due to differences in density. Warm gas in the chamber rises toward the outlet, drawing in colder, more dense gas through inlets near the bottom of the chamber.

natural frequency The frequency at which the free landing gear assembly will vibrate if disturbed from the normal position by external and upsetting forces.[AIR1489-88] See frequency, undamped; see also frequency, natural and frequency, resonant.

natural gas A mixture of gaseous hydrocarbons trapped in rock formations below the earth's surface. The mixture consists chiefly of methane and ethane, with smaller amounts of other low-molecular-weight combustible gases, and sometimes noncombustible gases such as nitrogen, carbon dioxide, helium and H_2S called "sour gas".

natural gas exploration Searching the geological features to identify locations for stimulating swells for recovery of natural gas.[NASA]

natural language A language, the rules of which reflect and describe current usage rather than prescribed usage. Contrast with artificial language.

natural language (computers) A computer language whose rules reflect and describe current rather than prescribed usage. The language is often loose and ambiguous in interpretation.[NASA]

natural lasers See lasers.

natural radioactivity Spontaneous radioactive decay of a naturally occurring nuclide.

nautical charts Charts and maps of oceans, coasts and harbors now compiled from satellite data for precision and correction of local errors.[NASA]

nautical mile A unit of distance equal to 6076.11549 ft or 1852 meters.[ARP 4107-88]

NAVAID Acronym for NAVigational AID Any visual or electronic device, either airborne or on the surface, which provides point-to-point guidance information or position data to aircraft in flight. NAVAIDs include VOR, VORTAC, and TACAN aids.[ARP4107-88]

NAVAID classes VOR, VORTAC, and TACAN aids are classed according to their operational use as follows: 1. T—Terminal, 2. L—Low altitude, 3. H—High altitude. These classifications assure serviceable signals within specified distances and altitudes from the particular station emitting the navigational signal.[ARP4107-88]

Navier-Stokes equation The equation of motion for a viscous fluid.[NASA]

navigable airspace Airspace at and above the minimum flight altitudes prescribed in FAR, including airspace needed for safe takeoff and landing. [ARP4107-88]

navigation The practice or art of directing the movement of a craft from one point to another. Navigation usually implies the presence of a human, a navigator, aboard the craft.[NASA]

navigation technology satellites Class of navigation satellites utilizing the global positioning system as well as a precise frequency and timing system.[NASA]

NDB Nondirectional Beacon.

near letter quality With computer printers, a dot- matrix character formation that resembles the print of earlier cloth-ribbon typewriters.

neck A reduced section of pipe or tubing between sections of larger diameter or between a pipe and a chamber.

needle valve A type of metering valve used chiefly for precisely controlling flow. Its essential design feature is a slender tapered rodlike control element which fits into a circular or conoidal seat. Operating the valve causes the rod to move into or out of the seat, gradually changing the effective cross-sectional area of the gap between the rod and its seat.

neg See negative.

negative alert A corrective or preventive alert that requires the pilot not to do something to resolve a conflict (for example, "don't climb").[ARP4153-88]

negative damping 1. The opposite of damping, i.e.: The propagation of oscillations or disturbances; a build up of energy with time. 2. Landing gear vibrations induced by the sensitivity of the torque developed between the rotating and nonrotating brake parts to slip velocity.[AIR1489-88]

negative feedback 1. Feedback which results in decreasing the amplification. [AIR1489-88] [NASA] 2. Returning part of an output signal and using it to reduce the value of an input signal.

negative G The opposite of positive G. Negative G occurs when, in a gravitational field or during an acceleration, the human body is so positioned that the force of gravity or inertia, or both, acts in foot-to-head direction, or on an aircraft to decrease or reverse the wing loading.[ARP4107-88]

negative-going edge The edge of a pulse going from a high to a low level.

negative ions Ions singly or in groups which acquire negative charges by gaining one or more electrons.[NASA]

negative thrust See reverse thrust.

negatron(s) 1. Negative electrons. Sometimes shortened to negatons.[NASA] 2. A negatively charged beta particle.

NEMA standard Consensus standards for electrical equipment approved by the majority of the members of the National Electrical Manufacturers Association.

nephelometer Instrument which measures, at more than one angle, the scattering function of particles suspended in a medium. Instrument for chemical analysis by measuring the light scattering properties of a suspension.[NASA]

neopheloscope An apparatus for making clouds in the laboratory by expanding moist air or by condensing water vapor.

neoprene A synthetic rubber made by polymerization of chloroprene (2-chlorobutadiene-1,3). Its color varies from amber to silver to cream. It exhibits excellent resistance to weathering, ozone, flames, various chemicals and oils.

neper A unit of measure determined by taking the natural logarithm of the scalar ratio of two voltages or two currents.

nephelometry The application of photometry to the measurement of the concentration of very dilute suspensions.

nephelometer A general term for instruments that measure the degree of cloudiness or turbidity.

nephoscope An instrument for determining the direction in which clouds move.

Neptune atmosphere The atmosphere of the planet Neptune which is primarily composed of hydrogen and methane. [NASA]

NESA Acronym for NonElectroStatically Activated. A term used to describe a particular type of aircraft window designed to prevent the buildup of ice. [ARP4107-88]

nest 1. To embed a subroutine or block of data into a larger routine or block of data. 2. To evaluate as nth degree polynomial by a particular algorithm which uses (n-1) multiply operations and (n-1) addition operations in succession. See also crimp anvil.

nested Used in relation to subroutines used or called with another subroutine.

nested DO loop A FORTRAN statement which directs the computer to perform a given sequence repeatedly.

nesting In computer software, a program that has loops within loops.

net fan requirements The calculated

operating conditions for a fan excluding tolerances.

net heat of combustion The energy released per unit mass of fuel due to its complete oxidation at constant pressure as measured by cooling the products to the initial temperature without condensation of the water vapor formed in the reaction.[AIR1533-91]

net positive suction head The minimum difference between the static pressure at the inlet to a pump and vapor pressure of the liquid being pumped. Below that pressure, fluid is not forced far enough into the pump inlet to be acted upon by the impeller.

net thrust The gross thrust of a jet engine minus the drag imposed by the momentum of the incoming air.[ARP 4107-88]

network 1. In data processing, any system consisting of an interconnection of computers and peripherals. Information is transferred between the devices in the network. 2. LAN (Local Area Network) is a system at one location linked by cables; WAN (Wide Area Network) is a widely dispersed system usually connected by telephone lines. 3. In an electric or hydraulic circuit, any combination of circuit elements.

network analyzer An apparatus which contains numerous electric-circuit elements that can be readily combined to form models of electric networks.

network control The management of acquisition, routing, and switching primarily in satellite communication. [NASA]

network management The facility by which network communication and devices are monitored and controlled.

network structure A type of alloy microstructure in which one phase occurs predominantly at grain boundaries, enveloping grains of a second phase.

neurological manifestations See decompression sickness.

neurology The study of the anatomy, physiology, and pathology of the nervous system.[NASA]

neuroscience See neurology.

neurotransmitters Chemical substances secreted by the terminal ends of axons, which stimulate a muscle fiber contraction or an impulse in other neurons.[NASA]

neutral atmosphere An atmosphere which tends neither to oxidize nor reduce immersed materials.

neutral atoms Atoms in which the number of electrons surrounding the nucleus equals the number of protons in the nucleus resulting in no net electric charge. [NASA]

neutral currents Weak interaction currents that carry zero electric charge. [NASA]

neutral density filter A filter which has uniform transmission throughout the part of the spectrum where it is used.

neutral filter A light-beam filter which exhibits constant transmittance at all wavelengths within a specified range.

neutral gases In astronomy, gas clouds of some nebulae which have not been ionized by hot stars.[NASA]

neutral geometry A propellant grain configured in such a manner that the exposed surface area remains constant as burning progresses.[AIR913-89]

neutral point Point on the titration curve where the hydrogen ion concentration equals the hydroxyl ion concentration.

neutral static stability (of an airplane) See stability.

neutral zone See deadband; zone, neutral.

neutrino beams Organized collections of neutrinos traveling outward from the source.[NASA]

neutrinos Sub atomic particles of zero, or near zero, rest mass, having no electric charge, postulated by Fermi (1934) in order to explain apparent contradictions in the law of conservation of energy in beta particle emission.[NASA]

neutron A nuclear particle with a mass number of one and exhibiting zero (neutral) charge.

neutron flux See flux (rate).

neutron radiography Nondestructive testing and inspection utilizing neutron beams from nuclear reactors, particle accelerators, and/or radioisotopes. Imagery displaying structural defects utilizes neutron image recorders or screens. [NASA]

neutrons Subatomic particles with no electric charge, and with a mass of 1.67482 times 10 to the minus 24 gram. [NASA]

newton 1. A unit of force in the SI system; that force which gives to a mass of 1 kilogram an acceleration of 1 meter per second squared.[NASA] 2. Metric unit for force.

Newtonian flow Fluid characteristics adhering to the linear relation between shear stress, viscosity and velocity distribution.

nexus The point in a computer system where interconnections occur.

nibble A word with four bits, or one-half a byte.

nibbling Contour cutting of sheet metal by a rapidly reciprocating punch which makes numerous, successive small cuts.

nichols plot A nichols plot is the logarithm of the output-to-input magnitude ratio plotted against the phase angle, as the frequency of a sinusoidal input is varied.[AIR1823-86]

nick An indentation on the surface of the nut, produced by forceful abrasion or impact.[MA1568-87]

nickel This metal offers a combination of corrosion resistance, formability, and tough physical properties. For these reasons, nickel is used for alloying purposes and as a coating for copper.[ARP1931-87]

nickel iron batteries Alkaline-type electric cells using potassium hydroxide as the electrolyte and anodes of steel wool substrate with active iron material and cathode of nickel plated steel wool substrate with active nickel material.[NASA]

nickel steels Steels containing nickel as a main alloying element.[NASA]

nine-light indicator A remote indicator used in conjunction with a contact anemometer and a wind vane. It consists of a center lamp surrounded by eight lamps, equally spaced and labeled to indicate compass points. Wind speed is indicated by the number of flashes the center lamp makes in a certain time interval; wind direction, by the position of an illuminated lamp in the outer ring.

nipple A short piece of pipe or tube, usually with an external thread at each end.

nitinol alloys Shape memory alloys of titanium and nickel.[NASA]

nitrogen lasers Stimulated emission devices in which the nitrogen molecule is the lasing medium.[NASA]

noble gases See rare gases.

noble metal thermocouple A thermocouple whose elements are made of platinum (Pt) or platinum-rhodium (Pt-Rh alloys), and that resist oxidation and corrosion at temperatures up to about 1550 °C (2800 °F); three standard alloy pairs are in common use—Pt vs Pt-10%Rh, Pt vs Pt-13%Rh, and Pt-6%Rh vs Pt-30%Rh.

noctilucence See luminescence.

noctilucent clouds Clouds of unknown composition which occur at great heights, 75 to 90 kilometers. They resemble thin cirrus clouds, but usually with a bluish or silverish color, although sometimes orange to red, standing out against a dark night sky. Sometimes called luminous clouds.[NASA]

node In data processing, one component of a computer network where interconnections occur.

nodes (standing waves) Points, lines,

or surfaces in standing waves where some characteristic of the wave field has essentially zero amplitude.[NASA]

nodular iron See ductile iron.

NOE navigation See nap-of-the-earth navigation.

NOESS The acronym for the National Operational Environmental Satellite System.[NASA]

no-flow valve pressure gain Ratio of change in controlled pressure to the corresponding change of a controlling variable at no flow.[AIR1916-88]

noise 1. Random variation in analyzer output not associated with characteristics of the sample to which the analyzer is responding and which is distinguishable from analyzer drift characteristics.[ARP1256-90] 2. Meaningless stray signals in a control system similar to radio static. Some types of noise interfere with the correctness of an output signal.

noise and vibration General term which covers such conditions as tire, and wheel unbalanced vibration, tire tread pattern resonance and noise, etc.[AIR1489-88]

noise equivalent power The amount of optical power which must be incident on a detector to produce an electrical signal equal to the r.m.s. level of noise inherent in the detector. It is the measure of the sensitivity of the sensor.

noise factor In an electronic circuit, the ratio of total noise in the output signal to the portion thereof in the input signal under the following conditions—a selected input frequency and its corresponding output frequency, an input termination whose noise temperature is a standard 290 K at all frequencies, a linear system, and noise expressed as power per unit bandwidth.

noise figure A calculated or measured mathematical figure that denotes the inherent noise in a unit, system, or link.

noise immunity A device's ability to discern valid data in the presence of noise.

noise pollution Objectional or harmful levels of noise.[NASA]

noise prediction Estimation of intensity and frequencies based on analyses of probable oscillation of vibration producing components.[NASA]

noise prediction (aircraft) Estimating or forecasting of aircraft noise.[NASA]

noise quantization Inherent noise that results from the quantization process.

noise temperature At a pair of terminals and at a specified frequency, the temperature of a passive system exhibiting the same noise power per unit bandwidth as the actual terminals.

no-light No ignition on a start attempt for whatever reason.[ARP906A-91]

no-load pressure Pressure required to maintain a system at a specified speed with no external load.[AIR1916-88]

no-load valve flow gain The increment of valve flow per change in valve input variable at no load across the output.[AIR1916-88]

no-load velocity The output velocity of the actuation system measured with no external load. Maximum no-load velocity is determined with a saturation input to the power modulator.[AIR1916-88]

nominal bandwidth The difference between upper and lower nominal cutoff frequencies of an acoustic, or electric or optical filter.

nominal pressure, operating pressure The maximum steady working pressure to which a fitting assembly or component may be subjected. The basic operating pressure without regard to operating pressure variations.[MA2005-88]

nominal pressure (system pressure) (rated pressure) The general pressure setting of the system.[AIR1916-88]

nominal size 1. The standard dimension closest to the central value of a toleranced dimension. 2. A size used for general identification.

nominal stress The stress calculated by dividing nominal load by nominal cross sectional area, ignoring the effect of stress raisers but taking into account localized variations due to general part design.

nomograms See nomographs.

nomographs On charts or graphs, lines of constant value of given quantities with respect to either space or time.[NASA]

nonactive maintenance time The time during which no maintenance can be accomplished on the item because of administrative or logistic reasons.[ARD 50010-91]

nonadiabatic conditions In thermodynamics, changes in volume, temperature, flow, etc. accompanied by a transfer of heat.[NASA]

nonadiabatic process See heat transfer.

non-blackbody A term used to describe the thermal emittance of real objects, which emit less radiation than blackbodies at the same temperature, which may reflect radiant energy from other sources, and which may have their emitted radiation modified by passing through the medium between the body and a temperature-measuring instrument.

noncondensable gas The portion of a gas mixture (such as vapor from a chemical processing unit or exhaust steam from a turbine) that is not easily condensed by cooling. It normally consists of elements or compounds that have very low, often subzero, boiling points and vapor pressures.

nonconforming material Contract material which has physical, functional and/or dimensional variation or deviation from the applicable specification, or is incomplete.[AS7200/1-90]

noncontact gaging A method of determining physical dimensions without actual contact between the measuring device and the object.

noncontacting tachometer Any of several devices for measuring rotational speed without physical contact between a sensor and the rotating element—for example, stroboscopes or eddy-current tachometers.

noncritical dimension Any dimension which can be altered without affecting the basic function of a device.

noncritical viewing sector Those geometric sectors of an instrument excluding the critical sectors.[ARP1161-91]

non-destructive read out A method of reading from memory where the stored value is left intact by the reading process; plated wire and modern semiconductor random-access memory (RAM) are examples of NDRO memory.

nondestructive testing Any testing method which does not damage or destroy the sample. Usually, it consists of stimulating the sample with electricity, magnetism, electromagnetic radiation or ultrasound, and measuring the sample's response.

non-detectable failure A failure that, upon occurrence, is not recognized by the failure detection scheme(s) of a fault tolerant flight control system. Unless stated otherwise, a flight control system must maintain its fail-operative status after the occurrence of a non-detectable failure.[ARP1181A-85]

nondirectional beacon/radio beacon (NDB) A low- to middle-frequency (L/MF) or ultra-high frequency (UHF) radio beacon transmitting nondirectional signals whereby the pilot of an aircraft equipped with direction finding equipment can determine his/her bearing to or from the radio beacon and "home" on, or track to or from, the station.[ARP4107-88]

nondispersive infrared analyzer (NDIR) An analyzer that by absorption of infrared energy selectively measures specific components.[ARP1256-90] [AIR1533-91]

non ESS failures The following failures are non-ESS failures: (a) Failures directly attributable to improper installation in the test facility. (b) Failures of test instrumentation or monitoring equipment (other than the BIT function), except where it is part of the delivered item. (c) Failures resulting from test operator error in setting up or in testing the equipment. (d) Failures attributable to an error in or interpretation of the test procedures. (e) Dependent failures. (f) Failures occurring during repair. (g) Failures clearly attributable to the environmental generation test equipment overstress condition.[ARD 50010-91]

non-hazardous area An area in which explosive gas/air mixtures are not expected to be present so that special precautions for the construction and use of electrical apparatus are not required.

noninherent failure causes Failures due to external factors not associated with intrinsic design, such as foreign object damage, corrosion, material defects and maintenance error.[AIR1872-88]

noninteracting control system A multi-element control system designed to avoid disturbances to other controlled variables due to the process input adjustments which are made for the purpose of controlling a particular process variable. See control system, non-interacting.

nonisothermal process In thermodynamics, compression or expansion of substances at nonuniform temperatures.[NASA]

nonisotropy See anisotropy.

nonlinear distortion A departure from a desired linear relationship between corresponding input and output signals of a system.

nonlinear effects Optical interactions which are proportional to the square or higher powers of electromagnetic field intensities. Nonlinear effects generate harmonics of optical frequencies, and sum and difference frequencies when two lightwaves are mixed.

nonlinearity A type of error in an FM system, where the input to a device does not relate to the output in a linear manner. See linearity.

nonlinear optics Study of the interaction of radiation with matter in which certain variables describing the response of the matter are not proportional to variables describing the radiation. [NASA]

nonlinear optimization See nonlinear programming.

nonlinear programming 1. In operations research a procedure for locating the maximum or minimum of a function of variables which are subject to constraints, when either the function or the constraints, or both, are nonlinear. Contrast with convex programming and dynamic programming. 2. Synonymous with nonlinear optimization.

nonlinear system Any system whose operation cannot be represented by a finite set of linear differential equations.

nonlocking Pertaining to code extension characters that change the interpretation of one or a specified number of characters. Contrast with locking.

nonmission effects Failure effects which do not prevent mission success or equipment operation, but whose correction may or may not be economically desirable due to added labor and material cost for repair (including loss-of-use cost due to maintenance downtime). [ARD50010-91]

non-operating The condition when the galley, with loaded equipment and inserts, is not used but remains installed in the airplane on the ground and the water system is filled, but without air conditioning or power, when the airplane is exposed outdoors with cabin closed for a period of time of 2 hours or more duration.[AS1426-80]

non-operating conditions See environmental conditions, non-operating.

nonperception See perception.

nonpoint sources Undetermined or general areas from which pollutants, contaminants, and/or other unwanted materials or wastes enter the environment. [NASA]

nonprecision approach procedure/nonprecision approach A standard instrument approach procedure in which no electronic glide slope is provided (for example, VOR, TACAN, NDB, LOC, ASR, LDA, or SDF approaches).[ARP 4107-88]

nonprocessor request The system for accomplishing data transfers between two devices without involving the CPU.

nonradar approach control An air traffic controlled (ATC) facility providing approach control service without the use of radar. Under this condition, the controller uses radio communications to keep track of the aircraft on approach to the airfield and issues further clearances to aircraft to approach the field based upon his/her knowledge of the location of the other aircraft as determined from these radio communications.[ARP4107-88]

nonreclosing pressure relief device A device for relieving internal pressure which remains open when actuated and must be replaced or reset before it can actuate again.

non-repeatability See repeatability.

non-return-to-zero (NRZ) Coding of digital data for serial transmission or storage whereby a logic ONE is represented by one signal level and a logic ZERO is represented by a different signal level.

nonrigidity See flexibility.

non-scheduled maintenance 1. Unscheduled maintenance specifically intended to eliminate an existing fault. 2. An urgent need for repair or upkeep that was unpredicted or not previously planned, and must be added to or substituted for previously planned work.

nontransferred arc In arc welding and cutting, an arc sustained between the electrode and a constricting nozzle rather than between the electrode and the work.

nontrip-free circuit breaker A breaker so designed that the circuit can be maintained closed when carrying overload current that would automatically trip the breaker to the open position.[ARP 1199A-90]

nonvolatile matter (NVM) test A test for the percentage of nonvolatile matter in a sealant. Generally, volatile solvents are used to produce the desired amount of flow in brushable and sprayable sealants in particular.[AS7200/1-90]

nonvolatile memory Computer memory that retains data when power is removed.

noon The instant at which a time reference is over the upper branch of the reference meridian.[NASA]

NOR 1. A logic operator having the property that if P is a statement, Q is a statement, R is a statement, ..., then the NOR of P, Q, R, ... is true if all statements are false, false if at least one statement is true. P NOR Q is often represented by a combination of OR and NOT symbols. P NOR Q is also called NEITHER P NOR Q. Synonymous with NOT-OR. 2. Logical negation of OR. Supplies a logic 0 when any input is at logic 1.

Nordel 1070 An ethylene propylene rubber by E. I. du Pont de Nemours Co.

NORDO Acronym for NO RaDiO. A term used to indicate that an aircraft is unable to communicate because of radio failure.[ARP4107-88]

normal axis See vertical axis.

normal electric-system operation All the functional electric-system operations required for aircraft operation, aircraft mission, and electric-system controlled continuity. These operations occur at any given instant and any number of times during ground operation, flight preparation, takeoff, airborne conditions, landing, and anchoring. Examples of such operations are switching of utilization equipment loads, engine speed changes, bus switching and synchronization, and paralleling of electric power sources. Switching of utilization equipment loads is a type of system operation which occurs the greatest number of times.[AS1212-71]

normal fluid temperature The stabilized fluid temperature normally reached during continuous operation.[AIR1916-88]

normality Concentration units defined as the number of gram-ions of replaceable hydrogen or hydroxyl groups per liter of solution. A shorter notation of gram-equivalents per liter is frequently used.

normalize 1. In programming, to adjust the exponent and fraction of a floating-point quantity such that the fraction lies in a prescribed normal standard range. 2. In mathematical operations, to reduce a set of symbols or numbers to a normal or standard form; synonymous with standardize. 3. In heat treating ferrous alloys, to heat 50 to 100 °F above the upper transformation temperature, then cool in still air.

normally aspirated Describes an engine that takes its combustion air directly from the atmosphere without the benefit of mechanical devices to increase the pressure of the air. See supercharger.[ARP4107-88]

normally closed (NC) 1. A switch position where the usual arrangement of contacts permits the flow of electricity in the circuit. 2. In a solenoid valve, an arrangement whereby the disk or plug is seated when the solenoid is deenergized. 3. A field contact that is closed for a normal process condition and open when the process condition is abnormal. 4. A valve with means provided to move to and/or hold in its closed position without actuator energy supply. See also field contact.

normally open (NO) 1. A switch position where the usual arrangement of contacts provides an open circuit (no current flowing). 2. In a solenoid valve, an arrangement whereby the disk or plug is seated when the solenoid is energized. 3. A field contact that is open for normal process condition and closed when the process conditions are abnormal. 4. A valve with means provided to move to and/or hold in its wide-open position without actuator energy supply. See also field contact.

normal mode voltage An extraneous voltage induced across the circuit path (transverse mode voltage). See also voltage, normal mode.

normal opening fuse (fast-acting) A fuse which opens the circuit without deliberate time-delay.[ARP1199A-90]

normal operating conditions See operating conditions, normal.

normal operation Intrinsically safe electrical apparatus or associated electrical apparatus is in normal operation when it complies electrically and mechanically with the requirements of its design specification and is used within the limits specified by the manufacturer.

normal output curve The locus of the midpoints of a complete input/output curve. This locus is the zero-hysteresis output curve.[AIR1916-88]

normal output gain The slope of the normal output curve in units of output/input.[AIR1916-88]

normal rated power The maximum horsepower or thrust an engine can deliver for a protracted period of time without damage; specified by the manufacturer or other qualified authority.[ARP 4107-88]

Normal Thermometric Scale The first international standard temperature scale, adopted in 1887, which was based on the fundamental interval of 100°

between the ice point of pure water and the condensing point of pure water vapor.

northern sky That part of the sky visible from the northern hemisphere. [NASA]

North Polar Spur (astronomy) One of the largest sources of diffuse radio emission outside the galactic plane. The Spur, a ridge of enhanced emission, may be the remnant of the shells of supermovae which exploded over 100,000 years ago.[NASA]

nose caps See nose cones.

nose cones The cone shaped leading end of rocket vehicles, consisting (a) of chambers in which satellites, instruments, animals, plants, or auxillary equipment may be carried, and (b) of outer surfaces built to withstand high temperatures generated by aerodynamic heating. [NASA]

nose gear See gear.

nose gear launch Launch of an aircraft by means of an apparatus including a launch bar attached to the nose gear and imparting tow loads to the aircraft through the nose gear system.[AIR1489-88]

noseheavy The condition of an airplane in which the nose tends to sink when the longitudinal control is released in any given attitude of normal flight. [ARP4107-88]

nose tips The foremost sharp points of bombs, rockets, missiles, and other symmetrical bodies.[NASA]

nose wheel The wheel(s) designed for use on nose gears. They differ from main wheels in that they usually make no provision for brakes within the wheel. They are also designed to take the dynamic vertical loads resulting from brake drag loads applied at the main wheels.[AIR1489-88]

nosewheel cycle A full cycle is that nosewheel motion from the center position to one extreme position, returning past the center position to the opposite extreme, and returning again to the center position.[ARP1595-82]

nosewheel, electrically controlled, electrically powered Use of all-electrical servos for NWS systems is usually limited to low speed (taxi) control, with notable exceptions on some executive type aircraft.[ARP1595-82]

nosewheel, electrically controlled, hydraulically powered Inherently less reliable than mechanically controlled servos, redundancy and/or failure detection features are normally required. Such systems are now in common use on military combat aircraft and some business jet aircraft.[ARP1595-82]

nosewheel, mechanically controlled, hydraulically powered Hydromechanical position servo mechanism, commonly used on transport and business jet aircraft and some smaller military aircraft. Normally very reliable, but dependent upon aircraft hydraulic system power.[ARP1595-82]

nosewheel, mechanically controlled, manually powered Highly reliable, independent of aircraft power generation, normally employed on small general aviation aircraft.[ARP1595-82]

nosewheel, powered steering angle The powered steering angle is the number of degrees (radians) the nosewheel is displaced from the center (straight ahead) position while powered by the NWS control system.[ARP1595-82]

nosewheel rated load The rated load is the maximum load needed to fulfill the output torque requirement.[ARP 1595-82]

nose wheel steering The system, apparatus, and/or controls for maneuvering the aircraft on the ground by rotating the nose wheels about an axis normal to the ground.[AIR1489-88]

nosewheel, steering rate Steering rate is the rate of change of nosewheel steering angle in degrees (or radians) per second.[ARP1595-82]

nosewheel, steering ratio Steering ratio describes the relationship between input control movement and output nosewheel angle change (i.e.; degrees nosewheel angle per inch of rudder pedal travel).[ARP1595-82]

NOT 1. A logic operator having the property that if P is a statement, then the NOT of P is true if P is false, false if P is true. 2. Logical negation symbol. Supplies the complement of any input.

NOTAM Acronym for NOtices to AirMen. Notices containing information (not known sufficiently in advance to publicize by other means) concerning the establishment of, condition of or change in any component of the National Airspace System, the timely knowledge of which is essential to personnel concerned with flight operations. These notices may be disseminated by teletypewriter or by voice.[ARP4107-88]

NOT-AND Same as NAND.

not-at-intermediate position A position that is either above or below the specified intermediate position.

notation 1. The act, process, or method of representing facts or quantities by a system or set of marks, signs, figures, or characters. 2. A system of such symbols or abbreviations used to express technical facts or quantities, as mathematical notation. 3. An annotation.

notch 1. A V-shaped indentation in an edge or surface. 2. An indentation of any shape that acts as a severe stress raiser.

notching Cutting out various shapes from the edge of a metal strip, blank or part.

notch sensitivity A measure of the sensitiveness of a material to the presence of stress concentration caused by notches in the form of threads or grooves, scratches and other stress raisers. Materials with low notch sensitivity are selected for the high fatigue/high stress applications in aircraft primary structure and in aircraft engines.[ARP700-65]

notch width The horizontal distance between opposite sides of the weir notch.

not-closed position A position that is more than zero-percent open. A device that is not closed may or may not be open.

NOTE Nap-Of-The-Earth (flight).

NOT-IF-THEN Same as exclusive OR.

not mission capable NMC is a material condition indicating that systems are not capable of performing any of their assigned missions because of unit level maintenance requirements. Recording of NMCM time shall start for: (a) unscheduled maintenance, when a malfunction is discovered, or at mission completion, whichever is later, and (b) scheduled maintenance, when the determination is made that a system cannot be returned to mission-capable status within 2 hours. Time stops when maintenance has been completed or is interrupted by work stoppage due to supply shortage. The period of work stoppage due to supply shall be measured as NMCS. NMCM shall resume when required supply items are delivered to the maintenance activity.[ARD50010-91]

not mission capable supply NMCS is a material condition indicating that systems and equipment are not capable of performing any of their assigned missions because of maintenance work stoppage due to a supply shortage. Recording of NMCS time shall start when work stoppage results from lack of parts, and the NMCS requisition is not satisfied 1 hour after the demand is initiated and remains unsatisfied. For Army and Marine Corps ground equipment, when both NMCM time and NMCS time are encountered in the same day and the sum is more than 12 hours, the whole day is carried against the condition status with the most hours.[ARD50010-91]

not-open position A position that is less

than 100 percent open. A device that is not open may or may not be closed.

not operating time That element of uptime during which an item is not required to operate.[ARD50010-91]

NOT-OR Same as NOR.

Nova computers A series of minicomputers built by Data General.[NASA]

Nova Laser System Laser fusion system utilizing large neodymium glass lasers for irradiating DT pellets.[NASA]

Nova satellites A second generation Navy navigation satellite which replaces the transit satellites.[NASA]

NOVRAM Nonvolatile random-access memory (one type of nonvolatile semiconductor computer memory).

nowcasting A self contained short period meteorological forecast for the immediate future covering a period of up to six hours.[NASA]

NOx Oxides of nitrogen, specifically, the sum of nitric oxide (NO) and nitrogen dioxide (NO_2).[ARP1256-90] [AIR1533-91]

nozzle 1. A device used to control the flow of air, gas, or liquids. May be designed with a constricting throat section and/or a divergent sections.[ARP 480A-87] 2. A short flanged or welded neck connection on a drum or shell for the outlet or inlet of fluids; also a projecting spout through which a fluid flows. 3. A streamlined device for accelerating and directing fluid flow into a region of lower fluid pressure. 4. A particular type of restriction used in flow system to facilitate flow measurement by pressure drop across a restriction.

nozzle efficiency The efficiency with which a nozzle converts potential energy into kinetic energy, commonly expressed as the ratio of the actual change in kinetic energy to the ideal change at the given pressure ratio.[NASA]

nozzle/flapper A fundamental part of pneumatic signal processing and pneumatic control operations. Basically, the device converts a displacement of the flapper to a pressure signal.

NPSH See net positive suction head.

NRZ See non-return-to-zero.

NSE (nth sequential) algorithm See compressor.

NTS See navigation technology satellites.

NTSB National Transportation Safety Board.

nuclear devices Devices whose explosive potency is derived from nuclear fission of atoms of fissionable material with the consequent conversion of part of their mass into energy.[NASA]

nuclear emulsion(s) 1. Very thick photographic emulsions used in the study of cosmic rays and other energetic particles. The paths of the particles through the thick emulsion are recorded in three dimensions.[NASA] 2. A photographic emulsion specially designed to record the tracks of ionizing particles.

nuclear fluorescence thickness gage A device for determining the weight of an applied coating by exciting the coated material with gamma rays and measuring low-energy fluorescent radiation that results.

nuclear fuel reprocessing Periodic chemical, physical, and metallurgical treatment of materials used as fuel elements in nuclear reactors to recover and purify residual fissionable and fertile materials.[NASA]

nuclear fuels Fissionable materials of reasonable long life, used or usable in producing energy in a nuclear reactor.[NASA]

nuclear medicine That branch of medicine dealing with the effect of radiation such as x-rays, gamma rays, and energetic particles on the body and with the prevention and cure of physiological injuries resulting from such radiation.[NASA]

nuclear power systems There are two basic types of nuclear auxiliary power systems. One type is powered by radioisotopes, and the other type is powered by a nuclear reactor. They differ only in the heat power source with power conversion subsystems and power regulation subsystems being similar for both types of nuclear auxiliary power sources. The heat power conversion to electrical power is accomplished using thermoelectric, thermionic, or dynamic power conversion devices. Only the radioisotope thermoelectric generators have seen practical application.[AIR744A-71]

nuclear pumped lasers Lasers in which the excitation is supplied by a nuclear reactor as a high flux source or by the kinetic energy of the fission fragments only.[NASA]

nuclear pumping Laser-like pumping produced by electrons generated in nuclear reactions or, in general, by beams of charged particles.[NASA]

nuclear radiation Corpuscular emissions, such as alpha and beta particles, or electromagnetic radiation, such as gamma rays, originating in the nucleus of the atom.[NASA]

nuclear reactors Apparatus in which nuclear fission may be sustained in a self supporting chain reaction.[NASA]

nuclear rocket engines Rocket engines in which nuclear reactors are used as power sources or as sources of thermal energy.[NASA]

nuclear vulnerability The resistance of structures or materials to nuclear radiation or explosions.[NASA]

nuclei (nuclear physics) The positively charged cores of atoms with which are associated practically the whole mass of each atom but only a minute part of its volume.[NASA]

nucleonics Technology involving atomic nuclei; includes nuclear reactors, particle accelerators, radiation detectors and radioisotope applications.

nucleons In the classification of subatomic particles according to mass, the second heaviest type of particles; their mass is intermediate between that of the meson and the hyperon.[NASA]

nucleus 1. The positively charged core of an atom; it contains almost all of the mass of the atom but occupies only a small fraction of its volume. 2. A number of atoms or molecules bound together with interatomic forces sufficiently strong to make a small particle of a new phase stable in a mass otherwise consisting of another phase; creating a stable nucleus is the first step in phase transformation by a nucleation-and-growth process. 3. That portion of the control program that must always be present in main storage. Also, the main storage area used by the nucleus and other transient control program routines.

nucleus counter An instrument that measures the number of condensation or ice nuclei in a sample volume of air.

nuclide(s) 1. Individual atoms of a given atomic number Z and mass number A.[NASA] 2. A species of atom characterized by a unique combination of charge, mass number and quantum state of its nucleus.

nude vacuum gage A hot-filament ionization gage mounted entirely within the vacuum system whose pressure is being measured.

nuisance alert 1. An alert that is given when an aircraft is in the TCAS operational envelope and a maneuver by the TCAS aircraft is not necessary to achieve satisfactory aircraft separation.[ARP 4153-88] 2. An alert which occurs at too low a level of windshear to warrant a windshear alert.[ARP4109-87]

null 1. A situation of no input to the controller. 2. A condition, such as of balance, which results in a minimum absolute value of output.

null-balance recorder An instrument that records a measured value by means of a pen or printer attached to a motor-driven slide, where the position of the slide is determined by continuously balancing current or voltage in the measuring circuit against current or voltage from a sensing element.

null indicator An indicating device such as a galvanometer used to determine when voltage or current in a circuit is zero; used chiefly in balancing bridge circuits. Also known as null detector.

null leakage Internal leakage of a valve when output flow is negligible.[AIR 1916-88]

null shift A shift for whatever reason of the input/output relationship with respect to reference coordinate axes. [ARP1281A-81]

number 1. A mathematical entity that may indicate quantity or amount of units. 2. Loosely, a numeral. 3. See binary number and random numbers.

Numbers, the 1. Digits written on the approach end of a runway to identify the orientation of the runway with respect to magnetic north to the nearest even 10 deg with the last zero deleted. For example, 02 painted on the approach end of a runway indicates that the magnetic orientation of the runway is approximately 020 deg and that an aircraft on final approach to that runway should be heading approximately 020 deg magnetic. 2. An expression used to indicate the receipt of the current ATIS broadcast. For example, if a pilot tells the controller that he/she has "the numbers," it signifies that he or she is aware of the current ATIS information being broadcast for the destination airfield. [ARP4107-88]

number system 1. A systematic method for representing numerical quantities in which any quantity is represented as the sequence of coefficients of the successive powers of a particular base with an appropriate point. Each succeeding coefficient from right to left is associated with and usually multiplies the next higher power of the base. 2. The following are names of the number systems with bases 2 through 20: 2, binary; 3, ternary; 4, quaternary; 5, quinary; 6, senary; 7, septenary; 8, octal or octonary; 9, novenary; 10, decimal; 11, undecimal; 12, duodecimal; 13, terdenary; 14, quaterdenary; 15, quindenary; 16, sexadecimal—or hexadecimal; 17, septendecimal; 18, octodenary; 19, novemdenary; 20, vicenary. Also 32, duosexadecimal—or duotricinary; and 60, sexagenary. The binary, octal, decimal, and sexadecimal systems are widely used in computers. See decimal number and binary number and related to positional notation and clarified by octal digit and binary digit.

numerical analysis 1. Study of approximation methods using arithmetic techniques.[NASA] 2. The study of methods of obtaining useful quantitative solutions to mathematical problems, regardless of whether an analytic solution exists or not, and the study of the errors and bounds on errors in obtaining such solutions.

numerical aperture The sine of the half-angle over which an optical fiber or optical system can accept light rays, multiplied by the index of refraction of the medium containing the rays.

numerical control Automatic control of a process performed by a device that makes use of all or part of numerical data generally introduced as the operation is in process.

numerical differentiation Approximate estimation of a derivative of a function by numerical techniques. [NASA]

numerical keypad Typical of a computer keyboard, a separate set of 0 through 9 keys arranged like numerical keys on a 10-key adding machine.

numeric word A word consisting of digits and possibly space characters and special characters.

Nusselt number A number expressing the ratio of convective to conductive heat transfer between a solid boundary and a moving fluid, defined as hI/k where h is the heat transfer coefficient. I is the characteristic length, and k is the thermal conductivity of the fluid. Named after Wilhelm Nusselt, a German engineer.[NASA]

nut A fastening device of various shapes having threads, lugs, or prongs designed to mate with a thread for the purpose of securely holding a mating part.[ARP 480A-87]

nutating-disk flowmeter A type of positive-displacement flowmeter in which the advancing volume of fluid causes a measuring disk to wobble (nutate), thereby passing a precise volume of fluid through the meter with each revolution of the disk.

nutation 1. The oscillation of the axis of any rotating body, as a gyroscope rotor. Specifically, in astronomy, irregularities in the processional motion of the equinoxes because of varying positions of the moon and, to a lesser extent, of other celestial bodies with respect to the ecliptic.[NASA] 2. Rocking back and forth, or periodically repeating a circular, elliptical, conical or spiral path, usually involving relatively small degrees of motion.

nutational oscillation See nutation.

N-value The exponent in the power function $V(T)=KT^N$, which is the calibration function for a ratio thermometer. The N-value and mean effective wavelength can be used to express operating characteristics of a given ratio thermometer.

nylon A plastics material used to make filaments, fibers, fabric, sheet and extrusions; a generic name for a type of long-chain polymer containing recurring amide groups within the main chain.

Nyquist frequency One-half of the sampling frequency in a sampled data system.

nyquist plot The nyquist plot provides magnitudes and phase relationships (in polar form) between system input and response for any excitation frequency. [AIR1823-86]

nystagmus An involuntary oscillation of the eyeballs, especially occurring as a result of eye fixations and stimulations of the inner ear during rotation of the body.[NASA]

Oberon A satellite of Uranus orbiting at a mean distance of 587,000 kilometers.[NASA]

object code 1. The machine code that can be directly executed by the computer. It is produced as a result of the translation of the source code. 2. A relocatable machine-language code.

object color stimulus The reference stimulus (Y_n, u'_n, v'_n) by which color differences are measured in the CIE LUV system.[ARP1782-89]

objective A hardware requirement established for design to achieve optimum performance, minimum weight or other technical criteria. The feasibility of meeting an objective, or modifying the technical criteria to establish a firm requirement is subject to review. Renegotiation may be accomplished during development after analyses, test reports, and cost data are available which support a review to determine viability for manufacturing production, cost of development, or other program factors.[AS1426-80]

objective variable A quantity or condition that is not measured directly for the purpose of controlling it, but rather is controlled through its relation to another, controlled variable.

object language A language which is the output of an automatic coding routine. Usually object language and machine language are the same, however, a series of steps in an automatic coding system may involve the object language of one step serving as a source language for the next step and so forth.

object machine The computer on which the object program is to be executed. Same as target computer.

object module The primary output of an assembler or compiler, which can be linked with other object modules and loaded into memory as a program. The object module is composed of the relocatable machine-language code, relocation information, and the corresponding symbol table defining the use of symbols within the module.

object program A fully compiled or assembled program that is ready to be loaded into the computer. See also target program.

object time system The collection of modules called by the compiled code to perform various utility or supervisory operations; for example, an object time system usually includes I/O and trap-handling routines.

oblate spheroids Ellipsoids of revolution, the shorter axis of which is the axis of revolution.[NASA]

oblique coordinates Magnitudes defining a point relative to two intersecting nonperpendicular lines, called axes.[NASA]

obliqueness The state of being neither perpendicular nor horizontal.[NASA]

obscuration See occultation.

observability (systems) The property of a system for which observations of the output variables always is sufficient to determine the initial values of all state variables.[NASA]

observed mean time between failure (THETA) The total operating time of the equipment divided by the number of relevant failures.[ARD50010-91]

obsolescent Lower in physical or functional value due to changes in technology rather than to deterioration.

obsolete No longer suitable for the intended use because of changes in technology or requirements.

obstacle avoidance The use of sensors utilizing laser triangulation as means of preventing collisions, especially in the operation of roving vehicles on planetary surfaces.[NASA]

obstacles See barriers.

occasional A metallographic feature that occurs in 10% or less of the microstructure.[AS1814-90]

occlusion Specifically, the trapping of

undissolved gas in a solid during solidification.[NASA]

occultation The disappearance of a body behind another body of larger apparent size.[NASA]

ocean color scanner A multispectral scanning radiometer which is geared to observe ocean features such as chlorophyll, sediments, and topography in the invisible and thermal ranges of radiation.[NASA]

ocean dynamics The study of the controlling forces in different ocean phenomena.[NASA]

ocean temperature Surface or subsurface temperature of an entire or specific region of an ocean.[NASA]

octal Pertaining to eight; usually describing a number system of base or radix eight, e.g., in octal notation, octal 214 is 2 times 64, plus 1 times 8, plus 4 times 1, and equals decimal 140.

octal digit The symbol 0, 1, 2, 3, 4, 5, 6, or 7 used as a digit in the system of notation which uses 8 as the base or radix. Clarified by number systems.

octal number A number of one or more figures, representing a sum in which the quantity represented by each figure is based on a radix of eight. The figures used as 0, 1, 2, 3, 4, 5, 6, and 7. Clarified by octal.

octave The interval between two frequencies with a ratio of 2:1.

octave-band analyzer A portable sound analyzer which amplifies a microphone signal, feeds it into one of several band-pass filters selected by a switch, and indicates signal amplitude on a logarithmic scale; except for the highest and lowest band, each band spans an octave in frequency.

octave-band filter A band-pass filter in which the upper and lower cutoff frequencies are in a fixed ratio of 2:1.

octaves The intervals between two frequencies having the ratio 1:2.[NASA]

octet A group of eight bits treated as a unit. See byte.

OD See outside diameter.

ODA/ODIF Office Document Architecture/Office Document Interchange Format.[AS4159-88]

odd-even check Same as parity check.

odograph An instrument mounted in a vehicle to automatically plot its course and distance traveled on a map.

odometer An instrument for measuring and indicating distance traveled.

Oersted The CGS unit of magnetic field strength; the SI unit, ampere-turn per metre, is preferred.

off Describing the nonoperating state of a device or circuit.

off-axis mirrors Mirrors in which the mechanical center of the mirror does not correspond to the axis of the optical figure of the mirror.

off equipment work For the purpose of maintenance data reporting, it includes all maintenance actions performed on removed, repairable components, usually at the IMA.[ARD50010-91]

offgassing The relative high mass loss characteristics of many nonmetallic materials upon initial vacuum exposure.[NASA]

off-line 1. Not being in continuous, direct communication with the computer. 2. Done independent of the computer (as in off-line storage). 3. Describing the state of a subsystem or piece of computer equipment which is operable, but currently bypassed or disconnected from the main system. 4. Pertaining to a computer that is not actively monitoring or controlling a process or operation, and pertaining to a computer operation performed while the computer is not monitoring or controlling a process or operation. 5. Describing lateral or angular deviation from the intended axis of a drilled or bored hole.

off-line diagnostics 1. Describing the state of a control system, subsystem or piece of computer equipment which is operable, but currently not actively monitoring or controlling the process. 2. A program to check out systems and subsystems, providing error codes if an error is detected. This diagnostic program is run while the system is off line.

off-line equipment The peripheral equipment or devices not in direct communication with the central processing unit of a computer.

off line maintenance information Data to indicate long term engine degradation through trend monitoring and tracking of engine usage history; e.g., engine hours, starts, LCF counts, and hot section usage.[ARP1587-81]

off-line memory Any media, capable of being stored remotely from the computer, which can be read by the computer when placed into a suitable reading device. Also see external storage.

off-line system That kind of system in which human operations are required between the original recording functions and the ultimate data processing function. This includes conversion operations as well as the necessary loading and unloading operations incident to the use of point-to-point or data-gathering systems. Compare on-line system.

off-on control Flicker control, especially as applied to rockets.[NASA]

off-route vector A vector by ATC which takes an aircraft off a previously assigned route. Altitudes assigned by ATC during such vectors provide required obstacle clearance.[ARP4107-88]

offset 1. A sustained deviation of the controlled variable from set point. This characteristic is inherent in proportional controllers that do not incorporate reset action. 2. Offset is caused by load changes. 3. The steady-state deviation The count value output from an A/D converter resulting from a zero input analog voltage. Used to convert subsequent nonzero measurements. 4. A short distance measured perpendicular to a principal line of measurement in order to locate a point with respect to the line. 5. A constant and steady state of deviation of the measured variable from the set point. 6. A printing process in which ink is transferred from the printing plate or master to a rubber covered roller, which in turn transfers the ink to the paper.

offset (programming) The difference between a base location and the location of an element related to the base location; the number of locations relative to the base of an array, string, or block.

offset tongue terminal A terminal whose tongue is forward of its conductor barrel and whose stud hole is offset from the centerline of the conductor barrel.[ARP914A-79]

OFT See space transportation system flights.

ogives Bodies of revolution formed by rotating a circular arc about an axis that intersects the arc; the shape of these bodies; also noses of projectiles or the like so shaped.[NASA]

ohm The metric unit for electrical resistance; it is the resistance (or impedance) of a conductor such that an electrical potential of one volt exists across the ends of the conductor when it carries a current of one ampere.

ohmmeter A device for measuring electrical resistance.

ohms per volt A standard rating of instrument sensitivity determined by dividing the instrument's electrical resistance by its full-scale voltage.

oil Any of various viscous organic liquids that are soluble in certain organic solvents such as naphtha or ether but are not soluble in water; may be of animal, vegetable, mineral or synthetic origin.

oil bath 1. Oil, in a container or chamber, which a part or mechanism is submerged, or into which it dips, during operation or manufacture. 2. Oil poured on a cutting tool or in which it is submerged during a machining operation.

oil burner A burner for firing oil.

oil cone The cone of finely atomized oil discharged from an oil atomizer.

oil dilution system The mechanism by which dilution is added to the lubrication system in order to reduce the viscosity and pour point of the lubrication oil.[S-4, 6-84]

oil gas A heating gas made by reacting petroleum oil vapors and steam.

oil heating and pumping set A group of apparatus consisting of a heater for raising the temperature of the oil to produce the desired viscosity, and a pump for delivering the oil at the desired pressure.

olemeter 1. A device for measuring the specific gravity of oil. 2. A device for measuring the proportion of oil in a mixture.

oleo (oleo strut) A telescoping landing gear strut consisting essentially of a piston that travels in an air-oil filled cylinder. The oil, upon compression of the strut, is forced through an orifice to provide a shock absorbing effect.[AIR-1489-88] [ARP4107-88]

oleo-pneumatic See shock absorber, air-oil.[AIR1489-88]

OLI (Out of Limits) Algorithm See compressor.

Olsen ductility test A method for determining relative formability of metal sheet. A sheet metal sample is deformed at the center by a steel ball until fracture occurs; the cup height at fracture indicates relative ease of forming deep-drawn or stamped parts.

OM Outer Marker

ombroscope An instrument for indicating when precipitation occurs. A heated water-sensitive surface is exposed to the weather; when it rains or snows, an electrical or mechanical output trips an alarm or records the occurrence on a time chart.

omega A nonequilibrium, submicroscopic phase which can be formed either athermally or isothermally preceding the formation of alpha from beta. It occurs in metastable beta alloys, alpha-plus-beta alloys rich in beta content, and unalloyed titanium, and leads to severe embrittlement. Athermal omega is believed to form without change in composition and is analogous to martensite. Isothermal omega is generally formed by aging a retained beta structure in the 392 to 932 °F (200 to 500 °C) temperature range.[AS1814-90]

omnidirectional With reference to a beacon or radio aid to air navigation, transmitting a signal throughout 360 deg of azimuth.[ARP4107-88]

omnidirectional (antenna) An antenna having equal gains in all directions.

omnigraph An automatic acetylene flame-cutting device that cuts several blanks simultaneously, duplicating the pattern traced by a mechanical pointer.

on Describing the operating state of a device or circuit.

onboard data processing Processing of acquired data aboard an aircraft, satellite, etc. rather than transmission to ground stations for processing.[NASA]

on condition (OC) 1. A term used to indicate maintenance based upon the functional, structural or other condition of the unit or part, as differentiated from time schedule maintenance.[ARP1587-81] 2. An OC unit has no fixed restoration limit, but must be given a specific check at a prescribed frequency or interval. This check must be against specific, published requirements which determine that the unit is serviceable and airworthy and is reasonably certain to remain so until its next scheduled check. The degree of certainty required depends on the importance of the unit to aircraft airworthiness, available backup, and redundancy.[AIR1916-88]

on-course indication An indication on an instrument which provides the pilot with a visual means of determining that the aircraft is located on the centerline of a given navigational track, or an indication on a radar scope that an aircraft is on a given track.[ARP4107-88]

on equipment work For the purpose of maintenance data reporting, it includes those maintenance actions accomplished on complete end items, for example, aircraft, drones, SE, removed engines.[ARD50010-91]

ones complement The radix-minus-one complement in binary notation. The ones complement of an octal 3516 is 4261. See also complement.

one watt-one ampere EED An EED which will not fire or be rendered inoperative when one ampere and/or one watt is passed through the bridge wire circuit for a specified period of time. [AIR913-89]

one-way Not recommended, except as a modifier of such items as restrictor valves.[ARP243B-65]

onisotropy See anisotropy.

on-line 1. Describing the state of a subsystem or piece of computer equipment which is operable and currently connected to the main system. 2. Pertaining to a computer that is actively monitoring or controlling a process or operation, or pertaining to a computer operation performed while the computer is monitoring or controlling a process or operation. 3. Describing coincidence of the axis of a drilled or bored hole and its intended axis, without measurable lateral or angular deviation. 4. Directly controlled by, or in continuous communication with, the computer (on-line storage). 5. Done in real time.

on-line data-reduction The processing of information as rapidly as information is received by the computing system or as rapidly as it is generated by the source.

on-line debugging The act of debugging a program while time sharing its execution with an on-line process program.

on-line diagnostics 1. Describing the state of a control system, subsystem or piece of computer equipment which is operable and actively monitoring or controlling the process. 2. A program to check out systems and subsystems, providing error codes and alarms if errors are detected. This diagnostic program runs in background while the control system is in the operating mode.

on-line equipment A computer system and the peripheral equipment or devices in the system in which the operation of such equipment is under control of the central processing unit, and in which information reflecting current activity is introduced into the data processing system as soon as it occurs. Thus, directly in-line with the main flow of transaction processing. Clarified by on-line.

on line maintenance Maintenance performed on a system or equipment without interrupting its operation. Synonymous with reliability with repair.[ARD 50010-91]

on-line memory Any media directly accessible by the computer system. Also see internal storage.

on-line processing Same as on-line.

on-line system(s) 1. Synonymous with on-line. 2. A system in which the input data enters the computer directly from the point of origin and/or in which output data is transmitted directly to where it is used. Compare off-line.

ON-OFF control 1. Also called two position or cycling control. This type of control can assume only a maximum or minimum position. The differential of the controller and the system characteristics determine the rate of amplitude of controlled variable cycles.[ARP 89C-70] 2. A simple form of control whereby the control variable is switched fully ON or fully OFF in response to the process variable rising above the set-point or falling below the set-point respectively. Cycling always occurs with this form of control.

on-off controller See controller, on-off.

on-off system A type of skid control system which utilizes a valve which opens or closes when input signal exceeds or reduces below specific threshold levels. Sometimes called a "bang-bang" system.[AIR1489-88]

on-off valve Controls the flow of air by opening or closing the flow passage. It is used in such a manner that it is not stopped in intermediate positions.[ARP 986-90]

Oort cloud A region of millions of comets between 30,000 and 100,000 A.U. from the sun. Comets are perturbed out of the Oort cloud by passing stars and fall into the inner solar system.[NASA]

opacity Of an optical path, the reciprocal of transmission.[NASA]

opcode The pattern of bits in an instruction that indicates the addressing mode.

open barrel terminal A terminal with an open conductor and/or insulation barrel which is designed to be crimped around a conductor or wire.[ARP914A-79]

open center In a hydraulic system, when no service is actuated the system is open to flow, completing the circuit through the control units back to reservoir. An open center system will normally employ a fixed displacement pump.[ARP 243B-65]

open circuit 1. An interruption in an electrical or hydraulic circuit, usually due to a failure or disconnection, which renders the circuit inoperable. 2. A nonrecirculating (once-through) system or process.

open circuit voltage The steady state of equilibrium potential of an electrode in absence of external current flow to or from the electrode.[NASA]

open-end protecting tube A tube extending from a physical boundary into the body of a medium to surround and protect a thermocouple yet allowing direct contact between the thermocouple's measuring junction and the medium.

open-flow nozzle See Kennison nozzle.

opening pressure The static inlet pressure which initiates a discharge.

open loop 1. A system operating without feedback, or with only partial feedback. See closed loop.[AIR1489-88] 2. Pertaining to a control system in which there is no self-correcting action for misses of the desired operational condition, as there is in a closed-loop system. See feed-forward control action.

open loop control 1. A control system which does not take any account of the error between the desired and actual values of the controlled variables. 2. An operation in which computer-evaluated control action is applied by an operator. See open loop and closed loop. 3. A system in which no comparison is made between the actual value and the desired value of a process variable.

open-loop control system A control system in which gain and power elements are used to provide an output in direct response to a command without feedback comparison.[AIR1916-88]

open-loop frequency response The frequency response between command input and control system feedback signal with no feedback loop closure.[AIR 1916-88]

open loop gain The equivalent output rate per unit of position error, or the product of the forward and feedback gains in in./sec/in.[ARP1281A-81]

open-loop numerical control A type of numerical control system in which the drive motor provides both actuation and measurement with no feedback to the control console.

openness of scale With respect to measuring instruments, the amount of change in a measured quantity that causes the pointer to move 1 mm (or in some instances, 1 in.) on the instrument scale.

open position A position that is 100 percent open.

open seal An impulse line filled with a seal fluid open to the process.

open system A system that complies with the requirements of the OSI reference model in its communication with other open systems.

open system interconnection (OSI) A connection between one communication system and another using a standard protocol.

open time See assembly time.

operable The state of being able to perform the intended function.[ARD50010-91]

operand The address of an instruction to be executed by the processor.

operating The condition when the galley and its equipment and inserts are used by flight crew for food and beverage service and cleanup.[AS1426-80]

operating characteristic curves The curve which shows the probability that a submitted lot with given mean life would meet the acceptability criterion on the basis of that sampling plan.[ARD 50010-91]

operating conditions, reference Conditions to which a device is subjected, not including the variable measured by the device. See also environmental condition.

operating control A control to start and stop the burner—must be in addition to the high limit control.

operating costs The price for operating a system exclusive of the cost of the system itself.[NASA]

operating level The nominal position or output at which a system or process operates. Typical examples are water level in a boiler, production rate of a manufacturing process, or acoustical output (volume) of a loudspeaker system.

operating pressure 1. The pressure available to a component or system during normal operation.[AIR1916-88] 2. The nominal pressure or pressure limits at which a system or process operates. 3. In a pneumatic or hydraulic system, the high and low values (range) of pressure that will produce full-range operation of an output device such as a motor operator, positioning relay or data-transmission device. See pressure, operating.

operating system(s) 1. Compute programs for expediting, controlling and/or recording computer use by other programs.[NASA] 2. An integrated collection of service routines for supervising the sequencing of programs by a computer. Operating systems may perform debugging, input-output, accounting, compilation, and storage assignment tasks. Synonymous with monitor system and executive system. 3. A group of programming systems operating under control of a data processing monitor program.

operating temperature The temperature at which a device may function on a continuous basis.[ARP914A-79]

operating time 1. The time during which a system or equipment is actually operating (in an "up" status). Operating time is usually divisible among several operating periods or conditions. These include "standby time", filament "on-time", pre-flight "checkout" time, flight time.[ARD50010-91] 2. That part of available time during which the hardware is operating and assumed to be yielding correct results. It includes development time, production time, and makeup time. Contrast with idle time.

operating weight Aircraft weight usually including the weight of the oil, fuel, crew, crew's baggage, emergency equipment, and payload.[ARP4107-88]

operation A set of tasks or processes,

usually performed at one location.

operational 1. Of, or pertaining to, the state of actual usage.[ARD50010-91]

operational availability The percent of time that a subsystem, line replaceable unit (LRU) or line replaceable module (LRM) is capable of satisfactorily performing in the operational environment. (AO does not depict ability of an item to continue to operate for a specific period of time. This characteristic is covered via the weapon system reliability term.)[ARD50010-91] 2. Describing a state of readiness for immediate use, as may be said of equipment or vehicles.

operational limit A pre-established reference for engine, engine component or sub-system operation.[ARP1587-81]

operationally ready Available and in condition for serving the functions for which it is designed.[AIR1916-88]

operational readiness A measure of the degree to which an item is in the operable and committable state at the start of the mission, when the mission is called for at unknown (random) point in time.[ARD50010-91]

operational suitability The degree to which a system can be satisfactorily operated in the field, with consideration being given to availability, safety, human factors, electromagnetic compatibility, logistic supportability, and training requirements.[ARD50010-91]

operational test See test.

operation analysis An evaluation process in industrial engineering that assesses design, materials, equipment, tools, working conditions, methods and inspection standards, usually for the purpose of improving production output or decreasing cost.

operation code The part of a computer instruction word which specified, in coded form, the operation to be performed.

operations analysis See operations research.

operations research The use of analytic methods adopted from mathematics for solving operational problems. The objective is to provide management with a more logical basis for making sound predictions and decisions. Among the common scientific techniques used in operations research are the following: linear programming, probability theory, information theory, game theory, Monte Carlo method, and queuing theory.

operator 1. Any corporate entity or person who causes or authorizes the operation of an aircraft, such as the owner, lessee, or bailee of an aircraft.[ARP 4107-88] 2. In a mishap sequence, the person in control of the aircraft at the point of the mishap. Other personnel involved in the mishap sequence of events are considered part of the operator's equipment.[ARP4107-88] 3. The person who initiates and monitors the operation of a computer. 4. The person who initiates and monitors the operation of a process. 5. A mathematical symbol which represents a mathematical process to be performed on an associated operand. 6. The portion of an instruction which tells the machine what to do.

operator command A statement to the control program, issued via a console device, which causes the control program to provide requested information, alter normal operations, initiate new operations, or terminate existing operations.

operator's console A device which enables the operator to communicate with the computer. It can be used to enter information into the computer, to request and display stored data, to actuate various preprogrammed command routines, etc. See also process engineer's console and programmer's console.

operator station 1. Serves as the interface between the operator and other devices on the data highway. 2. The operator can observe and control several devices.

opisometer An instrument incorporating a tracing wheel used for measuring the length of curved lines, such as those on a map.

Ophiuchi clouds Dense concentrations of interstellar gas near the stars Rho Ophiuchi and Zeta Ophiuchi.[NASA]

opposing load pressure Pressure acting to oppose operating pressure.[AIR 1916-88]

optical activity Ability to rotate the plane of vibration of polarized light to the right or left.[NASA]

optical ammeter An electrothermic instrument commonly employing a photoelectric cell and indicating device to determine the magnitude of electric current by measuring the light emitted by a lamp filament carrying the current; the instrument is calibrated by determining the amount of light emitted when known currents are carried by the same filament.

optical amplifier A type of amplifier in which an electric input signal is converted to light, amplified as light, then converted back to an electric output signal.

optical attenuation meter A device which measures the loss or attenuation of an optical fiber, fiber optic cable, or fiber optic system. Measurements are usually made in decibels.

optical bench A rigid horizontal bar or track for holding and supporting optical devices in fixed positions, yet allowing the positions to be changed or adjusted quickly and easily.

optical bistability A property of certain materials in which a nonlinear response is exhibited when under the influence of an external driving coherent light, thereby allowing these materials to behave like optical switches.[NASA]

optical character reader A scanning device that can recognize some typewritten characters.

optical comparator 1. Any comparator in which movement of a measuring plunger tilts a small mirror, which in turn reflects light in an optical system. 2. A type of comparator in which the silhouette of a part is projected on a graduated screen and the dimensions or contour evaluated from the image.

optical computers Computers which use light rather than electricity for all or part of their operation. They perform multiple tasks in parallel as opposed to electronic computers which would perform those tasks sequentially. Such increased processing capability makes them suited for aerospace problems which involve systems that have a large number of degrees of freedom, i.e., large space structures, pattern recognition activity, and robotics.[NASA]

optical countermeasures Equipment for exploiting the vulnerability of laser guided weapon systems.[NASA]

optical density A measurement of transmission equal to the base 10 logarithm of the reciprocal of transmittance. An object with optical density of zero is transparent; an optical density of one corresponds to 10% transmission.

optical depth See optical thickness.

optical disk A large electronic storage device that uses laser beam patterns to store data.

optical-emission spectrometry Measurement of the wavelength(s) and intensities of visible light emitted by a substance following stimulation.

optical encoder tachometer A type of instrument that combines a sensor (optical encoder) with a microprocessor to convert sensor impulses into a measurement of rotational velocity.

optical fiber Fine glass stands that transmit data using light signals.

optical filter A semitransparent device that selectively passes rays of light having predetermined wavelengths.

optical flat A transparent disk, usually made of fused quartz, having precisely parallel faces, one face polished for clear vision and the other face ground optically flat; when placed on a surface and illuminated under proper conditions, interference bands can be observed and used to either assess surface contour (relative flatness) or determine differences between a reference gage or gage block and a highly accurate part or inspection gage.

optical fluid-flow measurement Any method for measuring the density of a fluid in motion which depends on measuring refraction and phase shift among different rays of light as they pass through a flow field of varying density.

optical gage A gage that measures the image of an object without touching the object itself.

optical glass Glass free of imperfections, such as bubbles, chemical inhomogeneity or unmelted particles, which degrade its ability to transmit light.

optical grating 1. Diffraction grating usually employed with other appropriate optics to fabricate a monochromator. These gratings consist of a series of parallel grooves carefully and uniformly shaped in an optical surface either flat or concave depending upon the application. The number of grooves form and the shape of the grooves (its profile) determines in what region of the spectrum it is applicable. 2. Commonly referred to as a Ronchi grating. 3. A highly accurate device used in precision dimensional measurement which consists of a polished surface, commonly aluminum coating on a glass substrate, onto which close, equidistant and parallel grooves have been ruled; the distribution of grooves ranging from several hundred to many thousands of grooves per inch; gratings are used in conjunction with monochromatic light to produce interference patterns sometimes referred to as moiré patterns. They are used in optical testing as well as generating the dot matrix for picture reproduction from a photographic negative.

optical indicator An instrument which plots pressure variations as a function of time by making use of magnification in an optical system coupled with photographic recording.

optical mark reader Using light sensing, a device that reads marks made on special forms.

optical masers See lasers.

optical material Any material which is transparent to visible light or to x-ray, ultraviolet or infrared radiation.

optical paths Lines of sight or the paths followed by rays of light through optical systems.[NASA]

optical plastic Any plastics material which is transparent to light and can be used in optical devices and instruments to take advantage of physical or mechanical properties where the plastics material is superior to glass, or to take advantage of the lower cost of the plastics material.

optical pressure transducer Any of several devices that use optical methods to accurately measure the position of the sensitive element of the pressure transducer.

optical pyrometer An instrument that determines the temperature of an object by comparing its incandescent brightness with that of an electrically heated wire; the current through the wire is adjusted until the visual image of the wire blends into the image of the hot surface, and temperature is read directly from a calibrated dial attached to the current adjustment.

optical rangefinder An optical instrument for measuring distance, usually from the instrument's location to a target some distance away, by measuring the angle between rays of light from the target to separate windows on the range-finder body.

optical recording Making a record of an instrument reading by focusing a tiny beam of light on photosensitive paper, the position of the light along one axis of the resulting orthagonal plot being directly related to the value of the quantity being measured.

optical relay systems Systems using photocouplers in which the output device is a light sensitive switch that provides the same on and off operations as the contacts of a relay.[NASA]

optical rotation Rotation of the plane of polarization about the axis of a beam of polarized light.

optical scanners A light source and phototube combined as a single unit for scanning moving strips of paper or other materials in photoelectric side-register control systems.[NASA]

optical slant range The horizontal distance in a homogeneous atmosphere for which the attenuation is the same as that actually encountered along the true oblique path.[NASA]

optical spectrum See light (visible radiation).

optical storage disk A computer storage medium using lasers to form surface patterns that represent data. CD-ROM (Compact Disk Read-Only Memory) is an optical storage disk that stores data in digital form.

optical thickness Specifically, in calculations of the transfer of radiant energy, the mass of a given absorbing or emitting material lying in a vertical column of unit cross sectional area and extending between two specific levels. Also called optical depth.[NASA]

optical time domain reflectometer A device that sends a very short pulse of light down a fiber optic communication system and measures the time history of the pulse reflection. The reflection indicates fiber dispersion and discontinuities in the fiber path, such as breaks and connectors. The time it takes for the light pulse to travel to and from the discontinuity indicates how far it is from the test set.

optimal control Optimal control is when the values of the control parameters (transfer function) are chosen such that a selected performance index is maximized (or minimized).[AIR1823-86]

optimization 1. Theoretical analysis of a system, including all of the characteristics of the process, such as thermal lags, capacity of tanks or towers, length and size of pipes, etc. This analysis is made, sometimes with the aid of frequency response curves, to obtain the most desirable instrumentation and control. 2. Making a design, process or system as nearly perfect in function or effectiveness as possible. 3. Using a structured decision-making technique to select the best way of achieving a defined goal from a set of alternatives.

optimize 1. To establish control parameters so as to make control as effective as possible. 2. To rearrange the instructions or data in storage so that the program can be run in minimum time.

optimizing control See control, optimizing, steady-state optimization and dynamic optimization.

optimum altitude The altitude at a specific gross weight that results in the maximum miles per pound of fuel. Continuous optimum altitude can only be achieved by cruise climbing.[ARP1570-86]

optimum cruise altitude The level flight altitude adjusted by step climbs that results in the maximum miles per pound or fuel for the trip. The general procedure is to initiate cruise above optimum; then maintain that flight level until approximately an equal weight burn below optimum; then step climb above, and repeat. The steps may be 4000, 2000, 1000 or as dictated by ATC. The winds and trip distance or distance remaining would also influence the optimum cruise altitude.[ARP1570-86]

option approach An approach requested and conducted by a pilot which will result in either a touch-and-go, missed approach, low approach, stop-and-go, or full stop landing.[ARP4107-88]

option module Any additional device that expands a computer's capability.

optoelectronic amplifier An amplifier whose input and output signals and method of amplification may be either optical or electronic.

optogalvanic spectroscopy A method of obtaining absorption spectra of atomic and molecular species in flames and electrical discharges by measuring voltage and current changes upon laser irradiation.[NASA]

OR A logic operator having the property that if P is an expression, Q is an expression, R is an expression ..., then the OR of P, Q, R ... is true if at least one expression is true, false if all expressions are false. P OR Q is often represented by P + Q, PVQ. Synonymous with inclusive OR. Contrast with exclusive OR.

orbital elements A set of seven parameters defining the orbit of a body attracted by a central, inverse square force.[NASA]

orbital flight tests (shuttle) See space transportation system flights.

orbital lifetime The predicted lifetime of a satellite in orbit, usually based on such criteria as solar flux density, atmospheric density, the lessening of the eccentricity of elliptical orbits, or the gravitational effects of the sun or the moon.[NASA]

orbital motion See orbits.

orbital resonances (celestial mechanics) Systems of two or more satellites (including planets) that orbit the same primary and whose orbital mean motions are in a ratio of small whole numbers.[NASA]

orbital servicing The replenishing of propellants, pressurants, coolants, and the replacement of modules and experiments, during some phase of a spacecraft flight to extend the mission and lifetime, or change the payloads.[NASA]

orbital simulator See space simulator.

orbital transfer See transfer orbits.

orbital velocity The average velocity at which an earth satellite or other orbiting body travel around its primary. The velocity of such a body at any given point in its orbit, as in 'its orbital velocity at the apogee is less than at the perigee.'[NASA]

orbits The paths or bodies or particles around the influence of a gravitational or other force.[NASA]

orbit spectrum utilization Telecommunication techniques in spectrum conservation for reducing user costs.[NASA]

orbit transfer vehicles Concept of propulsive (velocity producing) rockets or stages for use with crew transfer modules, manned sortie modules, or other payloads.[NASA]

ordered structure The orderly or periodic arrangement of solute atoms on the lattice sites of the solvent.[AS1814-90]

organic Designating or composed of matter originating in plant or animal life or composed of chemicals of hydrocarbon origin, either natural or synthetic.[ARP1931-87]

organic charge transfer salts Organic compounds exhibiting temperature-dependent electrical magnetic, and heat transfer properties.[NASA]

organic fiber A fiber derived or composed of matter originating in plant or animal life, or composed of chemicals of hydrocarbon origin, either natural or synthetic.[ARP1931-87]

organic matter Compounds containing carbon often derived from living organisms.

organic peroxides Organic compounds containing radical groups combined with oxides in which two atoms of oxygen are linked together, e.g., diethyl peroxide.[NASA]

organic solids Solid materials composed of organic materials.[NASA]

organizational maintenance Maintenance which is the responsibility of and performed by a using organization on its assigned equipment.[ARD50010-91]

orient To place an instrument, particularly one for making optical measurements, so that its physical axis is aligned with a specific direction or reference line.

orientation Alignment with a specific direction or reference line.

orientation alpha A nonuniform alpha structure that results from colonies or domains of platelets or wormy alpha lying at different angles, and having no significance to crystallographic orientation, such that different areas exhibit different aspect ratios and alpha grain outlines.[AS1814-90]

orifice 1. A short fluid passage which produces a substantial reduction in flow by virtue of its cross-sectional area. A true orifice has zero length.[ARP243B-65] 2. An opening mouth; vent; in a shock absorber, a hole through which oil is forced in the energy dissipation process.[AIR1489-88] 3. The opening from the whirling chamber of a mechanical atomizer or the mixing chamber of a steam atomizer through which the liquid fuel is discharged. 4. A calibrated opening in a plate, inserted in a gas stream for measuring velocity of flow.

orifice coefficient A numerical constant defining the characteristics of a specific orifice.[AIR1489-88]

orifice control The method or the process by which flow of oil through an orifice is controlled. i.e., variation of orifice area by metering pin, variation of orifice shape, etc.[AIR1489-88]

orifice fitting 1. A specially designed orifice plate holding device.

orifice flange taps The 1/2 in. or 3/4 in. pipe taps in the edge of an orifice flange union.

orifice flange union Two unique flanges used to hold an orifice plate primary element with specific design dimensions established by the American Gas Association.

orifice meter A general term used to describe any recording differential pressure measuring instrument.

orifice mixer A piece of equipment for mixing two or more liquids by simultaneously directing them, under pressure, through a constriction where the resulting turbulence blends them together.

orifice plate A disc or platelike member, with a sharp-edged hole in it, used in a pipe to measure flow or reduce static pressure.

orifice run The differential pressure producing arrangement consisting of selected pipe, orifice flange union and orifice plate. An orifice run has rigid specifications defined by the American Gas Association.

orifice-type variable-area flowmeter A flow-measurement device consisting of a tube section containing an orifice and a guided conically tapered float that rides within the orifice; flow of a fluid through the meter positions the float in relation to flow rate, with float position being determined magnetically or by other indirect means.

O ring A toroidal sealing ring made of synthetic rubber or similar material. The cross section through the torus is usually round or oval, but may be rectangular or some other shape.

Orion (radio interferometry network) An operational radio interferometry observational network.[NASA]

Orion nebula An H 11 region about 500 pc distant and barely visible to the naked eye in the center of Orion's sword. [NASA]

orometer A barometer for measuring

elevation above sea level.

orsat A gas-analysis apparatus in which certain gaseous constituents are measured by absorption in separate chemical solutions.

orthicon A camera tube that utilizes a low-velocity electron beam to scan an image stored electrically on a photoactive mosaic panel.

orthogonality The degree of perpendicularity between the mutually perpendicular trace axes on CRT screens. [ARP1782-89]

orthometric correction A systematic correction that must be applied to a measured difference in elevation to compensate for the fact that level surfaces at different elevations are not exactly parallel.

OS See operating system.

oscillating-piston flowmeter A flow measurement device similar to a nutating-disk flowmeter but in which motion of the piston takes place in one plane only; rotational speed of the piston is directly related to the volume of fluid passing through the meter.

oscillation(s) 1. Fluctuations or vibrations on each side of a mean value or position. One oscillation is half an oscillatory cycle, consisting of a fluctuation or vibration in one direction; half a vibration. The variation, usually with time, of the magnitude of a quantity with respect to a specified reference when the magnitude is alternately greater and smaller than the reference. [NASA] 2. Fluctuation around the set point.

oscillator(s) A nonrotating device for producing alternating current; the output frequency is determined by characteristics of the device. In some cases the frequency is fixed, but in others it can be varied.

oscillator crystal A piezoelectric crystal device used chiefly to determine the frequency of an oscillator.

oscillator strengths A quantum mechanical analog of the number of dispersion electrons having a given natural frequency in an atom, used in an equation for the absorption coefficient of a spectral line.[NASA]

oscillatory circuit A circuit that produces a periodically reversing current when energized by a direct-current voltage; the circuit contains R, L and C elements, which may be varied to change the characteristics of the resultant a-c output.

oscillogram The permanent record created by an oscillograph. Alternatively, a permanent record of the trace on an oscilloscope, such as might be recorded photographically.

oscillograph A device for determining waveform by plotting instantaneous values of a quantity such as voltage as a function of time.

oscilloscope Instrument for producing visual representations of oscillations or changes in an electric current.[NASA] 2. A CRT device that can display instantaneous values of alternating-current voltages or currents with respect to time or with respect to other alternating-current voltages or currents; it also can be used to display instantaneous values of other quantities that vary rapidly with time (not necessarily oscillatory values) and which can be converted to suitable electrical signals by means of a transducer; the display is a graphical representation of electrical signals produced by varying the position of the focused spot where an electron beam strikes the fluorescent coating on the inside surface of the CRT face.

OSI See open system interconnection.

OSI reference model A seven layered model of communications networks defined by ISO. The seven layers are: Layer 7—Application: provides the interface for application to access the OSI environment. Layer 6—Presentation:

provides for data conversion to preserve the meaning of the data. Layer 5—Session: provides user-to-user connections. Layer 4—Transport: provides end-to-end reliability. Layer 3—Network: provides routing of data through the network. Layer 2—Data Link: provides link access control and reliability. Layer 1—Physical: provides an interface to the physical medium.

otolith organs Structures of the inner ear (utricle and saccule) which respond to linear acceleration and tilting.[NASA]

OTV See orbit transfer vehicles.

ounce A U.S. unit of weight; one ounce (avoirdupois) equals 1/16 pound, and is used for most commercial products; one ounce (troy) equals 1/12 pound, and is used for precious metals.

outdoor area See area, outdoor.

outer compass locator (LOM) See compass locator.

outer fix A fix in the destination terminal area, other than the approach fix, to which aircraft are normally cleared by an air route control center or a terminal area traffic control facility, and from which aircraft are cleared to the approach fix or final approach course. [ARP4107-88]

outer marker (OM) A marker beacon at or near the glide slope intercept altitude of an Instrument Landing System (ILS) approach. It is keyed to transmit a signal consisting of two dashes per second, which is received visually or aurally, or both, (on a 400 Hz tone) by compatible airborne equipment. The OM is normally located between four and seven miles from the runway threshold on the extended centerline of the runway. It serves as an aid to the flight personnel in that it marks a particular distance from the approach end of the runway.[ARP4107-88]

outgassing 1. The dissipation of gas from a dielectric evidencing decomposition.[ARP1931-87] 2. The evolution of gas from a material in a vacuum.[NASA] 3. The release of adsorbed or occluded gases and water vapor, usually during evacuation or subsequent heating of an evacuated chamber.

outlet pressure Pressure at the outlet of a component.[AIR1916-88]

outlet temperature Fluid temperature at the plane of the outlet port.[AIR1916-88]

outliers (statistics) In sets of data values so far removed from other values in the distribution that their presence cannot be attributed to the random combination of change causes.[NASA]

out-of-round A dimensional condition where diameters taken in different directions across a nominally circular object are unequal, the difference between them being the amount of out-of-roundness.

output 1. The controlled variable resulting from activity of the control system. [AIR1916-88] 2. The yield or product of an activity furnished by man, machine, or system.[NASA] 3. The information transferred from the internal storage of a computer to secondary or external storage or to any device outside of the computer. 4. The routines which direct 2. 5. The device or collective set of devices necessary for 2. 6. To transfer from internal storage on to external media.

output area An area of storage reserved for output.

output block 1. A block of computer words considered as a unit and intended or destined to be transferred from an internal storage medium to an external destination. 2. A section of internal storage reserved for storing data which are to be transferred out of the computer. Synonymous with output area. 3. A block used as an output buffer. See buffer.

output device The part of a machine which translates the electrical impulses representing data processed by the machine into permanent results such as printed forms, punched cards, magnetic writing on tape or into control signals for a process.

output indicator A device connected to a radio receiver to indicate variations in output signal without indicating a specific signal value; usually used for alignment or tuning.

output shaft The starter shaft which connects the starter to the engine accessory gearbox.[ARP906A-91]

output signal A signal delivered by a device, element, or system. See also signal, output.

output variable A variable delivered by a control algorithm, e.g., the signal going to a steam valve in a temperature control loop. See controlled variable.

outrigger gear See gear, outrigger.

outriggers Stabilizer devices used to improve the stability of vehicles, extending outside the normal envelope of that vehicle.[ARP1328-74]

outside caliper A caliper used to measure distances across two external opposing surfaces.

outside diameter The outer dimension of a circular member such as a rod, pipe or tube.

oval-shaped gear flowmeter A type of positive-displacement flowmeter that operates by trapping a precise volume of fluid between an oval, toothed rotor and the meter housing as the rotor revolves in mesh with a second rotor; volume flow of an incompressible fluid is indicated directly by determining rotor speed.

oven A heated enclosure for baking, drying or heating.

oven dry A term often used by papermakers to indicate paper from which all moisture has been removed by artificial evaporation using heat. See bone dry.

overcast (meteorology) In surface aviation weather observations, descriptive of sky cover of 1.0 (95 percent or more) when at least a portion of this amount is attributable to clouds or obscuring phenomena aloft; that is, when the total sky cover is not due entirely to surface-based obscuring phenomena; a predominant opaque cover is implied.[ARP4107-88]

overcharging The continued passage of electrical "charging" current after the cell plates are charged. It results in the generation of hydrogen and oxygen gases. The onset of overcharge is manifested in the form of a transition from negligible gassing to complete conversion of all (over) charge current to gas.[AS8033-88]

overconfidence See confidence.

overcontrol (by a pilot) To displace or move an aircraft's controls more than is necessary for the desired performance.[ARP4107-88]

overcurrent Any current exceeding the rated current of the protective device (exceeding the maximum ultimate trip current for circuit breakers). This includes both overload and short-circuit currents.[ARP1199A-90]

overdamped See damping.

overdue aircraft (IFR) An aircraft operating in accordance with IFR is considered missing or overdue when communications with such aircraft or radar identification cannot be established within 30 min after it: (a) fails to report over an ATC specified reporting point or over a compulsory reporting point along the route of flight, whichever is earlier; or (b) becomes overdue at the point of intended landing.[ARP4107-88]

overdue aircraft (VFR) A VFR aircraft is considered overdue if the flight plan it is operating on is not closed within one half hour after the aircraft's estimated time of arrival (ETA) at its destination. If a flight plan is not properly closed, search and rescue procedures will be started.[ARP4107-88]

overflow 1. The condition which arises

when the result of an arithmetic operation exceeds the capacity of the storage space allotted in a digital computer. 2. The digit arising from this condition if a mechanical or programmed indicator is included, otherwise the digit may be lost.

overflow pipe A pipe with its open end protruding above the liquid level in a tank; it limits the height of liquid in the tank by carrying away any liquid entering the open end, usually to a drain or sewage system.

overfractionation Operation of a distillation column to produce a purer product than required.

overhaul The process of disassembly sufficient to inspect all the operating components and the basic end article. It includes repair, replacement, or servicing as necessary, followed by reassembly and bench check/flight test. Upon completion of the overhaul process, the component/end article will be capable of performing its intended service life/service tour.[ARD50010-91]

overhaul, partial The reconditioning of a subassembly.[ARD50010-91]

overhaul, the controlled The reconditioning in accordance with a plan under which the time histories of individual items are monitored. The monitoring system is used to schedule the removal of items before they exceed a specified time limit.[ARD50010-91]

Overhauser effect In atomic physics, a radio frequency field applied to a substance in an external magnetic field, whose nuclei have spin 1/2 and which has unpaired electrons at the electron spin resonance frequency. This results in polarization of the nuclei as great as if the nuclei had the much larger electron magnetic moment.[NASA]

overheat To raise the temperature above a desired or safe limit; in metal heat treating, to reach a temperature that results in degraded mechanical or physical properties.

overlap A condition in which one portion of an item lays upon another portion of that same or some other item, such a "tape overlap".[ARP1931-87]

overlay 1. The technique of repeatedly using the same blocks of internal storage during different stages of a problem. When one routine is no longer needed in storage, another routine can replace all or part of it.

overlay pavement A pavement which is layed on top of an existing pavement to increase the load carrying capability or repair or seal the existing pavement. For instance, asphaltic pavement layed on top of concrete or old asphalt. [AIR1489-88]

overload capacity The force weight, power, pressure or other capacity factor, usually higher than the rated capacity, beyond which permanent damage occurs to a device or structure.

overload current An overcurrent in excess of the current rating. The overload range is considered to be greater than the rated current up to approximately ten times rated current.[ARP 1199A-90]

override To manually apply a control force which exceeds the authority of an automatic control system or to manually intervene in an otherwise automatic sequence.[ARP4107-88]

override control 1. Generally, two control loops connected to a common final control element—one control loop being normally in control with the second being switched in by some logic element when an abnormal condition occurs so that constant control is maintained. 2. A technique in which more than one controller manipulates a final control element. The technique is used when constraint control is important.

overrun 1. Refers to the relative speeds of the engine and starter after the start cycle has been completed. Power transmission has changed from starter-to-engine, to, engine-to-starter.[ARP906A-91] 2. Extended clear area located at the end of a runway for the purpose of run out of an aircraft in an abort or emergency situation.[AIR1489-88]

overrun barrier See barrier overrun.

overshoot 1. The increment by which the output exceeds the final value when responding to a step command, usually expressed as a percentage of the output.[AIR1916-88] 2. Movement of the aircraft whereby it coasts past the commanded reference altitude, attitude, or flight path before settling out.[ARP419-57] 3. A transient response to a step change in an input signal which exceeds the normal or expected steady-state response. 4. The maximum difference between the transient response and the steady-state response. See also transient overshoot.

overspeed switch A speed sensing device, usually located on the starter side of the engaging mechanism, used to terminate starter operation at the predetermined overspeed.[ARP906A-91]

overtones See harmonics.

overvoltage protection A protective device that interrupts power or reduces voltage supplied to an operating device in the event the incoming voltage exceeds a preset value.

overwriting In data processing, the elimination of data by writing new data over it.

Owen bridge A type of a-c bridge circuit in which one leg contains a fixed capacitor, the opposite leg contains an unknown inductance and resistance, the third leg contains a fixed resistor and the fourth leg contains a variable resistor and a variable capacitor; this type of bridge is especially useful for measuring wide ranges of inductances using reasonable ranges of standard capacitances, and can be used to measure permeability or core loss.

oxazole Compounds that contain a five-membered heterocyclic ring containing one nitrogen and one oxygen atom.[NASA]

oxidation Chemical combination with oxygen.

oxidation-reduction reactions An oxidizing chemical change, where an element's positive valence is increased (electron loss), accompanied by a simultaneous reduction of an associated element (electron gain).[NASA]

oxidizer(s) Specifically, substances (not necessarily containing oxygen) that support the combustion of a fuel or propellant.[NASA]

oxidizing atmosphere An atmosphere which tends to promote the oxidation of immersed materials.

oxygen 17 An isotope of oxygen.[NASA]

oxygen attack Corrosion or pitting in a boiler caused by oxygen.

oxygen deficiency See hypoxia.

oxygen toxicity See hyperoxia.

oxynitrides Base for a broad field of nitrogen ceramics utilizing silicon, aluminum, and other elements to produce high temperature refractory materials.[NASA]

ozone layer See ozonosphere.[NASA]

ozonosphere The general stratum of the upper atmosphere in which there is an appreciable ozone concentration and in which ozone plays an important part in the radiation balance of the atmosphere. This region lies roughly between 10 and 50 kilometers, with maximum ozone concentration at about 20 to 25 kilometers.[NASA]

p See pressure.

P See poise.

P&ID See piping and instrumentation drawing.

Pa See pascal.

PACE See physics and chemistry experiment in space.

pachymeter An instrument used to measure the thickness of material such as paper.

pack 1. In data processing, a method to condense data in order to increase storage capacity. 2. A removable disk.

package(s) 1. The smallest enclosure into which an item(s) is placed for protection during storage or shipment.[AS 478G-79] 2. Any assemblies or apparatus, complete in themselves or practically so, identifiable as units and readily available for use or installation.[NASA]

packaged boiler A packaged steam or hot water firetube boiler is defined as a modified Scotch unit engineered, built, fire tested before shipment, with material, workmanship, and performance warranteed by manufacturer as stated in the manufacturer's standard conditions of sale. Components include, but are not limited to, burner, boiler and controls.

packaged steam generator See packaged boiler.

packed column A distillation column filled with packing (commonly Raschig rings) to mix the descending liquid with the ascending vapors. Packing is often used instead of trays in columns for certain applications (such as gas adsorption) or very-low-pressure drop systems.

packed decimal A method of representing a decimal number by storing a pair of decimal digits in one eight-bit byte, which takes advantage of the fact that the numbers zero through nine can be represented by four bits.

packet Block of data assembled with other control bits as a basic chunk of information to be transmitted.

packets (communication) Digital data messages which are almost always preceded by headers (containing address information and other control characters) and followed by control characters which signify the end of a message. [NASA]

packet switching Switching circuit system for multiple access time division date transmission.[NASA]

packet switching system (PSS) In a wide area network, a method of sending data between computers.

packet transmission Transmission of bursts of digital data.[NASA]

packing 1. The flexible sealing element in a fluid seal which is subject to sliding motion.[ARP243B-65] 2. A method of sealing a mechanical joint in a fluid system. A material such as oakum or treated asbestos is compressed into the sealing area (known as a packing box or stuffing box) by a threaded seal ring. 3. In data processing, the compression of data to save storage space.

packing density The number of units of useful information contained within a given linear dimension, usually expressed in units per inch; for example, the number of binary digit magnetic pulses or numbers of characters stored on tape or drum per-linear-inch on a single track by a single head.

packing gland See packing follower.

packing nut See packing flange.

pad 1. A pad is larger than a boss and is attached to a pressure vessel to reinforce an opening. 2. A fixed-value attenuator.

paddle-wheel level detector A device for detecting the presence or absence of bulk solids at the device location; it consists of a motor that slowly rotates a paddle in the absence of material, and rotates itself against a momentary switch when material is at or above the paddle location.

page A block of information that can be stored as a complete unit in the computer memory.

paired or grouped wheel control The skid control system may be configured to control brake pressure to two or more rakes when any one of the braked wheels require skid control. Pairing or grouping of wheels is a function of landing gear arrangement, performance required, and aircraft directional stability required.[AS483A-88]

paired wheel control Control of two brakes/wheels utilizing one hydraulic antiskid valve and one control circuit. [AIR1489-88]

paleobiology The study of life and organisms that existed in the geologic past.[NASA]

paleoclimatology The study of climates in the geologic past, involving fossil, glacial, isotropic, or other data.[NASA]

palette In data processing, the range of display colors that will show on a screen.

PAM The process (or the results of the process) in which a series of pulses is generated having amplitudes proportional to the measured signal samples.

PAM/FM Frequency modulation of a carrier by pulse amplitude modulated information.

PAM/FM/FM Frequency modulation of a carrier by subcarriers that are frequency modulated by pulse amplitude modulated (PAM) information.

pan The international radio-telephony urgency signal. When repeated three times, it indicates uncertainty or alert, followed by nature of urgency.[ARP 4107-88]

panel 1. The structure or surface to which a device is mounted.[ARP914A-79] 2. A sheet of material held in a frame. 3. A section of an equipment cabinet or enclosure, or a metallic or nonmetallic sheet, on which operating controls, dials, instruments or subassemblies of an electronic device or other equipment are mounted.

panel method (fluid dynamics) Technique for analyzing and predicting the properties and characteristics of fluid flow, sometimes called the finite element method.[NASA]

panspermia The theory that holds that reproductive bodies of living organisms exist throughout the universe and develop wherever the environment is favorable.[NASA]

PANT program The passive nosetip technology (PANT) program is an investigation of flow phenomena over reentry vehicle nosetips by the Air Force.[NASA]

paper (material) Felted or matted sheets of cellulose fibers, bonded together and used for various purposes, especially involving printed language, artwork or diagrams.[NASA]

paper machine A synchronized series of mechanical devices such as screens and heated rolls that transforms a dilute suspension of cellulose fibers (digested pulp) into a dry sheet of paper.

paper tape punch A hardware device that punches digital data into a paper tape.

paper tape reader (PTR) A hardware device for accepting punched hole paper tapes and transmitting their information content to the computer in digital form.

PAR Precision Approach Radar.

parabolas Open curves where all points of which are equidistant from a fixed point called the focus, and a straight line. The limiting case occurs when the point is on the line, in which case the parabola becomes a straight line.[NASA]

parabolic bodies Surfaces of revolution generated by revolving sections of parabolas about their major axis.[NASA]

parabolic reflectors Reflecting surfaces having the cross section along the axis in the shape of a parabola. Parallel rays striking the reflector are brought to a focus at a point, or if the source of the rays is placed at the focus, the reflected rays are parallel.[NASA]

parabolic velocity See escape velocity.

paraboloids See parabolic bodies.

parabrake See drag chute.

paracone A system for recovering men and objects from great distances above the earth's surface and landing them safely onto the earth.[NASA]

paragraph text Grammatically punctuated word strings which express a thought or point.[AS4159-88]

parallax The apparent differences in spatial relations when objects in different planes are viewed from different directions; in making instrument readings, for instance, parallax will cause an error in the observed value unless the observer's eye is directly in line with the pointer.

parallelIn data transfer operations, a procedure that handles a multiple-bit code, working with all bits simultaneously, usually one word at a time.

parallel actuators Two or more actuators arranged in parallel to drive a single output or load. Usually, parallel actuators are physically separated, each with its own output connection, and are tied together by the load in a force or torque-summing fashion. Sometimes, referred to as side-by-side actuators.[ARP1181A-85]

parallel computer 1. A computer having multiple arithmetic or logic units that are used to accomplish parallel operations or parallel processing. Contrast with serial computer. 2. Historically, a computer, some specified characteristic of which is parallel, for example, a computer that manipulates all bits of a word in parallel.

parallel elements In an electric circuit, two or more two-terminal elements connected between the same pair of nodes.

parallel I/O The simultaneous input/output of all the bits. Eight lines are required for the simultaneous transmission of eight bits.

parallel linkage A linkage mechanism that amplifies reciprocating motion. A parallel linkage depending on the geometry of the drive crank, driven crank and connecting link can amplify, attenuate as well as characterize the relationship of output driven crank to the input driven crank.

parallel offset route A parallel track to the left or right of the designated or established airway/route. Normally associated with Area Navigation (RNAV) operations. Aircraft with RNAV equipment may ask for courses that parallel the airway routes so that they may avoid congestion on the established airway.[ARP4107-88]

parallel operation The performance of several actions, usually of a similar nature, simultaneously through provision of individual similar or identical devices for each such action. Particularly flow or processing of information. Parallel operation is performed to save time.

parallel output To send data simultaneously between interconnecting devices.

parallel processing Pertaining to the concurrent or simultaneous execution of two or more processes in multiple devices such as channels or processing units. Contrast with serial processing.

parallels Spacers or pressure pods used in molding equipment to regulate height and prevent mold parts from being crushed.

parallel search storage A storage device in which one or more parts of all storage locations are queried simultaneously. Contrast with associative storage.

parallel servo A servo located in a control system so that the servo output drives in parallel with the major input. This arrangement usually is used with actuators which perform an alternate function to that of the pilot. The parallel servo output will drive both the pilot controls and the flight control system. [ARP1181A-85] [AIR1916-88]

parallel splice A device for joining two or more conductors in which the conductors lie parallel and adjacent. See also lap joint and splice.[ARP914A-79]

parallel task execution Concurrent execution of two or more programs. Also, simultaneous execution of one program and I/O.

parallel transfer A method of data transfer in which the characters of an element of information are transferred simultaneously over a set of paths.

parallel transmission A method of transmitting digitally coded data in which a separate channel is used to transmit each bit making up a coded word. See also serial transmission.

parameter 1. A measurable or calculated quantity which varies over a set of values.[ARP1587-81] 2. A quantity in a subroutine whose value specifies or partly specifies the process to be performed; it may be given different values when the subroutine is used in different main routines, or in different parts of one main routine, but usually remains unchanged throughout any one such use. 3. A quantity used in a generator to specify machine configuration, designate subroutines to be included, or otherwise to describe the desired routine to be generated. 4. A constant or a variable in mathematics that remains constant during some calculation. 5. A definable characteristic of an item, device, or system. See also measurand.

parameter identification The estimation of the unknown parameters of models of physical plants or processes from their dynamic response.[NASA]

parameterize To set up for variable execution depending on run-time parameter.

parameters See independent variables.

parametric analysis Analysis of the impact on circuit performance of changes in the individual parameters, such as component values, process parameters, temperature, etc.

parametric oscillator A nonlinear device which, when pumped by light from a laser, can generate tunable output. The beam produced by the parametric oscillator relies on oscillation within the nonlinear material.

parametric variation A change in system properties—magnification, resistance or area, for example—which may affect performance of a control system that incorporates a feedback loop.

parasitic oscillations Unintended self-sustaining oscillations or transient pulsations.

parity 1. A symmetry property of a wave function.[NASA] 2. A code that is used to uncover data errors by making the sum of the "1" bits in a data unit either an odd or even number.

parity bit A binary digit appended to a group of bits to make the sum of all the bits always odd (odd parity) or always even (even parity); used to verify data storage and transmission.

parity check A check that tests whether the number of ones or zeroes in an array of binary digits is odd or even. Synonymous with odd-even check. See also check, parity.

park A computer routine that will disengage a hard disk as protection from possible damage.

parking brake The system and/or apparatus for applying brakes for parking the aircraft for an extended length of time. Usually utilizes the same brake assemblies as for normal braking but with a different control and power source.[AIR1489-88]

parking valve The hydraulic valve which applies or blocks pressure to or from the brake assembly to maintain pressure for parking the aircraft.[AIR 1489-88]

Parr turbidimeter A device for determining the cloudiness of a liquid by measuring the depth of the turbid suspension necessary to extinguish the image of a lamp filament of fixed intensity.

parse To break a command string into its elemental components for the purpose of interpretation.

Parshall flume A venturi-type device for measuring flow in an open channel at flow rates up to 1.5-billion gal/day (5.7-million m^3/day); it consists of a converging upstream section, a downward sloping throat and an upward sloping discharge section, and may be made of any suitable structural material, usually concrete.

parsing algorithms Computer routines for the syntactic and/or semantic analysis and restructuring of natural language instructions or data for internal processing.[NASA]

part An element of an assembly or subassembly that normally is of little use by itself and cannot be disassembled further for repair or maintenance.

partial mission capable Systems and equipment shall be considered PMC when they are safely usable and can perform one or more but not all assigned missions because one or more of their mission essential subsystems are inoperative for maintenance or supply reasons. This status code is not used for equipment with a single mission, such as ground launch missile systems and Army and Marine Corps round equipment. The Military Services may further subdivide PMC into maintenance and supply categories.[ARD 50010-91]

partial node A point, line or plane in a standing wave field where some attribute of the wave has a nonzero minimum value.

partial obscuration A designation of sky cover when (a) part of the sky (0.1 to 0.9) is completely hidden by surface-based phenomena, or (b) the sky is hidden by surfaced-based phenomena but the vertical visibility is not otherwise restricted.[ARP4107-88]

partial-panel flight Instrument flight in which one or more of the usual cockpit flight instruments are inoperative or missing.[ARP4107-88]

partial pressure 1. The pressure exerted by a designated component or components of a gaseous mixture.[NASA] 2. The portion of total pressure in a closed system containing a gas mixture that is due to a single element or compound.

partial pressure of air (P_a) The partial pressure of the air in equilibrium with the liquid. The absolute static pressure of the liquid and vapor phase in psia minus true vapor pressure of fuel at t° F. P_a - P - P_{TVP}.[AIR1326-91]

particle accelerator 1. Any of several different types of devices for imparting motion to charged atomic particles. 2. Specifically, devices for imparting large kinetic energy to charged particles, such as electrons, protons, deuterons, and helium ions.[NASA]

particle concentration The number of individual particles per unit volume of liquid.[ARP1192-87]

particle counters See radiation counters.

particle detectors See radiation counters.

particle flux See flux (rate).

particle laden jets Fluid, mainly issuing from a nozzle, that are turbulent and contain dispersed particles.[NASA]

particle precipitation The precipitation of particles other than electrons and protons.[NASA]

particles Elementary subatomic particles such as protons, electrons or neutrons. Very small pieces of matter. In celestial mechanics, hypothetical entities which respond to gravitational forces but which exert no appreciable gravitational force on other bodies, thus simplifying orbital computations. [NASA]

particle size A measure of dust size, expressed in microns or per cent passing through a standard mesh screen.

parting agent Also known as release agent. Material used to prevent sealant from sticking to a surface.[AIR4069-90]

parting tool See cutoff tool.

partition A contiguous area of computer memory within which tasks are loaded and executed.

part program In numerical control, an ordered set of instructions in a language and format required to cause operations to be effected under automatic control which then is either written in the form of a machine program on an input medium or stored as input data for processing in a computer to obtain a machine program.

parts See components.

parts management The control of life-limited engine parts to allow determination of the life usage, and hence the life remaining.[AIR1872-88]

parts per million (ppm) The unit volume concentration of a gas per million unit volumes of the gas mixture of which it is a part.[AIR1533-91] [ARP1256-90]

parts per million carbon (ppmC) The mole fraction of hydrocarbon multiplied by 10^6 measured on a (C_1H_n) equivalence basis. Thus, 1 ppm of methane is indicated as 1 ppmC. To convert ppm concentration of any hydrocarbon to an equivalent ppmC value, multiply ppm concentration by the number of carbon atoms per molecule of the gas. For example, 1 ppm propane translates as 3 ppmC hydrocarbon; 1 ppm hexane as 6 ppmC hydrocarbon.[AIR1533-91] [ARP 1256-90]

pascal Metric unit for pressure or stress.

PASCAL 1. High order computer programming language developed by Niklaus Wirth originally as an educational tool to foster structured programming. [NASA] 2. A programming language developed by Nicholas Wirth and named for the mathematician Blaise Pascal.

pass 1. A single circuit through a process, such as gases through a boiler, metal between forging rolls, or a welding electrode along a joint. 2. In data processing, the single execution of a loop. 3. The shaped open space between rolls in a metal-rolling stand. 4. A confined passageway, containing heating surface, through which a fluid flows in essentially one direction. 5. A single circuit of an orbiting satellite around the earth. 6. A transit of a metal-cutting tool across the surface of a workpiece with a single tool setting.

pass, aircraft An aircraft passing a given station or a runway or taxiway; a takeoff and a landing constitute one pass for determining ground flotation capabilities.[AIR1489-88]

pass, gear One movement of a gear assembly past a specific point on the runway or taxiway under consideration. (One aircraft pass may include more than one gear pass).[AIR1489-88]

passivating A process for the treatment of stainless steel in which the material is subjected to the action of an oxidizing solution which augments and strengthens the normal protective oxide film providing added resistance to corrosive attack.

passivation of metal The chemical treatment of a metal to improve its resistance to corrosion.

passive A general class of devices which operate on the signal power alone.[ARP 993A-69]

passive AND gate An electronic or fluidic device which generates an output signal only when both of two control signals appear simultaneously.

passive failure The type of failure wherein the failed device or system has no effect on the operational performance of a fault-tolerant system even when it is commanded to function. Usually associated with standby or inactive features of a fault-tolerant system.[ARP 1181A-85]

passive metal A metal which has a natural or artificially produced surface film that makes it resistant to electrochemical corrosion.

passive nosetip technology See PANT program.

passive paralleling The simplest and most common type of redundancy wherein two parallel functional devices are utilized such that if one fails the second is still available. This approach is limited to the more simple elements of the control system which can only fail passively, such as springs and linkages. When failure of one element occurs, there may be a change in performance or capability.[ARP1181A-85][AIR1916-88]

passive transducer A transducer that produces output waves without any direct interaction with the source of power that produces the actuating waves.

pass-through current See let-through current.

password In data processing, a series of characters needed to access a computer known only to those authorized to access the data stored in the computer or diskettes.

paste (consistency) Mixtures with characteristic soft or plastic consistencies. [NASA]

paste solder Finely divided solder alloy combined with a semisolid flux.

pasteurizing column A column that purges either a lighter-than-light key impurity through a purge stream at the top of the column or heavier-than-heavy key impurity through a purge stream at the bottom of the column.

patch 1. A section of coding inserted into a routine to correct a mistake or alter the routine. Often it is not inserted into the actual sequence of the routine being corrected, but placed somewhere else, with an exit to the patch and a return to the routine provided. 2. To insert corrected coding.

patent defect An inherent or induced weakness which can be detected by inspection, functional test, or other defined means without the need for stress screens.[ARD50010-91]

path In MS-DOS, the instructions to the computer as to how to locate a particular file.

path loss (radio) The signal loss between transmitting and receiving antennas.

pathogens Disease-producing agents, usually referring to living organisms. [NASA]

path profile (PROF) An imaginary earth referenced line in space connecting successive 3-dimensional points through which flight is desired and/or controlled (altitude, latitude and longitude). The flight path angle may be varying or constant, resulting in a curved or straight vertical profile. [ARP1570-86]

patterned screen An electronic optical display that consists of discrete elements such as phosphor dots or stripes on a shadowmask CRT.[ARP1782-89]

pattern failures The occurrence of two or more failures of the same part in identical or equivalent applications when the failures are caused by the same basic failure mechanism and the failures occur at a rate which is inconsistent with the parts predicted failure rate.[ARD 50010-91]

pattern flow The paths of fluid flow connecting various ports in a given valve position.[ARP243B-65]

pattern recognition 1. The identification of shapes, forms and configurations by automatic means.[NASA] 2. The recognition of shapes or other patterns by a machine system.

patriot missile Surface to air, antiaircraft missile.[NASA]

pavement, flexible Pavement of flexible construction such as asphaltic pavement.[AIR1489-88]

pavement, rigid Pavement of rigid construction such as concrete or steel reinforced concrete.[AIR1489-88]

PAW See plasma arc welding.

payload The revenue-producing or other useful load carried by an aircraft; also, the bomb load.[ARP4107-88]

payload assist module Rocket vehicle with a spinning solid-propellant motor to attain injection velocity to place payload into intended orbits from the parking orbits of the STS.[NASA]

payload control Execution of events involved in operating the payload and supporting systems.[NASA]

payload delivery (STS) The transport of payloads via the Space Transportation System including ground to earth orbit delivery by the Space Shuttle and orbit to orbit delivery via orbit transfer vehicles.[NASA]

payload deployment & retrieval system System of mechanical and control devices, with associated data systems, for payload handling in space.[NASA]

payload integration plan Procedures providing for compatibility of spaceborne experiments with the carrier spacecraft (e.g., shuttle orbiter).[NASA]

payloads Originally, the revenue producing portions of an aircraft's load, e.g., passengers, cargo, and mail. By extension, that which an aircraft, rocket, or spacecraft carries over and above which is necessary for the operation of the vehicle for its flight.[NASA]

payload transfer The in-space movement of payloads from point to point.[NASA]

P band In telemetry, the portion of the radio frequency spectrum from 215 to 260 MHz; generally a narrow section of that band near 225 MHz is available for telemetry application.

PBB See polybrominated biphenyls.

PBW See proportional bandwidth.

p chart A type of data display in quality control which charts the fraction defective in a sample or over a production period against time or number of units of production.

PCM (materials See phase change materials.

PCM (modulation) See pulse code modulation.

PCM serial recording The technique of recording a train of bits on a single track of magnetic tape.

P controller See controller, proportional.

PD control Proportional plus derivative control, used in processes where the controlled variable is affected by several different lag times. See both proportional control and derivative control.

PD controller See controller, proportional plus derivative.

PDES Product Data Exchange Specification.

PDM (modulation) See pulse duration modulation.

PDP-11 A family of sixteen-bit minicomputers manufactured by DEC.

PDU See protocol data unit.

peak pressure Maximum pressure, usually of short duration.[AIR1916-88]

peak-to-peak Pertains to the maximum amplitude excursion of a signal; for example, in the case of a pure sine wave, the maximum value between the 90° and 270° excursion points. See also double amplitude.

peak-to-peak amplitude In an oscillating or alternating function, the difference between maximum and minimum instantaneous values of the function.

peak to peak value Of an oscillating quantity, the algebraic difference between the extremes of the quantity.[AIR 1489-88]

pearlite An aggregate in steel of ferrite and cementite.[NASA]

peat Dark brown or black residuum produced from the partial decomposition and disintegration of mosses, hedges, trees, and other plants that grow in marshes and other wet places.[NASA]

Peclet number A nondimensional number arising in problems of heat transfer in fluids.[NASA]

peculiar ground support equipment Any system, subsystem, component, or equipment designed and used solely for maintenance task(s) performance on a specified end article of hardware.[ARD 50010-91]

peculiar stars Stars with spectra that cannot be conveniently fitted into any of the standard spectral classifications. They are denoted by a 'p' after their spectral type.[NASA]

peculiar support equipment Support equipment which is compatible with only one item.[ARD50010-91]

pedestal In PAM, an arbitrary minimum signal value assigned to provide for channel synchronization and decommutation.

pedometer 1. A device for determining the distance traveled by walking. 2. A device for determining birth weight of an infant.

PEEK A class of semicrystalline polymers called polyayrlene ethers for use as molding compounds and for use as composite matrix materials.[NASA]

peel test Sealant is cured on a selected substrate, then peeled from it using one of several types of testing machines (Scott, Instron, Tinius Olson). Two factors are of particular importance: the force required to peel it off and the site of the failure. If it is within the sealant, it is called "cohesive failure". If it fails at the bond line, it is called "adhesive failure". The former type of failure is preferred.[AS7200/1-90]

peep door A small door usually provided with a shielded glass opening through which combustion may be observed.

peep hole A small hole in a door covered by a movable cover.

peer entities Entities within the same layer.

peer pressure A motivating factor stemming from a person's perceived need to meet peer expectations.[ARP4107-88]

peer-to-peer protocol Communication protocol between peer entities.

peg count meter A meter that counts the number of trunks tested, the number of circuits passed busy, the number of tests failed, or the number of repeat tests completed.

pellicle An extremely thin, tough membrane which is stretched over a frame. Because of its thinness, it transmits some light and reflects other light, and hence can serve as a beamsplitter. Its thinness avoids the problem of ghost reflections sometimes produced by other beamsplitters. Usually found as beam splitters in interferometers.

Peltier effect 1. The effect which results in the production or absorption of heat at the junction of two metals on the passage of an electrical current.[NASA] 2. The principle in solid-state physics that forms the basis of thermocouples—if two dissimilar metals are brought into electrical contact at one point, the difference in electrical potential at some other point depends on the temperature difference between the two points.

pelvic restraint That portion of a torso restraint system intended to restrain movement of the pelvis, commonly referred to as a lap belt, safety belt, or seat belt.[AS8043-86]

pen 1. A device for writing with ink. 2. An ink-filled device for drawing a graphical record of an instrument reading.

penalty function In mathematics, a function used in treating maxima and minima problems subject to restraints. [NASA]

pencil An implement for making marks with graphite, carbon or a colored solid substance.

pendant, catapult A wire rope or synthetic fiber assembly designed to connect an aircraft to a catapult for the purpose of accelerating the aircraft for takeoff. The pendant connects to catapult hook(s) on the aircraft and to the catapult shuttle spreader. The pendant falls off the aircraft (sheds) as it overtakes the shuttle and is generally restrained by a bridle arrester system on the deck. See also Bridle, catapult.[AIR 1489-88]

pendant, cross deck See arresting cable.

pendant, holdback A line or link assembly suspended from the aircraft to the deck fitting for holdback of the aircraft until the specified time for release for catapult or to restrain the aircraft for engine runup.[AIR1489-88]

pendulum scale A type of weighing device in which the weight of the load is counterbalanced by rotation of a bent lever with a fixed weight at the free end.

penetrameter A stepped piece of metal used to assess density of exposed and developed radiographic film, and to determine relative ability of the radiographic technique to detect flaws in a workpiece.

penetrating particles See corpuscular radiation.

penetration 1. That portion of a published high altitude instrument approach procedure which prescribes a descent path from the fix on which the procedure is based to a fix or altitude from which an approach to the airport is made.[ARP4107-88] 2. Distance from the original base metal surface to the point where weld fusion ends. 3. A surface defect on a casting where molten metal filled surface voids in the sand mold.

penetration ballistics See terminal ballistics.

penetration number A measure of the consistency of materials such as waxes and greases expressed as the distance that a standard needle penetrates a sample under specified ASTM test conditions.

penetration rate The distance per unit time that a drill cuts into a material, measured along the drill axis.

penetrometer An instrument for determining penetration number.

pen-motor recorder A data-versus-time strip chart recorder where each trace is written by a motor-driven pen.

Penning discharge A direct current discharge where electrons are forced to oscillate between two opposed cathodes and are restrained from going to the surrounding anode by the presence of a magnetic field.[NASA]

Penning effect An increase in the effective ionization rate of a gas due to the presence of a small number of foreign metastable atoms.[NASA]

pen recorder See pen.

pentode An electron tube containing five electrodes—an anode, a cathode, a control electrode, and two others which usually are grids.

perceived special sortie See special sortie.

percent bit cannot duplicate % CND = 100 x (number of BIT CNDs)/(Total number of BIT indications). Total number of BIT indications excludes false alarms that do not generate maintenance actions. A BIT CND is an on-equipment, BIT indication of a malfunction that cannot be confirmed by subsequent troubleshooting by maintenance personnel.[ARD50010-91]

percent built in test false alarm % FA = 100 x (Number of BIT indications not resulting in maintenance actions)/(Total number of BIT indications). A BIT FA is an indication of a failure that is not accompanied by system degradation or failure and, in the opinion of the operator, does not require any maintenance action.[ARD50010-91]

percent built in test fault detection %

BIT FD = 100 x (# of confirmed failures detected by BIT)/(# of confirmed failures detected via all methods) where a confirmed failure is a condition when (a) equipment performance (including BIT performance) is less than that required to perform a satisfactory mission and (b) corrective action is required to restore equipment performance. This formula assumes that a requirement exists for 100 percent diagnostics capability.[ARD50010-91]

percent conductivity Conductivity of a material expressed as a percentage of that of copper.[ARP1931-87]

percent defective The number of defective pieces in a lot or sample, expressed as a percent.

percent fault isolation In defining this term, it is essential to recognize that it is just as operationally valuable for BIT to fault isolate an aircrew reported fault or manually detected fault as it is for BIT to fault isolate BIT detected faults. Hence the definition is: % FI = 100 x (# of fault isolation in which BIT effectively contributed)/(# of confirmed failures detected via all methods). Effective isolation should be defined, for example, to mean that the fault is unambiguously isolated to a single item node (driver, receiver, connector, wire), or to a specified maximum number of items (an ambiguity group of x items).[ARD50010-91]

percent of actual Same accuracy value applies over the entire flow rate range.

percent of dilution The amount of diluent in the mixture expressed in whole numbers, i.e., a mixture consisting of thirty parts of diluent and seventy parts of oil by volume is a thirty percent mixture and its dilution shall be known as thirty percent. Percentage of dilution shall not be measured as the amount of diluent with respect to the original amount of oil.[S-4, 6-84]

percent of span Accuracy value applies only at the maximum rated flow.

percent slip Percentage representing the ratio of reduction in wheel rotational velocity under the influence of braking force to the equivalent free rolling rotational velocity of an unbraked wheel. [AIR1489-88]

perception The detection and interpretation of transthreshold cues from the environment by one or more of the senses. The awareness, or the process of becoming aware, of extraorganic or intraorganic objects or relations or qualities, by means of sensory processes and under the influence of perceptual set and of prior experiences.[ARP4107-88]

perception, delayed Failure to detect cues in a timely manner due to an anomaly of attention of motivation.[ARP 4107-88]

perception (misperception) Failure to detect or correctly interpret cues due to an inappropriate perceptual set.[ARP 4107-88]

perception (nonperception) Inability to detect cues from the environment because of sensory limitations or the manner in which the cues are presented, or both.[ARP4107-88]

perceptual errors Deviations from accuracy in the perception of objects, shapes, colors, weights, etc. through the use of the senses.[NASA]

perceptual set A cognitive or attitudinal framework in which a person expects to perceive certain environmental cues and tends selectively to search for those cues more actively than others. One extreme of this anomaly is expectancy so strong that the person perceives cues that in fact are not there; the other extreme occurs when he/she does not expect cues to the extent that he/she does not detect cues that are there. Perceptual distortions in the form of illusions may also result.[ARP4107-88]

percussion A method of initiating an explosive item by striking sharply.[AIR

913-89]

perfect combustion The complete oxidation of all the combustible constituents of a fuel, utilizing all the oxygen supplied.

perfect diffusion That in which light is scattered uniformly in all directions by the diffusing medium.[ARP798-91]

perfect gas See ideal gas.

perfect vacuum A reference datum analogous to a temperature of absolute zero that is used to establish scales for expressing absolute pressures.

perfluoroalkoxy (PFA) A fluorocarbon resin which offers excellent electrical characteristics, high temperature resistance, chemical inertness and flame resistance.[ARP1931-87]

performance and energy management (PM) Those performance functions including thrust which provide guidance signals to control the aircraft in climb, cruise and descent flying optimum or desired performance (min cost, min fuel, select IAS/Mach, max gradient, etc.). These signals may be directed to the auto flight system pitch channel and auto throttles as well as pilot displays. [ARP1570-86]

performance characteristic A qualitative or quantitative measurement unique to a piece of equipment or a system that is evident only during its test or operation.

performance chart A graphic representation of some aspect of operation of a piece of equipment or a system.

performance curves Plots of the abilities of rotating equipment under various operating conditions.

performance data Information on the way a material or device behaves during actual use.

performance degradation The condition or status indicating impaired or deteriorated engine gas path performance as referenced to some established or predetermined condition.[ARP1587-81]

performance evaluation A comparison of performance data, usually taken by an automatic data logging system, with predetermined standards or estimates, for assessing operating experience or identifying any need for corrective action.

performance factors Factors computed or applied to basic performance representing individual aircraft variance with the standard. It may be expressed as a percentage drag factor, fuel factor or combined SFC. Performance factor may also represent an assigned minimum maneuver margin (buffet boundary) or minimum climb and cruise rate of climb. [ARP1570-86]

performance index 1. Ratio in percent of time integral of instantaneous brake pressure divided by the time integral of a series of straight lines connecting the peak brake pressure levels during which skidding initiates.[AIR1489-88] 2. In industrial engineering, the ratio of standard hours to hours of work actually used to produce a given output. A ratio greater than 1.00 (100%) indicates standard output is being exceeded.

performance number Any of a series of numbers used to rate aviation gasolines with octane values greater than 100; the PN (performance number) compares fuel antiknock values with those of a standard reference fuel in terms of an index which indicates relative engine performance.

perfusion See diffusion.

perigees Those orbital points nearest the earth when the earth is the center of attraction.[NASA]

perihelions Those points in solar orbits which are nearest the sun.[NASA]

period 1. Of a periodic function, the smallest increment of the independent variable that can be repeated to generate the function. 2. Of an undamped instrument, the time between two successive transits of the pointer through the rest position in the same direction following a step change in the measured quantity.

period doubling The bifurcation of a nonlinear system to two stable periodic cycles on its route to chaotic turbulence.[NASA]

periodic duty A type of intermittent duty in which the load conditions are regularly recurrent.[ARP1199A-90]

periodic function An oscillating quantity whose values repeatedly recur for equal increments of the independent variable.

periodic orbits See orbits.

periodic processes See cycles.

period, natural the reciprocal of a natural frequency of a landing gear. The time for one complete vibrational cycle in seconds.[AIR1489-88]

period, wear out failure rate That period during which the failure rate of a family of the items can be expected to increase due to deterioration processes.[ARD50010-91]

peripheral 1. A supplementary piece of equipment that puts data into, or accepts data from the computer (printers, floppy disc memory devices, videocopiers). 2. Any device, distinct from the central processor, that can provide input or accept output from the computer.

peripheral equipment (computers) Equipment that works in conjunction with a computer but is not part of the computer itself. Card or paper-tape readers or punches, magnetic tape handlers, or line printers are among items of peripheral equipment.[NASA]

peripheral speed See cutting speed.

peripheral visual cues Visual stimuli occurring outside of an approximately 60-degree cone from a person's normal sight line. Visual cues in this region are typically detected scotopically (with rods). Peripheral vision is mostly used for detecting gross movement and aids in maintaining ambient orientation. [ARP4107-88]

periscope Optical instrument which displaces the line of sight parallel to itself to permit a view which may otherwise be obstructed.[NASA]

permafrost Any soil, subsoil or other surficial deposit, or even bedrock, occurring in arctic or subarctic regions at a variable depth beneath the Earth's surface in which a temperature below freezing has existed continuously for a long time.[NASA]

permanent magnet A shaped piece of ferromagnetic material that retains its magnetic field strength for a prolonged period of time following removal of the initial magnetizing force.

permanent-magnet moving-coil instrument See moving-coil instrument.

permanent marking Marking which will ensure identification during the normal service life of the item.[AS478G-79]

permanent pressure drop The unrecoverable reduction in pressure that occurs when a fluid passes through a nozzle, orifice or other throttling device.

permeability Of a magnetic material, the ratio of the magnetic induction to the magnetic field intensity in the same region. The ability to permit penetrations or passage. In this sense the term is applied particularly to substances which permit penetration or passage of fluids.[NASA]

permeameter 1. A device for determining the average size or surface area of small particles; it consists of a powder bed of known dimensions and degree of packing through which the particles are forced under pressure. Particle size is determined from flow rate and pressure drop across the bed; surface area, from pressure drop. 2. A device for determining the coefficient of permeability by measuring the gravitational flow of fluid across a sample whose permeability is to be determined. 3. An instrument for determining magnetic permeability of a ferromagnetic material by measuring the magnetic flux or flux density in a specimen exposed to a magnetic field of a given intensity.

permissible dose The amount of ionizing radiation that a human being can absorb over a given period of time without harmful result.

permittivity Preferred term for dielectric constant.[ARP1931-87]

perovskites Minerals with a close-packed lattice and the general formula ABX3 where A and B are metals and X is a nonmetal, usually O.[NASA]

persistence The continuation of luminance of a phosphor after electron excitation has been removed.[ARP1782-89]

personal computer (PC) 1. A personal computer is generally used at an office desk for word processing, data bases and spreadsheets. A personal computer may be used to aid in the configuration of programmable logic controllers and distributed control systems or may be used for data acquisition and control of small processes. Ruggedized PC's have been used on the process floor for control and data acquisition. 2. The letters PC are sometimes used to signify a programmable (logic) controller.

personal flying Any use other than pleasure flying of an aircraft for purposes not associated with a business or profession, and not for hire. This includes maintaining pilot proficiency. [ARP4107-88]

personality variables Those traits of a person which characterize his/her behavior, predispose him/her to certain response patterns, and allow for some generalized predictions as to how he/she will respond in different situations. The pattern of motivation and of temperamental or emotional traits of the individual.[ARP4107-88]

person transported for compensation or hire A person who would not be transported unless there were some payment or other consideration, including monetary or services rendered, by or for the person, and who is not connected with the operation of the aircraft or its navigation, ownership, or business.[ARP 4107-88]

PERT See Program Evaluation and Review Technique.

perturbation Any departure introduced into an assumed steady state of a system, or a small departure from a nominal path such as a desired trajectory. Usually used as equivalent to small perturbation. Specifically, a disturbance in the regular motion of a celestial body, the result of a force additional to that which causes the regular motion, specifically a gravitational force.[NASA]

perturbation generator An instrument that simulates typical data link perturbations such as blanking, noise, bit rate jitter, baseline offset, and wow.

petri nets Abstract, formal models of the information flow in systems with discrete sequential or parallel events. The major use has been the modeling of hardware systems and software concepts of computers.[NASA]

petroleum Naturally occurring mineral oil consisting predominately of hydrocarbons.

petroleum engineering A branch of engineering that deals with drilling for and producing oil, natural gas and liquifiable hydrocarbons.

petroleum products Materials derived from petroleum, natural gas, and asphalt deposits. Includes gasolines, diesel and heating fuels, lubricants, waxes, greases, petroleum coke, petrochemicals, an sulfur.[NASA]

petrology That branch of geology dealing with the origin, occurrence, structure, and history of rocks, especially igneous and metamorphic rocks.[NASA]

PFM (modulation) See pulse frequency modulation.

pH The symbol for the measurement of acidity or alkalinity. Solutions with a pH reading of less than 7 are acid; solutions with a pH reading of more than 7 are alkaline on the pH scale of 0 to 14, where the midpoint of 7 is neutral.

phase 1. The relationship between voltage and current waveforms in a-c electrical circuits. 2. A microstructural constituent of an alloy that is physically distinct and homogeneous. 3. For a particular value of the dependent variable in a periodic function, the fractional part of a period that the independent variable differs from some arbitrary origin. 4. In batch processing, an independent process-oriented action within the procedural part of a recipe. The phase is defined by boundaries that constitute safe and logical points where processing can be interrupted.

phase angle 1. Phase angle is a measure of how a system's output response lags or leads a sinusoidal input to the system.[AIR1823-86] 2. The difference between the phase of current and the phase of voltage in an alternating-current signal, usually determined as the angle between current and voltage vectors plotted on polar coordinates. 3. A measure of the propagation of a sinusoidal wave in time or space from some reference instant or position on the wave. See also phase shift.

phase angle firing A method of operation for a SCR stepless controller in which power is turned on for the proportion of each half cycle in the a-c power supply necessary to maintain the desired heating level.

phase balance The percentage of unbalance of the phase voltage for a three-phase alternator is 100 times the maximum deviation of the phase voltage from the average of the three phase voltages, divided by the average.[AS8011A-85]

phase change materials Materials undergoing solid/liquid phase transformations and whose latent heat of fusion properties are used to store and deliver thermal energy, usually solar energy.[NASA]

phase conjugation Technique for the removal of phase distortions during propagation of laser beams through the atmosphere.[NASA]

phase crossover frequency See frequency, phase crossover.

phase crossover, gain margin frequency The point on the Bode plot of the transfer function at which the phase angle is - 180 degrees. The frequency at which phase crossover occurs is called the gain margin frequency.[AIR1916-88]

phase detectors Devices that continuously compare the phase of two signals and provide an output proportional to their difference in phase.[NASA]

phase deviation The peak difference between the instantaneous phase of the modulated wave and the carrier frequency.[NASA]

phase discriminator A device that detects the phase relationship of a signal to that of a reference.

phase margin 1. A measure of system stability defined as the phase lag to be added to achieve 180 deg of phase lag at the open loop frequency response corresponding to the 0 dB amplitude ratio. [AIR1916-88] 2. Phase margin is the degree in magnitude the actual phase angle would have to be increased to make the system unstable when the gain is unity (i.e. log 1 = 0; Bode plot).[ARP 1281A-81] 3. The difference between 180° and the absolute value of the open-loop phase angle for a stable feedback system at that frequency where the gain is unity.

phase matching Alignment of a nonlinear crystal with respect to the incident laser beam in the proper way to generate a harmonic of the laser frequency in the material.

phase meter An instrument for measuring electrical phase angles. Also known as phase-angle meter.

phase modulation Angle modulation in which the angle of a sine wave carrier is caused to depart from the carrier angle by an amount proportional to the instantaneous value of the modulation wave. Combinations of phase and frequency modulation are commonly referred to as frequency modulation. [NASA]

phase of flight, mishap See mishap.

phase response See frequency response; phase shift.

phase-sequence indicator A device that indicates the sequence in which the fundamental components of a polyphase set of voltages or currents reach some particular value—their maximum positive value, for example.

phase shift 1. The phase difference of two periodically recurring phenomena of the same frequency, expressed in angular measure. The angle between the lines connecting a celestial body and the sun and a celestial body and the earth.[NASA] 2. The time difference between the input and output signal or between any two synchronized signals, of a control unit, system, or circuit, usually expressed in degrees or radians. 3. A change in phase angle between the sinusoidal input to an element and its resulting output.

phase shift circuit An electronic network whose output voltage is shifted in phase when compared to a specified reference voltage.

phase shifter An electronic device whose output voltage (or current) differs from its input voltage (or current) by some desired phase relationship; in some devices the phase is shifted a fixed amount because of an inherent design feature, but in others the phase relationship can be adjusted.

phase shift keying (PSK) 1. The form of phase modulation in which the modulating function shifts the instantaneous phase of the modulated wave among predetermined discrete values.[NASA] 2. A form of PCM achieved by shifting the phase of the carrier; e.g., ± 90 degrees to represent "ones" and "zeros."

phase velocity Of a traveling plane wave at a single frequency, the velocity of an equiphase surface along the wave normal.[NASA]

Phelps vacuum gage A modified hot-filament ionization gage useful for measuring pressures in the range 10^{-5} to 1 torr.

phenology A branch of science dealing with the relations between climate and periodic biological phenomena.[NASA]

Philips gage An instrument that measures very low gas pressure (vacuum) indirectly by determining current flow from a glow discharge device.

pH meter An instrument for electronically measuring electrode potential of an aqueous chemical solution and directly converting the reading to pH (a measure of hydrogen ion concentration, or degree of acidity).

Phobos A satellite of Mars orbiting at a mean distance of 9,400 kilometers. [NASA]

Phoebe A satellite of Saturn orbiting at a mean distance of 12,960,000 kilometers.[NASA]

phon A unit of loudness level equivalent to a unit pressure level in decibels of a 1000-Hz tone.

phoneme The basic phonological element of speech consisting of a simple sound that, by itself, cannot differentiate one word from another; the American English language, for example, contains 38 to 40 phonemes (14 to 16 vowel sounds and 24 consonant sounds) that are used in conjunction with inflection, volume and emphasis to produce synthetic speech.

phonotelemeter A sophisticated stopwatch for estimating the distance from artillery by measuring the elapsed time from gun flash to arrival of the detonation sound.

phosphatizing Forming an adherent phosphate coating on metal by dipping or spraying with a solution to produce an insoluble, crystalline coating of iron phosphate which resists corrosion and serves for a base for paint.

phosphazene A ring or chain polymer that contains alternating phosphorus and nitrogen atoms, with two substituents on each phosphorus atom.[NASA]

phosphor 1. Material which gives off visual radiant energy when bombarded by electrons or ultraviolet light.[ARP 1782-89] 2. A phosphorescent material.

phosphor bronze A hard copper-tin alloy, deoxidized with phosphorus, and sometimes containing lead to enhance its machinability.

phosphorescence 1. Emission of light which continues after the exciting mechanism has ceased.[NASA] 2. Emission of radiant energy—often in the visible-light range—following excitation due to absorption of shorter wavelength radiation; phosphorescent emission may persist for a long time after the exciting radiation stops. Contrast with fluorescence; incandescence.

phosphoric acid fuel cells Long life fuel cells for the low to medium wattage range which use phosphoric acid as an electrolyte.[NASA]

phosphors Phosphorescent substances such as zinc sulfide, which emit light when excited by radiation, as on the scope of a cathode ray tube.[NASA]

phot The CGS unit of illuminance, which equals one lumen power cm^2; the SI unit, lux, is preferred.

photoacoustic spectroscopy An optical technique for investigating solid and semisolid materials, in which the sample is placed in a closed chamber filled with a gas and illuminated with monochromatic radiation of any desired wavelength, and with intensity modulated at some acoustic frequency. Absorption of radiation results in a periodic heat flow from the sample, which generates sound detectable with a sensitive microphone.[NASA]

photocathodes Electrodes used for obtaining photoelectric emission.[NASA]

photocell A device that alters its electrical resistance in proportion to the amount of light that impinges on it. See also photoelectric cells.

photochemical oxidants Any of the chemicals which enter into oxidation reactions in the presence of light or other radiant energy.[NASA]

photochemical reactions Chemical reactions which involve either the absorption or emission of radiation.[NASA]

photochemistry See photochemical reactions.

photoclinometry See photogrammetry.

photoconductive cell 1. Photoelectric cell whose electrical resistance varies with the amount of illumination falling upon the sensitive area of the cell. [NASA] 2. A transducer that converts the intensity of EM radiation, usually in the IR or visible bands, into a change of cell resistance.

photoconductor A type of conductor which changes its resistivity when illuminated by light; the changes in resistance can be measured to determine the amount of incident light.

photocurrents See photoelectric emission.

photodarlington A detector in which a phototransistor is fabricated on the same chip with a second transistor which amplifies the signal from the phototransistor. The circuit formed is a Darlington circuit—a simple and inexpensive type of detector with limited performance.

photodetectors See photometers.

photodiode A diode which detects light. Vacuum photodiodes are tubes in which detection relies on the photoelectric effect producing free electrons which are collected by a positively charged electrode.

photodissociation The dissociation (splitting) of a molecule by the absorption of a photon. The resulting components may be ionized in the process (photoionization).[NASA]

photodraft A photographic reproduction of a master layout or design on an emulsion-coated sheet of metal; it is used chiefly as a master in tool- and die-making.

photoelastic stress analysis A visual full field technique for measuring stresses in parts and structures. When a photoelastic material is subjected to forces and viewed under polarized light, the resulting stresses are seen as color fringe patterns. Interpretation of the colorful pattern will reveal the overall stress distribution, and accurate measurements can be made of the stress directions and magnitudes at any point. Three broad categories are embraced: (1) 2-dimensional model analysis. (2) 3-dimensional model analysis and (3) photoelastic coating analysis.[AIR1489-88]

photoelectric cell Transducer which converts electromagnetic radiation in the infrared, visible, and ultraviolet regions in to electrical quantities such as voltage, current, or resistance.[NASA]

photoelectric control Modifying a controlled variable in accordance with a control signal whose value is related to the intensity of a light-beam input signal.

photoelectric counter A counting device actuated when a physical object passes through an incident beam of light.

photoelectric effect 1. The emission of an electron from a surface as the surface absorbs a photon of electromagnetic radiation. Electrons so emitted are termed photoelectrons.[NASA] 2. A physical phenomenon whereby a so-called photoelectric material emits electrons when struck by light—one bound electron being emitted for each photon of light absorbed.

photoelectric hydrometer A device for measuring specific gravity of a continuously glowing liquid, in which a weighted float, similar to a hand hydrometer, rises or falls with changes in liquid density, changing the amount of light that is permitted to fall on a sensitive phototube whose output is calibrated in specific gravity units.

photoelectric photometer A device that uses a photocell, phototransistor or phototube to measure the intensity of light. Also known as electronic photometer.

photoelectric pyrometer An instrument that measures temperature by measuring the photoelectric emission that occurs when a phototube is struck by light radiating from an incandescent object.

photoelectric threshold The amount of energy in a photon of light that is just sufficient to cause photoelectric emission of one bound electron from a given substance.

photoelectrochemical devices Electrochemical devices powered by light or other incident radiation to produce electricity and/or chemical fuels. (e.g., hydrogen).[NASA]

photoelectrochemistry The study of the interaction between impinging light energy and the electropotential of the chemical changes in the electrode, electrolytic solution or a photosensitive membrane.[NASA]

photoelectronics See electronics.

photoelectrons Electrons which have been ejected from their parent atoms by interaction between those atoms and high energy photons.[NASA]

photoemission See photoelectric emission.

photoemissive tube photometer A device that uses a tube made of photoemissive material to measure the intensity of light; it is very accurate, but requires electronic amplification of the output current from the tube; it is considered chiefly a laboratory instrument.

photoemissivity See emissivity; photoelectric emission.

photogrammetry 1. The art or science of obtaining reliable measurements by means of photography.[NASA] 2. The science of making maps or accurately measuring features from aerial photographs.

photographic emulsion A light-sensitive coating—usually a silver halide compound in gelatine—used to capture and store the visual image in photography or radiography.

photographic recording Using a signal-controlled light beam or spot to record information—either by recording the position of the spot, its intensity, or both.

photo image A photographic print with shading (halftone).[AS4159-88]

photoionization The ionization of an atom or molecule by the collision of a high energy photon with the particle.[NASA]

photolithography The process of making a printing plate by exposing a design photographically on a sensitized emulsion and removing unwanted portions chemically.[NASA]

photoluminescence Nonthermal emission of electromagnetic radiation that occurs when certain materials are excited by absorption of visible light.

photomasks In the production of integrated circuit devices, repeated arrays of microphotographs of the circuit patters on glass substrates used to form successive patterns on single wafers often of submicrometer sizes.[NASA]

photometer Instrument for measuring the intensity of light or the relative intensity of a pair of lights.[NASA]

photometric brightness The intensity of illumination reflected or emitted by a surface as measured by a photoelectric device.[ARP1161-91]

photometry 1. The study of the measurement of the intensity of light.[NASA] 2. Any of several techniques for determining the properties of a material or for measuring a variable quantity by analyzing the spectrum or intensity, or both, of visible light.

photomultiplier A type of electron tube in which photons incident on a photocathode produce electrons by photoemission. These electrons are then amplified by passing them through an electron multiplier, which increases their numbers. Electrons passing through the multiplier are accelerated by high voltages and hit metal screens, from which they free more electrons.

photomultiplier tube Phototube with one or more dynodes between its photocathode and output electrode.[NASA] See also multiplier tube.

photon(s) 1. According to the quantum theory of radiation, the elementary quantities of radiant energy. They are regarded as discrete quantities having a momentum equal to hv/c, where h is the Planck constant, v is the frequency of the radiation, and c is the speed of light in a vacuum. Photons are never at rest, have no electric charges and no magnetic moments, but they have spin moments. The energy of a photon (the unit quantum of energy) is equal to hv.[NASA] 2. A quantum of electromagnetic radiation.

photon counting A measurement technique used for measuring low levels of radiation, in which individual photons generate signals which can be counted.

photophoresis Production of unidirectional motion in a collection of very fine particles, suspended in a gas or falling in a vacuum, by a powerful beam of light.[NASA]

photoreduction See photochemical reactions.

photosphere The intensely bright portion of the sun visible to the unaided eye.[NASA]

photosynthesis A process operating in green plants in which carbohydrates are formed under the influence of light with chlorophyll serving as a catalyst.[NASA]

phototheodolite A telescopic instrument or device incorporating one or more cameras (sometimes a motion-picture camera) for taking and recording horizontal and vertical angular measurements. In aeronautics, the phototheodolite (sometimes in conjunction with radar equipment) is used to track airborne craft and to measure and record attitude, altitude, azimuth, and elevation angles.[ARP4107-88]

photothermal conversion Conversion into thermal energy from optical radiation by a photoabsorptive or photoselective material.[NASA]

photothermotropism See anisotropy.

phototransistor A transistor in which one of the two junctions is illuminated by light, and electrons are released. The transistor treats this current as an input, which it amplifies, making it a simple detector-amplifier.

phototube An electron tube containing at least two electrodes, one that functions as a photoelectric emitter.

photovoltaic cell A transducer that converts the intensity of EM radiation, usually in the IR or visible bands into a voltage.

phugoid oscillation In a flight path, a long-period longitudinal oscillation consisting of shallow climbing and diving motions about a median flight path and involving little or no change in angle of attack.[ARP4107-88] See also oscillations, oscillators, pitch (inclination).

physical addressing A mode of addressing using a unique station address.[AIR 4271-89]

physical address space The set of computer memory locations where information can actually be stored for program execution; virtual memory addresses can be mapped, relocated, or translated to produce a final memory address that is sent to hardware memory units; the final memory address is the physical address.

physical block A physical record on a mass storage device.

physical condition The physical state of a person in terms of the extent of a regular rigorous exercise program or a physically active life style, or both. (a) Athletic: At least 6 hours of rigorous exercise per week and a physically active life style. (b) Active: At least 2 hours of rigorous exercise per week or a very active physical life. (c) Inactive: Less than 1 hour of rigorous exercise per week or an intermittently physically active life style (for example, occasional sports or yardwork). (d) Sedentary: No rigorous exercise and not physically active.[ARP4107-88]

physical fatigue See fatigue.

physical input See measurand.

physical properties Inherent characteristics of a substance—such as electrical conductivity, magnetic permeability, density or melting point—that can be determined without applying mechanical force.

physical record The largest unit of data that read/write hardware of an I/O device can transmit or receive in a single I/O operation; the length of a physical record is device-dependent; for example, a punched card can be considered the physical record for a card reader; it is eighty bytes long.

physical task saturation A situation in which the number or difficulty of tasks to be performed in a compressed time period exceeds a person's physical capacity to perform all of them.[ARP 4107-88]

physical workload The total physical task demands upon a person.[ARP4107-88]

physiological acceleration The acceleration experienced by a human or an animal test subject in an accelerating vehicle.[NASA]

physiological telemetry See biotelemetry.

physiology The science that treats of the functions of living organisms or their parts, as distinguished from morphology or anatomy.[NASA]

phytometer A device for determining the transpiration rate of plants; it consists of a soil-filled container in which one or more plants are rooted and sealed so that water can escape only through transpiration.

phytoplankton The aggregate of passively floating or drifting plant organisms in aquatic ecosystems.[NASA]

phytotrons Apparatus for the growth of plants under a variety of controlled environmental conditions.[NASA]

PI See proportional-integral derivative (PID) control.

P/I A pressure to current converter linearly converts a signal pressure range into a signal current range (for example, 3-15 psi into 4-20 mA).

piano wire Carbon steel wire (0.75 to 0.85% C) cold drawn to high tensile strength and uniform diameter.

pica A unit of measure used in printing; one-sixth of an inch. See point.

pick The open area left by the crossing of any two carriers in the weave of a braid axially along its length.[ARP1931-87]

picketing See mooring.

pickle liquor Spent pickling solution.

pickling (metallurgy) Preferential removal of oxide or mill scale from the surface of a metal by immersion usually in an acidic or alkaline solution.[NASA]

pickoffs See sensors.

picks per inch The number of carriers in either direction contained in one inch of the braid measured parallel to the axis of the wire or cable.[AS1198-81]

pickup 1. A transducer or other device that converts optical, acoustical, mechanical or thermal images or signals into electrical output signals. 2. Electrical noise or interference from a nearby device, system or circuit. 3. The minimum value of an input signal—voltage, current or power, for instance—needed to make a relay function as intended. See also transducer; sensors.

picofarad (pF) A measure of capacitance (10^{-12}) farads.[ARP1931-87]

PI control Proportional plus integral control, used in combination to eliminate offset. See both proportional control and integral control. Also called proportional plus reset control.

PI controller See controller, proportional plus integral.

picture elements See pixels.

PID See proportional integral derivative.

PID action A mode of controller action in which proportional integral, and derivative action are combined.

PID control Proportional plus integral plus derivative control, used in processes where the controlled variable is affected by long lag times. See proportional control, integral control, and derivative control.

PID controller See controller, proportional plus integral plus derivative.

piercing An operation in which a tool is forced through a metal part in order to cut a hole of a specific shape and size.

piezoelectric accelerometer A device for measuring variable forces associated with acceleration, such as from an earthquake or from vibration, by means of response of a piezoelectric crystal in physical contact with a mass that reacts to the accelerating forces.

piezoelectric ceramics Ceramic materials with piezoelectric properties similar to those of some natural crystals. [NASA]

piezoelectric transducers Transducers utilizing piezoelectric elements.[NASA]

piezoelectric detector A sensing element for detecting seismic disturbances which consists of a stack of piezoelectric crystals with an inertial mass on top of the stack; metal foil between the crystals collects the charges that develop when the crystals are strained.

piezoelectric effect The generation of an electric potential when pressure is applied to certain materials or conversely a change in shape when a voltage is applied to such materials. The changes are small, but piezoelectric devices can be used to precisely control small motions of optical components.

piezoelectric gage A pressure measuring device used to detect and measure blast pressures from explosives and internal pressure transients in guns; it uses a piezoelectric crystal to sense a pressure transient and develop an output voltage pulse in response.

piezoelectricity The property exhibited by some asymmetrical crystalline materials which when subjected to strain in suitable directions develop polarization proportional to the strain.[NASA]

piezoelectric pressure transducer Any of several sensor designs in which a force acting on the sensing element is converted to an electrical output by a piezoelectric crystal.

piezoid A piezoelectric crystal adapted for use by attaching electrodes to its surface or by other suitable processing.

piezometer 1. An instrument for measuring fluid pressure. 2. An instrument for measuring compressibility of materials.

piezoresistive accelerometer A device for measuring variable forces associated with acceleration, such as from an earthquake or from vibration, by means of changes in resistance of two or four semiconductor strain gages connected in a Wheatstone bridge circuit.

pig 1. An in-line scraper for removing scale and deposits from the inside surface of a pipeline; a holder containing brushes, blades, cutters, swabs, or a combination is forced through the pipe by fluid pressure. 2. A crude metal casting, usually of primary refined metal intended for remelting to make alloys.

pigtail 1. A conductor or wire extending from an electrical or electronic device, or from a cable shield, to serve as a connection.[ARP1931-87] [ARP914A-79] 2. A 270° or 360° loop in pipe or tubing to form a trap for vapor condensate. Used to prevent high temperature vapors from reaching the instrument. Used almost exclusively in static pressure measurement.

pile 1. An assemblage of thermoelectric elements, dissimilar-metal plates or fissile-material components so arranged to produce electrical or thermal power—as in a thermopile, storage battery or atomic reactor. 2. A heap of aggregate or other bulk material stored on a floor or on a flat area outdoors. 3. A long, heavy column made of timber, steel or reinforced concrete which has been driven or cast in place below grade to support another structure or to hold earth in place.

pilot 1. A mechanical control system, such as may be used to guide an aircraft in flight. 2. A bar extending in front of a reamer to guide the reamer and force it to cut concentric with the original borehole. 3. A flame which is utilized to ignite the fuel at the main burner or burners. See also ignitor.

pilotage Navigation by visual reference to landmarks.[ARP4107-88]

pilot balloon (meteorology) A small

free balloon tracked with a theodolite to determine the direction and speed of the wind at various altitudes.[ARP4107-88]

pilot circuit That portion of a control circuit or system which carries the control signal from the signal-generating device to the control device.

pilot, constant A pilot that burns without turndown throughout the entire time the boiler is in service.

pilot, continuous See pilot, constant.

pilot, expanding A pilot that normally burns at a low turndown throughout the entire time the burner is in service whether the main burner is firing or not. Upon a call for heat, the pilot is automatically expanded so as to reliably ignite the main burner. This pilot may be turned down at the end of the trial-for-ignition period for the main burner.

pilot flame establishing period The length of time fuel is permitted to be delivered to a proved pilot before the flame-sensing device is required to detect pilot flame.

pilot induced oscillation oscillations of a flying aircraft caused by transients and system changeovers, by pilot overreaction upon such transients, or by misleading pilot cues or excessive pilot gain in modern high-grain high order aircraft control systems.[NASA]

pilot plant A test facility, built to duplicate or simulate a planned process or full-scale manufacturing plant, used to gain operating experience or evaluate design alternatives before the full-scale plant is built.

pilot, proved A pilot flame which has been proved by flame failure controls.

pilot stabilization period A timed interval synonymous on most systems today with timed trial for pilot ignition. Today's programmers prevent main valve operation for a specified number of seconds after commencement of trial for pilot ignition even though the pilot is immediately proved.

pin 1. A cylindrical object used as a connector or as a guide in aligning mating parts.[ARP480A-87] 2. A cylindrical or slightly tapered fastener made of wood, metal or other material which joins two or more members yet allows free angular movement at the joint. 3. In a dot-matrix printer, the tiny cylinders that as a group form a character. Typical computer printers are 9-pin and 24-pin.

pinboard A type of control panel which uses pins rather than wires to control the operation of a computer. On certain small computers which use pinboards, a program is changed by the operator removing one pinboard and inserting another. Related to control panel. See plugboard.

pinch effect The result of an electromechanical force that constricts, and sometimes momentarily ruptures, a molten conductor carrying current at a high density. The self contradiction of a plasma column carrying large currents due to the interaction of this current with its own magnetic field.[NASA]

pi network A network consisting of three branches connected in series to form a closed mesh; one of the three junctions is an input terminal, one is an output terminal and the third is a common terminal connected to both input and output circuits.

pinhole A fault in a casting or coating resulting from small blisters that have burst or from small voids that formed during plating.

pinhole cameras Cameras which have no lenses, but consist essentially of a darkened box with a small hole in one side, so that an inverted image of outside objects is projected on the opposite side where it is recorded on photographic film.[NASA]

pinhole detector A photoelectric device that can detect small holes or other defects in moving sheets of material.

pinion The smaller of two gear wheels, or the smallest gear in a gear train.

pinning Sites within a superconducting material that are produced by localizing inclusions, dislocations, voids, etc., which provide a means of resisting flux motion (flux jumps) due to Lorenz forces.[NASA]

PIN photodiode A semiconductor diode light detector in which a region of intrinsic silicon separates the p and n type materials. It offers particularly fast response and is often used in fiber optic systems.

pipe 1. A pipe is a discontinuity in the center of a rolled bar. It is caused by internal cavities in the ingot formed during solidification and which have become elongated or stretched in the rolling operations.[AS3071A-77] 2. A tubular structural member used primarily to conduct fluids, gases or finely divided solids; it may be made of metal, clay, ceramic, plastic, concrete or other materials. 3. A general class of tubular mill products made to standard combinations of diameter and wall thickness. 4. An extrusion defect caused by the oxidized metal surface flowing toward the center of the extrusion at the back end.

pipe elbow meter A variable-head meter used to measure flow around the bend in a pipe.

pipe fitting A piece with an internal cavity for connecting lengths of pipe together or for connecting them to tanks or other process equipment; types include couplings, elbows, nipples, tees and unions.

pipelining 1. Processing techniques for improving the capability of computer systems by modeling, sequencing control, resource allocation, etc.[NASA] 2. The process of increasing data processing speed by simultaneously executing a number of basic instructions.

pipe saddle See leveling saddle.

pipe tap A small hole in the wall of a pipe for sampling its contents or for connecting a control device or pressure measuring instrument.

pipe tee A pipe fitting in the shape of the letter T; it is used to connect a branch line at 90° to the main run of pipe.

pipe thread A type of screw thread used chiefly to connect pipe and fittings; in the usual configuration, it is a 60° thread with flat roots and crests, and with a longitudinal taper of about 3/4 in. per foot (about 6.3%).

piping A system of pipes for carrying a fluid stream or gaseous material.

piping and instrumentation drawing (P&ID) 1. Show the interconnection of process equipment and the instrumentation used to control the process. In the process industry, a standard set of symbols is used to prepare drawings of processes. The instrument symbols used in these drawings are generally based on Instrument Society of America (ISA) Standard S5.1. 2. The primary schematic drawing used for laying out a process control installation.

pipping pressure 1. The pressure at which a safety valve opens. 2. The pressure the pipe cannot withstand without exceeding its design characteristics.

Pirani gage A pressure transducer used to measure very low gas pressure based on measurement of the resistance of a heated wire filament; resistance varies in accordance with thermal conduction of the gas, which in turn is related to gas pressure. Used primarily for pressures less than one atmosphere.

piston 1. A cylindrical part which slides in a cylinder or barrel and serves to transfer force to or from the enclosed fluid.[ARP243B-65] 2. A solid cylinder or disc that fits snugly into a larger cylinder and moves back and forth under fluid pressure.[AIR1489-88]

piston displacement The volume traversed by a piston in a single cycle, or stroke.

piston engines Engines, especially internal combustion engines, in which a piston or pistons moving back nd forth work upon a crankshaft or other device to create rotational movement.[NASA]

piston meter A type of fluid flow meter; it is a variable-area, constant-head device in which the flow rate is indicated by a pointer attached to a piston, which in turn is positioned by the buoyant force of the fluid.

pistonphone A device consisting of a small chamber and reciprocating piston of measurable displacement; it is used to establish a known sound pressure in the chamber.

piston ring seal A seal ring installed in a groove on the piston circumference to minimize the clearance flow between the piston outer diameter and the cylinder bore.

piston rod A coaxial column or rod, attached to or integral with a piston, which serves to transmit force between the piston and another mechanical member. [ARP243B-65]

piston-type variable-area flowmeter Any of several flowmeter designs in which fluid passing through the meter exerts force on a piston such that the piston moves against a counterbalancing force to expose a portion of an exit orifice, the amount exposed being directly related to volume flow.

pit 1. Void, hole in the surface as caused for example by corrosion.[MA2005-88] 2. A small surface cavity in a metal part or coating usually caused by corrosion or formed during electroplating.

pitch 1. Rotation or oscillation of an aircraft about its lateral axis.[ARP4107-88] 2. The distance a propeller would advance in one revolution if it were acting in a solid medium.[ARP4107-88] 3. Also known as phosphor dot pitch or triad pitch. On a patterned CRT screen, the distance from the center of the most fundamental screen element (for example, red phosphor dot or line) to the center of the nearest like element (for example, the nearest adjacent red phosphor dot or line).[ARP1782-89] [ARP4067-89] 4. An auditory sensation of tone that is directly related to sound-wave frequency. 5. A heavy, black or dark brown liquid or solid residue from distillation of tar or oil; it occurs naturally as asphalt. 6. The distance between similar mechanical elements in an array, such as gear teeth, screw threads or screen wires. 7. The distance between centerlines of tubes, rivets, staybolts, or braces. 8.In computer printers, a measure of the number of characters printed per inch. Typically 10, 12 or 17.

pitch angles See pitch (inclination).

pitch attitude The angle between the longitudinal axis of an aircraft and the horizontal plane.[ARP4107-88]

pitch diameter The pitch diameter of any layer of strands or wires of a conductor or cable is the diameter of the circle passing through their centers. [AS1198-81]

pitch-down See pitchunder.

pitch moment A moment about a lateral axis of an aircraft, rocket, airfoil, etc. This moment is positive when the angle of attack is increased, or the body is nosed upward.[ARP4107-88]

pitch setting The propeller blade setting as determined by the blade angle and measured in a manner and at a radius specified by the instruction manual for the propeller.[ARP4107-88]

pitchunder An act or instance of an aircraft pitching nose downward; a tendency of an aircraft to pitch nose downward. Also called tuck down or pitch-down.[ARP4107-88]

pitchup An act or instance of an aircraft pitching nose upward; a tendency of an aircraft to pitch nose upward.[ARP 4107-88]

pitot-static tube 1. A combination of a pitot tube and a static port arranged coaxially or otherwise parallel to one another and mounted externally on an aircraft (generally on the wing, the nose, or the vertical stabilizer) in a position to sense the air flow and pressure undisturbed by the flow over or around other structures of the aircraft. It is used principally to determine airspeed from the difference between impact and static pressures; also called pitot-static head. [ARP4107-88] 2. A combination of a pitot tube and a static tube—the two may be either parallel or concentric.

pitot tube 1. Tube which receives the total pressure including impact pressure created by the forward motion of the aircraft.[ARP4107-88] 2. Open ended tube or tube arrangement which, when pointed upstream, may be used to measure the stagnation pressure of the fluid for subsonic flow; or the stagnation pressure behind the tube's normal shock wave for supersonic flow.

pitot-venturi tube A combination of a venturi device and a pitot tube.

pitting A concentrated attack by oxygen or other corrosive chemicals in a boiler, producing a localized depression in the metal surface.

pivoting The procedure or technique of turning or pivoting an aircraft about a point such as to pivot about one main gear.[AIR1489-88]

pivots The paths followed by a point in a diameter of a circle as the circle rolls along in a straight line.[NASA]

PIX See plasma interaction experiment.

pixel In data processing, a portion of a CRT display screen.

PIXEL (PEL) 1. Shortened term of picture elements'. They are image resolution elements in vidicon-type detectors. [NASA] 2. Picture Element.[AS4159-88]

PL/1 See programming language/1.

placard A posted notice on or in an aircraft setting forth a requirement or limiting operational condition.[ARP4107-88]

plages (faculae) See faculae.

plain binding A weave in which each warp wire and each shute wire passes over one and under the next adjacent wire in each direction. Also square weave.[AIR888-89]

planar network An electronic network which can be drawn or sketched on a plane surface without having any of the branches cross each other.

plane of polarization In a plane polarized electromagnetic wave, the plane that contains both the direction of propagation and the electric field vector.

plane of symmetry The vertical fore-and-aft plane that divides an airplane into symmetrical halves.[ARP4107-88]

plane polarized wave An electromagnetic wave in a homogeneous isotropic medium that has been generated, or modified by the use of filters, so that the electric field vector lies in a fixed plane which also contains the direction of propagation.

plan equation An equation for determining horsepower: HP=plan/33,000, where p is mean effective pressure in psi, l is piston stroke in feet, a is net piston area in in.^{2}, and n is number of strokes per minute.

plane strain A deformation of a body in which the displacement of all points in the body are parallel to a given plane, and the displacement values are not dependent on the distance perpendicular to the plane.[NASA]

planetary boundary layer The layer of the atmosphere from the earth's surface to the geostrophic wind level, including the surface boundary layer and the Ekman layer.[NASA]

planetary cores The centers of planets. [NASA]

planetary craters collective term for craters on any of the planetary surfaces.[NASA]

planetary crusts The outermost layers of planets. The planetary crusts are on top of the mantle and are modified by various processes of weathering, sedimentation, metamorphosis, volcanism, and bombardment by meteorites.[NASA]

planetary entry See atmospheric entry.

planetary geology Study or science of a planet, its history, and its life as recorded in the rocks. Includes the study of the surface features, the geometry of rock formations, weather and erosion, and sedimentation.[NASA]

planetary limb In astronomy, the circular outer edge of a planet.[NASA]

planetary systems Systems consisting of a star and the planets and other objects in orbit around it.[NASA]

planetary twister A twisting machine whose payoff spools are mounted in rotating cradles that hold the axis of the spools in a fixed direction as the spools are revolved about one another, so the wire will not kink as it is twisted.[ARP 1931-87]

plane wave(s) 1. Waves on uniform currents in two-dimensional nondivergent fluid systems rotating with varying angular speeds about the local vertical (beta plane). These waves represent a special case of barotropic disturbance, conserving absolute vorticity. As applied to atmospheric flow, the planetary waves take into account the variability of the Coriolis parameter while assuming the motion to be two-dimensional.[NASA] 2. A wave whose equiphase surfaces form an array of parallel planes.

planetesimals See protoplanets.

planets celestial bodies of the solar system, revolving around the sun in nearly circular orbits, or similar bodies revolving around stars. The larger of such bodies are sometimes called principal planets to distinguish them from asteroids, planetoids, or minor planets, which are comparatively small.[NASA]

planigraphy See tomography.

planimeter A device for measuring area of a plane surface, usually of irregular shape, by tracing its perimeter.

plankton The aggregate of passively floating or drifting plant and animal organisms which provide the major source of sustenance for animal life in the aquatic ecosystem.[NASA]

plankton bloom See plankton.

plan-position indicator A type of radar display that gives a maplike presentation on a circular screen, indicating the range and azimuth of any discernible object with respect to a point on the screen representing the location of the transmitter. Also called plan view display (PVD).[ARP4107-88]

plant design Encompasses all design consideration of physical plants, i.e., airports, industrial plants, test facilities, etc. Structural is just one aspect of this design.[NASA]

plant stress Stimulus or a series of stimuli of such magnitude as to disrupt the growth and/or survival of plants.[NASA]

plan view display (PVD) See plan-position indicator.

plasma antennas An air plasma made by ionizing the atmosphere which acts as the conducting element of an RF antenna.[NASA]

plasma arc cutting Use of plasma torches for cutting hard materials at extremely high temperatures.[NASA]

plasma arc welding (PAW) Metals are heated with a constricted arc between an electrode and the workpiece (transferred arc), or the electrode and the constricting nozzle (non-transferred arc). Shielding is obtained from the hot, ionized gas issuing from the orifice which may be supplemented by an auxillary source of shielding gas.

plasma bubbles Pockets of very low electron density in the equatorial F region of the ionosphere in which the plasma density is lower than the ambient density.[NASA]

plasma clouds Specifically, a mass of ionized gas flowing out of the sun. [NASA]

plasma compression Decrease in volume and consequent increase in density of a plasma usually by the application of an intense magnetic field.[NASA]

plasma cooling Temperature control of plasmas in controlled fusion operations. [NASA]

plasma core reactors Nuclear reactors utilizing fissionable plasmas (such as uranium fluoride) for the fuel.[NASA]

plasma currents Electric currents induced in plasmas by injection of fast ion beams or some other means.[NASA]

plasma display devices Digital matrix flat panel devices in which small gas discharge plasma cells are used as light emitting sources.[NASA]

plasma drift Movement in the ionosphere of ion and plasma concentration by electric field variations in the upper atmosphere.[NASA]

plasmadynamic lasers Stimulated emission devices in which the lasing gas flow has been replaced with a lasing plasma flow of atoms or ions.[NASA]

plasma engines Reaction engines using magnetically accelerated plasma as a propellant. Plasma engines are types of electrical engines.[NASA]

plasma equilibrium Condition of plasma in which the constituent particles or fluid elements are unaccelerated or collectively at rest in steady flow.[NASA]

plasma etching Removal of material by use of a focused plasma beam.[NASA]

plasma focus A highly compressed plasma.

plasma generation See plasma generators.

plasma generators Machines, such as electric arc chambers, that will generate very high heat fluxes to convert neutral gases into plasmas. Devices which use the interaction of plasmas and electrical field to generate currents.[NASA]

plasma pumping Application of radiation of appropriate frequencies to plasma to increase the population of atoms or molecules in the higher energy states. [NASA]

plasma renin activity See immunoassay.

plasmas (physics) Electrically conductive gases comprised of neutral particles, ionized particles, and free electrons but which, taken as a whole, are electrically neutral. Plasmas are further characterized by relatively large intermolecular distances, large amounts of energy stored in the internal energy levels of the particles, and the presence of plasma sheaths at all boundaries of the plasma. Plasmas are sometimes referred to as a fourth state of matter. [NASA]

plasma sheaths The boundary layers of charged particles between plasmas and their surrounding walls, electrodes, or other plasmas. Envelopes of ionized gases that surround bodies moving through an atmosphere at hypersonic velocities.[NASA]

plasma sound waves See magnetohydrodynamic waves.

plasmasphere envelope of highly ionized gases surrounding the earth or another planet.[NASA]

plasma torches Burners which attain 50,000 degrees C temperatures by the use of plasma gas injected into an electric arc. Plasma torches are used for welding, spraying molten metal, and cutting hard rock or hard metals.[NASA]

plasmoids See plasmas (physics).

plastic An imprecise term generally referring to any polymeric material, natural or synthetic. Its plural, plastics, is the preferred term for referring to the industry and its products.

plastic clad silica A step index optical fiber in which a silica core is covered by a transparent plastic cladding of lower refractive index. The plastic cladding is usually a soft material, although hard-clad versions have been introduced.

plastic deformation The changes in dimensions of items caused by stress, that are retained after the stress is removed.[ARP700-65]

plastic fibers Optical fibers in which both core and cladding are made of plastic material. Typically their transmission is much poorer than that of glass fibers.

plasticity See plastic properties.

plasticizer 1. A material added to a propellant to increase flexibility or workability.[AIR913-89] 2. A chemical agent added in compounding plastics to make them softer and more flexible.[ARP 1931-87]

plasticorder A laboratory device for measuring temperature, viscosity and shear-rate in a plastics material which can be used to predict its performance.

plastic properties The tendency of a loaded body to assume a deformed state other than its original state when the load is removed.[NASA]

plastometer An instrument for determining flow properties of a thermoplastic resin by forcing molten resin through a fixed orifice at specified temperature and pressure.

plate 1. Any tool which has the general configuration of a smooth, flat piece of material of uniform thickness.[ARP 480A-87] 2. A rolled flat piece of metal; depending on the type of metal, the minimum thickness for the product to be called plate instead of sheet or strip may vary—for instance, plate steel is any hot-finished flat-rolled carbon or alloy steel product more than 8 in. wide and more than 0.230 in. thick, or more than 48 in. wide and more than 0.180 in. thick. See electroplating.

plateau 1. Generally, any portion of a function where the value of the dependent variable is essentially constant over a range of values for the independent variable. 2. Specifically, a portion of the output versus input characteristic of an instrument, electronic component or control device where the output signal level is essentially independent of the input signal level.

plate baffle A metal baffle.

platelet alpha A relatively coarse acicular alpha, usually with low aspect ratios. This microstructure arises from cooling alpha or alpha-beta alloys at a slow rate from temperatures at which a significant fraction of beta phase exists.[AS1814-90]

platen A plane surface receiving heat from both sides and constructed with a width of one tube and depth of two or more tubes, bare or with extended surfaces.

plate polarized light Light polarized by means of optical plates set at Brewster's angle to the optic axis. The more plates the greater the purity of the plane polarized exit beam.

plates (tectonics) Rigid divisions of the outer surface of the earth (lithosphere) which moves over a weaker layer (asthenosphere). The plates are about 100 km thick, and the continents, which are 40 km thick, rest on the plates and move with them.[NASA]

plating The electrolytic application of one metal over another.[ARP1931-87]

PLATO Acronym for Programmed Logic for Automatic Teaching Operations—a system developed at the University of Illinois for developing computer-aided instructional programs.[ARP4107-88]

PLC See programmable logic controller.

plenum 1. A condition where air pressure within an enclosure is greater than barometric pressure outside the enclosure. 2. An enclosure through which gas or air passes at relatively low velocities.

pliers A hand tool with scissor-type action having special jaws for the purpose of compressing, cutting, expanding, forming, or handling of small parts.

[ARP480A-87]

PL/M A block-structured high-level language for preparing software for Intel microprocessors.

plotter 1. A device for automatically graphing a dependent variable on a visual display or flat board, in which a movable pen or pencil is positioned by one or more instrument control signals. 2. Hardware device which plots on paper the magnitudes of selected data channels, as related to each other or to time.

plotter/printer A plotter that can also print alphanumeric data from the computer.

plug 1. (mechanical) A device which fits into a hole and serves as a stopper. [ARP480A-87] 2. (electrical) A male fitting for making electrical connections by insertion into a receptacle.[ARP 480A-87] 3. A rod or mandrel over which a pierced billet is drawn to form a tube or pipe, or that is inserted into a tube or pipe during cold reduction. 4. A punch or mandrel over which a cup is drawn. 5. A fluid-tight seal made to prevent flow through a leaking pipe or tube. 6. A projecting portion of a die intended to form a recess in a forged part. 7. A term frequently used to refer to the closure component.

plugboard A perforated board that accepts manually inserted plugs to control the operation of equipment, such as a removable panel containing an ordered array of terminals, which may be interconnected by short electrical leads (plugged in by hand) according to a prescribed pattern and thereby designating a specific program or sequence of specified program steps. See pinboard.

plug die See floating plug.

plug fuseholder A receptacle with female threads to accommodate a plug-type fuse.

plug gage A metal member used to check the dimension of a hole. The gaging element may be straight or tapered, plain or threaded, and of any shape cross section.

plugging Physically stopping the flow of fluid, either intentionally or unintentionally, especially by the buildup of material.

plug meter A device for measuring flow rate in which a tapered rod extends through an orifice; when the rod is positioned so that the effective area of the annulus is just sufficient to handle the fluid flow, the rate of flow is read directly from a scale.

plug, sealing An accessory used to fill open, nonwired cavities in a connector grommet so as to prevent the entry of moisture, fluids or foreign particulate contaminants.[ARP914A-79]

plug valve A type of shutoff valve consisting of a tapered rod with a lateral hole through it. As the rod is rotated 90° about its longitudinal axis, the hole is first aligned with the direction of flow through the valve and then aligned crosswise, interrupting the flow.

plumb Indicating a true vertical position with respect to the earth's surface; the condition is usually determined by a plumb bob, which consists of a weight (plummet) suspended on a string and positioned entirely by gravity.

plumb-bob gage 1. A device for determining liquid level in which a weighted plummet is lowered on a calibrated tape or cable until it just touches the liquid surface. 2. A device for detecting solids level in a storage bin or hopper by lowering a plummet until the lowering cable slackens, which is usually detected by an electrical or mechanical triggering device.

plummet gage See plumb-bob gage.

plunger-type instrument A moving-iron instrument in which a pointer is attached to a long, specially shaped piece of iron that moves along the axis of a coil by variable electromagnetic attraction depending on the current flowing in the coil.

ply orientation The arrangement of bonded layers comprising laminated materials to obtain optimal strength or other characteristics.[NASA]

ply rating A rating number which indicates the relative strength of a tire carcass. It is proportional to, but not necessarily the same as, the number of structural plies in the carcass.[AIR 1489-88]

PL/Z A high-level language for Zilog microprocessors.

pneudraulic See shock absorber, air-oil.

pneumatic 1. Pertaining to or operated by a gas, especially air. 2. Systems that employ gas, usually air, as the carrier of information and the medium to process and evaluate information.

pneumatically actuated valves Those valves that utilize gas as the primary power source to position the valve. This includes valves that are solenoid controlled, but pneumatically actuated. [ARP986-90]

pneumatic caster See trail, pneumatic.

pneumatic controller A device activated by air pressure to mechanically position another device, such as a valve stem. Also known as pneumatic positioner.

pneumatic control system 1. A control system that uses air or gas as the energy source. 2. A system which makes use of air for operating control valves and actuators.

pneumatic control valve A spring-loaded valve that regulates the area of a fluid-flow opening by changing position in response to variable pneumatic pressure opposing the spring force.

pneumatic extensions A system that converts float position to a proportional standard pneumatic signal. A magnetic coupling connects the internal float extension with an external mechanical system linked to a pneumatic transmitter.

pneumatics The branch of physics dealing with the mechanical properties of gases with particular emphasis on gas statics in closed systems.[NASA]

pneumatic signal line 1. An air (pneumatic) signal, usually 3-15 psig is used as the energy medium. 2. Applies to a signal using any gas as the signal medium. If a gas other than air is used, the gas may be identified by a note on the signal symbol or otherwise.

pneumatic system A system which makes use of air for operating control valves and actuators (cylinders, motors).

pneumatic telemetering Remote transmission of a signal from a primary sensing element to an indicator or recorder by means of a pneumatic pressure impulse sent through small-bore tubing; may be used to monitor temperature, pressure, flow rate or other variables in a process unit or system. Also known as pneumatic intelligence transmission.

pneumatic to current converter (P/I) A pressure to current converter linearly converts a signal pressure range into a signal current range (for example, 3-15 psi into 4-20 mA).

Pockel's cell A device in which the Pockel's effect is used to modulate light passing through the material. The modulation relies on rotation of beam polarization caused by the application of an electric field to a crystal; the beam then has to pass through a polarizer, which transmits a fraction of the light dependent on its polarization.

Pockel's effect See birefringence.

pocket chamber A small ionization chamber that can be charged, then carried in a person's pocket and periodically read to determine cumulative radiation dose received since the instrument was last charged. Also known as pocket dosimeter.

pocket meter A pocket-size direct-reading instrument for measuring radiation dose rate.

poidometer An automatic weighing device used in conjunction with a belt conveyer.

point 1. A process variable derived from an input signal or calculated in a process calculation. 2. A unit of measure used in printing; one-seventy second of an inch. 12 points equal one pica.

point drift See drift, point.

pointer 1. A needle-shaped or arrowhead-shaped element whose position over a scale indicates the value of a measured variable. 2. In data processing: A. A data string that tells the computer where to find a specific item. B. Similar to or the same as a cursor on a computer screen.

pointing 1. Reducing the diameter and tapering a short length at one end of a wire, rod or tube; usually done so the pointed end may be inserted through a reducing die and clamped in the moving element of a drawbench. 2. Finishing a mortar joint, or pressing mortar into a raked joint.

point matching method (mathematics) See boundary value problems.

point module See alarm module.

point of mishap See mishap.

point spread functions Mathematical functions involved in image processing. [NASA]

point test A means of safe access to permit a measurement that will facilitate maintenance, repair, calibration, alignment, or monitoring. Test points may be accessible in the normally installed position or may require disassembly of the equipment for accessibility. The former are called exposed test points; the latter, internal test points.[AIR 1916-88]

point-to-point numerical control A simple form of numerical control in which machine elements are moved between programmed positions without particular regard to path or speed control. Also known as positioning control.

poise The CGS unit of dynamic viscosity, which equals one dyne-second per cm^2; the centipoise (cP) is more commonly used.

Poiseuille flow Laminar flow of gases in long tubes at pressures and velocities such that the flow can be described by Poiseuille's equation. See also laminar flow.

Poisson process See stochastic processes.

POL See problem-oriented language.

polar auroras See auroras.

polar coordinates In a plane, a system of curvilinear coordinates in which a point is located by its distance r from the origin (or pole) and by the angle theta which a line (radius vector) joining the given point and the origin makes a fixed reference line, called the polar axis. In three dimensions, short for space polar coordinates.[NASA]

polar diagram 1. A diagram showing the relative effectiveness of an antenna system for either transmitting or receiving. Principally shows directional characteristics. 2. A schematic representation that relates events in the cycle of a piston engine to crankshaft position.

polarimeter Instrument for determining the degree of polarization of electromagnetic radiation, specifically the polarization of light.[NASA]

polarimetry Chemical analysis in which the amount of substance present in a solution is estimated from the amount of optical rotation (polarization) that occurs when a beam of light passes through the sample.

polariscope Instrument for detecting polarized radiation and investigating its properties.[NASA]

polarity The sign of the electric discharge associated with a given object as an electrode or an ion.[NASA]

polarization 1. The arrangement or orientation of connector inserts, jackscrews, polarizing pins/sockets, keys/keyways or housing configurations to prevent the mismating or cross mating of connectors.[ARP914A-79] 2. The state of electromagnetic radiation when transverse vibrations take place in some regular manner, e.g., all in one plane, in a circle, in an ellipse, or in some other definite curve. With respect to particles in an electric field, the displacement of the charge centers within a particle in response to the electric force acting thereon. The response of the molecules of a paramagnetic medium (such as iron) when subjected to a magnetic field. [NASA]

polarization maintaining fiber A single-mode optical fiber which maintains the polarization of the light which entered it, normally by including some birefrigence within the fiber itself. Normal single-mode fibers, and all other types, allow polarization to be scrambled in light transmitted through them.

polarized meter A meter with its zero point at the center of the scale so the direction the pointer deflects indicates electrical polarity and the distance it deflects indicates value of a measured voltage or current.

polarizer Device for polarizing radiant energy.[NASA] 2. A filter which transmits light of only a single polarization.

polarizing coating Coatings which influence the polarization of light passing through them, typically by blocking or reflecting light of one polarization and passing light that is orthogonally polarized.

polarizing pin, socket, key or keyway Devices incorporated in a connector to accomplish polarization.[ARP914A-79]

polarographic analysis A method of determining the amount of oxygen present in a gas by measuring the current in an oxygen-depolarized primary cell.

polarography A method of chemical analysis that involves automatically plotting the voltage-current characteristic between a large, non-polarizable electrode and a small polarizable electrode immersed in a dilute test solution; a curve containing a series of steps is produced, the potential identifying the particular cation involved and the step height indicating cation concentration; actual values are determined by comparing each potential and step height with plots generated from test solutions of known concentrations.

polar wandering (geology) Migration during geologic time of the earth's poles of rotation and magnetic poles. Also known as polar migration.[NASA]

pole-dipole array An electrode array for making resistivity or induced-polarization surveys in which one current electrode is placed far away from the area being surveyed while an assembly containing one current electrode and two potential electrodes is moved laterally across the area in a search pattern.

pole face On a magnetized part, the surface through which magnetic lines of flux enter or leave the part.

pole piece A shaped piece of ferromagnetic material, integral with or attached to one end of a magnet, whose function is to control the distribution of magnetic lines of flux.

pole-pole array An electrode array for making resistivity or induced-polarization surveys in which one current electrode and one potential electrode, in close proximity, are moved laterally across the area being surveyed.

polestar recorder An instrument used to determine the amount of cloudiness during the night. It consists of a fixed, long-focus camera positioned so that Polaris is permanently within its field of view; the apparent motion of Polaris is recorded as a circular pattern on the film, with approximate span of cloudiness indicated by interruptions in the arc due to clouds passing between the star and the camera.

poling A stage in fire-refining of copper

during which green-wood poles are thrust into the bath of molten metal; the wood decomposes and forms reducing gases that react with oxygen in the bath.

polished plate glass Glass whose surface irregularities have been removed by grinding and polishing, so that the surfaces are approximately plane and parallel.[ARP798-91]

polled access A media access method by which the node that has the right to use the network medium delegates that right to other stations on a per message basis. See master-slave.

polling 1. A method of sequentially observing each channel to determine if it is ready to receive data or requesting computer action. 2. The repetitive search of a LAN system to determine whether a work station is holding data for the main computer.

pollution transport Dispersing or diffusion of atmospheric or water pollutants.[NASA]

poloidal flux Plasma confinement concept with multipole magnetic fields. [NASA]

polyacetylene An aliphatic organic polymer that has high semiconductor properties which can be enhanced by doping.[NASA]

polyamide Nylon. A polymer containing a characteristic amide linkage which is derived from the condensation products of diacids and diamines, or amino acids.[ARP1931-87]

polybrominated biphenyls A group of 209 chemicals whose toxicity varies and includes principally one fire retardant called firemaster.[NASA]

polychloroprene Chemical name for neoprene. A synthetic elastomer rubber material.[ARP1931-87]

polydispersed A suspension containing a mix of particle sizes.[ARP1192-87]

polyetheretherketones See PEEK.

polyethylene A thermoplastic material composed of polymers of ethylene, and derived from the polymerization of ethylene gas.[ARP1931-87]

polyimide Kapton, liquid "H". A polymer containing a characteristic imide linkage which is usually based on the reaction between aromatic dianhydrides and aromatic diamines.[ARP1931-87]

polymer A material formed by the chemical union of one or more monomers. [ARP1931-87]

polymer matrix composites Materials consisting of reinforcing fibers, filaments, and/or whiskers embedded in polymeric bonding matrices for increased mechanical and physical properties. [NASA]

polynuclear organic compounds Hydrocarbon molecules with two or more nuclei and with or without oxygen, nitrogen, or other elements.[NASA]

polynucleotides Linear sequences of esters of nucleotides and phosphoric acid.[NASA]

polyolefin Any of the polymers and copolymers of the olefin family of hydrocarbons such as ethylene, propylene, butylene, etc.[ARP1931-87]

polypeptides In organic chemistry, chains of amino acids linked by peptide bonds but with lower molecular weights than proteins; obtained by synthesis or by partial hydrolysis of proteins.[NASA]

polyphase meter An instrument for measuring a quantity such as power factor or electric power in a polyphase electric circuit.

polypropylene A tough, light-weight, rigid plastic made by the polymerization of high-purity propylene gas.[ARP 1931-87]

polyvinylchloride (PVC) A family of insulating compounds whose basic ingredient is either polyvinylchloride or its copolymer with vinyl acetate.[ARP 1931-87]

polyvinyl fluoride DuPont's Tedlar,

unplasticized PVF films with outstanding resistance to ultraviolet radiation. [NASA]

polyvinylidene fluoride Thermoplastic resin, characterized by good mechanical, electrical, and chemical properties. Radiation cross-linking improves heat resistance.[ARP1931-87]

PONA analysis Determination of amounts of paraffins (P), olefins (O), naphthalenes (N) and aromatics (A) in gasoline in ASTM standard tests.

Pope cell A type of relative humidity sensor that employs a bifilar conductive grid on an insulating substrate whose resistance varies with relative humidity over a range of about 15 to 99% RH.

poppet A spring-loaded ball that engages a notch.

poppet valve A mushroom-shaped valve that controls the intake or exhaust of working fluid in a reciprocating engine; it may be cam operated or spring loaded, and its direction of movement is at right angles to the plane of its seat.

popping pressure In compressible fluid systems, the inlet pressure at which a safety relief valve opens.

porcelain enamel See enamel; see vitreous enamel.

porcupine boiler A boiler consisting of a vertical shell from which project a number of dead end tubes.

porosimeter A laboratory device for measuring the porosity of reservoir rock using compressed gas.

porosity The lack of soundness, usually in the form of gas holes or shrinkage voids that take on the character of gas holes.[AS3071A-77]

porpoise To oscillate about the lateral axis in the manner of a porpoise.[ARP 4107-88]

port 1. An opening at a surface of a component incorporating provisions for attachment of a fluid carrying passage, line, fitting or removable plug.[ARP 243B-65] 2. Threaded connection with a seal, component to pipe line, machined into the component.[MA2005-88] 3. The entry or exit point from a computer for connecting communications or peripheral devices. 4. An aperture for passage of steam or other fluids.

portable 1. Refers to a self-contained, battery-operated instrument that can be carried. 2. Capable of being carried, especially by hand, to any desired location.

portable standard meter A portable instrument used primarily as a reference standard for testing or calibrating other instruments.

ports Data channels dedicated to input or output.

positional notation A numeration system with which a number is represented by means of an ordered set of digits such that the value contributed by each digit depends on its position as well as upon its value.

positioner A device attached to the crimping tool to position the conductor barrel between the indentors.[ARP 914A-79]

positioner, amplifying A pneumatic positioner in which the input control signal is amplified to a proportionately higher pressure, needed to drive the actuator, e.g., 3-15 psig input/6-30 psig output.

positioner, characterized A positioner in which the valve position feedback is modified to produce a nonlinear response.

positioner, double acting A positioner with two outputs, suited to a double acting actuator.

positioner, electro-pneumatic A positioner which converts an electronic control signal input to a pneumatic output.

positioner, reversing A positioner which converts the input control signal into an output which is directionally opposite to the input.

position error An error in the reading

of an airspeed indicator owing to the difference between the pressure (especially the static pressure) at the pressure-measuring location and the free-stream pressure.[ARP4107-88] See also error, position.

positioner, single acting A positioner with one output, suited to a spring opposed actuator.

positioner, split range A positioner which drives an actuator full stroke in proportion to only a part of the input signal range.

position independent code (PIC) A code that can execute properly wherever it is loaded in memory, without modification or relinking; generally, this code uses addressing modes that form an effective memory address relative to the central processor's program counter (PC).

positioning Manipulating a workpiece in relation to working tools.

positioning action Controller action in which the final position of the control element has a predetermined relation to the value of the controlled variable.

positioning control See point-to-point numerical control.

positioning control system A system of control in which each controlled motion operates in accordance with instructions which only specify the next required position, the movement in the different axes of motion not being coordinated with each other and being executed, simultaneously or consecutively, at velocities which are not controlled by instructions on the input data medium.

position light Any of the three lights used on an aircraft to indicate its position and direction of flight (green on the right wing, red on the left wing, and white on the tail).[ARP4107-88]

position sensor Any device for measuring position and converting the measurement into an electrical, electromechanical or other signal for remote indication or recording.

position symbol A computer-generated indication on a radar display to indicate the mode of tracking.[ARP4107-88]

position telemeter A remote-reading instrument for indicating linear or angular position of an object or machine component.

position, valve The position of the valve mechanism which determines the flow pattern.[ARP243B-65]

positive control The separation of all air traffic, within designated airspace, by air traffic control.[ARP4107-88]

positive displacement Referring to any device that captures or confines definite volumes of fluid for purposes of measurement, compression or transmission.

positive-displacement flowmeter Any of several flowmeter designs in which volumetric flow through the meter is broken up into discrete elements and the flow rate is determined from the number of discrete elements that pass through the meter per unit time.

positive draft Pressure in a furnace, gas chamber or duct that is greater than ambient atmospheric pressure.

positive feedback 1. Feedback which results in increasing the amplification. [AIR1489-88] [NASA] 2. A closed loop in which any change is reinforced until a limit is eventually reached. 2. Returning part of an output signal and using it to increase the value of an input signal.

positive G In a gravitational field or during an acceleration, the effect that occurs when the human body is normally positioned so that the force of gravity or inertia, or both, acts on it in a head-to-foot direction or on an aircraft to increase the wing loading.[ARP4107-88]

positive-going edge The edge of a pulse going from a low to a high level.

positive ions Group of atoms which has acquired a positive electric charge by the loss of one or more electrons.[NASA]

positive meter Any of several devices for measuring fluid flow by alternately filling and emptying a container or chamber of known capacity; in such a device, fluid passes through it in a series of discrete amounts by weight or volume.

positive motion Motion transmitted from one machine part to another without slippage.

positron(s) 1. Subatomic particles which are identical to electrons in atomic mass, theoretical rest mass, and energy, but opposite in sign.[NASA] 2. A positively charged beta particle.

post 1. A structural brace or member which supports the pivot point for the trailing arm of a levered suspension gear system.[AIR1489-88] 2. A vertical support member resembling a pillar or column.

postassembly seal A seal that is applied after the tank structure and subassemblies have been assembled/attached.[AIR4069-90]

postconversion bandwidth The bandwidth presented to a detector.

post flight engine status An immediate GO/NO GO indication of engine availability for next flight and if in a NO GO status situation, indication of required maintenance action as appropriate.[ARP1587-81]

post guiding A design using guide bushing or bushings fitted into the bonnet or body to guide the plug's post.

post insulate To insulate an electrical connection after assembly.[ARP914A-79]

postlaunch reports Memoranda issued following spacecraft launchings to report launch data, the launch vehicle performance, orbital elements (expected and measured), and current status. [NASA]

postmission analysis (spacecraft) A broader term than postflight analysis which deals with the scientific aspects of a mission.[NASA]

post processor In numerical control, a computer program which adapts the output of a processor, applicable to a piece part, into a machine program for the production of that part on a particular combination of machine tool and controller.

potassium hydroxide The caustic material which is mixed with pure water to formulate the electrolyte solution which is used in nickel cadmium cells.[AS 803388]

potential energy 1. Energy possessed by a body by virtue of its position in a gravity field in contrast with kinetic energy, that possessed by virtue of its motion.[NASA] 2. Energy related to the position or height above a place to which fluid could possibly flow or a solid could fall or flow.

potential gradients In general, the local space rate of change of any potential, as the gravitational potential gradient or the velocity potential gradient. [NASA]

potential transformer An instrument transformer so connected in the circuit that its primary winding is in parallel with a voltage to be measured or controlled.

potentiometer 1. Instrument for measuring differences in electric potential by balancing the unknown voltage against a variable known voltage. If the balancing is accomplished automatically, the instrument is called a self balancing potentiometer. A variable electric resistor.[NASA] 2. A device for measuring an electromotive force by comparing it with a known potential difference.

potentiometric titration A technique of automatic titration where the end point is determined by measuring a change in the electrochemical potential of the sample solution.

potentiometer An instrument which provides and measures a controllable emf that can be made equal and opposite to that of the thermocouple.[AIR 46-90]

potometer A device for measuring transpiration from a leaf, twig or small plant; it consists of a small water-filled container, sealed so that moisture can escape only through the plant.

potting 1. The sealing of the cable-end of a connector with a material compound to exclude entry of fluids or contaminants and to provide strain relief to terminated wires or cable.[ARP914A-79] 2. A liquid thermosetting material used as a sealant.[ARP1931-87]

potting form See mold, potting.

pound The British or U.S. unit of mass or weight and is equal to .45 Kilograms.

poundal A unit of force in the English system of measurement; it is defined as the force necessary to impart an acceleration of one ft/s/s/ to a body having a mass of one pound.

pound of thrust A measurement unit of the reaction force generated and available for propulsion in jet or rocket engines.[ARP4107-88]

pour point 1. Lowest temperature at which a fluid will flow under specified conditions.[AIR1916-88] 2. Temperature at which molten metal is cast. 3. The temperature at which a petroleum-base lubricating oil becomes too viscous to flow, as determined in a standard ASTM test.

pour point depressant A substance which when added to an engine lubricating oil in relatively low concentrations will reduce its pour point by retarding the formation of a rigid wax structure.[S-4, 6-84]

pour test Chilling a liquid under specified conditions to determine its ASTM pour point.

powder coating A painting process in which finely ground dry plastic is applied to a part using electrostatic and compressed air transfer mechanisms. The applied powder is heated to its melting point and flows out, forming a smooth film and cures by means of a chemical reaction.

powder pattern The x-ray diffraction pattern consisting of a series of rings on a flat film or a series of lines on a circular strip film which results when a monochromatic beam of x-rays is reflected from a randomly oriented polycrystalline metal or from powdered crystalline material.

power absorption The amount of power which is absorbed by the tire in the rolling process. It results in a temperature rise in the tire carcass.[AIR1489-88]

power amplifier A component designed specifically for increasing signal power.[ARP993A-69]

power approach A landing approach during which the airplane is under power (contrasted to power-off gliding approach).[ARP4107-88]

power assurance The measurement and interpretation of the engine instrumentation data to determine whether the power or thrust delivered by the engine will be achieved within operating limits.[AIR1873-88]

power-boosted flight control systems Reversible control systems wherein the pilot effort, which is exerted through a set of mechanical linkages, is at some point in these linkages boosted by hydraulic power.[ARP578-69]

power-by-wire (PBW) actuation An integrated servoactuator which incorporates an electric motor to receive power from the aircraft main electric power system in lieu of an actuator connected directly to the aircraft main hydraulic system. Power conversion from electrical to mechanical may include mechanical devices such as ballscrews, gears, chains, cables, etc. or through a motor/pump and hydraulics. See also integrated actuator package.[ARP1181A-85]

power common 1. The reference point for power supplies and return currents from powering equipment. 2. Sometimes referred to simply as common. Power common or common is not to be confused with signal common.

power consumption The maximum amount of electrical power used by a device during normal steady-state operation.

power converter The component of the power system that transforms energy from the power source into a form compatible with the servoactuator requirements.[AIR1916-88]

power density (electromagnetic) See radiant flux density.

powered models Models that can be tested in complete force equilibrium, including propulsion.[NASA]

power/energy meter An instrument which measures the amount of optical power (watts) or energy (joules). It can operate in the visible, infrared, or ultraviolet region, and detect pulsed or continuous beams.

power factor The ratio of actual power to apparent power delivered to an electrical power circuit; in practice this is determined by dividing the resistive load, in watts, by the product of voltage and current, in volt-amperes, and expressing the result in percent.

power factor controllers A solid state electronic device that reduces excess energy waste in AC induction motors by providing only the amount of voltage required to satisfy a given load. [NASA]

power-factor meter An instrument for directly indicating power factor in a circuit.

power failure 1. Engine stoppage for a reason not directly attributable to the engine structure. For example, fuel exhaustion usually results in power failures, not engine failure. See engine failure.[ARP4107-88] 2. The removal of all power accidentally or intentionally.

power gain The ratio of the power that a transducer delivers to a specified load, under specified operating conditions, to the power absorbed by its input circuit. Of an antenna, in a given direction, 4 pi times the ratio of the radiation intensity in that direction to the total power delivered to the antenna.[NASA]

power input The energy required to drive a fan, expressed in brake horsepower delivered to fan shaft. See also excitation.

power interrupt A momentary loss of power to the engine monitoring equipment resulting in possible loss of data. [ARP1587-81]

power level At any given point in a system, the amount of power being delivered or used.

power level (dBm) The ratio of the power at a point to some arbitrary amount of power chosen as a reference. This ratio is usually expressed either in decibels based on 1 milliwatt (abbreviated dBm) or in decibels based on 1 watt (abbreviated dBW). See also decibel.

power limited A type of starter which can put out only a limited amount of power during a start cycle. Examples: any starter operating on a constant power supply such as pneumatic, electric, or hydraulic ground cart or airborne auxiliary power unit.[ARP906A-91]

power line protector A device used between a computer and a power outlet to absorb power surges or other interference that could damage the computer. See surge protector.

power loading The ratio of the gross weight of a propeller-driven aircraft to its power, usually expressed as the gross weight of the aircraft divided by the rated horsepower of the powerplant corrected for air of standard density; with turboprop engines, the equivalent shaft horsepower is used.[ARP4107-88]

power loss See also power absorption.

1. Power absorbed by the tires is a loss as far as useful work is concerned.[AIR 1489-88] 2. In a power transmission system or circuit, the difference between input power and output power, often expressed as a percent of input power. 3. In a current or voltage measuring instrument, the active power at its terminals when the pointer is at the upper end of the scale. 4. In any other electrical circuit, the difference between active power and electrical load at a stated value of current or voltage.

power modulator The component of the actuation system that regulates the potential energy of the power source to the actuator as a function of controller output. In a simple hydraulic power boost system, which has no separate controller, the power modulator controls the power to the actuator directly as a function of the command or error signal. [AIR1916-88]

power modules (STS) Modules for providing power for payloads for STS and mission dependent equipment.[NASA]

power-operated flight control system An irreversible control system wherein the pilot, through a set of mechanical linkages, actuates a power-control servomechanism. This mechanism actuates the primary control surface or corresponding device. A system of this type may have electrical or electronic pilot input modes.[ARP578-69]

power output The energy delivered by the fan, expressed in horsepower based on air or gas pressure and volume.

power run That part of the launching run after release in which the applied catapult towing force is accelerating the aircraft.[AIR1489-88]

power setting A value of the quantity used to rate the power output of an engine. The power setting of turbojet and similar engines should generally be expressed in terms of net thrust (corrected). For turboprop and similar engines, power setting should generally be in terms of shaft horsepower.[ARP 1179A-80]

power setting parameter (PS) That parameter used by the engine manufacturer to determine engine power or thrust. Various parameters are used for this purpose (for example, EPR, torque, N1, etc.).[AIR1873-88]

power source The component that supplies energy for load actuation.[AIR 1916-88]

power spectral density (PSD) A type of frequency analysis on data that can be done by computer using special software, or by an array processor, or by a special-purpose hardware device.

power splitter At the output of a telemetry radio transmitter, the device which splits the transmitter power between two or more antennas.

power stall A flight condition in which an airplane may be stalled despite the existence of a given amount of power. [ARP4107-88]

power supply The device within a computer that transforms external AC power to internal DC voltage.

power system The power system generates, conditions, and distributes power to the servoactuator. It is not a part of the servo actuation system.[AIR1916-88]

power system capacity The capacity of the power sources rated under the prescribed operating and environmental conditions in the aircraft. For parallel systems, this is the sum of the individual power source ratings with a paralleling factor applied. For nonparallel systems, this is the individual power source rating.[AS1212-71]

power transmission (lasers) Space-to-earth power transmission utilizing a laser (from solar power satellites). [NASA]

power-up test Test conducted automatically when power is provided to the system.[ARP4102/13-90]

Poynting-Robertson effect The gradual decrease in orbital velocity of a small particle such as a micrometeorite in orbit about the sun due to the absorption and remission of radiant energy by the particle.[NASA]

PPM (modulation) See pulse position modulation.

practice instrument approach An instrument approach procedure conducted by visual flight rules (VFR) or instrument flight rules (IFR) for the purpose of pilot training or proficiency demonstrations.[ARP4107-88]

Prandtl number A dimensionless number representing the ratio of momentum transport to heat transport in a flow.[NASA]

preamplifiers Amplifiers, the primary function of which is to raise the output of a low level source to an intermediate level so that the signal may be further processed without appreciable degradation in the signal-to-noise ratio. In radar amplifiers separated from the remainder of the receiver and located so as to provide the shortest possible input circuit path from the antenna so as to avoid deterioration of the signal-to-noise ratio.[NASA]

preassembly seal A seal that is applied during or prior to the assembly of the fuel tank structure (such as faying-surface seals and prepack seals).[AIR 4069-90]

precession Change in the direction of the axis of rotation of a spinning body, such as a gyro, when acted upon by a torque.[NASA]

precharged pressure (inflation pressure) The pressure at which a component is initially charged or inflated.[AIR1916-88] 2. That gas pressure to which a liquid type of accumulator is charged, prior to loading with the liquid to be stored.[ARP906A-91]

precipitate To separate materials from a solution by the formation of insoluble matter by chemical reaction. The material which is removed.

precipitation The removal of solid or liquid particles from a fluid.

precipitation (of defects) The process of transforming a latent defect into a patent defect through the application of stress screens.[ARD50010-91]

precipitator A fly ash separator and collector of the electrostatic type.

precision 1. The closeness with which a measurement upon a given, invariant sample can be reproduced in short-term repetitions of the measurement with no intervening analyzer adjustment.[ARP 1256-90] 2. The quality of being exactly or sharply defined or stated. A measure of the precision of a representation is the number of distinguishable alternatives from which it was selected, which is sometimes indicated by the number of significant digits it contains.[NASA] See also repeatability and stability.

precision approach procedure/precision approach A standard instrument approach procedure in which an electronic glide scope is provided (for example, ILS and PAR).[ARP4107-88]

precision approach radar Radar equipment in some ATC facilities operated by the FAA or the military services, or both, to detect and display azimuth, elevation, and range of aircraft on the final approach course to a runway. Primarily used to conduct a precision instrument approach to a runway.[ARP 4107-88]

precision control motor A reversible, two-phase induction type of motor specifically designed to be operated from two independent voltage sources of the same frequency with one voltage source maintained at a fixed voltage and phase direction and with the second source varied in voltage and phase direction to control speed, torque, and direction of rotation of the motor. A high torque to inertia ratio and an approximately straight line torque-speed curve are inherent characteristics of good control motors.[ARP497A-90]

precision depth recorder A machine that plots sonar depth soundings on electrosensitive paper.

precision guided projectiles Missiles guided by precise laser radiation.[NASA]

precoat sealing The application of a coat of brushable sealant to serve as a base for a fillet sealing.[AIR4069-90]

preconscious level See awareness, preconscious level.

predetection In instrumentation tape recorders, the process of recording a "low" intermediate frequency from the telemetry radio receiver (typically, 900 kHz center frequency) rather than the demodulated output of the receiver.

predicted preconditioning The operation of an item under stress to stabilize its characteristics, often referred to as "burn-in".[AIR1916-88]

predictive control 1. A type of automatic control in which the current state of a process is evaluated in terms of a model of the process and controller actions modified to anticipate and avoid undesired excursions. 2. Self-tuning. 3. Artificial intelligence.

predictive maintenance 1. A preventive maintenance program that anticipates failures which can be corrected before total failure. 2. A variation from a normal can indicate a system or equipment is approaching non-conformance. Vibration, eccentricity and noise monitoring are measurements that can predict failure. Also, an increase in diagnostic errors and retries can indicate a failure about to happen.

predisposing events, mishap See mishap.

pre-existing defect A hereditary or traumatic medical discrepancy, existing prior to the mishap sequence of events, which either has been resolved or has been medically waived for flying. This differs from an illness or disease in that it is asymptomatic.[ARP4107-88]

preface, mag tape The first few words of each tape record, which identify the record and document the status of the equipment.

pre-flight engine status Cockpit indications to aid in pre-takeoff checks; e.g., thrust check.[ARP1587-81]

preform 1. A cylinder of glass which is made to have a refractive index profile that would be desired for an optical fiber. The cylinder is then heated and drawn out to produce a fiber. 2. Brazing metal foil cut to the exact outline of the mating parts and inserted between the parts prior to placing in a brazing furnace.

preheater air Air at a temperature exceeding that of the ambient air.

preignition Spontaneous ignition of the explosive mixture in a cylinder of an internal combustion engine before the spark flashes.

pre-Imbrian period One of four stratigraphic classifications adopted for displaying (on maps) the geological ages of major features on the moon.[NASA]

pre-insulate To insulate an electrical connection before assembly.[ARP 914A-79]

prelaunch summaries Summaries prior to launch of the preparations and parameters of the mission.[NASA]

preliminary flight rating The Preliminary Flight Rating (PFR) provides an engine configuration which has demonstrated sufficient flight safety for limited use in experimental flight tests.[ARD 50010-91]

preload The final installation load expressed in pounds.[ARP700-65]

pre-main sequence stars Stars in which nuclear reactions that take place in its core have not yet occurred.[NASA]

premixing The mixing of ingredients prior to a specified action (mixing of fuel and air prior to ignition in combustion, for example).[NASA]

premodulator filter A lowpass filter at the input to a telemetry transmitter; its purpose is to limit modulation frequencies and thereby limit radiated frequencies outside the desired operating spectrum.

prepack seals A preassembly seal installed to fill voids or provide a support seal for subsequent fillet sealing.[AIR 4069-90]

prepolymers High molecular weight bifunctional molecules which, when catalyzed, produce chain extension.[AIR 4069-90]

prepregs The reinforcing materials containing or combined with the full complement or resin before moling operations in the production of composite materials.[NASA]

preprocessor 1. A hardware device in front of a computer, capable of making certain decisions or calculations more rapidly than the computer can make them. 2. The first of the two compiler stages. At this stage, the source program is examined for the preprocessor statements which are then executed, resulting in the alteration of the source program text. More generally, a program that performs some operation prior to processing by a main program.

pre-retracted gear A gear system which is not utilized in a takeoff mode and therefore is pre-retracted of placed in the stowed position. It is extended prior to landing the aircraft.[AIR1489-88]

pre-rotation The process of imparting rotation to the tires (wheels) prior to the landing touchdown to avoid the inertia loads on the gear imposed by rapid spin up and the wear on the tires caused by initial impact skidding.[AIR1489-88]

preselectors See preamplifiers.

present position (PPOS) The actual latitude and longitude coordinates the point on the earth's surface directly below the airplane at any given instant. Could also be 3-dimensional, including altitude.[ARP1570-86]

presintering See sintering.

press A mechanical device deriving its power from a drive screw or hydraulic plunger for the purpose of assembling, disassembling, cutting, straightening, etc. of parts when equipped with proper accessories.[ARP480A-87]

pressed density The density of a powder-metal compact after pressing and before sintering.

pressure 1. Normally used as a modifier to designate a portion of a system or unit which is normally exposed to system pressure. The term "pressure" is considered to mean gage pressure, except where otherwise specified.[ARP 243B-65] 2. Pressure is defined as fluid force per unit area. Pressure in a fluid at rest is equal in all directions.[AIR 1916-88] 3. Measure of applied force compared with the area over which the force is exerted, psia.

pressure, absolute True pressure, as related to a complete vacuum.[ARP 243B-65]

pressure altimeter A precision aneroid barometer that measures air pressure at the altitude a plane is flying and converts the reading to indicated height above sea level.

pressure amplifier A component designed specifically for amplifying pressure signals.[ARP993A-69]

pressure, burst The test pressure which a component or system must withstand without rupture.[ARP243B-65]

pressure breathing The breathing of oxygen or a suitable mixture of gases at a pressure higher than the surrounding pressure.[NASA]

pressure control A device or system that can raise, lower or maintain the internal pressure in a vessel or process equipment.

pressure curve Pressure variation, expressed graphically in relation to another variable, for example, time.[AIR1916-88]

pressure-demand oxygen system A demand oxygen system that furnishes oxygen at a pressure higher than atmospheric pressure above a certain altitude.[ARP4107-88]

pressure dependence Study of how a rate constant changes with pressure.[NASA]

pressure droop change in pressure from a higher level to a lower level with flow of a pressure-compensated pump or regulator.[AIR1916-88]

pressure drop 1. The reduction in fluid pressure due to flow. When applied to a fluid control unit, pressure drop is measured between given ports of the unit at a given flow and does not include the loss in fittings which are installed in ports (normally, the value applicable to a complete flow pattern at rated flow, unless otherwise stated).[ARP243B-65] 2. The difference in pressure between two points in a system, caused by resistance to flow.

pressure efficiency The ratio of the measured brake pressure integral to peak to peak straight line integral between the skid pressure levels from brake pressure time history instrumentation.[AIR 1489-88]

pressure elements The portions of a pressure-measuring gage that move or are temporarily deformed by the system pressure, the amount of movement or deformation being proportional to the pressure.

pressure energized seal A seal ring retained in the body bore with raised flexible lip which contacts an offset disk in the closed position yet is clear of the disk in other positions.

pressure-expanded joint A tube joint in a drum, header or tube sheet expanded by a tool which forces the tube wall outward by driving a tapered pin into the center of a sectional die.

pressure fluctuation Variation of pressure with time, usually occurring randomly.[AIR1916-88]

pressure gage An instrument for measuring pressure by means of a metallic sensing element or piezoelectric crystal.

pressure, gage Pressure as related to ambient atmospheric pressure.[ARP 243B-65]

pressure gradient The change in pressure value with distance in a steady-state flow.[AIR1916-88]

pressure head Equivalent height of a column of liquid required to produce a given pressure.[AIR1916-88]

pressure impulse A rapid rise and fall of pressure, or vice versa, of extremely short duration.[AIR1916-88]

pressure level In acoustic measurement, P=1 log (P_s/P_r), where P is the pressure level in bels, P_s is the sound pressure, and P_r is a reference pressure, usually taken as 0.002 $dyne/cm^2$.

pressure loss (pressure drop) Pressure loss is the reduction in pressure caused by resistance to flow or by an extraction of energy that is not converted into useful work.[AIR1916-88]

pressure measurement Any method of determining internal force per unit area in a process vessel, tank or piping system due to fluid or compressed gas; this includes measurement of static or dynamic pressure, absolute (total) or gage (total minus atmospheric), in any system of units.

pressure microphone An acoustic transducer which converts instantaneous sound pressure of impinging sound waves into an electrical signal that directly corresponds in both frequency and amplitude.

pressure, minimum operating That pressure below which a mechanism may not operate (as in a pilot operated valve, which requires a certain minimum pressure for operation).[ARP243B-65]

pressure modulator radiometer A cell containing a known quantity of a gas is placed in the single optical path of the radiometer and subjected to cyclical pressure changes which alter the absorption lines in the infrared spectrum of the gas. A narrow band signal results from the different voltages at the detector at high and low cell pressures. A wideband signal is generated by physically chopping a percentage of the input beam with a rotating chopper blade. [NASA]

pressure, output In a pressure control device, such as a pressure reducer or power brake valve, the pressure which will be produced at the outlet port. In a pressure modulating unit, as in a brake valve, the output pressure should be specified as maximum or the range stated.[ARP243B-65]

pressure, proof The pressure which a component must withstand as a production inspection test without damage, normally related to rated pressure.[ARP 243B-65]

pressure pulsation (pressure ripple) Periodical variation of pressure in synchronism with the operating speed of rotating equipment.[AIR1916-88]

pressure, rated The nominal maximum input or operating pressure.[ARP 243B-65]

pressure rating The maximum allowable internal force per unit area of a pressure vessel, tank or piping system during normal operation.

pressure ratio 1. The pressure ratio of a starter or APU is expressed as the ratio of inlet pressure to exhaust discharge pressure.[ARP906A-91] 2. Numerical ratio of the value of two pressures. [AIR1916-88] 3. The relationship of a force to the deformation of a system whose deformation varies in some proportion to the force.[NASA]

pressure-regulating valve A valve that can assume any position between fully open and fully closed, or that opens or remains closed against fluid pressure on a spring-loaded valve element, to release internal pressure or hold it and allow it to build up, as desired.

pressure regulator An in-line device that provides controlled venting from a high-pressure region to a lower-pressure region of a closed compressed gas system to maintain a preset pressure value in the lower-pressure region.

pressure relief device A mechanism that vents fluid from an internally pressurized system to counteract system overpressure; the mechanism may release all pressure and shut the system down (as does a rupture disc) or it may merely reduce the pressure in a controlled manner to return the system to a safe operating pressure (as does a spring-loaded safety valve).

pressure, reseat In a valve which closes itself against pressure, as in a check valve or relief valve, that pressure at which the valve will close itself so that flow rate recedes to a certain specified leakage.[ARP243B-65]

pressure rise 1. The rise of inflation pressure in a tire due to the rolling process and the energy absorption and temperature rise associated with the process.[AIR1489-88] 2. Change in pressure from a lower level to a higher level (due to energy addition or kinetic energy recovery).[AIR1916-88]

pressure suits Garments designed to provide pressure upon the body so that the respiratory and circulatory functions may continue normally, or nearly so, under low pressure conditions, such as occur at high altitudes or in space without benefit of pressurized cabins.[NASA]

pressure surge Pressure rise and fall over a period of time.[AIR1916-88]

pressure, surge The maximum magnitude of a transient pressure rise.[ARP 243B-65]

pressure switch A device that activates or deactivates an electrical circuit when a preselected pressure is exceeded in a process vessel or piping system.

pressure tap A small hole in the wall of an internally pressurized vessel or pipe so that the pressure element of an instrument can be attached to measure static pressure.

pressure transducer 1. A transducer which produces an output proportional to the imparted pressure.[AIR1489-88] 2. An instrument component that senses fluid pressure and produces an electrical output signal that is related to the magnitude of the pressure.

pressure transient Pressure rise or drop, or both, of extremely short duration with negligible energy.[AIR1916-88]

pressure under load (load pressure) The pressure reacting to a static or dynamic load.[AIR1916-88]

pressure-vacuum gage An instrument for measuring pressure both above and below atmospheric.

pressure vessel A metal container designed to withstand a specified bursting pressure; it is usually cylindrical with hemispherical end closures (but may be of some other shape, such as spherical) and is usually fabricated by welding.

pressure wave A cyclic variation of pressure with relative low amplitude and long period.[AIR1916-88]

pressurization, fuel Fuel control movement from cutoff to initiate fuel flow which causes the fuel manifold to become pressurized.[ARP906A-91]

pressurization, starter Opening the start valve to initiate starter operation.[ARP906A-91]

Preston tubes See pitot tubes.

pre-tinned The application of solder to a contact, conductor or other connecting device prior to soldering; the application of tin plating to the basis metal of connecting devices prior to fabrication.[ARP914A-79]

prevailing visibility The greatest horizontal visibility equaled or exceeded throughout at least half (not necessarily continuous) of the horizon from a particular observation point.[ARP 4107-88]

prevaporization The phase transformations of liquids to gases prior to some physical or chemical reaction.[NASA]

preventive alert An alert which provides the crew an escape maneuver (usually a vertical speed limit or a negative command) for which, because of their current flight path, they do not need to respond, for example, a "LIMIT CLIMB 500 fpm" alert when the aircraft is in level flight.[ARP4153-88]

preventive maintenance 1. The actions performed to retain an item at a specified level of performance by providing systematic inspection, detection, and prevention of impending failures.[ARD 50010-91] 2. Maintenance specifically intended to prevent faults from occurring during subsequent operation. Contrast with corrective maintenance. Corrective maintenance and preventive maintenance are both performed during maintenance time.

preventive maintenance time That part of the maintenance time during which preventive maintenance is performed on an item.[ARD50010-91]

primary air Air introduced with the fuel at the burners.

primary-air fan A fan to supply primary air for combustion of fuel.

primary alpha The allotrope of titanium with hexagonal, close-packed crystal structure which is retained from the last high temperature alpha-beta heating.[AS1814-90]

primary calibration Primary calibration is calibration performed using referee particles of known size and physical properties.[ARP1192-87]

primary colors Colors of constant hue and variable brilliance which can be mixed in varying proportions to produce or specify other colors. Also known as primaries.

primary color stimuli Color stimuli (x, y, z) by whose additive mixture (linear combination with positive coefficients) all other color stimuli may be matched. [ARP1782-89]

primary control surfaces Movable surfaces, such as ailerons, elevators, rudders, or spoilers, which move the aircraft about its three axes.[ARP4107-88]

primary detector The system element or device that first responds quantitatively to the attribute or characteristic being measured and performs the initial conversion or control of measurement energy.

primary device The part of a flowmeter which generates a signal responding to the flow from which the flow rate may be inferred.

primary element 1. Synonym for sensor. 2. Detector. 3. The first system element that responds quantitatively to the measured variable and performs the initial measurement operation. 4. A primary element performs the initial conversion of measurement energy. 5. Any device placed in a flow line to produce a signal for flow rate measurement. 6. The component of a measurement or control system that first uses or transforms energy from a given medium to produce an effect which is a function of the value of the measured variable. 7. The portion of the measuring means which first either utilizes or transforms energy from the controlled medium to produce an effect in response to change in the value of the controlled variable. 8. The effect produced by the primary element may be a change of pressure, force, position, electrical potential, or resistance. See also element, primary; sensing element; and sensor.

primary element (detector) The first system element that responds quantitatively to the measured variable and performs the initial measurement operation. A primary element performs the initial conversion of measurement energy. For transmitters not used with external primary elements, the sensing portion is the primary element.

primary explosive An explosive, very sensitive to initiation by impact or heat, usually one of the first elements in an explosive train.[AIR913-89]

primary failure A failure which is not a result of another failure.[ARP1587-81]

primary feedback A signal which is a function of the controlled variable and which is used to modify an input signal to produce an actuating signal.

primary flight controls Those used for controlling aircraft flight paths (e.g., ailerons, rudders, and elevators).[ARP 578-69]

primary flight display (PFD) The pilot's primary flight attitude display. The display may include attitude (pitch and roll), altitude, pitch steering commands, lateral-directional steering commands, expanded localizer, glide slope, radar height, decision height, heading reference, flight path angle, potential flight path, speed or speed reference, angle of attack.[ARP1570-86]

primary instrument An instrument that can be calibrated without reference to another instrument.

primary insulation The layer or layers of nonconducting material which is designed to act as electrical insulation, excluding cosmetic top coatings.[ARP 1931-87]

primary loop The outer loop in a cascade system.

primary luminous standard A standard by which the unit of light is established and from which the values of other standards are derived. A satisfactory primary standard must be reproducible from specifications.[ARP 798-91]

primary measuring element A component of a measuring or sensing device that is in direct contact with the substance whose attributes are being measured.

primary power system The electric system whose generators are driven by the aircraft propulsion engines. Power conversion systems (not part of utilization systems) powered by the primary generators are part of the primary power system.[AS1212-71]

primary seals A preassembly seal installed to fill voids or provide a support seal for subsequent fillet sealing.[AIR 4069-90]

prime To introduce fuel directly into the induction system or cylinders of an engine as an aid in starting the engine. [ARP4107-88]

prime (primer mixture) An explosive mixture containing a primary explosive, usually the first element in an explosive train.[AIR913-89]

primer 1. A "first" coating applied to a surface to provide a good bonding surface for the sealant or coating. The primer can be single or multifunctional. It can contain only an adhesion promoter, or only a metallic pacifier or a surface sealer, or other special agent, or can contain several of these substances. [AIR4069-90] 2. A primary initiating device to produce a hot flame.[AIR913-89]

priming The discharge of steam containing excessive quantities of water in suspension from a boiler, due to violent ebullition.

primitive equations Eulerian equations of fluid motion in which the primary dependent variables are the fluid's velocity components. The equations govern a wide variety of fluid motions and form the basis of most hydrodynamical analysis.[NASA]

printed circuit A system of conductors formed or deposited on a nonconducting substrate in a predetermined pattern to allow quick and repetitive construction of electronic devices.

printer In data processing, the device that produces printed paper copy of computer data.

print spooler A computer program that directs the computer to store certain data to be printed so that computer processing is not limited to printer speed.

prior beta grain size Size of beta grains established during the most recent beta field excursion. The grains may be distorted by subsequent subtransus deformation. The beta grain boundaries may be obscured by a superimposed alpha-beta microstructure and detectable only by special techniques.[AS1814-90]

prioritized significant events and conditions A dynamic hierarchy of environmental events and conditions which determines what tasks need to be performed, and in which order, to safely accomplish the assigned mission.[ARP 4107-88]

priority 1. The relative importance attached to different phenomena. 2.Level of importance of a program or device.

priority interrupt The temporary suspension of a program currently being executed in order to execute a program of higher priority. Priority interrupt functions usually include distinguishing the highest priority interrupt active, remembering lower priority interrupts which are active, selectively enabling or disabling priority interrupts, executing a jump instruction to a specific memory location, and storing the program counter register in a specific location. See hardware priority interrupt and software priority interrupt.

prismatic glass Clear glass into whose surface is fabricated a series of prisms, the function of which is to direct the incident light in desired directions.[ARP 798-91]

privacy Freedom from observation and/or intrusion. Applies to such things as communications, personal records, photographs.[NASA]

private aircraft Any civilian aircraft

not being used to transport persons or property for compensation or hire.[ARP 4107-88]

privilege A characteristic of a user, or program, that determines what kind of operations a user or program can perform; in general, a privileged user or program is allowed to perform operations that are normally considered the domain of the monitor or executive, or that can affect system operation as a whole.

probe 1. An instrument for remote inspection or sampling.[ARP480A-87] 2. A small, movable capsule or holder that allows the sensing element of a remote-reading instrument, usually an electronic instrument, to be inserted into a system or environment and then withdrawn after a series of instrument readings has been taken. 3. A small tube, movable or fixed, inserted into a process fluid to take physical samples or pressure readings.

probe-type consistency sensor A device in which forces exerted on a cylindrical body in the direction of flow are detected by a strain-gage bridge circuit; if the fluid is water, circuit output is a measure of flow rate, but if a solution or suspension is flowing at a steady flow rate, the output varies with changes in viscosity (or consistency).

problem definition The art of compiling logic in the form of general flow charts and logic diagrams which clearly explain and present the problem to the programmer in such a way that all requirements involved in the program are presented.

problem description In information processing, a statement of a problem. The statement also may include a description of the method of solution, the procedures and algorithms.

problem-oriented language (POL) Programming language designed for ease of problem definition and problem solution of specific classes and problems; for example, a language specifically convenient for expressing a specific problem in mathematical form, such as the ordinary algebraic languages or the symbolic notation of the Boolean algebra applied to a special problem; a special language for machine tool control.

procedural knowledge Knowledge of the capabilities and limitations of the system (and of the tactics used to employ it in various environmental conditions) that has been acquired through formal training.[ARP4107-88]

procedural language See procedural-oriented language.

procedure 1. A precise step-by-step method for effecting a solution to a problem. 2. In data processing, a smaller program that is part of a large program. 3. In batch processing, the part of a recipe that defines the generic strategy for producing a batch of material.

procedure-oriented language A programming language designed for convenience in expressing the technique or sequence of steps required to carry out a process or flow. Usually it is a source language and usually is not machine-oriented. Since many classes of problems involve similar procedures for their solution, the procedure-oriented language lends itself more readily to describing how a problem is to be solved. Flow diagrams, process control languages, and many of the common programming languages, such as COBOL, FORTRAN, and ALGOL are considered procedure-oriented languages.

procedure turn (PT) The maneuver prescribed when it is necessary to reverse direction to establish an aircraft on the intermediate approach segment or final approach course. The outbound course, direction of turn, distance within which the turn must be completed, and minimum altitude are specified in the procedure. However, unless otherwise restricted, the point at which the turn may be commenced, and the type and rate of the turn, are left to the discretion of the pilot.[ARP4107-88]

process 1. The collective functions performed in and by industrial equipment, exclusive of computer and/or analog control and monitoring equipment. 2. A series of continuous or regularly recurring steps or actions intended to achieve a predetermined result, as in refining oil, heat treating metal, or manufacturing paper. 3. A general term covering such terms as assemble, compile, generate, interpret, and compute. 4. The functions and operations utilized in the treatment of material. 5. A progressive course or series of actions. 6. Any operation or sequence of operations involving a change of energy, state, composition, dimension, or other properties that may define with respect to a datum. 7. An assembly of equipment and material that relates to some manufacturing sequence.

process alarm Alarm which occurs as a result of a process parameter exceeding some preset limit.

process and instrumentation diagram (P&ID) Show the interconnection of process equipment and the instrumentation used to control the process. In the process industry, a standard set of symbols is used to prepare drawings of processes.

process block valve The first valve off the process line or vessel used to isolate the measurement piping. See line class valve.

process calculations Installation-dependent calculations providing derived data to supplement the input signals. e.g., efficiencies, flows by material balance, etc.

process chart A graphical representation of the events in a process.

process computer A computer which, by means of inputs from and outputs to a process, directly controls or monitors the operation of elements in that process. See control computer and industrial computer. See also on-line.

process control 1. Descriptive of systems in which computers or controllers are used for automatic regulation of operations or processes. Typical are operations wherein the operation control is applied continuously and adjustments to regulate the operation are directed by the computer to keep the value of a controlled variable constant. Contrasted with numerical control. 2. An operation that regulates parameters by observation of the parameter, comparison with some desired value, and action to bring the parameter as close as possible to the desired value. 3. Adapting automatic regulatory procedures to the more efficient manufacture of products or the processing of material.

process control (industry) The ways and means by which continuous manufacturing and other industrial processes are monitored and maintained to create products of planned, uniform dimension and quality.[NASA]

process control chart A table or graph of test results or inspection data for each unit of production, arranged in chronological sequence for the entire assembly or production lot.

process control computer See process computer.

process control engineering A branch of engineering that deals with ways and means of keeping process variables as close as possible to desired values, or keeping them within specified ranges.

process control loop A system of control devices linked together to control one phase of a process.

process database Organized collection of data relating to the operation of a process.

process dynamics A set of dynamic interactions among process variables in a complex system, as in a petroleum refinery or chemical process plant.

process engineering An element of production engineering that involves selecting processes and equipment to be used, establishing the sequence and method of controlling all operations, and acquiring the tools needed to make a product.

process heat Increase in enthalpy accompanying chemical reactions or phase transformations at constant pressure (heat of crystallization and heat of sublimation are examples).[NASA]

process interrupt See interrupt, process.

process I/O Input and output operations directly associated with a process as contrasted with I/O operations not associated with the process. For example, in a process control system, analog and digital inputs and outputs would be considered process I/O whereas inputs and outputs to bulk storage would not be process I/O. See process.

process I/O bus 1. A circuit over which data or power is transmitted; often one which acts as a common connection among a number of locations. Synonymous with trunk. 2. A communications path between two switching parts.

process I/O device An apparatus for performing a prescribed function.

process I/O network A communication system. A set of OSI subnetworks interconnected by OSI intermediate systems and sharing a common network protocol.

processor(s) Abbreviated form for central processing unit. See also central processing units.

process parameter A characteristic of a process which can be monitored and measured to provide information on the process.

process pressure See pressure, process.

process reaction method A method of determination of optimum controller settings when tuning a process control loop. The method is based on the reaction of the open loop to an imposed disturbance.

process reaction rate The rate at which a process reacts to a step change.

process steam Steam used for industrial purposes other than for producing power.

process temperature See temperature, process.

process time 1. Elapsed time for the portion of the work cycle controlled by machines. 2. Elapsed time for an entire process.

process variable In the treatment of material, any characteristic or measurable attribute whose value changes with changes in prevailing conditions. Common variables are flow, level, pressure and temperature.

process variable alarm Alarm that is set whenever PV exceeds the limits set for a given input.

procuring activity A procuring activity is an organization or agency within or acting for the Government, including prime or lower tier contractors.[AS 1933-85]

producer gas Gaseous fuel obtained by burning solid fuel in a chamber where a mixture of air and steam is passed through the incandescent fuel bed. The process results in a gas, almost oxygen free containing a large percentage of the original heating value of the solid fuel in the form of CO and H_2.

product definition data A digital representation for product data that is required for the analysis, design, manufacture, test and support of a product. [AS4159-88]

production Output of a process or manufacturing facility.

production costs The process of fabrication, from raw materials through the finished products, including packaging and other prorated costs.[NASA]

production engineering An element of industrial engineering that deals with planning and control of manufacturing processes, especially for the purpose of improving efficiency and reducing costs associated with mechanical equipment.

production reliability acceptance test A test conducted under specified conditions, by, or on behalf of the government, using delivered or deliverable production items, to determine the producer's compliance with specified reliability requirements.[ARD50010-91]

productivity 1. Production output per unit of input, such as number of items per labor manhour. 2. Generically, the effectiveness with which labor, materials and equipment are used in a production operation.

products of combustion The gases, vapors, and solids resulting from the combustion of fuel.

proficiency flight A flight made by a pilot or other aircrew member or members to develop or improve proficiency in flying duties.[ARP4107-88]

profile descent An uninterrupted descent (except where level flight is required for speed adjustment) from cruising altitude/level to interception of a glide slope or to a minimum altitude specified for the initial or intermediate approach segment of a nonprecision instrument approach. The profile descent normally terminates at the final approach fix or where the glide slope or other appropriate minimum altitude is intercepted.[ARP4107-88]

prognosis The forecast of future functional status and condition based on current and accumulated inputs.[ARP 1587-81]

program 1. In data processing, a series of instructions that tell the computer how to operate. 2. Any series of actions proposed in order to achieve a certain result. 3. To design, write, and test a program. 4. A unit of work for the central processing unit from the standpoint of the executive program. See task.

program address counter Same as location counter.

program, block maintenance A program which divides major structural inspections and/or maintenance tasks into groups, or blocks, which permit convenient, economical and effective accomplishment.[ARD50010-91]

program, continuous maintenance A type of complete maintenance program which is expected to assure continuous availability of the airplane. Under this system the total maintenance effort is apportioned to each of the various and more frequent types of maintenance. NOTE: A complete overhaul at one point in time is not a part of a continuous maintenance plan.[ARD50010-91]

program control Descriptive of a system in which a computer is used to direct an operation or process and automatically to hold or to make changes in the operation or process on the basis of a prescribed sequence of events.

program controller See controller, program.

program counter A register which contains the address of the next instruction to be executed. At the end of each instruction execution the program counter is incremented by 1 unless a jump is to be carried out in which case the address of the jump label is entered.

program documentation The complete listing of a program's use, content and installation.

program, equalized maintenance A maintenance program whereby work packages are scheduled for accomplishment in such a manner that the required maintenance manpower will remain relatively constant. Portions of the heavier maintenance tasks are integrated into the lighter, or lesser, maintenance periods so that the workload fluctuations will be minimized.[ARD50010-91]

Program Evaluation and Review Technique A management control tool for managing complex projects; project milestones are defined and interrelated, then using a flowchart or computer progress is measured against the milestones; deviations from the integrated plan are used to trigger decisions or preplanned alternative actions to minimize adverse effects on the overall goal.

program generator Computer software that translates simple statements into program codes.

program library A collection of available computer programs and routines.

programmable data distributor (PDD) An optional module for the telemetry frame synchronizer that causes it to send certain predefined words from each telemetry frame out through a separate port (as to a special buffer area).

programmable logic controller (PLC) 1. A microcomputer-based control device used to replace relay logic. 2. A solid-state control system which has a user-programmable memory for storage of instructions to implement specific functions such as I/O control, logic, timing, counting, three mode (PID) control, communication, arithmetic, data and file manipulation. 3. A PLC consists of a central processor, input/output interface, and memory. 4. A PLC is designed as an industrial control system.

programmable read only memory (PROM) A hardware device that stores digital words; the computer can read the contents but cannot modify them.

programmer's console A man-machine interface, consisting of various information entry/retrieval devices, arranged as a packaged unit. It is used by the programmer of a computer control system for a manufacturing process, to monitor, modify, and control the internal behavior of the digital controller. See also operator's console, process engineer's console.

programming The design, the writing and testing of a program. See convex programming, dynamic programming, linear programming, mathematical programming, nonlinear programming, quadratic programming, macroprogramming, microprogramming, and multiprogramming.

programming language A language used to prepare computer programs.

programming language/1 (PL/1) A high-level programming language for general purpose scientific and commercial applications.

programming module A discrete identifiable set of instructions usually handled as a unit.

programming system A system consisting of a programming language and a computer program (the processor) to convert the language into absolute coding.

program parameter A parameter incorporated into a subroutine during computation. A program parameter frequently comprises a word stored relative to either the subroutine or the entry point and dealt with by the subroutine during each reference. It may be altered by the routine and/or may vary from one point of entry to another. Related to parameter.

program statement A source code instruction which translates into machine code instructions.

program storage A portion of the internal storage reserved for the storage of programs, routines, and subroutines. In many systems, protection devices are used to prevent inadvertent alteration of the contents of the program storage. Contrasted with working storage.

program timer A timing device which actuates a series of switches in programmed sequence.

progressive geometry A propellant grain configured in such a manner that the exposed surface area increases as burning progresses.[AIR913-89]

project engineering 1. Engineering activities associated with designing and constructing a manufacturing or processing facility. 2. Engineering activities related to a specific objective such as solving a problem or developing a product.

projectile penetration See terminal ballistics.

projectiles Objects, especially missiles, fired, thrown, launched, or otherwise projected in any manner, such as bullets, guided rocket missiles, sounding rockets, or pilotless airplanes. Originally, objects, such as bullets or artillery shells, projected by applied external forces.[NASA]

projection welding A welding process in which the arc is localized by projections, embossments or intersections.

Project SETI A program to search for extraterrestrial intelligence by means of radio communication.[NASA]

prolate spheroids Ellipsoids of revolutions, the longer axis of which is the axis of revolution.[NASA]

PROM See programmable read only memory.

PROM programmer A device which allows PROM's to be programmed.

prompt In data processing, instructions that appear on the CRT screen that requests response from the user.

proof-of-concept testing The evaluation of the feasibility of new concepts during the development of a system. [ARP4107-88]

proof pressure 1. The static pressure for testing an assembly, a prescribed multiple of the nominal or operating pressure.[MA2005-88] 2. The pressure above normal operating pressure which is used to non-destructively test a pressure container.[ARP906A-91] 3. Pressure above nominal system pressure of the component or system the application of which, at defined test conditions, must not lead to external leakage, permanent deformation, or any detrimental influence on function.[AIR1916-88]

proof test A high pressure test over a specific period of time for a shock absorber or other oil or air oil unit to insure against seal leakage or structural weakness.[AIR1489-88]

propagation The spreading abroad or sending forward, as of radiant energy. [NASA]

propagation constant A complex quality, characteristic of a radio frequency transmission line, which indicates the effect of the line on the transmitted wave. The real part indicates the attenuation, the imaginary part the phase shift.[ARP1931-87]

propagation delay The time period between the input of a logic signal to a device and a valid output from that signal at the output of the device.

propagation loss 1. Reduction in amplitude of the radio telemetry signal due to natural laws of attenuation. 2. Reduction in amplitude of optical signal in a fiber optic cable due to losses associated with scattering, absorption and reflection.

propagators See propagation.

propargyl groups Crosslinking agents for certain aromatic polyamides used as matrix resins in fiber composites. [NASA]

propellant An explosive substance or mixture of substances which when burned liberates gases.[AIR913-89]

propellant explosions Detonations of propellants as a result of motor malfunction.[NASA]

propellants Any agents used for consumption or combustion in rockets and from which the rockets derive their thrust, such as fuels, oxidizers, additives, catalysts, or any compounds of mixture of these. Specifically, fuels, oxidants, or a combination of mixture of fuels and oxidants used in propelling rockets. Propellants are commonly in either liquid or solid form.[NASA]

propeller, inboard component The component of a dual rotation or coaxial propeller which is mounted on the inboard shaft (nearest to power source).[ARP355-84]

propeller meter An instrument for measuring the quantity of fluid flowing past a given point; the flowing stream turns a propeller-like device, and the number of revolutions are related directly to the volume of fluid passed.

propeller, outboard component The component of a duel rotation or coaxial propeller which is mounted on the outboard shaft (farthest from the power source).[ARP355-84]

propeller wash See wake turbulence.

prop-fan technology Technology of a small diameter, highly loaded, many-bladed variable pitch advanced turboprop.[NASA]

proportional band 1. The total amount of change in the controlled variable required for the controller to move the controlled device through its complete stroke. That is, the amount that the control point of the system can be changed by varying the controlled device over its entire range. Also called throttling range.[ARP89C-70] 2. An expression of gain of an instrument (the wider the band, the lower the gain).

proportional bandwidth FM telemetry, where each subcarrier is deviated a fixed percentage of center frequency (and therefore an amount proportional to the center frequency) by data.

proportional control 1. Control of an aircraft, etc. in which control surface deflection (or output) is proportional to the movement of the remote control (or input).[AIR1489-88] 2. In proportional control, there is a definite value of controller output (and a definite position of the final controlled device) for every value of the error. The range of error which causes full range of position of the controlled device is called proportional band or throttling range. The proportional band is usually adjustable within the controller.[ARP89C-70] 3. A control mode in which there is a continual linear relationship between the deviation computer in the controller, the signal of the controller, and the position of the final control element.

proportional control action Corrective action which is proportional to the error, that is, the change of the manipulated variable is equal to the gain of the proportional controller multiplied by the error (the activating signal). See also control action, proportional.

proportional controller See controller, proportional.

proportional control mode 1. A controller mode in which the controller output is directly proportional to the controlled variable error. 2. Produces an output signal proportional to the magnitude of the input signal. 3. In a control system proportional action produces a value correction proportional to the deviation of the controlled variable from set point.

proportional control (Type 0) A control system that uses proportional forward and feedback control elements to provide an output in response to the error signal. This is referred to as a Type 0 control system.[AIR1916-88]

proportional counter An instrument whose primary element is a radiation counter tube or chamber operated in the range where the amplitude of each current pulse in its output is proportional to the energy of the quantum of radiation absorbed.

proportional gain See gain, proportional.

proportional, integral and derivative 1. Three mode controller. 2. Refers to a control method in which the controller output is proportional to the error, its time history, and the rate at which it is changing. The error is the difference between the observed and desired values of the variable that is under control action. 3. Proportional plus integral plus derivative control, used in processes where the controlled variable is affected by long lag times.

proportional integral derivative (PID) control A combination of proportional, integral and derivative control actions. Refers to a control method in which the controller output is proportional to the error, its time history, and the rate at which it is changing. The error is the difference between the observed and desired values of the variable that is under control action. Also called three-mode control.

proportional pitch In computer printers, a typeface in which each character is a different width, such as a W versus an I.

proportional region The range of operating voltage of a radiation counter tube where the gas amplification factor is greater than 1, usually about 10^3 to 10^5, and the output current pulse is proportional to the number of ions produced by the primary ionizing event, which in turn is proportional to the energy of the radiation quantum absorbed.

proportional speed floating control A variation of floating speed control is proportional speed floating. Here the motor speed is proportional to the error. The speed is high for large errors and low for small errors.[ARP89C-70] See controller, integral.

proportioning control or proportional control A control which has a unique stable position of the final control device for any given error. This may result in a difference between the control point and set point, known as droop. See also proportional band.[ARP89C-70]

proportioning probe A probe used in leak testing in which the ratio of air to tracer gas can be changed without changing the amount of flow transmitted to the detector.

propulsive efficiency The efficiency with which energy available for propulsion is converted into thrust by a rocket engine.[NASA]

PROSPRO IBM process-oriented language.

protected location A computer storage location, reserved for special purposes, in which data cannot be stored without undergoing a screening procedure to establish suitability for storage therein. May be indicated by a set guard bit.

protecting tube A closed-end tube that surrounds the measuring junction of a thermocouple and protects it from physical damage, corrosion or thermochemical interaction with the medium whose temperature is being measured. Also known as thermowell.

protective component A component, or an assembly of components, which is not liable to become defective, in service or in storage, in such a manner as to lower the intrinsic safety of the circuit. Such a component or assembly of components is considered as not subject to fault in that manner when tests of intrinsic safety are made.

protector A device which encloses a part or portion of a part to prevent mutilation. See also cap, cover, guard, mask. [ARP480A-87]

protein synthesis Process by which protein molecules are formed.[NASA]

protium See light water.

protocol 1. A formal set of conventions governing the mechanisms used for message transfer.[AIR4271-89] 2. A set of rules and formats which determine the communications behavior of an entity.

protocol data unit (PDU) Each of the seven OSI layers accepts data (SDUs) from the layer above, adds its own header (PCI), and passes the data to the layer below as a PDU. Conversely, each of the layers also accepts data from the layer below, strips off its header, and passes it up to the layer above.

protocols A set of conventions which

governs the way in which devices communicate with each other.

proton An elementary atomic particle of mass number 1 and a positive charge equal in magnitude but opposite in sign to the charge on an electron.

proton-proton reactions Thermonuclear reactions in which two protons collide at very high velocities and combine to form deuterons. The resultant deuterons may capture other protons to form tritium and the latter may undergo proton capture to form helium. The proton-proton reactions are now believed to be the principal sources of energy within the sun and other stars of its class. A temperature of 5 million degrees Kelvin and high hydrogen (proton) concentrations are required for these reactions to proceed at rates compatible with energy emission by such stars. [NASA]

protons Positively charged subatomic particles having a mass of 1.67252 times 10 to the minus 24 gram, slightly less than that of an electron.[NASA]

protoplanets Transition objects formed during primeval cloud condensation into stellar systems (stars, planets, etc.) which form the nucleus of planetary accretion.[NASA]

prototype A preproduction model suitable for evaluating a product's design, functionality, operability and form, but not necessarily its durability and reliability.

protractor A flat shaped piece of material graduated radially from a given point in units of angular measurement.-[ARP480A-87]

provider A stand rds issuing company or organization which delivers standards from their computing system to a user computer system.[AS4159-88]

proving Determination of flowmeter performance by establishing the relationship between the volume actually passed through the meter and the volume indicated by the meter.

PROWAY A standard for a process control highway based on IEEE 802.4 Token Bus immediate acknowledged MAC, a physical layer utilizing a phase continuous signaling technique.

proximity detector A sensor that produces an electric signal when the distance from the sensor to another object is less than a predetermined value.

proximity effect (electricity) Redistribution of current in a conductor caused by the presence of another conductor. [NASA]

proximity switch A device that senses the presence or absence of an object without physical contact and activates or deactivates an electrical circuit as a result.

PSAP address The fully qualified network address used to access application entities.

PSD See power spectral density.

pseudo code A code that requires translation prior to execution.

pseudo fly-by-wire An FBW control system having a means to mechanically control or override the servoactuators. [AIR1916-88]

pseudo instruction 1. A symbolic representation in a compiler or interpreter. 2. A group of characters having the same general form as a computer instruction, but never executed by the computer as an actual instruction. Synonymous with quasi instruction.

pseudo-operations A group of instructions which, although part of a program, do not perform any application related function. They generally provide information to the assembler.

pseudoplastic A material that exhibits flow (permanent deformation) at all values of shear stress, although in most cases the flow that occurs below some specific value (an apparent yield stress) is low and increases negligibly with increasing stress.

pseudopotentials Factors in an approximate method for calculation of energy bands in solids by the use of approximation which includes the many body effect.[NASA]

pseudo-random The property of satisfying one or more of the standard criteria for statistical randomness but being produced by a definite calculation process. Related to uniformly distributed random numbers.

pseudo-random number sequence A sequence of numbers, determined by some defined arithmetic process, that is satisfactorily random for a given purpose, such as by satisfying one or more of the standard statistical tests for randomness. Such a sequence may approximate any one of several statistical distributions, such as uniform distribution or normal (gaussian) distribution.

pseudo-variable A variable that requires manipulation prior to calculation or processing.

PSK See phase shift keying.

psophometer An instrument for measuring noise in electric circuits; its output is exactly one-half of the psophometric emf in a circuit when it is connected across a 600-ohm resistance in the circuit.

psycholinguistics Study of linguistic behavior such as conditioning by psychological factors including the speaker's and listener's culturally determined categories of expression and comprehension.[NASA]

psychology The science which studies the functions of the mind, such as sensation, perception, memory, thought, and more broadly the behavior of an organism in relation to its environment.[NASA]

psychometric chromaticities A color quality of a color stimulus definable by its coordinates in a uniform-chromaticity-scale such as the CIE UCS 1976 diagram. [ARP1782-89]

psychomotor performance Of or pertaining to muscular action ensuing directly from a mental process, as in the coordinated manipulation of aircraft or spacecraft controls.[NASA]

psychopharmacology The science that deals with the action of drugs on mental function.[NASA]

psychophysical color Specified by the tristimulus values of the radiant power (color stimulus) entering the eye.[ARP 1782-89]

psychrometer A device consisting of two thermometers, one of which is covered with a water-saturated wick, used for determining relative humidity; for a given set of wet-bulb and dry-bulb temperature readings, relative humidity is read from a chart. Also known as wet-and-dry-bulb thermometer.

PT Procedure Turn.

PTM (modulation) See pulse time modulation.

PTR See paper tape reader.

puck A pad or cup which carries brake lining and is attached to the rotors and/or stators of the brake assembly.[AIR 1489-88]

puff A minor combustion explosion within the boiler furnace or setting.

puller A mechanism which exerts a pulling force on a specific part in respect to another. May be of the hydraulic, knocker, lever, cam, or screw type. See also driver, extractor.[ARP480A-87]

pulley A form of wheel used to transmit and/or change direction of force applied by a flexible band, belt or the like.[ARP 480A-87]

pullout A maneuver in which an aircraft is brought to (or toward) level flight from a diving attitude or a steep glide.[ARP 4107-88]

pull-out force The axial force required to remove a terminated conductor from its attached contact or terminal; the axial force required to remove a contact from its retention member. See also force, contact retention.[ARP914A-79]

pullup A maneuver, in the vertical plane, in which the airplane is forced into a short climb, usually from approximately level flight.[ARP4107-88]

pulsating current Unidirectional current whose magnitude alternately rises and falls in a regularly recurring pattern.

pulsating flow 1. Irregular or repeating variations in fluid flow, often due to pressure variations in reciprocating pumps or compressors in the system. 2. A flow rate that varies with time, but for which the mean flow rate is constant when obtained over a sufficiently long period of time.

pulsating pressure Pressure whose magnitude alternately rises and falls in a regularly recurring pattern, and whose variation exceeds 1% per second, or 5% per minute, of the scale on the measuring instrument.

pulsation Rapid fluctuations in furnace pressure.

pulsation dampening A device installed in a gas or liquid piping system to smooth out fluctuations due to pulsating flow and/or pressure.

pulse 1. A significant and sudden change of short duration in the level of an electrical variable, usually voltage. 2. A regular or intermittent variation in a normally constant quantity characterized by a relatively rapid rise and subsequent decay within a finite time period.

pulse amplitude A general term indicating the magnitude of a pulse.[NASA]

pulse amplitude modulation (PAM) The process (or the results of the process) in which a series of pulses is generated with amplitudes proportional to the measured signal samples.

pulse-averaging discriminator In an FM system, a subcarrier demodulator which uses the width of each cycle of the subcarrier to derive a data output.

pulse band That range of controlled variable over which the final control device is pulsed to take corrective action. That is, it is the range of controlled variable that falls between a "maximum on" condition in one direction and a "maximum on" condition in the other direction. [ARP89C-70]

pulse charging Rapid and efficient method for charging electric batteries. [NASA]

pulse code 1. A code in which sets of pulses have been assigned particular meanings. 2. The binary representations of characters. 3. A series of energy pulses, or a pulse train, modulated in accordance with a data signal. 4. Generally, any data transmission scheme that utilizes pulsed energy to encode the transmitted values.

pulse code modulation (PCM) 1. Any modulation which involves a pulse code. [NASA] 2. The process of sampling a signal and encoding the height or amplitude of each sample into a series of uniform pulses. 3. In telemetry, serial data transmission (generally a series of binary-coded words).

pulse count telemetering A method of transmitting information that involves an "off-on" switching signal whose number of signal pulses per unit time represents the transmitted value.

pulse decay time The time between the instant when the amplitude of a pulse begins to drop from a specified upper limit and the instant when it reaches a specified lower limit.

pulse discriminator A device that detects pulses that have defined characteristics.

pulse Doppler radar A pulse radar system which utilizes the Doppler effect for obtaining information about the target (not including simple resolution from fixed targets).[NASA]

pulse duration 1. The time interval between the first and last instances at which the instantaneous amplitude reaches a stated fraction of the peak pulse amplitude.[NASA]

pulse duration modulation (PDM) 1. A form of pulse time modulation in which the duration of a pulse is varied. [NASA] 2. The process of sampling a signal and encoding each sample into a series of pulses whose duration or widths are proportional to the amplitude of the sample.

pulse duty factor The ratio of average pulse duration to average pulse spacing.

pulse-forming network Electrical circuitry used to generate high voltage pulses of particular shapes, and to modify the shapes of pulses generated by other sources.

pulse frequency modulation A form of pulse time modulation in which the pulse repetition rate is the characteristic varied.[NASA]

pulse height See pulse amplitude.

pulse-height discriminator An electronic circuit that selects and passes only those voltage pulses that exceed a given minimum amplitude.

pulse-height selector An electronic circuit that selects and passes only those voltage pulses whose peak amplitudes are within a specific range of values.

pulse input In process control systems, a type of input used to measure pulse or tachometer-type signals (speed, rpm, frequency, etc.).

pulse-interval modulation Modulation of a pulsed carrier wave in which the time interval between pulses is varied in accordance with the modulating signal.

pulsejet engines Compressorless jet engines in which combustion takes place intermittently, producing thrust by a series of explosions, commonly occurring at the approximate resonance frequency of the engine.[NASA]

pulse mode A type of sequential circuit in which inputs are nonperiodic pulses, as opposed to logic levels used in conjunction with clock pulses (clock mode).

pulse modulation 1. A method of varying the rate of corrective action as a function of error by moving the final control device in discrete steps. See also pulse band.[ARP89C-70] 2. Modulation of a carrier wave by a pulsed modulating wave. 3. Modulation of a carrier by a pulse train. Modulation of one or more characteristics of a pulse carrier.[NASA] 4. Modulation of one or more attributes of a pulsed carrier wave.

pulse motor See stepping motor.

pulse-position modulation 1. A form of pulse time modulation in which the position in time of a pulse is varied. Also called pulse phase modulation. [NASA] 2. A form of pulse-time modulation in which the position in time of a pulse varies in accordance with some attribute of the modulating wave.

pulse radar A type of radar, designed to facilitate range measurement, in which the transmitted energy is emitted in periodic short pulses.[NASA]

pulse repeater An electronic device that receives pulses from one circuit and transmits them into another circuit, with or without altering the frequency and waveform of the pulses.

pulse repetition frequency The rate at which pulses are repeated in a periodic pulse train.

pulse repetition period The reciprocal of pulse repetition frequency.

pulse repetition rate 1. The average rate at which pulses are repeated, whether or not the pulse train is periodic. 2. The number of electric pulses per unit of time experienced by a point in a computer, usually the maximum, normal, or standard pulse rate.

pulse rise time The time between the instant when the amplitude of a pulse reaches some specified fraction of its peak value and the instant when it reaches some specified higher fraction—unless otherwise stated, the two fractions are taken to be 10% and 90%.

pulse spacing The average time interval between corresponding locations on the waveforms of consecutive pulses.

pulse spectrum The frequency distribution of a series of sine waves that can be combined to yield a given periodic pulse train.

pulse switch A switch that provides one pulse of electric current for each cycle of operation.

pulse telemetering Any system for transmitting information in terms of electric pulses that are independent of electrical variations in the transmission channel; they can be classified as pulse duration, pulse count or pulse code systems.

pulse-time modulation Modulation of a pulsed carrier wave in which the time of occurrence of some specific point on each pulse waveform is varied from the unmodulated value in accordance with a modulating signal.

pulse transformer A type of transformer designed to convert an a-c input signal to a pulsed output signal.

pulse width See pulse duration.

pulse width modulation See pulse duration modulation.

pulse width modulation control A control scheme wherein the controller output is a train of pulses of variable duration is called pulse width modulation. The width of the "on" pulse is varied relative to the width of the "off" pulse as some function of the error.[ARP89C-70]

pultrusion Process of pulling continuous lengths of resin impregnated fiber through a shaped, heated die to produce lengths of reinforced plastic.[NASA]

pumice A light-colored, vesicular, glassy rock commonly having the composition of a rhyolite.[NASA]

pump 1. A device for converting mechanical energy into fluid energy.[ARP243B-65] 2. A machine that draws fluid into itself through an inlet and forces the fluid out through an exit port, often at higher pressure than at the inlet.

pump, fixed displacement A pump whose delivery at a fixed RPM and output pressure cannot be varied, and which delivers a relatively constant volume of fluid per cycle.[ARP243B-65]

pump, mechanical volume control A variable delivery pump whose output per cycle is controlled by external mechanical means.[ARP243B-65]

pump, pilot control A variable delivery pump whose output is controlled by the pressure at a control port.[ARP243B-65]

pump, pressure volume control A variable delivery pump whose output is controlled by its discharge pressure.[ARP 243B-65]

pump, variable delivery A pump whose delivery can be controlled independently of RPM by varying the output volume per cycle.[ARP243B-65]

punched card A durable paper-board card in which patterned punched holes can be read by a computer.

pure code A computer code that is never modified during execution; it is possible to allow many users to share the same copy of programs that are written as pure code.

purge 1. To cause a liquid or gas to flow from an independent source into the impulse line(s). 2. To introduce air into the furnace and the boiler flue passages in such volume and manner as to completely replace the air or gas-air mixture contained therein.

purge interlock A device so arranged that an air flow to the furnace above a minimum must exist for a definite time interval before the interlocking system will permit the automatic ignition torch to be placed in operation. See also purging classifications.

purge meter A device designed to measure small flow rates of liquids and gases used for purging measurement piping.

purge post An acceptable method of

scavenging the furnace and boiler passes to remove all combustible gases after flame failure controls have sensed pilot and main burner shutdown and safety shut-off valves are closed.

purge, pre-ignition An acceptable method of scavenging the furnace and boiler passes to remove all combustible gases before the ignition system can be energized.

purging Elimination of an undesirable gas or material from an enclosure by means of displacing the undesirable material with an acceptable gas or material.

purity The degree to which a substance is free of foreign materials.

push back The property of a braid of shield which allows the braid or shield to be pushed back along the cable core easily.[ARP1931-87]

pushbroom sensor modes Spacecraft instrument arrangements in which large numbers of detectors comprising linear arrays are swept by the forward motion of the spacecraft to attain increased fidelity and high sensitivity in the data captured.[NASA]

push-down list A list that is constructed and maintained so that the next item to be retrieved is the most recently stored item in the list, that is: last-in, first-out.

pusher A mechanism which exerts a pushing force on a specific part in respect to another. May be of the hydraulic, knocker, screw, lever, or cam type.[ARP 480A-87]

pusher propeller A propeller mounted aft of an engine; a propeller mounted behind the lateral axis or center of gravity of an aircraft.[ARP4107-88]

push-pull amplifiers Amplifiers in which there are two identical signal branch circuits so as to operate in phase opposition and with input and output connections each balanced to ground. [NASA]

push-pull circuit breaker Push-pull circuit breakers are those which may be manually actuated by a "push" to close and a "pull" to open.[ARP1199A-90]

push-up list A list that is constructed and maintained so the next item to be retrieved and removed is the oldest item still in the list, that is: first-in, first-out.

PVD Plan View Display.

PV tracking Set point automatically tracks the process variable when the controller is in manual.

PWM (modulation) See pulse duration modulation.

pycnometer A container of precisely known volume that is used to determine density of a liquid by weighting the filled container and dividing the weight by the known volume. Also spelled pyknometer.

pylon 1. A rigid structure protuding from the wing or fuselage to support an engine. 2. A tower that marks a turning point in a race or is used for precision flying exercises.[ARP4107-88]

pyranometer Actinometer which measures the combined intensity of incoming direct solar radiation and diffuse sky radiation. The pyranometer consists of a recorder and a radiation sensing element which is mounted so that it views the entire sky. Sometimes called solarimeter.[NASA]

pyrazines Compounds that contain a six-membered heterocyclic ring containing nitrogen atoms in the 1 and 4 positions.[NASA]

Pyrex (trademark) See borosilicate glass.

pyrgeometer An instrument for measuring radiation from the earth's surface into space.

pyrheliometer An instrument for measuring the intensity of direct solar radiation only.

pyridines Compounds that contain a six-membered heterocyclic ring containing one nitrogen atom.[NASA]

pyrimidines Compounds that contain a

six-membered heterocyclic ring containing nitrogen atoms in the 1 and 3 positions.[NASA]

pyroelectric detectors Detectors of visible, infrared, and ultraviolet radiation which rely on the absorption of radiation by pyroelectric materials. Heating of such materials by the absorbed radiation produces electric charges on opposite sides of the crystal, which can be measured to determine changes in the amount of radiation incident on the detector. These detectors usually also exhibit piezoelectric properties and may require isolation from acoustic or acceleration phenomena.

pyrogen A rocket ignition system containing a solid propellant grain as its main ignition material.[AIR913-89]

pyrographalloy See composite materials.

pyroheliometer Actinometer which measures the intensity of direct solar radiation, consisting of a radiation sensing element enclosed in a casing which is closed except for a small aperture, through which the direct solar rays enter, and a recorder unit.[NASA]

pyrolysis Chemical decomposition by the action of heat.[NASA]

pyrometer 1. Instrument that measures high temperature, e.g. of molten lavas, by electrical or optical means.[NASA] 2. Any of a broad class of temperature measuring instruments or devices. The term originally applied only to devices for measuring temperatures well above room temperature, but now it applies to devices for measuring temperatures in almost any range. Some typical pyrometers include thermocouples, radiation pyrometers, resistance pyrometers and thermistors, but usually not thermometers. It is a temperature transducer that measures temperatures by the EM radiation emitted by an object, which is a function of the temperature.

pyrophyllite A white, greenish, gray, or brown phyllosilicate mineral that resembles talc.[NASA]

pyrotechnic A mixture of chemicals designed to produce heat, light, noise, smoke, or gas.[AIR913-89]

pyroxenes A group of dark, rock-forming silicate minerals.[NASA]

pyrrhotite A common reddish-brown to bronze hexagonal mineral.[NASA]

pyrroles Compounds that contain a five-membered heterocyclic ring containing one nitrogen atom.[NASA]

q See volumetric flow rate.

QBC See queue control block.

QCD See quantum chromodynamics.

***Q* factor** 1. A rating factor for electronic components such as coils, capacitors and resonant circuits that equals reactance divided by resistance. 2. In a periodically repeating mechanical, electrical or electromagnetic process, the ratio of energy stored to energy dissipated per cycle.

Q flow Quota Flow Control.

Q meter A direct-reading instrument that measures the "Q" of an electric circuit at radio frequencies by determining the ratio of inductive reactance to resistance. Also known as quality-factor meter.

Q-switch An optical device which changes the Q (quality factor) of a laser cavity, typically raising it from a value below laser threshold to one well above threshold. This technique produces a short, intense pulse, known as a Q switched pulse. Q-switches can be based on acousto-optic or electro-optic devices, rotating mirrors, frustrated internal reflection, or saturation of absorption in a dye.

quadrant detectors Detectors which are divided up into four angularly symmetric sectors or quadrants. The amounts of radiation incident on each quadrant can be compared to one another for applications such as making sure that a beam is centered on the detector.

quadrant-edged orifice An orifice having a rounded contour at the inlet edge to yield more constant and predictable discharge coefficient at low flow velocity (Reynolds number less than 10,000).

quadratic programming In operations research, a particular case of nonlinear programming in which the function to be maximized or minimized and the constraints are quadratic functions of the controllable variables. Contrast with convex programming, dynamic programming, linear programming, and mathematical programming.

quadrature Elongation of 90 deg., usually specified as east or west in accordance with the direction of the body from the sun. The moon is a quadrature at first and last quarters. The situation of two periodic quantities differing by a quarter of a cycle.[NASA]

quadrature approximation See quadrature.

quadrilateral mechanism Also 4 bar linkage. A mechanical linkage composed of four elements or links, one of which may be the fixed base of the mechanism. Widely used in landing gear retraction and folding systems. [AIR1489-88]

quadruplex An adjective meaning fourfold, as used for a four-channel system. [AIR1916-88] [ARP1181A-85]

quadruplex system A control system containing four single paths so as to provide multiple failure capability, such as SFO/FS, DFO or DFO/FS.[AIR 1916-88] [ARP1181A-85]

quadrupole mass spectrometer A type of mass spectrometer employing a filter consisting of four conductive rods electrically connected in such a manner that, by varying the absolute potential applied to the rods, all ions except those possessing a specific mass-to-charge ratio are prevented from entering the detector.

quadrupoles A linear accelerator having four longitudinal vanes in its resonating cavity, which are shaped to create RF electric fields that simultaneously accelerate, bunch, and focus the charged particle beam.[NASA]

quad-slope converter An integrating analog-to-digital converter that goes through two cycles of dual slope conversion, once with zero input and once with the analog input being measured.

quad voter 1. A voter that performs in

a fashion to select one common output to represent the four signals or values. [AIR1916-88] 2. A voter that selects one common output to represent the four input signals or values.[ARP1181A-85]

qualification tests Those tests which are run on one or more units that are representative of the production article, to test the performance, endurance, environmental and special features of a design to demonstrate specification compliance and suitability of the unit for production and use in service.[ARP906A-91]

qualified 1. Competent, suited, or having met the requirements for a specific position or task. 2. To declare competent or capable.

quality assurance A set of systematic actions intended to provide confidence that a product or service will continually fulfill a defined need.

quality assurance system The combination of policy, organization, personnel, procedures, inspection and test plans which assure that from receipt of materials to shipment of product, the quality of the product will not be compromised and the performance requirements will be met.[AS7200/1-90]

quality control 1. An aggregate of functions designed to insure adequate quality in manufactured products by initial critical study of engineering design, materials, processes, equipment, and workmanship followed by periodic inspection and analysis.[NASA] 2. A set of systematic actions that make it possible to measure significant characteristics of a product or service and to control the characteristics within established limits.

quantity The value that marks frequency distribution interval boundaries that are determined by arranging a set of N observations in order of magnitude and marking off equal parts (N/P) of the total population P.[NASA]

quantity meter A flowmeter in which the flow is separated into known isolated quantities which are separately counted to determine the total volume passed through the meter.

quantity of light The product of the luminous flux by the time it is maintained. The unit is the lumen-hour.[ARP 798-91]

quantization The subdivision of the range of values of a variable into a finite number of nonoverlapping, and not necessarily equal, subranges or intervals, each of which is represented by an assigned value within the subrange. For example, a person's age is quantized for most purposes with the quantum of one year. See also measurement.

quantization distortion Inherent distortion introduced when a range of values for a wave attribute is divided into a series of smaller subranges.

quantization level A particular subrange for a quantized wave, or its corresponding symbol.

quantize To convert information from an analog pulse (as from a multiplexer) into a digital representation of that pulse. See encoder.

quantized pulse modulation Pulse modulation in which either the carrier wave or modulating wave is quantized.

quantum chromodynamics A gauge theory describing the interaction between quarks and gluons.[NASA]

quantum efficiency A measure of the efficiency of conversion or utilization of light or some other form of energy. [NASA]

quantum electronics The branch of electronics that essentially deals with lasers and laser devices which require quantum theory for their exact description.[NASA]

quantum noise Noise due to the discrete nature of light—i.e., its quantization into photons.

quantum theory The theory first stated

by Max Planck that all electromagnetic radiation is emitted and absorbed in quanta, each of magnitude hv, h being the Planck constant and v the frequency of the radiation.[NASA]

quantum wells Effective potential wells created by a minimum in the conduction band or a maximum in the valence band that arises when a smaller band-gap semiconductor is sandwiched between a larger band-gap semiconductor. [NASA]

quark parton model A theoretical model which summarizes our understanding of how protons and neutrons are made up of the fundamental subparticles called quarks.[NASA]

quarter amplitude A process-control tuning criteria where the amplitude of the deviation (error) of the controlled variable, following a disturbance, is cyclic so that the amplitude of each peak is one quarter of the previous peak.

quarter-wave plate A polarization retarder which causes light of one linear polarization to be retarded by one quarter wavelength (90°) relative to the orthogonal polarization.

quartz 1. Crystalline silica, an important rock-forming mineral.[NASA] 2. A natural transparent form of silica, which may be marketed in its natural crystalline state, or crushed and remelted to form fused quartz.

quasi instruction Same as pseudo instruction.

quefrencies In cepstral analysis, the frequency of periodic ripples in a spectra of a signal that contains echoes. Quefrencies are expressed in cycles per hertz or in seconds.[NASA]

quenching (atomic physics) Phenomena in which very strong electric fields cause the orbit of an electron or atom to precess rapidly so the average magnetic moment associated with its orbit angular momentum is reduced to zero. [NASA]

query language 1. Command language used to search and retrieve information. [NASA] 2. A means of getting information from a database without the need to write a program.

queue 1. Waiting line resulting from temporary delays in providing service. 2. In data processing, a waiting list of programs to be run next. 3. Any list of items; for example, items waiting to be scheduled or processed according to system-or user-assigned priorities.

queue control block (QCB) A control block that is used to regulate the sequential use of a programmer-defined facility among requesting tasks.

queued access method Any access method that automatically synchronizes the transfer of data between the program using the access method and input/output devices, thereby eliminating delays for input/output operations.

queuing An ordered progression of items into and through a system or process, especially when there is waiting time at the point of entry.

queuing discipline The rules or priorities for queue formation within a system of "customers" and "servers" as well as the rules for arrival time and service time.

queuing theory A form of probability theory useful in studying delays or lineups at servicing points.

Quevenne scale A specific gravity scale used in determining the density of milk; a difference of 1° Quevenne is equivalent to a difference of 0.001 in specific gravity, and therefore 20° Quevenne expresses a specific gravity of 1.020.

quick attach-detach (QAD) mounting flange A disconnect device that can attach the starter or accessory to the engine using simple tools on a single fastener. Usually incorporates a special flange which is pre-bolted to the mounting pad.[ARP906A-91]

quick disconnect A disconnect that can

be uncoupled or coupled without the use of tools, commonly employed in hydraulic and pneumatic lines. May be self-sealing when disconnected. [ARP906A-91] See also coupling.

quick engine change (QEC) A package or kit of hardware items not included on the engine as delivered by the engine producer but required in the build-up of the engine prior to installation in an aircraft.[ARP1587-81]

quick look 1. A feature of NAS Stage A and ARTS which provides the controller with the capability to display full data blocks of tracked aircraft from other controller positions.[ARP4107-88] 2. Essentially, an "instant replay" of data, generally at the same rate at which it was recorded.

quick return A device that makes the return stroke of a reciprocating machine element faster than the power stroke.

quiescent flow 1. A continuous leakage flow to return through hydraulic control interface.[AIR1489-88] 2. Total internal leakage of a complete hydraulic system or a branch of a hydraulic system at rest.[AIR1916-88]

quinoxalines A group of heterocyclic compounds consisting of a benzene ring condensed with a diazine ring.[NASA]

quota flow control (Q flow) A flow control procedure by which the central flow control function (CFCF) restricts traffic to an ARTC center area having an impacted airport, thereby avoiding sector/area saturation. May be implemented to prevent air traffic saturation in a sector/area serving an airport having weather or some facility problems that prevent it from accepting its normal flow of aircraft.[ARP4107-88]

R See rankine.

rack 1. A framework, stand, or grating on or in which articles are placed.[ARP 480A-87] 2. A housing for electronic equipment which permits convenient removal of portions of the equipment. [ARP914A-79] 3. A standardized steel framework designed to hold 19-in.-wide panels of various heights that have units of electronic equipment mounted on them. Also known as relay rack. 4. A frame for holding or displaying articles. 5. A bar having racon beacons. See radar beacons.

rad See radian.

RADAR Acronym for RAdio Detection And Ranging, (generally printed in lower case). A device which, by measuring the time interval between transmission of radio pulses and reception of their echoes and correlating the angular orientation of the radiated antenna beam or beams in azimuth or elevation, or both, provides information on range, azimuth, or elevation of objects in the path of the transmitted pulses. [ARP4107-88]

radar altimeter See radio altimeter.

radar altitude The altitude of an aircraft or spacecraft as determined by a radio altimeter; thus, the actual vertical distance from the terrain.[ARP4107-88]

radar approach control facility See RAPCON facility.

radar astronomy The study of celestial bodies within the solar system by means of radiation originating on earth but reflected from the body under observation.[NASA]

radar beacon Beacon transmitting characteristic signals on radar frequency, permitting crafts to determine their bearings and sometimes the range of the beacons.[NASA]

radar contact The term air traffic controllers use to indicate that an aircraft is identified on the radar display and that radar service can be provided until radar identification is lost or radar service is terminated. When the pilot is informed of radar contact, he/she is no longer required to report over compulsory reporting points.[ARP4107-88]

radar-controlled departure Use of surveillance radar (ASR) vectors to establish an aircraft on the en route track, expediting the departure by using radar traffic separation standards.[ARP4107-88]

radar cross sections The ratios of power returned in a radar echo to power received by the target reflecting the signal.[NASA]

radar direction finder See radio direction finder.

radar display See radarscope.

radar flight following The observation of the progress of a radar-identified aircraft, whose primary navigation is being provided by the pilot, wherein the controller retains and correlates the aircraft identity with the appropriate target or target symbol displayed on the radar scope.[ARP4107-88]

radar geology The application of imaging radar to geologic problems.[NASA]

radar handoff The action whereby radar identification of an aircraft is made known from one controller to another without interruption of the radar flight following.[ARP4107-88]

radar homing missiles Radar-following missiles designed to attack radar transmitters.[NASA]

radar identification The process of ascertaining that an observed radar target is the radar return from the particular aircraft.[ARP4107-88]

radar-identified aircraft An aircraft whose position has been correlated with an observed target or symbol on the radar display.[ARP4107-88]

radar-monitored departure Use of ASR to monitor departing aircraft, with advisories given concerning other radar-observed traffic, which might come into a conflict situation with the departing aircraft.[ARP4107-88]

radar monitoring See radar service.

radar navigation guidance The vectoring of aircraft to provide course guidance.[ARP4107-88]

radar networks A series of tracking stations each of which can individually or jointly track a target by utilizing an interchange of radar information used for multiradar tracking.[NASA]

radar point out/point out The term used between controllers to indicate radar handoff action where the initiating controller plans to retain communications with an aircraft penetrating the other controller's airspace and additional coordination is required.[ARP 4107-88]

radar range The distance from a radar to a target as measured by the radar. The maximum distance at which a radar set is effective in detecting targets. [NASA]

radar reflector Device capable of or intended for reflecting radar signals. [NASA]

radar scanning The action or process of moving or directing a searching radar beam.[NASA]

radarscope The viewing screen (generally a cathode-ray tube) portion of radar equipment upon which electronic pulses represent the distance and bearing of radar target returns. In modern radar systems, computer-generated displays represent the aircraft and necessary information about it (such as altitude or call sign).[ARP4107-88]

radar separation See radar service.

radar service A term which encompasses one or more of the following services based on the use of radar which can be provided by a controller to a pilot of a radar identified aircraft: (a) radar separation: The use of radar to maintain spacing between aircraft in accordance with established minimums. (b) radar navigational guidance: Radar vectoring of aircraft to provide course guidance. (c) radar monitoring: The radar flight following of an aircraft (whose primary navigation is being performed by the pilot) to observe and note deviations from its authorized flight path, airway, or route. Radar monitoring may include the monitoring of instrument approaches as well as en route radar flight following.[ARP4107-88]

radar target Objects which reflect a sufficient amount of a radar signal to produce an echo signal on the radar screen.[NASA]

radar traffic advisories See traffic advisories.

radar traffic information service See traffic advisories.

radar vectoring The use of radar to provide navigational or traffic separation guidance. In vectoring, the controller uses his/her radar display to ascertain the aircraft's position and track with respect to the target of concern and gives the pilot the heading to fly in order to make the desired track or separation from other aircraft.[ARP 4107-88]

radar weather echo intensity level Because radar echo intensities have a direct correlation with the degree of turbulence and other weather features associated with thunderstorms, the National Weather Service has categorized the intensities of radar weather echoes into the following levels: 1. Level 1 (WEAK) and level 2 (MODERATE). Light to moderate turbulence is possible with lightning. 2. Level 3 (STRONG). Severe turbulence possible, with lightning. 3. Level 4 (VERY STRONG). Severe turbulence likely, with lightning. 4. Level 5 (INTENSE). Severe turbulence, with lightning and organized wind gusts. Hail likely. 5. Level 6 (EXTREME). Severe turbulence with large hail, lightning, and extensive wind gusts.[ARP4107-88]

radial A magnetic bearing extending from a VOR/VORTAC/TACAN navigation facility.[ARP4107-88]

radial engine An internal-combustion reciprocating engine, the cylinders of which are disposed radially about the crankshaft in circular rows of five or more cylinders to a row, the row(s) being a plane perpendicular to the crankshaft, which rotates while the cylinders remain stationary. See also rotary engine.[ARP4107-88]

radial velocity In radar, that vector component of the velocity of a moving target that is directed away from or toward the ground station.[NASA]

radian Metric unit for a plane angle.

radiance 1. In radiometry, a measure of the intensic radiant intensity emitted by a radiator in a given direction. It is the irradiance (radiant flux density) produced by radiation from the source upon a unit surface area oriented normal to the line between source and receiver, divided by the solid angle subtended by the source at the receiving surface. It is assumed that the medium between the radiator and receiver is perfectly transparent, therefore radiance is independent of attenuation between source and receiver.[NASA] 2. Radiant flux per unit solid angle per unit of projected source area.

radiancy The rate of radiant energy emission from a unit area of a source in all the radial directions of the overspreading hemisphere.[NASA]

radiant flux The rate of flow of radiant energy with respect to time.

radiant fluxmeter A device which measures the amount of radiant flux emitted or absorbed. The units typically would be watts per unit area.

radiant intensity The energy emitted per unit time per unit solid angle along a specific linear direction.

radiation Transfer of heat by waves or particles.

radiation belts Envelopes of charged particles trapped in the magnetic field of a spatial body.[NASA]

radiation chemistry The branch of chemistry concerned with the chemical effects, including decomposition, of energetic radiation of particles of matter.[NASA]

radiation counter Instrument used for detecting or measuring moving subatomic particles by a counting process.[NASA]

radiation damage A general term for the deleterious effects of radiant energy-either electromagnetic radiation or particulate radiation—on physical substances or biological tissues.

radiation dosage The amount of radiation absorbed by a material, system, or tissue in a given amount of time; usually measured in units as roentgen.[NASA]

radiation exposure see radiation dosage.

radiation loss A comprehensive term used in a boiler-unit heat balance to account for the conduction, radiation, and convection heat losses from the settings to the ambient air.

radiation medicine See nuclear medicine.

radiation pressure Pressure exerted upon any material body by electromagnetic radiation incident upon it.[NASA]

radiation pyrometer An instrument that uses the radiant power emitted by a hot object in determining its temperature.

radiation sickness A syndrome following intense acute exposure to ionizing radiation. It is characterized by nausea and vomiting a few hours after exposure. Further symptoms include bloody diarrhea, hemorrhage under the skin (and internally), epilation (hair falling out), and a decrease in blood cell level.

[NASA]

radiation thermometer Any of several devices for determining the temperature of a body by measuring its emitted radiant energy, without physical contact between the sensor and the body.

radiation transport The study of radiation from emission to absorption. [NASA]

radiation trapping Confinement of radiation with a magnetic field.[NASA]

radiator Any source of radiant energy, especially electromagnetic radiation. Devices that dissipate the heat from something as from water or oil, not necessarily by radiation only.[NASA]

radioactive half life The time it takes for the radioactivity of a specific nuclide to be reduced by one half.

radioactive tracer A radioactive nuclide used at small concentration to follow the progress of some physical, chemical or biological process; typically, a radioactive isotope is substituted for a small proportion of the same element normally present, and the movement and behavior of the elemental substance or of a radioactively "labeled" compound is observed with radiation detection instruments.

radioactivity Spontaneous disintegration of atomic nuclei with emission of corpuscular or electromagnetic radiations. The number of spontaneous disintegrations per unit mass and per unit time of a given unstable (radioactive) element, usually measured in curies. [NASA]

radio altimeter A device that measures the altitude of a craft above the terrain by measuring the elapsed time between the transmission of radio waves from the craft and the reception of the same waves reflected from the terrain; also called radar altimeter.[ARP4107-88]

radio assisted detection and ranging See radar.

radio astronomy The study of celestial objects through observation of radio-frequency waves emitted or reflected by these objects. Specifically, the study of celestial objects by measurement of the radiation emitted by them in the radio-freuqency range of the electromagnetic spectrum.[NASA]

radio beacon(s) Transmitters, together with their associated equipment, that emit signals enabling the determination, by means of suitable receiving equipment, of direction, distance, or position with respect to the beacon.[NASA] See also nondirectional beacon.

radiobiology 1. The study of the effects produced on living organisms by radiation.[NASA] 2. A branch of biology that deals with the effect of radiation on biological processes and tissues.

radiocardiography The technique of recording of an intravenously injected radioisotope in the heart chambers. [NASA]

radio channel A band of frequencies wide enough to be used for radio communication.

radio control Remote control of a pilotless airplane, rocket, or spacecraft by means of radio signals that activate controlling devices.[NASA]

radio direction finder (RDF) A radio receiving set, together with its associated equipment, including a directional antenna that can be rotated. The position of the antenna when the peak signal is received indicates the direction to the transmitting station.[ARP4107-88]

radio-doppler Determining the radial component of relative velocity between an object and a fixed or moving point of observation by measuring the difference in frequency between transmitted radio-frequency waves and reflected-waves returning from the object.

radio engineering A discipline often considered part of electronic engineering that deals with generating, transmitting and receiving radio waves, and with the design, fabrication and testing of equipment for performing these functions.

radio frequencies Frequencies at which coherent electromagnetic radiation of energy is useful for communications purposes.[NASA]

radio-frequency heating See electronic heating.

radio frequency interference (RFI) A type of electrical noise that can affect electronic circuits adversely.

radio frequency ion thrustor engines See RIT engines.

radio frequency radiation see radio waves.

radiography A form of nondestructive testing that involves the use of ionizing radiation to detect flaws, characterize internal structure or measure thickness of metal parts or the human body; it may involve determining the attenuation of radiation passing through a test object.

radio horizons Loci or points at which direct rays from a radio transmitter become tangential to the earth's surface.[NASA]

radioimmunoassay A medical diagnostic procedure for the components (hormones and immunoglobulins primarily) as a well as pharmaceuticals in the blood. The RIA is based on the antigen antibody reactions.[NASA]

radio interferometer Interferometer operating at radio frequencies. Radio interferometers are used in radio astronomy and in satellite tracking.[NASA]

radioisotope 1. A radioactive isotope of a chemical element. 2. A nonpreferred synonym for radionuclide.

radioisotope tracer flowmeter A device for determining flow rate by injecting a radioactive substance into the fluid stream; two types are available—peak timing, in which flow rate is calculated from a measurement of the time it takes a peak concentration of tracer element to pass between two fixed points along the flow path, and dilution method, in which a known concentration of tracer is injected into the fluid stream and its concentration at a point downstream determined by analysis.

radio jets (astronomy) Jets of energetic particles occurring in radio galaxies and quasars usually emitted from the nuclear (active) region of the extragalactic radio source.[NASA]

radiology The application of radiation science to the study of medicine, especially the diagnosis and treatment of injury and disease.

radioluminescence Emission of light due to radioactive decay of a nuclide.

radio magnetic indicator (RMI) An aircraft navigational instrument coupled with a gyro compass or similar compass that indicates the direction of a selected NAVAID and indicates bearing with respect to the heading of the aircraft.[ARP4107-88]

radio meteors Meteors which have been detected by the reflection of radio signals from the meteor trails of relatively high ion density (ion columns).[NASA]

radiometer 1. Instrument for detecting and, usually, measuring radiant energy.[NASA] 2. An instrument which measures optical power in radiometric units (watts). It measures electromagnetic power linearly over its entire spectral range.

radiometric analysis A method of quantitative chemical analysis based on measuring the absolute disintegration rate of a radioactive substance of known specific activity.

radiometric correction An effort to correct the intensity range of an image.[NASA]

radiometric rectification See radiometric correction.

radiometric resolution The sensitivity of the sensor to distinguish between gray levels.[NASA]

radionuclide Any nuclide which undergoes spontaneous radioactive decay.

radio ranges See radio beacons.

radiosonde(s) A balloon-borne instrument for taking meteorological data. It consists of transducers for determining temperature, humidity and barometric pressure; a modulator for converting transducer outputs to radio signals; a selector switch for establishing transmission sequence; and a transmitter for generating a radio-frequency carrier wave.

radio spectra Frequencies of electromagnetic radiation usable for radio communication.[NASA]

radio telescope Device for receiving, amplifying, and measuring the intensity of radio waves originating outside the earth's atmosphere or reflected from the body outside the atmosphere.[NASA]

radio waves Waves produced by oscillation of an electric charge at a frequency useful for radio communication.[NASA]

radius of gyration Also radius of inertia. If the moment of inertia of a solid body with respect to a given axis is the limit of the sum of the products of the masses of each of the elementary particles into which the body may be conceived to be divided and the square of their distance from the given axis. If $I=K^2{}_m$, the quantity K is called the radius of gyration or radius of inertia.[AIR1489-88]

radius rod A rod or link in a mechanical system which serves to maintain a point in locus, equidistant from a given point. Used in landing gear retraction/extension system to predetermine and maintain the path of motion of the gear during the cycle.[AIR1489-88]

radix Also called the base number; the total number of distinct marks or symbols used in a numbering system. For example, since the decimal numbering system uses ten symbols (0, 1, 2, 3, 4, 5, 6, 7, 8, 9), the radix is ten. In the binary numbering system the radix is two, because there are only two marks or symbols (0, 1).

radix-50 A storage format in which three ASCII characters are packed into a sixteen-bit word.

radix complement A number obtained by subtracting each digit of the given number from one less than the radix, then adding to the least significant digit, executing all required carries; in radix two (binary), the radix complement of a number is used to represent the negative number of the same magnitude.

radix notation A positional representation in which the significances of any two adjacent digit positions have an integral ratio called the radix of the less significant of the two positions. Permissible values of the digit in any position range from zero to one less than the radix of that position.

radix number The quantity of characters for use in each of the digital positions of a numbering system. In the more common number systems the characters are some or all of the Arabic numerals. Unless otherwise indicated, the radix of any number is assumed to be 10. For positive identification of a radix 10 number, the radix is written in parentheses as a subscript to the expressed number. Synonymous with base and base number.

radix point Also called base point, binary point, decimal point, and others, depending on the numbering system; the index that separates the integral and fractional digits of the numbering system in which the quantity is represented.

radome(s) 1. Dielectric housings for antennas. (From RAdar DOME. Pronounced ray-domes).[NASA] 2. A weatherproof cover for a primary radar device constructed of a material which is transparent to radio waves.

RAIL Runway Alignment Indicator Light System.

railgun accelerator Linear dc motor consisting of a pair of rigid, field-producing rails, and a movable conducting armature.[NASA]

rails, optical Long, linear rods which are attached to an optical bench. Optical mounts can be affixed to the rail, creating an optical bench.

rake A mechanical device with projecting tubes, prongs, etc., set in such a manner as to pick up temperature, pressure, etc.[ARP480A-87]

RAM See random access memory.

ram The moving portion in the head of a crimping tool.[ARP914A-79]

Raman shifter A device which alters the wavelength of light by inducing Raman shifts in the light passing through it. Raman shifts are changes in photon energy caused by the transfer of vibrational energy to or from the molecule.

ram extrusion An extrusion process in which the insulation material (in a fine powder form), is mixed with a lubricant and forced through a circular die without heating.[ARP1931-87]

ramjet engine Jet engine with no mechanical compressor consisting of specially shaped tubes or ducts open at both ends, the air necessary for combustion being shoved into the duct and compressed by the forward motion of the engine, where the air passes through a diffuser and is mixed with fuel and burned, the exhaust gases issuing in a jet from the rear opening. Ramjet engines cannot operate under static conditions.[NASA]

ramp the area of an airport in which aircraft are loaded and unloaded.[ARP 4107-88]

ramp encoder An analog-to-digital conversion process whereby a binary counter is incremented during the generation of a ramp voltage; when the amplitude of the ramp voltage is equal to the amplitude of the voltage sample, the counter clock is inhibited. The counter contents therefore contain the binary equivalent of the sampled data.

ramp generator An electrical power supply which generates a voltage that increases at a constant rate. A plot of voltage vs. time shows a ramp-like waveform.

ramp response The total (transient plus steady-state) time response resulting from a sudden increase in the rate of change in the input from zero to some finite, constant value. See also response, ramp.

ramp response time See time, ramp response.

ramp weight The weight of an airplane before engine start; includes the takeoff weight plus a fuel allowance for engine start, taxi, runup, and takeoff ground roll to liftoff. See also takeoff weight. [ARP4107-88]

random access 1. The process of obtaining data from, or placing data into, storage when there is no sequential relation governing the access time to successive storage location.[NASA] 2. Synonymous with direct access. See also access, random.

random access device A device in which the access time is effectively independent of the location of the data. Synonymous with direct access device.

random access memory (RAM) This is memory that can be written into or read from and allows access to any address within the memory. RAM is volatile in that contents are lost when the power is switched off. A type of computer memory used for temporary storage of data; allows the CPU to have fast access to read or change any of its memory locations.

random access storage A storage technique in which the time required to obtain information is independent of the location most recently obtained.

random error(s) 1. Errors that are not systematic, are not erratic, and are not mistakes.[NASA] 2. Precision or repeatability data that deviate from a mean value in accordance with the laws of chance.

random file A collection of records, stored on random access devices such as discs. Algorithms are used to define the relationship between the record key and the physical location of the record.

randomized non-return-to-zero A coding scheme which provides the greatest packing density on instrumentation tape with low possibility of DC components.

randomizer A hardware device to inject a pseudo-random bit sequence into an NRZ wavetrain, thereby guaranteeing frequent data transitions so that the low-frequency component is not too low for transmission or recording. A "derandomizer" removes the sequence and restores data to the original form; a form of data enhancement.

random noise Oscillations whose instantaneous amplitudes occur, as a function of time according to a normal (Gaussian) curve.[NASA]

random number generator A special routine or hardware designed to produce a random number or series of random numbers according to specified limitations.

random numbers 1. Expressions formed by sets of digits selected from a sequence of digits in which each successive digit is equally likely to be any of the digits.[NASA] 2. A series of numbers obtained by chance. 3. A series of numbers considered appropriate for satisfying certain statistical tests. 4. A series of numbers believed to be free from conditions which might bias the result of a calculation. 5. See pseudo-random number sequence.

random sampling Selecting a small number of items for inspection or testing (the sample) from a much larger number of items (the lot or population) in a manner that gives each item in the population an equal probability of being included in the sample.

random sequence In welding, the technique of depositing a longitudinal weld bead in increments of random length and location.

random variables Variables characterized by random behavior in assuming their different possible values. Mathematically, they are described by their probability distribution, which specifies the possible values of a random variable together with the probability associated (in an appropriate sense) with each value. Random variables are said to be continuous if their possible values extend over a continuum and discrete if their possible values are separated by finite intervals.[NASA]

random walk The path followed by a particle which makes random scattering collisions with other particles in a gaseous or liquid medium.

random walk method In operations research, a variance-reducing method of problem analysis in which experimentation with probabilistic variables is traced to determine results of a significant nature. Uninteresting walks add only to the variance of the process and thus contribute nothing. An interesting walk tends to lead toward a predictive solution.

range 1. For instrumentation, the set of values over which measurements can be made without changing the instrument's sensitivity. 2. The extent of a measuring, indicating or recording scale. 3. An area within defined boundaries or landmarks used for testing vehicles, artillery or missiles, or for other test purposes. 4. The maximum distance a vehicle, aircraft or ship can travel without refueling. 5. The distance between a weapon and a target. 6. The maximum distance a radio, radar, sonar or television transmitter can send a signal without excessive attenuation. 7. The set of values that a quantity or function may assume. 8. The difference between the highest and lowest value that a quantity or function may assume. 9. The difference between the maximum and minimum values of physical output over which an instrument is designed to operate normally. See also error range.

rangeability 1. Describes the relationship between the range and minimum quantity that can be measured. 2. The ratio of the maximum flow rate to the minimum flow rate of a meter. 3. Installed rangeability may be defined as the ratio of maximum to minimum flow within which limits the deviation from a desired installed flow characteristic does not exceed some stated limits. 4. Inherent rangeability, a property of the valve alone, may be defined as the ratio of maximum to minimum flow coefficients between which the gain of the valve does not deviate from a specified gain by some stated tolerance.

range check In data processing, a validation that data is within certain limits.

range errors Errors in radar range measurement due to the propagation of radio energy through a nonhomogeneous atmosphere. These errors are due to the fact that the velocity of radio wave propagation varies with the index of refraction and that ray travel is not in straight lines through actual atmospheres. The resulting range errors are generally insignificant.[NASA]

range, wire The designation of wire/conductor sizes that a given conductor barrel, ferrule, grommet or accessory will accommodate.[ARP914A-79]

ranging The measurement of distance by timing how long it takes a light, radio frequency sound or ultrasound pulse to make a round trip from the source to a distant object.

rank 1. To arrange in an ascending or descending series according to importance. 2. Position in some ascending or descending series.

Rankine An absolute temperature scale where the zero point is defined as absolute zero (the point where all spontaneous molecular motion ceases) and the scale divisions are equal to the scale divisions in the Fahrenheit system; 0 °F equals approximately 459.69 °R.

Rankine cycle An ideal thermodynamic cycle consisting of heat addition at constant pressure, isentropic expansion, heat rejection at constant pressure, and isentropic compression; used as an ideal standard for the performance of heat-engine and heat-pump installations operating with a condensable vapor as the working fluid, such as a steam power plant.[NASA]

RAPCON Acronym for Radar APproach CONtrol.

RAPCON facility A terminal ATC facility that uses radar and nonradar capabilities to provide approach control services to aircraft arriving, departing, or transiting airspace controlled by the facility.[ARP4107-88]

rapid quenching (metallurgy) Rapid cooling of molten metals or alloys to achieve maximum uniformity in the crystal structure.[NASA]

rapid solidification See rapid quenching (metallurgy).

rapid traverse A machine tool mechanism that quickly moves the workpiece to a new position while the cutting tool is retracted.

rare gases Gases such as helium, neon, argon, krypton, xenon, and radon, all of whose shells of planetary electrons contain stable numbers of electrons so that the atoms are almost completely chemically inactive.[NASA]

rare gas-halide lasers A class of lasers in which the inert gases are used as the amplifying medium.[NASA]

RAST (Recover, Assist, Secure, and Traverse) A system, installed in the helicopter landing area of small ships, used to guide the helicopter to the landing area by means of a tensioned cable. Once landed, the helicopter is automatically secured to the deck. Further, the system assists in traversing the helicopter between the landing area and the hangar.[AIR1489-88]

raster A set of lines that provide essentially uniform coverage of a given area—for instance, the set of parallel lines on a television picture tube that are most easily seen when there is no picture.

raster pitch The distance between corresponding points of adjacent active scan lines of a raster scan.[ARP1782-89]

raster scan 1. A regular pattern of scanning lines.[ARP1782-89] 2. A form of display similar to television in which the signal is scanned backwards and forwards across the screen.

rate See control action, derivative.

rate action 1. Another name for the derivative control mode. 2. A control action which produces a corrective signal proportional to the rate at which the controlled variable is changing. 3. Rate action produces a faster corrective action than proportional action alone.

rate control See derivative control.

rate control action See control action, derivative.

rated capacity The manufacturers stated capacity rating for mechanical equipment, for instance, the maximum continuous capacity in pounds of steam per hour for which a boiler is designed.

rated conditions The inlet and outlet conditions at which the starter is designed to operate and performance calibration is desired. Usually expressed in terms of pressure, temperature, and flow rate, or current and volts.[ARP 906A-91]

rated continuous power The maximum power that can be delivered by the actuation system for an indefinite period without sustaining damage or reducing life.[AIR1916-88]

rated flow 1. Specified flow at steady-state conditions for a component or system.[AIR1916-88] 2. Design flow rate for a piping system or process vessel. 3. Normal operating flow rate for a fluid passing through a piping system.

rated horsepower The maximum or allowable power output of an engine, turbine or other prime mover under normal, continuous operating conditions.

rated load 1. The static linear impedance required to draw rated current under a steady state condition of rated units.[AS8011A-85] 2. A specified steady-state load applied to the actuator for determining rated velocity. Usually, rated load is an opposing load.[AIR1916-88] 3. The maximum design load for a machine, structure or vehicle.

rated power The power required at rated load and rated velocity, also called the power point. This point usually determines the maximum power required from the actuation system.[AIR1916-88]

rated pressure See nominal pressure.

rated temperature The maximum temperature at which an electric component can operate for extended periods without loss of its basic properties.[ARP1931-87]

rated velocity The maximum velocity the servoactuator is capable of under specified external load conditions and rated pressures.[ARP1281A-81]

rated voltage The voltage at which an electrical component can operate for extended periods of time without undue degradation.[ARP1931-87]

rate gain See gain, derivative action.

rate gyroscope A gyro wheel mounted

in a single gimbal ring in such a manner that rotation about an axis perpendicular to both the gimbal axis and gyro axis produces a precessional torque proportional to the rotation rate.

rate limit The maximum velocity of the control system output. Usually, rate limit is defined under aiding load condition with saturation signals applied to the servoactuator.[AIR1916-88]

rate of blowdown A rate normally expressed as a percentage of the water fed.

rate-of-climb indicator A navigation instrument that indicates the rate at which an aircraft gains or loses altitude.

rate response A relationship describing the output of a control system as a function of its input signal.

rate time In the action of a proportional-plus-rate or proportional-plus-reset-plus-rate controller, the time by which the rate action advances the proportional action on the controlled device.

rating See load.

ratio controller 1. A controller that maintains a predetermined ratio between two or more variables. 2. Maintains the magnitude of a controlled variable at a fixed ratio to another variable. See also controller, ratio.

ratio meter A measuring instrument whose pointer deflection is proportional to the ratio of the currents passing through two coils; it measures the quotient of two electrical qualities.

ratio of specific heats Specific heat at constant pressure divided by specific heat at constant volume.

ratio spectrofluorometer A type of instrument used in chemical assaying to determine proportions when two compounds are similar in bioassay and spectrophotometry but differ markedly in fluorescence; quantitative assays can be made either by proportionality or by use of a linearity curve.

ratio thermometer A device consisting essentially of two radiation thermometers in the same housing, the output of each thermometer having a separate wavelength response; the output is a ratio signal that is a function of temperature, but that is relatively insensitive to target size and therefore is as accurate for small radiating bodies as it is for larger ones.

ratio-type telemeter A telemeter that translates data in terms of the relative phase relation between two electrical quantities, or their relative magnitudes.

R.A.T.O. (Rocket assisted takeoff) A system and/or installation on an air vehicle or missile which utilizes a rocket motor assist during the launch or takeoff phase.[AIR1489-88]

raw data In data processing, information that has not been processed or analyzed by the computer.

rawinsondes Combinations of raob and rawin; observations of temperature, pressure, relative humidity, and winds-aloft by means of radiosonde and radio direction finding equipment of radar tracking.[NASA]

raw water Water supplied to the plant before any treatment.

ray acoustics See geometrical acoustics.

Rayleigh-Benard convection The flow of a fluid contained between horizontal thermally conducting plates and heated from below. The Rayleigh number is proportional to the temperature difference between the plates.[NASA]

Rayleigh disc A special form of acoustic radiometer used to measure particle velocities.

Rayleigh scattering Any scattering process produced by spherical particles whose radii are smaller than about one tenth the wavelength of the scattered radiation.[NASA]

Rayleigh waves Two dimensional barotropic disturbances in a fluid having one or more discontinuities in the vorticity profile. Surface waves associated with the free boundary of a solid, such that a surface particle describes an ellipse whose major axis is normal to the surface and whose center is at the undisturbed surface. At maximum particle displacement away from the solid surface the motion of the particle is opposite to that of the wave.[NASA]

ray optics See geometrical optics.

ray tracing A procedure used in the graphical determination of the path followed by a single ray of radiant energy as it travels through media of varying indices of refraction.[NASA]

RBN Radio Beacon.

RCAG Remote Communications Air/Ground Facility.

RCLM Runway Centerline Marking.

RCLS Runway Centerline Light System.

RCO Remote Communications Outlet.

RC oscillator Any oscillator whose frequency is determined by the interaction between resistors and capacitors in an electronic circuit.

RCR Runway Condition Reading.

RCTL See resistor-capacitor-transistor logic.

Re See Reynolds number.

reactance (X) 1.Opposition offered to the flow of alternating current by inductance or capacitance of a component or circuit. The imaginary portion (j factor) of the impedance.[ARP1931-87] 2. A component in an electrical circuit which is due to the presence of capacitative or inductive elements and not resistive elements, and which opposes the flow of electric current.

reactance drop The voltage drop 90° out of phase with the current.

reaction A chemical transformation or change brought about by the interaction of two substances.

reaction bonding Chemical combining of ingredients to produce silicon nitride ceramics.[NASA]

reaction factor See load factor.

reaction jets See jet thrust.

reaction products The substances formed in a chemical reaction—the desired items as well as the unwanted fumes, sludge, residues, etc.[NASA]

reaction time In human engineering, the interval between the input signal (physiological) or a stimulus (psychophysiological) and the response elicited by the signal. An inherent perceptual limitation which requires an individual specific amount of time for information processing before action is taken.[ARP 4107-88]

reactive volt-ampere meter See varmeter.

reactor 1. A circuit element that introduces capacitative or inductive reactance. 2. A vessel in which a chemical reaction takes place. 3. An enclosed vessel in which a nuclear chain reaction takes place.

reactor cores In nuclear reactors, the regions containing the fissionable material.[NASA]

reactor, endothermic A reactor which absorbs heat from the surroundings.

reactor, exothermic A reactor which generates heat.

reactor fuels See nuclear fuels.

reactor safety Theoretical and experimental investigations of the behavior of reactor types and designs under various real or hypothetical accidents. [NASA]

read 1. In data processing, to retrieve information from any memory medium. 2. To playback data from a tape, disk, and the like, or to obtain the contents of computer memory.

readability The smallest fraction of the scale on an instrument which can be easily read—either by estimation or by use of a vernier.

reading The indicated value determined from the scale of an indicating instrument, or from the position of the index on a recording instrument with respect to an appropriate indicating scale.

read-only memory (ROM) Storage containing data that cannot be changed by computer instruction, but requires alteration of construction circuits; therefore, data that is nonerasable and reusable, or fixed.

read-only memory devices Computer devices for storing data in permanent or nonerasable form.[NASA]

readout In data processing, the display of information on the CRT screen or other display unit.

read time See access time.

read write memory Memory whose contents can be written and read from. Read write memory is not necessarily random access memory.

real number Any number that can be represented by a point on a number line.

real time 1. In data processing, the actual time that is required to solve a problem. 2. Pertaining to the actual time during which a physical process transpires. 3. Pertaining to the performance of a computation during the actual time that the related physical process transpires in order that results of the computation can be used in guiding the physical process.

real-time clock A clock which indicates the passage of actual time, in contrast to a fictitious time set up by the computer program; such as, elapsed time in the flight of a missile, wherein a 60-second trajectory is computed in 200 actual milliseconds, or a 0.1 second interval is integrated in 100 actual microseconds. See clock, real time.

real-time input Input data inserted into a system at the time of generation by another system.

real-time interrupt process (RIP) Software within TELEVENT that responds to "events" (interrupts).

real-time language (RTL) A computer language designed to work on problems of a time-critical nature.

real-time operation 1. Time in which reporting on events or recording of events is simultaneous with the events. [NASA] 2. See also real-time processing

real-time output Output data removed from a system at time of need by another system.

real-time processing 1. The processing of information or data in a sufficiently rapid manner so that the results of the processing are available in time to influence the process being monitored or controlled. 2. Computation that is performed while a related or controlled physical activity is occurring so that the results of the computation can be used to guide the process.

real-time program A program which operates concurrently with an external process which it is monitoring or controlling, meeting the needs of that process with respect to time.

real-time system A system which responds within the time scale of the process being controlled, i.e. whose response time depends upon the process dynamics or time constants.

reamer Special type cutting device to size close tolerance holes.[ARP480A-87] See also drill.

rear insertion-front release The type of connector whose contacts are inserted from the rear, with the proper insertion tool, and released from the rear with the removal tool inserted from the face of the connector. This requires demating the plug and receptacle.[ARP914A-79]

rear insertion-rear release The type of connector whose contacts are both inserted and removed from the rear of the connector with the proper tools. This does not require demating of an electrical installation.[ARP914A-79]

rearward facing steps See backward facing steps.

reasonableness test A test providing a means for detecting a gross error in calculation by comparing results against upper and lower limits representing an allowable reasonable range.

reassociation The recombination of the products of dissociation.

Réaumur scale A temperature scale having 0° as the ice point and 80° as the steam point; the scale is little used outside the brewing, winemaking and distilling industries.

REB See relativistic electron beams. [NASA]

reboiler 1. The heat exchanger at the bottom of a distillation column. The reboiler generates the vapors that ascend through the column from liquid which comes down the column. The term is derived from "re-boil". 2. A closed heat exchanger which uses one medium to heat a second. The purpose is to maintain a separation between the two mediums due to noncompatibility or to prevent contamination between systems while transferring thermal energy.

rebound The phase of operation of a shock absorber after the compression and storing of energy within the unit. The stored energy forces the extension or rebound of the piston to release the energy. Sometimes causes the aircraft to bounce back into the air after initial impact unless the rebound is snubbed properly.[AIR1489-88]

receiver(s) 1. An apparatus near the outlet of a compressor which collects excess oil or moisture in the compressed air, and which reduces or eliminates pressure pulsations and stores compressed air for later use. 2. An electronic device which detects radio-wave signals and amplifies them into electrical signals of varying frequency and amplitude to drive an output device such as a loudspeaker. Also known as radio receiver. 3. A device which detects an optical signal, converts it into electronic form, then processes it further so it can be used by electronic equipment.

receiver gage A gage, calibrated in engineering units, which receives the output of a pneumatic transmitter.

receiving gage A fixed gage designed to inspect several dimensions on a part, and also inspect the relationships between dimensions.

receiving systems See receivers.

receptacles (containers) See containers.

receptors (physiology) Sensory nerve endings or organs in a living organism that is sensitive to physical or chemical stimuli.[NASA]

recharging The restoring of discharged electric storage batteries to a charged condition by passing direct current through them in a direction opposite to that of the discharging current.[NASA]

recipe The complete set of data and procedure that defines the control requirements of a particular product manufactured by a batch process. A recipe consists of a header, equipment requirements, procedure and formula.

reciprocal transducer A transducer in which the output signal is proportional to the reciprocal of the level of the stimulus.

reciprocating engines See piston engines.

reciprocity theorem Any theorem expressing reciprocal relations for the behavior of some physical system in which input and output can be interchanged without altering the response of the system to a given excitation.[NASA]

recirculation The reintroduction of part of the flowing fluid to repeat the cycle of circulation. See also circulation.

recognition The psychological process in which an observer so interprets the visual or auditory stimuli he receives from a distant object that he forms a correct conclusion as to the exact nature of that object or sound.[NASA]

recoil See rebound.

recoil control The process of control of the energy stored in a shock absorber after impact. The energy is dissipated in the extension phase usually by forcing hydraulic oil through orifices or passages.[AIR1489-88]

recoil snubber The valve, orifice(s), or snubbing device within a shock absorber which controls impact forces within the shock absorber upon extension of the landing gear after takeoff or catapulting.[AIR1489-88]

recombination The reaction between an ion and one or more electrons that returns the ionized element or molecule to the neutral state.

recombination coefficient A measure of the specific rate at which oppositely charged ions join to form neutral particles (a measure of ion recombination). [NASA]

recommended firing current (or energy) In an EED, the current (or energy) which must be applied to a bridge circuit to cause operation within a specified time.[AIR913-89]

recommended test current (or energy) In an EED, the current (or energy) that can be applied to a bridge circuit for extended periods without degrading the explosive material or firing the device. [AIR913-89]

reconditioning The programmed electrical exercise, consisting of a discharge/charge cycle(s), which is periodically utilized to restore the capability of full energy and power delivery to a battery which has electrically deteriorated while in service. It may also include water replacement and mechanical cleaning procedures.[AS8033-88]

reconfigurable That characteristic of a fault-tolerant device or system whereby continued functional operation, subsequent to a failure, is provided by rearranging or recombining the surviving control elements. Sometimes referred to as self-repairing.[ARP1181A-85]

record 1. A segment of a file consisting of an arbitrary number of words or characters. 2. In data processing, a group of data that contains all the information about a single item.

record (noun) On a computer tape, the smallest group of words which can be located and input by the computer. In telemetry systems, this is usually 1,000 to 4,000 data words.

record (verb) To store on some permanent (generally magnetic) medium, as on a magnetic tape or disk.

recorded value The value of a measured variable as determined from the position of a trace or mark on chart paper or as determined from a permanent or semipermanent effect on an alternative recording medium.

recorder 1. An instrument that makes and displays a continuous graphic, acoustic or magnetic record of a measured variable. 2. A measuring instrument in which the values of the measured variable are recorded.

record format The content and organization of a single record.

record gap An area between two consecutive records.

recording channel In a recording system that incorporates any means of producing multiple traces on a single chart or multiple tracks on an alternative recording medium, the circuit and associated apparatus needed to produce a single trace.

recording extensions A recorder is attached directly to the meter body with the recorder pen positioned by the metering float through a magnetic coupling.

recording instrument See instrument, recording.

recording stylus See stylus.

recoverable Items which may be rehabilitated to a serviceable condition one or more times before scrapping. Rehabilitation is by rework or servicing, such as welding, refinishing, recharging, etc.[ARD50010-91]

recovery procedure A flight path control technique used to escape from an inadvertent encounter with windshear. [ARP4109-87]

recovery time In a radiation counter tube, the time that must elapse after detection of a photon of radiation until the instrument can deliver another pulse of substantially full response level upon interaction with another photon of ionizing radiation.

recovery voltage The voltage impressed across the protective device after the circuit has been interrupted and after high frequency transients have subsided. [ARP1199A-90]

recrystallization In metals, the change from one crystal structure to another, as occurs on heating or cooling through a critical temperature. The formation of a new strain free grain structure from that existing in cold worked metal, usually accomplished by heating.[NASA]

recrystallization temperature The minimum temperature at which complete recrystallization of a cold worked metal occurs in a specified time, usually one hour.

rectangular coordinates See Cartesian coordinates.

rectangular terminal A terminal with a rectangular shaped tongue.[ARP914A-79]

rectennas Devices that convert microwave energy into direct-current power by utilizing a number of small diodes each with its own diode rectifier.[NASA]

rectifier(s) 1. Static devices having an asymmetrical conduction characteristic which is used to convert attending current into direct current.[NASA] 2. Any of several devices for converting alternating current to a relatively steady or pulsating direct-current signal.

rectifier antennas See rectennas.

rectifier instrument An instrument for measuring certain quantities of an alternating current circuit in which the a-c input is converted to direct current by a rectifier and the d-c signal is actually measured.

rectifier-type voltmeter An instrument that incorporates four semiconductor rectifier elements arranged in a square (full wave bridge configuration), with the a-c input connected across one diagonal and a permanent-magnet moving-coil detector connected across the other diagonal; the d-c output is proportional to the average current or voltage over any given half cycle of the a-c input, which makes the instrument useful for measuring nominal voltage or low-range (milliampere) current.

rectifying section The section of trays in a distillation column above the feed plate. In this section the vapor is enriched in the light components that are taken overhead.

recuperators See regenerators.

recursion The property which allows a callable program to call itself.

recursive Pertaining to a process which is inherently repetitive. The result of each repetition is usually dependent upon the result of the previous repetition.

recycle synchronizer A method of telemetry subframe recognition in which a specific subcommutator word (or words) contains a unique signal that marks the recycling of the subcommutator.

red data Data that require safeguards. [AIR4271-89]

red dwarf stars Red stars of low luminosity, so designated by E. Hertzsprung. Red Dwarf stars are commonly those main sequence stars fainter than an absolute magnitude of plus 1, and are the faintest and coolest of the dwarfs. [NASA]

red giant stars Stars whose evolution has progressed to the point where hydrogen core burning has been completed, the helium core has become denser and hotter than originally, and the envelope has expanded to perhaps 100 times its initial size.[NASA]

redout A temporary condition in which vision is obscured by a reddishness, or in which objects appear to have a reddish color. This condition, sometimes followed by unconsciousness (not considered a part of redout), is caused by blood rushing to the head. Redout is generally caused by the presence of negative G forces.[ARP4107-88]

Redox cells Cells for converting the energy of reactants to electrical energy, an intermediate reductant in the form of liquid electrolyte reacts at the anode in a conventional manner and is regenerated by reaction with a primary fuel. [NASA]

redox potential The electrochemical potential prevailing in a chemical reaction involving an exchange of electrons (reduction-oxidation potential).

red plague A powdery brown-red cuprous oxide deposit sometimes found on silver coated copper conductors and shield braids. It is fungus-like in appearance and will appear in random spots along the length of a conductor or shield.[ARP1931-87]

red shift In astronomy, the displacement of observed spectral lines toward the longer wavelengths of the red end of the spectrum.[NASA]

reduced gravity A condition in which the acceleration acting on a body is less than normal gravity, between 0 and 1 g.[NASA]

reducer A pipe fitting used to couple a pipe of one size to a pipe of a different size. When the flow is from the smaller pipe to the larger, increaser may be used to differentiate.

reducing atmosphere An atmosphere which tends to: a) promote the removal of oxygen from a chemical compound; b) promote the reduction of immersed materials.

reducing coupling A pipe fitting for connecting two pipes of different sizes.

reduction Removal of oxygen from a chemical compound.

redundancy 1. The existence of more than one means for accomplishing a given function. Each means of accomplishing the function need not necessarily be identical.[ARD50013-91][AIR 1916-88][ARD50010-91] 2. In the transmission of information, that fraction of the gross information content of a message which can be eliminated without loss of essential information. Any circuitry or program instructions present solely to handle faults or errors and not necessary for normal system operation. 3. A parallel or secondary system that takes over when the primary system fails so control can continue uninterrupted.

redundancy, active That redundancy wherein all redundant items are operating simultaneously rather than being activated when needed.[ARD50010-91] [AIR1916-88]

redundancy check An automatic or programmed check based on the systematic insertion of components or characters used especially for checking purposes.

redundancy management That portion of the system logic and control (hardware or software) which detects and isolates failures in a fault-tolerant system; and reconfigures the system after the failure is detected and isolated so as to maintain the same or a reduced level of operation.[ARP1181A-85][AIR 1916-88]

redundancy, standby That redundancy wherein the alternative means of performing the function is inoperative until needed and is activated upon failure of the primary means of performing the function.[ARD50010-91] [AIR1916-88]

redundant An adjective to denote the use of duplicate or alternate components for the purpose of improving mission safety or reliability.[ARP1181A-85][AIR

1916-88]

redundant character A character specifically added to a group of characters to insure conformity with certain rules which can be used to detect computer malfunction.

redundant check See redundancy check.

redundant design A technique of incorporating into a system two or more components that perform the same function(s) so that if one fails or malfunctions the other(s) will perform the necessary functions to enable the system to continue to operate safely. See fail safe and fail operational.[ARP4107-88]

redundant sealing Supplemental sealing that serves as a backup for the primary seal, giving a form of insurance.[AIR4069-90]

Redwood scale A time-based viscosity scale used predominantly in Great Britain; it is similar in concept to the Saybolt scale used in the United States.

reed A thin bar of metal, wood or cane that is clamped at one end and set into transverse elastic vibration by a stimulating force such as wind pressure; it is used to create sound in musical instruments and to act as a frequency standard in some meters.

re-engagement The act of the starter engaging the engine while the engine is at any speed condition between rest and starter cutoff.[ARP906A-91]

re-enterable load module A load module that can be used concurrently by two or more tasks.

re-entrancy The property which allows a callable program to be called and executed before it has completed the execution from a previous call. The results of the previous call are not affected.

reentrant The property of a program that enables it to be interrupted at any point by another program, and then resume from the point where it was interrupted.

re-entrant program A program which can be used for various tasks.

reentry The event occurring when a spacecraft or other object comes back into the sensible atmosphere after going to higher altitudes; the action involved in this event.[NASA]

reentry bodies See reentry vehicles.

reentry edge A negative contact angle between a fillet and its substrate caused by improper tooling (fairing, feathering) of the edges of the fillet, or failure to tool at all.[AIR4069-90]

re-entry point The instruction at which a program is re-entered from a subroutine.

reentry trajectories Those parts of rocket trajectories that begin at reentry and end at target or at the surface.[NASA]

refereed test A predetermined destructive or nondestructive test made by a regulatory body or a disinterested organization, often to fulfill a regulatory requirement; in some cases, the test may be done by the regulated organization and merely witnessed by an agent of the regulatory body.

reference accuracy In process instrumentation, a number or quantity that defines a limit that error will not exceed when a device is used under specified operating conditions. Error represents the difference between the measured value and the standard or ideal value. See accuracy rating.

reference dimension A dimension without tolerance on a mechanical drawing which is given for information only and is not to be used in making or inspecting the part or assembly.

reference input A signal from an independent reference source that is used as one of the inputs to an automatic controller.

reference junction A device used with a thermocouple transducer which couples it to copper wires without introducing an error.

reference junction compensation A

means of counteracting the effect of temperature variations of the reference junction, when allowed to vary within specified limits.

reference level The basis for comparison of an audio-frequency signal level given in decibels or volume units. See datum plane.

reference operating conditions See operating conditions, reference.

reference pressure The pressure established as a reference value.[AIR1916-88]

references (standards) See standards.

refinery gas The commercially non-condensible gas resulting from fractional distillation of crude oil, or the cracking of crude oil or petroleum distillates. Refinery gas is either burned at the refineries or supplied for mixing with city gas.

reflectance 1. The ratio of the radiant flux reflected by a body to that incident upon it.[NASA] 2. The fraction of incident light which is reflected by the surface.

reflected glare Glare resulting from specular reflections of high brightness in polished or glossy surfaces in the field of view. It usually is associated with reflections from within a visual task or areas in close proximity to the region being viewed.[AIR1151-70]

reflected radiation See reflected waves.

reflected rays See reflected waves.

reflected waves Shock waves, expansion waves, or compression waves reflected by another wave incident upon a wall or other boundary. In electronics, radio waves reflected from a surface or object.[NASA]

reflecting telescopes Telescopes which collect light by means of concave mirrors.[NASA]

reflection The process whereby a surface of discontinuity turns back a portion of the incident radiation into the medium through which the radiation approached.[NASA]

reflection coefficient See reflectance.

reflection loss That part of a signal which is lost due to reflection of power at a line discontinuity.[ARP1931-87]

reflection nebulae Any celestial body having a hazy cloudy appearance whose brightness results from the scattering by dust particles of light from nearby stars.[NASA]

reflectivity See reflectance.

reflectometer A photoelectric instrument for measuring the proportion of light reflected from a given surface.

reflector A device, the chief use of which is to redirect the light of a lamp by reflection in a desired direction or directions.[ARP798-91]

reflector antennas Antennas consisting of a reflecting surface and a feed.[NASA]

reflux The recycle stream that is returned to the top of the column. This stream supplies a liquid flow for the rectifying section that enriches the vapor stream moving up the column. Material in the stream is condensate from the overhead condenser. Reflux closes the energy balance by removing heat introduced at the reboiler.

reflux ratio A quantity usually expressed as the ratio of the reflux flow to the distillate flow. The ratio is used primarily in column design.

refracted radiation See refracted waves.

refracted rays See refracted waves.

refracted wave(s) 1. Waves that have had their direction of motion changed by refraction.[NASA] 2. The resultant wave train produced when an incident wave crosses the boundary between its original medium and a second medium; in many cases, only a portion of the wave crosses the boundary, with the remainder being reflected from the boundary.

refracting telescopes Telescopes which collect light by means of a lens or system of lenses.[NASA]

refraction The process in which the direction of energy propagation is changed as the result of a change within the propagating medium, or as the energy passes through the interface representing a density discontinuity between the two media. In the first instance the rays undergo a smooth bending over a finite distance. In the second case, the index of refraction changes through an interfacial layer that is thin compared to the wavelength of the radiation; thus, the refraction is abrupt, essentially discontinuous.[NASA]

refraction loss Reduction in amplitude or some other wave characteristic due to the refraction occurring in a nonuniform medium.

refractive index 1. The change in direction (angle) of a beam of light at the interface when progressing from one medium to another. This change is a function of the ratio of the wavelength (or phase velocity) of an electromagnetic wave in a vacuum to that of the substance.[ARP1192-87] 2. When a light ray is refracted at the surface of separation between air and glass, the ratio of the sine of the angle of incidence in the air to the sine of the angle of refraction in the glass is equal to the refractive index of the glass.[ARP924-91] 3. The ratio of phase velocity of a wave in free space to phase velocity of the same wave in the specific medium. See also refractivity.

refractivity The algebraic difference between an index of refraction and unity. [NASA]

refractometer 1. Instrument for measuring the index of refraction of a liquid, gas or solid.[NASA] 2. An instrument for measuring the index of refraction of a transparent substance; measurement can be accomplished in any of several ways, including measuring the critical angle, measuring refraction produced by a prism, observing interference patterns in transmitted light, and measuring the substance's dielectric constant.

refractor A device, usually of prismatic glass which redirects the light of a lamp in desired directions principally by refraction.[ARP798-91]

refractory baffle A baffle of refractory material.

refractory coating Pyrolytic material used for coating other materials exposed to high temperatures.[NASA]

refractory lined fire-box boiler A horizontal fire tube boiler, the front portion of which sets over a refractory or water cooled refractory furnace, the rear of the boiler shell having an integral or separately connected section containing the first pass tubes through which the products of combustion leave the furnace, then returning through the second-pass upper bank of tubes.

refractory metals Usually alloys of high-melting point, hard-to-work metals, but can also refer to certain unalloyed elements.[NASA]

refractory wall A wall made of refractory material.

refusal speed The highest speed attained during takeoff acceleration from which the aircraft may be decelerated to a stop within the remaining runway length.[ARP4107-88]

refuse The solid portions of the products of combustion.

regain moisture content Same as dry basis moisture content.

regenerative cooling The cooling of a part of an engine by the fuel or propellant being delivered to the combustion chamber; specifically, the cooling of a rocket engine combustion chamber or nozzle by circulating the fuel or oxidizer, or both, around the part to be cooled. [NASA]

regenerative feedback See positive feedback.

regenerator(s) 1. Devices used in a thermodynamic process for capturing and returning to the process heat that would otherwise be lost.[NASA] 2. A repeater, that is, a device which detects a weak signal in a fiber optic communication system, amplifies it, cleans it up, and retransmits it in optical form.

register(s) 1. (computers) Devices capable of retaining information, often that contained in a small subset (e.g. one word) of the aggregate information in a digital computer.[NASA] 2. Accurate matching or superimposition of two or more images. 3. Alignment with respect to a reference position or set of coordinates. 4. A subassembly of the burner on a furnace or oven that directs airflow into the combustion chamber. 5. The component of a meter which counts the revolutions of a rotor or individual pulses of energy and indicates the number of counts detected. 6. In data processing, the specific location of data in memory.

regolith The layer rock or blanket or unconsolidated rocky debris of any thickness that overlies bedrock and forms the surface of the land.[NASA]

regression analysis The statistical counterpart or analog of the functional expression, in ordinary mathematics, of one variable in terms of others. [NASA]

regressive geometry A propellant grain configured in such a manner that the surface area decreases as burning progresses.[AIR913-89]

regrowth alpha Alpha that grows on preexisting (primary) alpha during cooling from some temperature high in the alpha-beta field.[AS1814-90]

regular reflection factor The ratio of the regularly reflected light to the incident light.[ARP798-91]

regular transmission That in which the transmitted light is not diffused. In such transmission, the direction of the transmitted pencil of light has a definite geometrical relation to the corresponding incident pencil of light.[ARP 798-91]

regular transmission factor The ratio of regularly transmitted light to the incident light.[ARP798-91]

regulate The act of maintaining a controlled variable at or near its setpoint in the face of load disturbances.

regulating transformer A transformer used to adjust the voltage, phase relation, or both, in steps, without interrupting the load; it generally consists of one or more windings connected in series with the load circuit, and one or more windings excited from the load circuit or from a separate source.

regulation 1. Control of flow or of some other process variable. 2. The difference between maximum and minimum anode voltage drop over a range of anode current for a cold-cathode glow-discharge tube.

regulator A device for controlling pressure or flow in a process. See also controller, self-operated (regulator).

regulator, gas pressure A spring loaded, dead weighted or pressure balanced device which will maintain the gas pressure to the burner supply line.

regulatory control Maintaining the outputs of a process as close as possible to their respective setpoint values despite the influences of setpoint changes and disturbances.

regulatory mechanisms (biology) Specific processes by which living organisms control the rates of biochemical and physiological reactions involved in processes such as metabolism and cellular differentiation.[NASA]

reheated steam Superheated steam which derived its superheat from a reheater.

reheating The process of adding heat to steam to raise its temperature after it has done part of its intended work. This is usually done between the high pressure and low pressure sections of a compound turbine or engine.

reignition See ignition.

REIL Runway End Identification Lights.

reinforced-concrete stack A stack constructed of concrete reinforced by steel.

reinforcing materials Fibers, filaments, fabrics, and other substances used for strengthening of matrices in composite materials.[NASA]

Reissner-Nordstrom solution The unique solution of general relativity theory describing a nonrotating, charged black hole.[NASA]

REJ (reject) algorithm See compressor.

rejected takeoff (RTO) See also abort. The process of rejecting or aborting a takeoff of an aircraft because of engine or any other failure or command decision. This is one of the critical design conditions for aircraft brakes.[AIR1489-88]

relational data base In data processing, an information base that can draw data from another information base outside the original information base.

relational operator In data processing, a symbol used to determine a relationship to be tested, such as "greater than".

relative accuracy The maximum deviation from a straight line that passes through the end point of an ADC or a DAC transfer junction. Expressed as percent, ppm of the full-scale range or in LSBs.

relative address An address to which the base address must be added in order to find the machine address.

relative addressing Technique for addressing data in memory locations. A number contained in the address part of an instruction, when added to a base address, gives the actual address.

relative bearing See bearing.

relative code A code in which all addresses are specified or written with respect to an arbitrarily selected position, or in which all addresses are represented symbolically in a computable form.

relative damping See damping, relative.

relative density bottle See specific gravity bottle.

relative gain An open-loop gain determined with all other manipulated variables constant, divided by the same gain determined with all other controlled variables constant.

relative humidity The ratio, expressed as a percentage, of the amount of water vapor (grams of water vapor per kilogram of dry air) actually present in the air to the amount of water vapor which would be present if the air were saturated with respect to water at the same temperature and pressure.[AIR 1335-75]

relative luminosity The ratio of measured luminosity at a particular wavelength to measured luminosity at the wavelength of maximum luminosity.

relative response The ratio of the response of a device or system under some specific condition to its response under stated reference conditions.

relative wind 1. The direction and velocity in which the ambient air moves relative to an aircraft or airfoil.[ARP 4107-88] 2. The velocity of airflow with respect to a specific stationary or moving body, neglecting any localized disturbances due to the presence of the body in the airstream.

relativistic electron beams Beams of electrons traveling at approximately the speed of light.[NASA]

relativistic particles Particles with a velocity so large that their relativistic mass exceeds its rest mass by an amount which is significant for the computation or other considerations at hand.[NASA]

relativistic velocity A velocity sufficiently high that some properties of a particle of this velocity have values significantly different from those obtaining when the particle is at rest.[NASA]

relativity A principle that postulates the equivalence of the description of the universe, in terms of physical laws, by various observers, or for various frames of reference.[NASA]

relaxation The relief of stress with regard to time, temperature, and a specific strain.[ARP700-65]

relaxation method (mathematics) An iterative numerical method for solving elliptic partial differential equations, e.g. a Poisson equation.[NASA]

relaxation oscillator A device which generates a periodic nonsinusoidal electrical signal by gradually storing electrical energy and then rapidly releasing it.

relaxation time In general, the time required for a system, object, or fluid to recover to a specified condition or value after disturbance. Specifically, the time taken by an exponentially decaying quantity to decrease in amplitude by a factor of $1/e = 0.3679$.[NASA]

relay An electromechanical device which is operated by a change in a relatively low power electric signal to control the flow of electric current in one or more electric circuits generally not interconnected with the relay-control circuit.

relay-operated controller A control system or device in which the signal that operates the final control element or device is produced by supplementing the energy from the primary control element with energy from another source.

relay rack See rack.

release A device or catch for locking or releasing a mechanism. The point in the launching operation at which the final tensioning process has been completed, the catapult fired and the catapult tow force increased to such a magnitude as to rupture the release element or disengage the repeatable release holdback bar and effect release of the aircraft.[AIR1489-88]

releasc element The member in a catapult holdback which initiates the release process. Several types are (1) tension bar or frangible link (tension bar or ring) which fails at a predetermined load (2) hydromechanical device (repeatable release holdback bar) which releases at a given load and is reusable. [AIR1489-88]

reliability 1. The probability that an item shall perform its intended function for a specified interval under stated conditions.[ARD50010-91] [AIR1916-88] [ARD50013-91] 2. The closeness of agreement among a number of consecutive measurements of the output for the same value of the input under the same operating conditions, approaching from the same direction, for full range traverses.

reliability, assessed The reliability of an item determined within stated confidence limits from tests or failure data on nominally identical items. Results can only be accumulated (combined) when all the conditions are similar. [ARD50013-91] [ARD50010-91]

reliability assurance The exercise of positive and deliberate measures to provide confidence that a specified reliability will be obtained.[ARD50013-91]

reliability centered maintenance RCM is a disciplined logic or methodology used to identify preventive maintenance tasks to realize the inherent reliability of equipment at a minimum expenditure of resources.[ARD50010-91]

reliability control See quality control.

reliability development/growth test A series of tests conducted to disclose deficiencies and to verify that corrective actions will prevent recurrence in the operational inventory (also known as "TAAF" testing).[ARD50013-91] [ARD 50010-91]

reliability, dispatch The percentage of flights which depart without incurring a delay (technical) or cancellation (technical).[ARD50010-91]

reliability, enroute The probability of successfully completing a flight plan without incurring a failure that would cause deviation from flight plan.[ARD 50010-91]

reliability, estimated A reliability factor that is postulated for a system, subsystem, or equipment under specified conditions of test or use.[AIR1916-88]

reliability growth The positive improvement of the reliability of equipment through the systematic and permanent removal of failure mechanisms, regardless of their sources, by implementing corrective action.[ARD50010-91]

reliability, inherent The potential reliability of an item present in its design. [ARD50013-91]

reliability, mission The ability of an item to perform its required functions for the duration of a specified mission profile.[ARD50010-91]

reliability, operational The assessed reliability of an item based on field data.[ARD50010-91]

reliability qualification test A test conducted under specified conditions, by, or on behalf of, the government, using items representative of the approved production configuration, to determine compliance with specified reliability requirements as a basis for production approval. Also known as a "Reliability Demonstration", or "Design Approval", test.[ARD50010-91]

relic radiation Background radiation resulting from the primordial big bang. [NASA]

relief Clearance around the cutting edge of a tool, provided by tapering or contouring the adjacent surfaces.

relief valve 1. A modulating valve that limits supplied pressure by dumping from a system or compartment to a lower pressure region.[ARP986-90] 2. A device used to protect piping and components from overpressure.

relief valve (safety) An automatic pressure relieving device actuated by the pressure upstream of the valve and characterized by opening pop action with further increase in lift with an increase in pressure over popping pressure.

relocatable A program that can be moved about and located in any part of a system memory without affecting its execution.

relocatable coding Absolute coding containing relative addresses, which when derelativized, may be loaded into any portion of a computer's programmable memory and will execute the given action properly. The loader program normally performs the derelativization.

relocate In programming, to move a routine from one portion of storage to another and to adjust the necessary address references so that that routine, in its new location, can be executed.

relocation dictionary The part of an object or load module that identifies all relocatable address constants in the module.

reluctance Resistance of a substance to the passage of magnetic lines of force; it is the reciprocal of magnetic permeability. Also known as magnetic resistance.

reluctive pressure transducer A type of pressure sensor in which a moving armature attached to a pressure-sensitive element varies the reluctance of a magnetic circuit—either a permanent magnet or an electromagnet—thus producing an output current in a measuring coil.

REM See roentgen equivalent man.

remanence The magnetic flux density which remains in a magnetic circuit after the removal of an applied magnetomotive force.[NASA]

remedial maintenance The maintenance performed following equipment failure, as required, on an unscheduled basis. Contrasted with preventive maintenance.

remote In data processing, a term used to refer to any devices that are not located near the main computer.

remote access Pertaining to communication with a data processing facility by one or more stations that are distant from that facility.

remote communications air/ground facility (RCAG) An unmanned VHF/UHF transmitter/receiver facility which is used to expand ARTCC air/ground communications coverage and to facilitate direct contact between pilots and controllers.[ARP4107-88]

remote communications outlet (RCO) An unmanned air/ground communications facility remotely controlled by air traffic personnel. The RCO may be UHF or VHF and is intended to extend the communication range of the air traffic facility.[ARP4107-88]

remote control Control of an operation from a distance, especially by means of electricity or electronics; a controlling switch, lever, or other device used in this kind of control.[NASA]

remote manipulation Using electromechanical or hydromechanical equipment to enable a person to perform manual operations while remaining some distance from the work location; usually used for handling radioactive or otherwise hazardous materials.

remote manipulator system Devices used in space for deploying and retrieving payloads by remote control; also used for space maintenance and/or servicing of satellites and other spacecraft. [NASA]

remote processing unit (RPU) Field station with input/output circuitry and the main processor. These devices measure analog and discrete inputs, convert these inputs to engineering units, perform analog and logical calculations (including control calculations) on these inputs and provide both analog and discrete (digital) outputs.

remote sensing 1. The sensing of remote phenomena by whatever means. [NASA] 2. Detecting, measuring, indicating or recording information without actual contact between an instrument and the point of observation—for example, as in optical pyrometry.

removal, confirmed A removal where a failure or defect is found which substantiates the reason for removal.[ARD 50010-91] [AIR1916-88]

removal justified The total removals of some unit may include items removed for maintenance convenience, as well as those for which no problems are found, and those for which a shop check reveals a defect. The last group is the only one for which removals are justified.[AIR1916-88]

removal rate The number of removals of an item expressed in terms of a base period; usually per 1000 aircraft hours, 1000 items hours, 1000 engine hours, and sometimes per 100 or 1000 departures.[ARD50010-91]

removal, scheduled 1. Removal of an item when the hard time limit is reached. 2. Removal of an "on-condition" item during a scheduled periodic inspection or test.[AIR1916-88]

removal tool See extraction tool.

removal unconfirmed The removal of an item where no defect or failure is found, which substantiates the reason for removal, even though another defect or failure may be found.[AIR1916-88]

removal unscheduled The removal of an item brought about as a result of a known or suspect malfunction or defect or both at a time other than scheduled maintenance, inspection, or test. It includes: (a) all conditions monitored (CM) items, (b) only those on-condition (OC) items removed prior to their scheduled check, (c) only those hard-time (HT) items removed prior to their schedule. [AIR1916-88]

rendezvous The event of two or more

objects meeting with zero relative velocity at a preconceived time and place. The point in space at which such an event takes place, or is to take place. [NASA]

rene 95 High-strength nickel-base superalloy.[NASA]

REP See roentgen equivalent physical.

repair 1. The restoration of an item to a serviceable condition; same as corrective maintenance. Repair includes airworthiness documentation.[AIR1916-88] 2. Narrowly, restoring an item to serviceable condition, but not exactly to original design specifications.

repairability The probability that a failed system will be restored to operable condition within a specified active repair time.[ARD50010-91]

repairable item An item that can be restored to perform all of its required functions by corrective maintenance. [AIR1916-88]

repair cycle The period that elapses from the time the item is removed in a repairable condition to the time it is returned to stock in a serviceable condition.[AIR1916-88]

repair cycle time The elapsed time from failure of an item until the item is repaired and restored to an RFI condition and returned to the operational site. [ARD50010-91]

repair, essential Only those repairs necessary to ensure that the end item will fulfill its mission efficiently and safely.[AIR1916-88]

repair/overhaul, modular Application of maintenance procedures and techniques that concentrate attention to a defective subassembly or module and its repair or overhaul, in lieu of treating the complete assembly as an entity for all maintenance actions.[AIR 1916-88]

repair part, high mortality A repair part with failures anticipated or actual, of 30 or more per 100 end items per year.[AIR1916-88]

repeatable release holdback bar A type of catapult holdback which is designed to restrain the aircraft until a predetermined load is built up in the towing mechanism, then release the aircraft automatically, allowing the aircraft to be launched. The reusable release element may then be reset for the next launch.[AIR1489-88]

repeater A device that amplifies or regenerates data signals in order to extend the distance between data stations. Also called a regenerator.

replicated optics Optical components formed by transferring a master pattern to a roughly machined substrate, using an epoxy layer to form a final optical surface. The epoxy layer is then coated with a reflective layer to form the final component. The process allows mass production of complex surfaces much less expensively than conventional polishing techniques.

report generator A computer program that gives a less experienced user the ability to create reports from various files.

reporting point A geographical location in relation to which the position of an aircraft is reported.[ARP4107-88]

reproducibility The ability of an instrument to duplicate, with exactness, measurements of a given value. Usually expressed as a percent of span of the instrument. See also repeatability.

repulsion See force.

required time of arrival (RTA) A term associated with 4-D representing the required ETA or, the ETA to be programmed, at a designated downstream waypoint or waypoints.[ARP1570-86]

rerun In data processing, executing a program again.

resealing pressure The inlet pressure at which fluid no longer leaks past a relief valve after it is closed.

reseat pressure The pressure at which

the valve is reseated and flow is interrupted.[AIR1916-88]

reserved variable Any variable available only to specific programs in the system. Contrast with global variable.

reserved words Certain words in a language which may only appear in the context reserved for them.

reserve energy The capacity of a shock absorber to accept energy inputs over and above the design conditions; i.e., reserve for overload or over design landing conditions.[AIR1489-88]

reservoir 1. A container for operating fluid supply.[ARP243B-65] 2. A holding tank, cistern or pond for storing reserves of potable or make-up water.

reservoir, pneumatic A pressure storage chamber in which pneumatic pressure energy may be accumulated and from which it may be withdrawn.[ARP 243B-65]

reset 1. The process for returning a system to an operational state after a failure has occurred and has been corrected or isolated. Some systems have the capability for providing this function automatically.[AIR1916-88] 2. Another term for an integrating control function. Also used for any control characteristics which eliminate droop in a system.[ARP 89C-70] 3. To restore a storage device to a prescribed initial state, not necessarily that denoting zero. 4. To place a binary cell into the state denoting zero. 5. See integral control action (reset).

reset action 1. A control action which produces a corrective signal proportional to the length of time the controlled variable has been away from the set point. 2. Takes care of load changes. 3. Another name for the integral control mode.

reset control See integral control.

reset cycle To return a cycle index to its initial value.

reset rate See integral action rate.

reset windup Saturation of the integral mode of a controller developing during times when control cannot be achieved; this condition often causes the controlled variable to overshoot its setpoint when the obstacle to control is removed.

resident In data processing, a program that is permanently stored in the memory of the computer.

residual error The error remaining after attempts at correction.

residual fuels Products remaining from crude petroleum by removal of some of the water and an appreciable percentage of the more volatile hydrocarbons.

residual stress In structures, any stress in an unloaded body. These stresses arise from local yielding of the material due to machining, welding, quenching or cold working.[NASA]

residue check 1. Any modulo n check. 2. A check of numerical data or arithmetic operations in which the number A is divided by n and the remainder B accompanies A as a check digit.

resin 1. An organic substance of natural or synthetic origin characterized by being polymeric in structure. Most resins are of high molecular weight and consist of long chain or network molecular structure. Usually, resins are more soluble in their lower molecular weight forms. 2. A polymer or insulating compound used for metal extrusions or molding.[ARP1931-87]

resin matrix composites Composite materials utilizing a matrix of filaments and/or fibers of glass, metal, or other material bound with a polymer or resin. [NASA]

resistance 1. That property of an electric circuit which determines, for a given current, the rate at which electric energy is converted into heat.[ARP1931-87] 2. In electricity, the factor by which the square of the instantaneous conduction current must be multiplied to obtain the power lost by heat dissipation or other permanent radiation of energy away from the electrical current. In mechanics, the opposition by frictional effects to forces tending to produce motion. [NASA] 3. Impediment to gas flow, such as pressure drop or draft loss through a dust collector. Usually measured in inches water gage (w.g.).

resistance coefficients See resistance.

resistance drop The voltage drop in phase with the current.

resistance magnetometer A device for measuring magnetic field strength by means of a change in electrical resistance of a material immersed in the magnetic field.

resistance meter Any instrument for measuring electrical resistance. Also called an ohmmeter or megger.

resistance strain gage A fine wire or similar device whose electrical resistance changes in direct proportion to the amount of elastic strain it is subjected to.

resistance thermometer A temperature measuring device in which the sensing element is a resistor of a known variation in electrical resistance with temperature. See RTD.

resistance welding A group of welding processes in which heat is obtained from resistance of the work to electrical current in a circuit of which the work is one part. Spot welding is an example of resistance welding.

resistive flowmeter A device for measuring liquid flow rates in which an electrical output signal proportional to flow rate is determined from the rise and fall of a conductive differential-pressure manometer fluid in contact with a resistance-rod assembly.

resistivity Electrical resistance per unit length and unit cross section.

resistor 1. Passive fluidic element which because of viscous losses produces a pressure drop as a function of the flow through it and has a transfer function of essentially real components (i.e., negligible phase shift) over the frequency range of interest.[ARP993A-69] 2. An electrically conductive material shaped and constructed so that it offers a known resistance to the flow of electricity.

resistor-capacitor-transistor logic (RCTL) A type of computer circuit used to perform the "not or" logic function at speeds higher than can be achieved with RTL circuits. Similar to RTL except that capacitors reduce switching time.

resistor-capacitor-transistor logic (RCTL) A type of computer circuit used to perform the "not or" logic function at speeds higher than can be achieved with RTL circuits. Similar to RTL except that capacitors reduce switching time.

resistor-transistor logic (RTL) A form of logic circuit using resistors and transistors that performs not or nor logic.

resizing tool A tool designed to shape or size to a specific dimension. See also anvil.[ARP480A-87]

resolution 1. The accuracy with which the actuation system output can be positioned. Output accuracy is usually expressed in absolute terms; that is, a position resolution of 0.5 degrees.[AIR 1916-88] 2. A measure of the ability to delineate picture detail on a CRT, also, the smallest discernible or measurable detail in a visual presentation. It may be stated by a combination of the terms modulation transfer function, spot diameter, line width, raster or video lines. [ARP1782-89] 3. The smallest change in a measurement which can be detected.[ARP1256-90] 4. The ability of a film, a lens, a combination of both, or a vidicon system to render barely distinguishable a standard pattern of black and white lines. In radar, the minimum angular separation at the antenna at which two targets can be distinguished (a function of beamwidth); or the minimum range at which two targets at the same azimuth can be separated (equal to one half the pulse height). Of a gyro, a measure of response to small changes in input; the maximum value of the minimum input change that will cause a detectable change in the output for inputs greater than the threshold, expressed as a percent of one half the input range.[NASA]

resolution advisory Term used with the TCAS system for the display indication given to the crew recommending an escape maneuver to increase or maintain separation relative to an intruder aircraft.[ARP4153-88]

resolution sensitivity The smallest change in an input that produces a discernible response.

resolver Any means for determining the mutually perpendicular components of a vector quantity.

resolving power 1. A measure of the ability to respond to small changes in input. 2. The ability of an optical device to separate the images of two objects very close together. 3. The ability of a monochromator to separate two lines in a multi-line spectrum. See also resolution.

resolving time The minimum separation time between events that will enable a counting device to detect and respond to both events.

resonance The phenomena of amplification of a free wave or oscillation of a system by a forced wave or oscillation of exactly equal period. The forced wave may arise from an impressed force upon the system or from a boundary condition. The growth of the resonant amplitude is characteristically linear in time. Of a system in forced oscillation, the condition which exists when any change, however small, in the frequency of excitation causes a decrease in the response of the system.[NASA]

resonance bridge An electrical network used in measuring inductance, capacitance or frequency; it normally consists of four arms—one containing both inductance and capacitance and the other three containing only nonreactive resistances—and an adjustment device which balances the network by establishing resonance.

resonance fluorescence The emission of radiation by a gas or vapor as a result of excitation of atoms to a higher energy level by incident photons at the resonance frequency of the gas or vapor.[NASA]

resonance lines Spectral lines which occur either as absorption or emission lines.[NASA]

resonance radiation See resonance fluorescence.

resonant frequencies 1. Frequencies at which resonance exists.[NASA] 2. The wave frequency at which mechanical or electronic resonance is achieved. See also frequency, resonant.

resonator(s) 1. In radio and radar applications, circuits which will resonate at a given frequency, or over a range of frequencies, when properly excited.[NASA] 2. Generally a pair of mirrors located at either end of a laser medium, which cause light to bounce back and forth between them while passing through the laser medium.

resource Any facility of the computing system or operating system required by a job or task, and including main storage, input/output devices, the central processing unit, data sets, and control processing programs.

resource manager A general term for any control program function responsible for the allocation of a resource.

respiration The interchange of gases of living organisms and the gases of the medium in which they live.[NASA]

responders See transponders.

response(s) 1. Of a device or system, the motion (or other output) resulting from a excitation or input under specified conditions. Response characteristics, often presented graphically, give the response as a function of some independent variable such as frequency or direction. For such purposes it is customary to assume the other characteristics of the input are held constant.[AIR1489-88] 2. The execution of a selected course of action. This may include taking no action if that is the selected response. Any muscular or glandular process that depends upon stimulation.[ARP4107-88] 3. A quantitative expression of the output as a function of the input under conditions must be explicitly stated. [ARP89C-70] 4. The change in analyzer output signal that occurs with change in sample concentration. Also, the output signal corresponding to a given sample concentration.[ARP1256-90]

response-critical That aspect of controlling a process which implies the need to react to random disturbances in time to prevent impairment of yield, or dangerous conditions. Real time is often used synonymously.

response, delayed The execution of a selected course of action so long after the decision was made that the selected course of action is no longer appropriate. An adaptive or goal-seeking response evoked a considerable time after the disappearance of its usual stimulus. [ARP4107-88]

response, poor Ineffective execution of a selected course of action due to cognitive or physical task saturation, an anomaly of attention, an anomaly of motivation, or lack of sufficient procedural knowledge.[ARP4107-88]

response pressure The pressure at which a function is initiated.[AIR1916-88]

response set A cognitive framework of expectations which predisposes a person to a certain course of action regardless of the environmental cues perceived. [ARP4107-88]

response time 1. The time delay between the application of an input signal and the resulting output signal.[AIR1489-88] 2. The time between the initiation of an operation from a computer terminal and the receipt of results at the terminal. Response time includes transmission of data to the computer, processing, file, access and transmission of results to the terminal. 3. The time it takes for a controlled variable to react to a change in input.

responsiveness The ability of an instrument or control device to follow wide or rapid changes in the value of a dynamic measured variable.

restart In electronic computing, the process of recommencing a computing function from a known point in a program following computer system failure or other unusual event during execution of a task.

restart address The address at which a program can be restarted; normally, the address of the code required to initialize variables, counters, and the like.

restoring torque gradient The rate of change, with respect to deflection, of the resultant of electric and mechanical torques that tend to restore an instrument's moving element to any position of equilibrium.

restricted area See special use airspace.

restricted burning grain A solid propellant grain in which certain surfaces are restricted or inhibited to provide particular burning characteristics.[AIR 913-89]

resultants The sums of two or more vectors.[NASA]

retainer A device designed to position and/or restrain the movement of one or more parts.[ARP480A-87]

retarded elastic-chamber gage A pressure gage whose sensitive element is an elastic chamber that moves freely only through the lower portion of its indicating range.

retarder A straight or helical strip inserted in a fire tube primarily to increase the turbulence.

retarding ion mass spectrometer See mass spectrometer.

retarding magnet A magnet used in a motor-type meter to limit rotor speed to a value proportional to the quantity being measured.

reticle(s) 1. Systems of lines or wires placed in the focal plane of an optical instrument to serve as a reference.[NASA] 2. A glass window on which is etched or printed a pattern, typically for use in measurement or alignment. The simplest type of reticle is the cross-hairs of an alignment telescope.

reticulated foam A three-dimensional, net-like, open-cell material, made of a flexible polyurethane compound. It is useful for explosion suppression and slosh attenuation in aircraft fluid tanks.[AIR1664-90]

retirement for cause Procedure, primarily on aircraft, based on fracture mechanics, which allows safe utilization of the full life capacities of each component.[NASA]

retort processing One method for converting shale oil into oil similar to petroleum oils.[NASA]

retractable landing gear See landing gear.

retraction 1. To draw back or in. 2. The process of retracting, lifting, or folding a landing gear within the confines of the mold lines of the air vehicle to a stowed position for high speed, low aerodynamic drag configuration of the air vehicle.[AIR1489-88]

retraction curve A type of load-stroke curve which presents power requirements for the retraction cycle of operation for a landing gear.[AIR1489-88]

retraction geometry The geometric representation (mathematics) of the properties, measurement, and relationship of points, lines, angles, surfaces, and solids, of component parts of a landing gear system in the retraction process.[AIR1489-88]

retractor A device for storing webbing in a torso restraint system.[AS8043-86]

retreating blade On a rotary-wing aircraft in horizontal motion, any rotor blade or wing moving with the relative wind. See advancing blade.[ARP4107-88]

retrieve In data processing, to search for and extract data that is contained in a computer file.

retroaction See retrothrust.

retrofit Acronym for RETROactive FIT. 1. A modification of an aircraft, aircraft component, or other object that duplicates a change or modification made in later models of the same type.[ARP 4107-88] 2. Modification and upgrading of older control systems. 3. Parts, assembly, or kit that will replace similar components originally installed on equipment.

retrofitting Modification of equipment to incorporate changes made in later production of similar equipment; the changes may be performed in the factory or in the field.[NASA]

retroreflection Reflection wherein the reflected rays return along paths parallel to those of their corresponding incident rays.[NASA]

retroreflector(s) 1. Class of optical instruments which cause reflected radiation to return along paths parallel to those of their corresponding incident rays.[NASA] 2. An optical device which reflects an incident beam of light back to the source. The corner-cube prism is an example.

retro-rocket A rocket fired in a direction opposite to the line of flight of the vehicle to which it is attached.[AIR913-89]

retrorocket engines Rocket engines fitted on or in spacecraft, satellites, or the like to produce thrust opposed to forward motion.[NASA]

retrothrust Thrust used for a braking maneuver; reverse thrust.[NASA]

return alert See ringback.

return flow oil burner A mechanical atomizing oil burner in which part of the oil supplied to the atomizer is withdrawn and returned to storage or to the oil line supplying the atomizer.

return key In data processing, a frequently used key on a keyboard that activates a variety of instructions.

return pressure (back pressure) Pressure caused by resistance to flow in the return line or by precharged reservoirs, or both.[AIR1916-88]

return signal See signal, return.

reverberation The persistence of sound in an enclosed space, as a result of multiple reflections after the sound source has stopped. The sound that persists in an enclosed space, as a result of repeated reflection or scattering after the source of the sound has stopped.[NASA]

reverberation chambers Chambers designed to eliminate outside noise for accurate acoustic measurement.[NASA]

reverberation time The time in seconds it takes for average sound-energy density to decrease to one-millionth of its original steady-state value after sound from the source has stopped.

reverberation time meter An instrument for determining reverberation time of an acoustic enclosure.

reversal error See error.

reverse-acting controller A controller in which the value of the output signal decreases as the value of the input (measured variable or controlled variable) increases. See controller, reverse acting.

reverse drawing Drawing, especially a deep-drawn part, a second time in a direction opposite to the original draw direction.

reverse osmosis The application of pressure to stop or reverse the transport of solvent through a semipermeable membrane separating two solutions of different solute concentration. The applied pressure required to prevent the flow of solvent across a perfectly semipermeable membrane is called the osmotic pressure and is a characteristic of the solution.[NASA]

reverse pitch The angle of a propeller blade which produces reverse thrust; a negative pitch angle on the propeller blade.[ARP4107-88]

reverse polarity 1. An electrical circuit in which the positive and negative electrodes have been interchanged. 2. An arc welding circuit in which the electrode is electrically positive and the workpiece electrically negative.

reverse thrust The force developed by an air vehicle or a rocket engine in a reverse direction. This reverse thrust is utilized for deceleration of the air vehicle in ground operations such as landing rollout or rejected takeoff to assist the wheel brakes and/or other systems of the air vehicle.[AIR1489-88]

reverse time See reaction time.

reverse video A CRT screen display of dark characters on a light background—the opposite of the usual CRT screen display.

reversible transducer A transducer in which the transducer loss is independent of the direction of energy transmission through the transducer.

reversing switch An electrical switch whose function is to reverse connections, on demand, of one part of the circuit.

reversion The process for changing over control from an active to a standby channel or from a primary to a secondary channel.[AIR1916-88]

revolution (motion) See revolving.

revolutions per minute (rpm) A standard unit of measure for rotational speed.

revolving Moving in a path about an axis, usually external to the body accomplishing the motion.[NASA]

rework 1. The action taken to restore nonconforming materials to an acceptable configuration.[AS7200/1-90] 2. Restoring an item to a condition exactly conforming to original design specifications; usually applied to corrective action taken when an item has failed an inspection but requires a relatively simple operation such as replacing a part to enable the item to pass an identical inspection.

REX "A" hardness The hardness of a sealant as measured by a REX "A" hardness gauge.[AS7200/1-90] [AIR 4069]

Reynolds number A dimensionless criterion of the nature of flow in pipes. It is proportional to the ratio of dynamic forces to viscous forces: the product of diameter, velocity and density, divided by absolute viscosity.

Reynolds stress In the mathematical treatment of a viscous, incompressible, homogeneous fluid in turbulent motion, that represents the transfer of momentum due to turbulent fluctuations. [NASA]

RF See radio frequency.

RFI See radio frequency interference.

RFI protector A device that protects a computer from strong radio or television transmissions.

Rhea (astronomy) A natural satellite of the planet Saturn orbiting at a mean distance of 527,000 kilometers.[NASA]

rheocasting Use of partially solidified metal alloys (fractions solids) fed directly into a casting machine for forming into machine parts.[NASA]

rheology The study of the deformation and flow of matter.[NASA]

rheopectic substance A fluid whose apparent viscosity increases with time at any constant shear rate.

rheostat An adjustable variable resistor.

rhombic antennas Antennas composed of long wire radiators comprising the sides of the rhombus. The antenna usually is terminated in an impedance. The sides of the rhombus, the angle between the sides, the elevation, and the termination are proportioned to give the desired directivity.[NASA]

rhomboids Parallelograms whose adjacent sides are not equal.[NASA]

riblets Longitudinal striations forming v-shaped grooves on aerodynamic and hydrodynamic surfaces. The riblet devices act to reduce large-scale disturbances near the boundary layer. These grooves are dimensional on the order of the wall vortices and turbulent dimensions.[NASA]

ribbon cable A cable of individually insulated round conductors lying parallel and held together by means of films, adhesive, or woven textile yarn. See also flat cable.[ARP1931-87]

ribbon-in-the-sky An artist's concept sketch, used in briefings and discussions, that provides a pictorial representation of the total flight path for a mission. The sketch includes three-dimensional data and notations of specific task objectives (for example, orbits, refuel).[ARP4107-88]

ribbon parachutes Parachutes having a canopy consisting of an arrangement of closely spaced tapes. These parachutes have high porosity with attendant stability and slight opening shock. [NASA]

Richardson-Dushman equation See thermionic emission.

Richardson number A nondimensional number arising in the study of shearing flows of a stratified fluid.[NASA]

rich mixture A fuel-air mixture in which the ratio of fuel to air is higher than that required for efficient operation of the engine, generally with an air to fuel ratio from 12:1 to 15:1. See also lean mixture.[ARP4107-88]

rifled tube A tube which is helically grooved on the inner wall.

rifts See geological faults.

rigidity Resistance of a body to instantaneous change of shape.

rigid pavement See pavement.

rigid rotors (plasma physics) Ensembles of electrons moving in circular or nearly circular orbits at a constant angular frequency.[NASA]

rime icing/rime ice A white or milky and opaque granular deposit of ice formed by the rapid freezing of supercooled water drops as they impinge upon the exposed aircraft; formation is favored by small drop size, slow accretion, a high degree of supercooling and rapid dissipation of latent heat of fusion. The white appearance results from numerous relatively large air pockets that form because one particle freezes before the next one strikes. Rime ice weights less than clear ice, but may seriously distort airfoil shape and diminish aerodynamic efficiency.[ARP4107-88]

rim width The lateral dimension between the inside of the rim flanges of a wheel which restrain the installed tire in a lateral direction. Therefore, the rim width of the wheel is equal to the installed width between tire beads.[AIR 1489-88]

ring 1. A circular item(s) used for handling, assembling, fabricating, supporting, etc. of parts.[ARP480A-87] 2. A sequential network topology where each node is connected to exactly two nodes, and serves as a repeater when it is not sourcing data onto the network.

Ringelmann chart A series of four rectangular grids of black lines of varying widths printed on a white background, and used as a criterion of blackness for determining smoke density.

ringing An oscillating transient in an output signal that occurs following a sudden rise or fall in the input signal.

ringing time In ultrasonic testing, the length of time that a piezoelectric crystal continues to vibrate after the ultrasonic pulse has been generated.

ring spring A type of spring used in shock absorber design; circular rings with a double trapezoidal cross section. The springs are utilized in stacks and absorb energy by friction of the adjacent ring surfaces.[AIR1489-88]

ring-tongue terminal A terminal having a round-end tongue with a hole to accommodate a screw or stud.[ARP914A-79]

RIP See real-time interrupt process.

ripple 1. The cyclic variation of voltage about the mean level of the dc voltage during steady-state dc electric-system operation.[AS1212-71] 2. A small alternating-current signal superimposed on a larger direct-current signal; it usually results from imperfect filtering of the d.c. output from the rectifier.

ripstop A mechanical design technique for achieving mechanical separation of hydraulic systems where more than one pressure source exists. If material fracture occurs in one portion of a rip-stop design, the fracture cannot propagate from one hydraulic system containment to a second hydraulic system containment and cause loss of two hydraulic systems. Thus, if two hydraulic supplies are to be used, they can never enter the same piece of material anywhere in a system designed to provide rip-stop.[ARP1181A-85] [AIR1916-88]

riser See feedhead.

rise rate, initial instantaneous pressure The pressure rise rate, measured in the first 1/4 s of pressure rise, expressed in psi/s, developed from the tangent to a pressure time curve.[ARP 906A-91]

rise time The time required for the leading edge of a pulse to rise from one-tenth of its final value to nine-tenths of its final value. Rise time is proportional to time constant (see decay time).[AIR 1489-88] 2. The time required for the output voltage of a digital circuit to change from a logical low level (0) to a logical high level (1). See also time, rise.

RIT engines Radio frequency ion thrustors which generate thrust by converting electric energy into a reaction force by utilizing an electromagnetic field. [NASA]

riveter A manual or power driven tool, designed to perform the complete heading operation when setting rivets.[ARP 480A-87]

RMIRadio Magnetic Indicator.

RNAV Partial acronym for ARea NAVigation.

R-NAV-area navigation A term created by ARINC currently used interchangeably with L-NAV (lateral navigation). [ARP1570-86]

RNAV way point (W/P) A predetermined geographical position, used for route or instrument approach definition or progress reporting purposes, that is defined relative to a VORTAC station position.[ARP4107-88]

RNRZ See randomized non-return-to-zero.

roadway powered vehicles Surface vehicles utilizing a combination of an electrical power source embedded in a roadway and an inductive coupled power pickup.[NASA]

robot An intelligent multi-purpose device, usually programmable, which carries out pick and place, assembly or other manipulative operations.

robotics 1. The study and development of reprogrammable devices that do multifunctional tasks, conventionally done by humans, using manipulative functions and/or sensory feedback. The extension of human capabilities to manipulate, repair, service, construct and/or manufacture in space or on the ground is of primary interest to the aerospace community.[NASA] 2. The area of artificial computer intelligence as applied to the use of industrial robots.

robustness (mathematics) Insensitivity of systems to uncontrolled perturbations and independent of changes in environmental parameters as demonstrated mathematically.[NASA]

rocket A missile containing combustibles, independent of atmospheric oxygen, which on being ignited, liberate gases producing thrust.[AIR913-89]

rocket engines Reaction engines that contain within themselves, or carry along with themselves, all the substances necessary for their operation or for the consumption or combustion of their fuel, not requiring any outside substance and hence capable of operation in outer space.[NASA]

rocket launchers Devices for launching rockets.[NASA]

rocket lining In solid rockets, the layers of inhibitors applied to the inner surface of the chamber holding the grain.[NASA]

rocket nozzles The exhaust nozzles of rockets.[NASA]

rocket propellants Agents used for consumption or combustion in rockets and from which the rockets derive their thrust, such as fuels, oxidizers, additives, catalysts or any compounds or mixtures of these.[NASA]

rocket sondes See sounding rockets.

rocket thrust The thrust of a rocket engine usually expressed in pounds. [NASA]

rocket vehicles Vehicles propelled by rocket engines, used to place satellites in orbit, place missiles on target or carry passengers over rails as on rocket sleds. [NASA]

rock intrusions Vertical tabular bodies of rock that fill fissures in host rocks. [NASA]

rock mechanics The theoretical and applied science of the physical behavior of rocks, representing a branch of mechanics concerned with the response of rock to the force fields of its physical environment.[NASA]

rockoons High altitude sounding systems that consist of small solid propellant research rockets carried aloft by a large plastic balloons.[NASA]

rod-out The act of pushing a specially designed rod through a valve or opening to loosen deposits.

Roentgen A quantity of x-ray or gamma-ray radiation that produces, in air, ions carrying one electrostatic unit of electrical charge of either sign per 0.001293 gram of air.

roentgen equivalent man (REM) The unit of dose in radiation dosimetry; it equals the amount of radiation of any type that produces the same amount of biological damage in human beings as a dose of 1 roentgen of 200-kV x-rays.

roentgen equivalent physical (REP) A unit of radiation equal to the amount of radiation of any type that results in energy absorption of 93 ergs/g in soft tissue.

Roentgen rays An alternative term for x-rays.

roll 1. Angular displacement about the longitudinal axis of an airplane.[ARP 4107-88] 2. The act of rolling; rotational or oscillatory movement of an aircraft or similar body about a longitudinal axis through the body—called roll for any degree of such rotation. The amount of this movement, i.e., the angle of roll.[NASA]

rolled joint A joint made by expanding a tube into a hole by a roller expander.

roller A conical or cylindrical part, which functions by rolling on a surface such as a roller adapter, forming roller, spreading roller, etc.[ARP480A-87]

rolling moment A moment about the center of gravity of an airplane which tends to cause the airplane to roll about its longitudinal axis.[ARP4107-88]

rolling moments Moments that tend to rotate an aircraft, rocket or spacecraft about a longitudinal axis. These moments are considered positive when they tend to depress the starboard side of the body.[NASA]

rolling radius See tire rolling radius.

rolling resistance See tire rolling resistance.

rollout 1. The process of landing an aircraft, the time from touchdown through deceleration and stop: i.e., roll-out time, roll-out distance etc., also other applications such as rejected take off from brake application to stop. 2. The first showing of a prototype.[AIR1489-88][ARP4107-88] 3. An act or instance of recovering from a banked attitude.[ARP 4107-88]

rollout RVR See runway visual range.

roll-over error For an analog-to-digital converter with bipolar input range, the output difference for inputs of equal magnitude but opposite polarity. Specified in counts or LSBs.

roll test A type of test for a wheel and/or tire which consists of a series of landings, takeoffs, taxis or a continuous roll of the tire/wheel assembly against a rotating flywheel. The test is intended to be an index of the anticipated service fatigue life of the wheel or tire.[AIR1489-88]

ROM See read-only memory.

ROM devices See read-only memory devices.

Ronchi test An improvement on the Foucault knife-edge test for curved mirrors, in which the knife edge is replaced with a transmission grating with 15 to 80 lines per centimeter, and the pinhole source is replaced with a slit or a section of the same grating.[NASA]

root 1. The bottom of a screw thread. 2. The region of closest approach between two members of a weld joint. 3. The points at which the fusion zone of a weld intersect the base metal.

root directory In MS-DOS, the primary directory of a floppy or hard disk that contains subdirectories.

root locus analysis The root locus analysis is a graphical procedure for determining all possible roots of the characteristic equation for a control system. [AIR1823-86]

root mean square (rms) A means of expressing AC voltage in terms of the DC voltage, (peak AC voltage divided by the square root of two).[ARP1931-87]

root-mean-square errors In statistics, the square root of the arithmetic mean of the squares of the deviations of the various items from the arithmetic mean of the whole.[NASA]

root mean square value (RMS) The square root of the mean of the squares of sample values.

rope lay 1. A method of stranding or cabling in which the members themselves are stranded groups. The groups (either conductor strands or wires in a cable) are then combined in a concentric configuration to form the final conductor or cable.[ARP1931-87]

rosette strain gage See strain rosette.

rosette-type strain gage A type of resistance strain gage having three individual gage elements arranged to measure strain in three different directions simultaneously; typical arrangement has two elements oriented 90° to each other and the third at 45° to the first two.

Rossby waves See planetary waves.

rotameter A variable-area, constant-head, indicating type rate-of-flow volume meter in which fluid flows upward through a tapered tube, lifting a shaped plummet to a position where upward fluid force just balances the weight of the plummet.

rotary actuator A device that converts electric energy into controlled rotary force; it usually consists of an electric motor, a gear box, a control relay and one or more limit switches.

rotary engine 1. An engine in which the cylinders rotate around a stationary crankshaft. See also radial engine. [ARP4107-88] 2. A positive displacement engine consisting of a rotor and stator. The control volume which enclosed the working fluid during the thermodynamic cycle moves in a generally circular motion rather than a linear motion as in a piston engine.[NASA]

rotary oil burner A burner in which atomization is accomplished by feeding oil to the inside of a rapidly rotating cup.

rotary valve A valve for the admission or exhaust of working fluid, where the valve is a ported piston or disk that turns on its axis.

rotating See rotation.

rotating-cup viscometer A laboratory device for measuring viscosity in terms of the drag torque on a stationary element, such as a paddle or cylinder, immersed in a liquid contained in a cup that rotates at constant speed.

rotating meter See velocity-type flowmeter.

rotation The nose-up pitch rotation (pitch) of the aircraft by the pilot, increasing the angle of attack to generate additional lift for takeoff. See also V_r.[ARP4107-88]

rotational flow See vortices.

rotational transition A change in the rotational state of a molecule. Rotational transitions involve less energy than either electronic or vibrational transitions, and typically correspond to wavelengths in the far infrared, longer than about 20 micrometers.

rotational viscometer A device for measuring the apparent viscosity of non-Newtonian fluids by determining the torque required to rotate a spindle in a container filled with the substance; in some instruments, the container may rotate while the spindle does not.

rotifera A phylum of multicellular animals in the subkingdom Eumatazoa. [NASA]

rotor 1. The rotating member of a turbine, electric motor, compressor, pump or similar machine. 2. Any rotating assembly of vanes or airfoils.

rotor body interactions Aerodynamic interactions between a helicopter rotor and a body.[NASA]

rotorcraft An aircraft which in all of its usual flight attitudes is supported in the air wholly or in part by a rotor or rotors; that is, by airfoils rotating or revolving about an axis. The term rotorcraft is commonly applied to a helicopter, autogyro, or the like, in which the sustaining airfoils rotate about a substantially vertical axis.[ARP4107-88]

rotorcraft-load combination The combination of a rotorcraft and an external load, including the external-load-attaching means. Rotorcraft-load combinations are designed as Class A, Class B, and Class C, as follows: 1. Class A: A combination in which the external load cannot move freely, cannot be jettisoned, and does not extend below the landing gear. 2. Class B: A combination in which the external load is jettisonable and is lifted free of land or water during the rotorcraft operation. 3. Class C: A combination in which the external load is jettisonable and remains in contact with land or water during the rotorcraft operation.[ARP4107-88]

rotor disks See turbine wheels.

rotor-type vacuum gage A device for measuring low pressures, down to 10^{-7} torr, by sensing the deceleration of a rotor (usually a steel ball) levitated in a rotating magnetic field, the rotor being exposed directly to the evacuated space.

rotor wash See downwash and wake turbulence.

round-chart instrument A recording instrument whose output trace is written on a circular paper chart.

rounded orifice An orifice whose inlet side is rounded rather than sharp edged.

rounding error The error resulting from rounding off a quantity by deleting the less significant digits and applying some rule of correction to the part retained; e.g., 0.2751 can be rounded to 0.275 with a rounding error of .0001. Synonymous with round-off error and contrasted with truncation error.

round-off Synonymous with round. See also rounding error and half-adjust.

round-off error Same as rounding error.

route A defined path, consisting of one or more courses in a horizontal plane, which aircraft traverse over the surface of the earth.[ARP4107-88]

router A network device that interconnects two computer networks that have the same network architecture. A router requires OSI Level 1, 2 and 3 protocols. See bridge and gateway.

routine A subdivision of a program consisting of two or more instructions that are functionally related; therefore, a program. Clarified by subroutine and related to program.

rpm See revolutions per minute.

RS232 A logic level and connector specification for serial ASCII data transmission; sometimes called the "EIA interface".

RS-232C 1. An EIA standard, originally introduced by the Bell System, for the transmission of data over a cable less than 50 feet in length; it defines pin assignments, signal levels, etc. for receiving and transmitting devices. 2. A communications interface between a modem and other computer devices that complies with EIA Standard RS-232C.

RS422 Standard for serial data transmission.

RSX-11 Real-time resource sharing executive. A real-time, multiprogramming operating system that controls the sharing of system resources among any number of user-prepared tasks.

RT-11 A single-user foreground/background real-time disk-operating system; real-time does not mean that it is capable of accepting real-time data directly, but rather is an indication that it can support a real-time system such as TELEVENT.

RTD See resistance temperature detector.

RTL See real-time language; also see resistor-transistor logic.

rubber A material that is capable of recovering from large deformations quickly and forcibly in the vulcanized state. Originally applicable to a natural substance taken in liquid from trees, the term now includes synthetic rubber of many types and is synonymous with elastomer.[AS1933-85]

ruddevator/ruddervator One of a pair of control surfaces set in a V, each of which combines the functions of both a rudder and an elevator. See also stabilizer.[ARP4107-88]

ruled diffraction gratings Diffraction gratings in which the lines are mechanically ruled by a precision ruling engine. These gratings are normally replicated and the replicas are sold for most applications.

run 1. In data processing, to start a program on the computer. 2. The TELEVENT executive command to initiate a background program.

rung Group of program elements in a ladder diagram. The group controls a single output element (coil or function).

Runge-Kutta method A method for the numerical solution of an ordinary differential equation.[NASA]

run-in See break-in.

running fit Any of a class of clearance fits that allow assembled parts to run freely, especially shafts within their bearings or pinned joints in linkages. See also sliding fit.

run time The length of time between the beginning and the end of a program execution.

run-time error In data processing, an error that occurs during a program operation that may or may not cause the program to stop.

runway 1. A path, channel, or track over which something runs. 2. A long and narrow prepared strip or piece of ground, usually hard surfaced, for the takeoff and landing of aircraft. The term "runway" is rarely applied to the flight deck of an aircraft carrier. The term is applied to water area or strips only in a general sense, as, the natural runway provided by a river. See also "water lane".[AIR1489-88]

runway centerline light system (RCLS) Flush centerline lights spaced at 50-ft intervals beginning 75 ft from the landing threshold and extending to within 75 ft of the opposite end of the runway.[ARP4107-88]

runway condition reading (RCR) Numerical decelerometer readings relayed by air traffic controllers at USAF and certain civil bases for use by the pilot in determining runway braking action. See also braking action.[ARP4107-88]

runway edge light system Lights used to outline the edges of runways during periods of darkness and restricted visibility conditions. These light systems are classified according to the intensity or brightness they are capable of producing: high intensity runway lights (HIRL), medium intensity runway lights (MIRL), and low intensity runway lights (LIRL).[ARP4107-88]

runway end identification lights (REIL) Two synchronized flashing lights, one on each side of the runway threshold facing the approach area, which provide rapid and positive identification of the approach end of a particular runway. They are effective for: (a) identification of a runway surrounded by a preponderance of other lighting, (b) identification of a runway which lacks contrast with surrounding terrain, and (c) identification of a runway during reduced visibility.[ARP4107-88]

runway, grooved A runway which has

lateral grooves cut in the surface to provide for water drainage, and the prevention of aircraft skids or hydroplaning. [AIR1489-88]

runway loading index A factor which takes into account the runway load in terms of variations of pressure and single wheel load; it can be given with reasonable accuracy by the expression: RLI = the square root of PW where P = tire pressure and W = single wheel load.[AIR 1489-88]

runway markings See airport marking aids.

runway profile descent An instrument flight rules (IFR) air traffic control arrival procedure to a runway published for pilot use in graphic or textual form, or both. It may be associated with a standard terminal arrival route (STAR) procedure. Runway profile descents provide routings, and may depict crossing altitudes, speed restrictions, and headings to be flown from the en route structure to the point where the pilot will receive clearance for and execute an instrument approach procedure.[ARP4107-88]

runway visibility value (RVV) The visibility determined for a particular runway by a transmissometer. A meter provides a continuous indication of the visibility (reported in miles or fractions of miles) for the runway. RVV is used in lieu of prevailing visibility in determining minimums for a particular runway.[ARP4107-88]

runway visual range (RVR) An instrumentally derived value, based on standard calibrations, that represents the horizontal distance a pilot will see down the runway from the approach end. It is based on the sighting of either high intensity runway lights or on the visual contrast of other targets, whichever yields the greater visual range.[ARP 4107-88]

rupture In breaking strength or tensile strength tests, the point at which a material physically comes apart, as opposed to yield strength.[ARP1931-87]

rupture capacity Rupture capacity is applied to reusable protective devices. See also interrupting capacity.[ARP 1199A-90]

rupture disc A diaphragm designed to burst at a predetermined pressure differential.

rupture disc device A nonreclosing pressure relief device that relieves excessive static inlet pressure via a rupture disc.

rupture pressure See pressure, rupture.

Rutherford 1. 10^{-6} radioactive disintegration per second. 2. A quantity of a nuclide having an activity equal to one Rutherford.

RVR Runway Visual Range.

RVV Runway Visibility Value.

S

s See second; has largely replaced sec as the preferred abbreviation.

S See siemens.

S100 bus A hobbyist and small business user standard board and bus system which has become a de facto standard for microcomputers. The S100 bus uses 100 pins.

Sabin A unit of measure for sound absorption equivalent to one square foot of a perfectly absorptive surface.

sabotage Deliberate destructive action that may be directed against property, processes, systems, organizations, governments, or people and that is intended to prevent a process, undermine a group, or interfere with progress towards a goal.[NASA]

sabot projectiles Projectiles having devices fitted around or in back of the projectiles in gun barrels or launching tubes to support or protect the projectiles or to prevent the escape of gas ahead of it. The sabot separates from the projectile after launching.[NASA]

saddle 1. A device to support or cradle a part by engaging the contour of the part.[ARP480A-87] 2. A casting, fabricated chair, or member used for the purpose of support.

safe and arm A device to interrupting (safing) and aligning (arming) an explosive train.[AIR913-89]

safe area 1. Non-hazardous area. 2. An area in which explosive gas/air mixture are not expected to be present so that special precautions for the construction and use of electrical apparatus are not required. See non-hazardous area.

safety 1. The conservation of human life and its effectiveness, and the prevention of damage to items, consistent with mission requirements.[ARD50010-91] 2. To secure against loosening or rotating, as, to safety a bolt by passing a restraining wire through its head.[ARP 4107-88]

safety can A metal can with a special closure used for storing, handling and transporting flammable liquids.

safety control See control, safety.

safety ground 1. A connection between metal structures, cabinets, cases, etc. which is required to prevent electrical shock hazard to personnel. 2. Safety ground is not a signal reference point.

safety hoist A hoisting device that stops automatically when tension is released.

safety hook A lifting hook with a spring-loaded latch that prevents the lifting sling from accidentally slipping off.

safety of flight event An engine, engine component or sub-system functional status or condition that could seriously jeopardize flight integrity. [ARP1587-81]

safety of flight warning Timely crew warning to permit in-flight correction of problems seriously affecting flight integrity or leading to catastrophic engine failure. (The intent is to provide either real time crew warning or post-flight maintenance indication for all engine discrepancies that could cause an unsafe flight situation).[ARP1587-81]

safety plug A nonreclosable pressure-relief device containing a fusible element that melts at a predetermined temperature.

safety relief valve An automatic pressure relieving device actuated by the pressure upstream of the valve and characterized by opening pop action with further increase in lift with an increase in pressure over popping pressure.

safety rod See control rod.

safety shut down The action of shutting off all fuel and ignition energy to the burner by means of safety control or controls such that restart cannot be accomplished without operator action.

safety stop 1. A device on a hoisting apparatus to prevent the load from falling. 2. A device on a hoisting engine that automatically prevents it from overwinding. 3. A device that prevents mechanical over-travel on a piece of equipment.

safety valve A spring loaded valve that automatically opens when pressure attains the valve setting. Used to prevent excessive pressure from building up in a boiler.

safe working pressure See design pressure.

sag The "droop" of a sealant after it is applied to an overhead or vertical surface.[AIR4069-90]

SAGE satellite Spacecraft for the study of stratospheric aerosols and gases. [NASA]

Sagnac effect A phase shift (and consequent measurable rotation rate) caused by nonreciprocity (different optical path lengths) of two counterpropagating light waves traveling in the same coil in a fiber optic gyro or ring interferometer. [NASA]

salinometer An instrument for measuring water salinity; may utilize electrical conductivity measurement or a hydrometer calibrated to read percent salt content directly.

SALS Short Approach Light System.

salt beds Deposits of sodium chloride and other salts resulting from the evaporation and/or precipitation of ancient oceans.[NASA]

salt spray corrosion test This test is used to determine the relative abilities of thin coatings of greases to prevent corrosion on steel surfaces in the presence of salt spray.[S-5C,40-55]

SAMA See Scientific Apparatus Makers Association.

sample-and-hold per channel A method of sampling and holding analog data channels in advance of a normal multiplex sample-and-hold A/D encoding process.

sampled-data control That branch of automatic control theory concerned with the control of variables whose current values are not continuously available for comparison with the setpoint but instead are sampled only at given intervals.

sampled signal A signal which is updated only at given intervals by a new observation of the variable.

sample-hold A device that takes a "snapshot" of an analog signal so that it is held stationary for an A/D conversion.

sample/hold amplifier An amplifier which samples the input signal and holds the value for sufficient time before it is input to an analog to digital (A/D) converter.

sample inspection The monitoring or withdrawal, or both, of selected devices from service to establish realistic time limits.[AIR1916-88]

sample interval The time interval between measurements or observations of a variable.

sample plan The plan designed by a telemetry engineer to sample and encode data incrementally so that it may be accurately decoded and re-created.

sampling 1. The collection of exhaust sample under controlled conditions for the purpose of analysis.[ARP1179A-80] 2. Obtaining the values of a function for discrete, regularly or irregularly spaced values of the independent variable. 3. Selecting only part of a production lot or population for inspection, measurement or testing. 4. The removal of a portion of a material for examination or analysis. 5. In statistics, obtaining a sample from a population.

sampling action A type of controller action in which the value of the controlled variable is measured at intermittent intervals rather than continuously.

sampling controller A controller using intermittently observed values of a signal such as the setpoint signal, the actuating error signal, or the signal representing the controlled variable, to effect control action. See controller, sampling.

sampling rate For a given measurement, the number of times that it is sampled per second in a time-division-multiplexed system. Typically, it is at least five times the highest data frequency of the measurement.

sampling theorem Nyquist's result that equispaced data, with two or more points per cycle of highest frequency, allow reconstruction of band-limited functions; the theorem states: "If the rms spectrum/G(t), is identically zero at all frequencies above W cycles-per-second, then g(t) is uniquely determined by giving its ordinates at a series of points spaced 1/2W seconds apart, the series extending through the time domain."

sandblasting Grit blasting, especially when the abrasive is ordinary sand. See grit blasting.

sanding Smoothing a surface with abrasive cloth or paper; usually implies use of paper covered with adhesive-bonded flint or quartz fragments.

sandwich braze A joining technique for reducing thermal stress in a brazed joint, in which a shim is placed between the opposing surfaces to act as a transition layer.

sandwich construction A technique of producing composite materials that consists of gluing hard outer sheets onto a center layer, usually a foamed or honeycomb material.

sanitary engineering A field of civil engineering that deals with construction and operation of facilities that protect public health.

sans-serif In typesetting, a type style with no straight or curved decorative additions. See serif.

SAP See service access point.

Sargasso Sea A region in the Atlantic characterized by mixing ocean currents and a lack of winds. Located northeast of the West Indies.[NASA]

SARS System and Application Requirements Subcommittee.[AIR4271-89]

satellite atmospheres The atmospheres that are found on natural satellites. [NASA]

satellite communication Use of communication satellites, passive reflecting belts of dipoles or needles, or reflecting orbiting balloons to extend the range of radio communication by returning signals to earth from the orbiting object, with or without amplification. [NASA]

satellite defense See spacecraft defense.

satellites Any objects, man-made or natural, that orbit celestial bodies.[NASA]

satellite surfaces The crust and soil of natural satellites.[NASA]

Satellite Tracking and Data Acq Network See STDN (network).

satin finish A type of metal finish produced by scratch brushing a polished metal surface to produce a soft sheen.

"SATS" (Short Airfield for Tactical Support) A portable airfield design for use by one high-performance jet squadron. The field consists of a surfaced runway 2210 feet long and 96 feet wide, turn off areas at either end, a hot pad, and parking and maintenance. It also includes a CE1-3 catapult, two M-21 primary recovery systems, two Fresnel Lens Optical Landings Systems (FLOLS), and extensive field lighting and audio visual communications systems necessary to operate (launch and recovery) high performance jet aircraft. A STATS field has the capability of being readily expanded to an expeditionary airfield.[AIR1489-88]

saturable-core magnetometer A magnetometer whose output is derived from changes in magnetic permeability of a ferromagnetic core as a function of the strength of the magnetic field being measured.

saturable-core reactor A device for introducing inductive reactance into a circuit, where the effective reactance can be varied by varying auxiliary direct-current excitation of a ferromagnetic core.

saturant See lacquer.

saturated air Air which contains the maximum amount of water vapor that it can hold at its temperature and pressure.

saturated steam Steam at the temperature corresponding to its pressure.

saturated temperature The temperature at which evaporation occurs at a particular pressure.

saturated water Water at its boiling point.

saturation A characteristic curve exhibits saturation when further change of the input variable beyond a certain value results in a negligible additional change of the output variable.

saturation (chemistry) The state of a solution when it holds the maximum equilibrium quantity of dissolved matter at a given temperature.[NASA]

saturation current 1. In ionic conduction, the current obtained when the applied voltage is sufficient to collect all of the ions present. 2. In an electromagnet, the excitation current required to produce magnetic saturation.

saturation level The saturation level or maximum counting rate of the electronic counting circuitry is specified by the manufacturer. The APC shall be operated at the specified flow rate and at a particle concentration that yields counting rates below the saturation limit of the instrument.[ARP1192-87]

saturation voltage The minimum applied voltage that produces saturation current.

saturator A device, equipment or person that saturates one material with another.

sawtooth wave A type of cyclic direct-current waveform in which voltage and current rise gradually from zero to a peak value, then drop instantaneously to zero and repeat the cycle.

Saybolt color scale A standardized color scale used primarily in the petroleum and pharmaceutical industries to grade the yellowness of pale products; it is based on matching the color of a column of the sample liquid with one of a set of color-controlled glass disks, as described in ASTM standard D156.

Saybolt Furol viscosimeter An instrument similar to a Saybolt Universal viscosimeter, but with a larger diameter tube for measuring the viscosity of very thick oils.

Saybolt Universal viscosimeter An instrument for determining viscosity by measuring the time it takes an oil or other fluid to flow through a calibrated tube.

s-band In telemetry, the portion of the radio frequency spectrum between 2200 and 2300 MHz.

scab A surface defect on a casting or rolled metal product consisting of a thin, flat piece of metal partly detached from the substrate.

SCADA See supervisory control and data acquisition.

scaffold A movable or temporary platform that allows workers to perform tasks at considerable heights above the ground; it may be either supported from ground level on a framework or suspended from above on ropes or cables.

scalar quantity Any quantity that can be described by magnitude alone, as opposed to a vector quantity which can only be described by both magnitude and direction.

scalars Any physical quantity whose field can be described by a single numerical value at each point in space. [NASA]

scale 1. A device designed to weigh, count, or compute (or combination thereof) through an element which indicates the state of equilibrium between two opposite forces.[ARP480A-87] 2. A graduated series of markings, usually used in conjunction with a pointer to indicate a measured value. 3. A graduated measuring stick, such as a ruler. 4. A device for weighing objects. 5. A thick metallic oxide, usually formed by heating metals in air. 6. A hard coating or layer of materials on surfaces of boiler pressure parts.

scale effect Any variation in the nature of the flow and in the force coefficients associated with a change in value of the Reynolds number, i.e., caused by change in size without change in shape. [NASA]

scale factor 1. The coefficients used to multiply or divide quantities in a problem in order to convert them so as to have them lie in a given range of magnitude, e.g., plus one to minus one. 2. A constant multiplier which converts an instrument reading in scale divisions to a measured value in standard units. 3. In analog computing, a proportionality factor that relates the value of a specified variable to the circuit characteristic which represents it in the computer. 4. In digital computing, an arbitrary factor applied to some of the numerical quantities in the computer to adjust the position of the radix point so that the significant digits occupy specific positions.

scale height A measure of the relationship between density and temperature at any point in the atmosphere.[NASA]

scale length The distance that the pointer of an indicating instrument, or the marking device of a recording instrument, travels in moving from one end of the instrument scale to the other, measured along the baseline of the scale divisions.

scale models Three-dimensional representations of objects or structures containing all parts in the same proportion as their true size.[NASA]

scale-of-ten circuit A decade scaler.

scale-of-two circuit A binary scaler.

scaler(s) A measuring-circuit or control-circuit component that produces one output pulse each time a specific number of input pulses have been received.

scale span The algebraic difference, measured in scale units, between the highest value that can be read from the scale and the lowest value.

scale units The units of measure stated on an instrument scale.

scale-up Using data from an experimental model or pilot plant to design a larger (scaled-up) facility or device, usually of commercial size.

scaling 1. A misnomer for descaling. 2. Forming a thick layer of oxide on a metal, especially at high temperatures. 3. Depositing solid, adherent inorganic layers on the internal surface of a boiler tube, process pipe or vessel, usually from a very dilute water solution.

scaling circuit An electronic circuit that produces an output pulse whenever a predetermined number of input pulses has been received.

scaling factor A factor used in heat-exchange calculations to allow for reduced thermal conductivity across a tube or pipe wall due to scaling.

scalp To remove the surface layer of a billet, slab or ingot, thereby removing surface defects that might persist through later operations.

scan 1. Collection of data from process sensors by a computer for use in calculations, usually obtained through a multiplexer. 2. Sequential interrogation of devices or lists of information under program control. 3. A single sweep of PC applications program operation. The scan operates the program logic based on I/O status, and then updates outputs and input status. The time required for this is called the scan time. 4. To examine an area, volume or portion of the electromagnetic spectrum, point by point, in an ordered manner.

scan linearity Degree of uniformity of

deflection sensitivity at the face of the CRT.[ARP1782-89]

scanned image The representation of an image by successive scan lines of picture elements or PIXELS (PELS). Scanned images are often interchanged as facsimile data transmissions.[AS 4159-88]

scanner(s) 1. Radar mechanisms incorporating such things as rotatable antennas, radiators, motor drives, or mountings for directing a searching radar beam through space and imparting target information to an indicator. [NASA] 2. An instrument which automatically samples or interrogates the state of various processes, files, conditions, or physical states and initiates action in accordance with the information obtained. 3. Any device that examines as region or quantity point by point in a continuous systematic manner.

scanning In radar, the motion of the antenna assembly when searching for targets.[NASA]

scanning devices See scanners.

scanning laser acoustic microscope (SLAM) See acoustic microscopes.

scanning rate (or speed) The speed at which a computer can select and convert an analog input variable.

scan rate 1. A single sweep of PC applications program operation. The scan operates the program logic based on I/O status, and then updates output and inputs status. The time required for this is called scan time. 2. Sample rate, in a predetermined manner, each of a number of variables intermittently.

scars (geology) See erosion.

scatterers See scattering.

scattering 1. The process by which small particles suspended in a medium of a different index of diffraction diffuse a portion of the incident radiation in all directions. In scattering, no energy transformation results, only a change in the spatial distribution of the radiation.[NASA] 2. A collision or other interaction that causes a moving particle or photon of electromagnetic energy to change direction.

scattering coefficients Measures of the attenuation due to scattering of radiation as it traverses a medium containing scattering particles.[NASA]

scattering cross sections The hypothetical areas normal to the incident radiation that would geometrically intercept the total amount of radiation actually scattered by a scattering particle. They are also defined, equivalently, as the cross section areas of isotropic scatterers (spheres) which would scatter the same amount of radiation as the actual amount.[NASA]

scattering functions The intensities of scattered radiation in a given direction per lumen of flux incident upon the scattering material.[NASA]

scattering loss A reduction in the intensity of transmitted radiation due to internal scattering in the transmission medium or to roughness of a reflecting or transmitting surface.

scatter plates (optics) Holograms of diffusing screens for scattering incident light by the process of diffraction. [NASA]

scatter propagation Specifically, the long-range propagation of radio signals by scattering due to index of refraction inhomogeneities in the lower atmosphere.[NASA]

scavenger 1. A reactive metal added to molten metal to combine with and remove dissolved gases or other impurities. A chemical added to boiler water to remove oxygen.

scavenging 1. Removing spent gases from the cylinder of an internal combustion engine and replacing them with a fresh charge or with air. 2. Removing dissolved gases or other impurities from molten metal by reaction with an additive.

SCCF See solar cell calibration facility.

scenario An outline of a mission flight plan giving the particulars of each mission phase. See also design mission scenario and total mission scenario. [ARP4107-88]

Schach effect When a slowly or nonrotating stellite is heated on its sunward side, the photons of thermal radiation carry away more momentum from the hot sunward side than the cold shadowed side, thereby given the satellite a certain net acceleration in the direction away from the sun.[NASA]

schedule 160 A term used to define the wall thickness of pipe (schedule 40, 80, 160 and others).

scheduled maintenance 1. Periodic prescribed inspection and/or serving of equipment accomplished on a calendar, mileage or hours of operation basis. Included in preventive maintenance. [ARD50010-91] 2. Maintenance carried out in accordance with an established plan.

schematic A flow diagram of a fluid unit or system of units including all interconnections.[ARP243B-65]

Schering bridge A type of a-c bridge circuit particularly useful for measuring the combined capacitive and resistive qualities of insulating materials and high-quality capacitors.

schist A strongly foliated crystalline rock formed by dynamic metalmorphism which can be readily split into thin flakes or slabs due to the well developed parallelism of more than 50% of the minerals present.[NASA]

schlieren photography A method of photography for flow patterns that takes advantage of the fact that light passing through a density gradient in a gas is refracted as though it were passing through a prism.[NASA]

Schmitt trigger A bi-stable pulse generator in which an output pulse or constant amplitude exists only as the input voltage exceeds a certain DC value; the circuit can convert a slowly changing input waveform to an output waveform with sharp transitions.

Schuler tuning Adjusting a system performing the function of a pendulum so that it has a period of 84 minutes. [NASA]

Scientific Apparatus Makers Association (SAMA) An industrial association in the United States.

scintillation Generic term for rapid variations in apparent position, brightness, or color of a distant luminous object viewed through the atmosphere. A flash of light produced in a phosphor by an ionizing event. On a radar display, a rapid apparent displacement of the target from its mean position. [NASA]

scintillation counter(s) The combinations of phosphor, photomultiplier tube, and associated circuits for counting scintillations.[NASA]

scintillation spectrometer A scintillation system so designed that it can separate and determine the energy distribution in heterogeneous radiation.

scintillators See scintillation counters.

scissor jack A lifting jack whose operating mechanism consists of parallelogram linkages driven by a horizontal screw.

sclerometer An instrument that determines hardness of a material by measuring the force needed to scratch or indent the surface with a diamond point.

scleroscope An instrument that determines hardness of a material by measuring the height to which a standard steel ball rebounds when dropped from a standard height.

scoop proof A design feature whereby exposed contacts of a connector cannot be touched or damaged by any portion of the mating connector.[ARP914A-79]

scoring Deep scratches on the surface

of a metal.

scotch boiler A cylindrical steel shell with one or more cylindrical internal steel furnaces located (generally) in the lower portion and with a bank or banks (passes) of tubes attached to both end closures.

Scotch yoke A type of four-bar linkage used to convert uniform rotation into simple harmonic motion. Also converts linear motion to rotary motion.

scouring 1. Physical or chemical attack on internal surfaces of process equipment. 2. Mechanical finishing or cleaning using a mild abrasive and low pressure.

SCPC transmission See single channel per carrier transmission.

SCR See silicon-controlled rectifier.

scrap 1. Solid material or inspection rejects suitable for recycling as feedstock in a primary operation such as plastic molding, alloy production or glass remelting. 2. Narrowly, any unusable reject at final inspection.

scraper ring A piston ring that scrapes oil from the cylinder wall to prevent it from being burnt.

scrap rate The number of items beyond the limits of economical repair, expressed as a quantity per unit of time.[ARD 50010-91]

scratches Furrows or grooves in the surface of the glass caused by the removal of glass, usually made by coarse grit, fragments of glass, sharp tools, etc., rubbed over the surface.[ARP924-91]

scratch hardness 1. A measure of the resistance of minerals or metals to scratching; for minerals it is defined by comparison with 10 selected minerals comprising the Mohs scale. 2. A method of measuring metal hardness in which a cutting point is drawn across a metal surface under a specified pressure, and hardness is determined by the width of the resulting scratch.

scratch pad An intermediate work file which stores the location of an interrupted program, and retrieves the program when the interruption is complete.

scratchpad memory A high-speed, limited-capacity computer information store that interfaces directly with the central processor unit; it is used to supply the central processor with data for immediate computation, thus avoiding delays that would be encountered by interfacing with main memory. (The function of the scratchpad memory is analogous to that of a pad of paper used for jotting down notes.)

scratch register Addresses of scratch pad storage locations which can be referenced by the use of only one character.

screen 1. A device using meshed fabric, usually mounted on a frame, used to prevent ingestion of foreign objects into the air intake of jet engines.[ARP 480A-87] 2. In data processing, the plane surface of a CRT that is visible to the user. See also shield.

screen analysis A method of finding the particle size distribution of any loose, flowing aggregate by sifting it through a series of standard screens with holes of various sizes and determining the proportion that passes each screen.

screen grid An internal electron-tube element positioned between a control grid and an anode; it is usually maintained at a fixed positive potential so that electrostatic influence of the anode is reduced in the region between the screen grid and the cathode.

screening 1. A process for inspecting items to remove those that are unsatisfactory or those likely to exhibit early failure. Inspection includes visual examination, physical dimension measurement and functional performance measurement under specified environmental conditions.[ARD50010-91][AIR 1916-88] 2. Separation of an aggregate mixture into two or more portions according to particle size, by passing the mixture through one or more standard screens. 3. Removing solids from a liquid-solid mixture by means of a screen.

screw 1. A cylindrical machine element with a helical groove cut into its surface. 2. A type of fastener with threads cut into a cylindrical or conical shank and with a slitted, recessed, flat or rounded head that is usually larger in diameter than the shank.

screw conveyor A device for moving bulk material which consists of a helical blade or auger rotating within a stationary trough or casing; a screw elevator moves material in a vertical direction; a screw feeder moves material into a process unit. Also known as auger conveyor; spiral conveyor; worm conveyor.

screwed ends See end connections, threaded.

screw lock See jackscrew.

screw pinch A cylindrical plasma equilibrium in which the axial and azimuthal components of the vacuum field are of the same size.[NASA]

screw thread Any of several forms of helical ridges formed on or cut into the surface of a cylindrical or conical member; standard thread designs are used to connect pipes to fittings or to construct threaded fasteners.

scriber A sharp pointed tool for drawing lines on metal or plastic workpieces.

scroll In data processing, to move what is visible on the CRT screen up and down.

scrubber 1. A device for removing entrained dust or moisture from a process gas stream. 2. A device for washing out or otherwise removing an undesirable gaseous component from a process gas stream. 3. An apparatus for the removal of solids from gases by entrainment in water.

scuffing 1. A dull mark or blemish, sometimes due to abrasion, on a smooth or polished surface. 2. A form of mild adhesive wear generally exhibited as a dulling of the worn surface.

scuff patch

scuff pad An additional thickness of material designed to prevent tank wall damage from any anticipated source. The material can be rubber-coated fabric or molded rubber applied to either the inside or outside surface of the tank. Also called a wear patch.[AIR1664-90]

scum 1. A film of impurities floating on the surface of a liquid. 2. A slimy film on a solid surface.

S/D See synchro-to-digital converter.

SDF Simplified Directional Facility.

SDLC See synchronous data link control.

SDV See shuttle derived vehicles.

sea keeping Maintaining the stability of a surface vessel in linear response to wave height, pitch, heave, center of gravity, and bow acceleration.[NASA]

seal 1. The closure of a fuel tank or vessel to make it leak proof by the application of sealant to fasteners, seams, and any other possible leak path. Seals may also be accomplished by compressive interference (such as O-rings, plugs, elastomeric seals, interference fit fasteners.)[AIR4069-90] 2. Any device or system that creates a nonleaking union between two mechanical components. 3. A perfectly tight closure or joint.

sea law United Nations declaration regarding rights to minerals and other marine resources.[NASA]

seal chambers Enlarged pipe sections in measurement impulse lines to provide a) a high area-to-volume displacement ratio to minimize error from hydrostatic head difference when using large volume displacement measuring elements, and b) to prevent loss of seal fluid by displacement into the process.

seal coat 1. A layer of bituminous material flowed onto macadam or concrete to prevent moisture from penetrating the surface. 2. A preliminary coating to seal the pores in a material such as wood or unglazed ceramic.

seal drain A drain provision to remove seal leakage oil from the accessory drive cavity, and/or accessory housing.[ARP 906A-91]

sea level The level of the surface of the ocean; especially, the mean level halfway between high and low tide used as a standard in reckoning land elevation or sea depths.[NASA]

sealing 1. Impregnating castings with resins to fill regions of porosity. 2. Immersing anodized aluminum parts in boiling water to reduce porosity in the anodic oxide film.

sealing voltage The voltage required to move the armature of a magnetic relay from the position where contacts first touch to its fully closed position; a similar term applies to current.

seal, interfacial A seal provided at the interface of connectors to prevent the entry of fluids or contaminants across the interface of mated connectors.[ARP 914A-79]

seal leg The piping from the instrument to the top elevation of the seal fluid in the impulse line.

seal, peripheral A seal provided around the periphery of connector inserts to prevent the entry of fluids or contaminants at the perimeter of mated connectors. [ARP914A-79]

seal plane All surfaces of a tank that establish fuel seal continuity and are an immediate contact with fuel. These surfaces may be composed of structure, fastener, and/or sealing materials.[AIR 4069-90]

seal pot See seal chambers.

seal weld A weld used primarily to obtain tightness and prevent leakage.

seam 1. Usually a surface crack resulting from a defect obtained in casting or forging, also extraneous materials, stringer in the material, not homogeneous with base metal.[MA2005-88] 2. Open surface defect that is narrow and continuous, usually straight, running generally parallel to the nut axis. Seams are generally inherent in the bar from which the nut is made.[MA1568-87] 3. An extended length weld. 4. A mechanical joint, especially one made by folding edges of sheet metal together so that they interlock. 5. A mark on ceramic or glass parts corresponding to the mold parting line.

seamless tubing Tubular products made by piercing and drawing a billet, or by extrusion.

seamounts Elevations of the ocean flow rising to about 3000-1000 feet or more with the summit about 1000-6000 feet below sea level.[NASA]

seam welding A process for making a weld between metal sheets that consists of a series of overlapping spot welds; it is usually done by resistance welding but may be done by arc welding.

search In data processing, to seek out data meeting specific criteria.

season cracking A term usually reserved for describing stress-corrosion cracking of copper or copper alloys in an environment that contains ammonium ions.

seat The fixed area of a valve into which the moving part of a valve rests when the valve is closed to retain pressure and prevent flow.

seat belts Safety belts that fasten across the lap.[NASA]

seating, downstream Seating assisted by pressure differential across the closure component in the closed position, moving the closure component slightly downstream into tighter contact with the seat ring seal that is supported by the body.

seating, upstream A seat on the upstream side of a ball, designed so that the pressure of the controlled fluid causes the seat to move toward the ball.

seat ring A part that is assembled in the valve body and may provide part of the flow control orifice. The seat ring may have special material properties and may provide the contact surface for the closure member.

seat, spring loaded A seat utilizing a design that exerts a greater force at the point of closure component contact to improve the sealing characteristics, particularly at low pressure differential.

second A unit of time, metric and English systems.

secondary air Combustion air introduced into a combustion chamber over the burner flame to provide excess air and ensure complete combustion.

secondary calibration Secondary calibration (standardization) is an electronic calibration performed using the reference system built into the instrument.[ARP1192-87]

secondary circuit The part of an electrical circuit that conducts current output from a transformer to perform the circuit's function.

secondary combustion Combustion which occurs as a result of ignition at a point beyond the furnace. See also delayed combustion.

secondary cosmic rays Secondary emission in the atmosphere stimulated by primary cosmic rays.[NASA]

secondary damage Damage resulting from a primary failure.[ARP1587-81]

secondary device Part of a flowmeter which receives a signal from the primary device and displays, records, and/or transmits it as a measure of the flow rate.

secondary electron An energetic electron set in motion by the transfer of momentum from primary electromagnetic or particulate radiation.

secondary emission Emission of subatomic particles of photons stimulated by primary radiation; for example, cosmic rays impinging on other particles and causing them, by disruption of their electron configurations or even of their nuclei, to emit particles or photons or both in turn.[NASA]

secondary explosive A high explosive which is relatively insensitive to heat and shock, usually initiated by a primary explosive or by an exploding bridge wire.[AIR913-89]

secondary hardening Hardening of certain alloy steels by precipitation hardening during tempering; the hardening occurring during this stage supplements hardening achieved by controlled cooling from above the critical temperature in a step that precedes tempering.

secondary loop The inner loop of a cascade system.

secondary power All aircraft nonpropulsive power generation and transmission. This includes electric, hydraulic, pneumatic, mechanical, and auxiliary power. It is sometimes called flight vehicle power. See also auxiliary power. [ARP906A-91]

secondary radar A radar technique or mode of operation in which the return signals are obtained from a beacon, transponder, or repeater carried by the target, as contrasted with primary radar in which the return signals are obtained by reflection from the target.[NASA]

secondary seal A seal that alone, cannot provide a dependable absolute seal. It is sometimes used in conjunction with a primary seal.[AIR4069-90]

secondary (or Class B) steering system A secondary system is one that is normally in full time use during ground operation, but is not essential for safe ground operation of the aircraft. Fail-safety (fail passive) is an implied requisite.[ARP1595-82]

secondary storage The storage facilities not an integral part of the computer but directly connected to and controlled by the computer, e.g., magnetic drum and magnetic tapes.

secondary treatment Treatment of boiler feed water or internal treatment of boiler-water after primary treatment.

secondary waves See S waves.

secondary winding The output winding of a transformer or similar electrical device.

sectional conveyor A belt conveyor that can be made shorter or longer by removing or adding interchangeable sections.

sectional header boiler A horizontal boiler of the longitudinal or cross-drum type, with the tube bank comprised of multiple parallel sections, each section made up of a front and rear header connected by one or more vertical rows of generating tubes and with the sections or groups of sections having a common steam drum.

sector 1. In magnetic information storage medium, a defined area of a track or band. 2. The smallest addressable portion of storage on some disk and drum storage units.

sector gear A toothed machine component that looks like part of a gear wheel containing the bearing and part of the rim and its teeth.

sector link In an elastic-chamber pressure gage, the connecting link between the elastic chamber and the sector gear that positions the pointer.

sedentary See physical condition.

sediment 1. Matter in water which can be removed from suspension by gravity or mechanical means. 2. A non-combustible solid matter which settles out at the bottom of a liquid; a small percentage is present in residual fuel oils.

sedimentation 1. Classification of metal powders according to the rate at which they settle out of a fluid suspension. 2. Removal of suspended matter either by quiescent settling or by continuous flow at high velocity and extended retention time to allow the matter to deposit out.

sediment trap 1. A device for measuring the rate at which sediment accumulates on the floor of a body of water. 2. A device used for removing sediment from an instrument sensing line on a boiler.

see and avoid A visual procedure wherein pilots of aircraft flying in visual meteorological conditions (VMC), regardless of type of flight plan, are charged with the responsibility to observe the presence of other aircraft and to maneuver their aircraft as required to avoid collisions.[ARP4107-88]

Seebeck coefficient See Seebeck effect.

Seebeck effect The establishment of an electric potential difference tending to produce a flow of current in a circuit of two dissimilar metals the junctions of which are at different temperatures. [NASA]

seed(s) 1. Small bubbles of air or gaseous inclusions entrapped within the glass. [ARP924-91] 2. A small, single crystal of semiconductor material used to start the growth of a single large crystal from which semiconductor wafers are cut.

seek To position the access mechanism of a direct-access storage device at a specified location.

seek time The time taken to execute operation.

seepage 1. Occurrence of extremely slight fluid at the surface of a component or part of a hydraulic system, normally due to "breathing" of seals under cyclic pressure load. No drops are allowed over an extended period of observation.[AIR1916-88] 2. Not recommended; see leakage.[ARP243B-65]

segment In computer software programming, the division of a routine.

segmented circle A system of visual indicators laid upon the ground generally surrounding the wind sock or tetrahedron, designed to provide traffic pattern information at airports which do not have operating control towers. [ARP4107-88]

segments of an instrument approach procedure An instrument approach may have as many as four separate segments, depending on how the approach procedure is structured. (a) Initial Approach Segment: The segment between the initial approach fix and the intermediate fix or the point where the aircraft is established on the intermediate course or final approach course. (b) Intermediate Approach Segment: The segment between the intermediate fix or point and the final approach. (c) Final Approach Segment: The segment between the final approach fix or point and the runway, airport, or missed approach point. (d) Missed Approach Segment: The segment between the missed approach point, or point of arrival at decision height, and the missed approach fix at the prescribed altitude. [ARP4107-88]

segregation 1. Keeping process streams apart. 2. Nonuniform distribution of alloying elements and impurities in a cast metal microstructure. 3. A series of close, parallel, narrow and sharply defined wavy lines of color on the surface of a molded plastics part that differ in shade from surrounding areas and make it appear as if the components have separated. 4. The tendency of refuse of varying compositions to deposit selectively in different parts of the unit.

seismic detector An instrument that registers seismic impulses.

seismic profiler A continuous seismic reflection system used to study geologic structure beneath the oceans' floor to depths of 10,000 ft or more; the reflections are recorded on a rotating drum.

seismic waves The disturbance of earth tremors produced by a mechanical disturbance on the surface or underground. [NASA]

seismocardiography The measurement of the high frequency vibrations of the heart.[NASA]

seismochronograph A device for precisely determining the time at which an earthquake shock arrives at the instrument's location.

seismograph An instrument that detects and records vibrations in the earth, such as an earthquake.

seismology The study of earthquakes, by extension, the structure of the interior of the Earth via both natural and artificially generated seismic signals. [NASA]

seismoscope A device that records the occurrence or time of occurrence of an earthquake, but does not record the frequency or amplitude of earthquake shocks.

seizure See freeze-up.

selection Addressing a terminal and/or a component on a selective calling circuit. See also lockout and polling.

selective inattention See attention, anomalies of inattention.

selective-ion electrode A type of pH electrode that involves use of a metal-metal-salt combination as the measuring electrode, which makes the electrode particularly sensitive to solution activities of the anion in the metal salt.

selective plating Any of several methods of electrochemically depositing a metallic surface layer at only localized areas of a base metal, the remaining unplated areas being masked with a nonconductive material during the plating step.

selective surfaces Surfaces, often coated, for which the spectral optical properties, such as reflectance, absorptance, emittance, or transmittance vary significantly with wavelength. Such properties are of interest in solar energy applications.[NASA]

selective system A system in which the protective device closest to the faulted circuit opens and isolates that circuit without disturbing the remainder of the system.[ARP1199A-90]

selectivity The characteristic of an electronic receiver which determines the extent to which it can differentiate between a desired signal and electronic noise or undesired signals of other frequencies.

selector 1. A device for choosing objects or materials according to predetermined attributes. 2. A device for starting or stopping a mechanism at predetermined positions or locations. 3. A gearshift lever for operating an automatic transmission in a motor vehicle. 4. The part of a gearshift in a motor vehicle transmission that selects the required gearshift bar. 5. A converter that separates purified copper from residue in a single operation. 6. A device which selects one of a plurality of signals.

selenology That branch of astronomy that treats of the moon, its magnitude, motion, constitution, and the like. Selene is Greek for moon.[NASA]

self absorption Attenuation of radiation due to absorption within the substance that emits the radiation.

self-adapting Pertaining to the ability of a system to change its performance characteristics in response to its environment.

self adaptive control systems Particular types of stability augmentation systems which change the responses of given control inputs by constantly sampling responses and adjusting their gain, rather than having fixed or selective gain systems.[NASA]

self braking strut A folding brace or strut in a landing gear system which carries its own actuator.[AIR1489-88]

self-checking code Same as error detecting code.

self-cleaning A descriptor for any device fitted with a mechanism that removes accumulated deposits from its interior without disassembly.

self contained Used to describe independency from outside appliances, e.g., a self-contained lubrication system. [ARP906A-91]

self-contained apparatus Apparatus that is not necessarily connected to equipment in the non-hazardous area. (Usually therefore is self powered). Self-contained apparatus is normally portable e.g. walkie-talkie radios—but the term does not imply that it must be.

self-contained instrument An instrument that contains all of its component parts within a single case or enclosure, or has them incorporated into a single assembly.

self contained system Ignition systems meeting one or more of the other definitions herein which, in addition, are designed to operate from a power source which is engine supplied equipment. [AIR784-88]

self diffusion (solid state) The spontaneous movement of an atom to a new site in a crystal of its own species. [NASA]

self-discharge The decrease in the state of charge of a cell, not resulting from an applied useful load, over a period of time, normally due to internal electrochemical losses and accelerated by higher temperature.[AS8033-88]

self-documenting language Pertains to languages which permit comments to be interspersed with the commands. These comments become documentation. Also, the language is generally very readable.

self-extinguishing The quality of a material to stop burning once the source of a flame is removed.

self-indicting fuse or limiter That type of fuse or limiter that incorporates a device to visually denote severance of the fusible element as an integral part of the limiter is a self-indicating current limiter. Limiters or fuses that visually display the fusible element or incorporate a mechanical indicting device are classified as self-indicating current limiters or fuses.[ARP1199A-90]

self monitoring The capability of a component to monitor itself without additional sensors or other devices.[AIR 1916-88]

self-organizing environment A class of equipment which may be characterized loosely as containing a variable network in which the elements are organized by the equipment itself, without external intervention, to meet criteria of successful operation.

self overcommitment A response set in which a person commits to a task for which he/she is knowingly ill prepared and which presses him/her or his/her aircraft, or both, beyond reasonable limits.[ARP4107-88]

self-quenched counter tube A type of radiation counter tube in which internal interactions inhibit the reignition of electron discharge.

self regulating See automatic control.

self-repairing See reconfigurable.

self-sealing fastener A fastener that provides a fuel-tight seal without the need for sealant material nor the use of a mechanical seal. (An interference fit fastener is an example).[AIR4069-90]

self-sealing tank A tank in which construction is such that an automatic sealing action occurs when penetrated by a projectile. The weight and thickness of constructions is adjusted to the ballistic threat anticipated for the aircraft. [AIR1664-90]

self subtraction holography See holographic subtraction.

self sufficient A starting system independent of all sources of energy external to the aircraft.[ARP906A-91]

self-supporting steel stack A steel stack of sufficient strength to require no lateral support.

self-tapping screw A threaded fastener with specially designed and hardened threads that form internal threads in a hole in sheet metal or soft materials as the screw is driven.

self test(s) 1. A test or series of tests, performed by a device upon itself, which shows whether or not it is operating within design limits. This includes test programs on computers and ATE performing functional and diagnostic tests. [ARD50010-91] 2. Programmed functions performed by a machine, either automatically at start-up or on user demand, that test the working order of the machine. In particular, programs stored in read-only memory that test the integrity of a machine's integrated circuits and the connections between the circuits and the devices they control.[NASA] 3. A circuit used by a computer used to check its operation when power is first turned on.

self-tuning See adaptive control.

Selsyn A trade name for a synchro device.

semantics The relationships between symbols and their meanings.

semaphore A task synchronization mechanism. See synchronization of parallel computational processes.

semi-articulated A variation on the articulated shock strut configuration. The wheel is mounted on a lever which pivots about the lower end of the piston of a telescoping shock absorber. The front end of the lever is restrained by a link to the outer cylinder of the shock absorber.[AIR1489-88]

semiautomatic controller A control device in which some of the basic functions are performed automatically.

semicircular canals Structures of the inner ear, the primary function of which is to register movement of the body in space. They respond to change in the rate of movement.[NASA]

semi-conductor An electronic device such as an integrated circuit or transistor.

semiconductor devices Electron devices in which the characteristic distinguishing electronic conduction takes place within semiconductors.[NASA]

semiconductor diodes Two electrode semiconductor devices utilizing the rectifying properties of junctions or point contacts.[NASA]

semiconductor insulator semiconductors See SIS (semiconductors).[NASA]

semiconductors, II-VI Semiconductors composed of elements from group II and VI of the periodic table—sometimes extended to cover elements with valances of 2 and 6. Typical II-VI compounds are cadmium telluride and cadmium selenide.

semiconductors, III-V Semiconductors composed of atoms group II and V of the periodic table, such as gallium (III) and arsenic (V), which form gallium arsenide.

semiconductors (materials) Electronic conductors, with resistivity in the range between metals and insulators, in which the electrical charge carrier concentration increases with increasing temperature over some temperature range. Certain semiconductors possess two types of carriers, namely, negative electrons and positive holes.[NASA]

semiconductor strain gage A type of strain measuring device particularly well suited to use in miniature transducer elements; it consists of a piezoresistive element that is either bonded to a force-collecting diaphragm or beam or diffused into its surface.

semiconductor temperature sensor See thermistor.

semi-graded index An optical fiber with refractive index profile intermediate between step-index and graded-index. Strictly speaking, this might be considered a type of graded index fiber with refractive index profile somewhat steeper than normal.

semikilled steel Steel that is partly deoxidized during teeming so that only a small amount of dissolved gas is evolved as the metal solidifies.

semirigid plastic Any plastics material having an apparent modulus of elasticity of 10,000 to 100,000 psi.

senders See transmitters.

sense 1. To examine, particularly relative to a criterion. 2. To determine the present arrangement of some element of hardware, especially a manually set switch. 3. To read punched holes or other marks.

sense indicator An aircraft radio indicator that shows whether the aircraft is flying toward or away from an omnirange station. Sometimes called a to-from indicator.[ARP4107-88]

sense switch See alteration switch.

sensibility reciprocal A balance characteristic equal to the change in load required to vary the equilibrium position by one scale division at any load.

sensible heat Heat that causes a temperature change.

sensibility See sensitivity.

sensing element 1. A device which measures the value of a variable by providing an output that can be utilized by the controller.[ARP89C-70] 2. The portion of a device directly responsive to the value of the measured quantity. It may include the case protecting the sensitive portion. See element, sensing.

sensitive part A life-limited part, whose failure is not catastrophic but likely to seriously affect engine performance, reliability or operating cost (durability critical).[AIR1872-88]

sensitive time A characteristic of a cloud chamber equal to the amount of time after expansion when the degree of supersaturation is sufficient to allow a track to form and be detected.

sensitivity 1. The characteristic of an explosive component which expresses its susceptibility to initiation by externally applied stimuli.[AIR913-89] 2. In a sensing element, the ratio of change in output to a specified change in input.[ARP89C-70] 3. Response of a mathematical model to variations of the input parameters.[NASA] 4. Ratio of change of output to change of input. 5. Also defined as the least signal input capable of causing an output signal having desired characteristics. 6. The smallest change in actual value of a measured quantity that will produce an observable change in an instrument's indicated or recorded output. 7. The minimum value of an observed quantity that can be detected by a specific instrument. 8. The degree to which a process characteristic can be influenced or changed by a small change in some physical or chemical stimulus.

sensitized stainless steel Any austenitic stainless steel having chromium carbide deposited at the grain boundaries. This deprives the base alloy of chromium resulting in more rapid corrosion in aggressive media.

sensitometer An instrument for determining the sensitivity of light-sensitive materials.

sensitometry The measurement of the light response characteristics of photographic film under specified conditions of exposure and development.[NASA]

sensor 1. The component of a system that converts an input signal into a quantity which is measured by another part of the system. Also called sensing element.[AIR1489-88] 2. A component which senses variables and produces a signal in a medium compatible with fluidic devices; for example, a temperature or angular rate sensor.[ARP993A-69] 3. A mechanical, electrical, optical or fluidic device that provides data inputs; e.g., transducers, position indicators, discretes.[ARP1587-81] 4. That portion of a device directly responsive to the value of the measured quantity. It may include the case protection and the sensitive portion. 5. The portion of a channel which responds to changes in a plant variable or condition, and converts the measured process variable into an instrument signal. 6. A device that produces a voltage or current output representative of some physical property being measured (speed, temperature, flow, etc.).

sensor output format The form of the signal or waveform supplied by the sensor; e.g., analog, pulse frequency, digital.[ARP1587-81]

separation 1. Spacing of aircraft to achieve their safe and orderly movement in flight and while landing and taking off.[ARP4107-88] 2. An action that disunites a mixture of two phases into the individual phases. 3. Partition of aggregates into two or more portions of different particle size, as by screening. 4. The degree, in decibels, to which right and left channels of a stereophonic radio or sound system are isolated from each other. 5. The parting of two connected members of a structure or system, as occurs at preplanned times after launching a multistage rocket. 6. The removal of dust from a gas stream.

separation minima The minimum longitudinal, lateral, and vertical distances by which aircraft are spaced through the application of air traffic control procedures. See also composite separation.[ARP4107-88]

separator 1. The movable or flexible member in a fluid container, such as an accumulator or reservoir, the function of which is to prevent intermixture of fluids, such as air and hydraulic fluid.[ARP243B-65] 2. Layer(s) of insulating material (conventionally a fabric) which provide electronic separation between electrically conductive cell plates of opposing polarities. It is placed adjacent to the gas barrier and provides physical support for the gas barrier material.[AS8033-88] 3. Any machine for dividing a mixture of materials according to some attribute such as size, density or magnetic properties. 4. A device for separating materials of different specific gravity using water or air. 5. A cage in a ball-bearing or roller-bearing assembly. See cage.

separator-filter A piece of process equipment that removes solids and entrained liquid from a fluid stream by passing the fluid both through a set of baffles or a coalescer and through a screen.

sequence checking routine A routine which checks every instruction executed, and prints out certain data, e.g., to print out the coded instructions with addresses, and the contents of each of several registers, or it may be designed to print out only selected data, such as transfer instructions and the quantity actually transferred.

sequence control A system of control in which a series of machine movements occurs in a desired order, the completion of one movement initiating the next, and in which the extent of the movements is not specified by numerical input data.

sequence module See alarm module.

sequence monitor Computer monitoring of the step-by-step actions that should be taken by the operator during a startup and/or shutdown of a power unit. As a minimum, the computer would check that certain milestones had been reached in the operation of the unit. The maximum coverage would have the computer check that each required step is performed, that the correct sequence is followed, and that every checked point falls within its prescribed limits. Should an incorrect action or result occur, the computer would record the fault and notify the operator.

sequencer A mechanical or electronic control device that not only initiates a series of events but also makes them follow each other in an ordered progression.

sequence of events See mishap.

sequence report The weather report transmitted hourly to all teletype stations, and available at all flight service stations.[ARP4107-88]

sequencing 1. A following of one thing after another; an order of succession; a related or continuous series. 2. In landing gear system, the process of control of the gear retraction or extension in the proper order of succession. For example, the gear control is placed in the gear "up" position; a) heel doors open, b) gear retracts within the wheel well and locks in the stowed position, c) doors close and lock, and d) system hydraulic pressure is deenergized, thus leaving the gear in position for high speed flight.[AIR1489-88] 3. Planning a series of operations or tasks to optimize the use of available production facilities and resources.

sequential access A data access method in which records or files are read one at a time in the order in which they appear in the file or volume.

sequential control 1. Control by completion of a series of one or more events. [NASA] 2. A mode of computer operation in which instructions are executed in consecutive order by ascending or descending addresses of storage locations, unless otherwise specified by a jump. 3. A class of industrial process control functions in which the objective of the control system is to sequence the process units through a series of distinct states (as distinct from continuous control). See sequence monitor.

sequential files Collection of related records stored on secondary storage devices such as magnetic tapes and discs. The records are physically stored in the same order as the key number and it is not necessary to use any index or algorithms to locate a particular record.

sequential life test Sequential life test

is a life test sampling plan whereby neither the number of failures nor the time required to reach a decision are fixed in advance but instead decisions depend on the accumulated results of the life test. Information on the observed time to failure are accumulated over time and the results at any time determine the choice of one among three possible decisions: (a) the lot meets the acceptability criterion, (b) the lot does not meet the acceptability criterion, or (c) the evidence is insufficient for either decision (a) or (b) and the test must continue. [ARD50010-91]

sequential logic A logic circuit in which the output depends on the inputs to the circuit and the internal states of the circuit.

sequential sampling A method of inspection that involves testing an undetermined number of samples, one by one, until enough test results have been accumulated to allow an accept/reject decision to be made.

serial 1. In reference to digital data, the presentation of data as a time-sequential bit stream, one bit after another. 2. In PCM telemetry, the transfer of information on a bit by bit basis. 3. In data transfer operations, a procedure that handles the data one bit at a time in contrast to parallel operations.

serial access See access, serial.

serial computer 1. A computer having a single arithmetic and logic unit. 2. A computer, some specified characteristic of which is serial, for example, a computer that manipulates all bits of a word serially. Contrast with parallel processing.

serial I/O Method of data transmission in sequential mode, one bit at a time. Only one line is needed for the transmission. However, it takes longer to send/receive the data than parallel I/O.

serial operation A mode of computer operation in which information flows sequentially in time using only one digit, word, line or channel at any given time.

serial output In data processing, programming that instructs the computer to send only one bit at a time between interconnecting devices.

serial-to-parallel converter In PCM telemetry, the circuitry which converts a serial bit stream into bit-parallel data outputs, each transfer representing one measurement.

serial transmission In telemetering, sending bits of information from different sensors or devices over a single channel in sequence, with each bit of information coded to identify its source as well as its value.

series cascade action A type of control-system interaction whereby the output of each controller in a series (except the last one) serves as an input signal to the next controller.

series element Any of a number of two-terminal electronic elements that form a path from one node of a network to another in such a way that only elements of the path terminate at intermediate nodes along the path; alternatively, any of a number of two-terminal elements connected in such a way that any mesh containing one of the elements also contains the others.

series expansion In mathematics, a divergent series of terms the sum of which is asymptotic or ascending. [NASA]

series resistor A resistive element of the voltage circuit of an instrument that adapts the instrument to operate on some designated voltage.

series servo 1. A servo located in a control system so that the servo output adds algebraically to that of a major input. This arrangement is commonly used with SAS actuators to superimpose control on primary commands. The series servo output will not cause motion at the major input.[ARP1181A-85] [AIR

1916-88]

serif In typesetting, a type style that has decorative additions at the top and bottom of letters. Commonly called a roman typeface. See sans-serif.

serious injury Any injury which: (a) requires hospitalization for more than 48 hours, commencing within 7 days from the date of the injury was received; (b) results in a fracture of any bone (except simple fractures of fingers, toes, or nose); (c) causes severe hemorrhages, or severe nerve-, muscle-, or tendon-damage; (d) involves any internal organ; or (e) involves second- or third-degree burns, or any burns affecting more than 5 percent of the body surface.[ARP4107-88]

serrations Alteration of the inside surface of a conductor barrel to provide better gripping of the conductor or on the outside of a connector housing to provide better gripping of the connector, protusions on the rear of a connector housing for positive orientation of accessories.[ARP914A-79]

serve A filament, or group of filaments, such as wires or fibers, helically wound around a central core, normally unbonded.[ARP1931-87]

serviceable The condition of an end item in which all requirement for repair, bench check, overhaul, or modification, as applicable have been accomplished, making it capable of performing the function or requirements for which originally designed. The fact that signs of previous use are apparent does not necessarily mean that it is unserviceable. When appearance is not a primary consideration, and the condition of the item meets all safety and performance requirements, it will be processed as serviceable.[ARD50010-91]

service access point (SAP) The connection point between a protocol in one OSI layer and a protocol in the layer above. SAPs provide a mechanism by which a message can be routed through the appropriate protocols as it is passed up through the OSI layers.

service bulletin The only document issued by the manufacturer to notify the airline of recommended modification, substitution of parts, special inspections/checks, reduction of existing life limits or establishment of first time life limits and conversion from one engine model to another.[ARD50010-91]

service ceiling The height above sea level, under standard atmospheric conditions, at which a given airplane is unable to climb faster than 100 ft/min. [ARP4107-88]

service factor 1. For a facility such as a chemical processing plant or electric generating station, the proportion of time the facility is operating—actual operating time in hours divided by total elapsed time in hours, expressed as a percent. 2. In electric motors, a factor in which a motor can be operated above rated current without damage. For example, an electric motor with a service factor of 1.15 can be operated up to 115% of rated current without damage.

service life 1. Service life is the total time, specified in hours of engine operation with specified mission profiles and mission mix, that a part can function with repair, provided that after each repair the part can function for a minimum of one engine repair interval. Random failure events, such as FOD, are not considered in determining service life.[ARD50010-91] 2. A specified period of time during which a device is expected to function satisfactorily.[ARP914A-79] 3. The length of time a mechanism or piece of equipment can be used before it becomes either unreliable or economically impractical to maintain in good working order.

service rating The maximum voltage or current which a connector or electrical device is designed to function continuously at a specified temperature.

[ARP914A-79]

service test A test of an item, system, material, or technique conducted under simulated or actual operational conditions to determine whether the specified military requirements or characteristics are satisfied.[ARD50010-91]

service water General purpose water which may or may not have been treated for a special purpose.

servicing The performance of any act needed to keep an item in operating condition, i.e. lubricating, fueling, oiling, cleaning, etc., but not including preventative maintenance of parts or corrective maintenance tasks.[ARD50010-91] [AIR1916-88]

servicing point An access specifically provided to permit serving (lubrication, filling, draining, charging or cleaning) of equipment in the normal installed position.[ARD50010-91]

servo The actuating device which positions the aircraft control surfaces in response to a commanded signal from a controller, selector or sensor. Also called a servomotor or servo actuator.[ARP419-57]

servoactuator 1. An actuator that controls the magnitude of position outputs in response to the magnitude of the input signal.[ARP1281A-81] 2. An assembly of a controller, an actuator, and feedback element(s), typically packaged together as a single line replaceable unit (LRU).[AIR1916-88]

servoactuator flight control A mechanism capable of producing a controlled output position/motion of an aerodynamic control surface in response to a command input.[ARP1281A-81]

servo brake 1. A motor vehicle brake in which vehicle motion is used to increase the pressure on one of the brake shoes. 2. A power-assisted braking device.

servocylinder A servo actuator having such an integral follow-up mechanism that its final output may be made a function of the input signal to its control valve.[ARP243B-65]

servo engage/disengage The process of connecting or disconnecting the servo motor and gear train from the cable drum or sector through a mechanical or electrical clutch in the servo.[ARP 419-57]

servomechanism(s) 1. Control systems incorporating feedback in which one or more of the system signals represent mechanical motion.[NASA] 2. An automatic control system incorporating feedback that governs the physical position of an element by adjusting either the values of the coordinates or the values of their time derivatives. 3. Any feedback control system.

servo overpower The act of overpowering the servo by applying an appropriate pilot force at the control column or rudder pedals.[ARP419-57]

set point That value of the controlled temperature to which the selecting device is adjusted, representing the desired value of controlled variable.[ARP89C-70]

servo tab/servotab A tab directly actuated by the aircraft control system and which, when deflected, causes the control surface or other surface to which it is attached to be deflected or moved by the air forces acting on the tab. Sometimes called a Flettner tab or flying tab. [ARP4107-88]

SES See surface effect ships.

session Layer 5 of the OSI model.

set 1. A collection. 2. To place a storage device into a specified state, usually other than that denoting zero or blank. 3. To place a binary cell into the state denoting one. 4. In simulation theory, sets consist of entities with at least one common attribute. Additionally, entities may own any number of sets. Sets may be arranged (topologically ordered) on a "first-in, first-out," "last-in, last-out," or ranked basis. 5. A combination of units, assemblies, or parts connected together or used together to perform a single function, as in a television or radar set. 6. A group of tools, often with at least some of the individual tools differing from others only in size. 7. In plastics processing, conversion of a liquid resin or adhesive into a solid material. 8. Hardening of cement, plaster or concrete. 9. Permanent strain in a metal or plastics material.

setpoint The position at which the control point setting mechanism is set. This is the same as the desired value of the controlled variable.

setpoint control A control technique in which the computer supplies a calculated setpoint to a conventional analog instrumentation control loop.

set pressure 1. The pressure which a component is designed to provide for a defined operation.[AIR1916-88] 2. The inlet pressure at which a safety relief valve opens; usually a pressure established by specification or code.

set screw A small, headless machine screw used for holding a knob, gear or collar on a shaft; it usually has a sharp or cupped point on one end and a slot or recessed socket on the other end.

setting In an adjustable or calibrated unit (such as a relief valve, pressure switch or flow control device), operating characteristics which result from an adjustment or setting.[ARP243B-65]

settling Partial or complete separation of heavy materials from lighter ones by gravity.

settling time The time interval between the step change of an input signal and the instant when the resulting variation of the output signal does not deviate more than a specified tolerance from its steady-state value. See also time, settling.

setup 1. An arrangement of data or devices to solve a particular problem. 2. In a computer which consists of an assembly of individual computing units, the arrangement of interconnections between the units, and the adjustments needed for the computer to solve a particular problem. 3. Preliminary operations—such as control adjustments, installation of tooling or filling of process fluid reservoirs—that prepare a manufacturing facility or piece of equipment to perform specific work.

set-up driver A routine capable of accepting raw set-up and control information, converting this information to static stores, dynamic stores, or control words, and loading or transmitting the converted data to the associated module in order to achieve the desired effect.

severe weather avoidance plan (SWAP) A plan to minimize the effect of severe weather on traffic flows in impacted terminal or ARTCC areas, or both. SWAP is normally implemented to provide the least disruption to the ATC system when flight through portions of airspace is difficult or impossible due to severe weather.[ARP4107-88]

severity The consequences of a failure mode. Severity considers the worst potential consequence of a failure, determined by the degree of injury, property damage, or system damage that could ultimately occur.[ARD50010-91]

sexadecimal number A number, usually of more than one figure representing a sum in which the quantity represented by each figure is based on a radix of sixteen. Synonymous with hexadecimal number.

sextants Double reflecting instruments for measuring angles, primarily altitudes of celestial bodies.[NASA]

SFAR See sound fixing and ranging.

sferics See atmospherics.

SGEMP See system generated electromagnetic pulses.

SGML Standard Generalized Markup Language.[AS4159-88]

shackle An open or closed link having

extended arms, each with a hole to accommodate a single pin that spans the gap between the arms.

shade A device, the chief use of which is to diminish or intercept the light from the lamp in certain directions where such light is not desirable. Frequently the functions of a shade and a reflector are combined in the same unit.[ARP798-91]

shading Controlling the phase distribution and amplitude distribution of transducer action at the active face in order to control its directionality.

shadowgraph photography Photography in which steep density gradients in the flow about a body are made visible, the body itself being presented in silouette.[NASA]

shadowgraphs See shadowgraph photography.

shadows Darknesses in regions, caused by obstructions between the source of light and the regions.[NASA]

shaft A cylindrical metal rod used to position, and sometimes to drive or be driven by, rotating parts such as gears, pulleys or impellers, which transmit power and motion.

shaft balancing A method of reducing vibrations in rotating equipment by redistributing the mass to eliminate asymmetrical centrifugal forces.

shaft encoder A device for indicating the angular position of a cylindrical member. See gray code and cyclic code.

shaft horsepower 1. The horsepower delivered to the driving shaft of an engine as measured by a torsion meter. See also brake horsepower.[ARP4107-88] 2. The power output of an engine, turbine or motor. 3. The power input to a pump or compressor.

shakedown test An equipment test made during installation or prior to its initial production operation.

shakeout Removing sand castings from their molds.

shake table See vibration machine.

shake-table test A durability test in which a component or assembly is clamped to a table or platen and subjected to vibrations of predetermined frequencies and amplitudes.

shall function The instrument shall not exceed the given tolerances.[AIR818C-91]

shall not be adversely affected The instrument shall not exhibit characteristics or sustain damage which precludes proper functioning and/or use. [AIR818C-91]

shank The end of a tool that fits into a collet, chuck or other holding device.

shape control The control of large flexible platforms in orbit by means of actuators strategically located.[NASA]

shape memory alloys Martensitic alloys (titanium-nickel) which exhibit shape recovery characteristics by stress-induced transformation and reorientation. Reverse transformation during heating restores the original grain structure of the high temperature phase. [NASA]

shaping A machining process in which a reciprocating, single-point tool cuts a flat or simply contoured surface.

shareable program A (reentrant) program that can be used by several users at the same time.

shared display VDU, visual display unit.

shared time control See control, shared time.

shareware Computer software that can be freely copied. Additional payment to the program author is expected from those who find frequent use for the program.

sharpen To impart a keen edge or acute point to a cutting or piercing tool.

shatter cones Distinctively striated conical rock fragments along which fracturing has occurred, ranging in length from less than a centimeter to several meters, and generally found in nested or composite groups in rocks of cryptoexplosion structures and believed to be formed by shock waves generated by meteorite impact.[NASA]

shaving 1. Cutting a thin layer of material off the surface of a workpiece, such as to bring gear teeth to final shape. 2. Trimming thin layers or burrs from forgings, stampings or tubing to smooth parting lines, uneven edges or flash.

shear 1. A tool that cuts plate or sheet material by the action of two opposing blades that move along a plane approximately at right angles to the surface of the material being cut. 2. A type of stress tending to separate solid material by moving the portions on opposite sides of a plane through the material in opposite directions.

shear disturbances See S waves.

shear fatigue See shear stress.

shearing Separation of material by the cutting action of shears, or by similar action in a punch-and-die set.

shear lip A characteristic of ductile fractures in which the final portion of the fracture separation occurs along the direction of principal shear stress, as exhibited in the cup and cone fracture of a tensile-test specimen made of relatively ductile material.

shear pin 1. A pin or wire designed to hold parts in a fixed relative position until sufficient force to cut through the pin is applied to the assembly. 2. A pin through the hub and shaft of a powertrain member which is designed to fail in shear at a predetermined force, thereby protecting the mechanism from being overloaded.

shear section A special section, usually in the starter or accessory output shaft, that is designed to shear or fail at a specified torque range.[ARP906A-91]

shear spinning A metal forming process in which sheet metal or light plate is formed into a part having rotational symmetry by pressing a tool against a rotating blank and deforming the metal in shear until it comes in contact with a shaped mandrel. The resulting part has a wall thinner than the original blank thickness.

shear test 1. Two narrow strips of aluminum are overlapped at one end by approximately 1 inch and bonded in this area by sealant. After cure, the specimen is mounted in a testing machine and pulled in opposite directions, creating shear forces in the sealant. The force required for the sealant to fail in shear is a measure of its shear strength.[AS 7200/1-90] 2. Any of various tests intended to measure the shear strength of a solid.

shear wave(s) A wave in an elastic medium in which any element of the medium along the wave changes its shape without changing its volume. See also S waves.

sheath See jacket.

sheave A pulley or wheel with a grooved rim to guide a rope, cable or belt.

sheet Any flat material intermediate in thickness between film or foil and plate; specific thickness limits for sheet depend on the type of material involved, and sometimes also on other dimensions such as width.

sheet metal A flat-rolled metal product generally thinner than about 0.25 in.

sheld A metallic or electrically conductive braid, jacket, or other covering of a cable which is over the wire(s).[AS1198-82]

shelf life The storage time during which an explosive item remains serviceable. [AIR913-89] See also storage life.

shell 1. A thin metal cylinder. 2. The outer wall of a tank or pressure vessel. 3. A mold wall made of sand and a thermosetting plastics material used in certain casting processes. 4. A cast tube used as starting stock for certain types of drawn seamless tubing. 5. The metal tube that remains when a billet is extruded using a dummy block of smaller diameter than the billet. 6. A hollow, pierced forging. 7. The outer member of a pulley block surrounding the sheave.

shellac A flammable resinous material, produced by a species of insect found in India, that is used to make a water-resistant coating for wood by dissolving the resin in alcohol.

shell-and-tube heat exchanger A device for transferring heat from a hot fluid to a cooler one in which one fluid passes through the inside of a bundle of parallel tubes while the other fluid passes over the outside of the tubes but inside the vessel shell; heat is transferred by conduction across the walls of the tubes.

shell, electrical connector See housing, electrical connector.

sheltered area See area, sheltered.

shield 1. A conductive layer of material (usually wire, foil, or tape) applied over a wire or cable to provide electrical isolation from external interference. Also known as a screen.[ARP1931-87] 2. A protective device around any part(s), assembly or mechanism to protect persons from injury, parts from damage, etc.[ARP480A-87] 3. Any barrier to the passage of interference-causing electrostatic or electromagnetic fields. An electrostatic shield is formed by a conductive layer, like a foil, surrounding a cable core. 4. An electromagnetic shield is a ferrous metal cabinet or wireway. 5. An attenuating body that blocks radiation from reaching a specific location in space, or that allows only radiation of significantly reduced intensity to reach the specific location.

shielded cable 1. One or more wires enclosed within a conductive shield to minimize the interference effects of internal or external circuits.[ARP914A-79] 2. A cable surrounded by a shield.[ARP 1931-87]

shielded conductor An insulated conductor encased in one or more conducting envelopes, usually made of woven wire mesh or metal foil; similar products containing one or more insulated conductors are known as shielded cable, shielded conductor cable, or shielded wire.

shielded metal arc welding (SMAW) Metals are heated with an arc between a covered metal electrode and the work. Shielding is obtained from decomposition of the electrode covering. Pressure is not used and filler metal is obtained from the electrode.

shield, electrical connector A device placed around that portion of a connector which is used for attaching wires or cables so as to both shield against electromagnetic interference and/or protect the connector wires or cable from mechanical damage.[ARP914A-79]

shielding 1. The arrangement of shields used for any particular circumstance; the use of shields.[NASA] 2. Surrounding an electronic circuit or signal-transmission cable with a ground plane, such as a foil or woven-metal sheath, so that capacitive coupling between the circuit and ground plane remains stable. 3. Interposing a radiation absorbing material between a source of ionizing radiation and personnel or equipment to reduce or eliminate radiation damage.

shielding effectiveness A value equal to 20 times the log to the base ten, of the ratio between the voltage impressed on a given nonshielded wire and that impressed on an identical wire with a shield, when these wires are placed in a fixed magnetic field.[ARP1931-87]

shift register A data storage location in which data is stored and moved (shifted) from one position to another in the register. This shifting is usually to perform some logic or arithmetic operation on the stored data.

shim A thin piece of material, usually

metal, that is placed between two surfaces to compensate for slight variations in dimensions between two mating parts, and to bring about a proper alignment or fit.

shimmy 1. Abnormal vibration or wobbling. 2. A self-sustained vibration of a rolling wheel(s) about a real or theoretical swiveling axis. Sometimes classified as large angle and small angle, in which large angle consists of more or less violent oscillations of the wheel assembly about its caster axis; the tire contact area is sliding bodily sideways at each swing and derives energy to maintain the oscillation from the forward motion of the vehicle. Small angle or kinematic shimmy is a high frequency oscillation within the adhesion range of the tire.[AIR1489-88]

shimmy damper Any of various devices attached to a castering wheel to damp oscillations about the castering axis. See also damper.[AIR1489-88][ARP4107-88]

ship auger A wood boring tool consisting of a spiral body having a single cutting edge instead of two and without a spur at the outer end of the cutting edge; it may or may not have a central feed screw.

ship to shore communication Communication between a ship at sea and a shore station.[NASA]

shiva laser system High energy multiarm solid state (Nd doped ED-2 glass) infrared laser system used for laser driven fusion experiments.[NASA]

shock (physiology) Clinical manifestations of circulatory insufficiency, including hypotension, weak pulse, tachycardia, pallor, and diminished urinary output.[NASA]

shock absorber(s) 1. A device for absorbing the energy from impact of a mass or force input.[AIR1489-88] 2. Devices for the dissipation of energy used to modify the response of a mechanical system to applied shock.[NASA] 3. A component connected between a piece of equipment and its frame or support to damp out relative motion between them and reduce the effect of acceleration forces; it normally consists of a dashpot or a combination of a dashpot and a spring.

shock absorber, air-oil A shock absorber which utilizes the principle of forcing oil through an orifice for absorption of energy, and also provides a chamber with gas (air or more usually dry nitrogen) under pressure for the purpose of supporting the static load on the unit and by the gas pressure, extension of the shock absorber when the load is removed.[AIR1489-88]

shock absorber, constant A shock absorber which utilizes a fixed or constant orifice in metering the fluid.[AIR1489-88]

shock absorber, dash pot A shock absorber which utilizes fluid metering for shock absorption, and does not provide a gas charge for maintaining a static load.[AIR1489-88]

shock absorber, dual chamber A shock absorber which provides two separate chambers for gas charge and thereby a two stage spring rate as the shock absorber is compressed.[AIR1489-88]

shock absorber, frangible A shock absorber which utilizes material failure as a shock absorption medium i.e., a) buckling collapse of tubing, b) fragmenting of tubing, c) compression of honey comb material, d) compression of a fluid filled device, e) compression of a solid thermoplastic. These devices are "one shot" devices and must be replaced after each use.[AIR1489-88]

shock absorber, friction A shock absorber which utilizes friction as the energy absorption medium such as a ring spring configuration.[AIR1489-88]

shock absorber, honeycomb A shock absorber which utilizes a crushable honeycomb core material as the shock absorbing element.[AIR1489-88]

shock absorber, leaf spring A shock absorber which utilizes the deflection of a leaf spring for energy absorption capability. Common on light aircraft. [AIR1489-88]

shock absorber, liquid spring A shock absorber which utilizes the compressibility property of a liquid for the shock absorption medium. Generally provides a high spring rate short stroke, and requires special high pressure sealing techniques.[AIR1489-88]

shock absorber, pneumatic A shock absorber which utilizes air as the shock absorbing medium. Unusual on present day aircraft.[AIR1489-88]

shock absorber, rubber A shock absorber which utilizes rubber as the shock absorption medium. May be used with or without oil dampening.[AIR1489-88]

shock absorber, spring A shock absorber which utilizes helical coil springs as shock absorbing elements. May be used with or without oil damping.[AIR 1489-88]

shock absorber, variable orifice A shock absorber which utilizes a variable orifice for metering fluid which provides the shock absorption medium. The resistance or load on the unit can therefore be programmed versus stroke. A metering pin of variable diameter operating through the orifice is the most common method used for control. See also metering pin and metering tube. [AIR1489-88]

shock cord A strong, many stranded, rubber cord encased in a braided fabric sheath, used as a shock absorber element on light planes or as a launching element for gliders. Also known as bungee cord.[AIR1489-88]

shock diffusers See diffusers.

shock fronts Shock waves regarded as the forward surfaces of fluid regions having characteristics different from those of the region ahead of the wave. The front sides of shock waves.[NASA]

shock motion A sudden transient motion of large relative displacement.

shock mount A supporting structure that isolates sensitive equipment from the effects of mechanical shock or relatively high amplitude vibrations.

shock resistance The ability to absorb mechanical shock without cracking, breaking or excessively deforming.

shock spectra Plots of the maximum acceleration experienced by single degree of freedom systems as a function of their own natural frequency in response to applied shocks.[NASA]

shock strut A structural strut or link which embodies a shock absorbing element.[AIR1489-88]

shock tubes Relatively long tubes or pipes in which very brief high speed gas flows are produced by the sudden release of gas at very high pressure into low pressure portions of the tubes; the high speed flows move into the region of low pressure behind shock waves. [NASA]

shock tunnels Shock tubes used as wind tunnels.[NASA]

shock wave 1. The intense compressive wave produced by the detonation of an explosive charge.[AIR913-89] 2. Surfaces or sheets of discontinuity (i.e., abrupt changes in conditions) set up in a supersonic field of flow, through which the fluids undergo a finite decrease in velocity accompanied by a marked increase in pressure, density, temperature, and entropy, as occurs, e.g. in supersonic flows about bodies.[NASA]

shoe 1. A renewable friction element whose contour fits that of a drum and stops it from turning when lateral pressure is applied. Also known as brake shoe. 2. A metal block used as a form or support during bending of tubing, wire, rod or sheet metal. 3. A generic term for machine elements that provide support, or separate two members, while allowing relative sliding motion.

shop fabrication Making components and assemblies in a workshop for later transportation to the jobsite for installation.

shop replaceable assembly An item which is designated to be removed or replaced upon failure from a higher level assembly in the shop (intermediate or depot maintenance activity), and is to be tested as a separate entity.[ARD 50010-91]

shop replaceable unit An item which is designated to be removed or replaced upon failure from a higher level assembly in the shop (intermediate or depot maintenance activity), and is to be tested as a separate entity.[ARD50010-91]

shop weld A weld made during shop fabrication.

SHORAN An acronym for Short Range Navigation; a precision electronic position-fixing system using a pulse transmitter and receiver and two transponder beacons at fixed points.[ARP4107-88]

shoran A precision electronic position fixing system using a pulse transmitter and receiver and two transponder beacons at fixed points.[NASA]

Shore A hardness The hardness of a sealant as measured by a Shore "A" hardness gauge. It is essentially the same value as Rex "A" hardness.[AS 7200/1-90]

shore hardness An instrument measure of the surface hardness of an insulation or jacket material.[ARP1931-87]

short An electrical short circuit; an electrical circuit with nearly zero resistance.

short-circuit current(s) (fault current) 1. The maximum current that the system can produce at the point of application of the protective device.[ARP 1199A-90] 2. The steady value of the input alternating currents that flow when the output direct current terminals are short-circuited and rated line alternating voltage is applied to the line terminals.[NASA]

shortcoming A characteristic or operational deviation which does not prevent an item from functioning, but which should be corrected to achieve optimum efficiency and serviceability.

short finish Gray appearance of a polished glass surface resulting from this operation not being carried to the point where all traces of the previous grinding or smoothing operation are removed. [ARP924-91]

shortness A form of brittleness in alloys, usually brought about by grain boundary segregation; it may be referred to as hot, red or cold, depending on the temperature range in which it occurs.

short range clearance A clearance issued to a departing IFR flight which authorizes IFR flight to a specific fix short of the destination while air traffic control facilities are coordinating and obtaining the complete clearance. [ARP4107-88]

short range navigation See shoran.

short run Failure of molten metal to completely fill the mold cavity.

shorts Large particles remaining on a sieve after the finer portion of an aggregate has passed through the screen.

short takeoff and lading aircraft See STOL.

short-term memory See memory.

short term trending Tracking of engine, engine component or sub-system operational degradation by noting data on a particular flight or ground check and comparing with a pre-established trend.[ARP1587-81]

short time duty A requirement of service that demands operation at a substantially constant load for a short and definitely specified time.[ARP1199A-90]

shot 1. Small spherical particles of a metal. 2. Small, roughly spherical steel particles used in a blasting operation to remove scale from a metal surface. 3. An explosive charge.

shot effect In an electron tube, random variation in electron emission from the cathode, or random variation in electron distribution between electrodes.

shoulder A portion of a cylindrical machine element such as a shaft, screw or flange that is larger in diameter than the remainder.

showers Showery precipitation begins and ends abruptly in periods of short duration usually of the order of 15 minutes, although sometimes lasting half an hour or more.[AIR1335-75]

shrinkage Contraction of a metal or plastics material upon cooling, or in the case of plastics, upon curing (polymerization).

shrink fit A tight interference fit between mating parts where the amount of interference varies almost directly with diameter; parts are assembled by heating the outer member so that it expands, assembling the parts, and then allowing the outer member to cool and shrink onto the inner member. See also force fit.

shrink forming A process for forming metal parts that uses a combination of mechanical force and shrinkage of a heated blank to achieve final shape.

shrink link A link or rod element of a landing gear system which during retraction provides a kinematic constraint and shrinks or shortens a shock strut to reduce the required stowage volume within the aircraft or vehicle.[AIR1489-88]

shrink ring A heated ring or collar that is placed over an assembly and allowed to contract to hold the parts firmly in position.

shrink strut A strut which is shrunk or shortened before or during the retraction process to reduce the required stowage volume within the aircraft or vehicle. The shortening may be accomplished by means of a shrink link (kinematically) or hydraulically, electrically, mechanically or other methods.[AIR1489-88]

shrink wrapping A method of packaging that involves heating a plastics film, releasing internal strain and causing the film to shrink tightly over the object being packaged.

shroud 1. A device used to limit airflow through a rotor during balancing operations.[ARP480A-87] 2. A machine element used chiefly as a protective covering over other elements, especially in a rotating assembly.

shroud, insulation A part of a connector or device which provides physical protection to otherwise exposed contacts or terminals. See also insulation support.[ARP914A-79]

shroud line In a parachute, any one of a number of lines that attach the harness or load to the canopy. Also called a shroud, or a suspension line.[ARP 4107-88]

shunt(s) 1. In an electric circuit, a low-resistance element connected in parallel with another portion of the circuit and used chiefly to carry most of the current flowing in the circuit. 2. To divert all or part of a process flow away from the main stream and into a secondary operation, holding area, or bypass. See also circuits.

shunt diode barrier A network consisting of a fuse or resistor provided to protect voltage limiting shunt diodes and a current limiting resistor or other limiting components provided to limit current and voltage to intrinsically safe circuits.

shunt valve A valve that allows a fluid under pressure to escape into a passage that is of lower pressure or can accommodate higher flow rates than the normal passage.

shutdown 1. Cessation of engine operation for any reason other than training or normal operating procedures.[ARD50010-91] 2. The processes of decreasing engine thrusts to zero.[NASA]

shut-down circuit An electronic, electrical, hydraulic or pneumatic circuit that provides controlled steps for turning off or closing down process equipment; it is usually designed to automatically sequence shutdown actions and prevent equipment damage due to performing them out of sequence; it may be used for normal or emergency situations.

shute Sometimes called filler or "weft" wires. Run across the short way of the cloth as woven.[AIR888-89]

shutoff head The pressure developed by a centrifugal or axial-flow pump at its discharge when the discharge flow is zero.

shut off valve A device which is used to shut off or close down flow or pressure in a hydraulic or fluid system. Used in many landing gear system applications such as parking brake, skid control, etc.[AIR1489-88]

shuttle 1. A device which moves back and forth. See also catapult shuttle.[AIR 1489-88] 2. A back and forth motion in a machine, where the moving element continues to face one direction only. 3. The machine element that undergoes shuttle motion.

shuttle derived vehicles New configuration resulting from the production and operation of the Space Shuttle. [NASA]

shuttle engineering simulator Training equipment for crew members in mission operation procedures including various approach maneuvers, braking, final approach, etc.[NASA]

shuttle pallet satellites Reusable pallet type structures designed to be shuttle launched which will act as building blocks for larger platforms.[NASA]

shuttle valve A valve used in a brake or other fluid power system which serves to shift from one power source to an alternate source or other similar applications.[AIR1489-88]

SI See Systeme International d'Unites. See also International System of Units.

sialon Any composition containing silicon, aluminum, oxygen, and nitrogen and usually produced by the high-temperature reactions among the gradients. [NASA]

SID An acronym for Standard Instrument Departure (route). A preplanned instrument flight rule (IFR) air traffic control departure procedure printed for pilot use in graphic or textual form, or both.[ARP4107-88]

SID (ionospheric disturbances) See sudden ionospheric disturbances.

sidebands The frequency components on either side of center frequency (fo) which are generated when a carrier wave is modulated.

side brake A brake which is mounted at the side of the wheel.[AIR1489-88]

side-draw product A product stream removed from the column midway between the top and bottom trays. If the side draw is above the feed tray, a vapor is generally removed; if it is below the feed tray, it is usually a liquid.

side force coefficient A mathematical (numerical) measure of the physical force property that is constant in determination of side force on a tire.[AIR 1489-88]

side milling Using a milling machine and a cutter with teeth on one or both sides to machine a vertical surface.

side rake The angle between a reference plane and the tool face of a single-point turning tool.

sidereal time Time based upon the rotation of the earth relative to the vernal equinox.[NASA]

side relief angle (SRF) In a cutting tool, the angle between a plane normal to the base and the flanks of the tool below the cutting edge.

siderites A spathic iron ore; an iron carbonate.[NASA]

side rod 1. In a side-lever engine, one of the members linking the piston-rod crossheads to the side levers. 2. In a railroad locomotive, a large link joining the crankpins of adjacent drive wheels on one side of the engine.

siderograph An instrument combining

a clock and a navigation instrument which keeps a reference time equivalent to the time at 0° longitude (Greenwich meridian).

sideslip A slip in which the airplane's longitudinal axis remains parallel to the original flight path, but in which the actual flight path changes direction according to the steepness of the bank. A sideslip is used to make the airplane move sideways through the air to counteract the drift which results from a crosswind. See also slip.[ARP4107-88]

side-step maneuver A VFR maneuver accomplished by a pilot at the completion of an instrument approach to permit a straight-in landing on a parallel runway not more than 1200 ft to either side of the runway to which the instrument approach was conducted. Landing minimums for a side-step maneuver will be higher than those to the primary runway, but normally will be lower than the published circling minimums.[ARP 4107-88]

sidetone Feature which enables speaker to hear his/her own transmission in the headset.[ARP4107-88]

siemens Metric unit of conductance.

sieve 1. A meshed or perforated sheet, usually of metal, used for straining liquids, classifying particulate matter or breaking up masses of loosely adherent or softly compacted solids. 2. A meshed sheet with apertures of uniform standard size used as an element of a set of screens for determining particle size distribution of a loose aggregate.

sieve fraction The portion of a loose aggregate mass that passes through a standard sieve of given size number but does not pass through the next finer standard sieve; usually expressed in weight percent.

sight glass A glass tube, or a glass-faced section of a process line, used for sighting liquid levels or taking manometer readings.

sighting tube A tube, usually made of a ceramic material, that is used primarily for directing the line of sight for an optical pyrometer into a hot chamber.

sigma phase A brittle, nonmagnetic, intermetallic compound generally formed between iron and chromium during long periods of exposure at 1050 to 1800 °F.

SIGMET An acronym for SIGnificant METeorological information; a weather advisory concerning weather significant to the safety of all aircraft. SIGMET advisories cover severe and extreme turbulence, severe icing, and widespread dust or sandstorms that reduce visibility to less than 3 miles.[ARP4107-88]

sign 1. In arithmetic, a symbol which distinguishes negative quantities from positive ones. 2. An indication of whether a quantity is greater than zero or less than zero. The signs often are the marks, + and -, respectively, but other arbitrarily selected symbols may be used, such as 0 and 1, or 0 and 9; when used as codes at a predetermined location, they can be interpreted by a person or machine.

signal 1. A measure or quantity of the medium used to communicate a condition, effect, or other desired intelligence from one point in the system to another. In an electrical system this may be a voltage; in a hydraulic system, a pressure. A signal may be generated by a sensor, controller or selector or other such reference device.[ARP419-57] 2. Information conveyed from one point in a transmission or control system to another. Signal changes usually call for action or movement. 3. The event or phenomenon that conveys data from one point to another. 4. A time-dependent value attached to a physical phenomenon and conveying data.

signal attenuation The reduction in the strength of electrical signals.

signal common The reference point for all voltage signals in a system. Current flow into signal common is minimized to prevent IR drops which induce inaccuracy in the signal common reference.

signal conditioner (analog data) Hardware device which accepts data from some type of transducer and conditions it to a common scale for multiplexer input.

signal conditioner (PCM data) One of the functions of the bit synchronizer whereby serial data are accepted in the presence of perturbations (noise, jitter) and are reconstructed into coherent data.

signal conditioning To process the form or mode of a signal so as to make it intelligible to or compatible with a given device, including such manipulation as pulse shaping, pulse clipping, digitizing, and linearizing.

signal distance The path length which a signal is required to traverse.

signal generator An instrument used in testing and calibrating other electronic instruments that delivers an output wave-form at an accurately calibrated frequency any where in the audio to microwave range.

signal piping That piping interconnecting instruments, instrument devices or bulkhead fittings.

signal simulator A hardware device that generates a signal similar in most respects to actual data from a test vehicle.

signal-to-noise ratio 1. Ratios which measure the comprehensibility of a data source or transmission link, usually expressed as the root mean square signal amplitude divided by the root mean square noise amplitude.[NASA] 2. The ratio of the power of the signal conveying information to the power of the signal not conveying information.

signal validation Equipment "error" indications and program limit checks that detect faulty signals.

signature A signal or combination of data inputs that are characteristic of an individual engine, engine component or subsystem that can be used to indicate functional status and condition. [ARP1587-81]

signature analysis A process can be identified as having a particular signature when operating correctly. This can be noise spectrum or vibration spectrum. Signature analysis involves identifying departures from the reference signature and recognizing the source of the departure.

sign bit A single bit, usually the most significant bit in a word, which is used to designate the algebraic sign of the information contained in the remainder of the word.

sign-check indicator An error-checking device, indicating no sign or improper signing of a field used for arithmetic processes. The machine can, upon interrogation, be made to stop or enter into a correction routine.

sign digit In coded data, a digit incorporating 1 to 4 binary bits which is associated with an item of data to indicate its algebraic sign.

significance In positional representation, the factor, dependent on the digit position, by which a digit is multiplied to obtain its additive contribution in the representation of a number. Synonymous with weight.

significant digits A set of digits, usually from consecutive columns beginning with the most significant digit different from zero and ending with the least significant digit whose value is known and assumed relevant, e.g., 2300.0 has five significant digits; whereas, 2300 probably has two significant digits; however, 2301 has four significant digits and 0.0023 has two significant digits.

significant event An event or condition that indicates a change in system status in an event-driven system; a significant event is declared, for example, when an I/O operation completes. A declaration of a significant event indicates that the executive should review the eligibility of task execution since the event might unblock the execution of a higher-priority task. The following are considered to be significant events: I/O queuing, I/O request completion, a task request, a scheduled task execution, a mark time expiration, a task exit.

significant meteorological information See SIGMET.

sign position A position, normally located at one end of a numeral, that contains an indication of the algebraic sign of the number represented by the numeral.

silica See silicon dioxide.

silica gel A colloidal, highly absorbent silica used as a dehumidifying and dehydrating agent, as a catalyst carrier, and sometimes as a catalyst.[NASA]

silica glass A transparent or translucent material consisting almost entirely of fused silica (silicon dioxide). Also known as fused silica; vitreous silica.

silicon A basic material used to make semiconductors that has limited capacity for conductivity.

silicon bronze A corrosion-resistant alloy of copper and 1 to 5% silicon that has good mechanical properties.

silicon-controlled rectifier (SCR) A semiconductor device used to provide stepless control of an electric power circuit without the necessity of load matching; usually, two rectifiers are used in the circuit to provide full-wave control of the heater element, but in some instances two SCRs are used in a single package, known as a triac.

silicon dioxide The chemically resistant dioxide of silicon.[NASA]

silicone 1. Polymeric materials in which the recurring chemical group contains silicon and oxygen atoms as links in the main chain.[ARP1931-87] 2. A generic name for semiorganic polymers of certain organic radicals; they can exist as fluids, resins or elastomers, and are used in diverse materials such as greases, rubbers, cosmetics and adhesives.

siliconizing Producing a surface layer alloyed by diffusing silicon into the base metal at elevated temperature.

silicon-on-insulator semiconductors See SOI (semiconductors).

silicon solar cells See solar cells.

silky fracture A type of fracture surface appearance characterized by a fine texture, usually dull and nonreflective, typical of ductile fractures.

silver hydrogen batteries Secondary batteries having silver and hydrogen electrodes. They have good energy density and cycle life.[NASA]

silver solder A brazing alloy composed of silver, copper and zinc that melts at a temperature below that of silver but above that of lead-tin solder.

silviculture The theory and practice of controlling the establishment, composition, and growth of stands of trees for the harvesting of foliage limbs, and possibly the trees themselves for biomass.[NASA]

SIMD (computers) A type of parallel computer with multiple memories and an arithmetic logic unit for each memory. A single control unit allocates instruction execution according to the memory that holds the required operands.[NASA]

similitude Resembling something else—for example, a process that has been scaled up from a laboratory or pilot plant operation to commercial size.

simmer Detectable leakage from a safety relief valve at a pressure below the popping pressure.

simple balance A weighting device consisting of a bar resting on a knife edge and two pans, one suspended from each end of the bar; to determine precise weight, an unknown weight on one of the pans is approximately balanced by known weights placed in the other pan, and a precise balance is obtained by sliding a very small weight along the bar until a pointer attached to the bar at the balance point indicates a null position.

simple buffering A technique for controlling buffers in such a way that the buffers are assigned to a single data control block and remain so assigned until the data control block is closed.

simple electrical apparatus and components Those items (e.g. thermocouples, photocells, junction boxes) which do not generate or store more than 1.2V, 0.1A, 20μJ and 25 mW in the intrinsically safe system.

simple engine A machine for converting thermal energy to mechanical power by expanding a working fluid in a single stage, after which the fluid passes out of the engine through an exhaust port.

simple harmonic motion A motion such that the displacement is a sinusoidal function of time.[NASA]

simple machine Any of several elementary mechanical devices that form the basis for creating more complex devices—usually the lever, wheel, pulley, inclined plane and screw are the only devices considered simple machines.

simple sound source A sound source that radiates uniformly in all directions in an unrestrictive airspace.

simple systems Simple systems, in which all the electrical apparatus is certified intrinsically safe or certified associated electrical apparatus, do not require to be certified provided it is completely clear from the information given on the electrical apparatus certification documents that the system is intrinsically safe.

simplex An adjective meaning nonredundant.[AIR1916-88]

simplex method A finite iterative algorithm used in linear programming whereby successive solutions are obtained and tested for optimality.[NASA]

simplex mode Method of data transmission whereby the data is transmitted in one direction only, i.e. send or receive.

simplex pump A reciprocating pump with only one power cylinder and one pumping cylinder.

simplified directional facility (SDF) A NAVAID used for nonprecision instrument approaches. The final approach course is similar to that of an ILS localizer except that the SDF course may be offset from the runway, generally not more than 3 deg, and the course may be wider than the localizer, resulting in a lower degree of accuracy.[ARP 4107-88]

simplified short approach light system (SSALS) Similar to the approach light system but the installation consists of fewer light fixtures. See also approach light system.[ARP4107-88]

SIMSCRIPT A generic class of discrete, event-oriented simulation languages.

simulate 1. To artificially create behavior, environmental conditions or operating conditions pertaining to one system by using another, different system; usually done to accomplish testing, experimentation or training that would be difficult or hazardous to accomplish with the real system. 2. The representation of physical phenomena by use of mathematical formulas.

simulation 1. The representation of certain features of the behavior of a physical or abstract system by the behavior of another system. For example, the representation of physical phenomena by means of operations performed by a computer or the representation of operations of a computer by those of another computer. 2. Using computers, electronic circuitry, models or other imitative devices to gain knowledge about operations and interactions that take place in real physical systems.

simulation framework A simulation system capable of integrating and running two or more simulation algorithms in a single simulation environment.

simulator 1. A device or facility that provides a representation of the essential elements of a system out of their normal setting in such a manner that the representation is a valid analog of the system. A common type of simulator is a device, such as a link trainer or cockpit mock-up, that simulates flight or some other activity that is used in the training and maintenance of pilots' skills.[ARP4107-88] 2. A device which produces synthetic signals or readings similar to the actual operating condition.[ARP480A-87] 3. A program which simulates the operation of another device or system. In the case of microprocessors a simulator allows the execution of a microprocessor object program on a computer which is different from the microprocessor for which the program has been written. The simulator provides a range of debugging tools which allows the programmer to correct errors in the program. 4. A device, system, or computer program that represents certain features of the behavior of a physical or abstract system.

sine bar An accurately constructed layout aid consisting essentially of a straight bar with cylindrical rests at each end; one end of the sine bar is placed on a surface plate or gage block and the other end on a stack of gage blocks equal to the sine of a desired angle to the surface plate or another reference plane.

sine galvanometer A magnetometer whose measuring element consists of a small magnet suspended in the center of a pair of Helmholtz coils; the magnitude of a magnetic field is determined from the position of the magnet when various known currents are passed through the coils.

sine wave(s) 1. Waves which can be expressed as the sine of a linear function of time, or space, or both.[NASA] 2. A waveform in which the value of wave parameters—such as voltage and current in certain alternating-current circuits—vary directly as the sine of another variable—such as time.

sine-wave response See frequency response.

single acting Producing power or motion in one direction only.

single-axis tracking antenna A receiving antenna which tracks the transmitting station automatically in azimuth, but not in elevation.

single board computer (SBC) A complete computer, including memory, clock and input/output ports assembled on a single board.

single cascade action A type of control-system action whereby the input to the second of two automatic controllers is supplied by the first.

single channel per carrier transmission Voice and data transmission system for satellite communication featuring the use of a carrier frequency for each channel of communication.[NASA]

single compression A weave in which the shute wires are compressed tightly by the comb of the loom so that they are deformed in the machine direction. The shute wires are in contact with each other.[AIR888-89]

single density A computer diskette that can store approximately 3,400 bits per inch.

single-ended amplifier An electronic amplifier in which each stage operates asymmetrically with respect to ground; each stage contains one tube or amplifying transistor; alternatively, each contains two or more, connected in parallel.

single event upsets Radiation-induced errors in microelectronic circuits caused when charged particles (usually from the radiation belts or from cosmic rays) lose energy by ionizing the medium through which they pass, leaving behind a wave of electron-hole pairs. [NASA]

single-fail operative (SFO) A quality wherein a control device or system can sustain any single failure and remain operative. Unless specifically stated, it is understood that no nominal loss of performance occurs after the failure. [ARP1181A-85]

single frequency approach (SFA) A service provided to military single-piloted turbojet aircraft which permits use of a single UHF frequency during approach for landing so that pilots will not normally be required to change frequency from the beginning of the approach to touchdown.[ARP4107-88]

single instruction multiple data stream See SIMD (computers).

single-mode An optical fiber which can carry only a single waveguide mode of light. Components such as connectors used with such fiber are also labeled single-mode.

single-phase meter An instrument for determining power factor in a single-phase alternating-current circuit; it contains a fixed coil that carries the load current and crossed coils connected to the load voltage; the moving system is not restrained by a spring and therefore takes a position related directly to the phase angle between voltage and current.

single point failure The failure of an item which would result in failure of the system and is not compensated for by redundancy or alternative operational procedure.[ARD50010-91]

single pole A type of device such as a switch, relay or circuit breaker that is capable of either opening or closing one electrical path.

single sampling A type of inspection where an entire lot or production run (population) is accepted or rejected based on results of inspecting a single group of items (sample) selected from the population.

single-sideband modulation A type of modulation whereby the modulating wave's spectrum is translated in frequency by a specific amount, either with or without inversion.

single-sided A computer diskette that can record data on only one side.

single-speed floating control 1. A reversible motor drives the final controlled device at a given speed in one direction when the temperature at the sensing element is above the set point and reverses when the temperature is below the set point. A dead band in the controller causes the motor to remain stationary when there is no load change or other disturbance to the system.[ARP 89C-70]

single-stage compressor A machine that raises pressure in a compressible fluid in a single pass through a single set of machine elements.

single-stage pump A machine that develops pressure to drive a relatively incompressible fluid through a system by passing the fluid through a single set of machine elements.

single-stream batch A method of batch processing in which only one stream of batch commands is processed.

single wheel load The share of aircraft (or vehicle) or gear load as geometrically broken down and carried by a given wheel.[AIR1489-88]

sink A reservoir into which material or energy is rejected.

sinkhead See feedhead.

sinkholes Circular depressions in a Karst area. Their drainage is subterraneous, their size is measured in meters or tens of meters, and they are commonly funnel shaped.[NASA]

sink speed 1. The rate at which an aircraft loses altitude; especially the rate at which a heavier-than-air aircraft descends in a glide in still air under given conditions of equilibrium. 2. The vertical sink rate at which an aircraft (i.e., landing gear) contacts the ground.[AIR 1489-88]

sinter To thermally seal or fuse a material.[ARP1931-87]

sintering Heating a powder metal compact at a temperature below the melting point to form diffusion bonds between the particles.

sinus block See trapped gas effects.

sinuses A term used in anatomical nomenclature to designate a cavity or hollow space.[NASA]

sinusoids See sine waves.

sipes Lateral cuts in the tread of a tire. Under braking loads the tread sections deflect, thus exposing the cut edges and increase traction.[AIR1489-88]

siphon A tube, hose or pipe for moving liquid from a higher to a lower elevation by a combination of gravity acting on liquid in the longer leg and atmospheric pressure acting to keep the shorter leg filled.

siphoning The transfer of a liquid from a high to a lower level by atmospheric pressure forcing it up the shorter leg while the weight of the liquid in the longer leg causes continuous downward flow.[NASA]

SIS (semiconductors) Semiconductor devices consisting of an electrically insulating layer sandwiched between two semiconducting materials.[NASA]

site license In data processing, a software agreement that allows unlimited use of a program to a single organization at one location.

site selection Selecting the location for any physical plant (nuclear power, solar house, etc.) while considering the environmental impact, safety, etc. [NASA]

situational awareness Keeping track of the prioritized significant events and conditions in one's environment.[ARP 4107-88]

size 1. A specified value for some dimension that establishes an object's comparative bulk or magnitude. 2. One of a set of standard dimensions used to select an object from among a group of similar objects to obtain a correct fit. 3. In welding, the joint penetration of a groove weld or the nugget diameter of a spot weld or the length of the nominal legs of a fillet weld. 4. A material such as casein, gum, starch or wax used to treat the surface of leather, paper or textiles.

size distribution The study of the size of objects or features and their distribution.[NASA]

size effect The conditions which prevent the direct application of the same strength criteria over a wide range of size considerations.[ARP700-65]

size factor A factor used in tire design to compensate for variation in characteristics due to size of the tire.[AIR1489-88]

skew 1. Having an oblique position in relation to a specific reference plane, reference direction, or physical object. 2. A tape motion characterized by an angular velocity between the gap center line and a line perpendicular to the tape center line.

skid 1. A runner or slide used as an element of a landing gear of certain aircraft. See also wing skid, tail skid.[AIR 1489-88] 2. The act of sliding or slipping over a surface.[AIR1489-88] 3. To slide without revolving, as a wheel.[AIR1489-88] 4. Sometimes used to designate the portion of a tire tread which wears away in the normal service life of the tire; i.e. skid depth.[AIR1489-88] 5. A metal bar or runner on an object that provides support or wear resistance when the object contacts a floor, runway, apron or other flat areaway. 6. A wood or metal platform on legs, runners or wheels used to support parts or materials a short distance off the ground during storage or material handling. 7. A brake for a power machine. 8. A device placed under a wheel to prevent it from turning while a heavy wheeled object descends a steep hill.

skid control box A component of the skid control system containing the electrical and electronic components (black box).[AIR1489-88]

skid control system A group of interconnected components which interact to control excessive brake pressure and thus prevent inadvertent tire skidding and contribute to shorter aircraft stopping distances. Also called antiskid system.[AIR1489-88]

skid control valve A component of the skid control system which controls or modulates brake pressure in response to an electrical (or mechanical) signal from the skid control box.[AIR1489-88]

skid control system operating environment The skid control system operating environment is the environment which the skid control system feels in performing its function. The skid control operating environment includes the wheel braking system, the airframe elastic structure (including struts), the wheel dynamic loading tire elastic properties, runway friction levels available, runway roughness, and ambient and hydraulic fluid temperature.[AS483A-88]

skid depth The depth dimension of that part of a tire tread which wears away in the normal service life of a tire.[AIR 1489-88]

skid resistance The physical characteristic of a tire to resist or oppose skidding. The characteristic is affected by inflation pressure, tire shape, tread pattern, degree of wear, and other factors. [AIR1489-88]

skin A general term for a thin exterior covering—may be applied to the exterior walls of a building, the exterior covering of an airplane, a protective covering made of wood or plastics sheeting, or a thin layer on a mass of metal that differs in composition or some other attribute from the main mass of metal.

skis Long, flat runners attached to the landing gear of an aircraft for takeoff, landing, and operation over snow or ice; also in special cases, over water.[AIR 1489-88]

skived tape Tape formed by shaving thin layers from a cylindrical block of material.[ARP1931-87]

skyhook balloons Large free balloons having plastic envelopes, used especially for constant level meteorological observations at very high altitudes.[NASA]

sky waves In radio, radio energy that is received after having been reflected by the ionosphere.[NASA]

slab 1. A flat piece of concrete that spans beams, piers, columns or walls to make a floor, roof or platform. 2. A relatively thick piece of metal whose width is at least twice its thickness—generally used to describe a mill product intermediate between ingot and a flat rolled product such as sheet or plate.

slack Looseness or play in a mechanism, sometimes due to normal tolerances in the assembly and sometimes due to wear.

slag Molten or fused refuse.

slant perception See space perception.

slant-range distance The distance from an aircraft directly to an airfield, a navigational fix, or another aircraft which is at a different elevation. The slant range distance between two objects when they are at different elevations is greater than the horizontal distance.[ARP 4107-88]

slash An elongated mark (slash, virgule) on a radar presentation screen indicating the radar beacon reply of an aircraft.[ARP4107-88]

slat A section of the wing leading edge which is free to move fore and aft on tracks. At high angles of attack, the local suction at the wing leading edge creates a chordwise force forward and displaces the slat from the contour of the designed wing leading edge. A slat is a high lift device designed to increase the maximum lift coefficient for low-speed flight.[ARP4107-88]

slave A mechanical or electronic device that is under the control of a another device.

slave operated Pressure operated, so as to position in a manner equivalent to rigid mechanical inter-lock, by means of a master or control unit.[ARP243B-65]

sleek A scratch having boundaries which appear polished.[ARP924-91]

sleeve 1. In general, a hollow cylindrical member used to line a housing to impart different metallurgical properties to the rubbing surface than those inherent in the housing. In a slide valve, the hollow cylindrical member which directly affects the flow pattern through its relative position to an internal slide or spool.[ARP243B-65] 2. A tubular device designed to fit over another part; a hollow shaft.[ARP480A-87] 3. A tubular member through a wall to permit passage of pipe or other connections. 4. A cylindrical part that fits over another part.

sleeve bearing A cylindrical machine element that fits around a shaft and supports the shaft while it turns.

sleeve coupling A hollow cylinder that fits over the ends of two adjacent shafts or pipes to hold them together.

slewing 1. Of a gyro, the rotation of the spin axis caused by applying torque about the axis of rotation. In radar, changing the scale on the display. [NASA] 2. Rapidly moving a device to a new rotational position or a new elevation direction, or both.

slew rate 1. The steady-state output velocity of the actuation system in response to a step input command. Slew rate is usually determined by the average velocity between 10 and 90% of load position (so as to avoid actuator acceleration and deceleration effects) when both static and dynamic loads are imposed at the servoactuator output.[AIR 1916-88] 2. The maximum rate of change of an output signal from a device. 3. The limitation in a device or circuit in the rate of change of output voltage, usually imposed by some basic circuit considerations.

slide(s) 1. In a slide valve, the moving member which directly affects the flow pattern, usually cylindrical and internal to a sleeve. See also spool.[ARP243B-65] 2. Any mechanism that moves with predominantly sliding motion. 3. The main reciprocating member of a mechanical press, which moves up and down in the press frame and carries the punch or upper die. 4. A flat-bottomed chute.

slides (microscopy) Rectangular pieces of glass on which objects are mounted for microscopic examination.[NASA]

sliding-block linkage A mechanism for converting rotary motion into linear motion, or vice versa, which consists of a crank, a block that slides back and forth in a slot or on ways, and a link bar attached to the crank and block with pin joints.

sliding fit A type of clearance fit used to accurately locate parts that must assemble together without perceptible play (close-sliding fit) or used to allow assembled parts to move or turn easily but not run freely (sliding fit). See also running fit.

sliding gear A gear set whose speed can be changed by sliding gears along their axes to put them in or out of mesh with other gears of different sizes.

sliding-vane rotary flowmeter A type of positive-displacement flowmeter in which radial vanes slide in or out to trap and release discrete volumes of the metered fluid as a rotor containing the vanes revolves about a central cam surface which controls vane position.

slime 1. A soft, viscous or semisolid surface layer—often resulting from corrosion or bacterial action, and often having a foul appearance or odor. 2. A mudlike deposit in the bottom of a chemical process or electroplating tank. 3. A thick slurry of very fine solids. Also known as mud; pulp; sludge.

sling 1. An item made of strap, chain, rope, webbing, or the like, specifically designed to hold securely something to be hoisted, lowered, carried, rotated or suspended. See adapter.[ARP480A-87] 2. A length of rope, wire rope or chain used to support a load hanging from a crane hook.

sling psychrometer A device for determining relative humidity that consists of a wet-and-dry bulb thermometer mounted in a frame that can be whirled about, usually by means of a handle and short piece of chain or wire rope attached to the upper end of the frame.

slip 1. A descent with one wing lowered and the airplane's longitudinal axis at an angle to the flight path. See also forward slip and sideslip.[ARP4107-88] 2. A term commonly used to express leakage in positive-displacement flowmeters. 3. A suspension of ground flint or fine clay in water that is used in making porcelain or in decorating ceramic ware. Sometimes called slurry. See also tire.

slip clutch A clutch which may be incorporated into a starter to limit the maximum torque (impact and steady) that can be transmitted by the starter to the accessory drive train or engine. [ARP906A-91]

slip flow Rarefied gas flow in the region between Knudsen numbers 0.01 and 0.1.[NASA]

slip joint 1. A telescoping joint between two parts in an assembly. 2. A mechanical union that allows limited axial movement of one member, such as a pipe or duct, with respect to a mating member. 3. In civil engineering, a type of contraction joint consisting of a tongue and groove that allows independent movement between two members such as wall sections, slabs or precast structural units. 4. A type of scarf joint used in flexible-bag molding in which plastics veneers are laid up so that their beveled edges overlap.

slippage 1. Fluid leakage along the clearance between a reciprocating-pump piston and its bore. Also known as slippage loss. 2. Movement that unintentionally displaces two solid surfaces in contact with each other. 3. Movement of a gas phase through or past a gas-liquid interface instead of driving the interface forward; especially applicable to certain phenomena in petroleum engineering.

slipping turns A flight maneuver which consists of a turn in which the aircraft is allowed to slip.[ARP4107-88]

slip plane A crystallographic plane along which dislocations move under local shear stresses to produce permanent plastic strain in ductile metals.

slip ratio The quotient of actual advance of a screw propeller divided by the theoretical advance determined by blade pitch angle and number of revolutions in a specific period of time. See also tire.

slip seal A seal between members designed to permit movement of either member by slipping or sliding.

slipstream The flow of air pushed back by a revolving propeller or rotor. See also wake turbulence.[ARP4107-88]

slit 1. A long, narrow opening—often used for directing and shaping streams of radiation, fluids or suspended particulates. 2. To cut sheet metal, rubber, plastics or fabric into sheet or strip stock of precise width using rotary cutters, knives or shears.

sliver 1. Any thin, elongated, often sharp edged fragment of solid material. 2. A thin fragment attached at one end to the surface of flat-rolled metal and rolled into the surface during reduction.

slope control Electronically producing specific changes in a parameter with time, especially applied to a method of varying welding current.

sloshing See liquid sloshing.

slot 1. Any of certain apertures in an airfoil to improve aerodynamic behavior. 2. Any elongated opening in a machine part or structural member. 3. A special socket in a PC designed to accept an additional circuit board.

slotted nut A hexagon nut with slots cut across the flats of the hexagon so that a lockwire or cotter pin can be used to prevent the nut from turning.

slotted ring An access procedure that uses a pattern that circulates around the ring to determine access. The pattern is divided into sections called slots. When a node has data to transmit, it waits for an unused slot, writes its data into the slot, and marks the slot used. When the slot rotates to the destination node, it reads the data from the slot. When the slot rotates back to the original node, it marks the slot unused again.

slotted tongue terminal A terminal, having a bifurcated tongue, that allows attachment to a screw or stud without removal of the mounting hardware. [ARP914A-79]

slot washer 1. A washer with a slot extending to one edge so that the washer can be removed from a bolt without completely disassembling the fastened joint. 2. A type of lockwasher with an indentation on its rim so that it can be held in place with a nail or screw.

slow neutron A free (uncombined) neutron having a kinetic energy of about 100 eV or less. See also thermal neutrons.

slow (time) code A spread-out time code, modulated in amplitude and width, suitable for display on a relatively low-speed chart recorder.

sludge 1. A soft water-formed sedimentary deposit which normally can be removed by blowing down. 2. Fine sediment such as may be found in the bottom of an oil crankcase or boiler drum. See also slime.

slug 1. A large "dose" of chemical treatment applied internally to a steam boiler intermittently. Also used sometimes instead of "priming" to denote a discharge of water out through a boiler steam outlet in relatively large intermittent amounts. 2. A small, simply shaped piece of metal used as starting stock for forging, upsetting or extrusion. 3. The offal resulting from piercing a hole in sheet metal. 4. Liquid that completely fills the internal passage of a tube for a short distance.

slugging Producing a substandard weld joint by adding a separate piece of material which is not completely fused into the joint.

sluice 1. A waterway fitted with a vertical sliding gate for controlling the flow of water. 2. A channel for draining away excess water.

slump See sag.

slump test A quality control test for determining the consistency of concrete; the amount of slump is expressed as the decrease in height that occurs when a conical mold filled with wet concrete is inverted over a flat plate and then removed, leaving the concrete behind.

slurry 1. A suspension of fine solids in a liquid which can be pumped or can flow freely in a channel. 2. See slip. 3. A semiliquid refractory material used to repair furnace linings. 4. An emulsion of soluble oil and water used as a cutting fluid in certain machining operations.

slush drag The drag or resistance to forward motion of a tire(s) and the aircraft which is generated as it moves through melting snow, slush and water. [AIR1489-88]

small aircraft Aircraft of 12,500 lb or less, maximum certified takeoff weight.

[ARP4107-88]

small scale integration Low density of integrated circuits per unit area.

SMAW See shielded metal arc welding.

SME See CASA/SME.

smoke 1. Small gas-borne solid particles, including but not limited to black carbonaceous material from the burning of fuel, which in sufficient concentration creates visible opacity.[ARP1179A-80] 2. A dispersion of fine solid or liquid particles in a gas.

smoke detector A device that produces an alarm signal when the density of smoke in an area exceeds a preset value; usually, smoke density is detected photoelectrically.

smoke point The maximum flame height in a standard test that kerosene or jet fuel will burn without smoking.

SN Smoke Number, the dimensionless term quantifying smoke emission. SN increases with smoke density and is rated on a scale from 0 to 100. SN is evaluated for a sample size of 0.0162 grams of exhaust gas per sq. millimeter (0.0230 lb. per sq. in.) of filter area. [ARP1179A-80]

SNAP See sub-network access protocol.

snap fastener A type of fastening device used primarily to hold the edges of fabric articles together; it consists of a flange with a protruding ball attached to one edge of the fabric, and a mating flange with a socket attached to the opposing edge.

snap gage A device with two flat, parallel surfaces that are precisely spaced apart for checking one limit of tolerance on a diameter or length dimension. Sometimes, go-no-go tolerance limits are built into a single snap gage frame to permit checking both high and low limits of tolerance at the same time.

snap ring A type of retaining fastener in the shape of the letter C which is expanded across its diameter and allowed to snap back into a groove to hold parts in position, and especially keep them from sliding axially along a shaft.

snap roll A maneuver in which an airplane is made, by a quick movement of the controls, to complete a full rotation about its longitudinal axis while maintaining an approximately level line of flight.[ARP4107-88]

snapshot dump A selective dynamic dump performed at various points in a machine run.

snatch block A pulley whose side can be opened to allow a loop or rope to be inserted.

sneak circuit analysis 1. A procedure conducted to identify latent paths which cause occurrence of unwanted function or inhibit desired functions assuming all components are functioning properly. [ARD50010-91]

snorkel A tube that supplies air for an underwater operation.

snow 1. Precipitation of mainly hexagonal ice crystals, most of which are branched (star shaped).[AIR1335-75] 2. A speckled background on an intensity-modulated CRT display that is produced by electronic noise; the appearance is similar to the display on a television screen when the station is not broadcasting; snow may or may not make a transmitted image or data display unsuitable for its intended purpose.

snow grains Precipitation of very small white and opaque grains of ice. These grains are fairly flat or elongated; their diameter is generally less than 1/25 of an inch.[AIR1335-75]

snow pellets Precipitation of white and opaque particles of ice. These ice particles are spherical or sometimes conical; their diameter is about 1/2 to 1/5 of an inch. Precipitation of snow pellets generally occurs in showers together with precipitation of snow flakes or rain drops, when surface temperatures are around 32 °F (0 °C).[AIR1335-75]

snubber 1. A device which provides a

resisting force to "snub" or damp an imposed load; i.e., arresting hook snubber which resists the upward force imposed by the carrier deck or ground, prevents hook bounce, and maintains point contact with the deck.[AIR1489-88] 2. A mechanical or hydraulic device for restraining motion.

snubber valve A device within a snubber which provides resistance to fluid motion and thereby snubs or damps the force imposed.[AIR1489-88]

soaked sleeve A firesleeve, so saturated with fluid conveyed by the hose assembly the sleeve is mounted to that sweating or permeating of the fluid through the firesleeve results. (A soaked sleeve is usually indicative of a damaged or faulty hose assembly or spillage of fluid from a nearby port.)[ARP1658A-86]

soak test Sealants are soaked in fluids as specified in specifications to determine resistance to these materials. The conditions are somewhat harsher than are found in actual use.[AS7200/1-90]

soap bubble solution A noncorrosive soap solution used in the detection of leaks.[AIR4069-90]

soap bubble test A leak test consisting of applying soap solution to the external surface or joints of a system under internal pressure and observing the location, if any, where bubbles form indicating the existence of a gas leak.

soar To fly without propulsive power, as in a glider. It is called dynamic soaring unless it is done on ascending air currents; then it is called up-current soaring.[ARP4107-88]

Sobolev space A Banach space whose elements are functions defined in a domain in Euclidean space and whose norm measures the size and smoothness of the functions.[NASA]

socket weld An external weld joining the plain-ended male portion and the corresponding socket. Used here as a male valve inlet in a process line or vessel socket.

sodar Sound detection and ranging.[NASA]

sodium sulfates Sodium compounds containing the -SO4 group.[NASA]

sodium sulfur batteries One of several types of rechargeable batteries under consideration as power sources for electrically actuated vehicles. This battery uses a solid electrolyte as well as a sodium reservoir made of metal.[NASA]

SOFAR See sound fixing and ranging.

softening The act of reducing scale forming calcium and magnesium impurities from water.

softening agent A substance—often an organic chemical—that is added to another substance to soften it.

soft hammer A hammer with a head made of annealed copper, leather or plastic which keeps it from damaging finished surfaces.

soft landing The act of landing on the surface of a planet or natural satellite without damage to any portion of the vehicle or payload except possibly the landing gear.[NASA]

soft recovery See soft landing.

soft-sector disk In data processing, a disk that accepts magnetic patterns to define the boundaries of each sector.

software 1. A set of programs, procedures, rules, and possibly associated documentation concerned with the operation of a computer system, for example, compilers, library routines, manuals, circuit diagrams. 2. Contrast with hardware.

software engineering The systematic approach to the development, operation, maintenance, and retirement of software.[NASA]

software error failure A failure caused by an error in the computer program associated with the hardware.[ARD 50010-91]

software license An agreement between the seller and buyer of a computer program that usually limits the use of the program to one person at one time at one location. See site license.

software priority interrupt The programmed implementation of priority interrupt functions. See priority interrupt.

software tools Computer programs that aid in the specification, construction, testing, analysis, management, documentation, and maintenance of other computer programs.[NASA]

soft water Water which contains little or no calcium or magnesium salts, or water from which scale forming impurities have been removed or reduced.

SOI (semiconductors) Semiconductor devices consisting of a silicon layer coupled to an electrically insulating layer.[NASA]

soil mechanics 1. Mechanical properties of unconsolidated accumulations of particles produced by the disintegration and chemical decomposition of rocks.[NASA] 2. A branch of civil engineering that deals with the application of principles of solid and fluid mechanics to the design, construction and maintenance of earthworks and stable foundations.

soil pipe A vertical drain for carrying sewage from a building into a sewer or septic system.

solar activity Any type of variation in the appearance of energy output of the sun.[NASA]

solar atriums Open courts within buildings designed for passive solar heating.[NASA]

solar azimuth See azimuth.

solar backscatter UV spectrometer A spaceborne spectrometer that measures solar UV spectral irradiance incident on the earth and backscattered radiance from the earth and thereby estimates the total atmospheric ozone content of the atmosphere and the attitude distribution of ozone.[NASA]

solar blankets Large, high-temperature, low-mass solar arrays consisting of ultrathin silicon solar cells interconnected, welded, and bonded to flexible substances.[NASA]

solar blind A detector which contains filters to block sunlight, so the detector becomes essentially "blind" to the sun. In most cases, this involves blocking wavelengths longer than approximately 300 nm, simulating the absorptive effects of upper-atmosphere ozone.

solar cells 1. Solar cells are constructed from semiconductor materials. In metals the resistivity is inversely proportional to the number of free electrons available for the conduction of an electric current. Those materials with resistivities of less than 10^{-3} ohm-cm which conduct electricity well are classified as metals. Those elements whose resistivities are in excess of 10^{12} ohm-cm are classified as insulators. Those whose resistivities are in between are classified as semiconductors. Common semiconductors are germanium, silicon, cadmium, etc. Thus, some of the common solar cell materials are silicon and cadmium sulphide.[AIR744A-72] 2. Photovoltaic cells that convert sunlight into electrical energy.[NASA]

solar constant The rate at which solar radiation is received outside the earth's atmosphere on a surface normal to the incident radiation and at the earth's mean distance from the sun.[NASA]

solar cooling Conversion of solar energy into refrigeration energy.[NASA]

solar cosmic rays Cosmic rays supposedly originating in the sun.[NASA]

solar diameter Observable dimension of the sun.[NASA]

solar disk See sun.

solar dynamic power systems Electric power systems using a solar heated working fluid to drive a turboalternator. Primary applications are for space stations and spacecraft.[NASA]

solar eclipses Obscurations of the light of the sun by the moon.[NASA]

solar engine A device for converting thermal energy from the sun into electrical or mechanical energy or for using thermal energy from the sun to run a refrigeration system.

solar faculae See faculae.

solar furnace A device for producing high temperatures by focusing solar radiation.

solar heating The use of solar radiation to produce enough heat for cooking, industrial operations, or heating buildings.

solar houses Habitable buildings designed with large expanses of glass or other transparent materials to collect solar radiation for heating.[NASA]

solarimeter See pyranometer.

solar neighborhood The portion of the Milky Way Galaxy centering around the sun and containing the nearest neighboring stars.[NASA]

solar neutrinos Neutral particles originating from nuclear reactions in the core of the sun.[NASA]

solar noise See solar radio emission.

solar oscillations Irregular oscillations in the solar atmosphere.[NASA]

solar parallax The angle at the sun subtended by the equatorial diameter of the earth.[NASA]

solar planetary interactions The interactions and subsequent effects caused by the interactions of solar activity and/or wind with a planet, its magnetic field, its atmosphere, or natural satellites.[NASA]

solar plasma (radiation) See solar wind.

solar ponds (heat storage) Large, shallow ponds covered with thin, transparent plastic shields and used for collecting and storing solar heat for conversion to electric power.[NASA]

solar power Any of several methods of using energy from the sun to perform useful work.

solar prominences Filamentlike protuberances from the chromosphere of the sun.[NASA]

solar radiation The total electromagnetic radiation emitted by the sun.[NASA]

solar radio bursts Sudden increases in the flux from the sun at radio frequencies.[NASA]

solar radio emission Radiation at radio frequencies originating from the sun or its corona.[NASA]

solar radio waves See solar radio emission.

solar receivers See solar collectors.

solar selective coatings See selective surfaces.

solar simulators Devices which produce thermal energy, equivalent in intensity and spectral distribution to that from the sun, used in testing materials and space vehicles.[NASA]

solar system The sun and other celestial bodies within its gravitational influence, including planets, asteroids, satellites, comets, and meteors.[NASA]

solar thermal electric power plants The use of solar energy to generate steam for producing electricity.[NASA]

solar thermal propulsion Proposed energy source for spacecraft propulsion by passing hydrogen through a heat exchanger placed at the focal point of a large parabolic dish solar concentrator mirror.[NASA]

solar total energy systems Systems for converting solar energy directly into electrical and thermal energy.[NASA]

solar wind Streams of plasma flowing approximately radially outward from the sun.[NASA]

solder A joining alloy with a melting point below about 450 °F, such as certain lead-base or tin-base alloys.

solder cup A solder retaining cavity on the terminating end of solder contacts or terminals.[ARP914A-79]

solder eye A contact or terminal having a hole at its terminating end through which a conductor can be inserted prior to being soldered.[ARP914A-79]

solder glass A special glass that softens below about 900 °F, and that is used to join two pieces of higher melting glass without deforming them.

soldering A joining process which uses an alloy, melting below 800 °F, to form an electrical joint without alloying the base metals.[ARP1931-87]

soldering embrittlement Penetration by molten solder along grain boundaries of a metal with resulting loss of mechanical properties.

solderless wrap A technique of mechanically winding a solid conductor to a terminal post having a series of edges. Also called "wire wrap".[ARP914A-79]

solder sleeve A heat shrinkable tubing device containing a predetermined amount of solder and flux used for environmental resistant solder connections and shield termination.[ARP914A-79]

solenoid 1. An item consisting of one or more coils surrounding an iron core. The coil(s) and the core are movable in relation to each other. The axial or rotary movement is a result of the magnetic flux of the coil. It is designed to convert electrical energy into mechanical energy. It does not include switch contacts.[ARP480A-87]

solenoid valve A shutoff valve whose position is determined by whether or not electric current is flowing through a coil surrounding a moving-iron valve stem; the valve may be normally open, in which case gas or liquid flows through the valve when electricity to the coil is turned off; normally closed, in which case gas or liquid flows only when electricity is turned on; or three-way, in which gas or liquid flows in one path through the valve when electricity is off and in a different path when electricity is on.

soleplate 1. A flat member used as the supporting base of a machine. 2. A flat member in a machine's frame on which a bearing can be mounted and, if necessary, adjusted slightly.

solettas Orbiting solar mirrors (reflectors).[NASA]

solid coupling A device used to rigidly connect two shafts together and usually capable of transmitting full torque from one shaft to the other.

solid cyrogen cooling Cooling with solidified cryogenic fluids.[NASA]

solid die A one-piece tool with internal threads used for cutting screw threads on rod stock or small-diameter pipe.

solid electrolytes Single crystals, certain alloys, alkaline metals, and other compact compounds used in galvanic cells (batteries).[NASA]

solidification The change in state from liquid to solid in a material as its temperature passes through its melting temperature or melting range on cooling.

solid propellant combustion The burning of solid propellants by rapid oxidation and production of expanding gases, heat, and light.[NASA]

solid propellant gas generator Consists of three components: a self-sustaining combustible solid mixture called the propellant; an initiator to start the reaction upon command; and a chamber with the dual function of storing the above components and expelling the combustion gases at a rate determined by the propellant, its geometry and temperature, and the output or load orifice size.[AIR744A-72]

solid propellant rocket engines Rocket engines fueled with solid propellants. Such motors consist essentially of a combustion chamber containing the propellant, and a nozzle for the exhaust jet, although they often contain other components, as grids or liners.[NASA]

solid propellants Specifically, a rocket

propellant in the solid form, usually containing both fuel and oxidizer combined or mixed, and formed into a monolithic (not powdered or granulated) grain. [NASA]

solid state Pertaining to an electronic device or circuit whose operation is controlled by some combination of electrical, magnetic and optical phenomena within a circuit element consisting largely of a single piece of solid material, usually a crystalline semiconductor.

solid state devices Devices which utilize the electric, magnetic, and photic properties of the solid materials, e.g., binary magnetic cores or transistors. [NASA]

solid-state welding Any welding process that produces a permanent bond without exceeding the melting point of the base materials and without using a filler metal.

solid taper pin receptacles Solid receptacles are taper pin receptacles in which the tapered hole is formed by a suitable means and is seamless.[ARP 592-90]

solid taper pins Solid taper pins are pins machined from solid stock.[ARP 592-90]

solo flight time Flight time during which a pilot is the only occupant of the aircraft.[ARP4107-88]

solstices The two points of the ecliptic farthest from the celestial equator; two points on the celestial sphere occupied by the sun at maximum declination. [NASA]

solubility, S The air solubility in volume percent, measured at 32 °F and one atmosphere pressure, which will dissolve in a petroleum liquid when the air in equilibrium with a liquid is at a partial pressure of 760 mm of Hg. In the basic V/L formula, "S" is expressed in units of volume of air, measured at 32 °F and one atmosphere pressure, dissolved in 100 volumes of fuel at 60 °F.[AIR1326-91]

solubility coefficient, k The solubility of non-reactive gases in liquids generally follow Henry's Law which states that the mass of gas dissolved in a liquid at equilibrium is proportional to the partial pressure of the gas in the vapor phase with which the solution is in equilibrium.[AIR1326-91]

soluble oil An oil-based fluid that can form a stable emulsion or colloidal suspension with water; used principally as a cutting fluid or coolant.

solution A liquid, such as boiler water, containing dissolved substances.

solution heat treatment Heating an alloy into a temperature range where the principal alloying element(s) become dissolved in a single solid phase, then cooling the material rapidly enough to prevent precipitation of secondary phases.

solvation The process of swelling, getting, or dissolving of a material by a solvent; for resins, the solvent can be plasticized.[NASA]

solvent refined coal Low-sulfur distillate fuels from coal, plus the byproducts of methane, light hydrocarbons, and naphtha, all useful for making pipeline gas, ethylene, and high-octane unleaded gasoline.[NASA]

solvent retention The occurrence of solvent residues in chemical or material end products or intermediates. [NASA]

somatogravic illusion See illusion, vestibular.

somatogyral illusion See illusion, vestibular.

sonar 1. A method or system, analogous to radar used under water, in which high frequency sound waves are emitted so as to be reflected back from objects and used to detect the objects of interest. [NASA] 2. Any method or system that depends on underwater sound to passively listen for noises made by ships or sea creatures, to search for underwater objects by sending out sound pulses and listening for returning echoes, or to communicate by sending sound signals from one undersea location to another. An acronym derived from "SOund NAvigation and Ranging".

sonic barrier A popular term for the large increase in drag encountered when the speed of an aircraft or missile approaches the speed of sound in air; the speed at which this occurs is somewhat indefinite, and depends on altitude and general atmospheric conditions.

sonic booms Noises created by shock waves that emanate from aircraft or other objects traveling at or above sonic velocity.[NASA]

sonic fatigue stresses The fluctuating stresses induced in structure and fastener assemblies by oscillating sound waves which may be produced by pulse, ram or turbo jet, and rocket motors. [ARP700-65]

sonic flow See transonic flow.

sonic opacity A characteristic of a medium such as one containing a large quantity of particles or small bubbles that results in sound or ultrasound being reflected randomly from the discontinuities rather than being transmitted through the medium.

sonic speed The speed of sound in the specific medium of concern. See also acoustic velocity.

sonic vibration Refers to the dynamics of mechanical vibration waveforms in the frequency range up to 20KHz.[ARP 1587-81]

sonic waveguides See acoustic delay lines.

sonobuoy A combination of a passive sonar set and a radio transmitter mounted in a buoy that can be dropped by parachute from an airplane; underwater sounds, such as those from a submarine, are picked up by the sonar set and transmitted by radio to a receiver on the aircraft or a ship; the source of underwater sound can be determined by triangulation using several sonobuoys dropped in a known pattern, and comparing time-delay data from their signals by computer analysis.

sonograph 1. An instrument for recording sound or seismic vibration patterns. 2. An instrument for converting sound into percussive (seismic) vibrations.

soot A black deposit containing impure carbon and oily compounds resulting from the incomplete combustion of resinous materials, oils, wood and coal.

soot blower A mechanical device for discharging steam or air to clean heat absorbing surfaces.

sorbates Gas taken up by sorbents. [NASA]

sorbents The materials which take up gas by sorption.[NASA]

sorption The taking up of gas by absorption, adsorption, chemisorption, or any combination of these processes. [NASA]

sort In data processing, a routine that puts data in a specific order based on established criteria.

sortie 1. An operational flight by one aircraft.[ARD50010-91] 2. A term specifically used to refer to flight on a combat mission, but also commonly used to identify any flight, from takeoff to landing, that is part of a larger mission which may contain several flights. A flight from takeoff to landing that remains in the traffic pattern is not considered a sortie.[ARP4107-88]

sorting table Any horizontal conveyor where operators stationed along one or both sides manually sort bulk material, packages or individual items by selectively removing them from the conveyor.

sound 1. A pressure wave in an elastic or compressible material which exists in the form of alterations in pressure, stress, particle displacement or particle velocity; the term is usually restricted to such waves whose frequency is in the range of human hearing. 2. The human auditory sensation produced by such pressure waves. See also acoustics.

sound analyzer A device for measuring the band pressure level at various sound frequencies.

sound barrier See acoustic velocity.

sounders See sounding.

sound fields Regions containing sound waves.[NASA]

sound fixing and ranging A method for acoustically tracking submerged bodies or floats utilizing fixed hydrophones. [NASA]

sound generators Transducers which convert electrical, mechanical or other forms of energy into sound.[NASA]

sounding 1. Any penetration of the natural environment for scientific observation usually by sounding rockets or balloons.[NASA] 2. Determining the depth of a body of water, either by echo ranging or by taking depth readings with a weight attached to a line having knots or telltales regularly spaced along its length. 3. Measuring the depth of soil above bedrock by driving a steel rod into the soil. 4. Generically, any penetration of the natural environment for the purpose of taking scientific measurements.

sounding rockets Rockets designed primarily for routine upper air observation (as opposed to research) in the lower 250,000 feet of the atmosphere, especially that portion inaccessible to balloons, i.e., above 100,000.[NASA]

sound intensity In a specified direction at a point, the average rate of sound energy transmitted in the specified direction through a unit area normal to this direction at the point considered. [NASA]

sound-level meter An electronic instrument for measuring noise or sound levels in either decibels or volume units.

sound measurement See acoustic measurement.

sound-powered telephone A type of telephone usually used for emergency communications over short distances; electric current for transmitting the signal is generated by the speaker's voice in a specially designed microphone, and no external source of power is required.

sound pressure At a point, the total instantaneous pressure at that point in the presence of a sound wave minus the static pressure at that point.[NASA]

sound pressure level (SPL) The intensity of a sound wave which, in decibels, equals 20 log (P_s/P_r), where P_s is the pressure produced by the sound and P_r is a stated reference pressure.

sound reproduction Any process for detecting sound at one location and time and regenerating the same sound at the same location and time, or at another location, at another time, or both, and with any desired intensity.

sound velocity See acoustic velocity.

sound waves Mechanical disturbances advancing with infinite velocity through an elastic medium and consisting of longitudinal displacements of the medium, i.e., consisting of compressional and rarefactional displacements parallel to the direction of advance of the disturbance; a longitudinal wave. Sound waves are small amplitude adiabatic oscillations.[NASA]

source A reservoir from which material or energy is drawn.

source code 1. The program instructions written in high-level languages or assembly languages. The program must be translated into object code before it can be executed by the computer. 2. Software generated by a programmer in assembly language, generally with comments, headings, and other annotation.

source language The system of symbols and syntax, easily understood by people, used to describe a procedure a computer can execute.

source module A series of statements in the symbolic language of an assembler or compiler, which constitutes the entire input to a single execution of the assembler or compiler.

source program A program written in a source language.

sour crude Crude oil containing excessive amounts of sulfur, which liberate corrosive sulfur compounds during refining. Contrast with sweet crude.

sour gas Natural gas that contains corrosive sulfur-bearing compounds such as H_2S or mercaptans.

southern sky That portion of the celestial sphere between the celestial equator and the celestial south pole (and generally visible from areas in the earth's southern hemisphere).[NASA]

space In data processing, a unit of storage that is empty.

space based radar Radar systems installed on large space structures.[NASA]

space biology See exobiology.

spaceborne experiments A collective term designating the various experiments performed or planned in orbiting spacecraft and usually involving physical phenomena in space environments.[NASA]

space capsules Containers used for carrying out experiments in space. [NASA]

space charge The electric charge carried by a cloud or stream of electrons or ions in a vacuum or a region of low gas pressure when the charge is sufficient to produce local changes in the potential distribution. The net electric charge within a given volume.[NASA]

space commercialization For profit activities in space or prefatory to space activity.[NASA]

space cooling (buildings) The cooling of buildings with a solar energy system which incorporates water chillers controlled by thermostats and other devices to provide a comfortable living environment.[NASA]

spacecraft Devices, manned and unmanned, which are designed to be placed into an orbit about the earth or into a trajectory to another celestial body. [NASA]

spacecraft charging Electric charge induction upon the surface of a spacecraft by magnetopheric plasmas or other ion sources.[NASA]

spacecraft docking The act of coupling tow or more orbiting objects; the operation of mechanically connecting together, or in some manner bringing together orbital payloads.[NASA]

spacecraft survivability The ability of a spacecraft to survive adverse conditions including reentry problems. [NASA]

Spacecraft Tracking and Data Network See STDN (network).

space heating (buildings) Heating of living areas for the comfort of occupants (human and/or animals) by any means (electricity, fuels, solar radiation, etc.). [NASA]

Spacelab payloads A general, collective term for the diverse and numerous ESA payloads planned for space experiments.[NASA]

space observations (from earth) Surveillance of extraterrestrial phenomena from the earth's surface.[NASA]

space operation center A proposed NASA space station to be assembled in space that is designed for conducting space based operations such as satellite servicing, orbit transfer vehicle launch and recovery, and assembly of large space structures. Onboard capabilities could include space manufacturing and research experiments. When fully assembled it will be larger in size than the Space Shuttle.[NASA]

space perception The ability to estimate depth or distance between points in the field of vision. [NASA]

space plasmas Concentrations of free

electrons and protons in the ionosphere, plasmasphere, and beyond.[NASA]

space platforms Gimbal-mounted platforms equipped with gyros and accelerometers for maintaining a desired orientation in inertial space independent of spacecraft motion.[NASA]

space processing Forming and fabrication techniques aboard a spacecraft in a weightless or low-gravity environment and involving improved chemical and/or physical procedures for the creation of new or better products. [NASA]

spacer 1. An intermediate object which is used to provide a fixed dimension between two or more parts. See washer, shim.[ARP 480A-87] 2. A simple mechanical member designed to keep two other members in an assembly a specific distance apart.

space radiation See extraterrestrial radiation.

Space Shuttle ascent stage Shuttle takeoff configuration comprising the orbiter, solid rocket boosters, and external tank. [NASA]

Space Shuttle main engine Liquid propellant propulsion system using fuel drawn from external tanks to provide power for the orbiter to attain orbital speed. [NASA]

space simulators Devices used to simulate one or more parameters of the space environment used for testing space systems or components. Specifically, a closed chamber capable of approximating the vacuum and normal environments of space. [NASA]

space suits Pressure suits for wear in space or at very low ambient pressures within the atmosphere, designed to permit the wearer to leave the protection of a pressurized cabin.[NASA]

space technology The systematic application of science and engineering to the exploration and exploitation of outer space.

spacetennas The transmitting antennas of a solar power satellite transmission system which directs the high-power beam from space to a focus on the rectennas on earth.[NASA]

space-time continuum See relativity.

space transportation system flights Revised collective designation for all Space Shuttle flights.[NASA]

space vehicles See spacecraft.

spade bolt A bolt having a flattened head shaped like a spade with a transverse hole; it is used predominantly for fastening components such as shielded coils and capacitors to the chassis of heavy-duty electronic equipment.

spade drill A drill made from round or square stock by hammering one end, tapered to a thin edge, then sharpening a point on the thin edge with the finished piece looking like a pointed spade.

spade lug A device consisting of a body, that can be clamped or crimped onto the end of an electrical wire or cable, and a flat two-pronged projection, that can be slipped under a screw or nut on a terminal block to complete an electrical connection yet allow the connection to be disassembled without completely removing the screw or nut.

spade tongue terminal See slotted tongue terminal.[ARP914A-79]

spall To detach material from a surface in the form of thin chips whose major dimensions are in a plane approximately parallel to the surface.

spalling The breaking off of the surface of refractory material as a result of internal stresses.

span 1. A structural dimension measured in a straight line between two specific extremities, such as the ends of a beam or two columnar supports. 2. The dimension of an airfoil, such as the wings of an aircraft, from tip to tip, measured in a straight line. 3. The difference between maximum and minimum calibrated measurement values. Example: an instrument having a calibrated range of 20-120 has a span of 100.

span adjustment See adjustment, span.

span drift The time related change in response of the analyzer in repetition of a span gas measurement under identical conditions of flow and concentration.[ARP1256-90]

span error The difference between actual span and ideal span, usually expressed as a per cent of ideal span. See also error, span.

span gas A calibration gas to be used for routine verification and adjustment of analyzer response. [ARP1256-90]

spanloader aircraft Advanced distributed-load cargo aircraft configurations in which the payloads are distributed across the span of the wing for a close match between aerodynamic and inertial loading for minimal bending stresses.[NASA]

spanner 1. A wrench with a semicircular head having a projection or hole at one end. 2. A horizontal structural brace. 3. An attachment for a sextant that establishes an artificial horizon.

SPAR (rocket) See space processing applications rocket.

spare alarm point See alarm point.

spares Reparable components or assemblies used for maintenance replacement purposes in major end items of equipment.[ARD50010-91]

spares, provisioned Those spare parts that are procured under certain special procedures at a certain point in the system acquisition cycle.[AIR1916-88]

spark arrester 1. A device that reduces or prevents electric sparks at a point where a power circuit is opened or closed, such as at a circuit breaker or knife switch. 2. A device that prevents airborne embers from escaping from a chimney.

spark duration The length of time usually expressed in microseconds, required to dissipate the total energy of any one spark discharge occurring between the electrodes of a spark igniter.[AIR784-88]

spark energy The energy (joules) released between electrodes of the spark igniter.[AIR784-88]

spark igniter configuration The low tension type of spark igniter utilizes a material classified as a semi-conductor to bridge the gap from the center electrode to the outer or ground shell.[ARP 846-88]

sparking voltage The sparking voltage of a particular shunted surface gap spark igniter design is the lowest voltage applied to an igniter which will initiate a spark.[ARP846-88]

spark rate The number of spark discharges per unit time occurring at the spark igniter under a given set of conditions. (Example — 2 sparks per second minimum at room temperature and 24 V DC input.) "Usually the spark rate is specified as a minimum figure at the lowest input voltage to control the minimum number of sparks in the worst case condition and a maximum value at the highest input voltage to control the life of the spark igniter."[AIR784-88]

spark recorder A type of recorder where sparks passing between a metal pointer and an electrically grounded plate periodically burn small holes in recording paper as it moves slowly across the face of the plate; sparks are produced at regular intervals by a circuit powering an induction coil, and the varying lateral position of the moving pointer creates the trace.

spark shadowgraph photography See shadowgraph photography.

spark test A test used to locate pinholes in wire or cable insulation by the application of an electrical potential across the insulation while it is passing through an ionized field.[ARP1931-87]

spatial coherence The coherence of light over an area of the wavefront of a beam; where the beam hits the surface.

spatial disorientation (Type I) Unrecognized incorrect orientation in space. It may result from an illusion, an anomaly of attention, or an anomaly of motivation, but it is not accompanied by discomfort or confusion because it is not noticed. Also referred to as spatial misorientation.[ARP4107-88]

spatial disorientation (Type II) Recognized incorrect orientation in space typified by a discrepancy between sensory information and cognitive expectancy. The illusory sensory source may be visual, kinesthetic, or vestibular and the effect of the cognitive conflict may range from mild discomfort or confusion to incapacitation.[ARP4107-88]

spatial isotropy See isotropy.

spatial marching Techniques for solving partial differential equations that move along in a space direction.[NASA]

spatial misorientation See spatial disorientation (Type I).

spatial orientation See attitude (inclination).

spatial resolution The precision with which an optical instrument can produce separable images of different points on an object.[NASA]

spatial unorientation Lack of knowledge as to orientation in space due to the inability to detect orienting cues, as in a rapidly spinning or tumbling aircraft. In this situation, the lack of orientation is recognized but there are neither usable orienting cues nor a cognitive expectancy of true orientation.[ARP4107-88]

spatter 1. A term used to denote the condition resulting when large particles of coating material condense on the glass surface, and adhere there, in evaporated coating processes.[ARP924-91] 2. Particles of molten metal expelled during a welding operation and becoming adhered to an adjacent surface.

spatula A nonmetallic tool used to apply and/or to spread sealants.[AIR4069-90]

special IFR See fixed-wing special IFR.

special observation A category of aviation weather observations taken to report significant changes in one or more of the observed elements since the last preceding record or special observation.[ARP4107-88]

special sortie A sortie which is singularly urgent or a measure of capability.[ARP4107-88]

special sortie, actual A sortie which is objectively urgent or a measure of capability such as combat, medical evacuation, weather evacuation, or search and rescue.[ARP4107-88]

special sortie, perceived A sortie which is subjectively perceived to be urgent or is a measure of capability such as a checkride, an operational readiness inspection (ORI), or higher headquarters exercise.[ARP4107-88]

special test equipment Equipment developed for the principal purpose of maintaining quality assurance of end items development and production. Some STE may be used for depot repair.[ARD 50010-91]

special use airspace Airspace of defined dimensions identified by an area on the surface of the earth wherein activities must be confined because of their nature or wherein limitations may be imposed upon aircraft operations that are not part of those activities, or both. Special use airspace includes such areas as military operating areas, prohibited areas, and restricted areas.[ARP4107-88]

special VFR conditions Weather conditions in a control zone which are less than basic VFR but in which some aircraft are permitted flight under visual flight rules.[ARP4107-88]

special VFR operations aircraft operating in accordance with clearances within control zones in weather conditions less than the basic VFR weather minima. Such operations must be requested by the pilot and approved by ATC.[ARP4107-88]

specific acoustic impedance The complex ratio of sound pressure to particle velocity at a given point within the medium.

specific acoustic reactance The imaginary component of specific acoustic impedance.

specific acoustic resistance The real component of specific acoustic impedance.

specific address See absolute address.

specification 1. A detailed, precise description of a weapon system, its hardware, software, geometry, or other design parameter.[ARP4107-88] 2. A list of requirements that must be met when making a material, part, component or assembly; installing it in a system; or testing its attributes or functions. 3. A set of standard requirements applicable to any product or process within the jurisdiction of a given standards-making organization; an industry consensus standard.

specific code See absolute code.

specific fuel consumption The amount of fuel required to produce a given unit of power—expressed, for example, in pounds per horsepower-hour.

specific gravity (sp gr) The ratio of the density of a material to the density of the water at the same conditions. Specific gravity: G_f=liquid at flowing condition referred to water at 60 °F; G_g=gas referred to air, both at STP.

specific gravity bottle A small flask used to determine density; its precise weight is determined when empty, when filled with a reference liquid such as water, and when filled with a liquid of unknown density. Also known as density bottle; relative density bottle.

specific gravity, gas The density of a gas compared to the density of air.

specific gravity, liquid The density of a liquid compared to the density of water.

specific heat (sp ht) 1. The ratio of the heat absorbed (or released) by unit mass of a system to the corresponding temperature rise (or fall).[NASA] 2. The quantity of heat, expressed in Btu, required to raise the temperature of 1 lb of a substance 1 °F. 3. The ratio of the thermal capacity of a substance to that of water. The specific heat at constant pressure of a gas is designated c_p. The specific heat at constant volume of a gas is designated c_v. The ratio of the two (c_p/c_v), is called the ratio of specific heats, k.

specific humidity The weight of water vapor in a gas water vapor mixture per unit weight of dry gas.

speckle holography An imaging technique whereby a speckle pattern results from laser illumination of a diffusely reflecting surface when interference occurs between the fields passing through the various portions of lens aperture. Information about the motion of an object can then be obtained from the imaged fringes resulting from the translation of two speckle patterns. [NASA]

speckle interferometry An imaging process whereby the pattern on the image plane of an interferometer is the result of interference between two mutually coherent, but randomly speckled, fields of two lens formed images from laser illuminated, diffusely reflecting surfaces.[NASA]

spectral absorption See absorption spectra.

spectral analysis A frequency decomposition of the analog input signals. Identification of the frequency spectrum.

spectral density (As in PCM-coded data) the amount of a signal level at each frequency or portion of the spectrum.

spectral emissivity The ratio, at a specified wavelength, of thermal radiation emitted from a non-blackbody to that emitted from a blackbody at the same temperature.

spectral lines See line spectra.

spectral sensitivity In electronics, radiant sensitivity considered as a function of wavelength, or in physics, the response of a device or material to monochromatic light as a function of wavelength; also known as spectral response. [NASA]

spectral shift control Type of reactor moderator control in which the neutron spectrum is intentionally changed. [NASA]

spectrofluorometer An instrument for determining chemical concentration by fluorometric analysis using two monochromators—one to analyze the wavelength of strongest emission and the other to select the wavelength of best excitation in the sample.

spectroheliographs Instruments for taking photographs (spectroheliograms) of the image of the sun in monochromatic light. The wavelength of light chosen for this purpose corresponds to one of the Fraunhofer lines, usually the light of hydrogen or ionized calcium.[NASA]

spectrohelioscopes See spectroheliographs.

spectrometer A spectroscope which includes an angular scale for measurement of the angular deviation and wavelengths of the components of the spectrum.

spectrophotometric titration Instrumented titration in which the end point is determined by measuring a change in absorbed radiation with a spectrophotometer.

spectrophotovoltaics The enhancement of solar cell productivity by concentrating and subdividing the sunlight spectrum and focusing on specific spectrum efficient solar cells.[NASA]

spectropolarimeters See polarimeters.

spectroradiometer An instrument which measures power as a function of wavelength.

spectroreflectometer A device which measures the reflectance of a surface as a function of wavelength.

spectroscope A device which spreads out the spectrum for analysis. The simplest type is a prism or diffraction grating which spreads out the spectrum on a piece of paper or ground glass.

spectroscopic analysis Identification of chemical elements by characteristic emission and absorption of light rays.

spectrum Frequency band.

spectrum analyzer 1. An instrument for measuring the distribution of energy among the frequencies emitted by a pulse magnetron. 2. An electronic instrument for analyzing the output, amplitude and frequency of audio or radio frequency generators or amplifiers under normal or abnormal operating conditions.

spectrum display unit 1. An adjunct to a radio receiver that displays the radio spectrum in and on each side of the carrier being received. 2. On a telemetry receiver, a device which displays the spectrum at and on both sides of the frequency to which the receiver is tuned.

specular reflection Reflection in which the reflected radiation is not diffused; reflection as from a mirror.[NASA]

specular transmission density The value of photographic density obtained when only the normal component of transmitted flux is measured for source illumination whose rays are perpendicular to the plane of the film.

speech baseband compression Technique for reducing the bandwidth required to represent the human voice waveform.[NASA]

speed See velocity.

speed brake An air brake in the form of a flap or plate, which when exposed to the airstream provides a decelerating force. Also drag brake.[AIR1489-88] See also dive brakes.

speed climb/speed descent Vertical modes which maintain an assigned speed (or Mach) usually with the elevator or stabilizer. The throttle setting may be fixed or variable and may be applied independently. Such modes usually can be flown independent of waypoints.[ARP1570-86]

speed, critical engine failure For a multiengine aircraft, that speed at which during takeoff, the aircraft can fail the critical engine and a) elect to continue takeoff and just clear the ground at the end of the runway or b) elect to abort the takeoff and just stop the aircraft within the limit of the remaining runway.[AIR1489-88]

speed, dynamic hydroplaning That speed at which dynamic hydroplaning of a tire or complete air vehicle occurs. [AIR1489-88]

speed, engine idle The minimum normal operating speed of the engine.[ARP 906A-91]

speed, engine lite-off The rotor speed at which combustion is initiated.[ARP 906A-91]

speed, engine self-sustaining The engine rotor speed from which the engine is capable of accelerating to idle speed within specified engine limits without the assistance of a starter.[ARP906A-91]

speed, free running See speed, no load.

speed, landing gear operational limit The maximum airspeed of the aircraft at which the landing gear is designed to retract and/or extend.[AIR 1489-88]

speed, landing gear structural design The maximum airspeed of the aircraft at which the landing gear may be exposed to the airstream without danger of structural damage.[AIR1489-88]

speed management—time control (4D) The functions which provide guidance for adjusting speed and/or flight path so as to conform with air traffic flow control arrival time requirements at a designated downstream waypoint or waypoints. The requirement would normally also include a specified altitude and an arrival speed. These command signals may be directed to the auto flight system as well as pilot displays. (4D) means 4-dimensional—i.e., latitude, longitude, altitude and time. An assigned speed at the final (or designated) waypoints is also usually implied.[ARP1570-86]

speed, no load The peak speed to which the starter will nominally accelerate with the load disconnected and a given power available to the starter. In electrical starter-generator systems, this is the maximum speed to which the starter will assist the engine.[ARP906A-91]

speed of response See response time, time constant.

speed, overrunning The speed at which the engine side of the engaging mechanism overruns the starter side.[ARP 906A-91]

speed, pad Engine rotor speed as related to the accessory pad.[ARP906A-91]

speed reducer A gear train for transmitting power from a motor to the machinery it drives at a rotational speed less than that of the motor.

speed, rotor The speed of the engine rotor or starter turbine.[ARP906A-91]

speed, runaway In electrical starter-generator systems, this is the speed to which the starter-generator will go without load at maximum applied voltage. [ARP906A-91]

speed, starter cut-off The speed, up to which the starter provides starting assistance, and at which its power source is switched off.[ARP906A-91]

spent fuels Nuclear reactor fuels irradiated to the extent that they no longer can effectively sustain a chain reaction. [NASA]

spent liquor The liquid effluent from the pulping stage of papermaking; it consists of wood chemicals such as lignin and partly reacted digestion chemicals (caustic, sulfite or sulfate, depending on which pulping process was used).

sp gr See specific gravity.

sphalerite See zincblende.

spherical aberration A lens defect that makes rays from the peripheral part of the lens focus at a different point than do rays from the central portion of the lens, which produces an image lacking in contrast.

spherical candlepower The (mean) spherical candlepower of a lamp is the average candlepower of the lamp in all directions in space. It is equal to the total luminous flux of the lamp in lumens divided by 4 π.[ARP798-91]

spherical coordinates A system of curvilinear coordinates in which the position of a point in space is designated by its distance from the origin or pole (the radius vector), the angle phi between the radius vector and a vertically directed polar axis (the cone angle or coaltitude) and the angle theta between the plane of the phi and a fixed meridian plane through the polar axis (the polar angle or longitude).[NASA]

spherical plasmas Confined circular plasmas.[NASA]

spherical wave A wave whose equiphase surfaces form a series of concentric spheres.

spheroids Ellipsoids; figure resembling spheres.[NASA]

Spheromaks Toroidal fusion reactors.[NASA]

spherometer A device for measuring the spherical curvature of a surface.

spherulitic-graphite cast iron See ductile iron.

sp ht See specific heat.

spicules Bright spikes extending into the chromosome of the sun from below.[NASA]

SPIFR An acronym for Single Pilot IFR; this term is generally used to describe a general aviation aircraft operated in instrument flight conditions with only one pilot at the controls.[ARP4107-88]

spikes A variation from the surge level or from the controlled steady-state level of a characteristic which reaches its greatest amplitude in an extremely short time. It results from very high frequency currents of complex wave form produced when loads are switched. Transients so generated usually consist of a train of spikes.[AS1212-71]

spill The accidental release of a hazardous chemical or radioactive liquid from a process system or a container.

spin A maneuver, either deliberate or inadvertent, of a stalled airplane in which the airplane descends in a helical path at an angle of attack greater than the angle of maximum lift. The nose of the aircraft in a spin is usually, though not necessarily, pointed sharply downwards. In a normal spin, the longitudinal axis of the aircraft inclines downward at an angle greater than 45 deg. See also flat spin and inverted spin.[ARP4107-88]

spindle 1. An arbor, mandrel, axle, or shaft; esp., a revolving piece less in size than shaft; as in the live spindle of a lathe, that imparts motion to the work.[ARP480A-87] 2. An element of a landing gear which provides an axis for larger revolving parts.[AIR1489-88]

spin down 1. The process of angular deceleration of the wheels of an aircraft after takeoff. 2. The decrease in angular velocity of a wheel as measured by the skid control system. Also increase in slip ratio.[AIR1489-88]

spin glass A magnetic alloy in which the concentration of magnetic atoms is such that below a certain temperature their magnetic moments are no longer able to fluctuate thermally in time but are still directed at random in loose analogy to the atoms of ordinary glass.[NASA]

spinning 1. Production of plastics filament by extrusion through a spinneret. 2. Forming sheet metal into rotationally symmetrical shapes such as bowls or cones by pressing a round-ended tool against the flat stock and forcing it to conform to the shape of a rotating mandrel.

spinning solid upper stage Space shuttle upper stage designed for launching of satellites not requiring the full capacity of the interim upper stage; does not require inertial guidance system nor three-axis stabilization; can handle payloads of the class now launched by Delta or Atlas/Centaur.[NASA]

spin stabilization Directional stability of a spacecraft obtained by the action of gyroscopic forces which result from spinning the body about its axis of symmetry.[NASA]

spin up 1. The process of accelerating the wheel of an aircraft from zero to high angular velocity in a very small increment of time at touchdown and contact with the runway which induces high drag loads in the landing gear structure. 2. Increase in angular velocity of a wheel (decrease in slip ratio) as measured by the skid control system. [AIR1489-88]

spiral A maneuver or performance, especially of an airplane, in which the craft ascends or descends in a helical (corkscrew) path, distinguished from a spin in that the angle of attack is within the normal range of flight angles; also, the flight path of an aircraft so ascending or descending.[ARP4107-88]

spiral bevel gear A bevel gear with curved oblique teeth, which provide for gradual engagement and bring more teeth into contact with each other at any given time than for an equivalent straight bevel gear.

spiral flow test Determining the flow characteristics of thermoplastic resins by measuring the length and weight of resin that flow along a spiral cavity.

spiral gear A helical gear that transmits power from a driving shaft to a nonparallel driven shaft.

spiral welded pipe Pipe made by forming steel plate into long helical strips, fitting the strips together, and welding the spiral seams.

spiral wrap A helical wrap of material over a core. Also called a serve.[ARP 1931-87]

SPL See sound pressure level.

splash lubrication A method of lubricating a piston engine where the connecting-rod bearings dip into troughs filled with oil, splashing it onto other engine parts.

splice 1. A device with provision to accommodate and join two or more conductors to each other.[ARP914A-79] 2. To connect two pieces, forming a single longer piece, as in connecting the ends of wire, rope or tubing; the connection may be made by any of several methods including weaving and welding, and may be made with or without a connector. 3. A permanent junction between two optical fiber ends. It can be a mechanical splice, formed by gluing or otherwise attaching the ends together mechanically, or a fusion splice, formed by melting the ends together.

splice housing A housing designed to protect a splice in an optical fiber from damage such as from the application of stress on the fiber.

splice plate A piece of flat-rolled stock used to connect the webs or flanges of two girders together.

spline One of a set of axial keyways or gearlike ridges on the end of a shaft or the interior of a hub; in use, the splined shaft fits into a mating splined hub to transmit rotational power and motion, while permitting limited axial play between the two members.

split-beam colorimeter An instrument for determining the difference in radiation absorption by the sample at two wavelengths in the visible or ultraviolet region.

split-beam ultraviolet analyzer An instrument for monitoring the concentration of a specific chemical substance in a process stream or coating by measuring the amount of ultraviolet light absorbed at one wavelength and comparing it to the amount at a reference wavelength that is only weakly absorbed by the sample. Also known as a dual beam analyzer.

split bearing A journal bearing consisting of two semicylindrical pieces bolted together.

split flap See flap, split.

splitnut A nut that is cut in half lengthwise and hinged so that it can be rapidly engaged, on closing, and rapidly disengaged, on opening. Found on all thread cutting lathes.

split ranging See signal amplitude sequencing (split ranging).

split S A flight maneuver consisting of a half snap roll followed by a pullout and accomplishing a 180 deg change in direction accompanied by a loss of altitude.[ARP4107-88]

splits (geology) See geological faults.

splitter Plates spaced in an elbow of a duct so disposed as to guide the flow of fluid through the elbow with uniform distribution and to minimize pressure drop.

splitter vanes A set of curved, parallel strips of metal placed along the flow direction in a gas conduit to guide gas flow around a sharp bend in the conduit.

spoiler(s) 1. A plate, series of plates, comb, tube, bar, or other device that projects into the airstream about a body to break up or spoil the smoothness of the flow, especially such a device that projects from the upper surface of an airfoil, giving an increased drag and a decreased lift. Spoilers are normally movable and consist of two basic types: the flap spoiler, which is hinged along one edge and lies flush with the airfoil or body when not in use, and the retractable spoiler, which retracts edgewise into the body.[ARP4107-88]

spoilers, ground A type of drag brake and lift killing device designed to be operated after the aircraft has touched down on the runway. The intended purpose is to transfer the aircraft weight to the wheels as quickly as possible to enable braking and produce drag forces to decelerate the aircraft.[AIR1489-88]

spoke A bar, rod or wire connecting the hub of a wheel to its rim.

spokeshave A small tool for planning concave or convex surfaces.

sponge metal Any metal mass produced by decomposition or chemical reduction of a compound at a temperature below the metal's melting temperature.

sponging A phenomenon in sealants that is sometimes observed after fuel soak and temperature cycling. The sealant swells and a cross-section reveals many voids of various sizes. Sponging can also be caused by other factors but the occurrence is rare.[AIR4069-90]

spontaneous combustion Ignition of combustible materials following slow oxidation without the application of high temperature from an external source.

spool 1. The internal member of cylindrical slide valve which directly affects the flow pattern through its relative position to a surrounding sleeve.[ARP243B-65] 2. The drum of a hoist. 3. The movable member of a slide-type hydraulic valve. 4. A reel or drum for winding up thread or wire. 5. A relatively short transition member (also known as a spool piece) for making a welded connection between two lengths of pipe.

spooldown The process of rotor deceleration from part or full power following a blowout or fuel cutoff.[ARP906A-91]

spooldown start An engine airstart initiated during rotor spooldown, prior to reaching equilibrium windmilling speed.

[ARP906A-91]

spooling The technique by which output to low-speed devices is placed into queues on faster devices to await transmission to the slower devices.

spores The reproductive elements of the lower forms of living organisms, usually unicellular.[NASA]

spot The small luminescent area of the screen surface instantaneously excited by the impact of the electron beam.[ARP 1782-89]

spot check A type of random inspection in which only a very small percentage of total production is checked to verify that a process remains within its control limits.

spot drilling Drilling a small, shallow hole in a surface to act as a centering guide in a subsequent machining operation.

spot facing Producing a flat, machined surface concentric with a drilled hole to serve as a seat for a washer or bolthead, or to allow for flush mounting of mating parts.

spot welding A form of resistance welding where a weld nugget is produced along the interface between two pieces of metal, usually sheet metal, by passing electric current across the joint which is clamped between two small-diameter electrodes or between an electrode and an anvil or plate.

spray A mechanically produced dispersion of liquid drops in a gas stream; the larger the drops, the greater must be the gas velocity to keep the drops from separating out by gravity.

spray angle The angle included between the sides of the cone formed by liquid fuel discharged from mechanical, rotary atomizers and by some forms of steam or air atomizers.

sprayer plate A metal plate used to atomize the fuel in the atomizer of an oil burner.

spray nozzle A nozzle from which a liquid fuel is discharged in the form of a spray.

spray painting A process in which compressed air atomizes paint and carries the resulting spray to the surface to be painted.

spray tower A duct through which liquid particles descend countercurrent to a column of gas; a fine spray is used when the object is to concentrate the liquid, a coarse spray when the object is to clean the gas by entrainment of the solid particles in the liquid droplets.

spreader A device used to expand or open a part to facilitate assembly or disassembly of the part in respect to final position of part.[ARP480A-87]

spread or diffusing Surfaces (and media) which break up the incident light and distribute it as though the surface were incandescent, uniformly bright in all directions or approximately so. Examples are rough plaster, white glass, white plastic.[ARP798-91]

spread reflection Reflection of electromagnetic radiation from a rough surface with large irregularities.[NASA]

spreadsheet program In data processing, a program that will do a variety of calculations frequently needed by accountants and other businessmen.

spread spectrum transmission Communications technique with many different signal waveforms transmitted in a wide band; power is spread thinly over the band so that narrow-band radios can operate within the band without interference.[NASA]

spring A machine element whose chief purpose is to store mechanical energy or to induce mechanical force through elastic deformation of the element's material; the element may be shaped in the form of a plate, leaf, flat-wound helix, coil or washer; it may be made of almost any relatively hard metal or alloy; it may be stressed in tension, compression, bending or torsion; and in most spring designs the amount of deflection is directly proportional to applied load—if the load is released, the element returns to its normal, unstressed shape or position.

springback 1. A structural reaction in the landing gear system which follows "spin up". The energy stored in the gear structure by "spin up" causes a cyclic reaction to produce forward drag forces in the system.[AIR1489-88] 2. Movement of a part in the direction of recovering original size or shape upon release of elastic stress. 3. The amount of elastic deflection that occurs in cold-formed material upon release of the forming force; movement is in a direction opposite to the direction of plastic flow. 4. In flash, upset or pressure welding, the amount of deflection in the welding machine due to the upsetting pressure.

spring clip 1. A U-shaped fastener that attaches a leaf spring to an axle. 2. A fastener used chiefly in electrical connections that grips a part by elastic force.

spring coupling A flexible coupling with resilient parts.

spring hook A hook-shaped device with a spring-loaded member spanning the gap to form an eye; the spring-loaded member allows a bight of rope or cable to be quickly inserted into the eye and prevents the rope from slipping off the hook unless the member is deliberately depressed toward the center of the eye.

spring load A load that varies proportionally with load position. This load may be caused by aerodynamic forces on a movable surface, or by thrust deflection reaction forces on a movable nozzle element. The load can be either opposing or aiding, depending on output position and motion direction.[AIR 1916-88]

spring rate The force required to deflect a member a specified amount. Also fore and aft spring rate, lateral spring rate, torsional spring rate, etc.[AIR1489-88]

spring steel Carbon or low-alloy steel that is cold-worked or heat treated to give it the high yield strength normally required in springs; if it is a heat treatable composition, the springs may be formed prior to heat treatment (hardening).

spring temper A level of hardness and strength for nonferrous alloys and some ferrous alloys corresponding approximately to a cold worked state two-thirds of the way from full hard to extra spring temper.

sprocket A tooth on the periphery of a wheel or spool for engaging the links of a chain or the perforations in computer paper or motion picture film, or some other similar device, so that the chain or paper or film can be driven without slippage or will traverse the wheel without lateral movement, or both.

sprocket chain A flat chain, usually with pinned links, that meshes with the teeth of a sprocket for transmitting motion and mechanical power from one sprocket (the driving sprocket) to another (the driven sprocket).

sprocket hole Any of a series of perforations along the edge of motion picture film, paper tape, computer paper or continuous stationery which engage the teeth of a sprocket wheel or spool so the material can be driven through a mechanical device such as a camera, projector, printer, or recording instrument.

sprung arch An arch in the form of a segment of a circle supported by skew blocks at the two ends.

sprung mass The portion of a landing gear which is attached to fixed air vehicle structure as opposed to the unsprung mass, (tires, wheels, brakes, axle, etc.) separated from the A/V by means of the shock absorber.[AIR1489-88]

spur gear A toothed wheel whose teeth run parallel to the axis of the hub.

spurious error Errors due to instrument malfunction or to human goof-ups.
sputter-ion pump See getter-ion pump.
sputtering Dislocation of surface atoms of a material from bombardment of high energy atomic particles.[NASA]
Squama A scale or structure resembling a scale.[NASA]
square mesh A weave in wire cloth or textile fabric where the number of wires or threads per inch is the same both with the weave and in the cross-weave direction.
square mil The area of a square, one mil by one mil.[ARP1931-87]
square thread A machine thread with a square cross section; the widths of land and groove are each equal to one-half the pitch.
square wave(s) 1. Oscillations, the amplitudes of which show periodic discontinuities between two values, remaining constant between jumps. Specifically, in radar pulses initiated by a rapid rise to peak power, maintained at a constant peak power over the finite pulse length, and terminated by rapid decrease from peak power.[NASA] 2. A wave in which the dependent variable assumes one fixed value for one-half of the wave period, then assumes a second fixed value for the other half, with negligible time of transition between the two fixed values at each transition point.
square wells The impurity potential areas which bound an electron or hole in semiconducting crystals such as silicon.[NASA]
squat switch An electrical switch which is activated by virtue of presence of a load on the landing gear. This signal provides intelligence to many aircraft systems as to whether the A/C is airborne.[AIR1489-88]
squawk (mode, code, function) Request to a flight crew person to activate specific modes/codes/functions on the aircraft transponder: for example, "Squawk three/alpha, two one zero five, low". [ARP4107-88]
squeal See brake squeal.
squeegee A tool for spreading liquids onto a surface or scraping them off; it consists of a simple handle and a transverse blade with a flexible scraping edge usually made of rubber.
squeezed states (quantum theory) Single mode minimum uncertainty states for which the fluctuations in one quadrature phase of the field are smaller than would occur for a coherent state.[NASA]
squeeze films Thin viscoelastic fluid films squeezed between two usually planar structures to serve as sealants, load dampers, lubricants, etc.[NASA]
squeeze-out time A test performed on faying surface sealants at standard conditions. It is essentially the time from the start of mixing of a two-part kit (or the thawing of a frozen tube of sealant) to the tightening of two surfaces with sealant between them. That period of time which still allows the partially cured sealant to be squeezed out to a preselected thickness between the two surfaces is called the squeeze-out time. [AS7200/1-90]
squeeze roll One of two opposing rollers designed to exert pressure on a material passing between them.
squeeze time In resistance welding, the time from initial application of pressure until welding current begins to flow.
squib 1. Used as a general term to mean any of various small-size pyrotechnic or explosive devices containing deflagrating materials.[AIR913-89] 2. Various small explosive devices. Explosive devices used in the ignition of a rocket. [NASA]
squid (detectors) Superconducting quantum interference device magnetometers. [NASA]
SRF See side relief angle.
SSALS Simplified Short Approach Light System.

SSALSR Simplified Short Approach Light System with Runway Alignment Indicator Lights.

stability 1. In meteorology, a state in which the vertical distribution of temperature is such that a parcel of air will resist displacement from its initial level.[ARP4107-88] 2. In aerodynamics, the inherent flight characteristics of an aircraft tending to restore it to its original condition when disturbed by an unbalancing force or moment. See also stability, inherent and stability, aerodynamic.[ARP4107-88] 3. The ability of an explosive material to retain its original properties without degradation when exposed to various environmental conditions over a period of time.[AIR 913-89] 4. Freedom from undesirable deviation. 5. A measure of the controllability of a process. 6. The relative ability of a substance to retain its mechanical, physical and chemical properties during service. 7. The relative ability of a chemical to resist decomposition during storage. 8. The ability of an electronic device or circuit to maintain specified operating characteristics over extended periods of service. 9. The ability of a machine element to retain its original dimensions when exposed to heat, humidity or other environmental conditions. 10. The relative ability of a waterborne vessel to remain upright in a moving sea. 11. The state of a system if the magnitude of the response produced by an input variable, either constant or varied in time, is limited and related to the magnitude of the input variable.

stability, aerodynamic The stability of a body with respect to aerodynamic forces.[ARP4107-88]

stability augmentation Maintenance of aircraft stability in flight by means of automatic control devices which supplement a pilot's manipulation of the aircraft controls. The automatic controls are used to modify inherent aircraft handling problems.[NASA]

stability augmentation system (SAS) A function of the flight control system, including sensors, actuators, etc., that perform in such a manner so as to augment the basic dynamic stability of the aircraft. When considered as an entity, it is essentially a closed-loop regulator control system. The SAS generally has limited authority. To prevent undesirable coupling between the SAS signals and the pilot inputs, SAS signals are normally introduced by a series servo that does not cause stick motion or forces.[AIR1916-88]

stability, directional The stability of a vehicle about its yaw axis.[ARP4107-88]

stability, dynamic That characteristic of an aircraft that determines the nature of its subsequent motion when displaced from its normal flight attitude(s) by an external force or by movement of the control surfaces.[ARP4107-88]

stability, inherent Stability of an aircraft due solely to the disposition and arrangement of its fixed parts; that is, that characteristic which causes it, when displaced, to return to its normal attitude of flight without the use of controls or the interposition of any mechanical devices.[ARP4107-88]

stability, lateral The characteristic of an aircraft that causes it to remain stable or regain stability when caused to roll or sideslip.[ARP4107-88]

stability, longitudinal The characteristic of an aircraft that causes it to right itself or retain stability with respect to vertical displacement of the nose and tail of the aircraft about the center of lift (that is pitching motion).[ARP4107-88]

stability matrix A stability matrix is the eigenvalues of a system about an operating point.[AIR1823-86]

stability, negative dynamic The characteristic of an aircraft that causes the amplitude of an oscillatory motion to increase with time.[ARP4107-88]

stability, negative static The characteristic of an aircraft which causes it, when disturbed from equilibrium, to continue to change attitude in the direction of disturbance (also called static instability).[ARP4107-88]

stability, neutral static The characteristic of an aircraft that, when disturbed from equilibrium, neither causes it to continue in the direction of displacement nor causes it to return to its original attitude. A neutrally stable airplane is one which, if once disturbed from a state of steady flight, will not return to its original flight attitude but may seek any new flight attitude and state of steady flight. Dynamically, such an airplane is neither stable or unstable.[ARP4107-88]

stability of a linear system A linear system is stable if, having been displaced from its steady state by an external disturbance, it comes back to that steady state when the disturbance has ceased.

stability, positive dynamic The characteristic of an aircraft that causes the amplitude of an oscillatory motion to decrease with time.[ARP4107-88]

stability, positive static The characteristic of an aircraft which causes it, when disturbed, to return to its previous attitude of equilibrium.[ARP4107-88]

stability, static The initial tendency that an aircraft displays after its equilibrium has been disturbed.[ARP4107-88]

stabilizer(s) 1. Structural devices used to reduce the lateral deflection of vehicles, applied within the envelope of the vehicle.[ARP1328-74] 2. An airfoil or combination of airfoils, considered a single unit, with the principal function of maintaining stable flight for an aircraft or missile. 3. Any chemical added to a formulation for the chief purpose of maintaining mechanical or chemical stability throughout the useful life of the substance.

stabilizers The fixed surfaces (airfoils) at the rear of an aircraft that give it stability about the longitudinal axis. The vertical stabilizer supports the rudder, and the horizontal stabilizer supports the elevator. On some aircraft the functions of the vertical and horizontal stabilizers, together with those of the rudder and elevator, are integrated into two surfaces in the form of a shallow "V". See also ruddevator.[ARP4107-88]

stabilizing treatment Any of various treatments—mechanical or thermal—intended to promote dimensional or microstructural stability in a metal or alloy.

stable element Any device, such as a gyroscope, used to maintain a stable spatial position for devices such as instrumentation or ordnance mounted in a ship or aircraft.

stack 1. The portion of a chimney above roof level. 2. Any structure that contains flues for discharging waste gases to the atmosphere. 3. A vertical conduit, which due to the difference in density between internal and external gases, creates a draft at its base. 4. An area of memory set aside for temporary storage, or for procedure and interrupt linkages. A stack uses the last-in, first-out (LIFO) concept. As items are added to ("pushed on") the stack, the stack pointer decrements; as items are retrieved from ("popped off") the stack, the stack pointer increments.

stack draft The magnitude of the draft measured at inlet to the stack.

stack effect That portion of a pressure differential resulting from difference in elevation of the points of measurement.

stack effluent Gas and solid products discharged from stacks.

stacker A machine for lifting goods on a platform or fork and placing them in tiered storage, such as in a warehouse.

stacking The installation of two or more terminals on a single screw or stud.[ARP 914A-79]

stack pointer (SP) The SP contains the address of the top (lowest) address of the processor-defined stack.

STADAN (satellite tracking network) See STDN (network).

stadimeter Instrument for determining the distance to an object of known dimension by measuring the angle subtended at the observer by the object. The instrument is graduated directly in distance.[NASA]

stage In electronics, that portion of a circuit between the control tap of one tube or transistor and the control tap of another.

Stage I/II/III Service See terminal radar program.

staggered-intermittent fillet welding Welding a T joint on both sides of the tee in such a manner that the weld bead is segmented, with the segments on either side being opposite gaps between segments on the opposing side.

stagnant hypoxia See hypoxia.

stagnation The condition of being free from movement or lacking circulation.

stagnation point Point in a field of flow about a body where the fluid particles have zero velocity with respect to the body.[NASA]

stagnation pressure A theoretical pressure that could be developed if a flowing fluid could be brought to rest without loss of energy (isentropically).

stagnation region See stagnation point.[NASA]

stagnation temperature The temperature that would be attained if all of the kinetic energy of a moving stream of fluid were converted to heat.

stain 1. A stain is a discoloration of the glass surface, caused by the deposit of foreign matter on the surface or changes produced on the surface of the glass by chemical action of some substance with the glass.[ARP924-91] 2. A nonprotective liquid coloring agent used to bring out the grain in decorative woods. 3. A permanent or semipermanent discoloration on wood, metal, fabric or plastic caused by a foreign substance. 4. Any colored organic compound used to prepare biological specimens for microscopic examination.

stainless alloy Any member of a large and complex group of alloys containing iron, at least 5% chromium, and often other alloying elements, and whose principal characteristic is resistance to atmospheric corrosion or rusting. Also known as stainless steel.

stall A condition wherein the airflow separates from the airfoil surface, or the airflow around the airfoil becomes turbulent. The result of a stall is that the force of lift effect of the airfoil is lost or severely reduced.[ARP4107-88]

stall load The steady-state force or torque from the actuator at zero velocity, when the controller has a saturated input. It is the load that the servoactuator cannot overpower.[AIR1916-88]

stall speed Speed at which an aircraft will stall under given flight conditions.[ARP4107-88]

stamping Virtually any metal forming operation carried out in a press.

stand An item designed to mount and/or support a part of an assembly in a desired position. It may incorporate casters.[ARP480A-87]

standard air Dry air weighing 0.075 lb per cu ft at sea level (29.92 in. barometric pressure) and 70 °F.

standard atmospheric pressure A standard unit of atmospheric pressure defined as that pressure exerted by a 760 mm column of mercury at standard gravity (980.665 centimeters per second or 9.8066 cm/sec^2) at temperature zero deg Centigrade. One standard atmosphere = 760 mm of mercury or 29.9213 in of mercury of 1013.25 millibars.[ARP

4107-88]

standard cell A reference cell for electromotive force.

standard components In the EASY computer program, they are subroutines which define dynamic performance relations of ECS components and controls, dynamic performance relations of general basic controls, and miscellaneous analytical functions.[AIR1823-86]

standard conditions Temperature of 59 °F and pressure of 14.7 psia.[ARP906A-91]

standard control A measuring or controlling device used to provide assurance that allowable process variation has not been exceeded.[AS7200/1-90]

standard deviation The square root of the average squares of deviations from an arithmetic mean.[ARP1192-87]

standard fit Any fit between mating parts whose allowance and tolerance have been standardized.

standard flue gas Gas weighing 0.078 lb per cu ft at sea level (29.92[030] barometric pressure) and 70 °F.

standard gage 1. A highly accurate gage used only as a reference standard for checking or calibrating working gages. 2. A set span across tracks of a railroad that measures 4 ft 8-1/2 in. (1.44 m).

standard instrument arrival route See STAR.

standard instrument departure See SID.

standardization The act or process of reducing something to, or comparing it with, a standard. A measure of uniformity. A special case of calibration whereby a known input is applied to a device or system for the purpose of verifying the output of adjusting the output to a desired level or scale factor.[NASA]

standard leak A controlled finite amount of tracer gas allowed to enter a leak detector during adjustment and calibration.

standard pitch The geometrical pitch of a propeller taken at two-thirds of its radius. See also geometric pitch and pitch (propeller).[ARP4107-88]

standard rate turn A turn of 3 deg per second.[ARP4107-88]

standard, reference The most accurate company held device for assuring the accuracy of control and acceptance standards by periodic comparison.[AS7200/1-90]

standard sphere gap The maximum distance between the surfaces of two metal spheres, measured along a line connecting their centers, at which sparkover occurs when a dynamically variable voltage is applied across the spheres under standard atmospheric conditions; this value is a measure of the crest value of an alternating-current voltage.

standard terminal arrival Synonymous with standard instrument arrival route. See also STAR.[ARP4107-88]

standard tools Standard tools (normally hand tools) used for the assembly, disassembly, inspection, servicing, repair and maintenance of equipment, and which are manufactured by two or more recognized tool manufacturing companies and listed in those companies' catalogs.[ARD50010-91]

standard, transfer A standard device of reference- standard caliber to be used in calibrating control or acceptance standards which cannot readily be brought to the reference standard.[AS 7200/1-90]

standard volume A standard volume of gas is defined as the volume at 25 °C (77 °F) and 1001.32 kPa (29.921 in. Hg Abs.).[ARP1179A-80]

standard wire rope Wire rope made of six wire strands laid around a sisal core.

standby A term used to describe the normal status of a channel in a fault detection-correction fault-tolerant system when that channel may be switched into control in the event of a failure of a normally active operating channel. [ARP1181A-85] [AIR1916-88]

standing wave 1. When a tire is operated at high speeds, the centrifugal forces, internal stresses, external loads, and deflections imposed result in a departure from a round shape. A "standing wave" builds up in front of and behind the contact area. This condition represents a dangerous mode of operation for the tire.[AIR1489-88] 2. Periodic waves having fixed distribution in space which are the result of interference of progressive waves of the same frequency and kind. Such waves are characterized by the existence of nodes or partial nodes and antinodes that are fixed in space.[NASA] 3. A wave in which, for any of the dependent wave functions, the ratio of its instantaneous value at one point on the wave to its instantaneous value at any other point does not vary with time.

standing wave meter An instrument for measuring the standing-wave ratio in a radio frequency transmission line.

standpipe A vertical tube filled with a liquid such as water.

staple A fastener consisting of a U-shaped piece of wire with pointed ends; the fastener may be driven into a solid material such as wood as if it were a double-pointed nail, or it may be driven through thin sheets of paper or fabric and the ends folded over to hold the sheets together.

star A wiring technique where devices are inter-connected via a central hub or wiring closet.

STAR An acronym for STandard instrument Arrival Route, a preplanned instrument flight rule (IFR) air traffic control arrival procedure published for pilot use in graphic or textual form. STARs provide transitions from the en route structure to an outer fix or an instrument approach fix/arrival waypoint in the terminal area.[ARP4107-88]

star clusters Groups of stars physically close together.[NASA]

star coupler A coupler in which many fibers are brought together to a single optical element in which their signals are mixed. The mixed signals are then transmitted back through all the fibers.

star formation The collapse under gravity of molecular clouds of interstellar matter to form clusters of protostars, and the continuing collapse of the protostars to form main-sequence stars. [NASA]

star formation rate The rate at which stars are formed within a specified region or galaxy; sometimes expressed as the number of solar masses per year. [NASA]

star grain A solid propellant grain with an internal star-shaped cross section. [AIR913-89]

Stark effect The broadening or splitting of a spectral line observed when a luminous gas is acted upon by a strong electric field.[NASA]

starlan StarLAN is a proposed standard for 1 Mbit CSMA/CD on twisted pair medium. It is wired as a star with active hubs and is designed to use existing wiring. See TTP.

star network A set of three or more branches in an electronic network where one terminal of each branch is connected at a common node.

stars Self luminous celestial bodies exclusive of nebulas, comets, and meteors; suns seen in the heavens. Distinguished from planets or natural satellites that shine by reflected light.[NASA]

Starsat telescope An anastigmatic 3-mirror reflecting telescope for ultraviolet astronomy purposes aboard the Starsat satellite.[NASA]

starspots Temporary disturbed areas in the stellar photosphere that appear dark because they are colder than the surrounding areas.[NASA]

start bit The first bit in any asynchronous serial data transmission. Used to wake up the system; it carries none of the message information.

start cycle The events which take place between start initiation and the point of engine idle.[ARP906A-91]

starter 1. An electric motor and gear used to turn the crankshaft of an internal combustion engine until its operation becomes self-sustaining. 2. In some chemical processes, a reactive mixture used to initiate a reaction between less reactive chemicals. Also known as starting mix.

starter-assisted start An airstart in which the aircraft starter is engaged either during spooldown, during windmilling, or from zero rpm high pressure rotor speed.[ARP906A-91]

starter drive coupling The elements of the starter which engage the engine and through which torque to accelerate the engine is transmitted.[ARP906A-91]

starter envelope The external three-dimensional shape of the starter, or space allowance for the starter.[ARP 906A-91]

starter, fixed displacement A hydraulic starter which uses a constant flow of fluid per revolution.[ARP906A-91]

starter/generator An electrical starter that will function as a starting motor when external electrical energy is applied and will also function as an electric generator when driven by the engine.[ARP906A-91]

starter, hydraulic A device for converting fluid energy into rotary mechanical energy intended to provide continuous torque for engine starting purposes, and usually incorporating a suitable mechanism to connect the starter to the engine during starting cycles only.[ARP 243B-65]

starter power transform The relationship between starter power input, expressed in terms of voltage and current or pressure and flow rate, etc., and starter power output expressed in terms of torque and speed. Usually presented graphically on torque vs speed axis with input axes overlaid.[ARP906A-91]

starter/pump, hydraulic A hydraulic starter which will function as a motor when external hydraulic energy is applied and will also function as a hydraulic pump when driven by the engine. [ARP906A-91]

starter, variable displacement A starter whose displacement during the starting cycle is automatically controlled, usually to limit the fluid flow rate to a predetermined maximum.[ARP243B-65]

starting resistance The force needed to produce an oil film in a set of journal bearings supporting a shaft when the shaft first begins to turn.

star trackers Telescopic instruments on rockets or other flight borne vehicles that lock onto a celestial body and give guidance reference to the vehicles during flight.[NASA]

star tracking See star trackers.

start switch A switch or other device used to initiate a start cycle.[ARP906A-91]

start time, actual The measured length of engine starting time from initiation of the start cycle to engine idle speed. [ARP906A-91]

start time, calculated The calculated time of engine starting time from initiation of the start to engine idle speed, analytically obtained using graphical or tabulated data, or equations describing engine and starter characteristics. [ARP906A-91]

state 1. Condition of a circuit, system, etc., such as the condition at the output of a circuit that represents logic 0 or logic 1. 2. A description of the process in terms of its measured variables, or a description of the condition of a circuit or device as in "logic state 1".

state-dependent learning A learning anomaly in which a learned task is best remembered when the conditions exist that were present at the time of learning. Thus, procedural knowledge gained in a classroom setting may not be recalled in an operational setting if it was too dependent upon its environmental cues for retrieval.[ARP4107-88]

state equations See equations of state.

statement A software instruction to a computer telling it to perform some sequence of operations.

state variables 1. Variables which completely describe the behavior of a dynamic system.[AIR1823-86] 2. The output(s) of the memory element(s) of a sequential circuit.

static accuracy The degree to which the controlled temperature coincides with the specified or selected temperature after all transients have decayed. Static accuracy is usually specified as a deviation from nominal.[ARP89C-70]

static connection A pipe tap on a manifold used to connect process pressure to an instrument.

static efficiency The mechanical efficiency multiplied by the ratio of static pressure differential to the total pressure differential, from fan inlet to fan outlet.

static firing The firing of a rocket engine in a hold down position to measure thrust and accomplish other tests.[NASA]

static friction See stiction.

static gain See gain, static.

static-head liquid-level meter A pressure-sensing device, such as a gage, so connected in the piping system that any dynamic pressures in the system cancel each other and only the pressure difference due to liquid head above the gage position is registered.

static load That load which is imposed on a member when in a static state or 1 g condition.[AIR1489-88]

static model Set of equations of physical laws to determine a balance of systems at rest.[NASA] See also steady-state model.

static port A opening used as a source of ambient (static) pressure in the pitot-static system. One static port can generally be found on each side of an aircraft in an area where there is usually no dynamic (positive or negative) pressure due to the motion of the airplane through the air. Static air pressure is used to determine altitude and vertical velocity of an aircraft, and when compared with dynamic or impact pressure from the pitot tube, it is used in determining airspeed. See also pitot-static tube and static tube.[ARP4107-88]

static position The state of deflection or compression of a landing gear when the aircraft is in a static or 1g condition. Usually, further compression of the strut (taxi stroke) is available for dynamic taxi conditions. Decrease of the 1g static load permits extension of the strut (landing stroke region) to the free flight or full extended position.[AIR1489-88]

static pressure 1. The local pressure in a fluid that has no element due to velocity of the fluid.[AIR1916-88] 2. The pressure of a fluid that is independent of the kinetic energy of the fluid. 3. Pressure exerted by a gas at rest, or pressure measured when the relative velocity between a moving stream and a pressure-measuring device is zero. See also pressure, static.

static pressure gage An indicating instrument for measuring pressure.

static pressure tube See static tube.

static RAM Random access memory which requires continuous power but does not need to be refreshed as with dynamic RAM. Memory density is not as high as for dynamic RAM.

static register A computer register which retains information in static form.

static seal See gasket.

static stability The property of a physical system which maintains constancy in its static and dynamic responses despite changes in its internal conditions and variations in its environment. Compare with dynamic stability.

static stability, negative See stability, negative static.

static stability, neutral See stability, neutral static.

static stability, positive See stability, positive static.

static steering torque The torque required to turn the nose wheel(s) of an aircraft as it sits static on the runway surface (without the benefit of engine thrust or forward motion of the aircraft). [AIR1489-88]

static stores Digital registers in telemetry devices that hold set-up instructions from the computer.

static subroutine A subroutine which involves no parameters other than the addresses of the operands. Contrasted with dynamic subroutine.

static temperature The temperature of a fluid as measured under conditions of zero relative velocity between the fluid and the temperature-sensitive element, or as measured under conditions that compensate for any relative motion.

static test 1. Any measurement taken in a normally dynamic system under static conditions—for instance, a pressure test of a hydraulic system under no-flow conditions. 2. Specifically, a test to verify structural characteristics of a rocket, or to determine rocket-engine thrust, while a rocket is in a stationary or hold-down position.

static torque The torque which a brake is capable of resisting in a static (non rotating) state as opposed to dynamic torque.[AIR1489-88]

static tube 1. A tube vented to the atmosphere used in measuring ambient (static) air pressure for comparison with impact air pressure to determine airspeed. See also pitot-static tube and static port.[ARP4107-88] 2. A device used to measure static pressure in a stream of fluid. Normally, a static tube consists of a perforated, tapered tube with a branch tube for connecting it to a manometer; a related device called a static pressure tube consists of a smooth tube with a rounded nose that has radial holes in the tube behind the nose.

static unbalance That condition of unbalance for which the central principal axis is displaced only parallel to the shaft axis. NOTE: The quantitative measure of static unbalance can be given by the resultant of the two dynamic unbalance vectors.[ARP588A-89

static weighing A method in which the net mass of liquid collected is deduced from tare (empty tank) and gross (full tank) weighings respectively made before the flow is diverted into the weighing tank and after it is diverted to the by-pass.

station A device capable of originating to or utilizing data from the HSDB.[AIR 4271-89]

stationary orbits Orbits in which the satellite revolves about the primary at the angular rate at which the primary rotates on its axis. From the primary, the satellite thus appears to be stationary over a point on the primary.[NASA]

stationary wave A standing wave in which the energy flux is zero at all points on the wave.

stationkeeping The sequence of maneuvers that maintains a vehicle in predetermined orbit.[NASA]

station management The portion of network management that applies to the lowest two OSI layers.

statistical error 1. Generally, any error in measurement resulting from statistically predictable variations in measurement system response. 2. Specifically, an error in radiation-counter response resulting from the random time distribution of photon-detection events.

statistical quality control Any method

for controlling the attributes of a product or controlling the characteristics of a process that is based on statistical methods of inspection.

stator(s) 1. In machinery, parts or assemblies that remain stationary with respect to rotating or moving parts or assemblies such as the field frames of electric motors or generators, or the stationary casings and blades surrounding axial flow compressor rotors or turbine wheels; sator blades.[NASA] 2. The stationary portion of a machine that interacts with a rotor to produce power or motion.

statoscope 1. A barometer for recording small changes in atmospheric pressure. 2. An instrument for indicating small changes in altitude of an aircraft.

statuary bronze Any of several copper alloys used chiefly for casting ornamental objects such as statues; a typical composition is 90% Cu-6% Sn-3% Zn-1% Pb.

status words Sixteen-bit words, available for computer input, that tell the status of telemetry or magnetic tape equipment.

statute mile A unit of distance equal to 5280 feet.[ARP4107-88]

stay A tensile stress member to hold material or other members rigidly in position. See also brace.

staybolt A bolt threaded through or welded at each end, into two spaced sheets of a firebox or box header to support flat surfaces against internal pressure.

STC Supplemental Type Certificate.

STDN (network) Spaceflight Tracking and Data Network. Name changed from Space Tracking and Data Acquisition Network (STDAN).[NASA]

steady flow A flow in which the flow rate in a measuring section does not vary significantly with time.

steady state The condition of a substance or system whose local physical and chemical properties do not vary with time.[NASA]

steady-state deviation The system deviation after transients have expired. See offset; see also deviation, steady-state.

steady state flow See equilibrium flow.

steady-state model A mathematical model that represents the process at equilibrium (infinite time) conditions.

steady-state optimization A method of optimizing some criterion function of a process usually using a steady-state model of the process. Linear programming is frequently the optimization method used and a function approximating the profit of the process is a typical optimizing criterion. Contrast with dynamic optimization.

steady state response to steering The stable operating condition of a tire under lateral load inputs from the steering system.[AIR1489-88]

steady-state vibration A condition within a vibrating system where the velocity of each moving particle can be described by a periodic function.

steady state wind A wind which produces a constant force.[ARP1328-74]

steam The vapor phase of water substantially unmixed with other gases.

steam atomizing oil burner A burner for firing oil which is atomized by steam. It may be of the inside or outside mixing type.

steam attemperation Reducing the temperature of superheated steam by injecting water into the flow or passing the steam through a submerged pipe.

steam binding A restriction in circulation due to a steam pocket or a rapid steam formation.

steam cock A valve for admitting or releasing steam.

steam cure To hasten the curing cycle of concrete or mortar by the use of heated water vapor, at either atmospheric or higher pressure.

steam dryer A device for removing water droplets from steam. See steam scrubber.

steam-free water Water containing no steam bubbles.

steam gage A device for measuring pressure in a steam system.

steam generating unit A unit to which water, fuel, and air are supplied and in which steam is generated. It consists of a boiler furnace, and fuel burning equipment, and may include as component parts water walls, superheater, reheater, economizer, air heater, or any combination thereof.

steam jacket A casing around the cylinders and heads of a steam engine, or around some other mechanism or space, to keep the surfaces hot and dry.

steam jacketed valve See jacketed vlave.

steam-jet blower A device which utilizes the energy of steam flowing through a nozzle or nozzles to induce a flow of air to be supplied for combustion.

steam purity The degree of contamination. Contamination usually expressed in ppm.

steam quality The percent by weight of vapor in a steam and water mixture.

steam scrubber A series of screens, wires, or plates through which steam is passed to remove entrained moisture.

steam separator A device for removing the entrained water from steam.

steam trace The technique of preventing freezing in a pipe or tubing line with an adjacent steam line; usually 1/4 in. to 1/2 in. copper tubing. See heat tracing.

steam tracing An arrangement for heating a process line or instrument-air line to keep liquids from freezing or condensing—often, a piece of pipe or tubing carrying live steam is simply run alongside or coiled around the line to be heated.

steam trap A device that automatically collects condensate in a steam line and drains it away.

steel Any alloy of iron with up to 2% carbon that may or may not contain other alloying elements to enhance strength or other properties.

steep gradient aircraft See V/STOL aircraft.

steerable antennas Directional antennas whose major lobe can be readily shifted in direction.[NASA]

steer damper A device which provides power for steering (usually on a nose gear) and also damping to resist shimmy.[AIR1489-88]

steering angle The angle through which the landing gear or wheel(s) may be steered. Usually expressed as degrees left and right (one half the total angle). [AIR1489-88]

steering bars See command bars.

steering rockets See control rockets.

Stefan-Boltzmann law One of the radiation laws which states that the amount of energy radiated per unit time from a unit surface area of an ideal black body is proportional to the fourth power of the absolute temperature of the black body.[NASA]

stellar (star tracker) See CCD star tracker.

stellar activity A general term encompassing stellar phenomena such as stellar flares, starspot activity, magnetic activity, nuclear fusion, etc.[NASA]

stellarators Experimental thermonuclear devices where containment in a magnetic field is achieved by closing the field upon itself and thus allowing the particles to perform endless spiral motion.[NASA]

stellar color The particular wavelengths of optical radiation emitted by a star. [NASA]

stellar cores The central portion of the interior of stars.[NASA]

stellar coronas Ionized regions about stars formed by x-rays emitted during stellar flares.

stellar Doppler shift See Doppler effect; extraterrestrial radiation.

stellar flares Ejections of material from stars in eruptions that last from a few minutes to an hour or more.[NASA]

stellar interiors The subsurface portions of stars.[NASA]

stellar magnitude The measure of the relative brightness of a star. Stellar magnitudes are expressed in a variety of ways, according to the method or process of observation or determination. [NASA]

stellar mass accretion Process by which a star accumulates matter as it moves through dense clouds of interstellar gas.[NASA]

stellar oscillations Irregular fluctuations of the stellar atmospheres.[NASA]

stellar parallax The subtended angle at a star formed by the mean radius of the earth's orbit; it indicates distance to a star.[NASA]

stellar physics A term that encompasses the physical properties of stars, such as luminosity, size, mass, density, temperature, chemical composition, evolution, activity, etc.[NASA]

stellar systems Gravitationally bound groups of stars.[NASA]

stellite Any of a family of cobalt-containing alloys known for their wear resistance, corrosion resistance, and resistance to softening at high temperature.

stem A rod connecting a knob or handwheel to the moving part it operates.

stem rotation A phenomenon which occurs in linear motion valves when the hydraulic forces from the process fluid cause the closure component to rotate about the stem axis.

STEP Standard for the Exchange of Product Model Data.[AS4159-88]

step 1. One operation in a computer routine. 2. To cause a computer to execute one operation.

step bearing A bearing that supports the lower end of a vertical shaft. Also known as pivot bearing.

step brazing Making a series of brazed joints in a single assembly by sequentially making up individual joints and heating each one at a lower temperature than the previous joint to maintain joint integrity of earlier joints; the process requires a lower melting brazing alloy for each successive joint in the assembly.

step change The change from one value to another in a single increment in negligible time.

stepdown fix A fix permitting additional descent within a segment of an instrument approach procedure by identifying a point at which a controlling obstacle has been safely overflown.[ARP 4107-88]

step faults See geological faults.

step gage 1. A plug gage consisting of a series of cylindrical gages of increasing diameter mounted on the same axis. 2. A gage for measuring the height of a step or shoulder; it consists of a gage body and a sliding blade.

step-index fiber An optical fiber in which there is a discontinuous change in refractive index at the boundary between fiber core and cladding. Such fibers have a large numerical aperture (light accepting angle), and are simple to connect; but have lower bandwidth than other types of optical fibers.

stepping motor 1. Motor whose rotations are in short and essentially uniform angular movements rather than a continuous motion.[NASA] 2. A motor useful for low torque applications and suitable for computer interfacing. Pulse input results in a precise rotary step, typically 0.8° per pulse or 1.6° per pulse. It is often operated in open loop mode.

step recovery diodes Varactors in which forward voltage injects carriers across the junction, but before the carriers can combine, the voltage reverses and carriers return to their origin in a group. The result is an abrupt cessation of reverse current and a harmonic rich waveform.[NASA]

step response The time response of a device or process when subjected to an instantaneous change in input from one steady-state value to another. See also response, step.

step response time See time, step response.

step soldering Making a series of joints by soldering them sequentially at successively lower temperatures.

step turn A maneuver used to put a float plane in planing configuration prior to entering an active sea lane for takeoff. [ARP4107-88]

steradian The solid angle subtended at the center of a sphere by an area on the surface equal to a square with sides of length equal to the radius of the sphere.

stereochemistry Chemistry dealing with the arrangement of atoms and molecules in three dimensions.[NASA]

stereophonics 1. The use of two sound channels to mimic normal hearing. Stereophonic satellite broadcasting has now been developed.[NASA] 2. Reproducing or reinforcing sound by using two or more audio channels so that the sound gives three-dimensional sensations similar to those of the sound sources.

stereo route A routinely used route of flight established by users and ARTCCs identified by a coded name; for example, ALPHA 2. These routes minimize flight plan handling and communications.[ARP4107-88]

sterns See afterbodies.

stick gage A vertical rod or stick with a graduated scale or markings that is fixed in an open tank or vessel so that liquid level changes can be observed directly.

stick shaker A device that induces vibration felt in the pilot's yoke, warning of an approach to a stall; it is usually set to activate approximately 10 knots above the stall speed.[ARP4107-88]

stiffener A plate, angle, channel or similar structural element attached to a slender beam or column to prevent it from buckling by increasing its stiffness.

stiffness 1. Servoactuator stiffness is the ability of the output to resist motion when subjected to static and dynamic external loading. Measured in units of output motion/external force.[ARP 1281A-81] 2. Force or torque per unit displacement.[AIR1916-88] 3. The ratio of change of force (or torque) to the corresponding change in translational (or rotational) displacement of an elastic element.[NASA]

stiffness to ground The combined stiffness of all spring rates that determines the spatial reference of the load mass. This term is frequency dependent due to servoloop characteristics. Stiffness to ground in conjunction with the load inertia, determines the load natural frequency.[AIR1916-88]

Stilb A unit of luminescence equal to one candela per cm2; it is rarely used, as the candela per m2 is preferred.

stilling basin An area ahead of the weir plate large enough to pond the liquid so that it approaches the weir plate at low velocity, also called weir pond.

stimulate To cause an occurrence or action artificially, rather than waiting for it to occur naturally, as to stimulate an event.

stimulus See measurand.

Stirling cycle A theoretical heat engine cycle in which heat is added at constant volume, followed by isothermal expansion with heat addition. The heat is then rejected at constant volume, followed by isothermal compression with heat rejection.[NASA]

stishovite A mineral consisting essentially of silicon trioxide.[NASA]

stitch bonding A method of making wire connections on an integrated circuit board using impulse welding or heat and pressure to bond a connecting wire at two or more points while feeding the wire through a hole in the welding electrode.

stitching 1. Making a seam in fabric using a sewing machine. 2. Progressive welding of thermoplastics by successively pressing two small induction-heated electrodes against the material along a seam in a manner resembling the action of a sewing machine.

stitch welding Making a welded seam using a series of spot welds that do not overlap.

stochastic Pertaining to direct solution by trial-and-error, usually without a step-by-step approach, and involving analysis and evaluation of progress made, as in a heuristic approach to trial-and-error methods. In a stochastic approach to a problem solution, intuitive conjecture or speculation is used to select a possible solution, which is then tested against known evidence, observations or measurements. Intervening or intermediate steps toward a solution are omitted. Contrast with algorithmic and heuristic.

stochastic processes Ordered sets of observations in one or more dimensions, each being considered as a sample of one item from a probability distribution. [NASA]

stock Material, parts or components kept in storage until needed.

stockpile A reserve stock of supplies in excess of normal usage.

Stoddard solvent A specific type of petroleum naphtha used chiefly in dry cleaning, but also used in small quantities for cleaning soiled surfaces by hand.

stoichiometric conditions In chemical reactions, the point at which equilibrium is reached, as calculated from the atomic weights of the elements taking part in the reaction; stoichiometric equilibrium is rarely achieved in real chemical systems but, rather, empirically reproducible equivalence points are used to closely approximate stoichiometric conditions.

stoker A mechanized means of feeding coal or other solid combustibles into a furnace, burning them under controlled conditions, and carrying away solid combustion products.

Stokes A unit of kinematic viscosity (dynamic viscosity divided by sample density); the centistoke is more commonly used.

STOL Acronym for Short TakeOff and Landing aircraft; an aircraft which has the capability of operating from a short runway in accordance with applicable airworthiness and operating regulations. See also VTOL.[ARP4107-88]

stones (rocks) See rocks.

stoneware Glazed ceramic ware used in certain laboratory and industrial applications involving corrosive chemicals.

stop A device designed to limit movement of a part beyond a fixed point.[ARP 480A-87]

stop altitude squawk Command used by ATC to request an aircraft to turn off the automatic altitude reporting feature of its transponder. It is used when the verbally reported altitude varies 300 ft or more from the automatic altitude report.[ARP4107-88]

stop bit The last bit in an asynchronous serial transmission. Like the start bit, it is used for timing control and carries none of the message information.

stop cock A small valve for roughly controlling or shutting off the flow of fluid in a pipe.

stop nut 1. A nut positioned on an adjusting screw to restrict its travel. 2. A nut with an insert made of a compressible material that keeps the nut tight without requiring a lock washer.

stop plate See locator.

stop squawk Command used by ATC

to tell the pilot to turn off specified functions of the aircraft transponder. [ARP4107-88]

stopway An area beyond the takeoff runway, no narrower than the runway and centered upon the extended centerline of the runway, able to support the airplane during an aborted takeoff without causing structural damage to the airplane; designated by the airport authorities for use in decelerating the airplane during the aborted takeoff.[ARP 4107-88]

storage 1. Pertaining to a device in which data can be stored and from which it can be obtained at a later time. The means of storing data may be chemical, electrical or mechanical. 2. A device consisting of electronic electrostatic, electrical hardware, or other elements into which data may be entered and from which data may be obtained as desired. 3. The erasable storage in any given computer. Synonymous with memory.

storage address register A portion of core memory in a computer that contains the address of a storage location to be activated, either for reading the contents of the location or for storing information at the location.

storage allocation The process of reserving storage for specified information.

storage block A contiguous area of main or secondary storage.

storage buffer 1. A synchronizing element between two different forms of storage, usually between internal and external. 2. An input device in which information is assembled from external or secondary storage and stored, ready for transfer to internal storage. 3. An output device into which information is copied from internal storage and held for transfer to secondary or external storage. Computation continues while transfers between buffer storage and secondary or internal storage or vice versa take place. 4. Any device which stores information temporarily during data transfer. Clarified by buffer.

storage calorifier See cylinder.

storage capacity The amount of data that can be contained in a storage device.

storage cell An elementary unit of storage, for example, a binary cell, a decimal cell.

storage cycle A periodic sequence of events occurring when information is transferred to or from the storage device of a computer. Storing, sensing, and regeneration form parts of the storage sequence.

storage device A device into which data can be inserted, in which it can be retained, and from which it can be retrieved.

storage dump A listing of the contents of a storage device, or selected parts of it. Synonymous with memory dump. See also core dump.

storage key An indicator associated with a storage block or blocks, which requires that tasks have a matching protection key to use the blocks.

storage life The length of time an item can be stored under specified conditions and still meet specified requirements. [ARD50010-91] [AIR1916-88]

storage location A storage position holding one machine word and usually have a specific address.

storage, main See main storage.

storage protection An arrangement of preventing access to storage for either reading or writing, or both. See memory protect.

storage register A register in the storage of the computer, in contrast with a register in one of the other units of the computer.

store 1. To enter data into a storage device. 2. To retain data in a storage device.

stored energy The energy (joules) stored in the tank or storage capacitor of a capacitor discharge system (1/2 CE^2), or in the inductance coil of an inductive discharge system (1/2 LI^2).[AIR784-88]

stored program See stored routine.

stored program computer A computer controlled by internally stored instructions that can synthesize, store, and in some cases alter instructions as though they were data, and that can subsequently execute these instructions.

stored routine A series of instructions in storage to direct the step-by-step operation of the machine. See stored program.

straightening vanes Horizontal vanes inside a fluid conduit or pipe to reduce turbulent flow ahead of an orifice or venturi meter.

straight fittings Parts such as unions, machined out of bar stock, connecting to a port or tube-to-tube.[MA2005-88]

straight-in approach, IFR An instrument approach wherein final approach is begun without first having executed a procedure turn.[ARP4107-88]

straight-in approach, VFR Entry into the traffic pattern by interception of the extended runway centerline (final approach course) without executing any other portion of the traffic pattern.[ARP 4107-88]

straight-in landing A landing made on a runway aligned within 30 deg of the final approach course following completion of an instrument approach.[ARP 4107-88]

straight polarity Arc welding in which the electrode is connected to the negative terminal of the power supply.

straight-tube boiler See boilers.

strain aging A change in properties of a metal or alloy that occurs at room or slightly elevated temperature following cold working.

strainer A screen or porous medium positioned in a flowing stream of fluid (such as a water intake) to separate out harmful objects or particles before the fluid enters process equipment.

strain fatigue See fatigue (materials).

strain foil A type of strain gage made by photoetching a resistance element out of thin foil.

strain gage(s) 1. Instruments used to measure the strain of distortion in a member or test specimen (such as a structural part) subjected to a force.[NASA] 2. A device that can be attached to a surface, usually with an adhesive, and that indicates strain magnitude in a given direction by changes in electrical resistance of fine wire; it may be used to measure strain due to static or dynamic applied loading, in tension or compression, or both, depending on design of the gage, bonding technique, and type of instrumentation used to determine resistance changes in the strain element. 3. A high-resistance, fine-wire or thin-foil grid for use in a measuring bridge circuit. When the grid is securely bonded to a specimen, it will change its resistance as the specimen is stressed. These devices are used in many forms of transducers. 4. A transducer that converts information about the deformation of solid objects, called the strain, into a change of resistance.

strain hardening The increase in tensile and yield strengths, and the corresponding reduction in ductility, associated with plastic deformation of a metal at temperatures below its recrystallization range.

strain in contact area The internal strain within the tire carcass in the contact area which is deflected from a normal state.[AIR1489-88]

strain relief A technique, involving devices or methods of termination or installation, which reduces the transmission of mechanical stresses to the conductor termination.[ARP914A-79]

strain relief clamp See cable clamp.

strain rosette An assembly of two or

more strain gages used for determining biaxial stress patterns. Also known as rosette strain gage.

strain sensitivity A characteristic of a conductor that describes its resistance change in relation to a corresponding length change; it can be calculated as Δ R/R divided by Δ L/L; when referring to a specific strain gage material, strain sensitivity is commonly known as the gage factor.

strand 1. A strand is a rod or filament of metal or electrically conductive material.[AS1198-82] 2. An uninsulated monofilament conductor.[ARP1931-87]

stranded conductor 1. A conductor composed of more than one strand.[AS1198-82][ARP1931-87] 2. One of several wires that are twisted together to form wire rope, cable or electrical conductors. 3. One of the fibers or filaments used to produce yarn, thread, rope or cordage. 4. A piece of cable, rope, string, thread, wire or yarn of specified length. 5. A bar, billet, bloom or slab produced by continuous casting.

strand casting See continuous casting.

strand size The size of a strand that number corresponding to the AWG size of the strand.[AS1198-82]

strange attractors Abstract geometrical objects in theoretical physics that represent motion which is bounded but not periodic. Their detailed behavior is sensitive to external perturbations, but their overall qualitative behavior is stable. They are of particular interest in the study of turbulence.[NASA]

strap 1. A narrow nonwoven material used in a torso restraint system in place of webbing.[AS8043-86] 2. A preformed item of rigid or semi-rigid material, designed to partially, or completely, surround a part for the purpose of retaining and/or positioning the part close to a structure.[ARP480A-87]

strap bolt 1. A bolt with a hook or flattened extension instead of a head. 2. A double-ended bolt with a flattened, non-threaded center section that can be bent around an object to form a U-bolt.

strategic materials Critical raw materials whose foreign source of supply is uncertain and subject to potential cut-off. Examples of such materials are chromium, cobalt, manganese, and platinum group metals.[NASA]

strategy, frame synchronizer The procedure defined by an operator to emphasize rapid acquisition, or to emphasize accuracy of acquisition, or any point between those extremes.

stratification Non-homogeneity existing transversely in a gas stream.

stratigraphy That branch of geology which treats of the formation, composition, sequence, and correlation of the stratified rocks as part of the earth's crust.[NASA]

stratosphere radiation Any infrared radiation involved in the complex infrared exchange continually proceeding within the stratosphere.[NASA]

stratospheric warming A temperature rise in the global stratosphere.[NASA]

stray current corrosion Galvanic corrosion of a metal or alloy induced by electrical leakage currents passing between a structure and its service environment.

stray light Light emitted by components of the lighting assembly directly into the crew station area. Stray light is not visible to crew station members but contributes to the ambient light level; normally associated with flood lit integrally-lighted displays.[ARP1161-91]

streak cameras Cameras for measuring radiation pulses by deflection of an electron beam.[NASA]

streak photography The process of taking a time exposure photograph of a tracer particle in a fluid; the photograph reveals the motion of each tracer particle in the form of a streak which may be interpreted as a velocity vector.

[NASA]

stream(s) 1. Bodies of flowing water, great or small, contained within channels as well as uncontained fluids such as air.[NASA] 2. An input data path to the computer from a single telemetry source, as PCM, PAM, and so on.

streamline flow A type of fluid flow in which flow lines within the bulk of the fluid remain relatively constant with time. See also laminar flow.

streamlining Contouring the exterior shape of a body to reduce drag due to relative motion between the body and a surrounding fluid.

stream tube In the characterization of fluid flow, an imaginary tube whose wall is generated by streamlines passing through a closed curve.

street elbow A pipe elbow with an external thread at one end and an internal thread at the other end.

strength weld A weld capable of withstanding a design stress.

stress 1. The force per unit area of a body that tends to produce a deformation.[ARP4107-88] 2. The effect of a physiological psychological, or mental load on a biological organism, which causes fatigue and tends to degrade performance.[ARP4107-88]

stress amplitude One-half the algebraic difference between the maximum stress and minimum stress in one cycle of repeated variable loading.

stress concentration 1. The presence of shoulders, grooves, holes, keyways, threads, etc., results in a modification of the simple stress distribution so that localized high stresses occur. This localization of high stresses is known as stress concentration.[ARP700-65] 2. In structures, a localized area of high stress.[NASA]

stress concentration factor A measure of the severity of stress concentration is expressed as the ratio of the maximum local stress to the nominal simple stress, the latter of which is determined irrespective of stress concentration. [ARP700-65]

stress-corrosion cracking Deep cracking in a metal part due to the synergistic action of tensile stress and a corrosive environment, causing failure in less time than could be predicted by simply adding the effects of stress and the corrosive environment together. The tensile stress may be a residual or applied stress, and the corrosive environment need not be severe but only must contain a specific ion that the material is sensitive to.

stress cycles A variation of stress with time, repeated periodically and identically.[NASA]

stress intensity factors Load-induced variables in tension, compression and/or shear which are conducive to crack initiation and propagation and fatigue fracture in materials.[NASA]

stress raiser A discontinuity or change in contour that induces a local increase in stress in a structural member.

stress ratio The ratio of the minimum stress to the maximum stress in one cycle, considering tensile stresses as positive, compressive stresses as negative.[ARP700-65]

stress relieving Heating to a suitable temperature, holding long enough to reduce residual stress, and then cooling slowly enough to avoid inducing new residual stresses.

stress, residual The stress present in the body that is free of external forces or thermal gradients which is usually the result of fabrication methods. In fastener fabrication, the stresses resulting from heading, forming, machining, grinding and rolling are residual stresses and may be desirable or not depending upon their location and type. In centerless grinding, the residual stress is usually tensile, in thread rolling it is compressive; occasionally, as in heading, there are combinations of tensile and compressive residual stress. The compressive stress induced by fillet rolling is residual stress.[ARP700-65]

stress, rupture The tensile stress based on the original cross section in material or item at which failure occurs at a given temperature and time.[ARP700-65]

stress-strain relationship Relationship between the stress or load on a structure, structural member, or a specimen, and the strain or deformation that follows.[NASA]

stress tensors Complete sets of stress components in a solid or fluid medium. [NASA]

stretcher leveling Removing warp and distortion in a piece of metal by gripping it at both ends and subjecting it to tension loading at stresses higher than the yield strength.

stretch forming Shaping a piece of sheet metal or plastics sheet by applying tension and then wrapping the sheet around a die form; the process may be performed cold or the sheet may be heated first. Also known as wrap forming.

striae (cords) Striae are apparent streaks or veins in the glass which are the result of minor variations in the index of refraction within the body of the glass.[ARP924-91]

striation technique A method of making sound waves in air visible by using their individual ability to refract light.

strike 1. A thin electroplated film to be followed by other plated coatings. 2. A plating solution of high covering power and low efficiency used for electroplating very thin metallic films. 3. A local crater or remelted zone caused by accidental contact between a welding electrode and the surface of a metal object; also known as arc strike.

string 1. In data processing, a group of consecutive characters. 2. A linear sequence of entities, such as characters of physical elements.

stringer 1. A stringer is a solid nonmetallic impurity in the parent metal, often the result of an inclusion that has been stretched during a rolling process.[AS 3071A-77][MA1568-87] 2. Slender, lightweight, lengthwise fill-in structural members in a rocket body, or the like, serving to reinforce and give shape to the skin. [NASA]

string manipulation The handling of string data by various methods generally in terms of bits, characters, and sub-strings.

string-shadow instrument An indicating instrument in which the measured value is indicated by means of the shadow of a filamentary conductor whose position in an electric or magnetic field depends on the magnitude of the quantity being measured.

stringy alpha Platelet alpha that has been elongated and distorted by nondirectional metal working but not broken up or recrystallized. Also called "wormy alpha".[AS1814-90]

strip 1. The removal of insulation material from wire or cable.[ARP914A-79] 2. A flat-rolled metal product of approximately the same thickness range as sheet but having a width range narrower than sheet. 3. To mine stone, coal or ore without tunneling, but rather by removing broad areas of the earth's surface to relatively shallow depths.

strip chart A hardware device that records analog data (generally, six or eight channels) on a continuous chart.

strip-chart recorder Any instrument that produces a trace or series of data points, using one or more pens or a print wheel, on a grid printed on a continuous roll of paper that is moved at a uniform rate of travel in a direction perpendicular to the motion of the instrument's indicating mechanism. The resulting trace is a graph of the measured variable as a function of time.

stripper 1. A tool or chemical used to remove insulation material from wire or cable.[ARP914A-79] 2. A distillation column that has no rectifying section. In such a column, the feed enters at the top, and there is no other reflux.

stripping The process of removing insulation material from a wire or cable. [ARP1931-87]

stripping section That section of a distillation column below the feed. This section strips the light components from the liquid moving down the column.

strip printer A device that prints the output from a computer, telegraph or recording instrument on a very narrow, continuous length of paper tape.

strip terminal A contact or terminal supplied in some means of continuous form, for use in automatic or semiautomatic crimping machines.[ARP914A-79]

strobe pulse A pulse of light whose duration is less than the period of a recurring event or periodic function, and which can be used to render a specific event or characteristic visible so it can be closely observed.

stroboscope A device for intermittently viewing or illuminating moving bodies so that they appear to be motionless, either by placing an intermittent shutter between the object and an observer or by repeatedly flashing a brilliant light on the object. In this manner, a vibrating or rotating object can be made to appear stationary by adjusting the stroboscope's frequency; the indicated frequency of the stroboscope is equal to the object's vibrational or rotational frequency.

stroboscopic tachometer A stroboscopic lamp and variable-flashing-rate control circuit that enables the frequency to be adjusted until a rotating object appears to stand still; the frequency is read from a calibrated dial, and represents either the fundamental rotational speed in cycles per unit time or one of its harmonics; sometimes, a patterned disk centered on the axis of rotation is used to make it easier to determine fundamental frequency.

stroke The linear extent of movement of a reciprocating mechanical part. See also travel.

stroke-written symbology (vector-written) A type of system that allows the electron beams to create symbology by moving directly from one location to another on the CRT screen; the speed of the beam may be varied or the symbol may be retraced to increase luminance.[ARP1782-89]

strong interactions (field theory) One of the fundamental interactions of elementary particles, primarily responsible for nuclear forces and other interactions among hadrons.[NASA]

strongly coupled plasmas Highly compressed and collisional plasmas with electron densities in order 10 to the 24th power per cubic centimeter or more. The mean kinetic and potential energies of particles in the plasma are typically of the same order of magnitude.[NASA]

Strouhal number 1. A nondimensional number occurring in the study of periodic or quasiperiodic variations in the wake of objects immersed in the fluid stream.[NASA] 2. A nondimensional parameter defined as: $S=fh/V$, where f is frequency, V is velocity and h is reference length.

structural analysis Determination of the stresses and strains in a structural member due to combined gravitational and applied service loading.

structural fatigue See fatigue (materials).

structural steel Hot-rolled steel produced in standard sizes and shapes for use in constructing load-bearing structures, supports and frameworks; some of the standard shapes are angles, channels, I-beams, H-beams and Z-sections.

strut A structural support member for an aircraft for ground operation. Usually includes provisions for wheels, tires, axles, and shock absorber. Also, a bar or rod used to strengthen a framework by resisting longitudinal thrust.[AIR 1489-88]

stud 1. A post used for connecting conductors or terminals. It may be threaded, serrated or plain.[ARP914A-79] 2. A headless bolt threaded at both ends. 3. A threaded fastener with one end intended for welding to a metal surface. 4. A rivet, boss or nail with a large ornamental head. 5. A projecting pin serving as a support or means of attachment.

stud arc welding (SW) Metals are heated with an arc between a metal stud, or similar part, and the work. Once the surfaces to be joined are properly heated, they are brought together under pressure.

stud hole The hole or opening in the tongue of a terminal lug that is intended to accommodate a screw or stud.[ARP 914A-79]

stud-type board A terminal board used for connecting conductors or terminals by means of binding posts or stud terminations. See also terminal board. [ARP914A-79]

stud welding Producing a joint between the end of a rod-shaped fastener and a metal surface, usually by drawing an arc briefly between the two members then forcing the end of the fastener into a small weld puddle produced on a metal surface.

stuffing See packing.

stuffing box A cavity around a rod or shaft that penetrates a pump casing, valve body or other portion of a pressure boundary which can be filled with packing material and compressed to form a leak-tight seal while still permitting axial or rotary motion of the shaft.

stylus 1. Generically, any device that produces a recorded trace by direct contact with a chart or similar recording medium. 2. A needle-shaped device that follows the grooves in a phonograph record and converts the resulting mechanical vibrations into an audio-frequency signal.

subassembly 1. Two or more parts which form a portion of an assembly or component replaceable as a whole, but having a part or parts which are individually replaceable.[ARD50010-91] 2. Assemblies that are component parts of larger assemblies.[NASA] 3. An assembled group of parts intended for incorporation into a device or mechanism as a unit; often a subassembly performs a specific function independently or in conjunction with other subassemblies, and can be removed from the device for maintenance or repair without completely disassembling the device itself.

subcarrier A carrier applied as a modulating wave to another carrier or an intermediate subcarrier.

subcarrier band A band (of frequencies) associated with a given subcarrier and specified in terms of maximum subcarrier deviation.

subcarrier channel The channel required to convey telemetry information involving a subcarrier band.

subcarrier discriminator In FM telemetry, the device which is tuned to select a specific subcarrier and demodulate it to recover the data.

subcarrier oscillator The basic subcarrier frequency generator whose output frequency is used as the transmission or carrier medium of desired signal information; in telemetry, the desired signal information is most often used to frequency modulate the subcarrier for transmission.

subcarrier waves See carrier waves.

subcircuits See circuits; subassemblies.

subcommutation Commutation of a number of channels with the output applied to an individual channel of the primary commutator; subcommutation is synchronous if its rate is a submultiple of that of the primary commutator. Unique identification must be provided for the subcommutation frame pulse.

subcommutation frame In PCM systems, a recurring integral number of subcommutator words, which includes a single subcommutation frame synchronization word. The number of words in subcommutation frame is equal to an integral number of primary commutator frames. The length of a subcommutation frame is equal to the total number of words or bits generated as a direct output of the subcommutator.

subconscious level See awareness, subconscious level.

subdirectory In MS-DOS, a file that is stored in another directory. See root directory.

subduction (geology) Descent of one tectonic unit under another. Most commonly used for descent of a slab of lithosphere, but appropriate at any scale. [NASA]

subframe A multiplex generated at a slower rate than a frame, and input to the frame through one of the channels.

subgiant stars Celestial bodies whose position on the Hertzsprung-Russell (H-R) diagram is intermediate between that of the main-sequence stars and normal giants of the same spectral type.[NASA]

subgravity See reduced gravity.

subharmonic A sinusoidal function whose frequency is a submultiple of some other periodic function to which it is related.

subjective fatigue See fatigue, subjective.

sublayer A subdivision of an OSI layer; e.g. the IEEE 802 Standard divides the link layer into the LLC and MAC sublayers.

sublimation The transition of a substance directly from the solid state to the vapor state, or vice versa, without passing through the intermediate liquid state.[NASA]

submarine Any self-powered underwater craft or towed underwater barges and arrays.[NASA]

submerged-arc welding An electric-arc welding process in which the arc between a bare-wire welding electrode and workpiece is completely covered by granular flux during welding.

submergence The distance measured from the crest level to the downstream water surface when the flow is submerged, i.e., no air is contained beneath the nappe.

submersible pump A pump and electric motor housed together in a water-tight enclosure so that the unit may operate when submerged.

submultiplexer boundary See submiltiplexer group.

submultiplexing See block switching.

sub-network access protocol (SNAP) Provides a mechanism to uniquely identify private protocols above LLC.

sub-optimization The process of fulfilling or optimizing some chosen objective which is an integral part of a broader objective. Usually the broad objective and lower-level objective are different.

subprogram A part of a larger program which can be converted into machine language independently.

subroutine 1. A set of instructions necessary to direct a computer to carry out a well defined mathematical or logical operation; a subunit of a routine, usually coded in such a manner that it can be treated as a black box by the routine using it.[NASA] 2. A routine which is arranged so that control may be transferred to it from a master routine and so that, at the conclusion of the subroutine, control reverts to the master routine. Such a subroutine is usually called a closed subroutine. 3. A single routine may simultaneously be both a subroutine with respect to another routine and a master routine with respect to a third. Usually control is transferred to a single subroutine from more than one place in the master routine and the reason for using the subroutine is to avoid having to repeat the same sequence of instructions in different places in the master routine. Clarified by routine.

subroutine call The subroutine, in object coding, that performs the call functions.

subroutine library A set of standard and proven subroutines which is kept on file for use at any time.

subscale Subsurface oxides formed by reaction of a metal with oxygen that diffuses into the interior of the section rather than combining with metal in the surface layer.

subset 1. In data processing, any set of items that relate to a larger set. 2. A set within a set. 3. A subscriber apparatus in a communications network.

subsidence See damping, also subsidence ratio.

subsieve analysis Determination of particle-size distribution in a powdered material, none of which is retained on a standard 44-micrometer sieve.

subsieve fraction The portion of a powdered material that passes through a standard 44-micrometer sieve.

subsonic 1. In aerodynamics, speeds less than the speed of sound.[ARP4107-88] 2. A generic term roughly designating a speed less than the speed of sound in a given fluid medium. 3. For an aircraft, any speed from hovering (zero) up to about 85% of the speed of sound in the atmosphere at ambient temperature.

subsonic flow Flow of a fluid, as air over an airfoil, at speeds less than acoustic velocity.[NASA]

substantial damage Damage or failure which adversely affects the structural strength, performance, or flight characteristics of the aircraft, and which normally would require major repair or replacement of the affected component. [ARP4107-88]

substitutional element An alloying element with an atom size and other features similar to the titanium atom, which can replace or substitute for the titanium atoms in the lattice and form a significant region of solid solution in the phase diagram. Such elements used in alloying titanium include but are not limited to aluminum, vanadium, molybdenum, chromium, iron, tin, and zirconium.[AS1814-90]

substitution error See error, substitution.

substrate A surface underlying a coating such as paint, porcelain enamel or electroplate.

sub-subcommutation Commutation of a number of channels with the output applied to an individual channel of a subcommutator; unique identification must be provided for sub-subcommutation frame synchronization.

subsystem 1. A major functional portion of a system which contributes to operational completeness of the system. [ARD50010-91] 2. A major functional portion of a system or a combination of sets or groups that contribute to operational completeness of the system and which perform an operational function within that system.[AIR1916-88] 3. A portion of a larger system consisting of several components or process units which, together, have the characteristics of a system by themselves.

subthreshold See threshold.

successive approximation A type of analog-to-digital conversion that compares the unknown input with sums of accurately known binary fractions of full scale, starting with the largest, and rejecting any that changes the comparator's state. At the end of conversion, the output of the converter is a digital representation of the ratio of the input to full scale by a fractional binary code.

suction lift The pressure, in feet of fluid, that a pump must induce on the suction side to raise the fluid from the level in the supply well to the level of the pump. Also known as suction head.

suction line A tube, pipe or conduit that leads fluid from a reservoir or intake system to the intake port of a pump or compressor.

suction pressure Atmospheric pressure minus absolute pressure for values below atmospheric pressure.[AIR1916-88]

sudden ionospheric disturbances Complex combinations of sudden changes in the conditions of the ionosphere and the effects of these changes.[NASA]

sulfonated oil Mineral or vegetable oil treated with sulfuric acid to make an emulsifiable form of oil.

sulfurized oil Any of various oils containing active sulfur to increase film strength and load carrying ability.

sulphate-carbonate ratio The proportion of sulphates to carbonates, or alkalinity expressed as carbonates, in boiler water. The proper maintenance of this ratio has been advocated as a means of inhibiting caustic embrittlement.

sum The quantity resulting from the addition of an addend to an augend.

summation action A type of control-system action where the actuating signal is the algebraic sum of two or more controller output signals, or where it depends on a feedback signal which is the algebraic sum of two or more controller output signals.

summing point Any point in a control system where an algebraic summation of two or more control-loop variables is performed.[AIR1916-88]

sump A tank or pit for temporarily storing drainage.

sump pump A small, single-stage vertical pump used to remove drainage from a shallow well or pit.

sun The star at the center of the solar system, around which the planets, planetoids, and comets revolve. It is a G-type star.[NASA]

sunflowers Any of a number of tall related plants having yellow, daisylike flowers with yellow, brown, purple, or almost black disks containing seeds from which an oil is extracted.[NASA]

sunrise The crossing of the visible horizon by the upper limb of the ascending sun.[NASA]

sunset The crossing of the visible horizon by the upper limb of the descending sun.[NASA]

sunspot cycle A cycle with an average length of 11.1 years but varying between 7 and 17 years in the number and area of sunspots, as given by the relative sunspot number. This number rises from a minimum of 0 to 10 to a maximum of 50 to 140 about 4 years later, and then declines more slowly.[NASA]

sunspots Relatively dark areas on th surface of the sun consisting of dark central umbras surrounded by penumbras which are intermediate in brightness between the umbras and the surrounding photosphere.[NASA]

superalloys See heat resistant alloys.

supercalendered finish A shiny, smooth finish on paper obtained by subjecting the material to steam and pressure while passing it between alternating fiber-filled and steel rolls.

supercharger 1. A high-speed impeller driven by the engine or its exhaust gases to increase manifold pressure and thus enhance the performance of the engine. 2. A device such as an air pump or blower fitted into the intake of an internal combustion engine to raise the pressure of combustion air above the pressure that can be developed by natural aspiration.

supercommutation Commutation at a rate higher than once per commutator cycle, accomplished by connecting a single data input source to equally spaced contacts of the commutator (cross-patching); corresponding cross-patching is required at the decommutator.

supercompressibility The extent to which behavior of a gas departs from Boyle's law.

supercomputers Computers with very large capacity and very high speed. [NASA]

superconducting quantum interferometers. See squid (detectors).

superconductor A compound capable of exhibiting superconductivity—that is, an abrupt and large increase in electrical conductivity as the material's temperature approaches absolute zero.

superfines The portion of a metal powder whose particle size is less than 10 micrometers.

superfinishing Producing a finely honed surface by rubbing a metal with abrasive stones.

superheat To raise the temperature of steam above its saturation temperature. The temperature in excess of its saturation temperature.

superheated steam Steam at a higher temperature than its saturation temperature.

superheater A nest of tubes in the upper part of a steam boiler whose function is to raise the steam temperature above saturation temperature.

superhybrid materials Composites of polymers, boron-aluminum, and titanium.[NASA]

superlattices Crystals grown by depositing semiconductors in layers whose thickness is measured in atoms.[NASA]

superluminescent diode A compromise between a diode laser and LED, which is operated at the high drive currents characteristic of diode lasers, but lacks the cavity-mirror feedback mechanisms that produce stimulated emission. It is used when high power output is desired, but coherent emission is not wanted.

supermassive stars Stars with masses exceeding about 50 times that of the sun.[NASA]

supernatant liquor The liquid above settled solids, as in a gravity separator.

superplasticity The unusual ability of some metals and alloys to elongate uniformly by several thousand percent at elevated temperatures without separating.

superpressure balloons Meteorological balloons consisting of nonextensible envelopes designed to withstand higher internal pressure differentials than external ones. Such balloons will maintain constant elevations until sufficient gas diffuses from them to cause a change in buoyancy.[NASA]

superrotation The generally more rapid relative motions found in the very tenuous regions of the atmosphere at heights around 300 km. The density of the atmosphere decreases rapidly with height and more than 95% of the mass of the atmosphere is contained within the troposphere and lower stratosphere. These regions of the atmosphere rotate faster on average than the underlying solid earth.[NASA]

supersonic 1. A generic term roughly designating a speed that exceeds the speed of sound in a given fluid medium. 2. For an aircraft, any speed that exceeds Mach 1, which is about 650 to 750 mph depending on atmospheric conditions and altitude.

supersonic compressors Compressors in which supersonic velocity is imparted to the fluid relative to the rotor blades, the stator blades, or to both the rotor and the stator blades, producing oblique shock waves over th blades to obtain a high pressure rise.[NASA]

supersonic diffusers Diffusers designed to reduce the velocity and increase the pressure of fluid moving at supersonic velocities.[NASA]

supersonic flow In aerodynamics, flow of a fluid over a body at speeds greater than the acoustic velocity and in which the shock waves start at the surface of the body.[NASA]

supersonic nozzles Converging diverging nozzles designed to accelerate a fluid to supersonic speed.[NASA]

supersonics Specifically, the study of aerodynamics of supersonic speeds. [NASA]

supervisory control 1. A term used to imply that a controller output or computer program output is used as an input to other controllers, e.g., generation of setpoints in cascaded control systems. Used to distinguish from direct digital control. 2. An analog system of control in which controller setpoints can be adjusted remotely, usually by a supervisory computer. 3. Also known as a digitally-directed analog (DDA) control system. See also control, supervisory.

supervisory control system (SCS) Remote setpoint information to single loop analog controllers provided by a digital computer.

supervisory pressure A motivating factor stemming from a person's need to meet perceived supervisory expectations.[ARP4107-88]

supervisory program 1. A program used in supervisory control. 2. Same as executive program.

supplemental type certificate (STC) An FAA approval of a major change in an aircraft, engine, or subsystem which is not great enough to require a new application for a type certificate. The STC supplements the aircraft's type certificate data sheet to include the approved modification.[ARP4107-88]

supply pressure In a hydraulic or pneumatic system, the output pressure from the primary source of pressure, which is subsequently regulated or controlled to provide desired system functions. See also pressure, supply.

support A device or structure which serves to hold in position and/or act as a proper foundation by bearing the weight or stress of another part or parts. [ARP480A-87]

support and test equipment One of the nine principle elements of ILS. It consists of tools, metrology and calibration equipment, performance monitoring and fault isolation equipment, maintenance stands and handling devices required to support the operation and maintenance of systems. Items are categorized as peculiar (to the system under development) and common (commercially available) or currently in the defense inventory. Includes equipment categorized as Ground Support Equipment (GSE) or Aerospace Ground Equipment (AGE). [ARD50010-91]

support cost The total cost of ownership, excluding operating crews and using personnel, of an item during its operational life including the total impact of requirements for skill levels, technical data, test equipment, spares, spare parts, special tools, operational and maintenance equipment, facilities, levels and location of maintenance facilities, manpower, training and training equipment. [ARD50010-91]

support equipment Equipment required to support the operation and maintenance of the aircraft and all its airborne equipment.[ARD50010-91]

supporting electrode An electrode in a spectroscopic apparatus, other than a self electrode, that is designed to hold the analytical sample on or inside it.

support system A programming system used to support the normal translating functions of machine-oriented, procedural-oriented, and problem-oriented language processors.

suppressed range A suppressed range is an instrument range which does not include zero. The degree of suppression is expressed by the ratio of the value at the lower end of the scale to the span. See range, elevated-zero.

suppressed span See range, elevated-zero.

suppressed weir A rectangular weir in which the width of the approach channel is equal to the crest width, i.e., there are no end contractions.

suppressed-zero instrument Any indicating or recording instrument whose zero (no load) indicator position is off-scale, below the lower limit of travel for the pointer or marking device.

suppressed-zero range See range, supp. The individual surface of a part is that area which continues uninterrupted until it adjoins a fillet, corner, or another individual surface.[AS291D-64] 2. The exterior skin of a solid body, considered to have zero thickness.

suppression See range, elevated-zero.

suprathreshold See threshold.

surface 1. The individual surface of a part is that area which continues uninterrupted until it adjoins a fillet, corner, or another individual surface.[AS291D-64] 2. The exterior skin of a solid body, considered to have zero thickness.

surface-active agents See surfactants.

surface analyzer An instrument that measures irregularities in the surface of a body by moving a stylus across the surface in a predetermined pattern and producing a trace showing minute differences in height above a reference plane magnified as much as 50,000 times.

surface area 1. The total amount of exterior area on a solid body. 2. The sum of the individual surface areas of all the particles in a mass of particulate matter.

surface blowoff Removal of water, foam, etc. from the surface at the water level in a boiler. The equipment for such removal.

surface combustion The non-luminous burning of a combustible gaseous mixture close to the surface of a hot porous refractory material through which it has passed.

surface condenser Any of several designs for inducing a change of state from gas to liquid by allowing the gas phase to come in contact with a surface such as a plate or tube which is cooled on the opposite side, usually by being in direct contact with flowing cooled water.

surface density Any amount distributed over a surface, expressed as amount per unit area of surface.

surface effect ships Vessels using ground effect principle and having submerged rigid sidewalls (sealants). [NASA]

surface filter Porous materials that retain contaminants primarily on the influence face. The filtration holes for particle retention are on the same plane within the filtration media.[AIR888-89]

surface finish The roughness of a surface after finishing, measured either by comparing its appearance with a set of standards of different patterns and lusters or by measuring the height of surface irregularities with a profilometer or surface analyzer.

surface flaws Irregularities of any sort which occur at only one place or at relatively infrequent and widely varying random intervals on a surface. A flaw may be a scratch, ridge, hole, peak, crack, check, etc. Unless otherwise specified, the effect of flaws shall not be included in the roughness height measurement.[AS291D-64]

surface gage 1. A scribing tool in an adjustable stand that is used to check or lay out heights above a reference plane. 2. A gage for measuring height above a reference plane.

surface grinder A machine for grinding a plane surface; usually consists of a motor-driven wheel made of bonded abrasive and mounted on an arbor above a reciprocating table that holds the workpiece.

surface hardening Any of several processes for producing a surface layer on steel that is harder and more wear resistant than the softer, tougher core; the process usually involves some kind of heat treatment, and may or may not involve changing the chemical composition of the surface layer.

surface irregularity Nonconformity with general surface appearance, possible defect.[MA2005-88]

surface, nominal The imaginary true surface which would result if all surface irregularities (peaks, waves, ridges and hollows) were leveled off to zero value; or non-existent. It is this nominal surface or "mean line" from which the surface irregularities deviate.[AS 291D-64]

surface plate A table, usually made of granite or steel at least 2 ft square, that has a very accurate flat plane surface; it is used primarily in inspection and layout work as a reference plane for determining heights.

surface pressure See pressure.

surface resistivity The ratio of the potential gradient parallel to the current along the surface of a material to the current per unit width of the surface. [ARP1931-87]

surface roughness 1. Relatively finely spaced irregularities; the height, width, shape and direction of which, establish the predominant surface pattern.[AS 291D-64] 2. Minute pits, projections, scratches, grooves and the like which represent deviations from a true planar or contoured surface on solid material.

surface, roughness height rating Roughness height rating is a height rating of surface roughness over a length equal to the roughness width cutoff obtained by averaging the microinch deviations from the nominal surface.[AS 291D-64]

surface, roughness width The distance in inches between successive ridges which constitute the predominant pattern of the surface roughness.[AS291D-64]

surface tension See interfacial tension.

surface transfer impedance If a current is caused to flow on the outside of a cable shield, then an induced longitudinal voltage will result along the inside of the shield. The ratio of that induced voltage to the driving current is an impedance or surface transfer impedance.[ARP1931-87]

surface treating Any of several processes for altering properties of a metal surface, making it more receptive to ink, paint, electroplating, adhesives or other coatings, or making it more resistant to weathering or chemical attack.

surface, waviness Irregularities of the nominal surface evidenced by recurrent forms of waves. Waviness may be caused by factors such as machining deflections, vibration, heat treatment, or warping strains.[AS291D-64]

surface, waviness height value A physical measurement in inches which represents the maximum height of the waves; from wave peak to wave valley. [AS291D-64]

surface, waviness width value A physical measurement in inches which represents the width of the waves; from wave peak to wave peak.[AS291D-64]

surfacing Depositing filler metal on the surface of a part by welding or thermal spraying.

surge(s) 1. A variation from the controlled steady-state level of a characteristic, resulting from the inherent regulation of the electric power supply system and remedial action by the regulator. [AS1212-71] 2. A transient rise in power, pressure, etc., such as a brief rise in return pressure in a hydraulic system. [AIR1489-88]

surge, in compressors An unstable operating regime in which internal oscillations persist.

surge pressure See pressure, surge.

surge protector A device positioned between a computer and a power outlet designed to absorb power bursts that could damage the computer.

surge tank 1. A vessel used to absorb fluctuations in flow so that they are not passed on to other units. 2. A standpipe or storage reservoir in a downstream channel or conduit to absorb sudden rises in pressure and to prevent starving the conduit during sudden drops in pressure. 3. An open tank connected to the top of a surge line which maintains steady loading on a pump.

surveillance 1. Systematic observation of an area—usually by visual, electronic or photographic means—to gather intelligence for military or law-enforcement purposes. 2. Systematic observation of a process or operation while it is being performed to verify the use of proper equipment, materials, procedures and methods.

surveillance approach An instrument approach wherein the air traffic controller issues instructions for pilot compliance based on aircraft position in relation to the final approach course (azimuth) and the distance (range) from the end of the runway is displayed on the controller's radar scope.[ARP4107-88]

survivability The measure of the degree of capability to which a system will withstand a hostile environment without suffering an abortive impairment of its ability to accomplish its designated function.[AIR1916-88]

survivor curve A type of reliability curve that shows the average percent of total production of a given model or type of machine still in service after various lengths of service life in hours.

susceptibility meter An instrument for determining magnetic susceptibility at low magnitudes.

susceptometer A device for measuring the magnetic susceptibility of ferromagnetic, paramagnetic or diamagnetic materials.

suspend A halt in the process of attaching the aircraft to the catapult which results from the needs to ascertain whether the launch can be safely accomplished.[AIR1489-88]

suspended arch An arch in which the refractory blocks or shapes are suspended by metallic hangers.

suspended solids Undissolved solids in boiler water.

suspension 1. A fine wire or coil spring that supports the moving element of a meter or other instrument. 2. A system of springs, shock absorbers and other devices that support the chassis of a motor vehicle on its running gear. 3. A mixture of finely divided insoluble particles of solid or liquid in a carrier fluid (liquid or gas). 4. A system of springs or other devices that support an instrument or sensitive electronic equipment on a frame and reduce the intensity of mechanical shock or vibration transmitted through the frame to the instrument.

suspension line See shroud line.

sustainer grain A propellant or pyrotechnic grain used in a pressure cartridge or igniter to sustain burning.[AIR 913-89]

sustainer rocket engines Rocket engines that maintain the velocity of the rocket once it has achieved its programmed velocity by use of boosters or other engines.[NASA]

SW See stud arc welding.

swaging (swedging) 1. The mechanical reshaping of barrels; an obsolete term for crimping.[ARP914A-79] 2. Any of several methods of tapering or reducing the diameter of a rod or tube, most commonly involving hammering, forging, or squeezing between simple concave dies.

SWAP Severe Weather Avoidance Plan.

swapping The process of copying areas

of memory to mass storage, and back, in order to use the memory for two or more purposes. Data are swapped out when a copy of the data in memory is placed on a mass storage device; data are swapped in when a copy on a mass storage device is loaded in memory.

swapping device A mass storage device that is especially suited for swapping because of its fast transfer rate.

swath width The width of the area covered by an imaging sensor determined by the geometry of the instrument. [NASA]

S waves Waves in an elastic media which cause an element of the medium to change its shape without a change in volume. Mathematically, S waves are ones whose velocity field has zero divergence.[NASA]

sweat The condensation of moisture from a warm saturated atmosphere on a cooler surface. A slight weep in a boiler joint but not in sufficient amount to form drops.

sweat cooling A process by which a body having a porous surface is cooled by forced flow of coolant through the surface from the interior.[NASA]

sweet crude Crude petroleum containing very little sulfur.

sweet gas Natural gas containing no hydrogen sulfide or mercaptans.

swell The sudden increase in the volume of steam in the water steam mixture below the water level.

swinging load A load that changes at relatively short intervals.

swing joint A connection between two pipes that allows them to be repositioned with respect to each other.

swing pipe A discharge pipe whose inlet end can be raised or lowered within a tank.

switch 1. A device for controlling whether an electrical circuit is 'on' or 'off', usually by making or breaking an electrical connection between the circuit and its power supply. 2. Any electrical or mechanical device for placing another device or circuit into an operating or nonoperating condition. Also known as switching device; switching mechanism. 3. A device for re-routing signals from one optical fiber into another. 4. Any device for controlling which of two fixed paths a railway train, tram, subway car or conveyor line will follow.

switch, air flow proving A device installed in an air stream which senses air flow or loss thereof and electrically transmits the resulting impulses to the flame failure circuit.

switch, high pressure A device to monitor liquid, steam or gas pressure and arranges to open and/or close contacts when the pressure value is exceeded.

switching pressure The pressure at which a system or a component is activated, deactivated, or reversed.[AIR 1916-88]

switching time 1. The time interval in a switching device between the reference time, or time at which the leading edge of a switching or driving pulse occurs, and the last instant at which the instantaneous output response reaches a stated fraction of its peak value. 2. The time interval between the reference time and the first instant at which the instantaneous output response reaches a stated fraction of its peak value.

switch, low pressure A device to monitor liquid steam or gas pressure and arranged to open and/or close contacts when the pressure drops below the set value.

switch, oil temperature limit A device to monitor the temperature of oil between preset limits and arranged to open and/or close contacts should improper oil temperature be detected.

switch, pressure A device which opens or closes an electrical circuit at a give fluid pressure.[ARP243B-65]

swivel 1. A rotatable fluid connection.

[ARP243B-65] 2. A mechanical device that can move freely about a pin joint.

swivel angle 1. The angle through which a gear or wheel is free to swivel or caster.[AIR1489-88] 2. A transient variation in the current and/or potential at some point in the circuit. 3. An upheaval of liquid in a process system, which may result in carryover of liquid into vapor lines. 4. An unstable pressure buildup in a process system. 5. The peak system pressure. 6. The sudden displacement or movement of water in a closed vessel or drum.

swivel fitting A fitting, directionally adjustable without lateral movement, usually retained by a flange, and not free to rotate in service.[ARP243B-65]

swivel joint A fluid connection which is free to rotate in service, usually under pressure.[ARP243B-65]

symbiosis The intimate living together of two organisms of different species, for mutual benefit.[NASA]

symbol 1. A letter, figure, or other code used to represent something else.[ARP 4107-88] 2. Symbols are used on drawings to indicate devices, instruments, types of communication lines, connection points, valves, actuators, primary elements, and other graphic representations. 3. A name which represents a quantity of operation.

symbolic address A label, alphabetic or alphanumeric, that is used to specify a storage location in the context of a particular program; programs are often first written using symbolic addresses in some convenient code and are then translated into absolute addresses by an assembly program.

symbolic code A code which expresses programs in source language, i.e., by referring to storage locations and machine operations by symbolic names and addresses which are independent of their hardware determined names and addresses. Synonymous with pseudo code and contrasted with machine language code.

symbolic instruction An instruction in an assembly language directly translatable into a machine code.

symbolic logic The discipline that treats formal logic by means of a formalized artificial language or symbolic calculus whose purpose is to avoid the ambiguities and logical inadequacies of natural languages.

symbolic notation A method of representing a storage location by one or more figures, or labels.

symbolic number A numeral, used in writing routines, for referring to a specific storage location; such numerals are converted to actual storage addresses in the final assembling of the program.

symbolic programming The use of arbitrary symbols to represent addresses in order to facilitate programming.

symbology The use of one or more symbols which make up a format to portray/define information.[ARP4107-88]

symmetrical transducer A transducer in which all possible termination pairs may be interchanged without affecting transducer function.

symmetry breaking See broken symmetry.

sympathetic detonation (ignition) The explosion of a second charge or device caused by nearby detonation (ignition) of another.[AIR913-89]

synchronism The relationship between two or more periodic quantities of the same frequency when the phase difference between them is zero or constant at a predetermined value.[NASA]

synchronization Keeping one part of a process in a fixed relationship to another part of a process—for example, keeping the advance of strip stock through a stamping press timed to move ahead only when the press has raised the die free of the work. See also synchronism.

synchronization of parallel computa-

tional processes Controlling the execution of parallel computational processes so as to maintain some desired sequential relationship between programmed actions within the processes. See semaphore.

synchronization pattern A sequence of ones and zeros that signals the start of each frame of PCM data; the pattern generally chosen is a pseudo-random sequence, one that is unlikely to occur randomly in data.

synchronize To lock one element of a system into step with another. The term usually refers to locking a receiver to a transmitter, but it can refer to locking the data terminal equipment bit rate to the data set frequency.

synchronizer A hardware device that can recognize a predetermined pattern in a telemetry format and generate clock pulses that coincide with those occurrences.

synchronous 1. Electrically or mechanically in phase or in step—as applied to two or more circuits, motors, machines or other devices. 2. Describes data transmission or other logic events that use a timing signal or clock pulse to control the rate at which events occur. 3. The performance of a sequence of operations controlled by an external clocking device; implies that no operation can take place until the previous operation has been completed.

synchronous data link control (SDLC) A type of data link protocol.

synchronous detectors See correlators.

synchronous platforms Space platforms whose rotation is synchronized with that of earth.[NASA]

synchronous satellites Equatorial west-to-east satellites orbiting the earth at an altitude of approximately 35,900 kilometers at which altitude they make one revolution in 24 hours, synchronous with the earth's rotation.[NASA]

synchronous system trap (SST) A system condition that occurs as a result of an error or fault within the executing tasks.

synchroscope An instrument for indicating whether two periodic quantities are synchronous by means of a rotating pointer or cathode-ray oscilloscope; the position of the pointer or pattern on the oscilloscope tube indicates the instantaneous phase difference between the two quantities.

synchro-to-digital converter (S/D) An electronic device for converting the analog output signal of a rotary transformer (synchro) into a digital word for further processing.

synchrotrons Devices for accelerating particles, ordinarily electrons, in a circular orbit in an increasing magnetic field by means of an alternating field applied in a synchronism with the orbital motion.[NASA]

synergism An action where the total effect of two components or agents is greater than the individual effects of the components when simply added together; for instance, in stress-corrosion cracking, cracks form and propagate deep into a material in a much shorter time and at a much lower stress than could be predicted from known effects of stress and the corrosive environment.

synoptic meteorology The study and analysis of weather information gathered at the same time.[NASA]

syntax 1. The structure of expressions in a language. 2. The rules governing the structure of a language. 3. In data processing, grammatical rules for software programming that specify how instructions can be written.

syntectic alloys Metallic composite materials characterized by a reversible convertibility of their solid phases into two liquid phases by the application of heat.[NASA]

synthesis (chemistry) The application

of chemical reactions to obtain desired chemical products.[NASA]

synthetic aperture radar Active microwave sensors providing all-weather, high resolution imagery.[NASA]

synthetic apertures In radar technology, the simulations of large antennas by correcting the phase and magnitude of the return signals from smaller antennas, permitting the use of lower frequencies for airborne radars.[NASA]

synthetic metals Materials which do not occur in nature but have the appearance and physical properties of true metals.[NASA]

synthetic lubricant Any of a group of lubricating substances that can perform better than straight petroleum products in the presence of heat, chemicals or other severe environmental conditions.

synthetic relationship A relation existing between concepts which pertains to empirical observation. Such relationships are involved not in defining concepts or terms, but in reporting the results of observations and experiments.

syntony The situation of two or more oscillating circuits having the same resonant frequency.[NASA]

SYSGEN See system generation.

system 1. A combination of inter-related items arranged to perform a specific function.[ARD50010-91] 2. An organized arrangement in which each component part acts, reacts, or interacts in accordance with an overall design inherent in the arrangement.[AIR1489-88] 3. An assembly of procedures, processes, methods, routines, or techniques united by some form of regulated interaction to form an organized whole. 4. The complex or hardware and software utilized to affect the control of a process. 5. In data processing, any group of software and hardware that is connected to operate as a unit.

system, arresting A system comprising the gear and apparatus designed to arrest the aircraft in its landing roll. See also arresting gear.[AIR1489-88]

systematic error 1. An error in a set of measurements or control that can be predicted from scientific principles; individual errors from the same cause bias the value of the mean because they all act in the same direction (sense); the amount of error in each individual value may or may not have a direct mathematical relationship to true value of the quantity. 2. That which cannot be reduced by increasing the number of measurements if the equipment and conditions remain unchanged. 3. Any constant or reproducible error introduced into a measured or controlled value due to failure to control or compensate for a specific side effect. See error, systematic.

system board The control center of a computer.

system, brake control A system comprising the gear and apparatus for decelerating the aircraft by means of wheel brakes and includes: control pedals or input, cable, mechanical or other connection to brake pressure metering valves, skid control valves and subsystem, wheel brakes and other elements.[AIR1489-88]

system characteristic equation The characteristic equation is the denominator of the system transfer function.[AIR1823-86]

system check A check on the overall performance of the system, usually not made by built-in computer check circuits, e.g., control totals, hash totals, and record counts.

system control See control system.

system crash The sudden and complete failure of a computer system.

system designation A unique part number which identifies a torso restraint system and its separable sub-assemblies.[AS8043-86]

system deviation See deviation, system.

system device The device on which the

operating system is stored.

system effectiveness The probability that a system can successfully meet an operational demand within a given time when operated under specified conditions.[ARD50010-91]

Systeme Internationale d'Unites (SI) The current International System of Units.

system error In a control system, the difference between the value of the ultimately controlled variable and its ideal value.

system, gear positioning A system comprising the gear and apparatus for positioning the landing gear in its respective operational positions. Includes uplocks, downlocks, actuator, sequencing and indicating subsystems.[AIR 1489-88]

system generated electromagnetic pulses Electromagnetic fields generated by the emission of a large electronic current from a metallic body in space caused by the incidence on its surface of strong ionizing radiation pulses (usually x-ray) from space.[NASA]

system generation (SYSGEN) The process of building an operating system on or for a particular hardware configuration with software configuration modifications.

system generator A program that performs system-level functions; any program that is part of the basic operating system; a system utility program.

system, landing impact A system comprising the gear and apparatus for absorbing and dissipating the energy and loads imposed on the air vehicle by impact with the ground.[AIR1489-88]

system reliability The probability that a specific system will complete a specific mission without failure of mission essential system functions.[ARD50013-91]

system resonance The closed-loop resonant frequency observed with complete system operation. The system resonance is lower than the load natural frequency due to dynamic effects of system control elements, damping, and backlash. [AIR1916-88]

systems analysis The examination of an activity, procedure, method, technique, or a business to determine: a) behavioral relationships or, b) what must be accomplished and how.

systems engineering 1. Designing, installing and operating a system in a manner intended to achieve optimum output while conserving manpower, materials and other resources. 2. Designing and operating a system to achieve a predetermined level of performance, taking into consideration all of the factors that contribute to system performance, including manpower utilization and human factors.

systems integration The combining of subsystems each with numerous interfaces for the input and output of data and each with specified functions vital to the planned success of the main system.[NASA]

systems simulation The simulation of any dynamic system.[NASA]

system, steering control A system comprising the gear and apparatus for steering the aircraft in ground operations. Includes an input control (wheel or rudder pedals) mechanical or electrical connection to power control valves, a power unit and other elements.[AIR1489-88]

system synchronization A means by which transmitting and receiving stations establish a time reference.[AIR 4271-89]

system test Determining performance characteristics of an integrated, interconnected assemblage of equipment under conditions that evaluate its ability to perform as intended and that verify suitability of its interconnections.

T

t See temperature.

T See tesla; also see temperature.

tab A small auxiliary control surface attached to a larger control surface such as an aileron, rudder, or elevator, usually at the trailing edge. Deflection of the tab causes the larger surface to deflect in the opposite direction.[ARP 4107-88]

table 1. An orderly arrangement of data in rectangular form of rows and columns which is used to extract discrete data. [AS4159-88] 2. A flat plate, with or without legs, used primarily to support workpieces or other items at a given vertical height. 3. The flat portion of a machine tool such as a grinder that directly or indirectly supports and positions the work. 4. A collection of data in a form suitable for ready reference, frequently as stored in sequenced machine locations or written in the form of an array of rows and columns for ready entry and in which an intersection of labeled rows and columns serves to locate a specific piece of data or information. 5. In data processing, any group of data organized as an array.

table look up A procedure for obtaining the function value corresponding to an argument from a table of function values.

tableting A method of compacting powdered or granular solids using a punch and die; used to make certain food products, dyes, and pharmaceuticals.

TACAN Acronym for TACtical Air Navigation, an ultra-high frequency electronic air navigation aid which provides suitably equipped aircraft with a continuous indication of bearing and distance to the TACAN station.[ARP 4107-88]

tachometer An instrument designed to indicate the revolutions per minute of an engine crankshaft.[ARP4107-88]

tachometer generator A generator which is designed to be energized from a single phase voltage source and to deliver a generated output voltage essentially proportional to speed and energizing voltage. In practice residual voltages at zero speed exist. The frequency of this output voltage is the same as the fundamental energizing voltage frequency.[ARP667-82]

tack 1. A small, sharp nail with a broad head. 2. The quality of an adhesive, paint, varnish or lacquer to remain sticky to the touch for a prolonged period of time.

tack free time The time required at standard conditions for a curing sealant to lose its surface tackiness as determined by placing a small piece of polyethylene film on its surface and then peeling the film away. The sealant is tack free when no sealant is removed by the film.[AS7200/1-90] [AIR4069-90]

tackiness agent An additive that imparts adhesive qualities to a nonadhesive material.

tackle Any arrangement of ropes and pulleys used to produce a mechanical advantage.

tack weld 1. Any small, isolated arc weld especially one that does not bear load but rather merely holds two pieces in a fixed relationship. 2. A weld joint made by arc welding at small, isolated points along a seam.

tactical air navigation See TACAN.

tag 1. An item consisting of a flap, tab, strip, or the like, attached to a part by string, cord, wire, or similar material for the purpose of identification.[ARP 480A-87] 2. A unit of information whose composition differs from that of other members of the set so that it can be used as a marker or label; also called a flat or sentinel.

tag (from data compressor) A unique sixteen-bit word, preselected by the operator, that precedes each data output word and identifies it.

tag number Instrument loop identification number.

Tag-Robinson colorimeter A laboratory device used to compare shades of color in oil products by varying the thickness of a column of the oil until its color matches that of a standard.

tail assemblies The rear part of a body, as of an aircraft or a rocket. The tail surfaces of an aircraft or rocket.[NASA]

tail bumper A landing gear installation at the aft end of the fuselage or tail of the aircraft to protect it from contact with the ground. It may be a simple skid, a shock absorbing unit, or a complete retractable installation.[AIR 1489-88]

tail gear See gear.

tail mountings See tail assemblies.

tails (assemblies) See tail assemblies.

tail skid A tail bumper installation which consists of a simple skid or runner.[AIR 1489-88]

tail wheel A wheel at the tail used to support the tail section on the ground and/or protect that portion of the aircraft during takeoff and landing. A tail wheel may be retractable, steerable, fixed, castering, etc.[AIR1489-88]

takeoff The action of a rocket vehicle departing from its launch pad. The action of an aircraft as it becomes airborne.[NASA]

takeoff minimums The minimum weather conditions for aircraft to take off generally 1 mile visibility or RVR 50 for one- and two-engine aircraft and 1/2 mile visibility or RVR 24 for aircraft with three or more engines. See also runway visual range (RVR).[ARP 4107-88]

takeoff phase See mishap, phase of flight.

takeoff power 1. With respect to reciprocating engines, the brake horsepower that is developed under standard sea-level conditions, and under the maximum conditions of crankshaft rotational speed and engine manifold pressure approved for the normal takeoff, and limited in continuous use to the period of time shown in the approved engine specification. 2. With respect to turbine engines, the pounds of thrust that are developed under static conditions at a specified altitude and atmospheric temperature, and under the maximum conditions of rotor shaft rotational speed and gas temperature approved for the normal takeoff, and limited in continuous use to the period of time shown in the approved engine specification.[ARP4107-88]

takeoff roll The movement of the aircraft along the runway from the start or takeoff to rotation and liftoff.[ARP 4107-88]

takeoff thrust The equivalent of takeoff power.

takeoff weight The weight of an aircraft at liftoff. See also ramp weight.[ARP 4107-88]

talcum powder (red) A powder used in determining leak exit location; the powder turns bright red in fuel.[AIR4069-90]

tandem actuators Two or more coaxial actuators that are mechanically connected to move together. A tandem actuator usually has two pistons on the same rod, carried in dual actuator cylinder housings. Separate cylinders (with a common piston rod) can be used to give partial rip-stop protection.[ARP 1181A-85] [AIR1916-88]

tandem networks An arrangement of two-terminal-pair networks such that the output terminals of one network are directly connected to the input terminals of the other network.

tang 1. The slim, tapered end of a hand file that fits into a handle. 2. A tongue-like projection on the shank end of a drill that fits into the spindle of a drill press and ensures transmission of torque to the drill body.

tangent galvanometer A galvanometer consisting of a small compass mounted horizontally in the center of a large vertical coil of wire; the current through the coil is proportional to the tangent of the angle the compass needle makes with its rest (no current) position.

tank 1. A receptacle or structure, varying in design to contain a liquid or gas. [ARP480A-87] 2. A large container, covered or open, for holding, storing or transporting liquids. See also reservoir. [ARP243B-65]

tank circuit A resonant electronic circuit that consists of a capacitor and an inductor connected in parallel.

tank, barrier A film or material applied to the tank inner liner that resists fluid permeation. Used on most non-self-sealing and self-sealing fluid tanks containing aviation fuels.[AIR1664-90]

tank building form A male shape duplicating the shape of the finished tank. The various plies of rubber-coated fabric or spray-on polyurethane are fabricated over this shape. The form maintains the tank shape during vulcanization and is removed after vulcanization.[AIR 1664-90]

tank, convolution Generally, a recess projecting into the tank surface to clear structural ribs, formers, longerons, etc. Convolutions permit utilization of the maximum volume of the structure for fluid storage.[AIR1664-90]

tank, crash-resistant A variation of either a non-self-sealing tank or self-sealing flexible tank fabricated of very strong elastomer coated fabric and high-strength fittings. It is intended that this tank retain fluid without leakage under impact-survivable crash conditions. [AIR1664-90]

tank, fit check fixture A fixture accurately simulating the size and shape of the aircraft flexible tank cavity. It is used as a quality assurance acceptance test tool.[AIR1664-90]

tank fiber-lock fitting This is a fitting design developed under the auspices of the U.S. Army specifically for crash-resistant tanks. It uses numerous individual fibers fabricated readily into the tank wall as means of retaining the metal fitting in the tank. See also tank, wedge-lock fitting.[AIR1664-90]

tank fitting, flange Some fitting subassemblies are attached to the flexible tank wall by means of plies of rubber-coated fabric or molded rubber that are called flanges. These flanges are bonded to the metal rings of the fitting and also to the tank wall and become an integral part of the tank assembly during vulcanization.[AIR1664-90]

tank fittings, mounted Fittings are subassemblies fabricated into the tank wall for the purpose of: (a) attaching the tank to structure, (b) providing a sealing surface for and a means of attaching equipment items to the tank, (c) providing for the attachment and sealing of vent lines and fluid flow lines to the tank.[AIR1664-90]

tank, hanger fittings Various designs of fittings as single bolts, fabric loops, and formed wire rings which are fabricated into the flexible tank wall. Generally, they provide a means of externally attaching or supporting the flexible tank to the aircraft structure. However, the term is also applied to fittings used on the inside of the tank to support items of equipment and fluid flow lines.[AIR 1664-90]

tank, hydrodynamic ram Hydrodynamic ram refers to the pressure loads that develop in a fluid filled tank as a result of shock loading or the penetration and travel of a projectile through the fluid. The loads can be such as to cause structural failure.[AIR1664-90]

tank, lace line A high-strength, nylon cord used to support a flexible tank to the structure, It is used in conjunction with attach point on the external surface of the cell and on the structure. [AIR1664-90]

tank, liner (inner liner) The innermost

ply of a flexible fluid tank. It can be either a rubber layer, a rubber-coated fabric, or a spray-on polyurethane.[AIR 1664-90]

tank, molded in place seal A proprietary commercial sealing gasket in which a partial O-ring is molded and retained in grooves positioned on both sides of a thin metal ring. The metal ring is designed to be installed in the interface of the tank mounted fitting and the structure or item of equipment.[AIR 1664-90]

tank, nipple fitting A fitting generally molded entirely of rubber but can contain fabric. The fitting consists of a tubular section whose inside diameter is sized to match standard aluminum tubing. A seal is obtained using a hose clamp assembled to the outside of the molded tube. The molded tube may extend outwardly or inwardly from the tank external surface. The tubular section is adapted to the flexible tank by means of molded flanges. This type fitting is used primarily in general aviation non-self-sealing tanks.[AIR1664-90]

tank, non-self-sealing A tank having no ballistic sealing properties. It can, however, possess tear-resistant characteristics.[AIR1664-90]

tank-selector valve A valve used to select one of the separate gas tanks from which fuel is to be drawn; usually located in the flight station or cockpit of an aircraft.[ARP4107-88]

tank test See wet dielectric test.

tap 1. An internal threading tool with longitudinally separated cutting edges designed to form specific threads.[ARP 480A-87] 2. A threaded plug, where the threads are of accurate form and dimensions, and have cutting edges that form internal threads in a hole as the plug is screwed into the hole. 3. A small hole in the wall of a pipe or process vessel, usually threaded, where an instrument, control device or sampling device is attached. 4. A coupler in which part of the light carried by one fiber is split off and inserted into another fiber, essentially the same as a tee coupler. 5. To withdraw a quantity of molten metal from a refining or remelting furnace.

tap drill A drill used to make a small, precise hole for tapping.

tape 1. A graduated steel ribbon used to measure lengths, as in surveying. 2. A ribbon made of plastic, metal, paper or other flexible material suitable for data recording by means of electromagnetic imprinting; punching or embossing patterns; or printing. 3. An adhesive-backed ribbon used for sealing packages, attaching labels, and various other purposes.

tape-and-plumb-bob liquid-level gage See plumb-bob gage.

tape-controlled machine A machine tool operated automatically by means of control signals read off a length of magnetic or punched paper tape.

tape drive A device that moves magnetic tape past a head that can "read" the tape.

tape formatter A device, including buffers and controls, for recording ordered data on magnetic tape in gapped form, in a format recognized by a computer.

tape header data Several recorded characters at the beginning of a magnetic tape, used to identify the content of the tape.

taper A dimensional feature where thickness, height, diameter or some other measurement varies linearly with distance along a given axis.

tapered-roller bearing A roller bearing having tapered rollers that run in conical races; it can support both radial and thrust loads.

tapered-tube rotameter A type of variable-area flowmeter in which a float that has greater density than the fluid rides inside a tapered tube in such a manner that fluid flowing upward through the tapered section carries the float with it until the upward force exerted by the flowing fluid just balances the downward force due to float weight; as the float rides upward, the annular area around it becomes larger and force on the float decreases; if the tube is made of glass, it can be graduated so that flow rate is read directly by observing float position; otherwise flow rate must be determined from an indirect indication of float position.

tapered waveguide A waveguide section having a continuous change in cross section.

taper pin An electric solderless terminal used to connect a wire to a special receptacle. It has a round tapered shank on one end, and a wire barrel on the other end that is crimped to the wire. Formed and solid pins are available. [ARP592-90]

taper pin receptacles Electric terminals having round tapered holes that mate with taper pins. Formed and solid receptacles are available.[ARP592-90]

taper pin terminal blocks Electric connector blocks made from insulating material, and containing taper pin receptacles.[ARP592-90]

tape search A hardware process by which an operator or computer can cause instrumentation tape to be searched automatically for specific start and stop times for data reduction.

tape speed compensation signal A signal recorded on instrumentation tape along with the data (preferably on the same track as the data) to correct electrically for tape speed errors during playback.

tape terminal See strip terminal.

tape-type liquid-level gage A liquid-level gage consisting of a tape wound around a drum which is attached to a pointer or other level indicator, with one end of the tape attached to a float and the other counterweighted to keep the tape taut.

tape wrap A spirally or longitudinally applied insulating tape, wrapped around a conductor or cable, either insulated or uninsulated, and used as an insulation or mechanical barrier.[ARP1931-87]

taphole A hole in the side or bottom of a furnace or ladle for draining off molten metal.

tappet An oscillating part such as a lever, operated by a cam or push rod, and used to tap or push another machine element such as a valve.

tappet rod A pivot rod carrying one or more tappets and acting as a fulcrum for their motion.

tapping See dither.

TARE (data reduction) See data reduction.

tare weight In any weighing operation, the residual weight of any containers, scale components or residue that is included in total indicated weight and must be subtracted to determine weight of the live load.

target 1. A goal or standard against which some quantity such as productivity is compared. 2. A point of aim or object to be observed by visual means, electromagnetic imaging, radar, sonar or similar noncontact method. 3. In MS-DOS, the location where data is to be copied and stored.

target acquisition The process of optically, manually, mechanically, or electronically orienting tracking systems in the direction and range to lock on a target.[NASA]

target computer 1. The computer in which the target program is used. 2. A computer which has its programs prepared by a host processor. Same as object machine.

target flowmeter A device for measuring fluid flow rates by means of the drag force exerted on a sharp-edged disk centered in a circular flowpath due to differential pressure created by fluid flowing through the annulus; usually, the disk is mounted on a bar whose axis coincides with the tube axis, and drag force is measured by a secondary device attached to the bar.

target language The language into which some other language is to be translated.

target masking Technique used in vision contrast discrimination testing involving the ratio of the luminance of a target (object) to the luminance of the background, especially when light and dark adaptation are factors.[NASA]

target penetration See terminal ballistics.

targets Objects or points toward which something is directed. Objects which reflect a sufficient amount of a radiated signal to produce an echo signal on detection equipment.[NASA]

target program An object program which has been assembled or compiled by a host processor for a target computer.

target system The microcomputer system to be used in the final product.

target-type flowmeter An instrument for measuring fluid flow in which the fluid exerts force on a small circular disc suspended in the center of the flow conduit by means of a pivoted bar; the force exerted on the bar by a force-balance transmitter to counteract the fluid force on the target is an indication of the flow rate.

tarnish Discoloration of a finished surface by a thin film of corrosion products.

TAS True Airspeed.

task 1. A unit of work for the computer central processing unit from the standpoint of the executive program. 2. A specific "run time" execution of a program and its subprograms. 3. In the MULTICS sense, a virtual processor. (A single processor may be concurrently simulating many virtual processors.) 4. The execution of a segment on a virtual processor. See virtual processor. 5. In RSX-11 terminology, a load module with special characteristics; in general, any discrete operation performed by a program.

task control block (TCB) The consolidation of control information related to a task.

task dispatcher The control program that selects from the task queue the task that is to have control of the central processing unit and gives control to the task.

task management Those functions of the control program that regulate the use by tasks of the central processing unit and other resources, except for input/output devices.

task queue A queue of all the task control blocks present in the system at any one time.

taut-band ammeter An instrument for measuring electric current in which a moving coil is mounted on a taut metal band held rigidly at the ends; when current flows through the coil, it deflects within the gap of a permanent magnet, twisting the metal band; the magnitude of current is indicated by a pointer attached to the coil when the torque exerted by magnetic-field interaction is balanced by restoring torque in the twisted band.

taxi To move about on the ground, water, or other surface (as a carrier deck) under the craft's own power, except during the takeoff and landing rolls.[ARP 4107-88]

taxi phase See mishap, phase of flight.

taxiway A strip or area of the airfield set aside for taxi operations, from parking areas to the runway or interconnecting runways. Not intended to be used for takeoffs and landings.[AIR1489-88]

taxistrip See taxiway.

taxi stroke The stroke of a shock absorber which is used mainly during taxi operations. Generally the portion from "static" portion toward "full compressed".[AIR1489-88]

T-bolt A bolt shaped like the letter T, used primarily in conjunction with a dog or other holddown device to secure workpieces against a machine bed or table containing a number of T-shaped slots that the bolt head fits into.

T/C See thermocouple.

TCAS 1. Partial acronym (pronounced Tee-CAS) standing for Traffic Alert and Collision Avoidance System.[ARP 4153-88] 2. An avionics system designed to present real-time information about the close proximity of other aircraft to facilitate the pilot's responses to avoid mid-air collisions.[ARP4107-88]

TCAS-I The first version of TCAS; it provides the pilot with vertical (altitude) information about other aircraft in close proximity.[ARP4107-88]

TCAS-II Proposed second-generation TCAS that would give pilots left-right guidance information in addition to vertical (altitude) information about other aircraft in close proximity to their own aircraft so that they can take evasive measures to avoid mid-air collisions.[ARP4107-88]

TCH Threshold Crossing Height.

TCV program See terminal configured vehicle program.

TDM See time division multiplex.

TDMA See time division multiplex access.

TDZE Touchdown Zone Elevation.

TDZL Touchdown Zone Lights.

teardown inspection That inspection of the unit detail parts usually imposed as a part of the qualification testing of a starter or at the completion of a test program.[ARP906A-91]

tear-down time The amount of time required to disassemble a machine set-up following a production run and prior to setting up the jigs and fixtures for the next order.

teardrop procedure turn An instrument approach procedure turn starting from over the approach facility on an outbound course that is displaced to the left (to the right if a nonstandard approach) of the reciprocal of the inbound course and is followed by a turn to the right (left for nonstandard) to intercept the inbound course.[ARP4107-88]

tearing modes (plasmas) Explosive reconnections of energetic particle accelerations at high voltages in the magnetosphere during substorms.[NASA]

tears A sharp break or fissure in the surface of a hose cover, generally caused by strain and service conditions.[ARP 1658A-86]

tear strength Force required to initiate or continue a tear in a material under specified conditions.[ARP1931-87]

Technical and Office Protocol (TOP) A specification for a suite of communication standards for use in office automation developed under the auspices of Boeing Computer Services. The development of this specification is being taken over by the MAP/TOP Users Group under the auspices of CASA/SME.

technical characteristics Those attributes of equipment that pertain to the engineering principles governing its functions.

technical evaluation An investigation to determine the suitability of materials, equipment or systems to perform a specific function.

technical specifications A detailed description of the technical characteristics of an item or system in sufficient detail to form the basis for design, development, production and, in some cases, operation.

tectonic movement See tectonics.

tectonics A branch of geology dealing with the broad architecture of the upper part of the Earth's crust, that is, the regional assembling of structural or deformational features, a study of their mutual relations, their origin, and their historical evolution.[NASA]

TED See transferred electron devices.

Tediar (trademark) See polyvinyl fluoride.

tee coupler A fiber optic coupler in which the three fiber ends are joined together, and a signal transmitted from one fiber is split between the other two.

tee joint A junction, such as in piping or a weldment, where a branch member is connected at one end to a cross member running at right angles to the branch.

teeming Pouring molten metal into an ingot mold; most often used with reference to steel production.

tektites Small glassy bodies containing no crystals, composed of at least 65 percent silicon dioxide, bearing no relation to the geological formations in which they occur, and believed to be of extraterrestrial origin.[NASA]

teldata In these systems, the common parallel interface for telemetry data, as from a frame synchronizer to a buffered data channel.

telecommunications Pertaining to the transmission of signals over long distances, such as by telegraph, radio, or television.

teleconnections (meteorology) Statistically significant temporal correlations between meteorological parameters at widely separated points.[NASA]

telemetering 1. The transmission of a measurement over long distances, usually by electromagnetic means. 2. Using radio waves, wires or other means to transmit instrument readings to a remote location. Also known as remote metering; telemetry.

telemeters See telemetry.

telemetry The science of measuring quantities, transmitting the results to a distant station, and interpreting, indicating, and/or recording the quantities measured.[NASA]

telemetry front end (TFE) Hardware devices that accept multiplexed data and time, establish synchronization, convert to parallel data, and provide timing pulses, status, and the like for computer entry.

telemetry input channel A device which prepares telemetry data for input to a real-time computer.

telephone twisted pair (TTP) A network medium that uses existing telephone wiring. Standard work is in progress on a TTP standard for IEEE 802.3 StarLAN and IEEE 802.5 Token Ring.

telephotometer See telephotometry.

telephotometry The body of principles and techniques concerned with measuring atmospheric extinction using various types of telephotometers.[NASA]

teleran An aircraft navigation system that combines radar position information with a television image; ground plan position indicator, map and weather information are displayed together in the aircraft.

telescoping gage An adjustable gage for measuring inside dimensions such as hole diameters; it consists of a spring-loaded member that extends until it touches both sides of a hole, then is locked in place to prevent further extension when it is withdrawn from the hole; an outside micrometer or vernier calipers is used to measure the length of the locked gage member.

teletypewriter A popular hard copy device attached to the computer, which prints one character at a time.

telltale A marker on the outside of a tank that indicates water level on the inside of the tank.

telluric currents Large scale surges of electric charges within the earth's crust, associated with disturbances of the ionosphere.[NASA]

telluric lines Absorption lines in a solar spectrum produced by constituents of the atmosphere of the earth itself rather than by gases in the outer solar atmosphere such as those responsible for the Fraunhofer lines.[NASA]

TELSET In these systems, the common

parallel interface for telemetry set-up, as from a buffered data channel.

temper 1. The relative hardness and strength of flat-rolled steel or stainless steel that cannot be further hardened by heat treatment. 2. The relative hardness and strength of nonferrous alloys, produced by mechanical or thermal treatment (or both) and characterized by a specific structure, range of mechanical properties or reduction of area during cold working. 3. In the production of casting molds, to moisten mold sand with water. 4. In the heat treatment of ferrous alloys, to reheat after hardening for the purpose of decreasing hardness and increasing toughness without undergoing a eutectoid phase change. 5. In tool steels, an imprecise shop term sometimes used to denote carbon content. 6. In glass manufacture, to anneal or toughen by heating below the softening temperature. 7. To moisten and mix clay, mortar or plaster to a consistency suitable for use. 8. A master alloy added to tin to make the finest pewter.

temperature 1. In general, the intensity of heat as measured on some definite temperature scale by means of any of various types of thermometers. In statistical mechanics, a measure of translational molecular kinetic energy (with three degrees of freedom). In thermodynamics, the integrating factor of the differential equation referred to as the first law of thermodynamics.[NASA] 2. Indication of how hot or cold a substance is.

temperature coefficient The rate of change of some physical property—electrical resistivity, for instance—with temperature; the coefficient may be constant or nearly constant, or it may vary itself with temperature.

temperature coefficient of resistivity The amount of resistance change of a material per degree of temperature rise. [ARP1931-87]

temperature compensating circuit breaker A circuit breaker in which means may be inherent or otherwise provided for partially or completely neutralizing the effect which the ambient temperature may have upon the tripping characteristics of the circuit breaker.[ARP 1199A-90]

temperature compensation Any construction or arrangement that makes a measurement device or system substantially unaffected by changes in ambient temperature.

temperature dependence The characteristic of a material which is dependent on changes in the ambient temperature.[NASA]

temperature error An error in an instrument reading caused by a difference between the ambient temperature of the instrument and some desired standard temperature.

temperature range, compensated See temperature range, operating.

temperature range of the fluid Specified range of fluid temperature that should not be exceeded for satisfactory operation of a system.[AIR1916-88]

tempering 1. Heating hardened ferrous alloys below the transformation temperature to reduce hardness and improve toughness. 2. Adding moisture to molding sand, clay, mortar or plaster. 3. Heating glass below its softening temperature.

tempering air Air at a lower temperature added to a stream of pre-heated air to modify its temperature.

tempilstick A crayon made of a material having a sharp reaction at a specific temperature; in use, a crayon sensitive to a specific temperature is used to mark the surface of a metal to be heated; confirmation that the intended temperature was reached or exceeded is indicated by a change in color of the mark.

template 1. A device used to establish outline, form or location of a part.[ARP 480A-87] 2. A guide or pattern used in laying out parts to be manufactured. 3. A guide used in drawing standard shapes on an engineering or architectural drawing.

temporal coherence The coherence of light over time. Light is temporally coherent when the phase change during an interval T remains constant regardless of when the interval is measured.

temporal distribution The statistical distribution based on time of phenomena, occurrences or events.[NASA]

temporal resolution The precision with which an optical instrument or a system differentiates between time intervals.[NASA]

temporary marking Marking which will ensure identification during ordinary handling and storage of items prior to assembly and use. Under certain service conditions these markings may exhibit the characteristics of permanent markings.[AS478G-79]

tensile specimen A bar, rod or wire of specified dimensions used in a tensile test. Also known as tensile bar; test specimen.

tensile strength 1. The resistance to lengthwise stress, measured by the greatest load in weight per unit area pulling in the direction of length, that a given substance can bear without tearing apart.[ARP1931-87] 2. The maximum force per unit of the original cross-sectional area of the sample which results in the rupture of the sample. It is calculated by dividing the maximum force in pounds by the original cross-sectional area in square inches.[AS 1198-82]

tensile stress Force per unit cross-sectional area applied to elongate a material.[ARP1931-87]

tensile test A method of determining mechanical properties of a material by loading a machined, cast or molded specimen of specified cross-sectional dimensions in uniaxial tension until it breaks; the test is used principally to determine tensile strength, yield strength, ductility and modulus of elasticity. Also known as pull test.

tensioning That operation prior to firing the catapult during which a specified tension force is applied to the bridle, pendant or launch bar by moving the catapult tow fitting forward.[AIR1489-88]

tensor fields See tensors.

tensors Arrays of functions which obey certain laws of transformation. A one row or one column tensor array is a vector.[NASA]

ten to the minus sixth Written 10^{-6} and meaning one millionth. Used in the design of a system to define the probability of failure; that is, a failure probability of one in 1,000,000 flight hours (events). The likelihood of such a failure is expressed as "improbable".[ARP4107-88]

ten to the minus ninth Written 10^{-9} and meaning one billionth. Used in the design of a system to define the probability of failure; that is, a failure probability of one in 1,000,000,000 flight hours (events). The likelihood of such a failure is expressed as "extremely improbable".[ARP4107-88]

terminal 1. A device attached to the end of a conductor to provide both electrical and mechanical connection to a post, stud, chassis or another terminal.[ARP 1931-87] [ARP914A-79] 2. The contacts of a fuse or limiter, either blade or ferrule, for connecting the current-responsive element of a fuse or limiter to the fuse clips.[ARP1199A-90] 3. An I/O device that includes a keyboard and a display mechanism; a terminal is used as the primary communication device between a computer system and a person. A terminal can be dumb with no processing capability or intelligent when some processing capability is included within the terminal. 4. The connection points where the field wiring is brought to the I/O modules.

terminal area A general term used to describe airspace in which approach control service or airport traffic control service is provided.[ARP4107-88]

terminal ballistics That branch of ballistics dealing with the motion and behavior of projectiles at the termination of their flight, or in striking and penetrating a target.[NASA]

terminal-based conformity See conformity, terminal-based.

terminal-based linearity See linearity, terminal-based.

terminal block An assembly containing connection provisions to facilitate the connection of one or more conductors. [ARP914A-79]

terminal board 1. A board fabricated from an insulating material containing a single or multiple row or arrangement of termination points for the purpose of making connections.[ARP914A-79] 2. A structural component that provides one or more electrical terminals which are electrically insulated from the chassis or mounting, and almost always from each other.

terminal control area See controlled airspace.

terminal lug See terminal.

terminal pair A set of two associated terminals, so arranged as to be accessible for connecting a pair of associated leads.

terminal plate A conductive bushing bar or commoning bar (link, jumper bar).[ARP914A-79]

terminal radar service area (TRSA) Airspace surrounding designated airports wherein ATC provides radar vectoring, sequencing, and separation on a full-time basis for all IFR and participating VFR aircraft. Service in a TRSA is called Stage III service.[ARP4107-88]

terminal strip See terminal board.

terminal style The design or configuration of a terminal. See also terminal. [ARP914A-79]

terminal velocity The maximum velocity attainable, especially by a free falling body under given conditions.[NASA]

terminal very high frequency omnidirectional range station (T-VOR) A very high frequency terminal omnirange station located on or near an airport and used as an approach aid. [ARP4107-88]

termini Hardware to physically constrain and position optical fiber at a signal-path disconnect point. The optical equivalent of an electrical pin or socket contact.[ARD50020-91]

terrain clearance indicator An instrument for measuring absolute altitude; also known as absolute altimeter.

terrain following (TF) The flight of a military aircraft maintaining a constant altitude above the terrain or the highest obstruction. The altitude of the aircraft will change with the varying terrain or obstruction(s), or both.[ARP 4107-88]

terrestrial magnetism See geomagnetism.

terrestrial planets The four small planets near the sun (Earth, Mercury, Venus, and Mars).[NASA]

tertiary air Air for combustion supplied to the furnace to supplement the primary and secondary air.

tertiary (for Class C) steering systems A tertiary NWS system is one that is used primarily for taxi-parking-catapult spotting, and is not normally required or used for take-off or landing operations. A totally passive disengaged mode is an implied requisite.[ARP1595-82]

tesla Metric unit for magnetic flux density.

test A procedure or action taken to determine under real or simulated conditions the capabilities, limitations, characteristics, effectiveness, reliability or suitability of a material, device, system or method.[ARD50010-91]

test, analyze and fix A test concept designed to provide for positive reliability growth of equipment under simulated mission environments by removing all failure mechanisms, through failure analysis and positive corrective actions, until the required reliability level is achieved.[ARD50010-91]

test-block fan requirements The operating conditions for which a fan is designed which are to be proven by test, following the procedure outline by the Test Code of the National Association of Fan Manufacturers.

test chambers Places, sections, or rooms having special characteristics where a person or object is subjected to experiment, as an altitude chamber.[NASA]

tester A portable, self-contained unit designed to test the function of a component or system. Includes connecting adapters, harnesses, hoses, etc.[ARP 480A-87]

test firing The firing of a rocket engine, either live or static, with the purpose of making controlled observations of the engine or of an engine component. [NASA]

test, follow-on operational Tests conducted on a continuing basis to ensure that the established reliability and accuracy factors are preserved during the life of the vehicle, aircraft, etc.[AIR 1916-88]

test gage A pressure gage specially built for test service or other types of work requiring a high degree of accuracy and repeatability.

testing without replacement Life test sampling without replacement is a life test procedure whereby failed units are not replaced.[ARD50010-91]

testing with replacement Life test sampling with replacement is a life test procedure whereby the life test is continued with each failed unit of product replaced by a new one, drawn at random from the same lot, as soon as the failure occurred. In the case of complex unit of product, this may be interpreted to mean replacement of the component which caused the failure by a new component drawn at random from the same lot of components. When the "sample sizes" are the same in both instances, the expected waiting time required for decision when testing with replacement is less than when testing without replacement.[ARD50010-91]

test measurement and diagnostic equipment Any system or device used to evaluate the condition of an item to identify or isolate any actual or potential failures.[ARD50010-91]

test pattern generators Image-processing software.[NASA]

test ports Calibration connection points on the manifold between the manifold block valves and the instrument.

test report A report listing physical and chemical properties of a sealant. Test reports are generated for every batch of aerospace sealant manufactured. Generally, the tests are those acceptance tests called for by a given specification. [AS7200/1-90]

tests, acceptance A test program usually imposed on production articles run prior to delivery of a component or batch of components to demonstrate that the item meets certain specification requirements.[ARP906A-91]

test sequence A series of functionally related tests in which the test operation without interruption progresses systematically from one test mode to another.[ARP1256-90]

tests, preliminary flight rating (PFRT) A series of tests conducted prior to qualification tests run on a newly developed component which will require flight demonstration. The PFRT demonstrates that the component is safe for flight.[ARP 906A-91]

tests, qualification A series of tests run on a newly developed component to demonstrate compliance with specification requirements. Also called preproduction tests or certification tests.[ARP 906A-91]

test stand(s) 1. An equipment specifically designed to provide suitable mounting, connections, and controls for testing electrical, mechanical, or hydraulic equipment as an entire system.[ARD 50010-91] 2. A cabinet-like item usually with a front work area and consisting of the necessary instrumentation and mechanical and electrical facilities for component testing.[ARP480A-87] 3. A mobile unit usually equipped with a control booth, instrumentation, engine controls, auxiliary power unit, fuel, oil, electrical systems, and the like, primarily designed for testing aircraft engines.[ARP480A-87]

test to failure Testing conducted on one or more items until a predetermined number of failures have been observed. [ARD50010-91]

tethered satellites Concept for scientific payloads suspended at altitudes of 120 Km from Space Shuttle orbiters flying at 200-Km altitude, control system would permit deployment and retrieval of the tethered satellites.[NASA]

Tethys One of the natural satellites of Saturn orbiting at a mean distance of 295,000 kilometers.[NASA]

tetraethyl orthosilicate An oxidation inhibiting coating used on the wing leading edges and nose cap of the Space Shuttle.[NASA]

tetrahedron A device normally located on uncontrolled airports and used as a landing direction indicator. The small end of a tetrahedron points in the direction of landing. At controlled airports, the tetrahedron, if installed, should be disregarded because tower instructions supersede the indicator.[ARP4107-88]

tetrahydrofuran In organic chemistry, an intermediate and a solvent for polyvinyl chloride.[NASA]

tetroons See superpressure balloons.

text In data processing, any information that has a specific meaning.

text editor In data processing, a function used to process textual instructions rather than other program language.

TF Terrain Following.

TFE See telemetry front end.

theodolite(s) 1. Optical instruments which consist of a sighting telescope, mounted so that it is free to rotate around horizontal and vertical axes, and graduated scales so that the angle of rotation may be measured. The telescope is usually fitted with a right angle prism so that the observer continues to look horizontally into the eyepiece, whatever the variation of the elevation angle. [NASA] 2. A telescopic instrument used to measure horizontal and vertical angles simultaneously.[ARP4107-88]

theoretical air The amount of air required to completely burn a given amount of a combustible material.

theoretical cutoff frequency Disregarding any dissipation effects, the characteristic frequency at which the image attenuation constant of a transducer changes from zero to a positive value, or vice versa.

theoretical draft The draft which would be available at the base of a stack if there were no friction or acceleration losses in the stack.

theoretical flame temperature See adiabatic temperature.

theoretical plate A hypothetical device for bringing two streams of material into such perfect contact that they leave the device in equilibrium with each other.

therm A unit of heat applied especially to gas. One therm equals 100,000 Btu.

thermal accommodation coefficients See accommodation coefficient.

thermal aging Exposure to a given thermal condition or a series of conditions for prescribed periods of time.[ARP 1931-87]

thermal-agitation voltage The electrical potential difference induced in circuits by the agitated motion of electrons in the circuit conductors.

thermal analysis Determining transformation temperatures and other characteristics of materials or physical systems by making detailed observations of time-temperature curves obtained during controlled heating and cooling.

thermal-arrest calorimeter A device for measuring heats of fusion in which a sample is frozen under vacuum at subzero temperatures and thermal measurements are taken as the calorimeter warms to room temperature.

thermal bulb A device for measuring temperature in which the liquid in a bulb expands and contracts with changes in temperature, causing a Bourdon-tube element to elastically deform, thereby moving a pointer in direct relation to the temperature at the bulb.

thermal comfort That condition which expresses satisfaction with the thermal environment and which is measured by such factors as air temperature, relative humidity, air velocity, etc.[NASA]

thermal compensation See compensation.

thermal conductivity Heat flow per unit cross section per unit temperature gradient.

thermal conductivity gage A device for measuring pressure in a high-vacuum system by observing changes in thermal conductivity of an electrically heated wire that is exposed to the low-pressure gas in the system.

thermal converter A device consisting of one or more thermoelectric junctions in contact with or integral with an electric heater; the output of the thermoelectric component is directly related to the current flowing in the electric heater.

thermal cutout A device for protecting a circuit or electrical device from excessive current; it consists of a heater element and a replaceable fusible link which melts and opens the circuit when too much current flows through the heater element.

thermal decomposition The braking apart of complex molecules into simpler units by the application of heat.[NASA]

thermal degradation Impairment of properties caused by exposure to heat. [NASA]

thermal delay timer A timing device relying on the movement of a heated bimetal to actuate a set of contacts.

thermal detector See bolometer.

thermal diffusion Spontaneous movement of solvent atoms or molecules to establish a concentration gradient as a direct result of the influence of a temperature gradient.

thermal efficiency See thermodynamic efficiency.

thermal electromotive force The electromotive force developed across the free ends of a bimetallic couple when heat is applied to a physical junction between the opposite ends of the couple. Also known as thermal emf.

thermal emf The electrical potential generated in a conductor or circuit due to thermal effects, usually differences in temperature between one part of the circuit and another.

thermal emission The process by which a body emits electromagnetic radiation as a consequence of its temperature only.[NASA]

thermal energy Energy that flows between bodies because of a difference in temperature.

thermal expansion A physical phenomenon whereby raising the temperature of a body causes it to change dimensions (usually increasing) in a manner characteristic of the material of construction.

thermal fatigue Failure induced by clinical varying temperature gradients that create cyclic thermal stress and strain in the material. Because the tensile and compressive stresses occur in very localized areas, their calculation requires detailed knowledge of temperature distribution.[AIR1872-88]

thermal instability The conditions of temperature gradient, thermal conductivity, and viscosity which lead to the onset of convection in a fluid.[NASA]

thermal instrument Any instrument that measures a physical quantity by relating it to the heating effect of an electric current, such as in a hot-wire instrument.

thermal neutron(s) 1. Neutrons in thermal equilibrium with a medium in which they exist.[NASA] 2. A free (uncombined) neutron having a kinetic energy approximately equivalent to the kinetic energy of its surroundings.

thermal noise The noise at radiofrequency caused by thermal agitation in a dissipative body.[NASA]

thermal pollution Environmental temperature rise due to waste heat disposal. [NASA]

thermal power plant A facility or system for converting thermal energy into electric power.

thermal printer Prints characters on paper using a high speed heating element activating chemicals in the paper to form an image.

thermal radiation 1. The electromagnetic radiation emitted by any substance as the result of the thermal excitation of its molecules. Thermal radiation ranges in wavelength from the longest infrared radiation to the shortest ultraviolet radiation.[NASA] 2. Electromagnetic radiation that transfers heat out of a heated mass. 3. Electromagnetic radiation that is absorbed by a grey or black body from a source at a higher temperature than the absorbing body.

thermal rating The maximum or minimum temperature at which a material will perform its design function without undue degradation.[ARP1931-87]

thermal shielding See heat shielding.

thermal shock The development of a steep temperature gradient and accompanying high stresses within a structure.[NASA]

thermal spraying A method of coating a substrate by introducing finely divided refractory powder or droplets of atomized metal wire into a high temperature plasma stream from a special torch, which propels the coating material against the substrate.

thermal stresses Stresses in metal, resulting from nonuniform temperature distribution.[NASA]

thermal transducer Any device which converts thermal energy into electric power or other useful measuring medium. An example is a thermocouple.

thermal-type flowmeter An apparatus in which heat is injected into a flowing fluid stream and flow rate is determined from the rate of heat dissipation; either the rise in temperature at some point downstream of the heater or the amount of thermal or electrical energy required to maintain the heater at a constant temperature is measured.

thermal-type liquid-level meter Any of several devices which indicate the position of liquid level in a vessel by means of a thermally activated property such as an abrupt change in temperature, evaporation or condensation effects, or thermal expansion effects.

thermal variable A characteristic of a material or system that depends on its thermal energy—temperature, thermal expansion, calorific value, specific heat or enthalpy, for instance.

thermionic emission Direct ejection of electrons as the result of heating the material, which raises electron energy beyond the binding energy that holds the electron to the material.[NASA]

thermionic reactors See ion engines; nuclear rocket engines.

thermionics The study of the emission

of electrons by heat.[NASA]

thermionic tube An electron tube in which at least one of its electrodes is heated to induce electron or ion emission.

thermistor(s) 1. Electron devices employing the temperature dependent change of resistivity of a semiconductor.[NASA] 2. A temperature transducer constructed from semiconductor material and for which the temperature is converted into a resistance, usually with negative slope and highly nonlinear. Its usual applications are as a nonlinear circuit element (either alone or in combination with a heater), as a temperature compensator in a measurement circuit, or as a temperature-measurement element.

thermites Fire-hazardous mixtures of ferric oxide and powdered aluminum; upon ignition with a magnesium ribbon, the mixtures reach temperatures up to 4000 degrees F (sufficient to soften steel).[NASA]

thermit welding A fusion welding process in which a mixture of finely divided iron oxide and aluminum particles is ignited, reducing the iron oxide and producing a molten ferrous alloy that is then cast in a mold built up around the joint to be welded.

thermoammeter A device used chiefly to measure radio-frequency currents where the current is run through a wire of appropriate size; on this wire is mounted a thermocouple whose output is proportional to the temperature of the wire, which is a function of the R.F. current passing through the wire. Also called a hot wire ammeter.

thermochemistry A branch of chemistry that treats the relations of heat and chemical changes.[NASA]

thermocouple(s) 1. A junction of two dissimilar metals in which the energy is changed directly into electrical energy.[ARP480A-87] 2. A junction of dissimilar alloys which produce an electrical voltage due to the temperature change.[ARP1931-87]

thermocouple cable A two-conductor cable, each conductor employing a dissimilar metal, made up specifically for temperature measurements.[ARP1931-87]

thermocouple extension wire A matched pair of wires having specific temperature-emf properties that make the pair suitable for use with a thermocouple to extend the location of its reference junction (cold junction) to some remote location; alloys for such wires are specially designed and processed to make the pair suitable for use with only one type of thermocouple.

thermocouple instrument An electrothermic instrument having a direct-current mechanism, such as a permanent-magnet moving coil, which is driven by the output of one or more thermojunctions heated directly or indirectly by electric current.

thermocouple lead wire Similar to thermocouple cable used to transmit thermocouple information to remote indicators.[ARP1931-87]

thermocouple vacuum gage A device for measuring pressures in the range of about 0.005 to 0.02 torr by means of current generated by a thermocouple welded to the midpoint of a small heating element exposed to the vacuum chamber; alternatively, the instrument current may be generated in a specially constructed thermopile that serves as both heater and thermoelectric element.

thermodynamic efficiency In thermodynamics, the ratio of the work done by a heat engine to the total heat supplied by the heat source.[NASA]

thermodynamic equilibrium A very general result from statistical mechanics which states that if a system is in equilibrium, all processes which can exchange energy must be exactly balanced by the reverse process so that there is no net exchange of energy.[NASA]

thermodynamics The study of the flow of heat.[NASA]

thermoelasticity Dependence of the stress distribution of an elastic solid on its thermal state, or of its thermal conductivity on the stress distribution.[NASA]

thermoelectric cooling A method of cooling a chamber based on the Peltier effect, in which an electric current is circulated in a thermocouple whose cold junction is coupled to the chamber; the hot junction dissipates heat to the environment. Also known as thermoelectric refrigeration.

thermoelectric heating A method of heating involving a device similar to one used for thermoelectric cooling, except that the direction of current is reversed in the circuit.

thermoelectric hygrometer A condensation-type hygrometer in which the mirror element is chilled thermoelectrically.

thermoelectric series A tabulation of metals and alloys, arranged in order according to the magnitude and sign of their characteristic thermal emf.

thermoelectric thermometer A thermometer which uses a thermocouple or thermocouple array in direct contact with the body whose temperature is to be measured, and whose reading is given in relation to the reference junction whose temperature is known or automatically compensated for.

thermoforming A method of forming sheet plastic by heating it then pulling it over a contoured mold surface.

thermograd probe An instrument for recording temperature versus depth as it is lowered to the ocean floor; it records the flow of heat through the ocean floor.

thermograph An instrument for recording air temperature. Also known as recording thermometer.

thermography 1. Technique employing heat transfer transients.[NASA] 2. Either of two methods—contact thermography or projection thermography—for measuring surface temperature using thermoluminescent materials.

thermojunction Either of the two locations where the conductors of a thermocouple are in electrical contact; one, the measuring junction, is in thermal contact with the body whose temperature is being determined, and the other, the reference junction, is generally held at some known or controlled temperature.

thermomechanical treatment Combination of material-forming processes with heat treatments in order to obtain specific material properties.[NASA]

thermomechanics See thermodynamics.

thermometer(s) 1. Devices for measuring temperature.[NASA] 2. An instrument for measuring temperature—usually involving a change in a physical property such as density or electrical resistance of a temperature-sensitive material.

thermomigration A technique for doping semiconductors in which exact amounts of known impurities are made to migrate from the cool side of a wafer of pure semiconductor material to the hotter side when the wafer is heated in an oven.[NASA]

thermophone An electroacoustic transducer that produces sound waves when a conductor whose temperature varies in response to a varying electric current causes air adjacent to the conductor to expand and contract.

thermophoresis A process in which particles migrate in a gas under the influence of forces created by a temperature gradient.[NASA]

thermophysics See thermodynamics.

thermopile(s) 1. Transducers for converting thermal energy directly into electrical energy, composed of pairs of thermocouples which are connected either in series or in parallel. Batteries of thermocouples connected in series to form single compact units.[NASA] 2. An array of thermocouples used for measuring temperature or radiant energy, or for converting radiant energy into power; the thermocouples may be connected in series to give a higher-voltage output, or in parallel to give a higher-current output.

thermoplastic A classification of resin that can be readily softened and resoftened by heating.[ARP1931-87]

thermoplastic films Materials with a linear macromolecular structure that will repeatedly soften when heated and harden when cooled.[NASA]

thermoplastic resin An organic solid that will repeatedly soften when heated and harden when cooled; examples include styrene, acrylics, polyethylene, vinyl and nylon.

thermoregulation A mechanism by which mammals and birds balance heat gain and loss in order to maintain a constant body temperature.[NASA]

thermoregulator A highly accurate or highly sensitive thermostat—for instance, a mercury-in-glass thermometer with sealed-in electrodes, which turns an electric circuit on and off as the level of mercury rises and falls past the position of the electrodes.

thermosetting A classification of resin which cures by chemical reaction when heated, and when cured cannot be softened by heating.[ARP1931-87]

thermosetting resin An organic solid that sets up (solidifies) under heat and pressure, and cannot be softened and remolded readily; examples include phenolic, epoxy, melamine and urea.

thermostatic switch An electric switch whose contacts open and close in response to the amount of heat received by conduction or convection from the device whose power is regulated by the switch.

thermotropism See anisotropy.

thickener Equipment for removing free liquid from a slurry or other liquid-solid mixture to give a solid or semisolid mass without using filtration or evaporation; usually, the process involves centrifuging or gravity settling.

thick film metallization Conductive metallization applied to an insulating substrate by either: (a) silk screening an admixture of finely divided gold powder and base metal oxides which are suspended in an organic vehicle and firing in air to produce adherence, or (b) silk screening an admixture of finely divided base metal powder and base metal oxides which are suspended in an organic vehicle, firing in a controlled atmosphere to produce adherence, and electroplating a gold overlayer.[AS1346-74]

thickness 1. A physical dimension usually considered to represent the shortest of the three principal measurement axes—in flat products such as sheet or plate, for instance, it is the distance between top and bottom surfaces measured along an axis mutually perpendicular to them. 2. The distance from the external surface of a coating to the substrate, measured along a direction perpendicular to the coating-substrate interface. 3. The short transverse dimension of rolled, drawn or extruded stock.

thickness gage A device for measuring thickness of sheet material; it may involve physical gaging, but more often involves methods such as radiation absorption or ultrasonics.

thick plates Plates of steel or other material that are over two inches thick. The exact definition of dimensions that constitute thickness varies.[NASA]

thin film metallization Conductive metallization applied to an insulating substrate by evaporating, sputtering, or plating metallic layers, or by an equivalent process.[AS1346-74]

thin-film potentiometer A potentiometer in which the conductive element is a thin film of a cermet (metal mix), conductive plastic or deposited metal; usually thin-film potentiometers are useful when a stepless output is desired.

thin-film strain gage A strain gage in which the gage is produced by depositing an insulating layer (usually a ceramic) onto the structural element then depositing a metal gage element onto the insulation layer by sputtering or vacuum deposition through a mask which defines the strain gage configuration; thin-film techniques are used almost exclusively for transducer applications; for greatest sensitivity, four bridge elements, wiring between the elements, balance components and temperature-compensation components are deposited simultaneously.

thin-layer chromatography (TLC) A form of chromatography in which a small amount of sample is placed at one end of thin sorbent layer deposited on a metal, glass or plastic plate; after washing a solvent through the sorbent bed by capillary action, the individual components of the sample may be detected by visual means, ultraviolet analysis, radiochemistry or other suitable technique, or by a combination of techniques; thin-layer chromatography is a rapid and inexpensive method for screening and selecting solvent-stationary-phase systems for liquid chromatographic analysis.

thinner(s) An organic liquid added to a mixture such as paint to reduce its viscosity and make it more free-flowing. See also solvents.

thixotropic substance A substance whose flow properties depend on both shear stress and agitation; at a given shear stress, flow increases with increasing time of agitation; when agitation stops, internal shear stress exhibits hysteresis.

thioxotropy The property of a nonsag material that permits it to be moved (stirred or extruded) with less force than would be required with a Newtonian fluid. Non-Newtonian pseudoplastic materials (such as the nonsag sealants) stand like whipped cream without seeking their own level, but flow easily from a sealant gun under relatively low pressure.[AIR4069-90]

thread 1. A continuous helical rib used to provide interconnection by twisting the ribbed member into a mating ribbed member; used extensively in pipe-and-fitting connections and in threaded fasteners. 2. A thin single-strand or twisted filament of natural or synthetic fibers, plastics, metal, glass or ceramic.

threaded coupling See coupling.

threat aircraft An intruder that has been determined by the CAS threat detecting logic to warrant either a caution or a warning alert.[ARP4153-88]

threat evolution The evolution of the potential harm of an approaching aircraft or other objects.[NASA]

three axis stabilization Maintenance of a stable platform in a desired 3-axis orientation in inertial space by utilizing gyros and accelerometers and which is independent of vehicle motion.[NASA]

three body problem That problem in classical celestial mechanics which treats the motion of a small body, usually with negligible mass, relative to and under the gravitational influence of two other finite point masses.[NASA]

three-mode controller Another name for a PID controller.

three-position controller See controller, three-position.

three-quarters hard A temper of nonferrous alloys and some ferrous alloys that corresponds approximately to a hardness and tensile strength midway between those of half hard and full hard tempers.

three term control Proportional integral

derivative control.

three-way Not recommended; see valve, selector.[ARP243B-65]

threshold 1. The increment of input required to produce a change of actuation system output. Threshold is normally specified as the input increment required to reverse the direction of output motion when the input is cycled at a rate sufficiently low to minimize dynamic effects.[AIR1916-88] 2. The beginning of that portion of the runway usable for landing. 3. An inherent perceptual limitation which requires that stimuli be presented within a certain range of intensity and duration to ensure perception by the organism. (a) subthreshold stimuli: Stimuli that are presented below a detectable range or duration and thus are not perceived. (b) suprathreshold stimuli: Stimuli that are presented above a detectable range or duration and thus are perceived. (c) transthreshold stimuli: Stimuli that are presented within a detectable range and duration and thus are perceived.[ARP4107-88]

threshold crossing height (TCH) The theoretical height above the runway threshold at which the aircraft's glide slope antenna would be if the aircraft maintains the trajectory established by the mean ILS glidepath.[ARP4107-88]

threshold sensitivity 1. Of a transducer, the lowest level of the input signal which produces desired response at the output.[AIR1489-88] 2. The lowest value of a measured quantity that a given instrument or controller responds to effectively.

threshold shift See thresholds.

threshold voltage The threshold energy necessary to remove an electron from the bound position to the conduction band in solid state devices.[NASA]

throat 1. The narrowest portion of a constricted duct, as in a diffuser, or a venturi tube. 2. The shortest distance from the root of a fillet weld to its face. 3. The distance in a machine such as a resistance spot welder or arbor press from the centerline of the electrodes or punch to the nearest point of interference with the frame or other machine component; it establishes the maximum distance from the edge of flat work that the machine can perform an operation.

throat-injection amplifier An amplifier which utilizes auxiliary flow at a nozzle throat for a control signal to modulate the output flow. Pressure level of the control signal may either be above or below local throat pressure to result in a positive or negative (suction) quiescent control flow.[ARP993A-69]

throttle burst Instantaneous advancement of the throttle from any given setting to the full-open position.[ARP4107-88]

throttle chop Instantaneous retarding of the throttle from any given position to the full-closed position.[ARP4107-88]

throttle control A function or mode of operation which provides power or throttled control in response to pitch attitude, glide slope, or airspeed signals.[ARP419-57]

throttle valve A device for regulating flow of a fluid by alternatively opening up or closing down a restriction in a passage or inlet.

throttling calorimeter An instrument that determines the moisture content of steam by admitting steam to a well-insulated expansion chamber through an orifice and then measuring steam temperature; moisture content is found by referring to steam tables.

through connection See feed-thru.

throughput rate The net rate at which data can be received by a device, manipulated as specified, and output to some other specified device.

through stay A brace used in fire-tube boilers between the heads or tube sheets.

throwing power The relative ability of an electroplating solution or electrophoretic paint to cover irregularly shaped parts with a uniform coating. Contrast with covering power.

throwout The device for disengaging the driving and driven plates of a motor vehicle clutch.

thrust 1. The forward-directed force (positive thrust) developed by a jet engine or rocket, or by the rotation of a propeller. See also reverse thrust.[ARP4107-88] 2. Weight or pressure applied to a drill bit or other tool to make it cut. 3. Generically, the force any body exerts on another body—both can be stationary, both can be in motion, or one can be stationary and the other in motion.

thrust augmentation The increasing of the thrust of an engine or power plant, especially of a jet engine and usually for a short period of time, over the thrust normally developed.[NASA]

thrust bearing A bearing that supports axial load on a shaft and prevents the shaft from moving in an axial direction.

thrust distribution The location of areas of upward thrust (lift) on wings, airfoils, etc.[NASA]

thrust faults See geological faults.

thrust power See thrust.

thryatron An electronic switching tube filled with gas. Application of a voltage to the control grid turns on the current. Normally used for high voltage switches.

thumbscrew A threaded fastener with a head that is flattened along the axis of the threads so that it can be gripped between the thumb and forefinger for tightening or loosening.

thumbwheel A multiple position switch driven by a notched disc that can be rotated by thumb action. Frequently used to enter numerical information into a PC from a machine location.

thyratron A hot-cathode gas tube having one or more control electrodes which initiate anode current, but do not limit it except under specific operating conditions.

tidal oscillation See tides.

tide gage Any of several types of devices, ranging from a simple graduated rod observed visually to an elaborate recording instrument, used for determining the rise and fall of tides.

tide indicator The part of the tide gage—at water level or remote—that indicates the current height of a tide.

tides The periodic rising and falling of the earth's oceans and atmosphere. It results from the gravitational forces of the moon and sun acting upon the rotating earth. The disturbance actually propagates as a wave through the atmosphere and along the surface of the waters of the earth. Atmospheric tides are always so designated, whereas the term tide alone commonly implies the oceanic variety.[NASA]

tie down 1. The process of bridling or mooring an aircraft to the ground or deck to protect it from movement due to winds or deck roll. 2. An aircraft fitting or device to which the chain or cable is attached. 3. The chain, cable, or strap used to secure the aircraft (ground support equipment). See also mooring. [AIR1489-88]

tie plate A plate, through which a bolt or tie rod is passed, to hold brick in place.

tie rod 1. An item, long in relation to its cross-section, that connects and maintains relative position between two parts.[ARP480A-87] 2. A structural brace designed to bear tensile loads. 3. A part of the linkage in an automotive steering gear that connects the steering mechanism to the wheel supports and transmits steering forces to the wheels. 4. A rodlike member that connects mechanical or structural members of a machine together. 5. A tension member between buckstays or tie plates.

tiger start A type of start where the power control is set at part power or above at fuel manifold pressurization instead of at the usual idle detent.

[ARP906A-91]

tight 1. Inadequate clearance, or barest minimum clearance between moving parts. 2. In a pressurized or vacuum system, freedom from leaks. 3. A class of fit having slight negative allowance, which requires light to moderate levels of force to assemble mating parts together.

TIG welding See gas tungsten-arc welding.

tile A preformed refractory, usually applied to shapes other than standard brick.

tile baffle A baffle formed of preformed refractory shapes.

tilt See attitude (inclination).

tilting See attitude (inclination).

tiltmeter Instrument used to measure small changes in the tilt of the earth's surface usually in relation to a liquid-level surface or to the rest position of a pendulum.[NASA]

tilt rotor aircraft A type of convertible aircraft which takes off, hovers, and lands as a helicopter but is converted into a fixed wing aircraft by the 90-degree tilting of its rotor or rotors for use as a propeller for forward flight. [NASA]

tilt-switch level detector A relatively simple device for detecting high level in a bulk solids container by means of a free-hanging sensor that produces a switch action when the rising level of bulk material tilts the sensor from its normal vertical position.

TIMDATA In these systems, the common parallel interface for time data, as from a time code generator/translator to a buffered data channel.

time 1. Any appropriate unit of life measurement such as hours, cycles, landings, etc.[ARD50010-91] [AIR1916-88] 2. The hour of the day reckoned by the position of a celestial reference point relative to a reference celestial median. [NASA] 3. A fundamental measurement whose value indicates the magnitude of an interval between successive events.

time, active maintenance The time during which prevention and corrective maintenance work is actually being done on the item.[ARD50010-91]

time adjustment and calibration The time for recalibration returning, etc., when adjustments are necessary, either to compensate for performance degradation or to compensate for differences between the operating characteristics of the replacement item and those of the original item.[ARD50010-91]

time, administrative That element of delay time not included in supply delay time.[ARD50010-91]

time, awaiting maintenance That time during which a vehicle is not operationally ready because of maintenance (NORM) and no maintenance work is as yet being performed on either the vehicle or its related equipment.[AIR 1916-88]

time base In PC timer programming, refers to the basic increment of time used in the timer operation.

time base error In instrumentation tape recording and playback, the data error which results from a difference between tape recording speed and tape playback speed.

time between overhaul The maximum time that an item is permitted to operate between overhauls.[ARD50010-91]

time, block Block time is the time an aircraft is underway. It covers the time from pulling the chocks at the parked position until parked in the destination. Block time is entered on the aircraft log sheets and is used in computing airline statistical information. It includes taxi time, waiting on taxiways and at the end of runways, takeoff run, flight time, landing roll, and taxi-in. It excludes running of engines or systems, or both, while aircraft are parked or for strictly maintenance check/test purposes.

[AIR1916-88]

time, checkout That element of maintenance time during which performance of an item is verified to be a specified condition.[ARD50010-91]

time, cleanup The portion of total maintenance time following reassembly and checkout of an item in which tools, test equipment, and material not required for operation are removed from the equipment operating area.[ARD50010-91]

time code A serial BCD code, superimposed on a carrier so that it can be recorded on instrumentation tape, to annotate the time of day at which all data were recorded.

time code translator A hardware device to accept the serial time code (as from a separate track of an instrumentation tape), recognize synchronization, and prepare the time of day in parallel format for computer entry.

time constant Generally, the time required for an instrument to indicate a given percentage of the final reading result from an input signal; the relaxation time of an instrument. In the case of instruments such as thermometers, whose response to step changes in an applied signal is exponential in character, the time constant is equal to the time required for the instrument to indicate 63.2% of the total change, that is, when the transient error is reduced to 1/e of the original signal change. Also called lag coefficient.[AIR1489-88]

time, correction See time, settling.

time critical warning A warning condition in which the time to respond is extremely limited and the response to the alert is the most important action that the crew can take at that specific time (for example, ground proximity, windshear, collision avoidance).[ARP 4153-88]

time-delay A qualifying term indicating that there is purposely introduced a delayed action, which delay decreases as the magnitude of the current increases.[ARP1199A-90] See also time lag.

time-delay fuse A fuse that has its total clearing time deliberately delayed in the overload current range.[ARP 1199A-90]

time division multiple access Radio transmission method in which each station of a satellite communication network is assigned a time schedule for transmission (in lieu of frequency division); a multi-element antenna with an adaptive null steering array eliminates interference.[NASA]

time division multiplex (TDM) A system for the transmission of information about two or more quantities (measurands) over a common channel, by dividing available time intervals among the measurands to form a composite pulse train; information may be transmitted by variation of pulse duration, pulse amplitude, pulse position, or by a pulse code. Abbreviations of the codes used are PDM, PAM, PPM, and PCM.

time division multiplex access (TDMA) A type of protocol for access to communication system network.

time-division multiplexing 1. A system for the transmission of information about two or more quantities (measurands) over a common channel by dividing available time intervals among the measurands to form a composite pulse train.[NASA] 2. A digital technique for combining two or more signals into a single stream of data by interleaving bits from each signal. Bit one might be from signal one, bit two from signal two, etc.

time, down That element of active time during which an item is not in condition to perform its required function. Reduces availability and dependability. Synonymous with downtime.[ARD 50010-91]

time error The difference between the current ETA and the required time of arrival (RTA) at any designated downstream waypoint or waypoints.[ARP 1570-86]

time, fault correction Time required to perform corrective maintenance on a failed item (in place and on the equipment) or to replace the failed item with a replacement item drawn from storage. [ARD50010-91]

time, fault detection Time between the occurrence of a failure and the point at which it is recognized that the system or equipment does not respond to operational demand during the mission sequence.[ARD50010-91]

time, fault location The portion of active repair time required to test and analyze and isolate an equipment malfunction.[ARD50010-91]

time, final test The portion of active repair time required after completion of maintenance, adjustments, and calibration to verify by measurement that equipment performance is within specified or otherwise previously established limits.[ARD50010-91]

time, flight The airborne time elapsed from "wheels off" at takeoff to "touchdown" at landing.[AIR1916-88]

time gate A transducer that gives an output signal only during specific time intervals.

time, initial provisioning lead The maximum number of calendar days quoted by the supplier to cover the period of time from receipt of a customer's initial provisioning order to shipment. [ARD50010-91]

time-inverse A time-current relationship where the protective device opening time decreases as the current increases.[ARP1199A-90]

time lag The total time between the application of a signal to a measuring instrument and the full indication of that signal within the uncertainty of the instrument.[AIR1489-88]

time, logistics delay The portion of downtime during which repair is delayed solely because of the necessity to wait for a replacement part or other subdivisions of the system.[AIR1916-88]

time, maintenance An element of down time which excludes modification and delay time.[ARD50010-91]

time marching Techniques for solving a problem with partial differential equations that have a time derivation.[NASA]

time, mission The time period from "pulling the chocks" at the parked departure position to the end of the landing roll, with the aircraft safely on the ground. Taxi-in time is excluded.[AIR 1916-88]

time, operating The time period during which the equipment is performing its intended function.[AIR1916-88]

time, proportioning control See control, time proportioning.

timer 1. A device that automatically starts or stops a machine function, or series of functions, depending on either time of the day or elapsed time from an arbitrary starting point. 2. An instrument that measures elapsed time from some arbitrary starting point. 3. A device that fires the ignition spark in an internal combustion engine at a preset point in the engine cycle. 4. A device that opens or closes a set of contacts, and automatically returns them to their original position after a preset time interval has elapsed. Also known as interval timer. 5. A device providing the system with the ability to read elapsed time in splitsecond increments and to inform the system when a specified period of time has passed.

time, reaction That element of uptime needed to initiate a mission, measured from the time command is received. [ARD50010-91]

time response The variation of an output variable of an element or a system, produced by a specified variation of one of the input variables. See also response, time.

timers Hardware devices or PC program elements that can produce control actions based on elapsed time.

time-schedule controller A controller in which the setpoint (or reference input signal) automatically adheres to a predetermined time schedule. See also, controller, time schedule.

time series The discrete or continuous sequence of quantitative data assigned to specific moments in time, usually studied with respect to their distribution in time.

time sharing 1. A computer system in which CPU time and system resources are shared with a number of tasks or jobs under the direction of a scheduling formula or plan. 2. The use of a device for two or more purposes during the same overall time interval, accomplished by interspersing component actions in time. 3. Participation in available computer time by multiple users, via terminals. Characteristically, the response time is such that the computer seems dedicated to each user.

time signals Accurate signals marking specified times or time intervals. They are used primarily for determining errors of timepieces. Such signals are usually sent from an observatory by radio or telegraph.[NASA]

time slice The time allocated by the operating system for processing a particular program.

time slicing A method of scheduling programs in a multiprogramming environment in which specific fixed time periods or "time slices" are assigned to each task in a cyclic fashion.

time, supply delay That element of delay time which a needed replacement item is being obtained.[ARD50010-91]

time, task elapsed The calendar time from the commencement to the completion of a defined task.[ARD50010-91]

time, testing The time required to determine whether designated characteristics of a system are within specified values.[AIR1916-88]

time to go (TTG) The time in min (or min:sec) to fly the great circle course from present position to the next waypoint or the cumulative time between present position and a designated downstream waypoint or waypoints.[ARP 1570-86]

time, up That element of active time during which an item is in condition to perform its required functions. Increases availability and dependability. Synonymous with uptime.[ARD50010-91]

tinning 1. Coating with a thin layer of molten solder or tin to prevent corrosion or prepare a connection for soldering. 2. Covering or preserving a metal surface with tin. 3. A protective surface layer of tin or solder.

tipback angle The angle between a vertical line through the center of gravity (CG) of the aircraft (normal to the static ground line) and a line which passes through the CG and the ground contact point of the main landing gear of a tricycle gear aircraft. The angle to which an aircraft may rotate when the CG is vertically above the pivot point of the main landing gear and tipback is impending. The pivot point is the axle on a single axle configuration, or the bogie pivot axis on a multi-axle configuration. Generally considered to be the minimum angle achievable as determined by the combination of possible CG locations and landing gear positions corresponding to those locations.[AIR 1489-88]

tip path/tip path plane That circular plane described by the rotating tip (the furthermost rotating part from the axis of rotation) of a propeller or rotor.[ARP 4107-88]

tip speed The (rotational) velocity of that part of a propeller or rotor blade nearest the perimeter of its rotation. (Tip speed in miles per hour = 2 Pi x 1.894 x 10^{-4} r x RPH, when r = the length of the lade in feet from the center of rotation to the tip and RPH is rotational speed of the propeller or rotor in revolutions per hour.)[ARP4107-88]

tip vanes Wing mounted rotor tips with their spans oriented approximately parallel to the local free stream to increase the capture area and power output of the rotor.[NASA]

tip vortex A vortex flowing from the tip of a wing, owing to the flow of air around the tip from the high-pressure region below the surface to the low-pressure region above it.[ARP4107-88]

tire 1. A solid or air filled covering for a wheel, typically of rubber and/or fabric, fitted around the wheel's rim to absorb shock and provide traction. 2. A hoop of iron or heavy rubber fitted about the rim of a wheel.[AIR1489-88]

tire, anti-shimmy A tire with a channel tread cross section which provides contact points with the ground in two separate areas outboard (both sides) from the plane of symmetry of the tire. Often used for tail wheel tires or other applications to resist a tendency to shimmy.[AIR1489-88]

tire aspect ratio The ratio of the section height (outside radius minus bead seat radius) to the section width. See also aspect ratio.[AIR1489-88]

tire blister A thin swelling of the tread or outer ply of a tire due to air entrapped between plies of the carcass.[AIR1489-88]

tire bead See bead.

tire centrifugal growth See centrifugal growth.

tire, chafer See chafer strip.

tire, chevron cuts A pattern of cuts in a tire tread resembling chevrons or "Vs" due to a tearing of the surface rubber under traction.[AIR1489-88]

tire, chine A tire whose cross section has a flare near the shoulder similar to the chine of a boat for the purpose of modifying the normal spray or splash pattern as the tire runs through water. For engine inlet protection or other purposes.[AIR1489-88]

tire contact area See tire footprint.

tire cornering force See cornering force.

tire cornering power The power developed by a tire in producing turning capability in an aircraft. Resistance to lateral movement in a rolling tire.[AIR 1489-88]

tire, cut protected A tire with provisions for protection from cutting. May consist of inclusions (wire, fabric, etc.) in the tread or other design features.[AIR 1489-88]

tire-cut resistant A tire which has provisions for cut resistance. See also tire, cut protected.[AIR1489-88]

tire dynamic rating The load for which a tire is rated for dynamic loads such as load on a nose wheel tire induced by aircraft braking or a dynamic landing load (as opposed to static load).[AIR 1489-88]

tire, expandable A tire which is designed to be expanded or inflated after the landing gear is extended and prior to landing. After takeoff and prior to retraction into the wheel well, the tire is deflated and contracts to a smaller or natural state in order to require less stowage volume in the aircraft. See also, expandable tire.[AIR1489-88]

tire, fabric reinforced tread A tire whose tread is reinforced with layers of fabric to provide greater strength and resistance to centrifugal force at high rotational speeds.[AIR1489-88]

tire footprint, gross The total area within the boundary of a tire footprint. See also contact area.[AIR1489-88]

tire footprint, net The total area within the boundary of a tire footprint less the ineffective area which might be a result of tread pattern, pressure inefficiency in a low aspect ratio tire, etc.[AIR 1489-88]

tire, groove cracking The tendency of a tire to crack at the juncture of tread ribs to the basic carcass, or in the grooves of the tread.[AIR1489-88]

tire growth Increase in size of a tire for a number of reasons including: a) pressure growth, b) age growth, c) centrifugal growth etc.[AIR1489-88]

tire, ice grip A tire configuration which includes special provisions for service on ice such as steel wire in the tread to provide traction or grip on ice.[AIR1489-88]

tire loaded radius Generally given at the rated load of the tire. The normal dimension from the axle centerline to the surface against which the tire is loaded. Generally considered as a static condition as opposed to rolling radius which is a dynamic condition.[AIR1489-88]

tire ply One of the layers of fabric in the tire carcass.[AIR1489-88]

tire ply rating A number which represents a strength rating of the tire. It may or may not be the same as the number of plies or fabric layers of the carcass.[AIR1489-88]

tire, pneumatic A tire which contains a gas (i.e., air or dry nitrogen) as a shock absorbing and load carrying medium. [AIR1489-88]

tire rated pressure The service pressure at which a tire is rated to carry a given load at a specified deflection.[AIR1489-88]

tire rib undercutting The tendency of a tire to develop cuts at the base of a rib type tread due to bending of the tread from side loads and thus to undercut the tread.[AIR1489-88]

tire rolling and loaded radius The tire loaded radius is the instantaneous distance from the axle center to the tire-ground interface. This distance should be used in calculation of drag forces. In calculating the airplane synchronous velocity, however, use of rolling radius is recommended.[AS483A-88]

tire rolling radius The normal dimension from the axle centerline to the surface on which the tire is rolling. A dynamic condition as opposed to loaded radius which represents a static condition.[AIR1489-88]

tire rolling resistance The resistance which a tire exhibits to pure rolling movement. Generally a function of the air pressure, the radial load, the tire construction, and other factors.[AIR 1489-88]

tire self aligning torque See aligning torque.

tire sipes See sipes.

tire skid depth See skid depth.

tire slip In a braked wheel, the difference between actual surface speed of an unbraked or free rolling wheel.[AIR 1489-88]

tire slip angle The difference in angle between actual steered wheel angle and the direction which the wheel is traveling on the surface on which it is rolling. [AIR1489-88]

tire slip ratio In a braked wheel, the ratio of the difference in actual surface speed and the surface speed of a free rolling wheel to the actual surface speed of the free rolling wheel.[AIR1489-88]

tire, solid rubber A tire configuration which utilizes a solid rubber cross section. Sometimes used as tailwheels for light aircraft.[AIR1489-88]

tire standing wave See standing wave.

tire static rating The load for which a tire is rated for a static load. Generally given at a specified service pressure and deflection.[AIR1489-88]

tire to runway coefficient The coefficient of friction which a tire will develop against the runway surface. A function of tire design and construction, runway construction, climatic conditions, etc. [AIR1489-88]

tire traction force 1. The tractive force which a tire will develop. 2. Adhesive friction of a wheel (tire) to accelerate a load or provide braking.[AIR1489-88]

tire tread The outside diameter layer of a tire which contacts the surface on which a tire rolls and provides a specified depth for wear in service.[AIR 1489-88]

tire tread chunking A throwing off of the tread layer or sections of the tread of a tire due to abuse, improper construction, or high centrifugal forces in high speed operation.[AIR1489-88]

tire, tube type A tire which is designed to be used with an inner tube within for the purpose of air retention as opposed to a tubeless tire which is designed to operate without a tube.[AIR1489-88]

tire, tubeless A tire which is designed to retain air without use of an inner tube. Must be used with a wheel also designed to retain air in the rim section.[AIR1489-88]

tire, twin contact See tire, anti-shimmy.

tire undertread The layer of the tire carcass directly under the tread. It may contain shredded wire, fabric, or other construction for reinforcement.[AIR 1489-88]

tire vent hole Small holes, usually pierced, in the outer side walls of a tire (part way through only) for the purpose of relieving trapped air within the carcass plies and for prevention of blisters. [AIR1489-88]

tire wear indicator A means of indication of the amount of wear which a tire has used (or how much remains). May be in the form of colored fabric plies in the tread or carcass, molded dimples, or recesses in the tread, etc. [AIR1489-88]

Titan Centaur launch vehicle A Titan III rocket augmented with a Centaur rocket for launching spacecraft requiring high-velocity escape trajectories. [NASA]

titration curve A plot with pH as the ordinate and units of reagent added per unit of sample as the abscissa.

TLC See thin-layer chromatography.

T network A network consisting of three branches, one terminal of each branch being connected at a common node and the remaining terminals being connected, respectively, to an input junction, an output junction and a common input and output junction.

toe 1. The junction between the face of a weld and the adjacent base metal. 2. The portion of the base of a dam, earthwork or retaining wall opposite to the retained material.

toe crack A crack in the weldment that runs into the base metal from the toe of a weld.

to-from indicator See sense indicator.

toggle 1. A flip-flop. 2. Pertaining to a manually operated on-off switch, i.e., a two-position switch. 3. Pertaining to flip-flop, see-saw, or bistable action. 4. A pinned lever that can be used to amplify forces.

toggle circuit breaker A circuit breaker which has a toggle actuating means. [ARP1199A-90]

toggle switch A manually operated electric switch with a small projecting knob or arm that may be placed in either of two positions, "on" or "off," and will remain in that position until changed. 2. An electronically operated circuit that holds either of two states until changed.

tokamak devices Experimental toroidal magnetic confinement devices where toroidal current runs through the plasma in order to produce fusion reactor like plasma conditions. The name is a Russian acronym for toroidal magnetic current.[NASA]

token A token represents the right to use the network medium.

token bus An access procedure where the right to transmit is passed from device to device via a logical ring on a physical bus.

token passing Networking procedure in

which access to the bus for transmission is conditional on possession of a circulating token signal.

token ring An access procedure where the right to transmit is passed from device to device around the physical ring.

tol See tolerance.

tolerance (tol) 1. Permissible variation in the dimension of a part. 2. Permissible deviation from a specified value; may be expressed in measurement units or percent.

tolerance limits The extreme upper and lower boundaries of a specified range; it is computed from the nominal value and its tolerance.

tomography Technique of making radiographs of plane sections of a body or an object; its purpose is to show detail in a predetermined plane of the body, while blurring the images of structures in other planes.[NASA]

ton 1. A weight measurement equal to 2,000 lb (avoirdupois), short ton; 2,240 lb (avoirdupois), long ton; or 1,000 kg, metric ton. 2. A unit volume of sea freight equal to 40 cu ft. 3. A unit of measurement for refrigerating capacity equal to 200 Btu/min, or about 3517 W; derived from the capacity equal to the rate of heat extraction needed to produce a short ton of ice having a latent heat of fusion of 144 Btu/lb from water at the same temperature in 24 hr.

tong hold The end of a forging billet where the forger grips it with his tongs; the end is cropped off and scrapped after the operation is completed.

tongs Any of various tools designed to grip, hold or manipulate hot metal or other solid materials; they consist generally of two bars connected by a pivot joint near one end, and are used much like a large pliers.

tongue The protrusion, usually flat in configuration, of a terminal which is designed to be fastened to a stud, terminal block or chassis, or inserted in a receptacle.[ARP914A-79]

tool 1. Any device used to assist man in his work. 2. In manufacturing, any hand-held or machine-operated device for shaping material into a finished product—includes cutting, piercing or hammering devices; jigs and fixtures; dies and molds; shaped or flat rolls; and abrasives. 3. To equip a factory for production, or to design and build special devices needed to manufacture a specific product or model.

tool bit A piece of hardened metal with a sharpened edge or point used in a metal-cutting operation.

tooling a fillet After application of a fillet, the bead is shaped (tooled) to a feathered edge where it meets the substrate. The goal is to ensure good surface contact at the feathered edges, to eliminate voids, trapped air, and reentrant edges, and to produce a contour of the correct thickness and shape over the area being sealed.[AIR4069-90]

tool marks Marks on the surface of the nut resulting from machining, trimming, or forming of the locking feature.[MA 1568-87]

tool post A device attached to the tool slide on a lathe or similar machine tool clamping and positioning a tool holder.

tool steel Any of various steel compositions containing sufficient carbon and alloying elements to permit hardening to a level suitable for use in cutting tools, dies, molds, shear blades, metalforming rolls and other tooling applications.

tooth 1. One of the shaped projections on the rim or face of a gear. 2. A projection on a tool such as a comb, rake or saw.

tooth pain See trapped gas effects.

TOP Technical and Office Protocols.

top coat A material applied as a thin coating over the surface of applied sealant to protect it from the possible deleterious effects of fuel.[AIR4069-90]

top dead center The position of a piston and its connecting rod when at the extreme outer end of its stroke.

top of descent (TOD) The point at which the main descent from enroute cruise is initiated.[ARP1570-86]

topology The logical interconnection between devices. Local area networks typically use either a broadcast topology (bus) in which all stations receive all messages, or a sequential topology (ring) where each station receives messages from the station before them and transmits (repeats) messages to the station after them. Wide area networks typically use a mesh topology where each station is connected to one or more other stations and acts as a bridge to pass messages through the network.

topworks A nonstandard term for actuator.

torch A device used to control and direct a gas flame, such as in welding, brazing, flame cutting or surface heat treatment.

torching 1. The burning of fuel at the end of an exhaust pipe or stack of a reciprocating aircraft engine, the result of an excessive richness in the fuel-air mixture.[ARP4107-88] 2. The rapid burning of combustible material deposited on or near boiler-unit heating surfaces.

torn sleeve A firesleeve which evidences surface tear, or which has been stretched in two or fissured to the degree that it will no longer provide adequate fire protection to the hose assembly.[ARP 1658A-86]

toroidal wheels Doughnut-shaped wheels designed particularly for vehicles used in soft, granular soil (planetary surfaces).[NASA]

torque 1. About an axis, the product of a force and the distance of its line of action from the axis. Steering torque, for example, would be the torque available to rotate the steered wheel(s) about the spindle or steering axis.[AIR1489-88] 2. A rotary force, such as that applied by a rotating shaft at any point on its axis of rotation.

torque, accelerating The net torque (algebraic summation of applied torque, accessory torque and engine starting torque) which produces acceleration of the combined engine, accessory and starter, as typified by the starting cycle. [ARP906A-91]

torque, accessory The torque attributable to the accessories driven by the engine which must be overcome during the start cycle.[ARP906A-91]

torque, aerodynamic The portion of the engine drag torque which results from the engine compressor pumping.[ARP 906A-91]

torque amplifier A two-shaft device that supplies power to rotate the output shaft, maintaining corresponding position between output and input shafts, but without imposing any additional torque on the input shaft.

torque, applied Starter output torque applied to the engine.[ARP906A-91]

torque arms A contrivance consisting of two (usually metal) arms hinged to one another at their ends (appex), the other ends being hinged to two separate parts, between which rotation would be possible except for the constraint provided by the torque arms. Reciprocation of the two separate parts is permitted by virtue of the hinges of the torque arms.[AIR1489-88]

torque, breakaway Torque required to induce engine rotor rotation.[ARP906A-91]

torque-coil magnetometer An instrument for measuring properties of a magnetic field whose output is related to the torque developed in a coil that can turn within the field being measured.

torque converter 1. A hydraulic device consisting of a pump runner and turbine in a close coupled package. It is capable of torque multiplication and can be utilized in a turbine engine starting system in combination with an APU. [ARP906A-91] 2. Device for changing the torque speed or mechanical advantage between an input shaft and an output shaft.[NASA] 3. Any of several mechanisms designed to change or vary the torque, speed or mechanical advantage between an input shaft and an output shaft.

torque efficiency Torque efficiency is the ratio of measured brake torque integral to the integral of straight lines between peak torques taken from torque-time history instrumentation.[AIR1489-88]

torque, engine drag The sum total of torque loads developed by the engine that resist engine rotation, and whose absolute value is considered to be negative.[ARP906A-91]

torque, engine starting The sum total of torque developed by the engine during the start cycle. Its absolute value is negative when resisting acceleration and positive when assisting acceleration. [ARP906A-91]

torque error See mounting error.

torque, friction That portion of engine drag starting torque required to overcome the friction of mechanical and/or fluid components.[ARP906A-91]

torque, impact The highest transient torque which may occur when the starter first engages the engine during a start or during re-engagement, and which exceeds that torque nominally generated by the starter.[ARP906A-91]

torque, input See torque, applied.

torque link See torque arms.

torque-to-inertia or force-to-mass ratio The ratio of actuator shall load capability to rotary actuator inertia or linear actuator mass. This ratio determines the maximum actuator output acceleration.[AIR1916-88]

torque, net resisting See torque, engine drag.

torque, output See torque, applied.

torque, rated Starter torque developed at rated input and output conditions. [ARP906A-91]

torque, resisting See torque, engine drag.

torque, stall that torque developed by the starter at zero output speed.[ARP 906A-91]

torque, starter The torque produced by the starter during the start cycle. It is a function of starter, speed, and starter power supply.[ARP906A-91]

torque, steady state That torque developed by the starter or engine while operating with its output held to a constant speed.[ARP906A-91]

torque, transient That torque developed by the starter or engine while operating with its drive accelerating.[ARP 906A-91]

torque-tube flowmeter A device for measuring liquid flow through a pipe in which differential pressure due to the flow operates a bellows, whose motion is transmitted to a recorder arm by means of a flexible torque tube.

torque-type viscometer An instrument that can measure viscosity of Newtonian fluids, non-Newtonian fluids, and suspensions by determining the torque needed to rotate a vertical paddle or cylinder submerged in the fluid.

torque, valve The maximum actuating moment required at a given fluid pressure (usually rated pressure) to move the valve mechanism from one position to another.[ARP243B-65]

torque wrench 1. A hand or power tool that can be adjusted to deliver a preset rotary force to a nut or bolt. 2. A wrench that can measure the torque required to start rotary motion when tightening or loosening a bolt.

torr Also spelled tor. A unit of pressure equal to the pressure exerted by a column of mercury 1 mm high at 0 °C.

torsiometer An instrument consisting of angular scales mounted around a rotating shaft to determine the amount of twist in the loaded shaft, and thereby determine the power transmitted. Also known as torsionmeter.

torsional elastic moment The total torque moment which a tire is capable of producing through the footprint in contact with a surface.[AIR1489-88]

torsional spring rate The stiffness of a mechanical drive train expressed in terms of torque per angular displacement.[ARP906A-91]

torsion balance An instrument for measuring minute magnetic, electrostatic or gravitational forces by means of the rotational deflection of a horizontal bar suspended on a torsion wire whose other end is fixed.

torsion bar A type of spring that flexes by twisting about its axis rather than by bending.

torsion galvanometer A galvanometer whose reading is determined by the angle through which the moving system must be rotated to bring it to its zero position while under the influence of a specific force between the fixed and moving systems.

torsion hygrometer An instrument for measuring humidity in which a substance sensitive to humidity is twisted or spiraled under tension in such a manner that changes in length of the sensitive element will rotate a pointer in direct relation to atmospheric humidity.

torso restraint systems Consists of any strap, webbing, or similar device designed to secure a person in an aircraft with the intention of minimizing injury, including all buckles or other fasteners, and all integral hardware.[AS8043-86]

total absorption spectrometer An instrument that measures the total amount of x-rays absorbed by a sample and compares it to the amount absorbed by a reference sample; the sample may be solid, liquid or gas.

total accuracy See accuracy, total.

total adjusted error The maximum output deviation from the ideal expected values. Expressed as LSBs or percent of full-scale range at a fixed reference voltage.

total air The total quantity of air supplied to the fuel and products of combustion. Percent total air is the ratio of total air to theoretical air, expressed as percent.

total clearing time The total time measured from the beginning of the specified overcurrent condition until the interruption of the circuit. The total clearing time for a fuse or limiter is equal to the sum of the melting time and the arcing time.[ARP1199A-90]

total emissivity The ratio of the total amount of thermal radiation emitted by a non-blackbody to the total amount emitted by a blackbody at the same temperature.

total energy systems Energy systems which supply both electrical and heat requirements.[NASA]

total error band See error band.

total emissivity The ratio of the total amount of thermal radiation emitted by a non-blackbody to the total amount emitted by a blackbody at the same temperature.

total energy systems Energy systems which supply both electrical and heat requirements.[NASA]

total error band See error band.

total harmonic content The total harmonic content of a complex wave is the total rms value remaining when the fundamental component is removed. [ARP1212-71]

total hydrocarbons The total of hydrocarbon compounds of all classes and molecular weights.[ARP1256-90] [AIR 1533-91]

total mission scenario(s) All portions of all missions or sorties that the aircraft will accomplish. Typically many portions are repetitious, such as takeoff, climb, and cruise segments accomplished under similar environmental conditions. All mission scenarios are designed to assure that all aircraft capabilities are defined.[ARP4107-88]

total pressure The sum of the static and dynamic pressures at a location.[AIR 1916-88]

total radiation pyrometer See wide-band radiation thermometer.

total range The portion of a system of units that is between an instrument's upper and lower scale limits, and therefore defines the values of the measured quantity that can be indicated or recorded.

total solids concentration The weight of dissolved and suspended impurities in a unit weight of boiler water, usually expressed in ppm.

touch and go/touch and go landing A landing in which the aircraft touches down on the runway but does not come to a stop before taking off again.[AIR 1489-88]

touchdown The first point of contact of an air vehicle with the landing surface in the landing process.[AIR1489-88]

touchdown protection A skid control system feature which provides a full brake pressure release upon extending the landing gear prior to arming the system at touchdown.[AIR1489-88]

touchdown RVR See runway visual range.

touchdown zone The first 3000 ft of the runway beginning at the threshold. The area is used for determination of the touchdown zone elevation (TDZE) in the development of straight-in landing minimums for instrument approaches.[ARP 4107-88]

touchdown zone elevation (TDZE) The highest elevation, expressed in feet above MSL, in the first 3000 ft of the landing surface. TDZE is indicated on the instrument approach procedure chart when straight-in landing minimums are authorized.[ARP4107-88]

touchdown zone lights (TZL) Two rows of transverse light bars located symmetrically about the runway centerline in the runway touchdown zone. The system starts 100 ft from the landing threshold and extends to 3000 ft from the threshold or the midpoint of the runway, whichever is the lesser.[ARP4107-88]

touch feedback A type of interaction in a manipulator in which servos provide force feedback to the manipulator fingers, providing a sense of resistance so the operator does not crush the object.

touch panel See membrane switch.

toward targets See targets.

tow bar 1. A device or assembly for the purpose of connecting the air vehicle to a tow tractor and serving to carry the loads imposed in moving the vehicle in the towing process.[AIR1489-88] 2. A detachable bar or rigid linkage use to tow a vehicle.

tower en route control service/tower-to-tower control service The control of IFR en route traffic within delegated airspace between two or more adjacent approach control facilities. This service is designed to expedite traffic and reduce control and pilot communication requirements.[ARP4107-88]

tow fitting A structural fitting attached to the landing gear or airframe structure which serves for attachment to the tow bar for the aircraft towing process. [AIR1489-88]

towing The process of moving the air vehicle by an external power source provided by a tow tractor or vehicle and connected to the air vehicle by means of a tow bar.[AIR1489-88]

towing tank See model basin.

Townsend discharge A type of direct current discharge between two electrodes immersed in a gas and requiring electron emission from the cathode.[NASA]

trace 1. A graphical output from a recorder, usually in the form of an ink line on paper. 2. An interpretive diagnostic technique which provides an analysis of each executed instruction and writes it on an output device as each instruction is executed. See also sequence checking routine.

tracer A colored thread or filament visible in the insulation on an electrical wire so that the wire can be easily identified or traced between connections.

tracer gas A gas used in connection with a leak detecting instrument to find minute openings in a sealed vacuum system.

tracer milling Cutting a duplicate of a three-dimensional form by using the position of a stylus that traces across the form to operate the quill and table controls on a milling machine.

track 1. The actual flight path of an aircraft over the surface of the earth.[ARP 4107-88] 2. A pair of parallel metal rails for a railroad, tram, or similar wheeled vehicle. 3. A crawler mechanism for earth-moving equipment or military vehicles. 4. A band of data on recording tape or the spiral groove in a phonograph record. 5. An overhead rail for repositioning hoisting gear. 6. To follow the movement of an object—for instance, by continually repositioning a telescope or radar set so its line of slight is always on the object. 7. In data processing, a specific area on any storage medium that can be read by drive heads. See also tread.

track angle error (TKE) The angle between the airplane's actual ground track and the desired ground track or the angular difference between ground track angle and desired track angle. Track angle error is left when the actual track angle is less than the desired track angle and right when the actual track angle is greater than the desired track angle.[ARP1570-86]

track, disk The path on one disk plotter traversed by a head during one revolution.

tracked vehicle Land vehicle equipped with continuous roller belts over cogged wheels for moving over rough terrain.[NASA]

tracking In the nose gear launch catapult hookup process, the movement of the aircraft under its own power from entry into the aft section approach ramp, to the hookup condition.[AIR1489-88]

tracking accuracy The accuracy with which a channel output or combination of channel outputs match one another and/or the commanded output or outputs.[ARP1181A-85] [AIR1916-88]

tracking antennas See directional antennas.

tracking error In lateral mechanical recording equipment, the angle between the vibration axis of the pickup and a plane that is both perpendicular to the record surface and tangent to the unmodulated recording groove at the point where the needle rides in the groove.

tracking filter Electron device for attenuating unwanted signals while passing desired signals, by means of phase lock techniques which reduce the effective bandwidth of the circuit and eliminate amplitude variations.[NASA]

tracking problem The problem of controlling a system so that the output follows a given path.[NASA]

tracking radar A radar used for following a target.[NASA]

tracking stations Stations set up to track objects moving through the atmosphere or space, usually by means of radio or radar.[NASA]

tracking system Any device that continually repositions a mechanism or instrument to follow the movement of a target object.

track, landing gear test A track installation designed and built for the purpose of testing landing gears and having means provided for vertical sink speed and horizontal translation for imposing loads simulating use on an aircraft.[AIR 1489-88]

track, tape The path traversed by one head during the record or playback process.

track type landing gear A landing gear having endless belts running on bogies and rollers (as in a caterpillar tractor) instead of having wheels.[AIR1489-88]

tractor feed On a computer printer, a mechanical feed mechanism that uses gears with teeth that mesh with holes on the side of computer paper to pull the paper through the printer.

tractor (propeller) A propeller mounted forward of the engine, or an engine mounted forward of the main supporting surface. The opposite of pusher. [ARP4107-88]

traffic 1. A term used by a controller to transfer radar identification of an aircraft to another controller for the purpose of coordinating separation action. Traffic is normally issued: (a) in response to a handoff or point out, (b) in anticipation of a handoff or point out, or (c) in conjunction with a request for control of an aircraft. 2. A term used by ATC to refer to one or more aircraft. [ARP4107-88]

traffic advisories Advisories issued to alert a pilot to other known or observed air traffic which may be in such proximity to his/her aircraft's position or intended route of flight as to warrant his/her attention.[ARP4107-88]

traffic advisory Term used with the TCAS system for the display indication that there is a traffic situation that could subsequently require a resolution advisory. The information contains no suggested maneuver.[ARP4153-88]

traffic control control of vehicular traffic such as priority highway lanes, stoplight control, rapid-transit train control, or air traffic control.[NASA]

traffic information See traffic advisories.

traffic in sight Term used by pilots to inform a controller that traffic previously issued as a traffic advisory from the controller is in sight.[ARP4107-88]

traffic pattern The traffic flow that is prescribed for aircraft landing at, taxiing on, or taking off from an airport. The components of a typical traffic pattern are an upwind leg, a crosswind leg, a downwind leg, a base leg, and a final approach.[ARP4107-88]

trail angle, arresting hook The angle at which the arresting hook trails in a free flight configuration ready for touchdown. Usually given as the angle between a line normal to the aircraft horizontal reference line and through the hook pivot and a line through the hook pivot and extending through the throat of the hook point.[AIR1489-88]

trailer A wheeled vehicle, non-powered, with all of its weight supported by its wheels, which is towed by a self-propelled vehicle.[ARP480A-87]

trail, geometric The offset of a tire contact area centroid behind the caster axis by virtue of the caster axis being canted forward (wheel(s) mounted on the centerline of the caster axis).[AIR1489-88]

trailing edge The second transition of a pulse.

trailing vortex A vortex that trails from a wing or other lifting body; also, a vortex trailing from a wing tip.[ARP4107-88]

trail, mechanical The offset of a tire contact area centroid behind the caster axis by virtue of the axle being mounted dimensionally away from the centerline of the caster axis.[AIR1489-88]

trail, pneumatic The offset of a tire contact area centroid behind the caster axis by virtue of aft movement of the tire contact centroid as it rolls. Additive to geometric and/or mechanical trail.[AIR1489-88]

train A collection of one or more associated units and equipment modules, arranged in serial and/or parallel paths, used to make a complete batch of product.

training aid Any object or device designed, constructed or adapted chiefly for the purpose of instructing personnel.

training analysis Evaluation of all facets of instruction—presentation methods, instructors, effectiveness of training, and testing.[NASA]

training evaluation Procedures for determining the effectiveness of instruction.[NASA]

trajectories In general, paths traced by bodies moving as a result of an externally applied force, considered in three dimensions.[NASA]

transceiver A device that provides the electrical interface to the physical medium. Typically used to connect a node to a baseband network. See also MODEM. See also transmitter receivers.

transconductance The change in plate current divided by the change in control-grid voltage that causes it, when the plate voltage and all other voltages are kept constant.[NASA]

transducer 1. A hardware sensing device which measures a physical phenomenon (e.g., pressure, temperature, position) and outputs a calibrated signal.[ARP1587-81] 2. A component which converts a signal from one medium to an equivalent signal in a second medium, one of which is compatible with fluidic devices.[ARP993A-69] See also primary element, signal transducer, and transmitter.

transducer dissipation loss The ratio of the power delivered by a specified source to a transducer connected to a specified load, to the power available from the transducer connected to the same source.

transducer gain 1. The ratio of the power that a transducer delivers to a specified load under specified operating conditions to the available power of a specified source. Usually expressed in terms of decibels.[AIR1489-88] 2. The ratio of the power that a transducer delivers to a specified load under specific operating conditions to the power available from a specified source.

transducer loss The reciprocal of transducer gain.

transfer 1. The conveyance of control from one mode to another by means of instructions or signals. 2. The conveyance of data from one place to another. 3. An instruction for transfer. 4. To copy, exchange, read, record, store, transmit, transport, or write data. 5. An instruction which provides the ability to break the normal sequential flow of control. Synonymous with jump.

transfer admittance The complex ratio of current at a second pair of terminals of an electrical transducer to the emf across a given pair of terminals, at a specified frequency, both pairs being terminated in a specified manner.

transfer chamber In plastics molding, an intermediate chamber or vessel for softening a thermosetting resin with heat and pressure before admitting it to the mold for final curing.

transfer characteristic The current at one electrode expressed as a function of voltage at another electrode, with all other voltages held constant; the relationship is usually shown graphically.

transfer constant A transducer rating consisting of a complex number equal to 1/2 the natural logarithm of the complex ratio of the product of voltage and current entering a transducer to that leaving the transducer when the transducer is connected to its image impedance—the real part of the transfer coefficient is the image attenuation constant and the imaginary part is the image phase constant; transfer constants also can be determined for pressure and volume flow rate or force and velocity, instead of voltage and current. Also known as transfer factor.

transfer function 1. The transfer function for any component or system is defined as the ratio of the Laplace transform of the output to the Laplace transform of the input.[AIR1823-86] 2. A mathematical expression which describes the relationship between physical conditions at two different points in time or space in a given system, and perhaps, also describes the role played by the intervening time or space. 3. The response of an element of a process-control loop that specifies how the output of the device is determined by the input.

transfer impedance The complex ratio of applied a-c voltage, force or pressure at one point in a transducer to a-c current, velocity or volume velocity at another point in the same transducer, all inputs and outputs being connected to the system in some specified manner.

transfer instruction See branch instruction.

transfer lag See capacity lag.

transfer of control See branch and jump.

transfer of training See learning transfer.

transfer operation An operation which moves information from one storage location or one storage medium to another, e.g., read, record, copy, transmit, or exchange. Transfer is sometimes taken to refer specifically to movement between different storage media.

transfer orbit In interplanetary travel, elliptical trajectories tangent to the orbits of both the departure planet and the target planet.[NASA]

transfer ratio The ratio of the number of turns in the secondary winding to the number of turns in the primary winding.

transferred electron devices Electronic equipment utilizing diodes exhibiting negative conductance and susceptance.[NASA]

transferring controller/facility A controller/facility transferring control of an aircraft to another controller/facility.[ARP4107-88]

transfer switch A switch that controls whether a given conductor is connected to one circuit or to another.

transfer time The time interval between the instant the transfer of data to or from storage commences and the instant it is completed.

transfer vector A transfer table used to communicate between two or more programs. The table is fixed in relationship with the program for which it is the transfer vector. The transfer vector provides communication linkage between that program and any remaining subprograms.

transform To change the form of data according to specific rules.

transformation temperature The temperature at which a phase change occurs in a crystalline solid; sometimes, the term is applied to the upper or lower limit of a transformation range.

transformation tensors See tensors.

transformed beta A local or continuous structure comprised of decomposition products arising either by martensitic or by nucleation and growth processes during cooling from either above the beta transus or some temperature high in the alpha-beta phase field. The structure typically consists of alpha platelets which may or may not be separated by beta phase.[AS1814-90]

transformer An electrical device that uses electromagnetic induction to transfer power from one electric circuit to another at the same frequency, usually increasing voltage and decreasing current (or vice versa) in the process.

transformer voltage divider An inductive-type voltage divider used in some a-c bridge circuits to provide high accuracy, much as a KVVD is used in some d-c bridge circuits.

transgranular corrosion A slow mode

of failure that requires the combined action of stress and aggressive environment where the path of failure runs through the grains producing branched cracking.[NASA]

transient 1. A temporary deviation of the aircraft from its normal attitude produced by momentarily overpowering the automatic controls or by momentarily injecting a signal to cause a deviation. [ARP419-57] 2. A dynamic condition or characteristic—power level, voltage, magnetic field strength, force or pressure, for example—that is not periodically repeated; often, it implies an anomalous, temporary departure from a steady-state condition, the latter being either constant or cyclic. 3. Pertaining to rapid change. See also surges.

transient analyzer An electronic device used to capture a record of a transient event for later analysis.

transient deviation See deviation, transient.

transient digitizer A device which records a transient analog waveform and converts the information it has collected into digital form.

transient overshoot An excursion beyond the final steady-state value of the output as the result of a step-input change. It is usually referred to as the first excursion; expressed as a percentage of the steady-state output step.

transient overvoltage A momentary excursion in voltage occurring in a signal or supply line of a device which exceeds the maximum rated conditions specified for that device.

transient response The time response of the actuation system output, following a transient command input. Step response is usually specified as the time required to reach a particular percentage of the final output, along with limits on the percentage overshoot.[AIR1916-88]

transients The short term changing condition of a characteristic beyond the steady-state limits, returning to the steady-state limits within the specified time period.[ARP1212-71]

transilluminated systems Integrally lit displays in which the light source is behind the display and the light passes through the lit portions of the display. [ARP1161-91]

transistor A three terminal solid state semiconductor device which can be used as an amplifier, switch, detector or wherever a three terminal device with gain or switching action is required.

transistor/transistor logic (TTL) A type of digital circuitry.

transit A surveying instrument having a telescope mounted for measuring both horizontal and vertical angles. Also known as transit theodolite; theodolite.

transition 1. The general term that describes the change from one phase of flight or flight condition to another; for example, transition from en route flight to approach, or transition from instrument flight to visual flight. 2. A published procedure used to connect the basic SID to one of several en route airways/jet routes (SID transition), or a published procedure used to connect one of several en route airways/jet routes to the basic STAR (STAR transition). [ARP4107-88] 3. The switching from one state (for example, positive voltage) to another (negative) in a serial transmission.

transitional flow Flow between laminar and turbulent flow; generally between a pipe Reynolds number 2000 and 7000.

transition area See controlled airspace.

transition frequency See crossover frequency.

transition loss The ratio of signal power delivered to the portion of a transmission system following a discontinuity, after insertion of an ideal transducer, to the signal power delivered to the same portion prior to insertion of the ideal transducer.

transition points In aerodynamics, the points of change from laminar to turbulent flow.[NASA]

transition pressure The pressure at which phase transition occurs.[NASA]

transition temperature An arbitrarily defined temperature within the temperature range in which metal fracture characteristics determined usually by notched tests are changing rapidly such as from primarily fibrous (shear) to primarily crystalline (cleavage) fracture. The arbitrarily define temperature in a range in which the ductility of a material changes rapidly with temperature. [NASA]

transit time The time it takes for a particle, such as an electron or atom, to move from one point to another in a system or enclosure.

translate To convert from one language to another language.

translator 1. A program whose input is a sequence of statements in some language and whose output is an equivalent sequence of statements in another language. 2. A translating device.

transmeridian disrhythmia See circadian desynchronization.

transmission factor The ratio of the light transmitted to the incident light. [ARP798-91]

transmission line A continuous conductor or other pathway capable of transmitting electromagnetic power from one location to another while maintaining the power within a system of material boundaries.

transmission loss The reduction in the magnitude of some characteristic of a signal between two stated points in a transmission system.[NASA]

transmissions (machine elements) The gearing system by which power is transmitted from the engine to the live axle in an automobile.[NASA]

transmissometer 1. An instrument system which measures and indicates the transmissivity of light in the atmosphere. It compares the actual transmissivity of the light along a known baseline with the total possible, and computes either the (a) visibility, or (b) visual range. Since the instrument is located along a runway, it is the source of runway visibility and runway visual range data.[ARP4107-88] 2. An instrument for measuring the extinction coefficient of the atmosphere, and for determining visual range. Also known as hazemeter; transmittance meter.

transmittance 1. The ratio of the radiant flux transmitted by a medium or a body to the incident flux.[NASA] 2. The ratio of transmitted electromagnetic energy to incident electromagnetic energy impinging on a body that is wholly or partly transparent to the particular wavelength(s) involved.

transmitter 1. A device that translates the low-level output of a transmission to a site where it can be further processed. 2. In process control, a transmitter mounted together with a sensor or transducer in a single package designed to be used at or near the point of measurement. 3. A light source (LED or diode laser) which is combined with electronic circuitry to drive it. A transmitter operates directly from the signal generated by the other electronic equipment to produce the drive current needed for an LED or diode laser. 4. In process control, a device that converts a variable into a form suitable for transmission of information to another location (for example, resistance changed to current that is propagated on wires to a control installation).

transmitter receivers Combinations of transmitters and receivers in singe housings, with some components being used by both units.[NASA]

transmitters Devices used for the generation of signals of any type and form which are to be transmitted.[NASA]

transmutation A nuclear reaction that changes a nuclide into a nuclide of a different element.

transonic flow In aerodynamics, flow of a fluid over a body in the range just above and just below the acoustic velocity.[NASA]

transonic speed The speed of a body relative to the surrounding fluid at which the flow is in some places on the body subsonic and in other places supersonic.[ARP4107-88] [NASA]

transonics See transonic flow.

transparent 1. The quality of a substance that permits light, some other form of electromagnetic radiation, or particulate radiation to pass through it. 2. In data processing, a programming routine that allows other programs to operate identically regardless of whether the transparent instructions are installed or not installed.

transparent glass Glass having no apparent diffusing properties. Varieties of such glass are referred to as flint, crown, crystal, clear.[ARP798-91]

transpiration The passage of gas or liquid through a porous solid (usually under conditions of molecular flow). [NASA]

transponder(s) 1. The airborne radar beacon receiver/transmitter portion of the air traffic control radar beacon system (ATCRBS) which automatically receives radio signals from interrogators on the ground, and selectively replies with a specific reply pulse or pulse group only to those interrogations being received on the mode to which it is set to respond.[ARP4107-88] 2. Combined receiver and transmitter whose function is to transmit signals automatically when triggered by a interrogator.[NASA]

transponder codes See codes.

transponder mode C See mode C.

transport Layer 4 of the OSI.

transportability A measure of the ability to reuse computer programs on an industrial computer.

Transportation networks Networks of highways, railways, subways, etc. for the movement of passenger and cargo. [NASA]

transport delay The time from the initiation of an input signal until the first discernible change in the output, caused by the input signal.[AIR1489-88]

transport, tape A hardware device that moves magnetic tape past heads for recording or playback.

transthreshold See threshold.

transverse electric wave A type of electromagnetic wave having its electric field vector everywhere perpendicular to the direction of propagation in a homogeneous isotropic medium.

transverse electromagnetic wave A type of electromagnetic wave having both its electric field vector and its magnetic field vector everywhere perpendicular to the direction of propagation in a homogeneous isotropic medium.

transverse interference See interference, normal mode.

transverse oscillation Oscillation in which the direction of motion of the particles is perpendicular to the direction of advance of the oscillatory motion in contrast with longitudinal oscillation, in which the direction of motion is the same as that of advance.[NASA]

transverse response See transverse sensitivity.

transverse vibration See transverse oscillation.

transverse waves Waves in which the direction of displacement at each point of the medium is parallel to the wave front.[NASA]

trap 1. Conditional jump to a known location, automatically activated by hardware or software, with the location from which the jump occurred recorded. Often a temporary measure taken to determine the source of a computer bug. 2. A vertical S-, U- or J-bend in a soil pipe that always contains water to prevent sewer odors from backing up into the building. 3. A device on the intake, or high-vacuum side, of a diffusion pump to reduce backflow of oil or mercury vapors from the pumping medium into the evacuated chamber. 4. A receptacle for the collection of undesirable material.

trapped-air process A method of forming closed blow-molded plastics objects, in which sliding machine elements pinch off the top of the object after blowing to form a sealed, inflated product.

trapped fuel Any fuel in a fuel-delivery system, such as the fuel system of an internal combustion engine, that is not contained in the tanks.

trapped gas effects The physiological effects of the expansion of trapped gases in the body due to changes in barometric pressure. (a) ear block: Unequalized pressure of the gases between the middle ear and the atmosphere. (b) gastro-intestinal gas expansion: Unrelieved expansion of gases in the stomach or intestines. (c) sinus block: Unequalized pressure of the gases between the sinus cavities and the atmosphere. (d) tooth pain: Unrelieved expansion of gases beneath fillings and in periapical abscesses in the teeth.[ARP4107-88]

trapped vortexes Air flow in rotary motion but trapped relative to leading edge vortex separation, which increases not only lift but also drag. The trapped vortexes result in thrust and reduced drag.[NASA]

trapping A feature of some computers whereby an unscheduled jump is made to a predetermined location in response to a machine condition, e.g., a tagged instruction, or an anomalous arithmetic situation. Such a feature is commonly used by monitor routines to provide automatic checking or for communication between input-output routines and the programs using them.

trapping mode A scheme used mainly in program diagnostic procedures for certain computers. If the trapping mode flip-flop is set and the program includes any one of certain instructions, the instruction is not performed and the next instruction is taken from location 0. Program counter contents are saved in order to resume the program after executing the diagnostic procedure.

travel cycle Travel of the closure component from its closed position to the rated travel opening and its return to the closed position.

traveler A piece of documentation which travels with material in-process, identifying it, indicating the steps of processing necessary, the inspections required, the approvals as initialed by compounders and inspectors. Compounders indicate steps completed. Inspectors confirm quality standards have not been compromised.[AS7200/1-90]

traveling block In a block-and-tackle system, the portion of the hoisting apparatus—excluding any slings or special rigging—that is raised and lowered with the load; it usually consists of the sheaves, pulley frame, clevis and hook.

traveling wave A wave in which the ratio of the instantaneous value for any component of the wave field at one point to the instantaneous value at any other point varies with time; it also has the property of transmitting energy from one point to another along its direction of propagation.

traveling wave tubes Electron tubes in which streams of electrons interact continuously or repeatedly with guided electromagnetic waves moving substantially in synchronism with them, and in such a way that there is a net transfer of energy from the streams to the waves. [NASA]

travel time The time required for one-half a travel cycle at specified conditions.

traverse 1. To swivel a gun, antenna, tracking device or similar mechanism in a horizontal plane. 2. A survey consisting of a set of connecting lines of known length which meet each other at specific angles.

tray A horizontal plate in a distillation column that temporarily holds a pool of descending liquid until it flows into a vertical "downcomer" and onto the next tray. Each tray has openings to permit passage of ascending vapors.

tread 1. The lateral dimension between gears mounted in line (laterally) on an aircraft. Measured to the centerline of the wheel if a single wheel gear or the centroid of the wheel pattern if a multiwheel gear. Generally specified dimensionally.[AIR1489-88] 2. The outer surface of a wheel or tire that contacts the roadway or rails. 3. The horizontal portion of a stairstep. 3. The horizontal distance between successive risers in a stairway.

tread depth, tire The thickness of depth of tread which may be worn away before the tire becomes unserviceable.[AIR 1489-88]

treadle A bar or machine element that is pivoted at one end and connected to one or more other machine elements so that when it is stepped on, power or motion, or both, are transmitted to the other elements.

treated water Water which has been chemically treated to make it suitable for boiler feed.

tree 1. A decoder, the diagrammatic representation of which resembles the branching of a tree. 2. In microcomputing, the arrangement of DOS directories and subdirectories.

tree ring dating See dendrochronology.

trend analysis A technique to utilize deviation of recorded data and signature characteristics with respect to time to diagnose and prognosticate a malfunction or failure.[ARP1587-81]

trending A technique which presents deviation of recorded data from a baseline or signature characteristic with respect to a time scale.[AIR1873-88]

triac Semiconductor switching element. Commonly used in a-c output modules for PC's.

trial-for-ignition That period of time during which the programming flame failure controls permit the burner fuel valves to be open before the flame sensing device is required to detect the flame.

trial for main flame ignition A timed interval when with the ignition means proved, the main valve is permitted to remain open. If the main burner is not ignited during this period, the main valve and ignition means are cut off. A safety switch lockout follows.

trial for pilot ignition A timed interval when the pilot valve is held open and an attempt made to ignite and prove it. If the presence of the pilot is proved at the termination of the interval, the main valve is energized; if not, the pilot and ignition are cut off followed by a safety lock-out.

triangular wings See delta wings.

triangulation 1. In navigation, determining position by laying out lines of sight to three celestial bodies or landmarks, widely spaced around the horizon; if properly corrected for time of observation, for the ship's or aircraft's speed and heading, and for current or wind, the lines will meet at a point or will form a small triangle on a map that indicates position at the time of observation. 2. In surveying, a method of measuring a large land area by establishing a baseline, then building up a network of triangles each having at least one side common with an adjacent triangle.

triaxial A cable construction, having three coincident axes, such as conductor, first shield, and second shield, all insulated from one another.[ARP1931-87]

tribology Science of friction, wear, and lubrication.[NASA]

triboluminescence The emission of light caused by application of mechanical energy to a solid.[NASA]

trifilter hydrophotometer An instrument for measuring the transparency of water at three wavelengths using red, green and blue optical filters.

triggers See actuators.

trim, anti-cavitation A combination of control valve trim that by its geometry reduces the tendency of the controlled liquid to cavitate.

trim, anti-noise A combination of control valve trim that by its geometry reduces the noise generated by fluid flowing through the valve.

trim, balanced Control valve trim designed to minimize the net static and dynamic fluid flow forces acting on the trim.

trim condition data Data to indicate an out of trim engine condition and to implement a corrective action to return to scheduled trim.[ARP1587-81]

trim indicator An instrument used to indicate the relative amount of servo effort being used to maintain the aircraft trim.[ARP419-57]

trimming 1. Removing irregular edges from a stamped or deep drawn part. 2. Removing gates, risers and fins from a casting. 3. Removing parting-line flash from a forging. 4. Adding or removing small amounts of R, L or C from electronic circuits to cause minor changes in the circuit performance or to bring into specification.

trim, reduced Control valve trim which has a flow area smaller than the full flow area for that valve.

trim, soft seated Valve trim with an elastomeric, plastic or other readily deformable material used either in the closure component or seat ring to provide shutoff with minimal actuator forces.

triode An electron tube containing three electrodes—an anode, a cathode and a control electrode, or grid.

trip 1. To release a catch or free a mechanism. 2. An apparatus for automatically dumping minecars.

trip-free circuit breaker A breaker so designed that the pole(s) of the circuit breaker cannot be maintained closed when carrying overload currents that would automatically trip the breaker to the open position, and none of the circuit breaker poles would reclose while the operating mechanism is maintained in the closed position.[ARP1199A-90]

trip hammer A large power hammer that falls by gravity when released from its raised position by a cam or lever.

triple point A temperature at which all three phases of a pure substance—solid, liquid and gas—are in mutual equilibrium.

triple-tandem actuator A tandem actuator having three separate actuation sections.[ARP1181A-85]

triplex An adjective meaning threefold, as a triplex valve, a triplex actuator, etc.[ARP1181A-85] [AIR1916-88]

triplex system A fault-tolerant control system containing three signal paths. [ARP1181A-85] [AIR1916-88]

tripod A three-legged support for a transit, camera or other instrument which can be readily set up, collapsed and adjusted.

tripper A device that discharges a load from a conveyor by snubbing the conveyor belt.

tripropellants See liquid rocket propellants.[NASA]

trisonic wind tunnels Wind tunnels designed for subsonic, transonic, and supersonic flows.[NASA]

tri-state A type of logic device that has a high impedance state in addition to a high- and low-level output state. The high impedance state effectively disconnects the output of the device from the circuit; useful in the design of bus-oriented systems.

tri states A three-way switch, 1, 0 and a neutral state (effectively disconnected).

tristimulus values The amounts of the three primary color stimuli required to give, by additive mixture, a color match with the color stimulus considered.[ARP 1782-89]

tritium An isotope of hydrogen having atomic weight of 3 (one proton and two neutrons in the nucleus).

trochoids See pivots.

trochotron A multiple-electrode electron tube which generates an output signal proportional to an input signal by charging elements in sequence with an electron beam manipulated by a magnetic field.

trolley A wheeled car running on an overhead track or rail.

trombe walls Structures with passive solar collectors in the walls.[NASA]

tropical finish A coating applied to electronic equipment to protect it from insects, fungi and high humidity characteristic of tropical climates.

tropopause The boundary between the troposphere and the stratosphere, usually characterized by an abrupt change of lapse rate. The change is in the direction of increased atmospheric stability from regions below to regions above the tropopause. Its height varies from 15 to 20 kilometers in the tropics to about 10 kilometers in polar regions.[NASA]

troposphere That portion of the atmosphere from the earth's surface to the stratosphere; that is, the lowest 10 to 20 kilometers of the atmosphere. The troposphere is characterized by decreasing temperature with height, appreciable vertical wind motion, appreciable water vapor content, and weather. Dynamically, the troposphere can be divided into the following layers: surface boundary layer, Ekman layer, and free atmosphere.[NASA]

tropospheric waves Radio waves that are propagated by reflection from a place of abrupt change in the dielectric constant or its gradient in the troposphere. [NASA]

trouble contact See field contact.

trouble-shoot To search for the cause of a malfunction or erroneous problem behavior, in order to remove the malfunction. See debug.

troubleshooting Locating and diagnosing malfunctions or breakdowns in equipment by means of systematic checking or analysis.[ARD50010-91]

TRSA Terminal Radar Service Area.

truck 1. A wheeled conveyance with two or more wheels usually pushed by hand and having a superstructure for support of aircraft engines or major engine sections.[ARP480A-87] 2. On a multi-wheel gear, the assembly of more than one axle, wheels, tires, connecting structural beam and other equipment. See also bogie.[AIR1489-88]

true airspeed (TAS) 1. The velocity of the airplane with respect to the air mass around the airplane.[ARP1570-86] 2. Indicated airspeed adjusted for error from the installation of the sensing equipment, the compressibility of the air, and the density of the air.[ARP 4107-88]

true bearing See bearing.

true bond width On a bonded device, the maximum width of the beam lead in the bonded area directly over the conductor film metallization.[AS1346-74]

true complement See complement.

true concentric A true concentric stranding or twisted cable is when each successive layer has a reversed direction of lay from the preceding layer. [ARP1931-87]

true concentric stranding Composed of a central wire, surrounded by one or more layers of helically laid wires, with a reversed direction of lay, and an increased length of lay, for each successive layer.[ARP1931-87]

true hydrocarbon vapor pressure (P_{TVP})

Physical data in psia obtained from TVP vs. temperatures curves for various air free hydrocarbon fuel blends at a vapor-liquid ratio of zero.[AIR1326-91]

true mass flow A measurement that is a direct measurement of mass and independent of the properties and the state of the fluid.

true north The direction of the earth's north terrestrial pole (that is, the northern extremity of the earth's rotational axis). Also called geographic north. See also magnetic north.[ARP4107-88]

true ratio A characteristic of an instrument transformer equal to root-mean-square primary current (or voltage) divided by root-mean-square secondary current (or voltage) determined under specified conditions.

truncate 1. To terminate a computational process in accordance with some rule, e.g., to end the evaluation of a power series at a specified term. 2. To drop digits of a number of terms of a series thus lessening precision, e.g., the number 3.14159265 is truncated to five figures in 3.1415, whereas one may round off to 3.1416.

truncated address An operand address whose address field is shorter than the programmable memory's memory address register. The remaining fields in the instruction determine the algorithm for computing an effective address from the truncated address.

truncation error(s) The error resulting from the use of only a finite number of terms of an infinite series, or from the approximation of operations in the infinitesimal calculus for operations in the calculus of finite differences. It is frequently convenient to define truncation error, by exclusion, as an error generated in a computation not due to rounding, initial conditions, or mistakes. Contrasted with rounding error.

trunk See bus.

trunk communication system A communication link joining two telephone central offices or other large switching facilities. It is distinguished by its large capacity, and by the fact that all signals go from point to point, without branching off to many separate points except at the end points.

truth table A table that describes a logic function by listing all possible combinations of input values and indicating, for each combination, the true output values.

TTL See transistor/transistor logic.

TTP See telephone twisted pair.

tube 1. A hollow device, round, square or any other cross section, used to convey fluid gases and/or semi-solids.[ARP 480A-87] 2. A long hollow cylinder used for conveying fluids or transmitting pressure. Also known as tubing. 3. An evacuated glass-enveloped device used in electronic equipment to modify operating characteristics of a signal. Also known as electron tube.

tube cleaner A device for cleaning tubes by brushing, hammering, or by rotating cutters.

tube hole A hole in a drum, header, or tube sheet to accommodate a tube.

tube lasers Stimulated emission devices activated with shock tubes.[NASA]

tube plug A solid plug driven into the end of a tube.

tubercle A localized scab of corrosion products covering an area of corrosive attack.

tube seat That part of a tube hole with which a tube makes contact.

tube sheet A perforated plate for mounting an array of tubes so that fluid on one side of the plate is admitted to the interior of the tubes and is kept separate from fluid on the outside of the tubes, such as in a shell-and-tube heat exchanger.

tube turbining Passing a power-driven rotary device through a length of tubing to clean its interior surface.

tubular terminal A terminal manufactured from tubing rather than flat stock. [ARP914A-79]

tubular-type collector A collector utilizing a number of essentially straight-walled cyclone tubes in parallel.

tuck-down See pitchunder.

tumbling 1. A process for smoothing and polishing small parts by placing them in a barrel with wooden pegs, sawdust and abrasives, or with metal slugs, and rotating the barrel about its axis until the desired surface smoothness and lustre is obtained. 2. Loss of control in a two-frame free gyroscope due to a slowing of the wheel.

tumbling motion An attitude situation in which the vehicle continues on its flight, but turns end over end about its center of mass.[NASA]

tunable lasers Stimulated emission devices with selectable frequency output. [NASA]

tundish A pouring basin for molten metal.

tuner In a telemetry receiver, the input circuitry which selects and amplifies the desired frequency band.

tungsten inert-gas welding A nonpreferred term for gas tungsten-arc welding. Also known as TIG welding.

tuning The adjustment of control constants in algorithms or analog controllers to produce the desired control effect.

turbidity The optical obstruction to the passing of a ray of light through a body of water, caused by finely divided suspended matter.

turbine 1. A machine for converting thermal energy in a flowing stream of fluid into rotary mechanical power by expanding the working fluid through one or more sets of vanes on the periphery of a rotor. 2. A machine for converting fluid flow into mechanical rotary motion, such as steam, water or gasturbines of single or multiple stages.

turbine blades The blades of a turbine wheel.[NASA]

turbine engines Engines incorporating a turbine as a principal component; especially gas turbine engines.[NASA]

turbine meter A volumetric flow measuring device using the rotation of a turbine type element to determine flow rate.

turbine wheels Multivaned wheels or rotors, especially in gas turbine engines, rotated by the impulse from or reaction to a fluid passing across the vanes. [NASA]

turbining See tube turbining.

turboblower An axial-flow or centrifugal compressor.

turbofan(s) 1. A turbojet engine in which additional propulsive thrust is gained by extending a portion of the compressor or turbine blades outside the inner engine case. The extended blades propel bypass air which flows along the engine axis but between the inner and outer engine casing. This air is not combusted but does provide additional thrust (30 to 40 percent), caused by the propulsive effect imparted to it by the extended compressor blading.[ARP4107-88]

turbojet engine A jet engine incorporating a turbine-driven air compressor to take in and compress the air for the combustion of fuel, the gases of combustion being used both to rotate the turbine and to create a thrust-producing jet.[ARP4107-88] [NASA]

turbojet en route descent A procedure for effecting the descent of jet aircraft from an en route altitude to the final approach without execution of the maneuvers prescribed in a published high altitude instrument approach procedure. Its purpose is to expedite the movement of air traffic.[ARP4107-88]

turbopropeller engine An aircraft engine of the gas-turbine type in which the turbine power is used to drive both a compressor and a propeller. This type of engine usually delivers jet thrust in addition to its propeller thrust; it is often called a turboprop or turboprop engine and sometimes called a propeller turbine or propeller-turbine engine.[ARP 4107-88]

turborotors See turbine wheels.

turbosupercharger A gas-turbine-driven air compressor used to increase air-intake pressure of a reciprocating internal-combustion engine.

turbulence A state of fluid flow in which the instantaneous velocities exhibit irregular and apparently random fluctuations so that in practice only statistical properties can be recognized and subjected to analysis.[NASA]

turbulence amplifier An amplifier which utilizes control of the laminar-to-turbulent transition of a power jet to modulate the output.[ARP993A-69]

turbulent boundary layer The layer in which the Reynolds stresses are much larger than the viscous stresses. When the Reynolds number is sufficiently high, there is a turbulent layer adjacent to the laminar boundary layer.[NASA]

turbulent burner A burner in which fuel and air are mixed and discharged into the furnace in such a manner as to produce turbulent flow from the burner.

turbulent flow 1. Fluid motion in which random motions of parts of the fluid are superimposed upon a simple pattern of flow. All or nearly all fluid flow displays some degree of turbulence. The opposite is laminar flow.[AIR1489-88] [NASA]

turnaround (STS) The intervals between flights of the shuttle orbiters.[NASA]

turnaround time The particular amount of time required for a computation task to get from the requester, to the computer and back to the requester with desired results.

turnbuckle A device for tightening stays or tension rods in which a sleeve with a thread in one end and a swivel at the other (or with threads at both ends) is turned about its axis, drawing the ends of the device together.

turndown The ratio of the maximum plant design flow rate to the minimum plant design flow rate. See rangeability, inherent.

turner fluorometer A type of uv fluorometer in which primary filters pass only uv radiation to excite the sample and secondary filters pass only visible light to the photomultiplier tube; the intensity of emitted light is proportional to sample concentration even when exciting and measured light are not at optimum wavelengths.

turning clearance radius The dimensional radius required for clearance for turning an aircraft. May be determined by wing tip clearance, tail clearance, nose clearance or other. Generally specified.[AIR1489-88]

turning radius Usually specified as the minimum radius about which an aircraft will turn and must be specified, i.e., nose wheel turning radius, outside main gear turning radius, etc.[AIR1489-88]

turn-key system A system that includes all computer hardware and software, ready to operate.

turnover angle The angle formed between (a) a vertical line through the center of gravity of the aircraft and (b) a line through the center of gravity of the aircraft and normal to a line between the turnover points (for example nose gear and outside main gear). In a turn, the roll angle at which turnover is impending.[AIR1489-88]

turnover frequency See crossover frequency.

turnstile antennas Antennas composed of two dipole antennas, normal to each other, with their axes intersecting at their midpoints. Usually, the currents are equal and in phase quadrature. [NASA]

turret head A device attached to a crimping tool, having multiple positions that can be rotated to position a specific conductor barrel between the indentors.-[ARP914A-79]

T-VOR Partial acronym for Terminal Very High Frequency Omnidirectional Range.

Twaddle scale A specific gravity scale that attempts to simplify measurement of liquid densities heavier than water, such as industrial liquors; the range of density from 1.000 to 2.000 is divided into 200 equal parts, so that one degree Twaddle equals a difference in specific gravity of 0.005; on this scale, 40° Twaddle indicates a specific gravity of 1.200.

twilled binding A weave in which each shute wire passes successively over two and under two warp wires and each warp wire passes successively over two and under two shute wires.[AIR888-89]

twin Two portions of a crystal having a definite crystallographic relationship; one may be regarded as the parent, the other as the twin. The orientation of the twin is either a mirror image of the orientation of the parent about a "twinning plane" or an orientation that can be derived by rotating the twin portion about a "twinning axis".[AS1814-90]

twist 1. A hose or hose assembly subjected to internal pressures or improperly installed so that each end of the hose is turned awkwardly about the hose axis in opposing directions.[ARP1658A-86] 2. The number of turns per unit length in the lay of fiber, rope, thread, yarn or cord.

twist drill A sharpened cylindrical tool for cutting holes in solid material where the cutting edges run in a general radial direction at one end of the tool and helical grooves extend from the cutting edges along the length of the tool to eject chips and sometimes admit coolant.

twisted pair 1. A cable composed of two insulated conductors, twisted together, without common covering.[ARP1931-87] 2. Two insulated wires (signal and return) which are twisted around each other. Since both wires have nearly equal exposure to any electrostatic or electromagnetic interference, the differential noise is slight.

two body orbits See two body problem.

two body problem That problem in classical celestial mechanics which treats of the relative motion of two point masses under their mutual gravitational attraction.[NASA]

two component plastic injection kit A two component plastic injection kit packages the base compound and catalyst of a two-part sealant in separate sections until the time of use.[AS7200/1-90]

two-position action A type of control-system action that involves positioning the final control device in either of two fixed positions, without permitting it to stop at any intermediate position.

two-position controller See controller, two-position.

two's complement 1. A method of representing negative numbers in binary; formed by taking the radix complement of a positive number. 2. A form of binary arithmetic used in most computers to perform both addition and subtraction with the same circuitry where the representation of the numbers determines the operation to be performed.

two photon coherent states See squeezed states (quantum theory).[NASA]

two-sided sampling plan Any statistical quality control method whereby acceptability of a production lot is determined against both upper and lower limits.

two-stroke cycle An engine cycle for a reciprocating internal combustion engine that requires two strokes of the piston to complete.

two-way Not recommended, except as modifier in such items as valve, restrictor.[ARP243B-65]

Tyndall effect A physical phenomenon first observed by Sir John Tyndall, who noted that particles suspended in a fluid could be seen readily if illuminated by strong light and viewed from the side, even though they could not be seen when viewed from the front in the same light beam; this effect is the basis for nephelometry, which involves measurement of the intensity of side-reflected light, and is commonly used in such applications as analyzing for trace amounts of silver in solution, determining the concentration of small amounts of calcium in titanium alloys, measuring bacterial growth rates, and controlling the clarity of beverages, potable water and effluent discharges.

type 1. As used with respect to the certification, ratings, privileges, and limitations of aircraft: A specific make and basic model of aircraft, including modifications thereto that do not change its handling or flight characteristics. 2. As used with respect to the certification of aircraft: Those aircraft which are similar in design to each other. 3. As used with respect to the certification of aircraft engines: Those engines which are similar to one another in design. [ARP4107-88]

type of hydraulic system A classification standard for a military aircraft hydraulic system based on minimum and maximum allowable fluid temperatures, defined in ISO 6771 as: Type I -54 to +71 °C (-65 to +160 °F), Type II -54 to 135 °C (-65 to +275 °F), Type III -54 to 232 °C (-65 to +450 °F). NOTE: The upper limit of Type III was reduced from +232 °C (+450 °F) by the U.S. in recognition of hydraulic fluid limitations. [AIR1916-88]

U See velocity (ft/sec).

U bolt A rod threaded at both ends and bent into a U shape; it is most often used with a bar across the span of the U to provide a clamping force around a tubular or cylindrical object.

UHF Ultra High Frequency.

UL See Underwriters Laboratories.

ullage The empty volume of a propellant tank which is not occupied by fuel or oxidizer.[AIR913-89]

ultimate analysis See analysis, ultimate.

ultimate cycle method See Ziegler-Nichols method.

ultimately controlled variable See variable, ultimately controlled.

ultimate strength See tensile strength.

ultra high frequency (UHF) The frequency band between 300 and 3000 MHz. The band of radio frequencies used for military air/ground voice communications. In some instances, frequencies as low as 225 MHz may still be called UHF. [ARP4107-88]

ultralight aircraft An aircraft for one person weighing less than 254 pounds with a top speed of 55 knots and a maximum stalling speed of 24 knots.[NASA]

ultrasonically-assisted machining A machining method in which ultrasonic vibrations are imparted to a tool to make it cut with better quality or speed than would be possible using the same machining process but without vibrating the tool.

ultrasonic atomizer A type of atomizer that produces uniform droplets at low feed rates by flowing liquid over a surface which is vibrating at ultrasonic frequency.

ultrasonic bonding A method of joining two solid materials by subjecting a joint under moderate clamping pressure to vibratory shearing action at ultrasonic frequencies until a permanent bond is achieved; it may be used on both soft metals and thermoplastics.

ultrasonic cleaning Removing soil from a surface by the combined action of ultrasonic vibrations and a chemical solvent, usually with the part immersed.

ultrasonic coagulation A process that uses ultrasonic energy to bond small particles together, forming an aggregated mass.

ultrasonic delay line A constrained pathway for propagating sound so that the transit time along the pathway becomes a fixed time delay for another signal.

ultrasonic densimeter Density measuring instruments utilizing ultrasonic devices (sensors).[NASA]

ultrasonic density sensor A device for determining density from the attenuation of ultrasound beams passing through a liquid or semisolid; a typical application involves immersing an ultrasonic transducer in fully agitated lime slurry, thus avoiding coating and clogging which occurs with other devices.

ultrasonic detector Any of several devices for detecting ultrasound waves and measuring one or more wave attributes.

ultrasonic drilling A method of producing holes of almost any desired shape in very hard materials, such as tungsten carbide or gemstones, by causing a suitable tool which is pressed against the workpiece to vibrate axially under the driving force of an ultrasonic transducer.

ultrasonic flowmeter A device for measuring flow rates across fluid streams by either Doppler-effect measurements or time-of-transit determination; in both types of flow measurement, displacement of the portion of the flowing stream carrying the sound waves is determined and flow rate calculated from the effect on soundwave characteristics.

ultrasonic frequency Any frequency for compression waves resembling sound where the frequency is above the audible range—that is, above about 15-20 kHz.

ultrasonic generator A device for producing compression waves of ultrasonic frequencies.

ultrasonic level detector Any of several devices that use either time-of-transit or intensity attenuation of an ultrasonic beam to determine the position of the upper surface in a body of confined liquid or bulk solids.

ultrasonic light diffraction Forming optical diffraction patterns when light passing through a longitudinal ultrasound field is refracted by interaction with the sound waves.

ultrasonic machining A machining method in which an abrasive slurry is driven against a workpiece by a tool vibrating axially at high frequency to cut an exact shape in the workpiece surface.

ultrasonic material dispersion Using ultrasound waves to break up one component of a mixture and disperse it in another to create a suspension or emulsion.

ultrasonics The technology of sound at frequencies above the audio frequency range.[NASA] 2. Technology associated with the production and utilization of sound having a frequency higher than about 15 kHz.

ultrasonic stroboscope A device for producing pulsed light by using ultrasound to modulate a light beam.

ultrasonic testing A nondestructive testing method in which high frequency sound waves are projected into a solid to detect and locate flaws, to measure thickness, or to detect structural differences.

ultrasonic thickness gage Any of several devices that use either resonance or pulse-echo techniques to determine the thickness of metal parts—sheet or plate thickness, or pipe-wall thickness, for example; the technique also may be used to determine coating thickness in applications where a suitable reflection can be obtained from the coating-substrate interface.

ultrasonic transducer A device for converting high- frequency electric impulses into mechanical vibrations, or vice versa, usually through the use of a magnetostrictive or piezoelectric material.

ultrasonic vibration Refers to the dynamics of mechanical vibration waveforms in the frequency ranges of more than 20KHz.[ARP1587-81]

ultrasonic welding Same as ultrasonic bonding.

ultrasonoscope An instrument for displaying an echosonogram on an oscilloscope, and sometimes for providing auxiliary output to a chart recorder.

ultraviolet astronomy Use of special optical instruments for the observation of astronomical phenomena in the ultraviolet spectrum.[NASA]

ultraviolet degradation Degradation caused by long time exposure of a material to sunlight or other ultraviolet rays. [ARP1931-87]

ultraviolet-erasable read-only memory (UVROM) A type of computer memory that can be erased or changed only by exposure to ultraviolet light.

ultraviolet light See ultraviolet radiation.

ultraviolet radiation Electromagnetic radiation having wavelengths shorter than visible light and longer than low-frequency x-rays—that is, wavelengths of about 14 to 400 nanometers.

ultraviolet spectrophotometry Determination of the concentration of various compounds in a water solution or gas stream based on characteristic absorption of ultraviolet rays; uv absorption patterns are not as distinctive "fingerprints" as their ir counterparts, but in many cases the former are more selective and sensitive for use in process control applications.

ultraviolet telescopes Optical telescopes designed to collect ultraviolet light (wavelengths not capable of passing through earth's atmosphere) and as such must be used in space.[NASA]

umbilical connection Any flexible grouping of electrical, mechanical or hydraulic connections between a machine, vehicle or robotics device and a source of power, control signals and data acquisition or auxiliary services.

umbras The darkest parts of shadows in which light is completely cut off by intervening objects. Lighter parts surrounding the umbras, in which the light is only partly cut off, are called penumbras. The darker central portions of sun spots, surrounded by lighter penumbra. [NASA]

Umkehr effect Due to the presence of the ozone layer, an anomaly of the relative zenith intensities of scattered sunlight at certain wavelengths in the ultraviolet as the sun approaches the horizon. [NASA]

unacceptable message segment (UMS) A UMS condition occurs when a bus monitor determines that a message segment is not an AMS.[AS4116-90]

unaccounted-for loss That portion of a boiler heat balance which represents the difference between 100 percent and the sum of the heat absorbed by the unit and all the classified losses expressed as percent.

unarming Removal of normal skid control function upon reading a low predetermined wheel velocity threshold corresponding to a low aircraft forward velocity.[AIR1489-88]

unbalanced (to ground) Opposite of balanced (to ground); unbalanced (to ground) cable pairs can be susceptible to noise and crosstalk and can cause crosstalk to other pairs.

unbonded beam lead length On a bonded device, the distance from the edge of the dielectric layer to the apparent inner edge of the bonded area. [AS1346-74]

unbonded strain gage A type of wire strain gage sometimes used in transducer applications where strain is determined from elastic tension developed across the gage between mechanical end connections.

unburned combustible The combustible portion of the fuel which is not completely oxidized.

unburned combustible loss See combustible loss.

uncertainty The interval within which the true value of a measured quantity is expected to lie with a stated probability.

uncompensated integrating tachometer generator A generator characterized by a relatively high output-to-null ratio and, as such, is employed in high-gain rate and computing servomechanism applications. Such applications require a high degree of linearity over the range of speeds of the generator. [ARP667-82]

unconditional branch See unconditional transfer.

unconditional jump See unconditional transfer.

unconditional transfer An instruction which switches the sequence of control to some specified location. Synonymous with unconditional branch and unconditional jump. Loosely, jump.

uncontrolled airspace That portion of the airspace that has not been designated as continental control area, control area, control zone, terminal control area, or transition area, and within which ATC has neither the authority nor the responsibility for exercising control over air traffic.[ARP4107-88]

uncontrolled reentry (spacecraft) The descent into a denser atmosphere of a spacecraft in an elliptical orbit due to aerodynamic drag and other perturbation forces. The gradually increasing deceleration causes some kinetic energy to be converted into atmospheric heat.

The centrifugal force decreases and gravity pulls the spacecraft further into the atmosphere. The spacecraft eventually burns.[NASA]

uncouple To disengage a screwed, pinned or latched connection.

uncoupled modes Modes of vibration that can exist in systems concurrently with and independently of other modes. [NASA]

undamped frequency See frequency, undamped.

undamped natural frequency Of a landing gear or mechanical system, the frequency of free vibration resulting from only elastic and inertial forces of the system.[AIR1489-88]

underbead crack A crack in the heat-affected zone of a weldment that does not extend to the base metal surface.

undercarriage The landing gear of an aircraft. See also landing gear.[AIR 1489-88]

underconfidence See confidence.

undercut 1. An unfilled groove in the base metal along the toe of a weld. 2. A groove or recess along the transition zone from one cross-section to another, such as from a hub to a fillet, that leaves a portion of one cross-section undersized.

underdamped See damping.

underfill A condition whereby the face of a weld is lower than the position of an adjacent base metal surface.

underflow Pertaining to the condition that arises when a machine computation yields a nonzero result that is smaller than the smallest nonzero quantity that the intended unit of storage is capable of storing.

underground acoustics The sounding of subsoils, rocks, etc. for mineralogy and other exploratory purposes.[NASA]

underground structures Subterranean construction of tunnels, passageways, chambers, or excavations.[NASA]

undermotivation See motivation, under-.

undershoot Movement of the aircraft whereby it stops short of the commanded reference altitude, attitude or flight path, before settling out.[ARP419-57]

understressing Repeatedly stressing a part at a level below the fatigue limit or below the maximum service stress to improve fatigue properties.

under surface blowing Use of jets blowing on the underside of airfoils for variations in pressure distribution.[NASA]

under the hood Term indicating that the pilot is using a hood to restrict visibility outside the cockpit while simulating instrument flight. An appropriately rated pilot is required in the other control seat while this operation is being conducted.[ARP4107-88]

undertread See tire undertread.

underwater physiology The study of the bodily responses to the environmental stresses of the underwater milieu such as pressure, temperature and immersion effects.[NASA]

underwater resources Earth resources (minerals, petroleum, etc.) within or under the oceans.[NASA]

Underwriters Laboratories (UL) An independent testing and certifying organization.

undetected failure A failure that is not identifiable until a second and detectable failure has occurred.[AIR1916-88]

undocumented In data processing, the absence of instructions on how to use a program or computer.

unfired pressure vessel A vessel designed to withstand internal pressure, neither subjected to heat from products of combustion nor an integral part of a fired pressure vessel system.

unfired torque The torque required to rotate an engine during the start cycle before light off has occurred. It is the sum of compressor pumping torque, bearing friction and engine accessory drag.[ARP906A-91]

unformatted ASCII A mode of data transfer in which the low-order seven bits of each byte are transferred; no special formatting of the data occurs or is recognized.

unformatted binary A mode of data transfer in which all bits of a byte are transferred without regard to their contents.

uniaxial strain See axial strain.

UNIBUS DEC's common communications bus to the CPU, memory, and all peripheral devices along which address, data, and control are transferred on fifty-six lines; the form of communication is identical for all devices on this bus.

Unicom A nongovernment facility which may provide airport information at certain airports.[ARP4107-88]

unidirectional concentric stranding A stranding where each successive layer has a different lay length, thereby retaining a circular form without migration of strands from one layer to another.[ARP1931-87]

unidirectional lay That variation of concentric lay in which all the helical layers of strands comprising the concentric conductor have the same direction of lay. The construction includes normal unidirectional lay, in which each successive layer has a greater lay length than the preceding layer, and unidirectional equal lay (unilay), generally limited to 19 strands, in which all helical layers have the same length of lay. [AS1198-82]

unidirectional pulse A wave pulse in which intended deviations from the normally constant values occur in only one direction.

unidirectional stranding A term denoting that in a stranded conductor all layers have the same direction of lay. [ARP1931-87]

unified field theory Any theory which attempts to express gravitational theory and electromagnetic theory within a single unified framework; usually an attempt to generalize Einstein's general theory of gravitation alone to a theory of gravitation and classical electromagnetism.[NASA]

unified screw thread A system of standard 60° V threads that are classified coarse, fine and extra-fine (UNC, UNF and UNEF) to provide different levels of strength and clamping power.

uniform-chromaticity-scale diagram (UCS diagram) A chromaticity diagram in which the coordinate scales are chosen with the intention of making equal intervals represent, as nearly as possible, equal steps of discrimination for colors of the same luminance at all parts of the diagram.[ARP1782-89]

uniform corrosion Chemical reaction or dissolution of a metal characterized by uniform receding of the surface.

uniform screen An optical display such as a monochrome CRT that has a continuous light emitting surface that is uninterrupted by any physical structure.[ARP1782-89]

unilateral tolerance A method of dimensioning in which either the upper or lower limit of the allowable range is given as the stated size or location, and the permissible variation is given as a positive or negative tolerance from that size, but not both; the decision as to whether the upper or lower limit is given as the stated dimension depends on the critical value for the dimension, and should be chosen so that the tolerance is always away from the critical value and toward a less critical condition.

unilateral transducer A transducer that produces output waves related to input waves when connected in one direction, but that cannot produce such waves when input and output connections are reversed.

unilay strand A conductor constructed with a central core surrounded by more than one layer of helically-laid wires, with all layers having a common length and direction of lay.[ARP1931-87]

uninstalled Hose assemblies offered for acceptance against procurement contracts.[AS1933-85]

unintentional activation See error, unintentional activation.

uninterruptible power source The use of resident batteries in a device to phase in when external power is interrupted.

uninterruptible power supply (UPS) A type of power supply that can provide electrical power even when line power is lost.

union A threaded assembly for joining the ends of two pipes or tubes where neither can be rotated to complete the joint; it usually consists of flanged members that are threaded or soldered onto the pipe ends and a ring member that surrounds the flange edges and makes a leak-tight seal.

unit 1. An assembly or any combination of parts, subassemblies, and assemblies mounted together, normally capable of independent operation in a variety of situations.[ARD50010-91][ARD50013-91] 2. A collection of associated elements, loops, devices, and/or equipment modules that perform a coordinated function and which operates relatively independently. 3. A device having a special function. 4. A basic element. See arithmetic unit, binary unit, central processing unit, and control unit.

unit sensitivity The specific amount that a measured quantity must rise or fall to cause a pointer or other indicating element to move one scale division on a specific instrument.

universal development system A development system that, by means of personality modules, can be used to develop the software and hardware for a range of microcomputers.

universal instrument See altazimuth.

universal joint A linkage for transmitting rotational motion and power between two shafts whose axes do not coincide, especially when the axis of one must be allowed to pivot through a small angle with respect to the other during operation.

universal output transformer An output transformer for an electronic device having several taps on its secondary winding so that it can be connected to almost any loudspeaker system by choosing the proper tap.

universal ratio set Abbreviated URS. An arrangement of variable resistors used as a highly accurate continuously adjustable arm of a Wheatstone bridge with a resolution of 0.001 ohm and a total range of 2111.110 ohms; a URS is particularly well suited for measuring unknown resistances in terms of highly accurate fixed-decade standards within a 10:1 ratio of the unknown resistances.

universal time Time defined by the rotational motion of the earth and determined from the apparent diurnal motions which reflect this rotation; because of variations in the rate of rotation, universal time is not rigorously uniform. Also called Greenwich mean time. [NASA]

unmate The disengagement, disconnecting or uncoupling of mated connectors. [ARP914A-79]

unsafe condition Any condition within an aircraft that jeopardizes the safety of the aircraft or the personnel aboard. [AS1212-71]

unsaturation (chemistry) A state in which the atomic bonds of an organic compound's chain or ring are not completely satisfied (not saturated); unsaturation usually results in a double bond (as for olefins) or a triple bond (as for the acetylenes).[NASA]

unscheduled landing A landing at other than the scheduled destination, for mechanical or operational reasons.[ARP 4107-88]

unscheduled maintenance 1. Unscheduled maintenance is any maintenance other than scheduled preventative maintenance (e.g., corrective maintenance required as a result of a problem uncovered during scheduled preventative maintenance).[ARD50010-91] 2. Any urgent need for repair or upkeep that was unpredicted or not previously planned, and must be added to or substituted for previously planned work.

unsprung mass The portions of the landing gear (wheel, tires, axles, lower part of shock absorber, etc.) which are supported or partially supported by the ground as opposed to the spring mass which is separated by the shock absorber. See also sprung mass.[AIR1489-88]

unsprung weight That portion of a vehicle's gross weight that is comprised of the wheels, axles and various other components not supported by its springs.

unsteady flow A flow in which the flow rate fluctuates randomly with time and for which the mean value is not constant.

up-converters Parametric amplifiers characterized by the output signal frequencies being greater than the frequencies of the input signals.[NASA]

update 1. To put into a master file changes required by current information or transactions. 2. To modify an instruction so that the address numbers it contains are increased by a stated amount each time the instruction is performed.

upgrade To increase the value or quality of an operating system or commercial product by incorporating changes in design or manufacture without changing its basic function.

upgrading Major changes to older control systems.

uplink Message or data transmitted by the ground network to an aircraft by address (registration mark or flight number).[ARP4102/13-90]

uplinking The transmission of signals from ground terminals to satellites in telecommunication systems.[NASA]

uplock A locking device to hold a landing gear in the retracted or stowed position.[AIR1489-88]

upper air See upper atmosphere.

upper atmosphere The general term applied to the atmosphere above the troposphere.[NASA]

upper range-limit See range-limit, upper.

upper range-value See range-value, upper.

upper torso restraint That portion of a torso restraint system intended to restrain movement of the chest and shoulder region, commonly referred to as a shoulder harness.[AS8043-86]

UPS See uninterruptible power supply.

upset To cause a local increase in diameter or other cross-sectional dimension by applying an axial deforming force to a piece of rod or wire, such as is used to produce heads on nails or screws.

upsetting See cold heading.

upper surface blowing Use of jet blowing on the upper surface of airfoils to create variations in pressure distribution.[NASA]

uptake A conduit for exhaust gases connecting the outlet of a furnace or firebox to a chimney or stack.

up time The time during which equipment is either producing work or is available for productive work. Contrasted with downtime.

upward compatibility In data processing, the ability of a computer device or program to function on newer models.

upwelling See upwelling water.

Uranus rings Ring structures encircling the planet Uranus and similar to those of the planet Saturn.[NASA]

USASCII U.S. Standard Code for Information Exchange. The standard code, using a coded charter set consisting of 7-bit coded characters (8 bits including parity check), used for information exchange among data processing systems, communications systems, and associated equipment. The USASCII set consists of control characters and graphic characters. See also ASCII.

use factor The ratio of hours in operation to the total hours in that period.

user 1. A company or body of users which receives standards from external standards-making organizations and uses the standards via their computer system. [AS4159-88] 2. In data processing, the person or client who makes use of a computer system.

user-computer interface See man-computer interface.

user-friendly In data processing, a general term to describe programs that do not require extensive learning or technical skill to use successfully.

user interface The way a program communicates with an operator.

utility 1. Any general-purpose computer program included in an operating system to perform common functions. 2. Any of the systems in a process plant, manufacturing facility not directly involved in production; may include any or all of the following—steam, water, refrigeration, heating, compressed air, electric power, instrumentation, waste treatment and effluent systems.

utility program See utility routine.

utility routine A standard routine used to assist in the operation of the computer, e.g., a conversion routine, a sorting routine, a printout routine, or a tracing routine. Synonymous with utility program.

utility software A library of programs for general use in a computer system.

utilization equipment Utilization equipment will be considered as comprising either an individual unit, set, or a complete system to which the electrical power is applied.[AS1212-71]

utilization rate The planned or actual number of life units expended, or missions attempted during a stated interval of calendar time.[ARD50010-91] [AIR1916-88]

U tube See manometer.

U-tube manometer A device for measuring gage pressure or differential pressure by means of a U-shaped transparent tube partly filled with a liquid, commonly water; a small pressure above or below atmospheric is measured by connecting one leg of the U to the pressurized space and observing the height of liquid while the other leg is open to the atmosphere; similarly, a small differential pressure is measured by connecting both legs to pressurized space—for example, high- and low-pressure regions across an orifice or venturi.

UV Ceti stars See flare stars.

UV erasable PROM (EPROM) Memory whose contents can be erased by a period of intense exposure to UV radiation.

UVROM See ultraviolet-erasable read-only memory.

V

v See volt.

V 1. Velocity. The speed of an aircraft, expressed in knots or statute miles per hour.[ARP4107-88] 2. The measured sample size (volume).[ARP1179A-80]

V_1 Takeoff decision speed (formerly denoted as critical engine failure speed). The speed below which an aircraft can lose an engine and still stop on the remaining runway.[ARP4107-88]

V_2 Takeoff safety speed. The normal takeoff speed for a multiengine aircraft. If an aircraft loses an engine after V_1, the pilot should continue to accelerate to this airspeed and attempt to take off. [ARP4107-88]

V_a The design maneuvering speed; the maximum speed at which application of full aerodynamic control will not overstress the aircraft. V_a generally decreases as the gross weight of the aircraft decreases.[ARP4107-88]

vacuum 1. A given space filled with gas at pressures below atmospheric pressure.[NASA] 2. A low-pressure gaseous environment having an absolute pressure lower than ambient atmospheric pressure.

vacuum brake A type of power-assisted vehicle brake whose released position is maintained by maintaining a pressure below atmospheric in the actuating cylinder, and whose actuated position is obtained by admitting air at atmospheric pressure to one side of the cylinder.

vacuum breaker A device used in a water supply line to relieve a vacuum and prevent backflow. Also known as backflow preventer.

vacuum degassing Removing dissolved or trapped gases in a metal by melting or heating it under high vacuum.

vacuum deposition A process for coating a substrate with a thin film of metal by condensing it on the substrate in an evacuated chamber. See also vacuum plating.

vacuum filtration A process for separating solids from a suspension or slurry by admitting the mixture to a filter at atmospheric pressure (or higher) and drawing a vacuum on the outlet side to assist the liquid in passing through the filter element.

vacuum forming A method of forming sheet plastics by clamping the sheet to a stationary frame, then heating it and drawing it into a mold by pulling a vacuum in the space between the sheet and mold.

vacuum fusion A laboratory technique for determining dissolved gas content of metals by melting them in vacuum and measuring the amount of hydrogen, oxygen and sometimes nitrogen released during melting; the process can be used on most metals except reactive elements such as alkali and alkaline-earth metals.

vacuum gage Any of several devices for measuring pressures below ambient atmospheric.

vacuum-gage control circuit An electric circuit which energizes the tube of an electrically operated vacuum gage, controls and measures gage currents or voltages, and sometimes supplies and regulates power that degases tube elements.

vacuum-gage tube An enclosed portion of a pressure-measuring system connected to an evacuated chamber or system; its essential component is the pressure-sensing element, but it also includes the envelope and any support structure, plus the means for connecting the gage to the evacuated space.

vacuum jacketed valve See jacketed valve.

vacuum photodiode A vacuum tube in which light incident on a photoemissive surface (cathode) frees electrons, which are collected by the positively biased anode.

vacuum plating A process for producing a thin film of metal on a solid substrate by depositing a vaporized compound on the work surface, or by reacting a vapor with the surface, in an evacuated chamber. Also known as vapor deposition.

vacuum pump 1. A pump which maintains a vacuum in a line or system of lines. Commonly used for driving vacuum-driven flight instruments such as turn-and-bank indicators and gyro compasses.[ARP4107-88] 2. A device similar to a compressor whose inlet is attached to a chamber to remove noncondensible gases such as air and maintain the chamber at a pressure below atmospheric.

vacuum system(s) 1. Chambers having walls capable of withstanding atmospheric pressure and having an opening through which the gas can be removed through a pipe or manifold to a pumping system. The pumping system may or may not be considered as part of the vacuum system.[NASA] 2. A system consisting of one or more chambers that can withstand atmospheric pressure without completely collapsing, and having an opening for pumping gas out of the enclosed space.

vacuum tube(s) 1. Electron tubes evacuated to such a degree that their electrical characteristics are essentially unaffected by the presence or residual gas or vapor.[NASA] 2. A device for use in an electronic circuit to amplify d-c, audio, or microwave frequencies or rectify radio-frequency signals; it consists of an arrangement of metal emitters, grids and plates enclosed in a thin, evacuated glass envelope with a molded plastic base containing pin connectors that are attached to the tube internals.

validity The correctness, especially the degree of the closeness by which iterated results approach the correct result.

validity check A check based upon known limits or upon given information or computer results, e.g., a calendar month will not be numbered greater than 12, and a week does not have more than 168 hours. See also check, validity.

Valsalva exercise The procedure of raising the pressure in the nasapharynx by forcible expiration with the mouth closed and nostrils pinched, in order to clear the eustachian tubes.[NASA]

Valsalva maneuver See Valsalva exercise.

value engineering The systematic use of engineering principles to identify the functions of a product or service, and to provide these functions reliably at lowest cost. Also known as value analysis; value control.

valve 1. A device for directing, regulating or stopping flow or regulating pressure in a fluid system, usually through the operation of one or more movable members. [ARP243B-65] 2. An in-line device in a fluid-flow system that can interrupt flow, regulate the rate of flow, or divert flow to another branch of the system. 3. A term used by the British to denote a vacuum tube since valve action is the way a vacuum tube operates with a stream of electrons.

valve, automatic A valve in which operation is controlled entirely by action of fluid which passes through it.[ARP 243B-65]

valve, ball A valve with a rotary motion closure component consisting of a full ball or a segmented ball.

valve, blowoff A valve which releases fluid to overboard when its setting is exceeded.[ARP243B-65]

valve, brake A valve for control of wheel brake actuating pressure. A brake valve is usually a mechanically or pressure operated variable pressure reducing valve with provision for opening brake pressure outlet port to "return" when outlet pressure exceeds the pressure setting at the moment.[ARP243B-65]

valve, check A valve which allows free flow of fluid in one direction only and prevents flow in the opposite direction.

[ARP243B-65]

valve, control A valve in which flow pattern or rate is controlled by external means.[ARP243B-65]

valve, controllable check A two position manually operated valve which functions as a conventional check valve in one position, but permits free flow in either direction when actuated to the other operating position.[ARP243B-65]

valve, diaphragm type A valve with a flexible linear motion closure component which is moved into the fluid flow passageway of the body to modify the rate of flow through the valve by the actuator.

valve, directional control A selector valve having four working ports; "pressure", "return" and two "cylinder" or "load" ports; and a reversible flow pattern.[ARP243B-65]

valve, dump A two position control valve having two or more ports with all ports blocked in "off" position and all ports interconnected in "on" position.[ARP 243B-65]

valve, engine start A valve whose function is to initiate or terminate hydraulic starter operation by opening or closing the fluid inlet line in response to an external signal, and which usually provides automatic shutoff at the end of an engine starting cycle.[ARP243B-65]

valve flame tests The flame test requirement imposed on an aircraft valve depends on whether the valve must be fire proof or fire resistant. Fireproof valves are usually mounted on the engine side of a firewall or form part of the firewall. Fuel-actuated valves also fall within this category. Flame-resistant valves are valves mounted in similar regions where fires could occur, but the environment is not as severe.[ARP986-90]

valve, floating ball A valve with a full ball positioned within the valve that contacts either of two seat rings and is free to move toward the seat ring opposite the pressure source when in the closed position to effect shutoff.

valve, flow regulating (flow regulator) A valve which limits flow in a line to a predetermined valve, irrespective of variation in pressure differential caused by back-pressure or working against load.[ARP243B-65]

valve follower A linkage that transmits motion from a cam to the push rod of a valve, especially in an internal combustion engine.

valve, fuel control An automatically or manually operated device consisting essentially of a regulating valve and an operating mechanism. It is used to regulate fuel flow and is usually in addition to the safety shut-off valve. Such valve may be of the automatic or manually opened type.

valve, globe A valve with a linear motion closure component, one or more parts and a body distinguished by a globular shaped cavity around the port region.

valve, lock A pressure operated controllable check valve, sometimes known as a "counter balance valve" or "ratchet valve".[ARP243B-65]

valve, manual gas shutoff A manually operated valve in a gas line for the purpose of completely turning on or shutting off the gas supply.

valve, manual oil shutoff A manually operated valve in the oil line for the purpose of completely turning on or shutting off the oil supply to the burner.

valve, manual reset safety shut-off A manually opened, electrically latched, electrically operated safety shut-off valve designed to automatically shut off fuel when de-energized.

valve, motor driven reset safety shutoff An electrically operated safety shut-off valve designed to automatically shut off fuel flow upon being de-energized. The valve is opened and reset automatically by integral motor device only.

valve, overpressure relief A valve designed to relieve pressure at a predetermined level. Provides safety from explosive failure.[AIR1489-88]

valve plug An obsolete term. See closure component.

valve, pressure reducing (pressure reducer) A valve which reduces any inlet pressure to a pre-determined maximum outlet pressure regardless of flow or inlet pressure.[ARP243B-65]

valve, pressure regulating (unloader valve) A valve which directs flow from pump port to system port until pressure at the system port, or a control port, reaches cut-out pressure. The valve then opens and diverts flow from pump port to return port until system pressure recedes to cut-in pressure.[ARP243B-65]

valve, pressure relief A valve which limits maximum pressure in a circuit by releasing excess to return on the basis of differential pressure between pressure and return ports. A pressure relief valve must be able to accommodate full rated flow of the line size for which it is constructed.[ARP243B-65]

valve, priority A valve whose relief pressure is independent of downstream pressure and which opens and permits free flow when both upstream and downstream pressures exceed its setting. A priority valve also permits free reverse flow. Normal use of a priority valve is to provide pressure priority to one subcircuit over another.[ARP243B-65]

valve, relief A valve automatically releasing pressure higher than its setting; see also valve, pressure relief and valve, thermal relief.[ARP243B-65]

valve, restrictor A valve whose function is to produce a relatively high pressure drop in a fluid circuit by means of a reduced flow area or orifice.[ARP243B-65]

valve, restrictor, adjustable A restrictor valve having provisions for external adjustment.[ARP243B-65]

valve, restrictor, one-way (restrictor check valve) A restrictor valve which permits free flow in the reverse direction. [ARP243B-65]

valve, restrictor, two-way A restrictor valve which restricts flow in either direction, usually through a fixed orifice.[ARP243B-65]

valve, selector A control valve having more than one flow pattern. A selector valve will have a minimum of three working ports.[ARP243B-65]

valve, semi-automatic A valve incorporating automatic operation subject to external control.[ARP243B-65]

valve, servo A directional control valve which infinitely modulates flow or pressure as a function of its input signal. [ARP243B-65]

valve, servo, electro-hydraulic A hydraulic servo valve, the operation of which is controlled by an electrical input signal.[ARP243B-65]

valve, shutoff A two port, two position valve which opens or closes a fluid passage. A shutoff valve may be reversible operating equally well with pressure at either port or non-reversible, performing satisfactorily only with pressure entering at one of its ports.[ARP243B-65]

valve, shuttle A 3-port valve with one outlet and two inlet ports which automatically connects the "outlet" port to the "inlet" port having the higher pressure and blocks the "inlet" port having the lower pressure.[ARP243B-65]

valve, surge damping (surge damper) A valve whose function is to reduce surge pressures during intermittent flow.[ARP 243B-65]

valve, thermal relief A valve designed to bypass only the additional volume caused by thermal expansion of the fluid, i.e., no appreciable flow required.[ARP 243B-65]

van Allen radiation belts See radiation belts.

Van de Graaf generator An electrostatic device that uses a system of belts to generate electric charges and carry them to an insulated electrode, which becomes charged to a high potential.

vane 1. A flat or curved machine element attached to a hub or rotor that is acted upon by a flowing stream of fluid to produce rotary motion. 2. A fixed or adjustable plate inserted in a gas or air stream used to change the direction of flow.

vane control A set of movable vanes in the inlet of a fan to provide regulation of air flow.

vane guide A set of stationary vanes to govern direction, velocity and distribution of air or gas flow.

vane starter An engine starting device, incorporating a positive displacement vane type air/gas motor, driven by cold or heated compressed air or other gas. The starter may include speed increasing or reduction gears, and usually includes a jaw type engaging mechanism or an overrunning clutch.[ARP906A-91]

Van Stone nipples (flanges) A pipe nipple made with one enlarged integral end held against another face with a loose flange around the nipple.

vapor(s) 1. Gases whose temperatures are below their critical temperatures, so that they can be condensed to the liquid or solid state by increase of pressure alone. [NASA] 2. The gaseous product of evaporation.

vapor barrier A sheet or coating of low gas permeability that is applied to a structural wall to prevent condensation and absorption of moisture.

vapor barrier clothing Impermeable garments used with respirators as life support systems in toxic environments (caustic chemicals, etc.).[NASA]

vapor deposition See vacuum plating.

vapor-filled thermometer A type of filled-system thermometer in which temperature is determined from the vapor pressure developed from partial vaporization of a volatile liquid contained within the system.

vapor generator A container of liquid, other than water, which is vaporized by the absorption of heat.

vaporimeter 1. An apparatus in which the volatility of oils are estimated by heating them in a current of air. 2. An instrument used to determine alcohol content by measuring the vapor pressure of the substance.

vaporization The change from liquid or solid phase to the vapor phase.

vaporization cooling A method of cooling hot electronic equipment by spraying it with a volatile, nonflammable liquid of high dielectric strength; the liquid absorbs heat from the electronic equipment, vaporizes, and carries the heat to enclosure walls or to a radiator or heat exchanger. Also known as evaporative cooling.

vapor liquid ratio, V/L The equilibrium ratio of volume of vapor (actually air and fuel vapor) to volume of liquid, both at the same temperature.[AIR1326-91]

vapor phase epitaxy A crystal growth process whereby an element or a compound is deposited at a thin layer on a slice of substrate single crystal material by the vapor phase technique. [NASA]

vapor pressure 1. The pressure exerted by the molecules of a given vapor. For a pure confined vapor, it is that vapor's pressure on the walls of its containing vessel; and for a vapor mixed with other vapors or gases, it is that vapor's contribution to the total pressure (i.e., its partial pressure).[NASA] 2. The pressure (for a given temperature) at which a liquid is in equilibrium with its vapor. As a liquid is heated, its vapor pressure will increase until it equals the pressure above the liquid; at this point the liquid will begin to vaporize.

vapor pressure thermometer A temperature transducer for which the pressure of vapor in a closed system of gas and liquid is a function of temperature.

var A unit of measure for reactive power; it is calculated by taking the product of voltage, current, and the sine of the phase angle.

variable The symbolic representation of a logical storage location that can contain a value that changes during a discrete processing operation. See also measurand.

variable address See indexed address.

variable-area track A motion-picture sound track divided laterally into transparent and opaque areas, where the line of demarcation is an oscillographic trace that corresponds to the wave shape of the recorded sound.

variable-density track A motion-picture sound track of constant width, where the photographic density varies along the length of the track in accordance with a defined wave parameter of the recorded sound; the track is usually, but not always, of uniform density in the transverse direction at any point along the direction of travel.

variable-inductance accelerometer An instrument for measuring instantaneous acceleration of a body; it consists of a differential transformer with a center coil excited from an external a-c signal whose magnitude is proportional to displacement of a ferromagnetic core mass suspended on springs in the center of the three coils.

variable-inductance pickup A transducer that converts mechanical oscillations into audio-frequency electrical signals by varying the inductance of an internal coil.

variable-length record format A file format in which records are not necessarily the same length.

variable lift See lift.

variable lift devices A means of varying the lift of a basic airfoil through the use of various structures; for example, flaps, slots, and slats.[ARP4107-88]

variable pitch propeller See controllable pitch propeller.

variable position valve Controls flow by selective throttling of the flow passage. Intermediate areas may be infinitely or incrementally selectable.[ARP 986-90]

variable-reluctance pickup A transducer that converts mechanical oscillations into audio-frequency electrical signals by varying the reluctance of an internal magnetic circuit.

variable reluctance proximity sensor A device that senses the position (presence) of an actuating object by means of the voltage generated across the terminals of a coil surrounding a pole piece that extends from one end of a permanent magnet; coil voltage is proportional to the rate of change of magnetic flux as the object passes through the field near the pole piece.

variable reluctance tachometer A type of tachometer designed to measure rotational speeds of 10,000 to 50,000 rpm by detecting electrical pulses generated as an actuating element integral with the rotating body repeatedly passes through the magnetic field of a variable-reluctance sensor; the pulses are amplified and rectified, then used to control direct current to a milli-ammeter, which is calibrated directly in rpm.

variable-resistance accelerometer An instrument that measures acceleration by determining the change in electrical resistance in a measuring element such as a strain gage or slide wire whose dimensions are changed mechanically under the influence of acceleration.

variable-resistance pickup A transducer that converts mechanical oscillations into audio-frequency electrical signals by varying the electrical resistance of an internal circuit.

variable stream control engines Advanced, moderate bypass-ratio turbofan configurations that use duct burner thrust augmentation and coannular nozzles for jet noise reduction.[NASA]

variable word-length Having the property that a machine word may have a variable number of characters. It may be applied either to a single entry whose information content may be changed from time to time or to a group of functionally similar entries whose corresponding components are of different lengths.

variometer 1. Instrument for comparing magnetic forces, especially of the earth's magnetic field.[NASA] 2. A form of variable inductance, consisting of two coils connected in a series and arranged one inside the other, the inner equipped to rotate and, thereby, vary the mutual inductance between coils. This device was principally used in the early days of radio communications, but has found continued usefulness in electronics.

varistors Two electrode semiconductor devices having a voltage dependent nonlinear resistance.[NASA]

varmeter An instrument for measuring the electric power drawn by a reactive circuit. Also known as reactive volt-ampere meter.

varnish A transparent coating material consisting of a resinous substance dissolved in an organic liquid vehicle.

varying duty A requirement of service that demands operations at loads, and for intervals of time, both of which may be subject to wide variations.[ARP1199A-90]

vascular system See cardiovascular system.

VASI Acronym for Visual Approach Slope Indicator. An airport lighting facility providing vertical visual approach slope guidance to aircraft during approach to landing by radiating a directional pattern of high-intensity red and white focused light beams which indicate to the pilot that he/she is "on path" if red/white, "above path" if white/white, and "below path" if red/red. Some airports serving large aircraft have three-bar VASIs which provide two visual glide paths to the same runway.[ARP 4107-88]

VATOL aircraft Vertical attitude takeoff and landing aircraft.[NASA]

VAX 11-780 DEC's most powerful computer; uses thirty-two-bit words and has virtual memory.

V_b The design speed for maximum gust intensity; maximum turbulence penetration speed.[ARP4107-88]

V_c The design cruising speed used in calculation of structural strength in designing the aircraft; at this speed, the structure will sustain 30 ft/s gusts.[ARP 4107-88]

VCO See voltage controlled oscillators.

V_d The design diving speed to which the aircraft is carried in official certification tests; speed chosen by designer at which aircraft must be flown to ascertain that no adverse flight conditions exist at that speed.[ARP4107-88]

V_{df} Maximum demonstrated flight diving speed. Generally demonstrated only by test pilots. See also V_{ne}.[ARP4107-88]

VDU See video display unit.

vection illusion See illusions, visual.

vectopluviometer A rain gage, or a circular array of four or more rain gages, that measures the direction and inclination of falling rain.

vector(s) 1. Quantities such as force, velocity, or acceleration, which have both magnitude and direction at each point in space, as opposed to scalar which has magnitude only. Such quantities may be represented geometrically by an arrow of length proportional to its magnitude, pointing in the assigned direction.[NASA] 2. A one dimensional matrix. See matrix.

vectored interrupt An interrupt which carries the address of its service routine.

vectoring See radar vectoring.

vector quantity A property or characteristic which is completely defined only when both magnitude and direction are given.

vector voltmeter A two-channel, high-frequency sampling voltmeter that can be connected to two input signals of the same frequency to measure not only their voltages but also the phase angle between them.

vegetative index Linear combinations of spectral band responses in digital count, reflectance factor, or voltage to determine the vigor, greenness and/or biomass of the vegetation. Observations can be made by satelliteborne, aircraft-borne, truck mounted, or hand held spectrometers.[NASA]

vehicle 1. A body such as an aircraft or rocket designed to carry a payload aloft. 2. A self-propelled machine for transporting goods or personnel. 3. A solvent or other carrier for the resins and pigments in paint, lacquer, shellac or varnish.

vehicle instability Deflection of vehicle due to a wind force, causing unsafe working conditions.[ARP1328-74]

vehicle pivot point That point of the vehicle in contact with the ground on the opposite side of the vehicle from the side to which the wind force is applied and furthest from the wind's application point.[ARP1328-74]

vehicle tip point Maximum vehicle instability where the vehicle center of gravity has been rotated by a wind force to a point directly above the vehicle pivot point.[ARP1328-74]

veiling brightness Brightness superimposed on the retinal image which reduces its contrast. This veiling effect, produced by bright sources or areas in the visual field, results in decreased visual performance and visibility. [AIR1151-70]

velocimeter An instrument for measuring the speed of sound in gases, liquids or solids.

velocity (V) 1. Denotes the maximum design system operating velocity and is the takeoff speed under maximum gross weight standard hot day, 8000 ft (2432m) altitude, or as defined by the SPS.[AS483A-88] 2. Rate of motion. Rate of motion in a straight line is called linear speed, whereas change of direction per unit time is called angular speed. [NASA] 3. The rate of change of a position vector with respect to time at any given point in space; the first derivative of distance with respect to time.

velocity coupling The response of the burning propellant surface to the local velocity which would include both mean flow as well as acoustic velocity (both being parallel to the burning surface). [NASA]

velocity gain The proportional change of actuator velocity with error signal. This parameter is usually specified for no-load conditions and is usually measured open loop.[AIR1916-88]

velocity head The pressure, measured in height of fluid column, needed to create a fluid velocity. Numerically, velocity head is the square of the velocity divided by twice the acceleration of gravity ($U^2/2g$).

velocity limiting control See control, velocity limiting.

velocity meter A flowmeter that measures rate of flow of a fluid by determining the rotational speed of a vaned rotor inserted into the flowing stream; the vanes may or may not occupy the entire cross section of the flowpath.

velocity of approach A factor (F) determined by the ratio (m) of the valve orifice area to the inlet pipe area.

velocity of propagation In cable measurements, a function of dielectric constant. The transmission speed of an electric signal down a length of cable compared to its speed in free space.[ARP 1931-87]

velocity pressure The measure of the kinetic energy of a fluid.

velocity-type flowmeter A flow-measurement device in which the fluid flow causes a wheel or turbine impeller to turn, producing a volume-time readout. Also known as current meter; rotating meter.

velocity vs force or torque characteristic The curves of steady-state velocity vs steady-state force or torque when the actuator is driven by the power modulator with constant values of signal input.[AIR1916-88]

Venn diagram A graphical representation in which sets are represented by closed areas. The closed regions may bear all kinds of relations to one another, such as be partially overlapped, be completely separated from one another, or be contained totally one within another. All members of a set are considered to lie within or be contained within the closed region representing the set. The diagram is used to facilitate the determination of whether several sets include or exclude the same members.

vent Any opening or passage that allows gases to escape from a confined space to prevent the buildup of pressure or the accumulation of hazardous or unwanted vapors.

vent cap The check valve component on top of a cell which regulates the maximum internal pressure which the cell jar will hold while preventing the entrance of atmospheric gases under conditions of fluctuating ambient pressures. It is removable to allow water replacement to the electrolyte.[AS8033-88]

vented vs. closed amplifier A vented amplifier utilizes auxiliary ports to establish a reference pressure in a particular region of the amplifier geometry. A closed amplifier has no communication with an independent reference. [ARP993A-69]

venturi A constriction in a pipe, tube or flume consisting of a tapered inlet, a short straight constricted throat and a gradually tapered outlet; fluid velocity is greater and pressure is lower in the throat area than in the main conduit upstream or downstream of the venturi; it can be used to measure flow rate, or to draw another fluid from a branch into the main fluid stream.

venturi meter A type of flowmeter that measures flow rate by determining the pressure drop through a venturi constriction.

venturi tube(s) 1. Short tubes of smaller diameter in the middle than at the ends. When fluids flow through such tubes, the pressure decreases as the diameters become smaller, the amount of decrease being proportional to the speed of flow and the amount of restriction.[NASA] 2. A primary differential-pressure producing device having a cone section approach to a throat and a longer cone discharge section. Used for high volume flow at low pressure loss.

verify 1. To determine whether a transcription of data or other operation has been accomplished accurately. 2. To check the results of keypunching.

Verneuil process Method of single-crystal growth in which powder is dropped through an oxy-hydrogen flame, falling molten on crystal seed.[NASA]

vernier A short auxiliary scale which slides along a main instrument scale and permits accurate interpolation of fractional parts of the least division on the main scale.

vernier engines Rocket engines of small thrust used primarily to obtain a fine adjustment in the velocity and trajectory of a rocket vehicle just after the thrust cutoff of the last sustainer engine, and used secondarily to add thrust to a booster or sustainer engine.[NASA]

vernine See guanosines.

vertical attitude takeoff-landing aircraft See VATOL aircraft.

vertical axis An imaginary line passing through the center of gravity and lying in the lane of symmetry of an aircraft. It is perpendicular to both the longitudinal axis and the lateral axis. Angular movement about this axis is called yaw. Vertical axis is also referred to as the normal axis.[ARP4107-88]

vertical bearing (V/B) The angle in the vertical plane between the horizon and a line from the present position in space to the next 3-dimensional defined waypoint.[ARP1570-86]

vertical boiler A fire-tube boiler consisting of a cylindrical shell, with tubes connected between the top head and the tube sheet forms the top of the internal furnace. The products of combustion pass from the furnace directly through the vertical tubes.

vertical deviation The vertical distance or altitude between the existing flight altitude and the programmed path or profile altitude.[ARP1570-86]

vertical fins See fins.

vertical firing An arrangement of a burner such that air and fuel are discharged into the furnace, in practically a vertical direction.

vertical junction solar cells Solar cells made from wafers on which narrow grooves are formed using a preferential KOH etch. The grooved region is radiation tolerant.[NASA]

vertical motion simulators Vibration machines which produce mechanical oscillations parallel to the vertical axis. [NASA]

vertical navigation (VNAV) Those functions which provide guidance signals to control the aircraft during climb or descent.[ARP1570-86]

vertical orientation The attitude of an object in reference to a plane which is parallel to the direction of gravity (determined with a plumbline).[NASA]

vertical orifice installation, vertical orifice run, vertical meter run An orifice plate used in a vertical pipeline. See basic definitions.

vertical profile (VPROF) See path and profile beginning paragraphs. See also V-path.

vertical situation display (VSD) A display that represents the aircraft's situation on an imaginary vertical plane ahead of the aircraft. The basic dimensions of a VSD are azimuth, elevation, and attitude with respect to roll. Lateral translation of display elements signifies a change in aircraft heading, vertical translation of display elements represents a change in pitch or vertical flight path, and revolution of the display elements denotes rotation of the aircraft about the roll axis.[ARP4107-88]

vertical speed (VS) The rate of change of altitude. The term usually refers to baro altitude rate, but may also be radar altitude rate or inertial vertical velocity. [ARP1570-86]

vertical tails See tail assemblies.

vertical takeoff and landing aircraft (VTOL aircraft) See VTOL.

vertical velocity See sink speed.

vertigo The sensation that the outer world is revolving about the person (objective vertigo) or that he himself is moving in space (subjective vertigo). The word frequently is used erroneously as a synonym for dizziness or giddiness to indicate an unpleasant sensation of disturbed relations to surrounding objects in space.[NASA]

very fast-acting fuse A fuse that opens the circuit without deliberate time-delay and whose short-circuit opening time is faster than a normal opening fuse. [ARP1199A-90]

very high frequency (VHF) The frequency band between 30 and 300 MHz. [ARP4107-88]

very high frequency omnidirectional range (VOR) A ground-based electronic navigational aid transmitting very high frequency navigation signals, 360 deg in azimuth, oriented from magnetic north. Used as the basis for navigation in the National Airspace System. The VOR periodically identifies itself by Morse Code and may have an additional voice identification feature. Voice information/instructions to pilots.[ARP4107-88]

very high speed integrated circuits See VHSIC (circuits).

Very Large Array (VLA) A synthetic aperture radio telescope, consisting of 27 parabolic antennas each of which is 25 meters in diameter. The system when connected together is capable of arc-second resolution with high sensitivity resulting in the world's most powerful radio telescope. Operated by the National Radio Astronomy Observatory, it is located in Socorro, New Mexico. [NASA]

very large scale integration A very complex integrated circuit, which contains ten thousand or more individual devices, such as basic logic gates and transistors, placed on a single semiconductor chip.[NASA]

very long base interferometry The simultaneous observation of radio sources by two radio telescopes spaced very far apart to enhance angular resolution. The signals are recorded on magnetic tapes and combined electronically on a computer.[NASA]

Very Long Baseline Array (VLBA) A transcontinental radio telescope, being developed by the National Radio Astronomy Observatory to consist of ten dedicated and automated 25-meter (82 foot) diameter antennas distributed from Hawaii to St. Croix, Virgin Islands. [NASA]

vessel A container or structural enclosure in which materials—especially liquids, gases and slurries—are processed, stored or treated.

vestibular illusion See illusions, vestibular.

V/F (boilup-to-feed ratio) A quantity used to analyze the operation of a distillation column.

v_f The design maximum speed with the wing flaps in a prescribed extended position.[ARP4107-88]

V_{fe} The highest permissible speed at which wing flaps may be actuated.[ARP 4107-88]

VFR Visual Flight Rules.

VFR aircraft/VFR flight An aircraft conducting flight in accordance with visual flight rules.[ARP4107-88]

VFR conditions Weather conditions equal to or better than the minimum for flight under visual flight rules.[ARP 4107-88]

VFR conditions on-top/VFR-on-top ATC authorization for an IFR aircraft to operate in VFR conditions at any appropriate VFR altitude (as specified in FAR and as restricted by ATC) at or above the minimum en route altitude (MEA) which is at least 1000 ft above a cloud layer.[ARP4107-88]

VFR over-the-top The special permission for the operation of an aircraft over-the-top of an area of poor weather conditions under VFR when it is not being operated on an IFR flight plan.[ARP 4107-88]

V_h The maximum level flight airspeed with maximum continuous power applied.[ARP4107-88]

VHF Very High Frequency.

VHSIC (circuits) Chips being developed by a DOD program to provide high speed MIL spec VLSI device for us in military systems.[NASA]

vibrating density sensor Any of several devices in which a change in natural oscillating frequency of a device element—cylinder, single tube, twin tube, U-tube or vane—is detected and related to density of process fluid flowing through the system.

vibrating quartz-crystal moisture sensor A device for detecting the presence of moisture in a sample gas stream by dividing the stream into two portions, one of which is dried, then alternately passing the two streams across the face of a hygroscopically sensitized quartz crystal whose wet and dry vibrational frequencies are continuously monitored and compared to the frequency of an uncoated sealed reference crystal.

vibrating-reed electrometer An instrument that uses a vibrating capacitor to measure small electrical charges, often in combination with an ionization chamber.

vibrating-reed tachometer A device consisting of an extended series of reeds of various lengths mounted on the same base; the device is placed on a vibrating surface, such as the enclosure of rotating equipment, and the frequency determined by observing which of the reeds is vibrating at its natural frequency.

vibration Motion due to a continuous change in the magnitude of a given force which reverses its direction with time. Motion of an oscillating body during one complete cycle; two oscillations. [NASA]

vibrational frequencies (structural) See resonant frequencies.

vibration dampers See vibration isolators.

vibration damping Any method of converting mechanical vibrational energy into heat.

vibration isolators Resilient support that tend to isolate systems from steady state excitation.[NASA]

vibration machine A device for determining the effects of mechanical vibrations on the structural integrity or function of a component or system—especially electronic equipment. Also known as shake table.

vibration meter A device for measuring vibrational displacement, velocity and acceleration; it consists of a suitable pickup, electronic amplification circuits, and an output meter.

vibration mode In a system undergoing vibration, a characteristic pattern assumed by the system in which the motion of every particle is simple harmonic with the same frequency.[NASA]

vibration protection See vibration isolators.

vibration sensitivity See vibration error.

vibration-type level detector A device for detecting the level of solids in a bin or hopper, in which a tuning fork driven by a piezoelectric crystal vibrates freely when the level is below the sensor position and is inhibited from vibrating when bulk material surrounds the sensor.

vibratory separation A technique for separating or classifying particulate solids using screens that are subjected to vibratory or oscillating motion.

vibrograph An instrument for making an oscillographic recording of the amplitude and frequency of a mechanical vibration, such as by producing a trace on paper or film using a moving stylus.

vibrometer A device for measuring the amplitude of a mechanical vibration. Also known as vibration meter.

vibronic isolation Systems which minimize the transfer of vibrations from the floor and surrounding environment to the surface of an optical table or other equipment mounted on them.

vibronic transition A simultaneous change in both vibrational and electronic energy state of a molecule, with the amount of energy involved similar to that for electronic transitions.

Vickers hardness See diamond-pyramid hardness.

Victor airway Phonetic designation of VOR airways; for example, Victor 123. [ARP4107-88]

video In radio telemetry, this is the term generally applied to a telemetry multiplex output from a radio receiver.

video card In data processing, a plug-in circuit board that controls the display of data on the monitor.

video disks Disks, usually the size of long-playing stereo records, which store video data. The data is recorded by one of two techniques: the capacitance method, in which the disk has spiral grooves and is read by a contact stylus, and the optical method, which uses lasers in both the recording and playback of the data.[NASA]

video display unit (VDU) Any one of several types of shared human interface devices that use digital video technology.

video landmark acquisition and tracking Shuttle era system for earth-feature identification, acquisition, and tracking.[NASA]

video map An electronically displayed map on the radar display that may depict data such as airports, heliports, runway centerline extensions, hospital emergency landing areas, NAVAIDs and fixes, reporting points, airway/route centerlines, boundaries, handoff points, special-use tracks, obstructions, prominent geographic features, map alignment indicators, range accuracy marks, and minimum vectoring altitudes.[ARP 4107-88]

video receiver The data output of a telemetry receiver; the multiplex of telemetry measurements.

video terminal An operator terminal with a cathode-ray-tube (CRT) display instead of a printer; see CRT display.

video signals Signals with a bandwidth of over 20 kilohertz.[NASA]

vidicons Television pickup tubes utilizing photoconductors as the sensing elements.[NASA]

viewable screen size (useful screen size) The dimensions of the useable light emitting area of the CRT faceplate projected through the glass onto a plane perpendicular to the Z-axis of the faceplate.[ARP1782-89]

view effects Effects of change in angular size of field of view upon receptors of radiation.[NASA]

viewing envelope The envelope in space that is defined by horizontal and vertical angles measured from the perpendicular to the display face within which a viewer can perceive the symbology presented by the display.[ARP1782-89]

Virgo galactic cluster A cluster of galaxies nearest to the Milky Way Galaxy, centered in the constellation Virgo and about 16 mission light-years from earth.[NASA]

Virgo star cluster See Virgo galactic cluster.

virtual address space A set of memory addresses that are mapped into physical memory addresses by the paging or relocation hardware where a program is executed.

virtual block One of a collection of blocks comprising a file (or the memory image of that file). The block is virtual only in that its block number refers to its position relative to other blocks on the volume; that is, the virtual blocks of a file are numbered sequentially beginning with one, while their corresponding logical block numbers can be any random list of valid volume-relative block numbers.

virtual leak A gradual release of gas by desorption from the interior walls of a vacuum system in a manner that cannot be accurately predicted; its effect on system operation resembles that of an irregularly variable physical leak.

virtual memory The set of storage locations in physical memory, and on disk, that are referred to by virtual addresses. From the programmer's viewpoint, the secondary storage locations appear to be locations in physical memory. The size of virtual memory in any system depends on the amount of physical memory available and the amount of disk storage used for nonresident virtual memory.

virtual page number The virtual address of a page of virtual memory.

virtual processor Software which allows an individual user to consider a computer's resources to be entirely dedicated to him. A computer can simulate several virtual processors simultaneously.

viscoelastic damping The absorption of oscillatory motions by materials which are viscous while exhibiting certain elastic properties.[NASA]

viscoelastic flow See viscoelasticity.

viscometer An instrument that measures the viscosity of a fluid.

viscometer gage An instrument that determines pressure in a vacuum system by measuring the viscosity of residual gases.

viscosity 1. The property of a fluid relating to shearing stress and continued resistance to flow. The viscosity of sealants is measured at standard conditions by a device which has a spindle mounted on a gauge.[AS7200/1-91] 2. That molecular property of a fluid which enables it to support tangential stresses for a finite time and thus to resist deformation; the ratio of shear stress divided by shearing strain.[NASA] 3. Measure of the internal friction of a fluid or its resistance to flow.

viscous damping 1. The dissipation of energy that occurs when a particle in a vibrating system is resisted by a force that has a magnitude proportional to the magnitude of the velocity of the particle and direction opposite to the direction of the particle.[NASA] 2. A method of converting mechanical vibration energy into heat by means of a piston attached to the vibrating object which moves against the resistance of a fluid—usually a liquid or air—confined in a cylinder or bellows attached to a stationary support.

viscous-drag-type density meter A type of meter for determining gas density by comparing the drag force or linked impellers driven by flow of a standard gas and the test gas; the balance point is a function of gas density, and the instrument can be calibrated to read directly in density units.

viscous flow 1. The flow of a fluid through a duct under conditions such that the mean free path is very small in comparison with the smallest dimensions of a transverse section of the duct. This flow may be either laminar or turbulent.[NASA] See also laminar flow.

viscous fluids Fluids whose molecular viscosity is sufficiently large to make the viscous forces a significant part of the total force field in the fluid.[NASA]

viscous friction load A load opposing motion and proportional to load velocity.[AIR1916-88]

visibility 1. The quality or state of being perceivable by the eye.[AIR1151-70] 2. The ability, as determined by atmospheric conditions and expressed in units of distance, to see and identify prominent unlighted objects by day and prominent lighted objects by night. The term "visibility", as reported in statute miles, hundreds of feet, or meters, normally refers to horizontal visibility. [ARP4107-88]

visibility meter An instrument for directly or indirectly determining visual range in the earth's atmosphere.

visible infrared spin scan radiometer A radiometer used for satellite sounding of the atmosphere.[NASA]

visible radiation See light (visible radiation).

visible spectrum The range of wavelengths of visible radiation; display or graph of the intensity of visible radiation emitted or absorbed by a material as a function of wavelength or some related parameter.[NASA]

visual approach An approach wherein an aircraft on an IFR flight plan, operating in VFR conditions under the control of an air traffic control facility and having an air traffic control authorization, may proceed to the airport of destination in VFR conditions.[ARP 4107-88]

visual approach slope indicator See VASI.

visual flight rules (VFR) Rules that govern the procedures for conducting flight under visual conditions. The term "VFR" is also used in the United States to indicate weather conditions that are equal to or greater than minimum VFR requirements.[ARP4107-88]

visual illusion See illusions, visual.

visual line width The total width of the display line as seen by the human eye with or without the use of a magnifier. [ARP4067-89]

visual meteorological conditions (VMC) Meteorological conditions expressed in terms of visibility, distance from cloud, and ceiling equal to or better than specified minimums.[ARP4107-88]

visual photometry A subjective approach to the problem of photometry, wherein the human eye is used as the sensing instrument; to be distinguished from photoelectric photometry.[NASA]

visual separation A means employed by ATC to separate aircraft in terminal areas. There are two ways to effect this separation: 1. The tower controller may see the aircraft involved and issue instructions, as necessary, to ensure that the aircraft avoid each other. 2. A pilot may see the other aircraft involved and upon instructions from the controller provide his/her own separation by maneuvering the aircraft, as necessary, to avoid it.[ARP4107-88]

Viton A A fluorocarbon rubber by E.I. du Pont de Nemours Co.

vitreous enamel A coating applied to metal by covering the surface with powdered alkaliborosilicate glass frit and fusing it onto the surface by firing at a temperature of 800 to 1600 °F (425 to 875 °C). Also known as porcelain enamel.

vitreous silica See silica glass.

vitreous slag Glassy slag.

vitrification Formation of a glassy or noncrystalline material.[NASA]

vitrified wheel A grinding wheel made by compacting a mixture of abrasive particles and glass frit, then firing it to produce a bonded mass.

VLBI See very long base interferometry.

V_{le} The maximum safe speed for flight with the landing gear in the extended position.[ARP4107-88]

V_{lo} Maximum speed for safe extension or retraction of landing gear (landing gear operating speed).[ARP4107-88]

V_{lof} Liftoff speed; the velocity at which an aircraft separates from the ground during takeoff.[ARP4107-88]

VLSI See very large scale integration.

voice control Using the voice to activate devices which respond or operate by means of speech recognition.[NASA]

VMC Visual Meteorological Conditions. [ARP4107-88]

V_{mca} Minimum speed at which a multiengine aircraft is controllable in flight with the critical engine inoperative and the remaining engine(s) at takeoff power. It is not required that an airplane be able to climb or hold its altitude under these conditions, only that it be ale to maintain its heading.[ARP4107-88]

V_{mcg} The minimum speed at which a multiengine aircraft is controllable with the critical engine inoperative during the takeoff roll.[ARP4107-88]

V_{mo} Maximum operating limit speed. Limiting airspeed for turboprop and jet aircraft based on 80 percent of the flight-demonstrated diving speed. Flight at or beyond this speed has a high probability of causing serious structural damage to the aircraft. Similar to V_{ne}. [ARP4107-88]

V_{mu} Minimum unstick speed: that is, the calibrated airspeed at and above which the airplane can safely lift off the ground and continue flight.[ARP4107-88]

V_{ne} Never-Exceed speed. That airspeed which is 90 percent of the maximum flight-demonstrated diving speed. Flight at or beyond this airspeed entails a high risk of structural damage to the aircraft, especially if it is poorly rigged or a slight bit of turbulence is encountered. Use of this term is limited to piston-powered aircraft. Similar to V_{mo}.[ARP4107-88]

V_{no} The maximum structural cruising speed; the maximum speed for normal operation.[ARP4107-88]

vocabulary A list of operating codes or instructions available to the programmer for writing the program for a given problem, for a specific computer, or for a specific language.

voice print An acoustic spectrograph that can be used to analyze sound patterns, especially the harmonic patterns that distinguish one person's voice from another's.

void 1. Any opening, small crack, or crevice occurring at the juncture of structural members (such as chambers, reliefs, joggles, butt joints, or fasteners.) [AIR4069-90] 2. A shallow pocket or hollow on the surface of the nut due to nonfilling of metal during forging.[MA 1568-87]

void seal A seal used to fill holes, joggles, channels, and often other voids caused by the build-up of structure in a fuel tank. The void seal provides continuity of sealing where fillet seals are interrupted by such structure gaps.[AIR 4069-90]

vol See volume.

volatile 1. Of a liquid, having appreciable vapor pressure at room or slightly elevated temperature. 2. Of a computer, having memory devices that do not retain information if the power is interrupted.

volatile matter Those products given off by a material as gas or vapor, determined by definite prescribed methods.

volatile memory Memory whose contents are lost when the power is switched off.

volatile storage 1. A storage device in which stored data are lost when the applied power is removed, for example, an acoustic delay line. 2. A storage area for information subject to dynamic change.

volatization See vaporization.

volt A unit of electromotive force which when steadily applied to a conductor whose resistance is one ohm will produce a current of one ampere.

voltage See electric potential.

voltage amplification The ratio of voltage of an output signal to voltage of the corresponding input signal.

voltage breakdown test. Test to determine maximum voltage of insulted wire before electrical current leakage through the insulation.[ARP1931-87]

voltage, common mode (CMV) 1. A voltage of the same polarity on both sides of a deferential input relative to ground.

voltage controlled oscillators An oscillator whose frequency of oscillation can be varied by changing an applied voltage.[NASA]

voltage divider An electronic network that consists of multiple impedance elements connected in series; for a given voltage impressed across the network, one or more lower output voltages can be obtained by tapping across one or more node pairs in the network.

voltage drop The amount of voltage loss from original input in a conductor of given size and length.[ARP1931-87]

voltage modulation The cyclic variation or random variation, or both, about the mean level of the ac peak voltage during steady-state electric-system operation such as caused by voltage regulation and speed variations. The modulation envelope is formed by a continuous curve connecting the successive peaks of the basic voltage wave.[AS1212-71]

voltage modulation frequency characteristics The component frequencies which make up the modulation envelope wave form.[AS1212-71]

voltage-range multiplier A separate device installed externally to an instrument so its voltage range can be extended beyond the upper limit of the instrument scale; it consists principally of a special type of series resistance or impedance element.

voltage rating The maximum alternating current and/or the direct current voltage at which the protective device is designed to operate.[ARP1199A-90]

voltage standing-wave ratio In a waveguide, the ratio of the amplitude of the electric field at a voltage minimum to the amplitude at an adjacent voltage maximum.

voltage-type telemeter A system for transmitting information to a remote location using the amplitude of a single voltage as the telemeter signal.

volt-ampere meter An instrument for measuring apparent power—the product of voltage and current—in an a-c power circuit; in high-power applications the scale is usually graduated in kilovolt-amperes.

voltmeter An instrument for determining the magnitude of an electrical potential; it generally is constructed as a moving-coil instrument having high internal series resistance; if the high internal resistance is replaced with a low-resistance shunt connected in parallel with the instrument terminals, it can function as an ammeter.

voltmeter-ammeter An instrument consisting of a voltmeter and an ammeter in the same housing, but with separate electrical connections.

volt-ohm-milliammeter A test instrument having different ranges for measuring voltage, resistance and current flow (in the milliampere range) in electrical or electronic circuits. Also known as circuit analyzer; multimeter; multiple-purpose meter.

volume (vol) 1. The magnitude of a complex audio-frequency current measured in standard volume units on a graduated scale. 2. The three-dimensional space occupied by an object. 3. A measure of capacity for a tank or other container in standard units. 4. A mass storage media that can be treated as file-structured data storage.

volume control A device or system that regulates or varies the output-signal amplitude of an electronic circuit, such as for varying the loudness of reproduced sound.

volume flow rate Calculated using the area of the full closed conduit and the average velocity in the form, Q=VxA, to arrive at the total quantity of flow.

volume indicator A standard instrument for indicating the magnitude of a complex wave such as an electronic signal for reproducing speech or music; the magnitude in volume units equals the number of decibels above a reference level established by connecting the instrument across a 600-ohm resistor that is dissipating 1 mW of power at 100 Hz.

volume meter Any flowmeter in which actual flow of a fluid is determined by measuring a characteristic associated with the flow.

volume of air The number of cubic feet of air per minute expressed at fan outlet conditions.

volume of air dissolved (S_1) Volume of air initially dissolved in the fuel, measured at 32 °F and one atmosphere pressure in the vapor phase, per 100 volumes of the fuel at 60 °F.[AIR1326-91]

volume of air dissolved, (S_f) Volume of air measured at 32 °F and one atmosphere pressure that the fuel will dissolve under the lower or final partial pressure of air in the vapor phase, per 100 volumes of the fuel at 60 °F.[AIR-

1326-91]

volume of air evolved (S_g) Equals S_1-S_f; volume of air evolved, measured at 32 °F and one atmosphere pressure in the vapor phase per 100 volumes of the fuel at 60 °F.[AIR1326-91]

volumetric efficiency For a reciprocating engine or gas compressor, the ratio of volume of working fluid (at a specified temperature and pressure) admitted divided by piston displacement.

volumetric flow rate (q) The volumes of fluid moving through a pipe or channel within a given period of time.

volute A spiral casing for a centrifugal pump or fan; it allows the speed developed at the rotor vanes to be converted to pressure without hydraulic shock.

VOR Very-high-frequency Omnidirectional Range.

VORTAC Acronym for VHF Omnidirectional Range/TACtical Air Navigation. A combination of VOR (very-high-frequency omnidirectional range) and TACAN (tactical air navigation) ground-based radio navigation facilities.[ARP 4107-88]

vortex 1. The swirling motion of a liquid in a vessel at the entrance to a discharge nozzle. 2. The point in a cyclonic gas path where the outer spiral converge to form an inner spiral and where the two spirals change general direction by 180°.

vortex advisory system Display system which compares measured on-minute-average wind magnitudes and direction with the wind-rose criterion to predict wake vorticity and to indicate to the air traffic controller (with a red or green light) when the interarrival spacings for landings may be reduced to the 3 nautical mile limit.[NASA]

vortex alleviation The alteration of airfoil configurations to change the airflow patterns directly behind the wings to eliminate or inhibit the vertical motion which directly affects the aircraft immediately following, during closely spaced landings.[NASA]

vortex amplifier An amplifier which utilizes the pressure drop across a controlled vortex for modulating the output.[ARP993A-69]

vortex avoidance Schemes which involve airborne or ground-based equipment to track, monitor, and/or predict vortex behavior which might affect the approach and landing operations.[NASA]

vortex columns See vortices.

vortex disturbances See vortices.

vortex filaments The fine-scale structure of turbulent flow; the small non-energy containing eddies convected at mean freestream velocities.[NASA]

vortex flaps Leading edge flap designs for highly swept wings, in which the leading edge tabs, which are counter reflected, cause vortices to form on the flap. The trapped vortices cause significantly improved wind flow characteristics.[NASA]

vortex flow See vortices.

vortex shedding 1. Periodic separation of a fluid flowing past an unstreamlined body.[NASA] 2. A phenomenon that occurs when fluid flows past an obstruction; the shear layer near the obstruction has a high velocity gradient, which makes it inherently unstable; at some point downstream of the immediate vicinity of the obstruction, the shear layer breaks down into well-defined vortices, which are captured by the flowing stream and carried further downstream.

vortex streets Two parallel rows of alternately placed vortices along the wake of an obstacle in a fluid of moderate Reynolds number.[NASA]

vortex traps See trapped vortexes.

vortex tubes See vortices.

vortex-type flowmeter A device that uses differential-pressure variations associated with formation and shedding of vortices in a stream of fluid flowing past a standard flow obstruction—usually a circular element with a T-shaped cross section—to actuate a sealed detector at a frequency proportional to vortex shedding which, in turn, provides an output signal directly related to flow rate.

vortices In fluids, circulations drawing their energy from flows of much larger scale and brought about by pressure irregularities.[NASA]

vorticity equations Dynamic equations for the rate of change on the vorticity of a parcel, obtained by taking the curl of the vector equation of motion.[NASA]

votator A device used principally in food-processing industries for simultaneously chilling and mechanically working a continuous emulsified stream, such as in the production of margarine.

voter 1. A logic element or device that selects one signal to represent the output (of the voter) from 3 or more signals input to the device.[ARP1181A-85] 2. A logic element or device that compares the signal condition in two or more channels and changes state when a predetermined signal mismatch occurs.[AIR1916-88]

V-Path, path descent (climb), profile descent (climb) Terms used to designate modes which control the vertical trajectory to specific earth referenced paths or profiles which include one or more 3-dimensional waypoints (latitude, longitude and altitude). Usually the path (or profile) is controlled by elevator or stabilizer and the speed by throttles or drag. Other techniques (i.e. "backside") may be used. The path (or profile) may be linear or non-linear. Currently most usage involves descents but climbs are not excluded.[ARP1570-86]

V_r Rotation speed. The airspeed during takeoff roll at which the aircraft is rotated up in pitch to attain its takeoff attitude.[ARP4107-88]

V_{ref} Reference speed for final approach; generally 1.3 x V_s.[ARP4107-88]

V_s Stalling speed; the minimum steady flight speed at which the airplane is controllable.[ARP4107-88]

VSD Vertical Situation Display.

V_{so} Power-off stalling speed or minimum steady speed at which an airplane in the landing configuration is controllable.[ARP4107-88]

V_{ss} Stick shaker speed. That velocity at which the pilot's control stick (column) is caused to shake, warning the pilot of the nearness to stall speed.[ARP4107-88]

V/STOL (Vertical/Short Takeoff & Land) A type of aircraft designed to make a vertical or short run takeoff and landing.[AIR1489-88]

V/STOL aircraft A hybrid form of heavier-than-air aircraft that is capable, by virtue of one or more horizontal rotors or units acting as rotors, of taking off, hovering, and landing as, or in a fashion similar to a helicopter, and once aloft, and moving forward, capable, by means of a mechanical conversion of one sort or another, of flying as a fixed-wing aircraft, especially in its higher speed ranges.[NASA]

V/STOL forward operating facility A portable airfield capable of providing support by STOL fixed-wing aircraft as well as helicopters.[AIR1489-88]

VTOL Partial Acronym (pronounced VEE-TOL) for Vertical Takeoff and Landing (aircraft). Aircraft capable of vertical climbs and descents and of using very short runways or small areas for takeoff and landings. See also STOL.[ARP4107-88]

VTOL forward landing site A portable airfield of minimum size designed for operations dependent upon logistic or tactical support by helicopters.[AIR 1489-88]

vulcanization A complex reaction wherein the polymeric elastomer combines with additives, usually at an elevated temperature, to effect a change in state from an easily formed plasticity to a form with characteristic resilience and strength. The process is also referred to as curing or cross-linking. [AS1933-85]

vulcanizing Producing a hard, durable, flexible rubber product by steam curing a plasticized mixture of natural rubber, synthetic elastomers and certain chemicals.

V_x Best angle of climb speed. The airspeed which delivers the greatest gain of altitude for a given horizontal distance.[ARP4107-88]

V_{xse} Best angle of climb speed with one engine inoperative.[ARP4107-88]

V_y Best rate of climb speed. The airspeed which delivers the greatest gain in altitude for a given period of time.[ARP 4107-88]

V_{yse} Best rate of climb speed with one engine inoperative.[ARP4107-88]

w See flow rate (lb/hr).

W The calculated mass of the measured sample volume.[ARP1179A-80] See also watt.

wafer 1.A thin disc of a solid substance. 2. A thin part or component, such as a filter element.

wafer-type temperature detector A type of resistance-thermometer element designed with fine insulated wire of copper, nickel or platinum sandwiched between protective sheets of insulating material inside a sealed wafer-type enclosure; theses combine small mass with good thermal contact to give fast response times.

waist The center portion of a vessel, tank or container that is smaller in cross section than adjacent sections.

wait condition As applied to tasks, the condition of a task such that it is dependent on an event or events in order to enter the ready condition.

wait state In data processing, the elapsed time between the request for data from memory, and when the memory chip responds.

wake turbulence Phenomena resulting from the passage of an aircraft through the atmosphere. The term includes vortices, thrust stream turbulence, jet blast, jet wash, propeller wash, and rotor wash both on the ground and in the air.[ARP 4107-88]

walking beam landing gear A type of bogie landing gear configuration in which a forward and aft axle are carried on separate levers pinned to a structural post. A shock absorber linkage is attached to the levers and at the upper end to a swinging link or walking beam which is pivoted to the structural post. [AIR1489-88]

wall attachment amplifier An amplifier which utilizes control of the attachment of a free jet to a wall (Coanda effect) to modulate the output. Usually employed as a digital amplifier.[ARP 993A-69]

wall box A structure in a wall of a steam generator through which apparatus, such as sootblowers, extend into the setting.

wall thickness A term expressing the thickness of a layer of applied insulation.[ARP1931-87]

wandering sequence A welding technique in which increments of a weld bead are deposited along the seam randomly in both increment length and location.

warheads Originally the parts of the missile carrying the explosive, chemical, or other charge intended to damage the enemy. By extension, the term is sometimes used as synonymous with payload or nose cone.[NASA]

warning alert Emergency operational or aircraft system conditions that require immediate corrective or compensatory action by the crew.[ARP4153-88]

warning area See special use airspace.

warning indicating system A system which indicates to the pilot, or crew member, that a hazardous condition requiring immediate action exists, such as fire warning.[ARP1088-91]

warning, landing gear A device, mechanism, or system to provide a warning to the pilot when the landing gear is not in the landing configuration (extended) under specified conditions which may include aircraft altitude, engine power condition, and may provide visual or aural warning.[AIR1489-88]

warranty, failure-free A procurement requirement intended to have the manufacturer, or the design control agent, continuously upgrade the field reliability of designated equipment.[AIR1916-88]

wash 1. Airflow or turbulence induc ed by a propeller, rotor, jet engine, or airfoil.[ARP4107-88] 2. A stream of air or other fluid sent back along the axis of a propeller or jet engine. 3. A surface defect in castings caused by heat from the metal rising in the mold which induces expansion and shear of interface sand in the cope cavity. 4. A coating applied to the face of a mold prior to casting. Also known as mold wash. 5. To remove soil from parts, especially using a detergent or soap solution. 6. To remove cuttings or debris from a hole during drilling by introducing a liquid stream into the borehole and flushing it out.

washer 1. An intermediate object with a ratio of thickness to outside diameter of less than 25% which is used to provide a fixed dimension between two or more parts. See also spacer.[ARP480A-87] 2. A ring-shaped component used to distribute a fastener's holding force, insulate or cushion a nut or bolthead from its bearing surface, lock a nut in place, or improve tightness of a bolted joint. 3. A machine for mechanically agitating parts or materials in a detergent solution. Also known as washing machine.

waste 1. Rubbish from a building, or refuse from a manufacturing or process plant. 2. Dirty water from domestic or industrial uses. 3. The amount of excavated material remaining after the hole has been backfilled. 4. A relatively loose mass of threads or yarns used for wiping up spilled oil or other liquids.

waste fuel Any by-product fuel that is waste from a manufacturing process.

waste heat Sensible heat in noncombustible gases.

waste lubrication A method of delivering oil or other lubricant to a bearing surface by wicking action using cloth waste to absorb and transfer the lubricant.

waste treatment The processing of waste materials (liquid and solid) with chemicals high temperature, chopping, grinding, and filtering equipment. Bacterial action, dryers, separators, for conversion to useful products.[NASA]

watchdog In control systems, a combination of hardware and software which acts as an interlock scheme, disconnecting the system's output from the process in event of system malfunction.

watchdog timer An electronic internal timer which will generate priority interrupt unless periodically recycled by a computer. It is used to detect program stall or hardware failure conditions. See also timer, watchdog.

water 1. Dihydrogen oxide (molecular formula H_2O). The word is used ambiguously to refer to the chemical compound in general and to its liquid phase; when the former is meant, the term water substance is often used.[NASA] 2. A liquid composed of two parts of hydrogen and sixteen parts oxygen by weight.

water calorimeter A device for measuring radio-frequency power by determining the rise in temperature of a known volume of water in which the radio-frequency power is absorbed.

water column (w.c.) A vertical tubular member connected at its top and bottom to the steam and water space respectively of a boiler, to which the water gage, gage cocks, high and low level alarms and fuel cutoff may be connected.

water cooling Using a stationary or flowing volume of water to absorb heat and disperse it or carry it away.

water currents Net transport of water along a definable path.[NASA]

water-flow pyrheliometer A device for determining intensity of solar radiation in which the radiation sensor is a blackened water calorimeter; radiation intensity is calculated from the rise in temperature of water flowing through the calorimeter at a constant rate.

water gage The gage glass and its fittings for attachment.

water gas Gaseous fuel consisting primarily of carbon monoxide and hydrogen made by the interaction of steam and incandescent carbon.

water hammer 1. A sudden increase in pressure of water due to an instantaneous conversion of momentum to pressure. 2. A series of shocks, sounding like hammer blows, caused by suddenly reducing fluid-flow velocity in a pipe.

water heating The heating of water by any means including solar technology. [NASA]

water jacket A casing around a pipe, process vessel or operating mechanism for circulating cooling water.

water lane A lane or strip of water marked or set aside and maintained for the takeoff and landing of seaplanes. [AIR1489-88]

water level The elevation of the surface of the water in a boiler.

water path In ultrasonic testing, the distance from the search unit to the workpiece in a water column or immersion test setup.

waterproof Impervious to water. Compare with water resistant.

waterproof grease A viscous lubricant that does not dissolve in water and that resists being washed out of bearings or other moving parts.

waterproofing agent A substance used to treat textiles, paper, wood and other porous or absorbent materials to make them shed water rather than allow it to penetrate.

water resistant Slow to absorb water or to allow water to penetrate, often expressed as a maximum allowable immersion time. Compare with waterproof.

water rudder A small auxiliary rudder attached to the rear portion of a float or hull of a seaplane for the purpose of aiding directional control while the aircraft is on the water.[ARP4107-88]

water tube A tube in a boiler having the water and steam on the inside and heat applied to the outside.

water tube boiler A boiler in which the tubes contain water and steam, the heat being applied to the outside surface.

water vapor 1. Water (H_2O) in gaseous form. Also called aqueous vapor.[NASA] 2. A synonym for steam, usually used to denote steam of low absolute pressure.

waterways Navigable streams of canals; also channels for the passage or escape of water.[NASA]

watt (W) Metric unit of power. The rate of doing work or the power expended equal to 10^7 ergs/second, 3.4192 Btu/hour or 44.27 footpounds/minute.

watt-hour meter An integrating meter that automatically registers the integral of active power in a circuit with respect to time, usually providing a readout in kW-h.

wattmeter 1. Instrument for measuring the magnitude of the active power in an electric circuit. They are provided with a scale usually graduated in either watts, kilowatts, or megawatts. If the scale is graduated in kilowatts or megawatts, the instruments are usually designated as kilowattmeters or megawattmeters.[NASA] 2. An instrument for directly measuring average electric power in a circuit.

wave Variation of a physical attribute of a solid, liquid or gaseous medium in such a manner that some of its parameters vary with time at any position in the medium, while at any instant of time the parameters vary with position.

wave analyzer An electronic instrument for measuring magnitude and frequency of the various sinusoidal components of a complex electrical signal.

wave filter A transducer that separates waves by introducing relatively small insertion loss into waves of one or more frequency bands while introducing relatively large insertion loss into waves of other frequencies.

waveform digitizer A device which generates a digital signal corresponding to an analog waveform which it receives.

waveforms The graphical representations of waves, showing variation of amplitude with time.[NASA]

wave front 1. Of a wave propagating in a bulk medium, any continuous surface where the wave has the same phase at any given instant in time. 2. Of a wave propagating along a continuous surface, any continuous line where the wave has the same phase at any given instant in time.

wave gage A device for measuring the height of waves on the ocean or a large lake and for measuring the period between successive waves.

waveguide An elongated volume of air or other dielectric used in guided transmission of electromagnetic waves; it usually consists of a circular or rectangular tube of dimensions chosen for efficient propagation of a specific frequency or frequencies; the tube walls may be electrically conductive, or they may constitute surfaces where permittivity or permeability, or both, are discontinuous.

waveguide lasers Pump sources for deuterium oxide lasers.[NASA]

wave impedance In an electromagnetic wave, the ratio of the transverse electric field to the transverse magnetic field.

wave interference A pattern of varying wave amplitude caused by superimposing one wave on another in the same medium.

wavelength In any periodic wave, the distance from any point on the wave to a point having the same phases on the next succeeding cycle; the wavelength, λ, equals the phase velocity, v, divided by the frequency, f.

wavelength-division multiplexing Combination of two or more signals so they can be transmitted over a common optical path, usually through a single optical fiber, by a technique in which the signals are generated by light sources having different wavelengths.

wavelength division multiplexing The process in which each modulating wave modulates a separate subcarrier and the subcarriers are spaced in wavelengths. This term is used in optical communication where wavelength usage is preferred over frequency.[NASA]

wavelength meter An instrument which measures the wavelength of a laser beam, or other monochromatic source of light.

wavemeter An instrument for determining the wavelength of an a-c or high frequency signal, either directly or indirectly. The usual method is by measuring the frequency and converting to wavelength. Also called frequency meter.

wave motor A power conversion device for producing mechanical power from the lifting power of sea waves.

wave normal A unit vector perpendicular to a wave front and having a positive component coincident with the direction of propagation; in an isotropic medium, the wave normal lies along the direction of the propagation.

wave oscillators See oscillators.

wave radiation See electromagnetic radiation.

wave soldering A soldering technique used extensively to bond electronic components to printed circuit boards; soldering is precisely controlled by moving the assemblies across a flowing wave of solder in a molten soldering bath; the process also minimizes heating of the assemblies, and thus avoids one of the causes of early failure in electronic components. Also known as flow soldering.

wave tail The trailing portion of a signal-wave envelope, in time or distance, where the rms wave amplitude decreases from its steady-state value to the end of the wave train.

waypoint (WP) See area navigation, RNAV waypoint.

Wb See weber.

w.c. See water column.

weak interactions (field theory) One

class of the fundamental interactions among elementary particles responsible for beta decay of nuclei, and for the decay of elementary particles with lifetimes greater than about 10(-10) sec such as muons, K mesons, and lambda hypersons.[NASA]

weapons delivery Total requirements for locating the target, establishing the release conditions, and maintaining to the target (if required); includes the detection, recognition, and acquisition of the target, the weapons release as well as guidance.[NASA]

weapons replaceable assembly A generic term which includes all the replaceable packages of an avionic equipment, pod, or system as installed in an aircraft weapon system, with the exception of cables, mounts, and fuse boxes or circuit breakers.[ARD50010-91]

weapon system A weapon and those components/parts required for its operation.[ARD50010-91]

wear Progressive deterioration of a solid surface due to abrasive or adhesive action resulting from relative motion between the surface and another part or a loose solid substance.

wearout The process that results in an increase of the failure rate or probability of failure with an increasing number of life units.[AIR1916-88]

wear oxidation See fretting.

weathercocking/weathervaning The tendency of an aircraft to head into the wind.[ARP4107-88]

weathering The surface deterioration of a hose cover during outdoor exposure, such as checking, cracking, crazing, or chalking.[ARP1658A-86]

weatherometer A test apparatus used to estimate the resistance of materials and finishes to deterioration when exposed to climatic conditions; it subjects test surfaces to accelerated weathering conditions such as concentrated ultraviolet light, humidity, water spray and slat fog.

weatherproof Capable of being exposed to an outdoors environment without substantial degradation for an extended period of time.

weather resistance The relative ability of a material or coating to withstand the effects of wind, rain, snow and sun on its color, luster, and integrity.

web 1. In a grain of propellant, the minimum distance which can burn through as measured perpendicular to the burning surface (the distance between perforations in a grain).[AIR913-89] 2. The vertical plate connecting upper and lower flanges of a rail or girder. 3. The central portion of the tool body in a twist drill or reamer. 4. A thin section of a casting or forging connecting two regions of substantially greater cross section.

webbing A narrow fabric woven with continuous filling yarns and finished selvages.[AS8043-86]

weber (Wb) Metric unit for magnetic flux.

Weber-Fechner law An approximate psychological law relating the degree of response or sensation of a sense organ and the intensity of the stimulus. The law asserts that equal increments of sensation are associated with equal increments of the logarithm of the stimulus, or that the just noticeable difference in any sensation results from a change in the stimulus which bears a constant ratio to the value of the stimulus.[NASA]

wedge A wedge of wood, metal, plastic, etc., tapering to a thin edge.[ARP480A-87]

wedge lock fitting A fitting design developed under the auspices of the U.S. Army specifically for use in crash-resistant fuel tanks. The fitting gains its name from the use of a wedging action to retain the metal fitting rings to the flexible tank materials.[AIR1664-90]

weep A term usually applied to a minute leak in a boiler joint which forms droplets (or tears) of water very slowly.

weepage Occurrence of slight fluid at the surface of the sliding parts, such as piston rods, pump shafts, and valve spools, due to wiping of fluid from the wetted surface by seals or scrapers. Drops of fluids may be formed after a number of actuation cycles, to be defined. If there is no relative motion, such as at parting lines, weepage must not occur.[AIR1916-88]

Weibel instability An instability of collisionless plasmas characterized by the unstable growth of transverse electromagnetic waves and large magnetic field fluctuation brought about by an anisotropic distribution of electronic velocities.[NASA]

weight (wt) As in place weight. The multiplier value associated with a digit because of its position in a set of digits. The second place from the right has a place weight of 10 in decimal and 16 in hexadecimal, so a 2 in this location is worth 2x10 in decimal and 2x16 in hexadecimal. See also significance.

weight, carrier landing design gross The maximum weight at which an aircraft is designed to be landed aboard an aircraft carrier.[AIR1489-88]

weight, catapulting design gross The maximum weight at which an aircraft is designed to be catapulted from an aircraft carrier.[AIR1489-88]

weight, first landing The maximum weight at which the aircraft is designed to be landed after takeoff in a fully loaded condition.[AIR1489-88]

weighting Artificial adjustment of a measurement to account for factors peculiar to conditions prevailing at the time the measurement was taken.

weight, landing design gross The weight at which an aircraft is designed to be landed for a specified rate of sink velocity.[AIR1489-88]

weight, landplane landing design gross Applies to land based aircraft and to field landing of carrier based and amphibious types. The weight at which the aircraft is designed to be landed at a specified rate of sink velocity.[AIR1489-88]

weightlessness A condition in which no acceleration, whether of gravity or other force, can be detected by an observer within the system in question. [NASA]

weight, maximum design gross The weight of the aircraft with maximum internal and maximum external load for which provision is required.[AIR 1489-88]

weight, maximum landing design gross Applicable to land based operations. The maximum design gross weight of the aircraft less certain permissible reductions such as (a) assist takeoff fuel, (b) droppable fuel and tanks, (c) dumpable fuel, (d) other items expended during or immediately after takeoff as routine takeoff procedure.[AIR1489-88]

weight, maximum takeoff design gross The maximum taxi design gross weight less weight for fuel required in taxi out. [AIR1489-88]

weight, maximum taxi design gross The maximum weight at which the aircraft is designed to be taxied not necessarily the same as maximum design gross weight.[AIR1489-88]

weir An open-channel flow measurement device analogous to the orifice plate-flow constriction.

weir pond See stilling basin.

welded strain gage A type of foil strain gage especially designed to be attached to a metal substrate by spot welding; used almost exclusively in stress analysis.

welder A person or machine that makes a welded joint.

welding Producing a coherent bond between two similar or dissimilar metals by heating the joint, with or without pressure, and with or without filler metal, to a temperature at or above their melting point.

welding force See electrode force.

welding ground See work lead.

weldment A structure or assembly whose parts are joined together by welding.

weld metal The metal in the fusion zone of a welded joint.

well 1. The compartment, bay, or volume within the air vehicle for stowage of the retracted landing gear wheel assembly, strut and/or other associated equipment.[AIR1489-88] 2. A pressure-tight tube or similarly shaped chamber, closed at one end, and usually having external threads so it can be screwed into a tapped hole in a pressure vessel to form a pressure-tight means of inserting a thermocouple and other temperature-measuring element. Also known as thermowell. 3. A chamber at the bottom of a condenser, vacuum filter or other vessel where liquid droplets collect. Also known as hot well.

well-type manometer A type of double-leg, glass-tube manometer in which one leg is substantially smaller than the other; the large-diameter leg acts as a reservoir whose liquid level does not change appreciably with changes in pressure.

Westphal balance An instrument for determining the density of solids and liquids by direct reading, using a balance with movable weights; to measure liquid density, a plummet of known weight and volume is suspended in the liquid, whereas to measure solid density, a sample of the solid is suspended in a liquid of known density.

wet assay Determining the amount of recoverable mineral in an ore or metallurgical residue, or the amount of specific elements in an alloy, using flotation, dissolution and other wet-chemistry techniques.

wet back Baffle provided in a firetube boiler joining the furnace to the second pass to direct the products of combustion, that is completely water cooled.

wet-back boiler A baffle provided in a firetube boiler or water leg construction covering the rear end of the furnace and tubes, and is completely water cooled. The products of combustion leaving the furnace are turned in this area and enter the tube bank.

wet basis The more common basis for expressing moisture content in industrial measurement, in which moisture is determined as the quantity present per unit weight or volume of wet material; by contrast, the textile industry uses dry basis or regain moisture content as the measurement standard.

wet-bulb temperature The lowest temperature which a water wetted body will attain when exposed to an air current. This is the temperature of adiabatic saturation.

wet-bulb thermometer A thermometer whose bulb is covered with a piece of fabric such as muslin or cambric that is saturated with water; it is most often used as an element in a psychrometer.

wet classifier A device for separating solids in a liquid-solid mixture into fractions by making use of the difference in settling rates between small and large particles. The classifications can take place on a moving stream with appropriate spiral trays or in a tank or pond that is not agitated.

wet dielectric test A voltage dielectric test where the specimen to be tested is submerged in a liquid, and a voltage potential placed between the conductor and the liquid. Also called a tank test. [ARP1931-87]

wet grinding Any technique for removing surface layers of a metal part by abrasive action in the presence of a fluid, such as water or soluble oil, which cools and lubricates the work surface and carries away grinding debris.

wet installed fasteners Fasteners that are coated on the shank and under the head with a curing-type sealant to provide a corrosion barrier and a secondary seal.[AIR4069-90]

wet leg The liquid-filled low-side impulse line in a differential pressure level measuring system.

wetness A term used to designate the percentage of water in steam. Also used to describe the presence of a water film on heating surface interiors.

wet snow Snow existing at near freezing ambient temperatures. Wet snow tends to cling to exposed surfaces and may create an ice-like formation similar to the double horn shape, but is more likely to form a narrow hard ridge on the stagnation line.[AIR1667-89]

wet spinning The production of synthetic and man-made filaments by extruding the chemical solution through spinnerets into a chemical bath where they coagulate.[NASA]

wet steam Steam containing moisture.

wetting The process of supplying a water film to the water side of a heating surface.

wet-type differential pressure meter A design of differential-pressure instrument in which pressure difference is determined across an intermediate liquid in the instrument, by means of either an inverted-bell or float-type indicating mechanism.

Wheatstone bridge A four-arm resistance bridge, usually having three fixed resistances and one variable resistance.

wheel A simple machine consisting of a circular rim connected by a web or by spokes to a central hub or axle about which it revolves; in most applications, a wheel is used to support a load that rests on the axle while allowing the load to be moved easily from location to location.

wheelbarrowing A pilot-induced condition resulting when the nose wheel rather than the main gear is forced to support an abnormal share of the aircraft weight. This condition is aggravated by a crosswind and often results in loss of directional control.[ARP4107-88]

wheelbase The distance from the front gear axle to the aft gear axle(s) of an aircraft. If the aft gear axle is of multi wheel or bogie configuration, the dimension is to the pivot point of the assembly.[AIR1489-88]

wheel bead seat The portion of the rim of the wheel upon which the tire bead rests. The bead seat diameter is a basic dimension for a wheel.[AIR1489-88]

wheel bearing A device that allows a wheel to revolve about a stationary axle with low friction while supporting a heavy load; the chief types include journal, roller and ball bearings.

wheel, castering A wheel which is free to caster or swivel about an axis normal or near normal to the ground surface. Nose wheels are generally free to caster when not in a power steering mode. Tail wheels and outrigger wheels also are generally castering wheels. Most castering wheels require damping to avoid dynamic instability.[AIR1489-88]

wheel, convertible A wheel which is designed to be used either with a tube type tire or a tubeless tire.[AIR1489-88]

wheel, demountable flange A wheel which is designed to permit removal of one rim flange to allow tire installation and removal.[AIR1489-88]

wheel, detachable rim See wheel, demountable flange.

wheel diagram A method of representing a time-multiplexed format in the manner of a mechanical commutator or rotary switch.

wheel dresser A tool for refacing a grinding wheel to restore its dimensional accuracy and ability to cut work metal.

wheel, drop center divided A split type wheel with a drop center rim to aid in tire installation and removal.[AIR1489-88]

wheel flange A protruding rim, edge, etc., as on a wheel used to strengthen the wheel and to hold the tire in place.[AIR1489-88]

wheel, main The main wheels of an aircraft are those which carry the main portion of the load. In a standard tricycle configuration the main wheels are normally located just aft of the center of gravity of the air vehicle and carry 85-90% of the total aircraft weight. A main wheel is different from a nose wheel in that it is designed to house a brake(s) assembly.[AIR1489-88]

wheel, nonfrangible A wheel configuration whose design provides a means of continued load support and rolling capability in case of failure of the tire(s).[AIR1489-88]

wheel, nose The nose wheel(s) is located at the forward or nose portion of the aircraft and carry normally 10-15% of the aircraft weight in a standard tricycle gear configuration. Nose wheels normally are not designed to house a brake assembly.[AIR1489-88]

wheel, outrigger A wheel which is located in an outboard location such as a wing tip gear and carries little or no load. It provides stability for the air vehicle in ground operations.[AIR1489-88]

wheel rim flange The flange on the outboard edge of a wheel rim which holds the tire in place laterally. See also wheel flange.[AIR1489-88]

wheels, co-rotating Multiple (usually two) wheels rotating about a common axis and coupled together to rotate as a unit. Sometimes employed as an anti-shimmy measure in landing gear design.[AIR1489-88]

wheels, coupled See wheels, co-rotating.

wheel speed transducer A mechanism which generates an electrical signal with voltage or frequency proportional to wheel velocity.[AIR1489-88]

wheel, split A wheel configuration whose design provides a means of separating the wheel into two halves or parts for tire installation and removal.[AIR1489-88]

wheel, sprung A type of wheel whose design provides a shock absorber inside the wheel. Sometimes combined with a rigid structure landing gear.[AIR1489-88]

wheel, tail The tail wheel is located at the aft portion or tail of the air vehicle. On a tail wheel type gear configuration, it carries a small percentage of air vehicle weight. On a standard tricycle gear configuration, it is normally clear of the ground and carries no weight, but serves the purpose of tail protection in landing and takeoff operations.[AIR1489-88]

whip antennas Thin flexible monopole antennas.[NASA]

whirl See rotation.

whirling See rotation.

whispering gallery modes Electromagnetic (or elastic) waves that differ in frequency by more than an order of magnitude.[NASA]

whistlers Radiofrequency electromagnetic signals generated by some lightning discharges.[NASA]

white glass Highly diffusing glass having a nearly white, milky, or gray appearance. The diffusing properties are an inherent, internal characteristic of the glass.[ARP798-91]

white holes (astronomy) Time-reversed black holes, expanding sources with growing intensity and photon energy.[NASA]

white light A mixture of colors of visible light that appears white to the eye. A mixture of the three primary colors is sufficient to produce white light.

white noise 1. A sound or electromagnetic wave whose spectrum is continuous and uniform as a function of frequency.[NASA] 2. A noise whose power is distributed uniformly over all frequencies and has a mean noise power-per-unit bandwidth; since idealistic white noise is an impossibility, bandwidth restrictions have to be applied.

white radiation See Bremsstrahlung.

Whitworth screw thread A British standard screw threaded characterized by a 55° V form with rounded crests and roots.

wicking 1. Distribution of a liquid by capillary action into the pores of a porous solid or between the fibers of a material such as cloth. 2. Flow of solder under the insulation on wire, especially stranded wire.

wide angle diffusion That in which light is scattered over a wide angle. [ARP798-91]

wide-area network In data processing, the inter-connection of computers that can be miles apart in contrast to computers interconnected within one building.

wideband radiation thermometer A low-cost pyrometer that responds to a wide spectrum of the total radiation emitted by a target object; depending on the lens or window material used, the instrument responds to wavelengths from 0.3 μm to between 2.5 and 20 μm, which usually represents a significant fraction of the total radiation emitted. Also known as broadband pyrometer; total radiation pyrometer.

wide range metering The measurement of flow rates over a wide range of values at a defined accuracy.

Widmanstatten structure See basketweave.

Wien bridge A type of a-c bridge circuit that uses one leg containing a variable resistance and a variable capacitance to achieve balance with a leg containing a resistance and capacitance in parallel; although a Wien bridge can be used for measurements of this type, its more usual application is in determining frequency of an unknown a-c excitation signal.

wiggler magnets Components used in the production of coherent x-rays by the pumping of a gas with synchrotron radiation in combination with low energy photon beams.[NASA]

Wightman theory See quantum theory.

wild card In MS-DOS, a symbol used to search for any other symbol in a file.

winch A machine, usually power driven, consisting chiefly of a horizontal drum on which to wind cable, rope or chain and with which to apply a pulling force for hauling or hoisting; a winch with a vertical drum is usually known as a capstan.

Winchester disk A high-storage-capacity, small, moderately priced magnetic recording medium with a nonremovable storage element.

wind Movement of air which causes a force to be imposed on surfaces of aircraft ground support equipment.[ARP-1328-74]

windbox A chamber below the grate or surrounding a burner, through which air under pressure is supplied for combustion of the fuel.

windbox pressure The static pressure in the windbox of a burner or stoker.

wind circulation See atmospheric circulation.

wind direction (WD) The clockwise angle from true north to the wind velocity vector.[ARP1570-86]

wind gust A temporary increased wind force that exceeds the steady state wind force.[ARP1328-74]

windmill With reference to a propeller or rotor, to rotate without engine power, as from the force of the relative wind, or from prior existing power.[ARP4107-88]

windmilling Steady state engine rotation which is caused solely by air being forced through the engine by motion of flight (ram air).[ARP906A-91]

windmill start An airstart initiated after the engine has reached equilibrium windmilling speed.[ARP906A-91]

wind-on-nose (WON) The wind component acting along the longitudinal axis of the aircraft. A tailwind is positive and is sometimes designated as S by some navigators (acting on the stern). A headwind is negative and is sometimes designated as N (acting on the nose).[ARP1570-86]

window 1. An opening in the wall of a building or the sidewall of a vehicle body, usually covered with transparent material such as glass, which admits light and permits occupants to see out. 2. An interval of time when conditions are favorable for taking some predetermined action, such as launching a space vehicle. 3. A defect in thermoplastic sheet or film, similar to a fisheye but generally larger. 4. An aperture for passage of magnetic or particulate radiation. 5. An energy range or frequency range that is relatively transparent to the passage of waves. 6. A span of time when specific events can be detected, or when specific events can be initiated to produce a desired result. 7. In data processing, an area of a computer screen that can be temporarily opened to run a second program without disturbing the original program.

wind propagation The use of current present position instantaneous winds for all downstream leg ETE's and ETA's. The ETE's and ETA's are continuously revised as the present position winds change. The accuracy is increased if the wind is resolved into directional components corresponding to the downstream leg direction of flight as opposed to simply using present position ground speed for all downstream legs.[ARP1570-86]

wind shear A change in the wind speed or wind direction, or both, in a very short distance resulting in a tearing or shearing effect. It can exist in a horizontal or vertical direction and occasionally in both.[ARP4107-88]

windshear cautionary alert An alert which is set at a windshear level requiring pilot awareness and may require pilot action.[ARP4109-87]

windshear warning alert An alert which is set at a windshear level requiring immediate corrective or compensatory action by the pilot.[ARP4109-87]

wind speed The velocity of the wind with respect to a point on the earth's surface.[ARP1570-86]

wind tunnel(s) Tubelike structures or passages, sometimes continuous, together with their adjuncts, in which high speed movements of air or other gases are produced, as by fans, and within which objects such as engines or aircraft, airfoils, rockets (or models of these objects), are placed to investigate the airflow about them and the aerodynamic forces acting upon them.[NASA]

wind turbines Machines which convert wind energy into electricity.[NASA]

windup Saturation of the integral mode of a controller developing during times when control cannot be achieved, which causes the controlled variable to overshoot its setpoint when the obstacle to control is removed.

winglets In aerospace engineering, small nearly vertical, winglike surfaces mounted rearward above the wing tips to reduce the drag coefficients of lifting conditions.[NASA]

wing load(ing) Gross weight of an aircraft divided by the area of the wing.[ARP4107-88]

wing nacelle configurations Aerodynamic configurations involving various arrangements of wings and nacelles (over-the-wing, etc.).[NASA]

wing nut An internally threaded fastener having radial projections that allow it to be tightened or loosened by finger pressure.

wingover A flight maneuver or stunt in which the plane is put into a steep climbing turn until almost stalled after which the nose is allowed to drop while the turn is continued until normal flight is attained in a direction opposite to the original heading.[ARP4107-88]

wing root The base of a wing where it joins the fuselage.[ARP4107-88]

wing skid A skid or runner located in an outboard location near the wing tip to serve as protection to the air vehicle structure in case of tip over or instability.[AIR1489-88]

wingtip gear A landing gear assembly located at or near the wingtip usually to provide stability for ground operations when the main load carrying gear assemblies are located near the centerline of the aircraft.[AIR1489-88]

winning Recovering a metal from an ore or chemical compound.

winterize To prepare a measurement station for local winter conditions.

wiped joint A type of soldered joint in which molten filler metal is applied, then distributed between the faying surfaces by sliding mechanical motion.

wiping action The interaction of two electrical contacts which come in contact by their surfaces sliding against each other. See also contact wipe.[ARP 914A-79]

wire 1. A single metallic conductor of solid, stranded, or tensile construction designed to carry current in an electric circuit, but not having a metallic covering, sheath, or shield.[ARP1931-87] 2. A very long, thin length of metal, usually of circular cross section, and usually made by drawing through a die.

wire cloth Screening made of wires crimped or woven together.

wired program computer A computer in which the instructions that specify the operations to be performed are specified by the placement and interconnection of wires. The wires are usually held by a removable control panel, allowing flexibility of operation, but the term is also applied to permanently wired machines which are then called fixed program computers. Related to fixed program computer.

wiredrawing Reducing the cross section of wire or rod by pulling it through a die.

wire gage 1. A device for measuring the diameter of wire or thickness of sheet metal. 2. A standard series of sizes for wire diameter or sheet metal thickness, usually indicated by size codes consisting of a series of consecutive numbers.

wire insulation A flexible covering for electrical wire, which has relatively high dielectric strength to prevent inadvertently grounding the conductor, and which may have specific mechanical or chemical attributes that protect the conductor from damage or environmental effects in service.

wire rope A flexible length of stranded metal wire suitable for carrying substantial axial tension loads; it is most often used to haul or hoist loads or to act as a tensioned stationary support member in a structure, framework or truss.

wire size 1. A numerical designation for a conductor, usually expressed in terms of American Wire Gage (AWG), based on the approximate circular mil area of the conductor.[ARP914A-79][ARP1931-87] 2. The size of a wire is that number corresponding to its conductor size, such as "Size 20". See also conductor size.[AS1198-82]

wire stripper A tool for removing insulation from the end of insulated wire to prepare the wire for termination.

wire throw-out In braided hose, a broken end or ends in the wire reinforcements protruding from the surface of the braid.[ARP1658A-86]

wiring 1. (noun) Wires, cables, groups, harnesses, and bundles, and their terminations, associated hardware, and support, installed in the vehicle.[ARP1931-87] 2. (verb) When used as a verb, it is the act of fabricating and installing these items in the vehicle. [ARP1931-87]

wiring closet The room or location where the telecommunication wiring for a building, or section of a building, comes together to be interconnected.

wiring devices Wiring devices are the accessory parts and materials which are used in the installation of wiring, such as terminals, connectors, junction boxes, conduit, insulation, and supports.[ARP-1931-87]

Wobbe index The ratio of the heat of combustion of a gas to its specific gravity. For light hydrocarbon gases the Wobbe index is almost a linear function of the gas' specific gravity.

wobble switch Any of several designs of momentary or limit switches which are actuated by physical contact of an object with an extended wire, rod or cable projecting from the switch body; in most instances, wobble switches can be actuated by an object approaching from any direction in a plane perpendicular to the axis of the actuator.

Wolf-Rayet stars Very luminous, very hot (as high as 50,000K) stars whose spectra have broad emission lines (mainly He I and He II, which are presumed to originate from material ejected from the stars at very high velocities. Some W-R spectra show emission lines due to carbon CWC stars; others show emission lines due to nitrogen (WN stars).[NASA]

Wollaston wire Extremely fine platinum wire used in electroscopes, microfuses and hot wire instruments; it is made by coating platinum wire with silver, drawing the sheathed wire, and dissolving the silver away with acid.

Wood's glass A type of glass that is relatively opaque to visible light but relatively transparent to ultraviolet rays.

word 1. A group of bits that contain a measurement, command, tag or other information—typically sixteen bits. 2. A collection of bits (8, 12, 16, or 32) that represents the basic instruction or data in the computer. 3. In PCM telemetry terminology, one group of bits which represents a specific measurement; typically, a binary number. See also alphabetic word, computer word, machine word, and numeric word.

word length The number of characters or bits in a machine word. In a given computer, this number may be constant or variable.

word processing 1. The use of a computer, often with a CRT under full-screen control to facilitate the recording, storage, editing, updating, and organization of information in the form of words, especially sentential information. [NASA] 2. Software and computer hardware that can manipulate text in a variety of ways, such as formatting, moving, replacing, etc.

word selector A hardware device that can select a given word or words whenever they occur in the format, and present each to the user as a display and/or analog output.

word time In a storage device that provides serial access to storage locations, the time interval between the appearance of corresponding parts of successive words.

word wrap The ability of word processing programs to automatically start another line when a designated line-length is reached.

work-around An action required to complete the process run, even though all equipment is not working satisfactorily. You may have to run part of the process on manual, or you might jump out of an interlock until the maintenance can be scheduled.

work curve A graph which plots the pull-out force, indent force and relative conductivity of a crimp joint as a function of various depths of crimping.[ARP 914A-79]

work function An electrical potential difference corresponding to the amount of work that must be done to remove an electron from the surface of a metal.

work hardening See strain hardening.

working fluid(s) 1. Fluids (gas or liquid) used as the medium for the transfer of energy from one part of a system to another.[NASA] 2. Fluid that does the work for a system.

working load The maximum service load that an individual structural member, or an entire structure, is designed to carry.

working pressure The maximum allowable operating pressure for an internally pressurized vessel, tank or piping system, usually defined by applying the ASME Boiler and Pressure Vessel Code or the API piping code.

working space See working storage.

working standard Any standardized luminous source for daily use in photometry.[ARP798-91]

working storage A portion of the internal storage reserved for the data upon which operations are being performed. Synonymous with working space and contrasted with program storage.

working temperature The temperature of the fluid immediately upstream of a primary device.

working voltage See service rating.

work lead The electrical conductor that connects the workpiece to a welding machine or other source of welding power. Also known as ground lead; welding ground.

work life See assembly time.

workload The total of the combined physical and mental demands upon a person.[ARP4107-88]

work softening The phenomena of a drop in the yield strength of a metal when it has been strained or cold worked at low temperature and subsequently strained at an elevated temperature to cause the dislocations to become unstable.[NASA]

World Federation The joining together of three international regions: a) The Americas (Canadian MAP Interest Group and U.S. MAP/TOP Users Group) and Western Pacific (Australian MAP Interest Group), b) Asia (Japan MAP Users Group), and c) Europe (European MAP Users Group).

worm A shaft having at least one complete spiral tooth around the pitch surface, and used as the driving member for a worm gear or worm wheel.

worm gear A gear with teeth cut on an angle so it can be driven by a worm; it is used to transmit power and motion between two nonparallel, nonintersecting shafts.

worm wheel A gear wheel with curved teeth that mesh with a worm; it is usually used to transmit power and motion from the worm shaft to a nonintersecting shaft whose axis is at right angles to the worm shaft.

wormy alpha See stringy alpha.

wow In instrumentation tape recording and playback, high-frequency tape speed variations. See flutter.

wraparound contact solar cells See solar cells.

wrap forming See stretch forming.

wrapper sheet The outside plate enclosing the firebox in a firebox or locomotive boiler. Also the thinner sheet in the shell of a two thickness boiler drum.

wrench A manual or power operated tool, designed to impart a rotary movement of force, and is used for installing or removing nuts, bolts, and the like.[ARP 480A-87]

wringing fit A type of interference fit having zero to slightly negative allowance.

write To transfer information and store it. The information may be written in memory devices like RAM, or on storage media such as magnetic discs or tapes. Sometimes the storage element is absent, as in writing to a display.

write-protect notch A cut-out in the diskette envelope that prevents the computer from writing on the diskette, but does not prevent the computer from reading the diskette.

write time The amount of time it takes to record information. Related to access time.

wrought alloy A metallic material that has been plastically deformed, hot or cold, after casting to produce its final shape or an intermediate semifinished product.

W-R stars See Wolf-Rayet stars.

W stars See Wolf-Rayet stars.

wt See weight.

wye A pipe fitting similar to a tee, but in which the branch is at a 45° angle to the run.

wye-delta bridge A d-c bridge arrangement where resistors no larger than one megohm are used in wye configurations to simulate high-accuracy, high-stability resistors of one gigohm (10^9 ohms) or greater; these arrangements are used to accurately determine the value of very large resistances or to calibrate a bridge.

xenon chloride lasers Rare gas-halide lasers using XaCl as the active material. [NASA]

xenon fluoride lasers Lasers using XeF as the active material.[NASA]

xerography A dry copying process involving the photoelectric discharge of an electrostatically charged plate. The copy is made by tumbling a resinous powder over the plate, the remaining electrostatic charge discharged and the resin transferred to paper or an offset printing master.

XM-6 squib See squibs.

XM-8 squib See squibs.

x-ray binaries Bright galactic x-ray sources consisting of a compact star (neutron star or black hole) accreting matter from a close companion star. [NASA]

x-ray diffraction analyzer Any of several devices for detecting the positions of monochromatic x-rays diffracted from characteristic scattering planes of a crystalline material; used primarily in detecting and characterizing phases in crystalline solids.

x-ray diffractometer An instrument used in x-ray crystallography to measure the diffracted angle and intensity of x-radiation reflected from a powdered, polycrystalline or single-crystal specimen.

x-ray emission analyzer An apparatus for determining the elements present in an unknown sample (usually a solid) by bombarding it with electrons and using x-ray diffraction techniques to determine the wavelengths of characteristic x-rays emitted from the sample; wavelength is used to identify the specific atomic species responsible for the emission, and relative intensity at each strong emission line can be used to quantitatively or semiquantitatively determine composition.

x-ray fluorescence analyzer An apparatus for analyzing the composition of materials (solid, liquid or gas) by exciting them with strong x-rays and determining the wavelengths and intensities of secondary x-ray emissions.

x-ray goniometer An instrument for measuring the angle between incident and refracted beams of radiation in x-ray analysis.

x-ray imagery Reproduction of an object by means of focusing penetrating electromagnetic radiation (wavelengths ranging from 10-5 to 103 angstroms) coming from the object or reflected by the object. Analogous to infrared imagery, radar imagery and microwave imagery using the IR, radar and microwave frequencies.[NASA]

x-ray microscope An apparatus for producing greatly enlarged images by projection using x-rays from a special ultra-fine-focus x-ray tube, which acts essentially as a point source of radiation.

x-ray monochromator A device for producing an x-ray beam having a narrow range of wave-lengths; it usually consists of a single crystal of a selected substance mounted in a holder that can be adjusted to give proper orientation.

x-rays Short-wavelength electromagnetic radiation, having a wavelength shorter than about 15 nanometers, usually produced by bombarding a metal target with a stream of high-energy electrons; wavelengths are in the same range as gamma rays, longer than cosmic rays but shorter than ultraviolet; like gamma rays, x-rays are very penetrating and can damage human tissues, induce ionization, and expose photographic films.

x-ray stars Stars with strong emission in the x-ray portion of the electromagnetic spectrum.[NASA]

x-ray thickness gage A device used to continually measure the thickness of moving cold-rolled sheet or strip during the rolling process; it consists of an x-ray source on one side of the strip and a detector on the other—thickness is proportional to the loss in intensity as the x-ray beam passes through the moving material.

x-ray tubes Vacuum tubes designed to produce x-rays by accelerating electrons to a high velocity by means of an electrostatic field, then suddenly stopping them by collision with a target.[NASA]

XV-15 aircraft Experimental model of a tilt-rotor aircraft built by Bell Aircraft Company.[NASA]

X-value A term sometimes used to designate the inductive or capacitative reactance of an a-c electrical device or circuit.

x-wing rotors A new VTOL concept utilizing the stopped rotor x-wing aircraft.[NASA]

XY plotter A device used in conjunction with a computer to plot coordinate points in the form of a graph.

XY recorder A recorder for automatically drawing a graph of the relationship between two experimental variables; the position of a pen or stylus at any given instant is determined by signals from two different transducers that drive the pen-positioning mechanism in two directions at right angles to each other.

Y See expansion factor.

yagi antennas Directional antennas used on some types of radar and radio equipment consisting of an array of elemental, single wire dipole antennas and reflectors.[NASA]

Yang-Mills fields Types of fields based upon Yang-Mills Theory.[NASA]

Yang-Mills theory Mathematical idea for describing interactions among elementary particles which is based on the idea of gauge invariance under a non Abelian group.[NASA]

yaw The lateral rotational or oscillatory movement of an aircraft about the vertical axis.[ARP4107-88]

yawed angle or angle of yaw The amount of angular movement about a vertical axis. In a tire therefore, the yawed angle is that angle between the plane of symmetry of the wheel, and the direction line of roll of the wheel/tire assembly.[AIR1489-88]

yawed rolling The process of rolling in a yawed mode of operation.[AIR1489-88]

yawing moments Moments that tend to rotate aircraft, airfoils, rockets, or spacecraft about a vertical axis.[NASA]

yawing response to steer angle The tire response (yawed rolling angle) for a given input steering signal (angle).[AIR1489-88]

yield The quantity of a substance produced in a chemical reaction or other process from a specific amount of incoming material.

yield strength 1. The stress at which a material exhibits a specified deviation from proportionality of stress and strain.[ARP700-65] 2. The lowest stress at which a material undergoes plastic deformation. Below this stress the material is elastic; above it, viscous.[ARP 1931-87]

yield strength-offset The distance along the strain coordinate between the initial portion of the curve and a parallel line that intersects the stress-strain curve at a value of stress which is used as a measure of the yield strength. It is used for materials that have no obvious yield point. A value of 0.2% of the strain is normally used.[ARP700-65]

yield stress The force per unit area at the onset of plastic deformation, as determined in a standard mechanical-property test such as a uniaxial tension test.

yoke 1. A clamp or structure that attaches to two or more other parts to hold them in place.[ARP480A-87] 2. A slotted crosshead used in some steam engines instead of a connecting rod. 3. The framework surrounding the rotor of a d-c generator or motor which supports the field coils and provides magnetic linkage between them.

Young modulus See modulus of elasticity.

Z

Z See compressibility factor.

zap In data processing, a slang word meaning to erase or wipe-out data.

ZEL (Zero Launch) The launch of a rocket or aircraft by a zero length launcher.[AIR1489-88]

zenith That point of the celestial sphere vertically overhead. The point 180 deg. from the zenith is called the nadir. [NASA]

zero a device To erase all the data stored on a volume and reinitialize the format of the volume.

zero adjuster A mechanism for repositioning the pointer on an instrument so that the instrument reading is zero when the value of the measured quantity is zero.

zero adjustment See adjustment, zero.

zero-based conformity See conformity, zero-based.

zero-based linearity See linearity, zero-based.

zero bias A positive or negative adjustment to instrument zero to cause the measurement to read as desired.

zero code error A measure of the difference between the ideal (0.5 LSB) and the actual differential analog input level required to produce the first positive LSB code to transition (00...00 to 00...01).

zero defects A management program that encourages perfect performance in a manufacturing operation, and usually provides workers with rewards and incentives for achieving perfection.

zero drift Time related deviation of analyzer output from zero set point when it is operating on gas free of the component to be measured. This is not to be confused with "interference".[ARP1256-90]

zero elevation Biasing the zero output signal to raise the zero to a higher starting point. Usually used in liquid level measurement for starting measurement above the vessel connection point.

zero error See error, zero.

zero frequency gain See gain, zero frequency.

zero-fuel weight The ramp weight of an airplane minus the weight of usable fuel onboard.[ARP4107-88]

zero-g ACPL (spacelab) See atmospheric cloud physics lab (spacelab).

zero gas A gas to be used in establishing the zero, or no-response, adjustment of an analyzer.[ARP1256-90]

zero governor A regulating device which is normally adjusted to deliver gas at atmospheric pressure within its flow rating.

zero gravity See weightlessness.

zero length launcher A launcher that holds a vehicle in position and releases simultaneously at points such that a buildup of thrust (normally rocket thrust) is sufficient to take the missile or vehicle directly into the air without need of a takeoff run and without imposing a pitch rate upon release.[AIR1489-88]

zero level A reference level used for comparing signal intensities in electronic or sound-reproduction systems—in electronics, the zero level is usually taken as 0.006 W of power; in sound reproduction, it is usually taken as the threshold of hearing.

zero-lift angle of attack That angle between the chord plane of an airfoil and the relative wind at which the airfoil produces no lift.[ARP4107-88]

zero point energy Kinetic energy retained by molecules of a substance at a temperature of absolute zero.[NASA]

zero shift A shift in the instrument calibrated span evidenced by a change in the zero value. Usually caused by temperature changes, overrange, or vibration of the instrument.

zero suppression 1. The elimination of nonsignificant zeros in a numeral. 2. Biasing the zero output signal to produce the desired measurement. Used in level measurement to counteract the zero elevation caused by a wet-leg.

zero-zero Condition of no effective visibility in either a horizontal or a vertical direction, as in zero-zero weather. [ARP4107-88]

zero, zero out The procedure of adjusting the measuring instrument to the proper output value for a zero-measurement signal.

zeta pinch Type of plasma pinch produced by an electric current applied axially to a plasma cylinder in a controlled fusion reactor.[NASA]

Ziegler-Nichols method A method of determination of optimum controller settings when tuning a process-control loop (also called the ultimate cycle method). It is based on finding the proportional gain which causes instability in a closed loop.

zinc chlorides Reaction products of hydrochloric acid and zinc; white crystals soluble in water and alcohol and with a melting point of 290 degrees C. [NASA]

zincblende Zinc sulfide, ZnS; a cubic crystal.[NASA]

zinc-bromide batteries Electric cells in which during charge, zinc is plated on the anode and bromine is evolved at the cathode. The bromine is transferred to an external chamber for mixing and storing with an organic liquid complexing oil. During discharge, the zinc is oxidized at the anode and the complexed bromine is reduced at the cathode. [NASA]

zinc-chlorine batteries Candidate electric cells under development for electric vehicles.[NASA]

zinc plating An electroplating coating of zinc on a steel surface which provides corrosion protection in a manner similar to galvanizing.

zinc rich coating A single component zinc-rich coating which can be applied by brush, spray or dip and dries to a gray matte finish. ZRC is accepted by Underwriters Laboratories, Inc., as the equivalent to hot dip galvanizing.

zonal circulation See zonal flow (meteorology).

zonal flow (meteorology) The flow of air along a latitude circle; more specifically, the latitudinal (east or west) of existing flow.[NASA]

zone 1. A portion of internal storage allocated for a particular function or purpose. 2. The international method for specifying the probability that a location is made hazardous by the presence or potential presence, of flammable concentrations of gases or vapors.

Zone 0 A location in which a concentration of a flammable gas or vapor mixture is continuously present, or is present for long periods of time.

Zone 1 A location in which a concentration of a flammable gas or vapor mixture is likely to occur in normal operation.

Zone 2 A location in which a concentration of a flammable gas or vapor mixture is unlikely to occur in normal operation and if it does occur, will exist only for a short period of time.

zone bit 1. One of the two leftmost bits in a binary coding system in which six bits are used to define each character. 2. Any bit in a group of bit positions used to indicate the classification of the group—numeral, alpha character, special sign or command, for instance.

zone control A method of controlling temperature or some other process characteristic by dividing a physical area or process flowpath into several regions, or zones, and independently controlling the process characteristic in each zone.

zones Formerly called divisions. A zone is an area of similar probability of the presence and concentration of the potentially explosive mixture. It is part of the area classification.

zooplankton The aggregate of passively floating or drifting animal organisms in aquatic ecosystems.[NASA]

Z-value A term sometimes used to designate the impedance of a device or circuit, which is the vector sum of resistance and reactance (X-value).

Zyglo method A technique for liquid-penetrant testing to detect surface flaws in a metal using a special penetrant that fluoresces when viewed under ultraviolet radiation.